TD 196 .O73 M32 1997
Mackay, Donald, 1936-
Illustrated handbook of
physical-chemical

MHCC WITHDRAWN

D1737466

ILLUSTRATED HANDBOOK
OF
PHYSICAL-CHEMICAL PROPERTIES
AND
ENVIRONMENTAL FATE
FOR
ORGANIC CHEMICALS

Volume V
Pesticide Chemicals

ILLUSTRATED HANDBOOK
OF
PHYSICAL-CHEMICAL PROPERTIES
AND
ENVIRONMENTAL FATE
FOR
ORGANIC CHEMICALS

Volume V
Pesticide Chemicals

Donald Mackay
Wan-Ying Shiu
Kuo-Ching Ma

Boca Raton   New York

Library of Congress Cataloging-in-Publication Data

Mackay, Donald, Ph.D.
  Illustrated handbook of physical-chemical properties and environmental fate for organic
    chemicals/Donald Mackay, Wan Ying Shiu, and Kuo Ching Ma.
      p.   cm.
  Includes bibliographical references and index.
  ISBN 1-56670-255-0
    Contents: v. 1. Monoaromatic hydrocarbons, chlorobenzenes, and PCBs.
  1. Organic compounds - Environmental aspects - Handbooks, manuals, etc.
  2. Environmental chemistry - Handbooks, manuals, etc.
  I. Shiu, Wan Ying. II. Ma, Kuo Ching. III. Title.
  TD196.073M32   1997
  628.5'2—dc20   91-33888
                                                                                          CIP

This book contains information obtained from authentic and highly regarded sources. Reprinted material is quoted with permission, and sources are indicated. A wide variety of references are listed. Reasonable efforts have been made to publish reliable data and information, but the author and the publisher cannot assume responsibility for the validity of all materials or for the consequences of their use.

Neither this book nor any part may be reproduced or transmitted in any form or by any means, electronic or mechanical, including photocopying, microfilming, and recording, or by any information storage or retrieval system, without prior permission in writing from the publisher.

All rights reserved. Authorization to photocopy items for internal or personal use, or the personal or internal use of specific clients, may be granted by CRC Press LLC, provided that $.50 per page photocopied is paid directly to Copyright Clearance Center, 27 Congress Street, Salem, MA 01970 USA. The fee code for users of the Transactional Reporting Service is ISBN 1-56670-255-0/97/$0.00+$.50. The fee is subject to change without notice. For organizations that have been granted a photocopy license by the CCC, a separate system of payment has been arranged.

The consent of CRC Press LLC does not extend to copying for general distribution, for promotion, for creating new works, or for resale. Specific permission must be obtained in writing from CRC Press LLC for such copying.

Direct all inquiries to CRC Press LLC, 2000 Corporate Blvd., N.W., Boca Raton, Florida 33431.

© 1997 by CRC Press LLC
Lewis Publishers is an imprint of CRC Press LLC

No claim to original U.S. Government works
International Standard Book Number 1-56670-255-0
Library of Congress Card Number 91-33888
Printed in the United States of America   1 2 3 4 5 6 7 8 9 0
Printed on acid-free paper

ILLUSTRATED HANDBOOK OF PHYSICAL-CHEMICAL PROPERTIES AND ENVIRONMENTAL FATE FOR ORGANIC CHEMICALS

Volume I
Monoaromatic Hydrocarbons, Chlorobenzenes, and PCBs
Monoaromatic Hydrocarbons
Chlorobenzenes
Polychlorinated Biphenyls (PCBs)

Volume II
Polynuclear Aromatic Hydrocarbons, Polychlorinated Dioxins, Dibenzofurans
Polynuclear Aromatic Hydrocarbons (PAHs)
Chlorinated Dibenzo-p-Dioxins
Chlorinated Dibenzofurans

Volume III
Volatile Organic Chemicals
Hydrocarbons
Halogenated Hydrocarbons
Ethers

Volume IV
Oxygen, Nitrogen, and Sulfur Containing Compounds
Alcohols
Aldehydes and Ketones
Phenolic Compounds
Carboxylic Acids
Esters
Nitrogen and Sulfur Compounds

Volume V
Pesticide Chemicals
Herbicides
Insecticides
Fungicides

PREFACE

This series of Handbooks brings together physical-chemical data for similarly structured groups of chemical substances which influence their fate in the multimedia environment of air, water, soils, sediments and their resident biota. The task of assessing chemical fate locally, regionally and globally is complicated by the large (and increasing) number of chemicals of potential concern, by uncertainties in their physical-chemical properties, and by lack of knowledge of prevailing environmental conditions such as temperature, pH and deposition rates of solid matter from the atmosphere to water, or from water to bottom sediments. Further, reported values of properties such as solubility are often in conflict. Some are measured accurately, some approximately and some are estimated by various correlation schemes from molecular structure. In some cases, units or chemical identity are wrongly reported. The user of such data thus has the difficult task of selecting the "best" or "right" values. There is justifiable concern that the resulting deductions of environmental fate may be in substantial error. For example, the potential for evaporation may be greatly underestimated if an erroneously low vapor pressure is selected.

To assist the environmental scientist and engineer in such assessments, this Handbook contains compilations of physical-chemical property data for series of chemicals such as, in this case, the pesticide chemicals. It has long been recognized that within such series, properties vary systematically with molecular size, thus providing guidance about the properties of one substance from those of its homologs. Where practical, plots of these systematic property variations can be used to check the reported data and provide an opportunity for interpolation and even modest extrapolation to estimate unreported properties of other homologs. Most handbooks treat chemicals only on an individual basis, and do not contain this feature of chemical-to-chemical comparison which can be valuable for identifying errors and estimating properties.

The data are taken a stage further and used to estimate likely environmental partitioning tendencies, i.e., how the chemical is likely to become distributed between the various media which comprise our biosphere. The results are presented numerically and pictorially to provide a visual impression of likely environmental behavior. This will be of interest to those assessing environmental fate by confirming the general fate characteristics or behavior profile. It is, of course, only possible here to assess fate in a "typical" or "generic" or "evaluative" environment; thus, no claim is made that a chemical will behave in this manner in all situations, but this assessment should reveal the broad characteristics of behavior. These evaluative fate assessments are generated using simple fugacity models which flow naturally from the compilations of data on physical-chemical properties of relevant chemicals.

It is hoped that this series of Handbooks will be of value to environmental scientists and engineers and to students and teachers of environmental science. Its aim is to contribute to better assessments of chemical fate in our multimedia environment by serving as a reference source for environmentally relevant physical-chemical property data of classes of chemicals and by illustrating the likely behavior of these chemicals as they migrate throughout our biosphere.

Donald Mackay, born and educated in Scotland, received his degrees in Chemical Engineering from The University of Glasgow. After a period of time in the petrochemical industry he joined The University of Toronto, where he is now Professor Emeritus in the Department of Chemical Engineering and Applied Chemistry, and in the Institute for Environmental Studies. He is also the director of Environmental Modelling Centre, Trent University, Ontario since 1995. Professor Mackay's primary research is the study of organic environmental contaminants, their sources, fates, effects, and control, and particularly in understanding and modeling their behavior with the aid of the fugacity concept. His work has focused especially on the Great Lakes Basin and on cold northern climates.

Wan-Ying Shiu is a Research Associate in the Department of Chemical Engineering and Applied Chemistry, and the Institute for Environmental Studies, University of Toronto. She received her Ph.D. in Physical Chemistry from the Department of Chemistry, University of Toronto, M.Sc. in Physical Chemistry from St. Francis Xavier University and B.Sc. in Chemistry from Hong Kong Baptist College. Her research interest is in the area of physical-chemical properties and thermodynamics for organic chemicals of environmental concern.

Kuo-Ching Ma obtained his Ph.D. from The Florida State University, M.Sc. from The University of Saskatchewan and B.Sc. from The National Taiwan University; all in Physical Chemistry. After working many years in the Aerospace, Battery Research, Fine Chemicals and Metal Finishing industries in Canada as Research Scientist, Technical Supervisor/Director, he is now dedicating his time and interests to environmental research.

Quick Reference: A Guide to Readers

The reader requiring information on the properties of a specific chemical such as DDT should first consult the list of chemicals in the appropriate class (in this case organochlorines in Chapter 3, Section 3.1.3). The page number for the detailed data compilation is given in that section. Summary tables giving selected values are given in Section 3.2.

An index of the common names of compiled pesticide chemicals in alphabetical order is included at the end of this volume. Also included are indexes to the chemicals in Volume I to IV.

Illustrations of estimated environmental fate are given in Section 3.3, which in this case includes DDT, but not all pestcides are treated. Guidance about likely environmental fate can be obtained by examining the illustrations for a similar compound. Reference can be made to the Commentary in Section 3.4.

It is recommended that the reader consult the original references for verification, and determine if a more recent determination has been reported. Attempts have been made to make the compilation complete up to the end of 1996.

A diskette is provided with this volume which contains the programs used to calculate environmental fate. Three programs are provided which should yield identical results. One is a program written in simple BASIC or GWBASIC which regrettably are obsolete and are not available in the newer MS-DOS systems (MS-DOS 5.0 and up), but are still widely available. The program is saved in ASCII code and can be loaded and run in QBASIC. The other two are written for a spreadsheet program, in this case Lotus 123° (for both DOS and Windows versions) which not only performs fugacity calculations of the chemical of interest but is also able to produce distribution pie-charts. A "readme" file is also included.

TABLE OF CONTENTS

Chapter 1. Introduction 1
    1.1 The Incentive 1
    1.2 Physical-Chemical Properties 3
        1.2.1 The Key Physical-Chemical Properties 3
        1.2.2 Experimental Methods 6
    1.3 Quantitative Structure-Property Relationships (QSPRs) 10
        1.3.1 Objectives 10
        1.3.2 Examples 11
    1.4 Fate Models 15
        1.4.1 Evaluative Environmental Calculations 15
        1.4.2 Treatment of Dissociating Compounds 15
        1.4.3 Treatment of Water-Miscible Compounds 16
        1.4.4 Level I Fugacity Calculation 17
        1.4.5 Level II Fugacity Calculation 21
        1.4.6 Level III Fugacity Calculation 22
    1.5 Data Sources and Presentations 29
        1.5.1 Data Sources 29
        1.5.2 Data Format 29
        1.5.3 Explanation of Data Presentations 30
        1.5.4 Evaluative Calculations 32
    1.6 References 40
Chapter 2. Herbicides 55
    2.1 List of Chemicals and Data Compilations 57
    2.2 Summary Tables 251
    2.3 Illustrative Fugacity Calculations: Levels I, II and III 264
    2.4 Commentary on the Physical-Chemical Properties and Environmental Fate . . 304
    2.5 References 306
Chapter 3. Insecticides 334
    3.1 List of Chemicals and Data Compilations 336
    3.2 Summary Tables 559
    3.3 Illustrative Fugacity Calculations: Levels I, II and III 572
    3.4 Commentary on the Physical-Chemical Properties and Environmental Fate . . 600
    3.5 References 602
Chapter 4. Fungicides 649
    4.1 List of Chemicals and Data Compilations 651
    4.2 Summary Tables 727
    4.3 Illustrative Fugacity Calculations: Levels I, II and III 736
    4.4 Commentary on the Physical-Chemical Properties and Environmental Fate . . 752
    4.5 References 753
List of Symbols and Abbreviations 769
Appendices 773
    A1 BASIC Computer Program for Fugacity Calculations 773
    A2 Fugacity Calculations using Lotus 123° Spreadsheet Program 788
Index 795

Chapter 1. Introduction

1.1 THE INCENTIVE

It is alleged that there are some 60,000 chemicals in current commercial production, with approximately 1000 being added each year. Most are organic chemicals and many are pesticides designed to modify the biotic environment. Of these, perhaps 500 substances are of environmental concern because of their presence in detectable quantities in various components of the environment, their toxicity, their tendency to bioaccumulate, or their persistence. A view is emerging that some of these chemicals, including certain pesticides, are of such extreme environmental concern that all production and use should cease, i.e., as a global society we should elect not to synthesize or use these chemicals. They should be "sunsetted." PCBs, "dioxins" and DDT are examples. A second group consists of chemicals which are of concern because they are used or discharged in large quantities, or they are somewhat toxic or persistent. They are, however, of sufficient value to society that their continued use is justified, but only under conditions in which we fully understand their sources, fate and effects. This understanding is essential if society is to be assured that there are no adverse ecological or human health effects. Other groups of increasingly benign chemicals can presumably be treated with less rigor.

A key feature of this "cradle-to-grave" approach is that society must improve its skills in assessing chemical fate in the environment. We must better understand where chemicals originate, how they migrate in, and between, the various media of air, water, soils, sediments and their biota which comprise our biosphere. We must understand how these chemicals are transformed by chemical and biochemical processes and thus how long they will persist in the environment. We must seek a fuller understanding of the effects which they will have on the multitude of interacting organisms which occupy these media, including ourselves.

It is now clear that the fate of chemicals in the environment is controlled by a combination of three groups of factors. First are the prevailing environmental conditions such as temperatures, flows and accumulations of air, water and solid matter and the composition of these media. Second are the properties of the chemicals which influence partitioning and reaction tendencies, i.e., whether the chemical evaporates or associates with sediments, and how the chemical is eventually destroyed by conversion to other chemical species. Third are the patterns of use, into which compartments the substance is introduced, whether introduction is episodic or continuous and in the case of pestcides how, and, with which additives, the active ingredient is applied.

In recent decades there has emerged a discipline within environmental science concerned with increasing our understanding of how chemicals behave in our multimedia environment. It has been termed "chemodynamics." Practitioners of this discipline include scientists and engineers, students and teachers who attempt to measure, assess and predict how this large number of chemicals will behave in laboratory, local, regional and global environments. These individuals need data on physical-chemical and reactivity properties, as well as information on how these properties translate into environmental fate. This Handbook provides a compilation of such data and uses them to estimate the broad features of environmental fate. It does so for classes or groups of chemicals, instead of the usual approach of treating chemicals on an individual basis. This has the advantage that systematic variations in properties with molecular size can be revealed and used to check reported values, interpolate and even extrapolate to other chemicals of similar structure.

With the advent of inexpensive and rapid computation there has been a remarkable growth of interest in this general area of Quantitative Structure-Property Relationships (QSPRs). The ultimate goal is to use information about chemical structure to deduce physical-chemical properties, environmental partitioning and reaction tendencies, and even uptake and effects on biota. The goal is far from being fully realized, but considerable progress has been made. In earlier volumes of this Handbook series we adopted a simple, and well tried, approach of using molecular structure to deduce a molar volume, which in turn is related to physical-chemical properties. In the case of pesticides the application of QSPR approaches is complicated by the large number of chemical classes and the frequent complexity of the molecules. We do not treat a sufficient number of substances in each class or homologous series to permit useful QSPRs to be presented, although such QSPRs are available in the literature.

Regrettably, the scientific literature contains a great deal of conflicting data with reported values often varying over several orders of magnitude. There are some good, but more not-so-good reasons for this lack of accuracy. Many of these properties are difficult to measure because they involve analyzing very low concentrations of 1 part in 10^9 or 10^{12}. For many purposes an approximate value, for example, that a solubility is less than 1 mg/L, is adequate. There has been a mistaken impression that if a vapor pressure is low, as is the case with DDT, it is not important. DDT evaporates appreciably from solution in water, despite its low vapor pressure, because of its low solubility in water. In some cases the units are reported incorrectly or there are uncertainties about temperature or pH. In other cases the chemical is wrongly identified. One aim of this Handbook is to assist the user to identify such problems and provide guidance when selecting appropriate values.

The final aspect of chemical fate treated in this Handbook is the depiction or illustration of likely chemical fate. This is done using a series of multimedia "fugacity" models as is described in a later section. The authors' aim is to convey an impression of likely environmental partitioning and transformation characteristics, i.e., we seek to generate a "behavior profile." A fascinating feature of chemodynamics is that chemicals differ so greatly in their behavior. Some, such as chloroform, evaporate rapidly and are dissipated in the atmosphere. Others, such as DDT, partition into the organic matter of soils and sediments and the lipids of fish, birds and mammals. Phenols and carboxylic acids tend to remain in water where they may be subject to fairly rapid transformation processes such as hydrolysis, biodegradation and photolysis. By entering the physical-chemical data into a model of chemical fate in a generic or evaluative environment, it is possible to estimate the likely general features of the chemical's behavior and fate. The output of these calculations is presented numerically and pictorially.

In total, the aim of this series of Handbooks is to provide a useful reference work for those concerned with the assessment of the fate of existing and new chemicals, including pesticides, in the environment.

1.2 PHYSICAL-CHEMICAL PROPERTIES
1.2.1 The Key Physical-Chemical Properties

The major differences between behavior profiles of organic chemicals in the environment are attributable to physical-chemical properties. The key properties are believed to be solubility in water, vapor pressure, octanol-water partition coefficient, dissociation constant in water (when relevant) and susceptibility to degrading or transformation reactions. Other essential molecular descriptors are molecular mass and molar volume, with properties such as critical temperature and pressure and molecular area being occasionally useful for specific purposes.

Chemical identity may appear to present a trivial problem, but many chemicals have several names, and subtle differences between isomers (e.g., cis and trans) may be ignored. The most commonly accepted identifiers are the IUPAC name and the Chemical Abstracts System (CAS) number. More recently, methods have been sought of expressing the structure in line notation form so that computer entry of a series of symbols can be used to define a three-dimensional structure. The Wiswesser Line Notation is quite widely used, but it appears that for environmental purposes it will be superseded by the SMILES (Simplified Molecular Identification and Line Entry System, Anderson et al. 1987).

Molecular mass is readily obtained from structure. Also of interest are molecular volume and area, which may be estimated by a variety of methods.

Solubility in water and vapor pressure are both "saturation" properties, i.e., they are measurements of the maximum capacity which a phase has for dissolved chemical. Vapor pressure P (Pa) can be viewed as a "solubility in air," the corresponding concentration C (mol/m^3) being P/RT where R is the ideal gas constant (8.314 J/mol·K) and T is absolute temperature (K). Although most chemicals are present in the environment at concentrations well below saturation, these concentrations are useful for estimating air-water partition coefficients as ratios of saturation values. It is usually assumed that the same partition coefficient applies at lower sub-saturation concentrations. Vapor pressure and solubility thus provide estimates of air-water partition coefficients K_{AW} or Henry's law constants H (Pa·m^3/mol), and thus the relative air-water partitioning tendency.

The octanol-water partition coefficient K_{OW} provides a direct estimate of hydrophobicity or of partitioning tendency from water to organic media such as lipids, waxes and natural organic matter such as humin or humic acid. It is invaluable as a method of estimating K_{OC}, the organic carbon-water partition coefficient, the usual correlation invoked being that of Karickhoff (1981)

$$K_{OC} = 0.41 \cdot K_{OW}$$

K_{OC} is an important parameter which describes the potential for movement or mobility of pesticides in soil, sediment and groundwater. Because of the molecular structural complexity of these agrochemicals, the above simple relationship which considers only the chemical's hydrophobicity may fail for polar and ionic compounds. The effects of pH, soil properties, mineral surfaces and other factors influencing sorption become important. Other quantities, K_d (sorption partition coefficient), and K_{OM} (organic matter-water partition coefficient) are also commonly used to describe the extent of sorption.

K_{OW} is also used to estimate equilibrium fish-water bioconcentration factors K_B or BCF using a correlation similar to that of Mackay (1982)

$$K_B = 0.05 \, K_{ow}$$

where the term 0.05 corresponds to a 5% lipid content of the fish. If metabolism is appreciable, the effective K_B will be lower to an extent dictated by the relative rates of uptake and loss by metabolism and other clearance processes.

For dissociating chemicals it is essential to quantify the extent of dissociation as a function of pH using the dissociation constant pK_a. The parent and ionic forms behave and partition quite differently, thus pH and the presence of other ions may profoundly affect chemical fate.

Characterization of chemical reactivity presents a severe problem in environmental science in general and in Handbooks in particular. Whereas radioisotopes have fixed half-lives, the half-life of a chemical in the environment depends not only on the intrinsic properties of the chemical, but also on the nature of the surrounding environment. Factors such as sunlight intensity, hydroxyl radical concentration and the nature of the microbial community, as well as temperature, affect the chemical's half-life so it is impossible (and misleading) to document a single reliable half-life. The compilation by Howard et al. (1991) provides an excellent review of the existing literature for a large number of chemicals. It is widely used as a source document in this work. The best that can be done is to suggest a semi-quantitative classification of half-lives into groups, assuming average environmental conditions to apply. Obviously, a different class will generally apply in air and bottom sediment. In this compilation we use the following class ranges for chemical reactivity in a single medium such as water.

Class	Mean half-life (hours)	Range (hours)
1	5	< 10
2	17 (~ 1 day)	10-30
3	55 (~ 2 days)	30-100
4	170 (~ 1 week)	100-300
5	550 (~ 3 weeks)	300-1,000
6	1700 (~ 2 months)	1,000-3,000
7	5500 (~ 8 months)	3,000-10,000
8	17000 (~ 2 years)	10,000-30,000
9	55000 (~ 6 years)	> 30,000

These times are divided logarithmically with a factor of approximately 3 between adjacent classes. With the present state of knowledge it is probably misleading to divide the classes into finer groupings; indeed, a single chemical may experience half-lives ranging over three classes, depending on season.

When compiling the suggested reactivity classes, the authors have examined the available information as reaction rates of the chemical in each medium, by all relevant processes. These were expressed as an overall half-life for transformation. For example, a chemical may be subject to biodegradation with a half-life of 20 days (rate constant 0.0014 h^{-1}) and a photolysis with a rate constant of 0.0011 h^{-1} (half-life 630 hours). The overall rate constant is thus 0.0025 h^{-1} and the half-life 277 hours or 12 days. Data for homologous chemicals were also compiled, and insights into the reactivity of various functional groups considered. In most cases a single reaction class was assigned to the series; in this case, class 4 with a mean half-life of 170 hours was chosen. It must be appreciated that this chemical could fall into class 3 in summer and class 5 in winter.

These half-lives must be used with caution and it is wise to test the implications of selecting longer and shorter half-lives.

The pestcides treated in this volume are in many cases more polar and soluble in water than the more hydrophobic chemicals treated in Volumes I, II, and III. In many cases the substances are subject to dissociation, thus complicating the estimation of physical-chemical properties and environmental fate since both ionic and neutral species must be considered. There is thus a tendency for these chemicals to be more associated with water, to evaporate more slowly and to be less sorptive, and less bioaccumulative than the more common environmental contaminants.

Extensive research has been conducted into the atmospheric chemistry of organic chemicals because of air quality concerns. Recently, Atkinson and coworkers (1984, 1985, 1987, 1989, 1990, 1991), Altshuller (1980, 1991) and Sabljic and Güsten (1990) have reviewed the photochemistry of many organic chemicals of environmental interest for their gas phase reactions with hydroxyl radicals (OH), ozone (O_3) and nitrate radicals (NO_3) and have provided detailed information on reaction rate constants and experimental conditions, which allowed the estimation of atmospheric lifetimes. Klöpffer (1991) has estimated the atmospheric lifetimes for the reaction with OH radicals to range from 1 hour to 130 years, based on these reaction rate constants and an assumed constant concentration of OH radicals in air. As Atkinson (1985) has pointed out, the gas-phase reactions with OH radicals are the major tropospheric loss process for the alkanes, haloalkanes, the lower alkenes, the aromatic hydrocarbons, and a majority of the oxygen-containing organics. In addition, photooxidation reactions with O_3 and NO_3 radicals can result in transformation of these compounds. The night-time reaction with NO_3 radicals may also be important (Atkinson and Carter 1984, Sabljic and Güsten 1990).

There are fewer studies on direct or indirect photochemical degradation in the water phase; however, Klöpffer (1991) had pointed out that the rate constant or lifetimes derived from these studies "is valid only for the top layer or surface waters." Mill (1982, 1989, 1993) and Mill and Mabey (1985) have estimated half-lives of various chemicals in aqueous solutions from their reaction rate constants with singlet oxygen, as well as photooxidation with hydroxyl and peroxy radicals. Buxton et al. (1988) gave a critical review of rate constants for reactions with hydrated electrons, hydrogen atoms and hydroxyl radicals in aqueous solutions. Mabey and Mill (1978) also reviewed the hydrolysis of organic chemicals in water under environmental conditions. Recently, Ellington and coworkers (1987a,b, 1988, 1989) also reported the hydrolysis rate constants in aqueous solutions for a variety of organic chemicals.

Other processes, such as biodegradation, may be important, especially in the subsurface environment where there is widespread concern about groundwater contamination.

In most cases a review of the literature suggested that reaction rates in water by chemical processes are 1 to 2 orders of magnitude slower than in air, but with biodegradation often being significant, especially for hydrocarbons and oxygen-containing chemicals. Generally, the water half-life class is three more than that in air, i.e., a factor of about 30 slower. Chemicals in soils tend to be shielded from photolytic processes, and they are less bioavailable, thus the authors have frequently assigned a reactivity class to soil of one more than that for water. Bottom sediments are assigned an additional class to that of soils largely on the basis that there is little or no photolysis, there may be lack of oxygen, and the intimate sorption to sediments renders the chemicals less bioavailable.

Because of the requirements of pesticide regulation processes, extensive data usually exist on partitioning properties and reactivity or half-lives of active ingredients. In some cases these data have been peer-reviewed and published in the scientific literature, but often they are not generally available. A reader with interest in a specfic pesticide can often obtain additional data from manufacturers including accounts of chemical fate under field application conditions. Frequently these data are used as input to pesticide fate models, and the results of these modeling excercises may be available on request, and in some cases they are published in the scientific literature.

The chemical reactivity of these substances is a topic which continues to be the subject of extensive research; thus there is often detailed, more recent information about the fate of chemical species which is of particular relevance to air or water quality. The reader is thus urged to consult the original and recent references because it is impossible in a volume such as this, which considers the entire multimedia picture, to treat this subject in the detail it deserves.

When selecting physical-chemical properties or reactivity classes the authors have been guided by:
(1) the age of the data and acknowledgment of previous conflicting or supporting values,
(2) the method of determination,
(3) the perception of the objectives of the authors, not necessarily as an indication of competence, but often as an indication of the need of the authors for accurate values, and
(4) the reported values for structurally similar, or homologous compounds.

It is appropriate, therefore, to review briefly the experimental methods which are commonly used for property determinations and comment on their accuracy.

1.2.2. Experimental Methods
Solubility in Water

Most conventional organic contaminants are fairly hydrophobic and thus exhibit a measurable solubility in water. Solubility is often used to estimate the air-water partition coefficient or Henry's law constant, but this is not possible for miscible chemicals, indeed the method is suspect for chemicals of appreciable solubility in water, i.e., 1 g/100 g or more. Direct measurement of the Henry's law constant is thus required.

The conventional method of preparing saturated solutions for the determination of solubility is batch equilibration. An excess amount of solute chemical is added to water and equilibrium is achieved by shaking gently (generally referred as the "shake flask method") or slow stirring with a magnetic stirrer. The aim is to prevent formation of emulsions or suspensions and thus avoid extra experimental procedures such as filtration or centrifuging which may be required to ensure that a true solution is obtained. Experimental difficulties can still occur because of the formation of emulsion or microcrystal suspensions with the sparingly soluble chemicals such as higher normal alkanes and polycyclic aromatic hydrocarbons (PAHs). An alternative approach is to coat a thin layer of the chemical on the surface of the equilibration flask before water is added. An accurate "generator column" method has also been developed (Weil et al. 1974, May et al. 1978a,b) in which a column is packed with an inert solid support, such as glass beads or Chromosorb, and then coated with the solute chemical. Water is pumped through the column at a controlled, known flow rate to achieve saturation.

The method of concentration measurement of the saturated solution depends on the solute solubility and its chemical properties. Some common methods used for solubility measurement are listed below.

1. Gravimetric or volumetric methods (Booth and Everson 1948)
   An excess amount of solid compound is added to a flask containing water to achieve saturation solution by shaking, stirring, centrifuging until the water is saturated with solute and undissolved solid or liquid residue appears, often as a cloudy phase. For liquids, successive known amounts of solute may be added to water and allowed to reach equilibrium, and the volume of excess undissolved solute is measured.
2. Instrumental methods
   a. UV spectrometry (Andrews and Keffer 1950, Bohon and Claussen 1951, Yalkowsky and Valvani 1976);
   b. Gas chromatographic analysis with FID, ECD or other detectors (McAuliffe 1966, Mackay et al. 1975, Chiou et al. 1982, Bowman and Sans 1983);
   c. Fluorescence spectrophotometry (Mackay and Shiu 1977);
   d. Interferometry (Gross and Saylor 1931);
   e. High-pressure liquid chromatography (HPLC) with R.I., UV or fluorescence detection (May et al. 1978a,b, Wasik et al. 1983, Shiu et al. 1988, Doucette and Andren 1988a);
   f. Liquid phase elution chromatography (Schwarz 1980, Schwarz and Miller 1980);
   g. Nephelometric methods (Davis and Parke 1942, Davis et al. 1942, Hollifield 1979);
   h. Radiotracer or liquid scintillation counting (LSC) method (Banerjee et al. 1980, Lo et al. 1986).

For most organic chemicals the solubility is reported at a defined temperature in distilled water from completely miscible chemicals such as lower alcohols and amines to sparingly soluble hydrophobic chemicals such as PAHs and dioxins. For substances which dissociate (e.g., phenols, carboxylic acids and amines) it is essential to report the pH of the determination because the extent of dissociation affects the solubility. It is common to maintain the desired pH by buffering with an appropriate electrolyte mixure. This raises the complication that the presence of electrolytes modifies the water structure and changes the solubility. The effect is usually "salting-out." For example, many hydrocarbons have solubilities in seawater about 75% of their solubilities in distilled water. Care must thus be taken to interpret and use reported data properly when electrolytes are present.

The most common problem encountered with reported data is inaccuracy associated with very low solubilities, i.e., those less than 1.0 mg/L. Such solutions are difficult to prepare, handle and analyze, and reported data often contain appreciable errors.

Octanol-Water Partition Coefficient K_{ow}

The experimental approaches are similar to those for solubility, i.e., employing shake flask or generator-column techniques. Concentrations in both the water and octanol phases may be determined after equilibration. Both phases can then be analyzed by the instrumental methods discussed above and the partition coefficient is calculated from the concentration ratio C_o/C_w. This is actually the ratio of solute concentration in octanol saturated with water to that in water saturated with octanol.

As with solubility, K_{OW} is a function of the presence of electrolytes and for dissociating chemicals it is a function of pH. Accurate values can generally be measured up to about 10^6, but accurate measurement beyond this requires meticulous technique. A common problem is that the presence of small quantities of emulsified octanol in the water phase could create a high concentration of chemical in that emulsion which would cause an erroneously high apparent water phase concentration.

Considerable success has been achieved by calculating K_{OW} from molecular structure; thus, there has been a tendency to calculate K_{OW} rather than measure it, especially for "difficult" hydrophobic chemicals. These calculations are, in some cases, extrapolations and can be in serious error. Any calculated log K_{OW} value above 7 should be regarded as suspect, and any experimental or calculated value above 8 should be treated with extreme caution.

For many hydrophilic compounds such as the alcohols, K_{OW} is low and can indeed be less than 1.0, resulting in negative values of log K_{OW}. In such cases, care should be taken when using correlations developed for more hydrophobic chemicals since partitioning into biota or organic carbon phases may be primarily into aqueous rather than organic media.

Details of experimental methods are described by Fujita et al. (1964), Leo et al. (1971), Hansch and Leo (1979), Rekker (1977), Chiou et al. (1977), Miller et al. (1984), Bowman and Sans (1983), Woodburn et al. (1984), Doucette and Andren (1987), and De Bruijn et al. (1989).

Vapor Pressure

In principle, the determination of vapor pressure involves the measurement of the saturation concentration or pressure of the solute in a gas phase. The most reliable methods involve direct determination of these concentrations, but convenient indirect methods are also available based on evaporation rate measurement or chromatographic retention times. Some methods and approaches are listed below.

a. Direct measurement by use of pressure gauges: diaphragm gauge (Ambrose et al. 1975), Rodebush gauge (Sears and Hopke 1947), inclined-piston gauge (Osborn and Douslin 1975);
b. Comparative ebulliometry (Ambrose 1981);
c. Effusion methods, torsion and weight-loss (Balson 1947, Bradley and Cleasby 1953, Hamaker and Kerlinger 1969, De Kruif 1980);
d. Gas saturation or transpiration methods (Spencer and Cliath 1970, 1972, Sinke 1974, Macknick and Prausnitz 1979, Westcott et al. 1981, Rordorf 1985a,b, 1986);
e. Dynamic coupled-column liquid chromatographic method - a gas saturation method (Sonnefeld et al. 1983);
f. Calculation from evaporation rates and vapor pressures of reference compound (Gückel et al. 1974, 1982, Dobbs and Grant 1980, Dobbs and Cull 1982);
g. Calculation from GC retention time data (Hamilton 1980, Westcott and Bidleman 1982, Bidleman 1984, Kim et al. 1984, Foreman and Bidleman 1985, Burkhard et al. 1985a, Hinckley et al. 1990).

The greatest difficulty and uncertainty arise when determining the vapor pressure of chemicals of low volatility, i.e., those with vapor pressures below 1.0 Pa. Vapor pressures are strongly dependent on temperature, thus accurate temperature control is essential. Data are often regressed against temperature and reported as Antoine or Clapeyron constants. Care must be taken when using the Antoine or other equations to extrapolate data beyond the temperature range specified. It must be clear if the data apply to the solid or liquid phase of the chemical.

Henry's Law Constant

The Henry's law constant is essentially an air-water partition coefficient which can be determined by measurement of solute concentrations in both phases. This raises the difficulty of accurate analytical determination in two very different media which require different techniques. Accordingly, some effort has been devoted to devising techniques in which concentrations are measured in only one phase and the other concentration is deduced by a mass balance. These methods are generally more accurate. The principal difficulty arises with hydrophobic, low-volatility chemicals which can establish only very small concentrations in both phases.

Henry's law constant can be regarded as a ratio of vapor pressure to solubility, thus it is subject to the same effects which electrolytes have on solubility and temperature has on both properties. Some methods are as follows:

a. Volatility measurement of dilute aqueous solutions (Butler et al. 1935, Burnett 1963, Buttery et al. 1969);
b. Multiple equilibration method (McAuliffe 1971, Munz and Roberts 1987);
c. Equilibrium batch stripping (Mackay et al. 1979, Dunnivant et al. 1988, Betterton and Hoffmann 1988, Zhou and Mopper 1990);
d. GC-determined distribution coefficients (Leighton and Calo 1981);
e. GC analysis of both air/water phases (Vejrosta et al. 1982, Jönsson et al. 1982);
f. EPICS (Equilibrium Partioning In Closed Systems) method (Lincoff and Gossett 1984, Gossett 1987, Ashworth et al. 1988);
g. Wetted-wall column (Fendinger and Glotfelty 1988, 1990);
h. Headspace analyses (Hussam and Carr 1985);
i. Calculation from vapor pressure and solubility (Mackay and Shiu 1981);
j. GC retention volume/time determined activity coefficient at infinite dilution γ^∞ (Karger et al. 1971a,b, Sugiyama et al. 1975, Tse et al. 1992).

When using vapor pressure and solubility data, it is essential to ensure that both properties apply to the same chemical phase, i.e., both are of the liquid, or of the solid. Occasionally, a solubility is of a solid while a vapor pressure is extrapolated from higher temperature liquid phase data.

As was discussed earlier under solubility, for miscible chemicals it is necessary to determine the Henry's law constant directly, since solubilities are not measurable.

1.3 QUANTITATIVE STRUCTURE-PROPERTY RELATIONSHIPS (QSPRs)
1.3.1 Objectives

Because of the large number of chemicals of actual and potential concern, the difficulties and cost of experimental determinations, and scientific interest in elucidating the fundamental molecular determinants of physical-chemical properties, a considerable effort has been devoted to generating quantitative structure-property relationships (QSPRs). This concept of structure-property relationships or structure-activity relationships (QSARs) is based on observations of linear free-energy relationships, and usually takes the form of a plot or regression of the property of interest as a function of an appropriate molecular descriptor which can be obtained from merely a knowledge of molecular structure.

Such relationships have been applied to solubility, vapor pressure, K_{OW}, Henry's law constant, reactivities, bioconcentration data and several other environmentally relevant partition coefficients. Of particular value are relationships involving various manifestations of toxicity, but these are beyond the scope of this Handbook. These relationships are valuable because they permit values to be checked for "reasonableness" and (with some caution) interpolation is possible to estimate undetermined values. They may be used (with extreme caution!) for extrapolation.

A large number of descriptors have been, and are being, proposed and tested. Dearden (1990) and the compilations by Karcher and Devillers (1990) and Hermens and Opperhuizen (1991) give comprehensive accounts of descriptors and their applications.

Among the most commonly used molecular descriptors are molecular weight and volume, the number of specific atoms (e.g., carbon or chlorine), surface areas (which may be defined in various ways), refractivity, parachor, steric parameters, connectivities and various topological parameters. Several quantum chemical parameters can be calculated from molecular orbital calculations including charge, electron density and superdelocalizability. It is likely that existing and new descriptors will continue to be tested, and that eventually a generally preferred set of readily accessible parameters will be adopted for routine use for correlating purposes.

The usual approach is to compile data for the property in question for a series of structurally similar molecules and plot the logarithm of this property versus molecular descriptors, on a trial and error basis seeking the descriptor which best characterizes the variation in the property. It may be appropriate to use a training set to obtain a relationship and test this relationship on another set. Generally a set of five to ten data points is necessary before any reliable QSPR can be developed. In earlier volumes of this series of Handbooks these QSPRs were developed and displayed, but in this volume there are generally insufficient data points in any one chemical class to permit the development of reliable QSPRs. This is not to imply that such QSPRs can not be developed, indeed the literature contain examples of successful QSPRs, and it is believed that pesticide manufacturers routinely use this approach to suggest modifications to chemical structure which will improve efficacy. Accordingly, in this volume we treat a total of 161 pesticides comprising, i.e., herbicides (68), insecticides (60) and fungicides (33) which are then subdivided into classes based on their molecular structure. Since there is an average of only 4 to 5 pesticides per class, we are unable to present QSPRs. However, we have included the relevant information concerning properties useful for developing QSPRs, correlation methods and examples being listed in the following section.

1.3.2 Examples

Recently, there have been efforts to extend the long-established concept of Quantitative-Structure-Activity-Relationships (QSARs) to Quantitative-Structure-Property Relationships (QSPRs) to compute all relevant environmental physical-chemical properties (such as aqueous solubility, vapor pressure, octanol-water partition coefficient, Henry's law constant, bioconcentration factor (BCF), sorption coefficient and envrionmental reaction rate constants from molecular structure. Examples are Burkhard (1984) and Burkhard et al. (1985a) who calculated solubility, vapor pressure, Henry's law constant, K_{OW} and K_{OC} for all PCB congeners. Hawker and Connell (1988) also calculated log K_{OW}; Abramowitz and Yalkowsky (1990) calculated melting point and solubility for all PCB congeners based on the correlation with total surface area (planar TSAs). Doucette and Andren (1988b) used six molecular descriptors to compute the K_{OW} of some chlorobenzenes, PCBs and PCDDs. Mailhot and Peters (1988) employed seven molecular descriptors to compute physical-chemical properties of some 300 compounds. Isnard and Lambert (1988, 1989) correlated solubility, K_{OW} and BCF for a large number of organic chemicals. Nirmalakhandan and Speece (1988a,b, 1989) used molecular connectivity indices to predict aqueous solubility and Henry's law constants for 300 compounds over 12 logarithmic units in solubility. Kamlet and co-workers (1987, 1988) have developed the solvatochromic parameters with the intrinsic molar volume to predict solubility, log K_{OW} and toxicity of organic chemicals. Warne et al. (1990) correlated solubility and K_{OW} for lipophilic organic compound with 39 molecular descriptors and physical-chemical properties. Atkinson (1987, 1988) has used the structure-activity relationship (SAR) to estimated gas-phase reaction rate constants of hydroxyl radicals for organic chemicals. Mabey et al. (1984) have reviewed the estimation methods from SAR correlation for reaction rate constants and physical-chemical properties in environmental fate assessment. Other correlations are reviewed by Lyman et al. (1982) and Yalkowsky and Banerjee (1992). As Dearden (1990) has pointed out, "new parameters are continually being devised and tested, although the necessity of that may be questioned, given the vast number already available." It must be emphasized, however, that regardless of how accurate these predicted or estimated properties are claimed to be, utimately they have to be confirmed or verified by experimental measurement.

A fundamental problem encountered in these correlations is that the molecular descriptors can be calculated with relatively high precision, usually within a few percent. The accuracy may not always be high, but for empirical correlation purposes precision is more important than accuracy. The precision and accuracy of the experimental data are often poor, frequently ranging over a factor of two or more. Certain isomers may yield identical descriptors, but have different properties. There is thus an inherent limit to the applicability of QSPRs imposed by the quality of the experimental data, and further efforts to improve descriptors, while interesting and potentially useful, are unlikely to yield demonstrably improved QSPRs.

For correlation of **solubility**, the correct thermodynamic quantities for correlation are the activity coefficient γ, or the excess Gibbs free energy ΔG, as discussed by Pierotti et al. (1959) and Tsonopoulos and Prausnitz (1971). Examples of such correlations are given below.

1. Carbon number or carbon plus chlorine number (Tsonopoulos and Prausnitz 1971, Mackay and Shiu 1977);
2. Molar volume cm^3/mol
   a. Liquid molar volume - from density (McAuliffe 1966, Lande and Banerjee 1981, Chiou et al. 1982, Abernethy et al. 1988, Wang et al. 1992);
   b. Molar volume by additive group contribution method, e.g., LeBas method, Schroeder method (Reid et al. 1987, Miller et al. 1985);

    c. Intrinsic molar volume, V_I, cm^3/mol - from van der Waals radius with solvatochromic parameters α and β (Leahy 1986, Kamlet et al. 1987, 1988);
    d. Characteristic molecular volume, m^3/mol (McGowan and Mellors 1986);
3. Group contribution method (Irmann 1965, Korenman et al. 1971, Polak and Lu 1973, Klopman et al. 1992);
4. Molecular volume - Å^3/mol (cubic Angstrom per mole)
    a. van der Waals volume (Bondi 1964);
    b. Total Molecular Volume (TMV) (Pearlman et al. 1984, Pearlman 1986);
5. Total Surface Area (TSA) - Å^2/mol (Hermann 1971, Amidon et al. 1975, Yalkowsky and Valvani 1976, Yalkowsky et al. 1979, Pearlman 1986, Andren et al. 1987, Hawker and Connell 1988, Dunnivant et al. 1992);
6. Molecular Connectivity Indices (MCI) or χ (Kier and Hall 1976, Andren et al. 1987, Nirmalakhandan and Speece 1988b, 1989);
7. Boiling point (Almgren et al. 1979);
8. Melting point (Amidon and Williams 1982);
9. Melting point and TSA (Abramowitz and Yalkowsky 1990);
10. High Pressure Liquid Chromatography (HPLC) - retention data (Locke 1974, Whitehouse and Cooke 1982, Brodsky and Ballschmiter 1988);
11. Adsorbability Index (AI) (Okouchi et al. 1992);
12. Fragment solubility constants (Wakita et al. 1986).

Several workers have explored the linear relationship between octanol-water partition coefficient and solubility as a means of estimating solubility.

Hansch et al. (1968) established the linear free-energy relationship between aqueous and octanol-water partition of organic liquid. Others, such as Tulp and Hutzinger (1978), Yalkowsky et al. (1979), Mackay et al. (1980), Banerjee et al. (1980), Chiou et al. (1982), Bowman and Sans (1983), Miller et al. (1985), Andren et al. (1987) and Doucette and Andren (1988b) have all presented similar but modified relationships.

The UNIFAC (UNIQUAC Functional Group Activity Coefficient) group contribution (Fredenslund et al. 1975, Kikic et al. 1980, Magnussen et al. 1981, Gmehling et al. 1982 and Hansen et al. 1991) is widely used for predicting the activity coefficient in nonelectrolyte liquid mixtures by using group-interaction parameters. This method has been used by Kabadi and Danner (1979), Banerjee (1985), Arbuckle (1983, 1986), Banerjee and Howard (1988) and Al-Sahhaf (1989) for predicting solubility (as a function of the infinite dilution acitivity coefficient, γ^∞) in aqueous systems. Its performance is reviewed by Yalkowsky and Banerjee (1992).

HPLC retention time data have been used as a psuedo-molecular descriptor by Whitehouse and Cooke (1982), Hafkenscheid and Tomlinson (1981), Tomlinson and Hafkenscheid (1986) and Swann et al. (1983).

The **octanol-water partition coefficient** K_{OW} is widely used as a descriptor of hydrophobicity. Variation in K_{OW} is primarily attributable to variation in activity coefficient in the aqueous phase (Miller et al. 1985); thus, the same correlations used for solubility in water are applicable to K_{OW}. Most widely used is the Hansch-Leo compilation of data (Leo et al. 1971, Hansch and Leo 1979) and related predictive methods. Examples of K_{OW} correlations are:

1. Molecular descriptors
   a. Molar volumes: LeBas method; from density; intrinsic molar volume; characteristic molecular volume (Abernethy et al. 1988, Chiou 1985, Kamlet et al. 1988, McGowan and Mellors 1986);
   b. TMV (De Bruijn and Hermens 1990);
   c. TSA (Yalkowsky et al. 1979, 1983, Pearlman 1980, 1986, Pearlman et al. 1984, Hawker and Connell 1988);
   d. Molecular connectivity indices (Doucette and Andren 1988b);
   e. Molecular weight (Doucette and Andren 1988b).
2. Group contribution methods
   a. π-constant or hydrophobic substituent method (Hansch et al. 1968, Hansch and Leo 1979, Doucette and Andren 1988b);
   b. Fragmental constants or f-constant (Rekker 1977, Yalkowsky et al. 1983);
   c. Hansch and Leo's f-constant (Hansch and Leo 1979; Doucette and Andren 1988b).
3. From solubility - K_{ow} relationship
4. HPLC retention data
   a. HPLC-k' capacity factor (Könemann et al. 1979, McDuffie 1981);
   b. HPLC-RT retention time (Veith et al. 1979, Rapaport and Eisenreich 1984, Doucette and Andren 1988b);
   c. HPLC-RV retention volume (Garst 1984);
   d. HPLC-RT/MS HPLC retention time with mass spectrometry (Burkhard et al. 1985c).
5. Reversed-phase thin-layer chromatography (TLC) (Ellgehausen et al. 1981, Bruggeman et al. 1982).
6. Molar refractivity (Yoshida et al. 1983).
7. Combination of HPLC retention data and molecular connctivity indices (Finizio et al. 1994).
8. Molecular orbital methods (Reddy and Locke 1994).

As with solubility and octanol-water partition coefficient, **vapor pressure** can be estimated with a variety of correlations as discussed in detail by Burkhard et al. (1985a) and summarized as follows:
1. Interpolation or extrapolation from equation for correlating temperature relationships, e.g., the Clausius-Clapeyron, Antoine equations (Burkhard et al. 1985a);
2. Carbon or chlorine numbers (Mackay et al. 1980, Shiu and Mackay 1986);
3. LeBas molar volume (Shiu et al. 1987, 1988);
4. Boiling point T_B and heat of vaporization ΔH_v (Mackay et al. 1982);
5. Group contribution method (Macknick and Prausnitz 1979);
6. UNIFAC group contribution method (Jenson et al. 1981, Yair and Fredenslund 1983, Burkhard et al. 1985a, Banerjee et al. 1990);
7. Molecular weight and Gibbs' free energy of vaporization ΔG_v (Burkhard et al. 1985a);
8. TSA and ΔG_v (Amidon and Anik 1981, Burkhard et al. 1985a, Hawker 1989);
9. Molecular connectivity indices (Kier and Hall 1976, 1986, Burkhard et al. 1985a);
10. Melting point T_M and GC retention index (Bidleman 1984, Burkhard et al. 1985a);
11. Solvatochromic parameters and intrinsic molar volume (Banerjee et al. 1990).

As described earlier, **Henry's law constants** can be calculated from the ratio of vapor pressure and aqueous solubility. Henry's law constants do not show a simple linear pattern as solubility, K_{ow} or vapor pressure when plotted against simple molecular descriptors, such as numbers of chlorine or LeBas molar volume, e.g., PCBs (Burkhard et al. 1985b), pesticides

(Suntio et al. 1988), and chlorinated dioxins (Shiu et al. 1988). Henry's law constants can be estimated from:

1. UNIFAC-derived infinite dilution activity coefficients (Arbuckle 1983);
2. Group contribution and bond contribution methods (Hine and Mookerjee 1975, Meylan and Howard 1991);
3. Molecular connectivity indices (Nirmalakhandan and Speece 1988b, Sabljic and Güsten 1989, Dunnivant et al. 1992);
4. Total surface area - planar TSA (Hawker 1989);

For water-miscible compounds the use of aqueous solubility data is obviously impossible.

Bioconcentration Factors:
1. Correlation with K_{ow} (Neely et al. 1974, Könemann and van Leeuwen 1980, Veith et al. 1980, Chiou et al. 1977, Mackay 1982, Briggs 1981, Garten and Trabalka 1983, Davies and Dobbs 1984, Zaroogian et al. 1985, Oliver and Niimi 1988, Isnard and Lambert 1988);
2. Correlation with solubility (Kenaga 1980, Kenaga and Goring 1980, Briggs 1981, Garten and Trabalka 1983, Davies and Dobbs 1984, Isnard and Lambert 1988);
3. Correlation with K_{oc} (Kenaga 1980, Kenaga and Goring 1980, Briggs 1981);
4. Calculation with HPLC retention data (Swann et al. 1983);
5. Calculation with solvatochromic parameters (Hawker 1989, 1990b).

Sorption Coefficients:
1. Correlation with K_{ow} (Karickhoff et al. 1979, Schwarzenbach and Westall 1981, Mackay 1982, Oliver 1984);
2. Correlation with solubility (Karickhoff et al. 1979);
3. Molecular connectivity indices (Gerstl 1984; Sabljic 1984, 1987, Bahnick and Doucette 1988, Sabljic et al. 1989, Meylan et al. 1992);
4. Estimation from molecular connectivity index/fragment contribution method (Meylan et al. 1992, Lohninger 1994);
5. From HPLC retention data (Swann et al. 1983, Szabo et al. 1990).
6. Molecular orbital method (Reddy and Locke 1994).

1.4 FATE MODELS
1.4.1 Evaluative Environmental Calculations

As was discussed earlier, since pesticides are designed to be discharged into ecosystems and induce toxic effects in organisms, there are incentives to build up a detailed understanding of their environmental fate. An obvious economic incentive exists to apply the active ingredient in minimal, yet effective quantities. Another incentive is to ensure that the substance does not adversely affect non-target species or neighboring ecosystems, by for example, excessive evaporation or run-off. A considerable literature exists on mass balance models which describe the detailed fate of pesticides following application. This fate is profoundly affected by the method of application, the nature of the receiving environment (which is usually a soil) and the weather following application, especially temperature and rainfall. In this volume, we make no attempt to characterize this short term, localized behavior. Rather, we focus on the longer term fate of the substance in a wider environment.

When conducting such "field" assessments, there is incentive to standardize environments using "evaluative" environmental models. The nature of these calculations has been described in a series of papers, notably Mackay (1979), Paterson and Mackay (1985), Mackay and Paterson (1990, 1991), and a recent text (Mackay 1991). Only the salient features are presented here. Three calculations are completed for each chemical, namely the Level I, II and III fugacity calculations. The results of these calculations are presented in illustrated form for representative chemicals throughout this Volume.

1.4.2 Treatment of Dissociating Compounds
Corrections for Dissociation

In the case of dissociating or ionizing organic chemicals such as organic acids and bases, e.g., phenols, carboxylic acids and amines, it is desirable to calculate the concentrations of ionic and non-ionic species, and correct for this effect. Recently, Westall et al. (1985), Schwarzenbach et al. (1988), Jafvert et al. (1990), Johnson and Westall (1990) and the text by Schwarzenbach, Gschwend and Imboden (1993) have discussed and reviewed the effect of pH and ionic strength to the distribution of these types of chemicals in the environment.

In the following section, an approach is suggested for estimating the effect of pH on properties and environmental fate using the phenols as an example. A similar approach can be used for bases. The extent of dissociation is characterized by the acid dissociation constant, K_a, expressed as its negative logarithm, pK_a, which for most chloro-phenolic compounds range between 4.75 for pentachlorophenol and 10.2 for phenol, and between 10.0 and 10.6 for the alkylphenols. The dissolved concentration in water is thus the sum of the undissociated, parent or protonated compound and the dissociated phenolate ionic form. When the pK_a exceeds pH by 2 or more units, dissociation is 1% or less and for most purposes is negligible. The ratio of ionic to non-ionic or dissociated to undissociated species is given by,

$$\text{ionic/non-ionic} = 10^{(pH-pKa)} = I$$

The fraction ionic X_I is $I/(1 + I)$. The fraction non-ionic X_N is $1/(1 + I)$. For compounds such as pentachlorophenol in which pH generally exceeds pK_a, I and X_I can be appreciable and there is an apparently enhanced solubility (IUPAC 1985, Yoshida et al. 1987, Acrand et al. 1995). There are other reports of pH effects on octanol-water partition coefficient (Kaiser and Valdmanis 1982, Westall et al. 1985, Lee et al. 1990, Smejtek and Wang 1993), soil sorption behavior (Choi and Amoine 1974, Lee et al. 1990, Schellenberg et al. 1984, Yoshida et al. 1987), bioconcentration and uptake kinetics to goldfish (Stehly and Hayton 1990) and toxicity to algae (Shigeoka et al. 1988).

The following treatment has been suggested by Shiu et al. (1994) and is reproduced briefly below. It is suggested that the simplest, "first order" approach is to take into account the effect of dissociation by deducing the ratio of ionic to non-ionic species I, the fraction ionic X_I and fraction non-ionic X_N for the chemical at both the pH of experimental data determination (I_D, X_{ID}, X_{ND}) and at the pH of the desired environmental simulation (I_E, X_{IE}, X_{NE}). It is assumed that dissociation takes place only in aqueous solution, not in air, organic carbon, octanol or lipid phases. Some ions and ion pairs are known to exist in the latter two phases, but there are insufficient data to give a general procedure for estimating the quantities. No correction is made for the effect of cations other than H^+. This approach must be regarded as merely a first correction for the dissociation effect, and not a full evaluation. This should preferably be based on direct experimental determinations. The reported solubility S mol/m³ and K_{OW} presumably refer to the total of ionic and non-ionic forms, i.e., S_T and $K_{OW,T}$, at the pH of experimental determination, i.e.,

$$S_T = S_N + S_I$$

The solubility and K_{OW} of the non-ionic forms can be estimated as

$$S_N = S_T \cdot X_{ND}; \quad K_{OW,N} = K_{OW,T}/X_{ND}$$

Vapor pressure P^S is not affected; an apparent Henry's law constant, H_T, must also be adjusted to H_T/X_N, being P^S/S_N or $P^S/(S_T \cdot X_N)$.

The Z values are calculated using the conventional equations at the pH of the experimental data (i.e., the system pH). The total Z value in water is then separated into its ionic and non-ionic contributions, i.e., fractions of $I/(I + 1)$ and $1/(I + 1)$. The Z value for the non-ionic form in water is assumed to apply at all pHs, i.e., including the environmental pH, but an additional and possibly different ionic Z value in water is deduced at the environmental pH using I calculated at that pH. The total Z values in water are then calculated. Z values in other media are unaffected.

The calculation was illustrated for pentachlorophenol in Volume IV, Chapter 4, the Phenolic Compounds section. Since pentachlorophenol has been used not only as a wood preservative but also as an insecticide, herbicide and fungicide, it is recommended that the reader refer to Volume IV for a more detailed discussion.

Similar treatment can be applied to other dissociating compounds such as the carboxylic acids, nitrophenols. For bases such as amines, the basicity constant, pK_b is defined and the extent of dissociation I estimated as $10^{-(pH-pKb)}$.

1.4.3 Treatment of Water-Miscible Compounds

In the multimedia models used in this series of volumes, an air-water partition coefficient K_{AW} or Henry's law constant (HLC) is required, and is calculated from the ratio of the pure substance vapor pressure and aqueous solubility. This method is widely used for hydrophobic chemicals but is inappropriate for water-miscible chemicals for which no solubility can be measured (although there are some reported "calculated solubilities" which have been derived from correlations with molecular descriptors for alcohols, aldehydes and amines by Leahy 1986, Kamlet et al. 1987, 1988 and Nirmalakhandan and Speece 1988). The obvious option is to input the HLC directly either as such or as the air-water partition coefficient. If the chemical's activity coefficient γ in water is known, then H can be estimated as $v_w \gamma P_L^S$ where v_w is the molar volume of water and P_L^S is the liquid vapor pressure. Since H can be regarded as P_L^S/C_L^S where C_L^S is the solubility, it is apparent that $(1/v_w\gamma)$ is a "pseudo-solubility." Correlations for γ are available as has been discussed earlier. For example, if γ is 5.0, the pseudo-solubility is 11,100 mol/m³ since v_w is 18×10^{-6} m³/mol or 18 cm³/mol. Chemicals with γ less than about 20 are usually miscible in water. In this case if the liquid vapor pressure is 1000 Pa, H will be 1000/11,100 or 0.090 Pa·m³/mol and K_{AW} will be H/RT or 3.6×10^{-5} at 25°C. Alternatively, if H or K_{AW} is known, C_L^S can be calculated. It is possible to apply existing models to hydrophilic chemicals if this pseudo-solubility is calculated from the activity coefficient or from a known H, i.e., C_L^S is P_L^S/H or $P_L^S/(K_{AW} \cdot RT)$. This is the approach used in this volume. In the fugacity illustrations all pseudo-solubilities are so designated and should not be regarded as real experimentally accessible quantities.

1.4.4 Level I Fugacity Calculation

The Level I calculation describes how a given amount of chemical partitions at equilibrium between six media: air, water, soil, bottom sediment, suspended sediment and fish. No account is taken of reactivity. Whereas most early evaluative environments have treated a one square kilometer region with about 70% water surface (simulating the global proportion of ocean surface), it has become apparent that a more useful approach is to treat a larger, principally terrestrial area similar to a jurisdictional region such as a U.S. state. The area selected is 100,000 km² or 10^{11} m², which is about the area of Ohio, Greece or England.

The atmospheric height is selected as a fairly arbitrary 1000 m reflecting that region of the troposphere which is most affected by local air emissions. A water surface area of 10% or 10,000 km² is used, with a water depth of 20 m. The water volume is thus 2×10^{11} m³. The soil is viewed as being well mixed to a depth of 10 cm and is considered to be 2% organic carbon. It has a volume of 9×10^9 m³. The bottom sediment has the same area as the water, a depth of 1 cm and an organic carbon content of 4%. It thus has a volume of 10^8 m³.

For the Level I calculation both the soil and sediment are treated as simple solid phases with the above volumes, i.e., the presence of air or water in the pores of these phases is ignored.

Two other phases are included for interest. Suspended matter in water is often an important medium when compared in sorbing capacity to that of water. It is treated as having 20% organic carbon and being present at a volume fraction in the water of 5×10^{-6}, i.e., it is about 5 to 10 mg/L. Fish is also included at an entirely arbitrary volume fraction of 10^{-6} and are assumed to contain 5% lipid, equivalent in sorbing capacity to octanol. These two phases are small in volume and rarely contain an appreciable fraction of the chemical present, but it is in these phases that the highest concentration of chemical often exists.

Another phase which is introduced later in the Level III model is aerosol particles with a volume fraction in air of 2×10^{-11}, i.e., approximately 30 μg/m³. Although negligible in volume, an appreciable fraction of the chemical present in the air phase may be associated with aerosols. Aerosols are not treated in Level I or II calculations because their capacity for chemical is usually negligible when compared with soil.

These dimensions and properties are summarized in Table 1.1. The user is encouraged to modify these dimensions to reflect conditions in a specific area of interest.

Table 1.1a Compartment Dimensions and Properties for Level I and II Calculations

Compartment	Air	Water	Soil	Sediment	Suspended Sediment	Fish
Volume, V (m^3)	10^{14}	2×10^{11}	9×10^9	10^8	10^6	2×10^5
Depth, h (m)	1000	20	0.1	0.01	-	-
Area, A (m^2)	100×10^9	10×10^9	90×10^9	10×10^9	-	-
Org. Fraction (ϕ_{OC})	-	-	0.02	0.04	0.2	-
Density, ρ (kg/m^3)	1.2	1000	2400	2400	1500	1000
Adv. Residence Time, t (hours)	100	1000	-	50,000	-	-
Adv. flow, G (m^3/h)	10^{12}	2×10^8	-	2000	-	-

Table 1.1b Bulk Compartment Dimensions and Volume Fraction (v) for Level III Calculations

Compartment		Volume	
Air	Total volume	10^{14} m^3	(as above)
	Air phase	10^{14} m^3	
	Aerosol phase	2000 m^3	(v = 2×10^{-11})
Water	Total volume	2×10^{11} m^3	
	Water phase	2×10^{11} m^3	(as above)
	Suspended sediment phase	10^6 m^3	(v = 5×10^{-6})
	Fish phase	2×10^5 m^3	(v = 1×10^{-6})
Soil	Total volume	18×10^9 m^3	
	Air phase	3.6×10^9 m^3	(v = 0.2)
	Water phase	5.4×10^9 m^3	(v = 0.3)
	Solid phase	9.0×10^9 m^3	(v = 0.5) (as above)
Sediment	Total volume	500×10^6 m^3	
	Water phase	400×10^6 m^3	(v = 0.8)
	Solid phase	100×10^6 m^3	(v = 0.2) (as above)

The amount of chemical introduced in the Level I calculation is an arbitrary 100,000 kg or 100 tons. If dispersed entirely in the air, this amount yields a concentration of 1 $\mu g/m^3$ which is not unusual for ubiquitous contaminants such as hydrocarbons. If dispersed entirely in the water, the concentration is a higher 500 $\mu g/m^3$ or 500 ng/L, which again is reasonable for a well-used chemical of commerce. The corresponding value in soil is about 0.0046 $\mu g/g$. It is believed that this amount is a reasonable common value for evaluative purposes. Clearly for restricted chemicals such as PCBs, this amount is too large, but it is preferable to adopt a common evaluative amount for all substances. No significance should, of course, be attached to the absolute values of the concentrations which are deduced from this arbitrary amount. Only the relative values have significance.

The Level I calculation proceeds by deducing the fugacity capacities, Z values for each medium (see Table 1.2), following the procedures described by Mackay (1991). These working equations show the necessity of having data on molecular mass, water solubility, vapor pressure, and octanol-water partition coefficient. The fugacity f (Pa) common to all media is deduced as

$$f = M / \sum V_i Z_i$$

where M is the total amount of chemical (mol), V_i is the medium volume (m³) and Z_i is the corresponding fugacity capacity for the chemical in each medium.

The molar concentration C (mol/m³) can then be deduced as Zf mol/m³ or as WZf g/m³ or 1000 WZf/ρ μg/g where ρ is the phase density (kg/m³) and W is the molecular mass (g/mol). The amount m_i in each medium is $C_i V_i$ mol, and the total in all media is M mol. The BASIC computer program for undertaking this calculation is appended. For those who prefer a spreadsheet format, an identical Lotus 123® program is also provided.

The information obtained from this calculation includes the concentrations, amounts and distribution. In the figures, a pie-chart illustrates the distribution between the four primary compartments of air, water, soil and sediment, the amount in fish and suspended sediment being ignored. This information is useful as an indication of the relative concentrations.

Note that this simple treatment assumes that the soil and sediment phases are entirely solid, i.e., there are no air or water phases present to "dilute" the solids. Later in the Level III calculation these phases and aerosols are included.

Table 1.2a Equations for Phase Z Values Used in Levels I and II and the Bulk Phase Values Used in Level III

Compartment	Z values
Air	$Z_1 = 1/RT$
Water	$Z_2 = 1/H = C^S/P^S$
Soil	$Z_3 = Z_2 \cdot \rho_3 \cdot \phi_3 \cdot K_{OC}/1000$
Sediment	$Z_4 = Z_2 \cdot \rho_4 \cdot \phi_4 \cdot K_{OC}/1000$
Suspended Sediment	$Z_5 = Z_2 \cdot \rho_5 \cdot \phi_5 \cdot K_{OC}/1000$
Fish	$Z_6 = Z_2 \cdot \rho_6 \cdot L \cdot K_{OW}/1000$
Aerosol	$Z_7 = Z_1 \cdot 6 \times 10^6 / P_L^S$

where
- R = gas constant (8.314 J/mol·K)
- T = absolute temperature (K)
- C^S = solubility in water (mol/m³)
- P^S = vapor pressure (Pa)
- H = Henry's law constant (Pa·m³/mol)
- P_L^S = liquid vapor pressure (Pa)
- K_{OW} = octanol-water partition coefficient
- ρ_i = density of phase i (kg/m³)
- ϕ_i = mass fraction organic-carbon in phase i (g/g)
- L = lipid content of fish

Note for solids $P_L^S = P_S^S / \exp\{6.79(1 - T_M/T)\}$ where T_M is melting point (K) of the solute.

Table 1.2b Bulk Phase Z Values, Z_{Bi} Deduced as $\Sigma\ v_i Z_i$, in which the Coefficients, e.g., 2×10^{-11}, are the Volume Fractions v_i of Each Pure Phase as Specified in Table 1.1b

Compartment	Bulk Z values	
Air	$Z_{B1} = Z_1 + 2 \times 10^{-11} Z_7$	(approximately 30 μg/m³ aerosols)
Water	$Z_{B2} = Z_2 + 5 \times 10^{-6} Z_5 + 1 \times 10^{-6} Z_6$	(5 ppm solids, 1 ppm fish by volume)
Soil	$Z_{B3} = 0.2 Z_1 + 0.3 Z_2 + 0.5 Z_3$	(20% air, 30% water, 50% solids)
Sediment	$Z_{B4} = 0.8 Z_2 + 0.2 Z_4$	(80% water, 20% solids)

1.4.5 Level II Fugacity Calculation

The Level II calculation simulates a situation in which a chemical is continuously discharged into the multimedia environment and achieves a steady-state equilibrium condition at which input and output rates are equal. The task is to deduce the rates of loss by reaction and advection.

The reaction rate data developed for each chemical in the tables are used to select a reactivity class as described earlier, and hence a first-order rate constant for each medium. Often these rates are in considerable doubt, thus the quantities selected should be used with extreme caution because they may not be widely applicable. The rate constants k_i h^{-1} are used to calculate reaction D values for each medium D_{Ri} as $V_i Z_i k_i$. The rate of reactive loss is then $D_{Ri}f$ mol/h.

For advection, it is necessary to select flow rates. This is conveniently done in the form of advective residence times, t in hour (h), thus the advection rate G_i is V_i/t m³/h for each medium. For air, a residence time of 100 hours is used (approximately 4 days), which is probably too long for the geographic area considered, but shorter residence times tend to cause air advective loss to be a dominant mechanism. For water, a figure of 1000 hours (42 days) is used, reflecting a mixture of rivers and lakes. For sediment burial (which is treated as an advective loss), a time of 50,000 hours or 5.7 years is used. Only for very persistent, hydrophobic chemicals is this process important. No advective loss from soil is included. The D value for loss by advection D_{Ai} is $G_i Z_i$ and the rates are $D_{Ai}f$ mol/h. These rates are listed in Table 1.1a.

There may thus be losses caused by both reaction and advection D values for the four primary media. These loss processes are not included for fish or suspended matter. At steady state, and equilibrium conditions, the input rate E mol/h can be equated to the sum of the output rates, from which the common fugacity can be calculated as follows

$$E = f \cdot \sum D_{Ai} + f \cdot \sum D_{Ri}$$

thus,

$$f = E/(\sum D_{Ai} + \sum D_{Ri})$$

The common assumed emission rate is 1000 kg/h or 1 ton/h. To achieve an amount equivalent to the 100 tons in the Level I calculation requires an overall residence time of 100 hours. Again, the concentrations and amounts m_i and $\sum m_i$ or M can be deduced, as well as the reaction and advection rates. These rates obviously total to give the input rate E. Of particular interest are the relative rates of these loss processes, and the overall persistence or residence time which is calculated as

$$t_O = M/E$$

where M is the total amount present. It is also useful to calculate a reaction and an advection persistence t_R and t_A as

$$t_R = M/\sum D_{Ri}f \qquad t_A = M/\sum D_{Ai}f$$

Obviously,
$$1/t_O = 1/t_R + 1/t_A$$

These persistences indicate the likelihood of the chemical being lost by reaction as distinct from advection. The percentage distribution of chemical between phases is identical to that in Level I. A pie-chart depicting the distribution of losses is presented.

1.4.6 Level III Fugacity Calculation

Whereas the Levels I and II calculations assume equilibrium to prevail between all media, this is recognized as being excessively simplistic and even misleading. In the interests of algebraic simplicity only the four primary media are treated for this level. The task is to develop expressions for intermedia transport rates by the various diffusive and nondiffusive processes as described by Mackay (1991). This is done by selecting values for 12 intermedia transport velocity parameters which have dimensions of velocity (m/h or m/year), are designated as U_i m/h and are applied to all chemicals. These parameters are used to calculate seven intermedia transport D values.

It is desirable to calculate new "bulk phase" Z values for the four primary media which include the contribution of dispersed phases within each medium as described by Mackay and Paterson (1991) and as listed in Tables 1.2a and 1.2b. The air is now treated as an air-aerosol mixture, water as water plus suspended particles and fish, soil as solids, air and water, and sediment as solids and porewater. The Z values thus differ from the Level I and Level II "pure phase" values. The necessity for introducing this complication arises from the fact that much of the intermedia transport of the chemicals occurs in association with the movement of chemical in these dispersed phases. To accommodate this change the same volumes of the soil solids and sediment solids are retained, but the total phase volumes are increasd. These Level III volumes are also given in Table 1.1b. The reaction and advection D values employ the generally smaller bulk phase Z values but the same residence times, thus the G values are increased and the D values are generally larger.

Intermedia D Values

The justification for each intermedia D value follows. It is noteworthy that, for example, air-to-water and water-to-air values differ because of the presence of one-way nondiffusive processes. A fuller description of the background to these calculations is given by Mackay (1991).

1. Air to Water (D_{12})

Four processes are considered: diffusion (absorption), dissolution in rain of gaseous chemical, and wet and dry deposition of particle-associated chemical.

For diffusion, the conventional two-film approach is taken with water-side (k_W) and air-side (k_A) mass transfer coefficients (m/h) being defined. Values of 0.05, 5 m/h for k_W and k_A are used. The absorption D value is then

$$D_{VW} = 1/[1/(k_A A_W Z_1) + 1/(k_W A_W Z_2)]$$

where A_W is the air-water area (m²) and Z_1 and Z_2 are the pure air and water Z values. The velocities k_A and k_W are designated as U_1 and U_2.

For rain dissolution, a rainfall rate of 0.876 m/year is used, i.e., U_R or U_3 is 10^{-4} m/h. The D value for dissolution D_{RW} is then

$$D_{RW} = U_R A_W Z_2 = U_3 A_W Z_2$$

For wet deposition, it is assumed that the rain scavenges Q (the scavenging ratio) or about 200,000 times its volume of air. Using a particle concentration (volume fraction) v_Q of 2×10^{-11}, this corresponds to the removal of Qv_Q or 4×10^{-6} volumes of aerosol per volume of rain. The total rate of particle removal by wet deposition is then $Qv_QU_RA_W$ m³/h, thus the wet "transport velocity" Qv_QU_R is 4×10^{-10} m/h.

For dry deposition, a typical deposition velocity U_Q of 10 m/h is selected yielding a rate of particle removal of $U_Qv_QA_W$ or $2 \times 10^{-10}A_W$ m³/h corresponding to a transport velocity of 2×10^{-10} m/h. Thus,

$$U_4 = Qv_QU_R + U_Qv_Q = v_Q(QU_R + U_Q)$$

The total particle transport velocity U_4 for wet and dry deposition is thus 6×10^{-10} m/h and the total D value D_{QW} is

$$D_{QW} = U_4A_WZ_7$$

where Z_7 is the aerosol Z value.

The overall D value is given by

$$D_{12} = D_{VW} + D_{RW} + D_{QW}$$

2. Water to Air (D_{21})

Evaporation is treated as the reverse of absorption; thus D_{21} is simply D_{VW} as before.

3. Air to Soil (D_{13})

A similar approach is adopted as for air-to-water transfer. Four processes are considered with rain dissolution (D_{RS}) and wet and dry deposition (D_{QS}) being treated identically except that the area term is now the air-soil area A_S.

For diffusion, the approach of Jury et al. (1983, 1984a,b,c) is used as described by Mackay and Stiver (1991) and Mackay (1991) in which three diffusive processes are treated. The air boundary layer is characterized by a mass transfer coefficient k_S or U_7 of 5 m/h, equal to that of the air-water mass transfer coefficient k_A used in D_{12}.

For diffusion in the soil air-pores, a molecular diffusivity of 0.02 m²/h is reduced to an effective diffusivity using a Millington-Quirk type of relationship by a factor of about 20 to 10^{-3} m²/h. Combining this with a path length of 0.05 m gives an effective air-to-soil mass transfer coefficient k_{SA} of 0.02 m/h which is designated as U_5.

Similarly, for diffusion in water a molecular diffusivity of 2×10^{-6} m²/h is reduced by a factor of 20 to an effective diffusivity of 10^{-7} m²/h, which is combined with a path length of 0.05 m to give an effective soil-to-water mass transfer coefficient of k_{SW} 2×10^{-6} m/h.

It is probable that capillary flow of water contributes to transport in the soil. For example, a rate of 7 cm/year would yield an equivalent water velocity of 8×10^{-6} m/h which exceeds the water diffusion rate by a factor of four. For illustrative purposes we thus select a water transport velocity or coefficient U_6 in the soil of 10×10^{-6} m/h, recognizing that this may be in error by a substantial amount, and will vary with rainfall characteristics and soil type.

The soil processes are in parallel with boundary layer diffusion in series, so the final equation is

$$D_{VS} = 1/[1/D_S + 1/(D_{SW} + D_{SA})]$$

where

$$D_S = U_7 A_S Z_1 \quad (U_7 = 5 \text{ m/h})$$
$$D_{SW} = U_6 A_S Z_2 \quad (U_6 = 10 \times 10^{-6} \text{ m/h})$$
$$D_{SA} = U_5 A_S Z_1 \quad (U_5 = 0.02 \text{ m/h})$$

where A_S is the soil horizontal area.

Air-soil diffusion thus appears to be much slower than air-water diffusion because of the slow migration in the soil matrix. In practice, the result will be a nonuniform composition in the soil with the surface soil (which is much more accessible to the air than the deeper soil) being closer in fugacity to the atmosphere.

The overall D value is given as

$$D_{13} = D_{VS} + D_{QS} + D_{RS}$$

4. Soil to Air (D_{31})

Evaporation is treated as the reverse of absorption, thus the D value is simply D_{VS}.

5. Water to Sediment (D_{24})

Two processes are treated, diffusion and deposition.

Diffusion is characterized by a mass transfer coefficient U_8 of 10^{-4} m/h which can be regarded as a molecular diffusivity of 2×10^{-6} m²/h divided by a path length of 0.02 m. In practice, bioturbation may contribute substantially to this exchange process, and in shallow water current-induced turbulence may also increase the rate of transport. Diffusion in association with organic colloids is not included. The D value is thus given as $U_8 A_W Z_2$.

Deposition is assumed to occur at a rate of 5000 m³/h which corresponds to the addition of a depth of solids of 0.438 cm/year; thus 43.8% of the solids resident in the accessible bottom sediment is added each year. This rate is about 12 cm³/m²·day which is high compared to values observed in large lakes. The velocity U_9, corresponding to the addition of 5000 m³/h over the area of 10^{10} m², is thus 5×10^{-7} m/h.

It is assumed that of this 5000 m³/h deposited, 2000 m³/h or 40% is buried (yielding the advective flow rate in Table 1.1a), 2000 m³/h or 40% is resuspended (as discussed later) and the remaining 20% is mineralized organic matter. The organic carbon balance is thus only approximate.

The transport velocities are thus:

deposition U_9 5.0×10^{-7} m/h or 0.438 cm/y

resuspension U_{10} 2.0×10^{-7} m/h or 0.175 cm/y

burial U_B 2.0×10^{-7} m/h or 0.175 cm/y
(included as an advective residence time of 50,000 h)

The water-to-sediment D value is thus

$$D_{24} = U_8 A_W Z_2 + U_9 A_W Z_5$$

where Z_5 is the Z value of the particles in the water column.

6. Sediment to Water (D_{42})

This is treated similarly to D_{24} giving:

$$D_{42} = U_8 A_W Z_2 + U_{10} A_W Z_4$$

where U_{10} is the sediment resuspension velocity of 2.0×10^{-7} m/h and Z_4 is the Z value of the sediment solids.

7. Sediment Advection (D_{A4})

This D value is $U_B A_W Z_4$ where U_B, the sediment burial rate, is 2.0×10^{-7} m/h. It can be viewed as $G_B Z_{B4}$ where G_B is the total burial rate specified as V_S/t_B where t_B (residence time) is 50,000 h, and V_S (the sediment volume) is the product of sediment depth (0.01 cm) and area A_W. Z_4, Z_{B4} are the Z values of the sediment solids and of the bulk sediment respectively. Since there are 20% solids, Z_{B4} is about $0.2\ Z_4$. There is a slight difference between these approaches because in the advection approach (which is used here) there is burial of water as well as solids.

8. Soil to Water (D_{32})

It is assumed that there is run-off of water at a rate of 50% of the rain rate, i.e., the D value is

$$D = 0.5\ U_3 A_S Z_2 = U_{11} A_S Z_2$$

thus the transport velocity term U_{11} is $0.5 U_3$ or 5×10^{-5} m/h.

For solids run-off it is assumed that this run-off water contains 200 parts per million by volume of solids; thus the corresponding velocity term U_{12} is $200 \times 10^{-6} U_{11}$, i.e., 10^{-8} m/h. This corresponds to the loss of soil at a rate of about 0.1 mm per year. If these solids were completely deposited in the aquatic environment (which is about 1/10th the soil area), they would accumulate at about 0.1 cm per year, which is about a factor of four less than the deposition rate to sediments. The implication is that most of this deposition is of naturally generated organic carbon and from sources such as bank erosion.

Summary

The twelve intermedia transport parameters are listed in Table 1.3 and the equations are summarized in Table 1.4.

Table 1.3 Intermedia Transport Parameters

U		m/h	m/year
1	Air side, air-water MTC*, k_A	5	43,800
2	Water side, air-water MTC, k_W	0.05	438
3	Rain rate, U_R	10^{-4}	0.876
4	Aerosol deposition	6×10^{-10}	5.256×10^{-6}
5	Soil-air phase diffusion MTC, k_{SA}	0.02	175.2
6	Soil-water phase diffusion MTC, k_{SW}	10×10^{-6}	0.0876
7	Soil-air boundary layer MTC, k_S	5	43,800
8	Sediment-water MTC	10^{-4}	0.876
9	Sediment deposition	5.0×10^{-7}	0.00438
10	Sediment resuspension	2.0×10^{-7}	0.00175
11	Soil-water run-off	5.0×10^{-5}	0.438
12	Soil-solids run-off	10^{-8}	8.76×10^{-5}

* Mass transfer coefficient with. Scavenging ratio $Q = 2 \times 10^5$, dry deposition velocity $U_Q = 10$ m/h and sediment burial rate $U_B = 2.0 \times 10^{-7}$ m/h

Table 1.4 Intermedia Transport D Value Equations

Air-Water	$D_{12} = D_{VW} + D_{RW} + D_{QW}$
	$D_{VW} = A_W/(1/U_1Z_1 + 1/U_2Z_2)$
	$D_{RW} = U_3 A_W Z_2$
	$D_{QW} = U_4 A_W Z_7$
Water-Air	$D_{21} = D_{VW}$
Air-Soil	$D_{13} = D_{VS} + D_{RS} + D_{QS}$
	$D_{VS} = 1/(1/D_S + 1/(D_{SW} + D_{SA}))$
	$D_S = U_7 A_S Z_1$
	$D_{SA} = U_5 A_S Z_1$
	$D_{SW} = U_6 A_S Z_2$
	$D_{RS} = U_3 A_S Z_2$
	$D_{QS} = U_4 A_S Z_7$
Soil-Air	$D_{31} = D_{VS}$
Water-Sediment	$D_{24} = U_8 A_W Z_2 + U_9 A_W Z_5$
Sediment-Water	$D_{42} = U_8 A_W Z_2 + U_{10} A_W Z_4$
Soil-Water	$D_{32} = U_{11} A_S Z_2 + U_{12} A_S Z_3$

Algebraic Solution

Four mass balance equations can be written, one for each medium, resulting in a total of four unknown fugacities, enabling simple algebraic solution as shown in Table 1.5. From the four fugacities, the concentration, amounts and rates of all transport and transformation processes can be deduced, yielding a complete mass balance.

Table 1.5 Level III Solutions to Mass Balance Equations

Compartment	Mass balance equations
Air	$E_1 + f_2 D_{21} + f_3 D_{31} = f_1 D_{T1}$
Water	$E_2 + f_1 D_{12} + f_3 D_{32} + f_4 D_{42} = f_2 D_{T2}$
Soil	$E_3 + f_1 D_{13} = f_3 D_{T3}$
Sediment	$E_4 + f_2 D_{24} = f_4 D_{T4}$

where E_i is discharge rate, E_4 usually being zero.

$$D_{T1} = D_{R1} + D_{A1} + D_{12} + D_{13}$$
$$D_{T2} = D_{R2} + D_{A2} + D_{21} + D_{23} + D_{24}, \quad (D_{23} = 0)$$
$$D_{T3} = D_{R3} + D_{A3} + D_{31} + D_{32}, \quad (D_{A3} = 0)$$
$$D_{T4} = D_{R4} + D_{A4} + D_{42}$$

Solutions:

$$f_2 = [E_2 + J_1 J_4/J_3 + E_3 D_{32}/D_{T3} + E_4 D_{42}/D_{T4}]/(D_{T2} - J_2 J_4/J_3 - D_{24} \cdot D_{42}/D_{T4})$$
$$f_1 = (J_1 + f_2 J_2)/J_3$$
$$f_3 = (E_3 + f_1 D_{13})/D_{T3}$$
$$f_4 = (E_4 + f_2 D_{24})/D_{T4}$$

where

$$J_1 = E_1/D_{T1} + E_3 D_{31}/(D_{T3} \cdot D_{T1})$$
$$J_2 = D_{21}/D_{T1}$$
$$J_3 = 1 - D_{31} \cdot D_{13}/(D_{T1} \cdot D_{T3})$$
$$J_4 = D_{12} + D_{32} \cdot D_{13}/D_{T3}$$

The new information from the Level III calculations are the intermedia transport data, i.e., the extent to which chemical discharged into one medium tends to migrate into another. This migration pattern depends strongly on the proportions of the chemical discharged into each medium; indeed, the relative amounts in each medium are largely a reflection of the locations of discharge. It is difficult to interpret these mass balance diagrams because, for example, chemical depositing from air to water may have been discharged to air, or to soil from which it evaporated, or even to water from which it is cycling to and from air.

To simplify this interpretation, it is best to conduct three separate Level III calculations in which unit amounts (1000 kg/h) are introduced individually into air, soil and water. Direct discharges to sediment are unlikely and are not considered here. These calculations show clearly the extent to which intermedia transport occurs. If, for example, the intermedia D values are small compared to the reaction and advection values, the discharged chemical will tend to remain in the discharge or "source" medium with only a small proportion migrating to other media. Conversely, if the intermedia D values are relatively large the chemical becomes very susceptible to intermedia transport. This behavior is observed for persistent substances such as PCBs, which have very low rates of reaction.

A direct assessment of multimedia behavior is thus possible by examining the proportions of chemical found at steady state in the "source" medium and in other media. For example, when discharged to water, an appreciable fraction of the benzene is found in air, whereas for atrazine, only a negligible fraction of atrazine reaches air.

Linear Additivity
Because these equations are entirely linear, the solutions can be scaled linearly. The concentrations resulting from a discharge of 2000 kg/h are simply twice those of 1000 kg/h. Further, if discharge of 1000 kg/h to air causes 500 kg in water and discharge of 1000 kg/h to soil causes 100 kg in water, then if both discharges occur simultaneously, there will be 600 kg in water. If the discharge to soil is increased to 3000 kg/h, the total amount in the water will rise to (500 + 300) or 800 kg. It is thus possible to deduce the amount in any medium arising from any combination of discharge rates by scaling and adding the responses from the unit inputs. This "linear additivity principle" is more fully discussed by Stiver and Mackay (1989).

In the diagrams presented later, these three-unit (1000 kg/h) responses are given. Also, an illustrative "three discharge" mass balance is given in which a total of 1000 kg/h is discharged, but in proportions judged to be typical of chemical use and discharge to the environment. For example, benzene is believed to be mostly discharged to air with minor amounts to soil and water.

Also given in the tables are the rates of reaction, advection and intermedia transport for each case.

The reader can deduce the fate of any desired discharge pattern by appropriate scaling and addition. It is important to re-emphasize that because the values of transport velocity parameters are only illustrative, actual environmental conditions may be quite different; thus, simulation of conditions in a specific region requires determination of appropriate parameter values as well as the site specific dimensions, reaction rate constants and the physical-chemical properties which prevail at the desired temperature.

In total, the aim is to convey an impression of the likely environmental behavior of the chemical in a readily assimilable form.

1.5. DATA SOURCES AND PRESENTATIONS
1.5.1 Data Sources

Most physical properties such as molecular weight (MW, g/mol), melting point (M.P., °C), boiling point (B.P., °C), and density have been obtained from commonly used handbooks such as the CRC Handbook of Physics and Chemistry (Weast 1972, 1982), Lange's Handbook of Chemistry (Dean 1979, 1985), Dreisbach's Physical Properties of Chemical Compounds, Vol. I, II and III (1955, 1959, 1961), Organic Solvents, Physical Properties and Methods of Purification (Riddick et al. 1986), the Merck Index (1983, 1989) and several Handbooks and compilations of chemical property data for pesticides. Notable are the text by Hartley and Graham-Bryce (1980), the Agrochemicals Handbook (Hartley and Kidd 1987), The Pesticide Manual (Worthing 1983, 1987, 1991, Tomlin 1994), the Agrochemicals Desk Reference (Montgomery 1993) and the SCS/ARS/CES Pesticide Properties Database by Wauchope and coworkers (Wauchope et al. 1992, Augustij-Beckers et al. 1994, Hornsby et al. 1996). Other physical-chemical properties such as aqueous solubility, vapor pressure, octanol-water partition coefficient, Henry's law constant, bioconcentration factor and sorption coefficient have been obtained from scientific journals or other environmental handbooks, notably Verschueren's Handbook of Environmental Data on Organic Chemicals (1977, 1983) and Howard and coworkers' Handbook of Environmental Fate and Exposure Data, Vol. I, II, III and IV (1989, 1990, 1991 and 1993). Other important sources of vapor pressure are the CRC Handbook of Physics and Chemistry (Weast 1972, 1982), Lange's Handbook of Chemistry (Dean 1985), the Handbook of Vapor Pressures and Heats of Vaporization of Hydrocarbons and Related Compounds (Zwolinski and Wilhoit 1971), the Vapor Pressure of Pure Substances (Boublik et al. 1973, 1984), the Handbook of the Thermodynamics of Organic Compounds (Stephenson and Malanowski 1987). For aqueous solubilities, valuable sources include the IUPAC Solubility Data Series (1984, 1985, 1989a,b) and Horvath's Halogenated Hydrocarbons, Solubility-Miscibility with Water (Horvath 1982). Octanol-water partition coefficients are conveniently obtained from the compilation by Leo et al. (1971) and Hansch and Leo (1979), or can be calculated from molecular structure by the methods of Hansch and Leo (1979) or Rekker (1977). Lyman et al. (1982) also outline methods of estimating solubility, K_{OW}, vapor pressure, and bioconcentration factor for organic chemicals. The recent Handbook of Environmental Degradation Rates by Howard et al. (1991) is a valuable source of rate constants and half-lives for inclusion in subsequent fugacity calculations.

The most reliable sources of data are the original citations in the reviewed scientific literature. Particularly reliable are those papers which contain a critical review of data from a number of sources as well as independent experimental determinations. Calculated or correlated values are reported in the tables but are viewed as being less reliable. A recurring problem is that a value is frequently quoted, then requoted and the original paper may not be cited. The aim in this work has been to gather and list the citations, interpret them and select a "best" or "most likely" value.

1.5.2 Data Format
Each data sheet lists the following properties, although not all quantities are included for all chemicals. In all cases citations are provided.

Common Name:
Synonym:
Chemical Name:
Uses:
CAS Registry No:
Molecular Formula:
Molecular Weight (g/mol):

Melting Point (°C):
Boiling Point (°C):
Density (g/cm^3 at 20°C):
Molar Volume (cm^3/mol):
Molecular Volume, TMV (Å^3):
Dissociation Constant: pK_a or pK_b:
Total Surface Area, TSA (Å^2):
Heat of Fusion, ΔH_{fus}, kcal/mol:
Entropy of Fusion, ΔS_{fus}, cal/mol·K (or e.u.):
Fugacity Ratio at 25°C:
Water Solubility (g/m^3 or mg/L at 25°C):
Vapor Pressure (Pa at 25°C):
Henry's Law Constant (Pa·m^3/mol) or Air-Water Partition Coefficient:
Octanol-Water Partition Coefficient K_{OW} or log K_{OW}:
Bioconcentration Factor K_B or BCF (or log K_B):
Sorption Partition Coefficient to Organic Carbon K_{OC} or to Organic Matter K_{OM}:
Half-lives in the Environment:
    Air:
    Surface water:
    Groundwater:
    Soil:
    Sediment:
    Biota:
Environmental Fate Rate Constants or Half-Lives:
    Volatilization/Evaporation:
    Photolysis:
    Oxidation or Photooxidation:
    Hydrolysis:
    Biotransformation/Biodegradation:
    Bioconcentration, Uptake (k_1) and Elimination (k_2) Rate Constants:

1.5.3 Explanation of Data Presentations

Example: DDT (data sheets presented in Chapter 3)

Chemical Properties.
    The names, formula, melting and boiling point and density data are self-explanatory.
    The molar volumes are in some cases at the stated temperature and in others at the normal boiling point. Certain calculated molecular volumes are also used; thus the reader is cautioned to ensure that when using a molar volume in any correlation, it is correctly selected. In the case of polynuclear aromatic hydrocarbons, the LeBas molar volume is regarded as suspect because of the compact nature of the multi-ring compounds. It should thus be regarded as merely an indication of relative volume, not absolute volume.

    The total surface areas (TSAs) are calculated in various ways and may contain the hydration shell, thus giving a much larger area. Again, the reader is cautioned to ensure that values are consistent.

    Heats of fusion, ΔH_{fus}, are generally expressed in kcal/mol or kJ/mol and entropies of fusion, ΔS_{fus} in cal/mol·K (e.u. or entropy unit) or J/mol·K. The fugacity ratio F, defined below, is used to calculate the supercooled liquid vapor pressure or solubility for correlation purposes. In the case of liquids such as benzene, it is 1.0. For solids it is a fraction representing the ratio of solid to liquid solubility or vapor pressure. It is generally assumed that for a rigid organic molecule, the entropy of fusion is 13.5 e.u. or 56 J/mol·K, which is an average value of a number of organic compounds (Yalkowsky 1979, Miller et al. 1984). The fugacity ratio, F, given is calculated using ΔS_{fus} = 56 J/mol·K in the following expression

$$F = \exp\{(\Delta H_{fus}/RT)(1 - T_M/T)\} = \exp\{(\Delta S_{fus}/R)(1 - T_M/T)\}$$

where R is the ideal gas constant (8.314 J/mol·K or 1.987 cal/mol·K) and T_M is the melting point and T is the system temperature (K).

As is apparent, a wide variety of solubilities (in units of g/m³ or the equivalent mg/L) have been reported. Experimental data have the method of determination indicated. In other compilations of data the reported value has merely been quoted from another secondary source. In some cases the value has been calculated. The abbreviations are generally self-explanatory and usually include two entries, the method of equilibration followed by the method of determination. From these values a single value is selected for inclusion in the summary data table. In the case of 1,1,1-trichloro-2,2-bis-(4-chlorophenyl)ethane (p,p'-DDT), the insecticide was banned in mny countries in the 1970s but has persisted in the environment with its parent form and metabolites p,p'-DDD, p,p'-DDE. The selected solubility is 0.0055 g/m³ at 25°C. From an examination of the data it is judged that the true value almost certainly lies between 0.002 and 0.015 g/m³.

The vapor pressure data are treated similarly with a value of 2×10^{-5} Pa being selected. The true value is judged to lie between 1.0×10^{-5} and 5×10^{-5} Pa. Vapor pressures are, of course very temperature-dependent.

The Henry's law constant data are reported experimentally to be 1.31 and 1.20 Pa·m³/mol determined by fog chamber and wetted-wall column methods respectively (Fendinger et al. 1989). These values agree well with the value of 1.29 Pa·m³/mol estimated from the ratio of selected vapor pressure and solubility. The range of calculated values is from 1.31 to 7.3 Pa·m³/mol.

The octanol-water partition coefficient data are a combination of experimental and calculated values. A range of log K_{OW} value of about 3.98 to 6.91 is reported, and a value of 6.19 is selected. The authors believe that the actual value lies between 5.90 to 6.40.

Several (log) bioconcentration factors are listed, most lying in the range from 2.0 to 5.0. For a log K_{OW} of 6.19, the expected BCF is about 5% of K_{OW}, i.e., log BCF of 4.89.

The (log) organic-carbon partition coefficients listed range mostly from 5.2 to 6.0. It is expected that K_{OC} usually lies in the range of 20 to 80% of K_{OW}, i.e., K_{OC} will range from 200,000 to 800,000. There is clustering of experimental values around 250,000, i.e., log K_{OC} is 5.4. Organic matter partition coefficients are also reported. Since organic carbon accounts for some 50 to 60% of the content of organic matter, K_{OM} the organic matter-water partition coefficient, is expected to be about half K_{OC}.

The reader is advised to consult the original reference when using these values of BCF, K_{OC} and K_{OM} to ensure that conditions are as close as possible to those of specific interest.

The "Half-life in the Environment" data reflect observations of the rate of disappearance of the chemical from a medium, without necessarily identifying the cause of mechanism of loss. For example, loss from water may be a combination of evaporation, biodegradation and photolysis. Clearly these times are highly variable and depend on factors such as temperature, meteorology and the nature of the media. Again, the reader is urged to consult the original reference.

The "Environmental Fate Rate Constants" refer to specific degradation processes rather than media. As far as possible the original numerical quantities are given and thus there is a variety of time units with some expressions being rate constants and others half-lives.

The conversion is
$$k = 0.693/t_{1/2}$$

where k is the first-order rate constant (h^{-1}) and $t_{1/2}$ is the half-life (h).

From these data a set of medium-specific degradation reaction half-lives was selected for use in Level II and III calculations. Emphasis was based on the fastest and the most plausible degradation process for each of the environmental compartments considered. Instead of assuming an equal half-life for both the water and soil compartment, as suggested by Howard et al. (1991), a slower active class (in the reactivity table described earlier) was assigned for soil and sediment compared to that of the water compartment. This is in part because the major degradation processes are often photolysis (or photooxidation) and biodegradation. There is an element of judgement in this selection and it may be desirable to explore the implications of selecting other values. The selected values of the insecticides are given in Table 3.3 at the end of Chapter 3.

In summary, the physical-chemical and environmental fate data listed result in the selection of values of solubility, vapor pressure, K_{OW} and reaction half-lives which are used in the evaluative environmental calculations.

1.5.4. Evaluative Calculations

The illustrative evaluative environmental calculations discussed here and presented later for a number of chemicals have been modified from those in Volume I of this series to give more data in a more spacious layout. Level I and II diagrams are assigned to separate pages and the physical-chemical properties are included in the Level I diagram. The Level III diagram is identical to that in Volume I, but additional pie-chart diagrams are included to show how the mass of chemical is distributed between compartments, and how loss processes are distributed as a function of the compartment of discharge. For dissociating compounds the environmental pH is specified and the calculation of Z values has been modified to include ionic species as discussed in Section 1.4.2. For water-miscible compounds, a "pseudo-solubility" is calculated from the measured Henry's law constant. Generally, if discharge is into a compartment such as water, most chemical will be found in that compartment and will react there, but a quantity does migrate to other compartments and is lost from these media. Three pie-charts corresponding to discharges of 1000 kg/h into air, water and soil are included. The percentage emission in each medium is different from the previous Volumes, i.e., 5, 25 and 70% will be discharged to air, water and soil respectively (instead of 60, 30 and 10% to air, water and soil compartments). A fourth pie-chart with discharges to all three compartments is also given. This latter chart is, in principle, the linear sum of the first three, but since the overall residence times differ, the diagram with the longest residence time, and greatest resident mass, tends to dominate.

Figures 1.1 and 1.2 show the mass distributions obtained in Level I calculations and the removal distribution from Level II fugacity calculation of p,p'-DDT in the generic environment. Figure 1.3 shows the Level III fugacity calculation. Both mass and removal distributions are shown in Figure 1.4 for the four scenarios of discharges into air, water, soil, and mixed compartments.

Level I

The Level I calculations illustrated in Figure 1.1 suggest that if 100,000 kg (100 tons) of DDT are introduced into the 100,000 km² environment, most DDT will tend to be associated with soil. Only 0.019% will partition into air at a concentration of 1.85×10^{-10} g/m³ or about 185 pg/m³. The water will contain 0.07% at a low concentration of 0.36 μg/m³ or equivalently 0.36 ng/L. Soil will contain 97.7% of the DDT at 4.5×10^{-3} μg/g and sediments about 2.17% at 9.0×10^{-3} μg/g. The fugacity is calculated to be 1.3×10^{-9} Pa. The dimensionless soil-water and sediment-water partition coefficients or ratios of Z values are 30,500 and 61,000 as a result of a K_{OC} of about 635,000 and a few percent organic carbon in these media. There is evidence of bioconcentration with a rather high fish concentration of 2.76×10^{-2} μg/g corresponding to a BCF of 77,400. The pie-chart in Figure 1.1 (which is the same as the Level I diagram for DDT in Chapter 3) clearly shows that soil is the primary medium of accumulation. Note that only four media (air, water, soil and bottom sediment) are depicted in the pie-chart, therefore the sum of the percent distribution figures is slightly less than 100%. The air-water partition coefficient is very low at about 5.2×10^{-4}. Partitioning to air is always slight, but as is discussed later the small amount represents a real potential for atmospheric transport.

Level II

The Level II calculation illustrated in Figure 1.2 includes the reaction half-lives of 170 h in air, 5500 h in water, 17,000 h (about 2 years) in soil and 55,000 h in sediment. No reaction is included for suspended sediment or fish. The steady-state input of 1000 kg/h results in an overall fugacity of 2.95×10^{-7} Pa which is about 230 times the Level I value. The concentrations and amounts in each medium are thus about 230 times the Level I values. The relative mass distribution is identical to Level I. The primary loss mechanism is reaction in soil which accounts for 906 kg/h or 90% of the input. Most of the remainder is lost by reaction and advection in air. The water and sediment loss processes are less important largely because little of the DDT is present in these media. The overall residence time is 22,800 h, i.e., 2.6 years; thus, there is an inventory of DDT in the system of 22,800 × 1000 or 23×10^6 kg. The pie-chart in Figure 1.2 shows the dominance of soil reaction as a removal process.

The primary loss mechanism of soil reaction has a D value of 8.7×10^9, thus, for any other process to compete with this would require a D value of at least 10^9 mol/Pa·h. The next largest D values are 1.64×10^8 and 4.03×10^8 for reaction and advection in air, which are over a factor of 10 smaller. It is noteworthy that 4% of the DDT is lost by advection in air, thus confirming the potential for long range atmospheric transport.

Chemical name: p,p'-DDT

Level I calculation: (six-compartment model)

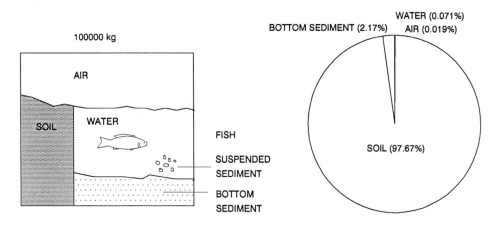

Distribution of mass

Physical-chemical properties:
molecular wt., g/mol	354.50
melting point, °C	109
solubility, g/m³	0.0055
vapor pressure, Pa	2.0E−05
log K_{OW}	6.19
fugacity ratio, F	0.1476
dissoc. const. pK_a	

Partition coefficients:
H, Pa·m³/mol	1.2891
K_{AW}	5.20E−04
K_{OC}	635015
BCF	77441
K_{SW}	30481
$K_{SD/W}$	60961
$K_{SSD/W}$	190504
$K_{AR/W}$	4.43E+10

COMPARTMENT	Z	CONCENTRATION			AMOUNT	AMOUNT
	mol/m³·Pa	mol/m³	mg/L, or g/m³	µg/g	kg	%
AIR	4.03E−04	5.22E−13	1.85E−10	1.56E−07	18.515	0.0185
WATER	7.76E−01	1.00E−09	3.56E−07	3.56E−07	71.205	0.0712
SOIL	2.36E+04	3.06E−05	1.09E−02	4.52E−03	97667	97.667
BIOTA (FISH)	6.01E+04	7.78E−05	2.76E−02	2.76E−02	5.514	5.51E−03
SUSPENDED SEDIMENT	1.48E+05	1.91E−04	6.78E−02	4.52E−02	67.82	6.78E−02
BOTTOM SEDIMENT	4.73E+04	6.12E−05	2.17E−02	9.04E−03	2170.4	2.170
Total					100000	100

Fugacity, f = 2.816E−12 Pa

Figure 1.1 Level I fugacity calculation of DDT in a generic environment.

Chemical name: p,p'-DDT
Level II calculation: (six-compartment model)

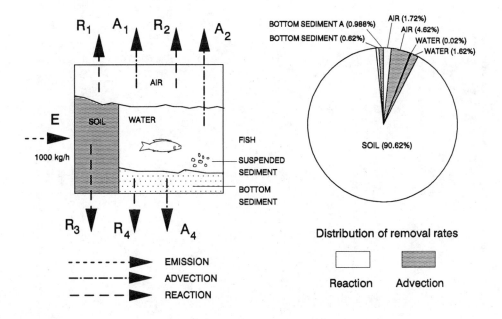

COMPARTMENT		D VALUES		CONC'N	LOSS	LOSS	REMOVAL
	Half-Life	Reaction	Advection		Reaction	Advection	%
	h	mol/Pa·h	mol/Pa·h	mol/m^3	kg/h	kg/h	
AIR	170	1.64E+08	4.03E+08	1.19E−10	17.181	42.147	5.933
WATER	5500	1.95E+07	1.55E+08	2.29E−07	2.042	16.209	1.825
SOIL	17000	8.67E+09		6.97E−03	906.31		90.631
BIOTA (FISH)				1.77E−02			
SUSPENDED SEDIMENT				4.36E−02			
BOTTOM SEDIMENT	55000	5.96E+07	9.46E+07	1.39E−02	6.225	9.881	1.611
Total		8.86E+09	5.59E+08		931.76	68.24	100
Reaction + Advection			9.42E+09			1000	
Fugacity, f = 2.947E−07 Pa				Overall residence time = 22764 h			
Total amount = 22763998 kg				Reaction time = 24431 h			
				Advection time = 333601 h			

Figure 1.2   Level II fugacity calculation of DDT in a generic environment.

Fugacity Level III calculations: (four-compartment model)
Chemical name: p,p'-DDT

Phase Properties and Rates:

Compartment	Bulk Z mol/m3 Pa	Half-life h	D Values Reaction mol/Pa h	Advection mol/Pa h
Air (1)	7.608E-04	170	3.10E+08	7.61E+08
Water (2)	1.575E+00	5500	3.97E+07	3.15E+08
Soil (3)	1.182E+04	17000	8.68E+09	
Sediment (4)	9.459E+03	55000	5.96E+07	9.46E+07

	E(1)=1000	E(2)=1000	E(3)=1000	E(1,2,3)
Overall residence time =	12047.48	17530.83	24509.07	22141.43 h
Reaction time =	19874.56	63896.78	24561.38	27766.74 h
Advection time =	30591.01	24159.18	11507952	109290.95 h

Legend:
- - - - EMISSION (E)
-·-·- REACTION (R)
····· ADVECTION (A)
——— TRANSFER D VALUE mol/Pa h

Phase Properties, Compositions, Transport and Transformation Rates:

Emission, kg/h			Fugacity, Pa				Concentration, g/m3			
E(1)	E(2)	E(3)	f(1)	f(2)	f(3)	f(4)	C(1)	C(2)	C(3)	C(4)
1000	0	0	1.301E-06	2.035E-07	1.455E-07	6.031E-07	3.509E-07	1.136E-04	6.099E-01	2.022E+00
0	1000	0	3.003E-08	3.400E-06	3.358E-09	1.008E-05	8.098E-09	1.898E-03	1.408E-02	3.379E+01
0	0	1000	2.968E-10	9.712E-09	3.242E-07	2.879E-08	8.004E-11	5.422E-06	1.359E+00	9.654E-02
50	250	700	7.277E-08	8.670E-07	2.351E-07	2.570E-06	1.962E-08	4.840E-04	9.852E-01	8.617E-01

Emission, kg/h			Loss, Reaction, kg/h				Loss, Advection, kg/h			
E(1)	E(2)	E(3)	R(1)	R(2)	R(3)	R(4)	A(1)	A(2)	A(3)	A(4)
1000	0	0	1.430E+02	2.862E+00	4.48E+02	1.274E+01	3.509E+02	2.272E+01	6.099E-01	2.022E+00
0	1000	0	3.301E+00	4.783E+01	1.03E+01	2.129E+02	8.098E+00	3.796E+02	1.408E-02	3.379E+01
0	0	1000	3.263E-02	1.366E-01	9.97E+02	6.082E-01	8.004E-02	1.084E+00	1.359E+00	9.654E-02
50	250	700	8.000E+00	1.220E+01	7.23E+02	5.429E+01	1.962E+01	9.679E+01	9.852E-01	8.617E-01

Amounts, kg				Total Amount, kg
m(1)	m(2)	m(3)	m(4)	
3.509E+04	2.272E+04	1.098E+07	1.011E+06	1.205E+07
8.098E+02	3.796E+05	2.534E+05	1.690E+07	1.753E+07
8.004E+00	1.084E+03	2.446E+07	4.827E+04	2.451E+07
1.962E+03	9.679E+04	1.773E+07	4.309E+06	2.214E+07

Intermedia Rate of Transport, kg/h

T12	T21	T13	T31	T32	T24	T42
air-water	water-air	air-soil	soil-air	soil-water	water-sed	sed-water
5.865E+01	1.383E+00	4.489E+02	7.291E+02	1.278E+00	5.335E+07	2.039E+01
1.353E+00	2.311E+01	1.036E+01	1.683E-03	2.949E-02	8.915E+02	3.407E+02
1.338E-02	6.602E-02	1.024E-01	1.624E-01	2.847E+00	2.547E+00	9.732E-01
3.280E+00	5.893E+00	2.511E+01	1.178E-01	2.064E+00	2.273E+02	8.687E+01

Figure 1.3 Fugacity Level III calculation for DDT in a generic environment.

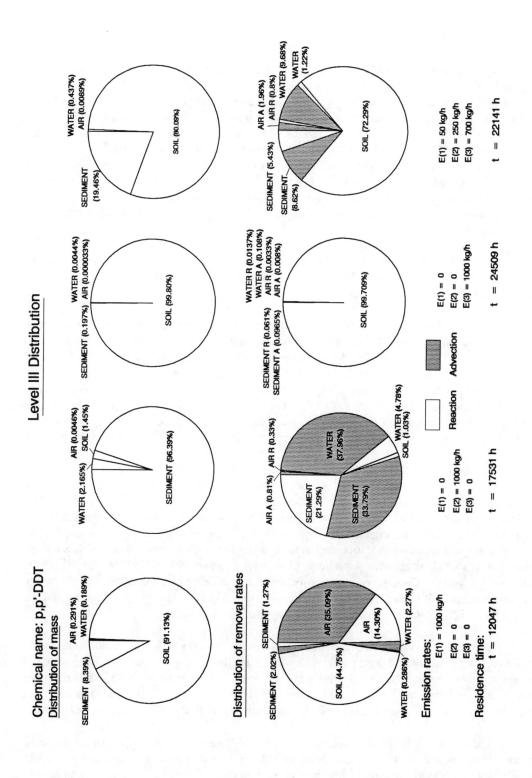

Figure 1.4 Fugacity Level III distributions of DDT for four emission scenarios in a generic environment.

Level III

The Level III diagrams (Figure 1.3 and Figure 1.4) are regarded as the most realistic depictions of chemical fate.

This calculation includes an estimation of intermedia transport. Examination of the magnitude of the intermedia D values given in the fate diagrams (Figure 1.3 and Figure 1.4, which are the same as the Level III calculations and distribution graphs for DDT in Chapter 3) suggest that water-sediment, air-soil and air-water transport are most important, with soil-air exchange being slower. The magnitude of these larger intermedia transport D values (approximately 1×10^8 to 9×10^8 mol/Pa·h) compared to the soil reaction value of 87×10^8 and the air reaction and advection values of some 3.1×10^8 and 7.6×10^8 suggests that reaction in soil will remain dominant and advection or reaction from air will be comparable in water to transport from water to sediment and from air to soil and water. This chemical tends to be fairly immobile in terms of intermedia transport when it is introduced to soil, but it is fairly mobile when introduced to air or water.

The bulk Z values are similar for air and water to the values for the "pure" phases in Levels I and II, but they are lower for soil and sediment because of the "dilution" of the solid soil and sediment phases with air or water.

These tabulated data are given in illustrated or pictorial form in Figure 1.3. The first row of values at the foot of Figure 1.3 describes the condition if 1000 kg/h is emitted into the air. The result is 35,000 kg in air, 23,000 kg in water, 11,000,000 kg in soil and 1,000,000 kg in sediment. It can be concluded that DDT discharged to the atmosphere has a fairly high potential to enter other media, especially soil. The rate of transfer from air to water (T_{12}) is about 59 kg/h and that from air to soil (T_{13}) 449 kg/h. The reason for this is the value of the mass transfer coefficients which control this transport process. The overall residence time is 12,000 hours or 1.4 years.

If 1000 kg/h of DDT is discharged to water, as in the second row, there is, as expected a much higher concentration in water. There is reaction of 48 kg/h in water, advective outflow of 380 kg/h and transfer to air (T_{21}) of 23 kg/h with substantial loss of 892 kg/h to sediment. The amount in the water is 380,900 kg, thus the residence time in the water is approximately 380 hours and the overall environmental residence time is a longer 17,500 hours or 2.5 years. The key processes are thus sedimentation and advective outflow. The evaporation half-life can be calculated as (0.693 × mass in water)/rate of transfer, i.e., (0.693 × 380,000)/23 = 11,400 hours or 1.3 years. Clearly, competition between advection and sedimentation in the water determines the overall fate. Of the DDT discharged, 96.4% is now found in the sediment and the concentration is fairly high, namely 34 g/m³ or approximately 23 μg/g.

The third row shows the fate if DDT is discharged into soil. The amount in soil is 24,500,000 kg, with only 8 kg in air. The overall residence time is 24,500 hours, which is largely controlled by the reaction rate in soil of 997 kg/h. There is no advection, thus the other loss mechanism is transfer to air (T_{31}) at a rate of 0.16 kg/h, 2.8 kg/h to water by run-off. The soil concentration of 1.36 g/m³ is controlled almost entirely by the rate at which the DDT reacts.

The net result is that DDT behaves entirely differently when discharged to the three media. If discharged to air, it is rapidly deposited to soil and it also advects and reacts fairly rapidly with a residence time of 12,000 h or about 500 days, controlled by the accumulation in soil with substantial transport to soil or air. If discharged to water, it is primarily subject to deposition to sediment. If discharged to soil, it mostly remains there and reacts with an overall residence time of about 24,500 hours or 2.8 years. It is clearly a persistent compound regardless of how it is discharged.

The final scenario is a combination of discharges, 50 kg/h to air, 250 kg/h to water, and 700 kg/h to soil (which are different proportions from the previous conditions in Volumes I, II, and III of 600 kg/h to air, 200 kg/h to water and 100 kg/h to soil but the same as Volume VI). The concentrations, amounts and transport and transformation rates are merely linearly combined versions of the three initial scenarios. For example, the rate of reaction in air is now 8.0 kg/h. This is 0.05 of the first (air emission) rate of 143 kg/h, i.e., 7.15 kg/h, plus 0.25 of the second (water emission) rate of 3.3 kg/h, i.e., 0.83 kg/h and 0.7 of the third (soil emission) rate of 0.032 kg/h, i.e., 0.022 kg/h yielding a total of (7.15 + 0.83 + 0.022) or 8.0 kg/h. It is also apparent that the amount in the air of 1962 kg causing a concentration of 1.96×10^{-8} g/m^3 or 19.6 ng/m^3 is attributable to emissions to air (0.05 × 351 or 17.5 ng/m^3), emissions to water (0.25 × 8.0 or 2.0 ng/m^3) and emissions to soil (0.7 × 0.08 or 0.006 ng/m^3). The concentration in water of 4.84×10^{-4} g/m^3 or 484 ng/L is largely attributable to the discharges to water which alone cause (0.25 × 1.9×10^{-3} g/m^3) or 4.75×10^{-4} g/m^3 or 475 µg/m^3 or 475 ng/L. Although more is emitted to soil it contributes only about 3.8 ng/L to the water with air emissions accounting for about 5.7 ng/L. Similarly, the prevailing soil concentration is controlled by the rate of discharge to the soil.

In this multimedia discharge scenario the overall residence time is 22,100 hours, which can be viewed as 5% of the air emission residence time of 12,000 h, 25% of the water emission residence time of 17,500 h and 70% of the soil emission residence time of 24,500 h. The overall amount in the environment of 22.1×10^6 kg is thus largely controlled by the discharges to soil and water.

Finally, it is interesting to note that the fugacities in this final case (in units of nPa) for the four media are 73 (air), 867 (water), 235 (soil) and 2570 (sediment). The water, soil and sediment media are fairly close to equilibrium, i.e., within a factor of about 5 of the average value of 1200 nPa, but the air is at a lower fugacity because of the relatively short residence time in the air.

Examining these three behavior profiles, and their combination in the fourth, illustrates and explains the environmental fate characteristics of this and other chemicals. Important intermedia transport processes and levels in various media arise from discharges into other media became clear. It is believed that the broad fate characteristics of environmental fate as described in the generic environment are applicable to other environments.

1.6 REFERENCES

Abernethy, S., Mackay, D., McCarty, L. S. (1988) Volume fraction correlation for narcosis in aquatic organisms: The key role of partitioning. *Environ. Toxicol. Chem.* 7, 469-481.

Abramowitz, R., Yalkowsky, S. H. (1990) Estimation of aqueous solubility and melting point of PCB congeners. *Chemosphere* 21, 1221-1229.

Almgren, M., Grieser, F., Powell, J. R., Thomas, J. K. (1979) A correlation between the solubility of aromatic hydrocarbons in water and micellar solutions, with their normal boiling points. *J. Chem. Eng. Data* 24, 285-287.

Al-Sahhaf, T. A. (1989) Prediction of the solubility of hydrocarbons in water using UNIFAC. *J. Environ. Sci. Health* A24, 49-56.

Altshuller, A. P. (1980) Lifetimes of organic molecules in the troposphere and lower stratosphere. *Adv. Environ. Sci. Technol.* 10, 181-219.

Altshuller, A. P. (1991) Chemical reactions and transport of alkanes and their products in the troposphere. *J. Atmos. Chem.* 12, 19-61.

Ambrose, D. (1981) Reference value of vapor pressure. The vapor pressures of benzene and hexafluorobenzene. *J. Chem. Thermodyn.* 13, 1161-1167.

Ambrose, D., Lawrenson, L. J., Sprake, C. H. S. (1975) The vapour pressure of naphthalene. *J. Chem. Thermodyn.* 7, 1173-1176.

Amidon, G. L., Yalkowsky, S. H., Anik, S. T., Leung, S. (1975) Solubility of nonelectrolytes in polar solvents. V. Estimation of the solubility of aliphatic monofunctional compounds in water using a molecular surface area approach. *J. Phys. Chem.* 9, 2239-2245.

Amidon, G. L., Anik, S. T. (1981) Application of the surface area approach to the correlation and estimation of aqueous solubility and vapor pressure. Alkyl aromatic hydrocarbons. *J. Chem. Eng. Data* 26, 28-33.

Amidon, G. L., Williams, N. A. (1982) A solubility equation for non-electrolytes in water. *Intl. J. Pharm.* 11, 249-156.

Anderson, E., Veith, G. D., Weininger, D. (1987) *SMILES: A Line Notation and Computerized Interpreter for Chemical Structures.* EPA Environmental Research Brief, U.S. EPA, EPA/600/M-87/021.

Andren, A. W., Doucette, W. J., Dickhut, R. M. (1987) Methods for estimating solubilities of hydrophobic organic compounds: Environmental modeling efforts. In: *Sources and Fates of Aquatic Pollutants.* Hites, R. A., Eisenreich, S. J., Eds., pp. 3-26, Advances in Chemistry Series 216, American Chemical Society, Washington, D.C.

Andrews, L. J., Keefer, R. M. (1950a) Cation complexes of compounds containing carbon-carbon double bonds. IV. The argentation of aromatic hydrocarbons. *J. Am. Chem. Soc.* 72, 3644-3647.

Andrews, L. J., Keefer, R. M. (1950b) Cation complexes of compounds containing carbon-carbon double bonds. VII. Further studies on the argentation of substituted benzenes. *J. Am. Chem. Soc.* 72, 5034-5037.

Arbuckle, W. B. (1983) Estimating activity coefficients for use in calculating environmental parameters. *Environ. Sci. Technol.* 17, 537-542.

Arbuckle, W. B. (1986) Using UNIFAC to calculate aqueous solubilities. *Environ. Sci. Technol.* 20, 1060-1064.

Arcand, Y., Hawari, J., Guiot, S. R. (1995) Solubility of pentachlorophenol in aqueous solutions: the pH effect. *Water Res.* 29, 131-136.

Ashworth, R. A., Howe, G. B., Mullins, M. E., Roger, T. N. (1988) Air-water partitioning coefficients of organics in dilute aqueous solutions. *J. Hazard. Materials* 18, 25-36.

Atkinson, R. (1985) Kinetics and mechanisms of the gas phase reaction of hydroxyl radicals with organic compounds under atmospheric conditions. *Chem. Rev.* 85, 69-201.

Atkinson, R. (1987) A structure-activity relationship for the estimation of the rate constants for the gas phase reactions of OH radicals with organic compounds. *Int. J. Chem. Kinetics* 19, 790-828.

Atkinson, R. (1988) Estimation of gas-phase hydroxyl radical rate constants for organic chemicals. *Environ. Toxicol. Chem.* 7, 435-442.

Atkinson, R. (1989) Kinetics and mechanisms of the gas-phase reactions of the hydroxyl radical with organic compounds. *J. Phys. Chem. Ref. Data Monograph No. 1.* 1-246.

Atkinson, R. (1990) Gas-phase tropospheric chemistry of organic compounds, a review. *Atmos. Environ.* 24A, 1-41.

Atkinson, R. (1991) Kinetics and mechanisms of the gas-phase reactions of the NO_3 radicals with organic compounds. *J. Phys. Chem. Data* 20, 450-507.

Atkinson, R., Carter, W. L. (1984) Kinetics and mechanisms of the gas-phase reactions of ozone with organic compounds under atmospheric conditions. *Chem. Rev.* 84, 437-470.

Augustijn-Becker, P. W. M., Hornsby, A. G., Wauchope, R. D. (1994) The SCS/ARS/CES pesticide properties database for environmental decision making. II. Additional compounds. *Rev. Environ. Contam. Toxicol.* 137, 1-82.

Bahnick, D. A., Doucette, W. J. (1988) Use of molecular connectivity indices to estimate soil sorption coefficients for organic chemicals. *Chemosphere* 17, 1703-1715.

Balson, E. W. (1947) Studies in vapour pressure measurement. Part III. An effusion manometer sensitive to 5×10^{-6} millimetres of mercury: vapour pressure of D.D.T. and other slightly volatile substances. *Trans. Farad. Soc.* 43, 54-60.

Banerjee, S. (1985) Calculation of water solubility of organic compounds with UNIFAC-derived parameters. *Environ. Sci. Technol.* 19, 369-370.

Banerjee, S., Howard, P. H. (1988) Improved estimation of solubility and partitioning through correction of UNIFAC-derived activity coefficients. *Environ. Sci. Technol.* 22, 839-841.

Banerjee, S., Howard, P. H., Lande, S. S. (1990) General structure-vapor pressure relationships for organics. *Chemosphere* 21, 1173-1180.

Banerjee, S., Yalkowsky, S. H., Valvani, S. C. (1980) Water solubility and octanol/water partition coefficients of organics. Limitations of the solubility-partition coefficient correlation. *Environ. Sci. Technol.* 14, 1227-1229.

Betterton, E. A., Hoffmann, M. R. (1988) Henry's law constants of some enviromentally important aldehydes. *Environ. Sci. Technol.* 22, 1415-1418.

Bidleman, T. F. (1984) Estimation of vapor pressures for nonpolar organic compounds by capillary gas chromatography. *Anal. Chem.* 56, 2490-2496.

Bohon, R. L., Claussen, W. F. (1951) The solubility of aromatic hydrocarbons in water. *J. Am. Chem. Soc.* 73, 1571-1576.

Bondi, A. (1964) van der Waals volumes and radii. *J. Phys. Chem.* 68, 441-451.

Booth, H. S., Everson, H. E. (1948) Hydrotropic solubilities: solubilities in 40 percent sodium xylenesulfonate. *Ind. Eng. Chem.* 40, 1491-1493.

Boublik, T., Fried, V., Hala, E. (1973) *The Vapor Pressure of Pure Substances*, Elsevier, Amsterdam, The Netherlands.

Boublik, T., Fried, V., Hala, E. (1984) *The Vapor Pressure of Pure Substances*, 2nd revised Edition, Elsevier, Amsterdam, The Netherlands.

Bowman, B. T., Sans, W. W. (1983) Determination of octanol-water partitioning coefficient (K_{ow}) of 61 organophosphorus and carbamate insecticides and their relationship to respective water solubility (S) values. *J. Environ. Sci. Health* B18, 667-683.

Bradley, R. S., Cleasby, T. G. (1953) The vapour pressure and lattice energy of some aromatic ring compounds. *J. Chem. Soc.* 1953, 1690-1692.

Briggs, G. G. (1981) Theoretical and experimental relationships between soil adsorption, octanol-water partition coefficients, water solubilities, bioconcentration factors, and the parachor. *J. Agric. Food Chem.* 29, 1050-1059.

Brodsky, J., Ballschmiter, K. (1988) Reversed phase liquid chromatography of PCBs as a basis for the calculation of water solubility and log K_{OW} for polychlorobiphenyls. *Frensenius Z. Anal. Chem.* 331, 295-301.

Bruggeman, W. A., van der Steen, J., Hutzinger, O. (1982) Reversed-phase thin-layer chromatography of polynuclear aromatic hydrocarbons and chlorinated biphenyls. Relationship with hydrophobicity as measured by aqueous solubility and octanol-water partition coefficient. *J. Chromatogr.* 238, 335-346.

Budavari, S., Editor (1989) *The Merck Index. An Encyclopedia of Chemicals, Drugs and Biologicals.* 11th Edition, Merck & Co. Inc., Rahway, New Jersey.

Burkhard, L.P. (1984) Physical-Chemical Properties of the Polychlorinated Biphenyls: Measurement, Estimation, and Application to Environmental Systems. Ph.D. Thesis, University of Wisconsin-Madison, Wisconsin.

Burkhard, L. P., Andren, A. W., Armstrong, D. E. (1985a) Estimation of vapor pressures for polychlorinated biphenyls: A comparison of eleven predictive methods. *Environ. Sci. Technol.* 19, 500-507.

Burkhard, L. P., Armstrong, D. E., Andren, A. W. (1985b) Henry's law constants for polychlorinated biphenyls. *Environ. Sci. Technol.* 590-595.

Burkhard, L. P., Kuehl, D. W., Veith G. D. (1985c) Evaluation of reversed phase liquid chromatograph/mass spectrometry for estimation of n-octanol/water partition coefficients of organic chemicals. *Chemosphere* 14, 1551-1560.

Burnett, M. G. (1963) Determination of partition coefficients in infinite dilution by the gas chromatographic analysis of the vapor above dilute solutions. *Anal. Chem.* 35, 1567-1570.

Butler, J. A. V., Ramchandani, C. N., Thomson, D. W. (1935) The solubility of non-electrolytes. Part I. The free energy of hydration of some aliphatic alcohols. *J. Chem. Soc.* 280-285.

Buttery, R. B., Ling, L. C., Guadagni, D. G. (1969) Volatilities of aldehydes, ketones, and esters in dilute water solution. *J. Agric. Food Chem.* 17, 385-389.

Buxton, G. V., Greenstock, G. L., Helman, W. P., Ross, A. B. (1988) Critical review of rate constants for reactions of hydrated electrons, hydrogen atoms and hydroxyl radicals in aqueous solutions. *J. Phys. Chem. Ref. Data* 17, 513-886.

Chiou, C. T. (1981) Partition coefficient and water solubility in environmental chemistry. In: *Hazard Assessment of Chemicals Current Developments.* Vol. 1, pp. 117-153. Academic Press, New York.

Chiou, C. T. (1985) Partition coefficients of organic compounds in lipid-water systems and correlations with fish bioconcentration factors. *Environ. Sci. Technol.* 19, 57-62.

Chiou, C. T., Freed, V. H., Schmedding, D. W. (1977) Partition coefficient and bioaccumulation of selected organic chemicals. *Environ. Sci. Technol.* 11, 475-478.

Chiou, C. T., Schmedding, D. W., Manes, M. (1982) Partitioning of organic compounds in octanol-water system. *Environ. Sci. Technol.* 16, 4-10.

Choi, J., Amoine, S. (1974) Adsorption of pentachlorophenol by soils. *Soil Sci. Plant Nutr.* 20, 135-144.

Coates, M., Connell, D. W., Barron, D. M. (1985) Aqueous solubility and octan-1-ol to water partition coefficients of aliphatic hydrocarbons. *Environ. Sci. Technol.* 19, 628-632.

Davies, R. P., Dobbs, A. J. (1984) The prediction of bioconcentration in fish. *Water Res.* 18, 1253-1262.

Davis, W. W., Krahl, M. E., Clowes, G. H. (1942) Solubility of carcinogenic and related hydrocarbons in water *J. Am. Chem. Soc.* 64, 108-110.

Davis, W. W., Parke, Jr., T. V. (1942) A nephelometric method for determination of solubilities of extremely low order. *J. Am. Chem. Soc.* 64, 101-107.

Dean, J. D., Ed. (1979) *Lange's Handbook of Chemistry.* 12th Edition, McGraw-Hill, Inc., New York, New York.

Dean, J. D., Ed. (1985) *Lange's Handbook of Chemistry.* 13th Edition, McGraw-Hill, Inc., New York, New York.

Dearden, J. C. (1990) Physico-chemical descriptors. In: *Practical Applications of Quantitative Structure-Activity Relationships (QSAR) in Environmental Chemistry and Toxicology.* Karcher, W. and Devillers, J., Eds., pp. 25-60. Kluwer Academic Publisher, Dordrecht, The Netherlands.

De Bruijn, J., Busser, G., Seinen, W., Hermens, J. (1989) Determination of octanol/water partition coefficient for hydrophobic organic chemicals with the "slow-stirring" method. *Environ. Toxicol. Chem.* 8, 499-512.

De Bruijn, J., Hermens, J. (1990) Relationships between octanol/water partition coefficients and total molecular surface area and total molecular volume of hydrophobic organic chemicals. *Quant. Struct.-Act. Relat.* 9, 11-21.

De Kruif, C. G. (1980) Enthalpies of sublimation and vapour pressures of 11 polycyclic hydrocarbons. *J. Chem. Thermodyn.* 12, 243-248.

Dobbs, A. J., Grant, C. (1980) Pesticide volatilization rate: a new measurement of the vapor pressure of pentachlorophenol at room temperature. *Pestic. Sci.* 11, 29-32.

Dobbs, A. J., Cull, M. R. (1982) Volatilization of chemical relative loss rates and the estimation of vapor pressures. *Environ. Pollut. Ser.B.,* 3, 289-298.

Doucette, W. J., Andren, A. W. (1987) Correlation of octanol/water partition coefficients and total molecular surface area for highly hydrophobic aromatic compounds. *Environ. Sci. Technol.* 21, 521-524.

Doucette, W. J., Andren, A. W. (1988a) Aqueous solubility of selected biphenyl, furan, and dioxin congeners. *Chemosphere* 17, 243-252.

Doucette, W. J., Andren, A. W. (1988b) Estimation of octanol/water partition coefficients: Evaluation of six methods for highly hydrophobic aromatic hydrocarbons. *Chemosphere* 17, 345-359.

Dreisbach, R. R. (1955) *Physical Properties of Chemical Compounds.* No. 15 of the Adv. in Chemistry Series, American Chemical Society, Washington, DC.

Dreisbach, R. R. (1959) *Physical Properties of Chemical Compounds-II.* No. 22, Adv. in Chemistry Series, American Chemical Society, Washington, DC.

Dreisbach, R. R. (1961) *Physical Properties of Chemical Compounds-III.* No. 29, Adv. in Chemistry Series, American Chemical Society, Washington, DC.

Dunnivant, F. M., Coate, J. T., Elzerman, A. W. (1988) Experimentally determined Henry's law constants for 17 polychlorobiphenyl congeners. *Environ. Sci. Technol.* 22, 448-453.

Dunnivant, F. M., Elzerman, A. W., Jurs, P. C., Hansen, M. N. (1992) Quantitative structure-property relationships for aqueous solubilities and Henry's law constants of polychlorinated biphenyls. *Environ. Sci. Technol.* 26, 1567-1573.

Ellgehausen, H., D'Hondt, C., Fuerer, R. (1981) Reversed-phase chromatography as a general method for determining octan-1-ol/water partition coefficients. *Pestic. Sci.* 12, 219-227.

Ellington, J. J. (1989) *Hydrolysis Rate Constants for Enhancing Property-Reactivity Relationships.* U.S. EPA, EPA/600/3-89/063, Athens, Georgia.

Ellington, J. J., Stancil, Jr., F. E., Payne, W. D. (1987a) *Measurements of Hydrolysis Rate Constant for Evaluation of Hazardous Land Disposal: Volume I. Data on 32 Chemicals* U.S. EPA, EPA/600/3-86/043, Athens, Georgia.

Ellington, J. J., Stancil, Jr., F. E., Payne, W. D., Trusty, C. D. (1987b) *Measurements of Hydrolysis Rate Constant for Evaluation of Hazardous Land Disposal: Volume II. Data on 54 Chemicals*. U.S. EPA, EPA/600/3-88/028, Athens, Georgia.

Ellington, J. J., Stancil, Jr., F. E., Payne, W. D., Trusty, C. D. (1988) *Interim Protocol for Measurement Hydrolysis Rate Constants in Aqueous Solutions*. USEPA, EPA/600/3-88/014, Athens, Georgia.

Fendinger, N. J., Glotfelty, D. E. (1988) A laboratory method for the experimental determination of air-water Henry's law constants for several pesticides. *Environ. Sci. Technol.* 22, 1289-1293.

Fendinger, N. J., Glotfelty, D. E. (1989) A comparison of two experimental techniques for determining air-water Henry's laws constants. *Environ. Sci. Technol.* 23, 1828-1831.

Fendinger, N. J., Glotfelty, D. E. (1990) Henry's law constants for selected pesticides, PAHs and PCBs. *Environ. Toxicol. Chem.* 9, 731-735.

Finizio, A., Di Guardo, A., Vighi, M. (1994) Improved RP-HPLC determination of K_{ow} for some chloroaromatic chemicals using molecular connectivity indices. *SAR & QSAR in Environ. Res.* 2, 249-260.

Foreman, W. T., Bidleman, T. F. (1985) Vapor pressure estimates of individual polychlorinated biphenyls and commercial fluids using gas chromatographic retention data. *J. Chromatogr.* 330, 203-216.

Fredenslund, A., Jones, R. L., Prausnitz, J. M. (1975) Group-contribution estimation of activity coefficients in nonideal liquid mixtures. *AIChE J.* 21, 1086-1099.

Fujita, T., Iwasa, J., Hansch, C. (1964) A new substituent constant, *"pi"* derived from partition coefficients. *J. Am. Chem. Soc.* 86, 5175-5180.

Garst, J.E. (1984) Accurate, wide-range, automated, high-performance chromatographic method for the estimation of octanol/water partition coefficients. II: Equilibrium in partition coefficient measurements, additivity of substituent constants, and correlation of biological data. *J. Pharm. Sci.* 73, 1623-1629.

Garten, C. T., Trabalka, J. R. (1983) Evaluation of models for predicting terrestrial food chain behavior of xenobiotics. *Environ. Sci. Technol.* 17, 590-595.

Gerstl, Z., Helling, C. S. (1984) Evaluation of molecular connectivity as a predictive method for the adsorption of pesticides by soils. *J. Environ. Sci. Health* B22, 55-69.

Gmehling, J., Rasmussen, P., Fredenslund, A. (1982) Vapor-liquid equilibria by UNIFAC group contribution. Revision and extension. 2 *Ind. Eng. Chem. Process Des. Dev.* 21, 118-127.

Gossett, R. (1987) Measurement of Henry's law constants for C_1 and C_2 chlorinated hydrocarbons. *Environ. Sci. Technol.* 21, 202-208.

Gross, P. M., Saylor, J. H. (1931) The solubilities of certain slightly soluble organic compounds in water. *J. Am. Chem. Soc.* 1931, 1744-1751.

Gückel, W., Rittig, R., Synnatschke, G. (1974) A method for determining the volatility of active ingredients used in plant protection. II. Application to formulated products. *Pestic. Sci.* 5, 393-400.

Gückel, W., Kästel, R., Lawerenz, J., Synnatschke, G. (1982) A method for determining the volatility of active ingredients used in plant protection. Part III: The temperature relationship between vapour pressure and evaporation rate. *Pestic. Sci.* 13, 161-168.

Hafkenscheid, T. L., Tomlinson, E. (1981) Estimation of aqueous solubilities of organic non-electrolytes using liquid chromatographic retention data. *J. Chromatogr.* 218, 409-425.

Hamaker, J. W., Kerlinger, H. O. (1969) Vapor pressures of pesticides. *Adv. Chem. Ser.* 86, 39-54.

Hamilton, D. J. (1980) Gas chromatographic measurement of volatility of herbicide esters. *J. Chromatogr.* 195, 75-83.

Hansch, C., Leo, A. (1979) *Substituent Constants for Correlation Analysis in Chemistry and Biology*. Wiley-Interscience, New York, New York.

Hansch, C., Quinlan, J. E., Lawrence, G. L. (1968) The linear-free energy relationship between partition coefficient and aqueous solubility of organic liquids. *J. Org. Chem*. 33, 347-350.

Hansen, H. K., Schiller, M., Gmehling, J. (1991) Vapor-liquid equlibria by UNIFAC group contribution. 5. Revision and extension. *Ind. Eng. Chem. Res*. 30, 2362-2356.

Hartley, G. S., Gram-Bryce, I. J. (1980) *Physical Principles of Pesticide Behavior, the Dynamics of Applied Pesticides in the Local Environments in Relation to Biological Reponse*. Vol. 2, Academic Press, London, United Kingdom.

Hartley, D., Kidd, H., Eds., (1987) *The Agrochemical Handbook*, 2nd ed., The Royal Society of Chemistry, The University of Nottingham, England.

Hawker, D. W. (1989) The relationship between octan-1-ol/water partition coefficient and aqueous solubility in terms of solvatochromic parameters. *Chemosphere* 19, 1586-1593.

Hawker, D. W. (1990a) Vapor pressures and Henry's law constants of polychlorinated biphenyls. *Environ. Sci. Technol*. 23, 1250-1253.

Hawker, D. W. (1990b) Description of fish bioconcentration factors in terms of solvatochromic parameters. *Chemosphere* 20, 267-477.

Hawker, D. W., Connell, D.W. (1988) Octanol-water partition coefficients of polychlorinated biphenyl congeners. *Environ. Sci. Technol*. 22, 382-387.

Hermann, R. B. (1971) Theory of hydrophobic bonding. II. The correlation of hydrocarbon solubility in water with solvent cavity surface area. *J. Phys. Chem*. 76, 2754-2758.

Hermens, J. L. M., Opperhuizen, A., Eds. (1991) *QSAR in Environmental Toxicology IV*. Elsevier, Amsterdam, The Netherlands. Also published in *Sci. Total Environ*. vol. 109/110.

Hinckley, D. A., Bidleman, T.F., Foreman, W.T. (1990) Determination of vapor pressures for nonpolar and semipolar organic compounds from gas chromatographic retention data. *J. Chem. Eng. Data* 35, 232-237.

Hine, J., Mookerjee, P. K. (1975) The intrinsic hydrophilic character of organic compounds. Correlations in terms of structural contributions. *J. Org. Chem*. 40, 292-298.

Hollifield, H. C. (1979) Rapid nephelometric estimate of water solubility of highly insoluble organic chemicals of environmental interest. *Bull. Environ. Contam. Toxicol*. 23, 579-586.

Hornsby, A. G., Wauchope, R. D., Herner, A. E. (1996) *Pesticide Properties in the Environment*. Springer-Verlag, Inc., New York, New York.

Horvath, A. L. (1982) *Halogenated Hydrocarbons, Solubility - Miscibility with Water*. Marcel Dekker, Inc., New York, New York.

Howard, P. H., Ed. (1989) *Handbook of Fate and Exposure Data for Organic Chemicals. Vol. I. Large Production and Priority Pollutants*. Lewis Publishers, Inc., Chelsea, Michigan.

Howard, P. H., Ed. (1990) *Handbook of Fate and Exposure Data for Organic Chemicals. Vol. II. Solvents*. Lewis Publishers, Inc., Chelsea, Michigan.

Howard, P. H., Ed. (1991) *Handbook of Fate and Exposure Data for Organic Chemicals. Vol. III. Pesticides*. Lewis Publishers, Inc., Chelsea, Michigan.

Howard, P. H., Ed. (1993) *Handbook of Fate and Exposure Data for Organic Chemicals. Vol. IV. Solvents 2*. Lewis Publishers, Inc., Chelsea, Michigan.

Howard, P. H., Boethling, R. S., Jarvis, W. F., Meylan, W. M., Michalenko, E. M. (1991) *Handbook of Environmental Degradation Rates*. Lewis Publishers, Inc., Chelsea, Michigan.

Hussam, A., Carr, P. W. (1985) A study of a rapid and precise methodology for the measurement of vapor liquid equilibria by headspace gas chromatography. *Anal. Chem.* 57, 793-801.

Irmann, F. (1965) Eine einfache korrelation zwishen wasserlöslichkeit und strucktur von kohlenwasserstoffen und halogenkohlenwasserstoffen. *Chem.-Ing.-Techn.* 37, 789-798.

Isnard, P., Lambert, S. (1988) Estimating bioconcentration factors for octanol-water partition coefficient and aqueous solubility. *Chemosphere* 17, 21-34.

Isnard, P., Lambert, S. (1989) Aqueous solubility/n-octanol-water partition coefficient correlations. *Chemosphere* 18, 1837-1853.

IUPAC Solubility Data Series (1984) Vol. 15: *Alcohols with Water.* Barton, A. F. M., Ed., Pergamon Press, Oxford, England.

IUPAC Solubility Data Series (1985) Vol. 20: *Halogenated Benzenes, Toluenes and Phenols with Water.* Horvath, A. L., Getzen, F. W., Eds., Pergamon Press, Oxford, England.

IUPAC Solubility Data Series (1989a) Vol. 37: *Hydrocarbons (C_5 - C_7) with Water and Seawater.* Shaw, D. G., Ed., Pergamon Press, Oxford, England.

IUPAC Solubility Data Series (1989b) Vol. 38: *Hydrocarbons (C_8 -C_{36}) with Water and Seawater.* Shaw, D. G., Ed., Pergamon Press, Oxford, England.

Iwase, K., Komatsu, K., Hirono, S., Nakagawa, S., Moriguchi, I. (1985) Estimation of hydrophobicity based on the solvent-accessible surface area of molecules. *Chem. Pharm. Bull.* 33, 2114-2121.

Jafvert, C. T., Westall, J. C., Grieder, E., Schwarzenbach, P. (1990) Distribution of hydrophobic ionogenic organic compounds between octanol and water: organic acids. *Environ. Sci. Technol.* 24, 1795-1803.

Jaworska, J. S., Schultz, T. W. (1992) Quantitive relationships of structure-activity and volume fraction for selected nonpolar and polar narcotic chemicals. *SAR. & QSAR in Environ. Res.* 1, 3-19.

Jensen, T., Fredenslund, A., Rasmussen, P. (1981) Pure-compound vapor pressures using UNIFAC group contribution. *Ind. Eng. Chem. Fundam.* 20, 239-246.

Johnson, C. A., Westall, J. C. (1990) Effect of pH and KCl concentration on the octanol-water distribution of methylanilines. *Environ. Sci. Technol.* 24, 1869-1875.

Jönsson, J. A., Vejrosta, J., Novak, J. (1982) Air/water partition coefficients for normal alkanes (n-pentane to n-nonane). *Fluid Phase Equil.* 9, 279-286.

Jury, W. A., Spencer, W. F., Farmer, W. J. (1983) Behavior assessment model for trace organics in soil: I. Model description. *J. Environ. Qual.* 12, 558-566.

Jury, W. A., Farmer, W. J., Spencer, W. F. (1984a) Behavior assessment model for trace organics in soil: II. Chemical classification and parameter sensitivity. *J. Environ. Qual.* 13, 567-572.

Jury, W. A., Farmer, W. J., Spencer, W. F. (1984b) Behavior assessment model for trace organics in soil: III. Application of screening model. *J. Environ. Qual.* 13, 573-579.

Jury, W. A., Spencer, W. F., Farmer, W. J. (1984c) Behavior assessment model for trace organics in soil: IV. Review of experimental evidence. *J. Environ. Qual.* 13, 580-587.

Kabadi, V. N., Danner, R. P. (1979) Nomograph solves for solubilities of hydrocarbons in water. *Hydrocarbon Processing*, 68, 245-246.

Kaiser, K. L. E., Valdmanis, I. (1982) Apparent octanol/water partition coefficients of pentachlorophenol as a function of pH. *Can. J. Chem.* 61, 2104-2106.

Kamlet, M. J., Doherty, R. M., Veith, G. D., Taft, R. W., Abraham, M. H. (1986) Solubility properties in polymers and biological media. 7. An analysis toxicant properties that influence inhibition of bioluminescence in *Photobacterium phosphoreum* (the Microtox test). *Environ. Sci. Technol.* 20, 690-695.

Kamlet, M. J., Doherty, R. M., Abraham, M. H., Carr, P. W., Doherty, R. F., Raft, R. W. (1987) Linear solvation energy relationships. Important differences between aqueous solublity relationships for aliphatic and aromatic solutes. *J. Phys. Chem.* 91, 1996.

Kamlet, M. J., Doherty, R. M., Carr, P. W., Mackay, D., Abraham, M. H., Taft, R. W. (1988) Linear solvation energy relationships. 44. Parameter estimation rules that allow accurate prediction of octanol/water partition coefficients and other solubility and toxicity properties of polychlorinated biphenyls and polycyclic aromatic hydrocarbons. *Environ. Sci. Technol.* 22, 503-509.

Karcher, W., Devillers, J., Eds., (1990) *Practical Applications of Quantitative-Structure-Activity Relationships (QSAR) in Environmental Chemistry and Toxicology.* Kluwer Academic Publisher, Dordrecht, The Netherlands.

Karger, B. L., Castells, R. C., Sewell, P. A., Hartkopf, A. (1971a) Study of the adsorption of insoluble and sparingly soluble vapors at the gas-liquid interface of water by gas chromatography. *J. Phys. Chem.* 75, 3870-3879.

Karger, B. L., Sewell, P. A., Castells, R. C., Hartkopf, A. (1971b) Gas chromatographic study of the adsorption of insoluble vapors on water. *J. Colloid Interface Sci.* 35(2), 328-339.

Karickhoff, S. W. (1981) Semiempirical estimation of sorption of hydrophobic pollutants on natural sediments and soil. *Chemosphere* 10, 833-846.

Karickhoff, S. W., Brown, D. S., Scott, T. A. (1979) Sorption of hydrophobic pollutants on natural water sediments. *Water Res.* 13, 241-248.

Kenaga, E. E. (1980) Predicted bioconcentration factors and soil sorption coefficients of pesticides and other chemicals. *Ecotox. Environ. Saf.* 4, 26-38.

Kenaga, E. E., Goring, C. A. I. (1980) Relationship between water solubility, soil sorption, octanol-water partitioning, and concentration of chemicals in biota. In: *Aquatic Toxicology.* ASTM STP 707, Eaton, J. G., Parrish, P. R., Hendrick, A. C., Eds., pp. 78-115, Am. Soc. for Testing and Materials, Philadelphia, Pennsylvania.

Kier, L. B., Hall, L. H. (1976) Molar properties and molecular connectivity. In: *Molecular Connectivity in Chemistry and Drug Design.* Medicinal Chem. Vol. 14, pp. 123-167, Academic Press, New York.

Kier, L. B., Hall, L. H. (1986) *Molecular Connectivity in Structure-Activity Analysis.* Wiley, New York.

Kikic, I., Alesse, P., Rasmussen, P., Fredenslunds, A. (1980) On the combinatorial part of the UNIFAC and UNIQUAC models. *Can. J. Chem. Eng.* 58, 253-258.

Kim, Y.-H., Woodrow, J. E., Seiber, J. N. (1984) Evaluation of a gas chromatographic method for calculating vapor pressures with organophosphorus pesticides. *J. Chromotagr.* 314, 37-53.

Klöpffer, W. (1991) Photochemistry in environmental research: Its role in abiotic degradation and exposure analysis. *EPA Newsletter* 41, 24-39.

Klopman, G., Wang, S., Balthasar, D. M. (1992) Estimation of aqueous solubility of organic molecules by the group contribution approach. Application to the study of biodegradation. *J. Chem. Inf. Comput. Sci.* 32, 474-482.

Könemann, H., van Leeuewen, K. (1980) Toxicokinetics in fish: accumulation of six chlorobenzenes by guppies. *Chemosphere* 9, 3-19.

Könemann, H., Zelle, R., Busser, F. (1979) Determination of log P_{oct} values of chloro-substituted benzenes, toluenes and anilines by high-performance liquid chromatography on ODS-silica. *J. Chromatogr.* 178, 559-565.

Lande, S. S., Banerjee, S. (1981) Predicting aqueous solubility of organic nonelectrolytes from molar volume. *Chemosphere* 10, 751-759.

Leahy, D. E. (1986) Intrinsic molecular volume as a measure of the cavity term in linear solvation energy relationships: octanol-water partition coefficients and aqueous solubilities. *J. Pharm. Sci.* 75, 629-636.

Lee, L. S., Rao, P. S. C., Nkedi-Kizza, P., Delfino, J. (1990) Influence of solvent and sorbent characteristics on distribution of pentachlorophenol in octanol-water and soil-water systems. *Environ. Sci. Technol.* 24, 654-661.

Leighton, Jr., D. T., Calo, J. M. (1981) Distribution coefficients of chlorinated hydrocarbons in dilute air-water systems for groundwater contamination applications. *J. Chem. Eng. Data* 26, 382-385.

Leo, A., Hansch, C., Elkins, D. (1971) Partition coefficients and their uses. *Chem. Rev.* 71, 525-616.

Lincoff, A. H., Gossett, J. M. (1984) The determination of Henry's law constants for volatile organics by equilibrium partitioning in closed systems. In: *Gas Transfer at Water Surfaces*. Brutsaert, W., Jirka, G. H., Eds., pp. 17-26, D. Reidel Publishing Co., Dordrecht, The Netherlands.

Lo, J. M., Tseng, C. L., Yang, J. Y. (1986) Radiometric method for determining solubility of organic solvents in water. *Anal. Chem.* 58, 1596-1597.

Locke, D. (1974) Selectivity in reversed-phase liquid chromatography using chemically bonded stationary phases. *J. Chromatogr. Sci.* 12, 433-437.

Lohninger, H. (1994) Estimation of soil partition coefficients of pesticides from their chemical structure. *Chemosphere* 29, 1611-1626.

Lyman, W. J., Reehl, W. F., Rosenblatt, D. H. (1982) *Handbook of Chemical Property Estimation Methods*. McGraw-Hill, New York.

Mabey, W., Mill, T. (1978) Critical review of hydrolysis of organic compounds in water under environmental conditions. *J. Phys. Chem. Ref. Data* 7, 383-414.

Mabey, W. J., Mill, T., Podoll, R. T. (1984) *Estimation Methods for Process Constants and Properties used in Fate Assessment*. USEPA, EPA-600/3-84-035, Athens, Georgia.

Mackay, D. (1979) Finding fugacity feasible. *Environ. Sci. Technol.* 13, 1218-1223.

Mackay, D. (1982) Correlation of bioconcentration factors. *Environ. Sci. Technol.* 16, 274-278.

Mackay, D. (1991) *Multimedia Environmental Models. The Fugacity Approach*. Lewis Publishers, Inc., Chelsea, Michigan.

Mackay, D., Bobra, A. M., Shiu, W.-Y., Yalkowsky, S. H. (1980) Relationships between aqueous solubility and octanol-water partition coefficient. *Chemosphere* 9, 701-711.

Mackay, D., Bobra, A. M., Chan, D. W., Shiu, W.-Y. (1982) Vapor pressure correlation for low-volatility environmental chemicals. *Environ. Sci. Technol.* 16, 645-649.

Mackay, D., Paterson, S. (1990) Fugacity models. In: *Practical Applications of Quantitative Structure-Activity Relationships (QSAR) in Environmental Chemistry and Toxicology*. Karcher, W., Devillers, J., Eds., pp. 433-460, Kluwer Academic Publishers, Dordrecht, The Netherlands.

Mackay, D., Paterson, S. (1991) Evaluating the multimedia fate of organic chemicals: A Level III fugacity model. *Environ. Sci. Technol.* 25, 427-436.

Mackay, D., Shiu, W. Y. (1977) Aqueous solubility of polynuclear aromatic hydrocarbons. *J. Chem. Eng. Data* 22, 339-402.

Mackay, D., Shiu, W. Y. (1981) A critical review of Henry's law constants for chemicals of environmental interest. *J. Phys. Chem. Ref. Data* 11, 1175-1199.

Mackay, D., Shiu, W. Y., Sutherland, R.P. (1979) Determination of air-water Henry's law constants for hydrophobic pollutants. *Environ. Sci. Technol.* 13, 333-337.

Mackay, D., Shiu, W. Y., Wolkoff, A. W. (1975) Gas chromatographic determination of low concentration of hydrocarbons in water by vapor phase extraction. In: *Water Quality Parameters. ASTM STP 573*, pp. 251-258, American Society for Testing and Materials, Philadelphia, Pennsylvania.

Mackay, D., Stiver, W. H. (1991) Predictability and environmental chemistry. In: *Environmental Chemistry of Herbicides*. Vol. II, Grover, R., Cessna, A. J., Eds., pp. 281-297, CRC Press, Inc., Boca Raton, Florida.

Macknick, A. B., Prausnitz, J. M. (1979) Vapor pressure of high-molecular weight hydrocarbons. *J. Chem. Eng. Data* 24, 175-178.

Magnussen, T., Rasmussen, P., Fredenslund, A. (1981) UNIFAC parameter table for prediction of liquid-liquid equilibria. *Ind. Eng. Chem. Process Des. Dev.* 20, 331-339.

Mailhot, H., Peters, R. H. (1988) Empirical relationships between the 1-octanol/water partition coefficient and nine physicochemical properties. *Environ. Sci. Technol.* 22, 1479-1488.

May, W. E., Wasik, S. P., Freeman, D. H. (1978a) Determination of the aqueous solubility of polynuclear aromatic hydrocarbons by a coupled-column liquid chromatographic technique. *Anal. Chem.* 50, 175-179.

May, W. E., Wasik, S. P., Freeman, D. H. (1978b) Determination of the solubility behavior of some polycyclic aromatic hydrocarbons in water. *Anal. Chem.* 50, 997-1000.

McAuliffe, C. (1966) Solubility in water of paraffin, cycloparaffin, olefin, acetylene, cycloolefin and aromatic hydrocarbons. *J. Phys. Chem.* 76, 1267-1275.

McAuliffe, C. (1971) GC determination of solutes by multiple phase equilibration. *Chem. Tech.* 1, 46-51.

McDuffie, B. (1981) Estimation of octanol/water partition coefficient for organic pollutants using reversed phase HPLC. *Chemosphere* 10, 73-83.

McGowan, J.C., Mellors, A. (1986) *Molecular Volumes in Chemistry and Biology-Applications including Partitioning and Toxicity*. Ellis Horwood Limited, Chichester, England.

The Merck Index (1989) *An Encyclopedia of Chemicals, Drugs and Biologicals*. 11th Edition, Budavari, S., Ed., Merck and Co., Inc., Rahway, New Jersey.

Meylan, W. M., Howard, P. H. (1991) Bond contribution method for estimating Henry's law constants. *Environ. Toxicol. Chem.* 10, 1283-1293.

Meylan, W. M., Howard, P. H., Boethling, R. S. (1992) Molecular topology/fragment contribution for predicting soil sorption coefficient. *Environ. Sci. Technol.* 26, 1560-1567.

Mill, T. (1982) Hydrolysis and oxidation processes in the environment. *Environ. Toxicol. Chem.* 1, 135-141.

Mill, T. (1989) Structure-activity relationships for photooxidation processes in the environment. *Environ. Toxicol. Chem.* 8, 31-45.

Mill, T. (1993) Environmental chemistry. In: *Ecological Risk Assessment*. Suter, II, G.W., Ed., pp. 91-127, Lewis Publishers, Inc., Chelsea, Michigan.

Mill, T., Mabey, W. (1985) Photodegradation in water. In: *Environmental Exposure from Chemicals*. Vol. 1. Neely, W. B., Blau, G. E., Eds., pp. 175-216, CRC Press, Inc., Boca Raton, Florida.

Miller, M. M., Ghodbane, S., Wasik, S. P., Tewari, Y. B., Martire, D. E. (1984) Aqueous solubilities, octanol/water partition coefficients and entropies of melting of chlorinated benzenes and biphenyls. *J. Chem. Eng. Data* 29, 184-190.

Miller, M. M., Wasik, S. P., Huang, G.-L., Shiu, W.-Y., Mackay, D. (1985) Relationships between octanol-water partition coefficient and aqueous solublity. *Environ. Sci. Technol.* 19, 522-529.

Montgomery, J. H. (1993) *Agrochemicals Desk Reference. Environmental Data*. Lewis Publishers, Inc., Chelsea, Michigan.

Munz, C., Roberts, P. V. (1987) Air-water phase equilibria of volatile organic solutes. *J. Am. Water Works Assoc.* 79, 62-69.

Neely, W. B., Branson, D. R., Blau, G. E. (1974) Partition coefficient to measure bioconcentration potential of organic chemicals in fish. *Environ. Sci. Technol.* 8, 1113-1115.

Nirmalakhandan, N. N., Speece, R. E. (1988a) Prediction of aqueous solubility of organic chemicals based on molecular structure. *Environ. Sci. Technol.* 22, 328-338.

Nirmalakhandan, N. N., Speece, R. E. (1988b) QSAR model for predicting Henry's law constant. *Environ. Sci. Technol.* 22, 1349-1357.

Nirmalakhandan, N. N., Speece, R. E. (1989) Prediction of aqueous solubility of organic chemicals based on molecular structure. 2. Application to PNAs, PCBs, PCDDs, etc. *Environ. Sci. Technol.* 23, 708-713.

Okouchi, H., Saegusa, H., Nojima, O. (1992) Prediction of environmental parameters by adsorbability index: water solubilities of hydrophobic organic pollutants. *Environ. Intl.* 18, 249-261.

Oliver, B. G. (1984) The relationship between bioconcentration factor in rainbow trout and physical-chemical properties for some halogenated compounds. In: *QSAR in Environmental Toxicology*. Kaiser, K. L. E., Ed., pp. 300-317, D. Reidel Publishing Co., Dordrecht, The Netherlands.

Oliver, B. G., Niimi, A. J. (1988) Trophodynamic analysis of polychlorinated biphenyl congeners and other chlorinated hydrocarbons in the Lake Ontario ecosystem. *Environ. Sci. Technol.* 22, 388-397.

Osborn, A. G., Douslin, D. R. (1975) Vapor pressures and derived enthalpies of vaporization of some condensed-ring hydrocarbons. *J. Chem. Eng. Data* 20, 229-231.

Paterson, S., Mackay, D. (1985) The fugacity concept in environmental modelling. In: *The Handbook of Environmental Chemistry*. Vol. 2/Part C, Hutzinger, O., Ed., pp. 121-140, Springer-Verlag, Heidelberg, Germany.

Pearlman, R. S. (1986) Molecular surface area and volume: Their calculation and use in predicting solubilities and free energies of desolvation. In: *Partition coefficient, Determination and Estimation*. Dunn, III, W. J., Block, J. H., Pearlman R. S., Eds., pp. 3-20, Pergamon Press, New York, New York.

Pearlman, R. S., Yalkowsky, S. H., Banerjee, S. (1984) Water solubilities of polynuclear aromatic and heteroaromatic compounds. *J. Phys. Chem. Ref. Data* 13, 555-562.

Pierotti, C., Deal, C., Derr, E. (1959) Activity coefficient and molecular structure. *Ind. Eng. Chem. Fundam.* 51, 95-101.

Polak, J., Lu, B. C. Y. (1973) Mutual solubilities of hydrocarbons and water at 0° and 25°C. *Can. J. Chem.* 51, 4018-4023.

Rapaport, R. A., Eisenreich, S. J. (1984) Chromatographic determination of octanol-water partition coefficients (K_{ow}'s) for 58 polychlorinated biphenyl congeners. *Environ. Sci. Technol.* 18, 163-170.

Reddy, K. N., Locke, M. A. (1994) Relationships between molecular properties and log p and soil sorption (K_{oc}) of substituted phenylureas: QSAR models. *Chemosphere* 28, 1929-1941.

Reid, R. C., Prausnitz, J. M., Polling, B. E. (1987) *The Properties of Gases and Liquids*. 4th Edition, McGraw-Hill, Inc., New York, New York.

Rekker, R. F. (1977) *The Hydrophobic Fragmental Constant*. Elsevier, Amsterdam/New York, New York.

Riddick, J. A., Bunger, W. B., Sakano, T. K. (1986) *Organic Solvents, Physical Properties and Methods of Purification*. 4th Edition, Wiley-Science Publication, John Wiley & Sons, New York, New York.

Rordorf, B. F. (1985a) Thermodynamic and thermal properties of polychlorinated compounds: the vapor pressures and flow tube kinetic of ten dibenzo-*p*-dioxins. *Chemosphere* 14, 885-892.

Rordorf, B. F. (1985b) Thermodynamic properties of polychlorinated compounds: the vapor pressures and enthalpies of sublimation of ten dibenzo-*p*-dioxins. *Thermochimica Acta* 85, 435-438.

Rordorf, B. F. (1986) Thermal properties of dioxins, furans and related compounds, *Chemosphere* 15, 1325-1332.

Sabljic, A. (1984) Predictions of the nature and strength of soil sorption of organic pollutants by molecular topology. *J. Agric. Food Chem.* 32, 243-246.

Sabljic, A. (1987) On the prediction of soil sorption coefficients of organic pollutants from molecular structure: Application of molecular topology model. *Environ. Sci. Technol.* 21, 358-366.

Sabljic, A., Lara, R., Ernst, W. (1989) Modelling association of highly chlorinated biphenyls with marine humic substances. *Chemosphere* 19, 1665-1676.

Sabljic, A., Güsten, H. (1989) Predicting Henry's law constants for polychlorinated biphenyls, *Chemosphere* 19, 1503-1511.

Sabljic, A., Güsten, H. (1990) Predicting the night-time NO_3 radical reactivity in the troposphere. *Atmos. Environ.* 24A, 73-78.

Schellenberg, K., Leuenberger, C., Schwarzenbach, R. P. (1984) Sorption of chlorinated phenols by natural sediments and aquifer materials. *Environ. Sci. Technol.* 18, 652-657.

Schwarz, F. P. (1980) Measurement of the solubilities of slightly soluble organic liquids in water by elution chromatography. *Anal. Chem.* 52, 10-15.

Schwarz, F. P., Miller, J. (1980) Determination of the aqueous solubilities of organic liquids at 10.0, 20.0, 30.0 °C by elution chromatography. *Anal. Chem.* 52, 2162-2164.

Schwarzenbach, R. P., Gschwend, P. M., Imboden, D. M. (1993) *Environmental Organic Chemistry*. John Wiley & Sons, Inc., New York, New York.

Schwarzenbach, R. P., Stierli, R., Folsom, B. R., Zeyer, J. (1988) Compound properties relevant for assessing the environmental partitioning of nitrophenols. *Environ. Sci. Technol.* 22, 83-92.

Schwarzenbach, R. P., Westall, J. (1981) Transport of nonpolar compounds from surface water to groundwater. Laboratory sorption studies. *Environ. Sci. Technol.* 11, 1360-1367.

Sears, G. W., Hopke, E. R. (1947) Vapor pressures of naphthalene, anthracene and hexachlorobenzene in a low pressure region. *J. Am. Chem. Soc.* 71, 1632-1634.

Shigeoka, T., Sato, Y., Takeda, Y. (1988) Acute toxicity of chlorophenols to green algae, *Selenastrum capricornutum* and *Chlorella vulgaris*, and quantitative structure-activity relationships. *Environ. Toxicol. Chem.* 7, 847-854.

Shiu, W.-Y., Ma, K.-C., Verhanickova, D., Mackay, D. (1994) Chlorophenols and alkylphenols: A review and correlation of environmentally relevant properties and fate in an evaluative environment. *Chemosphere* 29(6), 1155-1224.

Shiu, W.-Y., Mackay, D. (1986) A critical review of aqueous solubilities, vapor pressures, Henry's law constants, and octanol-water partition coefficients of the polychlorinated biphenyls. *J. Phys. Chem. Ref. Data* 15, 911-929.

Shiu, W.-Y., Gobas, F. A. P. C., Mackay, D. (1987) Physical-chemical properties of three congeneric series of chlorinated aromatic hydrocarbons. In: *QSAR in Environmental Toxicology II*. Kaiser, K. L. E., Ed., pp. 347-362, D. Reidel Publishing Co., Dordrecht, The Netherlands.

Shiu, W.-Y., Doucette, W., Gobas, F. A. P. C., Mackay, D., Andren, A. W. (1988) Physical-chemical properties of chlorinated dibenzo-*p*-dioxins. *Environ. Sci. Technol.* 22, 651-658.

Sinke, G. C. (1974) A method for measurement of vapor pressures of organic compounds below 0.1 torr. Naphthalene as reference substance. *J. Chem. Thermodyn.* 6, 311-316.

Smejtek, P., Wang, S. (1993) Distribution of hydrophobic ionizable xenobiotics between water and lipid membranes: pentachlorophenol and pentachlorophenate. A comparison with octanol-water partition. *Arch. Environ. Contam. Toxicol.* 25, 394-404.

Sonnefeld, W. J., Zoller, W. H., May, W. E. (1983) Dynamic coupled-column liquid chromatographic determination of ambient temperature vapor pressures of polynuclear aromatic hydrocarbons. *Anal. Chem.* 55, 275-280.

Spencer, W. F., Cliath, M. M. (1969) Vapor density of dieldrin. *Environ. Sci. Technol.* 3, 670-674.

Spencer, W. F., Cliath, M. M. (1970) Vapor density and apparent vapor pressure of lindane (γ-BHC). *J. Agric. Food Chem.* 18, 529-530.

Spencer, W. F., Cliath, M. M. (1972) Volatility of DDT and related compounds. *J. Agric. Food Chem.* 20, 645-649.

Stehly, G. R., Hayton, W. L. (1990) Effect of pH on the accumulation kinetics of pentachlorophenol in goldfish. *Arch. Environ. Contam. Toxicol.* 19, 464-470.

Stephenson, R. M., Malanowski, A. (1987) *Handbook of the Thermodynamics of Organic Compounds*. Elsevier, New York.

Stiver, W., Mackay, D. (1989) The linear additivity principle in environmental modelling: Application to chemical behaviour in soil. *Chemosphere* 19, 1187-1198.

Sugiyama, T., Takeuchi, T., Suzuki, Y. (1975) Thermodynamic properties of solute molecules at infinite dilution determined by gas-liquid chromatography. I. Intermolecular energies of n-alkane solutes in C_{28} - C_{36} n-alkane solvents. *J. Chromatogr.* 105, 265-272.

Suntio, L. R., Shiu, W.-Y., Mackay, D. (1988) Critical review of Henry's law constants for pesticides. *Rev. Environ. Contam. Toxicol.* 103, 1-59.

Swann, R. L., Laskowski, D. A., McCall, P. J., Vander Kuy, K., Dishburger, H. J. (1983) A rapid method for the estimation of the environmental parameters octanol/water partition coefficient, soil sorption constant, water to air ratio, and water solubility. *Res. Rev.* 85, 17-28.

Szabo, G., Prosser, S., Bulman, R. A. (1990) Determination of the adsorption coefficient (K_{OC}) of some aromatics for soil by RP-HPLC on two immobilized humic acid phases. *Chemosphere* 21, 777-788.

Tomlin, C., Ed. (1994) *The Pesticide Manual (A World Compendium)*, 10th Ed., Incorporating the Agrochemicals Handbook, The British Crop Protection Council and The Royal Society of Chemistry, England.

Tomlinson, E., Hafkenscheid, T. L. (1986) Aqueous solution and partition coefficient estimation from HPLC data. In: *Partition Coefficient, Determination and Estimation*. Dunn, III, W. J., Block, J. H., Pearlman, R. S., Eds., pp. 101-141, Pergamon Press, New York, New York.

Tse, G., Orbey, H., Sandler, S. I. (1992) Infinite dilution activity coefficients and Henry's law coefficients for some priority water pollutants determined by a relative gas chromatographic method. *Environ. Sci. Technol.* 26, 2017-2022.

Tsonopoulos, C., Prausnitz, J. M. (1971) Activity coefficients of aromatic solutes in dilute aqueous solutions. *Ind. Eng. Chem. Fundam.* 10, 593-600.

Tulp, M. T. M., Hutzinger, O. (1978) Some thoughts on the aqueous solubilities and partition coefficients of PCB, and the mathematical correlation between bioaccumulation and physico-chemical properties. *Chemosphere* 7, 849-760.

Veith, G. D., Austin, N. M., Morris, R. T. (1979) A rapid method for estimating log P for organic chemicals. *Water Res.* 13, 43-47.

Veith, G. D., Macek, K. J., Petrocelli, S. R., Caroll, J. (1980) An evaluation of using partition coefficients and water solubilities to estimate bioconcentration factors for organic chemicals in fish. In: *Aquatic Toxicology.* ASTM ATP 707, Eaton, J. G., Parrish, P. R., Hendrick, A.C., Eds, pp. 116-129, Am. Soc. for Testing and Materials, Philadelphia, Pennsylvania.

Vejrosta, J., Novak, J., Jönsson, J. (1982) A method for measuring infinite-dilution partition coefficients of volatile compounds between the gas and liquid phases of aqueous systems. *Fluid Phase Equil.* 8, 25-35.

Verschueren, K. (1977) *Handbook of Environmental Data on Organic Chemicals.* Van Nostrand Reinhold, New York, New York.

Verschueren, K. (1983) *Handbook of Environmental Data on Organic Chemicals.* 2nd Eition, Van Nostrand Reinhold, New York, New York.

Wakita, K., Yoshimoto, M., Miyamoto, S., Watsnabe, H. (1986) A method for calculations of the aqueous solubility of organic compounds by using new fragment solubility constants. *Chem. Pharm. Bull.* 34, 4663-4681.

Wang, L., Zhao, Y., Hong, G. (1992) Predicting aqueous solubility and octanol/water partition coefficients of organic chemicals from molar volume. *Environ. Chem.* 11, 55-70.

Warne, M., St. J., Connell, D. W., Hawker, D. W. (1990) Prediction of aqueous solubility and the octanol-water partition coefficient for lipophilic organic compounds using molecular descriptors and physicochemical properties. *Chemosphere* 16, 109-116.

Wasik, S. P., Miller, M. M., Tewari, Y. B., May, W. E., Sonnefeld, W. J., DeVoe, H., Zoller, W. H. (1983) Determination of the vapor pressure, aqueous solubility, and octanol/water partition coefficient of hydrophobic substances by coupled generator column/liquid chromatographic methods. *Res. Rev.* 85, 29-42.

Wauchope, R. D., Buttler, T. M., Hornsby, A. G., Augustijn-Beckers, P. W. M., Burt, J. P. (1992) The SCS/ARS/CES pesticide properties database for environmental decision making. *Rev. Environ. Contam. Toxicol.* 123, 1-156.

Weast, R., Ed. (1972-73) *Handbook of Chemistry and Physics.* 53th Edition, CRC Press, Inc., Cleveland, Ohio.

Weast, R., Ed. (1982-83) *Handbook of Chemistry and Physics.* 64th Edition, CRC Press, Inc., Boca Raton, Florida.

Weil, L., Dure, G., Quentin, K. L. (1974) Solubility in water of insecticide, chlorinated hydrocarbons and polychlorinated biphenyls in view of water pollution. *Z. Wasser Abwasser Forsch.* 7, 169-175.

Westall, J. C., Leuenberger, C., Schwarzenbach, R. P. (1985) Influence of pH and ionic strength on the aqueous-nonaqueous distribution of chlorinated phenols. *Environ. Sci. Technol.* 19, 193-198.

Westcott, J. W., Bidleman, T. F. (1982) Determination of polychlorinated biphenyl vapor pressures by capillary gas chromatography. *J. Chromatogr.* 210, 331-336.

Westcott, J. W., Simon, J. J., Bidleman, T. F. (1981) Determination of polychlorinated biphenyl vapor pressures by a semimicro gas saturation method. *Environ. Sci. Technol.* 15, 1375-1378.

Whitehouse, B. G., Cooke, R. C. (1982) Estimating the aqueous solubility of aromatic hydrocarbons by high performance liquid chromatography. *Chemosphere* 11, 689-699.

Windholz, M., Ed. (1983) *The Merck Index, An Encyclopedia of Chemicals, Drugs and Biologicals.* 10th Edition, Merck & Co. Inc. Rahway, New Jersey.

Woodburn, K. B., Doucette, W. J., Andren, A. W. (1984) Generator column determination of octanol/water partition coefficients for selected polychlorinated biphenyl congeners. *Environ. Sci. Technol.* 18, 457-459.

Worthing, C. R., Ed. (1983) *The Pesticide Manual (A World Compendium)*, 7th Edition, The British Crop Protection Council, Croydon, England.

Worthing, C. R., Ed. (1987) *The Pesticide Manual (A World Compendium)*, 8th Edition, The British Crop Protection Council, Croydon, England.

Worthing, C. R., Ed. (1991) *The Pesticide Manual (A World Compendium)*, 9th Edition, The British Crop Protection Council, Croydon, England.

Yair, O. B., Fredenslund, A. (1983) Extension of the UNIFAC group-contribution method for the prediction of pure-component vapor pressure. *Ind. Eng. Chem. Fundam. Des. Dev.* 22, 433-436.

Yalkowsky, S. H. (1979) Estimation of entropies of fusion of organic compounds. *Ind. Eng. Chem. Fundam.* 18, 108-111.

Yalkowsky, S. H., Banerjee, S. (1992) *Aqueous Solubility, Methods of Estimation for Organic Compounds*. Marcel Dekker, Inc., New York, New York.

Yalkowsky, S. H., Valvani, S. C. (1976) Partition coefficients and surface areas of some alkylbenzenes. *J. Med. Chem.* 19, 727-728.

Yalkowsky, S. H., Valvani, S. C. (1979) Solubility and partitioning. I: Solubility of nonelectrolytes in water. *J. Pharm. Sci.* 69, 912-922.

Yalkowsky, S. H., Orr, R. J., Valvani, S. C. (1979) Solubility and partitioning. 3. The solubility of halobenzenes in water. *Ind. Eng. Chem. Fundam.* 18, 351-353.

Yalkowsky, S. H., Valvani, S. S., Mackay, D. (1983) Estimation of the aqueous solubility of some aromatic compounds. *Res. Rev.* 85, 43-55.

Yoshida, K., Shigeoka, T., Yamauchi, F. (1983) Relationship between molar refraction and n-octanol/water partition coefficient. *Ecotox. Environ. Saf.* 7, 558-565.

Yoshida, K., Shigeoka, T., Yamauchi, F. (1987) Evaluation of aquatic environmental fate of 2,4,6-trichlorophenol with a mathematical model. *Chemosphere* 16, 2531-2544.

Zhou, X., Mopper, K. (1990) Apparent partition coefficients of 15 carbonyl compounds between air and seawater and between air and freshwater: Implications for air-sea exchange. *Environ. Sci. Technol.* 24, 1864-1869.

Zaroogian, G.E., Heltshe, J. F., Johnson, M. (1985) Estimation bioconcentration in marine species using structure-activity models. *Environ. Toxicol. Chem.* 4, 3-12.

Zwolinski, B. J., Wilhoit, R. C. (1971) *Handbook of Vapor Pressures and Heats of Vaporization of Hydrocarbons and Related Compounds*. API-44, TRC Publication No. 101, Texas A&M University, College Station, Texas.

Chapter 2. Herbicides

2.1 List of Chemicals and Data Compilations:
- 2.1.1 Aliphatic acids:
  - Dalapon 111
- 2.1.2 Aromatic Acids:
  - Chloramben 93
  - Dicamba 119
  - Picloram 196
- 2.1.3 Amides:
  - Alachlor 57
  - Butachlor 87
  - Diphenamid 135
  - Metolachlor 178
  - Pronamide 209
  - Propachlor 212
  - Propanil 215
- 2.1.4 Benzonitriles:
  - Bromoxynil 85
  - Dichlobenil 123
- 2.1.5 Carbamates:
  - Barban 76
  - Chlorpropham 95
  - Propham 221
- 2.1.6 Dinitroanilines:
  - Benefin 78
  - Butralin 89
  - Dinitramine 130
  - Fluchloralin 150
  - Isopropalin 163
  - Oryzalin 192
  - Profluralin 200
  - Trifluralin 243
- 2.1.7 Diphenylethers:
  - Bifenox 80
  - Fluorodifen 155
- 2.1.8 Phenols:
  - Dinoseb 132
  - PCP (Pentachlorophenol) 525
- 2.1.9 Phenoxyalkanoic acids:
  - 2,4-D 106
  - 2,4-DB 114
  - Dichlorprop 126
  - MCPA 171
  - MCPB 174
  - Mecoprop 176
  - 2,4,5-T 231
- 2.1.10 Thiocarbamates:
  - Butylate 91
  - Diallate 116

EPTC	144
Molinate	181
Pebulate	194
Triallate	240
Vernolate	249

2.1.11 Triazines:

Ametryn	61
Atrazine	67
Cyanazine	103
Prometon	203
Prometryn	206
Propazine	218
Simazine	226
Terbutryn	237

2.1.12 Uracils:

Bromacil	82
Terbacil	235

2.1.13 Ureas:

Chlorsulfuron	98
Chlortoluron	101
Diuron	139
Fenuron	147
Fluometuron	152
Isoproturon	165
Linuron	167
Monolinuron	184
Monuron	186
Neburon	190

2.1.14 Miscellaneous:

Amitrole (Triazole)	64
Diclofop-methyl (Chlorophenoxy acid ester)	128
Diquat (Bipyridyl)	137
Fluridone (Fluoro-phenyl pyridinone)	157
Glyphosate (Phosphate)	160
Pyrazon (Pyridazinone)	224

2.2	Summary Tables	251
2.3	Illustrative Fugacity Calculations: Levels I, II and III	264
	Atrazine	264
	2,4-D	268
	Diuron	272
	EPTC	276
	Glyphosate	280
	Linuron	284
	Metolachlor	288
	2,4,5-T	292
	Triallate	296
	Trifluralin	300
2.4	Commentary on the Physical-Chemical Properties and Environmental Fate	304
2.5	References	306

Common Name: Alachlor
Synonym: alachlore, Alanex, Bronco, Bullet, Cannon, Lasso, Lazo, metachlor, Pillarzo
Chemical Name: 2-chloro-2,6-diethyl-N-methoxymethylacetanilide; 2-chloro-N-(2,6-diethylphenyl)-N-(methoxymethyl)acetamide
Uses: pre-emergence, early post-emergence or soil-incorporated herbicide to control most annual grasses and many annual broadleaf weeds in beans, corn, cotton, milo, peanuts, peas, soybeans, sunflower, and certain woody ornamentals.
CAS Registry No: 15972-60-8
Molecular Formula: $C_{14}H_{20}ClNO_2$
Molecular Weight: 269.77
Melting Point (°C):
    39.5-41.5 (Khan 1980; Herbicide Handbook 1989; Montgomery 1993; Tomlin 1994)
    41.0 (Suntio et al. 1988)
    39.5 (Kühne et al. 1995)
    40-41 (Milne 1995)
Boiling Point (°C):
    100 (at 0.02 mmHg, Ashton & Crafts 1981; Herbicide Handbook 1989; Worthing 1991; Montgomery 1993; Tomlin 1994; Milne 1995)
    135 (at 0.30 mmHg, Herbicide Handbook 1989; Milne 1995)
Density (g/cm³ at 20°C):
    1.133 (25°C, Agrochemical Handbook 1987; Montgomery 1993; Tomlin 1994)
Molar Volume (cm³/mol):
    240.7 (calculated-LeBas method, Suntio et al. 1988)
Molecular Volume (Å³):
Total Surface Area, TSA (Å²):
Dissociation Constant pK_a: 0.62
Heat of Fusion, ΔH_{fus}, kcal/mol:
    7.0 (DSC method, Plato 1972)
Entropy of Fusion, ΔS_{fus}, cal/mol·K (e.u.):
Fugacity Ratio at 25°C (assuming ΔS_{fus} = 13.5 e.u.), F: 0.695

Water Solubility (g/m³ or mg/L at 25°C):
    242 (20°C, Weber 1972; Worthing 1987; quoted, Muir 1991)
    200 (Bailey & White 1965; quoted, Shiu et al. 1990)
    242 (Herbicide Handbook 1974, 1978, 1983, 1989; Martin & Worthing 1977; quoted, Kenaga 1980; Kenaga & Goring 1980; Nash 1988; Taylor & Glotfelty 1988; Glotfelty et al. 1989; Isensee 1991)
    240 (Hartley & Graham-Bryce 1980; Beste & Humburg 1983; Shiu et al. 1990)
    148 (Khan 1980; quoted, Suntio et al. 1988; Shiu et al. 1990)
    242 (Weber et al. 1980; quoted, Willis & McDowell 1982)
    242 (Ashton & Crafts 1981; Worthing 1987, 1991; Di Guardo et al. 1994)
    242 (Agrochemicals Handbook 1983,87; Tomlin 1994)
    130 (20°C, selected, Suntio et al. 1988; quoted, Majewski & Capel 1995)
    148, 242 (literature data variability, Heller et al. 1989)
    242 (Herbicide Handbook 1989; quoted, Pait et al. 1992)
    140 (23°C, Merck Index 1989)
    240 (Wauchope 1989; quoted, Shiu et al. 1990)
    240 (20-25°C, selected, Wauchope et al. 1992; Hornsby et al. 1996)
    242 (quoted, Montgomery 1993; Wienhold & Gish 1994)
    240 (selected, Lohninger 1994; Gish et al. 1995)

148   (quoted, Kühne et al. 1995)
23.5   (calculated-group contribution fragmentation method, Kühne et al. 1995)
140   (23°C, Milne 1995)
240   (selected, Halfon et al. 1996)
174, 512 (quoted, predicted-AQUAFAC, Lee et al. 1996)

Vapor Pressure (Pa at 25°C):
  0.00293 (20°C, Weber 1972; Worthing 1987, 1991; quoted, Muir 1991; Di Guardo et al. 1994)
  0.00293 (Herbicide Handbook 1974,83,89; quoted, Nash 1988; Suntio et al. 1988; Glotfelty et al. 1989; Taylor & Spencer 1990)
  0.00293 (20-25°C, Weber et al. 1980; quoted, Willis & McDowell 1982)
  0.00293 (Ashton & Crafts 1981; Schnoor & McAvoy 1981; Schnoor 1992)
  0.00290 (Beste & Humburg 1983; quoted, Taylor & Glotfelty 1988; Taylor & Spencer 1990)
  0.00290 (Agrochemicals Handbook 1987)
  0.00300 (20°C, selected, Suntio et al. 1988; quoted, Majewski & Capel 1995)
  0.00290 (Worthing 1991; Tomlin 1994)
  0.00187 (20-25°C, selected, Wauchope et al. 1992; Hornsby et al. 1996; quoted, Halfon et al. 1996)
  0.00413 (quoted, Montgomery 1993)

Henry's Law Constant (Pa·m^3/mol):
  6.20×10^{-3} (20°C, calculated-P/C, Suntio et al. 1988; quoted, Findinger & Glotfelty 1988; Fendinger et al. 1989; Müller et al. 1994; Majewski & Capel 1995)
  8.43×10^{-4} (wetted-wall-GC/ECD, Findinger & Glotfelty 1988; quoted, Fendinger et al. 1989)
  3.26×10^{-3} (calculated-P/C, Taylor & Glotfelty 1988)
  1.12×10^{-3} (fog chamber-GC/ECD, Findinger et al. 1989)
  8.38×10^{-4} (23°C, known LWAPC of Findinger et al. 1989, Meylan & Howard 1991)
  1.21×10^{-5} (bond-estimated LWAPC, Meylan & Howard 1991)
  3.26×10^{-3} (20°C, calculated-P/C, Muir 1991)
  6.20×10^{-3} (calculated-P/C, Montgomery 1993)
  3.22×10^{-3} (Gish et al. 1995)
  2.20×10^{-3} (calculated-P/C, this work)

Octanol/Water Partition Coefficient, log K_{ow}:
  2.92   (Leo et al. 1971; quoted, Kenaga & Goring 1980)
  2.30   (Kenaga 1980; quoted, Nash 1988)
  2.64   (quoted, Rao & Davidson 1980; Suntio et al. 1988; Bintein & Devillers 1994; Di Guardo et al. 1994)
  3.087 (shake flask, Dubelman & Bremer 1983; quoted, Sicbaldi & Finizio 1993)
  2.92   (selected, Dao et al. 1983)
  3.52   (Medchem Database 1988; quoted, Müller et al. 1994)
  2.80   (selected, Suntio et al. 1988)
  2.64, 2.90 (quoted values, Montgomery 1993)
  3.27   (RP-HPLC, Sicbaldi & Finizio 1993)
  3.52   (selected, Hansch et al. 1995)
  3.27, 2.39, 2.95 (RP-HPLC, ClogP, calculated-S, Finizio et al. 1997)

Bioconcentration Factor, log BCF:
- 1.45 (calculated-S, Kenaga 1980; quoted, Isensee 1991)
- 0.954 (calculated-K_{OC}, Kenaga 1980)
- 1.88 (selected, Schnoor & McAvoy 1981; quoted, Schnoor 1992)
- 0.778 (freshwater fish, Call et al. 1984; quoted, Isensee 1991)
- 1.70 (quoted, Pait et al. 1992)

Sorption Partition Coefficient, log K_{OC}:
- 2.28 (soil, Beestman & Demming 1976; quoted, Kenaga 1980; Kenaga & Goring 1980)
- 2.32 (soil, calculated, Kenaga & Goring 1978)
- 2.30 (soil, Kenaga 1980; quoted, Nash 1988)
- 1.70 (selected, sediment/water, Schnoor & McAvoy 1981; quoted, Schnoor 1992)
- 1.91 (soil, av. for soils 2-7, Weber & Peter 1982; quoted, Glotfelty 1989)
- 2.08 (soil, screening model calculations, Jury et al. 1987b)
- 2.28 (Carsel 1989)
- 2.18, 2.23, 2.28, 2.53 (quoted values, Bottoni & Funari 1992)
- 2.23 (soil, 20-25°C, selected, Wauchope et al. 1992; Hornsby et al. 1996; quoted, Richards & Baker 1993; Lohninger 1994)
- 1.63-2.28 (quoted values, Montgomery 1993)
- 2.21 (selected, Wienhold & Gish 1994)

Half-Lives in the Environment:

Air:

Surface water: half-life of 23 days for 0.244 µg/mL to biodegrade in river water at 23°C with biodegradation rate of 0.030 d^{-1} (Schnoor et al. 1982; quoted, Muir 1991); half-life >6 weeks for 0.01-1.0 µg/ml to biodegrade in sewage effluent lake water at 28°C (Novick & Alexander 1985; quoted, Muir 1991).

Groundwater: half-lives of <6 months for 0.07 µg/mL to biodegrade in groundwater, and >15 months for 10.0 µg/mL to biodegrade in groundwater both at 25°C (Weidner 1974; quoted, Muir 1991).

Sediment:

Soil: measured dissipation rate of 0.077 d^{-1} (Zimdahl & Clark 1982; quoted, Nash 1988); half-life of 18 d from screening model calculations (Jury et al. 1987b); estimated dissipation rate of 0.020 and 0.036 d^{-1} (Nash 1988); field half-life of <1.5 week by using field lysimeters (Bowman 1990); degradation rate constant of $(4.52 \pm 0.192) \times 10^{-2}$ d^{-1} with half-life of 15.3 d in control soil and $(7.27 \pm 0.772) \times 10^{-2}$ d^{-1} with half-life of 9.53 d in pretreated soil in the field; $(2.77 \pm 0.226) \times 10^{-2}$ d^{-1} with half-life of 25 d in control soil and $(14.1 \pm 1.75) \times 10^{-2}$ d^{-1} with half-life of 4.93 d in pretreated soil once only in the laboratory (Walker & Welch 1991); selected field half-life of 15 d (Wauchope et al. 1992; Hornsby et al. 1996; quoted, Richards & Baker 1993); soil half-life of 30 d (quoted, Pait et al. 1992); soil half-life of 14-28 d (Di Guardo et al. 1994); dissipation half-life of 42 d from soil surface (Gish et al. 1995); 15 d (selected, Halfon et al. 1996).

Biota: biochemical half-life of 18 d from screening model calculations (Jury e tal. 1987b).

Environmental Fate Rate Constants or Half-Lives:
- Volatilization: 9000 d^{-1} and 49000 d^{-1} (measured and estimated values, Glotfelty et al. 1989); estimated half-life of 2444 d from 1 m depth of water at 20°C (Muir 1991).
- Photolysis: half-life of 2.25 h in distilled water (Tanaka et al. 1981; quoted, Cessna & Muir 1991); 640 ppb contaminated water in the presence of TiO_2 and H_2O_2 photodegraded to 3.5 ppb by 15 h solar irradiation with complete degradation after 75 h (Muszkat et al. 1992).
- Oxidation: calculated rate constant of 7×10^9 M^{-1} s^{-1} for the reaction with hydroxyl radicals in aqueous solutions at (24 ± 1)°C (Haag & Yao 1992).
- Hydrolysis: alkaline chemical hydrolysis half-life of >365 d (Schnoor & McAvoy 1981; quoted, Schnoor 1992).
- Biodegradation: half-lives of <6 months for 0.07 µg/mL to biodegrade in ground water, >15 months for 10.0 µg/mL to biodegrade in groundwater both at 25°C and <12 weeks for 3.2 µg/mL to biodegrade in soil-water suspension at 35°C (Weidner 1974; quoted, Muir 1991); half-life of 23 d for 0.244 µg/mL to biodegrade in river water at 23°C with biodegradation rate of 0.030 d^{-1} (Schnoor et al. 1982; quoted, Muir 1991); half-life of 18 d from screening model calculations (Jury et al. 1987b); half-life >6 weeks for 0.01-1.0 µg/mL to biodegrade in sewage effluent lake water at 28°C (Novick & Alexander 1985; quoted, Muir 1991).

Biotransformation:

Bioconcentration, Uptake (k_1) and Elimination (k_2) Rate Constants:

Common Name: Ametryn
Synonym: Amephyt, Ametrex, Evik, Gesapax
Chemical Name: 6-methylthio-2-(ethylamino)-4-(isopropylamino)-1,3,5-triazine; N-ethyl-N'-(1-methylethyl)-6-(methylthio)-1,3,5-triazine-2,4-diamine
Uses: herbicide to control broadleaf and grass weeds in corn, sugarcane, some citrus fruits, and in noncropland; also used as pre-harvest and post-harvest dessicant in potatoes to control crop and weeds.
CAS Registry No: 834-12-8
Molecular Formula: $C_9H_{17}N_5S$
Molecular Weight: 227.35
Melting Point (°C):
    84-85   (Khan 1980; Suntio et al. 1988; Herbicide Handbook 1989; Tomlin 1994)
    88-89   (Suntio et al. 1988; Milne 1995)
    84-86   (Worthing 1991; Montgomery 1993)
    84   (Kühne et al. 1995)
Boiling Point (°C):
Density (g/cm³ at 20°C):
    1.19   (Agrochemicals Handbook 1987; Worthing 1991; Montgomery 1993; Tomlin 1994)
Molar Volume (cm³/mol):
    277.5   (calculated-LeBas method, Suntio et al. 1988)
Molecular Volume (Å³):
Total Surface Area, TSA (Å²):
Dissociation Constant:
    4.00   (pK_a, Weber 1970; quoted, Bintein & Devillers 1994)
    4.10   (pK_a, Worthing 1991; Montgomery 1993)
    10.07   (pK_b, Wauchope et al. 1992; Hornsby et al. 1996)
    9.90   (pK_b, Tomlin 1994)
Heat of Fusion, ΔH_{fus}, kcal/mol:
Entropy of Fusion, ΔS_{fus}, cal/mol·K (e.u.):
Fugacity Ratio at 25°C (assuming ΔS_{fus} = 13.5 e.u.), F: 0.255

Water Solubility (g/m³ or mg/L at 25°C):
    700   (Woodford & Evans 1963; quoted, Shiu et al. 1990)
    405   (26°C, shake flask-UV at pH 3, Ward & Weber 1968; quoted, Shiu et al. 1990)
    195   (26°C, shake flask-UV at pH 7, Ward & Weber 1968; quoted, Shiu et al. 1990)
    192   (26°C, shake flask-UV at pH 10, Ward & Weber 1968; quoted, Shiu et al. 1990)
    185   (Martin & Worthing 1977; Herbicide Handbook 1978; quoted, Kenaga 1980; Kenaga & Goring 1980; Isensee 1991)
    185   (20°C, Khan 1980)
    194   (Weber et al. 1980; quoted, Willis & McDowell 1982)
    185   (20°C, Ashton & Crafts 1981; quoted, Shiu et al. 1990)
    185   (20°C, Agrochemicals Handbook 1987)
    185   (20°C, Verschueren 1983; quoted, Suntio et al. 1988; Shiu et al. 1990)
    185   (selected, Gerstl & Helling 1987)
    185   (20°C, selected, Suntio et al. 1988; quoted, Majewski & Capel 1995)

185   (Herbicide Handbook 1989; quoted, Shiu et al. 1990)
185   (20°C, Worthing 1991)
185   (20-25°C, selected, Wauchope et al. 1992; Hornsby et al. 1996)
185   (20°C, quoted, Montgomery 1993)
200   (Tomlin 1994)
207   (quoted, Kühne et al. 1995)
134   (calculated-group contribution fragmentation method, Kühne et al. 1995)
185   (20°C, Milne 1995)

Vapor Pressure (Pa at 25°C):
1.12×10^{-4} (20°C, Khan 1980)
1.12×10^{-4} (20°C, Ashton & Crafts 1981; quoted, Shiu et al. 1990)
1.12×10^{-4} (20°C, Verschueren 1983; quoted, Suntio et al. 1988)
1.12×10^{-4} (20°C, Agrochemicals Handbook 1987; Worthing 1991)
1.00×10^{-4} (20°C, selected, Suntio et al. 1988; quoted, Majewski & Capel 1995)
1.12×10^{-4} (20°C, Herbicide Handbook 1989)
4.40×10^{-4} (30°C, Herbicide Handbook 1989)
3.65×10^{-4} (20-25°C, selected, Wauchope et al. 1992; Hornsby et al. 1996)
1.12×10^{-4} (20°C, quoted, Montgomery 1993)
3.65×10^{-4} (Tomlin 1994)

Henry's Law Constant (Pa·m³/mol):
1.20×10^{-4} (20°C, calculated, Suntio et al. 1988; quoted, Majewski & Capel 1995)
1.38×10^{-4} (calculated-P/C, Montgomery 1993)
1.23×10^{-4} (calculated-P/C, this work)

Octanol/Water Partition Coefficient, log K_{ow}:
2.69   (Kenaga & Goring 1980; quoted, Finizio et al. 1991)
2.58   (selected, Dao et al. 1983)
2.58   (Gerstl & Helling 1987; quoted, Bintein & Devillers 1994)
2.82   (Worthing & Hance 1990; quoted, Finizio et al. 1991)
2.98   (shake flask, Biagi et al. 1991; quoted, Sicbaldi & Finizio 1993)
3.07, 3.07 (RP-HPLC, calculated, Finizio et al. 1991)
2.98   (selected, Magee 1991)
2.83   (Worthing 1991)
2.63-3.07 (quoted, Montgomery 1993)
2.88   (RP-HPLC, Sicbaldi & Finizio 1993)
2.63   (Tomlin 1994)
2.61   (shake flask-UV, Liu & Qian 1995)
2.58   (calculated-RP-HPLC-k', Liu & Qian 1995)
2.83   (Milne 1995)
2.98   (selected, Hansch et al. 1995)
2.88, 3.08, 2.82 (RP-HPLC, ClogP, calculated-S, Finizio et al. 1997)

Bioconcentration Factor, log BCF:
- 1.52 (calculated-S, Kenaga 1980; quoted, Isensee 1991)
- 1.32 (calculated-K_{OC}, Kenaga 1980)

Sorption Partition Coefficient, log K_{OC}:
- 2.59 (soil, Hamaker & Thompson 1972; quoted, Kenaga 1980; Kenaga & Goring 1980)
- 2.40 (soil, calculated, Kenaga & Goring 1978; quoted, Kenaga 1980)
- 2.59 (soil, Kenaga & Goring 1980; quoted, Liu & Qian 1995)
- 2.59 (Rao & Davidson 1980)
- 2.59, 2.86 (quoted, calculated-χ, Gerstl & Helling 1987)
- 2.59 (reported as log K_{OM}, Magee 1991)
- 2.51 (estimated as log K_{OM}, Magee 1991)
- 2.40-2.59, 2.58 (quoted values, Buttoni & Funari 1992)
- 2.48 (soil, 20-25°C, selected, Wauchope et al. 1992; Hornsby et al. 1996)
- 2.23-2.44 (quoted, Montgomery 1993)
- 2.48 (Tomlin 1994)
- 2.42 (calculated-K_{OW}, Liu & Qian 1995)

Half-Lives in the Environment:
Air:
Surface water:
Groundwater:
Sediment:
Soil: half-life of 6.0 months at 15°C and 4.5 months at 30°C in soils (Freed & Haque 1973); selected field half-life of 60 d (Wauchope et al. 1992; Hornsby et al. 1996); half-life of 70-129 d in soil (Tomlin 1994).
Biota:

Environmental Fate Rate Constants or Half-Lives:
Volatilization:
Photolysis: half-life of 10 h for 10 µg/mL to degrade in distilled water under >290 nm light and 3.3 h in 1% acetone solution (Burkhard & Guth 1976; quoted, Cessna & Muir 1991); half-lives of 2.25 h for 17% of 33 µg/mL to degrade in 0.2% aqueous solutions of the surfactant Triton X-100 and for 8% of 33 µg/mL to degrade in distilled water both under 300 nm light (Tanaka et al. 1981; quoted, Cessna & Muir 1991).
Oxidation:
Hydrolysis: half-life of 32 d at pH 1 and more than 200 d at pH 13 (Montgomery 1993).
Biodegradation:
Biotransformation:
Bioconcentration, Uptake (k_1) and Elimination (k_2) Rate Constants:

Common Name: Amitrole
Synonym: Amazole, Amitrol, Amizole, aminotriazole, Azolan, Azole, cytrol, Diurol
Chemical Name: 3-amino-1,2,4-triazole; 3-amino-s-triazole; 1H-1,2,4-triazol-3-amine
Uses: nonselective, foliage-applied herbicide in uncropped land and orchards to control perennial weeds in certain grasses.
CAS Registry No: 61-82-5
Molecular Formula: $C_2H_4N_4$
Molecular Weight: 84.08
Melting Point (°C):
    159   (Khan 1980; Herbicide Handbook 1989)
    157-159 (Worthing 1991; Montgomery 1993; Tomlin 1994)
Boiling Point (°C):
Density (g/cm^3 at 20°C):
    1.138   (Agrochemicals Handbook 1987; Herbicide Handbook 1989; Montgomery 1993; Tomlin 1994)
Molar Volume (cm^3/mol):
    85.1   (calculated-LeBas method, this work)
Molecular Volume (Å^3):
Total Surface Area, TSA (Å^2):
Dissociation Constant:
    9.83   (pK_b, Wauchope et al. 1992; Hornsby et al. 1996)
Heat of Fusion, ΔH_{fus}, kcal/mol:
    5.90   (DSC method, Plato 1972)
Entropy of Fusion, ΔS_{fus}, cal/mol·K (e.u.):
Fugacity Ratio at 25°C (assuming ΔS_{fus} = 13.5 e.u.), F: 0.048

Water Solubility (g/m^3 or mg/L at 25°C):
    252000   (Freed & Burschel 1957; quoted, Shiu et al. 1990)
    280000   (Martin 1961; Spencer 1981; quoted, Shiu et al. 1990)
    280000   (Bailey & White 1965; quoted, Shiu et al. 1990)
    soluble (Wauchope 1978)
    280000   (Khan 1980; Weber et al. 1980; Willis & McDowell 1982)
    280000   (Ashton & Crafts 1981)
    280000   (selected, Dao et al. 1983)
    280000   (Worthing 1983,91; quoted, Shiu et al. 1990)
    280000   (Agrochemicals Handbook 1987)
    278415   (selected, Gerstl & Helling 1987)
    280000   (Herbicide Handbook 1989)
    280000   (Reinert 1989; quoted, Howard 1991)
    360000   (20-25°C, selected, Wauchope et al. 1992; Lohninger 1994; Hornsby et al. 1996)
    280000   (20°C at pH 7, quoted, Montgomery 1993)
    280000   (23°C, Tomlin 1994)

Vapor Pressure (Pa at 25°C):
    <0.001   (Agrochemicals Handbook 1983; quoted, Howard 1991)
    <0.001   (Agrochemicals Handbook 1987)
    5.50×10^{-8} (20°C, Worthing 1991; Tomlin 1994)
    5.87×10^{-5} (20-25°C, selected, Wauchope et al. 1992; Hornsby et al. 1996)
    5.51×10^{-7} (20°C, quoted, Montgomery 1993)

Henry's Law Constant (Pa·m³/mol):
    <3.04x10⁻⁷   (calculated-P/C, Howard 1991)
    1.650x10⁻¹⁰   (20°C, calculated-P/C, Montgomery 1993)
    1.650x10⁻¹⁰   (calculated-P/C, this work)

Octanol/Water Partition Coefficient, log K_{ow}:
    0.52   (selected, Dao et al. 1983)
    0.52   (selected, Gerstl & Helling 1987)
    −0.15   (Reinert 1989; quoted, Howard 1991)
    −0.15   (quoted, Montgomery 1993)
    −0.87, −0.84 (pH 7)   (quoted, Hansch et al. 1995)

Bioconcentration Factor, log BCF:
    −0.301   (estimated-S, Lyman et al. 1982; quoted, Howard 1991)
    −0.347   (estimated-log K_{ow}, Lyman et al. 1982; quoted, Howard 1991)

Sorption Partition Coefficient, log K_{oc}:
    2.04   (soil, estimated-molecular topology & QSAR, Sabljic 1984; quoted, Howard 1991)
    0.23   (calculated-χ, Gerstl & & Helling 1987)
    1.26   (Reinert 1989)
    2.00   (soil, 20-25°C, selected, Wauchope et al. 1992; Hornsby et al. 1996)
    1.73-2.31   (quoted, Montgomery 1993)
    2.00   (estimated-chemical structure, Lohninger 1994)

Half-Lives in the Environment:
    Air: 3.8 d, based on a theoretical calculation for the vapor-phase reaction with hydroxyl radicals in the atmosphere at 25°C (GEMS 1986; quoted, Howard 1989); 3.2-32 h, based on estimated rate constant for the vapor-phase reaction with hydroxyl radicals in air (Atkinson 1987; quoted, Howard et al. 1991).
    Surface water: 672-4320 h, based on estimated aqueous aerobic biodegradation half-life (Howard et al. 1991).
    Groundwater: 1344-8640 h, based on estimated aqueous aerobic biodegradation half-life (Howard et al. 1991).
    Sediment:
    Soil: half-lives: 1.4, 1.6, 1.3, 92, 36, and 56 d with disappearance rates: 0.495, 0.433, 0.533, 0.0075, 0.0193, and 0.124 d⁻¹ at pH 6.0, 7.0, 8.0, 5.3, 6.5, and 7.5 (Hamaker 1972; quoted, Nash 1988); half-life of 1.5 month at 15°C and 1.0 month at 30°C in soils (Freed & Haque 1973); persistence of one month in soil (Wauchope 1978); persistence in soil for ca. 2-4 weeks (Herbicide Handbook 1989; Tomlin 1994); 672-4320 h, based on estimated aqueous aerobic biodegradation half-life (Howard et al. 1991); selected field half-life of 14 d (Wauchope et al. 1992; Hornsby et al. 1996).
    Biota:

Environmental Fate Rate Constants or Half-Lives:
- Volatilization:
- Photolysis:
- Oxidation: photooxidation half-life of 3.2-32 h in air, based on estimated rate constant for the vapor-phase reaction with hydroxyl radicals in air (Atkinson 1987; quoted, Howard et al. 1991).
- Hydrolysis:
- Biodegradation: aqueous aerobic half-life of 672-4032 h, based on reported half-lives in soil and water (Freed & Haque 1973; Reinert & Rogers 1987; quoted, Howard et al. 1991); aqueous anaerobic half-life of 2688-16128 h, based on estimated aqueous aerobic biodegradation half-life (Howard et al. 1991).
- Biotransformation:
- Bioconcentration, Uptake (k_1) and Elimination (k_2) Rate Constants:

Common Name: Atrazine
Synonym: Aatrex, Akikon, Aktikon, Aktinit, Atratol, Atred, Atrex, Candex, Fenamine, Gesaprim, Hungazin, Inakor, Primatol, Primaze, Radazine, Strazine, Triazine A, Vectal, Weedex A, Wonuk, Zeazine
Chemical Name: 2-chloro-4-(ethylamino)-6-(isopropylamino)-1,3,5-triazine; 6-chloro-N-ethyl-N'-(1-methylethyl)-1,3,5-triazine-diamine
Uses: pre-emergence and post-emergence herbicide to control some annual grasses and broadleaf weeds in corn, fallow land, rangeland, sorghum, noncropland, certain tropical plantations, evergreen nurseries, fruit crops, and lawns.
CAS Registry No: 1912-24-9
Molecular Formula: $C_8H_{14}ClN_5$
Molecular Weight: 215.68
Melting Point (°C):
    173-175 (Khan 1980; Herbicide Handbook 1989)
    172 (Karickhoff 1981)
    176 (Nkedi-Kizza et al. 1985; Kühne et al. 1995)
    174 (Suntio et al. 1988; Riederer 1990; Yalkowsky & Banerjee 1992)
    175-177 (Worthing 1991; Montgomery 1993)
    175.8 (Tomlin 1994)
    171-174 (Milne 1995; Montgomery 1993)
Boiling Point (°C):
Density (g/cm³ at 20°C):
    1.187 (Worthing 1991; Montgomery 1993; Tomlin 1994)
Molar Volume (cm³/mol):
    250.6 (calculated-LeBas method, Suntio et al. 1988; Riederer 1990)
Molecular Volume (Å³):
Total Surface Area, TSA (Å²):
    219.0 (Nkedi-Kizza et al. 1985)
Dissociation Constant:
    1.68 (pK_a, Weber 1970; Somasundaram et al. 1991; Bintein & Devillers 1994)
    1.70 (pK_a, Weber et al. 1980; Willis & McDowell 1982; Worthing 1991; Francioso et al. 1992; Montgomery 1993; Tomlin 1994)
    1.60 (pK_a, Yao & Haag 1991)
    12.32 (pK_b, Wauchope et al. 1992; Hornsby et al. 1996)
    1.62 (pK_a, 20°C, Montgomery 1993)
Heat of Fusion, ΔH_{fus}, kcal/mol:
    9.70 (DSC method, Plato 1972)
Entropy of Fusion, ΔS_{fus}, cal/mol·K (e.u.):
Fugacity Ratio at 20°C (assuming ΔS_{fus} = 13.5 e.u.), F: 0.0336

Water Solubility (g/m³ or mg/L at 25°C):
    70.0 (26°C, Bailey & White 1965; quoted, Shiu et al. 1990)
    50.0 (Günther et al. 1968; quoted, Davies & Dobbs 1984)
    31.1 (26°C, shake flask-UV at pH 3, Ward & Weber 1968; quoted, Freed 1976; Shiu et al. 1990)
    34.9 (26°C, shake flask-UV at pH 7, Ward & Weber 1968; quoted, Freed 1976; Shiu et al. 1990)
    36.8 (26°C, shake flask-UV at pH 10, Ward & Weber 1968; quoted, Freed 1976; Shiu et al. 1990)

98.0	(50°C, Getzen & Ward 1971; quoted, Shiu et al. 1990)
33.0	(shake flask-GC, Hörmann & Eberle 1972; quoted, Freed 1976; Shiu et al. 1990)
29.9	(shake flask-UV, Hurle & Freed 1972; quoted, Freed 1976; Shiu et al. 1990)
30.0	(20°C, Weber 1972; Worthing 1987; quoted, Muir 1991)
32.0	(Freed 1976; quoted, Jury et al. 1984; Glotfelty et al. 1989)
31.5	(Spencer 1976; quoted, Suntio et al. 1988)
33.0	(Wauchope 1978)
33.0	(selected, Ellgehausen et al. 1980; Nkedi-Kizza et al. 1985)
33.0	(Kenaga 1980; Kenaga & Goring 1980; quoted, Bysshe 1982; Isensee 1991)
33.0	(27°C, Khan 1980)
35.0	(Weber et al. 1980; quoted, Willis & McDowell 1982)
33.0	(27°C, Ashton & Crafts 1981; Herbicide Handbook 1989; Pait et al. 1992)
30.0	(20°C, Burkhard & Guth 1981; quoted, Howard 1991)
30.0	(shake flask-HPLC, Ellgehausen et al. 1981; quoted, Shiu et al. 1990)
33.0	(selected, Schnoor & McAvoy 1981; quoted, Schnoor 1992)
33.0	(Lyman et al. 1982,90; quoted, Nash 1988; Hemond & Fechner 1994)
24.0	(quoted, Thomas 1982)
32.0	(Beste & Humburg 1983; Jury et al. 1983; quoted, Taylor & Glotfelty 1988; Grover 1991)
70.0	(Windholz 1983; quoted, Somasundaram et al. 1991)
28.0	(20°C, Agrochemicals Handbook 1987)
33.0	(selected, Gerstl & Helling 1987)
30.0	(20°C, selected, Suntio et al. 1988; quoted, Riederer 1990; Majewski & Capel 1995)
32.0	(Taylor & Glotfelty 1988; quoted, Shiu et al. 1990)
29.9, 33, 70	(literature data variability, Heller et al. 1989)
30.0	(20°C, Worthing 1991)
30.0	(20°C, selected, Francioso et al. 1992)
33.0	(MAFF 1992a; quoted, Meakins et al. 1994)
33.0	(20-25°C, selected, Wauchope et al. 1992; Hornsby et al. 1996)
60.8	(selected, Yalkowsky & Banerjee 1992)
28.0	(20°C, Montgomery 1993)
33.0	(27°C, Montgomery 1993)
33.0	(selected, Sieber et al. 1994; Wienhold & Gish 1994; Halfon et al. 1996)
33.0	(20°C, Tomlin 1994; quoted, Sanchez-Camazano et al. 1995)
32.0	(selected, Gish et al. 1995)
28.0	(Milne 1995)
30.5	(quoted, Kühne et al. 1995)
24.2	(calculated-group contribution fragmentation method, Kühne et al. 1995)
29.0	(selected, Pinsuwan et al. 1995)
30.0	(selected, Iglesias-Jimenez et al. 1996)
36.5, 114	(quoted, predicted-AQUAFAC, Lee et al. 1996)

Vapor Pressure (Pa at 25°C):
9.00×10^{-5} (gas saturation, Friedrich & Stammbach 1964; quoted, Jury et al. 1983; Glotfelty et al. 1989; Taylor & Spencer 1990; Grover 1991)
4.00×10^{-5} (20°C, Weber 1972; Worthing 1987; quoted, Muir 1991)

3.99×10^{-5} (20°C, gas saturation, extrapolated from Friedrich & Stammbach 1964, Spencer 1976; quoted, Suntio et al. 1988)
4.00×10^{-5} (20°C, Worthing 1979, 1991; quoted, Dobbs et al. 1984)
4.00×10^{-5} (20°C, Hartley & Graham-Bryce 1980; Beste & Humburg 1983; quoted, Taylor & Glotfelty 1988)
4.00×10^{-5} (20°C, Khan 1980)
4.00×10^{-5} (20-25°C, Weber et al. 1980; quoted, Willis & McDowell 1982)
4.00×10^{-5} (20°C, Ashton & Crafts 1981)
1.33×10^{-4} (selected, Schnoor & McAvoy 1981; quoted, Schnoor 1992)
3.7×10^{-5} (20°C, extrapolated from gas saturation measurement, ln P(Pa) = 36.8 − 13778/T, for temp range 51-81.5°C, Grayson & Fosbraey 1982)
1.13×10^{-4} (Thomas 1982; quoted, Nash 1988)
3.99×10^{-5} (selected, Nkedi-Kizza et al. 1985)
4.00×10^{-5} (20°C, Agrochemicals Handbook 1987)
4.00×10^{-5} (20°C, selected, Suntio et al. 1988; quoted, Riederer 1990; Majewski & Capel 1995)
8.70×10^{-5} (selected, Nash 1989)
3.99×10^{-5} (20°C, Herbicide Handbook 1989)
18.6×10^{-5} (30°C, Herbicide Handbook 1989)
4.05×10^{-5} (Riederer 1990; quoted, Hemond & Fechner 1994)
9.00×10^{-5} (selected, Taylor & Spencer 1990)
4.00×10^{-5} (20°C, selected, Francioso et al. 1992)
3.85×10^{-5} (20-25°C, selected, Wauchope et al. 1992; Hornsby et al. 1996)
4.00×10^{-5} (20°C, Montgomery 1993)
2.00×10^{-5} (selected, Sieber et al. 1994)
3.90×10^{-5} (Tomlin 1994)
3.90×10^{-5} (selected, Halfon et al. 1996)

Henry's Law Constant (Pa·m^3/mol):
6.20×10^{-4} (calculated-P/C, Jury et al. 1983, 1984, 1987a; Jury & Ghodrati 1989; quoted, Grover 1991)
2.90×10^{-4} (20°C, calculated-P/C, Suntio et al. 1988; quoted, Majewski & Capel 1995)
6.19×10^{-4} (calculated-P/C, Taylor & Glotfelty 1988)
5.70×10^{-4} (calculated-P/C, Nash 1989)
3.04×10^{-4} (Riederer 1990; quoted, Hemond & Fechner 1994)
2.66×10^{-4} (calculated-P/C, Howard 1991)
2.89×10^{-4} (20°C, calculated-P/C, Muir 1991)
3.08×10^{-4} (20°C, calculated-P/C, Montgomery 1993)
1.00×10^{-3} (calculated-P/C, Sieber et al. 1994)
6.20×10^{-4} (Gish et al. 1995)
2.88×10^{-4} (calculated-P/C, this work)

Octanol/Water Partition Coefficient, log K_{ow}:
2.68 (Brown 1978; quoted, Kenaga & Goring 1980; Lyman 1982, 1990; Hodson & Williams 1988; Finizio et al. 1991; Hemond & Fechner 1994)
2.75 (shake flask-GC, Erkell & Walum 1979; quoted, Sicbaldi & Finizio 1993)
2.63 (Veith et al. 1979, 1980; quoted, Bysshe 1982; Klein et al. 1988)

2.75   (selected, Ellgehausen et al. 1980)
2.35   (quoted, Rao & Davidson 1980; Karickhoff 1981; Suntio et al. 1988)
2.71   (selected, Brown & Flagg 1981; quoted, Karickhoff 1981; McDuffie 1981)
2.21   (HPLC-k', McDuffie 1981)
2.75   (shake flask, Ellgehausen et al. 1981)
2.52   (calculated-S, Ellgehausen et al. 1981; quoted, Sicbaldi & Finizio 1993)
2.63   (Veith & Kosian 1982; quoted, Saito et al. 1992)
2.68   (selected, Dao et al. 1983; Gerstl & Helling 1987)
2.80   (Elgar 1983; quoted, Suntio et al. 1988)
2.05   (RP-HPLC-k', Braumann et al. 1983)
2.64   (shake flask-GC, Geyer et al. 1984; quoted, Sicbaldi & Finizio 1993)
2.75   (Hansch & Leo 1985; quoted, Howard 1991)
2.64   (OECD method 1981, Kerler & Schönherr 1988; quoted, Riederer 1990)
2.60   (selected, Klein et al. 1988)
2.40   (selected, Suntio et al. 1988; quoted, Finizio et al. 1991)
2.33   (quoted, Bintein & Devillers 1994)
2.65   (selected, Travis & Arms 1988)
2.68   (Lopez-Avila et al. 1989; quoted, Somasundaram et al. 1991)
2.61, 2.61 (RP-HPLC, calculated, Finizio et al. 1991)
2.75   (selected, Geyer et al. 1991; Magee 1991)
2.34   (Worthing 1991; selected, Francioso et al. 1992; Sieber et al. 1994)
2.70   (MAFF 1992a; quoted, Meakins et al. 1994)
2.33-2.80 (quoted values, Montgomery 1993)
2.42   (RP-HPLC, Sicbaldi & Finizio 1993)
2.58   (Aquasol Database 1994; quoted, Pinsuwan et al. 1994)
2.50   (Tomlin 1994; quoted, Sanchez-Camazano et al. 1995)
2.27   (shake flask-UV, Liu & Qian 1995)
2.34   (Milne 1995)
2.58, 2.60 (selected, calculated-f const., Pinsuwan et al. 1995)
2.61   (selected, Hansch et al. 1995; Devillers et al. 1996)
2.50   (selected, Iglesias-Jimenez et al. 1996)
2.43, 2.40, 3.45 (RP-HPLC, ClogP, calculated-S, Finizio et al. 1997)

Bioconcentration Factor, log BCF:
        −2.00   (vegetation, correlated-K_{ow}, Beynon et al. 1972; quoted, Travis & Arms 1988)
        1.04   (Metcalf & Sanborn 1975; quoted, Kenaga & Goring 1980; Isensee 1991)
        1.00   (Isensee 1976; quoted, Isensee 1991)
        0.50   (whitefish, Burkhard & Guth 1976; quoted, Howard 1991)
        0.90   (fathead minnows, Veith et al. 1979; quoted, Howard 1991)
        0.30   (catfish, Ellgehausen et al. 1980; quoted, Howard 1991)
        0.26   (*Daphnia magna*, wet wt. basis, Ellgehausen et al. 1980; quoted, Geyer et al. 1991)
        0.48   (*Corygonus fera.* at 12°C, Gunkel & Streit 1980; quoted, Davies & Dobbs 1984)
        1.93, 0.845 (calculated-S, K_{OC}, Kenaga 1980; quoted, Isensee 1991)
        <0.90   (Veith et al. 1980; quoted, Bysshe 1982)
        1.90   (selected, Schnoor & McAvoy 1981; quoted, Schnoor 1992)
        1.93   (estimated-S, Bysshe 1982)

1.77   (estimated-K_{OW}, Bysshe 1982)
0.90   (fathead minnows, Veith & Kosian 1982; quoted, Saito et al. 1992)
2.00   (mottled sculpin, Lynch et al. 1982; quoted, Howard 1991)
1.60   (activated sludge, Freitag et al. 1984)
1.00   (golden ide, Freitag et al. 1985; quoted, Howard 1991)
0.477, 0.954, 0.845, 0.778 (zebrafish: egg, embryo, yolk sac fry, juvenile; Görge & Nagel 1990)
0.78   (*Brachydanio rerio*, Görge & Nagel 1990; quoted, Devillers et al. 1996)
0.983   (*Hydrilla*, Hinman & Klaine 1992)
0.845   (quoted, Pait et al. 1992)
1.98   (*Scenedesmus acutus*, Wang et al. 1996)
0.748   (*Ictalurus melas*, Wang et al. 1996)
0.230   (*Daphnia magna*, Wang et al. 1996)

Bioaccumulation Factor, log BF:
1.710   (algae, Ellgehausen et al. 1980; quoted, Wang et al. 1996)
0.329   (catfish, Ellgehausen et al. 1980; quoted, Wang et al. 1996)
0.261   (daphnids, Ellgehausen et al. 1980)
1.72, 0.477, 1.60 (algae, fish, sludge, Klein et al. 1984)
1.70, <1.00, 1.60 (algae, fish, sludge, Freitag et al. 1985)

Sorption Partition Coefficient, log K_{OC}:
2.17   (soil, Hamaker & Thompson 1972; quoted, Kenaga & Goring 1980; quoted, Karickhoff 1981; Bysshe 1982; Hodson & Williams 1988)
2.09   (av. of 4 soils, Rao & Davidson 1979; Davidson et al. 1980; quoted, Howard 1991)
2.17, 2.81 (exptl, calculated, Kenaga & Goring 1980; Kenaga 1980)
2.20   (av. soils/sediments, Rao & Davidson 1980; quoted, Lyman 1982)
2.21   (av. of 56 soils from lit. review, Rao & Davidson 1980; quoted, Karickhoff 1981; Jury et al. 1983; Glotfelty et al. 1989; Howard 1991)
2.33   (a Georgia pond sediment, Brown & Flagg 1981; quoted, Howard 1991; Muir 1991)
1.59   (a Swiss soil, Burkhard & Guth 1981; quoted, Howard 1991)
3.11, 2.31 (estimated-S, S & M.P., Karickhoff 1981)
1.94, 2.42 (estimated-K_{OW}, Karickhoff 1981)
0.7-1.48 (selected, sediment/water, Schnoor & McAvoy 1981; quoted, Schnoor 1992)
2.18   (soil, Thomas 1982; quoted, Nash 1988, 1989)
2.29-3.18 (Wolf & Jackson 1982; quoted, Muir 1991)
3.23-4.13 (Means & Wijayaratne 1982; quoted, Muir 1991)
2.21   (soil average, Jury et al. 1983; quoted, Grover 1991)
1.63-3.29 (Wauchope & Myers 1985; quoted, Muir 1991)
2.24, 2.46 (quoted, calculated-χ, Gerstl & Helling 1987)
2.20   (soil, screening model calculations, Jury et al. 1987a,b; Jury & Ghodrati 1989)
2.08-2.17 (Carsel 1989)
2.17, 2.21 (reported, estimated as log K_{OM}, Magee 1991)
2.0, 2.18, 2.17-2.81, 2.26 (quoted values, Bottoni & Funari 1992)

2.27, 2.41, 2.59, 2.16 (soils, no. 1, 2, 3, 4; Francioso et al. 1992)
2.00 (soil, 20-25°C, selected, Wauchope et al. 1992; quoted, Dowd et al. 1993; Richards & Baker 1993; Wienhold & Gish 1994; Hornsby et al. 1996)
1.95-2.71 (quoted values, Montgomery 1993)
2.40 (estimated-chemical structure, Lohninger 1994)
1.95-2.19 (Tomlin 1994)
1.81 (soil, HPLC-ring test, Kördel et al. 1995a)
1.81 (soil, HPLC-screening method, Kördel et al. 1993, 1995b)
2,17, 2.23 (quoted, calculated-K_{ow}, Liu & Qian 1995)
1.00 (sediment/water, Chung et al. 1996)

Half-Lives in the Environment:
  Air: 2.6 h, based on estimated rate constant of 147.2×10^{-12} cm^3/molecule·sec at 25°C for the vapor-phase reaction with hydroxyl radicals in air (Atkinson 1987; quoted, Howard 1991).
  Surface water: estimated half-life of 3.21 d in aqueous solution from river die-away tests (Furmidge & Osgerby 1967; quoted, Scow 1982); half-life of 1-4 weeks in esturine systems (Jones et al. 1982; quoted, Meakins et al. 1994); under laboratory conditions in distilled water and river water was completely degraded after 21.3 and 7.3 h, respectively (Mansour et al. 1989; quoted, Montgomery 1993); half-lives of 3.2 d to 7-8 months in aquatic environments (Eisler 1985; quoted, Day 1991); half-lives of 35.6-168 h in surface water system of a small stream in Iowa by water quality analyses (Kolpin & Kalkhoff 1993); half-life of 235 d at 6°C, 164 d at 22°C in darkness, 59 d under sunlight conditions for river water at pH 7.3; 130 d at 22°C in darkness for filtered river water at pH 7.3 and half-life of 200 d at 22°C in darkness, 169 d under sunlight conditions for seawater, pH 8.1 (Lartiges & Garrigues 1995).
  Groundwater: half-lives of 6-15 months for 0.72-10 μg/mL to biodegrade slowly at 25°C (Weidener 1974; quoted, Muir 1991).
  Sediment: half-life of 145 d in a Wisconsin Lake sediment (Armstrong et al. 1967; quoted, Jones et al. 1982; Means et al. 1983) and half-life of approximately 30 d for Chesapeake Bay sediment (Ballantine et al. 1978; quoted, Jones et al. 1982); half-life of 7-28 d for 0.1 μg/mL to rapid degrade in both aerobic and low oxygen systems in estuarine sediment/water at 12-35°C (Jones et al. 1982; quoted, Muir 1991); aerobic half-life of >35 d for 0.1-1.0 μg/mL to slowly biodegrade in sediment/water at 25°C (Wolf & Jackson 1982; quoted, Muir 1991).
  Soil: half-lives in aqueous buffered solutions in soil at 25°C and pH 1, 2, 3, 4, 11, 12, and 13 were reported to be 3.3, 14, 58, 240, 100, 12.5, and 1.5 d, respectively (Armstrong et al. 1967; quoted, Montgomery 1993); half-life of 3-5 yr in agricultural soils (Armstrong et al. 1967; quoted, Jones et al. 1982); estimated persistence of 10 months in soil (Kearney et al. 1969; quoted, Jury et al. 1987a); half-lives at 25°C and pH 4 with and without fulvic acid (2%) were 1.73 and 244 d (Li & Felbeck 1972; quoted, Montgomery 1993); persistence of 10 months in soil (Edwards 1973; quoted, Morrill et al. 1982); half-life of 6.0 months at 15°C and 2.0 months at 30°C in soils (Freed & Haque 1973); persistence of 12 months (Wauchope 1978); correlated half-lives: 37 d at pH 5.1-7.0, and 28 d at pH

7.7-8.2 (Boddington Barn soil, Hance 1979), about 30 d at pH 4.6-5.3 and 40 d at pH 6.3-8.0 (Triangle soil, Hance 1979); half-life of 37 d in agricultural soils (Dao et al. 1979; quoted, Jones et al. 1982); estimated first-order half-life of 36.5 d from biodegradation rate constant of 0.019 d^{-1} by soil incubation die-away studies (Rao & Davidson 1980; quoted, Scow 1982); half-lives in a Hatzenbühl soil at pH 4.8 and Neuhofen soil at pH 6.5 at 22°C were 53 and 113 d, respectively (Burkhard & Guth 1981; quoted, Montgomery 1993); half-life of 1-6 months (Jones et al. 1982; quoted, Meakins et al. 1994); moderately persistent in soils with half-life of 20-100 d (Willis & McDowell 1982); biodegradation half-life of 71 d from screening model calculations (Jury et al. 1984; 1987a,b; Jury & Ghodrati 1989); half-lives of about 6-10 weeks (Hartley & Kidd 1987; quoted, Montgomery 1993); field half-life of 4 weeks by using lysimeters (Bowman 1990); half-lives at 20 ± 1°C from soil surfaces: 655 to >1000 d in peat soil and 143-939 d in sandy soil (Dörfler et al. 1991); degradation rate constant of $(1.20 \pm 0.097) \times 10^{-2}$ d^{-1} with half-life of 57.8 d in control soil and $(1.01 \pm 0.034) \times 10^{-2}$ d^{-1} with half-life of 68.6 d in pretreated soil once only in the laboratory (Walker & Welch 1991); half-life based on extractable residues, was approximately 21 d in the microcosm studies, compared to 14 d in surface field soil (Winkelmann & Klaine 1991); selected field half-life of 60 d (Wauchope et al. 1992; quoted, Dowd et al. 1993; Richards & Baker 1993; Hornsby et al. 1996); soil half-life of 130 days (quoted, Pait et al. 1992); field half-life of 35-50 days in soil and water but may be longer under cold or dry conditions; 105 to >200 d under groundwater conditions, depending on test system (Wood et al. 1991; quoted, Tomlin 1994); first-order rate constants of -0.017 to -0.003 d^{-1} with corresponding half-lives of 41 to 231 d in Ames, Iowa, soil (Kruger et al. 1993); dissipation half-life of 71 d from soil surface (Gish et al. 1995); 60 d (selected, Halfon et al. 1996).

Biota: 0.03 h in algae, 1.52 d in catfish and 9.5 h in daphnids (Ellgehausen et al. 1980); biochemical half-life of 64 d from screening model calculations (Jury et al. 1987b); half-lives at 20 ± 1°C from plant surfaces: 25.6 d in bean, 24.3 d in turnips and 14.6 d in oats (Dörfler et al. 1991).

Environmental Fate Rate Constants or Half-Lives:
Volatilization: initial rate constant of 6.4×10^{-4} h^{-1} and predicted rate constant of 4.2×10^{-4} h^{-1} from soil with a half-life of 1650 h (Thomas 1982); half-life of 97 d (Jury et al. 1983; quoted, Grover 1991); rate constants of 1100 d^{-1} (measured) and 6000 d^{-1} (estimated) (Glotfelty et al. 1989); half-lives at 20 ± 1°C from soil surfaces: 655->1000 d in peat soil and 143-939 d in sandy soil; half-lives at 20 ± 1°C from plant surfaces: 25.6 d in bean, 24.3 d in turnips and 14.6 d in oats (Dörfler et al. 1991).
Photolysis: half-lives of 10 ppm aqueous solutions: 19 ± 9 h under summer sunlight of 9.1 h/d exposure and 61 ± 29 h under spring sunlight of 3.7 h/d exposure (Burkhard et al. 1975); half-lives of 4.9 h for 10 μg/mL to degrade in 1% acetone solution and 25 hours for 10 μg/mL to degrade in distilled water both under >290 nm light (Burkhard & Guth 1976); nearsurface direct sunlight photolysis rate constant of 9×10^{-6} d^{-1} with half-life of 81,000 d (Schnoor & McAvoy 1981; quoted, Schnoor 1992); half-life of 2.25 h for 17-27% of 100 μg/mL to degrade in distilled water under

300 nm light (Tanaka et al. 1981; quoted, Cessna & Muir 1991); rate of photolytic degradation was slightly higher in water (half-life of 3-12 d) than in sediments (half-life of 1-4 weeks) (Jones et al. 1982; quoted, Montgomery 1993); 40 ppb contaminated water in presence of TiO_2 and H_2O_2 degraded to 4 ppb after 15 h by solar irradiation with complete degradation after 75 h (Muszkat et al. 1992).

Oxidation: photooxidation half-life of 2.6 h in air, based on estimated rate constant of 147.2×10^{-12} cm^3/molecule·s at 25°C for the vapor-phase reaction with hydroxyl radicals in air (Atkinson 1987; quoted, Howard et al. 1991); rate constant of 5.9×10^9 M^{-1} s^{-1} for the reaction (photo-Fenton with reference to acetophenone) with hydroxyl radicals in aqueous solutions at pH 3.6 and (24 ± 1)°C (Buxton et al. 1988; quoted, Faust & Hoigné 1990; Haag & Yao 1992); rate constant of $(2.6 \pm 0.4) \times 10^9$ M^{-1} s^{-1} for the reaction (photo-Fenton with reference to acetophenone) with hydroxyl radicals in aqueous solutions at pH 3.6 and (24 ± 1)°C (Haag & Yao 1992).

Hydrolysis: half-lives in aqueous buffered solutions at 25°C and pH 3.1 (citrate buffer), 11.1 (carbonate buffer), 3.9 (phosphate buffer + sterile lake sediment) ~70, ~75, and ~2 days, respectively (Armstrong et al. 1967; quoted, Muir 1991); half-lives in aqueous buffered solutions in soil at 25°C and pH 1, 2, 3, 4, 11, 12, and 13 were reported to be 3.3, 14, 58, 240, 100, 12.5, and 1.5 d, respectively (Armstrong et al. 1967; quoted, Montgomery 1993); rate of hydrolysis was found to drastically increase upon small additions of humic materials, indicating hydrolysis could be catalyzed: for example, half-life at pH 4 and 25°C of 244 d without an additive while it could be drastically decreased to 1.37 days with the presence of 2% humic acid (Li & Felbeck 1972; quoted, Howard 1991; Montgomery 1993); rate constants of 3.9×10^{-5} M^{-1} s^{-1} and 7.6×10^{-5} M^{-1} s^{-1} with half-lives of 66 and 81 d in aqueous solutions of pH 3.1 and 11.1 (Wolfe et al. 1976; quoted, Muir 1991); rate constants: 19.9 d^{-1} at pH 2.9, 3.99 d^{-1} at pH 4.5, 1.74 d^{-1} at pH 6.0, and 0.934 d^{-1} at pH 7.0 with corresponding half-lives of 34.8, 174, 398, and 742 d all at 25°C in 0.5 mg/mL concn. of aqueous fulvic acid (Khan 1978; quoted, Howard 1991; Montgomery 1993); rate constants: 28.4 d^{-1} at pH 2.8, 12.6 d^{-1} at pH 4.5, 3.16 d^{-1} at pH 6.0, and 1.23 d^{-1} at pH 7.0 with corresponding half-lives of 24.4, 55.0, 219, and 563 d all at 25°C in 1.0 mg/mL concn. of aqueous fulvic acid (Khan 1978); rate constants: 151 d^{-1} at pH 2.4, 43.7 d^{-1} at pH 4.5, 13.2 d^{-1} at pH 6.0, and 7.93 d^{-1} at pH 7.0 with corresponding half-lives of 4.60, 15.9, 52.5 and 87.3 d all at 25°C in 5.0 mg/mL concn. of aqueous fulvic acid (Khan 1978); calculated rate constant of 9.30×10^{-6} s^{-1} with half-life of 86 d at 20°C in a buffer at pH 5 (Burkhard & Guth 1981; quoted, Muir 1991); half-lives of >3 months (in sterile buffer solution at pH 7.2) and >14 d (in sterile mineral salt solution at pH 7.2) for 20 μg/mL to hydrolyze at 23°C (Geller 1980; quoted, Muir 1991); alkaline chemical hydrolysis rate constant of 1×10^{-16} M^{-1} s^{-1} with half-life of 742 d (Schnoor & McAvoy 1981; quoted, Schnoor 1992); 1771 years at pH 7 and 25°C (Montgomery 1993).

Biodegradation: half-life of 64 d in soil (Armstrong et al. 1967; Dao et al. 1979; quoted, Means et al. 1983); estimated half-life of 3.21 d in aqueous solution from river die-away tests (Furmidge & Osgerby 1967; quoted, Scow 1982); aerobic half-life of >90 d for 10-20 μg/mL to degrade in soil-water suspension (Goswami & Green 1971; quoted, Muir 1991); rate constant of 0.019 d^{-1} by soil incubation die-away studies (Rao & Davidson 1980; quoted, Scow 1982); aerobic half-life of >35 d for 0.1-1.0 µg/mL to slowly biodegrade in sediment/water at 25°C (Wolf & Jackson 1982; quoted, Muir 1991); half-lives of 36 and 110 d in soil (Jones et al. 1982; quoted, Means et al. 1983); half-life of 71 d for a 100 d leaching and screening test in 0-10 cm depth of soil (Jury et al. 1983, 1984, 1987a; Jury & Ghodrati 1989; quoted, Grover 1991); 0.22 d^{-1} of aerobic degradation rate observed in incubations of river water samples (Lyman et al. 1990; quoted, Hemond & Fechner 1994); half-lives of aqueous atrazine using first-order dacay rate: 201 d with 12 mM methanol, 289 d with 6 mM sodium acetate, 164 days with 6 mM acetic acid and 200 d with 2 mM glucose; however the half-life without any organic amendments was 224 d in the sample reactors (Chung et al. 1996).

Biotransformation:

Bioconcentration, Uptake (k_1) and Elimination (k_2) Rate Constants:

k_2: 0.0248, 1.26 h^{-1} (algae, daphnids, Ellgehausen et al. 1980)

k_2: 27.2 d^{-1} (catfish, Ellgehausen et al. 1980)

k_1: 2.4, 30, 19.0 h^{-1} (zebrafish: egg, yolk sac fry, juvenile; Görge & Nagel 1990)

k_1: 227.0 h^{-1} (*Scenedesmus acutus*, Wang et al. 1996)

k_2: 2.354 h^{-1} (*Scenedesmus acutus*, Wang et al. 1996)

k_1: 0.412 h^{-1} (*Ictalurus melas*, Wang et al. 1996)

k_2: 0.073 h^{-1} (*Ictalurus melas*, Wang et al. 1996)

k_1: 2.027 h^{-1} (*Daphnia magna*, Wang et al. 1996)

k_2: 1.161 h^{-1} (*Daphnia magna*, Wang et al. 1996)

Common Name: Barban
Synonym: Barbamate, Barbane, Carbine, Carbyne, CBN, Chlorinat
Chemical Name: carbamic acid, (3-chlorophenyl)-, 4-chloro-2-butynyl ester; 4-chlorobut-2-ynyl 3-chloro-carbanilate; 4-chloro-2-butynyl 3-chlorophenylcarbamate
Uses: herbicide for post-emergence control of wild oats in wheat, barley, broad beans, field beans, soybeans, peas, sugar beet, flax, lucerne, lentils, mustard, oilseed rape, sunflowers, etc.
CAS Registry No: 101-27-9
Molecular Formula: $C_{11}H_9Cl_2NO_2$
Molecular Weight: 258.1
Melting Point (°C):
    75   (Khan 1980; Herbicide Handbook 1989)
    75-76   (Milne 1995)
Boiling Point (°C):
Density (g/cm³ at 20°C):
    1.403   (25°C, Agrochemicals Handbook 1987)
Molar Volume (cm³/mol):
    262.8   (calculated-LeBas method, this work)
Molecular Volume (Å³):
Total Surface Area, TSA (Å²):
Dissociation Constant pK_a:
Heat of Fusion, ΔH_{fus}, kcal/mol:
Entropy of Fusion, ΔS_{fus}, cal/mol·K (e.u.):
Fugacity Ratio at 25°C (assuming ΔS_{fus} = 13.5 e.u.), F: 0.313

Water Solubility (g/m³ or mg/L at 25°C):
    15.0   (Swezey & Nex 1961; quoted, Shiu et al. 1990)
    11.0   (20°C, Weber 1972; Worthing 1987; quoted, Muir 1991)
    11.0   (Martin & Worthing 1977; quoted, Kenaga 1980)
    11.0   (Khan 1980; Weber et al. 1980; Willis & McDowell 1982)
    11.0   (Ashton & Crafts 1981)
    11.0   (Agrochemicals Handbook 1987)
    11.0   (selected, Gerstl & Helling 1987)
    11.0   (Worthing 1987; quoted, Majewski & Capel 1995)
    11.0   (Herbicide Handbook 1989, Merck Index 1989; quoted, Milne 1995)
    11.0   (20-25°C, selected, Augustijn-Beckers et al. 1994; Hornsby et al. 1996)

Vapor Pressure (Pa at 25°C):
    1.33×10^{-3} (20°C, Weber 1972; Worthing 1987; quoted, Muir 1991)
    5.00×10^{-5} (Agrochemicals Handbook 1987)
    1.60×10^{-4} (Worthing 1987; quoted, Majewski & Capel 1995)
    5.05×10^{-5} (Herbicide Handbook 1989)
    5.07×10^{-5} (20-25°C, selected, Augustijn-Beckers et al. 1994; Hornsby et al. 1996)

Henry's Law Constant (Pa·m³/mol):
    0.00117   (20°C, calculated-P/C, Muir 1991)
    1.17   (calculated-P/C as per Worthing 1987, Majewski & Capel 1995)
    0.00117   (calculated-P/C, this work)

Octanol/Water Partition Coefficient, log K_{OW}:
    2.68   (selected, Gerstl & Helling 1987)

Bioconcentration Factor, log BCF:
    2.20   (calculated-S, Kenaga 1980)

Sorption Partition Coefficient, log K_{OC}:
    3.06   (soil, calculated-S, Kenaga 1980)
    2.66   (calculated-χ, Gerstl & Helling 1987)
    3.00   (20-25°C, estimated, Augustijn-Beckers et al. 1994; Hornsby et al. 1996)

Half-Lives in the Environment:
    Air:
    Surface water:
    Groundwater:
    Sediment:
    Soil: estimated persistence of 2 months (Kearney et al. 1969; quoted, Jury et al. 1987); persistence of 2 weeks in soil (Edwards 1973; quoted, Morrill et al. 1982); persistence of about 3 weeks in soil (Herbicide Handbook 1989); selected field half-life of 5 days (Augustijn-Beckers et al. 1994; Hornsby et al. 1996).
    Biota:

Environmental Fate Rate Constants or Half-Lives:
    Volatilization: estimated half-life of 6690 days from 1 m depth of water at 20°C (Muir 1991).
    Photolysis: half-life of 2.25 hours for 22-99% of 10 μg/ml to degrade in distilled water under 300 nm light (Tanaka et al. 1981; quoted, Cessna & Muir 1991).
    Oxidation:
    Hydrolysis:
    Biodegradation:
    Biotransformation:
    Bioconcentration, Uptake (k_1) and Elimination (k_2) Rate Constants:

Common Name: Benefin
Synonym: Balan, Bonalan, benfluralin
Chemical Name: N-butyl-N-ethyl-α,α,α-trifluoro-2,6-dinitro-p-toluidine
Uses: as pre-emergence herbicide for the control of annual grasses and broadleaf weeds in chicory, cucumbers, endive, groundnuts, lettuce, lucerne, and other foliage crops.
CAS Registry No: 1861-40-1
Molecular Formula: $C_{13}H_{16}F_3N_3O_4$
Molecular Weight: 335.3
Melting Point (°C): 65-66.5
Boiling Point (°C):
    121-122 (0.5 mmHg), 148-149 at 7 mmHg (Tomlin 1994)
Density (g/cm^3 at 20°C):
    1.28   (tech., Tomlin 1994)
Molar Volume (cm^3/mol):
    295.9   (calculated-LeBas method, Suntio et al. 1988)
Molecular Volume (Å^3):
Total Surface Area, TSA (Å^2):
Dissociation Constant pK_a:
Heat of Fusion, ΔH_{fus}, kcal/mol:
    9.25   (DSC method, Plato & Glasgow 1969)
Entropy of Fusion, ΔS_{fus}, cal/mol·K (e.u.):
Fugacity Ratio at 25°C (assuming ΔS_{fus} = 13.5 e.u.), F: 0.389

Water Solubility (g/m^3 or mg/L at 25°C):
    < 1.0   (Ashton & Crafts 1973)
    0.50   (Weber et al. 1980; quoted, Willis & McDowell 1982)
    1.0   (20°C, selected, Suntio et al. 1988)
    0.10   (Herbicide Handbook 1983; Tomlin 1994)
    0.10   (20-25°C, selected, Hornsby et al. 1996)

Vapor Pressure (Pa at 25°C):
    0.00519   (30°C, Ashton & Crafts 1973)
    0.0104   (Herbicide Handbook 1983)
    0.0040   (20°C, estimated, Suntio et al. 1988)
    0.0087   (quoted, Tomlin 1994)
    0.0088   (20-25°C, selected, Hornsby et al. 1996)

Henry's Law Constant (Pa·m^3/mol):
    1.34   (20°C, calculated-P/C, Suntio et al. 1988)

Octanol/Water Partition Coefficient, log K_{OW}:
    5.34   (selected, Magee 1991)
    5.29   (20°C, pH 7, Tomlin 1994)
    5.29   (pH 7, selected, Hansch et al. 1995)

Bioconcentration Factor, log BCF:
    3.36   (calculated-S per Kenaga 1980, this work)

Sorption Partition Coefficient, log K_{OC}:
    4.03, 3.75   (quoted, estimated; Magee 1991)
    3.95   (soil, Hornsby et al. 1996)

Half-Lives in the Environment:
    Air: half-life is 0.782-7.82 hours based on estimated reaction with OH radicals in the gas-phase (Howard et al. 1991).
    Surface water: half-life of 288-864 hours based on observed photolysis by sunlight (Howard et al. 1991).
    Groundwater: half-life of 144-5760 hours based on unacclimated aqueous aerobic and anaerobic biodegradation half-lives (Howard et al. 1991).
    Sediment:
    Soil: half-life of 504-2880 hours based on aerobic solid die-away test data (Howard et al. 1991); field half-life of 40 days (Hornsby et al. 1996).
    Biota:

Environmental Fate Rate Constants or Half-Lives:
    Volatilization:
    Photolysis: atmospheric and aqueous photolysis half-lives were estimated to be 288-864 hours (Howard et al. 1991).
    Oxidation: photooxidation half-life was estimated to be between 0.782-7.82 hours based on reaction with OH radicals in air (Howard et al. 1991).
    Hydrolysis: no hydrolyzable group (Howard et al. 1991).
    Biodegradation: aerobic half-life was estimated to be 504-2880 hours in soil, and anaerobic soil half-life is 144-480 hours (Howard et al. 1991).
    Biotransformation:
    Bioconcentration, Uptake (k_1) and Elimination (k_2) Rate Constants:

Common Name: Bifenox
Synonym: MC-4379, Modown
Chemical Name: benzoic acid, 5-(2,4-dichlorophenoxy)-2-nitro-, methyl ester; methyl-5-(2,4-dichlorophenoxy)-2-nitrobenzoate
Uses: selective pre-emergence and post-emergence herbicide to effectively control a wide variety of broadleaf weeds in corn, grain, sorghum, maize, rice, and soybeans.
CAS Registry No: 42576-02-3
Molecular Formula: $C_{14}H_9Cl_2NO_5$
Molecular Weight: 342.1
Melting Point (°C):
    84-86 (Herbicide Handbook 1989; Worthing 1991; Montgomery 1993; Tomlin 1994; Milne 1995)
Boiling Point (°C):
Density (g/cm³ at 20°C):
    1.155 (Ashton & Crafts 1981; Herbicide Handbook 1989; Montgomery 1993)
Molar Volume (cm³/mol):
    305.5 (calculated-LeBas method, this work)
Molecular Volume (Å³):
Total Surface Area, TSA (Å²):
Dissociation Constant pK_a:
Heat of Fusion, ΔH_{fus}, kcal/mol:
Entropy of Fusion, ΔS_{fus}, cal/mol·K (e.u.):
Fugacity Ratio at 25°C (assuming ΔS_{fus} = 13.5 e.u.), F: 0.255

Water Solubility (g/m³ or mg/L at 25°C):
    0.35   (20°C, Weber 1972; Worthing 1987; quoted, Muir 1991)
    0.35   (Martin & Worthing 1977; Herbicide Handbook 1978; quoted, Kenaga 1980; Kenaga & Goring 1980; Isensee 1991)
    0.35   (Ashton & Crafts 1981)
    0.35   (30°C, Worthing 1987; quoted, Shiu et al. 1990)
    0.35   (Agrochemicals Handbook 1987)
    0.35   (Herbicide Handbook 1989; quoted, Shiu et al. 1990)
    0.35   (Merck Index 1989)
    0.35   (Worthing 1991; Tomlin 1994)
    0.398   (20-25°C, selected, Wauchope et al. 1992; Lohninger 1994; Hornsby et al. 1996)
    0.35   (Montgomery 1993)
    0.35   (Milne 1995)

Vapor Pressure (Pa at 25°C):
    0.00032 (20°C, Weber 1972; Worthing 1987; quoted, Muir 1991)
    0.00032 (30°C, Ashton & Crafts 1981; Worthing 1991; Tomlin 1994)
    0.00032 (30°C, Agrochemicals Handbook 1987)
    0.00032 (30°C, Merck Index 1989)
    0.00032 (20-25°C, selected, Wauchope et al. 1992; Hornsby et al. 1996)
    0.00032 (30°C, Montgomery 1993)

Henry's Law Constant (Pa·m^3/mol):
- 0.321 (20°C, calculated-P/C, Muir 1991)
- 0.011 (calculated-P/C, Montgomery 1993)
- 0.313 (calculated-P/C, this work)

Octanol/Water Partition Coefficient, log K_{OW}:
- 5.63 (selected, Dao et al. 1983)
- 4.50 (Worthing 1991)
- 4.48 (Montgomery 1993)
- 4.48 (Tomlin 1994)
- 4.48 (selected, Hansch et al. 1995)

Bioconcentration Factor, log BCF:
- 2.30 (static water, Metcalf & Sanborn 1975; quoted, Kenaga & Goring 1980; Isensee 1991)
- 3.05 (calculated-S, Kenaga 1980; quoted, Isensee 1991)

Sorption Partition Coefficient, log K_{OC}:
- 3.89 (soil, calculated per Kenaga & Goring, Kenaga 1980)
- 4.0 (soil, 20-25°C, estimated, Wauchope et al. 1992; Hornsby et al. 1996)
- 2.24-4.39 (Montgomery 1993)
- 4.0 (estimated-chemical structure, Lohninger 1994)
- 2.70-4.36 (Tomlin 1994)

Half-Lives in the Environment:
  Air:
  Surface water:
  Groundwater:
  Sediment:
  Soil: half-life of 2-5 days for 10 μg/ml to biodegrade in flooded soil at 30°C (Ohyama & Kuwatsuka 1978; quoted, Muir 1991); average half-life in soils is 7-14 days (Hartley & Kidd 1987; Herbicide Handbook 1989; quoted, Montgomery 1993); selected field half-life of 7.0 days (Wauchope et al. 1992; Hornsby et al. 1996); average half-life of 7-14 days (Herbicide Handbook 1989); half-life of ca. 5-7 days in soil (Tomlin 1994).
  Biota:

Environmental Fate Rate Constants or Half-Lives:
  Volatilization: estimated half-life of 29.8 days from 1 m depth of water at 30°C (Muir 1991).
  Photolysis: with <5% degradation by UV light of 290-400 nm in 48 hours (Worthing 1991).
  Oxidation:
  Hydrolysis: stable in aqueous solution at pH 5.0-7.3 but rapidly hydrolyzed at pH 9.0 both at 22°C (Worthing 1991).
  Biodegradation: half-life of 2-5 days for 10 μg/mL to biodegrade in flooded soil at 30°C (Ohyama & Kuwatsuka 1978; quoted, Muir 1991).
  Biotransformation:
  Bioconcentration, Uptake (k_1) and Elimination (k_2) Rate Constants:

Common Name: Bromacil
Synonym: Borea, Bromax, Bromazil, Cynogan, Hyvar, Hyvarex, Krovar I or II, Nalkil, Uragan, Urox B, Uron HX, Weed Blast
Chemical Name: 5-bromo-3-*sec*-butyl-6-methyluracil; 5-bromo-6-methyl-3-(1-methylpropyl)-2,4-(1H,3H)pyrimidinedione
Uses: Herbicide applied to soil to control annual and perennial grasses, broadleaf weeds, and general vegetation on uncropped land; also used for selective weed control in apple, asparagus, cane fruit, hops, and citrus crops.
CAS Registry No: 314-40-9
Molecular Formula: $C_9H_{13}BrN_2O_2$
Molecular Weight: 261.1
Melting Point (°C):
    158.0-159 (Khan 1980; Suntio et al. 1988; Agrochemical Handbook 1987; Herbicide Handbook 1989; Worthing 1991; Milne 1995)
    158.0 (Karickhoff 1981; Patil 1994)
    157.5-160 (Montgomery 1993)
    149.0-153 (Tomlin 1994)
Boiling Point (°C):
Density (g/cm³ at 20°C):
    1.55 (25°C, Agrochemicals Handbook 1987; Herbicide Handbook 1989; Worthing 1991; Montgomery 1993)
    1.59 (23°C, Tomlin 1994)
    1.55 (Milne 1995)
Molar Volume (cm³/mol):
    193.1 (calculated-LeBas method, Suntio et al. 1988)
Molecular Volume (Å³):
Total Surface Area, TSA (Å²):
Dissociation Constant pK_a:
    9.10 (Wauchope et al. 1992; Hornsby et al. 1996)
    <7.0 (Montgomery 1993)
    9.27 (Tomlin 1994)
Heat of Fusion, ΔH_{fus}, kcal/mol:
Entropy of Fusion, ΔS_{fus}, cal/mol·K (e.u.):
Fugacity Ratio at 25°C (assuming ΔS_{fus} = 13.5 e.u.), F: 0.0470

Water Solubility (g/m³ or mg/L at 25°C):
    815 (Bailey & White 1965; quoted, Karickhoff 1981)
    815 (Melnikov 1971; Spencer 1973; quoted, Shiu et al. 1990)
    815 (20°C, Weber 1972; Worthing 1987, 1991; quoted, Muir 1991)
    815 (Herbicide Handbook 1978; quoted, Kenaga & Goring 1980; Jury et al. 1983, 1984; Isensee 1991)
    815 (Khan 1980; quoted, Suntio et al. 1988)
    820 (Beste & Humburg 1983; Jury et al. 1983; quoted, Taylor & Glotfelty 1988)
    1064 (shake flask-GC or LSC, Gerstl & Mingelgrin 1984; quoted, Shiu et al. 1990)
    626, 775, 1043 (4, 25, 40°C, shake flask-LSS, Madhun et al. 1986)
    815 (Agrochemicals Handbook 1987; Herbicide Handbook 1989)
    1064 (selected, Gerstl & Helling 1987)
    670 (20°C, selected, Suntio et al. 1988; quoted, Majewski & Capel 1995)

815, 1024 (literature data variability, Heller et al. 1989)
700 (20-25°C, selected, Wauchope et al. 1992; Lohninger 1994; Hornsby et al. 1996)
815 (quoted, Montgomery 1993)
807, 3522 (quoted, calculated, Patil 1994)
700, 807, 1287 (at pH 7, 5, 9, Tomlin 1994)
815 (Milne 1995)

Vapor Pressure (Pa at 25°C):
5×10^{-5} (20°C, Weber 1972; Worthing 1987; quoted, Muir 1991)
3×10^{-5} (estimated, USEPA 1975; quoted, Jury et al. 1983)
0.107 (100°C, Khan 1980; quoted, Suntio et al. 1988)
2.9×10^{-5} (Jury et al. 1983; quoted, Taylor & Glotfelty 1988; Taylor & Spencer 1990)
0.00033 (Agrochemicals Handbook 1987; Worthing 1991)
0.005 (20°C, selected, Suntio et al. 1988; quoted, Majewski & Capel 1995)
4×10^{-5} (20-25°C, selected, Wauchope et al. 1992)
3.3×10^{-5} (Montgomery 1993)

Henry's Law Constant (Pa·m^3/mol):
9.17×10^{-6} (Beste & Humburg 1983; Jury et al. 1983; quoted, Taylor & Glotfelty 1988)
9.17×10^{-5} (calculated-P/C, Jury et al. 1984, 1987a; Jury & Ghodrati 1989)
0.0019 (20°C, selected, Suntio et al. 1988; quoted, Majewski & Capel 1995)
1.06×10^{-5} (20°C, calculated-P/C, Muir 1991)
1.06×10^{-5} (calculated-P/C, Montgomery 1993)

Octanol/Water Partition Coefficient, log K_{OW}:
2.02 (Rao & Davidson 1980; quoted, Karickhoff 1981; Suntio et al. 1988)
1.33 (selected, Dao et al. 1983)
1.84 (shake flask-GC or LSC, Gerstl & Mingelgrin 1984)
1.84, 1.87, 1.90 (4, 25, 40°C, shake flask-LSS, Madhun et al. 1986)
1.85 (selected, Gerstl & Helling 1987)
2.00 (selected, Suntio et al. 1988)
2.11 (selected, Magee 1991; Devillers et al. 1996)
1.84-2.04 (Montgomery 1993)
2.02, 2.11; 2.11 (quoted values, selected, Sangster 1993)
2.02, 0.29 (quoted, calculated, Patil 1994)
1.87, 1.88, 1.63 (at pH 7, 5, 9, Tomlin 1994)
2.11 (selected, Hansch et al. 1995)

Bioconcentration Factor, log BCF:
0.505 (measured, Kenaga 1980; quoted, Isensee 1991)
2.27 (calculated-S, Kenaga 1980; quoted, Isensee 1991)
0.477 (calculated-K_{OC}, Kenaga 1980)
0.51 (*Pimephales promelas*, Call et al. 1987; quoted, Devillers et al. 1996)

Sorption Partition Coefficient, log K_{OC}:

    1.86   (soil, Hamaker & Thompson 1972; quoted, Kenaga & Goring 1980; Karickhoff 1981; Bahnick & Doucette 1988)

    3.13   (soil, calculated as per Kenaga & Goring 1978, Kenaga 1980)

    1.86   (Rao & Davidson 1980; quoted, Jury et al. 1983)

    2.33, 1.34, 1.63 (estimated-S, S & M.P., K_{OW}, Karickhoff 1981)

    1.61   (sediments average-Freundlich adsorption, Corwin & Farmer 1984)

    1.41-2.46 (California Lake sediments, Corwin & Farmer 1984; quoted, Muir 1991)

    1.98, 1.88 (4, 25°C, Semiahmoo soil, in μmol/kg OC, batch equilibrium method-LSS, Madhun et al. 1986)

    2.11, 1.88 (4, 25°C, Adkins soil, in μmol/kg OC, Madhun et al. 1986)

    1.90, 1.66, 1.75; 1.86, 1.89, 1.34 (estimated-K_{OW}; S, Madhum et al. 1986)

    1.53, 2.73 (quoted, calculated-χ, Gerstl & Helling 1987)

    1.86   (soil, screening model calculations, Jury et al. 1987a,b; Jury & Ghodrati 1989; Carsel 1989)

    2.56   (calculated-χ, Bahnick & Doucette 1988)

    1.86, 1.80 (reported, estimated as log K_{OM}, Magee 1991)

    1.53, 1.86, 3.13 (quoted values, Bottoni & Funari 1992)

    1.51   (soil, 20-25°C, selected, Wauchope et al. 1992; Hornsby et al. 1996)

    1.51   (Montgomery 1993)

    2.09   (estimated-chemical structure, Lohninger 1994)

Half-Lives in the Environment:

    Air:

    Surface water:

    Groundwater:

    Sediment:

    Soil: half-life of 7.0 months at 15°C and 4.5 months at 30°C in soils (Freed & Haque 1973); rate constant of 0.0038 d^{-1} with biodegradation half-life of 350 d under field conditions (Rao & Davidson 1980; quoted, Jury et al. 1984); half-life of 350 d from screening model calculations (Jury et al. 1987a,b; Jury & Ghodrati 1989); half-life of >100 d (Willis & McDowell 1982); selected field half-life of 60 d (Wauchope et al. 1992; Hornsby et al. 1996).

    Biota: biochemical half-life of 350 d from screening model calculations (Jury et al. 1987a,b; Jury & Ghodrati 1989).

Environmental Fate Rate Constants or Half-Lives:

    Volatilization: estimated half-life of 10,000 d from 1 m depth of water at 20°C (Muir 1991).

    Photolysis: 115 ppb contaminated water in the presence of TiO_2 and H_2O_2 photodegraded to 6 ppb by 15 h solar irradiation with complete degradation after 75 h (Muszkat et al. 1992).

    Oxidation:

    Hydrolysis:

    Biodegradation: half-life of 350 d for 100 d leaching and screening test in 0-10 cm depth of soil (Rao & Davidson 1980; quoted, Jury et al. 1983, 1984, 1987a).

    Biotransformation:

    Bioconcentration, Uptake (k_1) and Elimination (k_2) Rate Constants:

Common Name: Bromoxynil
Synonym: Brittox, Brominal, Brominex, Brominil, Broxynil, Buctril, Chipco crab-kleen, ENT 20852, Nu-lawn weeder, Oxytril M, Partner
Chemical Name: 3,5-dibromo-4-hydroxybenzonitrile; 4-cyano-2,6-dibromophenol
Uses: herbicide for post-emergence control of annual broadleaf weeds and it is often used in combination with other herbicides to extend the spectrum of control.
CAS Registry No: 1689-84-5
Molecular Formula: $C_7H_3Br_2NO$
Molecular Weight: 276.9
Melting Point (°C):
    190   (Khan 1980; Herbicide Handbook 1989; Montgomery 1993)
    194-195   (Worthing 1991; Montgomery 1993; Tomlin 1994; Milne 1995)
Boiling Point (°C):
Density (g/cm³ at 20°C):
Molar Volume (cm³/mol):
    176.7   (calculated-LeBas method, this work)
Molecular Volume (Å³):
Total Surface Area, TSA (Å²):
Dissociation Constant pK_a:
    4.20   (radiometer/pH meter, Cessna & Grover 1978)
    4.06   (Herbicide Handbook 1989; Montgomery 1993)
    4.06   (Merck Index 1989; Worthing 1991)
    3.86   (Tomlin 1994)
Heat of Fusion, ΔH_{fus}, kcal/mol:
    7.60   (DSC method, Plato 1972)
Entropy of Fusion, ΔS_{fus}, cal/mol·K (e.u.):
Fugacity Ratio at 25°C (assuming ΔS_{fus} = 13.5 e.u.), F: 0.021

Water Solubility (g/m³ or mg/L at 25°C):
    130   (20-25°C, Spencer 1973; quoted, Shiu et al. 1990)
    131   (Kenaga 1980; quoted, Isensee 1991)
    <200   (Khan 1980)
    130   (20-25°C, Ashton & Crafts 1981)
    130   (Agrochemicals Handbook 1987; Tomlin 1994)
    130   (Worthing 1987,91; quoted, Shiu et al. 1990)
    130   (20-25°C, Herbicide Handbook 1989)
    130   (Montgomery 1993, Milne 1995)

Vapor Pressure (Pa at 25°C):
    <0.0010   (20°C, Agrochemicals Handbook 1987; Tomlin 1994)
    0.00064   (Herbicide Handbook 1989)
    0.00064   (Montgomery 1993)

Henry's Law Constant (Pa·m³/mol):
    0.14180   (20-25°C, calculated-P/C, Montgomery 1993)
    1.36×10^{-3} (calculated-P/C, this work)

Octanol/Water Partition Coefficient, log K_{OW}:
    2.60   (selected, Dao et al. 1983)
    <2.00   (Herbicide Handbook 1989)
    <2.00   (Montgomery 1993)

Bioconcentration Factor, log BCF:
    1.60   (calculated, Kenaga 1980; quoted, Isensee 1991)

Sorption Partition Coefficient, log K_{OC}:
    2.48   (quoted from Kenaga 1980, Bottoni & Funari 1992)
    2.48   (calculated, Montgomery 1993)

Half-Lives in the Environment:
    Air:
    Surface water: half-life of ~24 hours for 0.03 µg/mL to biodegrade in runoff water at 20-25°C (Brown et al. 1984; quoted, Muir 1991).
    Groundwater:
    Sediment:
    Soil: half-life in soil is approximately 10 days (Hartley & Kidd 1987; quoted, Montgomery 1993; Tomlin 1994).

Environmental Fate Rate Constants or Half-Lives:
    Volatilization:
    Photolysis: rate constant of degradation in water of 1.04×10^{-3} s^{-1} at pH 8.3 and 1.08×10^{-3} s^{-1} at pH 11.6 (Kochany 1992).
    Oxidation:
    Hydrolysis:
    Biodegradation: half-life of ~24 hours for 0.03 µg/mL to biodegrade in runoff water at 20-25°C (Brown et al. 1984; quoted, Muir 1991).
    Biotransformation:
    Bioconcentration, Uptake (k_1) and Elimination (k_2) Rate Constants:

Common Name: Butachlor
Synonym: Butanex, Butanox, CP 53619, Lambast, Machete, Pillarsete
Chemical Name: N-butoxymethyl-2-chloro-2'6'-diethylacetanilide; N-(butoxymethyl)-2-chloro-N-(2,6-diethylphenyl)acetamide
Uses: herbicide for pre-emergence control of most annual grasses, some broadleaf weeds, and many aquatic weeds in both seeded and transplanted rice.
CAS Registry No: 23184-66-9
Molecular Formula: $C_{17}H_{26}ClNO_2$
Molecular Weight: 311.9
Melting Point (°C):
    < −5.0 (Agrochemicals Handbook 1987; Worthing 1991; Tomlin 1994; Milne 1995)
    < −10 (Herbicide Handbook 1989)
Boiling Point (°C):
    156 (at 0.5 mmHg, Ashton & Crafts 1981; Herbicide Handbook 1989; Tomlin 1994; Milne 1995)
Density (g/cm³ at 20°C):
    1.07 (25°C, Ashton & Crafts 1981; Agrochemicals Handbook 1987; Herbicide Handbook 1989; Tomlin 1994; Milne 1995)
Molar Volume (cm³/mol):
    387.8 (calculated-LeBas method, this work)
Molecular Volume (Å³):
Total Surface Area, TSA (Å²):
Dissociation Constant pK_a:
Heat of Fusion, ΔH_{fus}, kcal/mol:
Entropy of Fusion, ΔS_{fus}, cal/mol·K (e.u.):
Fugacity Ratio at 25°C (assuming ΔS_{fus} = 13.5 e.u.), F: 1.0

Water Solubility (g/m³ or mg/L at 25°C):
    23 (20°C, Weber 1972; Worthing 1987; quoted, Muir 1991)
    20 (Martin & Worthing 1977; quoted, Kenaga 1980)
    23 (24°C, Ashton & Crafts 1981; Herbicide Handbook 1989)
    20 (20°C, Agrochemicals Handbook 1987; Tomlin 1994)
    23 (24°C, Worthing 1987,91; quoted, Shiu et al. 1990)
    23 (20-25°C, selected, Augustijn-Beckers et al. 1994; Hornsby et al. 1996)
    20 (20°C, Milne 1995)

Vapor Pressure (Pa at 25°C):
    0.0007 (20°C, Weber 1972; Worthing 1987; quoted, Muir 1991)
    0.0006 (Ashton & Crafts 1981; Herbicide Handbook 1989)
    0.0006 (Agrochemicals Handbook 1987)
    0.0006 (Worthing 1991; Tomlin 1994)
    0.0006 (20-25°C, selected, Augustijn-Beckers et al. 1994; Hornsby et al. 1996)

Henry's Law Constant (Pa·m³/mol):
    0.00817 (20°C, calculated-P/C, Muir 1991)
    0.00814 (calculated-P/C, this work)

Octanol/Water Partition Coefficient, log K_{OW}:
    4.50   (quoted and selected, Hansch et al. 1995)
    4.09   (calculated-S as per Chiou et al. 1977; Chiou 1981; this work)

Bioconcentration Factor, log BCF:
    2.06   (calculated-S, Kenaga 1980)
    1.03, 0.756 (18, 9 µg/L; carp, Wang et al. 1992)
    0.38, 0.845 (10, 1 µg/L; tilapia, Wang et al. 1992)
    0.447, 0.845 (10, 1 µg/L; loach, Wang et al. 1992)
    1.71, 1.90 (5, 2.5 µg/L; eel, Wang et al. 1992)
    0.041, 0.778 (100, 10 µg/L; freshwater clam, Wang et al. 1992)

Sorption Partition Coefficient, log K_{OC}:
    2.92   (calculated-S, Kenaga 1980)
    2.85   (20-25°C, selected, Augustijn-Beckers et al. 1994; Hornsby et al. 1996)

Half-Lives in the Environment:
    Air:
    Surface water:
    Groundwater:
    Sediment:
    Soil: persists for 6-10 weeks in soil (Agrochemicals Handbook 1987; Tomlin 1994); 4 to 8 days depending upon soil type (Herbicide Handbook 1989); persists in soil 42-70 days (Worthing 1991); selected field half-life of 12 days (Augustijn-Beckers et al. 1994; Hornsby et al. 1996).

Environmental Fate Rate Constants or Half-Lives:
    Volatilization: estimated half-life of 1049 days from 1 m depth of water at 20°C (Muir 1991).
    Photolysis: half-life of 0.8-5.4 hours in distilled water (Chen et al. 1982; quoted, Cessna & Muir 1991).
    Oxidation:
    Hydrolysis: half-life of >2.5 months for 2 µg/mL to hydrolyze in phosphate buffer at pH 6 and borate buffer at pH 9 both at 25°C (Chen & Chen 1979; quoted, Muir 1991).
    Biodegradation:
    Biotransformation:
    Bioconcentration, Uptake (k_1) and Elimination (k_2) Rate Constants:

Common Name: Butralin
Synonym: Amex, Butalin, Rutralin, Sector, Tamex
Chemical Name: N-*sec*-butyl-4-*tert*-butyl-2,6-dinitroaniline; 4-(1,1-dimethylethyl)-N-(1-methylpropyl)-2,6-dinitrobenzenamine
Uses: herbicide for pre-emergence control of annual broadleaf weeds and grasses in cotton, beans, barley, rice, soybeans, alliums, vines, ornamentals and orchards of fruit and nut trees; also to control suckers on tobacco.
CAS Registry No: 33629-47-9
Molecular Formula: $C_{14}H_{21}N_3O_4$
Molecular Weight: 295.3
Melting Point (°C):
    60-61 (Khan 1980; Worthing 1991; Tomlin 1994; Milne 1995)
Boiling Point (°C):
    134-136 (at 0.5 mmHg, Ashton & Crafts 1981; Agrochemicals Handbook 1987; Worthing 1991; Tomlin 1994; Milne 1995)
Density (g/cm³ at 20°C):
Molar Volume (cm³/mol):
    313.6 (calculated-LeBas method, this work)
Molecular Volume (Å³):
Total Surface Area, TSA (Å²):
Dissociation Constant pK_a:
Heat of Fusion, ΔH_{fus}, kcal/mol:
Entropy of Fusion, ΔS_{fus}, cal/mol·K (e.u.):
Fugacity Ratio at 25°C (assuming ΔS_{fus} = 13.5 e.u.), F: 0.440

Water Solubility (g/m³ or mg/L at 25°C):
    1.0 (Herbicide Handbook 1978; quoted, Kenaga 1980; Kenaga & Goring 1980; Isensee 1991)
    1.0 (Khan 1980)
    10 (24°C, Ashton & Crafts 1981)
    1.0 (24°C, Agrochemicals Handbook 1987; Tomlin 1994)
    1.0 (24-26°C, Worthing 1987,91; quoted, Shiu et al. 1990)
    1.0 (Merck Index 1989)
    1.0 (24°C, Milne 1995)

Vapor Pressure (Pa at 25°C):
    0.002 (Ashton & Crafts 1981)
    0.0017 (Agrochemicals Handbook 1987; Worthing 1991; Tomlin 1994)
    0.0017 (Merck Index 1989)

Henry's Law Constant (Pa·m³/mol):
    0.502 (calculated-P/C, this work)

Octanol/Water Partition Coefficient, log K_{OW}:
    4.54 (selected, Dao et al. 1983)

Bioconcentration Factor, log BCF:
- 2.79 (calculated-S, Kenaga 1980; quoted, Isensee 1991)
- 2.80 (calculated-K_{OC}, Kenaga 1980)

Sorption Partition Coefficient, log K_{OC}:
- 3.64 (calculated, Kenaga & Goring 1978; quoted, Kenaga 1980)
- 3.91 (soil, Kenaga & Goring 1980; quoted, Bahnick & Doucette 1988)
- 3.75 (calculated-χ, Bahnick & Doucette 1988)

Half-Lives in the Environment:
  Air:
  Surface water:
  Groundwater:
  Sediment:
  Soil: half-life of 24 days for 0.5 μg/mL to biodegrade in soil at 20-42°C (Savage 1978; quoted, Muir 1991).

Environmental Fate Rate Constants or Half-Lives:
  Volatilization:
  Photolysis: half-life of 8 hours for 25% of 2000 μg/mL to degrade in methanol under sunlight (Plimmer & Klingebiel 1974; quoted, Cessna & Muir 1991).
  Oxidation:
  Hydrolysis:
  Biodegradation: half-life of 24 days for 0.5 μg/mL to biodegrade in soil at 20-42°C (Savage 1978; quoted, Muir 1991).
  Biotransformation:
  Bioconcentration, Uptake (k_1) and Elimination (k_2) Rate Constants:

Common Name: Butylate
Synonym: Butilate, diisocarb, Genate, R 1910, Sutan
Chemical Name: S-ethyldiisobutylthiocarbamate; S-ethyl-bis(2-methylpropylcarbamothioate
Uses: herbicide to control annual grass weeds in maize, by pre-plant soil incorporation; also to control some broadleaf weeds.
CAS Registry No: 2008-41-5
Molecular Formula: $C_{11}H_{23}NOS$
Molecular Weight: 217.38
Melting Point (°C): liquid
Boiling Point (°C):
    137.5-138 (at 21 mmHg, Agrochemicals Handbook 1987; Merck Index 1989; Worthing 1991; Tomlin 1994; Milne 1995)
    71.0 (at 10 mmHg, Herbicide Handbook 1989)
Density (g/cm³ at 20°C):
    0.9402 (25°C, Agrochemicals Handbook 1987; Herbicide Handbook 1989; Worthing 1991; Tomlin 1994; Milne 1995)
    0.9417 (Milne 1995)
Molar Volume (cm³/mol):
    280.9 (calculated-LeBas method, Suntio et al. 1988)
Molecular Volume (Å³):
Total Surface Area, TSA (Å²):
Dissociation Constant pK_a:
Heat of Fusion, ΔH_{fus}, kcal/mol:
Entropy of Fusion, ΔS_{fus}, cal/mol·K (e.u.):
Fugacity Ratio at 25°C (assuming ΔS_{fus} = 13.5 e.u.), F: 1.0

Water Solubility (g/m³ or mg/L at 25°C):
    45.0 (Kenaga 1980; Nash 1988)
    45.0 (Weber et al. 1980; quoted, Willis & McDowell 1982)
    45.0 (22°C, Ashton & Crafts 1981; Agrochemicals Handbook 1987; Herbicide Handbook 1989; quoted, Suntio et al. 1988; Pait et al. 1992)
    46.0 (20°C, Worthing 1987,91; quoted, Shiu et al. 1990)
    40.0 (20°C, selected, Suntio et al. 1988; quoted, Majewski & Capel 1995)
    44.0 (20-25°C, selected, Wauchope et al. 1992; Hornsby et al. 1996)
    46.0 (selected, Lohninger 1994)
    36.0 (20°C, Tomlin 1994)
    45.0 (22°C, Milne 1995)
    44.0 (selected, Halfon et al. 1996)

Vapor Pressure (Pa at 25°C):
    1.73 (Ashton & Crafts 1973; quoted, Suntio et al. 1988)
    0.096 (20°C, Hartley & Graham-Bryce 1980; quoted, Suntio et al. 1988)
    1.733 (Herbicide Handbook 1983, 1989; quoted, Nash 1988)
    0.10 (20°C, selected, Suntio et al. 1988; quoted, Majewski & Capel 1995)
    0.17 (Worthing 1991)
    1.733 (20-25°C, selected, Wauchope et al. 1992; Hornsby et al. 1996)
    1.73 (Tomlin 1994)
    1.733 (selected, Halfon et al. 1996)

Henry's Law Constant (Pa·m³/mol):
    0.560   (20°C, calculated-P/C, Suntio et al. 1988; quoted, Majewski & Capel 1995)

Octanol/Water Partition Coefficient, log K_{OW}:
    4.15   (Worthing 1991; Tomlin 1994; Milne 1995)
    4.15   (selected, Hansch et al. 1995)
    4.17, 4.01, 3.45   (RP-HPLC, ClogP, calculated-S, Finizio et al. 1997)

Bioconcentration Factor, log BCF:
    1.86   (calculated-S, Kenaga 1980; quoted, Pait et al. 1992)
    3.06   (calculated-K_{OW} as per Kenaga 1980, this work)

Sorption Partition Coefficient, log K_{OC}:
    2.73   (soil, Kenaga 1980; quoted, Nash 1988)
    2.73, 4.09   (quoted values, Bottoni & Funari 1992)
    2.60   (soil, 20-25°C, selected, Wauchope et al. 1992; quoted, Richards & Baker 1993; Hornsby et al. 1996)
    2.60   (estimated-chemical structure, Lohninger 1994)

Half-Lives in the Environment:
    Air:
    Surface water:
    Groundwater:
    Sediment:
    Soil: measured dissipation rate of 3.6 d^{-1} (Nash 1983; quoted, Nash 1988); estimated dissipation rate of 23 and 0.61 d^{-1} (Nash 1988); half-life of 1.5-3.0 weeks in several soils under crop growing conditions (Herbicide Handbook 1989); selected field half-life of 13 days (Wauchope et al. 1992; quoted, Richards & Baker 1993; Hornsby et al. 1996); soil half-life of 12 days (quoted, Pait et al. 1992); half-life of 1.5-10 weeks in soil and water (Tomlin 1994); soil half-life of 13 days (selected, Halfon et al. 1996).
    Biota: disappear from the stems and leaves of corn plants 7 to 14 days after application (Herbicide Handbook 1989).

Environmental Fate Rate Constants or Half-Lives:
    Volatilization:
    Photolysis:
    Oxidation:
    Hydrolysis:
    Biodegradation:
    Biotransformation:
    Bioconcentration, Uptake (k_1) and Elimination (k_2) Rate Constants:

Common Name: Chloramben
Synonym: ACP-M-728, Amiben, Amoben, Chlorambed, Chlorambene, M-728, NCI-C00055, Ornamental weeder, Vegaben, Vegiben
Chemical Name: 3-amino-2,5-dichlorobenzoic acid
Uses: pre-emergence or pre-plant herbicide used in many vegetable and field crops to control annual broadleaf weeds and grasses.
CAS Registry No: 133-90-4
Molecular Formula: $C_7H_5Cl_2NO_2$
Molecular Weight: 206.0
Melting Point (°C):
    201   (Khan 1980; Herbicide Handbook 1989)
    200-201 (Worthing 1991; Montgomery 1993; Tomlin 1994; Milne 1995)
Boiling Point (°C):
Density (g/cm³ at 20°C):
Molar Volume (cm³/mol):
    190.8   (calculated-LeBas method, this work)
Molecular Volume (Å³):
Total Surface Area, TSA (Å²):
Dissociation Constant pK_a:
    3.40   (Hornsby et al. 1996)
Heat of Fusion, ΔH_{fus}, kcal/mol:
    9.30   (DSC method, Plato 1972)
Entropy of Fusion, ΔS_{fus}, cal/mol·K (e.u.):
Fugacity Ratio at 25°C (assuming ΔS_{fus} = 13.5 e.u.), F: 0.018

Water Solubility (g/m³ or mg/L at 25°C):
    700   (Spencer 1973; quoted, Shiu et al. 1990)
    700   (Martin & Worthing 1977; Herbicide Handbook 1978; quoted, Kenaga 1980; Kenaga & Goring 1980; Isensee 1991)
    700   (Khan 1980; Weber et al. 1980; quoted, Willis & McDowell 1982)
    700   (Ashton & Crafts 1981; Herbicide Handbook 1989)
    700   (Agrochemicals Handbook 1987)
    700   (selected, Gerstl & Helling 1987)
    700   (Worthing 1987, 1991; quoted, Shiu et al. 1990; Tomlin 1994; Majewski & Capel 1995)
    700   (Merck Index 1989; quoted, Milne 1995)
    700   (Montgomery 1993)
    700   (selected, Lohninger 1994)

Vapor Pressure (Pa at 25°C):
    0.93   (100°C, Agrochemicals Handbook 1987)
    52.7   (Worthing 1987; quoted, Majewski & Capel 1995)
    0.93   (100°C, Worthing 1991; Tomlin 1994)

Henry's Law Constant (Pa·m³/mol):
    0.274   (calculated-P/C as per Worthing 1987; quoted, Majewski & Capel 1995)

Octanol/Water Partition Coefficient, log K_{OW}:
- 1.11 (quoted, Rao & Davidson 1980)
- 1.46 (selected, Dao et al. 1983)
- −2.64 (selected, Gerstl & Helling 1987)
- 1.11 (Magee 1991)
- 1.11 (Montgomery 1993)
- 1.11 (quoted, Sangster 1993)

Bioconcentration Factor, log BCF:
- 1.18 (calculated-S, Kenaga 1980; quoted, Isensee 1991)
- −0.097 (calculated-K_{OC}, Kenaga 1980)

Sorption Partition Coefficient, log K_{OC}:
- 1.32 (soil, Harris & Warren 1964; Farmer 1976; quoted, Kenaga 1980; Kenaga & Goring 1980)
- 2.08 (soil, calculated as per Kenaga & Goring 1978, Kenaga 1980)
- 1.78 (calculated-χ, Gerstl & Helling 1987)
- 1.32 (reported as log K_{OM}, Magee 1991)
- 2.28 (Montgomery 1993)
- 1.56 (selected, Lohninger 1994)

Half-Lives in the Environment:
Air:
Surface water:
Groundwater:
Sediment:
Soil: estimated persistence of 3 months (Kearney et al. 1969; quoted, Jury et al. 1987); half-lives: 36, 38, 41, and 20 d with disappearance rates: 0.0193, 0.0182, 0.0169 and 0.0347 d^{-1} at pH 4.3, 5.3, 6.5 and 7.5 (Hamaker 1972; quoted, Nash 1988); persistence in soil is of 6-8 weeks (Hartley & Kidd 1987; Herbicide Handbook 1989; quoted, Montgomery 1993).
Biota:

Environmental Fate Rate Constants or Half-Lives:
Volatilization:
Photolysis: half-life of 6 h for 206 μg/mL to degrade in distilled water under sunlight (Sheets 1963; quoted, Cessna & Muir 1991); half-life of <2 d for 16 μg/mL to degrade in distilled water under sunlight (Hahn et al. 1969; quoted, Cessna & Muir 1991).
Oxidation:
Hydrolysis:
Biodegradation: half-life of >70 d for 50 μg/mL to degrade in incubated soil with nutrient medium of 3 g/L (Schliebe et al. 1965; quoted, Muir 1991).
Biotransformation:
Bioconcentration, Uptake (k_1) and Elimination (k_2) Rate Constants:

Common Name: Chlorpropham
Synonym: Beet-Kleen, Bud-nip, Chlor-IFC, Chloro-IPC, CIPC, Ebanil, ENT 18060, Fasco Wy-hoe, Furloe, Nexoval, Prevenol, Preweed, Sprout-nip, Taterpex
Chemical Name: isopropyl N-(3-chlorophenyl) carbamate; isopropyl 3-chlorocarbanilate
Uses: pre-emergent and post-emergent herbicide used to regulate plant growth and control weeds in carrot, onion, garlic, and other crops.
CAS Registry No: 101-21-3
Molecular Formula: $C_{10}H_{12}ClNO_2$
Molecular Weight: 213.65
Melting Point (°C):
    38-40   (Khan 1980; Herbicide Handbook 1989)
    40.7-41.1   (Suntio et al. 1988)
    40.7-41.4   (Montgomery 1993)
    41.4   (Patil 1994; Tomlin 1994; Kühne et al. 1995)
Boiling Point (°C):
    149   (at 2 mmHg, Merck Index 1989)
Density (g/cm^3 at 20°C):
    1.180   (30°C, Agrochemicals Handbook 1987; Herbicide Handbook 1989; Worthing 1991; Montgomery 1993; Tomlin 1994)
    1.5388   (Merck Index 1989)
Molar Volume (cm^3/mol):
    232.4   (calculated-LeBas method, Suntio et al. 1988)
Molecular Volume (Å^3):
Total Surface Area, TSA (Å^2):
Dissociation Constant pK_a:
Heat of Fusion, ΔH_{fus}, kcal/mol:
    4.90   (DSC method, Plato & Glasgow 1969)
Entropy of Fusion, ΔS_{fus}, cal/mol·K (e.u.):
Fugacity Ratio at 25°C (assuming ΔS_{fus} = 13.5 e.u.), F: 0.6946

Water Solubility (g/m^3 or mg/L at 25°C):
    0.470   (Brust 1966; quoted, Shiu et al. 1990)
    102.3   (shake flask-GC, Freed et al. 1967; quoted, Freed 1976)
    108   (20°C, Günther et al. 1968; quoted, Suntio et al. 1988)
    89   (20°C, Weber 1972; Martin & Worthing 1977; Worthing 1987; quoted, Kenaga 1980; Muir 1991; Wauchope et al. 1992; Hornsby et al. 1996)
    2.0   (Spencer 1973; quoted, Shiu et al. 1990)
    88   (Martin & Worthing 1977; Herbicide Handbook 1978, 1989; quoted, Kenaga 1980; Kenaga & Goring 1980; Karickhoff 1981; Isensee 1991)
    0.70   (19°C, shake flask-GC, Bowman & Sans 1979; quoted, Shiu et al. 1990)
    0.73   (20°C, shake flask-GC, Bowman & Sans 1983a,b; quoted, Shiu et al. 1990)
    88   (Khan 1980; Ashton & Crafts 1981)
    80-102   (Weber et al. 1980; quoted, Willis & McDowell 1982)
    89   (Agrochemicals Handbook 1987; Worthing 1991; Tomlin 1994)
    89   (selected, Gerstl & Helling 1987; Montgomery 1993; Lohninger 1994)
    2.0   (20°C, Worthing 1987; quoted, Shiu et al. 1990)
    100   (20°C, selected, Suntio et al. 1988; quoted, Majewski & Capel 1995)
    107, 74   (quoted, calculated, Patil 1994)
    89, 355   (quoted, calculated-group contribution fragmentation method, Kühne et al. 1995)

Vapor Pressure (Pa at 25°C):
    0.00050 (20°C, Weber 1972; Worthing 1987; quoted, Muir 1991)
    0.00133 (extrapolated, Spencer 1976; quoted, Suntio et al. 1988)
    0.00133 (Khan 1980; quoted, Suntio et al. 1988)
    0.00133 (Ashton & Crafts 1981; Herbicide Handbook 1989)
    0.00100 (20°C, selected, Suntio et al. 1988; quoted, Majewski & Capel 1995)
    0.00130 (selected, Taylor & Spencer 1990)
    0.00107 (20-25°C, selected, Wauchope et al. 1992; Hornsby et al. 1996)
    0.00133 (estimated, Montgomery 1993)

Henry's Law Constant (Pa·m^3/mol):
    0.0021 (20°C, calculated-P/C, Suntio et al. 1988; quoted, Majewski & Capel 1995)
    0.0032 (20°C, calculated-P/C, Muir 1991)
    0.0021 (20-25°C, calculated-P/C, Montgomery 1993)

Octanol/Water Partition Coefficient, log K_{OW}:
    3.06 (Rao & Davidson 1980; quoted, Suntio et al. 1988)
    3.06 (Karickhoff 1981; quoted, Patil 1994)
    3.42 (selected, Dao et al. 1983)
    3.42 (selected, Gerstl & Helling 1987)
    3.10 (selected, Suntio et al. 1988)
    3.06, 3.09 (quoted, calculated, Patil 1994)
    3.51 (selected, Hansch et al. 1995)

Bioconcentration Factor, log BCF:
    1.70 (calculated-S, Kenaga 1980; quoted, Isensee 1991)
    1.52 (calculated-K_{OC}, Kenaga 1980)

Sorption Partition Coefficient, log K_{OC}:
    2.77 (soil, Hamaker & Thompson 1972; quoted, Kenaga 1980; Kenaga & Goring 1980; Karickhoff 1981)
    2.57 (soil, calculated-S as per Kenaga & Goring 1978, Kenaga 1980)
    2.85, 2.80 (estimated-S, Karickhoff 1981)
    3.17, 3.08 (estimated-S & M.P., Karickhoff 1981)
    2.67 (estimated-K_{OW}, Karickhoff 1981)
    3.06, 2.31 (quoted, calculated-χ, Gerstl & Helling 1987)
    2.80 (quoted exptl., Meylan et al. 1992)
    2.32 (calculated-χ & fragment contribution, Meylan et al. 1992)
    2.60 (soil, 20-25°C, estimated, Wauchope et al. 1992; Hornsby et al. 1996)
    2.77, 2.91 (Montgomery 1993)
    2.60 (estimated-chemical structure, Lohninger 1994)

Half-Lives in the Environment:
    Air:
    Surface water: rate constant of $3.6\text{-}6.7 \times 10^{-10}$ mL(cell)$^{-1}$ d^{-1} from measurements of different river water samples (Paris et al. 1978; quoted, Scow 1982); half-life of 120 hours for 2-25 µg/mL to biodegrade in stream water with biodegradation rate of 2.5×10^{-4} mg M^{-1} h^{-1} at 28°C (Wolfe et al. 1978; quoted, Muir 1991); $1.6\text{-}1.8 \times 10^{-8}$ mL(cell)$^{-1}$ d^{-1} from measurements of different river water samples (Steen et al. 1979; quoted, Scow 1982);

aerobic half-life of 190 hours for 0.1-1.0 µg/mL to biodegrade in lake water with biodegradation rate of 1.3-4.9x10^{-4} L org^{-1} h^{-1} at 22°C (Schnoor et al. 1982; quoted, Muir 1991); aerobic half-life of >4 months for 6-7 µg/ml to biodegrade in river water at 25°C (Stepp et al. 1985; quoted, Muir 1991).

Groundwater:

Sediment: aerobic half-life of 10-75 days for 0.1-5.4 µg/mL to biodegrade in activated sludge (Schwartz 1967; quoted, Muir 1991).

Soil: half-lives in soil at 15 and 29°C are 65 and 30 days, respectively (Hartley & Kidd 1987; Herbicide Handbook 1989; quoted, Montgomery 1993; Tomlin 1994); selected field half-life of 30 days (Wauchope et al. 1992; Hornsby et al. 1996).

Biota:

Environmental Fate Rate Constants or Half-Lives:

Volatilization: estimated half-life of 2220 days from 1 m depth of water at 20°C (Muir 1991).

Photolysis: half-life of 130 hours for 4 µg/mL to degrade in distilled water under >280 nm light (Guzik 1978; quoted, Cessna & Muir 1991); calculated half-life of 121 days in distilled water (Wolfe et al. 1978; quoted, Cessna & Muir 1991); half-life of 2.25 hours for 21-76% of 80 µg/mL to degrade in distilled water under 300 nm light (Tanaka et al. 1981; quoted, Cessna & Muir 1991).

Oxidation:

Hydrolysis: half-life of >4 months for 4274 µg/mL to hydrolyze in phosphate buffer at pH 5-9 and 20°C (El-Dib & Aly 1976; quoted, Muir 1991); half-life of >1 week for 2.10 µg/mL to hydrolyze in natural waters at 67°C (Schnoor et al. 1982; quoted, Muir 1991).

Biodegradation: aerobic half-life of 10-75 days for 0.1-5.4 µg/mL to biodegrade in activated sludge (Schwartz 1967; quoted, Muir 1991); rate constant of 3.6-6.7×10^{-10} mL(cell)$^{-1}$ d^{-1} from measurements of different river water samples (Paris et al. 1978; quoted, Scow 1982); half-life of 120 hours for 2-25 µg/mL to biodegrade in stream water with biodegradation rate of 2.5×10^{-4} mg M^{-1} h^{-1} at 28°C (Wolfe et al. 1978; quoted, Muir 1991); 1.6-1.8×10^{-8} mL(cell)$^{-1}$ d^{-1} from measurements of different river water samples (Steen et al. 1979; quoted, Scow 1982); primary degradation rate constant of $(2.6 \pm 0.72) \times 10^{-14}$ L(cell)$^{-1}$ h^{-1} in North American waters (Paris et al. 1981; quoted, Battersby 1990); aerobic half-life of 190 hours for 0.1-1.0 µg/mL to biodegrade in lake water with biodegradation rate of 1.3-4.9×10^{-4} L org^{-1} h^{-1} at 22°C (Schnoor et al. 1982; quoted, Muir 1991); 75 µg/mL to biodegrade with degradation rate of 1.4-4.2x10^{-13} L org^{-1} h^{-1} at 28°C in natural and sediment waters (Steen et al. 1982; quoted, Muir 1991); aerobic half-life of >4 months for 6-7 µg/mL to biodegrade in river water at 25°C (Stepp et al. 1985; quoted, Muir 1991).

Biotransformation:

Bioconcentration, Uptake (k_1) and Elimination (k_2) Rate Constants:

Common Name: Chlorsulfuron
Synonym: DPX 4189, Finesse, Glean, Telar
Chemical Name: 2-chloro-N-(((4-methoxy-6-methyl-1,3,5-triazin-2-yl)amino)-carbonyl)-benzenesulfonamide; 1-((o-chlorophenyl)-3-(4-methoxy-6-methyl-s-triazin-2-yl)urea
Uses: herbicide to control broadleaf weeds and some grass weeds.
CAS Registry No: 64902-72-3
Molecular Formula: $C_{12}H_{12}ClN_5O_4S$
Molecular Weight: 357.80
Melting Point (°C):
    174-178 (Agrochemicals Handbook 1987; Herbicide Handbook 1989; Worthing 1991; Montgomery 1993; Tomlin 1994; Milne 1995)
Boiling Point (°C):
    192 (dec., Herbicide Handbook 1989; Montgomery 1993)
Density (g/cm³ at 20°C):
Molar Volume (cm³/mol):
Molecular Volume (Å³):
Total Surface Area, TSA (Å²):
Dissociation Constant pK_a:
    3.60 (Herbicide Handbook 1989; Worthing 1991; Montgomery 1993; Tomlin 1994; Hornsby et al. 1996)
Heat of Fusion, ΔH_{fus}, kcal/mol:
Entropy of Fusion, ΔS_{fus}, cal/mol·K (e.u.):
Fugacity Ratio at 25°C (assuming ΔS_{fus} = 13.5 e.u.), F: 0.0321

Water Solubility (g/m³ or mg/L at 25°C):
    300 (at pH 5, Agrochemicals Handbook 1987; Worthing 1991; Tomlin 1994; Milne 1995)
    27900 (at pH 7, Agrochemicals Handbook 1987; Worthing 1991; Tomlin 1994; Milne 1995)
    28000 (at pH 7 with ionic strength 0.05, Herbicide Handbook 1989)
    7000 (20-25°C, at pH 7, selected, Wauchope et al. 1992; quoted, Majewski & Capel 1995)
    7000 (20-25°C, at pH 7, selected, Hornsby et al. 1996)
    60, 7000 (at pH 5, pH 7, Montgomery 1993)

Vapor Pressure (Pa at 25°C):
    6.10×10^{-4} (Agrochemicals Handbook 1987)
    6.13×10^{-4} (Herbicide Handbook 1989)
    3.00×10^{-9} (Worthing 1991; Tomlin 1994)
    1.98×10^{-2} (20-25°C, Wauchope et al. 1992; quoted, Majewski & Capel 1995)
    3.11×10^{-9} (Montgomery 1993)
    6.13×10^{-4} (20-25°C, selected, Hornsby et al. 1996)

Henry's Law Constant (Pa·m³/mol):
    3.60×10^{-11} (calculated-P/C, Montgomery 1993)
    1.98×10^{-5} (20-25°C, calculated-P/C as per Wauchope et al. 1992, Majewski & Capel 1995)

Octanol/Water Partition Coefficient, log K_{OW}:
    −0.84, 0.17, 1.09 (pH 8.4, pH 7.1, pH 4.5, UV, Ribo 1988)
    −0.88, 1.05 (pH 8.4, pH 4.5, HPLC, Ribo 1988, quoted, Sangster 1993)
    −1.34, 0.74 (pH 7, pH 4.5, Hay 1990, quoted, Sangster 1993)
    2.20   (Grayson & Kleier 1990, quoted, Sangster 1993)
    −0.10   (Montgomery 1993)
    −1.00   (at pH 7, Tomlin 1994)
    0.74, −1.34 (quoted, Hansch et al. 1995)

Bioconcentration Factor, log BCF:
    0.622 (calculated-S as per Kenaga 1980, this work)

Sorption Partition Coefficient, log K_{OC}:
    1.02 (Flanagan silt loam, Montgomery 1993)
    1.60 (Tomlin 1994)
    1.60 (at pH 7, selected, Hornsby et al. 1996)

Half-Lives in the Environment:
    Air:
    Surface water:
    Groundwater:
    Sediment:
    Soil: hydrolysis rates will be increased by warm soil temperatures at low pH and in the presence of moisture with an average half-life of 4-6 weeks under growing conditions (Agrochemicals Handbook 1987; Herbicide Handbook 1989); degrades in soil via hydrolysis followed by microbial degradation with half-life of 4-6 weeks (Hartley & Kidd 1987; quoted, Montgomery 1993; Tomlin 1994); degradation rate constants: 0.033 d^{-1} (depth 0-20 cm with half-life of 21 days), 0.0315 d^{-1} (depth 20-40 cm with half-life of 22 days) and for depth 40-60 cm with half-life of >150 days (Soakwaters soil, Walker et al. 1989); degradation rate constants: 0.0116 d^{-1} (depth 0-20 cm with half-life of 60 days), 0.0120 d^{-1} (depth 20-40 cm with half-life of 58 days) and 0.0076 d^{-1} (depth 40-60 cm with half-life of 91 days) (Wharf ground soil, Walker et al. 1989); degradation rate constants: 0.0126 d^{-1} (depth 0-20 cm with half-life of 55 days), 0.0073 d^{-1} (depth 20-40 cm with half-life of 95 days) and 0.0056 d^{-1} (depth 40-60 cm with half-life of 124 days) (Cottage Field soil, Walker et al. 1989); degradation rate constants: 0.0147 d^{-1} (depth 0-20 cm with half-life of 47 days), 0.0116 d^{-1} (depth 20-40 cm with half-life of 60 days) and 0.0047 d^{-1} (depth 40-60 cm with half-life of 147 days) (Hunts Mill soil, Walker et al. 1989); degradation rate constants: 0.0094 d^{-1} (depth 0-20 cm with half-life of 74 days), 0.0096 d^{-1} (depth 20-40 cm with half-life of 72 days) and 0.0082 d^{-1} (depth 40-60 cm with half-life of 85 days) (Bottom Barn soil, Walker et al. 1989); degradation rate constants: 0.0141 d^{-1} (depth 0-20 cm with half-life of 49 days), 0.0126 d^{-1} (depth 20-40 cm with half-life of 55 days) and 0.0089 d^{-1} (depth 40-60 cm with half-life of 78 days) (Long Ashton soil, Walker et al. 1989); degradation rate constants: 0.0144 d^{-1} (depth 0-20 cm with half-life

of 48 days), 0.0126 d^{-1} (depth 20-40 cm with half-life of 55 days) and 0.0124 d^{-1} (depth 40-60 cm with half-life of 56 days) (Norfolk Agricultural Station soil, Walker et al. 1989); degradation rate constants: 0.0248 d^{-1} (depth 0-20 cm with half-life of 28 days), 0.0289 d^{-1} (depth 20-40 cm with half-life of 24 days) and 0.0347 d^{-1} (depth 40-60 cm with half-life of 20 days) (Norfolk Agricultural Station soil, Walker et al. 1989); selected field half-life of 40 days (Hornsby et al. 1996).

Environmental Fate Rate Constants or Half-Lives:
    Volatilization:
    Photolysis: assuming first-order kinetics, calculated half-lives of ~186 hours for 33 µg/mL to degrade in distilled water, 31 hours for creek water, 136 hours for silica gel and 115 hours for montmorillonit under sunlight (Herrmann et al. 1985; quoted, Cessna & Muir 1991); however under indoor conditions half-lives of 92 hours in methanol, 78 hours in distilled water but only 18 hours in natural creek water (Herrmann et al. 1985); reported half-life in distilled water at >290 nm was 18 hours (Montgomery 1993).
    Oxidation:
    Hydrolysis: half-life of 4-8 weeks at 20°C and pH 5.7-7.0 (Agrochemicals Handbook 1987; Worthing 1991; Montgomery 1993; Tomlin 1994).
    Biodegradation:
    Biotransformation:
    Bioconcentration, Uptake (k_1) and Elimination (k_2) Rate Constants:

Common Name: Chlortoluron
Synonym: C 2242, Clortokem, Deltarol, Dicuran, Highuron, Higaluron, Tolurex
Chemical Name: 3-(3-chloro-p-tolyl)-1,1-dimethylurea; N'-(3-chloro-4-methylphenyl)-N,N-dimethylurea
Uses: herbicide to control pre- and post-emergent annual grasses and broadleaf weeds in winter cereals, particularly wheat and barley.
CAS Registry No: 15545-48-9
Molecular Formula: $C_{10}H_{13}ClN_2O$
Molecular Weight: 212.7
Melting Point (°C):
    147-148 (Khan 1980; Agrochemicals Handbook 1987)
    147.5 (Patil 1994)
    148.1 (Tomlin 1994)
Boiling Point (°C):
Density (g/cm³ at 20°C):
    1.40 (Tomlin 1994)
Molar Volume (cm³/mol):
    192 (modified LeBas method, Spurlock & Biggar 1994)
Molecular Volume (Å³):
Total Surface Area, TSA (Å²):
Dissociation Constant pK_a:
Heat of Fusion, ΔH_{fus}, kcal/mol:
Entropy of Fusion, ΔS_{fus}, cal/mol·K (e.u.):
Fugacity Ratio at 25°C (assuming ΔS_{fus} = 13.5 e.u.), F: 0.061

Water Solubility (g/m³ or mg/L at 25°C):
    10.0 (20°C, Spencer 1973; quoted, Shiu et al. 1990)
    70.0 (Martin & Worthing 1977; quoted, Kenaga 1980; Briggs 1981; Patil 1994)
    10.0 (20°C, Khan 1980)
    70.0 (20°C, Ashton & Crafts 1981)
    56.4, 80.6, 99.1 (4, 25, 40°C, shake flask-LSS, Madhun et al. 1986)
    70.0 (20°C, Agrochemicals Handbook 1987)
    70.0 (20°C, Worthing 1987; quoted, Shiu et al. 1990)
    70.0 (selected, Chaumat et al. 1991; Halfon et al. 1996)
    70.0 (20°C, selected, Evelyne et al. 1992)
    90.0 (Spurlock 1992; quoted, Spurlock & Biggar 1994)
    70.0, 10660 (quoted, calculated, Patil 1994)
    74.0 (Tomlin 1994)
    70.0, 49.3 (quoted, predicted-AQUAFAC, Lee et al. 1996)

Vapor Pressure (Pa at 25°C):
    4.8×10^{-6} (20°C, Khan 1980)
    1.7×10^{-5} (20°C, Ashton & Crafts 1981)
    1.7×10^{-5} (20°C, Agrochemicals Handbook 1987)
    1.7×10^{-5} (Tomlin 1994)
    1.7×10^{-5} (selected, Halfon et al. 1996)

Henry's Law Constant (Pa·m³/mol):
    5.17×10^{-5} (20°C, calculated-P/C, this work)

Octanol/Water Partition Coefficient, log K_{OW}:
- 2.41 (shake flask-UV, Briggs 1981; quoted, Sicbaldi & Finizio 1993; Patil 1994)
- 2.54 (Dao et al. 1983; Spurlock 1992; quoted, Spurlock & Biggar 1994)
- 2.33, 2.34, 2.32 (4, 25, 40°C, shake flask-LSS, Madhun et al. 1986; quoted, Sicbaldi & Finizio 1993)
- 2.241 (calculated, Evelyne et al. 1992)
- 2.25 (RP-HPLC, Sicbaldi & Finizio 1993)
- 0.26 (calculated, Patil 1994)
- 2.50 (Tomlin 1994)
- 2.41 (selected, Hansch et al. 1995)
- 2.38, 2.44 (shake flask-UV, calculated-RPHPLC-k', Liu & Qian 1995)
- 2.25, 2.49, 2.42 (RP-HPLC, ClogP, calculated-S, Finizio et al. 1997)

Bioconcentration Factor, log BCF:
- 1.75 (calculated-S, Kenaga 1980)
- 1.11 (calculated-K_{OW} as per Kenaga 1980, this work)
- 2.09, 2.16 (cuticle/water 24 hr: tomato, pepper, Chaumat et al. 1991)
- 2.01, 2.15 (cuticle/water 24 hr: box tree, pear, Chaumat et al. 1991)
- 1.30 (cuticle/water 24 hr: vanilla, Chaumat et al. 1991)
- 2.09, 2.16 (cuticle/water: tomato, pepper, Evelyne et al. 1992)

Sorption Partition Coefficient, log K_{OC}:
- 2.62 (soil, calculated-S, Kenaga 1980)
- 1.78 (reported as log K_{OM}, Briggs 1981)
- 2.75, 2.62 (4, 25°C, Semiahmoo soil, in μmol/kg OC, batch equilibrium method-LSS, Madhun et al. 1986)
- 2.57, 2.43 (4, 25°C, Adkins soil, in μmol/kg OC, batch equilibrium method-LSS, Madhun et al. 1986)
- 2.48, 2.18; 2.54, 2.50 (estimated-K_{OW}; S, Madhun et al. 1986)
- 2.81, 2.58 (exptl., calculated-K_{OW}, Liu & Qian 1995)

Half-Lives in the Environment:
    Air:
    Surface water:
    Groundwater:
    Sediment:
    Soil: half-life of 4 weeks in the moist silty loam at 25 ± 1°C (Smith & Briggs 1978); half-life of 30-40 days in soil (Tomlin 1994); 135 days (selected, Halfon et al. 1996).
    Biota:

Environmental Fate Rate Constants or Half-Lives:
    Volatilization:
    Photolysis:
    Oxidation:
    Hydrolysis: calculated half-life of >200 days at pH 5,7,9 and 30°C (Tomlin 1994).
    Biodegradation:
    Biotransformation:
    Bioconcentration, Uptake (k_1) and Elimination (k_2) Rate Constants:

Common Name: Cyanazine
Synonym: Bladex, 90DF, DW 3418, Fortrok, Fortrol, Payze, SD 15418, WL 19805
Chemical Name: 2-(4-chloro-6-ethylamino-1,3,5-triazin-2-ylamino)-2-methyl-propionitrile
Uses: herbicide to control annual grasses and broadleaf weeds in cereals, cotton, maize, onions, peanuts, peas, potatoes, soybeans, sugar cane, and wheat fallow.
CAS Registry No: 21725-46-2
Molecular Formula: $C_9H_{13}ClN_6$
Molecular Weight: 240.7
Melting Point (°C):
    167.0   (Karickhoff 1981)
    166.5-167   (Spencer 1982; Herbicide Handbook 1989; Worthing 1991)
    166.6   (Patil 1994)
    166.5   (Kühne et al. 1995)
    167.5-169   (Tomlin 1994; Milne 1995)
Boiling Point (°C):
Density (g/cm³ at 20°C):
Molar Volume (cm³/mol):
Molecular Volume (Å³):
Total Surface Area, TSA (Å²):
Dissociation Constant:
    1.00   (pK_a, Weber et al. 1980; Willis & McDowell 1982)
    12.9   (pK_b, Wauchope et al. 1992; Hornsby et al. 1996)
    0.63, 1.1   (pK_a, Montgomery 1993)
Heat of Fusion, ΔH_{fus}, kcal/mol:
Entropy of Fusion, ΔS_{fus}, cal/mol·K (e.u.):
Fugacity Ratio at 25°C (assuming ΔS_{fus} = 13.5 e.u.), F: 0.039

Water Solubility (g/m³ or mg/L at 25°C):
    171   (Melnikov 1971; Wauchope 1978; quoted, Shiu et al. 1990)
    171   (Martin & Worthing 1977; Herbicide Handbook 1978; quoted, Kenaga 1980; Kenaga & Goring 1980; Karickhoff 1981; Worthing 1987, 1991; Shiu et al. 1990; Isensee 1991; Majewski & Capel 1995)
    171   (Ashton & Crafts 1981; Herbicide Handbook 1989)
    171   (Weber et al. 1980; quoted, Willis & McDowell 1982)
    150   (selected, Schnoor & McAvoy 1981; quoted, Schnoor 1992)
    171   (Agrochemicals Handbook 1987; Montgomery 1993; Tomlin 1994)
    160   (23°C, Herbicide Handbook 1989; quoted, Shiu et al. 1990)
    171   (Merck Index 1989; quoted, Milne 1995)
    170   (20-25°C, selected, Wauchope et al. 1992; Hornsby et al. 1996)
    170   (selected, Lohninger 1994)
    170, 6046   (quoted, calculated, Patil 1994)
    179, 45 (quoted, calculated-group contribution fragmentation method, Kühne et al. 1995)

Vapor Pressure (Pa at 25°C):
    2.13×10^{-7} (20°C, Ashton & Crafts 1973; 1981; Spencer 1982; Herbicide Handbook 1989)
    2.67×10^{-7} (20-25°C, Weber et al. 1980; quoted, Willis & McDowell 1982)

5.33×10^{-7} (selected, Schnoor & McAvoy 1981; quoted, Schnoor 1992)
1.00×10^{-5} (20°C, extrapolated from gas saturation measurement, ln P (Pa) = 25.7-10913/T, for temp range 65.7-92°C, Grayson & Fosbraey 1982)
2.00×10^{-7} (20°C, Agrochemicals Handbook 1987; Worthing 1991; Tomlin 1994; Majewski & Capel 1995)
5.21×10^{-6} (Worthing 1987; quoted, Majewski & Capel 1995)
1.33×10^{-6} (30°C, Herbicide Handbook 1989)
2.13×10^{-7} (20°C, Merck index 1989)
2.13×10^{-7} (20-25°C, selected, Wauchope et al. 1992; Hornsby et al. 1996)
2.13×10^{-7} (20°C, Montgomery 1993)

Henry's Law Constant (Pa·m^3/mol):
    2816 (20-25°C, calculated-P/C, Montgomery 1993)
    2.87×10^{-7} (calculated-P/C as per Worthing 1987, Majewski & Capel 1995)
    3.00×10^{-7} (calculated-P/C, this work)

Octanol/Water Partition Coefficient, log K_{OW}:
    2.18 (quoted, Kenaga & Goring 1980)
    2.24 (selected, Brown & Flagg 1981; quoted, Karickhoff 1981; Madhun et al. 1986)
    1.80, 1.66 (RP-HPLC, calculated, Finizio et al. 1991)
    2.22 (selected, Magee 1991)
    1.80, 2.24 (Montgomery 1993)
    2.24, 0.79 (quoted, calculated, Patil 1994)
    2.10 (Tomlin 1994)
    2.22 (selected, Hansch et al. 1995)
    2.04 (shake flask-UV, Liu & Qian 1995)
    1.64, 1.29, 3.02 (RP-HPLC, ClogP, calculated-S, Finizio et al. 1997)

Bioconcentration Factor, log BCF:
    1.53 (calculated-S, Kenaga 1980; quoted, Isensee 1991)
    1.00 (calculated-K_{OC}, Kenaga 1980)
    1.48 (selected, Schnoor & McAvoy 1981; quoted, Schnoor 1992)

Sorption Partition Coefficient, log K_{OC}:
    2.30 (quoted, Kenaga 1980; Kenaga & Goring 1980; Karickhoff 1981; Bahnick & Doucette 1988)
    2.41 (soil, calculated-S as per Kenaga & Goring 1978, Kenaga 1980)
    2.26 (Georgia's Hickory Hill pond sediment, Brown & Flagg 1981; quoted, Karickhoff 1981; Muir 1991)
    2.71, 1.75, 1.85 (estimated-S, S & M.P., K_{OW}, Karickhoff 1981)
    0.48-1.48 (selected, sediment/water, Schnoor & McAvoy 1981; quoted, Schnoor 1992)
    2.57, 2.26 (soil, quoted, Madhun et a. 1986)
    2.36, 2.09; 2.33, 1.75 (estimated-K_{OW}; S, Madhun et al. 1986)
    2.23 (soil, screening model calculations, Jury et al. 1987b)

2.35 (calculated-χ, Bahnick & Doucette 1988)
2.30, 2.16 (reported, estimated as log K_{OM}, Magee 1991)
2.23, 2.26, 2.30 (quoted values, Bottoni & Funari 1992)
2.28 (soil, 20-25°C, selected, Wauchope et al. 1992; quoted, Richards & Baker 1993; Hornsby et al. 1996)
1.58-2.63 (Montgomery 1993)
2.54 (selected, Lohninger 1994)
2.05, 2.11 (exptl., calculated-K_{OW}, Liu & Qian 1995)

Half-Lives in the Environment:
    Air:
    Surface water: aerobic half-life of 14 d for 0.06 µg/mL to degrade in pond water at 10-20°C (Roberts 1974; quoted, Muir 1991).
    Groundwater:
    Sediment: aerobic half-life of >28 d for 0.06 µg/mL to slowly degrade in pond sediment at 10-20°C (Roberts 1974; quoted, Muir 1991).
    Soil: half-life in soil ca. 2 weeks (Beynon et al. 1972; quoted, Tomlin 1994); persistence of 12 months in soil (Wauchope 1978); half-life of 13.5 d from screening model calculations (Jury et al. 1987b); half-life is 12-15 d in sandy loam soils and 20-25 d in silt and clay loam soils (Herbicide Handbook 1989; quoted, Montgomery 1993); disapperance half-lives from the upper 15 cm on a clay loam Ontario soil were 181 d in 1987 and 90 days in 1988 with calculated half-lives of 27 and 12 d respectively (Frank et al. 1991); selected field half-life of 14 d (Wauchope et al. 1992; quoted, Richards & Baker 1993; Hornsby et al. 1996); soil half-life of 19 d (Pait et al. 1992).
    Biota: biochemical half-life of 13.5 d from screening model calculations (Jury et al. 1987b).

Environmental Fate Rate Constants or Half-Lives:
    Volatilization:
    Photolysis:
    Oxidation:
    Hydrolysis: alkaline chemical hydrolysis half-life of >365 d (Schnoor & McAvoy 1981; quoted, Schnoor 1992).
    Biodegradation: aerobic half-lives of 14 d for 0.06 µg/mL to degrade in pond water and >28 d in pond sediment both at 10-20°C (Roberts 1974; quoted, Muir 1991).
    Biotransformation:
    Bioconcentration, Uptake (k_1) and Elimination (k_2) Rate Constants:

Common Name: 2,4-D
Synonym: 2,4-Dichlorophenoxyacetic acid
Chemical Name: 2,4-dichlorophenoxyacetic acid
Uses: post-emergence control of annual and perennial broadleaf weeds in cereals, maize, sorgum, grassland, established turf, grass seed crops, orchards, cranberries, asparagus, sugar cane, rice, forestry, and on noncrop land, etc.
CAS Registry No: 94-75-7
Molecular Formula: $C_8H_6Cl_2O_3$, $Cl_2C_6H_3OCH_2COOH$
Molecular Weight: 221.04
Melting Point (°C):
    136-140 (Verschueren 1983)
    138 (Dean 1985; Lee et al. 1993)
    135-138 (Suntio et al. 1988; Riederer 1990)
    140.5 (Agrochemicals Handbook 1987; Howard 1991; Tomlin 1994)
Boiling Point (°C):
    160 (at 0.4 mmHg, Dean 1985)
    215 (Neely & Blau 1985)
Density (g/cm³ at 25°C):
    1.565 (30°C, Neely & Blau 1985; Tomlin 1994)
    1.416 (Montgomery 1993)
Molar Volume (cm³/mol):
    209.8 (calculated-LeBas method, Suntio et al. 1988; Riederer 1990)
    206.2 (calculated-LeBas method)
Molecular Volume (Å³):
Total Surface Area, TSA (Å²):
Dissociation Constant, pK_a:
    2.73 (potentiometric, Nelson & Faust 1969)
    2.87 (spectrophotometric, Cessna & Grover 1978; McCall et al. 1978; Somasundaram et al. 1991)
    2.80 (Reinert & Rogers 1984; selected, Wauchope et al. 1992)
    2.64 (Dean 1985; Lee et al. 1993)
    2.61-3.31 (Howard 1991)
    2.97 (Sangster 1993)
    3.10 (Kollig 1993)
    2.64-3.31 (Montgomery 1993)
Heat of Fusion ΔH_{fus}, kcal/mol:
    9.10 (DSC method, Plato & Glasgow 1969)
Entropy of Fusion ΔS_{fus}, cal/mol·K (e.u.):
Fugacity Ratio at 25°C (assuming ΔS_{fus} = 13.5 e.u.), F: 0.0720

Water Solubility (g/m³ or mg/L at 25°C):
    890 (Hodgman 1952; quoted, Freed et al. 1977)
    522 (shake flask-UV, Leopold et al. 1960; quoted, Shiu et al. 1990)
    725 (Bailey & White et al. 1965; quoted, Shiu et al. 1990)
    725, 400, 900, 550 (quoted, Gunther et al. 1968)
    900 (Herbicide Handbook 1974; quoted, Jury et al. 1983)
    890 (Hamaker 1975)
    900 (Wauchope 1978; quoted, Pait et al. 1992)
    900 (Kenaga 1980a,b; Kenaga & Goring 1980; selected, Isensee 1991)

600	(20°C, Khan 1980)
620-900	(Weber et al. 1980; quoted, Willis & McDowell 1982)
470	(20-25°C, pH 5.6, Geyer et al. 1981)
620	(20°C, Hartley & Kidd 1983, 1987)
890	(Verschueren 1983; quoted, Isnard & Lambert 1988; Shiu et al. 1990; Somasundaram et al. 1991; Lee et al. 1993; Montgomery 1993)
620	(Worthing 1983; quoted, Shiu et al. 1990)
690	(recommended, Neely & Blau 1985)
609	(Gerstl & Helling 1987)
400	(20°C, selected, Suntio et al. 1988; quoted, Riederer 1990)
703	(Gustafson 1989)
682	(Yalkowsky et al. 1987; quoted, Howard 1991)
540-890	(Nyholm et al. 1992; quoted, Meakins et al. 1994)
900, 600, 890, 703, 1072	(quoted, Wauchope et al. 1992)
890	(20-25°C, selected, Wauchope et al. 1992)
311	(pH 1, Tomlin 1994)

Vapor Pressure (Pa at 25°C):
- 8.0×10^{-5} (Hamaker 1975)
- 0.180-1.69 (transpiration method, Spencer 1976; quoted, Suntio et al. 1988)
- 53.0 (160°C, Hartley & Kidd 1983, 1987)
- 8.0×10^{-5} (recommended, Neely & Blau 1985; Lyman 1985)
- 1.0 (20°C, selected, Suntio et al. 1988; quoted, Riederer 1990)
- 6.0×10^{-6} (selected, Nash 1989)
- 0.20, 0.0032 (quoted, estimated from Henry's law constant, Howard 1991)
- 5.6×10^{-5} (selected, Mackay & Stiver 1991)
- 1.40, 3.2×10^{-3} (quoted, estimated from HLC, Howard 1991)
- 1.33×10^{-5}, 8.0×10^{-5}, 1.07×10^{-3} (quoted, Wauchope et al. 1992)
- 0.00107 (20-25°C, seleted, Wauchope et al. 1992)
- 0.627 (Montgomery 1993)
- 0.011 (Tomlin 1994)

Henry's Law Constant (Pa·m^3/mol):
- 1.36×10^{-5} (calculated-P/C, Jury et al. 1983)
- 1.39×10^{-5} (calculated-P/C, Jury et al. 1987a, Jury & Ghodrati 1989)
- 0.55 (20°C, calculated-P/C, Suntio et al. 1988; quoted, Riederer 1990)
- 0.0015 (calculated, Nash 1989)
- 1.38×10^{-5}, 1.03×10^{-3} (quoted, calculated-bond contribution, Howard 1991)

Octanol/Water Partition Coefficient, log K_{ow}:
- 2.81 (shake flask-AS, Fujita et al. 1964; quoted, Freed et al. 1977; Sangster 1993; Howard 1991; Somasundaram et al. 1991; Nyholm et al. 1992; Meakins et al. 1994; Finizio et al. 1997)
- 2.59 (electrometric titration, Freese et al. 1979; quoted, Sangster 1993)
- 1.57 (Kenaga & Goring 1980; Kenaga 1980b; quoted, Finizio et al. 1997)
- 2.49 (correlated-S, Mackay et al. 1980; quoted, Sangster 1993)
- 2.74 (selected, Dao et al. 1983)
- 2.81 (20°C, quoted, Verschueren 1983)
- 1.57, 4.88 (shake flask-OECD 1981 Guidelines, Geyer et al. 1984)

2.81   (recommended, Neely & Blau 1985)
−1.36   (Gerstl & Helling 1987)
2.65   (shake flask, Hansch & Leo 1987; quoted, Sangster 1993)
3.00   (quoted, Isnard & Lambert 1988)
2.50   (OECD 1981 method, Kerler & Schönherr 1988)
2.649   (liquid/liquid-countercurrent-chromatography, Ilchmann et al. 1993; quoted, Sangster 1993)
2.81   (recommended, Sangster 1993)
2.68   (calculated-QSAR, Kollig 1993)
1.44-4.18 (Montgomery 1993)
2.58-2.83 (pH 1, Tomlin 1994)
2.81   (selected, Hansch et al. 1995)
0.59, 2.77, 2.64 (RP-HPLC, CLOGP, Calculated-S, Finizio et al. 1997)

Bioconcentration Factor, log BCF:
    1.11, −0.097 (calculated-S, K_{ow}, Kenaga 1980a)
    −2.46, 1.30 (beef fat, fish, Kenaga 1980b)
    0.778, 1.94 (alga *Chlorella*: exptl. 24 h exposure, calculated-S, Geyer et al. 1981)
    0.778 (algae, Freitag et al. 1982)
    <1.00 (golden orfe, Freitag et al. 1982)
    1.23 (activated sludge, Freitag et al. 1982)
    0.0 (fish, microcosm conditions, Garten & Trabalka 1983)
    0.778, 1.23 (algae, calculated-K_{ow}, Geyer et al. 1984)
    1.23 (algae, Geyer et al. 1984)
    0.85 (quoted, Isnard & Lambert 1988)
    1.11 (calculated, Isensee 1991)
    −5.00 (bluegill sunfish and channel catfish, Howard 1991)
    −2.70 (frog tadpoles, Howard 1991)
    −3.0, −2.52 (pH 7.8, seaweeds, Howard 1991)
    0.778, 0.85 (quoted: alga, fish, Howard 1991)
    0.0 (quoted, Pait et al. 1992)
    0.0, 0.505 (*Ictalurus melas*, *Daphnia magna*, Wang et al. 1996)

Sorption Partition Coefficient, log K_{OC}:
    1.51 (Hamaker 1975; quoted, Nash 1989)
    1.30, 2.0 (quoted, calculated, Kenaga 1980a)
    1.30, 2.11 (quoted, Kenaga & Goring 1980)
    1.30 (quoted, Kenaga 1980b)
    1.76 (quoted, average value of 3 soils, McCall et al. 1980)
    1.29 (soil, Neely & Blau 1985)
    1.30 (soil, screening model calculations, Jury et al. 1987a,b; Jury & Ghodrati 1989)
    1.61 (soil, quoted, Sabljic 1987)
    1.75, 2.00 (quoted, calculated-χ, Gerstl & Helling 1987)
    1.30 (selected, Mackay & Stiver 1991)
    1.00, 1.23, 2.29 (sediment, Alfisol soil, Podzol soil, von Oepen et al. 1991)
    1,30, 1.78, 1.51, 1,26, 1.72, 1.75, 1.76 (soil, quoted, Wauchope et al. 1992)
    1.30 (soil, selected, Wauchope et al. 1992)
    0.68 (calculated-K_{ow}, Kollig 1993)
    1.68-2.73 (Montgomery 1993)

Half-Lives in the Environment:
- Air: 1.8-18 h, based on estimated rate constant for the vapor-phase reaction with hydroxyl radicals in air (Howard et al. 1991); photooxidation half-life of 23.9 h for reactions with hydroxyl radicals in air (Howard 1991).
- Surface water: 48-96 h, based on reported photolysis half-lives for aqueous solution irradiated at UV wavelength of 356 nm (Baur & Bovey 1974; selected, Howard et al. 1991); degradation half-life of 14 d in sensitized, filtered and sterilized river water, based on sunlight photolysis test of 1 μg mL^{-1} in distilled water (Zepp et al. 1975; quoted, Cessna & Muir 1991); typical biodegradation half-lives of 10 to <50 d with longer expected in oligotrophic waters, photolysis half-life of 29-43 d for water solutions irradiated at sunlight (Howard 1991); degraded relatively slowly when incubated in natural waters or in soil/sediment suspensions, with half-lives ranging from about 6 to 170 d (Muir 1991); calculated rate constant of 5×10^9 M$^{-1}\cdot$s^{-1} for the reaction with hydroxyl radicals in aqueous solution (Haag & Yao 1992); half-life of 2-4 d when irradiated at λ = 356 nm in aqueous solution (Montgomery 1993).
- Groundwater: 480-4320 h, based on estimated unacclimated aqueous aerobic and anaerobic biodegradation half-lives (Howard et al. 1991).
- Sediment: half-life of <1 d for degradation in sediments and lake muds (Howard 1991); degraded relatively slowly when incubated in natural waters or in soil/sediment suspensions, with half-lives ranging from about 6 to 170 d (Muir 1991).
- Soil: Lab. half-life of 4.0 d in Quachita grassland (Altom & Stritzke 1973; quoted, Nash 1983); field half-life of 5.2 d in Arid range (Lane et al. 1977; quoted, Nash 1983); field half-life of 19 d in Dykland soil (Stewart & Gaul 1977; quoted, Nash 1983); Lab. half-life of 5.5 d in Naff soil (Wilson & Chen 1978; quoted, Nash 1983); microagroecosystem half-life of 11 d for granular application to bluegrass turf (Nash & Beall 1980; quoted, Nash 1983); non-persistent in soil with half-life <20 d (Willis & McDowell 1982); microagroecosystem half-life of 3 d in moist fallow soil (Nash 1983); half-life of 15 d in soil (Jury et al. 1983, 1987a,b; Jury & Ghodrati 1989); persistence of one month in soil (Jury et al. 1987); 240-1200 h, based on estimated unacclimated aqueous aerobic biodegradation half-life (Howard et al. 1991); biodegradation half-lives ranging from <1 d to several weeks, of 3.9 and 11.5 d in 2 moist soils and 9.4 to 254 d in the same soils under dry conditions (Howard 1991); degraded relatively slowly when incubated in natural waters or in soil/sediment suspensions, with half-lives ranging from about 6 to 170 d (Muir 1991); field half-lives range from 2-16 d, with a selected value of 10 d (Wauchope et al. 1992); soil half-life of 18 days (Pait et al. 1992); rate constants for Amsterdam silt loam at soil depth 0-30 cm: 0.0053 d^{-1} at 10°C, 0.0046 d^{-1} at 17°C and 0.0127 d^{-1} at 24°C with corresponding first-order half-lives: 7, 7, and 2 d and at soil depth 30-60 cm: 0.00012 d^{-1} at 10°C, 0.0044 d^{-1} at 17°C and 0.0077 d^{-1} at 24°C with corresponding first-order half-lives: 273, 8, and 4 days; and at soil depth 60-120 cm: 0.00005 d^{-1} at 10°C, 0.0013 d^{-1} at 17°C and 0.0022 d^{-1} at 24°C with corresponding first-order half-lives: 593, 25, and 12 d (Veeh et al. 1996).
- Biota: depuration half-life of 13.8 h in daphnids, 1.32 d in catfish (Ellgehausen et al. 1980).

Environmental Fate Rate Constants or Half-Lives:
- Volatilization: volatilization from water is negligible, calculated volatilization half-lives from soil of 660 d (from 1 cm) and 7.1 years (from 10 cm) (Howard 1991).
- Photolysis: aqueous photolysis half-lives of 2-4 d when irradiated at 356 nm, a half-life of 50 minutes in water when irradiated at 254 nm and half-life of 29-43 d when exposed to September sunlight (Howard 1991).
- Oxidation: photooxidation half-life of 1.8-18 h, based on estimated rate constant for the vapor-phase reaction with hydroxyl radicals in air (Howard et al. 1991).
- Hydrolysis: no hydrolyzable groups and rate constant at neutral pH is zero (Kollig et al. 1987; selected, Howard et al. 1991); generally resistant to hydrolysis, may become important at pH >8 (Howard 1991).
- Biodegradation: degradation half-lives range from 10 to >50 d in clear to murky river water with lag time ranges from 6-12 d and degradation rate of 0.7-14.0 d^{-1} (Nesbitt & Watson 1980a); half-life of 4 d in river with nutrient and suspended sediments and 10 d with a lag time of 5 d for filtered river water (Nesbitt & Watson 1980b); degradation kinetics not first-order, time for 50% decomposition in six soils: Commerce 5 d, Catlin 1.5 d, Keith 3.9 d, Cecil 3.0 d, Walla-Walla 2.5 d and Fargo 8.5 d, with an average time of 4 d (McCall et al. 1981); easily degraded under aerobic conditions with a half-life of 1.8 and 3.1 d for cometabolism and metabolism respectively, under anaerobic conditions the degradation rate decreases and the half-lives are 69 and 135 d (Liu et al. 1981; quoted, Muir 1991); second-order rate constants of $(3.6-28.8) \times 10^{-6}$ mL·cell^{-1}·d^{-1} in natural water (Paris et al. 1981; quoted, Klečka 1985); first-order rate constant of <0.14-0.07 d^{-1} in river water at 25°C (Nesbitt & Watson 1980; quoted, Klečka 1985); rate constant of 0.058 ± 0.006 d^{-1} in lake water at 29°C (Subba-Rao et al. 1982; quoted, Klečka 1985); rate constant of 0.08-0.46 d^{-1} in soil at 25°C (McCall et al. 1981; quoted, Klečka 1985); aqueous aerobic half-life of 240-1200 h, based on unacclimated aerobic river die-away test data (Nesbitt & Watson 1980; selected, Howard et al. 1991); aqueous anaerobic half-life of 672-4320 h, based on unacclimated aqueous screening test data (Liu et al. 1981; selected, Howard et al. 1991); first-order biodegradation rate constant of 0.035 d^{-1} (die-away test), 0.029 d^{-1} (CO_2 evolution test) in soil and 6.9×10^{-1} mL·(g bacteria)$^{-1}$·d^{-1} by activated sludge cultures (Scow 1982); biodegradation half-lives in river water of 18 to over 50 d (clear water) and 10 to 25 d (muddy water) with lag times of 6 to 12 d; degradation with a mixture of microorganisms from activated sludge, soil, and sediments lead to half-lives of 1.8-3.1 d under aerobic conditions and 69-135 d under anaerobic conditions (Howard 1991).
- Biotransformation:
- Bioconcentration, Uptake (k_1) and Elimination (k_2) Rate Constants:
  - k_1: 0.0092 h^{-1} (catfish, Wang et al. 1996)
  - k_2: 0.0092 h^{-1} (catfish, Wang et al. 1996)
  - k_1: 0.8560 h^{-1} (*Daphnia magna*, Wang et al. 1996)
  - k_2: 0.2690 h^{-1} (*Daphnia magna*, Wang et al. 1996)

Common Name: Dalapon
Synonym: Alatex, Basinex P, Crisapon, D-Granulat, Dawpon-Rae, Ded-Weed, Dowpon, DPA, Gramevin, Kenapon, Liropon, Proprop, Radapon
Chemical Name: 2,2-dichloropropanoic acid; 2,2-dichloropropionic acid; α-dichloropropanoic acid; α,α-dichloropropionic acid
Uses: selective systemic herbicide to control perennial and annual grasses on noncrop land, fruits, vegetables, and some aquatic weeds.
CAS Registry No: 75-99-0
Molecular Formula: $C_3H_4Cl_2O_2$
Molecular Weight: 143
Melting Point (°C): liquid
Boiling Point (°C):
    185-190 (Kenaga 1974; Spencer 1976; Khan 1980; Herbicide Handbook 1989; Worthing 1991; Tomlin 1994)
    98-99 (sodium salt at 20 mmHg, Merck Index 1989)
    190 (Howard 1991)
Density (g/cm³ at 20°C):
    1.389 (Nelson & Faust 1969; quoted, Kenaga 1974; Montgomery 1993)
    1.389 (22.8°C, Herbicide Handbook 1989)
    1.4014 (Merck Index 1989; Milne 1995)
Molar Volume (cm³/mol):
Molecular Volume (Å³):
Total Surface Area, TSA (Å²):
Dissociation Constant pK_a:
    1.84 (Nelson & Faust 1969; Freed 1976; Hornsby et al. 1996)
    1.74 (Kenaga 1974; quoted, Howard 1991)
    1.74-1.84 (Worthing 1991; Tomlin 1994)
    2.06 (Yao & Haag 1991)
    1.84 (free acid, Montgomery 1993)
Heat of Fusion, ΔH_{fus}, kcal/mol:
Entropy of Fusion, ΔS_{fus}, cal/mol·K (e.u.):
Fugacity Ratio at 25°C (assuming ΔS_{fus} = 13.5 e.u.), F: 1.0

Water Solubility (g/m³ or mg/L at 25°C):
    900000 (Woodford & Evans 1963; quoted, Shiu et al. 1990)
    900000 (Bailey & White 1965; quoted, Shiu et al. 1990)
    >800000 (Kenaga 1974)
    502000 (Martin & Worthing 1977; quoted, Kenaga 1980; Kenaga & Goring 1980; Lyman 1982; Spencer 1982; Shiu et al. 1990; Howard 1991; Isensee 1991; Majewski & Capel 1995)
    450000 (Weber et al. 1980; quoted, Willis & McDowell 1982)
    501200 (Garten & Trabalka 1983; quoted, Shiu et al. 1990)
    431850 (selected, Gerstl & Helling 1987)
    900000 (sodium salt, Worthing 1987, 1991)
    500000 (Reinert 1989)
    450000 (Merck Index 1989)
    450000-900000 (Montgomery 1993)
    900000 (20-25°C, selected, Hornsby et al. 1996)

Vapor Pressure (Pa at 25°C):
- 16.0 (calculated from high temp., Foy 1976; quoted, Howard 1991; Majewski & Capel 1995)
- 1×10^{-5} (Worthing 1991; Tomlin 1994)
- 0.0 (20-25°C, selected, Hornsby et al. 1996)

Henry's Law Constant (Pa·m^3/mol):
- 6.50×10^{-3} (Hine & Mookerjee 1975; quoted, Howard 1991)
- 0.608 (calculated, Montgomery 1993)
- 4.56×10^{-3} (calculated-P/C as per Howard 1991, Majewski & Capel 1995)

Octanol/Water Partition Coefficient, log K_{ow}:
- 0.76 (Kenaga 1974; quoted, Rao & Davidson 1980; Howard 1991; Montgomery 1993; Sangster 1993)
- 0.78 (Kenaga 1980, quoted, Lyman 1982; Sangster 1993)
- 1.34 (selected, Dao et al. 1983)
- −2.76 (selected, Gerstl & Helling 1987)
- 1.48 (Reinert 1989)
- 0.78 (selected, Hansch et al. 1995)

Bioconcentration Factor, log BCF:
- 0.477 (dalapon sodium salt in fish, Kenaga 1974; quoted, Howard 1991)
- −0.444 (calculated-S, Kenaga 1980; quoted, Isensee 1991)
- 0.301 (estimated-K_{ow}, Lyman et al. 1982; quoted, Howard 1991)

Sorption Partition Coefficient, log K_{oc}:
- 0.477 (soil, calculated-S as per Kenaga & Goring 1978, Kenaga 1980)
- 0.97 (calculated-χ, Gerstl & Helling 1987)
- 2.13 (Reinert 1989)
- 0.48, 2.13 (quoted values, Bottoni & Funari 1992)
- 0.27-2.18 (calculated, Montgomery 1993)
- 0.0 (soil, 20-25°C, selected, Hornsby et al. 1996)

Half-Lives in the Environment:
- Air: 289-2893 h, based on an estimated rate constant for the vapor-phase reaction with hydroxyl radicals in air (Atkinson 1987; quoted, Howard et al. 1991).
- Surface water: 336-1440 h, based on estimated aqueous aerobic biodegradation half-life (Howard et al. 1991).
- Groundwater: 672-2880 h, based on estimated aqueous aerobic biodegradation half-life (Howard et al. 1991).
- Sediment:
- Soil: half-life of 7-8 d in soil (Kaufman 1966; quoted, Kaufman 1976); persistence across 43 soils from <2 weeks to >8 weeks (Day et al. 1963; quoted, Kaufman 1976); 336-1440 h, based on unacclimated aerobic soil grab sample data (Corbin & Upchurch 1967; Kaufman & Doyle 1977; quoted,

Howard et al. 1991); estimated persistence of 8 months (Kearney et al. 1969; quoted, Jury et al. 1987); persistence of 8 weeks in soil (Edwards 1973; quoted, Morrill et al. 1982); persistence of about 2 weeks in growing season in most agricultural soils (Herbicide Handbook 1974; quoted, Kaufman 1976); estimated first-order half-life of 15 d from biodegradation rate constant of 0.047 d^{-1} by soil incubation die-away studies (Rao & Davidson 1980; quoted, Scow 1982); non-persistent in soil with half-life of <20 d (Willis & McDowell 1982); field half-life of 30 d (20-25°C, selected, Hornsby et al. 1996).

Biota:

Environmental Fate Rate Constants or Half-Lives:
 Volatilization:
 Photolysis:
 Oxidation: photooxidation half-life of 289-2893 h in air, based on an estimated rate constant for the vapor-phase reaction with hydroxyl radicals in air (Atkinson 1987; quoted, Howard et al. 1991); 4.6×10^8 M^{-1} s^{-1} for the reaction (photo-Fenton with reference to acetophenone) with hydroxyl radicals in aqueous solutions at pH 3.4 and at (24 ± 1)°C (Buxton et al. 1988; quoted, Faust & Hoigné 1990; Haag & Yao 1992); rate constant of $(7.3 \pm 0.3) \times 10^7$ M^{-1} s^{-1} for the reaction (photo-Fenton with reference to acetophenone) with hydroxyl radicals in aqueous solutions at pH 3.4 and at (24 ± 1)°C (Haag & Yao 1992).
 Hydrolysis:
 Biodegradation: aqueous aerobic half-life of 336-1440 h, based on unacclimated aerobic soil grab sample data (Corbin & Upchurch 1967; Kaufman & Doyle 1977; quoted, Howard et al. 1991); rate constant of 0.047 d^{-1} by soil incubation die-away studies (Rao & Davidson 1980; quoted, Scow 1982); aqueous anaerobic half-life of 1344-5760 h, based on estimated aqueous aerobic biodegradation half-life (Howard et al. 1991).
 Biotransformation:
 Bioconcentration, Uptake (k_1) and Elimination (k_2) Rate Constants:

Common Name: 2,4-DB
Synonym: Butoxon, Butyrac, Butyrac 118, Embutox, Legumex D
Chemical Name: 4-(2,4-dichlorophenoxy)butanoic acid; 4-(2,4-dichlorophenoxy)butyric acid
Uses: herbicide for post-emergence control of many annual and perennial broadleaf weeds in lucerne, clovers, undersown cereals, grassland, forage legumes, soybeans, and groundnuts.
CAS Registry No: 94-82-6
Molecular Formula: $C_{10}H_{10}Cl_2O_3$
Molecular Weight: 249.1
Melting Point (°C):
    117-119 (Spencer 1982; Milne 1995)
    119-119.5 (Worthing 1987, 1991)
Boiling Point (°C):
Density (g/cm³ at 20°C):
Molar Volume (cm³/mol):
    254.2 (calculate-LeBas method, this work)
Molecular Volume (Å³):
Total Surface Area, TSA (Å²):
Dissociation Constant pK_a:
    5.95 (Bailey & White 1965; Que Hee et al. 1981)
    4.80 (Worthing 1987; Hornsby et al. 1996)
Heat of Fusion, ΔH_{fus}, kcal/mol:
Entropy of Fusion, ΔS_{fus}, cal/mol·K (e.u.):
Fugacity Ratio at 25°C (assuming ΔS_{fus} = 13.5 e.u.), F: 0.120

Water Solubility (g/m³ or mg/L at 25°C):
    82.3 (Bailey & White 1965; quoted, Que Hee et al. 1981; Shiu et al. 1990)
    53 (rm. temp., Melnikov 1971; quoted, Shiu et al. 1990)
    46 (Martin & Worthing 1977; Kenaga 1980)
    46 (Weber et al. 1980; quoted, Willis & McDowell 1982)
    46 (Agrochemicals Handbook 1987)
    46 (Worthing 1987, 1991; quoted, Shiu et al. 1990)
    46 (Merck Index 1989; quoted, Milne 1995)
    46 (20-25°C, selected, Hornsby et al. 1996)

Vapor Pressure (Pa at 25°C):
    negligible (Agrochemicals Handbook 1987)

Henry's Law Constant (Pa·m³/mol):

Octanol/Water Partition Coefficient, log K_{OW}:
    3.53 (shake flask-HPLC/UV, Jafvert et al. 1990)
    3.53 (selected, Hansch et al. 1995)

Bioconcentration Factor, log BCF:
    1.85 (calculated-S, Kenaga 1980)
    2.21 (calculated-log K_{OW} as per Mackay 1982, this work)

Sorption Partition Coefficient, log K_{OC}:
- 2.72 (calculated-S, Kenaga 1980; quoted, Bottoni & Funari 1992)
- 1.30 (organic carbon, Wauchope et al. 1991; quoted, Dowd et al. 1993)
- 2.64 (20-25°C, estimated, Hornsby et al. 1996)

Half-Lives in the Environment:
Air: 6-60 h, based on an estimated rate constant for the vapor-phase reaction with hydroxyl radicals in air (Atkinson 1987; Howard et al. 1991).
Surface water: 24-168 h, based on estimated unacclimated aqueous aerobic biodegradation half-life (Howard et al. 1991).
Groundwater: 48-336 h, based on estimated unacclimated aqueous aerobic biodegradation half-life (Howard et al. 1991).
Sediment:
Soil: 24-168 h, based on unacclimated soil grab sample data (Smith 1978; quoted, Howard et al. 1991); selected half-life of 10 d (Wauchope et al. 1991; quoted, Dowd et al. 1993); half-life <7 d (Worthing 1991); field half-life of 5 d (20-25°C, selected, Hornsby et al. 1996).
Biota:

Environmental Fate Rate Constants or Half-Lives:
Volatilization:
Photolysis:
Oxidation: photooxidation half-life of 6-60 h in air, based on an estimated rate constant for the vapor-phase reaction with hydroxyl radicals in air (Atkinson 1987; quoted, Howard et al. 1991).
Hydrolysis: stable in distilled water for 40 d (Chau & Thomson 1978; quoted, Howard et al. 1991).
Biodegradation: aqueous aerobic half-life of 24-168 h, based on unacclimated soil grab sample data (Smith 1978; quoted, Howard et al. 1991); aqueous anaerobic half-life of 96-672 h, based on estimated unacclimated aqueous aerobic biodegradation half-life (Howard et al. 1991).
Biotransformation:
Bioconcentration, Uptake (k_1) and Elimination (k_2) Rate Constants:

Common Name: Diallate
Synonym: Avadex, CP 15336, DATC, Pyradex
Chemical Name: S-(2,3-dichloroallyl)diisopropyl(thiocarbamate); S-(2,3-dichloro-2-propenyl)bis(1-methylethyl)carbamothioate
Uses: pre-emergent and selective herbicide to control wild oats and blackgrass in barley, corn, flax, lentils, peas, potatoes, soybeans, and sugar beets.
CAS Registry No: 2303-16-4
Molecular Formula: $C_{10}H_{17}Cl_2NOS$
Molecular Weight: 270.24
Melting Point (°C):
    25-30   (Suntio et al. 1988; Herbicide Handbook 1989; Montgomery 1993)
Boiling Point (°C):
    97   (at 0.15 mmHg, Herbicide Handbook 1989)
    108   (at 0.25 mmHg, Herbicide Handbook 1989; Montgomery 1993)
    150   (at 9 mmHg, Howard 1991; Milne 1995; Montgomery 1993)
Density (g/cm^3 at 20°C):
    1.188   (25°C, Agrochemicals Handbook 1987; Montgomery 1993)
Molar Volume (cm^3/mol):
    305.1   (calculated-LeBas method, Suntio et al. 1988)
Molecular Volume (Å^3):
Total Surface Area, TSA (Å^2):
Dissociation Constant pK_a:
Heat of Fusion, ΔH_{fus}, kcal/mol:
Entropy of Fusion, ΔS_{fus}, cal/mol·K (e.u.):
Fugacity Ratio at 25°C (assuming $\Delta S_{fus} = 13.5$ e.u.), F: 0.8920

Water Solubility (g/m^3 or mg/L at 25°C):
    40.0   (Günther et al. 1968; quoted, Suntio et al. 1988)
    14.0   (Ashton & Crafts 1973; quoted, Suntio et al. 1988)
    40.0   (rm. temp., Spencer 1973; quoted, Shiu et al. 1990)
    40.0   (Martin & Worthing 1977; quoted, Shiu et al. 1990)
    14.0   (Herbicide Handbook 1978; quoted, Kenaga 1980; Kenaga & Goring 1980; Suntio et al. 1988; Isensee 1991)
    40.0   (22°C, Khan 1980; quoted, Shiu et al. 1990)
    68.8   (22°C, shake flask-GC, Bowman & Sans 1979, 1983a,b; quoted, Shiu et al. 1990)
    14.0   (Weber et al. 1980; quoted, Willis & McDowell 1982)
    14.0   (Ashton & Crafts 1981; Herbicide Handbook 1989)
    14.0   (Agrochemicals Handbook 1987)
    40.5   (20-25°C, shake flask-GC, Kanazawa 1981; quoted, Shiu et al. 1990)
    14.0   (IARC 1983; quoted, Howard 1991)
    52.5   (Garten & Trabalka 1983; quoted, Shiu et al. 1990)
    14.0   (20°C, selected, Suntio et al. 1988; quoted, Majewski & Capel 1995)
    40.0   (Taylor & Glofelty 1988; quoted, Shiu et al. 1990)
    14.0   (Montgomery 1993)
    14.0   (20-25°C, selected, Augustijn-Beckers et al. 1994; Hornsby et al. 1996)
    14.0   (selected, Lohninger 1994)
    40.0   (Milne 1995)

Vapor Pressure (Pa at 25°C):
- 0.020   Ashton & Crafts 1973; quoted, Suntio et al. 1988; Herbicide Handbook 1989)
- 0.0117   (20°C, Hartley & Graham-Bryce 1980; quoted, Suntio et al. 1988)
- 0.020   (Agrochemicals Handbook 1987)
- 0.013   (20°C, selected, Suntio et al. 1988; quoted, Majewski & Capel 1995)
- 0.020   (IARC 1983; quoted, Howard 1991)
- 0.020   (20°C, Montgomery 1993)
- 0.020   (20-25°C, selected, Augustijn-Beckers et al. 1994; Hornsby et al. 1996)

Henry's Law Constant (Pa·m^3/mol):
- 0.250   (20°C, calculated-P/C, Suntio et al. 1988; quoted, Majewski & Capel 1995)
- 0.385   (calculated-P/C, Howard 1991)
- 0.253   (20-25°C, calculated-P/C, Montgomery 1993)
- 0.108   (calculated-P/C, this work)

Octanol/Water Partition Coefficient, log K_{ow}:
- 5.23   (estimated, USEPA 1988; quoted, Howard 1991)
- 3.29   (calculated, Montgomery 1993)

Bioconcentration Factor, log BCF:
- 2.15   (calculated-S, Kenaga 1980; quoted, Howard 1991; Isensee 1991)
- 2.08   (calculated-K_{oc}, Kenaga 1980)

Sorption Partition Coefficient, log K_{oc}:
- 3.28   (soil, Grover 1974; quoted, Kenaga 1980; Kenaga & Goring 1980)
- 2.96   (Melfort loam, Grover et al. 1979; quoted, Howard 1991)
- 2.46   (Weyburn sandy loam, Grover et al. 1979; quoted, Howard 1991)
- 2.59   (Regina clay, Grover et al. 1979; quoted, Howard 1991)
- 2.49   (Indian Head sandy loam, Grover et al. 1979; quoted, Howard 1991)
- 2.65   (Asquith loamy sand, Grover et al. 1979; quoted, Howard 1991)
- 3.28   (soil, measured value, Kenaga 1980; quoted, Howard 1991)
- 3.28   (Kenaga & Goring 1980; quoted, Lyman 1982; Bahnick & Doucette 1988)
- 3.00   (soil, calculated-S as per Kenaga & Goring 1978, Kenaga 1980)
- 2.77   (calculated-χ, Bahnick & Doucette 1988)
- 2.28   (Montgomery 1993)
- 3.52   (selected, Lohninger 1994)
- 2.70   (20-25°C, selected, Augustijn-Beckers et al. 1994; Hornsby et al. 1996)

Half-Lives in the Environment:
   Air: 0.58-5.8 h, based on an estimated rate constant for the vapor-phase reaction with hydroxyl radicals in air (Atkinson 1987; quoted, Howard et al. 1991).
   Surface water: 252-2160 h, based on aerobic soil die-away test data (Anderson & Domsch 1976; Smith 1970; quoted, Howard et al. 1991).

Groundwater: 504-4320 h, based on aerobic soil die-away test data (Anderson & Domsch 1976; Smith 1970; quoted, Howard et al. 1991).
Sediment:
Soil: 252-2160 h, based on aerobic soil die-away test data (Anderson & Domsch 1976; Smith 1970; quoted, Howard et al. 1991; Montgomery 1993); half-life of 30 d (Hartley & Kidd 1987; quoted, Montgomery 1993); selected field half-life of 30 d (Augustijn-Beckers et al. 1994; Hornsby et al. 1996).

Environmental Fate Rate Constants or Half-Lives:
Volatilization:
Photolysis: half-life of 4 h <1 % of 135 µg/mL to degrade in distilled water under >300 nm light (Ruzo & Casida 1985; quoted, Cessna & Muir 1991).
Oxidation: photooxidation half-life of 0.58-5.8 h, based on an estimated rate constant for the vapor-phase reaction with hydroxyl radicals in air (Atkinson 1987; quoted, Howard et al. 1991).
Hydrolysis: neutral hydrolysis rate constant of $(1.2 \pm 0.7) \times 10^{-5}$ h^{-1} with a calculated first-order half-life of 6.6 yr at pH 7 (Ellington et al. 1987, 1988); first-order half-life of 6.6 yr, based on measured first-order base catalytized hydrolysis rate constant at pH 7 (Ellington et al. 1987; quoted, Howard et al. 1991).
Biodegradation: aqueous aerobic half-life of 252-2160 h, based on aerobic soil die-away test data (Anderson & Domsch 1976; Smith 1970; quoted, Howard et al. 1991); aqueous anaerobic half-life of 1008-8640 h, based on aerobic soil die-away test data (Anderson & Domsch 1976; Smith 1970; quoted, Howard et al. 1991).
Biotransformation:
Bioconcentration, Uptake (k_1) and Elimination (k_2) Rate Constants:

Common Name: Dicamba
Synonym: Banex, Banvel, Banvel D, Brush buster, Dianat, MDBA, Mediben
Chemical Name: 3,6-dichloro-2-methoxybenzoic acid; 3,6-dichloro-o-anisic acid
Uses: systemic pre-emergent and post-emergent herbicide to control both annual and perennial broadleaf weeds.
CAS Registry No: 1918-00-9
Molecular Formula: $C_8H_6Cl_2O_3$
Molecular Weight: 221.04
Melting Point (°C):
    114-116 (Khan 1980; Spencer 1982; Suntio et al. 1988; Herbicide Handbook 1989; Worthing 1991; Montgomery 1993; Tomlin 1994; Milne 1995)
    115 (Lee et al. 1993; Kühne et al. 1995)
Boiling Point (°C):
Density (g/cm³ at 20°C):
    1.570 (25°C, Agrochemicals Handbook 1987; Worthing 1991; Montgomery 1993; Caux et al. 1993; Tomlin 1994; Milne 1995)
Molar Volume (cm³/mol):
    207.9 (calculated-LeBas method, Suntio et al. 1988)
Molecular Volume (Å³):
Total Surface Area, TSA (Å²):
Dissociation Constant pK_a:
    1.94 (Kearney & Kaufman 1975; Spencer 1982; Lee et al. 1993)
    1.90 (Cessna & Grover 1978; Weber et al. 1980; Willis & McDowell 1982; Howard 1991; Montgomery 1993)
    1.95 (Worthing 1991; Montgomery 1993; Caux et al. 1993)
    1.87 (Tomlin 1994)
    1.91 (Hornsby et al. 1996)
Heat of Fusion, ΔH_{fus}, kcal/mol:
    5.40 (DSC method, Plato & Glasgow 1969)
Entropy of Fusion, ΔS_{fus}, cal/mol·K (e.u.):
Fugacity Ratio at 25°C (assuming ΔS_{fus} = 13.5 e.u.), F: 0.1288

Water Solubility (g/m³ or mg/L at 25°C):
    7900 (Freed 1966; quoted, Suntio et al. 1988; Shiu et al. 1990)
    4500 (Martin & Worthing 1977; quoted, Kenaga 1980; Kenaga & Goring 1980; Shiu et al. 1990; Isensee 1991)
    4500 (Hartley & Graham-Bryce 1980; quoted, Taylor & Glotfelty 1988)
    4500 (Khan 1980; quoted, Suntio et al. 1988; Shiu et al. 1990)
    4500 (Weber et al. 1980; Willis & McDowell 1982)
    4500 (Ashton & Crafts 1981; quoted, Suntio et al. 1988)
    4470 (Garten & Trabalka 1983; quoted, Shiu et al. 1990)
    7900 (Verschueren 1983; quoted, Lee et al. 1993)
    6500 (Agrochemicals Handbook 1987; Herbicide Handbook 1989)
    4500 (selected, Gerstl & Helling 1987)
    6500 (Worthing 1987, 1991; quoted, Shiu et al. 1990)
    5600 (20°C, selected, Suntio et al. 1988; quoted, Howard 1991; Majewski & Capel 1995)
    4500 (selected, USDA 1989; quoted, Neary et al. 1993)
    4500 (Reinert 1989)

6500   (Caux et al. 1993)
6500   (Montgomery 1993; Tomlin 1994; Milne 1995)
4410, 221 (quoted, calculated-group contribution fragmentation method, Kühne et al. 1995)

Vapor Pressure (Pa at 25°C):
0.00454   (Ashton & Crafts 1973; quoted, Suntio et al. 1988)
0.00267   (Baur & Bovey 1974; quoted, Spencer 1976; Suntio et al. 1988)
0.49   (20°C, Hartley & Graham-Bryce 1980; quoted, Taylor & Glotfelty 1988; Taylor & Spencer 1990)
0.493   (Khan 1980; quoted, Suntio et al. 1988)
<0.00013 (20-25°C, Weber et al. 1980; Willis & McDowell 1982)
0.0045   (Ashton & Crafts 1981; Worthing 1991; quoted, Caux et al. 1993)
0.00453   (Herbicide Handbook 1983, 1989; quoted, Howard 1991)
0.0045   (Agrochemicals Handbook 1987; Tomlin 1994)
0.003   (20°C, selected, Suntio et al. 1988; quoted, Majewski & Capel 1995)
0.50   (100°C, Merck Index 1989)
0.50   (20°C, selected, Taylor & Spencer 1990)
0.0045   (20°C, Montgomery 1993)

Henry's Law Constant (Pa·m^3/mol):
0.00012   (20°C, calculated-P/C, Suntio et al. 1988)
0.0248   (calculated-P/C, Taylor & Glotfelty 1988)
0.0918   (Suntio et al. 1988; quoted, Howard 1991; Majewski & Capel 1995)
2.2×10^{-5} (calculated-P/C, Nash 1989)
1.22×10^{-4} (20-25°C, calculated-P/C, Montgomery 1993)
0.00012, 0.000154 (20, 25°C, quoted, Caux et al. 1993)

Octanol/Water Partition Coefficient, log K_{OW}:
0.477   (quoted, Rao & Davidson 1980; Suntio et al. 1988; Caux et al. 1993)
2.41   (selected, Dao et al. 1983)
2.21   (Hansch & Leo 1985; quoted, Howard 1991)
−1.69   (selected, Gerstl & Helling 1987)
0.50   (selected, Suntio et al. 1988)
3.01   (selected, Travis & Arms 1988)
2.46   (Reinert 1989)
2.49   (shake flask-HPLC/UV, Jafvert et al. 1990)
2.46   (quoted from EPA Environmental Fate one-liner database Version 3.04, Lee et al. 1993)
0.48, 2.21; 2.21 (quoted values, selected, Sangster 1993)
0.48   (Montgomery 1993)
-0.80   (pH 7, Tomlin 1994)
2.21   (selected, Hansch et al. 1995)

Bioconcentration Factor, log BCF:
0.699   (calculated-S, Kenaga 1980; quoted, Isensee 1991)
−2.00   (calculated-K_{OC}, Kenaga 1980)
−4.58   (beef biotransfer factor logB_b, correlated-K_{OW}, Oehler & Ivie 1980; quoted, Travis & Arms 1988)

-4.60 (milk biotransfer factor logB_m, correlated-K_{OW}, Oehler & Ivie 1980; quoted, Travis & Arms 1988)
1.450 (estimated-K_{OW} per Hansch & Leo 1985, Lyman et al. 1982; quoted, Howard 1991)
0.903 (estimated-S per Suntio et al. 1988, Lyman et al. 1982; quoted, Howard 1991)

Sorption Partition Coefficient, log K_{OC}:
-0.398 (soil, quoted exptl., Kenaga 1980)
1.63 (soil, calculated-S as per Kenaga & Goring 1978, Kenaga 1980)
0.342 (av. soils/sediments, Rao & Davidson 1980; quoted, Lyman 1982)
-0.40, 2.08 (quoted, calculated-χ, Gerstl & Helling 1987)
0.34 (soil, screening model calculations, Jury et al. 1987b)
-0.38 (quoted, Nash 1989)
2.67 (K_{OC} = 470 reported, Reinert 1989)
0.643 (soil, estimated, Shirmohammadi et al. 1989; quoted, Howard 1991)
-1.00 (selected, USDA 1989; quoted, Neary et al. 1993)
0.30 (organic carbon, Wauchope et al. 1991; quoted, Dowd et al. 1993)
-0.4-1.62, 0.18, 0.34 (quoted values, Bottoni & Funari 1992)
1.50 (soil, quoted exptl., Meylan et al. 1992)
1.46 (soil, calculated-χ & fragment contribution, Meylan et al. 1992)
-0.40, 0.34 (Montgomery 1993)
0.34 (quoted, Caux et al. 1993)
0.30 (Tomlin 1994)

Half-Lives in the Environment:
Air: half-life was estimated to be 2.42 d for reaction with hydroxyl radicals (Eisenreich et al. 1981; quoted, Caux et al. 1993); half-life of 2.42-6.0 d, based on estimated rate constant for the vapor-phase reaction with hydroxyl radicals in the atmosphere (Atkinson 1985; quoted, Howard 1991).
Surface water:
Groundwater: half-lives determined under batch conditions were 23.5 d at 28°C, 38 d at 20°C, and 151 d at 12°C and were all higher than estimated half-life of 13.5 d from the decrease in column effluent concentrations over time (Comfort et al. 1992); half-life <7 d in surface water (Caux et al. 1993).
Sediment:
Soil: estimated persistence of 2 months (Kearney et al. 1969; quoted, Jury et al. 1987a); half-lives: 59, 19, and 17 d with disappearance rates: 0.0117, 0.036 and 0.041 d^{-1} at pH 4.3, 5.3 and 6.5 (Hamaker 1972; quoted, Nash 1988); persistence of 2 months in soil (Edwards 1973; quoted, Morrill et al. 1982); estimated first-order half-life of 31.5 d in soil from biodegradation rate constant of 0.022 d^{-1} by soil incubation die-away studies (Rao & Davidson 1980; quoted, Scow 1982); nonpersistent in soils with half-life less than 20 days (Willis & McDowell 1982); from review of persistence literature, the mean half-life under lab. conditions was 14 days while the mean half-life under field conditions was 8 d (Rao & Davidson 1982; quoted, Howard 1991); non-persistent with half-life <20 d in soil (Willis & McDowell 1982); half-life of 14 d from screening model calculations (Jury et al. 1987b); half-life of less than 14 days under conditions amenable to rapid

metabolism (Herbicide Handbook 1989); selected half-life of 14 d (Wauchope et al. 1991; quoted, Dowd et al. 1993); half-life < 14-25 d (Worthing 1991; quoted, Montgomery 1993); half-life 4-555 d with a mean of 24 d (Caux et al. 1993); half-life < 14 d (Tomlin 1994).

Biota: biochemical half-life of 14 d from screening model calculations (Jury et al. 1987b); average half-life of 25 d in the forest (USDA 1989; quoted, Neary et al. 1993); biological half-life 0.64 h (Caux et al. 1993).

Environmental Fate Rate Constants or Half-Lives:
Volatilization:
Photolysis:
Oxidation: photooxidation half-life of 2.42-6.0 d, based on estimated rate constant for the vapor-phase reaction with hydroxyl radicals in the atmosphere (Atkinson 1985; quoted, Howard 1991).
Hydrolysis: half-life of > 133 d for 2 μg/mL to hydrolyze in dark sterile pond water at 37-39°C (Scifres et al. 1973; quoted, Muir 1991).
Biodegradation: half-lives of 60 d to > 160 d for 100 μg/mL to degrade in pond sediment/water under lighted conditions at 20-30°C (Scifres et al. 1973; quoted, Muir 1991); under lab. conditions using nonsterile sandy loam, silty clay, or heavy clay soil, 50% of applied dicamba degraded within two weeks; however in sterilized (via heating) soil, over 90% of applied dicamba was recovered after four weeks, suggesting that microbes were responsible for the decomposition (Smith 1973; quoted, Howard 1991); half-life of > 25 d for 5.85 mg to plants to degrade following washoff from plants and sands in model ecosystem (Yu et al. 1975; quoted, Muir 1991); rate constant of 0.022 d^{-1} by soil incubation die-away studies (Rao & Davidson 1980; quoted, Scow 1982); the rate of biodegradation in soil generally increases with temperature and soil moisture (up to 50%) and tends to be faster when the soil is slightly acidic (Herbicide Handbook 1983; quoted, Howard 1991).
Biotransformation:
Bioconcentration, Uptake (k_1) and Elimination (k_2) Rate Constants:

Common Name: Dichlobenil
Synonym: Barrier 2G, Barrier 50W, Casoron, DBN, DCB, Decabane, Du-Sprex, Dyclomec, NIA 5996, Niagara 5006, Niagara 5996, Norosac
Chemical Name: 2,6-dichlorobenzonitrile
Uses: soil applied herbicide to control many annual and perennial broadleaf weeds.
CAS Registry No: 1194-65-6
Molecular Formula: $C_7H_3Cl_2N$
Molecular Weight: 172.01
Melting Point (°C):
    145-146 (Verloop 1972; Spencer 1982; Worthing 1991; Tomlin 1994; Milne 1995)
    144-145 (Suntio et al. 1988; Montgomery 1993; Milne 1995)
Boiling Point (°C):
    270 (Verloop 1972; Khan 1980; Worthing 1991; Tomlin 1994; Milne 1995)
    270-270.1 (Montgomery 1993)
Density (g/cm³ at 20°C):
    > 1.0 (Milne 1995)
Molar Volume (cm³/mol):
    148.9 (calculated-LeBas method, Suntio et al. 1988)
Molecular Volume (Å³):
Total Surface Area, TSA (Å²):
Dissociation Constant pK_a:
Heat of Fusion, ΔH_{fus}, kcal/mol:
    6.20 (DSC method, Plato & Glasgow 1969)
    8.205 (Verloop 1972)
Entropy of Fusion, ΔS_{fus}, cal/mol·K (e.u.):
Fugacity Ratio at 25°C (assuming ΔS_{fus} = 13.5 e.u.), F: 0.065

Water Solubility (g/m³ or mg/L at 25°C):
18	(20°C, Günther et al. 1968; quoted, Suntio et al. 1988; Shiu et al. 1990)
25	(Günther et al. 1968; quoted, Suntio et al. 1988; Shiu et al. 1990)
18	(20°C, Verloop 1972; Spencer 1982)
18	(20°C, Weber 1972; quoted, Muir 1991)
18	(Martin & Worthing 1977; Herbicide Handbook 1978; quoted, Kenaga 1980; Kenaga & Goring 1980; Isensee 1991)
18	(Wauchope 1978; Burkhard & Guth 1981)
18	(Khan 1980; quoted, Suntio et al. 1988; Shiu et al. 1990)
18	(Weber et al. 1980; quoted, Willis & McDowell 1982)
18	(20°C, Ashton & Crafts 1981; Herbicide Handbook 1989)
18	(20°C, Verschueren 1983; quoted, Shiu et al. 1990)
18	(20°C, Agrochemicals Handbook 1987)
18	(selected, Gerstl & Helling 1987)
18	(20°C, Worthing 1987, 1991; quoted, Shiu et al. 1990; Tomlin 1994)
18	(20°C, selected, Suntio et al. 1988; quoted, Howard 1991; Majewski & Capel 1995)
18	(Reinert 1989)
21.2	(20-25°C, selected, Wauchope et al. 1992; Lohninger 1994; Hornsby et al. 1996)
25	(Montgomery 1993)
18, 25	(20°C, 25°C, Milne 1995)

Vapor Pressure (Pa at 25°C):
- 0.072 (20°C, effusion manometer tech. with calculation, Barnsley & Rosher 1961; quoted, Martin 1971; Ashton & Crafts 1973; Spencer 1976)
- 0.0733 (20°C, Verloop 1972)
- 0.0667 (20°C, Weber 1972; Worthing 1987; quoted, Muir 1991)
- 0.0004 (20°C, Spencer 1976; quoted, Suntio et al. 1988)
- 0.0666 (20°C, effusion method, Spencer 1976; quoted, Suntio et al. 1988)
- 0.0733 (20°C, Khan 1980; quoted, Suntio et al. 1988; quoted, Howard 1991; Majewski & Capel 1995)
- 0.0733 (20-25°C, Weber et al. 1980; quoted, Willis & McDowell 1982)
- 0.0733 (20°C, Ashton & Crafts 1981; Herbicide Handbook 1989)
- 0.0732 (quoted, Burkhard & Guth 1981)
- 0.073 (20°C, Agrochemicals Handbook 1987)
- 0.070 (20°C, selected, Suntio et al. 1988)
- 0.133 (20-25°C, selected, Wauchope et al. 1992; Hornsby et al. 1996)
- 0.0733 (Montgomery 1993)
- 0.088 (20°C, gas saturation, Tomlin 1994)

Henry's Law Constant (Pa·m^3/mol):
- 0.669 (20°C, calculated-P/C, Suntio et al. 1988; quoted, Howard 1991; Majewski & Capel 1995)
- 0.637 (20°C, calculated-P/C, Muir 1991)
- 0.669 (20-25°C, calculated-P/C, Montgomery 1993)

Octanol/Water Partition Coefficient, log K_{OW}:
- 2.90 (quoted, Rao & Davidson 1980; selected, Suntio et al. 1988, Magee 1991)
- 2.57, 2.65 (RP-HPLC, shake flask, Edasforth & Moser 1983; quoted, Sangster 1993)
- 3.06 (Geyer et al. 1984, quoted, Sangster 1993)
- 2.94 (Hansch & Leo 1985; quoted, Howard 1991)
- 1.63 (Reinert 1989)
- 2.98 (selected, Dao et al. 1983, Gerstl & Helling 1987)
- 2.90 (Montgomery 1993)
- 2.90 (quoted and selected, Sangster 1993)
- 2.70 (Tomlin 1994)
- 2.74 (selected, Hansch et al. 1995)

Bioconcentration Factor, log BCF:
- 1.74 (fish in static water, Kenaga 1975; Kenaga & Goring 1980)
- 2.08 (calculated-S, Kenaga 1980; quoted, Isensee 1991)
- 1.08 (calculated-K_{OC}, Kenaga 1980)
- 1.18-1.60 (fish, Freitag et al. 1982; quoted, Howard 1991)
- 1.30 (algae, Freitag et al. 1982; quoted, Howard 1991)
- 1.72 (estimated-S, Lyman et al. 1982; quoted, Howard 1991)
- 2.03-2.32 (Montgomery 1993)

Sorption Partition Coefficient, log K_{OC}:
- 2.91 (potting soil with 22% organic content, Massini 1961; quoted, Howard 1991)
- 2.08 (sandy loam with 5% organic content, Massini 1961; quoted, Howard 1991)
- 2.37 (soil, Hamaker & Thompson 1972; quoted, Kenaga 1980; Kenaga & Goring 1980)
- 2.95 (soil, calculated-S as per Kenaga & Goring 1978, Kenaga 1980)
- 2.35 (Rao & Davidson 1980)
- 2.94 (soil, estimated-S, Lyman et al. 1982; quoted, Howard 1991)
- 2.37, 1.45 (quoted, calculated-χ, Gerstl & Helling 1987)
- 2.96 (Reinert 1989)
- 2.37 (reported as log K_{OM}, Magee 1991)
- 2.31 (estimated as log K_{OM}, Magee 1991)
- 2.21, 2.57-2.96 (quoted values, Bottoni & Funari 1992)
- 2.60 (soil, 20-25°C, estimated, Wauchope et al. 1992; Hornsby et al. 1996)
- 2.60 (estimated-chemical structure, Lohninger 1994)

Half-Lives in the Environment:
Air: 92 d, based on estimation for the vapor-phase reaction with hydroxyl radicals in atmosphere (Atkinson 1987; quoted, Howard 1991).
Surface water:
Groundwater:
Sediment: half-life of ~7 d for 5 μg/mL to biodegrade in sediment suspension at 30°C (Miyazaki et al. 1975; quoted, Muir 1991).
Soil: estimated persistence of 4 months (Kearney et al. 1969; quoted, Jury et al. 1987); half-life in soil may vary between 1 and 6 months depending on soil type (Beynon & Wright 1972; Verloop 1972; quoted, Tomlin 1994); persistence of 4 months in soil (Edwards 1973; quoted, Morrill et al. 1982); persistence of 4 months (Wauchope 1978); 1.5 to 12 months depending upon soil type (Herbicide Handbook 1989); selected field half-life of 60 d (Wauchope et al. 1992; Hornsby et al. 1996).

Environmental Fate Rate Constants or Half-Lives:
Volatilization: estimated half-life of 7.4 d, based on Henry's law constant for a model river 1 m deep with a wind velocity of 3 m/s and flowing at 1 m/s (Lyman et al. 1982; quoted, Howard 1991); estimated half-life of 11 days from 1 m depth of water at 20°C (Muir 1991).
Photolysis: photolytic half-life of 15 d in water (Tomlin 1994).
Oxidation: photooxidation half-life of 92 d in air, based on estimation for the vapor-phase reaction with hydroxyl radicals in atmosphere (Atkinson 1987; quoted, Howard 1991).
Hydrolysis:
Biodegradation: half-life of ~7 d for 5 μg/mL to biodegrade in sediment suspension at 30°C (Miyazaki et al. 1975; quoted, Muir 1991).
Biotransformation:
Bioconcentration, Uptake (k_1) and Elimination (k_2) Rate Constants:

Common Name: Dichlorprop
Synonym: Cornox RK, dichloroprop, Dikofag DP, 2,4-DP, Hedonal DP, Polymone
Chemical Name: (±)-2-(2,4-dichlorophenoxy) propanoic acid; (±)-2-(2,4-dichlorophenoxy) propionic acid
Uses: herbicide and growth regulator to control annual broadleaf and grass weeds; also to control aquatic weeds and chemical maintenance of embankments and roadside verges.
CAS Registry No: 120-36-5
Molecular Formula: $C_9H_8Cl_2O_3$
Molecular Weight: 235.1
Melting Point (°C):
    117.5-118.1 (Agrochemicals Handbook 1987)
    116-117.5 (Herbicide Handbook 1989; Tomlin 1994)
Boiling Point (°C):
Density (g/cm³ at 20°C):
    1.64 (25°C, Bailey & White; quoted, Que Hee et al. 1981)
    1.42 (Herbicide Handbook 1989; Tomlin 1994)
Molar Volume (cm³/mol):
    232.0 (calculated-LeBas method, this work)
    165.6 (calculated-density, this work)
Molecular Volume (Å³):
Total Surface Area, TSA (Å²):
Dissociation Constant pK_a:
    2.855 (Cessna & Grover 1978)
    2.86 (Wauchope et al. 1992; Hornsby et al. 1996)
    3.00 (Tomlin 1994)
Heat of Fusion, ΔH_{fus}, kcal/mol:
    8.20 (DSC method, Plato 1972)
Entropy of Fusion, ΔS_{fus}, cal/mol·K (e.u.):
Fugacity Ratio at 25°C (assuming ΔS_{fus} = 13.5 e.u.), F: 0.123

Water Solubility (g/m³ or mg/L at 25°C):
    350 (20°C, Woodford & Evans 1963; quoted, Que Hee et al. 1981)
    350 (20°C, Spencer 1973; quoted, Shiu et al. 1990)
    350 (Martin & Worthing 1977; quoted, Kenaga 1980)
    350 (20°C, Agrochemicals Handbook 1987)
    350 (20°C, Worthing 1987, 1991; quoted, Shiu et al. 1990; Tomlin 1994)
    710 (28°C, Herbicide Handbook 1989; quoted, Shiu et al. 1990)
    350 (20°C, Merck Index 1987)
    50 (ester, 20-25°C, estimated, Wauchope et al. 1992; Lohninger 1994; Hornsby et al. 1996)

Vapor Pressure (Pa at 25°C):
    4.50×10^{-4} (20°C, Agrochemicals Handbook 1987)
    4.00×10^{-4} (20-25°C, estimated, Wauchope et al. 1992; Hornsby et al. 1996)
    $< 1.0 \times 10^{-5}$ (20°C, Tomlin 1994)

Henry's Law Constant (Pa·m^3/mol):
    2.69×10^{-4} (calculated-P/C, this work)

Octanol/Water Partition Coefficient, log K_{OW}:
    2.75   (RP-HPLC-k', Braumann et al. 1983)
    3.43   (shake flask-CC, Ilchmann et al. 1993; quoted, Sangster 1993)
    1.77   (Tomlin 1994)
    3.43   (selected, Hansch et al. 1995)

Bioconcentration Factor, log BCF:
    1.36   (calculated-S, Kenaga 1980)

Sorption Partition Coefficient, log K_{OC}:
    2.23   (soil, calculated-S, Kenaga 1980)
    3.00   (soil, 20-25°C, estimated, Wauchope et al. 1992; Lohninger 1994; Hornsby et al. 1996)
    1.08-1.60 (Tomlin 1994)

Half-Lives in the Environment:
    Air:
    Surface water:
    Groundwater:
    Sediment:
    Soil: selected field half-life of 10 days (Wauchope et al. 1992; Hornsby et al. 1996); half-life of ca. 8 days in soil (Tomlin 1994).
    Biota:

Environmental Fate Rate Constants or Half-Lives:
    Volatilization:
    Photolysis:
    Oxidation:
    Hydrolysis:
    Biodegradation:
    Biotransformation:
    Bioconcentration, Uptake (k_1) and Elimination (k_2) Rate Constants:

Common Name: Diclofop-methyl
Synonym: dichlordiphenoprop, Hoegrass, Illoxan
Chemical Name: propanoic acid, 2-[4-(2,4-dichlorophenoxy)]-, methyl ester; 2-[4-(2,4-dichlorophenoxy)]methylpropionate
Uses: herbicide to control post-emergent wild oats, wild millets, and other annual grass weeds in wheat, barley, rye, red fescue, and broadleaf weeds in crops such as soybeans, sugar cane, fodder beet, flax, legumes, oilseed rape, sunflowers, clover, lucerne, groundnuts, brassicas, carrots, celery, beet root, parsnips, lettuce, spinach, potatoes, tomatoes, fennel, alliums, herbs, etc.
CAS Registry No: 51338-27-3
Molecular Formula: $C_{16}H_{14}Cl_2O_4$
Molecular Weight: 341.2
Melting Point (°C):
    39-41 (Agrochemical Handbook 1987; Herbicide Handbook 1989; Worthing 1991; Milne 1995)
Boiling Point (°C):
    175-176 (at 0.1 mmHg, Agrochemical Handbook 1987; Herbicide Handbook 1989)
    175-177 (at 0.1 mmHg, Milne 1995)
Density (g/cm^3 at 20°C):
    1.30 (40°C, Agrochemical Handbook 1987; Herbicide Handbook 1989)
    1.035 (Herbicide Handbook 1989)
Molar Volume (cm^3/mol):
    349.6 (calculated-LeBas method, this work)
    329.7 (calculated-density, this work)
Molecular Volume (Å^3):
Total Surface Area, TSA (Å^2):
Dissociation Constant pK_a:
    3.1 (Wauchope et al. 1992; Hornsby et al. 1996)
Heat of Fusion, ΔH_{fus}, kcal/mol:
Entropy of Fusion, ΔS_{fus}, cal/mol·K (e.u.):
Fugacity Ratio at 25°C (assuming ΔS_{fus} = 13.5 e.u.), F: 0.711

Water Solubility (g/m^3 or mg/L at 25°C):
    3.0 (22°C, Agrochemicals Handbook 1987; Worthing 1987,91)
    3000 (22°C, Herbicide Handbook 1989)
    0.8 (20-25°C, selected, Wauchope et al. 1992; Hornsby et al. 1996)
    3.0 (Lohninger 1994; Milne 1995)
    0.8 (selected, Halfon et al. 1996)

Vapor Pressure (Pa at 25°C):
    3.4×10^{-5} (20°C, Agrochemical Handbook 1987)
    3.4×10^{-5} (20°C, Worthing 1987,91)
    3.4×10^{-5} (20°C, Herbicide Handbook 1989)
    1.5×10^{-4} (30°C, Herbicide Handbook 1989)
    4.7×10^{-4} (20-25°C, selected, Wauchope et al. 1992; Hornsby et al. 1996)
    4.7×10^{-4} (selected, Halfon et al. 1996)

Henry's Law Constant (Pa·m^3/mol):
    0.199   (calculated-P/C, this work)

Octanol/Water Partition Coefficient, log K_{OW}:
    4.58   (Worthing 1991)

Bioconcentration Factor, log BCF:
    2.74   (calculated-S as per Kenaga 1980, this work)

Sorption Partition Coefficient, log K_{OC}:
    4.15-4.39   (quoted values, Buttoni & Funari 1992)
    4.20   (soil, 20-25°C, selected, Wauchope et al. 1992; Hornsby et al. 1996)
    4.20   (estimated-chemical structure, Lohninger 1994)
    4.25   (soil, HPLC-screening method, Kördel et al. 1993,95b)
    3.69   (soil, calculated-S as per Kenaga 1980, this work)

Half-Lives in the Environment:
    Air:
    Surface water:
    Groundwater:
    Sediment:
    Soil: under aerobic conditions it decomposes in a matter of days to 2-[4-(2',4'-dichlorophenoxy)phenoxy]propionic acid which in turn is metabolized relatively quickly with a half-life of 10 days in sandy soils and about 30 days in sandy clay soils while under anaerobic conditions, results were similar except that the very rapid cleavage of the ester bond by hydrolysis within one hour to propionic acid derivatives was experienced and within two days, up to 86% of the parent compound was metabolized into various free acid metabolites and up to 3.7% of phenol metabolites (Herbicide Handbook 1989); selected field half-life of 30 days at pH 7.0 (Wauchope et al. 1992; Hornsby et al. 1996); 30 days (selected, Halfon et al. 1996).

Environmental Fate Rate Constants or Half-Lives:
    Volatilization:
    Photolysis:
    Oxidation:
    Hydrolysis:
    Biodegradation: first-order rate constants of -0.0883, -0.225, -0.266 h^{-1} in nonsterile sediment and -0.0158, -0.0139, -0.0134 h^{-1} in sterile sediment by shake-tests at Davis Bayou and of -0.0457, -0.103, -0.120 h^{-1} in nonsterile water and -0.00233, -0.00722, -0.00785 h^{-1} in sterile water by shake-tests at Davis Bayou (Walker et al. 1988).
    Biotransformation:
    Bioconcentration, Uptake (k_1) and Elimination (k_2) Rate Constants:

Common Name: Dinitramine
Synonym: Cobex, Cobexo
Chemical Name: N,N-diethyl-2,6-dinitro-4-trifluromethyl-m-phenylenediamine
Uses: herbicide for selective pre-plant soil incorporating control of many annual grass and broadleaf weeds in cotton, soybeans, peas, groundnuts, beans, sunflowers, safflowers, carrots, turnips, fennel, chicory, etc. and in transplanted tomatoes, capsicums, aubergines, and brassicas.
CAS Registry No: 29091-05-2
Molecular Formula: $C_{11}H_{13}N_4O_4F_3$
Molecular Weight: 322.2
Melting Point (°C):
    98-99   (Khan 1980; Agrochemicals Handbook 1987; Suntio et al. 1988; Worthing 1991; Tomlin 1994; Milne 1995)
Boiling Point (°C):
Density (g/cm³ at 20°C):
    1.50   (25°C, Ashton & Crafts 1981; Agrochemicals Handbook 1987)
Molar Volume (cm³/mol):
    265.7   (calculated-LeBas method, Suntio et al. 1988)
Molecular Volume (Å³):
Total Surface Area, TSA (Å²):
Dissociation Constant pK_a:
Heat of Fusion, ΔH_{fus}, kcal/mol:
Entropy of Fusion, ΔS_{fus}, cal/mol·K (e.u.):
Fugacity Ratio at 25°C (assuming ΔS_{fus} = 13.5 e.u.), F: 0.185

Water Solubility (g/m³ or mg/L at 25°C):
    1.1   (Martin & Worthing 1977; quoted, Kenaga 1980; Kenaga & Goring 1980; Isensee 1991)
    1.0   (Wauchope 1978)
    1.1   (Khan 1980)
    1.0   (20°C, Ashton & Crafts 1981; Agrochemicals Handbook 1987)
    1.0   (Verschueren 1983; quoted, Suntio et al. 1988; Shiu et al. 1990)
    1.1   (Worthing 1987, 1991; quoted, Shiu et al. 1990)
    0.8   (20°C, selected, Suntio et al. 1988)
    1.1   (20-25°C, selected, Augustijn-Beckers et al. 1994; Hornsby et al. 1996)
    1.1   (selected, Lohninger 1994)
    1.0   (20°C, Tomlin 1994; Milne 1995)
    1.1, 76.6   (quoted, predicted-AQUAFAC, Lee et al. 1996)

Vapor Pressure (Pa at 25°C):
    0.00048   (Khan 1980; Ashton & Crafts 1981)
    0.00048   (Verschueren 1983; quoted, Suntio et al. 1988)
    0.000479   (Agrochemicals Handbook 1987; Worthing 1991; Tomlin 1994)
    0.00040   (20°C, selected, Suntio et al. 1988)
    0.00048   (20-25°C, selected, Augustijn-Beckers et al. 1994; Hornsby et al. 1996)

Henry's Law Constant (Pa·m³/mol):
    0.160   (20°C, calculated-P/C, Suntio et al. 1988)

Octanol/Water Partition Coefficient, log K_{OW}:
- 4.31 (selected, Dao et al. 1983)
- 4.30 (Worthing 1991; Tomlin 1994)
- 4.30 (Milne 1995)
- 4.30 (selected, Hansch et al. 1995)

Bioconcentration Factor, log BCF:
- 2.77 (calculated-S, Kenaga 1980; quoted, Isensee 1991)
- 2.45 (calculated-K_{OC}, Kenaga 1980)

Sorption Partition Coefficient, log K_{OC}:
- 3.60 (soil, Harvey 1974; quoted, Kenaga 1980; Kenaga & Goring 1980)
- 3.61 (soil, calculated-S as per Kenaga & Goring 1978, Kenaga 1980)
- 3.60 (20-25°C, estimated, Augustijn-Beckers et al. 1994; Hornsby et al. 1996)
- 3.84 (estimated-chemical structure, Lohninger 1994)

Half-Lives in the Environment:
  Air:
  Surface water:
  Groundwater:
  Sediment:
  Soil: half-life of 22 d for 0.5 µg/mL to biodegrade in flooded soil with approximately 1 cm of water on top of the soil (Savage 1978; quoted, Muir 1991); persistence of 3 months in soil (Wauchope 1978); selected field half-life of 30 d (Augustijn-Beckers et al. 1994; Hornsby et al. 1996); half-life of 10-66 d (Tomlin 1994).
  Biota:

Environmental Fate Rate Constants or Half-Lives:
  Volatilization:
  Photolysis: half-life of < 1 h in distilled water, river water and ocean water under sunlight (Newsom & Woods 1973; quoted, Cessna & Muir 1991).
  Oxidation:
  Hydrolysis:
  Biodegradation: half-life of 22 d for 0.5 µg/mL to biodegrade in flooded soil with approximately 1 cm of water on top of the soil (Savage 1978; quoted, Muir 1991).
  Biotransformation:
  Bioconcentration, Uptake (k_1) and Elimination (k_2) Rate Constants:

Common Name: Dinoseb
Synonym: Anatox, Aretit, Basanite, Butaphene, Caldon, Chemox, Dibutox, Dinitrall, DNBP, DN-289, DNOSAP, DNOSBP, DNSBP, Dow General, Dyanap, Dytop
Chemical Name: 2-*sec*-butyl-4,6-dinitrophenol
Uses: herbicides/insecticides; pre- or post-emergence control of broadleaf weeds in cereals, maize, lucerne, clover, trefoil, grass leys, potatoes, peas, onions, garlics, peas, leeks, soya beans, orchards, groundnuts, strawberries, vineyards and other crops; for control of strawberry runners and raspberry suckers and overwintering forms of insect pests on fruit trees; also used as a desiccant for leguminous seed crops; destruction of potato haulms; as a pre-harvest hop defoliant, etc.
CAS Registry No: 88-85-7
Molecular Formula: $C_{10}H_{12}N_2O_5$
Molecular Weight: 240.22
Melting Point (°C):
    32 (Khan 1980; Herbicide Handbook 1989)
    42 (Spencer 1982)
    38-42.0 (Suntio et al. 1988; Milne 1995)
    30-40.0 (Milne 1995)
Boiling Point (°C):
    362 (estimated, Grain 1982)
Density (g/cm³ at 20°C):
    1.265 (45°C, Agrochemicals Handbook 1987; Herbicide Handbook 1989; Milne 1995)
Molar Volume (cm³/mol):
    218.0 (calculated-LeBas method, Suntio et al. 1988)
Molecular Volume (Å³):
Total Surface Area, TSA (Å²):
Dissociation Constant pK_a:
    4.62 (radiometer/pH meter, Cessna & Grover 1978; Hornsby et al. 1996)
    4.61 (radiometer/pH meter, Cessna & Grover 1978)
    4.62 (Worthing 1987, 1991; quoted, Howard 1991)
    4.50 (Yao & Haag 1991)
    4.62 (Haderlein et al. 1996)
Heat of Fusion, ΔH_{fus}, kcal/mol:
Entropy of Fusion, ΔS_{fus}, cal/mol·K (e.u.):
Fugacity Ratio at 25°C (assuming ΔS_{fus} = 13.5 e.u.), F: 0.7110

Water Solubility (g/m³ or mg/L at 25°C):
    50 (Günther 1968; quoted, Suntio et al. 1988; Shiu et al. 1990)
    52 (Kearney & Kaufman 1975; quoted, Howard 1991)
    50 (Martin & Worthing 1977; Kenaga 1980; quoted, Isensee 1991)
    52 (Khan 1980; quoted, Suntio et al. 1988; Shiu et al. 1990)
    52 (Ashton & Crafts 1981; Herbicide Handbook 1989)
    50 (Spencer 1982; Thomas 1982; quoted, Nash 1988)
    100 (rm. temp., Worthing 1983, 1991; quoted, Shiu et al. 1990)
    52 (20°C, Agrochemicals Handbook 1987; Milne 1995)
    47 (20°C, selected, Suntio et al. 1988; quoted, Majewski & Capel 1995)
    100 (selected, Lohninger 1994)
    52 (20-25°C, selected, Hornsby et al. 1996)

Vapor Pressure (Pa at 25°C):
- 130 (151.5°C, Khan 1980; quoted, Suntio et al. 1988)
- 133 (151.1°C, Ashton & Crafts 1981)
- 0.0067; 0.0008, 0.0067 (supercooled liq.: quoted; estimated, Grain 1982)
- 0.0067 (Thomas 1982; quoted, Dobbs et al. 1984; Nash 1988)
- 0.0023 (30°C, Spencer 1982)
- 10 (20°C, selected, Suntio et al. 1988; quoted, Howard 1991; Majewski & Capel 1995)
- 0.183 (60°C, Worthing 1991)
- 0.0067 (20-25°C, selected, Hornsby et al. 1996)

Henry's Law Constant (Pa·m³/mol):
- 51.1 (20°C, calculated-P/C, Suntio et al. 1988; quoted, Howard 1991; Majewski & Capel 1995)

Octanol/Water Partition Coefficient, log K_{OW}:
- 3.59 (Hansch & Leo 1979; quoted, Haderlein et al. 1996)
- 3.69 (calculated, Zitko et al. 1976; quoted, Kenaga & Goring 1980, Sangster 1993)
- 3.69 (Hansch & Leo 1985; quoted, Howard 1991, Magee 1991)
- 4.10 (RPPHLC-RT, Klein et al. 1988)
- 3.14 (shake flask-GC, De Bruijn et al. 1989)
- 3.0, 3.57 (pH 7, pH 2, shake flask, Brook et al. 1990)
- 3.56 (selected, Hansch et al. 1995)

Bioconcentration Factor, log BCF:
- 1.83 (calculated-S, Kenaga 1980a; quoted, Howard 1991)
- 0.778 (calculated-K_{OC}, Kenaga 1980)
- 1.51 (measured, Kenaga 1980; quoted, Isensee 1991)

Sorption Partition Coefficient, log K_{OC}:
- 2.85 (soil, Thomas 1982; quoted, Nash 1988)
- 2.09 (soil, Kenaga 1980a; Kenaga & Goring 1980; quoted, Howard 1991)
- 2.71 (soil, calculated-S as per Kenaga & Goring 1978, Kenaga 1980a)
- 3.82 (buffered soil at pH 3, Hodson & Williams 1988; quoted, Howard 1991)
- 2.09, 2.68 (reported, estimated as log K_{OM}, Magee 1991)
- 1.80, 2.04, 2.08 (quoted values, Bottoni & Funari 1992)
- 2.70 (selected, Lohninger 1994)
- 1.48 (soil, 20-25°C, estimated, Hornsby et al. 1996)

Adsorption coefficient, K_d (L·kg^{-1}):
- 6.4, 64 (homoionic K$^+$-kaolinite, K$^+$-montmorillonite clay minerals, Haderlein et al. 1996)

Half-Lives in the Environment:
- Air: 12.2-122 h, based on estimated rate constant for the reaction with hydroxyl radicals in air (Atkinson 1987; quoted, Howard et al. 1991).
- Surface water: 1032-2952 h, based on aerobic soil mineralization data for one soil (Doyle et al. 1978; quoted, Howard et al. 1991).
- Groundwater: 96-5904 h, based on estimated unacclimated aqueous aerobic and anaerobic biodegradation half-lives (Howard et al. 1991).
- Sediment:
- Soil: 1032-2952 h, based on aerobic soil mineralization data for one soil (Doyle et al. 1978; quoted, Howard et al. 1991); field half-life of 30 d (20-25°C, estimated, Hornsby et al. 1996).
- Biota:

Environmental Fate Rate Constants or Half-Lives:
- Volatilization: initial rate constant of 1.1×10^{-3} h^{-1} and predicted rate constant of 2.6×10^{-3} h^{-1} from soil with a half-life of 266.5 h (Thomas 1982).
- Photolysis:
- Oxidation: photooxidation half-life of 12.2-122 h in air, based on estimated rate constant for the reaction with hydroxyl radicals in air (Atkinson 1987; quoted, Howard et al. 1987); calculated rate constant of 4×10^9 M^{-1} s^{-1} for the reaction with hydroxyl radicals in aqueous solutions at (24 ± 1)°C (Haag & Yao 1992).
- Hydrolysis:
- Biodegradation: aqueous aerobic half-life of 1032-2952 h, based on aerobic soil mineralization data for one soil (Doyle et al. 1978; quoted, Howard et al. 1991) and aqueous anaerobic half-life of 96-360 h, based on anaerobic soil die-away test data for isopropalin (Gingerich & Zimdahl 1976; quoted, Howard et al. 1991).
- Biotransformation:
- Bioconcentration, Uptake (k_1) and Elimination Constants (k_2):

Common Name: Diphenamid
Synonym: Difenamide, Dimid, Dymid, Enide, Fenam, Rideon
Chemical Name: N,N-dimethyldiphenylacetamide; N,N-dimethyl-α-phenyl-benzeneacetamide
Uses: herbicide for pre-emergence control of annual grasses and some broadleaf weeds in cotton, sweet potatoes, tomatoes, vegetables, capsicums, okra, soybeans, groundnuts, tobacco, pome fruit, stone fruit, citrus fruit, bush fruit, strawberries, forestry nurseries, and ornamental plants, shrubs, and trees.
CAS Registry No: 957-51-7
Molecular Formula: $C_{16}H_{17}NO$
Molecular Weight: 239.3
Melting Point (°C):
    132.0-135.5 (Khan 1980)
    134.5-135.5 (Spencer 1982; Agrochemicals Handbook 1987; Herbicide Handbook 1989; Worthing 1991; Tomlin 1994; Milne 1995)
Boiling Point (°C):
Density (g/cm^3 at 20°C):
    1.17 (23.3°C, Agrochemicals Handbook 1987; Tomlin 1994; Milne 1995)
Molar Volume (cm^3/mol):
    284.2 (calculated-LeBas method, this work)
Molecular Volume (Å^3):
Total Surface Area, TSA (Å^2):
Dissociation Constant pK_a:
Heat of Fusion, ΔH_{fus}, kcal/mol:
    6.55 (DSC method, Plato & Glasgow 1969)
Entropy of Fusion, ΔS_{fus}, cal/mol·K (e.u.):
Fugacity Ratio at 25°C (assuming ΔS_{fus} = 13.5 e.u.), F: 0.081

Water Solubility (g/m^3 or mg/L at 25°C):
    240 (Melnikov 1971; quoted, Shiu et al. 1990)
    260 (27°C, Spencer 1973, 1982; Khan 1980; Worthing 1987; quoted, Shiu et al. 1990; Howard 1991; Lohninger 1994)
    260 (Martin & Worthing 1977; Weber et al. 1980; quoted, Kenaga 1980; Willis & McDowell 1982)
    260 (27°C, Agrochemicals Handbook 1987; Herbicide Handbook 1989; Tomlin 1994)
    280 (20-25°C, selected, Hornsby et al. 1996)

Vapor Pressure (Pa at 25°C):
    $< 1.33 \times 10^{-4}$ (Weber et al. 1980; quoted, Willis & McDowell 1982)
    negligible (20°C, Agrochemicals Handbook 1987; Tomlin 1994)
    4.0×10^{-6} (20-25°C, selected, Hornsby et al. 1996)

Henry's Law Constant (Pa·m^3/mol):

Octanol/Water Partition Coefficient, log K_{OW}:
    3.36 (calculated-S as per Chiou et al. 1977 & Chiou 1981, this work)

Bioconcentration Factor, log BCF:
    1.43   (calculated-S, Kenaga 1980)

Sorption Partition Coefficient, log K_{OC}:
    2.32   (soil, calculated-S, Kenaga 1980)
    2.32   (selected, Lohninger 1994)
    2.32   (soil, 20-25°C, selected, Hornsby et al. 1996)

Half-Lives in the Environment:
    Air:
    Surface water:
    Groundwater:
    Sediment:
    Soil: estimated persistence of 8 months (Kearney et al. 1969; Edwards 1973; quoted, Morrill et al. 1982; Jury et al. 1987); persistence under warm damp conditions is ca. 3-6 months (Herbicide Handbook 1989; Tomlin 1994); field half-life of 30 d (20-25°C, selected, Hornsby et al. 1996).
    Biota:

Environmental Fate Rate Constants or Half-Lives:
    Volatilization:
    Photolysis: half-life of 2.25 h in distilled water (Tanaka et al. 1981; quoted, Cessna & Muir 1991);
    Oxidation:
    Hydrolysis:
    Biodegradation:
    Biotransformation:
    Bioconcentration, Uptake (k_1) and Elimination (k_2) Rate Constants:

Common Name: Diquat
Synonym: Aquacide, Deiquat, Dextrone, Ortho, Pathclear, Preeglone, Reglone, Weedol, Weedtrine-D
Chemical Name: 1,1'-ethylene-2,2'-dipyridine
Uses: nonselective contact herbicide to control broadleaf weeds in fruit and vegetable crops.
CAS Registry No: 2764-72-9
Molecular Formula: $C_{12}H_{14}N_2$
Molecular Weight: 186.26
Melting Point (°C): 335-340 (Spencer 1982)
Boiling Point (°C):
Density (g/cm³ at 20°C):
    1.22-1.27 (Ashton & Crafts 1981; Herbicide Handbook 1989; Montgomery 1993; Tomlin 1994)
Molar Volume (cm³/mol):
    230.6 (calculated-LeBas method, this work)
    149.6 (calculated-density, this work)
Molecular Volume (Å³):
Total Surface Area, TSA (Å²):
Dissociation Constant pK_a:
Heat of Fusion, ΔH_{fus}, kcal/mol:
Entropy of Fusion, ΔS_{fus}, cal/mol·K (e.u.):
Fugacity Ratio at 25°C (assuming ΔS_{fus} = 13.5 e.u.), F: 0.00077

Water Solubility (g/m³ or mg/L at 25°C):
    700000 (Khan 1980; Spencer 1982)
    670000 (Weber et al. 1980; quoted, Willis & McDowell 1982)
    700000 (Verschueren 1983; quoted, Howard 1991)
    570 (Reinert 1989)
    700000 (Worthing 1991; Tomlin 1994)
    700000 (Montgomery 1993)

Vapor Pressure (Pa at 25°C):
    <0.00533 (Agrochemicals Handbook 1983; quoted, Howard 1991)
    <1.3×10^{-5} (Worthing 1991; Tomlin 1994)
    <1.3×10^{-5} (20°C, Montgomery 1993)

Henry's Law Constant (Pa·m³/mol):
    <6.38×10^{-9} (20-25°C, calculated-P/C, Montgomery 1993)
    <3.42×10^{-9} (calculated-P/C, this work)

Octanol/Water Partition Coefficient, log K_{OW}:
    −3.05 (Garten & Trabalka 1983; quoted, Howard 1991)
    2.78 (Reinert 1989)
    −4.60 (20°C, Worthing 1991; Tomlin 1994)
    −4.60 (Montgomery 1993)

Bioconcentration Factor, log BCF:
    −2.84   (calculated-S as per Kenaga 1980, this work)
    −5.92   (calculated-log K_{ow} as per Mackay 1982, this work)

Sorption Partition Coefficient, log K_{OC}:
    2.84   (Reinert 1989)
    0.420   (calculated, Montgomery 1993)
    0.425   (calculated-S as per Kenaga 1980, this work)

Half-Lives in the Environment:
    Air:
    Surface water: half-life of about 50 days to biodegrade in lake water (Hiltibran 1972; quoted, Muir 1991).
    Groundwater:
    Sediment: half-life of >158 days for 1.5 µg/mL of infested sediment-water microcosm to biodegrade (Simsiman & Chesters 1976; quoted, Muir 1991).
    Soil:
    Biota:

Environmental Fate Rate Constants or Half-Lives:
    Volatilization:
    Photolysis: half-life of 192 h for 100% of 10 µg/mL to degrade in distilled water under 240-260 nm light (Funderburk et al. 1960; quoted, Cessna & Muir 1991); half-life of <5 weeks for 4 µg/ml to degrade in distilled water under sunlight (Slade & Smith 1967; quoted, Cessna & Muir 1991); dry diquat photodecomposed by UV light with half-life of 48 h (Funderburk & Bozarth 1967; quoted, Montgomery 1993); half-life about 48 h when associated with aerosols (Howard 1991); half-life of 3 weeks for 3% of 5 µg/mL to degrade in distilled water under sunlight (Smith & Grove 1969; quoted, Cessna & Muir 1991).
    Oxidation: rate constant of 5.9×10^9 M^{-1} s^{-1} for the reaction (Fenton with reference to acetophenone) with hydroxyl radicals in aqueous solutions at pH 3.1 and at $(24 \pm 1)°C$ (Buxton et al. 1988; quoted, Faust & Hoigné 1990; Haag & Yao 1992); rate constant of $(8.0 \pm 1.8) \times 10^8$ M^{-1} s^{-1} for the reaction (Fenton with reference to acetophenone) with hydroxyl radicals in aqueous solutions at pH 3.1 and at $(24 \pm 1)°C$ (Haag & Yao 1992).
    Hydrolysis: half-life of 74 d under simulated sunlight at pH 7 (Montgomery 1993; Tomlin 1994).
    Biodegradation: half-life of about 50 d to biodegrade in lake water (Hiltibran 1972; quoted, Muir 1991); half-lives of >158 d for 1.5 µg/mL of infested sediment-water microcosm to biodegrade in sediment and only about 2 d in water both at 25°C (Simsiman & Chesters 1976; quoted, Muir 1991).
    Biotransformation:
    Bioconcentration, Uptake (k_1) and Elimination (k_2) Rate Constants:

Common Name: Diuron
Synonym: AF 101, Cekiuron, Crisuron, Dailon, DCMU, Diater, dichlorofonidim, Di-on, Direx, DMU, Drexel, Duran, Dynex, Herbatox, Karmex, Marmer, NA 2767, Telvar, Unidron, Urox D, Vonduron
Chemical Name: 3-(3,4-dichlorophenyl)-1,1-dimethylurea; N'-(3,4-dichlorophenyl)-N,N-dimethylurea
Uses: pre-emergence herbicide in soils to control germinating broadleaf grasses and weeds in crops such as apples, cotton, grapes, pears, pineapple, and alfalfa; also used as sugar cane flowering depressant.
CAS Registry No: 330-54-1
Molecular Formula: $C_9H_{10}Cl_2N_2O$
Molecular Weight: 233.1
Melting Point (°C):
    158-159 (Khan 1980; Spencer 1982; Suntio et al. 1988; Herbicide Handbook 1989; Worthing 1991; Tomlin 1994; Milne 1995)
    155 (Karickhoff 1981)
    158 (Swann et al. 1983; Nkedi-Kizza et al. 1985; Kühne et al. 1995)
    159 (Yalkowsky & Banerjee 1992)
    150-155 (Montgomery 1993)
    158.5 (Patil 1994)
Boiling Point (°C):
    180 (decomposes, Montgomery 1993)
Density (g/cm³ at 20°C):
Molar Volume (cm³/mol):
    223.8 (calculated-LeBas method, Suntio et al. 1988)
    188.0 (modified LeBas method, Spurlock & Biggar 1994a)
Molecular Volume (Å³):
Total Surface Area, TSA (Å²):
    219.0 (Nkedi-Kizza et al. 1985)
Dissociation Constant pK_a:
    −1 to −2 (Montgomery 1993)
Heat of Fusion, ΔH_{fus}, kcal/mol:
    8.10 (DSC method, Plato & Glasgow 1969)
Entropy of Fusion, ΔS_{fus}, cal/mol·K (e.u.):
Fugacity Ratio at 25°C (assuming ΔS_{fus} = 13.5 e.u.), F: 0.047

Water Solubility (g/m³ or mg/L at 25°C):
    40.0 (Günther et al. 1968; quoted, Suntio et al. 1988; Shiu et al. 1990)
    42.0 (Melnikov 1971; Khan 1980; quoted, Suntio et al. 1988; Shiu et al. 1990)
    42.0 (20°C, Weber 1972; Weber et al. 1980; quoted, Willis & McDowell 1982)
    42.0 (Spencer 1973. 1982; quoted, Shiu et al. 1990; Muir 1991)
    37.3 (shake flask-UV, Freed et al. 1976; Freed 1976; quoted, Spencer 1976; Jury et al. 1983,84; Suntio et al. 1988; Shiu et al. 1990)
    42.0 (Martin & Worthing 1977; quoted, Kenaga 1980; Kenaga & Goring 1980; Karickhoff 1981; Worthing 1987, 1991; Isensee 1991)
    42.0 (quoted, Wauchope 1978; Geyer et al. 1980; Thomas 1982; Nash 1988)
    42.0 (Ashton & Crafts 1981; Herbicide Handbook 1989)
    42.4 (shake flask, Briggs 1981)
    22.0 (shake flask-HPLC, Ellgehausen et al. 1981; quoted, Shiu et al. 1990)

37.0   (Jury et al. 1983; quoted, Taylor & Glotfelty 1988)
38.7   (gen. col.-HPLC/RI, Swann et al. 1983; quoted, Shiu et al. 1990)
120   (RP-HPLC-RT, Swann et al. 1983)
42.0   (selected, Nkedi-Kizza et al. 1985; Gerstl & Helling 1987)
19.6, 40.1, 53.4 (4, 25, 40°C, shake flask-LSS, Madhun et al. 1986)
42.0   (Agrochemicals Handbook 1987; Tomlin 1994)
40.0   (20°C, selected, Suntio et al. 1988; quoted, Mejewski & Capel 1995)
37.3   (Yalkowsky 1989; quoted, Howard 1991)
42.0   (selected, Chaumat et al. 1991; Evelyne et al. 1992; Lohninger 1994)
38.0   (Spurlock 1992; quoted, Spurlock & Biggar 1994b)
42.0   (20-25°C, selected, Wauchope et al. 1992; Hornsby et al. 1996)
40.5   (selected, Yalkowsky & Banerjee 1992)
40.0   (20°C, Montgomery 1993)
42.0, 307 (quoted, calculated, Patil 1994)
36.9   (quoted, Kühne et al. 1995)
97.2   (calculated-group contribution fragmentation method, Kühne et al. 1995)
42.0   (Milne 1995)
22.0   (selected, Pinsuwan et al. 1995)
35, 84   (quoted, predicted-AQUAFAC, Lee et al. 1996)

Vapor Pressure (Pa at 25°C):
   1.6×10^{-5} (estimated, Nex & Swezey 1954; quoted, Jury et al. 1983)
   3.8×10^{-6} (20°C, Johnson & Julin 1974; quoted, Muir 1991)
   4.1×10^{-4} (50°C, Khan 1980; quoted, Suntio et al. 1988)
   $< 1.3 \times 10^{-4}$ (20-25°C, Weber et al. 1980; quoted, Willis & McDowell 1982)
   4.1×10^{-6} (50°C, Ashton & Crafts 1981)
   2.5×10^{-4} (Thomas 1982; quoted, Nash 1988)
   2.1×10^{-5} (Jury et al. 1983; quoted, Taylor & Glotfelty 1988; Taylor & Spencer 1990)
   3.6×10^{-4} (Jury et al. 1983; quoted, Howard 1991)
   2.7×10^{-4} (selected, Nkedi-Kizza et al. 1985
   4.1×10^{-4} (50°C, Agrochemicals Handbook 1987; Worthing 1991)
   2.0×10^{-4} (20°C, selected, Suntio et al. 1988; quoted, Mejewski & Capel 1995)
   4.1×10^{-4} (50°C, Herbicide Handbook 1989)
   2.1×10^{-5} (selected, Taylor & Spencer 1990)
   9.2×10^{-6} (20-25°C, selected, Wauchope et al. 1992; Hornsby et al. 1996)
   1.1×10^{-6} (Tomlin 1994)

Henry's Law Constant (Pa·m^3/mol):
   1.4×10^{-4} (calculated-P/C, Jury et al. 1984, 1987a,b; Jury & Ghodrati 1989)
   1.2×10^{-4} (20°C, calculated-P/C, Suntio et al. 1988; quoted, Mejewski & Capel 1995)
   1.3×10^{-4} (calculated-P/C, Taylor & Glotfelty 1988)
   0.274   (calculated-P/C, Howard 1991)
   2.1×10^{-5} (20°C, calculated-P/C, Muir 1991)
   1.5×10^{-4} (20-25°C, calculated-P/C, Montgomery 1993)

Octanol/Water Partition Coefficient, log K_{OW}:
- 1.97 (Briggs 1969; quoted, Kenaga & Goring 1980; Karickhoff 1981; Suntio et al. 1988)
- 2.60 (calculated-f const., Rekker 1977; quoted, Ellgehausen et al. 1981)
- 2.81 (quoted, Rao & Davidson 1980; Karickhoff 1981; Suntio et al. 1988)
- 2.68 (shake flask-UV, Briggs 1981; quoted, Sicbaldi & Finizio 1993)
- 2.89 (calculated-S, Ellgehausen et al. 1981; quoted, Sicbaldi & Finizio 1993)
- 2.57 (selected, Dao et al. 1983)
- 2.60 (Elgar 1983; quoted, Suntio et al. 1988)
- 2.77 (Hansch & Leo 1985; quoted, Howard 1991)
- 2.69, 2.65, 2.63 (4, 25, 40°C, shake flask-LSS, Madhun et al. 1986; quoted, Sicbaldi & Finizio 1993)
- 2.49 (selected, Gerstl & Helling 1987)
- 2.60 (selected, Suntio et al. 1988; quoted, Bintein & Devillers 1994)
- 2.464 (calculated, Evelyne et al. 1992)
- 2.81 (Spurlock 1992; quoted, Spurlock & Biggar 1994b)
- 1.97-2.81 (Montgomery 1993)
- 2.45 (RP-HPLC, Sicbaldi & Finizio 1993)
- 2.80 (Aquasol Database 1994; quoted, Pinsuwan et al. 1995)
- 1.97, 1.95 (quoted, calculated, Patil 1994)
- 2.85 ± 1.70 (Tomlin 1994)
- 2.58, 2.73 (shake flask-UV, calculated-RPHPLC-k', Liu & Qian 1995)
- 2.80, 2.70 (quoted, calculated-f const., Pinsuwan et al. 1995)
- 2.68 (selected, Hansch et al. 1995; Devillers et al. 1996)
- 2.45, 2.68, 3.41 (RP-HPLC, ClogP, calculated-S, Finizio et al. 1997)

Bioconcentration Factor, log BCF:
- 1.40 (measured, Isensee 1976; quoted, Isensee 1991)
- 1.88 (calculated-S, Kenaga 1980; quoted, Howard 1991; Isensee 1991)
- 1.34 (calculated-K_{OC}, Kenaga 1980; quoted, Howard 1991)
- 2.16 (*Pimephales promelas*, Call et al. 1987; quoted, Devillers et al. 1996)
- 2.41, 2.48 (cuticle/water: tomato, pepper, Chaumat et al. 1991)
- 2.41, 2.51 (cuticle/water: box tree, laurel, Chaumat et al. 1991)
- 2.55, 2.28 (cuticle/water: pear, ivy, Chaumat et al. 1991)
- 1.18, 1.64 (cuticle/water: cleavers, vanilla, Chaumat et al. 1991)
- 2.45, 2.48 (cuticle/water: tomato, pepper, Evelyne et al. 1992)

Bioaccumulation Factor, log BF:
- −1.70 (adipose tissue in both male & female Albino rats, Hodge et al. 1967; quoted, Geyer et al. 1980)

Sorption Partition Coefficient, log K_{OC}:
- 2.60 (soil, Hamaker & Thompson 1972; Farmer 1976; Hane 1976; quoted, Kenaga 1980; Kenaga & Goring 1980; Karickhoff 1981; Liu & Qian 1995)
- 2.75 (soil, calculated-S as per Kenaga & Goring 1977, Kenaga 1980)
- 2.59 (average of 3 soils, HPLC-RT, McCall et al. 1980)
- 2.15-2.52 (Peck et al. 1980; quoted, Muir 1991)
- 1.97 (reported as log K_{OM}, Briggs 1981)

3.06, 2.41 (estimated-S, S & M.P., Karickhoff 1981)
1.58, 2.42 (estimated-K_{ow}, Karickhoff 1981)
2.58 (average of 84 soils, Rao & Davidson 1982; quoted, Jury et al. 1983, 1984; Howard 1991)
2.18 (soil, Thomas 1982; quoted, Nash 1988)
2.83 (Webster soil, Nkedi-Kizza 1983; quoted, Howard 1991)
2.49 (soil slurry method, Swann et al. 1983)
2.48 (RP-HPLC-RT, Swann et al. 1983)
3.03, 2.94 (4, 25°C, Semiahmoo soil, in μmol/kg OC, batch equilibrium method-LSS, Madhun et al. 1986)
2.82, 2.68 (4, 25°C, Adkins soil, in μmol/kg OC, Madhun et al. 1986)
2.86, 2.44, 2.48; 2.81, 2.74, 2.44 (estimated-K_{ow}; S, Madhun et al. 1986)
2.46, 2.50 (quoted, calculated-χ, Gerstl & Helling 1987)
2.58 (soil, screening model calculations, Jury et al. 1987a,b; Jury & Ghodrati 1989)
2.35, 2.57 (2 subsurface soils from Oklahoma, Bouchard & Wood 1988; quoted, Howard 1991)
2.94, 2.68 (mucky peat soil, loam sand soil, quoted, Howard 1991)
2.18, 2.48-2.49, 2.59, 2.66 (quoted values, Bottoni & funari 1992)
2.68 (soil, 20-25°C, selected, Wauchope et al. 1992; Hornsby et al. 1996)
2.21-2.87 (Montgomery 1993)
2.68 (selected, Lohninger 1994)
2.60 (Tomlin 1994)
2.70 (calculated-K_{ow}, Liu & Qian 1995)

Half-Lives in the Environment:

Air: 0.12 d, based on estimation for the vapor-phase reaction with hydroxyl radicals in the atmosphere (Atkinson 1987; quoted, Howard 1991).

Surface water: should be photolyzed within a few days (Howard 1991).

Groundwater:

Sediment: half-life of 3-10 d for 40 μg/mL to biodegrade in pond sediment of anaerobic media at 30°C (Attaway et al. 1982a; quoted, Muir 1991); half-life of <17 d for 40 μg/mL to biodegrade in pond sediment at 30°C (Attaway et al. 1982b; quoted, Muir 1991); half-life of ~5 d for 0.22 μg/mL to biodegrade in pond sediment of anaerobic media (Stepp et al. 1985; quoted, Muir 1991).

Soil: estimated persistence of 10 months in soil (Kearney et al. 1969; quoted, Jury et al. 1987); persistence of 8 months in soil (Edwards 1973; quoted, Morrill et al. 1982); half-life of 7.0 months at 15°C and 5.5 months at 30°C in soils (Freed & Haque 1973); persistence of 10 months (Wauchope 1978); rate constant of 0.0031 d^{-1} with half-life of 328 d under field conditions (Rao & Davidson 1980); the calculated half-life due to volatilization from soil when incorporated into 1 cm of soil is 1918 d (Jury et al. 1983; quoted, Howard 1991); half-lives in an Adkins loamy sand are 705, 414, and 225 d at 25, 30, and 35°C, respectively; however, in a Semiahoo mucky peat, the half-lives were considerable higher: 3991, 2164, and 1165 d at 25, 30, and 35°C, respectively (Madhun & Freed 1987; quoted, Montgomery 1993); half-life of 328 d from screening model calculations (Jury et al. 1987a,b;

Jury & Ghodrati 1989); selected field half-life of 90 d (Wauchope et al. 1992; Hornsby et al. 1996).

Biota: biochemical half-life of 328 d from screening model calculations (Jury et al. 1987a,b; Jury & Ghodrati 1989).

Environmental Fate Rate Constants or Half-Lives:

Volatilization: 2.5×10^{-3} h^{-1} (initial) and 5.3×10^{-4} h^{-1} (predicted) from soil with a half-life of 1307 h (Thomas 1982); the calculated half-life due to volatilization from soil when incorporated into 1 cm of soil is 1918 d (Jury et al. 1983; quoted, Howard 1991).

Photolysis: half-life of 2.25 h for 80-84% of 40 µg/mL to degrade in distilled water under 300 nm light (Tanaka et al. 1981; quoted, Cessna & Muir 1991); in surface waters should be photolyzed within a few days (Howard 1991).

Oxidation: photooxidation half-life of 0.12 d in air, based on estimation for the vapor-phase reaction with hydroxyl radicals in the atmosphere (Atkinson 1987; quoted, Howard 1991).

Hydrolysis: half-life of >4 months for 4660 µg/mL to hydrolyze in phosphate buffer at pH 5-9 and 20°C (El-Dib & Aly 1976; quoted, Muir 1991).

Biodegradation: half-life of 328 d for a 100 d leaching and screening test in 0-10 cm depth of soil (Rao & Davidson 1980; quoted, Jury et al. 1983, 1984, 1987a); half-life of 3-10 d for 40 µg/mL to biodegrade in pond sediment of anaerobic media at 30°C (Attaway et al. 1982a quoted, Muir 1991); half-life of <17 d for 40 µg/mL to biodegrade in pond sediment at 30°C (Attaway et al. 1982b; quoted, Muir 1991); 67-99% will be degraded in 10 weeks under aerobic conditions by mixed cultures isolated from pond water and sediments forming 6-7 products (Ellis & Camper 1982; quoted, Howard 1991; Muir 1991); half-life was <70 d at 30°C (Ellis & Camper 1982; quoted, Muir 1991; Montgomery 1993); half-life of ~5 d for 0.22 µg/mL to biodegrade in pond sediment of anaerobic media (Stepp et al. 1985; quoted, Muir 1991); aerobic half-life of ~20 d for 0.0005-10 µg/mL to biodegrade in filtered sewage water at 20°C (Wang et al. 1985; quoted, Muir 1991).

Biotransformation:

Bioconcentration, Uptake (k_1) and Elimination (k_2) Rate Constants:

Common Name: EPTC
Synonym: Eptam, Eradicane, FDA 1541, R 1608, Torbin
Chemical Name: carbamic acid, dipropylthio-, S-ethyl ester; S-ethyldipropyl(thiocarbamate); S-ethyldipropylcarbamothioate
Uses: selective systemic herbicide for pre-emergence control of perennial and annual grasses, broadleaf weeds.
CAS Registry No: 759-94-4
Molecular Formula: $C_9H_{19}NOS$
Molecular Weight: 189.31
Melting Point (°C): liquid
Boiling Point (°C):
    235.0   (Khan 1980; Herbicide Handbook 1989)
    127.0   (at 20 mmHg, Agrochemicals Handbook 1987; Merck Index 1989; Worthing 1991; Montgomery 1993; Tomlin 1994; Milne 1995)
Density (g/cm³ at 20°C):
    0.9546 (30°C, Spencer 1982; Agrochemicals Handbook 1987; Montgomery 1993; Tomlin 1994; Milne 1995)
    0.960   (25°C, Herbicide Handbook 1989; Montgomery 1993)
Molar Volume (cm³/mol):
    236.5   (calculated-LeBas method, Suntio et al. 1988)
Molecular Volume (Å³):
Total Surface Area, TSA (Å²):
Dissociation Constant pK_a:
Heat of Fusion, ΔH_{fus}, kcal/mol:
Entropy of Fusion, ΔS_{fus}, cal/mol·K (e.u.):
Fugacity Ratio at 25°C (assuming ΔS_{fus} = 13.5 e.u.), F: 1.0

Water Solubility (g/m³ or mg/L at 25°C):
    375   (shake flask-GC, Freed et al. 1967; quoted, Freed 1976; Shiu et al. 1990)
    365   (Martin & Worthing 1977; quoted, Kenaga 1980; Kenaga & Goring 1980; Suntio et al. 1988; Shiu et al. 1990; Isensee 1991)
    370   (20°C, Khan 1980; quoted, Suntio et al. 1988; Shiu et al. 1990)
    370-375 (Weber et al. 1980; quoted, Willis & McDowell 1982)
    370   (20°C, Ashton & Crafts 1981; Burkhard & Guth 1981; Herbicide Handbook 1989)
    375   (20°C, Spencer 1982)
    370   (Beste & Humburg 1983; Jury et al. 1983, 1984; quoted, Taylor & Glotfelty 1988; Grover 1991)
    375   (Agrochemicals Handbook 1987)
    375   (24°C, Worthing 1987, 1991; quoted, Shiu et al. 1990)
    370   (20°C, selected, Suntio et al. 1988; quoted, Majewski & Capel 1995)
    365   (20°C, Merck Index 1989; quoted, Milne 1995)
    344   (20-25°C, selected, Wauchope et al. 1992; Hornsby et al. 1996)
    375   (Montgomery 1993; Tomlin 1994)
    344   (selected, Lohninger 1994)
    370   (selected, Wienhold & Gish 1994)

Vapor Pressure (Pa at 25°C):
- 4.666 (extrapolated, Patchett et al. 1964; quoted, Freed 1976)
- 1.84 (20°C, effusion method, Hamaker & Kerlinger 1969; quoted, Suntio et al. 1988)
- 4.53 (Khan 1980; quoted, Suntio et al. 1988; quoted, Majewski & Capel 1995)
- 4.532 (Ashton & Crafts 1981; Herbicide Handbook 1989)
- 2.62 (quoted, Burkhard & Guth 1981)
- 2.80 (Patchett et al. 1983; quoted, Jury et al. 1983; Taylor & Glotfelty 1988; Taylor & Spencer 1990; Grover 1991)
- 4.70 (Agrochemicals Handbook 1987)
- 2.00 (20°C, selected, Suntio et al. 1988)
- 4.532 (35°C, Merck Index 1989)
- 4.50 (Worthing 1991)
- 4.532 (20-25°C, selected, Wauchope et al. 1992; Hornsby et al. 1996)
- 4.532 (20°C, Montgomery 1993)
- 0.00001 (Tomlin 1994)

Henry's Law Constant (Pa·m^3/mol):
- 1.463 (calculated-P/C, Jury et al. 1983, 1984, 1987a,b; Jury & Ghodrati 1989; quoted, Grover 1991)
- 1.02 (20°C, calculated-P/C, Suntio et al. 1988; quoted, Majewski & Capel 1995)
- 1.463 (calculated-P/C, Taylor & Glotfelty 1988)
- 1.013 (20-25°C, calculated-P/C, Montgomery 1993)
- 1.023 (calculated-P/C, this work)

Octanol/Water Partition Coefficient, log K_{ow}:
- 1.76 (selected, Dao et al. 1983)
- 3.20 (Worthing 1991; Montgomery 1993; Tomlin 1994; Milne 1995)
- 3.21 (quoted and selected, Sangster 1993)
- 3.21 (selected, Hansch et al. 1995)
- 3.45, 3.21, 2,77 (RP-HPLC, ClogP, calculated-S, Finizio et al. 1997)

Bioconcentration Factor, log BCF:
- 1.34 (calculated-S, Kenaga 1980; quoted, Isensee 1991)
- 1.08 (calculated-K_{oc}, Kenaga 1980)

Sorption Partition Coefficient, log K_{oc}:
- 2.38 (soil, Hamaker & Thompson 1972; quoted, Kenaga 1980; Kenaga & Goring 1980)
- 2.45 (soil, Hamaker & Thompson 1972; quoted, Jury et al. 1983; Grover 1991)
- 2.23 (soil, calculated-S as per Kenaga & Goring 1978, Kenaga 1980)
- 2.58 (soil, screening model calculations, Jury et al. 1987a,b; Jury & Ghodrati 1989)
- 2.23-2.38, 2.45 (quoted values, Bottoni & Funari 1992)
- 2.30 (soil, 20-25°C, selected, Wauchope et al. 1992; quoted, Richards & Baker 1993; Hornsby et al. 1996)
- 2.38 (Montgomery 1993)
- 2.30 (selected, Lohninger 1994)
- 2.45 (selected, Wienhold & Gish 1994)

Half-Lives in the Environment:
- Air: calculated lifetime of 6 h for the vapor-phase reaction with OH radicals in the troposphere (Atkinson et al. 1992; Kwok et al. 1992).
- Surface water:
- Groundwater:
- Sediment:
- Soil: estimated persistence of 4 months in soil (Kearney et al. 1969; quoted, Jury et al. 1987a); half-life of 30 d from screening model calculations (Jury et al. 1987a,b; Jury & Ghodrati 1989); half-life in moist loam soil at 21 to 27°C is approximately one week (Herbicide Handbook 1974, 1989); selected field half-life of 6 d (Wauchope et al. 1992; quoted, Richards & Baker 1993; Hornsby et al. 1996).
- Biota: biochemical half-life of 30 d from screening model calculations (Jury et al. 1987a,b; Jury & Ghodrati 1989).

Environmental Fate Rate Constants or Half-Lives:
- Volatilization: half-life of 3.7 d (Jury et al. 1983; quoted, Grover 1991).
- Photolysis:
- Oxidation: calculated lifetime of 6 hours for the vapor-phase reaction with OH radicals in the troposphere (Atkinson et al. 1992; Kwok et al. 1992).
- Hydrolysis:
- Biodegradation: half-life of 30 d for a 100 d leaching and screening test in 0-10 cm depth of soil (Nash 1980; quoted, Jury et al. 1983, 1984, 1987a; quoted, Grover 1991).
- Biotransformation:
- Bioconcentration, Uptake (k_1) and Elimination (k_2) Rate Constants:

Common Name: Fenuron
Synonym: Dozer, Dybar, Falisilvan, Fenidim, Fenulon, Urab
Chemical Name: 1,1-dimethyl-3-phenylurea; N,N-dimethyl-N'-phenylurea
Uses: herbicide to control woody plants and deep-rooted perennial weeds, often used in combination with chlorpropham to extend its weed control spectrum and range of crops.
CAS Registry No: 101-42-8
Molecular Formula: $C_9H_{12}N_2O$
Molecular Weight: 164.2
Melting Point (°C):
    128   (Karickhoff 1981)
    131-133   (Suntio et al. 1988)
    133-134   (Worthing 1991; Tomlin 1994)
    131   (Kühne et al. 1995)
Boiling Point (°C):
Density (g/cm³ at 20°C):
    1.13   (25°C, Agrochemicals Handbook 1987)
    1.08   (Worthing 1991; Tomlin 1994)
Molar Volume (cm³/mol):
    182.0   (calculated-LeBas method, Suntio et al. 1988)
    159.0   (modified LeBas method, Spurlock & Biggar 1994a)
Molecular Volume (Å³):
Total Surface Area, TSA (Å²):
Dissociation Constant pK_a:
Heat of Fusion, ΔH_{fus}, kcal/mol:
    5.80   (DSC method, Plato & Glasgow 1969)
Entropy of Fusion, ΔS_{fus}, cal/mol·K (e.u.):
Fugacity Ratio at 25°C (assuming ΔS_{fus} = 13.5 e.u.), F: 0.087

Water Solubility (g/m³ or mg/L at 25°C):
    2600   (Freed 1966; quoted, Suntio et al. 1988; Shiu et al. 1990)
    2400   (Günther et al. 1968; quoted, Suntio et al. 1988; Shiu et al. 1990)
    3850   (Martin & Worthing 1977; quoted, Kenaga 1980; Kenaga & Goring 1980; Briggs 1981; Karickhoff 1981; Isensee 1991)
    3850   (Khan 1980; Weber et al. 1980; Willis & McDowell 1982)
    3850   (Ashton & Crafts 1981)
    3700   (shake flask-HPLC, Ellgehausen et al. 1981; quoted, Shiu et al. 1990; Pinsuwan et al. 1995)
    3850   (Verschueren 1983; Worthing 1987, 1991; quoted, Suntio et al. 1988; Shiu et al. 1990; Tomlin 1994)
    3850   (Agrochemicals Handbook 1987)
    3850   (selected, Gerstl & Helling 1987; Evelyne et al. 1992)
    3000   (20°C, selected, Suntio et al. 1988)
    3900   (Spurlock 1992; quoted, Spurlock & Biggar 1994b)
    3850   (20-25°C, selected, Augustijn-Beckers et al. 1994; Hornsby et al. 1996)
    3850, 8421   (quoted, calculated, Patil 1994)
    4125, 1929   (quoted, calculated-group contribution fragmentation method, Kühne et al. 1995)
    3576, 1805   (quoted, predicted-AQUAFAC, Lee et al. 1996)

Vapor Pressure (Pa at 25°C):
    0.0213   (60°C, Khan 1980; Verschueren 1983; quoted, Suntio et al. 1988)
    0.0210   (60°C, Agrochemicals Handbook 1987)
    0.0050   (20°C, selected, Suntio et al. 1988)
    0.0210   (60°C, Worthing 1991; Tomlin 1994)
    0.0267   (20-25°C, selected, Augustijn-Beckers et al. 1994; Hornsby et al. 1996)

Henry's Law Constant (Pa·m^3/mol):
    0.00027   (20°C, calculated-P/C, Suntio et al. 1988)

Octanol/Water Partition Coefficient, log K_{OW}:
    1.00   (Leo et al. 1971; quoted, Kenaga & Goring 1980; McDuffie 1981; Suntio et al. 1988)
    0.72   (calculated-f const., Rekker 1977; quoted, Ellgehausen et al. 1981)
    0.96   (shake flask-UV, Briggs 1981; Karickhoff 1981; quoted, Sicbaldi & Finizio 1993; Aquasol Database 1994; Pinsuwan et al. 1995)
    0.88   (calculated-S, Ellgehausen et al. 1981; quoted, Sicbaldi & Finizio 1993)
    0.62   (HPLC-k', McDuffie 1981)
    0.70   (Elgar 1983; quoted, Suntio et al. 1988)
    0.20   (Jacob & Neumann 1983; quoted, Suntio et al. 1988)
    1.18   (RP-HPLC-k', Braumann et al. 1983)
    0.99   (selected, Gerstl & Helling 1987)
    0.50   (selected, Suntio et al. 1988)
    1.00   (Thor 1989; quoted, Connell & Markwell 1990)
    1.00   (Spurlock 1992; quoted, Spurlock & Biggar 1994b)
    1.18   (RP-HPLC, Sicbaldi & Finizio 1993)
    1.00, 0.48   (quoted, calculated, Patil 1994)
    0.977, 0.96 (selected, calculated-f const., Pinsuwan et al. 1995)
    0.98   (selected, Hansch et al. 1995)
    1.18, 0.98, 2.10 (RP-HPLC, ClogP, calculated-S, Finizio et al. 1997)

Bioconcentration Factor, log BCF:
    0.778   (calculated-S, Kenaga 1980; quoted, Isensee 1991)
    0.0   (calculated-K_{OC}, Kenaga 1980)
    1.34   (earthworms, Lord et al. 1980; quoted, Connell & Markwell 1990)
    0.699, 0.602 (cuticle/water: tomato, pepper, Evelyne et al. 1992)

Sorption Partition Coefficient, log K_{OC}:
    1.43   (soil, Hamaker & Thompson 1972; quoted, Kenaga 1980; Kenaga & Goring 1980; Karickhoff 1981)
    1.67   (soil, calculated-S as per Kenaga & Goring 1978, Kenaga 1980)
    0.88   (reported as log K_{OM}, Briggs 1981)
    0.61   (estimated-K_{OW}, Karickhoff 1981)
    1.80, 1.86 (estimated-S, Karickhoff 1981)
    0.72, 0.84 (estimated-S & M.P., Karickhoff 1981)
    1.39, 1.74 (quoted, calculated-χ, Gerstl & Helling 1987)
    1.62   (20-25°C, selected, Augustijn-Beckers et al. 1994; Hornsby et al. 1996)

Half-Lives in the Environment:
    Air:
    Surface water: aerobic half-life of ~10 d for 0.01 µg/mL to biodegrade in river water (Eichelberger & Lichtenberg 1971; quoted, Muir 1991); persistence of up to 4 weeks in river water (Eichelberger & Lichtenberg 1971).
    Groundwater:
    Sediment:
    Soil: half-life of 4.5 months at 15°C and 2.2 months at 30°C in soils (Freed & Haque 1973); persistence of 8 months in soil (Edwards 1973; quoted, Morrill et al. 1982); selected field half-life of 60 d (Augustijn-Beckers et al. 1994; Hornsby et al. 1996).
    Biota:

Environmental Fate Rate Constants or Half-Lives:
    Volatilization:
    Photolysis:
    Oxidation:
    Hydrolysis:
    Biodegradation: aerobic half-life of ~10 d for 0.01 µg/mL to biodegrade in river water (Eichelberger & Lichtenberg 1971; quoted, Muir 1991).
    Biotransformation:
    Bioconcentration, Uptake (k_1) and Elimination (k_2) Rate Constants:

Common Name: Fluchloralin
Synonym: BAS-392H, Basalin
Chemical Name: N-(2-chloroethyl)-2,6-dinitro-N-propyl-4-(trifluoromethyl)benzenamine; N-(2-chloroethyl)α,α,α-trifluoro-2,6-dinitro-N-propyl-p-toluidine
Uses: herbicide for pre-plant or pre-emergence control of annual grass and broadleaf weeds in cotton, groundnuts, jute, potatoes, rice soybeans, and sunflowers, etc.
CAS Registry No: 33245-39-5
Molecular Formula: $C_{12}H_{13}ClF_3N_3O_4$
Molecular Weight: 355.7
Melting Point (°C):
    42-43   (Khan 1980; Spencer 1982; Herbicide Handbook 1989; Worthing 1991; Tomlin 1994; Milne 1995)
Boiling Point (°C):
Density (g/cm^3 at 20°C):
Molar Volume (cm^3/mol):
    326.1   (calculated-LeBas method, this work)
Molecular Volume (Å^3):
Total Surface Area, TSA (Å^2):
Dissociation Constant pK_a:
Heat of Fusion, ΔH_{fus}, kcal/mol:
Entropy of Fusion, ΔS_{fus}, cal/mol·K (e.u.):
Fugacity Ratio at 25°C (assuming ΔS_{fus} = 13.5 e.u.), F: 0.664

Water Solubility (g/m^3 or mg/L at 25°C):
    1.00   (20°C, Weber 1972; Ashton & Crafts 1981; quoted, Muir 1991)
    1.00   (Edwards 1977; quoted, Nash 1988)
    <1.0   (Martin & Worthing 1977; Herbicide Handbook 1978,89; quoted, Kenaga 1980; Kenaga & Goring 1980; Shiu et al. 1990)
    0.70   (20°C, Spencer 1982)
    <1.0   (Worthing 1987, 1991; quoted, Shiu et al. 1990; Tomlin 1994)
    10   (Merck Index 1989; quoted, Milne 1995)
    0.90   (20-25°C, selected, Augustijn-Beckers et al. 1994; Hornsby et al. 1996)
    0.50   (selected, Lohninger 1994)

Vapor Pressure (Pa at 25°C):
    0.0033   (20°C, Weber 1972; Worthing 1987; quoted, Muir 1991)
    0.373   (20°C, Ashton & Crafts 1981)
    0.0008, 0.0033, 0.0133, 0.533   (20, 30, 40, 50°C, Spencer 1982)
    0.0035   (Herbicide Handbook 1983; quoted, Nash 1988)
    0.0037   (20°C, Herbicide Handbook 1989)
    0.0033   (30°C, Herbicide Handbook 1989)
    0.004   (20°C, Worthing 1991; Tomlin 1994)
    0.004   (20-25°C, selected, Augustijn-Beckers et al. 1994; Hornsby et al. 1996)

Henry's Law Constant (Pa·m^3/mol):
    1.174   (20°C, calculated-P/C, Muir 1991)
    1.343   (calculated-P/C, this work)

Octanol/Water Partition Coefficient, log K_{OW}:
    4.63   (selected, Magee 1991)
    4.33   (calculated-S as per Chiou et al. 1977 & Chiou 1981, this work)

Bioconcentration Factor, log BCF:
    >2.79   (calculated-S, Kenaga 1980)
    2.40   (calculated-K_{OC}, Kenaga 1980)

Sorption Partition Coefficient, log K_{OC}:
    3.56   (soil, Harvey 1974; quoted, Kenaga & Goring 1980; Bahnick & Doucette 1988)
    3.60   (soil, Kenaga 1980; quoted, Nash 1988)
    >3.64   (soil, calculated-S as per Kenaga & Goring 1978, Kenaga 1980)
    4.25   (calculated-χ, Bahnick & Doucette 1988)
    3.56   (reported as log K_{OM}, Magee 1991)
    3.58   (estimated as log K_{OM}, Magee 1991)
    3.48   (20-25°C, estimated, Augustijn-Beckers et al. 1994; Hornsby et al. 1996)
    3.80   (estimated-chemical structure, Lohninger 1994)

Half-Lives in the Environment:
    Air:
    Surface water:
    Groundwater:
    Sediment:
    Soil: half-life of 8 days for 0.5 µg/mL to biodegrade in soil at 20-42°C (Savage 1978; quoted, Muir 1991); measured dissipation rate of 0.099-0.13 d^{-1} (Savage & Jordan 1980; quoted, Nash 1988); half-lives of 28.7 weeks at 4°C, 10.5 weeks at 25°C for soils of field capcity moiture and 20.8 weeks at 4°C, 8.4 weeks at 25°C for flooded soil, Crowley silt loam; 29.3 weeks at 4°C, 10.5 weeks at 25°C for soil of field capacity moiture and 20.8 weeks at 4°C and 4.3 weeks at 25°C for flooded soil, Sharkey silty clay (Brewer et al. 1982); half-life of 3.6 weeks for 2.0 µg/ml to biodegrade in soil at 25°C (Brewer et al. 1982; quoted, Muir 1991); estimated dissipation rate of 0.29, and 0.120 d^{-1} (Nash 1988); estimated field half-life of 60 d (Augustijn-Beckers et al. 1994; Hornsby et al. 1996).
    Biota:

Environmental Fate Rate Constants or Half-Lives:
    Volatilization: estimated half-life of 1 d from 1 m depth of water (20°C, Muir 1991).
    Photolysis: half-life of 13 d for 84% of 5 µg/mL to degrade in distilled water under sunlight (Nilles & Zabik 1974; quoted, Cessna & Muir 1991); half-life of 8 h for 50% of 2000 µg/mL to degrade in methanol under sunlight (Plimmer & Klingebiel 1974; quoted, Cessna & Muir 1991).
    Oxidation:
    Hydrolysis:
    Biodegradation: half-life of 8 d for 0.5 µg/mL to biodegrade in soil at 20-42°C (Savage 1978; quoted, Muir 1991); half-life of 3.6 weeks for 2.0 µg/mL to biodegrade in soil at 25°C (Brewer et al. 1982; quoted, Muir 1991).
    Biotransformation:
    Bioconcentration, Uptake (k_1) and Elimination (k_2) Rate Constants:

Common Name: Fluometuron
Synonym: CIBA 2059, Cotoran, Cottonex, Higalcoton, Lanex, Meturon, Pakhtaran
Chemical Name: 1,1-dimethyl-3-(α,α,α-trifluoro-m-tolyl)urea; N,N-dimethyl-N'-[3-(trifluoromethyl)phenyl]urea
Uses: herbicide to control many annual broadleaf weeds in sugar cane and cotton.
CAS Registry No: 2164-17-2
Molecular Formula: $C_{10}H_{11}F_3N_2O$
Molecular Weight: 232.2
Melting Point (°C):
    163-164.5 (Khan 1980; Spencer 1982; Agrochemical Handbook 1987; Herbicide Handbook 1989; Worthing 1991; Montgomery 1993; Tomlin 1994; Milne 1995)
    163.8 (Patil 1994)
    163.0 (Kühne et al. 1995)
Boiling Point (°C):
Density (g/cm³ at 20°C):
    1.390 (Agrochemical Handbook 1987; Worthing 1991; Montgomery 1993; Tomlin 1994; Milne 1995)
Molar Volume (cm³/mol):
    229.7 (calculated-LeBas method, this work)
    167.1 (calculated-density, this work)
Molecular Volume (Å³):
Total Surface Area, TSA (Å²):
Dissociation Constant pK_a: -1.00 (Sangster 1993)
Heat of Fusion, ΔH_{fus}, kcal/mol:
    7.10 (DSC method, Plato 1972)
Entropy of Fusion, ΔS_{fus}, cal/mol·K (e.u.):
Fugacity Ratio at 25°C (assuming ΔS_{fus} = 13.5 e.u.), F: 0.042

Water Solubility (g/m³ or mg/L at 25°C):
    90 (Melnikov 1971; Spencer 1973, 1982; quoted, Shiu et al. 1990)
    90 (20°C, Martin & Worthing 1977; Herbicide Handbook 1978,89; quoted, Kenaga 1980; Kenaga & Goring 1980; Shiu et al. 1990; Isensee 1991)
    90 (Wauchope 1978; Khan 1980)
    90 (Weber et al. 1980; quoted, Willis & McDowell 1982)
    106 (shake flask-UV, Briggs 1981)
    90 (Herbicide Handbook 1983; quoted, Nash 1988)
    105 (20°C, Agrochemical Handbook 1987)
    8.05 (selected, Gerstl & Helling 1987)
    105 (20°C, Worthing 1987,91; quoted, Shiu et al. 1990)
    110 (20-25°C, selected, Wauchope et al. 1992; Hornsby et al. 1996)
    90 (quoted, Pait et al. 1992)
    80 (Montgomery 1993)
    110 (selected, Lohninger 1994)
    88, 13360 (quoted, calculated, Patil 1994)
    110 (Tomlin 1994)
    86, 207 (quoted, calculated-group contribution fragmentation method, Kühne et al. 1995)
    90, 283 (quoted, predicted-AQUAFAC, Lee et al. 1996)

Vapor Pressure (Pa at 25°C):
    6.70×10^{-5} (20-25°C, Weber et al. 1980; quoted, Willis & McDowell 1982)
    6.70×10^{-5} (Herbicide Handbook 1983; quoted, Nash 1988)
    6.70×10^{-5} (20°C, Agrochemical Handbook 1987)
    6.70×10^{-5} (20°C, Herbicide Handbook 1989)
    6.60×10^{-5} (20°C, Worthing 1991)
    1.25×10^{-4} (20-25°C, selected, Wauchope et al. 1992; Hornsby et al. 1996)
    6.70×10^{-5} (20°C, Montgomery 1993)
    1.25×10^{-4} (Tomlin 1994)

Henry's Law Constant (Pa·m^3/mol):
    <0.283   (20-25°C, calculated-P/C, Montgomery 1993)
    1.73×10^{-4} (calculated-P/C, this work)

Octanol/Water Partition Coefficient, log K_{OW}:
    1.34   (Briggs 1969; quoted, Kenaga & Goring 1980)
    2.42   (shake flask-UV, Briggs 1981; quoted, Sicbaldi & Finizio 1993)
    3.44   (selected, Dao et al. 1983)
    2.40   (selected, Gerstl & Helling 1987)
    2.23   (Worthing 1991; Tomlin 1994)
    2.23, 2.38   (Montgomery 1993)
    2.03   (RPHPLC-RT, Sicbaldi & Finizio 1993)
    1.88, 2.20, 2.42, 2.03; 2.20   (quoted values; selected, Sangster 1993)
    1.34, 0.33   (quoted, calculated, Patil 1994)
    2.23   (Milne 1995)
    2.42   (selected, Hansch et al. 1995)
    2.03, 2.39, 3.18   (RP-HPLC, ClogP, calculated-S, Finizio et al. 1997)

Bioconcentration Factor, log BCF:
    1.67   (calculated-S, Kenaga 1980; quoted, Isensee 1991; Pait et al. 1992)
    0.954   (calculated-K_{OC}, Kenaga 1980)

Sorption Partition Coefficient, log K_{OC}:
    2.24   (soil, Abernethy & Davidson 1971; Davidson & McDougal 1973; Savage & Wauchope 1974; Carringer et al. 1975; Wood & Davidson 1975; quoted, Kenaga 1980; Kenaga & Goring 1980)
    2.30   (soil, Kenaga 1980; quoted, Nash 1988)
    2.57   (soil, calculated-S as per Kenaga & Goring 1978, Kenaga 1980)
    1.58   (reported as log K_{OM}, Briggs 1981)
    2.03, 2.30   (quoted, calculated-χ, Gerstl & Helling 1987)
    2.00   (soil, 20-25°C, selected, Wauchope et al. 1992; Hornsby et al. 1996)
    1.46-2.08   (Montgomery 1993)
    2.00   (estimated-chemical structure, Lohninger 1994)
    1.49-2.07   (Tomlin 1994)

Half-Lives in the Environment:
: Air:
: Surface water: half-life of 730-1010 d at pH 5-9 and 20°C in aqueous solutions (Herbicide Handbook 1989).
: Groundwater:
: Sediment:
: Soil: measured dissipation rate of 0.023-0.043 d^{-1} (Horowitz & Herzlinger 1974: quoted, Nash 1988); estimated dissipation rate of 0.0012, and 0.011 d^{-1} (Nash 1988); persistence of 4 months in soil (Wauchope 1978); selected field half-life of 85 d (Wauchope et al. 1992; Hornsby et al. 1996); soil half-life of 30 d (Pait et al. 1992); median half-life of ca. 30 d in soil (Herbicide Handbook 1989; Tomlin 1994).
: Biota:

Environmental Fate Rate Constants or Half-Lives:
: Volatilization:
: Photolysis: half-lives in 10 ppm aqueous solutions: 11 ± 2 h under summer sunlight of 9.1 h/d exposure and 33 ± 16 h under spring sunlight of 3.7 h/d exposure (Burkhard et al. 1975).
: Oxidation:
: Hydrolysis: half-lives at 20°C: 1.6 yr at pH 1, 2.4 yr at pH 5, and 2.8 yr at pH 9 (Montgomery 1993).
: Biodegradation:
: Biotransformation:
: Bioconcentration, Uptake (k_1) and Elimination (k_2) Rate Constants:

Common Name: Fluorodifen
Synonym: Preforan, Soyex
Chemical Name: 4-nitrophenyl α,α,α-trifluoro-2-nitro-p-tolyl ether
Uses: herbicide.
CAS Registry No: 15457-05-3
Molecular Formula: $C_{13}H_7F_3N_2O_5$
Molecular Weight: 328.2
Melting Point (°C):
    94   (Spencer 1982; Milne 1995)
Boiling Point (°C):
Density (g/cm³ at 20°C):
Molar Volume (cm³/mol):
    282.6   (calculated-LeBas method, this work)
Molecular Volume (Å³):
Total Surface Area, TSA (Å²):
Dissociation Constant pK_a:
Heat of Fusion, ΔH_{fus}, kcal/mol:
Entropy of Fusion, ΔS_{fus}, cal/mol·K (e.u.):
Fugacity Ratio at 25°C (assuming ΔS_{fus} = 13.5 e.u.), F: 0.208

Water Solubility (g/m³ or mg/L at 25°C):
    2.0   (20°C, Spencer 1973, 1982; quoted, Shiu et al. 1990)
    <2.0   (Weber et al. 1980; quoted, Willis & McDowell 1982)
    2.0   (shake flask-HPLC, Ellgehausen et al. 1981; quoted, Shiu et al. 1990)
    2.0   (selected, Gerstl & Helling 1987; Pinsuwan et al. 1995)
    2.0   (20°C, Worthing 1987, 1991)

Vapor Pressure (Pa at 25°C):
    9.33×10^{-6} (20°C, Spencer 1982)

Henry's Law Constant (Pa·m³/mol):

Octanol/Water Partition Coefficient, log K_{OW}:
    3.30   (selected, Ellgehausen et al. 1980; Geyer et al. 1991)
    4.40   (20 ± 2°C, shake flask-UV, Briggs 1981)
    3.70   (calculated-S, Ellgehausen et al. 1981)
    3.65   (shake flask, Ellgehausen et al. 1981)
    4.57   (selected, Gerstl & Helling 1987)
    3.55   (Aquasol database 1994; quoted, Pinsuwan et al. 1995)
    3.55, 4.79   (selected, calculated, Pinsuwan et al. 1995)

Bioconcentration Factor, log BCF:
    2.019   (log BF-bioaccumulation factor, algae, Ellgehausen et al. 1980)
    2.386   (log BF-bioaccumulation factor, catfish, Ellgehausen et al. 1980)
    1.178   (log BF-bioaccumulation factor, daphnids, Ellgehausen et al. 1980; quoted, Geyer et al. 1991)

Sorption Partition Coefficient, log K_{OC}:
    3.13   (calculated-χ, Gerstl & Helling 1987)

Half-Lives in the Environment:
    Air:
    Surface water:
    Groundwater:
    Sediment:
    Soil:
    Biota:

Environmental Fate Rate Constants or Half-Lives:
    Volatilization:
    Photolysis:
    Oxidation:
    Hydrolysis:
    Biodegradation:
    Biotransformation:
    Bioconcentration, Uptake (k_1) and Elimination (k_2) Rate Constants:

Common Name: Fluridone
Synonym: Brake, EL-171, Fluridon, Pride, Sonar
Chemical Name: 1-methyl-3-phenyl-5-[3-(trifluoromethyl)phenyl]4(1H)-pyridinone; 1-methyl-3-phenyl-5-(α,α,α-trifluoro-m-tolyl)-4-pyridone
Uses: herbicide to control annual grass and broadleaf weeds and certain perennial species in cotton; also used to control aquatic weeds and plants in lakes, ponds, ditches, etc.
CAS Registry No: 59756-60-4
Molecular Formula: $C_{19}H_{14}F_3NO$
Molecular Weight: 328.2
Melting Point (°C):
    154-155   (Agrochemical Handbook 1987; Herbicide Handbook 1989; Worthing 1991; Tomlin 1994; Milne 1995)
    152   (Kühne et al. 1995)
Boiling Point (°C):
Density (g/cm³ at 20°C):
Molar Volume (cm³/mol):
    333.5   (calculated-LeBas method, this work)
Molecular Volume (Å³):
Total Surface Area, TSA (Å²):
Dissociation Constant:
    12.3   (pK_b, Wauchope et al. 1992)
Heat of Fusion, ΔH_{fus}, kcal/mol:
Entropy of Fusion, ΔS_{fus}, cal/mol·K (e.u.):
Fugacity Ratio at 25°C (assuming ΔS_{fus} = 13.5 e.u.), F: 0.052

Water Solubility (g/m³ or mg/L at 25°C):
    12.0   (20°C, Weber 1972; Worthing 1987; quoted, Muir 1991)
    12.0   (Kenaga 1980; quoted, Isensee 1991)
    12.0   (Herbicide Handbook 1983,89; quoted, Nash 1988)
    12.0   (Agrochemical Handbook 1987)
    12.0   (Worthing 1987, 1991; quoted, Shiu et al. 1990; Tomlin 1994)
    12.0   (Merck Index 1989; Milne 1995)
    12   (quoted, Reinert 1989)
    10.0   (20-25°C, selected, Wauchope et al. 1992; Hornsby et al. 1996)
    10.0   (selected, Lohninger 1994)
    11.9, 0.78 (quoted, calculated-group contribution fragmentation method, Kühne et al. 1995)

Vapor Pressure (Pa at 25°C):
    1.31×10^{-5} (20°C, Weber 1972; Worthing 1987; quoted, Muir 1991)
    1.00×10^{-5} (Herbicide Handbook 1983; quoted, Nash 1988)
    0.013   (Agrochemical Handbook 1987)
    1.33×10^{-5} (Herbicide Handbook 1989)
    0.013   (Worthing 1991)
    1.33×10^{-5} (20-25°C, selected, Wauchope et al. 1992; Hornsby et al. 1996)
    1.30×10^{-5} (Tomlin 1994)

Henry's Law Constant (Pa·m^3/mol):
    3.59×10^{-4} (20°C, calculated-P/C, Muir 1991)

Octanol/Water Partition Coefficient, log K_{ow}:
    1.87   (Reinert 1989)
    1.87   (Worthing 1991; Tomlin 1994; Milne 1995)
    2.98   (shake flask, Takahashi et al. 1993; quoted, Sangster 1993)

Bioconcentration Factor, log BCF:
    2.18   (calculated-S, Kenaga 1980; quoted, Isensee 1991)
    0.778   (measured, West et al. 1983; quoted, Isensee 1991)

Sorption Partition Coefficient, log K_{OC}:
    1.60   (soil, Kenaga 1980; quoted, Nash 1988)
    2.97-3.39 (pond sediment, Muir et al. 1980)
    3.36, 2.95 (lake and river sediment, Muir et al. 1980)
    2.94   (quoted, Reinert 1989)
    2.90, 3.81, 3.03 (Norfolk sand pH 6.0, Norfold with montmorillonite pH 5.9, Norfolk sand with added organic matter pH 5.3, Reinert 1989)
    3.43, 2.57, 2.43 (California soil at pH 6, 7, 7.3, Reinert 1989)
    3.00   (20-25°C, selected, Wauchope et al. 1992; Hornsby et al. 1996)
    3.00   (selected, Lohninger 1994)

Half-Lives in the Environment:
    Air: 0.359-3.20 h, based on estimated rate constant for reaction with hydroxyl radicals (Atkinson 1987; quoted, Howard et al. 1991) and ozone (Atkinson & Carter 1984; quoted, Howard et al. 1991).
    Surface water: half-life of ca. 21 d in water (Agrochemicals Handbook 1987); 288-864 h, based on estimated photolysis half-life in water (Howard et al. 1991); anaerobic half-life of 9 months and aerobic half-life of ca. 20 d (Tomlin 1994).
    Groundwater: 2112-9216 h, based on estimated unacclimated aqueous aerobic biodegradation half-life (Howard et al. 1991).
    Sediment: half-lives of 12 months for 5 µg/mL to biodegrade in static sediment and water, and ~9 months in aerobic and anaerobic sediment and water all at 25°C (Muir & Grift 1982; quoted, Muir 1991).
    Soil: measured dissipation rate of 0.0041 d^{-1} (Banks et al. 1979; quoted, Nash 1988) with estimated half-life of 44-192 d (Banks et al. 1979; quoted, Howard et al. 1991); estimated dissipation rate of 0.0067 and 0.025 d^{-1} (Nash 1988); selected field half-life of 21 d (Wauchope et al. 1992; Hornsby et al. 1996); half-life in the hydrosoil ca. 90 d (Tomlin 1994).
    Biota:

Environmental Fate Rate Constants or Half-Lives:
- Volatilization: estimated half-life of 10,000 d from 1 m depth of water at 20°C (Muir 1991).
- Photolysis: half-life of ~23 h to degrade in distilled water under >290 nm light (West et al. 1979; quoted, Cessna & Muir 1991); half-life of ~6 h for 5 μg/mL to degrade in nonsterile pond water under sunlight (Muir & Grift 1982; quoted, Cessna & Muir 1991); half-life of 27 d for 85% of 10 μg/mL to degrade in distilled water and for 85% of 10 μg/mL to degrade in lake water at pH 8.4 both under sunlight (Saunders & Mosier 1983; quoted, Cessna & Muir 1991; Howard et al. 1991); resistance to decomposition by UV light with a half-life of 23 h in deionized water (Herbicide Handbook 1989).
- Oxidation: photooxidation half-life of 0.359-3.20 h, based on estimated rate constant for reaction with hydroxyl radicals (Atkinson 1987; quoted, Howard et al. 1991) and ozone (Atkinson & Carter 1984; quoted, Howard et al. 1991).
- Hydrolysis: half-life of >113 d for 1 μg/mL to hydrolyze in pond water at 4°C (Ghassemi et al. 1981; quoted, Muir 1991); half-life of 23 h in water (Tomlin 1994).
- Biodegradation: aqueous aerobic half-life of 44-192 d, based on soil die-away test data and field study soil persistence (Banks et al. 1979; quoted, Howard et al. 1991); half-lives of 12 months for 5 μg/mL to biodegrade in static sediment and water, and ~9 months in aerobic and anaerobic sediment and water all at 25°C (Muir & Grift 1982; quoted, Muir 1991); aqueous anaerobic half-life of 176 d to 2.1 yr, based on estimated unacclimated aqueous aerobic biodegradation half-life (Howard et al. 1991); microbial degradation half-life of >343 d at pH 7.3 with 2.6% organic matter in a silt loam soil (Tomlin 1994).

Biotransformation:

Bioconcentration, Uptake (k_1) and Elimination (k_2) Rate Constants:

Common Name: Glyphosate
Synonym: Mon-0573, 0468, 2139; Polado, Roundup
Chemical Name: N-(phosphoromethyl)glycine
Uses: nonselective, post-emergent, broad spectrum herbicide to control annual and perennial grasses, sedges, broadleaf, and emerged aquatic weeds; also used to control insects on fruit trees.
CAS Registry No: 1071-83-6
Molecular Formula: $C_3H_8NO_5P$
Molecular Weight: 169.1
Melting Point (°C):
    200   (dec., Herbicide Handbook 1989; Worthing 1991; Tomlin 1994)
    230   (dec., Montgomery 1993; Milne 1995)
    200   (Milne 1995)
Boiling Point (°C):
Density (g/cm³ at 20°C):
    1.74   (Herbicide Handbook 1989; Montgomery 1993)
Molar Volume (cm³/mol):
Molecular Volume (Å³):
Total Surface Area, TSA (Å²):
Dissociation Constant pK_a:
    5.70   (Worthing 1991)
    2.60, 5.90, 10.40   (pK_1, pK_2, pK_3, Yao & Haag 1991)
    2.32, 5.86, 10.86   (pK_1, pK_2, pK_3, Montgomery 1993; Hornsby et al. 1996)
Heat of Fusion, ΔH_{fus}, kcal/mol:
Entropy of Fusion, ΔS_{fus}, cal/mol·K (e.u.):
Fugacity Ratio at 25°C (assuming ΔS_{fus} = 13.5 e.u.), F: 0.019

Water Solubility (g/m³ or mg/L at 25°C):
    10000   (Spencer 1973, 1982; quoted, Shiu et al. 1990)
    12000   (Martin & Worthing 1977; Herbicide Handbook 1978; quoted, Kenaga 1980; Kenaga & Goring 1980; Isensee 1991)
    12000   (Kenaga 1980; quoted, Shiu et al. 1990)
    12000   (Ashton & Crafts 1981)
    12000   (Worthing 1987, 1991; quoted, Shiu et al. 1990; Tomlin 1994)
    15700   (Herbicide Handbook 1989; quoted, Shiu et al. 1990)
    12000   (Merck Index 1989; Milne 1995)
    12000   (selected, USDA 1989; quoted, Neary et al. 1993)
    12000   (quoted, Reinert 1989)
    12000   (Montgomery 1993; quoted, Majewski & Capel 1995)
    12000   (selected, Halfon et al. 1996)

Vapor Pressure (Pa at 25°C):
    2.59×10^{-5} (45°C, Herbicide Handbook 1989)
    4.00×10^{-5} (Worthing 1991)
    0.001   (Montgomery 1993; quoted, Majewski & Capel 1995)
    negligible   (Tomlin 1994)
    0.0   (selected, Halfon et al. 1996)

Henry's Law Constant (Pa·m³/mol):
   1.41×10^{-5} (calculated-P/C, Montgomery 1993; quoted, Majewski & Capel 1995)

Octanol/Water Partition Coefficient, log K_{OW}:
   −3.25   (Reinert 1989)
   −4.59   (Worthing 1991)
   −1.60   (Montgomery 1993)
   −4.10, −1.70   (pH 2.5, pH 5.3, quoted, Sangster 1993)
   −1.70   (pH 5.3, selected, Hansch et al. 1995)
   0.94, 1.67   (RP-HPLC, calculated-S, Finizio et al. 1997)

Bioconcentration Factor, log BCF:
   0.477   (calculated-S, Kenaga 1980; quoted, Isensee 1991)
   2.26   (calculated-K_{OC}, Kenaga 1980)

Sorption Partition Coefficient, log K_{OC}:
   3.42   (soil, Sprankle et al. 1975; Hance 1976; Nomura & Hilton 1977; quoted, Kenaga 1980; Kenaga & Goring 1980)
   1.40   (soil, calculated-S as per Kenaga & Goring 1978, Kenaga 1980)
   1.22   (selected, USDA 1989; quoted, Neary et al. 1993)
   −0.43   (Reinert 1989)
   3.69, 3.53, 3.42 (3 agricultural soils: Houston clay loam at pH 7.5, Muskingum silt loam at pH 5.8, Sassafras sandy loam at pH 5.6, Reinert 1989)
   4.38   (organic carbon, Wauchope et al. 1991; quoted, Dowd et al. 1993)
   3.43-3.69 (Montgomery 1993)

Half-Lives in the Environment:
   Air:
   Surface water: half-life of >9 weeks for 2 µg/mL to biodegrade in polluted lake water (Rueppel et al. 1977; quoted, Muir 1991); half-lives of 70 days in pond water at pH 7.2, 63 d in swamp water at pH 6.3 and 49 d in Sphagnum bog water at pH 4.2 (Ghossemi et al. 1981; quoted, Muir 1991).
   Groundwater:
   Sediment:
   Soil: half-life of <28 d for 10 µg/mL to biodegrade in soil-water suspension (Rueppel et al. 1977; quoted, Muir 1991); estimated first-order half-life of 7 d from biodegradation rate constant of 0.1 d^{-1} from soil incubation die-away studies (Rao & davidson 1980; quoted, Scow 1982); moderately persistent in soil with half-life of 20-100 d (Willis & McDowell 1982); average half-life is less than 60 d (Hartley & Kidd 1987; Herbicide Handbook 1989; quoted, Montgomery 1993); selected half-life of 47 d (Wauchope et al. 1991; quoted, Dowd et al. 1993; Halfon et al. 1996).
   Biota: average half-life of 60 d in the forest (USDA 1989; quoted, Neary et al. 1993).

Environmental Fate Rate Constants or Half-Lives:
: Volatilization:
: Photolysis: half-life of 48 h for 0% of 168 µg/mL to degrade in distilled water under >290 nm light (Rueppel et al. 1977; quoted, Cessna & Muir 1991); half-life of 9 weeks for >90% of 2 µg/mL to degrade in distilled water under sunlight (Lund-Høie & Friestad 1986; quoted, Cessna & Muir 1991); half-lives of 4.0 d and 3-4 weeks for aqueous solutions of 1.0 and 2000 ppm under indoor UV light (Lund-Høie & Friestad 1986; quoted, Montgomery 1993).
: Oxidation: rate constant of 7.3×10^8 M^{-1} s^{-1} for the reaction (photo-Fenton with reference to glycolic acid) with hydroxyl radicals in aqueous solutions at pH 3.8 and at $(24 \pm 1)°C$ (Buxton et al. 1988; quoted, Faust & Hoigné 1990; Haag & Yao 1992); rate constant of $(1.8 \pm 0.5) \times 10^8$ M^{-1} s^{-1} for the reaction (photo-Fenton with reference to glycolic acid) with hydroxyl radicals in aqueous solutions at pH 3.8 and at $(24 \pm 1)°C$ (Haag & Yao 1992).
: Hydrolysis: half-life of 7 d for 10 µg/mL to hydrolyze in sterile water + soil (Rueppel et al. 1977; quoted, Muir 1991); half-life of 32 d for 25 and 250 µg/mL to hydrolyze in sterile distilled water at pH 3, 6 and 9 in the dark at 5 and 35°C (Ghassemi et al. 1981; quoted, Muir 1991).
: Biodegradation: half-life of <28 d for 10 µg/mL to biodegrade in soil-water suspension (Rueppel et al. 1977; quoted, Muir 1991); half-life of >9 weeks for 2 µg/mL to biodegrade in polluted lake water (Rueppel et al. 1977; quoted, Muir 1991); rate constant of 0.1 d^{-1} from soil incubation die-away studies (Rao & Davidson 1980; quoted, Scow 1982); half-lives of 70 d in pond water at pH 7.2, 63 d in swamp water at pH 6.3 and 49 d in Sphagnum bog water at pH 4.2 (Ghassemi et al. 1981; quoted, Muir 1991).
: Biotransformation:
: Bioconcentration, Uptake (k_1) and Elimination (k_2) Rate Constants:

Common Name: Isopropalin
Synonym: EL 179, Isopropaline, Isopropalin solution, Paarlan
Chemical Name: 4-ispopropyl-2,6-dinitro-N,N-dipropylaniline; 4-(1-methylethyl)-2,6-dinitro-N,N-dipropylbenzenamine; 2,6-dinitro-N,N-dipropylcumidine
Uses: herbicide used pre-planting and incorporated with soil preparation to control broadleaf weeds and grasses in transplanted tobacco, and in direct-seeded tomatoes and capsicums.
CAS Registry No: 33820-53-0
Molecular Formula: $C_{15}H_{23}N_3O_4$
Molecular Weight: 309.4
Melting Point (°C): liquid
Boiling Point (°C):
Density (g/cm³ at 20°C):
Molar Volume (cm³/mol):
    361.3   (calculated-LeBas method, this work)
Molecular Volume (Å³):
Total Surface Area, TSA (Å²):
Dissociation Constant pK_a:
Heat of Fusion, ΔH_{fus}, kcal/mol:
Entropy of Fusion, ΔS_{fus}, cal/mol·K (e.u.):
Fugacity Ratio at 25°C (assuming ΔS_{fus} = 13.5 e.u.), F: 1.0

Water Solubility (g/m³ or mg/L at 25°C):
    0.11   (Martin & Worthing 1977; Herbicide Handbook 1978; quoted, Kenaga 1980; Kenaga & Goring 1980; Isensee 1991)
    1.10   (Ashton & Crafts 1981)
    0.10   (Spencer 1982)
    0.10   (Agrochemicals Handbook 1987)
    0.10   (Worthing 1987, 1991; quoted, Shiu et al. 1990)
    0.08   (Herbicide Handbook 1989; quoted, Shiu et al. 1990)
    0.10   (Merck Index 1989; quoted, Milne 1955)
    0.10   (20-25°C, selected, Wauchope et al. 1992; Hornsby et al. 1996)
    0.10   (selected, Lohninger 1994)
    0.10, 0.02   (quoted, predicted-AQUAFAC, Lee et al. 1996)

Vapor Pressure (Pa at 25°C):
    0.0019   (30°C, Ashton & Crafts 1981)
    0.0019   (30°C, Agrochemicals Handbook 1987)
    0.0040   (25.6°C, Herbicide Handbook 1989)
    0.0012   (20-25°C, selected, Wauchope et al. 1992; Hornsby et al. 1996)

Henry's Law Constant (Pa·m³/mol):
    5.34   (calculated-P/C, this work)

Octanol/Water Partition Coefficient, log K_{OW}:
    4.71   (calculated-log BCF as per Kenaga 1980, this work)

Bioconcentration Factor, log BCF:
- 3.50 (calculated-S, Kenaga 1980; quoted, Isensee 1991)
- 3.88 (calculated-K_{OC}, Kenaga 1980)

Sorption Partition Coefficient, log K_{OC}:
- 4.88 (soil, Harvey 1974; quoted, Kenaga 1980; Kenaga & Goring 1980)
- 4.17 (soil, calculated-S as per Kenaga & Goring 1978, Kenaga 1980)
- 4.17-4.88 (quoted values, Bottoni & Funari 1992)
- 4.00 (20-25°C, selected, Wauchope et al. 1992; Hornsby et al. 1996)
- 4.00 (selected, Lohninger 1994)

Half-Lives in the Environment:
Air: 0.743-74.3 h, based on estimated rate constant for the vapor-phase reaction with hydroxyl radicals in air (Atkinson 1987; quoted, Howard et al. 1991).
Surface water: 288-864 h, based on observed photolysis on soil TLC plates under summer sunlight (Helling 1976; quoted, Howard et al. 1991) and adjusted for relative winter sunlight intensity (Lyman et al. 1982; quoted, Howard et al. 1991).
Groundwater: 96-5040 h, based on estimated unacclimated aqueous aerobic and anaerobic degradation half-lives (Howard et al. 1991).
Sediment:
Soil: 408-2520 h, based on aerobic soil die-away test data for one soil at 15°C and 30°C (Gingerich & Zimdahl 1976; quoted, Howard et al. 1991); selected field half-life of 100 d (Wauchope et al. 1992; Hornsby et al. 1996).

Environmental Fate Rate Constants or Half-Lives:
Volatilization:
Photolysis: atmosphere photolysis half-life of 288-864 h, based on observed photolysis on soil TLC plates under summer sunlight (Helling 1976; quoted, Howard et al. 1991) and adjusted for relative winter sunlight intensity (Lyman et al. 1982; quoted, Howard et al. 1991); aqueous photolysis half-life of 288-864 h, based on observed photolysis on soil TLC plates under summer sunlight (Helling 1976; quoted, Howard et al. 1991) and adjusted for relative winter sunlight intensity (Lyman et al. 1982; quoted, Howard et al. 1991).
Oxidation: photooxidation half-life of 0.743-74.3 h in air, based on estimated rate constant for the vapor-phase reaction with hydroxyl radicals in air (Atkinson 1987; quoted, Howard et al. 1991).
Hydrolysis:
Biodegradation: aqueous aerobic half-life of 408-2520 h, based on aerobic soil die-away test data for one soil at 15°C and 30°C (Gingerich & Zimdahl 1976; quoted, Howard et al. 1991); aqueous anaerobic half-life of 96-360 h, based on anaerobic soil die-away test which tested one soil (Gingerich & Zimdahl 1976; quoted, Howard et al. 1991).
Biotransformation:
Bioconcentration, Uptake (k_1) and Elimination (k_2) Rate Constants:

Common Name: Isoproturon
Synonym: Alon, Arelon, CGA 18731, Gramion, Graminon, Hoe 16410, Hytane, IP 50, IP flo, Tolkan
Chemical Name: 3-(4-isopropylphenyl)-1,1-dimethylurea; 3-p-cumenyl-l-1-dimethylurea
Uses: herbicide used for pre- and post-emergence control of annual grasses and broadleaf weeds in spring and winter wheat (except durum wheat), spring and winter barley, winter rye, and triticale.
CAS Registry No: 34123-59-6
Molecular Formula: $C_{12}H_{18}N_2O$
Molecular Weight: 206.3
Melting Point (°C):
    158-159 (Spencer 1982)
    155-156 (Worthing 1991)
    158 (Tomlin 1994)
Boiling Point (°C):
Density (g/cm³ at 20°C):
    1.16 (Agrochemicals Handbook 1987; Tomlin 1994)
Molar Volume (cm³/mol):
    259.1 (calculated-LeBas method, this work)
Molecular Volume (Å³):
Total Surface Area, TSA (Å²):
Dissociation Constant pK_a:
Heat of Fusion, ΔH_{fus}, kcal/mol:
Entropy of Fusion, ΔS_{fus}, cal/mol·K (e.u.):
Fugacity Ratio at 25°C (assuming ΔS_{fus} = 13.5 e.u.), F: 0.051

Water Solubility (g/m³ or mg/L at 25°C):
    60 (Martin & Worthing 1977; quoted, Kenaga 1980)
    70 (20°C, Spencer 1982)
    72 (20°C, Agrochemicals Handbook 1987)
    55 (Worthing 1987, 1991; quoted, Shiu et al. 1990; Evelyne et al. 1992)
    55.9 (Chaumat et al. 1991)
    65 (22°C, Tomlin 1994; quoted, Otto et al. 1997)
    65 (20°C, selected, Traub-Eberhard et al. 1994)
    61, 25 (quoted, predicted-AQUAFAC, Lee et al. 1996)

Vapor Pressure (Pa at 25°C):
    3.3×10^{-6} (20°C, Spencer 1982; Agrochemicals Handbook 1987)
    3.3×10^{-6} (20°C, Worthing 1991)
    3.3×10^{-6}, 3.15×10^{-2}, 0.172 (20, 77, 150°C, Tomlin 1994)
    3.3×10^{-6} (quoted, Otto et al. 1997)

Henry's Law Constant (Pa·m³/mol):
    1.05×10^{-5} (calculated-P/C, Otto et al. 1997)
    1.24×10^{-5} (calculated-P/C, this work)

Octanol/Water Partition Coefficient, log K_{OW}:
    2.25 (Worthing 1991)
    2.537 (calculated, Evelyne et al. 1992)
    2.50 (pH 7, 22°C, Tomlin 1994; quoted, Otto et al. 1997)

Bioconcentration Factor, log BCF:
    1.79   (calculated-S, Kenaga 1980)
    1.76, 1.82   (cuticle/water: tomato, pepper; Chaumat et al. 1991)
    1.71, 1.90   (cuticle/water: box tree, pear; Chaumat et al. 1991)
    1.52, 1.20   (cuticle/water: ivy, vanilla; Chaumat et al. 1991)
    1.76, 1.82   (cuticle/water: tomato, pepper; Evelyne et al. 1992)

Sorption Partition Coefficient, log K_{OC}:
    2.66   (soil, calculated-S, Kenaga 1980; quoted, Bottoni & Funari 1992)
    1.86   (soil, HPLC-screening method, Kördel et al. 1993)
    2.11   (soil, quoted from Kördel et al. 1993, Traub-Eberhard et al. 1994)

Half-Lives in the Environment:
    Air: 0.743-74.3 h, based on estimated rate constant for the vapor-phase reaction with hydroxyl radicals in air (Atkinson 1987; quoted, Howard et al. 1991).
    Surface water: 288-864 h, based on observed photolysis on soil TLC plates under summer sunlight (Helling 1976; quoted, Howard et al. 1991) and adjusted for relative winter sunlight intensity (Lyman et al. 1982; quoted, Howard et al. 1991).
    Groundwater: 96-5040 h, based on estimated unacclimated aqueous aerobic and anaerobic degradation half-lives (Howard et al. 1991).
    Sediment:
    Soil: 408-2520 h, based on aerobic soil die-away test data for one soil at 15°C and 30°C (Gingerich & Zimdahl 1976; quoted, Howard et al. 1991); 15-21 d at 20°C in soil (Traub-Eberhard et al. 1994); estimated half-lives of 14.6 d under conventional tillage, 7.99 d under ridge tillage and 12.17 d with no tillage (Otto et al. 1997).

Environmental Fate Rate Constants or Half-Lives:
    Volatilization:
    Photolysis: atmosphere photolysis half-life of 288-864 h, based on observed photolysis on soil TLC plates under summer sunlight (Helling 1976; quoted, Howard et al. 1991) and adjusted for relative winter sunlight intensity (Lyman et al. 1982; quoted, Howard et al. 1991); aqueous photolysis half-life of 288-864 h, based on observed photolysis on soil TLC plates under summer sunlight (Helling 1976; quoted, Howard et al. 1991) and adjusted for relative winter sunlight intensity (Lyman et al. 1982; quoted, Howard et al. 1991); half-life of 1.5 h for 215 μg/mL to degrade in distilled water under 254 nm light (Kulshrestha & Mukerjee 1986; quoted, Cessna & Muir 1991).
    Oxidation: photooxidation half-life of 0.743-74.3 h in air, based on estimated rate constant for the vapor-phase reaction with hydroxyl radicals in air (Atkinson 1987; quoted, Howard et al. 1991).
    Hydrolysis:
    Biodegradation: aqueous aerobic half-life of 408-2520 h, based on aerobic soil die-away test data for one soil at 15°C and 30°C (Gingerich & Zimdahl 1976; quoted, Howard et al. 1991); aqueous anaerobic half-life of 96-360 h, based on anaerobic soil die-away test which tested one soil (Gingerich & Zimdahl 1976; quoted, Howard et al. 1991).
    Bioconcentration, Uptake (k_1) and Elimination (k_2) Rate Constants:

Common Name: Linuron

Synonym: Afalon, Cephalon, Garnitan, Herbicide 326, Hoe 2810, Linex 4L, Linorox, Linurex, Lorox, Methoxydiuron, Premalin, Scarclex, Sinuron

Chemical Name: 3-(3,4-dichlorophenyl)-1-methoxy-1-methylurea; N'-(3,4-dichlorophenyl)-N-methoxy-N-methylurea

Uses: selective pre-emergence and post-emergence herbicide used on a wide variety of food crops to control many annual broadleaf and grass weeds.

CAS Registry No: 330-55-2

Molecular Formula: $C_9H_{10}Cl_2N_2O_2$

Molecular Weight: 249.11

Melting Point (°C):
- 93-94 (Khan 1980; Spencer 1982; Suntio et al. 1988; Herbicide Handbook 1989; Montgomery 1993; Milne 1995)
- 93.0 (Karickhoff 1981; Kühne et al. 1995)
- 93-95 (Worthing 1991)
- 93.5 (Patil 1994)

Boiling Point (°C):

Density (g/cm³ at 20°C):

Molar Volume (cm³/mol):
- 232.9 (calculated-LeBas method, Suntio et al. 1988)

Molecular Volume (Å³):

Total Surface Area, TSA (Å²):

Dissociation Constant pK_a:

Heat of Fusion, ΔH_{fus}, kcal/mol:
- 6.85 (DSC method, Plato & Glasgow 1969)

Entropy of Fusion, ΔS_{fus}, cal/mol·K (e.u.):

Fugacity Ratio at 25°C (assuming ΔS_{fus} = 13.5 e.u.), F: 0.2078

Water Solubility (g/m³ or mg/L at 25°C):
- 75 (Woodford & Evans 1963; Bailey & White 1965; quoted, Shiu et al. 1990)
- 75 (Melnikov 1971; Spencer 1973, 1982; quoted, Shiu et al. 1990)
- 75 (Martin & Worthing 1977; Worthing 1983, 1987; quoted, Kenaga 1980)
- 75 (Hartley & Graham-Bryce 1980; Beste & Humburg 1983; quoted, Taylor & Glotfelty 1988)
- 75 (Herbicide Handbook 1978, 1983, 1989; quoted, Kenaga 1980; Kenaga & Goring 1980; Karickhoff 1981; Isensee 1991; Pait et al. 1992)
- 75 (Wauchope 1978; Khan 1980; quoted, Suntio et al. 1988; Shiu et al. 1990)
- 75 (Weber et al. 1980; quoted, Willis & McDowell 1982)
- 75 (Ashton & Crafts 1981; selected, Gerstl & Helling 1987)
- 75 (Briggs 1981)
- 81 (Agrochemicals Handbook 1987)
- 65 (20°C, selected, Suntio et al. 1988; quoted, Majewski & Capel 1995)
- 81 (24°C, Worthing 1991; quoted, Di Guardo et al. 1994)
- 75 (20-25°C, selected, Wauchope et al. 1992; Hornsby et al. 1996; quoted, Lohninger 1994; Halfon et al. 1996)
- 75-81 (Montgomery 1993)
- 75, 207 (quoted, calculated, Patil 1994)
- 75.2 (quoted, Kühne et al. 1995)
- 168 (calculated-group contribution fragmentation method, Kühne et al. 1995)
- 81 (Milne 1995)

Vapor Pressure (Pa at 25°C):
    0.00147 (20°C, Quellette & King 1977; quoted, Suntio et al. 1988)
    0.0012 (20°C, Hartley & Graham-Bryce 1980; quoted, Taylor & Glotfelty; Taylor & Spencer 1990)
    0.002 (24°C, Khan 1980; quoted, Suntio et al. 1988)
    0.002 (20-25°C, Weber et al. 1980; quoted, Willis & McDowell 1982)
    0.002 (24°C, Agrochemicals Handbook 1987)
    0.0014 20°C, selected, Suntio et al. 1988; quoted, Majewski & Capel 1995)
    0.0011 (20°C, selected, Taylor & Spencer 1990)
    0.002 (24°C, Worthing 1991; quoted, Di Guardo et al. 1994)
    0.0023 (20-25°C, selected, Wauchope et al. 1992; Hornsby et al. 1996)
    0.002 (20°C, Montgomery 1993)
    0.0027 (selected, Halfon et al. 1996)

Henry's Law Constant (Pa·m^3/mol):
    0.0054 (20°C, calculated-P/C, Suntio et al. 1988; quoted, Majewski & Capel 1995)
    0.004 (Taylor & Glotfelty 1988)
    0.0062 (20-25°C, calculated-P/C, Montgomery 1993)
    0.00465 (calculated-P/C, this work)

Octanol/Water Partition Coefficient, log K_{ow}:
    2.19 (Briggs 1969; quoted, Kenaga & Goring 1980; Karickhoff 1981; Suntio et al. 1988)
    2.76 (shake flask-UV, Briggs 1981; quoted, Braumann et al. 1983; Madhun et al. 1986; Sicbaldi & Finizio 1993)
    2.19 (selected, Dao et al. 1983)
    2.48 (selected, Gerstl & Helling 1987)
    2.20 (selected, Suntio et al. 1988)
    3.00 (Worthing 1991; quoted, Di Guardo et al. 1994)
    2.19, 3.00 (Montgomery 1993)
    2.75 (RP-HPLC-RT, Sicbaldi & Finizio 1993)
    2,76, 2.75, 3.11, 3.20; 3.20 (quoted values; selected, Sangster 1993)
    2.19, 2.19 (quoted, calculated, Patil 1994)
    3.00 (Milne 1995)
    3.20 (selected, Hansch et al. 1995)
    2.75, 3.0, 3.26 (RP-HPLC, ClogP, calculated-S, Finizio et al. 1997)

Bioconcentration Factor, log BCF:
    1.73 (calculated-S, Kenaga 1980; quoted, Isensee 1991)
    1.68 (calculated-K_{OC}, Kenaga 1980)
    1.73 (calculated, Pait et al. 1992)

Sorption Partition Coefficient, log K_{OC}:
    2.91 (soil, Hamaker & Thompson 1972; quoted, Kenaga 1980; Kenaga & Goring 1980; Karickhoff 1981)
    2.61 (soil, calculated-S as per Kenaga & Goring 1978, Kenaga 1980)

2.93   (av. soils/sediments, Rao & Davidson 1980; quoted, Karickhoff 1981; Lyman 1982)
2.19   (reported as log K_{OM}, Briggs 1981)
2.93, 2.80, 1.80 (estimated-S, S & M.P., K_{OW}, Karickhoff 1981)
3.83   (Means & Wijayaratne 1982; quoted, Muir 1991)
2.96, 2.90 (soil, quoted, Madhun et al. 1986)
2.99, 2.58; 2.62, 2.80 (estimated-K_{OW}, S, Madhun et al. 1986)
2.76, 2.64 (quoted, calculated-χ, Gerstl & Helling 1987)
2.94   (screening model calculations, Jury et al. 1987b)
2.61-2.91, 2.83, 2.93 (quoted values, Bottoni & Funari 1992)
2.60   (soil, 20-25°C, selected, Wauchope et al. 1992; quoted, Richards & Baker 1993; Lohninger 1994; Hornsby et al. 1996)
2.70-2.78 (Montgomery 1993)
2.59   (soil, HPLC-ring test, Kördel et al. 1995a)
2.59   (soil, HPLC-screening method, Kördel et al. 1993,95b)

Half-Lives in the Environment:
   Air: 0.49-4.90 h, based on an estimated rate constant for the vapor-phase reaction with hydroxyl radicals in air (Atkinson 1987; quoted, Howard et al. 1991).
   Surface water: 672-4272 h, based on estimated unacclimated aqueous aerobic biodegradation half-life (Howard et al. 1991).
   Groundwater: 1344-8544 h, based on estimated unacclimated aqueous aerobic biodegradation half-life (Howard et al. 1991).
   Sediment: degradation half-life of 12 d in estuarine sediment (12°/∞) system (Cunningham et al. 1981; quoted, Means et al. 1983); degradation half-life of 6 d in estuarine sediment (18°/∞) system (Means et al. 1983).
   Soil: estimated persistence of 4 months (Kearney et al. 1969; Edwards 1973; quoted, Morrill et al. 1982; Jury et al. 1987a); 672-4272 h, based on soil die-away test data (Walker 1978; Walker & Zimdahl 1981; quoted, Howard et al. 1991); persistence of 4 months (Wauchope 1978); correlated half-lives: 57 d at pH 5.1-5.8, 22 d at pH 6.3-7.0 and 19 d at pH 7.7-8.2 (Boddington Barn soil, Hance 1979) and 67 d at pH 4.6-5.2, 53 days at pH 5.3-6.1, and about 20 d at pH 6.3-8.0 (Triangle soil, Hance 1979); estimated first-order half-life of 72 d from biodegradation rate constant of 0.0096 d^{-1} by soil incubation die-away studies (Rao & Davidson 1980; quoted, Scow 1982); decomposition half-lives: 11 d in fresh soil and 12 d in air dried soil both in polyethylene bags, 49 d in undisturbed cores and 40 d in perfusion (Hance & Haynes 1981); moderately persistent in soil with half-life of 20-100 d (Willis & McDowell 1982); half-life under field conditions is 2 to 5 months (Hartley & Kidd 1987; Herbicide Handbook 1989; quoted, Montgomery 1993); half-life of 75 d from screening model calculations (Jury 1987b); degradation rate constant of $(3.48 \pm 0.156) \times 10^{-2}$ d^{-1} with half-life of 19.9 d in control soil and $(23.2 \pm 2.07) \times 10^{-2}$ d^{-1} with half-life of 2.99 d in pretreated soil in the field; $(3.73 \pm 0.208) \times 10^{-2}$ d^{-1} with half-life of 18.6 d in control soil and $(18.8 \pm 2.76) \times 10^{-2}$ d^{-1} with half-life of 3.68 d in pretreated soil once only in the laboratory (Walker & Welch 1991); selected field half-life of 60 d (Wauchope et al. 1992; quoted, Richards & Baker 1993; quoted, Halfon et al. 1996; Hornsby et al. 1996);

soil half-life of 60 d (Pait et al. 1992); soil half-life of 29-67 d (Di Guardo et al. 1994).

Biota: biochemical half-life of 75 d from screening model calculations (Jury et al. 1987b).

Environmental Fate Rate Constants or Half-Lives:
Volatilization:
Photolysis: half-life of 2 months for 31% of 55 µg/mL to degrade in distilled water under sunlight (Rosen et al. 1969; quoted, Cessna & Muir 1991); half-life of 2.25 h for 67-75% of 75 µg/mL to degrade in distilled water under 300 nm light (Tanaka et al. 1981; quoted, Cessna & Muir 1991); atmosphere photolysis half-life of 1344-4032 h, based on measured rate constant for summer sunlight photolysis in distilled water (Rosen et al. 1969; quoted, Howard et al. 1991) and adjusted to relative winter sunlight intensity (Lyman et al. 1982; quoted, Howard et al. 1991); aqueous photolysis half-life of 1344-4032 h, based on measured rate constant for summer sunlight photolysis in distilled water (Rosen et al. 1969; quoted, Howard et al. 1991) and adjusted to relative winter sunlight intensity (Lyman et al. 1982; quoted, Howard et al. 1991).

Oxidation: photooxidation half-life of 0.49-4.90 h in air, based on an estimated rate constant for the vapor-phase reaction with hydroxyl radicals in air (Atkinson 1987; quoted, Howard et al. 1991).

Hydrolysis: half-life of >4 months for 4980 µg/mL to hydrolyze in phosphate buffer at pH 5-9 and 20°C (El-dib & Aly 1976; quoted, Muir 1991).

Biodegradation: half-lives in soil: 78 d (Moyer et al. 1972; quoted, Means et al. 1983), 87 d (Hance 1974; quoted, Means et al. 1983), 58 and 180 d in soil (Urosol & Hance 1974; quoted, Means et al. 1983); aqueous aerobic half-life of 672-4272 h, based on soil die-away test data (Walker 1978; Walker & Zimdahl 1981; quoted, Howard et al. 1991); rate constant of 0.0096 d^{-1} by soil incubation die-away studies (Rao & Davidson 1980; quoted, Scow 1982); aerobic half-lives of ~40 d for 1 µg/mL to biodegrade in lake sediment and ~60 d for 4 µg/mL to biodegrade in lake sediment and water (Huber & Gemes 1971; quoted, Muir 1991); aerobic half-lives of ~20 d for 0.22 µg/mL to biodegrade in pond sediment plus aerobic salts medium of 34 g/L (Stepp et al. 1985; quoted, Muir 1991); aqueous anaerobic half-life of 2688-17088 h, based on estimated unacclimated aqueous aerobic biodegradation half-life (Howard et al. 1991).

Biotransformation:

Bioconcentration, Uptake (k_1) and Elimination (k_2) Rate Constants:

Common Name: MCPA
Synonym: Agritox, Agroxohe, Agroxone, Anicon Kombi, Bordermaster, Chiptox, Chwastox, Cornox, Ded-weed, Dicopur-M, Dicotex, Dikotes, Emcepan, Empal, Hedapur M 52, Hederax M, Herbicide M, Hedonal, Hormotuho, Kilsem, Krezone, Legumex DB, Leuna M, Leyspray, Linormone, MCP, metaxon, Methoxone, Netazol, Okultin M, Phenoxylene Plus, Raphone, Razol dock killer, Rhomenc, Rhonox, Shamrox, Seppic MMD, Trasan, Ustinex, Vacate, Verdone, Weedar, Weed-rhap, Zelan
Chemical Name: (4-chloro-2-methylphenoxy)acetic acid; 4-chloro-o-tolyloxyacetic acid
Uses: systemic post-emergence herbicide to control annual and perennial weeds in cereals, rice, flax, vines, peas, potatoes, asparagus, grassland and turf.
CAS Registry No: 94-74-6
Molecular Formula: $C_9H_9ClO_3$
Molecular Weight: 200.6
Melting Point (°C):
    118-119 (Khan 1980; Seiber et al. 1986; Herbicide Handbook 1989; Worthing 1991; Milne 1995)
    120 (Montgomery 1993; Milne 1995)
    119-120.5 (Tomlin 1994)
Boiling Point (°C):
Density (g/cm^3 at 20°C):
    1.56 (25°C, Que Hee et al. 1981; Herbicide Handbook 1989; Montgomery 1993)
Molar Volume (cm^3/mol):
    211.1 (calculated-LeBas method, this work)
Molecular Volume (Å^3):
Total Surface Area, TSA (Å^2):
Dissociation Constant pK_a:
    3.05 (Nelson & Faust 1969; Que Hee et al. 1981)
    3.125 (Cessna & Grover 1978)
    3.07 (Worthing 1991)
    3.05-3.13 (Montgomery 1993)
    3.12 (Hornsby et al. 1996)
Heat of Fusion, ΔH_{fus}, kcal/mol:
Entropy of Fusion, ΔS_{fus}, cal/mol·K (e.u.):
Fugacity Ratio at 25°C (assuming ΔS_{fus} = 13.5 e.u.), F: 0.118

Water Solubility (g/m^3 or mg/L at 25°C):
    1605 (shake flask-UV, Leopold et al. 1960; quoted, Shiu et al. 1990)
    1605 (Bailey & White 1965; quoted, Que Hee et al. 1981)
    < 1000 (Khan 1980)
    630 (20°C, Melnikov 1971; quoted, Shiu et al. 1990)
    825 (Martin & Worthing 1977; quoted, Kenaga 1980)
    825 (Weber et al. 1980; quoted, Willis & McDowell 1982)
    1500 (selected, Seiber et al. 1986)
    825 (rm. temp., Agrochemicals Handbook 1987; Worthing 1991)
    817 (selected, Gerstl & Helling 1987)
    835 (rm. temp., Worthing 1987; quoted, Shiu et al. 1990; Majewski & Capel 1995)
    730-825 (Montgomery 1993)
    825 (Milne 1995; Halfon et al. 1996)

Vapor Pressure (Pa at 25°C):
- 7.9×10^{-4} (measured-volatilization rate, Seiber et al. 1986)
- 2.0×10^{-4} (20°C, Agrochemicals Handbook 1987)
- 2.0×10^{-4} (21°C, Worthing 1987, 91; quoted, Majewski & Capel 1995)
- 2.3×10^{-5} (20°C, Tomlin 1994)
- 2.0×10^{-4} (20°C, Milne 1995)
- 2.0×10^{-4} (selected, Halfon et al. 1996)

Henry's Law Constant (Pa·m³/mol):
- 1.0×10^{-4} (calculated-P/C, Seiber et al. 1986)
- $4.86 \times^{-4}$ (calculated-P/C as per Worthing 1987, Majewski & Capel 1995)
- 2.5×10^{-4} (calculated-P/C, this work)

Octanol/Water Partition Coefficient, log K_{OW}:
- 2.69 (selected, Dao et al. 1983)
- 2.30 (RP-HPLC-k', Braumann et al. 1983)
- −1.41 (selected, Gerstl & Helling 1987)
- 2.86; 3.252 (quoted; countercurrent LC, Ilchmann et al. 1993)
- 1.37-1.43 (calculated, Montgomery 1993)

Bioconcentration Factor, log BCF:
- 1.15 (calculated-S, Kenaga 1980)

Sorption Partition Coefficient, log K_{OC}:
- 2.04 (soil, calculated-S, Kenaga 1980)
- 1.95 (calculated-χ, Gerstl & Helling 1987)
- 2.04 (quoted from Kenaga 1980, Bottoni & Funari 1992)
- 2.03-2.07 (calculated, Montgomery 1993)
- −0.57, 3.25 (quoted, Sangster 1993)

Half-Lives in the Environment:
  Air:
  Surface water: dissipation half-life of ~4 days in rice field (Soderquist & Crosby 1975; quoted, Muir 1991); degraded rapidly with a half-life of 9 d in rice paddy water held under darkened conditions (Muir 1991).
  Groundwater:
  Sediment: half-lives ranging from 80 to 400 d of MCPA at low concentrations in marine sediments (Muir 1991).
  Soil: degradation half-life of 50 d in Finland sandy clay soil at room temp. from first-order rate constant obtained by linear regression and 41 days at 35°C; degradation half-life of 23 d in Bangladesh loam soil at room temp. from first-order rate constant obtained by linear regression and 20 days at 35°C (Sattar & Paasivirta 1980); persistence of 3 months in soil (Edwards 1973; quoted, Morrill et al. 1982); half-lives of 15 to 25 d in flooded soils (Muir 1991); 15 d (selected, Halfon et al. 1996).
  Biota:

Environmental Fate Rate Constants or Half-Lives:
  Volatilization: 9.78×10^{-7} h^{-1} at pH 3.5 (Seiber et al. 1986).
  Photolysis: half-life of 71 h for <10% of 50 µg/mL to degrade in NaOH solution at pH 9.8 under >290 nm light (Soderquist & Crosby 1975; quoted, Cessna & Muir 1991); half-life of 245 h for 17-98.5% of 9 µg/mL to degrade in distilled water under sunlight (Draper & Crosby 1984; quoted, Cessna & Muir 1991); half-life of 4.6 d for 14,700 µg/mL to degrade in droplets of spray solution suspended in air under sunlight (Freiberg & Crosby 1986; quoted, Cessna & Muir 1991).
  Oxidation:
  Hydrolysis:
  Biodegradation: half-life of >168 h for 1 µg/mL to degrade in activated sludge (Schmidt 1975; quoted, Muir 1991); aerobic half-life of ~9 d for 1 µg/mL to degrade in natural water in absence of sunlight (Soderquist & Crosby 1975; quoted, Muir 1991); half-life of >12 d for 0.045-0.156 µg/mL to degrade in water after application to model crop and washoff (Virtanen et al. 1979; quoted, Muir 1991); half-life of 15-25 d for 10 µg/mL to degrade in flooded soils (Duah-Yentumi & Kuwatsuka 1980; quoted, Muir 1991); half-life of >25 d for 10 µg/mL to degrade in flooded soils (Ursin 1985; quoted, Muir 1991).
  Biotransformation:
  Bioconcentration, Uptake (k_1) and Elimination (k_2) Rate Constants:

Common Name: MCPB
Synonym: Bexane, Can-Trol, Legumex, Thistrol, Thitrol, Trifolex, Tropotox
Chemical Name: 4-(4-chloro-2-methylphenoxy)butanoic acid; 4-(4-chloro-2-methylphenoxy)-butyric acid
Uses: herbicide for post-emergence control of annual and perennial broadleaf weeds in cereals, clovers, sainfoin, groundnuts, peas, etc. and also used to control broadleaf and woody weeds in forestry.
CAS Registry No: 94-81-5
Molecular Formula: $C_{11}H_{13}ClO_3$
Molecular Weight: 228.7
Melting Point (°C):
    100 (Agrochemicals Handbook 1987; Worthing 1991; Tomlin 1994; Milne 1995)
    100-101 (Herbicide Handbook 1989)
Boiling Point (°C):
    >280 (Tomlin 1994)
Density (g/cm³ at 22°C):
    1.254 (Tomlin 1994)
Molar Volume (cm³/mol):
    255.5 (calculated-LeBas method, this work)
Molecular Volume ($Å^3$):
Total Surface Area, TSA ($Å^2$):
Dissociation Constant pK_a:
    4.80 (Nelson & Faust 1969; Que Hee et al. 1981)
    4.84 (Worthing 1991; Tomlin 1994)
Heat of Fusion, ΔH_{fus}, kcal/mol:
    8.20 (DSC method, Plato 1972)
Entropy of Fusion, ΔS_{fus}, cal/mol·K (e.u.):
Fugacity Ratio at 25°C (assuming ΔS_{fus} = 13.5 e.u.), F: 0.181

Water Solubility (g/m³ or mg/L at 25°C):
    41 (shake flask-UV, Leopold et al. 1960; quoted, Shiu et al. 1990)
    44 (rm. temp., Melnikov 1971; quoted, Shiu et al. 1990)
    44 (Bailey & White 1965; quoted, Que Hee et al. 1981)
    44 (Martin & Worthing 1977; quoted, Kenaga 1980)
    44 (Agrochemicals Handbook 1987)
    44 (rm. temp., Worthing 1987, 1991; Tomlin 1994)
    44 (Milne 1995)

Vapor Pressure (Pa at 25°C):
    5.77×10^{-5}, 9.83×10^{-5} (20, 25°C, Tomlin 1994)

Henry's Law Constant (Pa·m³/mol):
    3.22×10^{-4} (calculated-P/C, this work)

Octanol/Water Partition Coefficient, log K_{OW}:
    4.60   (selected, Dao et al. 1983)
    3.53   (RP-HPLC-k', Braumann et al. 1983)
    3.473   (countercurrent LC, Ilchmann et al. 1993)
    2.79   (Tomlin 1994)
    3.43   (selected, Hansch et al. 1995)

Bioconcentration Factor, log BCF:
    1.86   (calculated-S, Kenaga 1980)

Sorption Partition Coefficient, log K_{OC}:
    2.73   (soil, calculated-S, Kenaga 1980)

Half-Lives in the Environment:
    Air:
    Surface water:
    Groundwater:
    Sediment:
    Soil: duration of residual activity in soil is ca. 3-4 months (Agrochemicals Handbook 1987; Tomlin 1994).
    Biota:

Environmental Fate Rate Constants or Half-Lives:
    Volatilization:
    Photolysis:
    Oxidation:
    Hydrolysis:
    Biodegradation:
    Biotransformation:
    Bioconcentration, Uptake (k_1) and Elimination (k_2) Rate Constants:

Common Name: Mecoprop
Synonym: Compitox, Duplosan, Hedonal, Iso-Cornox, Kilprop, MCPP, Mecopex, Mepro, Methoxone, Propal
Chemical Name: (±)-2-(4-chloro-2-methylphenoxy)propanoic acid; (±)-2-(4-chloro-o-tolyloxy)propionic acid
Uses: herbicide for post-emergence control of broadleaf weeds in wheat, barley, rye, herbage seed crops, grassland, and under fruit trees and vines, etc.
CAS Registry No: 7085-19-0
Molecular Formula: $C_{10}H_{11}ClO_3$
Molecular Weight: 214.6
Melting Point (°C):
- 92-93 (Spencer 1982)
- 94-95 (Agrochemicals Handbook 1987; Herbicide Handbook 1989; Worthing 1991)

Boiling Point (°C):
Density (g/cm^3 at 20°C):
Molar Volume (cm^3/mol):
- 233.3 (calculated-LeBas method, this work)

Molecular Volume (Å^3):
Total Surface Area, TSA (Å^2):
Dissociation Constant pK_a:
- 3.75 (Bailey & White; quoted, Que Hee et al. 1981)
- 3.105 (Cessna & Grover 1978)
- 3.78 (Worthing 1991)

Heat of Fusion, ΔH_{fus}, kcal/mol:
- 6.90 (DSC method, Plato 1972)

Entropy of Fusion, ΔS_{fus}, cal/mol·K (e.u.):
Fugacity Ratio at 25°C (assuming ΔS_{fus} = 13.5 e.u.), F: 0.203

Water Solubility (g/m^3 or mg/L at 25°C):
- 895 (Martin 1961; Bailey & White 1965; quoted, Shiu et al. 1990)
- 891 (Bailey & White 1965; quoted, Que Hee et al. 1981)
- 620 (20°C, Melnikov 1971; quoted, Shiu et al. 1990)
- 620 (Martin & Worthing 1977; quoted, Kenaga 1980)
- 620 (20°C, Ashton & Crafts 1981; Herbicide Handbook 1989)
- 620 (20°C, Agrochemicals Handbook 1987)
- 620 (20°C, Worthing 1987; quoted, Shiu et al. 1990)
- 620 (20°C, Worthing 1991)
- 620 (selected, Lohninger 1994)
- 734 (Tomlin 1994)
- 660000 (selected, Halfon et al. 1996)

Vapor Pressure (Pa at 25°C):
- $<1.0\times10^{-5}$ (20°C, Agrochemicals Handbook 1987)
- 3.10×10^{-4} (20°C, Worthing 1991)
- 0.0 (selected, Halfon et al. 1996)

Henry's Law Constant (Pa·m^3/mol):
- 7.43×10^{-5} (calculated-P/C, this work)

Octanol/Water Partition Coefficient, log K_{OW}:
    3.94   (selected, Dao et al. 1983)
    2.83   (RP-HPLC-k', Braumann et al. 1983)
    0.10   (Worthing 1991)
    0.09; 3.126 (quoted; countercurrent LC, Ilchmann et al. 1993)
    3.13   (selected, Hansch et al. 1995)

Bioconcentration Factor, log BCF:
    1.20   (calculated-S, Kenaga 1980)

Sorption Partition Coefficient, log K_{OC}:
    2.11   (quoted from Kenaga 1980, Bottoni & Funari 1992)
    1.30   (selected, Lohninger 1994)

Half-Lives in the Environment:
    Air: 3.8-37.8 h, based on an estimated rate constant for the vapor-phase reaction with hydroxyl radicals in air (Atkinson 1987; quoted, Howard et al. 1991).
    Surface water: 168-240 h, based on estimated aqueous aerobic biodegradation half-life (Howard et al. 1991).
    Groundwater: 336-4320 h, based on estimated aqueous aerobic and anaerobic biodegradation half-lives (Howard et al. 1991).
    Sediment:
    Soil: 168-240 h, based on aerobic soil grab sample data (Kirkland & Fryer 1972; Smith & Hayden 1981; quoted, Howard et al. 1991); 21 days (selected, Halfon et al. 1996).
    Biota:

Environmental Fate Rate Constants or Half-Lives:
    Volatilization:
    Photolysis:
    Oxidation: photooxidation half-life of 3.8-37.8 h in air, based on an estimated rate constant for the vapor-phase reaction with hydroxyl radicals in air (Atkinson 1987; quoted, Howard et al. 1991).
    Hydrolysis:
    Biodegradation: aqueous aerobic half-life of 168-240 h, based on aerobic soil grab sample data (Kirkland & Fryer 1972; Smith & Hayden 1981; quoted, Howard et al. 1991); aqueous anaerobic half-life of 672-4320 h, based on anaerobic digestor sludge data (Battersby & Wilson 1989; quoted, Howard et al. 1991).
    Biotransformation:
    Bioconcentration, Uptake (k_1) and Elimination (k_2) Rate Constants:

Common Name: Metolachlor

Synonym: Bicep, CGA 24705, Codal, Cortoran multi, Dual, Metetilachlor, Milocep, Ontrack 8E, Pennant, Primagram, Primextra

Chemical Name: 2-chloro-6'-ethyl-*N*-(2-methoxy-1-methylethyl)acet-*o*-toluidide; 2-chloro-*N*-(2-ethyl-6-methylphenyl)-*N*-(2-methoxy-1-methylethyl)acetamide

Uses: pre-emergence herbicide to control most annual grasses and weeds in beans, chickpeas, corn, cotton, milo, okra, peanuts, peas, potatoes, sunflower, soybeans and some ornamentals.

CAS Registry No: 51218-45-2

Molecular Formula: $C_{15}H_{22}ClNO_2$

Molecular Weight: 283.8

Melting Point (°C): liquid

Boiling Point (°C):
    100   (at 0.001 mmHg, Herbicide Handbook 1989; Merck Index 1989; Worthing 1991; Montgomery 1993; Milne 1995)

Density (g/cm³ at 20°C):
    1.12   (Agrochemicals Handbook 1987; Worthing 1991; Montgomery 1993; Milne 1995)
    1.085   (Herbicide Handbook 1989)

Molar Volume (cm³/mol):
    340.0   (calculated-LeBas method, this work)
    258.0   (calculated-density, this work)

Molecular Volume (Å³):

Total Surface Area, TSA (Å²):

Dissociation Constant pK_a:

Heat of Fusion, ΔH_{fus}, kcal/mol:

Entropy of Fusion, ΔS_{fus}, cal/mol·K (e.u.):

Fugacity Ratio at 25°C (assuming ΔS_{fus} = 13.5 e.u.), F: 1.0

Water Solubility (g/m³ or mg/L at 25°C):
    530   (Martin & Worthing 1977; quoted, Kenaga 1980; Burkhard & Guth 1981)
    440   (selected, Ellgehausen et al. 1980)
    520   (20°C, Ashton & Crafts 1981; Spencer 1982)
    530   (shake flask-HPLC, Ellgehausen et al. 1981; quoted, Shiu et al. 1990)
    530   (20°C, Agrochemicals Handbook 1987)
    530   (Hartley & Graham-Bryce 1980; Beste & Humburg 1983; quoted, Taylor & Glofelty 1988)
    530   (20°C, Worthing 1987, 1991; quoted, Di Guardo et al. 1994; Majewski & Capel 1995)
    530   (20°C, Herbicide Handbook 1989; Pait et al. 1992)
    530   (20°C, Merck Index 1989; quoted, Milne 1995)
    530   (20°C, Montgomery 1993)
    530   (selected, Lohninger 1994; Pinsuwan et al. 1995; Halfon et al. 1996)
    488   (Tomlin 1994; quoted, Otto et al. 1997)
    530   (20-25°C, selected, Hornsby et al. 1996)

Vapor Pressure (Pa at 25°C):
    0.00170 (20°C, Hartley & Graham-Bryce 1980; quoted, Taylor & Glotfelty 1988; Taylor & Spencer 1990)
    0.00173 (20°C, Ashton & Crafts 1981)
    0.00173 (quoted, Burkhard & Guth 1981)
    0.00170 (20°C, Agrochemicals Handbook 1987)
    0.00170 (20°C, Worthing 1987, 1991; quoted, Di Guardo et al. 1994; Majewski & Capel 1995)
    0.00173 (20°C, Herbicide Handbook 1989; Merck index 1989)
    0.00170 (20°C, selected, Taylor & Spencer 1990)
    0.00173 (20°C, Montgomery 1993)
    0.00420 (Tomlin 1994; quoted, Otto et al. 1997)
    0.00418 (selected, Halfon et al. 1996)
    0.00418 (20-25°C, selected, Hornsby et al. 1996)

Henry's Law Constant (Pa·m^3/mol):
    0.00092 (Hartley & Graham-Bryce 1980; quoted, Taylor & Glotfelty 1988)
    0.00091 (20°C, calculated-P/C as per Worthing 1987, Majewski & Capel 1995)
    0.00093 (20°C, calculated-P/C, Montgomery 1993)
    0.00244 (calculated-P/C, Otto et al. 1997)
    0.00110 (calculated-P/C, this work)

Octanol/Water Partition Coefficient, log K_{OW}:
    3.13 (shake flask-HPLC, Ellgehausen et al. 1980; Geyer et al. 1991)
    3.28 (shake flask-HPLC, Ellgehausen et al. 1981)
    3.45 (Worthing 1991)
    2.93, 3.45 (Montgomery 1993)
    3.13, 3.28 (quoted, Sangster et al. 1993)
    3.13 (quoted, Di Guardo et al. 1994)
    3.31 (Aquasol Database 1994; quoted, Pinsuwan et al. 1994)
    2.90 (Tomlin 1994; quoted, Otto et al. 1997)
    3.45 (Milne 1995)
    3.31, 2.95 (selected, calculated-f const., Pinsuwan et al. 1995)
    3.13 (selected, Hansch et al. 1995)

Bioconcentration Factor, log BCF:
    1.813 (log BF-bioaccumulation of algae, Ellgehausen et al. 1980)
    0.733 (log BF-bioaccumulation of daphnids, Ellgehausen et al. 1980; quoted, Geyer et al. 1991)
    0.851 (log BF-bioaccumulation of catfish, Ellgehausen et al. 1980)
    1.26 (calculated-S, Kenaga 1980)

Sorption Partition Coefficient, log K_{OC}:
    2.15 (soil, calculated-S, Kenaga 1980)
    2.26 (soil, screening model calculations, Jury et al. 1987b)
    2.00, 2.15, 2.28, 2.30 (quoted values, Bottoni & Funari 1992)

    2.46   (soil, quoted exptl., Meylan et al. 1992)
    2.46   (soil, calculated-χ and fragment contribution, Meylan et al. 1992)
    2.30   (20-25°C, selected, Wauchope et al. 1992; quoted, Richards & Baker 1993; Lohninger 1994; Hornsby et al. 1996)
    2.08-2.49   (Montgomery 1993; Tomlin 1994)

Half-Lives in the Environment:
    Air:
    Surface water:
    Groundwater: degradation time 500-1000 d (Tomlin 1994).
    Sediment:
    Soil: half-lives in clay loam soils and sandy loam soils were 15-38 and 33-100 d, respectively (Zimdahl & Clark 1982; quoted, Montgomery 1993); half-life of 42 d from field half-life of 3-4 weeks by using lysimeters (Bowman 1990); calculated half-lives of 80, 99 and 142 d for the disapperance from upper 15 cm on an Ontario clay loam soil while the decline was followed for 332, 364 and 370 d respectively in 1987, 1988 and 1989 (Frank et al. 1991); half-life in soil is approximately 6 d (Worthing 1991; quoted, Montgomery 1993); field half-life of 90 d (20-25°C, selected, Wauchope et al. 1992; quoted, Richards & Baker 1993; Halfon et al. 1996; Hornsby et al. 1996); soil half-life of 40 d (Pait et al. 1992); soil half-life of 28-46 d (Di Guardo et al. 1994); half-life about 30 (Tomlin 1994); estimated half-lives of 28.3 days under coventrional tillage, 25.61 days under ridge tillage and 8.63 days with no tillage (Otto et al. 1997).
    Biota: half-life of 1.15 d in catfish (Ellgehausen et al. 1980); biochemical half-life of 42 d from screening model calculations (Jury et al. 1987b).

Environmental Fate Rate Constants or Half-Lives:
    Volatilization:
    Photolysis: under optimum exposure conditions to natural sunlight, half-life is approximately 8 d (Herbicide Handbook 1989).
    Oxidation:
    Hydrolysis: half-life of >200 d at 20°C and $1 \leq pH \leq 9$ (Montgomery 1993); calculated half-life >200 d ($2 \leq pH \geq 10$) (Tomlin 1994).
    Biodegradation:
    Biotransformation:
    Bioconcentration, Uptake (k_1) and Elimination (k_2) Rate Constants:
        k_2:   9.11 d^{-1} (catfish, Ellgehausen et al. 1980)

Common Name: Molinate
Synonym: Felan, Higalnate, Hydram, Jalan, Molmate, Ordram, Stauffer R 4572, Sakkimok, Yalan, Yulan
Chemical Name: 1H-azepine-1-carbothioic acid, hexahydro, S-ethyl ester; ethyl 1-hexamethyleneiminecarbothioate
Uses: selective herbicide to control the germination of annual grasses and broadleaf weeds in rice crops.
CAS Registry No: 2212-67-1
Molecular Formula: $C_9H_{17}NOS$
Molecular Weight: 187.32
Melting Point (°C): <25 (Montgomery 1993)
Boiling Point (°C):
    202   (at 10 mmHg, Agrochemicals Handbook 1987; Herbicide Handbook 1989; Worthing 1991; Milne 1995)
    117   (at 10 mmHg, Montgomery 1993)
Density (g/cm³ at 20°C):
    1.064   (Agrochemicals Handbook 1987)
    1.0643   (Herbicide Handbook 1989; Montgomery 1993)
    1.063   (Worthing 1991; Milne 1995)
Molar Volume (cm³/mol):
    220.6   (calculated-LeBas method, this work)
    176.1   (calculated-density, this work)
Molecular Volume (Å³):
Total Surface Area, TSA (Å²):
Dissociation Constant pK_a:
Heat of Fusion, ΔH_{fus}, kcal/mol:
Entropy of Fusion, ΔS_{fus}, cal/mol·K (e.u.):
Fugacity Ratio at 25°C (assuming ΔS_{fus} = 13.5 e.u.), F: 1.0

Water Solubility (g/m³ or mg/L at 25°C):
    880   (20°C, Weber 1972; Worthing 1987; quoted, Muir 1991)
    800   (Martin & Worthing 1977; quoted, Kenaga 1980)
    800   (Wauchope 1978; Khan 1980)
    800-912 (Weber et al. 1980; quoted, Willis & McDowell 1982)
    912   (21°C, Spencer 1982)
    800   (20°C, Herbicide Handbook 1983, 1989; quoted, Seiber et al. 1986,89)
    880   (20°C, Agrochemicals Handbook 1987)
    880   (20°C, Worthing 1987, 1991; quoted, Majewski & Capel 1995)
    870   (Kanazawa 1989)
    800   (quoted, Pait et al. 1992)
    970   (20-25°C, selected, Wauchope et al. 1992; quoted, Lohninger 1994; Halfon et al. 1996; Hornsby et al. 1996)
    880   (20°C, Montgomery 1993)
    88   (20°C, Tomlin 1994)
    880   (20°C, Milne 1995)

Vapor Pressure (Pa at 25°C):
- 0.748 (20°C, Weber 1972; Worthing 1987; quoted, Muir 1991; Majewski & Capel 1995)
- 0.746 (20°C, Khan 1980)
- 0.413 (quoted, Seiber et al. 1986,89)
- 0.746 (Agrochemicals Handbook 1987; Tomlin 1994)
- 0.746 (Herbicide Handbook 1989; Worthing 1991)
- 0.746 (20-25°C, selected, Wauchope et al. 1992; quoted, Halfon et al. 1996)
- 0.746 (Montgomery 1993)

Henry's Law Constant (Pa·m^3/mol):
- 0.097 (calculated-P/C, Seiber et al. 1986,89)
- 0.159 (20°C, calculated-P/C as per Worthing 1987; quoted, Majewski & Capel 1995)
- 0.159 (20°C, calculated-P/C, Muir 1991)
- 0.095 (20°C, calculated-P/C, Sagebiel et al. 1992)
- 0.460 (20°C, gas-stripping method, Sagebiel et al. 1992)
- 0.390 (20°C, headspace-GC method, Sagebiel et al. 1992)
- 0.162 (calculated-P/C, Montgomery 1993)
- 0.145 (calculated-P/C, this work)

Octanol/Water Partition Coefficient, log K_{OW}:
- 3.21 (shake flask-GC, Kanazawa 1981; quoted, Kanazawa 1989; Sicbaldi & Finizio 1993; Devillers et al. 1996)
- 2.88 (Worthing 1991; Tomlin 1994)
- 2.88 (Montgomery 1993)
- 3.13 (RP-HPLC, Saito et al. 1993)
- 3.26 (RP-HPLC, Sicbaldi & Finizio 1993)
- 3.21, 3.26; 3.21 (quoted values; selected, Sangster 1993)
- 2.88 (Milne 1995)
- 3.21 (selected, Hansch et al. 1995)
- 3.25, 2.85, 2.49 (RP-HPLC, ClogP, calculated-K_{OW}, Finizio et al. 1997)

Bioconcentration Factor, log BCF:
- 1.15 (calculated-S, Kenaga 1980; quoted, Pait et al. 1992)
- 1.41 (*Peudorasbora parva*, Kanazawa 1981; quoted, Devillers et al. 1996)

Sorption Partition Coefficient, log K_{OC}:
- 2.04 (soil, calculated-S, Kenaga 1980)
- 1.92 (average of 2 soils, Kanazawa 1989)
- 1.92, 2.04 (quoted values, Bottoni & Funari 1992)
- 1.92 (soil, quoted exptl., Meylan et al. 1992)
- 2.46 (soil, calculated-χ and fragments contribution, Meylan et al. 1992)
- 2.28 (soil, 20-25°C, selected, Wauchope et al. 1992; Hornsby et al. 1996)
- 1.93-1.97 (Montgomery 1993)
- 2.28 (selected, Lohninger 1994)

Half-Lives in the Environment:
- Air: calculated lifetime of 6 hours for the vapor-phase reaction with OH radicals in the troposphere (Atkinson et al. 1992; Kwok et al. 1992).
- Surface water: half-life of 84 h from dissipation from flooded rice fields (Seiber & McChesney 1987; quoted, Seiber et al. 1989).
- Groundwater:
- Sediment:
- Soil: persistence of 2 months in soil (Wauchope 1978); half-life is approximately 3 weeks in moist loam soils at 21-27°C (Herbicide Handbook 1989); selected field half-life of 21 d (Wauchope et al. 1992; quoted, Halfon et al. 1996; Hornsby et al. 1996); soil half-life of 21 d (Pait et al. 1992).

Environmental Fate Rate Constants or Half-Lives:
- Volatilization: 0.0150 h^{-1} (average of 2 runs, Seiber et al. 1986); 1.1 kg/ha (1st 4 day) from flooded rice fields (Seiber et al. 1986; Seiber & McChesney 1987; quoted, Seiber et al. 1989); estimated half-life of 43 d from 1 m depth of water at 20°C (Muir 1991).
- Photolysis: half-life of 7-10 d for 8-10 µg/mL to degrade in distilled water under >290 nm light (Soderquist et al. 1977; quoted, Cessna & Muir 1991); half-life of 96 h for <5% of 0.2 µg/mL to degrade in distilled water under sunlight (Deuel et al. 1978; quoted, Cessna & Muir 1991); half-life of 245 h for 2-54% of 10 µg/mL to degrade in distilled water under sunlight (Draper & Crosby 1984; quoted, Cessna & Muir 1991).
- Oxidation: calculated life-time of 6 h for the vapor-phase reaction with OH radicals in the troposphere (Atkinson et al. 1992; Kwok et al. 1992).
- Hydrolysis: half-life of >10 d in aqueous buffer at pH 5-9 in the dark (Soderquist et al. 1977; quoted, Muir 1991).
- Biodegradation: half-life of ~16 d for 0.2 µg/mL to biodegrade in flooded soils (Deuel et al. 1978; quoted, Muir 1991); half-life of 10 weeks for 4.2 µg/mL to biodegrade in flooded soil and <2 weeks in water both at 21-26°C (Thomas & Holt 1980; quoted, Muir 1991).
- Biotransformation:
- Bioconcentration, Uptake (k_1) and Elimination (k_2) Rate Constants:

Common Name: Monolinuron
Synonym: Afesin, Aresin, Arresin, Hoe 02747
Chemical Name: 3-(4-chlorophenyl)-1-methoxy-1-methylurea; N'-(4-chlorophenyl)-N-methoxy-N-methylurea
Uses: herbicide for pre- or post-emergence control of annual broadleaf weeds and annual grasses in asparagus, berry fruit, cereals, maize, field beans, vines, leeks, onions, potatoes, herbs, lucerne, flowers, ornamental shrubs and trees, etc.
CAS Registry No: 1746-81-2
Molecular Formula: $C_9H_{11}ClN_2O_2$
Molecular Weight: 214.6
Melting Point (°C):
    79-80   (Khan 1980)
    75-78   (Spencer 1982)
    76   (Karickhoff 1981)
    80-83   (Agrochemicals Handbook 1987; Worthing 1991)
    81.5   (Patil 1994; Kühne et al. 1995)
Boiling Point (°C):
Density (g/cm³ at 20°C):
Molar Volume (cm³/mol):
    224.0   (calculated-LeBas method, this work)
Molecular Volume (Å³):
Total Surface Area, TSA (Å²):
Dissociation Constant pK_a:
Heat of Fusion, ΔH_{fus}, kcal/mol:
Entropy of Fusion, ΔS_{fus}, cal/mol·K (e.u.):
Fugacity Ratio at 25°C (assuming ΔS_{fus} = 13.5 e.u.), F: 0.273

Water Solubility (g/m³ or mg/L at 25°C):
    735   (20°C, Melnikov 1971; quoted, Shiu et al. 1990)
    735   (Spencer 1973, 1982; quoted, Shiu et al. 1990)
    580   (Martin & Worthing 1977; Khan 1980; quoted, Kenaga & Goring 1980; Briggs 1981; Karickhoff 1981; Shiu et al. 1990)
    735   (Agrochemicals Handbook 1987; Tomlin 1994)
    735   (Worthing 1987, 1987, 1991; quoted, Shiu et al. 1990)
    735   (20-25°C, selected, Augustijn-Beckers et al. 1994; Hornsby et al. 1996)
    578, 332   (quoted, calculated, Patil 1994)
    578, 7420   (quoted, calculated-group contribution fragmentation method, Kühne et al. 1995)

Vapor Pressure (Pa at 25°C):
    0.02   (22°C, Khan 1980; Agrochemicals Handbook 1987)
    0.0015   (20°C, Spencer 1982)
    6.40   (65°C, Worthing 1991)
    0.02   (20-25°C, selected, Augustijn-Beckers et al. 1994; Hornsby et al. 1996)
    0.0013, 0.10   (20, 50°C, Tomlin 1994)

Henry's Law Constant (Pa·m³/mol):
    0.0058   (calculated-P/C, this work)

Octanol/Water Partition Coefficient, log K_{OW}:
    1.60   (Briggs 1969; quoted, Kenaga & Goring 1980; Karickhoff 1981)
    2.30   (shake flask-UV, Briggs 1981; quoted, Sicbaldi & Finizio 1993)
    1.60   (selected, Dao et al. 1983)
    1.99   (RP-HPLC-k', Braumann et al. 1983)
    2.20   (Worthing 1991; Tomlin 1994)
    2.16   (RP-HPLC, Sicbaldi & Finizio 1993)
    1.60, 2.30, 2.22, 2.16; 2.30 (quoted values; selected, Sangster 1993)
    1.60, 2.21 (quoted, calculated, Patil 1994)
    2.30   (selected, Hansch et al. 1995)
    2.16, 2.31, 2.59 (RP-HPLC, ClogP, calculated-S, Finizio et al. 1997)

Bioconcentration Factor, log BCF:
    1.23   (calculated-S, Kenaga 1980)
    1.00   (calculated-K_{OC}, Kenaga 1980)
    1.85   (activated sludge, Freitag et al. 1982, 1984, 1985)
    1.52, <1.0 (algae, golden orfe, Freitag et al. 1982)
    1.60, 1.30 (algae, golden ide, Freitag et al. 1985)

Sorption Partition Coefficient, log K_{OC}:
    2.30   (soil, Hamaker & Thompson 1972; quoted, Kenaga 1980; Kenaga & Goring 1980; Karickhoff 1981)
    2.11   (soil, calculated-S as per Kenaga & Goring 1978, Kenaga 1980)
    1.60   (reported as log K_{OM}, Briggs 1981)
    2.36, 2.08, 1.21 (estimated-S, S & M.P., K_{OW}, Karickhoff 1981)
    2.40-2.70 (soil, Worthing 1991)
    2.26-2.30, 2.40-2.70 (quoted values, Bottoni & Funari 1992)
    2.30   (20-25°C, estimated, Augustijn-Beckers et al. 1994; Hornsby et al. 1996)
    1.78   (soil, HPLC-screening method, Kördel et al. 1993,95b)

Half-Lives in the Environment:
    Air:
    Surface water:
    Groundwater:
    Sediment:
    Soil: estimated field half-life of 60 d (Augustijn-Beckers et al. 1994; Hornsby et al. 1996).
    Biota:

Environmental Fate Rate Constants or Half-Lives:
    Volatilization:
    Photolysis: half-life of 23 h for 66% of 286 μg/mL to degrade in distilled water under >300 nm light (Kotzias et al. 1974; quoted, Cessna & Muir 1991).
    Oxidation:
    Hydrolysis:
    Biodegradation:
    Biotransformation:
    Bioconcentration, Uptake (k_1) and Elimination (k_2) Rate Constants:

Common Name: Monuron
Synonym: Chlorfenidim, CMU, Karmex, Lirobetarex, Monurex, Monurox, Rosuran, Telvar, Urox
Chemical Name: N'-(4-chlorophenyl)-N-N-dimethylurea; 1,1-dimethyl-3-(p-chlorophenyl)urea
Uses: herbicide; also as sugar cane flowering suppressant.
CAS Registry No: 150-68-5
Molecular Formula: $C_9H_{11}ClN_2O$
Molecular Weight: 198.65
Melting Point (°C):
    174-175 (Khan 1980; Spencer 1982; Montgomery 1993)
    170 (Karickhoff 1981)
    176-177 (Suntio et al. 1988)
    174.5 (Patil 1994)
    170.5 (Kühne et al. 1995)
Boiling Point (°C):
    185-200 (decomposes, Montgomery 1993)
Density (g/cm³ at 20°C):
    1.27 (Spencer 1982; Agrochemicals Handbook 1987; Montgomery 1993)
Molar Volume (cm³/mol):
    202.9 (calculated-LeBas method, Suntio et al. 1988)
    173.0 (modified LeBas method, Spurlock & Biggar 1994a)
Molecular Volume (Å³):
Total Surface Area, TSA (Å²):
Dissociation Constant pK_a:
Heat of Fusion, ΔH_{fus}, kcal/mol:
Entropy of Fusion, ΔS_{fus}, cal/mol·K (e.u.):
Fugacity Ratio at 25°C (assuming ΔS_{fus} = 13.5 e.u.), F: 0.031

Water Solubility (g/m³ or mg/L at 25°C):
    203 (Freed 1966; quoted, Suntio et al. 1988; Shiu et al. 1990)
    230 (Günther et al. 1968; quoted, Suntio et al. 1988; Shiu et al. 1990)
    262 (shake flask-UV, Hurle & Freed 1972; quoted, Freed 1976; Jury et al. 1983; Suntio et al. 1988; Shiu et al. 1990)
    230 (20°C, Weber 1972; Worthing 1987; quoted, Shiu et al. 1990; Muir 1991)
    230 (Martin & Worthing 1977; quoted, Kenaga 1980; Kenaga & Goring 1980; Karickhoff 1981; Isensee 1991)
    230 (Sanborn et al. 1977; quoted, Belluck & Felsot 1981)
    228 (selected, Ellgehausen et al. 1980)
    203-230 (Weber et al. 1980; quoted, Willis & McDowell 1982)
    230 (Khan 1980; Ashton & Crafts 1981; quoted, Burkhard & Guth 1981)
    233 (quoted, Briggs 1981)
    200 (shake flask-HPLC, Ellgehausen et al. 1981; quoted, Shiu et al. 1990; Pinsuwan et al. 1995)
    260 (Jury et al. 1983, 1984; quoted, Taylor & Glotfelty 1988)
    230 (Agrochemicals Handbook 1987; selected, Gerstl & Helling 1987)
    200 (20°C, selected, Suntio et al. 1988)
    230 (selected, Evelyne et al. 1992)
    275 (Spurlock 1992; quoted, Spurlock & Biggar 1994b)
    230 (at pH 6.26, Montgomery 1993)

230   (20-25°C, selected, Augustijn-Beckers et al. 1994; Hornsby et al. 1996)
230, 3850 (quoted, calculated, Patil 1994)
256, 308 (quoted, calculated-group contribution fragmentation method, Kühne et al. 1995)
230, 202 (quoted, predicted-AQUAFAC, Lee et al. 1996)

Vapor Pressure (Pa at 25°C):
 6.72×10^{-5} (20°C, Weber 1972; Worthing 1987; quoted, Muir 1991)
 6.67×10^{-5} (Khan 1980; quoted, Suntio et al. 1988)
 6.67×10^{-5} (Ashton & Crafts 1981)
 6.65×10^{-5} (quoted, Burkhard & Guth 1981)
 6.70×10^{-5} (OECD 1981; quoted, Dobbs et al. 1984)
 2.30×10^{-5} (calculated, Jury et al. 1983; quoted, Taylor & Glotfelty 1988; Taylor & Spencer 1990)
 6.00×10^{-5} (Agrochemicals Handbook 1987)
 3.00×10^{-3} (20°C, selected, Suntio et al. 1988)
 6.67×10^{-5} (Merck Index 1989)
 2.30×10^{-5} (selected, Taylor & Spencer 1990)
 6.00×10^{-5} (20°C, Montgomery 1993)
 6.67×10^{-5} (20-25°C, selected, Augustijn-Beckers et al. 1994; Hornsby et al. 1996)

Henry's Law Constant (Pa·m^3/mol):
 1.88×10^{-5} (calculated-P/C, Jury et al. 1984, 1987a; Jury & Ghodrati 1989)
 3.0×10^{-3} (20°C, calculated-P/C, Suntio et al. 1988)
 1.91×10^{-5} (calculated-P/C, Taylor & Glotfelty 1988)
 5.60×10^{-5} (20°C, calculated-P/C, Muir 1991)
 3.0×10^{-3} (20°C, calculated-P/C, Montgomery 1993)
 6.6×10^{-5} (calculated-P/C, this work)

Octanol/Water Partition Coefficient, log K_{ow}:
 1.46 (Briggs 1969; quoted, Kenaga & Goring 1980; Karickhoff 1981; Suntio et al. 1988)
 1.66 (calculated-f const., Rekker 1977; quoted, Ellgehausen et al. 1981)
 2.08 (selected, Ellgehausen et al. 1980; Geyer et al. 1991)
 2.12 (quoted, Rao & Davidson 1980; Suntio et al. 1988)
 2.95 (calculated as per Chiou et al. 1977, Belluck & Felsot 1981)
 1.98 (shake flask-UV, Briggs 1981; quoted, Madhun et al. 1986; Sicbaldi & Finizio 1993)
 1.95 (calculated-S, Ellgehausen et al. 1981; quoted, Sicbaldi & Finizio 1993)
 1.66 (shake flask, Ellgehausen et al. 1981; quoted, Sicbaldi & Finizio 1993)
 1.86 (selected, Dao et al. 1983)
 1.91 (RP-HPLC-k', Braumann et al. 1983)
 1.86 (selected, Gerstl & Helling 1987)
 1.80 (selected, Suntio et al. 1988)
 2.12 (Spurlock 1992; quoted, Spurlock & Biggar 1994b)
 1.46, 2.12 (Montgomery 1993)
 1.86 (RP-HPLC, Sicbaldi & Finizio 1993)

1.94   (selected, Sangster 1993)
1.96   (Aquasol Database 1994; quoted, Pinsuwan et al. 1995)
1.46, 0.89 (quoted, calculated, Patil 1994)
1.89, 1.88 (shake flask-UV, calculated-RPHPLC-k', Liu & Qian 1995)
1.96, 1.99 (quoted, calculated-f const., Pinsuwan et al. 1995)
1.94   (selected, Hansch et al. 1995)
1.86, 1.99, 2.92 (RP-HPLC, ClogP, calculated-S, Finizio et al. 1997)

Bioconcentration Factor, log BCF:
1.786   (log BF bioaccumulation factor for algae, Ellgehausen et al. 1980)
0.32   (log BF bioaccumulation factor for daphnids, Ellgehausen et al. 1980; quoted, Geyer et al. 1991)
0.245   (log BF bioaccumulation factor for daphnids, Ellgehausen et al. 1980)
1.46   (calculated-S, Kenaga 1980; quoted, Isensee 1991)
0.699   (calculated-K_{OC}, Kenaga 1980)
0.0   (*Triaenodes tardus*, Belluck & Felsot 1981)
1.58, 1.67 (cuticle/water: tomato, pepper; Eveline et al. 1992)

Sorption Partition Coefficient, log K_{OC}:
2.00   (soil, Hamaker & Thompson 1972; quoted, Kenaga 1980; Kenaga & Goring 1980; Karickhoff 1981)
2.34   (soil, calculated-S as per Kenaga & Goring 1978, Kenaga 1980)
2.26   (av. of 18 soils, Rao & Davidson 1980; quoted, Karickhoff 1981; Jury et al. 1983; Liu & Qian 1995)
1.46   (reported as log K_{OM}, Briggs 1981)
2.58, 1.51 (estimated-S, S & M.P., Karickhoff 1981)
1.07, 1.73 2.58 (estimated-K_{OW}, Karickhoff 1981)
2.23, 2.36 (soil, quoted, Madhun et al. 1986)
2.03, 1.85; 2.17, 1.52 (estimated-K_{OW}, S, Madhun et al. 1986)
1.99, 2.12 (quoted, calculated-χ, Gerstl & Helling 1987)
2.26   (screening model calculations, Jury et al. 1987a,b; Jury & Ghoodrati 1989)
1.99, 2.33 (Montgomery 1993)
2.18   (20-25°C, estimated, Augustijn-Beckers et al. 1994; Hornsby et al. 1996)
1.99   (soil, HPLC-ring test, Kördel et al. 1995a)
1.99   (soil, HPLC-screening method, Kördel et al. 1993,95b)
2.29   (calculated-K_{OW}, Liu & Qian 1995)

Half-Lives in the Environment:
Air:
Surface water: persistence of up to 8 weeks in river water (Eichelberger & Lichtenberg 1971).
Groundwater:
Sediment:
Soil: half-life of 5.0 months at 15°C and 4.1 months at 30°C in soils (Freed & Haque 1973); reported half-life of 166 d from screening model calculations (Jury et al. 1987a,b; Jury & Ghodrati 1989; quoted, Montgomery 1993);

estimated field half-life of 170 d (Augustijn-Beckers et al. 1994; Hornsby et al. 1996).

Biota: 0.45 d in catfish (Ellgehausen et al. 1980); biochemical half-life of 166 days from screening model calculations (Jury et al. 1987a,b; Jury & Ghodrati 1989).

Environmental Fate Rate Constants or Half-Lives:
Volatilization:
Photolysis: half-life of 14 d for 6% of 200 µg/mL to degrade in distilled water under sunlight (Crosby & Tang 1969; quoted, Cessna & Muir 1991); half-life of 2.25 h for 44% of 200 µg/mL to degrade in distilled water under 300 nm light (Tanaka et al. 1977; quoted, Cessna & Muir 1991); half-life of 2.25 s for 75% of 100 µg/mL to degrade in 0.2% Triton X-100 aqueous solution under 300 nm light (Tanaka et al. 1981; quoted, Cessna & Muir 1991); half-life of 2.25 h for >70% of 200 µg/mL to degrade in aqueous solutions of nonionic surfactants at concns. in excess of critical micelle concn. under 300 nm light (Tanaka et al. 1979; quoted, Cessna & Muir 1991); half-life of 45 h for 69% of 165 µg/mL to degrade in distilled water under >280 nm light (Tanaka et al. 1982; quoted, Cessna & Muir 1991).

Oxidation:
Hydrolysis: half-life of >4 months for 3974 µg/mL to hydrolyze in phosphate buffer at pH 5-9 and 20°C (El-Dib & Aly 1976; quoted, Muir 1991).

Biodegradation: aerobic half-life of ~7 d for 0.01 µg/mL to biodegrade in river water (Eichelberger & Lichtenberg 1971; quoted, Muir 1991); half-life of 166 d for a 100 d leaching and screening test in 0-10 cm depth of soil (Jury et al. 1983, 1984, 1987a,b; Jury & Ghodrati 1989); aerobic half-life of ~10-15 d for 0.0005-10 µg/mL to biodegrade in filtered sewage water at 20°C (Wang et al. 1985; quoted, Muir 1991).

Biotransformation:
Bioconcentration, Uptake (k_1) and Elimination (k_2) Rate Constants:
k_2: 21.05 d^{-1} (catfish, Ellgehausen et al. 1980)

Common Name: Neburon
Synonym: Kloben, Neburea, Neburex
Chemical Name: 1-butyl-3-(3,4-dichlorophenyl)-1-methylurea; N-butyl-N'(3,4-dichlorophenyl)-N-methylurea
Uses: pre-emergence herbicide to control grasses and broadleaf weeds in peas, beans, lucerne, garlic, beets, cereals, strawberries, ornamentals and forestry.
CAS Registry No: 555-37-3
Molecular Formula: $C_{12}H_{16}Cl_2N_2O$
Molecular Weight: 275.2
Melting Point (°C):
    102-103   (Khan 1980; Spencer 1982; Worthing 1991; Tomlin 1994)
    101.5-103 (Montgomery 1993)
    101.5   (Kühne et al. 1995)
Boiling Point (°C):
Density (g/cm³ at 20°C):
Molar Volume (cm³/mol):
    236.0   (modified LeBas method, Spurlock & Biggar 1994a)
Molecular Volume (Å³):
Total Surface Area, TSA (Å²):
Dissociation Constant pK_a:
Heat of Fusion, ΔH_{fus}, kcal/mol:
    7.10   (DSC method, Plato & Glasgow 1969)
Entropy of Fusion, ΔS_{fus}, cal/mol·K (e.u.):
Fugacity Ratio at 25°C (assuming ΔS_{fus} = 13.5 e.u.), F: 0.169

Water Solubility (g/m³ or mg/L at 25°C):
    4.8   (24°C, Bailey & White 1965; quoted, Shiu et al. 1990)
    4.8   (24°C, Melnikov 1971; quoted, Shiu et al. 1990)
    4.8   (Martin & Worthing 1977; quoted, Kenaga 1980; Kenaga & Goring 1980; Isensee 1991)
    4.8   (28°C, Khan 1980)
    5.0   (Agrochemicals Handbook 1987; Tomlin 1994)
    4.8   (selected, Gerstl & Helling 1987)
    4.8   (24°C, Worthing 1987, 1991; quoted, Shiu et al. 1990)
    5.2   (Spurlock 1992; quoted, Spurlock & Biggar 1994b)
    4.8   (24°C, Montgomery 1993)
    5.0   (20-25°C, selected, Augustijn-Beckers et al. 1994; selected, Hornsby et al. 1996)
    4.8   (selected, Lohninger 1994)
    4.67, 9.99 (quoted, calculated-group contribution fragmentation method, Kühne et al. 1995)

Vapor Pressure (Pa at 25°C):
    negligible (rm. temp., Montgomery 1993)

Henry's Law Constant (Pa·m³/mol):

Octanol/Water Partition Coefficient, log K_{OW}:
- 4.59 (selected, Dao et al. 1983; Gerstl & Helling 1987)
- 4.31 (RP-HPLC-k', Braumann et al. 1983)
- 4.22 (Spurlock 1992; quoted, Spurlock & Biggar 1994b)
- 3.80, 4.22; 3.80 (quoted values; selected, Sangster 1993)
- 3.80 (calculated, Montgomery 1993)
- 4.10 (shake flask-UV, Liu & Qian 1995)
- 3.99 (calculated-RPHPLC-k', Liu & Qian 1995)
- 3.80 (selected, Hansch et al. 1995)
- 3.40, 4.02, 4.13 (RP-HPLC, ClogP, calculated-S, Finizio et al. 1997)

Bioconcentration Factor, log BCF:
- 2.41 (calculated-S, Kenaga 1980; quoted, Isensee 1991)
- 1.85, 2.18 (calculated-S, K_{OC}, Kenaga 1980)

Sorption Partition Coefficient, log K_{OC}:
- 3.36 (soil, Hamaker & Thompson 1972; quoted, Kenaga 1980; Kenaga & Goring 1980)
- 3.26, 2.72 (soil, calculated-S, Kenaga 1980)
- 3.49 (av. soils/sediments, Rao & Davidson 1980; quoted, Lyman 1982; Liu & Qian 1995)
- 3.36, 3.23 (quoted, calculated- χ , Gerstl & Helling 1987)
- 3.40 (soil, quoted exptl., Meylan et al. 1992)
- 2.95 (soil, calculated-χ and fragment contribution, Meylan et al. 1992)
- 3.49 (Montgomery 1993)
- 3.40 (20-25°C, selected, Augustijn-Beckers et al. 1994; Hornsby et al. 1996)
- 3.40 (selected, Lohninger 1994)
- 3.60 (calculated-K_{OW}, Liu & Qian 1995)

Half-Lives in the Environment:
Air:
Surface water:
Groundwater:
Sediment:
Soil: residual activity in soil is limited to approximately 3-4 months (Hartley & Kidd 1987; quoted, Montgomery 1993); selected field half-life of 120 d (Augustijn-Beckers et al. 1994; Hornsby et al. 1996).
Biota:

Environmental Fate Rate Constants or Half-Lives:
Volatilization:
Photolysis:
Oxidation:
Hydrolysis: half-life of >4 months for 5500 µg/mL to hydrolyze in phosphate buffer at pH 5-9 and 20°C (El-dib & Aly 1976; quoted, Muir 1991).
Biodegradation:
Biotransformation:
Bioconcentration, Uptake (k_1) and Elimination (k_2) Rate Constants:

Common Name: Oryzalin
Synonym: Dirimal, EL 119, Rycelan, Rycelon, Ryzelan, Surflan
Chemical Name: 4-(dipropylamino)-3,5-dinitrobenzene-sulfonamide; 3,5-dinitro-N^4, N^4-dipropylsulfanilamide
Uses: herbicide for pre-emergence control of many annual grasses and broadleaf weeds in cotton, fruit trees, vines, nut trees, soybeans, groundnuts, oilseed rape, sunflowers, lucerne, peas, sweet potatoes, mint, ornamentals and also used in noncrop areas.
CAS Registry No: 19044-88-3
Molecular Formula: $C_{12}H_{18}N_4O_6S$
Molecular Weight: 346.4
Melting Point (°C):
    137-138 (Khan 1980; Milne 1995)
    141-142 (Spencer 1982; Agrochemicals Handbook 1987; Herbicide Handbook 1989; Worthing 1991; Milne 1995)
    137 (Kühne et al. 1995)
Boiling Point (°C): 265 (dec. Tomlin 1994)
Density (g/cm³ at 20°C):
Molar Volume (cm³/mol):
    351.1 (calculated-LeBas method, this work)
Molecular Volume (Å³):
Total Surface Area, TSA (Å²):
Dissociation Constant pK_a:
    9.40 (Worthing 1991; Tomlin 1994)
    8.60 (Wauchope et al. 1992; Hornsby et al. 1996)
Heat of Fusion, ΔH_{fus}, kcal/mol:
Entropy of Fusion, ΔS_{fus}, cal/mol·K (e.u.):
Fugacity Ratio at 25°C (assuming ΔS_{fus} = 13.5 e.u.), F: 0.070

Water Solubility (g/m³ or mg/L at 25°C):
    2.4 (Martin & Worthing 1977; quoted, Kenaga 1980; Spencer 1982)
    2.6 (Weber et al. 1980; quoted, Willis & McDowell 1982)
    2.4 (Ashton & Crafts 1981)
    2.5 (Agrochemicals Handbook 1987)
    2.4 (Worthing 1987, 1991; quoted, Shiu et al. 1990)
    2.6 (Herbicide Handbook 1989; quoted, Shiu et al. 1990)
    2.5 (Merck Index 1989; quoted, Milne 1995)
    2.5 (20-25°C, selected, Wauchope et al. 1992; Hornsby et al. 1996)
    2.5 (selected, Lohninger 1994)
    2.6 (Tomlin 1994)
    2.4, 13.5 (quoted, calculated-group contribution fragmentation method, Kühne et al. 1995)

Vapor Pressure (Pa at 25°C):
    <1.33×10^{-5} (30°C, Ashton & Crafts 1981)
    <1.30×10^{-5} (30°C, Agrochemicals Handbook 1987)
    <1.33×10^{-6} (Herbicide Handbook 1989; Tomlin 1994)
    <1.33×10^{-5} (30°C, Merck Index 1989)
    <1.30×10^{-6} (Worthing 1991)
    <1.30×10^{-6} (20-25°C, selected, Wauchope et al. 1992; Hornsby et al. 1996)

Henry's Law Constant (Pa·m³/mol):
    0.000188   (calculated-P/C, this work)

Octanol/Water Partition Coefficient, log K_{OW}:
    4.13   (selected, Dao et al. 1983)
    3.73   (Worthing 1991)
    3.72   (pH 7, Tomlin 1994)
    3.73   (Milne 1995)
    3.73   (selected, Hansch et al. 1995)

Bioconcentration Factor, log BCF:
    2.58   (calculated-S, Kenaga 1980)

Sorption Partition Coefficient, log K_{OC}:
    3.43   (soil, calculated-S, Kenaga 1980)
    2.78   (soil, 20-25°C, selected, Wauchope et al. 1992; Hornsby et al. 1996)
    2.78   (estimated-chemical structure, Lohninger 1994)
    2.85-3.04 (Tomlin 1994)

Half-Lives in the Environment:
    Air:
    Surface water:
    Groundwater:
    Sediment:
    Soil: selected field half-life of 20 d (Wauchope et al. 1992; Hornsby et al. 1996); half-life of 2.1 months for aerobic degradation and 10 d for anaerobic degradation (Tomlin 1994).
    Biota:

Environmental Fate Rate Constants or Half-Lives:
    Volatilization:
    Photolysis:
    Oxidation:
    Hydrolysis:
    Biodegradation: in soil, microbial degradation occurs rapidly, half-life of 2.1 month for aerobic and 10 d for anaerobic metabolism (Tomlin 1994).
    Biotransformation:
    Bioconcentration, Uptake (k_1) and Elimination (k_2) Rate Constants:

Common Name: Pebulate
Synonym: PEBC, R-2061, Stauffer 2061, Tillam, Timmam-6-E
Chemical Name: S-propyl butylethyl(thiocarbamate); S-propyl butylethylcarbamothioate
Uses: selective pre-emergence herbicide to control annual grasses and broadleaf weeds in tomatoes, sugar beet, and tobacco.
CAS Registry No: 1114-71-2
Molecular Formula: $C_{10}H_{21}NOS$
Molecular Weight: 203.36
Melting Point (°C): liquid
Boiling Point (°C):
    142   (at 20 mmHg, Agrochemicals Handbook 1987; Merck Index 1989; Montgomery 1993; Milne 1995)
    142   (at 21 mmHg, Herbicide Handbook 1989)
Density (g/cm³ at 20°C):
    0.956   (Agrochemicals Handbook 1987; Worthing 1991; Tomlin 1994; Milne 1995)
    0.9555   (Herbicide Handbook 1989; Montgomery 1993)
Molar Volume (cm³/mol):
    258.7   (calculated-LeBas method, Suntio et al. 1988)
Molecular Volume (Å³):
Total Surface Area, TSA (Å²):
Dissociation Constant pK_a:
Heat of Fusion, ΔH_{fus}, kcal/mol:
Entropy of Fusion, ΔS_{fus}, cal/mol·K (e.u.):
Fugacity Ratio at 25°C (assuming ΔS_{fus} = 13.5 e.u.), F: 1.0

Water Solubility (g/m³ or mg/L at 25°C):
    92   (21°C, Woodford & Evans 1963; quoted, Shiu et al. 1990)
    92   (21°C, Spencer 1973, 1982)
    60   (Herbicide Handbook 1978, 1989; quoted, Kenaga 1980; Kenaga & Goring 1980; Shiu et al. 1989; Isensee 1991; Lohninger 1994)
    60   (Ashton & Crafts 1973, 1981; quoted, Suntio et al. 1988)
    60   (20°C, Khan 1980; Agrochemicals Handbook 1987; Tomlin 1994)
    60   (20°C, Worthing 1987, 1991; quoted, Shiu et al. 1990)
    60   (20°C, selected, Suntio et al. 1988; quoted, Majewski & Capel 1995)
    100   (20-25°C, selected, Wauchope et al. 1992; Hornsby et al. 1996)
    60   (20°C, Montgomery 1993; Milne 1995)
    90.8, 162   (quoted, calculated-group contribution fragmentation method, Kühne et al. 1995)

Vapor Pressure (Pa at 25°C):
    4.67   (Ashton & Crafts 1973, 1981; quoted, Suntio et al. 1988; Herbicide Handbook 1989)
    3.60   (20°C, Hartley & Graham-Bryce 1980; quoted, Suntio et al. 1988)
    9.06   (30°C, Khan 1980)
    9.00   (30°C, Agrochemicals Handbook 1987; Tomlin 1994)
    3.50   (20°C, selected, Suntio et al. 1988; quoted, Majewski & Capel 1995)
    4.70   (Worthing 1991; Tomlin 1994)
    1.186   (20-25°C, selected, Wauchope et al. 1992; Hornsby et al. 1996)
    9.064   (20°C, Montgomery 1993)

Henry's Law Constant (Pa·m³/mol):
- 11.67 (20°C, calculated-P/C, Suntio et al. 1988; quoted, Majewski & Capel 1995)
- 11.65 (20°C, calculated-P/C, Montgomery 1993)

Octanol/Water Partition Coefficient, log K_{OW}:
- 3.78 (selected, Magee 1991)
- 3.84 (Worthing 1991; Montgomery 1993; Milne 1995)
- 3.83 (Tomlin 1994)
- 3.84 (selected, Hansch et al. 1995)
- 4.19, 3.74, 3.27 (RP-HPLC, ClogP, calculated-S, Finizio et al. 1997)

Bioconcentration Factor, log BCF:
- 1.79 (calculated-S, Kenaga 1980; quoted, Isensee 1991)
- 1.54 (calculated-K_{OC}, Kenaga 1980)

Sorption Partition Coefficient, log K_{OC}:
- 2.80 (soil, Hamaker & Thompson 1972; quoted, Kenaga 1980; Kenaga & Goring 1980)
- 2.66 (soil, calculated-S as per Kenaga & Goring 1978, Kenaga 1980)
- 2.80 (reported as log K_{OM}, Magee 1991)
- 2.65 (estimated as log K_{OM}, Magee 1991)
- 2.63 (soil, 20-25°C, selected, Wauchope et al. 1992; Hornsby et al. 1996)
- 2.80 (Montgomery 1993)
- 2.63 (selected, Lohninger 1994)

Half-Lives in the Environment:
- Air:
- Surface water: half-life of 11 d, at pH 4 and pH 10, 12 d at pH 7 (40°C, Tomlin 1994).
- Groundwater:
- Sediment:
- Soil: half-life in moist loam soil is approximately 2 weeks at 21-27°C (Herbicide Handbook 1989; Montgomery 1993); selected field half-life of 14 d (Wauchope et al. 1992; Hornsby et al. 1996); half-life of 2-3 weeks (Tomlin 1994); .
- Biora:

Environmental Fate Rate Constants or Half-Lives:
- Volatilization:
- Photolysis:
- Oxidation:
- Hydrolysis:
- Biodegradation: in soil, microbial degradation half-life is 2-3 weeks (Tomlin 1994).
- Biotransformation:
- Bioconcentration, Uptake (k_1) and Elimination (k_2) Rate Constants:

Common Name: Picloram
Synonym: Amdon, ATCP, Borolin, Grazon, K-Pin, Tordon
Chemical Name: 4-amino-3,5,6-trichloropicolinic acid; 4-amino-3,5,6-trichloro-2-pyridinecarboxylic acid
Uses: systemic herbicide to control most broadleaf weeds on grassland and noncropped land.
CAS Registry No: 1918-02-1
Molecular Formula: $C_6H_3Cl_3N_2O_2$
Molecular Weight: 241.48
Melting Point (°C):
- 215 (dec., Spencer 1982; Tomlin 1994; Milne 1995)
- 218-219 (Suntio et al. 1988)
- 215-219 (dec., Montgomery 1993)

Boiling Point (°C):
Density (g/cm³ at 20°C):
Molar Volume (cm³/mol):
- 204.2 (calculated-LeBas method, Suntio et al. 1988)

Molecular Volume ($Å^3$):
Total Surface Area, TSA ($Å^2$):
Dissociation Constant pK_a:
- 1.90 (Weber et al. 1980; Willis & McDowell 1982)
- 3.60 (Merck Index 1983; quoted, Howard 1991; Yao & Haag 1991; Montgomery 1993)
- 2.3 (22°C, Worthing 1991; Montgomery 1993; Tomlin 1994)
- 1.94 (Hornsby et al. 1996)

Heat of Fusion, ΔH_{fus}, kcal/mol:
Entropy of Fusion, ΔS_{fus}, cal/mol·K (e.u.):
Fugacity Ratio at 25°C (assuming ΔS_{fus} = 13.5 e.u.), F: 0.0123

Water Solubility (g/m³ or mg/L at 25°C):
- 430 (Bailey & White 1965; Freed 1966; Khan 1980; Ashton & Crafts 1981; quoted, Taylor & Glofelty 1988; Suntio et al. 1988; Shiu et al. 1990)
- 546 (20°C, shake flask-colorimetric, Cheung & Biggar 1974; quoted, Spencer 1976; Suntio et al. 1988; Shiu et al. 1990)
- 685 (30°C, shake flask-colorimetric, Cheung & Biggar 1974; quoted, Shiu et al. 1990)
- 430 (Martin & Worthing 1977, Worthing 1991; quoted, Kenaga 1980; Kenaga & Goring 1980; Isensee 1991; Howard 1991)
- 430 (Hartley & Graham-Bryce 1980; quoted, Taylor & Glotfelty 1988)
- 430 (Weber et al. 1980; Spencer 1982; Willis & McDowell 1982)
- 427 (Garten & Trabalka 1983; quoted, Shiu et al. 1990)
- 430 (Agrochemicals Handbook 1987; Herbicide Handbook 1989; Tomlin 1994)
- 430 (selected, Gertl & Helling 1987; Milne 1995)
- 430 (20°C, selected, Suntio et al. 1988)
- 400-430 (Montgomery 1993)

Vapor Pressure (Pa at 25°C):
- 7.30×10^{-7} (20°C, Hartley & Graham-Bryce 1980; quoted, Taylor & Glotfelty 1988; Taylor & Spencer 1990)
- 8.20×10^{-5} (35°C, Khan 1980; quoted, Suntio et al. 1988)

8.26×10^{-5} (20-25°C, Weber et al. 1980; Willis & McDowell 1982)
  8.20×10^{-5} (35°C, Ashton & Crafts 1981; Herbicide Handbook 1989)
  9.70×10^{-9} (Dobbs 1982; quoted, Howard 1991)
  7.30×10^{-6} (20°C, quoted from Hartlet & Graham-Bryce 1980, Dobbs et al. 1984)
  8.20×10^{-5} (35°C, Agrochemicals Handbook 1987)
  6.00×10^{-5} (20°C, selected, Suntio et al. 1988)
  1.40×10^{-4} (45°C, Herbicide Handbook 1989)
  4.50×10^{-8} (quoted, Nash 1989)
  7.40×10^{-7} (20°C, selected, Taylor & Spencer 1990)
  8.20×10^{-5} (35°C, Worthing 1991; Tomlin 1994)
  8.20×10^{-5} (35°C, Montgomery 1993)

Henry's Law Constant (Pa·m³/mol):
  3.40×10^{-5} (20°C, calculated-P/C, Suntio et al. 1988)
  4.20×10^{-7} (calculated-P/C, Taylor & Glotfelty 1988)
  2.50×10^{-5} (calculated-P/C, Nash 1989)
  4.10×10^{-6} (calculated-P/C, Howard 1991)
  3.40×10^{-5} (20-35°C, calculated-P/C, Montgomery 1993)
  3.17×10^{-5} (calculated-P/C, this work)

Octanol/Water Partition Coefficient, log K_{OW}:
  0.30   (Kenaga 1975; quoted, Kenaga & Goring 1980; Rao & Davidson 1980; Suntio et al. 1988)
  0.63   (selected, Dao et al. 1983)
  0.30   (Hansch & Leo 1985; quoted, Howard 1991)
  −3.47   (selected, Gerstl & Helling 1987)
  0.30   (selected, Suntio et al. 1988)
  1.166   (calculated as per Broto et al. 1984, Karcher & Devillers 1990)
  0.30   (Magee 1991)
  0.26, 0.30 (quoted, Sangster 1993)
  0.30   (Montgomery 1993)

Bioconcentration Factor, log BCF:
  −1.70   (fish in static water, quoted from Dow Chemical data, Kenaga & Goring 1980)
  1.30   (calculated-S, Kenaga 1980; quoted, Isensee 1991)
  −0.222 (calculated-K_{OC}, Kenaga 1980)
  0.0   (estimated-K_{OW}, Lyman et al. 1982; quoted, Howard 1991)
  1.49   (fish in flowing water, Garten & Trabalka 1983; quoted, Howard 1991)

Sorption Partition Coefficient, log K_{OC}:
  1.23   (soil, Hamaker & Thompson 1972; quoted, Kenaga 1980; Kenaga & Goring 1980)
  1.10   (average in soil, Hamaker & Thompson 1972; quoted, Howard 1991)
  1.10   (average in soil, Reinhold et al. 1979; quoted, Howard 1991)
  1.23   (Kenaga & Goring 1980; quoted, Bahnick & Doucette 1988)
  2.20   (soil, calculated-S as per Kenaga & Goring 1978, Kenaga 1980)

1.41   (av. of 26 soils, Rao & Davidson 1980; quoted, Lyman 1982)
1.40   (soil, Rao & Davidson 1982; quoted, Howard 1991)
1.31   (Catlin soil, McCall & Agin 1985; quoted, Brusseau & Rao 1989)
1.05   (Commerce soil, McCall & Agin 1985; quoted, Brusseau & Rao 1989)
1.34   (Fargo soil, McCall & Agin 1985; quoted, Brusseau & Rao 1989)
1.0   (Holdredge soil, McCall & Agin 1985; quoted, Brusseau & Rao 1989)
1.26   (Norfolk soil, McCall & Agin 1985; quoted, Brusseau & Rao 1989)
1.1   (Kawkawlin soil, McCall & Agin 1985; quoted, Brusseau & Rao 1989)
1.05   (Walla-Walla soil, McCall & Agin 1985; quoted, Brusseau & Rao 1989)
1.23, 2.11 (quoted, calculated-χ, Gerstl & Helling 1987)
1.68   (screening model calculations, Jury et al. 1987b)
1.47   (calculated-χ, Bahnick & Doucette 1988)
1.88   (quoted, Nash 1989)
−0.222 (selected, USDA 1989; quoted, Neary et al. 1993)
1.23   (reported as log K_{OM}, Magee 1991)
1.20   (organic carbon, Wauchope et al. 1991; quoted, Dowd et al. 1993)
1.11, 1.41, 1.68 (quoted values, Bottoni & Funari 1992)
1.41   (Montgomery 1993)

Half-Lives in the Environment:
   Air: 12.21 d, based on estimated rate constant for the vapor-phase reaction with photochemically produced hydroxyl radicals in the atmosphere (GEMS 1986; quoted, Howard 1991).
   Surface water: decomposed by u.v. irradiation with a half-life of 2.6 d (Tomlin 1994).
   Groundwater: half-life of >15 months for 0.07, 0.72 and 10 µg/mL to biodegrade in ground water (Wiedner 1974; quoted, Muir 1991).
   Sediment:
   Soil: estimated persistence of 18 months (Kearney et al. 1969; Edwards 1973; quoted, Morrill et al. 1982; Jury et al. 1987b); persistent in soils with half-life >5 years (Alexander 1973; quoted, Howard 1991); estimated first-order half-life of 95 d in soil from biodegradation rate constant of 0.0073 d^{-1} by soil incubation die-away test studies (Rao & Davidson 1980; quoted, Scow 1982); persistent in soil with half-life > 100 d (Willis & McDowell 1982); half-life 100 d from screening model calculations (Jury et al. 1987b); selected half-life of 90 d (Wauchope et al. 1991; quoted, Dowd et al. 1993); half-life of 3-330 d (Tomlin 1994).
   Biota: biochemical half-life of 100 d from screening model calculations (Jury et al. 1987b); average half-life of 60 d in the forest (USDA 1989; quoted, Neary et al. 1993).

Environmental Fate Rate Constants or Half-Lives:
   Volatilization:
   Photolysis: half-life of 200 h for 40% of 4,840 µg/mL to degrade in dilute NaOH solution under sunlight (Hall et 1968; quoted, Cessna & Muir 1991); half-lives for direct photolysis by sunlight under various conditions ranged from 2.3 d to 9.58 d at depths of 2.54 cm-3.65 m at various times of the year; one result at 3.65 m during Sept.-Oct. gave a half-life of 41.3 d; distilled water and canal water gave essentially the same results in one set of

experiments (Hedlund & Youngson 1972; quoted, Cessna & Muir 1991; Howard 1991); half-life of 72 h for 99% of 548 µg/mL to degrade in Na salt solution under 300-380 nm light (Mosier & Guenzi 1973; quoted, Cessna & Muir 1991); half-life of 0.5 h for 38% of 265 µg/mL to degrade in distilled water under 254 nm light (Glass 1975; quoted, Cessna & Muir 1991); half-life of 2.2 d for <2.4 µg/mL to degrade in distilled water under sunlight (Skurlatov et al. 1983; quoted, Cessna & Muir 1991); half-life of 16 h in surface water estimated from direct midday sunlight photolysis in mid-summer at 40°N (Zepp 1991).

Oxidation: photooxidation half-life of 12.21 d in air, based on estimated rate constant for the vapor-phase reaction with photochemically produced hydroxyl radicals in the atmosphere (GEMS 1986; quoted, Howard 1991); rate constant of 5.9×10^9 M^{-1} s^{-1} for the reaction (Fenton with reference to acetophenone) with hydroxyl radicals in aqueous solutions at pH 2.1-3.7 and at (24 ± 1)°C (Buxton et al. 1988; quoted, Haag & Yao 1992); rate constant of $(3.4 \pm 0.3) \times 10^9$ M^{-1} s^{-1} for the reaction (Fenton with reference to acetophenone) with hydroxyl radicals in aqueous solutions at pH 2.1-3.7 and at (24 ± 1)°C (Haag & Yao 1992).

Hydrolysis:

Biodegradation: half-lives in mixture of 5 gram soil and 1-4 mL water, 128-144 h; in mixture of 1 mL water with 0.25-10 gram soil, 90-1000 h (Hance 1969; quoted, Howard 1991); half-life of >15 months for 0.07, 0.72 and 10 µg/mL to biodegrade in groundwater (Weidner 1974; quoted, Muir 1991); rate constant of 0.0073 d^{-1} by soil incubation die-away test studies (Rao & Davidson 1980; quoted, Scow 1982); biochemical half-life of 100 from screening model calculations (Jury et al. 1987b); degraded slowly by soil micro- organisms with a half-life of 30-300 d (Tomlin 1994).

Biotransformation:

Bioconcentration, Uptake (k_1) and Elimination (k_2) Rate Constants:

Common Name: Profluralin
Synonym: CGA 10832, Pregard, Tolban
Chemical Name: N-(cyclopropylmethyl)-2,6-dinitro-N-propyl-4-trifluoromethylaniline; N-(cyclopropylmethyl)-2,6-dinitro-N-propyl-4-(trifluoromethyl)benzenamine
Uses: herbicide for pre-planting by soil incorporation to control annual and perennial broadleaf and grass weeds in cotton, soybeans, brassicas, capsicums, tomatoes and other crops.
CAS Registry No: 26399-36-0
Molecular Formula: $C_{14}H_{16}F_3N_3O_4$
Molecular Weight: 347.3
Melting Point (°C):
    32 (Spencer 1982)
    32.1-32.5 (Agrochemicals Handbook 1987)
    33-36 (Suntio et al. 1988; Milne 1995)
    32-33 (Worthing 1991)
Boiling Point (°C):
Density (g/cm³ at 20°C):
    1.45 (25°C, Ashton & Crafts 1981)
    1.38 (Agrochemicals Handbook 1987; Worthing 1991)
Molar Volume (cm³/mol):
    304.7 (calculated-LeBas method, Suntio et al. 1988)
Molecular Volume (Å³):
Total Surface Area, TSA (Å²):
Dissociation Constant pK_a:
Heat of Fusion, ΔH_{fus}, kcal/mol:
Entropy of Fusion, ΔS_{fus}, cal/mol·K (e.u.):
Fugacity Ratio at 25°C (assuming ΔS_{fus} = 13.5 e.u.), F: 0.778

Water Solubility (g/m³ or mg/L at 25°C):
    0.10 (20°C, Weber 1972; quoted, Muir 1991)
    0.10 (Spencer 1973, 1982; quoted, Shiu et al. 1990)
    0.10 (Wauchope 1978; Kenaga 1980; quoted, Isensee 1991; Pinsuwan et al. 1995)
    0.10 (27°C, Ashton & Crafts 1973, 1981; quoted, Suntio et al. 1988; Shiu et al. 1990)
    0.10 (shake flask-HPLC, Ellgehausen et al. 1981; quoted, Shiu et al. 1990)
    0.10 (20°C, Agrochemicals Handbook 1987)
    0.10 (20°C, Worthing 1987, 1991; quoted, Shiu et al. 1990; Muir 1991)
    0.08 (20°C, selected, Suntio et al. 1988; quoted, Majewski & Capel 1995)
    0.10 (20-25°C, selected, Augustijn-Beckers et al. 1994; Hornsby et al. 1996)
    0.10 (selected, Lohninger 1994)
    0.10 (20°C, Milne 1995)

Vapor Pressure (Pa at 25°C):
    0.0092 (20°C, Weber 1972; Worthing 1987; quoted, Muir 1991)
    0.0092 (20°C, Ashton & Crafts 1973,81; quoted, Suntio et al. 1988)
    0.0084 (20°C, Agrochemicals Handbook 1987)
    0.009 (20°C, selected, Suntio et al. 1988; quoted, Majewski & Capel 1995)
    0.0084 (20°C, Worthing 1991)
    0.0084 (20-25°C, selected, Augustijn-Beckers et al. 1994; Hornsby et al. 1996)

Henry's Law Constant (Pa·m³/mol):
    39.07   (20°C, calculated-P/C, Suntio et al. 1988; quoted, Majewski & Capel 1995)
    31.91   (20°C, calculated-P/C, Muir 1991)

Octanol/Water Partition Coefficient, log K_{OW}:
    5.16   (selected, Dao et al. 1983)
    6.34   (shake flask-HPLC/UV, Ellgehausen et al. 1981)
    5.58   (calculated-fragmental const., Ellgehausen et al. 1981)
    5.44   (selected, Magee 1991)
    6.34   (quoted and selected, Sangster 1993)
    6.34, 4.46   (quoted, calculated-f const., Pinsuwan et al. 1995)
    6.34   (selected, Hansch et al. 1995)

Bioconcentration Factor, log BCF:
    3.35   (calculated-S, Kenaga 1980; quoted, Isensee 1991)
    2.83   (calculated-K_{OC}, Kenaga 1980)

Sorption Partition Coefficient, log K_{OC}:
    3.93   (soil, exptl., Kenaga 1980)
    4.19   (soil, calculated-S as per Kenaga & Goring 1978, Kenaga 1980)
    3.93, 3.83   (reported, estimated as log K_{OM}, Magee 1991)
    3.93   (soil, quoted exptl., Meylan et al. 1992)
    4.26   (soil, calculated-χ and fragment contribution, Meylan et al. 1992)
    4.00   (20-25°C, estimated, Augustijn-Beckers et al. 1994; Hornsby et al. 1996)
    4.16   (selected, Lohninger 1994)

Half-Lives in the Environment:
    Air:
    Surface water: biodegradation half-life < 20 d in water and sediment with flooded soils and terrestrial-aquatic model ecosystems (Miur 1991).
    Groundwater:
    Sediment: biodegradation half-life < 20 d in water and sediment with flooded soils and terrestrial-aquatic model ecosystems (Miur 1991).
    Soil: half-life of 12 d for 0.5 µg/mL to biodegrade in flooded soils at 20-42°C (Savage 1978; quoted, Muir 1991); persistence of 12 months in soil (Wauchope 1978); half-life of less than one month for 1 µg/mL to biodegrade in flooded soils at 25°C (Camper et al. 1980; quoted, Muir 1991); half-lives of 19.9 weeks at 4°C, 6.7 weeks at 25°C for for soil of field capcity moisture, 20.4 weeks at 4°C, 4.8 weeks at 25°C for flooded soils, Crowley silt loam; 25.8 weeks at 4°C, 8.6 weeks at 25°C for soil of field capacity moisture, 21.3 weeks at 4°C and 6.2 weeks at 25°C for flooded soils, Sharkey silty clay (Brewer et al. 1982); selected field half-life of 110 d (Augustijn-Beckers et al. 1994; Hornsby et al. 1996).
    Biota:

Environmental Fate Rate Constants or Half-Lives:
    Volatilization: estimated half-life of 1.2 d from 1 m depth of water at 20°C (Muir 1991).
    Photolysis:
    Oxidation:
    Hydrolysis:
    Biodegradation: half-life of 12 d for 0.5 µg/mL to biodegrade in flooded soils at 20-42°C (Savage 1978; quoted, Muir 1991); half-life of less than 1 month for 1 µg/mL to biodegrade in flooded soils at 25°C (Camper et al. 1980; quoted, Muir 1991); biodegradation half-life <20 d in water and sediment with flooded soils and terrestrial-aquatic model ecosystems (Miur 1991)..
    Biotransformation:
    Bioconcentration, Uptake (k_1) and Elimination (k_2) Rate Constants:

Common Name: Prometon
Synonym: G 31435, Gesafram, Gesagram, Methoxypropazine, Ontracic 800, Ontrack, Pramitol, Prometone
Chemical Name: 6-methoxy-N,N'-bis(methylethyl)-1,3,5-triazine-2,4-diamine; 2,4-bis(isopropylamino)-6-methoxy-1,3,5-triazine
Uses: nonselective pre-emergence and post-emergence herbicide to control most annual and broadleaf weeds, grasses, and brush weeds on noncrop land.
CAS Registry No: 1610-18-0
Molecular Formula: $C_{10}H_{19}N_5O$
Molecular Weight: 225.3
Melting Point (°C):
    91-92 (Khan 1980; Spencer 1982; Herbicide Handbook 1989; Worthing 1991; Montgomery 1993; Tomlin 1994)
Boiling Point (°C):
Density (g/cm³ at 20°C):
    1.088 (Agrochemicals Handbook 1987; Worthing 1991; Montgomery 1993)
Molar Volume (cm³/mol):
    280.2 (calculated-LeBas method, this work)
Molecular Volume (Å³):
Total Surface Area, TSA (Å²):
Dissociation Constant:
    4.28 (pK_a, Weber 1970; quoted, Bintein & Devillers 1994)
    4.30 (pK_a, 21°C, Worthing 1991; Montgomery 1993)
    9.73 (pK_b, Wauchope et al. 1992; Hornsby et al. 1996)
    9.7 (21°C, pK_b, Tomlin 1994)
Heat of Fusion, ΔH_{fus}, kcal/mol:
    5.30 (DSC method, Plato & Glasgow 1969)
Entropy of Fusion, ΔS_{fus}, cal/mol·K (e.u.):
Fugacity Ratio at 25°C (assuming ΔS_{fus} = 13.5 e.u.), F: 0.220

Water Solubility (g/m³ or mg/L at 25°C):
    750 (20°C, Bailey & White 1965; quoted, Shiu et al. 1990)
    750 (Martin & Worthing 1977; Herbicide Handbook 1978; quoted, Kenaga 1980; Kenaga & Goring 1980; Isensee 1991)
    750 (20°C, Khan 1980)
    677 (Weber et al. 1980; quoted, Willis & McDowell 1982)
    750 (20°C, Ashton & Crafts 1981; Herbicide Handbook 1989)
    620 (20°C, Spencer 1982)
    750 (20°C, Verschueren 1983; quoted, Shiu et al. 1990)
    750 (quoted, Jury et al. 1984; Spencer et al. 1988; Spencer & Cliath 1990)
    750 (20°C, Agrochemicals Handbook 1987; Gerstl & Helling)
    620 (20°C, Worthing 1987, 1991; Tomlin 1994)
    720 (20-25°C, selected, Wauchope et al. 1992; Hornsby et al. 1996)
    750 (20°C, Montgomery 1993)
    720 (selected, Lohninger 1994)
    750, 597 (quoted, predicted-AQUAFAC, Lee et al. 1996)

Vapor Pressure (Pa at 25°C):
- 0.00030 (20°C, Khan 1980)
- 0.00031 (20°C, Ashton & Crafts 1981; Worthing 1991)
- 0.00083 (Jury et al. 1984; quoted, Spencer et al. 1988; Spencer & Cliath 1990; Taylor & Spencer 1990)
- 0.00031 (20°C, Agrochemicals Handbook 1987)
- 0.00031 (20°C, Herbicide Handbook 1989)
- 0.00105 (30°C, Herbicide Handbook 1989)
- 0.00083 (selected, Taylor & Spencer 1990)
- 0.00103 (20-25°C, selected, Wauchope et al. 1992; Hornsby et al. 1996)
- 0.00031 (20°C, Montgomery 1993)
- 0.000306 (20°C, Tomlin 1994)

Henry's Law Constant (Pa·m^3/mol):
- 2.50×10^{-4} (calculated-P/C, Jury et al. 1984; quoted, Spencer et al. 1988; Spencer & Cliath 1990)
- 9.02×10^{-5} (20°C, calculated-P/C, Montgomery 1993)
- 9.01×10^{-5} (calculated-P/C, this work)

Octanol/Water Partition Coefficient, log K_{OW}:
- 1.94 (selected, Dao et al. 1983)
- 1.94 (Gerstl & Helling 1987; quoted, Bintein & Devillers 1994)
- 2.99, 3.10 (RP-HPLC, calculated, Finizio et al. 1991)
- 2.85 (selected, Magee 1991)
- 2.99 (quoted and selected, Sangster 1993)
- 2.55 (shake flask-UV, Liu & Qian 1995)
- 2.69, 2.99 (Montgomery 1993)
- 2.99 (selected, Hansch et al. 1995)
- 2.82, 2.98, 2.58 (RP-HPLC, ClogP, calculated-S, Finizio et al. 1997)

Bioconcentration Factor, log BCF:
- 1.18 (calculated-S, Kenaga 1980; quoted, Isensee 1991)
- 1.28 (calculated-K_{OC}, Kenaga 1980)

Sorption Partition Coefficient, log K_{OC}:
- 2.54 (soil, Hamaker & Thompson 1972; quoted, Kenaga 1980; Kenaga & Goring 1980; Liu & Qian 1995)
- 2.04 (soil, calculated-S as per Kenaga & Goring 1978, Kenaga 1980)
- 2.61 (Jury et al. 1984; quoted, Spencer & Cliath 1990)
- 3.93 (selected, Gerstl & Helling)
- 2.40 (calculated-χ, Gerstl & Helling)
- 2.48 (Spencer et al. 1988)
- 2.54 (reported as log K_{OM}, Magee 1991)
- 2.35 (estimated as log K_{OM}, Magee 1991)
- 2.60 (soil, quoted exptl., Meylan et al. 1992)
- 2.20 (soil, calculated-χ and fragment contribution, Meylan et al. 1992)

2.18　　(soil, 20-25°C, selected, Wauchope et al. 1992; Hornsby et al. 1996)
　　1.92-2.24　(Montgomery 1993)
　　2.77　　(selected, Lohninger 1994)
　　2.39　　(calculated-K_{OW}, Liu & Qian 1995)

Half-Lives in the Environment:
　　Air:
　　Surface water:
　　Groundwater:
　　Sediment:
　　Soil: selected field half-life of 500 d (Wauchope et al. 1992; Hornsby et al. 1996).
　　Biota:

Environmental Fate Rate Constants or Half-Lives:
　　Volatilization: estimated half-life of 100 d (Spencer & Cliath 1990).
　　Photolysis: half-life of 2.25 h for 1% of 100 μg/mL to degrade in distilled water
　　　　under 300 nm light (Tanaka et al. 1981; quoted, Cessna & Muir 1991).
　　Oxidation:
　　Hydrolysis:
　　Biodegradation:
　　Biotransformation:
　　Bioconcentration, Uptake (k_1) and Elimination (k_2) Rate Constants:

Common Name: Prometryn
Synonym: Caparol, Cotton-Pro, Gesagard, G-34161, Mercasin, Mercazin, Polisin, Primatol, Prometrex, Prometrin, Selectin, Sesagard, Uvon
Chemical Name: N,N'-bis(1-methylethyl)-6-(methylthio)-1,3,5-triazine-2,4-diamine; 2,4-bis(isopropylamino)-6-(methylthio)-1,3,5-triazine
Uses: selective herbicide to control many annual grass and broadleaf weeds in celery, cotton and peas.
CAS Registry No: 7287-19-6
Molecular Formula: $C_{10}H_{19}N_5S$
Molecular Weight: 241.37
Melting Point (°C):
    118-120   (Khan 1980; Spencer 1982; Suntio et al. 1988; Herbicide Handbook 1989; Worthing 1991; Montgomery 1993; Milne 1995)
    72.5   (Kühne et al. 1995)
Boiling Point (°C):
Density (g/cm³ at 20°C):
    1.157   (Agrochemicals Handbook 1987; Worthing 1991; Montgomery 1993; Milne 1995)
Molar Volume (cm³/mol):
    299.7   (calculated-LeBas method, Suntio et al. 1988)
Molecular Volume (Å³):
Total Surface Area, TSA (Å²):
Dissociation Constant:
    4.05   (pK_a, Weber 1970; Pacakova et al. 1988; Somasundaram et al. 1991; Bintein & Devillers 1994)
    4.10   (pK_a, 21°C, Weber et al. 1980; Willis & McDowell 1982; Worthing 1991)
    9.95   (pK_b, Wauchope et al. 1992; Hornsby et al. 1996)
    4.05   (pK_a, 21°C, Montgomery 1993)
Heat of Fusion, ΔH_{fus}, kcal/mol:
    6.30   (DSC method, Plato & Glasgow 1969)
Entropy of Fusion, ΔS_{fus}, cal/mol·K (e.u.):
Fugacity Ratio at 25°C (assuming ΔS_{fus} = 13.5 e.u.), F: 0.115

Water Solubility (g/m³ or mg/L at 25°C):
    48   (20°C, Woodford & Evans 1963; quoted, Shiu et al. 1990)
    48   (20°C, Bailey & White 1965)
    206   (26°C, shake flask-UV at pH 3, Ward & Weber 1968)
    40.3   (26°C, shake flask-UV at pH 7, Ward & Weber 1968)
    41.8   (26°C, shake flask-UV at pH 10, Ward & Weber 1968)
    48   (20°C, Ashton & Crafts 1973, 1981; Khan 1980; quoted, Suntio et al. 1988; Shiu et al. 1990)
    48   (Martin & Worthing 1977; Herbicide Handbook 1978; quoted, Kenaga 1980; Kenaga & Goring 1980; Jury et al. 1983, 1984; Isensee 1991)
    48   (Wauchope 1978)
    40   (Weber et al. 1980; quoted, Willis & McDowell 1982)
    48   (20°C, Windholz 1983; quoted, Somasundaram et al. 1991)
    48   (quoted, Jury et al. 1984; Spencer & Cliath 1990)
    48   (20°C, Agrochemicals Handbook 1987; Gerstl & Helling)
    33   (20°C, Worthing 1987, 1991; quoted, Shiu et al. 1990)

48   (20°C, selected, Suntio et al. 1988; quoted, Majewski & Capel 1995)
48   (20°C, Herbicide Handbook 1989; quoted, Shiu et al. 1990)
33   (20-25°C, selected, Wauchope et al. 1992; Hornsby et al. 1996)
48   (20°C, Montgomery 1993)
33   (Tomlin 1994; selected, Lohninger 1994)
19.2, 1 (quoted exptl., calculated, Kühne et al. 1995)
48   (20°C, Milne 1995)

Vapor Pressure (Pa at 25°C):
   0.00028 (gas saturation, Friedrich & Stammbach 1964; quoted, Jury et al. 1983)
   0.00013 (20°C, Ashton & Crafts 1973,81; quoted, Agrochemicals Handbook 1987; Suntio et al. 1988)
   0.00013 (20°C, Khan 1980)
   0.00013 (20-25°C, Weber et al. 1980; quoted, Willis & McDowell 1982)
   0.00028 (quoted, Jury et al. 1984; Spencer & Cliath 1990)
   0.00010 (20°C, selected, Suntio et al. 1988; quoted, Majewski & Capel 1995)
   0.00013 (20°C, Herbicide Handbook 1989)
   0.00053 (30°C, Herbicide Handbook 1989)
   0.00013 (20°C, Worthing 1991)
   0.00017 (20-25°C, selected, Wauchope et al. 1992; Hornsby et al. 1996)
   0.00013 (20°C, Montgomery 1993)
   0.000169 (Tomlin 1994)

Henry's Law Constant (Pa·m^3/mol):
   0.00139 (calculated-P/C, Jury et al. 1984; quoted, Spencer & Cliath 1990)
   0.00139 (calculated-P/C, Jury et al. 1987a,b; Jury & Ghodrati 1989)
   0.00050 (20°C, calculated-P/C, Suntio et al. 1988; quoted, Majewski & Capel 1995)
   0.00050 (20°C, calculated-P/C, Montgomery 1993)

Octanol/Water Partition Coefficient, log K_{ow}:
   2.99   (selected, Dao et al. 1983; quoted, Liu & Qian 1995)
   1.91   (RP-HPLC-k', Braumann et al. 1983)
   3.46   (selected, Yoshioka et al. 1986; quoted, Somasundaram et al. 1991)
   3.51   (shake flask, Mitsutake et al. 1986)
   2.99   (Gerstl & Helling 1987; quoted, Bintein & Devillers 1994)
   3.34, 3.48 (RP-HPLC, calculated, Finizio et al. 1991)
   3.43   (selected, Magee 1991)
   3.34   (Worthing 1990, 1991; quoted, Finizio et al. 1991)
   3.34, 3.46 (Montgomery 1993)
   3.34, 3.43, 3.51; 3.51 (quoted values; selected, Sangster 1993)
   2.93   (calculated-RPHPLC-k', Liu & Qian 1995)
   3.34   (Milne 1995)
   3.51   (selected, Hansch et al. 1995)
   3.25, 3.39, 3.46 (RP-HPLC, ClogP, calculated-S, Finizio et al. 1997)

Bioconcentration Factor, log BCF:
   1.85, 1.67 (calculated-S, K_{oc}, Kenaga 1980; quoted, Isensee 1991)

Sorption Partition Coefficient, log K_{OC}:
- 2.91 (soil, Hamaker & Thompson 1972; quoted, Kenaga 1980; Kenaga & Goring 1980; Liu & Qian 1995)
- 2.72 (soil, calculated-S as per Kenaga & Goring 1978, Kenaga 1980)
- 2.79 (Rao & Davidson 1980; quoted, Jury et al. 1983)
- 2.79 (Jury et al. 1984; quoted, Spencer & Cliath 1990)
- 2.99, 3.17 (quoted, calculated- χ, Gerstl & Helling)
- 2.78 (screening model calculations, Jury et al. 1987a,b; Jury & Ghodrati 1989)
- 2.92, 2.75 (reported, estimated as log K_{OM}, Magee 1991)
- 2.72-2.91, 2.79, 2.83 (quoted values, Bottoni & Funari 1992)
- 2.60 (soil, 20-25°C, selected, Wauchope et al. 1992; Hornsby et al. 1996)
- 2.28-2.79 (Montgomery 1993)
- 3.15 (estimated-chemical structure, Lohninger 1994)
- 2.60 (soil, Tomlin 1994)
- 2.38 (soil, HPLC-screening method, Kördel et al. 1993, 1995b)
- 2.63 (calculated-K_{OW}, Liu & Qian 1995)

Half-Lives in the Environment:
  Air:
  Surface water: completely decomposed when exposed to UV light for 3 h (Montgomery 1993).
  Groundwater:
  Sediment:
  Soil: estimated persistence of 3 months (Kearney et al. 1969; Edwards 1973; quoted, Morrill et al. 1982; Jury et al. 1987a,b; Jury & Ghodrati 1989); half-life of about 6 months to biodegrade in flooded soils (Plimmer et al. 1970; quoted, Muir 1991); persistence of 2 months in soil (Wauchope 1978); selected field half-life of 60 days (Wauchope et al. 1992; Hornsby et al. 1996); half-life of 70 d for microbial degradation in soil (Tomlin 1994).
  Biota: biochemical half-life of 60 d from screening model calculations (Jury et al. 1987a,b; Jury & Ghodrati 1989).

Environmental Fate Rate Constants or Half-Lives:
  Volatilization: half-life of 60 d (Jury et al. 1984; quoted, Spencer & Cliath 1990).
  Photolysis:
  Oxidation:
  Hydrolysis: half-life of 22 d in 0.1 N hydrochloric acid solution, 500 yr at pH 7 in distilled water and 30 yr in 0.01 sodium hydroxide solution all at 25°C (Montgomery 1993).
  Biodegradation: half-life of 60 d (Wauchope 1978; quoted, Jury et al. 1983, 1984); half-life of 60 d for a 100 d leaching and screening test in 0-10 cm depth of soil (Jury et al. 1987a,b; Jury & Ghodrati 1989); soil microbial degradation half-life is 70 d (Tomlin 1994).
  Biotransformation:
  Bioconcentration, Uptake (k_1) and Elimination (k_2) Rate Constants:

Common Name: Pronamide
Synonym: Kerb, Promamide, Propyzamide, RH-315
Chemical Name: 3,5-dichloro-N-(1,1-dimethylpropynyl)benzamide
Uses: herbicide.
CAS Registry No: 23950-58-5
Molecular Formula: $C_{12}H_{11}Cl_2NO$
Molecular Weight: 256.13
Melting Point (°C):
    154-156 (Khan 1980)
    154   (Herbicide Handbook 1989)
    155-156 (Milne 1995)
Boiling Point (°C): 321
Density (g/cm³ at 20°C):
Molar Volume (cm³/mol):
    270.4 (calculated-LeBas method, this work)
Molecular Volume (Å³):
Total Surface Area, TSA (Å²):
Dissociation Constant pK_a:
Heat of Fusion, ΔH_{fus}, kcal/mol:
Entropy of Fusion, ΔS_{fus}, cal/mol·K (e.u.):
Fugacity Ratio at 25°C (assuming ΔS_{fus} = 13.5 e.u.), F: 0.053

Water Solubility (g/m³ or mg/L at 25°C):
    15   (Martin & Worthing 1977; Herbicide Handbook 1978, 1983; quoted, Kenaga 1980; Kenaga & Goring 1980; Steen & Collette 1989)
    15   15°C, Khan 1980)
    15   (Ashton & Crafts 1981)
    15   (Worthing 1987; quoted, Shiu et al. 1990; Howard 1991; Majewski & Capel 1995)
    15   (24°C, Herbicide Handbook 1989; quoted, Shiu et al. 1990)
    15   (20-25°C, selected, Wauchope et al. 1992; Hornsby et al. 1996)
    15   (selected, Lohninger 1994; Milne 1995)

Vapor Pressure (Pa at 25°C):
    0.0113 (Khan 1980)
    0.0113 (Ashton & Crafts 1981; Herbicide Handbook 1989)
    0.0536 (Dixon & Rissman 1985; quoted, Howard 1991)
    0.227 (Worthing 1987; quoted, Majewski & Capel 1995)
    0.0113 (20-25°C, selected, Wauchope et al. 1992)
    0.0113 (20-25°C, selected, Hornsby et al. 1996)

Henry's Law Constant (Pa·m³/mol):
    0.912 (Dixon & Rissman 1985; quoted, Howard 1991)
    0.193 (calculated-P/C as per Worthing 1987, Majewski & Capel 1995)
    0.188 (calculated-P/C, this work)

Octanol/Water Partition Coefficient, log K_{ow}:
- 3.26 (estimated, Lyman et al. 1982; quoted, Howard 1991)
- 3.36 (selected, Magee 1991)
- 3.26 (selected, Dao et al. 1983)
- 2.95 (estimated-QSAR & SPARC, Kollig et al. 1993)
- 3.09-3.28 (Milne 1995)

Bioconcentration Factor, log BCF:
- 2.13 (calculated-S, Kenaga 1980)
- 1.00 (calculated-K_{OC}, Kenaga 1980)
- 2.25 (estimated-K_{ow}, Lyman et al. 1982; quoted, Howard 1991)
- 2.13 (estimated-S, Lyman et al. 1982; quoted, Howard 1991)

Sorption Partition Coefficient, log K_{OC}:
- 2.30 (soil, Leistra et al. 1974; Carlson et al. 1975; quoted, Kenaga 1980; Kenaga & Goring 1980)
- 2.30 (measured for single soil, Kenaga 1980; quoted, Howard 1991)
- 3.00 (soil, calculated-S as per Kenaga & Goring 1978, Kenaga 1980)
- 2.99 (soil, estimated-S, Lyman et al. 1982; quoted, Howard 1991)
- 2.30 (reported as log K_{OM}, Magee 1991)
- 2.42 (estimated as log K_{OM}, Magee 1991)
- 2.30 (soil, quoted exptl., Meylan et al. 1992)
- 3.20 (soil, calculated-χ and fragment contribution, Meylan et al. 1992)
- 2.90 (soil, 20-25°C, selected, Wauchope et al. 1992; Hornsby et al. 1996)
- 2.63 (estimated-QSAR & SPARC, Kollig 1993)
- 2.54 (selected, Lohninger 1994)

Half-Lives in the Environment:
- Air: 4.2 h, based on an estimated rate constant for the vapor-phase reaction with photochemically produced hydroxyl radicals in the atmosphere (Atkinson 1985; quoted, Howard 1991).
- Surface water:
- Groundwater:
- Sediment:
- Soil: depending on soil and climatic conditions, the degradation half-life may range from 10 to 112 d, but a half-life of 40 d may be more common under field conditions (Walker 1976,78; Zandvoort et al. 1979; quoted, Howard 1991); selected field half-life of 60 d (Wauchope et al. 1992; Hornsby et al. 1996).
- Biota:

Environmental Fate Rate Constants or Half-Lives:
- Volatilization: based on a Henry's law constant of 0.9118 Pa·m³/mol, half-life from a river one meter deep flowing 1 m/s with a wind velocity of 3 m/sec is estimated to be 6.6 d (Lyman et al. 1982; quoted, Howard 1991).
- Photolysis:

Oxidation: photooxidation half-life of 4.2 h in air, based on an estimated rate constant for the vapor-phase reaction with photochemically produced hydroxyl radicals in the atmosphere (Atkinson 1985; quoted, Howard 1991).

Hydrolysis: neutral hydrolysis rate constant of $<1.5\times10^{-5}$ h^{-1} with a calculated half-life of more than 700 d in neutral solution and with faster hydrolysis rates in acidic and basic solutions to be expected (Ellington et al. 1987,88; quoted, Howard 1991).

Biodegradation: depending on soil and climatic conditions, the degradation half-life may range from 10 to 112 d, but a half-life of 40 d may be more common under field conditions (Walker 1976,78; Zandvoort et al. 1979; quoted, Howard 1991).

Biotransformation: second-order rate constant for microbial degradation in natural water was estimated to be 5×10^{-14} L/organisms·h with an estimated half-life of 580 d (Steen & Collette 1989; quoted, Howard 1991).

Bioconcentration, Uptake (k_1) and Elimination (k_2) Rate Constants:

Common Name: Propachlor
Synonym: Albrass, Bexton, CIPA, CP 31393, Niticid, Propachlore, Prolex, Ramrod, Satecid
Chemical Name: 2-chloro-*N*-(1-methylethyl)-*N*-phenylacetamide; 2-chloro-*N*-isopropyl-acetanilide
Uses: selective pre-emergence herbicide to control most annual grasses and some broadleaf weeds in brassicas, corn, cotton, flax, leeks, maize, milo, onions, peas, roses, ornamental trees and shrubs, soybeans, and sugar cane.
CAS Registry No: 1918-16-7
Molecular Formula: $C_{11}H_{14}ClNO$
Molecular Weight: 211.69
Melting Point (°C):
- 67-76 (Khan 1980; Spencer 1982; Suntio et al. 1988; Montgomery 1993)
- 77 (Herbicide Handbook 1989; Worthing 1991; Tomlin 1994; Kühne et al. 1995; Milne 1995)

Boiling Point (°C):
- 110 (at 0.03 mmHg, Ashton & Crafts 1981; Agrochemicals Handbook 1987; Herbicide Handbook 1989; Worthing 1991; Montgomery 1993; Milne 1995)

Density (g/cm^3 at 20°C):
- 1.13 (25°C, Ashton & Crafts 1981)
- 1.242 (25°C, Agrochemicals Handbook 1987; Worthing 1991; Montgomery 1993; Tomlin 1994; Milne 1995)
- 1.134 (25°C, Herbicide Handbook 1989)

Molar Volume (cm^3/mol):
- 231.6 (calculated-LeBas method, Suntio et al. 1988)

Molecular Volume (Å^3):

Total Surface Area, TSA (Å^2):

Dissociation Constant pK_a:

Heat of Fusion, ΔH_{fus}, kcal/mol:
- 6.60 (DSC method, Plato 1972)

Entropy of Fusion, ΔS_{fus}, cal/mol·K (e.u.):

Fugacity Ratio at 25°C (assuming ΔS_{fus} = 13.5 e.u.), F: 0.313

Water Solubility (g/m^3 or mg/L at 25°C):
- 700 (Melnikov 1971; quoted, Khan 1980; Suntio et al. 1988; Shiu et al. 1990)
- 614 (20°C, Weber 1972; Worthing 1987; quoted, Muir 1991)
- 693 (Spencer 1973, 1982; quoted, Shiu et al. 1990)
- 580 (20°C, Ashton & Crafts 1973; quoted, Suntio et al. 1988; Shiu et al. 1990)
- 580 (Martin & Worthing 1977; Herbicide Handbook 1978, 1983; quoted, Kenaga 1980; Kenaga & Goring 1980; Steen & Collette 1989; Shiu et al. 1990; Isensee 1991)
- 580 (Wauchope 1978; Weber et al. 1980; quoted, Willis & McDowell 1982)
- 700 (20°C, Khan 1980)
- 839 (gen. col.-HPLC-RI, Swann et al. 1983; quoted, Shiu et al. 1990)
- 2300 (HPLC-RT, Swann et al. 1983; quoted, Shiu et al. 1990)
- 613 (Agrochemicals Handbook 1987; Herbicide Handbook 1989)
- 700 (selected, Gerstl & Helling 1987)
- 613 (Worthing 1987, 1991; quoted, Shiu et al. 1990)
- 600 (20°C, selected, Suntio et al. 1988; quoted, Majewski & Capel 1995)
- 613 (20-25°C, selected, Wauchope et al. 1992; Hornsby et al. 1996)

613-700　(Montgomery 1993)
　　　613　　(selected, Lohninger 1994)
　　　700, 71.7 (quoted, calculated-group contribution fragmentation method, Kühne et al. 1995)
　　　613　　(Tomlin 1994; Milne 1995)

Vapor Pressure (Pa at 25°C):
　　　0.032　(20-25°C, Weber et al. 1980; quoted, Willis & McDowell 1982)
　　　0.0307　(24°C, Beestman & Deming 1974; quoted, Muir 1991)
　　　0.0307　(Ashton & Crafts 1981; Herbicide Handbook 1989; quoted, Suntio et al. 1988)
　　　0.03　(Agrochemicals Handbook 1987)
　　　0.03　(20°C, selected, Suntio et al. 1988; quoted, Majewski & Capel 1995)
　　　0.0306　(Worthing 1991; Tomlin 1994)
　　　0.0307　(20-25°C, selected, Wauchope et al. 1992; Hornsby et al. 1996)
　　　0.03　(Montgomery 1993)

Henry's Law Constant (Pa·m^3/mol):
　　　0.011　(20°C, calculated-P/C, Suntio et al. 1988; quoted, Majewski & Capel 1995)
　　　0.011　(20°C, calculated-P/C, Muir 1991)
　　　0.011　(calculated-P/C, Montgomery 1993)

Octanol/Water Partition Coefficient, log K_{OW}:
　　　2.75　(Leo et al. 1971; quoted, Kenaga & Goring 1980)
　　　1.61　(quoted, Rao & Davidson 1980; Suntio et al. 1988)
　　　2.80　(selected, Gerstl & Helling 1987)
　　　1.60　(selected, Suntio et al. 1988)
　　　2.18　(selected, Magee 1991)
　　　1.61　(Montgomery 1993)
　　　2.18　(quoted and selected, Sangster 1993)
　　　1.62-2.30 (Tomlin 1994)
　　　2.18　(selected, Hansch et al. 1995)
　　　2.36, 2.06, 2.47 (RP-HPLC, ClogP, calculated-S, Finizio et al. 1997)

Bioconcentration Factor, log BCF:
　　　1.23　(calculated-S, Kenaga 1980; quoted, Isensee 1991)
　　　1.15　(calculated-K_{OC}, Kenaga 1980)

Sorption Partition Coefficient, log K_{OC}:
　　　2.42　(soil, Beestman & Demming 1976; quoted, Kenaga 1980; Kenaga & Goring 1980)
　　　2.11　(soil, calculated-S as per Kenaga & Goring 1978, Kenaga 1980)
　　　2.42, 2.43 (quoted, calculated-χ, Gerstl & Helling 1987)
　　　2.62　(screening model calculations, Jury et al. 1987b)
　　　2.42, 2.31 (reported, estimated as log K_{OM}, Magee 1991)
　　　2.42　(soil, quoted exptl., Meylan et al. 1992)
　　　2.45　(soil, calculated-χ and fragment contribution, Meylan et al. 1992)
　　　1.90　(soil, 20-25°C, selected, Wauchope et al. 1992; Hornsby et al. 1996)
　　　2.07-2.11 (Montgomery 1993)
　　　2.62　(estimated-chemical structure, Lohninger 1994)

Half-Lives in the Environment:
: Air:
: Surface water: half-life of ~ 10-14 d for 0.001-1.0 µg/mL to biodegrade in sewage effluent lake water (Novick & Alexander 1985; quoted, Muir 1991).
: Groundwater:
: Sediment:
: Soil: persistence of 2 months (Wauchope 1978); half-life of 7 d from screening model calculations Jury et al. 1987b); persists in soil for 28-42 d (Worthing 1991); selected field half-life of 6.3 d (Wauchope et al. 1992; Hornsby et al. 1996).
: Biota: biochemical half-life of 7 d from screening model calculations (Jury et al. 1987b).

Environmental Fate Rate Constants or Half-Lives:
: Volatilization: estimated half-life of 671 d from 1 m depth of water at 20°C (Muir 1991).
: Photolysis: half-life of 2.25 h in distilled water (Tanaka et al. 1981; quoted, Cessna & Muir 1991); 1 ppb contaminated water in the presence of TiO_2 and H_2O_2 completely photodegraded after 3 h by solar irradiation (Muszkat et al. 1992).
: Oxidation:
: Hydrolysis:
: Biodegradation: half-life of ~ 10-14 d for 0.001-1.0 µg/mL to biodegrade in sewage effluent lake water (Novick & Alexander 1985; quoted, Muir 1991); biochemical half-life of 7 d from screening model calculations (Jury et al. 1987b).
: Biotransformation: second-order microbial rate constant of 1.1×10^{-9} L·organisms^{-1} h^{-1} (Steen & Collette 1989).
: Bioconcentration, Uptake (k_1) and Elimination (k_2) Rate Constants:

Common Name: Propanil
Synonym: Bay 30130, Chem rice, Crystal Propanil-4, DCPA, Dipram, DPA, DPA, Erban, Erbanil, Farmco propanil, FW-734, Grascide, Herbax technical, Prop-Job, Propanex, Propanid, Riselect, Rogue, Rosanil, S 10165, Stam F-34, Stampede, Stam Supernox, Strel, Supernox, Surcopur, Surpur, STAM, Synpran N, Vertac, Wham EZ
Chemical Name: N-(3,4-dichlorophenyl)propionamide; N-(3,4-dichlorophenyl)propanamide
Uses: selective emergence and post-emergence herbicide to control many grasses and broadleaf weeds in potatoes, rice and wheat.
CAS Registry No: 709-98-8
Molecular Formula: $C_9H_9Cl_2NO$
Molecular Weight: 218.09
Melting Point (°C):
 92-93 (Khan 1980; Spencer 1982; Montgomery 1993; Tomlin 1994)
 91-93 (Suntio et al. 1988; Milne 1995)
 85-89 (Herbicide Handbook 1989)
Boiling Point (°C):
Density (g/cm^3 at 20°C):
 1.25 (25°C, Ashton & Crafts 1981; Agrochemicals Handbook 1987; Herbicide Handbook 1989; Montgomery 1993; Milne 1995)
 1.41 (22°C, Tomlin 1994)
Molar Volume (cm^3/mol):
 220.1 (calculated-LeBas method, Suntio et al. 1988)
Molecular Volume (Å^3):
Total Surface Area, TSA (Å^2):
Dissociation Constant pK_a:
Heat of Fusion, ΔH_{fus}, kcal/mol:
 4.80 (DSC method, Plato & Glasgow 1969)
Entropy of Fusion, ΔS_{fus}, cal/mol·K (e.u.):
Fugacity Ratio at 25°C (assuming ΔS_{fus} = 13.5 e.u.), F: 0.217

Water Solubility (g/m^3 or mg/L at 25°C):
 225 (Woodford & Evans 1963; Khan 1980; quoted, Shiu et al. 1990)
 500 (Bailey & White 1965; quoted, Shiu et al. 1990)
 268 (Freed 1966; quoted, Suntio et al. 1988; Shiu et al. 1990)
 500 (Ashton & Crafts 1973; quoted, Suntio et al. 1988; Shiu et al. 1990)
 225 (Martin & Worthing 1977; quoted, Kenaga 1980)
 268-500 (Weber et al. 1980; quoted, Willis & McDowell 1982)
 130 (20°C, Spencer 1982)
 225 (Herbicide Handbook 1983; quoted, Steen & Collette 1989; Pait et al. 1992)
 225 (Agrochemicals Handbook 1987; quoted, Trevisan et al. 1991)
 225 (selected, Gerstl & Helling 1978; Milne 1995)
 225 (rm. temp., Worthing 1987; quoted, Shiu et al. 1990)
 300 (20°C, selected, Suntio et al. 1988; quoted, Majewski & Capel 1995)
 500 (Herbicide Handbook 1989)
 130 (20°C, Worthing 1991)
 200 (20-25°C, selected, Wauchope et al. 1992; Lohninger 1994; Hornsby et al. 1996; quoted, Halfon et al. 1996)
 130 (20°C, Montgomery 1993)
 225 (Montgomery 1993)

130   (Tomlin 1994)
218, 18.6 (quoted, calculated-group contribution fragmentation method, Kühne et al. 1995)

Vapor Pressure (Pa at 25°C):
- 0.012   (60°C, Khan 1980)
- 0.012   (60°C, Verschueren 1983; quoted, Suntio et al. 1988)
- 0.012   (60°C, Agrochemicals Handbook 1987; quoted, Trevisan et al. 1991)
- 0.005   (20°C, selected, Suntio et al. 1988; quoted, Majewski & Capel 1995)
- 2.60×10^{-5} (20°C, Worthing 1991)
- 0.00533   (20-25°C, selected, Wauchope et al. 1992; Hornsby et al. 1996; quoted, Halfon et al. 1996)
- 2.67×10^{-5} (20°C, Montgomery 1993)

Henry's Law Constant (Pa·m^3/mol):
- 0.0036   (20°C, calculated-P/C, Suntio et al. 1988; quoted, Majewski & Capel 1995)
- 0.0036   (20°C, calculated-P/C, Montgomery 1993)
- 0.00545   (calculated-P/C, this work)

Octanol/Water Partition Coefficient, log K_{ow}:
- 2.02   (quoted, Rao & Davidson 1980; Suntio et al. 1988)
- 2.80   (20 ± 2°C, shake flask-UV, Briggs 1981)
- 3.12   (selected, Dao et al. 1983)
- 2.99   (selected, Gerstl & Helling 1987)
- 2.00   (selected, Suntio et al. 1988)
- 2.53   (selected, Trevisan et al. 1991)
- 2.29   (Worthing 1991)
- 2.34   (quoted from Kenaga 1980, Bottoni & Funari 1992)
- 2.73   (RP-HPLC, Sicbaldi & Finizio 1993)
- 2.03, 2.29 (Montgomery 1993)
- 2.73, 3.07; 3.07 (quoted values; selected, Sangster 1993)
- 2.80   (RP-HPLC, Saito et al. 1993)
- 3.30   (Tomlin 199)
- 2.29   (Milne 1995)
- 3.07   (selected, Hansch et al. 1995)
- 2.73, 3.33, 2.81 (RP-HPLC, CLOGP, calculated-S, Finizio et al. 1997)

Bioconcentration Factor, log BCF:
- 1.46   (calculated-S, Kenaga 1980)
- 1.34   (calculated, Pait et al. 1992)

Sorption Partition Coefficient, log K_{oc}:
- 2.34   (calculated-S, Kenaga 1980)
- 2.23   (calculated-χ, Gerstl & Helling 1987)
- 2.33   (selected, Trevisan et al. 1991)
- 2.19   (Montgomery 1993)

2.17 (soil, 20-25°C, selected, Wauchope et al. 1992)
2.17 (soil, 20-25°C, selected, Hornsby et al. 1996)
2.17 (selected, Lohninger 1994)
2.38-2.90 (Tomlin 1994)

Half-Lives in the Environment:
Air:
Surface water: hydrolysis half-life >> 1 yr (pH 4, 7 ,9) at 22°C and photolysis half-life of 12-13 h in aqueous solution (Tomlin 1994).
Groundwater:
Sediment: half-life of ~10 d for 40 µg/mL to biodegrade in pond sediment (Stepp et al. 1985; quoted, Muir 1991).
Soil: half-life of 1-2 d for 30 µg/mL to biodegrade in flooded soil at 30°C (Kuwatsuka 1972; quoted, Muir 1991); selected field half-life of 1.0 d (Wauchope et al. 1992; Hornsby et al. 1996; quoted, Halfon et al. 1996); soil half-life of 15 d (Pait et al. 1992).
Biota:

Environmental Fate Rate Constants or Half-Lives:
Volatilization:
Photolysis: half-life of 34 d for 82% of 200 µg/mL to degrade in distilled water under sunlight (Moilanen & Crosby 1972; quoted, Cessna & Grover 1991); half-life of 2.25 h for 37-51% of 100 µg/mL to degrade in distilled water under >300 nm light (Tanaka et al. 1981; quoted, Cessna & Grover 1991); half-life of 245 h for 14-81% of 15 µg/mL to degrade in distilled water under sunlight (Draper & Crosby 1984; quoted, Cessna & Grover 1991); photolysis half-life of 12-13 h in water (Tomlin 1994).
Oxidation:
Hydrolysis: half-life of >4 months for 4360 µg/mL to hydrolyze in phosphate buffers pH 5-9 at 20°C (El-dib & Aly 1976; quoted, Muir 1991); hydrolysis half-life >> 1 year at pH 4, 7, 9 at 22°C (Tomlin 1994).
Biodegradation: half-life of 1-2 d for 30 µg/mL to biodegrade in flooded soil at 30°C (Kuwatsuka 1972; quoted, Muir 1991); half-life of ~10 d for 40 µg/mL to biodegrade in pond sediment (Stepp et al. 1985; quoted, Muir 1991).
Biotransformation: second-order microbial degradation rate constant of 5×10^{-10} L·organisms^{-1} h^{-1} (Steen & Collette 1989).
Bioconcentration, Uptake (k_1) and Elimination (k_2) Rate Constants:

Common Name: Propazine

Synonym: G-30028, Geigy 30028, Gesamil, Maax, Milogard, Plantulin, Primatol P, Propasin, Prozinex

Chemical Name: 6-chloro-N,N'-bis(1-methylethyl)-1,3,5-triazine-2,4-diamine; 2-chloro-4,6-bis(isopropylamino)-1,3,5-triazine

Uses: selective pre-emergence herbicide to control annual grasses and broadleaf weeds in milo and sweet sorghum.

CAS Registry No: 139-40-2

Molecular Formula: $C_9H_{16}ClN_5$

Molecular Weight: 229.7

Melting Point (°C):
- 212-214 (Khan 1980; Spencer 1982; Herbicide Handbook 1989; Worthing 1991; Tomlin 1994; Milne 1995)
- 213 (Suntio et al. 1988)
- 212 (Montgomery 1993)
- 211 (Kühne et al. 1995)

Boiling Point (°C):

Density (g/cm³ at 20°C):
- 1.162 (Agrochemicals Handbook 1987; Worthing 1991; Montgomery 1993; Tomlin 1994; Milne 1995)

Molar Volume (cm³/mol):
- 272.8 (calculated-LeBas method, Suntio et al. 1988)

Molecular Volume (Å³):

Total Surface Area, TSA (Å²):

Dissociation Constant:
- 1.85 (pK_a, Weber 1970; quoted, Bintein & Devillers 1994)
- 1.80 (pK_a, Weber et al. 1980; Willis & McDowell 1982)
- 1.85 (pK_a, Herbicide Handbook 1989)
- 1.70 (pK_a, 21°C, Worthing 1991)
- 12.15 (pK_b, Wauchope et al. 1992)
- 1.85 (pK_a, 22°C, Montgomery 1993)
- 12.3 (pK_b, 21°C, Tomlin 1994)

Heat of Fusion, ΔH_{fus}, kcal/mol:
- 10.0 (DSC method, Plato 1972)

Entropy of Fusion, ΔS_{fus}, cal/mol·K (e.u.):

Fugacity Ratio at 25°C (assuming ΔS_{fus} = 13.5 e.u.), F: 0.0141

Water Solubility (g/m³ or mg/L at 25°C):
- 10 (Gysin 1962; quoted, Shiu et al. 1990)
- 8.6 (20-22°C, Bailey & White 1965; quoted, Shiu et al. 1990)
- 4.82 (26°C, shake flask-UV at pH 3, Ward & Weber 1968)
- 4.60 (26°C, shake flask-UV at pH 7, Ward & Weber 1968)
- 5.05 (26°C, shake flask-UV at pH 10, Ward & Weber 1968)
- 8.60 (Spencer 1973; quoted, Shiu et al. 1990)
- 8.60 (Martin & Worthing 1977; quoted, Kenaga 1980; Kenaga & Goring 1980; Shiu et al. 1990; Isensee 1991)
- 8.60 (20°C, Quellette & King 1977; quoted, Suntio et al. 1988; Shiu et al. 1990)
- 8.60 (Wauchope 1978; Weber et al. 1980; quoted, Willis & McDowell 1982)
- 8.60 (20°C, Khan 1980; Ashton & Crafts 1981; selected, Suntio et al. 1988)

4.8-8.6   (Weber et al. 1980; quoted, Willis & McDowell 1982)
5.0   (20°C, Spencer 1982)
8.60   (20°C, Agrochemicals Handbook 1987; quoted, Majewski & Capel 1995)
8.53   (selected, Gerstl & Helling 1987)
8.60   (20°C, Herbicide Handbook 1989)
5.0   (20°C, Worthing 1987, 1991; quoted, Shiu et al. 1990)
8.60   (20-25°C, selected, Wauchope et al. 1992; Hornsby et al. 1996)
8.50   (20°C, Montgomery 1993)
8.60   (selected, Lohninger 1994)
8.53, 9.36 (quoted, calculated-group contribution fragmentation method, Kühne et al. 1995)
5.0   (20°C, Tomlin 1994; Milne 1995)
8.6, 16   (quoted, predicted-AQUAFAC, Lee et al. 1996)

Vapor Pressure (Pa at 25°C):
3.90×10^{-6} (20°C, Quellette & King 1977; quoted, Suntio et al. 1988)
3.90×10^{-6} (20°C, Khan 1980)
4.00×10^{-6} (20-25°C, Weber et al. 1980; quoted, Willis & McDowell 1982)
3.90×10^{-6} (20°C, Ashton & Crafts 1981; Herbicide Handbook 1989)
4.00×10^{-6} (20°C, Agrochemicals Handbook 1987)
3.90×10^{-6} (20°C, selected, Suntio et al. 1988; quoted, Majewski & Capel 1995)
2.10×10^{-5} (30°C, Herbicide Handbook 1989)
3.90×10^{-6} (20°C, Worthing 1991)
1.75×10^{-5} (20-25°C, selected, Wauchope et al. 1992; Hornsby et al. 1996)
3.90×10^{-6} (20°C, Montgomery 1993)

Henry's Law Constant (Pa·m³/mol):
1.00×10^{-4} (20°C, selected, Suntio et al. 1988; quoted, Majewski & Capel 1995)
1.00×10^{-3} (20°C, calculated-P/C, Montgomery 1993)
1.04×10^{-3} (calculated-P/C, this work)

Octanol/Water Partition Coefficient, log K_{ow}:
2.89   (quoted, Kenaga & goring 1980)
2.94   (shake flask, Brown & Flagg 1981)
3.09   (selected, Dao et al. 1983)
2.59   (RP-HPLC-k', Braumann et al. 1983)
2.89   (selected, Yoshioka et al. 1986)
3.09   (Gerstl & Helling 1987; quoted, Bintein & Devillers 1994)
2.93   (shake flask, Biagi et al. 1991; quoted, Sicbaldi & Finizio 1993)
2.91, 3.02 (RP-HPLC, calculated, Finizio et al. 1991)
2.93   (selected, Magee 1991)
2.91, 2.94 (Montgomery 1993)
2.77   (RP-HPLC, Sicbaldi & Finizio 1993)
2.93   (selected, Sangster 1993)
2.93   (selected, Hansch et al. 1995)
2.89   (shake flask-UV, Liu & Qian 1995)
2.77, 2.70, 3.98 (RP-HPLC, ClogP, calculated-S, Finizio et al. 1997)

Bioconcentration Factor, log BCF:
    2.26   (calculated-S, Kenaga 1980; quoted, Isensee 1991)
    0.903   (calculated-K_{OC}, Kenaga 1980)

Sorption Partition Coefficient, log K_{OC}:
    2.20   (soil, Hamaker & Thompson 1972; Brown 1978; quoted, Kenaga 1980; Kenaga & Goring 1980)
    3.11   (soil, calculated-S per Kenaga & Goring 1978; Kenaga 1980)
    2.56   (Georgia's Hickory Hill pond sediment, Brown & Flagg 1981)
    2.29, 2.78 (quoted, calculated- χ, Gerstl & Helling 1987)
    2.20, 2.34 (reported, estimated as log K_{OM}, Magee 1991)
    2.18   (quoted, Bottoni & Funari 1992)
    2.19   (soil, 20-25°C, selected, Wauchope et al. 1992; Hornsby et al. 1996)
    1.69-2.56 (Montgomery 1993)
    2.44   (selected, Lohninger 1994)
    1.90, 2.0 (Tomlin 1994)
    1.94   (soil, HPLC-screening method, Kördel et al. 1993, 1995b)
    2.20, 2.57 (quoted, calculated-K_{OW}, Liu & Qian 1995)

Half-Lives in the Environment:
    Air:
    Surface water:
    Groundwater:
    Sediment:
    Soil: persistence of 18 months (Edwards 1973; quoted, Morrill et al. 1982); persistence of 12 months in soil (Wauchope 1978); under lab. conditions, half-lives in a Hatzenbühl soil at pH 4.8 and Neuhofen soil at pH 6.5 at 22°C were 62 and 127 d (Burkhard & Guth 1981; quoted, Montgomery 1993); selected field half-life of 135 d (Wauchope et al. 1992; Hornsby et al. 1996).
    Biota:

Environmental Fate Rate Constants or Half-Lives:
    Volatilization:
    Photolysis: 1 ppb contaminated water in presence of TiO_2 and H_2O_2 completely photodegraded after 3.3 h by solar irradiation (Muszkat et al. 1992).
    Oxidation:
    Hydrolysis: calculated rate constant of 9.70×10^{-6} s^{-1} with half-life of 83 d at 20°C in a buffer at pH 5 (Burkhard & Guth 1981).
    Biodegradation:
    Biotransformation:
    Bioconcentration, Uptake (k_1) and Elimination (k_2) Rate Constants:

Common Name: Propham

Synonym: Agermin, Ban-Hoe, Beet-Kleen, Birgin, Chem-Hoe, Collavin, IFC, IFK, INPC, IPC

Chemical Name: carbanilate acid isopropyl ester; isopropyl carbanilate; isopropyl-N-phenyl carbamate; 1-methylethyl phenylcarbamate

Uses: pre-emergence and post-emergence herbicide to control annual grass weeds in peas, beans, sugar beet, lettuce, lucerne, clover, flax, sunflowers and lentils.

CAS Registry No: 122-42-9

Molecular Formula: $C_{10}H_{13}NO_2$

Molecular Weight: 179.2

Melting Point (°C):
    87   (Spencer 1982)
    87-88.0 (Herbicide Handbook 1989; Montgomery 1993)
    87-87.6 (Worthing 1991; Tomlin 1994)
    87.3   (Patil 1994)

Boiling Point (°C):
    >150   (sublimes but dec., Montgomery 1993)

Density (g/cm³ at 20°C):
    1.09   (Spencer 1982; Tomlin 1994; Agrochemicals Handbook 1987; Milne 1995)
    1.09   (30°C, Ashton & Crafts 1981; Herbicide Handbook 1989; Montgomery 1993)

Molar Volume (cm³/mol):
    213.6 (calculated-LeBas method, this work)

Molecular Volume (Å³):

Total Surface Area, TSA (Å²):

Dissociation Constant pK_a:

Heat of Fusion, ΔH_{fus}, kcal/mol:

Entropy of Fusion, ΔS_{fus}, cal/mol·K (e.u.):

Fugacity Ratio at 25°C (assuming ΔS_{fus} = 13.5 e.u.), F: 0.241

Water Solubility (g/m³ or mg/L at 25°C):
    100   (Freed 1953; quoted, Shiu et al. 1990)
    250   (Nex & Sweezey 1954; quoted, Shiu et al. 1990)
    22.5-32 (Bailey & White 1965; quoted, Shiu et al. 1990)
    250   (20°C, Spencer 1973, 1982; ; quoted, Shiu et al. 1990)
    250   (Martin & Worthing 1977; Herbicide Handbook 1978, 1983, 1989; quoted, Kenaga 1980; Kenaga & Goring 1980; Shiu et al. 1990; Isensee 1991)
    250-254 (Weber et al. 1980; quoted, Willis & McDowell 1982)
    250   (Ashton & Crafts 1981)
    250   (Agrochemicals Handbook 1987)
    127   (selected, Gerstl & Helling 1987)
    32-250 (20-25°C, Worthing 1991; Montgomery 1993)
    250   (20-25°C, selected, Wauchope et al. 1992; Hornsby et al. 1996)
    250   (selected, Lohninger 1994)
    250   (20°C, Tomlin 1994; Milne 1995)

Vapor Pressure (Pa at 25°C):
    sublimes (rm. temp., Herbicide Handbook 1989)
    sublimes (rm. temp., Montgomery 1993; Tomlin 1994)

Henry's Law Constant (Pa·m^3/mol):

Octanol/Water Partition Coefficient, log K_{OW}:
- 2.60 (20 ± 2°C, shake flask-UV, Briggs 1981)
- 2.16 (selected, Dao et al. 1983; Gerstl & Helling 1987)
- 1.93 (calculated, Montgomery 1993)
- 2.60, 1.98 (quoted, calculated, Patil 1994)

Bioconcentration Factor, log BCF:
- 1.43 (calculated-S, Kenaga 1980; quoted, Isensee 1991)
- 0.301 (calculated-K_{OC}, Kenaga 1980)

Sorption Partition Coefficient, log K_{OC}:
- 1.71 (Hamaker & Thompson 1972; quoted, Kenaga 1980; Kenaga & Goring 1980)
- 2.32 (soil, calculated-S as per Kenaga & Goring 1978, Kenaga 1980)
- 1.71 (20 ± 2°C, reported as log K_{OM}, Briggs 1981)
- 1.95, 1.93 (quoted, calculated- χ, Gerstl & Helling 1987)
- 2.30 (soil, 20-25°C, estimated, Wauchope et al. 1992; Hornsby et al. 1996)
- 1.71 (Montgomery 1993)
- 2.30 (estimated-chemical structure, Lohninger 1994)

Half-Lives in the Environment:

Air:

Surface water: half-life of 190 h for 3-40 µg/mL to biodegrade in stream water with rate of 1.5×10^{-4} mg M^{-1} h^{-1} at 28°C (Wolfe et al. 1978; quoted, Muir 1991); half-lives of ~30-40 d for 1-0.0004 µg/mL to biodegrade in filtered sewage water at 20-22°C and ~20 to 50 d at 29°C in filtered lake water (Wang et al. 1984; quoted, Muir 1991); aerobic half-lives of >4 months for 6-7 µg/mL to biodegrade in river water at 25°C (Stepp et al. 1985; quoted, Muir 1991).

Groundwater:

Sediment:

Soil: half-life in soil is approximately 15 d and 5 d at 16 and 29°C (Hartley & Kidd 1987; Herbicide Handbook 1989; quoted, Montgomery 1993; Tomlin 1994); selected field half-life of 10 d (Wauchope et al. 1992; Hornsby et al. 1996).

Biota:

Environmental Fate Rate Constants or Half-Lives:

Volatilization:

Photolysis: calculated half-life of 254 d in distilled water (Wolfe et al. 1978; quoted, Cessna & Muir 1991); half-life of 2.25 h for 1% of 100 µg/ml to degrade in distilled water under 300 nm light (Tanaka et al. 1981; quoted, Cessna & Muir 1991).

Oxidation:

Hydrolysis: half-life of >4 months for 3550 µg/mL to hydrolyze in phosphate buffer at pH 5-9 and 20°C (El-Dib & Aly 1976; quoted, Muir 1991).

Biodegradation: rate constant of 0.003-2.1 mg (g bacteria)$^{-1}$ d^{-1} in aquatic systems (Wolfe et al. 1978; quoted, Scow 1982); half-life of 190 hours for 3-40 µg/mL to biodegrade in stream water with rate of 1.5×10^{-4} mg M^{-1} h^{-1} at 28°C (Wolfe et al. 1978; quoted, Muir 1991); half-lives of ~30-40 d for 1-0.0004 µg/mL to biodegrade in filtered sewage water at 20-22°C and ~20 to 50 d at 29°C in filtered lake water (Wang et al. 1984; quoted, Muir 1991); aerobic half-lives of >4 months for 6-7 µg/mL to biodegrade in river water at 25°C (Stepp et al. 1985; quoted, Muir 1991).

Biotransformation:

Bioconcentration, Uptake (k_1) and Elimination (k_2) Rate Constants:

Common Name: Pyrazon
Synonym: chloridazon, chloridazone, Blurex, Burex, Dazon, Phenosane, Piramin, Pyramin
Chemical Name: 5-amino-4-chloro-2-phenylpyridazin-3(2H)-one
Uses: as pre- and post-emergence herbicide to control of annual broadleaf weeds in sugar-beet, fodder beet and beet root; and also used in combination with other herbicides, etc.
CAS Registry No: 1698-60-8
Molecular Formula: $C_{10}H_8ClN_3O$
Molecular Weight: 221.6
Melting Point (°C):
    205-206 (Spencer 1982; Worthing 1987; Milne 1995)
    205-206 (with dec., Agrochemicals Handbook 1987)
    207   (Suntio et al. 1988)
    206   (Tomlin 1994)
Boiling Point (°C):
Density (g/cm³ at 20°C):
    1.54   (Tomlin 1994)
Molar Volume (cm³/mol):
    205.7 (calculated-LeBas method, Suntio et al. 1988)
    143.9 (calculated-density, this work)
Molecular Volume (Å³):
Total Surface Area, TSA (Å²):
Dissociation Constant pK_a:
Heat of Fusion, ΔH_{fus}, kcal/mol:
    6.35   (DSC method, Plato & Glasgow 1969)
Entropy of Fusion, ΔS_{fus}, cal/mol·K (e.u.):
Fugacity Ratio at 25°C (assuming ΔS_{fus} = 13.5 e.u.), F: 0.0162
    0.013 (20°C, Suntio et al. 1988)

Water Solubility (g/m³ or mg/L at 25°C):
    400   (20°C, Ashton & Crafts 1973, 1981)
    300   (20°C, Khan 1980)
    400   (20°C, Spencer 1982)
    400   (selected, Gerstl & Helling 1987)
    400   (20°C, Worthing 1987; Agrochemicals Handbook 1987)
    360   (20°C, selected, Suntio et al. 1988)
    400   (20-25°C, selected, Wauchope et al. 1992; Hornsby et al. 1996)
    340   (20°C, Tomlin 1994)
    400   (20°C, Milne 1995)

Vapor Pressure (Pa at 25°C):
    9.86   (40°C, Ashton & Craft 1973; Spencer 1982)
    <0.00001 (20°C, Worthing 1987)
    <0.00001 (20°C, Agrochemicals Handbook 1987)
    7.0   (20°C, estimated, Suntio et al. 1988)
    6.67   (20-25°C, selected, Wauchope et al. 1992; Hornsby et al. 1996)
    <0.00001 (20°C, Tomlin 1994)

Henry's Law Constant (Pa·m³/mol):
    4.31   (calculated-P/C, Suntio et al. 1988)

Octanol/Water Partition Coefficient, log K_{ow}:
    1.14   (22°C, shake flask-AS, Braumann & Grimme 1981; quoted, Sangster 1993; Sicbaldi & Finizio 1993)
    1.50   (selected, Gerstl & Helling 1987)
    1.12   (RP-HPLC, Sicbaldi & Finizio 1993)
    1.19   (pH 7, Tomlin 1994)
    2.20   (Milne 1995)
    1.14   (selected, Hansch et al. 1995)
    1.12, 1.65, 2.79   (RP-HPLC, ClogP, calculated-S, Finizio et al. 1995)

Bioconcentration Factor, log BCF:
    1.32   (calculated-S per Kenaga 1980, this work)

Sorption Partition Coefficient, log K_{oc}:
    2.12, 2.18 (selected, calculated- χ , Gerstl & Helling 1987)
    2.08   (soil, Wauchope et al. 1992; Hornsby et al. 1996)
    1.95-2.53 (soil, Tomlin 1994)

Half-Lives in the Environment:
    Air:
    Surface water:
    Groundwater:
    Sediment:
    Soil: field half-life of 21 d (Wauchope et al. 1992; Hornsby et al. 1996)
    Biota:

Environmental Fate Rate Constants or Half-Lives:
    Volatilization:
    Photolysis: half-life was 150 h (pH 7) in simulated sunlight and 47.7 h by 80000 lux, xenon lamp (Tomlin 1994).
    Oxidation:
    Hydrolysis:
    Biodegradation:
    Biotransformation:
    Bioconcentration, Uptake (k_1) and Elimination (k_2) Rate Constants:

Common Name: Simazine

Synonym: A 2079, Aktinit S, Amizine, Aquazine, Batazina, Bitemol, Cekusan, CAT, CET, DCT, Framed, G 27692, Gesapun, Gesaran, Gesatop, Herbazin, Herbex, Herboxy, Premazine, Primatol, Primcep, Princep, Printop, Radocon, Radokor, Simadex, Simanex, Sim-Trol, Tafazine, Weedex, Zeapur

Chemical Name: 6-chloro-N,N'-diethyl-1,3,5-triazine-2,4-diamine; 2-chloro-4,6-bis(ethylamino)-1,3,5-triazine

Uses: selective pre-emergence systemic herbicide to control many broadleaf weeds and annual grasses in deep-rooted fruit and vegetable crops.

CAS Registry No: 122-34-9

Molecular Formula: $C_7H_{12}ClN_5$

Molecular Weight: 201.67

Melting Point (°C):
    226   (Karickhoff 1981)
    225-227 (Spencer 1982; Suntio et al. 1988; Herbicide Handbook 1989; Worthing 1991; Montgomery 1993; Tomlin 1994; Milne 1995)
    227   (Kühne et al. 1995)

Boiling Point (°C):

Density (g/cm^3 at 20°C):
    1.302   (Agrochemicals Handbook 1987; Milne 1995)
    1.203   (Montgomery 1993)

Molar Volume (cm^3/mol):
    228.4   (calculated-LeBas method, Suntio et al. 1988)

Molecular Volume (Å^3):

Total Surface Area, TSA (Å^2):

Dissociation Constant:
    1.65   (pK_a, Weber 1970; quoted, Bintein & Devillers 1994)
    1.60   (pK_a, Weber et al. 1980; Willis & McDowell 1982)
    1.70   (pK_a, 21°C, Worthing 1991; Montgomery 1993)
    2.00   (pK_a, Yao & Haag 1991)
    12.35   (pK_b, Wauchope et al. 1992; Hornsby et al. 1996)
    12.3   (pK_b, Tomlin 1994)

Heat of Fusion, ΔH_{fus}, kcal/mol:
    10.500   (DSC method, Plato 1972)

Entropy of Fusion, ΔS_{fus}, cal/mol·K (e.u.):

Fugacity Ratio at 25°C (assuming ΔS_{fus} = 13.5 e.u.), F: 0.01028

Water Solubility (g/m^3 or mg/L at 25°C):
    5.0   (Bailey & White 1965; quoted, Karickhoff 1981)
    5.8   (26°C, shake flask-UV at pH 3, Ward & Weber 1968; quoted, Freed 1976; Shiu et al. 1990)
    5.0   (26°C, shake flask-UV at pH 7, Ward & Weber 1968; quoted, Freed 1976; Shiu et al. 1990)
    5.0   (26°C, shake flask-UV at pH 10, Ward & Weber 1968; quoted, Freed 1976; Shiu et al. 1990)
    15.1   (26°C, Getzen & Ward 1971; quoted, Freed 1976; Shiu et al. 1990)
    5.0   (20°C, Weber 1972; Worthing 1987; quoted, Muir 1991)
    5.0   (20°C, Spencer 1973; quoted, Shiu et al. 1990)

5.0   (Freed 1976; Wauchope 1978; quoted, Jury et al. 1983, 1984; Taylor & Glotfelty 1988; Glotfelty et al. 1989; Grover 1991)
5.0   (20°C, Martin & Worthing 1977; Khan 1980; quoted, Suntio et al. 1988; Shiu et al. 1990)
3.5   (Herbicide Handbook 1978; quoted, Kenaga 1980; Kenaga & Goring 1980; Isensee 1991)
5.0   (Weber et al. 1980; Willis & McDowell 1982)
3.5   (20°C, Ashton & Crafts 1981)
5.0, 3.50, 7.4 (20°C, quoted, exptl., calculated-Parachor & M.P., Briggs 1981)
3.5   (20°C, Agrochemicals Handbook 1987)
4.95   (selected, Gerstl & Helling 1987)
5.0   (20°C, Worthing 1987, 1991; quoted, Shiu et al. 1990)
5.0   (20°C, selected, Suntio et al. 1988; quoted, Majewski & Capel 1995)
3.5   (20°C, Herbicide Handbook 1989; quoted, Pait et al. 1992)
5.0   (MAFF 1992b, quoted, Meakins et al. 1994)
6.2   (20-25°C, selected, Wauchope et al. 1992; Lohninger 1994; Hornsby et al. 1996; quoted, Halfon et al. 1996)
3.5-5.0 (20°C, Montgomery 1993)
5.68   (quoted, Kühne et al. 1995)
20   (calculated-group contribution fragmentation method, Kühne et al. 1995)
5.0   (20°C, Milne 1995)

Vapor Pressure (Pa at 25°C):
8.10×10^{-7} (20°C, gas saturation, calculated-Antoine eqn., Friedrich & Stammbach 1964; quoted, Jorden et al. 1970; Spencer 1976; Glotfelty et al. 1989)
8.10×10^{-7} (20°C, Weber 1972; Worthing 1987; quoted, Muir 1991)
2.00×10^{-6} (gas saturation, Spencer & Cliath 1974; quoted, Jury et al. 1983; Grover 1991)
8.10×10^{-7} (20°C, Khan 1980; quoted, Suntio et al. 1988)
8.00×10^{-7} (20-25°C, Weber et al. 1980; Willis & McDowell 1982)
8.10×10^{-7} (20°C, Ashton & Crafts 1981; Herbicide Handbook 1989)
2.00×10^{-6} (Jury et al. 1983; quoted, Taylor & Glotfelty 1988; Taylor & Spencer 1990)
8.10×10^{-7} (20°C, Agrochemicals Handbook 1987)
8.50×10^{-6} (20°C, selected, Suntio et al. 1988; quoted, Majewski & Capel 1995)
4.80×10^{-6} (30°C, Herbicide Handbook 1989)
8.10×10^{-7} (20°C, Worthing 1991)
2.95×10^{-6} (20-25°C, selected, Wauchope et al. 1992; Hornsby et al. 1996; quoted, Halfon et al. 1996)
8.10×10^{-7} (20°C, Montgomery 1993)
2.94×10^{-6} (OECD 104, Tomlin 1994)

Henry's Law Constant (Pa·m³/mol):
8.40×10^{-5} (calculated-P/C, Jury et al. 1983, 1984, 1987a; Jury & Ghodrati 1989; quoted, Grover 1991)
3.40×10^{-4} (20°C, calculated-P/C, Suntio et al. 1988; quoted, Montgomery 1993; Majewski & Capel 1995)
8.40×10^{-5} (calculated-P/C, Taylor & Glotfelty 1988)
3.30×10^{-5} (20°C, calculated-P/C, Muir 1991)
3.40×10^{-4} (20°C, calculated-P/C, Montgomery 1993)

Octanol/Water Partition Coefficient, log K_{OW}:
- 2.19 (quoted, Kenaga & Goring 1980)
- 1.94 (quoted, Rao & Davidson 1980; Suntio et al. 1988)
- 1.51 (20 ± 2°C, shake flask-UV, Briggs 1981; quoted, Braumann et al. 1983; Sicbaldi & Finizio 1993)
- 2.19, 1.75 (quoted exptl., calculated-Parachor, Briggs 1981)
- 2.16 (shake flask, Brown & Flagg 1981; quoted, Karickhoff 1981; McDuffie 1981)
- 1.96 (HPLC-k', McDuffie et al. 1981)
- 2.27 (selected, Dao et al. 1983)
- 2.14 (shake flask, Mitsutake et al. 1986)
- 2.27 (Gerstl & Helling 1987; quoted, Bintein & Devillers 1994)
- 1.90 (selected, Suntio et al. 1988; quoted, Finizio et al. 1991)
- 2.18 (Thor 1989; quoted, Connell & Markwell 1990; Magee 1991)
- 2.18 (shake flask, Biagi et al. 1991; quoted, Sicbaldi & Finizio 1993)
- 2.26, 2.20 (RP-HPLC, calculated, Finizio et al. 1991)
- 1.96 (Worthing 1991; Milne 1995)
- 2.30 (MAFF 1992b, quoted, Meakins et al. 1994)
- 1.94-2.26 (Montgomery 1993)
- 1.51, 2.07 (quoted, RP-HPLC, Sicbaldi & Finizio 1993)
- 2.18 (selected, Sangster 1993)
- 2.10 (Tomlin 1994)
- 2.18 (shake flask-UV, Liu & Qian 1995)
- 2.18 (selected, Hansch et al. 1995)
- 2.07, 2.09, 4.04 (RP-HPLC, ClogP, calculated-S, Finizio et al. 1997)

Bioconcentration Factor, log BCF:
- 2.48 (calculated-S, Kenaga 1980; quoted, Isensee 1991)
- 0.778 (calculated-K_{OC}, Kenaga 1980)
- 0.0 (quoted, Kenaga & Goring 1980)
- 2.16 (earthworms, Lord et al. 1980; quoted, Connell & Markwell 1990)
- 0.0 (quoted exptl., Briggs 1981)
- 0.699 (calculated-K_{OW}, Briggs 1981)

Sorption Partition Coefficient, log K_{OC}:
- 2.13 (soil, Hamaker & Thompson 1972; Brown 1978; quoted, Kenaga 1980; Kenaga & Goring 1980; Karickhoff 1981; Liu & Qian 1995)
- 3.34 (soil, calculated-S as per Kenaga & Goring 1978, Kenaga 1980)
- 2.15 (av. soils/sediments, Rao & Davidson 1980; quoted, Karickhoff 1981; Lyman 1982; Jury et al. 1983; Glotfelty et al. 1989; Sukop & Cogger 1992)
- 1.43 (20 ± 2°C, reported as log K_{OM}, Briggs 1981)
- 1.89 (quoted exptl., reported as log K_{OM}, Briggs 1981)
- 1.54 (calculated, reported as log K_{OM}, Briggs 1981)
- 2.33 (Georgia's Hickory Hill pond sediment, Brown & Flagg 1981; quoted, Karickhoff 1981; Muir 1991)
- 3.66, 2.53 (estimated-S, S & M.P., Karickhoff 1981)

1.77   (estimated-K_{OW}, Karickhoff 1981)
2.14   (soil average, Jury et al. 1983; quoted, Grover 1991)
2.20, 2.15   (selected, calculated- χ, Gerstl & Helling 1987)
2.15   (screening model calculations, Jury et al. 1987a,b; Jury & Ghodrati 1989)
1.60-2.20   (Carsel 1989)
2.13, 1.92   (reported, estimated as log K_{OM}, Magee 1991)
2.13-3.34, 2.15, 2.45, 2.70   (quoted values, Bottoni & Funari 1992)
2.11   (soil, 20-25°C, selected, Wauchope et al. 1991, 1992; quoted, Dowd et al. 1993; Richards & Baker 1993; Hornsby et al. 1996)
3.02   (av. 12 soils, calculated-linearized Freundlich Isotherm, Sukop & Cogger 1992)
2.14   (Montgomery 1993)
2.37   (selected, Lohninger 1994)
2.01-2.58   (Tomlin 1994)
1.78   (soil, HPLC-screening method, Kördel et al. 1995b)
2.18   (calculated-K_{OW}, Liu & Qian 1995)

Half-Lives in the Environment:
  Air:
  Surface water: half-life of >32 d for 3 µg/mL to biodegrade in pond water at 25°C (Tucker & Boyd 1981; quoted, Muir 1991); half-life of 1-4 weeks in esturine systems (Jones et al. 1982; quoted, Meakins et al. 1994); half-life in ponds of about 30 d (Herbicide Handbook 1989).
  Groundwater:
  Sediment: half-life of 8-27 d for 3 µg/mL to biodegrade in pond sediment/water at 25°C (Tucker & Boyd 1981; quoted, Muir 1991).
  Soil: estimated persistence of 12 months (Kearney et al. 1969; Edwards 1973; quoted, Morrill et al. 1982; Jury et al. 1987a); persistence of 12 months (Wauchope 1978); estimated first-order half-life of 49.5 d from biodegradation rate constant of 0.014 d^{-1} by soil incubation die-away studies (Rao & Davidson 1980; quoted, Scow 1982); under lab. conditions, half-lives of 45 d in Hatzenbühl soil at pH 4.8 and 100 d in Neuhofen soil at pH 6.5 both at 22°C, respectively (Burkhard & Guth 1981; quoted, Montgomery 1993); half-life of 1-6 months (Jones et al. 1982; quoted, Meakins et al. 1994); half-life of 75 d from screening model calculations (Jury et al. 1987a,b; Jury & Ghodrati 1989); moderately persistent in soils with half-life of 20-100 d (Willis & McDowell 1982); degradation rate constant of $(1.51 \pm 0.086) \times 10^{-2} d^{-1}$ with half-life of 45.9 d in control soil and $(1.76 \pm 0.177) \times 10^{-2} d^{-1}$ with half-life of 39.4 d in pretreated soil in the field; $(0.943 \pm 0.047) \times 10^{-2} d^{-1}$ with half-life of 73.5 d in control soil and $(0.864 \pm 0.048) \times 10^{-2} d^{-1}$ with half-life of 80.2 d in pretreated soil once only in the laboratory (Walker & Welch 1991); selected field half-life of 60 d (Wauchope et al. 1991, 1992; quoted, Dowd et al. 1993; Richards & Baker 1993; quoted, Halfon et al. 1996; Hornsby et al. 1996); soil half-life of 75 d (Pait et al. 1992); degradation half-life of 70-110 d (Tomlin 1994).
  Biota: biochemical half-life of 75 d from screening model calculations (Jury et al. 1987a,b; Jury & Ghodrati 1989).

Environmental Fate Rate Constants or Half-Lives:
- Volatilization: half-life of 276 d (Jury et al. 1983; quoted, Grover 1991); measured rate constant of 600 d^{-1} and estimated rate constant of 1000 d^{-1} (Glotfelty et al. 1989).
- Photolysis:
- Oxidation: rate constant of 5.9×10^9 M^{-1} s^{-1} for the reaction (photo-Fenton with reference to acetophenone) with hydroxyl radicals in aqueous solutions at pH 3.5 and (24 ± 1)°C (Buxton et al. 1988; quoted, Faust & Hoigné 1990; Haag & Yao 1992); rate constant of $(2.8 \pm 0.2) \times 10^9$ M^{-1} s^{-1} for the reaction (photo-Fenton with reference to acetophenone) with hydroxyl radicals in aqueous solutions at pH 3.5 and (24 ± 1)°C (Haag & Yao 1992).
- Hydrolysis: calculated rate constant of 8.32×10^{-6} s^{-1} with half-life of 96 d at 20°C in a buffer at pH 5 (Burkhard & Guth 1981).
- Biodegradation: rate constant of 0.014 d^{-1} by soil incubation die-away studies (Rao & Davidson 1980; quoted, Scow 1982); half-lives of 8-27 d for 3 μg/ml to biodegrade in pond sediment/water and >32 d in pond water both at 25°C (Tucker & Boyd 1981; quoted, Muir 1991); half-life of 75 d for a 100 day leaching and screening test in 0-10 cm depth of soil (Rao & Davidson 1980; quoted, Jury et al. 1983, 1984, 1987a,b; Jury & Ghodrati 1989; Grover 1991); microbial degradation half-life in soil is 70-110 d (Tomlin 1994).
- Biotransformation:
- Bioconcentration, Uptake (k_1) and Elimination (k_2) Rate Constants:

Common Name: 2,4,5-T
Synonym: Amine 2,4,5-T for rice, BCF-bushkiller, Brush rhap, Brush-Khap, Brushtox, Dacamine, Ded-Weed, Dinoxol, Envert-T, Estercide T-2 & T-245, Esterone 245, Fence rider, Forron, Fortex, Fruitone A, Gesatop, Inverton 245, Line rider, Phortox, Reddon, Reddox, Spontox, Super D Weedone, Tippon, Tormona, Transamine, Tributon, Trinoxol, Trioxone, Veon, Weddar, Weedone
Chemical Name: 2,4,5-trichlorophenoxyacetic acid
Uses: herbicide to control undesirable brush and woody plants; also used as plant hormone, defoliant.
CAS Registry No: 93-76-5
Molecular Formula: $C_8H_5Cl_3O_3$
Molecular Weight: 255.49
Melting Point (°C):
    158   (Spencer 1982)
    153   (Suntio et al. 1988; Howard 1991; Riederer 1990; Milne 1995)
    153-156   (Shiu et al. 1990; Worthing 1991)
    157-158   (Montgomery 1993)
Boiling Point (°C):
    >200   (dec., Howard 1991)
Density (g/cm^3 at 20°C):
    1.80   (Ashton & Crafts 1981; Agrochemicals Handbook 1987; Montgomery 1993)
    1.80   (25°C, Que Hee et al. 1981)
    1.80   (Spener 1982; Merck Index 1989; Milne 1995)
Molar Volume (cm^3/mol):
    226.1   (calculated-LeBas method, Suntio et al. 1988; Riederer 1990)
Molecular Volume (Å^3):
Total Surface Area, TSA (Å^2):
Dissociation Constant pK_a:
    2.88   (Nelson & Faust 1969; quoted, Que Hee et al. 1981)
    2.85   (Cessna & Grover 1978; Somasundaram et al. 1991; Augustijn-Beckers et al. 1994)
    2.88   (Que Hee et al. 1981; quoted, Howard 1991)
    2.80-2.88   (Montgomery 1993)
Heat of Fusion, ΔH_{fus}, kcal/mol:
    8.35   (DSC method, Plato & Glasgow 1969)
Entropy of Fusion, ΔS_{fus}, cal/mol·K (e.u.):
Fugacity Ratio at 25°C (assuming ΔS_{fus} = 13.5 e.u.), F: 0.0540

Water Solubility (g/m^3 or mg/L at 25°C):
    268.3   (shake flask-UV, Leopold et al. 1960; quoted, Freed 1976; Shiu et al. 1990)
    268   (Loos 1975; quoted, Suntio et al. 1988; Shiu et al. 1990)
    238   (20°C, Loos 1975; quoted, Suntio et al. 1988; Riederer 1990; Shiu et al. 1990)
    238   (Martin & Worthing 1977; quoted, Kenaga 1980; Kenaga & Goring 1980; Isensee 1991)
    238-280   (Weber et al. 1980; quoted, Willis & McDowell 1982)
    238   (30°C, Ashton & Crafts 1981)
    268   (quoted, Que Hee et al. 1981)
    278   (Spencer 1982)

278   (Verschueren 1983; quoted, Suntio et al. 1988; Shiu et al. 1990)
278   (20°C, Agrochemicals Handbook 1987; quoted, Howard 1991)
280   (selected, Gerstl & Helling 1987)
150   (Worthing 1987, 1991; quoted, Shiu et al. 1990; Lohninger 1994)
220   (20°C, selected, Suntio et al. 1988)
238   (30°C, Merck Index 1989; quoted, Somasundaram et al. 1991)
220   (20°C, Montgomery 1993)
278   (20-25°C, selected, Augustijn-Beckers et al. 1994)
238   (20°C, Milne 1995)

Vapor Pressure (Pa at 25°C):
   0.0063   (effusion method, Hamaker & Kerlinger 1969; quoted, Spencer 1976; Suntio et al. 1988)
   $<1.0 \times 10^{-6}$ (20°C, Hartley & Kidd 1983; quoted, Howard 1991)
   0.0050   (20°C, selected, Suntio et al. 1988; Riederer 1990)
   $<1.0 \times 10^{-5}$ (20°C, Agrochemicals Handbook 1987)
   7.00×10^{-7} (Worthing 1991)
   0.0040   (20°C, Montgomery 1993)
   0.0   (20-25°C at pH 7, selected, Augustijn-Beckers et al. 1994)

Henry's Law Constant (Pa·m³/mol):
   8.79×10^{-4} (Hine & Mookerjee 1975; quoted, Howard 1991)
   0.0058   (20°C, calculated-P/C, Suntio et al. 1988; Riederer 1990)
   0.0049   (20°C, calculated-P/C, Montgomery 1993)

Octanol/Water Partition Coefficient, log K_{OW}:
   0.60   (pH dependent quoted from Dow Chemical data, Kenaga & Goring 1980)
   0.85   (Rao & Davidson 1980; quoted, Suntio et al. 1988)
   2.99   (RP-HPLC-k', Braumann et al. 1983)
   3.13   (Hansch & Leo 1985; quoted, Howard 1991; Somasundaram et al. 1991)
   −0.58   (selected, Gerstl & Helling 1987)
   3.40   (OECD 81 method, Kerler & Schönherr 1988; quoted, Riederer 1990)
   0.80   (selected, Suntio et al. 1988)
   3.36   (selected, Travis & Arms 1988)
   3.31   (shake flask-HPLC/UV, Jafvert et al. 1990)
   0.60-3.40 (Montgomery 1993)
   3.13   (countercurrent LC, Ilchmann et al. 1993)
   −0.28, 3.13; 3.13 (quoted values; selected, Sangster 1993)
   3.13   (selected, Hansch et al. 1995)

Bioconcentration Factor, log BCF:
   −4.55   (milk biotransfer factor log B_m, correlated-K_{OW}, Bjerke et al. 1972; quoted, Travis & Arms 1988)
   1.18   (measured, Isensee 1976; quoted, Isensee 1991)
   −4.82   (beef biotransfer factor log B_b, correlated-K_{OW}, Kenaga 1980; quoted, Travis & Arms 1988)

1.63 (fish under flowing water conditions, Kenaga & Goring 1980; quoted, Howard 1991)
1.45 (calculated-S, Kenaga 1980; quoted, Isensee 1991)
0.301 (calculated-K_{OC}, Kenaga 1980)
1.36-1.40 (fish under static ecosystem tests, Kenaga & Goring 1980; Garten & Tralbalka 1983; quoted, Howard 1991)
1.41 (mosquitofish 32 days under unspecified conditions, Ang et al. 1989; quoted, Howard 1991)

Sorption Partition Coefficient, log K_{OC}:
1.72 (soil, Hamaker & Thompson 1972; quoted, Kenaga 1980; Kenaga & Goring 1980)
2.34 (soil, calculated-S as per Kenaga & Goring 1978, Kenaga 1980)
1.93 (sand soil, Nkedi-Kizza et al. 1983; quoted, Howard 1991)
2.27 (whole soil, Nkedi-Kizza et al. 1983; quoted, Howard 1991)
2.31 (fine soils, Nkedi-Kizza et al. 1983; quoted, Howard 1991)
2.31 (coarse clay soil, Nkedi-Kizza et al. 1983; quoted, Howard 1991)
2.45 (medium silt soil, Nkedi-Kizza et al. 1983; quoted, Howard 1991)
2.31 (coarse silt soil, Nkedi-Kizza et al. 1983; quoted, Howard 1991)
1.72, 2.38 (quoted, calculated- χ , Gerstl & Helling 1987)
1.90 (soil, screening model calculations, Jury 1987b)
1.72 (soil, Sabljic 1987; quoted, Howard 1991)
1.77 (Alfisol soil, von Oepen et al. 1991)
2.63 (Podzol soil, von Oepen et al. 1991)
1.94 (sediment, von Oepen et al. 1991)
1.72, 2.27 (Montgomery 1993)
1.90 (20-25°C at pH 7, selected, Augustijn-Beckers et al. 1994)
1.72 (estimated-chemical structure, Lohninger 1994)

Half-Lives in the Environment:
Air: 12.2-122 h, based on an estimated rate constant for the vapor-phase reaction with hydroxyl radicals in the atmosphere (Atkinson 1987; quoted, Howard et al. 1991).
Surface water: estimated first-order half-life of 693 d from biodegradation rate constant of 0.001 d^{-1} by river die-away test in aquatic systems (Lee & Ryan 1979; quoted, Scow 1982); 240-480 h, based on estimated unacclimated aqueous aerobic biodegradation half-life (Howard et al. 1991); extremely resistant degradation in natural water with half-lives from 580 d for static sediment-sea water to 1400 d for static estuarine river water (Miur 1991).
Groundwater: 480-4320 h, based on estimated unacclimated aqueous aerobic and anaerobic biodegradation half-life (Howard et al. 1991).
Sediment: estimated first-order half-life of 23-69.3 d from biodegradation rate constant of 0.01-0.03 d^{-1} at 9-21°C by river die-away test in slurry sediment of aquatic systems (Lee & Ryan 1979; quoted, Scow 1982); half-life of 27 d for sediment-water microcosm under aerobic conditions (quoted, Miur 1991).
Soil: 240-480 h, based on unacclimated soil grab sample data (Smith 1978, 1979; quoted, Howard et al. 1991); estimated first-order half-life of 19.8 d from

rate constant of 0.035 d^{-1} by soil incubation die-away studies (Rao & Davidson 1980; quoted, Scow 1982); half-life of 33 d from screening model calculations (Jury et al. 1987b); aerobic degradation half-life from >25 d (29°C) to 58 d (21°C) in soil suspension from pre-incubated soil (Muir 1991); selected field half-life of 30 d (Augustijn-Beckers et al. 1994).

Biota: biochemical half-life of 33 d from screening model calculations (Jury et al. 1987b).

Environmental Fate Rate Constants or Half-Lives:
Volatilization:
Photolysis: half-life of 48 h for 17-80% of 1 µg/mL to degrade in buffered aqueous solution at pH 7.8 under sunlight (Crosby & Wong 1973; quoted, Cessna & Muir 1991); half-life of 15 d for <2.6 µg/mL to degrade in distilled water under sunlight (Skurlatov et al. 1983; quoted, Cessna & Muir 1991); half-life of 8.7 d for <2.6 µg/mL to degrade in aqueous fulvic acid solution (17 mg/L) and 3.5 d for <2.6 µg/mL to degrade in aqueous fulvic acid solution (55 mg/L) under sunlight (Skurlatov et al. 1983; quoted, Cessna & Muir 1991).
Oxidation: photooxidation half-life of 12.2-122 h in air, based on an estimated rate constant for the vapor-phase reaction with hydroxyl radicals in the atmosphere (Atkinson 1987; quoted, Howard et al. 1991).
Hydrolysis: will not hydrolyze to any reasonable extent; however, it may undergo other abiotic transformation processes (Kollig 1993).
Biodegradation: aerobic half-life of 27 d for 50 µg/mL in sediment-water microcosm by long lag phase degradation (Alexander 1974; quoted, Muir 1991); aqueous aerobic half-life of 240-480 h, based on unacclimated soil grab sample data (Smith 1978, 1979; quoted, Howard et al. 1991); rate constant of 0.001 d^{-1} by river die-away test in aquatic systems (Lee & Ryan 1979; quoted, Scow 1982); rate constant of 0.035 d^{-1} by soil incubation die-away studies (Rao & Davidson 1980; quoted, Scow 1982); biodegradation rate constant of 0.01-0.03 d^{-1} at 9-21°C by river die-away test in slurry sediment of aquatic systems (Lee & Ryan 1979; quoted, Scow 1982); half-life of 33 d from screening model calculations (Jury et al. 1987b); aqueous anaerobic half-life of 672-4320 h, based on anaerobic digestor sludge data (Batterby & Wilson 1989; quoted, Howard et al. 1991).
Biotransformation:
Bioconcentration, Uptake (k_1) and Elimination (k_2) Rate Constants:

Common Name: Terbacil
Synonym: Sinbar, Turbacil
Chemical Name: 3-*tert*-butyl-5-chloro-6-methyluracil
CAS Registry No: 5902-51-2
Uses: control of most annual grasses and broadleaf weeds, and some perennial weeds in established apples, asparagus, blueberries, citrus, lucerne, mint, peaches, pecans, strawberries, and sugar cane, etc.
Molecular Formula: $C_9H_{13}ClN_2O_2$
Molecular Weight: 216.7
Melting Point (°C):
    175-177 (Spencer 1982; Agrochemicals Handbook 1987; Montgomery 1993; Tomlin 1994)
Boiling Point (°C):
    sublime (below M.P., Agrochemicals Handbook 1987; Tomlin 1994)
Density (g/cm³ at 20°C):
    1.34 (25°C, Agrochemicals Handbook 1987; Montgomery 1993; Tomlin 1994)
Molar Volume (cm³/mol):
    217.7 (LeBas method, Suntio et al. 1988)
Molecular Volume (Å³):
Total Surface Area, TSA (Å²):
Dissociation Constant pK_a:
    9.0 (Wauchope et al. 1992)
Heat of Fusion, ΔH_{fus}, kcal/mol:
Entropy of Fusion, ΔS_{fus}, cal/mol·K (e.u.):
Fugacity Ratio at 25°C (assuming ΔS_{fus} = 13.5 e.u.), F: 0.0321
    0.027 (20°C, Suntio et al. 1988)

Water Solubility (g/m³ or mg/L at 25°C):
    710   (Ashton & Craft 1973; 1981)
    710   (Martin & Worthing 1977; quoted, Kenaga 1980, Kenaga & Goring 1980; Montgomery 1993; Tomlin 1994)
    710   (Spencer 1982)
    710   (Agrochemicals Handbook 1987)
    600   (20°C, selected, Suntio et al. 1988)
    19330   (20-25°C, supercooled liquid value, quoted, Majewski & Capel 1995)
    710   (20-25°C, selected, Wauchope et al. 1992; Hornsby et al. 1996)

Vapor Pressure (Pa at 25°C):
    6.40×10^{-5} (29.5°C, Ashton & Craft 1973; 1981)
    6.00×10^{-5} (30°C, Agrochemicals Handbook 1987)
    5.00×10^{-5} (20°C, selected, Suntio et al. 1988)
    6.00×10^{-5} (20°C, Montgomery 1993)
    6.25×10^{-5} (29.5°C, Tomlin 1994)
    1.91×10^{-3} (20-25°C, supercooled liquid value, quoted, Majewski & Capel 1995)
    4.13×10^{-5} (20-25°C, selected, Wauchope et al. 1992; Hornsby et al. 1996)

Henry's Law Constant (Pa·m^3/mol):
- 1.80×10^{-5} (20°C, calculated-P/C, Suntio et al. 1988: quoted, Majewski & Capel 1995)
- 1.82×10^{-5} (20-25°C, calculated, Montgomery 1993)
- 1.53×10^{-5} (calculated-P/C, this work)

Octanol/Water Partition Coefficient, log K_{OW}:
- 1.89 (Karickhoff et al. 1979; quoted, Madhun et al. 1986)
- 1.89 (Rao & Davidson 1980; quoted and selected, Sangster 1993)
- 1.90 (selected, Suntio et al. 1988)
- 1.89, 1.90 (quoted, Montgomery 1993)
- 1.91 (Tomlin 1994)
- 1.89 (selected, Hansch et al. 1995)

Bioconcentration Factor, log BCF:
- 1.18 (calculated-S, Kenaga 1980)
- 1.74 (Montgomery 1993)
- 1.32-2.20 (soil, quoted, Montgomery 1993)

Sorption Partition Coefficient, log K_{OC}:
- 1.71, 2.08 (soil: quoted exptl., calculated, Kenaga 1980; Kenaga & Goring 1980)
- 1.62, 1.98 (soil, quoted, Madhun et al. 1986)
- 1.89, 1.76; 1.82, 1.04 (estimated-K_{OW}; S, Madhun et al. 1986)
- 1.62 (soil, screening model calculations, Jury et al. 1987b)
- 1.74 (soil, Wauchope et al. 1992; Hornsby et al. 1996)

Half-Lives in the Environment:
- Air:
- Surface water:
- Groundwater:
- Sediment:
- Soil: moderately persistent in soil with half-life of 20-100 d (Willis & McDowell 1982); half-life of about 5-7 months (Agrochemicals Handbook 1987); half-life of 50 d from screening model calculations (Jury et al. 1987b); field half-life reported to be between 50-175 d and the selected value is 120 d (Wauchope et al. 1992; Hornsby et al. 1996).
- Biota: biochemical half-life of 50 d (Jury et al. 1987b).

Environmental Fate Rate Constants or Half-Lives:
- Volatilization:
- Photolysis:
- Oxidation:
- Hydrolysis:
- Biodegradation: half-life of 50 d (Jury et al. 1987b).
- Biotransformation:
- Bioconcentration, Uptake (k_1) and Elimination (k_2) Rate Constants:

Common Name: Terbutryn
Synonym: Clarosan, GS 14260, Igran, Prebane, Shortstop, Terbutrex, Terbutrin, Terbutryn
Chemical Name: N-(1,1-dimethylethyl)-N'-ethyl-6-(methylthio)-1,3,5-triazine-2,4-diamine;
2-(*tert*-butylamino)-4-(ethylamino)-6-(methylthio)-*s*-triazine
Uses: selective herbicide to control annual broadleaf and grass weeds in wheat.
CAS Registry No: 886-50-0
Molecular Formula: $C_{10}H_{19}N_5S$
Molecular Weight: 241.37
Melting Point (°C):
    104-105 (Agrochemicals Handbook 1987; Worthing 1991; Montgomery 1993; Milne 1995)
    104 (Herbicide Handbook 1989; Suntio et al. 1988; Kühne et al. 1995)
Boiling Point (°C):
    154-160 (at 0.06 mmHg, Agrochemicals Handbook 1987; Worthing 1991; Montgomery 1993; Milne 1995)
Density (g/cm³ at 20°C):
    1.115 (Agrochemicals Handbook 1987; Worthing 1991; Montgomery 1993; Milne 1995)
Molar Volume (cm³/mol):
    273.8 (calculated-LeBas method, Suntio et al. 1988)
Molecular Volume (Å³):
Total Surface Area, TSA (Å²):
Dissociation Constant:
    4.30 (pK_a, Worthing 1991)
    9.70 (pK_b, Wauchope et al. 1992; Hornsby et al. 1996)
    4.07 (pK_a, Montgomery 1993)
Heat of Fusion, ΔH_{fus}, kcal/mol:
Entropy of Fusion, ΔS_{fus}, cal/mol·K (e.u.):
Fugacity Ratio at 25°C (assuming ΔS_{fus} = 13.5 e.u.), F: 0.162

Water Solubility (g/m³ or mg/L at 25°C):
    25 (20°C, Ashton & Crafts 1973, 1981; quoted, Suntio et al. 1988; Shiu et al. 1990)
    25 (20°C, Weber 1972; Worthing 1987; quoted, Muir 1991)
    58 (20°C, Quellette & King 1977; quoted, Suntio et al. 1988; Shiu et al. 1990)
    25 (Martin & Worthing 1977; Herbicide Handbook 1978; quoted, Kenaga 1980; Kenaga & Goring 1980; Isensee 1991)
    25 (shake flask-HPLC, Ellgehausen et al. 1981)
    25 (20°C, Agrochemicals Handbook 1987)
    25 (20°C, Worthing 1987, 1991; quoted, Shiu et al. 1990)
    25 (20°C, selected, Suntio et al. 1988; quoted, Majewski & Capel 1995)
    25 (20°C, Herbicide Handbook 1989; quoted, Shiu et al. 1990)
    22 (20-25°C, selected, Wauchope et al. 1992; Hornsby et al. 1996)
    25 (20°C, Montgomery 1993)
    22 (selected, Lohninger 1994)
    24, 63.5 (quoted, calculated-group contribution fragmentation method, Kühne et al. 1995)

Vapor Pressure (Pa at 25°C):
    0.00013 (20°C, Weber 1972; Worthing 1987; quoted, Muir 1991)
    0.00013 (20°C, Ashton & Crafts 1973; quoted, Suntio et al. 1988)
    0.00013 (20°C, Agrochemicals Handbook 1987)
    0.00013 (20°C, selected, Suntio et al. 1988; quoted, Majewski & Capel 1995)
    0.00013 (20°C, Herbicide Handbook 1989)
    0.00013 (20°C, Worthing 1991)
    0.00128 (20°C, Montgomery 1993)
    0.00028 (20-25°C, selected, Wauchope et al. 1992; Hornsby et al. 1996)

Henry's Law Constant (Pa·m^3/mol):
    0.0013 (20°C, calculated-P/C, Suntio et al. 1988; quoted, Majewski & Capel 1995)
    0.0012 (20°C, calculated-P/C, Muir 1991)
    0.0120 (20°C, calculated-P/C, Montgomery 1993)
    0.0014 (calculated-P/C, this work)

Octanol/Water Partition Coefficient, log K_{OW}:
    3.74 (shake flask-GC, Elkell & Walum 1979; quoted, Sicbaldi & Finizio 1993; Sangster 1993)
    3.72, 3.74 (shake flask, Ellgehausen et al. 1981; quoted, Sicbaldi & Finizio 1993, Sangster 1993)
    4.64 (selected, Dao et al. 1983)
    2.56 (RP-HPLC-k', Braumann et al. 1983)
    3.34 (Worthing 1987; quoted, Finizio et al. 1991)
    3.43, 3.48 (RP-HPLC, calculated, Finizio et al. 1991)
    3.74, 2.56 (selected, Geyer et al. 1991)
    3.49 (Worthing 1991)
    3.43-3.73 (Montgomery 1993)
    3.34 (RP-HPLC, Sicbaldi & Finizio 1993)
    3.74 (selected, Sangster 1993)
    3.38, 3.36 (shake flask-UV, calculated-RPHPLC-k', Liu & Qian 1995)
    3.49 (Milne 1995)
    3.74 (selected, Hansch et al. 1995)
    3.34, 3.48, 3.50 (RP-HPLC, ClogP, calculated-S, Finizio et al. 1997)

Bioconcentration Factor, log BCF:
    1.17 (*Daphania magna*, wet wt. basis, Ellgehausen et al. 1980; quoted, Geyer et al. 1991)
    2.00 (calculated-S, Kenaga 1980; quoted, Isensee 1991)
    2.00 (calculated-K_{oc}, Kenaga 1980)

Sorption Partition Coefficient, log K_{oc}:
    2.85 (soil, Colbert et al. 1975; Gaillardon et al. 1977; quoted, Kenaga 1980; Kenaga & Goring 1980; Liu & Qian 1995)
    2.87 (soil, calculated-S as per Kenaga & Goring 1978, Kenaga 1980)
    2.85-2.87 (quoted values, Bottoni & Funari 1992)

3.30   (soil, 20-25°C, selected, Wauchope et al. 1992; Hornsby et al. 1996)
3.21-4.07 (Montgomery 1993)
3.30   (estimated-chemical structure, Lohninger 1994)
2.68   (soil, HPLC-screening method, Kördel et al. 1995b)
2.84   (calculated-K_{OW}, Liu & Qian 1995)

Half-Lives in the Environment:
    Air:
    Surface water:
    Groundwater:
    Sediment: aerobic half-life of 80-240 d for 1 µg/mL to biodegrade in sediment-water and anaerobic half-life of >650 d for 1 µg/mL to biodegrade in sediment-water both at 25°C (Muir & Yarechewski 1982; quoted, Muir 1991).
    Soil: estimated half-life of 42 d (Wauchope et al. 1992; Hornsby et al. 1996).
    Biota:

Environmental Fate Rate Constants or Half-Lives:
    Volatilization:
    Photolysis: 4 ppb contaminated water in the presence of TiO_2 and H_2O_2 completely photodegraded after 15 h by solar irradiation (Muszkat et al. 1992).
    Oxidation:
    Hydrolysis:
    Biodegradation: aerobic half-life of 80-240 d for 1 µg/mL to biodegrade in sediment-water and anaerobic half-life of >650 d for 1 µg/mL to biodegrade in sediment-water both at 25°C (Muir & Yarechewski 1982; quoted, Muir 1991).
    Biotransformation:
    Bioconcentration, Uptake (k_1) and Elimination (k_2) Rate Constants:

Common Name: Triallate
Synonym: Avadex BW, Buckle, CP 23426, Dipthal, Far-Go
Chemical Name: 2,3,3-trichloro-2-propene-1-thiol diisopropylcarbamate; S-(2,3,3-trichloro-allyl)diisopropyl(thiocarbamate); S-(2,3,3-trichloro-2-propenyl) bis(1-methylethyl)-carbamothioate
Uses: herbicide to control wild oats in lentils, barley, peas, and winter wheat.
CAS Registry No: 2303-17-5
Molecular Formula: $C_{10}H_{16}Cl_3NOS$
Molecular Weight: 304.74
Melting Point (°C):
    29-30   (Suntio et al. 1988; Herbicide Handbook 1989; Worthing 1991; Montgomery 1993; Tomlin 1994; Milne 1995)
Boiling Point (°C):
    148-149   (Khan 1980; Spencer 1982)
    117   (at 40 mPa, Herbicide Handbook 1989: Montgomery 1993; Milne 1995)
Density (g/cm³ at 20°C):
    1.273   (25°C, Agrochemicals Handbook 1987; Herbicide Handbook 1989; Worthing 1991; Montgomery 1993; Tomlin 1994; Milne 1995)
Molar Volume (cm³/mol):
    314.0   (calculated-LeBas method, Suntio et al. 1988)
Molecular Volume (Å³):
Total Surface Area, TSA (Å²):
Dissociation Constant pK_a:
Heat of Fusion, ΔH_{fus}, kcal/mol:
Entropy of Fusion, ΔS_{fus}, cal/mol·K (e.u.):
Fugacity Ratio at 25°C (assuming ΔS_{fus} = 13.5 e.u.), F: 0.9130

Water Solubility (g/m³ or mg/L at 25°C):
    4.0   (20°C, Weber 1972; Worthing 1987; quoted, Muir 1991)
    4.0   (Ashton & Crafts 1973; quoted, Suntio et al. 1988; Shiu et al. 1990)
    4.0   (Spencer 1973, 1982; Khan 1980; quoted, Shiu et al. 1990)
    4.0   (Martin & Worthing 1977, Worthing 1983, 1991; Herbicide Handbook 1978, 1989; quoted, Kenaga 1980; Kenaga & Goring 1980; Shiu et al. 1990; Isensee 1991)
    4.0   (Hartley & Graham-Bryce 1980; Jury et al. 1980, 1983, 1984; quoted, Taylor & Glotfelty 1988; Spencer & Cliath 1990; Grover 1991)
    4.0   (Weber et al. 1980; quoted, Willis & McDowell 1982)
    <1.0   (27°C, Ashton & Crafts 1981)
    4.0   (Agrochemicals Handbook 1987; Tomlin 1994)
    3.0   (20°C, selected, Suntio et al. 1988; quoted, Majewski & Capel 1995)
    4.0   (20-25°C, selected, Wauchope et al. 1992; Hornsby et al. 1996)
    4.0   (selected, Lohninger 1994)
    4.0   (Montgomery 1993, Milne 1995)

Vapor Pressure (Pa at 25°C):
    0.016   (20°C, Weber 1972; Worthing 1987; quoted, Muir 1991)
    0.016   (Ashton & Crafts 1973; Spencer 1982; quoted, Suntio et al. 1988)
    0.0141   (gas saturation, Grover et al. 1978; quoted, Suntio et al. 1988)

0.0257 (gas saturation, Grover et al. 1978; quoted, Jury et al. 1983; Grover 1991)
0.0257 (calculated-regression eqn., Grover et al. 1978)
0.0276 (calculated-vapor density, Grover et al. 1978; quoted, Spencer & Cliath 1983)
0.0265 (29.5°C, Ashton & Crafts 1981)
0.0276 (gas saturation method, Spencer & Cliath 1983)
0.026 (Jury et al. 1983; quoted, Taylor & Glotfelty 1988; Taylor & Spencer 1990)
0.0257 (quoted, Jury et al. 1984; Spencer & Cliath 1990)
0.016 (Agrochemicals Handbook 1987)
0.010 (20°C, selected, Suntio et al. 1988; quoted, Majewski & Capel 1995)
0.015 (Herbicide Handbook 1989)
0.026 (selected, Taylor & Spencer 1990)
0.016 (Worthing 1991; Tomlin 1994)
0.0147 (20-25°C, selected, Wauchope et al. 1992; Hornsby et al. 1996)
0.016 (20°C, Montgomery 1993)

Henry's Law Constant ($Pa \cdot m^3/mol$):
1.96 (calculated-P/C, Jury et al. 1983, 1984, 1987a, 1990; Jury & Ghodrati 1989; quoted, Spencer & Cliath 1990; Grover 1991)
1.02 (20°C, calculated-P/C, Suntio et al. 1988; quoted, Majewski & Capel 1995)
1.983 (calculated-P/C, Taylor & Glotfelty 1988)
1.226 (20°C, calculated-P/C, Muir 1991)
1.013 (20-25°C, calculated-P/C, Montgomery 1993)
0.762 (calculated-P/C, this work)

Octanol/Water Partition Coefficient, log K_{ow}:
4.29 (Montgomery 1993)

Bioconcentration Factor, log BCF:
2.45 (calculated-S, Kenaga 1980; quoted, Isensee 1991)
2.18 (calculated-K_{oc}, Kenaga 1980)

Sorption Partition Coefficient, log K_{oc}:
3.56 (Guenzi & Beard 1974; quoted, Jury et al. 1983, 1984, 1990)
3.34 (soil, Grover 1974; Beestman & Demming 1976; quoted, Kenaga 1980; Kenaga & Goring 1980)
3.32 (soil, calculated-S as per Kenaga & Goring 1978, Kenaga 1980)
3.56 (Jury et al. 1984; quoted, Spencer & Cliath 1990)
3.56 (soil, screening model calculations, Jury et al. 1987a,b; Jury & Ghodrati 1989)
3.35 (soil, quoted exptl., Meylan et al. 1992)
3.22 (soil, calculated-χ and fragments contribution, Meylan et al. 1992)
3.38 (soil, 20-25°C, selected, Wauchope et al. 1992; Hornsby et al. 1996)
3.31 (calculated, Montgomery 1993)
3.38 (selected, Lohninger 1994)

Half-Lives in the Environment:
- Air: calculated life-time of 5 h for the vapor-phase reaction with OH radicals in the troposphere (Atkinson et al. 1992; Kwok et al. 1992).
- Surface water: half-lives of 680 d at pH 6.8 and 1170 d at pH 7.0, both at 25°C for biodegradation in aquatic systems (Smith 1969; quoted, Scow 1982).
- Groundwater:
- Sediment:
- Soil: biodegradation half-life of 100 d from screening model calculations (Jury et al. 1984, 1987a,b; 1990; Jury & Ghodrati 1989; quoted, Montgomery 1993); selected field half-life of 82 d (Wauchope et al. 1992; Hornsby et al. 1996).
- Biota: biochemical half-life of 100 d from screening model calculations (Jury et al. 1987a,b; Jury & Ghodrati 1989).

Environmental Fate Rate Constants or Half-Lives:
- Volatilization: half-life of 26 d (Jury et al. 1983; quoted, Grover 1991); half-life of 100 d (Jury et al. 1984; quoted, Spencer & Cliath 1990); estimated half-life of 8 d from 1 m depth of water at 20°C (Muir 1991).
- Photolysis:
- Oxidation: calculated lifetime of 5 h for the vapor-phase reaction with OH radicals in the troposphere (Atkinson et al. 1992; Kwok et al. 1992).
- Hydrolysis: half-lives of >24 weeks for 1 μg/mL to hydrolyze in aqueous buffer at pH 4, 7, and 9 in the dark at 25°C (Smith 1969; quoted, Muir 1991).
- Biodegradation: estimated half-lives of 680 d at pH 6.8 and 1170 d at pH 7.0, both at 25°C from biodegradation rate constants in aquatic systems (Smith 1969; quoted, Scow 1982); half-life of 100 d for a 100 d leaching and screening test in 0-10 cm depth of soil (Jury et al. 1983, 1984, 1987a,b; 1990; Jury & Ghodrati 1989; Grover 1991).
- Biotransformation:
- Bioconcentration, Uptake (k_1) and Elimination (k_2) Rate Constants:

Common Name: Trifluralin
Synonym: Agreflan, Crisalin, Digermin, Elancolan, L-36352, Nitran, Nitrofor, Olitref, Treflan, Trifluoramine, Trifurex, Trikepin, Trim
Chemical Name: 2,6-dinitro-N,N-dipropyl-4-trifluoromethylaniline;2,6-dinitro-N,N-dipropyl-4-(trifluoromethyl)benzenamine
Uses: pre-emergence herbicide to control many grass and broadleaf weeds.
CAS Registry No: 1582-09-8
Molecular Formula: $C_{13}H_{16}F_3N_3O_4$
Molecular Weight: 335.5
Melting Point (°C):
    48.5-49   (Khan 1980; Spencer 1982; Herbicide Handbook 1989; Worthing 1991; Tomlin 1994; Milne 1995)
    49.0   (Swann et al. 1983)
    46-47   (Suntio et al. 1988; Montgomery 1993)
    48.7   (Patil 1994)
    48.5   (Kühne et al. 1995)
Boiling Point (°C):
    362   (estimated, Grain 1982)
    139-140   (at 4.2 mmHg, Agrochemicals Handbook 1987; Montgomery 1993; Milne 1995)
    96-97   (at 0.18 mmHg, Herbicide Handbook 1989)
    139-149   (at 4.20 mmHg, Herbicide Handbook 1989)
Density (g/cm^3 at 20°C):
    1.294   (25°C, Montgomery 1993)
    1.36   (22°C, Tomlin 1994)
Molar Volume (cm^3/mol):
    295.9   (calculated-LeBas method, Suntio et al. 1988)
Molecular Volume (Å^3):
Total Surface Area, TSA (Å^2):
Dissociation Constant pK_a:
Heat of Fusion, ΔH_{fus}, kcal/mol:
    5.70   (DSC method, Plato & Glasgow 1969)
Entropy of Fusion, ΔS_{fus}, cal/mol·K (e.u.):
Fugacity Ratio at 25°C (assuming ΔS_{fus} = 13.5 e.u.), F: 0.5789

Water Solubility (g/m^3 or mg/L at 25°C):
    24   (27°C, Woodford & Evans 1963; quoted, Shiu et al. 1990)
    24   (27°C, Günther et al. 1968; quoted, Suntio et al. 1988; Shiu et al. 1990)
    40   (29.5°C, Melnikov 1971; quoted, Shiu et al. 1990)
    0.35   (20°C, Weber 1972; Worthing 1987; quoted, Muir 1991)
    24   (27°C, Spencer 1973; quoted, Shiu et al. 1990)
    0.1-0.5 (Probst et al. 1975; quoted, Suntio et al. 1988)
    0.60   (Herbicide Handbook 1978; quoted, Kenaga 1980; Kenaga & Goring 1980; Thomas 1982; Nash 1988; Isensee 1991)
    0.05   (Wauchope 1978)
    <1.0   (20°C, Khan 1980)
    0.05   (Weber et al. 1980; quoted, Willis & McDowell 1982)
    0.60, 1.50 (exptl., calculated-MP & Parachor, Briggs 1981)
    8.11   (20-25°C, Kanazawa 1981; quoted, Shiu et al. 1990)

4.0 (Spencer 1982)
0.30 (Beste & Humburg 1983; Jury et al. 1983,84; quoted, Taylor & Glotfelty 1988; Grover 1991)
0.32 (gen. col.-HPLC-RI, Swann et al. 1983; quoted, Jury et al. 1983; Shiu et al. 1990)
0.70 (HPLC-RT, Swann et al. 1983; quoted, Shiu et al. 1990)
4.0 (27°C, Verschueren 1983; quoted, Suntio et al. 1988)
0.75 (shake flask-GC or LSC, Gerstl & Mingelgrin 1984; quoted, Shiu et al. 1990)
<1.0 (27°C, Agrochemicals Handbook 1987; Worthing 1991)
0.60 (quoted, Isnard & Lambert 1988; Kanazawa 1989)
0.50 (20°C, selected, Suntio et al. 1988; quoted, Majewski & Capel 1995)
0.30 (Taylor & Glofelty 1988; quoted, Shiu et al. 1990)
0.30 (Herbicide Handbook 1989; quoted, Shiu et al. 1990)
<1.0 (27°C, selected, Francioso et al. 1992)
0.30 (20-25°C, selected, Wauchope et al. 1992; Lohninger 1994; Hornsby et al. 1996; quoted, Halfon et al. 1996)
4.0 (27°C, Montgomery 1993)
0.184, 0.221, 0.189 (at pH 5, 7, 9, Tomlin 1994)
0.60, 6.70 (quoted, calculated, Patil 1994)
0.7. 1.61 (quoted, calculated-group contribution fragmentation method, Kühne et al. 1995)
<1.0 (27°C, Milne 1995)
1.16, 0.145 (quoted, predicted-AQUAFAC, Lee et al. 1996)

Vapor Pressure (Pa at 25°C):
0.0292 (29°C, effusion method, Hamaker & Kerlinger 1971; quoted, Spencer 1976)
0.0138 (20°C, Weber 1972; Worthing 1987; quoted, Muir 1991)
0.0323 (30°C, quoted from Spencer & Cliath 1973 unpublished data, Spencer et al. 1973)
0.0148 (gas saturation, Spencer & Cliath 1974; quoted, Spencer 1976; Jury et al. 1983; Nash 1989)
0.0065 (20°C, gas saturation, Spencer & Cliath 1974)
0.0323 (30°C, gas saturation, Spencer & Cliath 1974)
0.0065 (20°C, gas saturation-GC, Spencer 1976; quoted, Suntio et al. 1988)
0.0323 (30°C, gas saturation-GC, Spencer 1976; quoted, Suntio et al. 1988)
0.0137 (Worthing 1979,91; quoted, Dobbs et al. 1984)
0.0265 (29.5°C, Khan 1980)
0.0029 (20-25°C, Weber et al. 1980; quoted, Willis & McDowell 1982)
0.027; 0.008, 0.040 (30°C, quoted; estimated values, Grain 1982)
0.0173 (Thomas 1982)
0.0173 (Herbicide Handbook 1983; quoted, Nash 1988)
0.015 (20°C, Jury et al. 1983; quoted, Taylor & Glotfelty 1988; Taylor & Spencer 1990; Grover 1991)
0.0137 (Agrochemicals Handbook 1987)
0.006 (20°C, selected, Suntio et al. 1988; quoted, Majewski & Capel 1995)
0.011 (20°C, P_L, Suntio et al. 1988; quoted, Bacci et al. 1990)
0.0147 (Herbicide Handbook 1989)
0.0147 (Carsel 1989)
0.015 (20°C, selected, Taylor & Spencer 1990)

0.0137 (selected, Francioso et al. 1992)
0.0147 (20-25°C, selected, Wauchope et al. 1992; Hornsby et al. 1996; quoted, Halfon et al. 1996)
0.0147 (20°C, Montgomery 1993)
0.0095 (Tomlin 1994)

Henry's Law Constant (Pa·m^3/mol):
16.61 (calculated-P/C, Jury et al. 1983, 1984, 1987a,b; Jury & Ghodrati 1989; quoted, Grover 1991)
4.02 (20°C, calculated-P/C, Suntio et al. 1988; quoted, Findinger et al. 1989; Bacci et al. 1990; Müller et al. 1994; Majewski & Capel 1995)
16.36 (calculated-P/C, Taylor & Glotfelty 1988)
5.206 (fog chamber-GC/ECD, Findinger et al. 1989)
5.95 (wetted-wall col.-GC/ECD, Findinger et al. 1989)
16.0 (calculated-P/C, Nash 1989)
13.27 (20°C, calculated-P/C, Muir 1991)
4.903 (23°C, calculated-P/C, Montgomery 1993)
4.026 (calculated-P/C, this work)

Octanol/Water Partition Coefficient, log K_{ow}:
5.34 (quoted, Kenaga & Goring 1980; Isnard & Lambert 1988; Kanazawa 1989)
3.06 (quoted, Rao & Davidson 1980; Suntio et al. 1988; Haderlein et al 1996)
5.34, 5.16 (shake flask-UV, calculated-Parachor, Briggs 1981)
5.30 (selected, Briggs 1981; quoted, Sicbaldi & Finizio 1993)
5.28 (shake flask, Brown & Flagg 1981; quoted, McDuffie 1981)
3.97 (shake flask-GC, Kanazawa 1981; quoted, Sicbaldi & Finizio 1993)
4.94 (HPLC-k', McDuffie 1981)
5.05 (selected, Dao et al. 1983)
4.86 (shake flask, Dubelman & Bremer 1983; quoted, Sicbaldi & Finizio 1993)
4.19 (shake flask-GC or LSC, Gerstl & Mingelgrin 1984)
5.34 (Medchem Database 1988; quoted, Müller et al. 1994)
3.0 (selected, Suntio et al. 1988; quoted, Bacci et al. 1990)
5.33 (selected, Travis & Arms 1988)
5.07 (Herbicide Handbook 1989; Worthing 1991; Milne 1995)
5.34 (selected, Magee 1991; Devillers et al. 1996)
5.07 (selected, Francioso et al. 1992; quoted, Bintein & Devillers 1994)
5.07, 5.28, 5.34 (Montgomery 1993)
4.88 (RP-HPLC, Saito et al. 1993)
4.82 (RP-HPLC, Sicbaldi & Finizio 1993)
5.34 (selected, Sangster 1993; Hansch et al. 1995)
5.34, 4.19 (quoted, calculated, Patil 1994)
5.27 (pH 7.7-8.9, Tomlin 1994)
4.82, 5.0, 4.45 (RP-HPLC, ClogP, calculated-S, Finizio et al. 1997)

Bioconcentration Factor, log BCF:
3.97, 3.66 (measured, Metcalf & Sanborn 1975; quoted, Isensee 1991)
3.51 (fathead minnow, kinetic test, Spacie & Hamelink 1979)
3.03 (fathead minnow, chronic exposure, Spacie & Hamelink 1979)
3.11 (mosquitofish, correlated-S, Spacie & Hamelink 1979)
3.01 (rainbow trout, correlated-K_{ow}, Spacie & Hamelink 1979)

3.26-3.76   (Spacie & Hamelink 1979; quoted, Isensee 1991)
3.66   (quoted exptl., Kenaga 1980)
3.04   (calculated-K_{OC}, Kenaga 1980)
2.92   (calculated-S, Kenaga 1980; quoted, Isensee 1991)
3.68, 2.95   (exptl., calculated-K_{OW}, Briggs 1981)
3.50   (*Pseudorasbora parva*, Kanazawa 1981; quoted, Devillers et al. 1996)
3.26-3.76   (selected, Schnoor & McAvoy 1981; quoted, Schnoor 1992)
−0.37   (vegetation, correlated-K_{OW}, U. Oklahoma 1986; quoted, Travis & Arms 1988)
3.76   (quoted, Isnard & Lambert 1988)
2.67, 5.02   (dry leaf, wet leaf, Bacci et al. 1990)

Sorption Partition Coefficient, log K_{OC}:
4.14   (soil, Harvey 1974; quoted, Kenaga 1980; Kenaga & Goring 1980; Hodson & Williams 1988)
3.76   (soil, calculated-S as per Kenaga & Goring 1978, Kenaga 1980; quoted, Nash 1988)
3.64   (av. 3 soils, McCall et al. 1980)
4.49   (Georgia's Hickory Hill pond sediment, Brown & Flagg 1981; quoted, Muir 1991)
2.70   (selected, sediment/water, Schnoor & McAvoy 1981; quoted, Schnoor 1992)
3.78   (soil, Thomas 1982; quoted, Nash 1988)
3.87   (soil average, Jury et al. 1983; quoted, Grover 1991)
3.63   (soil slurry method, Swann et al. 1983)
3.98   (RP-HPLC-RT, Swann et al. 1983)
3.86   (screening model calculations, Jury et al. 1987a,b; Jury & Ghodrati 1989)
3.59   (quoted, Nash 1988)
2.94   (average of 2 soils, Kanazawa 1989)
3.64-4.15, 3.76-4.14   (quoted values, Bottoni & Funari 1992)
4.71, 4.44, 4.59   (No. 1 & 2 soil, No. 3 soil & No. 4 soil; Francioso et al. 1992)
3.90   (soil, 20-25°C, selected, Wauchope et al. 1992)
4.37   (selected, Lohninger 1994)
3.94   (soil, HPLC-ring test, Kördel et al. 1995a)
3.94   (soil, HPLC-screening method, Kördel et al. 1993, 1995b)

Sorption Partition Coefficient, log K_{OM}:
3.87   (Grover et al. 1978; quoted, Jury et al. 1983)
3.63   (experimental, Grover et al. 1979; quoted, Briggs 1981)
4.14   (av. soils/sediments, Kenaga & Goring 1980; quoted, Lyman 1982)
3.90   (experimental, Briggs 1981)
3.32   (calculated, Briggs 1981)
1.36, 2.08, 2.98   (log K_P: with first-order rate 0.52, 0.2, 8.3x10^{-3} hr^{-1}, Karickhoff & Morris 1985; quoted, Brusseau & Rao 1989)
4.14, 3.75   (selected, estimated, Magee 1991)
2.94-4.49   (Montgomery 1993)
4.37   (selected, Lohninger 1994)
3.90   (soil, 20-25°C, selected, Hornsby et al. 1996)

Adsorption Coefficient K_d (L kg^{-1}):
8.1   (homoionic K^+-montmorillonite clay minerals, Haderlein et al. 1996)

Half-Lives in the Environment:
   Air:
   Surface water: calculated half-life of 21 min from midday direct sunlight photolysis rate constant of 2.0 h^{-1} (Zepp 1978; Zepp & Cline 1979; quoted, Zepp et al. 1984); calculated half-life of 0.94 h for disappearance via direct sunlight photolysis in aqueous media (Zepp & Baughman 1978; quoted, Harris 1982); half-lives of <20 d for 2.5-5 cm water over flooded soils, ~20 h in water above sediment in estuarine sediment-water microcosm (Muir 1991).
   Groundwater:
   Sediment: degradation half-life of 9 d in estuarine sediment (18°/∞) system (Means et al. 1983).
   Soil: half-lives of 4-5 d for 4 μg/mL to biodegrade in flooded soils at 24.5°C and >21 d at 3.3°C (Probst et al. 1967; quoted, Means et al. 1983; Muir 1991); estimated persistence of 6 months in soil (Kearney et al. 1969; Edwards 1973; quoted, Morrill et al. 1982; Jury et al. 1987a); degradation half-life of 93 d in soil (Parr & Smith 1973; quoted, Means et al. 1983); half-life of >20 d for 0.33 μg/mL to biodegrade in soil suspension at 25°C (Willis et al. 1974; quoted, Muir 1991); degradation half-lives of 1 days (Kearney et al. 1976; quoted, Means et al. 1983); and 54 d in soil (Zimdahl & Gwynn 1977; quoted, Means et al. 1983); half-life of 20 d for 0.5 μg/ml to biodegrade in flooded soil with 0.5-1.0 cm of water on top of the soil at 20-42°C (Savage 1978; quoted, Muir 1991); persistence of more than 6 months (Wauchope 1978); half-life of <1 month for 1.0 μg/mL to biodegrade in flooded soils at 25°C (Camper et al. 1980; quoted, Muir 1991); estimated first-order half-life of 86.6 d in soil from biodegradation rate constant of 0.008 d^{-1} by soil incubation die-away test and 27.7 d in anaerobic systems from rate constant of 0.025 d^{-1} by flooded soil incubation die-away test (Rao & Davidson 1980; quoted, Scow 1982); field half-life of 0.1-0.3 d in moist fallow soil (Glotfelty 1981; quoted, Nash 1983); half-lives of 19.6 weeks at 4°C, 7.1 weeks at 25°C for for soil of field capcity moisture, 16.2 weeks at 4°C, 3.9 weeks at 25°C for flooded soils, Crowley silt loam; 27.0 weeks at 4°C, 8.1 weeks at 25°C for soil of field capacity moisture, 18.6 weeks at 4°C and 5.4 weeks at 25°C for flooded soils, Sharkey silty clay (Brewer et al. 1982); microagroecosystem half-life of 3-4 d in moist fallow soil (Nash 1983); half-life of 46 weeks for 2.0 μg/mL to biodegrade in flooded soils at 25°C (Brewer et al. 1982; quoted, Muir 1991); very persistent in soils with half-life >100 d (Willis & McDowell 1982); half-life of 20 h for 0.36 μg/mL to biodegrade in sediment-water microcosm at 20°C (Spain & Van Veld 1983; quoted, Muir 1991); measured dissipation rate of 0.69 d^{-1} (Nash 1983; quoted, Nash 1988); estimated dissipation rate of 1.6 and 0.24 d^{-1} (Nash 1988); first-order adsorption rate constants: 0.52, 0.2, 8.3x10^{-3} h^{-1} (Karickhoff & Morris 1985; quoted, Brusseau & Rao 1989); half-life approximately 22 d in submerged soils in a model ecosystem (Muir 1991); selected field half-life of 60 d (Wauchope et al. 1992; Hornsby et al. 1996; quoted, Halfon et al. 1996).
   Biota: half-lives: 22-31 d in river sauger, 17-57 d in river shorthead redhorse, 23 d in river golden redhorse, 3 d in lab. fathead minnow (Spacie & Hamelink

1979); biochemical half-life of 132 d (Jury et al. 1987a,b; Jury & Ghodrati 1989).

Environmental Fate Rate Constants or Half-Lives:
    Volatilization: initial rate constant of 2.6×10^{-2} h^{-1} and predicted rate constant of 6.6×10^{-2} h^{-1} from soil with a half-life of 10.5 h (Thomas 1982); half-life of 18 d (Jury et al. 1983; quoted, Grover 1991); measured rate constant of 2-6 d^{-1} (Glotfelty et al. 1984; quoted, Glotfelty 1989); estimated rate constant of 0.7 d^{-1} (Glotfelty et al. 1989); estimated half-life of 1.6 d from 1 m depth of water at 20°C (Muir 1991).
    Photolysis: half-life of 22 min for direct sunlight photolysis near surface water at 40°N in the summer (Zepp & Cline 1977; quoted, Zepp 1980); calculated half-life of 0.94 h for disappearance via direct sunlight photolysis in aqueous media (Zepp & Baughman 1978; quoted, Harris 1982); near surface direct sunlight photolysis rate constant of 0.03 d^{-1} with half-life of 22 d (Schnoor & McAvoy 1981; quoted, Schnoor 1992); observed vapor photolysis half-lives ranged from 25-60 min for July, midday sunlight in an outdoor chamber (Mongar & Miller 1988); half-life of 0.5 h estimated from photolysis reaction rate by direct sunlight of midday in mid-summer at 40°N near surface water (Zepp 1991).
    Oxidation:
    Hydrolysis:
    Biodegradation: half-lives of 4-5 d for 4 μg/mL to biodegrade in flooded soils at 24.5°C and >21 d at 3.3°C (Probst et al. 1967; quoted, Means et al. 1983; Muir 1991); half-life of >20 d for 0.33 μg/ml to biodegrade in soil suspension at 25°C (Willis et al. 1974; quoted, Muir 1991); half-life of 20 days for 0.5 μg/mL to biodegrade in flooded soil with 0.5-1.0 cm of water on top of the soil at 20-42°C (Savage 1978; quoted, Muir 1991); half-life of <1 month for 1.0 μg/mL to biodegrade in flooded soils at 25°C (Camper et al. 1980; quoted, Muir 1991); rate constant of 0.008 d^{-1} by soil incubation die-away test and 0.025 d^{-1} by flooded soil incubation die-away test (Rao & Davidson 1980; quoted, Scow 1982); half-life of 132 days for a 100 d leaching and screening test in 0-10 cm depth of soil (Rao & Davidson 1980; quoted, Jury et al. 1983, 1984, 1987a,b; Jury & Ghodrati 1989; Grover 1991); half-life of 46 weeks for 2.0 μg/mL to biodegrade in flooded soils at 25°C (Brewer et al. 1982; quoted, Muir 1991); half-life of 20 h for 0.36 μg/ml to biodegrade in sediment-water microcosm at 20°C (Spain & Van Veld 1983; quoted, Muir 1991); rate constants of -0.00504 to -0.00730 h^{-1} in nonsterile sediment and -0.00160 to -0.00651 h^{-1} in sterile sediment by shake-tests at Range Point and rate constants of -0.00827 to -0.01140 h^{-1} in nonsterile water and -0.00499 to -0.00712 h^{-1} in sterile water by shake-tests at Range Point (Walker et al. 1988); rate constants of -0.00621, -0.0121 h^{-1} in nonsterile sediment and -0.00476, -0.00409 h^{-1} in sterile sediment by shake-tests at Davis Bayou and first-order rate constants of -0.00439, -0.00349 h^{-1} in nonsterile water and -0.00299, -0.00598 h^{-1} in sterile water by shake-tests at Davis Bayou (Walker et al. 1988).
    Biotransformation:
    Bioconcentration, Uptake (k_1) and Elimination (k_2) Rate Constants:

Common Name: Vernolate
Synonym: PPTC, R1607, Vanalate, Vernam, Vernnolaolate
Chemical Name: S-propyldipropylthiocarbamate; S-propyldipropylcarbamothioate
Uses: herbicide incorporated with soil for pre-planting or pre-emergence control of broadleaf and grass weeds in groundnuts, soybeans, maize, tobacco, and sweet potatoes.
CAS Registry No: 1929-77-7
Molecular Formula: $C_{10}H_{21}NOS$
Molecular Weight: 203.35
Melting Point (°C): liquid
Boiling Point (°C):
    150   (at 30 mmHg, Khan 1980; Herbicide Handbook 1989; Worthing 1991; Tomlin 1994; Milne 1995)
    149-150   (at 30 mmHg, Merck Index 1989)
Density (g/cm³ at 20°C):
    0.954   (Ashton & Crafts 1981; Herbicide Handbook 1989; Worthing 1991)
    0.952   (Agrochemicals Handbook 1987; tomlin 1994; Milne 1995)
Molar Volume (cm³/mol):
    269.2   (calculated-LeBas method, this work)
Molecular Volume (Å³):
Total Surface Area, TSA (Å²):
Dissociation Constant pK_a:
Heat of Fusion, ΔH_{fus}, kcal/mol:
Entropy of Fusion, ΔS_{fus}, cal/mol·K (e.u.):
Fugacity Ratio at 25°C (assuming ΔS_{fus} = 13.5 e.u.), F: 1.0

Water Solubility (g/m³ or mg/L at 25°C):
    107   (Martin & Worthing 1977; quoted, Kenaga 1980)
    90   (20°C, Khan 1980; Spencer 1982; quoted, Suntio et al. 1988; Shiu et al. 1990)
    90   (20°C, Ashton & Crafts 1981; Herbicide Handbook 1989; Pait et al. 1992)
    107   (21°C, Verschueren 1983; quoted, Suntio et al. 1988; Shiu et al. 1990)
    90   (20°C, Agrochemicals Handbook 1987)
    90   (20°C, Worthing 1987, 1991; quoted, Shiu et al. 1990)
    90   (20°C, selected, Suntio et al. 1988; quoted, Majewski & Capel 1995)
    107   (Merck Index 1989; quoted, Milne 1995)
    95   (Wauchope 1989; quoted, Shiu et al. 1990)
    108   (20-25°C, selected, Wauchope et al. 1992; Hornsby et al. 1996)
    108   (selected, Lohninger 1994)

Vapor Pressure (Pa at 25°C):
    0.84   (20°C, Hartley & Graham-Bryce 1980; quoted, Suntio et al. 1988)
    1.386   (Khan 1980; Spencer 1982; quoted, Suntio et al. 1988; Herbicide Handbook 1989)
    1.333   (Ashton & Crafts 1981)
    1.39   (Agrochemicals Handbook 1987)
    0.9   (20°C, selected, Suntio et al. 1988; quoted, Majewski & Capel 1995)
    1.386   (Merck Index 1989)
    1.39   (Worthhing 1991; Tomlin 1994)
    1.293   (20-25°C, selected, Wauchope et al. 1992; Hornsby et al. 1996)

Henry's Law Constant (Pa·m^3/mol):
- 2.05 (20°C, calculated-P/C, Suntio et al. 1988; quoted, Majewski & Capel 1995)
- 2.034 (calculated-P/C, this work)

Octanol/Water Partition Coefficient, log K_{OW}:
- 3.84 (20°C, Worthing 1991; Tomlin 1994)
- 3.84 (20°C, Milne 1995)
- 3.84 (selected, Hansch et al. 1995)
- 3.86, 3.74, 3.19 (RP-HPLC, ClogP, calculated-S, Finizio et al. 1997)

Bioconcentration Factor, log BCF:
- 1.64 (calculated-S, Kenaga 1980)
- 1.70 (calculated, Pait et al. 1992)

Sorption Partition Coefficient, log K_{OC}:
- 2.52 (calculated-S, Kenaga 1980)
- 2.41 (soil, 20-25°C, selected, Wauchope et al. 1992; Hornsby et al. 1996)
- 2.41 (estimated-chemical structure, Lohninger 1994)

Half-Lives in the Environment:
- Air:
- Surface water:
- Groundwater:
- Sediment:
- Soil: half-life in moist loam soil is approximately 1.5 week at 21-27°C (Herbicide Handbook 1989); selected field half-life of 12 d (Wauchope et al. 1992; Hornsby et al. 1996); soil half-life of 11 d (Pait et al. 1992); microbial degradation half-life are 8-16 d at 27°C, >64 d at 4°C (Tomlin 1994).
- Biota:

Environmental Fate Rate Constants or Half-Lives:
- Volatilization:
- Photolysis:
- Oxidation:
- Hydrolysis:
- Biodegradation: microbial degradation half-lives are 8-16 d at 27°C, >64 d at 4°C in soil (Tomlin 1994).
- Biotransformation:
- Bioconcentration, Uptake (k_1) and Elimination (k_2) Rate Constants:

Table 2.2.1 Common names, chemicals names and physical properties of herbicides

Name	Synonym	Chemical Name	Formula	MW	mp, °C	Fug. ratio at 25°C	pK_a	pK_b
Alachlor [15972-60-8]	Lasso, Metachlor	α-chloro-2,6-diethyl-N-methoxy-methylacetanilide	$C_{14}H_{20}ClNO_2$	269.77	41	0.695	0.62	
Ametryn [834-12-8]	Evik, Gesapax	2-methylthio-4-(ethylamino)-6-(isopropylamino)-s-triazine	$C_9H_{17}N_5S$	227.35	84-85 88-89	0.255	4.00 4.10	10.07
Amitrole [61-82-5]	Amerol, Aminotriazole	3-amino-1H-1,2,4-triazole	$C_2H_4N_4$	84.08	157-159	0.048		9.83
Atrazine [1912-24-9]	Gesaprim	2-chloro-4-(ethylamino)-6-(isopropylamino)-s-triazine	$C_8H_{14}ClN_5$	215.68	174	0.034	1.68 1.70	12.32
Barban [101-27-9]	Carbyne	4-chlorobut-2-ynyl-3-chlorocarbanilate	$C_{11}H_9Cl_2NO_2$	258.1	75-76	0.313		
Benefin [1861-40-1]	Balan, Bonalan Benfluralin	N-butyl-N-ethyl-α,α,α-trifluoro-2,6-dinitro-p-toluidine	$C_{13}H_{16}N_3O_4F_3$	335.3	65-66.5	0.389		
Bifenox [42576-02-3]	Modown	methyl-5-(2,4-dichlorophenoxy)-2-nitrobenzoate	$C_{14}H_9Cl_2NO_5$	342.1	84-86	0.255		
Bromacil [314-40-9]	Borea, Hyvar X	5-bromo-3-sec-butyl-6-methyl-uracil	$C_9H_{13}BrN_2O_2$	261.11	158-159	0.047	9.10 <7.0	
Bromacil lithium salt			$C_9H_{12}N_2O_3Li$	267.0			9.27	
Bromoxynil [1689-84-5]	Brominal, Buctril	3,5-dibromo-4-hydroxybenzonitrile	$C_7H_3Br_2NO$	276.9	194-195	0.021	4.06 4.20	
Bromoxynil butyrate ester [3861-41-4]			$C_{13}H_9BrNO_3$	347.01			4.10	
Bromoxynil octanoate [1689-99-2]			$C_{15}H_{17}Br_2NO_2$	403	45-46	0.620	4.08	
Butachlor [23184-66-9]	Machete	N-butoxymethyl-2-chloro-2',6'-diethylacetanilide	$C_{17}H_{26}ClNO_2$	311.9	<-5	1		
Butralin [33629-47-9]	Amex, Tamex	N-sec-butyl-4-tert-butyl-2,6-dinitroaniline	$C_{14}H_{21}N_3O_4$	295.3	60-61	0.440		
Butylate [2008-41-5]	Sutan	S-ethyl bis(2-methylpropyl)carbamothioate	$C_{11}H_{23}NOS$	217.36	liquid	1		

Name	Synonym	Chemical Name	Formula	MW	mp, °C	Fug. ratio at 25°C	pK$_a$	pK$_b$
Chloramben [133-90-4]	Amiben, Amoben	3-amino-2,5-dichlorobenzoic acid	$C_7H_5Cl_2NO_2$	206	200-201	0.018	3.40	
Chloramben salts [133-90-4]	Amiben	ammonium or sodium salt of chloramben	$C_7H_5Cl_2NO_2$	206				
Chlorbromuron [13360-45-7]	Maloran	3-(4-bromo-3-chlorophenyl)-1-methoxy-1-methylurea	$C_9H_{10}BrClN_2O_2$	293.5	95-97	0.199		
Chlorfenac [85-34-7]	Fenac	(2,3,6-trichlorophenyl)acetic acid	$C_8H_5Cl_3O_2$	239.5	156	0.051		
Chlorpropham [101-21-3]	Furloe	isopropyl 3-chlorocarbanilate	$C_{10}H_{12}ClNO_2$	213.65	40.7-41.1	0.695		
Chlorsulfuron [64902-72-3]	Glean, Telar	1-(2-chlorophenylsulfonyl)-3-(4-methoxy-6-methyl-1,3,5-trizaon-2-yl)urea	$C_{12}H_{12}ClN_5O_4S$	357.8	174-178	0.032	3.60	
Chlortoluron [15545-48-9]	Dicuran	3-(3-chloro-p-tolyl)-1,1-dimethylurea	$C_{10}H_{13}ClN_2O$	212.7	147-148	0.061		
Cyanazine [21725-46-2]	Bladex, Fortrol	2-(4-chloro-6-ethylamino-1,3,5-triazin-2-ylamino)-2-methylpropionitrile	$C_9H_{13}ClN_6$	240.7	166.5-167	0.039	1 0.63	12.9
2,4-D [94-75-7]	Agratect, Farmco, Weed Tox	2-(2,4-dichlorophenoxy)acetic acid	$C_8H_6Cl_2O_3$	221.04	140.5	0.072 0.072	2.64 3.31	
2,4-D dimethylamine salt [2008-39-1] 2,4-D esters			$C_{10}H_{13}Cl_2NO_3$ $C_8H_6Cl_2O_3$ (a)	266.1 221.04	85-87	0.249	2.80	
Dalapon [75-99-0]	Dowpon, Radapon	2,2-dichloropropionic acid	$C_3H_4Cl_2O_2$	143	liquid	1	1.74 1.84	
Dalapon sodium salt [120-20-8]		sodium 2,2-dichloropropionate	$C_3H_3Cl_2Na$	164.95	166.5 dec			
2,4-DB [94-82-6]	Embutox	4-(2,4-dichlorophenoxy)butyric	$C_{10}H_{10}Cl_2O_3$	249.1	119-119.5	0.120	4.80	
2,4-DB butoxyethyl ester			$C_{16}H_{22}O_4Cl_2$	349.3			4.80	
2,4-DB dimethylamine salt			$C_{10}H_{10}Cl_2O_3$	294.1			4.80	
Diallate [2303-16-4]	Avadex	S-(2,3-dichloroallyl)diisopropylthiocarbamate	$C_{10}H_{17}Cl_2NOS$	270.24	25-30	0.892		
Dicamba [1918-00-9]	Banvel, Dianat, Mediben	3,6-Dichloro-o-anisic acid	$C_8H_6Cl_2O_3$	221.04	114-116	0.129	1.91 1.95	
Dicamba-dimethylammonium salt [2300-66-5]			$C_{10}H_{13}Cl_2NO_3$	266.1				

Name	Synonym	Chemical Name	Formula	MW	mp, °C	Fug. ratio at 25°C	pK_a	pK_b
Dicamba-sodium salt [1982-69-0]			$C_8H_5Cl_2NaO_3$	243				
Dichlobenil [1194-65-6]	Casoron	2,6-dichlorobenzonitrile	$C_7H_3Cl_2N$	172.02	144-145	0.065		
Dichlorophen (F.A.B) [97-23-4]	Super Mosstox	4,4'-dichloro-2,2'-methylenediphenol	$C_{13}H_{10}Cl_2O_2$	269.1	177-178	0.031		
Dichloroprop [120-36-5]	2,4-DP	(RS)-2-(2,4-dichlorophenoxy)-propionic acid	$C_9H_8Cl_2O_3$	235.1	116-117.5	0.123	3.0	
Dichlorprop-P [15165-67-0]	Cornox RK	(R)-2-(2,4-dichlorophenoxy)propionic acid	$C_9H_8Cl_2O_3$	235.1	122	0.110	3.0	
Dichlorprop ester	Weedone	butoxyethyl ester of (R-2-(2,4-dichloro-	$C_9H_8Cl_2O_3$	335.3			2.86	
Diclofop [40843-25-2]		2-(4-aryloxyphenoxy)propionic acid	$C_{15}H_{12}Cl_2O_4$	327.2			2.85	
Diclofop-methyl [51338-27-3]	Hoelon	2-(4-(2,4-dichlorophenoxy)phenoxyl)-propanoic acid methyl ester	$C_{16}H_{14}Cl_2O_4$	341.2	39-41	0.711	3.10	
Dinitramine [29091-05-2]	Cobexo	N,N-diethyl-2,6-dinitro-4-trifluoro-methyl-m-phenyenediamine	$C_{11}H_{13}N_4O_4F_3$	322.2	98-99	0.185		
Dinoseb [88-85-7]	Antox, Aretit, BNP 30, DNBP	2-sec-butyl-4,6-dinitrophenol	$C_{10}H_{12}N_2O_5$	240.2	38-42	0.711	4.62	
Dinoseb salts [88-85-7]	Premerge, Dinitro	2-sec-butyl-4,6-dinitrophenyl ammonium, amine, acetate salts	$C_{10}H_{12}N_2O_5$#	240.2#			4.50	
Diphenamid [957-51-7]	Dymid, Enide	N,N-dimethyldiphenylacetamide	$C_{16}H_{17}NO$	239.3	134.5-135.	0.081		
Diquat [2764-72-9]	Reglone, Pathclear Cleansweep, Weedol	1,1'-ethylene-2,2'-dipyridine	$C_{12}H_{14}N_2$	186.2	300 dec.	0.00077		10
Diquat dibromide salt [85-00-7]			$C_{12}H_{12}Br_2N_2$	344.1				
Diuron [330-54-1]	DMU, Karmex DCMU	3-(3,4-dichlorophenyl)-1,1-dimethylurea	$C_9H_{10}Cl_2N_2O$	233.1	158-159	0.047		
EPTC [759-94-4]	Eptam, Eradicane	S-ethyl dipropylthiocarbamate	$C_9H_{19}NOS$	189.31	liquid	1		
Fenoprop (G.R.) [93-72-1]	Silvex, 2,4,5-TP	(±)-2-(2,4,5-trichlorophenoxy)-propionic acid	$C_9H_7Cl_3O_3$	269.5	179-181	0.029		

Name	Synonym	Chemical Name	Formula	MW	mp, °C	Fug. ratio at 25°C	pK$_a$	pK$_b$
Fenuron [101-42-8]	Dybar, Urab	1,1-dimethyl-3-phenylurea	C$_9$H$_{12}$N$_2$O	164.2	131-133	0.087		
Fenuron-TCA [4482-55-7]		1,1-dimethyl-3-phenyluronium trichloroacetate	C$_{11}$H$_{13}$Cl$_3$N$_2$O$_3$	327.6	65-68	0.384		
Fluchloralin [33245-39-5]	Basalin, BAS-392H	N-(2-chloroethyl) α,α,α-trifluoro-2,6-dinitro-N-propyl-p-toluidine	C$_{12}$H$_{13}$ClF$_3$N$_3$O$_4$	355.7	42-43	0.664		
Fluometuron [2164-17-2]	Cotoran, Cottonex, Meturon	N,N-dimethyl-N'-[3-(trifluoromethyl)-phenyl]urea	C$_{10}$H$_{11}$F$_3$N$_2$O	232.2	163-164.5	0.042		
Fluorodifen [15457-05-3]	Soyex	4-nitrophenyl α,α,α-trifluoro-2-nitro-p-tolyl ether	C$_{13}$H$_7$F$_3$N$_2$O$_5$	328.2	94	0.208		
Fluridone [59756-60-4]	Fluridon, Pride, Sonar	1-methyl-3-phenyl-5-[3-(trifluoromethyl)-phenyl]-4-(1H)-pyridinone	C$_{19}$H$_{14}$F$_3$NO	329.2	154-155	0.052		12.3
Glyphosate [1071-83-6]	Roundup, Polado	N-(phosphoromethyl)glycine	C$_3$H$_8$NO$_5$P	169.1	200	0.019	5.70	
Glyphosate-mono(iso-propylammonium) [38641-94-0]			C$_6$H$_{17}$N$_2$O$_5$P	228.2	200	0.019		
Ioxynil [1689-83-4]	Actril, Totril	4-hydroxy-3,5-di-iodobenzonitrile	C$_7$H$_3$I$_2$NO	370.9	209	0.015	3.96	
Ioxynil-octanoate [3681-47-0]		4-cyano-2,6-iodophenyl octanoate	C$_{15}$H$_{17}$I$_2$NO$_2$	497.1	59-60	0.451		
Ioxynil-sodium salt [2961-62-8]			C$_7$H$_2$I$_2$NNaO	392.9	360	0.00049		
Isopropalin [33820-53-0]	Paralan	4-isopropyl-2,6-dinitro-N-dipropylaniline	C$_{15}$H$_{23}$N$_3$O$_4$	309.4	liquid	1		
Isoproturon [34123-59-6]	Alon, Arelon, Graminon	3-p-cumenyl-1,1-dimethylurea	C$_{12}$H$_{18}$N$_2$O	206.3	155-156	0.051		
Linuron [330-55-2]	Afalon, Lorox	3-(3,4-dichlorophenyl)-1-methoxy-1-methylurea	C$_9$H$_{10}$Cl$_2$N$_2$O$_2$	249.1	93-94	0.208		
MCPA (G.R., H) [94-74-6]	Metaxon, Agroxone, Agritox	4-chloro-2-methylphenoxyacetic acid	C$_9$H$_9$ClO$_3$	200.6	118-119	0.118	3.05 3.13	
MCPA dimethylamine salt		dimethylamine salt of MCPA	C$_{11}$H$_{16}$ClNO$_3$	243.7			3.12	
MCPA ester	Weedone, Weedar	2-butoxyethyl, isooctyl esters of MCPA	C$_9$H$_9$ClO$_3$#	200.6#				
MCPA sodium salt		sodium salt of MCPA	C$_9$H$_8$ClNaO$_3$	222.6				
MCPA-thioethyl [25319-90-8]	Herbit, Zero One	S-ethyl 4-chloro-o-tolyoxythioacetate	C$_{11}$H$_{13}$ClO$_2$S	244.7	41-42	0.679		

Name	Synonym	Chemical Name	Formula	MW	mp, °C	Fug. ratio at 25°C	pK$_a$	pK$_b$
MCPB [94-81-5]	Tropotox	4-(4-chloro-2-methylphenoxy)butyric acid	C$_{11}$H$_{13}$ClO$_3$	228.7	100	0.181	4.84 4.80	
MCPB sodium salt	Thistrol	sodium salt of MCPB	C$_{11}$H$_{12}$ClO$_3$Na	250.6				
Mecoprop [7085-19-0]	Iso-Cornox, MCPP	($\pm$)-2-(4-chloro-o-tolyloxy)-propionic acid	C$_{10}$H$_{11}$ClO$_3$	214.6	94-95	0.203	3.78 3.75	
Mecoprop-P [16484-77-8]		(R)-2-(4-chloro-o-tolyoxy)propionic acid	C$_{10}$H$_{11}$ClO$_3$	214.6	95	0.203	3.78	
Metobromuron [3060-89-7]	Patoran	3-(4-bromophenyl)-1-methoxy-1-methylurea	C$_9$H$_{11}$BrN$_2$O$_2$	259.1	95.5-96	0.199		
Metolachlor [51218-45-2]	Codal, Dual, Primagram	2-chloro-6′-ethyl-N-(2-methoxy-1-methylethyl)acet-o-toluidide	C$_{15}$H$_{22}$ClNO$_2$	283.8	liquid	1		
Metoxuron [19937-59-8]	Dosanex	3-(3-chloro-4-methoxyphenyl)-1,1-dimethylurea	C$_{10}$H$_{13}$ClN$_2$O$_2$	228.7	126-127	0.098		
Molinate [2212-67-1]	Ordram	S-ethyl azepane-1-carbothioate	C$_9$H$_{17}$NOS	187.3	liquid	1		
Monolinuron [1746-81-2]	Aresin	3-(4-chlorophenyl)-1-methoxy-1-methylurea	C$_9$H$_{11}$ClN$_2$O$_2$	214.6	80-83	0.273		
Monuron [150-68-5]	Telvar, Urox	1,1-dimethyl-3-(p-chloro-phenyl)-urea	C$_9$H$_{11}$ClN$_2$O	198.6	176-177	0.031		
Neburon [555-37-3]	Kloben	1-butyl-3-(3,4-dichlorophenyl)-1-methylurea	C$_{12}$H$_{16}$Cl$_2$N$_2$O	275.2	102-103	0.169		
Nitralin [4726-14-1]	Planavin	4-(methylsulfonyl)-2,6-dinitro-N,N-dipropylaniline	C$_{13}$H$_{19}$N$_3$O$_6$S	345.4	150-151	0.057		
Oryzalin [19044-88-3]	Rycelan, Rycelon, Surflan	4-(dipropylamino)-3,5-dinitro-benzenesulfonamide	C$_{12}$H$_{18}$N$_4$O$_6$S	346.4	141-142	0.070	9.40 8.60	
Paraquat [4685-14-7]	Cyclone, Gramoxone	1,1′-dimethyl-4,4′-pyridinium	C$_{12}$H$_{14}$N$_2$	186.3	dec.			
Paraquat dichloride salt [1910-42-5]			C$_{12}$H$_{14}$Cl$_2$N$_2$	257.2				<4
Pebulate [1114-71-2]	Tillam	s-propyl butylethylcarbamothioate	C$_{10}$H$_{21}$NOS	203.4	liquid	1		
Pentachlorophenol [87-86-5]	PCP	pentachlorophenol	C$_6$Cl$_5$OH	266.3	190	0.0336	4.74	
Pentachlorophenol sodium salt (Pentacon)								

Name	Synonym	Chemical Name	Formula	MW	mp, °C	Fug. ratio at 25°C	pK$_a$	pK$_b$
Pentanochlor [2307-68-8]	Solan	3'-chloro-2-methylvaler-p-toluidide	C$_{13}$H$_{18}$ClNO	239.7	85-86	0.249		
Picloram [1918-02-1]	Tordon	4-amino-3,5,6-trichloro-picolinic acid	C$_6$H$_3$Cl$_3$N$_2$O$_2$	241.5	218-219	0.012	1.90 3.60	
Picloram-potassium salt [2425-60-0]			C$_6$H$_2$Cl$_3$KN$_2$O$_2$	279.6				
Profluralin [26399-36-0]	Pregard	N-(cyclopropylmethyl)-2,6-dinitro-N-propyl-4-(trifluoromethyl)-benzenamine	C$_{14}$H$_{16}$F$_3$N$_3$O$_4$	347.3	33-36	0.778		
Prometon [1610-18-0]	Primatol, Gesagram	2,4-bis(isopropylamino)-6-methoxy-s-triazine	C$_{10}$H$_{19}$N$_5$O	225.3	91-92	0.217	4.28 4.30	9.73
Prometryn [7287-19-6]	Caparol, Gesagard	N,N'-bis(isopropylamino)-6-(methylthio)-1,3,5-triazine-2,4-diamine	C$_{10}$H$_{19}$N$_5$S	241.4	118-120	0.115	4.05 4.10	9.95
Pronamide [23950-58-5]	Kerb, Promamide	3,5-dichloro-N-(1,1-dimethyl-propyny)-benzamide	C$_{12}$H$_{11}$Cl$_2$NO	256.1	154	0.053		
Propachlor [1918-16-7]	Ramrod	2-chloro-N-(1-methylethyl)-N-phenylacetamide	C$_{11}$H$_{14}$ClNO	211.7	67-76	0.313		
Propanil [709-98-8]	Propanex, Riselect, Stampede 3E	N-(3,4-dichlorophenyl)-propionamide	C$_9$H$_9$Cl$_2$NO	218.1	91-93	0.217		
Propazine [139-40-2]	Gesamil, Milogard	2-chloro-4,6-bis(isopropylamino)-s-triazine	C$_9$H$_{16}$ClN$_5$	230.0	212-214	0.014	1.85 1.80	12.15
Propham [122-42-9]	IPC	isopropyl carbanilate	C$_{10}$H$_{13}$NO$_2$	179.2	87-87.6	0.240		
Propyzamide [23950-58-5]	Kerb	3,5-dichloro-N-(1,1-dimethyl propynylbenzamide	C$_{12}$H$_{11}$Cl$_2$NO	256.1	155-156	0.051		
Pyrazon [1698-60-8]	Chloridazon	5-amino-4-chloro-2-phenyl-3(2H)-pyridazinone	C$_{10}$H$_8$ClON$_3$	221.6	207	0.016		
Secbumeton [26259-45-0]	Etazine, Sumitol	N-ethyl-6-methoxy-N'-(1-methyl-propyl)-1,3,5-triazine-2,4-diamine	C$_{10}$H$_{19}$N$_5$O	225.0	86-88	0.244	4.40	
Simazine [122-34-9]	Gesatop, Weedex, Aquazine	2-chloro-4,6-di(ethylamino)-s-triazine	C$_7$H$_{12}$ClN$_5$	201.7	225-227	0.010	1.65 1.60	12.35
Simetryne [1014-70-6]	Gy-bon	N,N'-diethyl-6-methylthio-1,3,5-triazine-2,4-diyldiamine	C$_8$H$_{15}$N$_5$S	213.3	82-83	0.267		11

Name	Synonym	Chemical Name	Formula	MW	mp, °C	Fug. ratio at 25°C	pKa	pKb
2,4,5-T [93-76-5]	Gesatop	2,4,5-trichlorophenoxyacetic acid	$C_8H_5Cl_3O_3$	255.5	153	0.054	2.80	
2,3,6-TBA [50-31-7]	Trysben, Cambilene	2,3,6-trichlorobenzoic acid	$C_7H_3Cl_3O_2$	225.5	125-126	0.10	2.88	
Terbacil [5902-51-2]	Sinbar	3-tert-butyl-5-chloro-6-methyluracil	$C_9H_{13}ClN_2O_2$	216.7	175-177	0.032	9.0	
Terbumeton [33693-04-8]	Caragard	N-tert-butyl-N'-ethyl-6-methoxy-1,3,5-triazine	$C_{10}H_{19}N_5O$	225.3	123-124	0.105		9.41
Terbuthylazine [5915-41-3]	Gardoprim	N-tert-butyl-6-chloro-N'-ethyl-1,3,5-triazine-2,4-diamine	$C_9H_{16}ClN_5$	229.7	177-179	0.031		12
Terbutryne [886-50-0]	Igran, Clarosan, Prebane	N-tert-butyl-N'-ethyl-6-methyl-thio-1,3,5-triazine-2,4-diamine	$C_{10}H_{19}N_5S$	241.4	104-105	0.162	4.30	9.7
Thiobencarb [28249-77-6]	Benthiocarb, Bolero, Saturno	S-4-chlorobenzyl-diethyl-thiocarbamate	$C_{12}H_{16}ClNOS$	257.8	3.3	1	4.07	
Triallate [2303-17-5]	Avadex BW, Far-Go	S-(2,3,3-trichloro-2-propenyl)-bis(1-methylethyl)carbamothioate	$C_{10}H_{16}Cl_3NOS$	304.7	29-30	0.913		
Trifluralin [1582-09-8]	Treflan, Triflurex, Elancolan	2,6-dinitro-N,N-dipropyl-4-trifluoromethylaniline	$C_{13}H_{16}F_3N_3O_4$	335.5	48.5-49	0.579		
Vernolate [1929-77-7]	Surpass, Vernam	S-propyldipropylthiocarbamate	$C_{10}H_{21}NOS$	203.4	liquid	1		

Note: F.A.B. - fungicide algicide bactericide; G.R. - growth regulator
Fug. ratio - fugacity ratio; pK_a - acid dissociation constant; pK_b - basicity constant
(a) ester is quickly converted to parent acid.
value of parent acid

Table 2.2.2 Summary of physical-chemical properties of herbicides at 25°C

Name	Selected properties						H, Pa·m³/mol		log K_{OC}
	P^S, Pa	P_L, Pa	S, g/m³	C^S, mol/m³	C_L, mol/m³	log K_{OW}	calcd P/C		
Alachlor	0.0020	2.88×10^{-3}	240	0.890	1.281	2.8	0.0022		2.23
Ametryn	0.0001	3.92×10^{-4}	185	0.814	3.191	2.58	1.23×10^{-4}		2.59
Amitrole	5.50×10^{-7}*	1.14×10^{-5}	280000	3330	68850	0.52	1.65×10^{-10}		2.04
Atrazine	4.00×10^{-5}	1.19×10^{-3}	30	0.139	4.140	2.75	2.88×10^{-4}		2.00
Barban	5.00×10^{-5}*	1.60×10^{-4}	11	0.043	0.1362	2.68	1.17×10^{-3}		2.66
Benefin	0.0088	0.0226	0.1	0.003	0.0077	5.29	29.4		3.95
Bifenox	3.20×10^{-4}	1.25×10^{-3}	0.35	0.0010	0.0040	4.48	0.313		
Bromacil	4.00×10^{-5}	8.46×10^{-4}	815	3.121	66.018	2.11	1.28×10^{-5}		1.86
Bromacil lithium salt	4.13×10^{-5}								1.51
Bromoxynil	6.40×10^{-4}	0.0307	130	0.469	22.54	<2.0	1.36×10^{-3}		
Bromoxynil octanoate	6.40×10^{-4}	1.03×10^{-3}				5.4			4.25
Butachlor	6.0×10^{-4}	6.00×10^{-4}	23	0.074	0.074	4.50	8.14×10^{-3}		2.8
Butralin	0.0017	3.86×10^{-3}	1	0.0034	7.69×10^{-3}	4.54	0.502		3.75
Butylate	1.73	1.73	45	0.182	0.182	4.15	8.36		2.60
Chloramben	0.93	51.19	700	3.398	187.05	1.11	0.274		1.32
Chloramben salts	0		900000						1.18
Chlorbromuron	5.33×10^{-5}	2.69×10^{-4}	50	0.170	0.858		3.13×10^{-4}		2.7
Chlorfenac	1	19.75	200	0.835	16.50		1.20		
Chlorpropham	0.001	0.001	89*	0.417	0.600	3.51	2.40×10^{-3}		2.85, 2.8
Chlorsufuron	6.13×10^{-4}*	0.019	7000	19.56	609.4	−1.0	3.13×10^{-5}		1.6
Chlortoluron	1.70×10^{-5}	2.80×10^{-4}	70	0.329	5.418	2.38	5.17×10^{-5}		2.81
Cyanazine	2.13×10^{-7}	5.41×10^{-6}	171	0.710	18.03	2.22	3.00×10^{-7}		2.3
2,4-D	8.0×10^{-5}*	1.11×10^{-3}	400	1.810	25.12	2.81	4.42×10^{-5}		1.68-2.73
2,4-D (a)	0.001	0.0139	890	4.026	55.92	2.81	2.48×10^{-4}		
2,4-D DMA salt	0		796000	2991	12000				1.3
2,4-D esters			100						2.00
Dalapon	1.0×10^{-5}*	1.0×10^{-5}	502000	3510	3510	0.78	2.85×10^{-9}		0.48, 2.13
Dalapon sodium salt			900000	5455					0

Name	Selected properties						H, Pa·m³/mol		
	P^S, Pa	P_L, Pa	S, g/m³	C^S, mol/m³	C_L, mol/m³	log K_{ow}	calcd P/C		log K_{oc}
2,4-DB	0		46	0.185	1.539	3.53			2.64
2,4-DB butoxyethyl ester	0.00001		8						2.7
2,4-DB DMA salt	0		709000						1.30
Diallate	0.02	0.0224	50*	0.185	0.207	5.23	0.108		2.70
Dicamba	0.0045*	0.0349	4500	20.36	158.1	2.21	2.21×10^{-4}		0.342, −0.4
Dicamba salt			400000	1646					0.301
Dicamba-dimethylammonium									
Dicamba-potassium salt									
Dicamba-sodium salt									
Dichlobenil	0.07*	1.076	18	0.105	1.609	2.74	0.669		2.91
Dichlorophen (F.A.B)	1.30×10^{-8}	4.24×10^{-7}	30	0.111	3.635		1.17×10^{-7}		
Dichloroprop	0.0004	3.25×10^{-3}	350	1.489	12.099	3.43	2.69×10^{-4}		3.0
Dichlorprop-P	6.20×10^{-5}	5.65×10^{-4}	590	2.510	22.855	1.95	2.47×10^{-5}		2.23
Dichlorprop(2,4-DP)ester	1.0×10^{-5}		50	0.149					3.00
Diclofop									
Diclofop-methyl	4.67×10^{-4}*	6.57×10^{-4}	0.8	2.34×10^{-3}	3.30×10^{-3}	4.58	0.199		4.2
Dinitramine	0.00048	2.59×10^{-3}	1	0.0031	0.017	4.30	0.155		3.6
Dinoseb	0.01*	0.0141	50	0.208	0.293	3.56	0.048		2.85
Dinoseb acetate									
Diphenamid	4.0×10^{-6}*	4.95×10^{-5}	260	1.087	13.46	1.92	3.68×10^{-6}		2.31
Diquat	1.30×10^{-5}	0.0170	700000	3800	4.96×10^6	−3.05*	3.42×10^{-9}		
Diquat dibromide	0		718000	2087					
Diuron	9.2×10^{-5}*	1.9×10^{-3}	40	0.172	3.630	2.78	6.83×10^{-4}		2.6
EPTC	2*	2.0	370	1.954	1.954	3.2	1.023		2.3
Fenopro (H., G.R.)	1.33×10^{-5}*	4.54×10^{-4}	140	0.519	17.73		2.56×10^{-5}		2.48
Fenuron	0.0267	0.305	3800	23.14	264.7	0.98	1.15×10^{-3}		1.43
Fenuron-TCA	0		4800	14.65	38.13				
Fluchloralin	0.004	6.03×10^{-3}	1	0.00281	0.0042	4.60*	1.343		3.50
Fluometuron	6.70×10^{-5}	1.61×10^{-3}	90	0.388	9.292	2.42	1.73×10^{-4}		2.24
Fluridone	1.3×10^{-5}	2.5×10^{-4}	12	0.036	0.7039	2.98	0.357		2.544-3.04
Fluorodifen	9.5×10^{-6}	4.47×10^{-6}	2	0.0061	0.0293	3.65			3.13

Name	Selected properties						H, Pa·m³/mol	
	P^s, Pa	P_L, Pa	S, g/m³	C^s, mol/m³	C_L, mol/m³	log K_{ow}	calcd P/C	log K_{oc}
Glyphosate	4.0×10^{-5}*	2.15×10^{-3}	12000	70.96	3818.4	−1.6	5.64×10^{-7}	3.43-3.69
Glyphosate-mono(iso-propylammonium)								
Ioxynil	0.001	0.066	50	0.135	8.904		7.42×10^{-3}	
Ioxynil-octanoate	0.0037	8.21×10^{-3}						
Ioxynil-sodium								
Isopropalin	0.0019	0.0019	0.11	3.56×10^{-4}	3.56×10^{-4}	4.71	5.34	4.0
Isoproturon	3.30×10^{-6}	6.52×10^{-5}	55	0.267	5.266	2.25	1.24×10^{-5}	1.86
Linuron	0.023*	6.74×10^{-2}	75	0.301	1.449	3.0	7.54×10^{-2}	2.91
MCPA (H., G.R.)	0.0002	1.70×10^{-3}	1605	8.001	68.05	2.69*	2.50×10^{-5}	2.03-2.07
MCPA dimethylamine salt								1.30
MCPA ester	0.0002		5.0					3.00
MCPA-thioethyl	0.021	0.0309	2.3	0.0094	0.0138		2.234	
MCPB	5.77×10^{-5}*	3.18×10^{-4}	41	0.179	0.989	3.43	3.22×10^{-4}	1.30
MCPB sodium salt	0		200000	798				
Mecoprop	3.1×10^{-4}	1.53×10^{-3}	620	2.89	14.23	3.94	7.43×10^{-5}	
Mecoprop-P	4.0×10^{-4}	1.97×10^{-3}	860	4.007	19.73		9.98×10^{-5}	
Metobromuron	4.0×10^{-4}	2.02×10^{-3}	330	1.274	6.416	2.41	3.14×10^{-4}	2.26
Metolachlor	0.0042*	4.20×10^{-3}	430	1.80	1.80	3.13	2.33×10^{-3}	
Metoxuron	0.0043	0.0439	678	2.965	30.26	1.6	1.45×10^{-3}	1.92
Molinate	0.75	0.750	970	5.179	5.179	3.21	0.145	2.3
Monolinuron	0.02	0.0732	735	3.425	12.54	2.30	5.84×10^{-3}	2.00
Monuron	6.66×10^{-5}	2.12×10^{-3}	230	1.007	32.08	1.94	6.62×10^{-5}	3.36
Neburon			4.8	0.017	0.1031	3.8		
Nitralin	0.2	3.526	0.5	0.0014	0.0255		138.2	
Oryzalin	1.30×10^{-6}	1.87×10^{-5}	2.4	0.0069	0.0995	3.73	1.88×10^{-4}	2.78
Paraquat	<0.0001		~700000					
Paraquat dichloride salt	0		620000					6.0
Pebulate	1.2*	1.20	92*	0.452	0.452	3.84	2.653	
Pentanochlor			8	0.033	0.134			2.63

Name	Selected properties						H, Pa·m³/mol	log K_{oc}
	P^s, Pa	P_L, Pa	S, g/m³	C^s, mol/m³	C_L, mol/m³	log K_{ow}	calcd P/C	
Picloram	6.0×10^{-5}*	4.98×10^{-3}	430	1.781	147.7	0.3	3.37×10^{-5}	1.23
Picloram-potassium salt			400000	1430.6				1.20
Profluralin	0.009	0.0116	0.10	2.88×10^{-4}	3.7×10^{-4}	6.34	31.35	4.0
Prometon	0.0003	1.38×10^{-3}	750	3.329	15.31	2.99	9.01×10^{-5}	2.54
Prometryn	0.0001	8.70×10^{-4}	48	0.199	1.730	3.51	5.03×10^{-4}	2.60
Pronamide	0.011	0.208	15	0.059	1.105	3.26	0.188	2.90
Propachlor	0.03	0.0958	600	2.834	9.055	2.18	0.011	1.90
Propanil	0.005*	0.0230	200	0.917	4.218	3.07	5.45×10^{-3}	2.17
Propazine	3.90×10^{-6}	2.89×10^{-4}	8.6	0.037	2.766	2.90	1.04×10^{-3}	2.19
Propham	sublime		250	1.395	5.804	2.60		1.71
Propyzamide	5.80×10^{-5}	1.15×10^{-3}	15	0.059	1.157	3.28	9.90×10^{-4}	2.90
Pyrazon (chloridazon)	7*	441.8	360	1.625	102.5	1.14	4.309	2.08
Secbumeton	0.00097	3.98×10^{-3}	620	2.756	11.31	2.18	3.52×10^{-4}	2.30
Simazine	8.50×10^{-6}	8.27×10^{-4}	5	0.025	2.412	2.18	3.43×10^{-4}	2.11
Simetryne	9.47×10^{-5}	3.55×10^{-4}	450	2.110	7.904	3.13	4.49×10^{-5}	2.30
2,4,5-T	0.005*	0.0922	220	0.861	15.89	4.34	5.81×10^{-3}	1.72
2,3,6-TBA			7700	34.15	340.6			
Terbacil	5.0×10^{-5}	1.56×10^{-3}	710	3.277	102.1	1.89	1.53×10^{-5}	1.74
Terbumeton	2.70×10^{-4}	2.57×10^{-3}	130	0.577	5.500	3.04	4.68×10^{-4}	
Terbuthylazine	1.50×10^{-4}	4.89×10^{-3}	8.5	0.037	1.206	3.04	4.05×10^{-3}	2.21-2.44
Terbutryne	0.00013*	8.04×10^{-4}	22	0.091	0.564	3.74	1.43×10^{-3}	2.85
Thiobencarb	2.2	2.20	19.1	0.074	0.0741	3.42	29.69	
Triallate	0.015	0.0164	4	0.013	0.0144	4.29	1.14	3.38
Trifluralin	0.026*	0.0259	0.5*	1.49×10^{-3}	2.57×10^{-3}	5.34	10.08	4.14
Vernolate	0.90	0.90	90	0.443	0.443	3.84	2.034	2.414

Note: F.A.B. - fungicide algicide bactericide; G.R. - growth regulator, H - herbicide
2,4-D(a) physical-chemical properties modified from values used in Vol. IV.
* The reported values for this quantity vary considerably; whereas this selected value represents the best judgement of the authors, the reader is cautioned that it may be subject to large error.

Table 2.2.3 Suggested half-life classes of herbicides in various environmental compartments

Compounds	Air class	Water class	Soil class	Sediment class
Atrazine	1	8	6	6
2,4-D	2	3	5	6
2,4-DB	3	4	4	5
Dalapon	5	6	6	6
Diallate	2	6	6	7
Dicamba	3	5	5	6
Diuron	2	5	6	7
EPTC	2	4	4	6
Glyphosate	4	6	6	7
Isopropalin	2	5	6	7
Linuron	2	5	6	7
Mecoprop	2	4	4	6
Methomyl	5	7	5	7
Metolachlor	4	6	6	7
Simazine	3	5	6	7
2,4,5-T	3	5	5	6
Triallate	4	6	4	5
Trifluralin	4	6	6	7

where,

Class	Mean half-life (hours)	Range (hours)
1	5	< 10
2	17 (~ 1 day)	10-30
3	55 (~ 2 days)	30-100
4	170 (~ 1 week)	100-300
5	550 (~ 3 weeks)	300-1,000
6	1700 (~ 2 months)	1,000-3,000
7	5500 (~ 8 months)	3,000-10,000
8	17000 (~ 2 years)	10,000-30,000
9	55000 (~ 6 years)	> 30,000

2.3 Illustrative Fugacity Calculations: Levels I, II, III

Chemical name: Atrazine

Level I calculation: (six-compartment model)

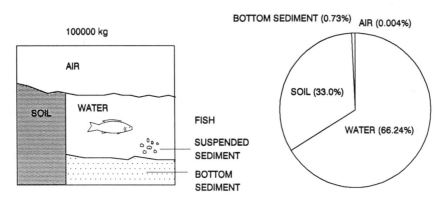

Distribution of mass

Physical-chemical properties:
molecular wt., g/mol	215.68
melting point, °C	174
solubility, g/m³	30.0
vapor pressure, Pa	4.0E−05
log K_{OW}	2.75
fugacity ratio, F	0.0336
dissoc. const. pK_b	12.32

Partition coefficients:
H, Pa·m³/mol	2.88E−04
K_{AW}	1.16E−07
K_{OC}	230.56
BCF	28.117
K_{SW}	11.067
$K_{SD/W}$	22.134
$K_{SSD/W}$	69.168
$K_{AR/W}$	5.04E+09

COMPARTMENT	Z	CONCENTRATION			AMOUNT	AMOUNT
	mol/m³·Pa	mol/m³	mg/L, or g/m³	µg/g	kg	%
AIR	4.03E−04	1.78E−13	3.84E−11	3.24E−08	3.843	3.84E−03
WATER	3.58E+03	1.54E−06	3.31E−04	3.31E−04	66247	66.247
SOIL	3.85E+04	1.70E−05	3.67E−03	1.53E−03	32992	32.992
BIOTA (FISH)	9.78E+04	4.32E−05	9.31E−03	9.31E−03	1.863	1.86E−03
SUSPENDED SEDIMENT	2.41E+05	1.06E−04	2.29E−02	1.53E−02	22.91	2.29E−02
BOTTOM SEDIMENT	7.70E+04	3.40E−05	7.33E−03	3.05E−03	733.14	0.733
Total					100000	100

Fugacity, f = 4.416E−10 Pa

Chemical name: Atrazine
Level II calculation: (six-compartment model)

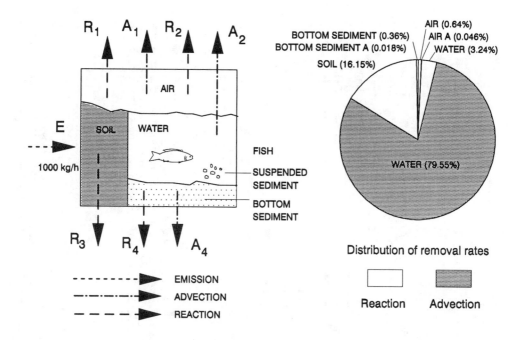

COMPARTMENT	Half-Life	D VALUES		CONC'N	LOSS	LOSS	REMOVAL
		Reaction	Advection		Reaction	Advection	%
	h	mol/Pa·h	mol/Pa·h	mol/m^3	kg/h	kg/h	
AIR	5	5.59E+09	4.03E+08	2.14E−12	6.395	0.462	0.686
WATER	17000	2.84E+10	6.95E+11	1.84E−05	32.427	795.46	82.789
SOIL	1700	1.41E+11		2.04E−04	161.49		16.149
BIOTA (FISH)				5.19E−04			
SUSPENDED SEDIMENT				1.28E−03			
BOTTOM SEDIMENT	1700	3.14E+09	1.54E+08	4.08E−04	3.589	0.1761	0.3765
Total		1.75E+11	6.96E+11		203.90	796.10	100
Reaction + Advection			8.71E+11			1000	

Fugacity, f = 5.303E−09 Pa

Total amount = 1200758 kg

Overall residence time = 1200.76 h

Reaction time = 5888.98 h

Advection time = 1508.30 h

Fugacity Level III calculations: (four-compartment model)
Chemical name: Atrazine

Phase Properties and Rates:

Compartment	Bulk Z mol/m3 Pa	Half-life h	D Values Reaction mol/Pa h	Advection mol/Pa h
Air (1)	4.441E-04	5	6.15E+09	4.44E+08
Water (2)	3.479E+03	17000	2.84E+10	6.96E+11
Soil (3)	2.029E+04	1700	1.49E+11	
Sediment (4)	1.818E+04	1700	3.70E+09	1.82E+08

	E(1)=1000	E(2)=1000	E(3)=1000	E(1,2,3)
Overall residence time =	1832.79	965.44	2310.80	1950.56 h
Reaction time =	2183.20	22859.34	2543.16	2832.74 h
Advection time =	11419.16	1008.02	25291	6263.37 h

↑ EMISSION (E)
↑ REACTION (R)
↑ ADVECTION (A)
↑ TRANSFER D VALUE mol/Pa h

Diagram shows four-compartment model:
- AIR (1): E_1, R_1, A_1 = 1.716E08, D_{13} = 2.017E07
- WATER (2): E_2, R_2, A_2, D_{12}, D_{21} = 3.510E09, D_{24} = 3.631E09
- SOIL (3): E_3, R_3, D_{31} = 3.158E10, D_{32} = 1.568E10
- SEDIMENT (4): R_4, A_4, D_{42} = 4.680E09

Phase Properties, Compositions, Transport and Transformation Rates:

Emission, kg/h

E(1)	E(2)	E(3)	f(1)	f(2)	f(3)	f(4)
1000	0	0	1.113E-07	9.984E-10	2.134E-08	6.215E-10
0	1000	0	3.090E-12	6.382E-09	5.925E-13	3.973E-09
0	0	1000	1.163E-10	6.087E-10	2.817E-10	3.789E-10
50	250	700	5.648E-09	2.071E-09	2.079E-08	1.290E-09

Concentration, g/m3

C(1)	C(2)	C(3)	C(4)
1.066E-08	7.491E-04	9.337E-02	2.436E-03
2.960E-13	4.788E-03	2.592E-06	1.557E-01
1.114E-11	4.567E-04	1.233E-01	1.485E-03
5.409E-10	1.554E-03	9.095E-02	5.055E-03

Emission, kg/h

E(1)	E(2)	E(3)	R(1)	R(2)	R(3)	R(4)
1000	0	0	1.478E+02	6.107E+00	6.85E+02	4.966E-01
0	1000	0	4.102E-03	3.904E+01	1.90E-02	3.174E+00
0	0	1000	1.543E-01	3.723E+00	9.04E+02	3.028E-01
50	250	700	7.497E+00	1.267E+01	6.67E+02	1.030E+00

Loss, Advection, kg/h

A(1)	A(2)	A(4)
1.066E+01	1.498E+02	2.436E-02
2.960E-04	9.576E+02	1.557E-01
1.114E-02	9.134E+01	1.485E-02
5.409E-01	3.108E+02	5.055E-02

Amounts, kg

m(1)	m(2)	m(3)	m(4)
1.066E+03	1.498E+05	1.681E+06	1.218E+03
2.960E-02	9.576E+05	4.666E+01	7.787E+03
1.114E+00	9.134E+04	2.219E+06	7.427E+02
5.409E+01	3.108E+05	1.637E+06	2.528E+03

Intermedia Rate of Transport, kg/h

T12	T21	T13	T31	T32	T24	T42
air-water	water-air	air-soil	soil-air	soil-water	water-sed	sed-water
8.426E+01	4.343E-03	7.581E+02	7.898E-01	7.219E+01	1.008E+00	4.868E-01
2.339E-03	2.776E-02	2.105E-02	2.193E-05	2.004E-03	6.442E+00	3.112E+00
8.801E-02	2.648E-03	7.918E-01	1.043E+00	9.530E+01	6.144E-01	2.968E-01
4.275E+00	9.012E-03	3.846E+01	7.694E-01	7.032E+01	2.091E+00	1.010E+00

Total Amount, kg

1.833E+06
9.655E+05
2.311E+06
1.951E+06

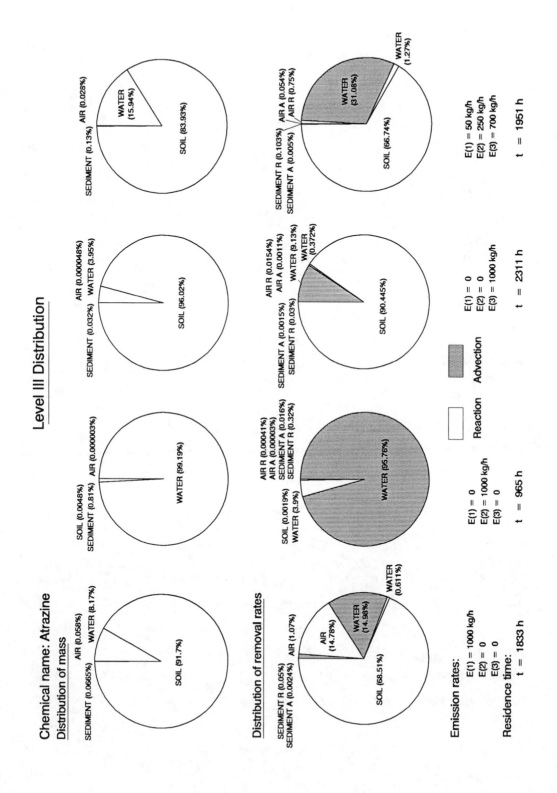

Chemical name: 2,4-D

Level I calculation: (six-compartment model)

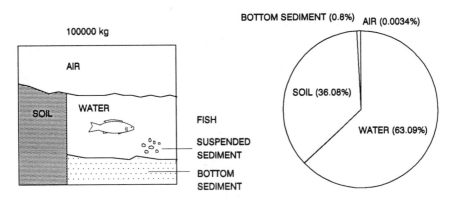

Distribution of mass

Physical-chemical properties:
molecular wt., g/mol	221.04
melting point, °C	140.5
solubility, g/m^3	890
vapor pressure, Pa	1.07E−03
log K_{OW}	2.81
fugacity ratio, F	0.0721
dissoc. const. pK_a	2.80

Partition coefficients:
H, Pa·m^3/mol	2.66E−04
K_{AW}	1.07E−07
K_{OC}	264.718
BCF	32.283
K_{SW}	12.706
$K_{SD/W}$	25.413
$K_{SSD/W}$	79.415
$K_{AR/W}$	4.04E+08

COMPARTMENT	Z	CONCENTRATION			AMOUNT	AMOUNT
	mol/m^3·Pa	mol/m^3	mg/L, or g/m^3	µg/g	kg	%
AIR	4.03E−04	1.53E−13	3.38E−11	2.85E−08	3.382	3.38E−03
WATER	3.76E+03	1.43E−06	3.15E−04	3.15E−04	63092	63.092
SOIL	4.78E+04	1.81E−05	4.01E−03	1.67E−03	36076	36.076
BIOTA (FISH)	1.21E+05	4.61E−05	1.02E−02	1.02E−02	2.037	2.04E−03
SUSPENDED SEDIMENT	2.99E+05	1.13E−04	2.51E−02	1.67E−02	25.053	2.51E−02
BOTTOM SEDIMENT	9.56E+04	3.63E−05	8.02E−03	3.34E−03	801.68	0.802
Total					100000	100

Fugacity, f = 3.793E−10 Pa

Chemical name: 2,4-D

Level II calculation: (six-compartment model)

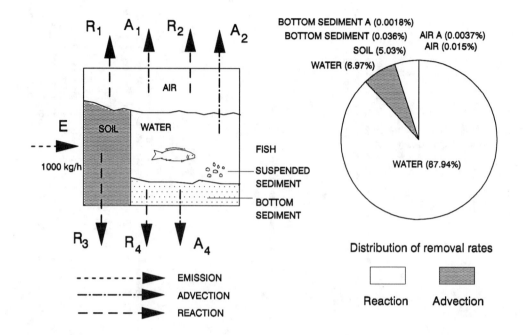

COMPARTMENT	Half-Life h	D VALUES Reaction mol/Pa·h	D VALUES Advection mol/Pa·h	CONC'N mol/m³	LOSS Reaction kg/h	LOSS Advection kg/h	REMOVAL %
AIR	17	1.64E+09	4.03E+08	1.69E−13	0.153	0.0374	0.0190
WATER	55	9.48E+12	7.53E+11	1.58E−06	879.36	69.790	94.915
SOIL	550	5.41E+11		2.01E−05	50.281		5.028
BIOTA (FISH)				5.10E−05			
SUSPENDED SEDIMENT				1.25E−04			
BOTTOM SEDIMENT	1700	3.90E+09	1.91E+08	4.01E−05	0.361	0.0177	0.0379
Total		1.03E+13	7.53E+11		930.15	69.846	100
Reaction + Advection			1.08E+13			1000	

Fugacity, f = 4.195E−10 Pa

Total amount = 110616 kg

Overall residence time = 110.62 h

Reaction time = 118.92 h

Advection time = 1583.73 h

Fugacity Level III calculations: (four-compartment model)
Chemical name: 2,4-D

Phase Properties and Rates:

Compartment	Bulk Z mol/m³ Pa	Half-life h	D Values Reaction mol/Pa h	Advection mol/Pa h
Air (1)	4.067E-04	17	1.66E+09	4.07E+08
Water (2)	3.765E+03	55	9.49E+12	7.53E+11
Soil (3)	2.504E+04	550	5.68E+11	
Sediment (4)	2.214E+04	1700	4.51E+09	2.21E+08

	E(1)=1000	E(2)=1000	E(3)=1000	E(1,2,3)
Overall residence time =	667.51	74.16	772.73	592.83 h
Reaction time =	680.43	80.05	774.39	605.44 h
Advection time =	35146.17	1008.73	361210	28463.69 h

- - - - - EMISSION (E)
– – – REACTION (R)
-·-·- ADVECTION (A)
——▶ TRANSFER D VALUE mol/Pa h

Phase Properties, Compositions, Transport and Transformation Rates:

Emission, kg/h			Fugacity, Pa				Concentration, g/m³			
E(1)	E(2)	E(3)	f(1)	f(2)	f(3)	f(4)	C(1)	C(2)	C(3)	C(4)
1000	0	0	1.134E-07	5.285E-11	6.602E-09	3.198E-11	1.020E-08	4.397E-05	3.653E-02	1.565E-04
0	1000	0	2.234E-13	4.417E-10	1.300E-14	2.673E-10	2.008E-14	3.675E-04	7.195E-08	1.308E-03
0	0	1000	3.342E-11	1.283E-11	7.736E-09	7.766E-12	3.004E-12	1.068E-05	4.281E-02	3.800E-05
50	250	700	5.695E-09	1.220E-10	5.745E-09	7.386E-11	5.119E-10	1.016E-04	3.179E-02	3.614E-04

Emission, kg/h			Loss, Reaction, kg/h				Loss, Advection, kg/h			
E(1)	E(2)	E(3)	R(1)	R(2)	R(3)	R(4)	A(1)	A(2)	A(3)	A(4)
1000	0	0	4.156E+01	1.108E+02	8.29E+02	3.189E-02	1.020E+01	8.795E+00		1.565E-03
0	1000	0	8.185E-05	9.262E+02	1.63E-03	2.666E-01	2.008E-05	7.351E+01		1.308E-02
0	0	1000	1.224E-02	2.691E+01	9.71E+02	7.745E-03	3.004E-03	2.136E+00		3.800E-04
50	250	700	2.087E+00	2.559E+02	7.21E+02	7.366E-02	5.119E-01	2.031E+01		3.614E-03

Emission, kg/h			Amounts, kg				Total Amount, kg
E(1)	E(2)	E(3)	m(1)	m(2)	m(3)	m(4)	
1000	0	0	1.020E+03	8.795E+03	6.576E+05	7.824E+01	6.675E+05
0	1000	0	2.008E-03	7.351E+04	1.295E+00	6.539E+02	7.416E+04
0	0	1000	3.004E-01	2.136E+03	7.706E+05	1.900E+01	7.727E+05
50	250	700	5.119E+01	2.031E+04	5.723E+05	1.807E+02	5.928E+05

Emission, kg/h			Intermedia Rate of Transport, kg/h						
E(1)	E(2)	E(3)	T12	T13	T21	T31	T32	T24	T42
			air-water	water-air	air-soil	soil-air	soil-water	water-sed	sed-water
1000	0	0	9.487E+01	2.356E-04	8.536E+02	2.514E-01	2.477E+01	6.141E-02	2.795E-02
0	1000	0	1.868E-04	1.969E-03	1.681E-03	4.952E-07	4.879E-05	5.133E-01	2.336E-01
0	0	1000	2.795E-02	5.722E-05	2.515E-01	7.706E+05	2.903E+01	1.491E-02	6.788E-03
50	250	700	4.763E+00	5.442E-04	4.286E+01	2.188E-01	2.156E+01	1.418E-01	6.455E-02

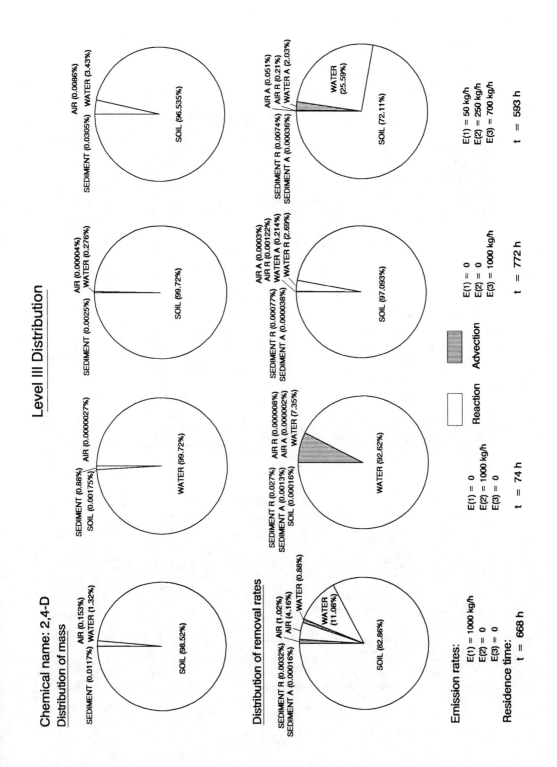

Chemical name: Diuron

Level I calculation: (six-compartment model)

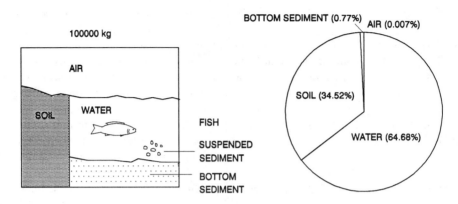

Distribution of mass

Physical-chemical properties:
molecular wt., g/mol	233.10
melting point, °C	158
solubility, g/m^3	40
vapor pressure, Pa	9.20E−05
log K_{OW}	2.78
fugacity ratio, F	0.0484
dissoc. const. pK_a	

Partition coefficients:
H, Pa·m^3/mol	5.36E−04
K_{AW}	2.16E−07
K_{OC}	247.05
BCF	30.13
K_{SW}	11.86
$K_{SD/W}$	23.72
$K_{SSD/W}$	74.11
$K_{AR/W}$	3.15E+09

COMPARTMENT	Z	CONCENTRATION			AMOUNT	AMOUNT
	mol/m^3·Pa	mol/m^3	mg/L, or g/m^3	µg/g	kg	%
AIR	4.03E−04	3.00E−13	6.99E−11	5.90E−08	6.995	0.0070
WATER	1.87E+03	1.39E−06	3.23E−04	3.23E−04	64683	64.68
SOIL	2.21E+04	1.65E−05	3.84E−03	1.60E−03	34517	34.517
BIOTA (FISH)	5.62E+04	4.18E−05	9.74E−03	9.74E−03	1.949	1.95E−03
SUSPENDED SEDIMENT	1.38E+05	1.03E−04	2.40E−02	1.60E−02	23.97	2.40E−02
BOTTOM SEDIMENT	4.42E+04	3.29E−05	7.67E−03	3.20E−03	767.04	0.767
Total					100000	100

Fugacity, f = 7.439E−10 Pa

Chemical name: Diuron

Level II calculation: (six-compartment model)

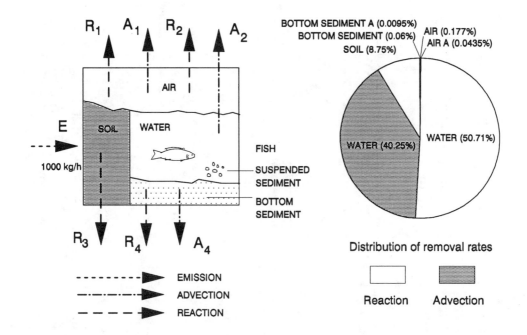

Distribution of removal rates

Reaction Advection

COMPARTMENT	Half-Life h	D VALUES Reaction mol/Pa·h	D VALUES Advection mol/Pa·h	CONC'N mol/m³	LOSS Reaction kg/h	LOSS Advection kg/h	REMOVAL %
AIR	17	1.64E+09	4.03E+08	1.87E−12	1.774	0.435	0.221
WATER	550	4.70E+11	3.73E+11	8.63E−06	507.09	402.46	90.955
SOIL	1700	8.11E+10		1.02E−04	87.546		8.755
BIOTA (FISH)				2.60E−04			
SUSPENDED SEDIMENT				6.40E−04			
BOTTOM SEDIMENT	5500	5.57E+08	8.85E+07	2.05E−04	0.6013	0.0954	0.0697
Total		5.53E+11	3.73E+11		697.01	402.99	100
Reaction + Advection			9.26E+11			1000	

Fugacity, f = 4.628E−09 Pa

Total amount = 622192 kg

Overall residence time = 622.19 h

Reaction time = 1042.17 h

Advection time = 1543.96 h

Fugacity Level III calculations: (four-compartment model)
Chemical name: Diuron

Phase Properties and Rates:

Compartment	Bulk Z mol/m3 Pa	Half-life h	D Values Reaction mol/Pa h	Advection mol/Pa h
Air (1)	4.289E-04	17	1.75E+09	4.29E+08
Water (2)	1.866E+03	550	4.70E+11	3.73E+11
Soil (3)	1.162E+04	1700	8.53E+10	
Sediment (4)	1.034E+04	5500	6.51E+08	1.03E+08

	E(1)=1000	E(2)=1000	E(3)=1000	E(1,2,3)
Overall residence time =	1876.00	447.92	2272.30	1796.39 h
Reaction time =	2066.44	803.03	2366.52	2096.21 h
Advection time =	20356.59	1012.91	57076	12559.63 h

Arrows legend:
- EMISSION (E)
- REACTION (R)
- ADVECTION (A)
- TRANSFER D VALUE mol/Pa h

Diagram values:
- E_1, R_1, A_1 (Air)
- AIR (1): 1.638E08, 2.017E07
- D_{31}, D_{13}, D_{12}, D_{21}
- WATER (2): 1.893E09
- R_2, A_2
- E_3, E_2
- 1.702E10, D_{32} 8.413E09
- SOIL (3)
- R_3, R_4, A_4
- D_{24} 1.954E09, D_{42} 2.556E09
- SEDIMENT (4)

Phase Properties, Compositions, Transport and Transformation Rates:

Emission, kg/h — Fugacity, Pa — Concentration, g/m3

E(1)	E(2)	E(3)	f(1)	f(2)	f(3)	f(4)	C(1)	C(2)	C(3)	C(4)
1000	0	0	2.037E-07	8.251E-10	3.695E-08	7.787E-10	2.036E-08	3.589E-04	1.001E-01	1.877E-03
0	1000	0	4.868E-12	5.082E-09	8.829E-13	4.797E-09	4.866E-13	2.211E-03	2.391E-06	1.156E-02
0	0	1000	3.561E-10	4.571E-10	4.578E-08	4.315E-10	3.560E-11	1.988E-04	1.240E-01	1.040E-02
50	250	700	1.044E-08	1.632E-09	3.390E-08	1.540E-09	1.043E-09	7.097E-04	9.180E-02	3.712E-03

Loss, Reaction, kg/h — Loss, Advection, kg/h

E(1)	E(2)	E(3)	R(1)	R(2)	R(3)	R(4)	A(1)	A(2)	A(3)	A(4)
1000	0	0	8.301E+01	9.044E+01	7.34E+02	1.182E-01	2.036E+01	7.177E+01		1.877E-03
0	1000	0	1.984E-03	5.571E+02	1.75E+02	7.284E-01	4.866E-04	4.421E+02		1.156E-02
0	0	1000	1.451E-01	5.010E+01	9.10E+02	6.551E-02	3.560E-02	3.977E+01		1.040E-02
50	250	700	4.253E+00	1.789E+02	6.74E+02	2.339E-01	1.043E+00	1.419E+02		3.712E-03

Amounts, kg — Total Amount, kg

m(1)	m(2)	m(3)	m(4)	Total
2.036E+03	7.177E+04	1.801E+06	9.385E+02	1.876E+06
4.866E-02	4.421E+05	4.304E+01	5.781E+03	4.479E+05
3.560E+00	3.977E+04	2.232E+06	5.199E+02	2.272E+06
1.043E+02	1.419E+05	1.652E+06	1.856E+03	1.796E+06

Intermedia Rate of Transport, kg/h

T12	T21	T13	T31	T32	T24	T42
air-water	water-air	air-soil	soil-air	soil-water	water-sed	sed-water
8.989E+01	3.879E-03	8.081E+02	1.411E+00	7.246E+01	4.917E-01	3.546E-01
2.148E-03	2.390E-02	1.931E-02	3.372E-05	1.731E-03	3.028E+00	2.185E+00
1.571E-01	2.149E-03	1.413E+00	1.748E+00	8.979E+02	2.724E+00	1.965E+00
4.605E+00	7.672E-03	4.140E+01	1.294E+00	6.648E+01	9.724E-01	7.014E-01

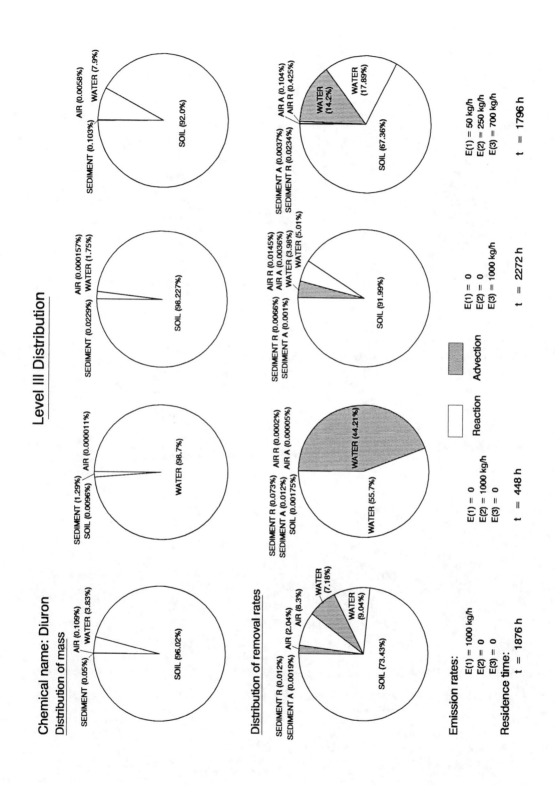

Chemical name: EPTC

Level I calculation: (six-compartment model)

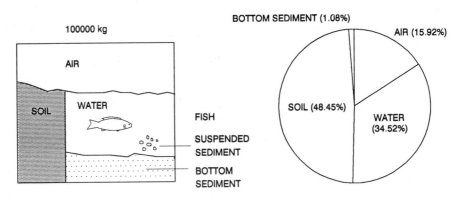

Distribution of mass

Physical-chemical properties:
molecular wt., g/mol	189.31
melting point, °C	< −30
solubility, g/m^3	375
vapor pressure, Pa	4.53
log K_{OW}	3.20
fugacity ratio, F	1.0
dissoc. const. pK_a	

Partition coefficients:
H, Pa·m^3/mol	2.2869
K_{AW}	9.23E−04
K_{OC}	649.81
BCF	79.245
K_{SW}	31.191
$K_{SD/W}$	62.381
$K_{SSD/W}$	194.942
$K_{AR/W}$	1.32E+06

COMPARTMENT	Z	CONCENTRATION			AMOUNT	AMOUNT
	mol/m^3·Pa	mol/m^3	mg/L, or g/m^3	µg/g	kg	%
AIR	4.03E−04	8.41E−10	1.59E−07	1.34E−04	15922	15.922
WATER	4.37E−01	9.12E−07	1.73E−04	1.73E−04	34517	34.517
SOIL	1.36E+01	2.84E−05	5.38E−03	2.24E−03	48448	48.448
BIOTA (FISH)	3.47E+01	7.22E−05	1.37E−02	1.37E−02	2.735	2.74E−03
SUSPENDED SEDIMENT	8.52E+01	1.78E−04	3.36E−02	2.24E−02	33.644	3.36E−02
BOTTOM SEDIMENT	2.73E+01	5.69E−05	1.08E−03	4.49E−03	1076.62	1.0766
Total					100000	100

Fugacity, f = 2.085E−06 Pa

Chemical name: EPTC

Level II calculation: (six-compartment model)

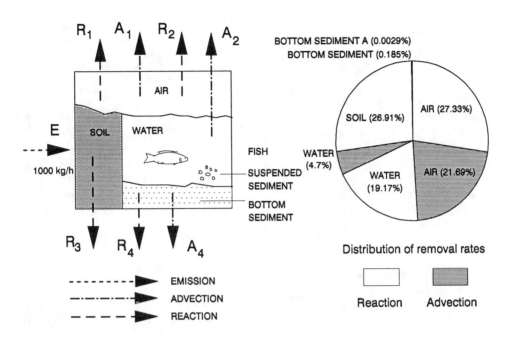

COMPARTMENT	Half-Life	D VALUES		CONC'N	LOSS	LOSS	REMOVAL
		Reaction	Advection		Reaction	Advection	%
	h	mol/Pa·h	mol/Pa·h	mol/m³	kg/h	kg/h	
AIR	55	5.08E+08	4.03E+08	1.15E−09	273.35	216.94	49.029
WATER	170	3.57E+08	8.75E+07	1.24E−06	191.72	47.030	23.875
SOIL	170	5.00E+08		3.87E−05	269.09		26.909
BIOTA (FISH)				9.84E−05			
SUSPENDED SEDIMENT				2.42E−04			
BOTTOM SEDIMENT	550	3.44E+06	5.46E+04	7.75E−05	1.8483	0.029	0.1878
Total		1.37E+09	4.91E+08		736.00	293.97	100
Reaction + Advection			1.86E+09			1000	

Fugacity, $f = 2.841E-06$ Pa

Total amount = 136251 kg

Overall residence time = 136.25 h

Reaction time = 185.12 h

Advection time = 516.10 h

Fugacity Level III calculations: (four-compartment model)
Chemical name: EPTC

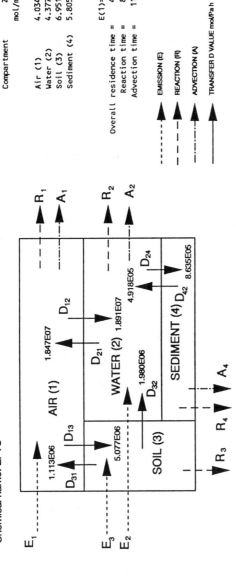

Phase Properties and Rates:

Compartment	Bulk Z mol/m3 Pa	Half-life h	D Values Reaction mol/Pa h	Advection mol/Pa h
Air (1)	4.034E-04	55	5.08E+08	4.03E+08
Water (2)	4.377E-01	170	3.57E+08	8.75E+07
Soil (3)	6.951E+00	170	5.10E+08	
Sediment (4)	5.805E+00	550	3.66E+06	5.81E+04

	E(1)=1000	E(2)=1000	E(3)=1000	E(1,2,3)
Overall residence time =	48.32	192.03	244.68	221.70
Reaction time =	85.57	241.90	245.10	239.55
Advection time =	111.00	931.35	140642	2974.65

- - - - EMISSION (E)
- - - - REACTION (R)
- - - - ADVECTION (A)
──── TRANSFER D VALUE mol/Pa h

Phase Properties, Compositions, Transport and Transformation Rates:

Emission, kg/h

E(1)	E(2)	E(3)	f(1)	f(2)	f(3)	f(4)	C(1)	C(2)	C(3)	C(4)
1000	0	0	5.650E-06	2.306E-07	5.590E-08	4.733E-08	4.315E-07	1.911E-05	7.356E-05	5.202E-05
0	1000	0	2.250E-07	1.140E-05	2.227E-09	2.340E-06	1.719E-08	9.448E-04	2.930E-06	2.572E-03
0	0	1000	1.312E-08	4.450E-08	1.029E-05	9.133E-09	1.002E-09	3.687E-06	1.355E-02	1.004E-05
50	250	700	3.479E-07	2.893E-06	7.210E-06	5.938E-07	2.657E-08	2.397E-04	9.487E-03	6.526E-04

Fugacity, Pa | Concentration, g/m3

Emission, kg/h

E(1)	E(2)	E(3)	R(1)	R(2)	R(3)	R(4)	A(1)	A(2)	A(3)	A(4)
1000	0	0	5.437E+02	1.558E+01	5.40E+00	3.277E-02	4.315E+02	3.822E+00		5.202E-04
0	1000	0	2.165E+01	7.703E+02	2.15E-01	1.620E+00	1.719E+01	1.890E+02		2.572E-02
0	0	1000	1.263E+00	3.006E+00	9.94E+02	6.323E-03	1.002E+00	7.375E-01		1.004E-04
50	250	700	3.348E+01	1.955E+02	6.96E+02	4.111E-01	2.657E+01	4.795E+01		6.526E-03

Loss, Reaction, kg/h | Loss, Advection, kg/h

Amounts, kg

m(1)	m(2)	m(3)	m(4)	Total Amount, kg
4.315E+04	3.822E+03	1.324E+03	2.601E+01	4.832E+04
1.719E+03	1.890E+05	5.274E+01	1.286E+03	1.920E+05
1.002E+02	7.375E+02	2.438E+05	5.018E+00	2.447E+05
2.657E+03	4.795E+04	1.708E+05	3.263E+02	2.217E+05

Intermedia Rate of Transport, kg/h

T12	T21	T13	T31	T32	T24	T42
air-water	water-air	air-soil	soil-air	soil-water	water-sed	sed-water
2.022E+01	8.063E-01	5.430E+00	1.178E-02	2.095E-02	3.770E-02	4.407E-03
8.054E-01	3.986E+01	2.163E-01	4.691E-04	8.346E-04	1.864E+00	2.179E-01
4.697E-02	1.556E+00	1.261E-02	2.169E+00	3.859E+00	7.274E-03	8.503E-04
1.245E+00	1.011E+01	3.344E-01	1.519E+00	2.702E+00	4.729E-01	5.529E-02

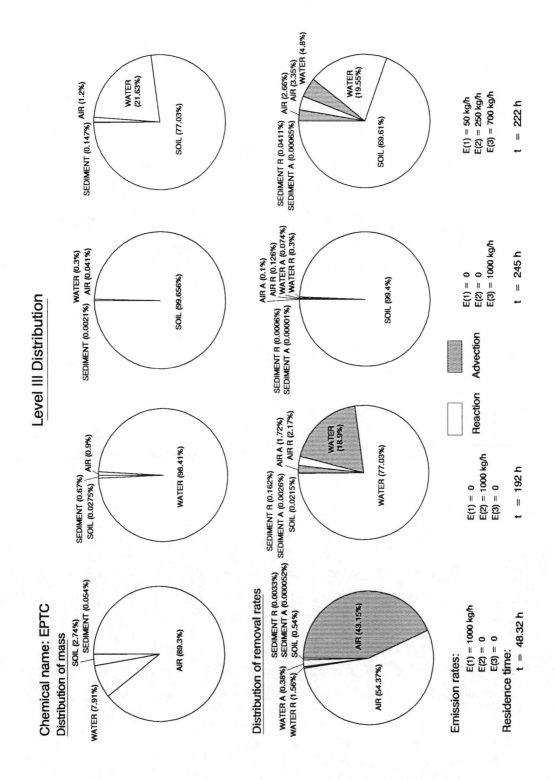

Chemical name: Glyphosate

Level I calculation: (six-compartment model)

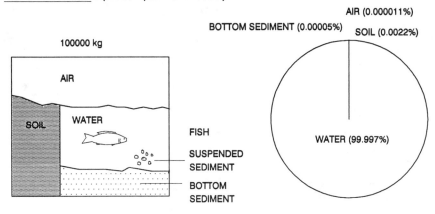

Distribution of mass

Physical-chemical properties:
molecular wt., g/mol	169.40
melting point, °C	200
solubility, g/m³	12000
vapor pressure, Pa	4.0E−05
log K_{OW}	−1.60
fugacity ratio, F	0.0186
dissoc. const. pK_a	2.32

Partition coefficients:
H, Pa·m³/mol	5.65E−07
K_{AW}	2.27E−10
K_{OC}	1.03E−02
BCF	1.26E−03
K_{SW}	4.94E−04
$K_{SD/W}$	9.89E−04
$K_{SSD/W}$	3.09E−03
$K_{AR/W}$	2.79E+09

COMPARTMENT	Z	CONCENTRATION				AMOUNT	AMOUNT
	mol/m³·Pa	mol/m³	mg/L, or g/m³	μg/g		kg	%
AIR	4.03E−04	6.72E−16	1.13E−13	9.61E−11		0.0114	1.14E−05
WATER	1.77E+06	2.95E−06	5.00E−04	5.00E−04		99998	99.998
SOIL	8.75E+02	1.46E−09	2.47E−07	1.03E−07		2.224	2.22E−03
BIOTA (FISH)	2.22E+03	3.71E−09	6.28E−07	6.28E−07		1.26E−04	1.26E−07
SUSPENDED SEDIMENT	5.47E+03	9.12E−09	1.54E−06	1.03E−06		1.54E−03	1.54E−06
BOTTOM SEDIMENT	1.75E+03	2.92E−09	4.94E−07	2.06E−07		4.94E−02	4.94E−05
Total						100000	100

Fugacity, f = 1.667E−12 Pa

Chemical name: Glyphosate
Level II calculation: (six-compartment model)

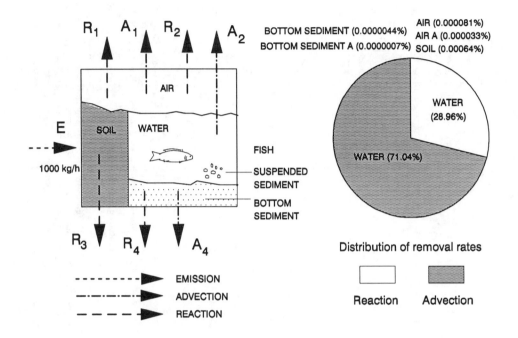

COMPARTMENT	Half-Life h	D VALUES Reaction mol/Pa·h	D VALUES Advection mol/Pa·h	CONC'N mol/m³	LOSS Reaction kg/h	LOSS Advection kg/h	REMOVAL %
AIR	170	1.64E+08	4.03E+08	4.78E−15	3.30E−04	8.09E−04	1.14E−04
WATER	1700	1.44E+14	3.54E+14	2.10E−05	289.59	710.40	99.999
SOIL	1700	3.21E+09		1.04E−08	6.44E−03		6.44E−04
BIOTA (FISH)				2.63E−08			
SUSPENDED SEDIMENT				6.48E−08			
BOTTOM SEDIMENT	5500	2.21E+07	3.50E+06	2.07E−08	4.42E−05	7.02E−06	5.13E−06
Total		1.44E+14	3.54E+14		289.60	710.40	100
Reaction + Advection			4.99E+14			1000	

Fugacity, $f = 1.184E-11$ Pa

Total amount = 710416 kg

Overall residence time = 710.42 h

Reaction time = 2453.1 h

Advection time = 1000.0 h

Fugacity Level III calculations: (four-compartment model)
Chemical name: Glyphosate

Phase Properties and Rates:

Compartment	Bulk Z mol/m3 Pa	Half-life h	D Values Reaction mol/Pa h	Advection mol/Pa h
Air (1)	4.259E-04	170	1.74E+08	4.26E+08
Water (2)	1.771E+06	1700	1.44E+14	3.54E+14
Soil (3)	5.317E+05	1700	3.90E+12	
Sediment (4)	1.417E+06	5500	8.93E+10	1.42E+10

	E(1)=1000	E(2)=1000	E(3)=1000	E(1,2,3)
Overall residence time =	1226.71	711.61	1283.98	1138.03 h
Reaction time =	2454.37	2456.30	2454.29	2454.61 h
Advection time =	2452.46	1001.85	2693	2121.71 h

▲ EMISSION (E)
▲ REACTION (R)
▲ ADVECTION (A)
▲ TRANSFER D VALUE mol/Pa h

Phase Properties, Compositions, Transport and Transformation Rates:

Emission, kg/h

E(1)	E(2)	E(3)	f(1)	f(2)	f(3)	f(4)	C(1)	C(2)	C(3)	C(4)
1000	0	0	3.333E-10	8.336E-12	4.475E-10	7.876E-12	2.405E-11	2.501E-03	4.031E-02	1.891E-03
0	1000	0	1.348E-17	1.184E-11	1.810E-17	1.119E-11	9.729E-19	3.552E-03	1.631E-09	2.685E-03
0	0	1000	5.106E-15	7.947E-12	4.973E-10	7.509E-12	3.684E-16	2.384E-03	4.479E-02	1.802E-03
50	250	700	1.667E-11	8.939E-12	3.705E-10	8.446E-12	1.203E-12	2.682E-03	3.337E-02	2.028E-03

Fugacity, Pa | Concentration, g/m3

Emission, kg/h

E(1)	E(2)	E(3)	R(1)	R(2)	R(3)	R(4)	A(1)	A(2)	A(4)
1000	0	0	9.803E-03	2.039E+02	2.96E+02	1.191E-01	2.405E-02	5.002E+02	1.891E-02
0	1000	0	3.966E-10	2.896E+02	1.20E-05	1.692E-01	9.729E-10	7.103E+02	2.685E-02
0	0	1000	1.502E-07	1.944E+02	3.29E+02	1.136E-01	3.684E-07	4.768E+02	1.802E-02
50	250	700	4.903E-04	2.186E+02	2.45E+02	1.277E-01	1.203E-03	5.364E+02	2.028E-02

Loss, Reaction, kg/h | Loss, Advection, kg/h

Amounts, kg

m(1)	m(2)	m(3)	m(4)	Total Amount, kg
2.405E+02	5.002E+05	7.256E+05	9.453E+02	1.227E+06
9.729E-08	7.103E+05	2.935E+02	1.343E+03	7.117E+05
3.684E-05	4.768E+05	8.063E+05	9.012E+02	1.284E+06
1.203E-01	5.364E+05	6.007E+05	1.014E+03	1.138E+06

Intermedia Rate of Transport, kg/h

T12	T21	T13	T31	T32	T24	T42
air-water	water-air	air-soil	soil-air	soil-water	water-sed	sed-water
1.000E+02	2.848E-05	9.000E+02	1.376E-02	6.042E+02	2.501E+02	2.363E+00
4.045E-06	4.045E-05	3.641E-05	5.567E-10	2.444E-05	3.552E+02	3.356E+00
1.532E-03	2.715E-05	1.379E-02	1.529E-02	6.713E+02	2.384E+02	2.253E+00
5.001E+00	3.054E-05	4.501E+01	1.139E-02	5.001E+02	2.682E+02	2.534E+00

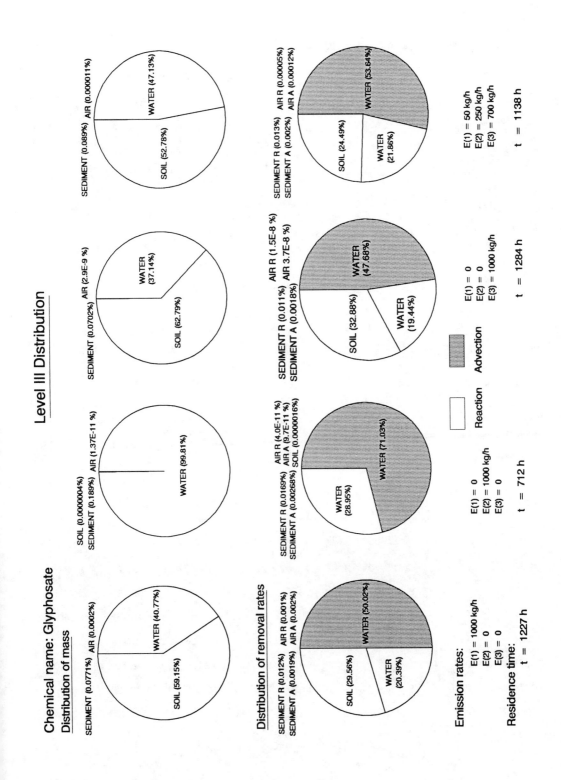

Chemical name: Linuron

Level I calculation: (six-compartment model)

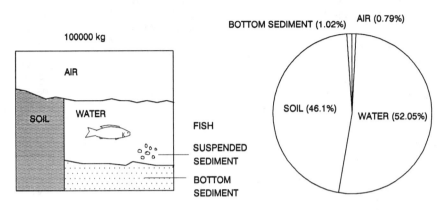

Distribution of mass

Physical-chemical properties:
molecular wt., g/mol	249.11
melting point, °C	94.0
solubility, g/m³	75
vapor pressure, Pa	2.27E−02
log K_{OW}	3.0
fugacity ratio, F	0.2078
dissoc. const. pK_a	

Partition coefficients:
H, Pa·m³/mol	0.0754
K_{AW}	3.04E−05
K_{OC}	410.0
BCF	50.00
K_{SW}	19.68
$K_{SD/W}$	39.36
$K_{SSD/W}$	123.0
$K_{AR/W}$	5.49E+07

COMPARTMENT	Z	CONCENTRATION			AMOUNT	AMOUNT
	mol/m³·Pa	mol/m³	mg/L, or g/m³	µg/g	kg	%
AIR	4.03E−04	3.18E−11	7.92E−09	6.68E−06	791.62	0.792
WATER	1.33E+01	1.04E−06	2.60E−04	2.60E−04	52052	52.052
SOIL	2.61E+02	2.06E−05	5.12E−03	2.13E−03	46097	46.097
BIOTA (FISH)	6.63E+02	5.22E−05	1.30E−02	1.30E−02	2.603	2.60E−03
SUSPENDED SEDIMENT	1.63E+03	1.29E−04	3.20E−02	2.13E−02	32.013	3.20E−02
BOTTOM SEDIMENT	9.56E+04	4.11E−05	1.02E−02	4.27E−03	1024.38	1.024
Total					100000	100

Fugacity, f = 7.877E−08 Pa

Chemical name: Linuron
Level II calculation: (six-compartment model)

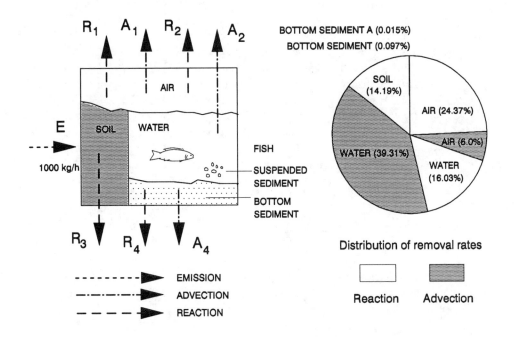

COMPARTMENT	Half-Life	D VALUES		CONC'N	LOSS	LOSS	REMOVAL
		Reaction	Advection		Reaction	Advection	%
	h	mol/Pa·h	mol/Pa·h	mol/m³	kg/h	kg/h	
AIR	17	1.64E+09	4.03E+08	2.40E−10	243.74	59.79	30.353
WATER	1700	1.08E+09	2.65E+09	7.89E−06	160.27	393.15	55.341
SOIL	1700	9.58E+08		1.55E−04	141.93		14.193
BIOTA (FISH)				3.95E−04			
SUSPENDED SEDIMENT				9.71E−04			
BOTTOM SEDIMENT	5500	6.58E+06	1.04E+06	3.11E−04	0.975	0.155	0.1130
Total		3.68E+09	3.06E+09		546.91	433.093	100
Reaction + Advection			6.74E+09			1000	

Fugacity, f = 5.950E−07 Pa

Total amount = 755296 kg

Overall residence time = 755.30 h

Reaction time = 1381.03 h

Advection time = 1666.98 h

Fugacity Level III calculations: (four-compartment model)
Chemical name: Linuron

Phase Properties and Rates:

Compartment	Bulk Z mol/m³ Pa	Half-life h	D Values Reaction mol/Pa h	D Values Advection mol/Pa h
Air (1)	4.039E-04	17	1.65E+09	4.04E+08
Water (2)	1.327E+01	1700	1.08E+09	2.65E+09
Soil (3)	1.345E+02	1700	9.87E+08	
Sediment (4)	1.150E+02	5500	7.25E+06	1.15E+06

	E(1)=1000	E(2)=1000	E(3)=1000	E(1,2,3)
Overall residence time =	168.29	720.43	2329.48	1819.16
Reaction time =	209.17	2454.17	2432.04	2320.07
Advection time =	861.09	1019.79	55237	8425.74

Legend:
- ---- EMISSION (E)
- — REACTION (R)
- -·-· ADVECTION (A)
- → TRANSFER D VALUE mol/Pa h

Diagram values:
- E_1, E_2, E_3 emissions
- R_1, R_2, R_3, R_4 reactions; A_1, A_2, A_4 advections
- AIR (1): 2.011E07
- WATER (2): 3.351E07
- SOIL (3): 5.992E07
- SEDIMENT (4): 2.142E07
- D_{31} 1.184E07, D_{13} 1.324E08
- D_{12}, D_{21}
- D_{24} 1.431E07, D_{42}
- D_{32}

Phase Properties, Compositions, Transport and Transformation Rates:

Emission, kg/h | Fugacity, Pa | | | | Concentration, g/m³ | | | | Loss, Advection, kg/h | | | |

E(1)	E(2)	E(3)	f(1)	f(2)	f(3)	f(4)	C(1)	C(2)	C(3)	C(4)	A(1)	A(2)	A(3)	A(4)
1000	0	0	1.813E-06	1.974E-08	2.267E-07	1.863E-08	1.824E-07	6.528E-05	7.596E-03	5.337E-04	1.824E+02	1.306E+01		5.337E-04
0	1000	0	9.684E-09	1.066E-06	1.211E-09	1.006E-06	9.743E-10	3.526E-03	4.058E-05	2.883E-02	9.743E-01	7.052E+02		2.883E-02
0	0	1000	2.082E-08	6.059E-08	3.795E-06	5.716E-08	2.095E-09	2.003E-04	1.271E-01	1.638E-03	2.095E+00	4.006E+01		1.638E-03
50	250	700	1.076E-07	3.100E-07	2.668E-06	2.925E-07	1.083E-08	1.025E-03	8.938E-02	8.380E-03	1.083E+01	2.050E+02		8.380E-02

Emission, kg/h | Loss, Reaction, kg/h | | | | Amounts, kg | | | | Total Amount, kg |

E(1)	E(2)	E(3)	R(1)	R(2)	R(3)	R(4)	m(1)	m(2)	m(3)	m(4)	
1000	0	0	7.435E+02	5.322E+01	5.57E+01	3.362E-02	1.824E+04	1.306E+04	1.367E+05	2.669E+02	1.683E+05
0	1000	0	3.972E+00	2.875E+02	2.98E-01	1.816E+00	9.743E+01	7.052E+05	7.304E+02	1.441E+04	7.204E+05
0	0	1000	8.538E+00	1.633E+01	9.33E+02	1.032E-01	2.095E+02	4.006E+04	2.288E+06	8.189E+02	2.329E+06
50	250	700	4.414E+01	8.356E+01	6.56E+02	5.280E-01	1.083E+03	2.050E+05	1.609E+06	4.190E+03	1.819E+06

Intermedia Rate of Transport, kg/h

T12	T21	T13	T31	T32	T24	T42
air-water	water-air	air-soil	soil-air	soil-water	water-sed	sed-water
1.513E+01	9.891E-02	5.979E+01	6.686E-01	3.384E+00	1.054E-01	6.639E-02
8.083E-02	5.343E+00	3.194E-01	3.572E-03	1.808E-02	5.691E-01	3.586E+00
1.738E-01	3.035E-01	6.867E-01	1.119E+01	5.664E+01	3.233E-01	2.037E-01
8.984E-01	1.553E+00	3.550E+00	7.867E+00	3.982E+01	1.654E+00	1.042E+00

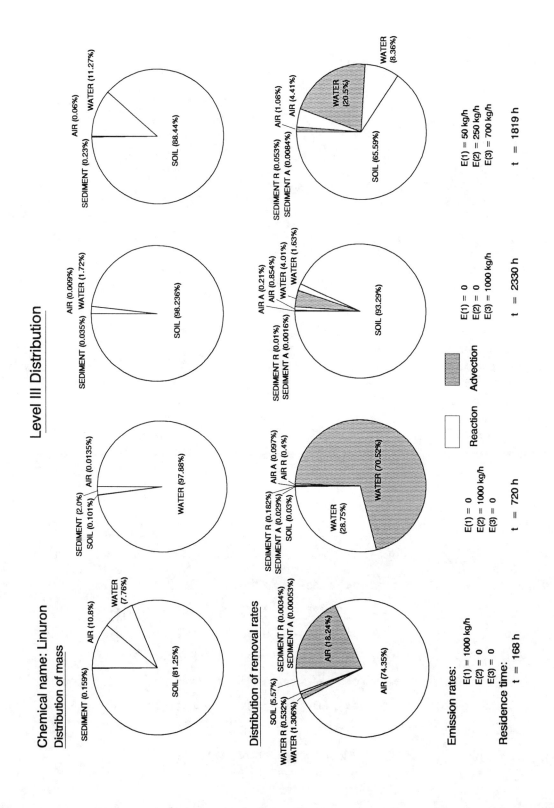

Chemical name: Metolachlor

Level I calculation: (six-compartment model)

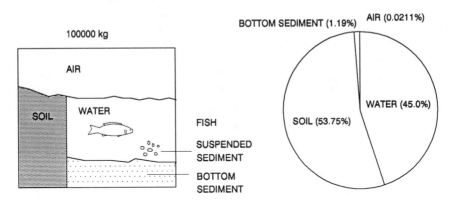

Distribution of mass

Physical-chemical properties:
molecular wt., g/mol	238.80
melting point, °C	25
solubility, g/m³	430
vapor pressure, Pa	4.20E−03
log K_{OW}	3.13
fugacity ratio, F	1.0
dissoc. const. pK_a	

Partition coefficients:
H, Pa·m³/mol	2.33E−03
K_{AW}	9.41E−07
K_{OC}	553.07
BCF	67.45
K_{SW}	26.55
$K_{SD/W}$	53.10
$K_{SSD/W}$	165.92
$K_{AR/W}$	1.43E+09

COMPARTMENT	Z	CONCENTRATION			AMOUNT	AMOUNT
	mol/m³·Pa	mol/m³	mg/L, or g/m³	µg/g	kg	%
AIR	4.03E−04	8.86E−13	2.12E−10	1.79E−07	21.168	0.0212
WATER	4.29E+02	9.42E−07	2.25E−04	2.25E−04	44993	44.99
SOIL	1.14E+04	2.50E−05	5.97E−03	2.49E−03	53751	53.75
BIOTA (FISH)	2.89E+04	6.35E−05	1.52E−02	1.52E−02	3.035	3.03E−03
SUSPENDED SEDIMENT	7.11E+04	1.56E−04	3.73E−02	2.49E−02	37.327	3.73E−02
BOTTOM SEDIMENT	2.28E+04	5.00E−05	1.19E−02	4.98E−03	1194.5	1.1945
Total					100000	100

Fugacity, f = 2.197E−09 Pa

Chemical name: Metolachlor
Level II calculation: (six-compartment model)

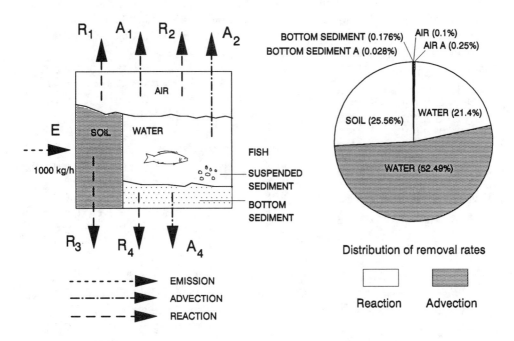

COMPARTMENT	Half-Life h	D VALUES Reaction mol/Pa·h	D VALUES Advection mol/Pa·h	CONC'N mol/m³	LOSS Reaction kg/h	LOSS Advection kg/h	REMOVAL %
AIR	170	1.64E+08	4.03E+08	1.03E−11	1.007	2.470	0.348
WATER	1700	3.50E+10	8.57E+10	1.10E−05	213.97	524.9	73.887
SOIL	1700	4.18E+10		2.92E−04	255.62		25.562
BIOTA (FISH)				7.41E−04			
SUSPENDED SEDIMENT				1.82E−03			
BOTTOM SEDIMENT	5500	2.87E+08	4.55E+07	5.84E−04	1.756	0.279	0.203
Total		7.69E+10	8.61E+10		472.36	527.64	100
Reaction + Advection			1.63E+11			1000	

Fugacity, f = 2.563E−08 Pa

Total amount = 1166612 kg

Overall residence time = 1166.61 h

Reaction time = 2469.78 h

Advection time = 2210.98 h

Fugacity Level III calculations: (four-compartment model)
Chemical name: Metolachlor

Phase Properties and Rates:

Compartment	Bulk Z mol/m3 Pa	Half-life h	D Values Reaction mol/Pa h	D Values Advection mol/Pa h	E(1,2,3) mol/Pa h
Air (1)	4.149E-04	170	1.69E+08	4.15E+08	
Water (2)	4.291E+02	1700	3.50E+10	8.58E+10	
Soil (3)	5.820E+03	1700	4.27E+10		
Sediment (4)	4.896E+03	5500	3.08E+08	4.90E+07	

	E(1)=1000	E(2)=1000	E(3)=1000
Overall residence time =	1962.52	727.69	2376.99	1943.94 h
Reaction time =	2365.73	2498.26	2453.48	2453.01 h
Advection time =	11514.50	1026.77	76241	9367.22 h

EMISSION (E)
REACTION (R)
ADVECTION (A)
TRANSFER D VALUE mol/Pa h

Phase Properties, Compositions, Transport and Transformation Rates:

Emission, kg/h

E(1)	E(2)	E(3)	f(1)	f(2)	f(3)	f(4)
1000	0	0	8.311E-07	4.296E-09	7.451E-08	4.052E-09
0	1000	0	1.383E-10	3.456E-08	1.240E-11	3.260E-08
0	0	1000	2.300E-09	1.509E-09	9.375E-08	1.424E-09
50	250	700	4.320E-08	9.912E-09	6.936E-08	9.349E-09

Fugacity, Pa — Concentration, g/m3

C(1)	C(2)	C(3)	C(4)
8.235E-08	4.402E-04	1.035E-01	4.737E-03
1.371E-11	3.542E-03	1.724E-05	3.811E-02
2.279E-10	1.547E-04	1.303E-01	1.664E-03
4.281E-09	1.016E-03	9.638E-02	1.093E-02

Loss, Advection, kg/h

A(1)	A(2)	A(3)	A(4)
8.235E+01	8.804E+01	1.035E-01	4.737E-03
1.371E-02	7.083E+02	1.724E-05	3.811E-02
2.279E-01	3.093E+01	1.303E-01	1.664E-03
4.281E+00	2.031E+02	9.638E-02	1.093E-02

Loss, Reaction, kg/h

E(1)	E(2)	E(3)	R(1)	R(2)	R(3)	R(4)
1000	0	0	3.357E+01	3.589E+01	7.60E+02	2.984E-01
0	1000	0	5.588E-03	2.887E+02	1.26E-01	2.401E+00
0	0	1000	9.288E-02	1.261E+01	9.56E+02	1.049E-01
50	250	700	1.745E+00	8.281E+01	7.07E+02	6.886E-01

Amounts, kg

m(1)	m(2)	m(3)	m(4)
8.235E+03	8.804E+04	1.864E+06	2.368E+03
1.371E+00	7.083E+05	3.103E+02	1.906E+04
2.279E+01	3.093E+04	2.345E+06	8.322E+02
4.281E+02	2.031E+05	1.735E+06	5.465E+03

Total Amount, kg

| 1.963E+06 |
| 7.277E+05 |
| 2.377E+06 |
| 1.944E+06 |

Intermedia Rate of Transport, kg/h

T12	T21	T13	T31	T32	T24	T42
air-water	water-air	air-soil	soil-air	soil-water	water-sed	sed-water
8.978E+01	2.069E+02	7.965E+02	2.198E+00	3.451E+01	8.046E-01	4.589E-01
1.494E-02	1.665E-01	1.326E-01	3.659E-04	5.745E-03	6.474E+00	3.692E+00
2.484E-01	7.269E-03	2.204E+00	2.766E+00	4.342E+01	2.827E-01	1.612E-01
4.667E+00	4.774E-02	4.140E+01	2.046E+00	3.212E+01	1.857E+01	1.059E+00

Compartment diagram values:

AIR (1): D_{31} 1.235E08, D_{13} 2.017E07, D_{21} 4.013E09
WATER (2) 4.524E08, D_{12}, 4.743E08
SOIL (3): D_{32} 1.940E09
SEDIMENT (4) D_{42} 7.844E08, D_{24}

290

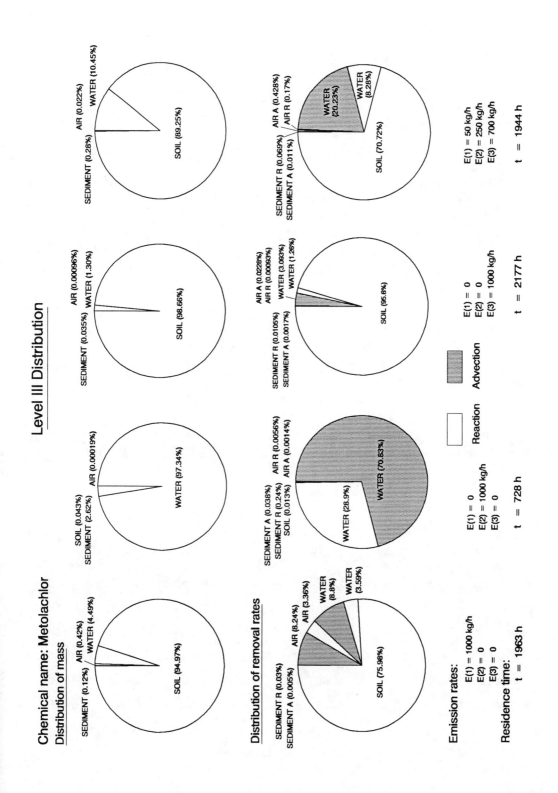

Chemical name: 2,4,5-T

Level I calculation: (six-compartment model)

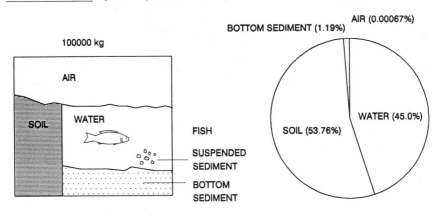

Distribution of mass

Physical-chemical properties:
molecular wt., g/mol	255.49
melting point, °C	153
solubility, g/m^3	278
vapor pressure, Pa	8.0E−05
log K_{OW}	3.13
fugacity ratio, F	0.0542
dissoc. const. pK_a	2.85

Partition coefficients:
H, Pa·m^3/mol	7.35E−05
K_{AW}	2.97E−08
K_{OC}	533.07
BCF	67.45
K_{SW}	26.55
$K_{SD/W}$	53.10
$K_{SSD/W}$	165.92
$K_{AR/W}$	4.07E+09

COMPARTMENT	Z	CONCENTRATION			AMOUNT	AMOUNT
	mol/m^3·Pa	mol/m^3	mg/L, or g/m^3	µg/g	kg	%
AIR	4.03E−04	2.61E−14	6.67E−12	5.63E−09	0.6674	6.67E−05
WATER	1.36E+04	8.81E−07	2.25E−04	2.25E−04	45002	45.002
SOIL	3.61E+05	2.34E−05	5.97E−03	2.49E−03	53762	53.762
BIOTA (FISH)	9.17E+05	5.94E−05	1.52E−02	1.52E−02	3.035	3.04E−03
SUSPENDED SEDIMENT	2.26E+06	1.46E−04	3.73E−02	2.49E−02	37.33	3.73E−02
BOTTOM SEDIMENT	7.22E+05	4.68E−05	1.19E−02	4.98E−03	1194.7	1.195
Total					100000	100

Fugacity, f = 6.475E−11 Pa

Chemical name: 2,4,5-T

Level II calculation: (six-compartment model)

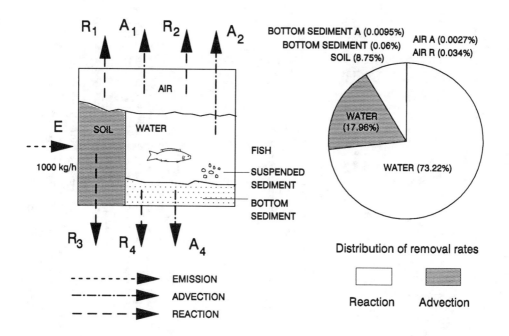

COMPARTMENT	Half-Life	D VALUES		CONC'N	LOSS	LOSS	REMOVAL
		Reaction	Advection		Reaction	Advection	%
	h	mol/Pa·h	mol/Pa·h	mol/m³	kg/h	kg/h	
AIR	55	1.64E+08	4.03E+08	1.04E−13	3.36E−02	2.66E−02	0.00602
WATER	170	1.11E+13	2.72E+12	3.51E−06	732.17	179.61	91.178
SOIL	1700	1.32E+12		9.33E−05	87.468		8.747
BIOTA (FISH)				2.37E−04			
SUSPENDED SEDIMENT				5.83E−04			
BOTTOM SEDIMENT	5500	9.10E+09	1.44E+09	1.87E−04	0.6008	0.0954	0.0696
Total		1.24E+13	2.72E+12		820.27	179.635	100
Reaction + Advection			1.51E+13			1000	

Fugacity, f = 2.584E−10 Pa

Total amount = 399108 kg

Overall residence time = 399.11 h

Reaction time = 486.56 h

Advection time = 2220.59 h

Fugacity Level III calculations: (four-compartment model)
Chemical name: 2,4,5-T

Phase Properties and Rates:

Compartment	Bulk Z mol/m3 Pa	Half-life h	D Values Reaction mol/Pa h	D Values Advection mol/Pa h
Air (1)	4.362E-04	55	5.50E+08	4.36E+08
Water (2)	1.361E+04	170	1.11E+13	2.72E+12
Soil (3)	1.846E+05	1700	1.35E+12	
Sediment (4)	1.553E+05	5500	9.78E+09	1.55E+09

	E(1)=1000	E(2)=1000	E(3)=1000	E(1,2,3)
Overall residence time =	2124.91	202.13	2355.27	1805.47
Reaction time =	2191.48	251.70	2375.61	1914.09
Advection time =	69948.03	1026.37	275171	31816.94

- ▲ EMISSION (E)
- ▲ REACTION (R)
- ▲ ADVECTION (A)
- ▲ TRANSFER D VALUE mol/Pa·h

Diagram: AIR (1) with D_{13}, D_{31} (1.789E08, 1.227E11), D_{12} (2.017E07), D_{21}; WATER (2) with D_{32} (6.153E10), D_{24} (1.505E10); SOIL (3); SEDIMENT (4) with D_{42} (1.362E10), (2.488E10); Emissions E_1, E_2, E_3; Reactions R_1–R_4; Advections A_1, A_2, A_4.

Phase Properties, Compositions, Transport and Transformation Rates:

Emission, kg/h — Fugacity, Pa

E(1)	E(2)	E(3)	f(1)	f(2)	f(3)	f(4)
1000	0	0	2.851E-08	3.908E-11	2.469E-09	3.686E-11
0	1000	0	4.158E-14	2.830E-10	3.601E-11	2.669E-10
0	0	1000	3.603E-12	1.230E-11	2.764E-09	1.160E-11
50	250	700	1.428E-09	8.130E-11	2.058E-09	7.669E-11

Concentration, g/m3

C(1)	C(2)	C(3)	C(4)
3.178E-09	1.359E-04	1.165E-01	1.463E-03
4.634E-15	9.842E-04	1.699E-07	1.059E-02
4.015E-13	4.277E-05	1.304E-01	4.603E-04
1.592E-10	2.828E-04	9.708E-02	3.043E-03

Emission, kg/h — Loss, Reaction, kg/h

E(1)	E(2)	E(3)	R(1)	R(2)	R(3)	R(4)
1000	0	0	4.004E+00	1.108E+02	8.55E+02	9.215E-02
0	1000	0	5.839E-06	8.024E+02	1.25E-02	6.672E-01
0	0	1000	5.059E-04	3.487E+01	9.57E+02	2.900E-02
50	250	700	2.005E-01	2.305E+02	7.12E+02	1.917E-01

Loss, Advection, kg/h

A(1)	A(2)	A(3)	A(4)
3.178E+00	2.719E+01		1.463E-02
4.634E-06	1.968E+02		1.059E-01
4.015E-04	8.554E+00		4.603E-03
1.592E-01	5.656E+01		3.043E-02

Amounts, kg

m(1)	m(2)	m(3)	m(4)
3.178E+02	2.719E+04	2.097E+06	7.314E+02
4.634E-04	1.968E+05	3.058E+00	5.295E+03
4.015E-02	8.554E+03	2.346E+06	2.301E+02
1.592E+01	5.656E+04	1.747E+06	1.521E+03

Intermedia Rate of Transport, kg/h

T12	T21	T13	T31	T32	T24	T42
air-water	water-air	air-soil	soil-air	soil-water	water-sed	sed-water
9.930E+01	2.014E-04	8.936E+02	1.129E-01	3.882E+01	2.485E-01	1.417E-01
1.448E-04	1.458E-03	1.303E-03	1.646E-07	5.661E-05	1.799E+00	1.026E+00
1.255E-02	6.338E-05	1.129E-01	1.263E-01	4.345E+01	7.819E-02	4.459E-02
4.974E+00	4.190E-04	4.476E+01	9.406E-02	3.235E+01	5.169E-01	2.948E-01

Total Amount, kg

2.125E+06
2.021E+05
2.355E+06
1.805E+06

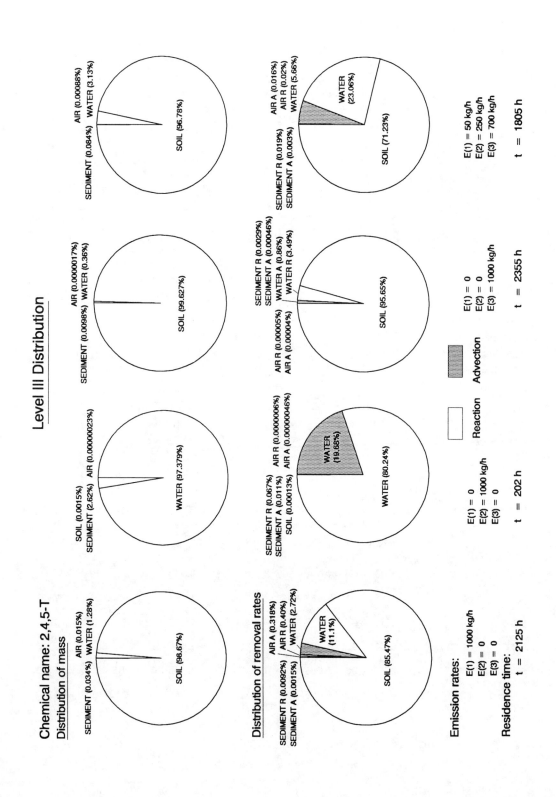

Chemical name: Triallate

Level I calculation: (six-compartment model)

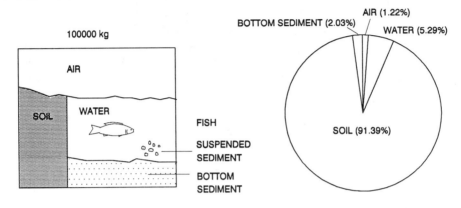

Distribution of mass

Physical-chemical properties:
molecular wt., g/mol	304.70
melting point, °C	29
solubility, g/m³	4.0
vapor pressure, Pa	1.50E−02
log K_{OW}	4.29
fugacity ratio, F	0.913
dissoc. const. pK_a	

Partition coefficients:
H, Pa·m³/mol	1.14
K_{AW}	4.61E−06
K_{OC}	7994.36
BCF	974.92
K_{SW}	383.73
$K_{SD/W}$	767.46
$K_{SSD/W}$	2398.31
$K_{AR/W}$	3.65E+08

COMPARTMENT	Z	CONCENTRATION			AMOUNT	AMOUNT
	mol/m³·Pa	mol/m³	mg/L, or g/m³	μg/g	kg	%
AIR	4.03E−04	4.00E−11	1.22E−08	1.03E−05	1219.78	1.2198
WATER	8.75E−01	8.68E−08	2.65E−05	2.65E−05	5292	5.292
SOIL	3.36E+02	3.33E−05	1.02E−02	4.23E−03	91388	91.388
BIOTA (FISH)	8.53E+02	8.47E−05	2.58E−02	2.58E−02	5.160	5.16E−03
SUSPENDED SEDIMENT	2.10E+03	2.08E−04	6.35E−02	4.23E−02	63.46	6.35E−02
BOTTOM SEDIMENT	6.72E+02	6.67E−05	2.03E−02	8.46E−03	801.6820	2.031
Total					100000	100

Fugacity, f = 9.923E−08 Pa

Chemical name: Triallate
Level II calculation: (six-compartment model)

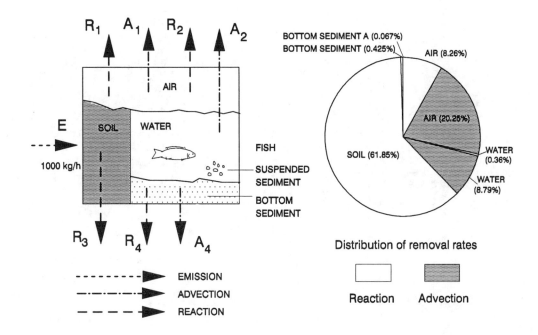

COMPARTMENT	Half-Life h	D VALUES Reaction mol/Pa·h	Advection mol/Pa·h	CONC'N mol/m³	LOSS Reaction kg/h	LOSS Advection kg/h	REMOVAL %
AIR	170	1.64E+08	4.03E+08	6.65E−10	82.558	202.524	28.508
WATER	17000	7.14E+06	1.75E+08	1.44E−06	3.582	87.871	9.145
SOIL	1700	1.23E+09		5.53E−04	618.54		61.854
BIOTA (FISH)				1.41E−03			
SUSPENDED SEDIMENT				3.46E−03			
BOTTOM SEDIMENT	5500	8.46E+06	1.34E+06	1.11E−03	4.2486	0.6744	0.4923
Total		1.40E+09	5.78E+08		708.93	291.07	100
Reaction + Advection			1.98E+09			1000	

Fugacity, f = 1.648E−06 Pa

Total amount = 1660328 kg

Overall residence time = 1660.33 h

Reaction time = 2342.02 h

Advection time = 5704.23 h

Fugacity Level III calculations: (four-compartment model)
Chemical name: Triallate

Phase Properties and Rates:

Compartment	Bulk Z mol/m3 Pa	Half-life h	D Values Reaction mol/Pa h	D Values Advection mol/Pa h
Air (1)	4.064E-04	170	1.66E+08	4.06E+08
Water (2)	8.865E-01	17000	7.23E+06	1.77E+08
Soil (3)	1.682E+02	1700	1.23E+09	
Sediment (4)	1.350E+02	5500	8.51E+06	1.35E+06

	E(1)=1000	E(2)=1000	E(3)=1000	E(1,2,3)
Overall residence time =	175.54	1146.24	2445.87	2007.45 h
Reaction time =	579.18	11586.37	2455.51	2723.30 h
Advection time =	251.88	1272.09	623001	7636.91 h

EMISSION (E)
REACTION (R)
ADVECTION (A)
TRANSFER D VALUE mol/Pa h

Phase Properties, Compositions, Transport and Transformation Rates:

Emission, kg/h			Fugacity, Pa				Concentration, g/m3			
E(1)	E(2)	E(3)	f(1)	f(2)	f(3)	f(4)	C(1)	C(2)	C(3)	C(4)
1000	0	0	5.394E-06	5.341E-07	7.541E-08	5.029E-07	6.679E-07	1.443E-04	3.864E-03	2.069E-02
0	1000	0	4.881E-07	1.545E-05	6.823E-09	1.455E-05	6.043E-08	4.173E-03	3.497E-04	5.985E-01
0	0	1000	8.201E-09	5.349E-08	2.647E-06	5.036E-08	1.015E-09	1.445E-05	1.357E-01	2.072E-03
50	250	700	3.974E-07	3.927E-06	1.859E-06	3.697E-06	4.921E-08	1.061E-03	9.524E-02	1.521E-01

Emission, kg/h			Loss, Reaction, kg/h				Loss, Advection, kg/h			
E(1)	E(2)	E(3)	R(1)	R(2)	R(3)	R(4)	A(1)	A(2)	A(3)	A(4)
1000	0	0	2.722E+02	1.176E+00	2.84E+01	1.303E+00	6.679E+02	2.885E+01		2.069E-01
0	1000	0	2.463E+01	3.402E+01	2.57E+00	3.771E+01	6.043E+01	8.347E+02		5.985E+00
0	0	1000	4.139E-01	1.178E-01	9.95E+02	1.305E-01	1.015E+00	2.890E+00		2.072E-02
50	250	700	2.006E+01	8.647E+00	6.99E+02	9.583E+00	4.921E+01	2.121E+02		1.521E+00

	Amounts, kg				Total Amount, kg
	m(1)	m(2)	m(3)	m(4)	
	6.679E+04	2.885E+04	6.956E+04	1.034E+04	1.755E+05
	6.043E+03	8.347E+05	6.294E+03	2.992E+05	1.146E+06
	1.015E+02	2.890E+03	2.442E+06	1.036E+03	2.446E+06
	4.921E+03	2.121E+05	1.714E+06	7.605E+04	2.007E+06

Intermedia Rate of Transport, kg/h							
T12	T21	T13	T31	T32	T24	T42	
air-water	water-air	air-soil	soil-air	soil-water	water-sed	sed-water	
3.458E+01	3.138E+01	2.849E+01	3.450E-02	9.744E-02	1.850E+00	3.399E-01	
3.129E+00	9.077E+01	2.578E+00	3.121E-03	8.817E-03	5.352E+01	9.833E+00	
5.258E-02	3.143E-01	4.331E-02	1.211E+00	3.421E+00	1.853E-01	3.404E-02	
2.548E+00	2.307E+01	2.099E+00	8.502E-01	2.401E+00	1.360E+01	2.499E+00	

Figure/Diagram: Four-compartment box model showing AIR (1), WATER (2), SOIL (3), SEDIMENT (4) with emissions E_1, E_2, E_3; reactions R_1, R_2, R_3, R_4; advections A_1, A_2, A_4; transfer D-values: D_{12}=1.928E07, D_{21}=2.114E07, D_{13}=4.241E06, D_{31}=1.501E06, D_{24}=2.219E06, D_{42}=1.137E07, D_{32}=9.002E06.

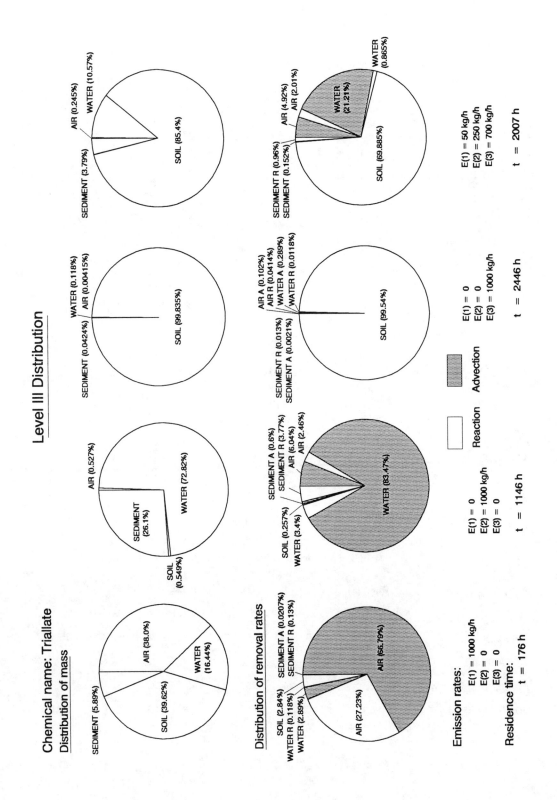

Chemical name: Trifluralin

Level I calculation: (six-compartment model)

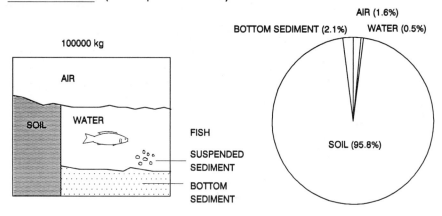

Distribution of mass

Physical-chemical properties:
molecular wt., g/mol   355.50
melting point, °C   48.5
solubility, g/m^3 0.32
vapor pressure, Pa   1.50E−02
log K_{OW} 5.34
fugacity ratio, F   0.5856
dissoc. const. pK_a

Partition coefficients:
H, Pa·m^3/mol   17.727
K_{AW} 6.34E−03
K_{OC} 89698
BCF   10939
K_{SW} 4306
$K_{SD/W}$ 8611
$K_{SSD/W}$ 26909
$K_{AR/W}$ 2.34E+08

COMPARTMENT	Z	CONCENTRATION			AMOUNT	AMOUNT
	mol/m^3·Pa	mol/m^3	mg/L, or g/m^3	μg/g	kg	%
AIR	4.03E−04	4.67E−11	1.57E−08	1.32E−05	1567.5	1.5675
WATER	6.36E−02	7.36E−09	2.47E−06	2.47E−06	494.14	0.494
SOIL	2.74E+02	3.17E−05	1.06E−02	4.43E−03	95739	95.74
BIOTA (FISH)	6.96E+02	8.06E−05	2.70E−02	2.70E−02	5.405	5.41E−03
SUSPENDED SEDIMENT	1.71E+03	1.98E−04	6.65E−02	4.43E−02	66.49	6.65E−02
BOTTOM SEDIMENT	5.48E+02	6.34E−05	2.12E−02	8.87E−03	2127.5	2.128
Total					100000	100

Fugacity, f = 1.158E−07 Pa

Chemical name: Trifluralin
Level II calculation: (six-compartment model)

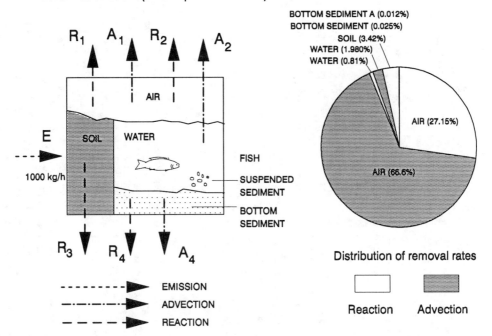

COMPARTMENT	Half-Life h	D VALUES Reaction mol/Pa·h	D VALUES Advection mol/Pa·h	CONC'N mol/m³	LOSS Reaction kg/h	LOSS Advection kg/h	REMOVAL %
AIR	170	1.64E+08	4.03E+08	1.34E−09	182.81	448.46	63.128
WATER	1700	5.18E+06	1.27E+07	2.11E−07	5.763	14.137	1.990
SOIL	5500	3.10E+08		9.07E−04	345.12		34.512
BIOTA (FISH)				2.30E−03			
SUSPENDED SEDIMENT				5.67E−03			
BOTTOM SEDIMENT	17000	2.34E+06	1.10E+06	1.81E−03	2.481	1.217	0.370
Total		4.80E+08	4.16E+08		536.18	463.82	100
Reaction + Advection			8.96E+08			1000	

Fugacity, $f = 3.313E-06$ Pa

Total amount = 2860989 kg

Overall residence time = 2861.0 h

Reaction time = 5335.85 h

Advection time = 6168.4 h

Fugacity Level III calculations: (four-compartment model)
Chemical name: Trifluralin

Phase Properties and Rates:

Compartment	Bulk Z mol/m3 Pa	Half-life h	D Values Reaction mol/Pa h	Advection mol/Pa h
Air (1)	4.053E-04	170	1.65E+08	4.05E+08
Water (2)	7.284E-02	1700	5.94E+06	1.46E+07
Soil (3)	1.369E+02	5500	3.11E+08	
Sediment (4)	1.096E+02	17000	2.23E+06	1.10E+06

Overall residence time = E(1)=1000   E(2)=1000   E(3)=1000   E(1,2,3)
Reaction time =   223.90   3122.99   7908.99   6328.24 h
Advection time =   749.42   8807.11   7931.65   7894.62 h
                        319.30   4838.84   2768855   31894.51 h

EMISSION (E)
REACTION (R)
ADVECTION (A)
TRANSFER D VALUE mol/Pa h

Phase Properties, Compositions, Transport and Transformation Rates:

Emission, kg/h

E(1)	E(2)	E(3)	f(1)	f(2)	f(3)	f(4)	C(1)	C(2)	C(3)	C(4)
1000	0	0	5.088E-06	1.683E-06	1.053E-07	3.233E-06	6.918E-07	4.113E-05	4.837E-03	1.188E-01
0	1000	0	1.600E-06	7.649E-05	3.312E-08	1.469E-04	2.176E-07	1.869E-03	1.521E-03	5.400E+00
0	0	1000	1.546E-08	1.348E-07	9.559E-06	2.590E-07	2.102E-09	3.295E-06	4.391E-01	9.520E-03
50	250	700	6.652E-07	1.930E-05	6.705E-06	3.707E-05	9.046E-08	4.716E-04	3.080E-01	1.363E+00

| | Fugacity, Pa | | | | Concentration, g/m3 | | | |

Emission, kg/h

E(1)	E(2)	E(3)	R(1)	R(2)	R(3)	R(4)	A(1)	A(2)	A(3)	A(4)
1000	0	0	2.820E+02	3.354E+00	1.10E+01	2.422E+00	6.918E+02	8.227E+02	4.837E+00	1.188E+00
0	1000	0	8.869E+01	1.524E+02	3.45E+00	1.101E+02	2.176E+02	3.738E+02	1.521E+00	5.400E+01
0	0	1000	8.570E-01	2.686E-01	9.96E+02	1.940E-01	2.102E+00	6.590E-01	4.391E+02	9.520E-02
50	250	700	3.687E+01	3.845E+01	6.98E+02	2.777E+01	9.046E+01	9.433E+01	3.080E+02	1.363E+01

| | Loss, Reaction, kg/h | | | | Loss, Advection, kg/h | | | |

Emission, kg/h

E(1)	E(2)	E(3)	m(1)	m(2)	m(3)	m(4)	Total Amount, kg
1000	0	0	6.918E+04	8.227E+03	8.707E+04	5.942E+04	2.239E+05
0	1000	0	2.176E+04	3.738E+05	2.738E+04	2.700E+06	3.123E+06
0	0	1000	2.102E+02	6.590E+02	7.903E+06	4.760E+03	7.909E+06
50	250	700	9.046E+03	9.433E+04	5.544E+06	6.813E+05	6.328E+06

| | Amounts, kg | | | | |

T12	T21	T13	T31	T32	T24	T42
air-water	water-air	air-soil	soil-air	soil-water	water-sed	sed-water
2.214E+01	6.969E+00	1.102E+01	2.756E-02	1.882E+00	4.867E+00	1.257E+00
6.963E+00	3.167E+02	3.465E+00	8.668E-03	5.918E-03	2.212E+02	5.711E+01
6.728E-02	5.583E-01	3.348E-02	2.502E+00	1.708E+00	3.899E-01	1.007E-01
2.895E+00	7.991E+01	1.441E+00	1.755E+00	1.198E+00	5.581E+01	1.441E+01

Intermedia Rate of Transport, kg/h

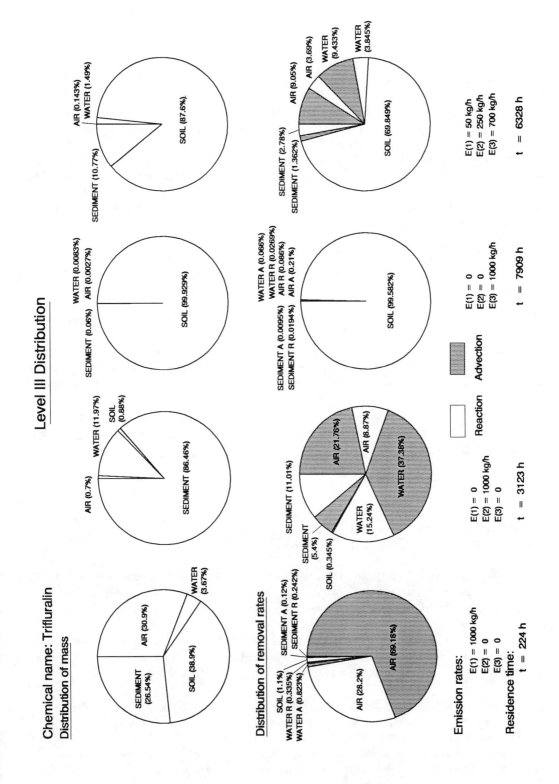

2.4 COMMENTARY ON PHYSICAL-CHEMICAL PROPERTIES AND ENVIRONMENTAL FATE

Properties and Reactivity

Selected physical-chemical properties of the herbicides are given in Tables 2.2.1 and 2.2.2 Most herbicides are fairly soluble in water, very few having solubilities less than 10 g/m^3. They also have relatively low vapor pressures, thus the Henry's law constants tend to be small, few having values exceeding 1 Pa·m^3/mol, corresponding to an air-water partition coefficient of 0.0004. The octanol-water partition coefficients (log K_{OW}) in most cases range from 1.5 to 3.4, i.e., they are slightly to moderately hydrophobic. These properties enhance their efficacy as herbicides because their relatively high solubility in water facilitates uptake into plants and transport in the transpiration stream to foliage. The low vapor pressure and Henry's law constant retard evaporation, and the moderate hydrophobicity prevents excessive sorption to soil which would result in reduced availability for uptake.

The half-lives of selected herbicides in soil in Table 2.2.3 range from Class 4 (1 week) to Class 6 (2 months), thus most are fairly rapidly degraded. Atrazine stands out as particularly persistent in water. It is emphasized that these half-lives must not be interpreted as universal values because rates of degradation are sensitive to environmental conditions, including temperature.

In many cases, the herbicides are highly water-soluble and they may dissociate as organic acids. Extreme care must be exercised in these cases because reported data on solubility may be for the undissociated acid at low pH, or for a salt at higher pH. A fundamental understanding of the solution chemistry of these substances is thus essential as a basis for performing partitioning calculations and interpreting the results. In many cases the assumption that sorption to soil is controlled by K_{OC} and organic carbon content may be invalid, there being potential for appreciable sorption to mineral surfaces.

Evaluative Calculations

Illustrative calculations are given for selected herbicides, namely atrazine, 2,4-D (with modified physical-chemical properties updated from those used in Vol. 4), diuron, EPTC, glyphosate, linuron, metolachlor, 2,4,5-T, triallate and trifluralin.

As is expected from their hydrophobic nature the Level I distributions show a strong tendency to partition into water, most of the remainder being sorbed to soil. The more hydrophobic triallate partitions more strongly to soil.

The Level II calculations suggest that degradation in water and soil and advective loss from water predominate, the proportions depending on hydrophobicity and relative half-lives. Atmospheric processes are unimportant except for a modest contribution in the case of triallate.

The Level III calculations are the most illuminating and realistic. The case of discharge to air is only of theoretical interest since application conditions are usually designed to avoid evaporation or spray drift to non-target sites. Similarly discharge entirely to water is rarely of practical importance. The third and fourth scenarios are regarded as most realistic. Not surprisingly most herbicides remain associated with the soil, with some present in water. Examination of the intermedia transport rates gives some indication of the potential for evaporation and runoff, and thus indirectly by leaching to groundwater. For example, when 1000 kg/h of atrazine is discharged to soil, the losses are 904 kg/h by reaction, 1 kg/h by evaporation and 95 kg/h by runoff. For the more hydrophobic triallate, the corresponding losses are 555 kg/h (reaction), 1 kg/h (evaporation) and 3 kg/h (runoff). For metolachlor, the quantities are 956 kg/h (reaction), 3 kg/h (evaporation) and 43 kg/h (runoff), there being return of 2 kg/h from air to soil. The data can thus be interpreted to give an indication of relative differences in behavior. Readers are encouraged to use the model to probe the sensitivities of these results to changes in input data, notably K_{ow} and the half-lives.

It is hoped that the behavior characteristics revealed in these illustrative calculations will be of value for interpreting the observed behavior of herbcides, for predicting the fate of existing herbicides in new conditions, and the fate of new pesticides.

2.5 REFERENCES:

Abernethy, J. R., Davidson, J. M. (1971) Effect of calcium chloride on prometryne and fluometuron adsorption in soil. *Weed Sci.* 19, 517-521.

Agrochemicals Handbook (1983) *The Agrochemicals Handbook*, The Royal Society of Chemistry, Nottingham, England.

Agrochemicals Handbook (1987) *The Agrochemicals Handbook*, 2nd Edition, Hartley, D., Kidd, H., Eds., The Royal Society of Chemistry, Nottingham, England.

Agrochemicals Handbook (1989) *The Agrochemicals Handbook*, The Royal Society of Chemistry, Nottingham, England.

Alexander, M. (1973) Nonbiodegradable and other recalcitrant molecules. *Biotech. Bioeng.* 15, 611-647.

Alexander, M. (1974) Microbial formation of environmental pollutants. *Appl. Environ. Microbiol.* 15, 611-647.

Altom, J. D., Strittzke, J. F. (1973) Degradation of dicamba, picloram, and few phenoxy herbicides in soils. *Weed Sci.* 21, 556-560.

Anderson, J. P. E., Domsch, K. H. (1976) Microbial degradation of thiocarbamate herbicide Diallate in soils and by pure cultures of soil microorganisms. *Arch. Environ. Contam. Toxicol.* 4, 1-7.

Ang, C., Meleady, K., Wallace L. (1989) Pesticide residues in drinking water in the north coast region of New South Wales, Australia. *Bull. Environ. Contam. Toxicol.* 42, 595-602.

Aquasol Database (1994) *Aquasol Database*. 5th Edition, Yalkowsky, S. H., Dannenfelzer, R. M., Editors, University of Arizona, Arizona.

Armstrong, D. E., Chesters, G., Harris, R. R. (1967) Atrazine hydrolysis in soil. *Soil Sci. Am. Proc.* 31, 61-66.

Ashton, F. M., Crafts, A. S. (1973) *Mode of Action of Herbicides*. John Wiley & Sons, New York, New York.

Ashton, F. M., Crafts, A. S. (1981) *Mode of Action of Herbicides*. John Wiley & Sons, New York, New York.

Atkinson, R. (1985) Kinetics and mechanisms of gas-phase reactions of hydroxyl radicals with organic compounds under atmospheric conditions. *Chem. Rev.* 85, 69-201.

Atkinson, R. (1987) Structure-activity relationship for estimation of rate constants for the gas-phase reactions of OH radicals with organic compounds. *Int'l. J. Chem. Kinetics* 19, 799-828.

Atkinson, R., Carter, W. P. L. (1984) Kinetics and mechanisms of the gas-phase reactions of ozone with organic compounds under atmospheric conditions. *Chem. Rev.* 84, 437-470.

Atkinson, R., Kwok, E. S. C., Arey, J. (1992) Photochemical processes affecting the fate of pesticides in the atmosphere. *Brighton Crop Prot. Conf.-Pests Dis.* (2), 469-476.

Atkinson, R., Tuazon, E. C., Wallington, T. J., Aschmann, S. M., Arey, J., Winer, A. M., Pitts, Jr., J. N. (1987) Atmospheric chemistry of aniline, N,N-dimethylaniline, pyridine, 1,3,5-triazine and nitrobenzene. *Environ. Sci. Technol.* 21, 64-72.

Attaway, H. H., Camper, N. D., Paynter, M. J. B. (1982a) Anaerobic microbial degradation of diuron by pond sediment. *Pest. Biochem. Physiol.* 17, 96.

Attaway, H. H., Paynter, M. J. B., Camper, N. D. (1982b) Degradation of selected phenylurea herbicides by anaerobic pond sediment. *J. Environ. Sci. Health* B17, 683-689.

Augustijn-Beckers, P. W. M., Hornsby, A. G., Wauchope, R. D. (1994) The SCS/ARS/CES pesticides database for environmental decision-making. II. Additional compounds. *Rev. Environ. Contam. Toxicol.* 137, 1-82.

Bacci, E., Calamari, D., Gaggi, C., Vighi, M. (1990) Bioconcentration of organic chemical vapors in plant leaves: Experimental measurements and correlation. *Environ. Sci. Technol.* 24, 885-889.

Bahnick, D. A., Doucette, W. J. (1988) Use of molecular connectivity indices to estimate soil sorption coefficients for organic chemicals. *Chemosphere* 17, 1703-1715.

Bailey, G. W., White, J. L. (1965) Herbicides - A compilation of their physical, chemical and biological properties. *Res. Rev.* 10, 97.

Ballatine, L. G., Newby, L. C., Simoneaux, B. J. (1978) Fate of atrazine in a marine environment. 4th International Congress of Pesticide Chemistry. Abstract No. V-528. IUPAC, Zurich, Switzerland.

Banks, P. A., Ketchersid, M. L., Merkle, M. G. (1979) The persistence of fluridone in various soils under field conditions and controlled conditions. *Weed Sci.* 27, 631.

Barnsley, G. E., Rosher, P. H. (1961) The relationship between the herbicidal effect of 2,6-dichlorobenzonitrile and its persistence in soil. *Weed Res.* 1, 147-158.

Battersby, N. S. (1990) A review of biodegradation kinetics in the aquatic environments. *Chemosphere* 21(10/11), 1243-1284.

Battersby, N. S., Wilson, V. (1989) Survey of the anaerobic biodegradation potential of organic chemicals in digesting sludge. *Appl. Environ. Microbiol.* 55, 433-439.

Baur, J. R., Bovey, R. W. (1974) Ultraviolet and volatility loss of herbicides. *Arch. Environ. Contam. Toxicol.* 2, 275-288.

Beestman, G. B., Demming, J. M. (1974) Dissipation of acetamide herbicides from soils. *Agron. J.* 66, 308-544.

Beestman, G. B., Demming, J. M. (1976) Triallate mobility in soils. *Weed Sci.* 24, 541-544.

Belluck, D., Felsot, A. (1981) Bioconcentration of pesticides by egg masses of the caddisfly, *Triaenodes tardus* milne. *Bull. Environ. Contam. Toxicol.* 26, 299-306.

Benoit-Guyod, J. L., Crosby, D. G., Bowers, J. B. (1986) Degradation of MCPA by ozone and light. *Water Res.* 20, 67-72.

Beste, C. E., Humburg, N. E. (1983) *Herbicide Handbook of the Weed Science Society of America.* 5th Edition, Weed Science Society, Champaign, Illinois.

Beynon, K. I., Stoydin, G., Wright, A. N. (1972) The breakdown of the triazine herbicide cyanazine in soils and maize. *Pest. Sci.* 3, 293-305; 379-401.

Beynon, K. I., Stoydin, G., Wright, A. N. (1972) Comparison of the breakdown of the triazine herbicides cyanazine, atrazine and simazine in soils and in maize. *Pest. Biochem. Physiol.* 2, 153-161.

Beynon, K. I., Wright, A. N. (1972) The fates of herbicides chlorthiamid and dichlobenil in relation to residues in crops, soils and animals. *Res. Rev.* 43, 23-53.

Biagi, G. L., Guerra, M. C., Barbaro, A. M., Recanatini, M., Borea, P. A., Sapone, A. (1991) Lipophilicity for *s*-triazine herbicides. In: *QSAR in Environmental Toxiocology IV.* Hermens, J. L., Opperhuizen, A., Editors, Elsevier, Amsterdam, The Netherlands.

Bintein, S., Devillers, J. (1994) QSAR for organic chemical sorption in soils and sediments. *Chemosphere* 28(6), 1171-1188.

Bjerke, E. L., Herman, J. L., Miller, P. W., Wetters, J. H. (1972) Residue study of phenoxy herbicides in milk and cream. *J. Agric. Food Chem.* 20, 963-967.

Bottoni, P., Funari, E. (1992) Criteria for evaluating the impact of pesticides on groundwater quality. *Sci. Total Environ.* 123/124, 581-590.

Bouchard, D. C., Wood, A. L. (1988) Pesticides sorption on geologic material of varying organic carbon content. *Toxicol. Ind. Health* 4, 341-349.

Bowman, B. T. (1990) Mobility and persistence of alachlor, atrazine and metolachlor in plain field sand, and atrazine and isazofos in honeywood silt loam, using lysimeters. *Environ. Toxicol. Chem.* 9, 453-461.

Bowman, B. T., Sans, W. W. (1979) The aqueous solubility of twenty-seven insecticides and related compounds. *J. Environ. Sci. Health* B14(6), 625-634.

Bowman, B. T., Sans, W. W. (1983a) Further water solubility determination of insecticidal compounds. *J. Environ. Sci. Health* B18(2), 221-227.

Bowman, B. T., Sans, W. W. (1983b) Determination of octanol-water partitioning coefficients (K_{OW}) of 61 organophosphorous and carbamate insecticides and their relationship to respective water solubility (S) values. *J. Environ. Sci. Health* B18(6), 667-683.

Braumann, T., Grimme, L. H. (1981) Determination of hydrophobic parameters for pyridazinone herbicides by liquid-liquid partition and reversed-phase high-performance liquid chromatography. *J. Chromatogr.* 206(1), 7-15.

Braumann, T., Weber, G., Grimme, H. (1983) Quantiative structure-activity relationship for herbicides. Reversed-phase liquid chromatographic retention parameter log k_w versus liquid-liquid partition coefficient as a model of the hydrophobicity of phenylureas s-trizines and phenoxycarbonic acid derivatives. *J. Chromatogr.* 261, 329-343.

Brewer, F., Lavy, T. L., Talbert, R. E. (1982) Effects of flooding on dinitroaniline persistence in soybean (*Glycine max*)-rice (*Oryza sativa*) rotation. *Weed Sci.* 30, 531-539.

Briggs, G. G. (1969) Molecular structure of herbicides and their sorption by soils. *Nature* 223, 1288.

Briggs, G. G. (1981) Theoretical and experimental relationships between soil adsorption, octanol-water partition coefficients, water solubilities, bioconcentration factors, and the parachor. *J. Agric. Food Chem.* 29, 1050-1059.

Broto, P., Moreau, G., Vandycke, C. (1984) Molecular structure: Perception, autocorrelation descriptor and SAR studies. System of atomic contribution for the calculation of the n-octanol/water partition coefficients. *Eur. J. Med. Chem. Chim. Term.* 19, 71-78.

Brown, D. F., McDonough, L. M., McCool, D. K., Papendick, R. I. (1984) High performance liquid chromatographic determination of bromoxynil octanoate and metribuzin in runoff water from wheat fields. *J. Agric. Food Chem.* 32, 195-200.

Brown, D. S., Flagg, E. W. (1981) Empirical prediction of organic pollutant sorption in natural sediments. *J. Environ. Qual.* 10, 382-386.

Brusseau, M. L., Rao, P. S. C. (1989) The influence of sorbate-organic matter interactions of sorption nonequilibrium. *Chemosphere* 18, 1691-1706.

Brust, H. F. (1966) A summary of chemical and physical properties of Dursban. *Down to Earth* 22(3), 21-22.

Budavari, S., Editor (1989) *The Merck Index. An Encyclopedia of Chemicals, Drugs and Biologicals.* 11th Edition, Merck and Co., Inc., Rahway, New Jersey.

Burkhard, N., Eberle, D. O., Guth, J. A. (1975) Model systems for studying the environmental behaviour of pesticides. *Environ. Quality & Safety* Supplement VIII, 204-213.

Burkhard, N., Guth, J. A. (1976) Photodegradation of atrazine, altraton and ametryne in aqueous solution with acetone as a photosensitizer. *Pest. Sci.* 7, 65-71.

Burkhard, N., Guth, J. A. (1981) Rate of volatilization of pesticides from soil surfaces; comparison of calculated results with those determined in a laboratory model system. *Pest. Sci.* 12, 37-44.

Burkhard, N., Guth, J. A. (1981) Chemical hydrolysis of 2-chloro-4,6-bis(alkylamino)-1,3,5-triazine herbicides and their breakdown in soil under the influence of adsorption. *Pest. Sci.* 12(1), 45-52.

Buxton, G. V., Greenstock, C. L., Helman, W. P., Ross, A. B. (1988) Critical review of rate constants for reactions of hydrated electrons, hydrogen atoms and hydroxyl radicals ($\cdot$OH/$\cdot$O$^-$) in aqueous solution. *J. Phys. Chem. Ref. Data* 17, 513-886.

Bysshe, S. E. (1982) Chapter 5, Bioconcentration factor in aquatic organisms. In: *Handbook on Chemical Property Estimation Methods, Environmental Behavior of Organic Compounds*. Lyman, W. J., Reehl, W. F., Rosenblatt, D. H., Editors, McGraw-Hill, Inc., New York.

Call, D. J., Brooke, L. T., Kent, R. J., Poirier, S. H., Knuth, M. L., Shubat, P. J., Slick, E. J. (1984) Toxicity, uptake, and elimination of the herbicides alachlor and dinoseb in freshwater fish. *J. Environ. Qual.* 13, 493-498.

Call, D. J., Brooke, L. T., Kent, R. J., Knuth, M. L., Poirier, S. H., Huot, J. M., Lima, A. R. (1987) Bromacil and diuron herbicides: Toxicity, uptake, and elimination in freshwater fish. *Arch. Environ. Contam. Toxicol.* 16, 607-613.

Camper, N. D., Stralka, K., Skipper, H. D. (1980) Aerobic and anaerobic degradation of profluralin and trifluralin. *J. Environ. Sci. Health* B15, 457-473.

Carlson, W. C., Lignowski, E. M., Hopen, H. J. (1975) Mode of action of pronamide. *Weed Sci.* 23, 155-161.

Carringer, R. D., Weber, J. B., Monaco, T. J. (1975) Adsorption-desorption of selected pesticides by organic matter and montmorillonte. *J. Agric Food Chem.* 23, 568-572.

Carsel, R. F. (1989) Hydrologic processes affecting the movement of organic chemicals in soils. In: *Reactions and Movement of Organic Chemicals in Soils*. SSSA Special Publication No. 22, Sawhney, B. L., Brown, K., Eds., pp. 439-445, Soil Science Society of America and Society of Agronomy, Madison, Wisconsin.

Caux. P.-Y., Kent, R. A., Taché, M., Grande, C., Fan, G. T., MacDonald, D. D. (1993) Environmental fate and effects of dicamba: A Canadian perspective. *Rev. Environ. Contam. Toxicol.* 133, 1-58.

Cessna, A. J., Grover, R. (1978) Spectrophotometric determination of dissociation constants of selected acidic herbicides. *J. Agric. Food Chem.* 26, 289-293.

Cessna, A. J., Muir, D. C. G. (1991) Chapter 6, Photochemical transformations. In: *Environmental Chemistry of Herbicides*. Vol. II, Grover, R., Cessna, A. J., Editors, pp. 199-264, CRC Press, Inc., Boca Raton, Florida.

Chaumat, E., Chamel, A., (1991) Sorption and permeation to phenylurea herbicides of isolated cuticles of fruit and leaves. Effect of cuticular characteristics and climatic parameters. *Chemosphere* 22, 85-97.

Chau, A. S. Y., Thomson, K. (1978) Investigation of the integrity of seven herbicidal acids in water samples. *J. Assoc. Off. Anal. Chem.* 61, 481-485.

Chen, Y-L., Chen, J-S. (1979) Degradation and dissipation of herbicide butachlor in paddy fields. *J. Pest. Sci.* 4, 431.

Chen, Y.-L., Lo, C.-C., Wong, Y.-S. (1982) Photodecomposition of herbicide butachlor in aqueous solution. *J. Pest. Sci.* 7, 41.

Cheng, H. H., Editor (1990) *Pesticides in the Soil Environment: Processes, Impacts, and Modeling*. Soil Science Society of America, Inc., Madison, Wisconsin.

Cheung, M. W., Biggar, J. W. (1974) Solubility and molecular structure of 4-amine-3,5,6-trichloropicolinic acid in relation to pH and temperature. *J. Agric. Food Chem.* 22, 202-206.

Chiou. C. T. (1981) Partition coefficient and water solubility in environmental chemistry. In: *Hazard Assessment of Chemicals. Current Development*. Vol. 1, 117-153. Academic Press, Inc., New York, New York.

Chiou. C. T., Freed, V. H., Peters, L. J., Kohnert, R. L. (1980) Evaporation of solutes from water. *Environ. Int'l* 3, 231-236.

Chiou. C. T., Freed, V. H., Schmedding, D. W., Kohnert, R. L. (1977) Partition coefficient and bioaccumulation of selected organic chemicals. *Environ. Sci. Technol.* 11, 475-478.

Chung, K. H., Ro, K. S., Roy, D. (1996) Fate and enhancement of atrazine biotransformation in anaerobic wetland sediment. *Water Res.* 30, 341-346.

Colbert, F. O., Volk, V. V., Appleby, A. P. (1975) Sorption of atrazine, terbutryn and GS-14254 on natural and lime-amended soils. *Weed Sci.* 23, 390-394.

Comfort, S. D., Inskeep, W. P., Macut, R. E. (1992) Degradation and transport of dicamba in a clay soil. *J. Environ. Qual.* 21, 653-658.

Connell, D. W., Markwell, R. D. (1990) Bioaccumulation in the soil to earthworm system. *Chemosphere* 20, 91-100.

Corbin, F. T., Upchurch, R. P. (1967) Influence of pH on detoxification of herbicides in soils. *Weeds* 15, 370-377.

Corwin, D. L., Farmer, W. J. (1984) Non-single-valued adsorption-desorption of bromacil and diquat by freshwater sediments. *Environ. Sci. Technol.* 18, 507-514.

Crosby, D. G., Leitis, E. (1973) The photodecomposition of trifluralin in water. *Bull. Environ. Contam. Toxicol.* 10, 237.

Crosby, D. G., Tang, C.-S. (1969) Photodecomposition of 3-(*p*-chlorophenyl)-1,1-dimethylurea (monuron). *J. Agric. Food. Chem.* 17, 1041-1043.

Crosby, D. G., Wong, A. S. (1973) Photodecomposition of 2,4,5-trichlorophenoxyacetic acid (2,4,5-T) in water. *J. Agric. Food. Chem.* 21, 1052.

Cunningham, J. J., Kemp, W. M., Stevenson, J. C., Boynton, W. R., Means, J. C. (1981) Stress effects of agricultural herbicides on submerged macrophytes in estuarine microcosms. pp. 147-182. In: Submerged aquatic vegetation in Chesapeake Bay. Annual Report to USEPA, UMCEES, Horn Point Environmental Laboratories, Cambridge, Maryland.

Dao, T. H., Lavy, T. L., Sorensen, R. C. (1979) Atrazine degradation and residue distribution in soil. *Soil Sci. Soc. Am. J.* 43, 1129-1134.

Dao, T. H., Lavy, T. L., Dragun, J. (1983) Rationale of the solvent selection for soil extraction of pesticide residues. *Res. Rev.* 87, 91-104.

Davidson, J. M., McDougal, J. R. (1973) Experimental and predicted movement of three herbicides in a water-saturated soil. *J. Environ. Qual.* 2, 428-433.

Davidson, J. M. et al. (1980) Adsorption, Movement and Biological Degradation of Large Concentration of Selected Pesticides in Soils. U.S. EPA-600/2-80-124.

Davies, R. P., Dobbs, A. J. (1984) The prediction of bioconcentration in fish. *Water Res.* 18(10), 1253-1262.

Day, B. E., Jordon, L. S., Russell, R. C. (1963) Persistence of dalapon residues in California soils. *Soil Sci.* 95, 326-330.

Day, K. E. (1991) Chapter 16, Pesticide transformation products in surface waters. Effects on aquatic biota. pp.217-241. In: *Pesticide Transformation Products. Fate and Significance in the Environment.* ACS Sym. series 457, Somasundaram, L., Coats, J. R., Editors, American Chemical Society, Washington DC.

Dean, J. D., Editor (1985) *Lange's Handbook of Chemistry.* 13th Edition, McGraw-Hill, Inc., New York.

Deuel, L. E., Turner, F. T., Brown, K. W., Price, J. D. (1978) Persistence and factors affecting dissipation of molinate under flooded rice culture. *J. Environ. Quality* 7, 373.

Devillers, J., Bintein, S., Domine, D. (1996) Comparison 0f BCF models based on log P. *Chemosphere* 33(6), 1047-1065.

Di Guardo, A., Calamari, D., Zanin, G., Consalter, A., Mackay, D. (1994) A fugacity model of pesticide runoff to surface water: development and validation. *Chemosphere* 28, 511-531.

Dixon, D., Rissman, E. (1985) Physical-Chemical Properties and Categorization of RCRA Wastes According to Volatility. U.S. EPA Report No. 450/3-85-007. NTIS PB 85-404527. Springfield, Virginia.

Dobbs, A. J., Cull, M. R. (1982) Volatilization of chemicals-relative loss rates and the estimation of vapor pressures. *Environ. Pollut.* (series B) 3, 289-298.

Dobbs, A. J., Hart, G. F., Parsons, A. H. (1984) The determination of vapour pressures from relative volatilization rates. *Chemosphere* 13, 687-692.

Dörfler, U., Alder-Köhler, R., Schneider, P., Scheunert, I., Korte, F. (1991) A laboratory model system for determining the volatility of pesticides from soil and plant surfaces. *Chemosphere* 23(4), 485-496.

Dowd, J. F., Bush, P. B., Neary, D. G., Taylor, J. W., Berisford, Y. C. (1993) Modeling pesticide movement in forested watersheds: Use of PRZM for evaluating pesticide options in loblolly pine stand management. *Environ. Toxicol. Chem.* 12, 429-439.

Doyle, R. C., Kaufman, D. D., Burt, G. W. (1978) Effect of dairy manure and sewage sludge on ^{14}C-pesticide degradation in soil. *J. Agric. Food Chem.* 26, 987-989.

Draper, W. M., Crosby, D. G. (1984) Solar photooxidation of pesticides in dilute hydrogen peroxide. *J. Agric. Food Chem.* 32, 231.

Duah-Yentumi, S., Kuwatsuka, S. (1980) Effect of organic matter and chemical fertilizers on the degradation of benthiocarb and MCPA herbicides in the soil. *Soil Sci. Plant Nutr.* 26, 541.

Dubelman, S., Bremer, M. J. (1983) Determination of the octanol/water partition coefficient of MAPC products. Report No. MSL-3219, Monsanto Company Agricultural Research Division, St. Louis.

Edwards, C. A. (1973) *Persistent Pesticides in the Environment.* 2nd Edition, CRC Press, Cleveland, Ohio.

Edwards, C. A. (1977) Nature and origins of pollution of aquatic systems by pesticides. In: *Pesticides in Aqatic Environments.* Khan, M. A. Q., Editor, Plenum Press, New York.

Eichelberger, J. W., Lichtenberg, J. J. (1971) Persistence of pesticides in river water. *Environ. Sci. Technol.* 5, 541-544.

Eisler, R. (1985) *Atrazine Hazards to Fish, Wildlife, and Invertebrates: A Synoptic Review.* U.S. Fish and Wildlife Service Biological Rep. 53pp.

El-Dib, M. A., Aly, O. A. (1976) Persistence of some phenylamide pesticides in the aquatic environment. I. Hydrolysis. *Water Res.* 10, 1047.

Elgar, K. E. (1983) Pesticide residues in water-An appraisal. In: *International Union of Pure and Applied Chemistry. Pesticide Chemistry: Human Welfare and the Environment.* Vol. 4, Miyamoto, J., Kearney, P. C., Editors, Pergamon Press, Oxford, England.

Ellgehausen, H., Guth, J. A., Esser, H. O. (1980) Factors determining bioaccumulation potential of pesticides in the individual compartments of aquatic food chains. *Ecotoxicol. Environ. Saf.* 4, 134-157.

Ellgehausen, H., D'Hondt, C., Fuerer, R. (1981) Reversed-phase chromatography as a general method for determining octanol/water partition coefficients. *Pest. Sci.* 12, 219.

Ellington, J. J., Stancil, F. E., Payne, W. D. (1987) Measurement of Hydrolysis Rate Constants for Evaluation of Hazardous Waste Land Disposal. Volume 1, Data on 32 chemicals. U.S. EPA-600/3-86/043, Washington DC.

Ellington, J. J., Stancil, F. E., Payne, W. D. (1987) Measurement of Hydrolysis Rate Constants for Evaluation of Hazardous Waste Land Disposal. Volume 2, Data on 54 chemicals. U.S. EPA, EPA-600/53-87/019, Washington DC.

Ellington, J. J., Stancil, F. E., Payne, W. D., Trusty, C. D. (1988) Measurement of Hydrolysis Rate Constants for Evaluation of Hazardous Waste Land Disposal. Volume 3, Data on 70 chemicals. U.S. EPA, EPA-600/3-88/028, NTIS PB 88-234042, Springfield, Virginia.

Ellis, P. A., Camper, N. D. (1982) Aerobic degradation of diuron by aquatic microorganisms. *J. Environ. Sci. Health* B17, 277-290.

Erkell, L., Walum, E. (1979) Differentiation of cultured neuroblastoma cells by urea derivatives. *Febs Letters* 104, 401.

Evelyne, C., André, C., Georges, T., Michel, T. (1992) Quantitative relationships between structure and penetration of phenylurea herbicides through isolated plant cuticles. *Chemosphere* 24(2), 189-200.

Farmer, W. J. (1976) *A Literature Survey of Benchmark Pesticides.* Science Communication Division of Dept. of Medical and Public Affairs, Medical Center of George Washington University, Washington DC.

Faust, B. C., Hoigné, J. (1990) Photolysis of Fe(III)-hydroxy complexes as sources of OH radicals in clouds, fog and rain. *Atmos. Environ.* 24A, 79-89.

Findinger, N. J., Glotfelty, D. E. (1988) A laboratory method for the determination of air-water Henry's law constants for several pesticides. *Environ. Sci. Technol.* 22, 1289-1293.

Findinger, N. J., Glotfelty, D. E., Freeman, H. P. (1989) Comparison of two experimental techniques for determining air-water Henry's law constants. *Environ. Sci. Technol.* 23(12), 1528-1531.

Finizio, A., Di Guardo, A., Arnoldi, A., Vighi, M., Fanelli, R. (1991) Different approaches for the evaluation of K_{ow} for *s*-triazine herbicides. *Chemosphere* 23, 801-812.

Foy, C. L. (1976) The chlorinated aliphatic acid. In: *Herbicides: Chemistry, Degradation and Mode of Action.* Marcel Dekker, Inc., New York.

Francioso, O., Bak, E., Rossi, N., Sequi, P. (1992) Sorption of atrazine and trifluralin in relation to the physico-chemical characteristics of selected soils. *Sci. Total Environ.* 123/124, 503-512.

Frank, R., Clegg, B. S., Patni, N. K. (1991) Dissipation of cyanazine and metolachlor on a clay loam, Ontario, Canada, 1987-1990. *Arch. Environ. Contam. Toxicol.* 21, 253-262.

Freed, V. H. (1953) Herbicides mechanisms - Mode of action other than aryloxyalkyl acids. *J. Agric. Food Chem.* 1, 47-51.

Freed, V. H. (1966) Chemistry of herbicides. In: *Pesticides and Their Effects on Soils and Water.* Breth, S. A., Editor, Soil Science Society of America, Inc., pp. 28-39.

Freed, V. H. (1976) Solubility, hydrolysis, dissociation constants and other constants of benchmark pesticides. In: A Literature Survey of Benchmark Pesticides. pp.1-18, Medical Center of George Washington University, Washington DC.

Freed, V. H., Burschel, P. (1957) The relationship of water solubility to dosage of herbicides. *Z. Pflanzenkrankh, µ. Pflanzenschutz* 64, 477.

Freed, V. H., Chiou, C. T., Haque, R. (1977) Chemodynamics: Transport and behaviour of chemicals in the environment-A problem in environmental health. *Environ. Health Prospect.* 20, 55-70.

Freed, V. H., Haque, R. (1973) Chapter 10, Adsorption, movement, and distribution of pesticides in soil. In: *Pesticide Formulations.* Van Valkenburg, Editor, pp. 441-459, Marcel Dekker, Inc., New York.

Freed, V. H., Haque, R., Vernetti, J. (1967) Thermodynamic properties of some carbamates and thiocarbamates in aqueous solutions. *J. Agric. Food Chem.* 15, 1121-1123.

Freese, E., Levin, B. C., Pearce, R., Sreevalson, T., Kaufman, J. J., Koski, W. S., Semo, N. M. (1979) Correlation between the growth inhibitory effects, partition coefficients and teratogenic effects of lipophilic acids. *Teratology* 20(3), 413-440.

Freiberg, M. B., Crosby, D. G. (1986) Loss of MCPA from simulated spray droplets. *J. Agric. Food Chem.* 34, 92-95.

Freitag, D., Balhorn, L., Geyer, H., Körte, F. (1985) Environmental hazard profile of organic chemicals. An experimental method for the assessment of the behaviour of chemicals in the ecosphere by simple laboratory tests with C-14 labelled chemicals. *Chemosphere* 14, 1589-1616.

Freitag, D., Geyer, H., Kraus, A., Viswanathan, R., Kotzias, D., Attar, A., Klein, W., Körte, F. (1982) Ecotoxicological profile analysis. VII. Screening chemicals for their environmental behavior by comparative evaluation. *Ecotoxicol. Environ. Saf.* 6, 60-81.

Freitag, D., Lay, J. P., Körte, F. (1984) Environmental hazard profile - Test results to structure and translation into the environment. In: *QSAR in Environmental Toxicology*. Kaiser, K. L. E., Editor, pp. 111-136, D. Reidel Publishing Company, Dordrecht, The Netherlands.

Friedrich, K., Stammbach, K. (1964) Gas chromatographic determination of small vapour pressures. Determination of the vapour pressures of some triazine herbicides. *J. Chromatogr.* 16, 22-28.

Fujita, T., Iwasa, J., Hansch, C. (1964) A new substituent constant derived from partition coefficients. *J. Am. Chem. Soc.* 86(23), 5175-5180.

Funderburk, Jr., H. H., Bozarth, G. A. (1967) Review of the metabolism and decomposition of diquat and paraquat. *J. Agric. Food Chem.* 15(4), 563-567.

Funderburk, Jr., H. H., Negi, N. S., Lawrence, J. M. (1960) Photochemical decomposition of diquat and paraquat. *Weeds*. 14, 240.

Furmidge, C. G., Osgerby, J. M. (1967) Persistence of herbicides in soil. *J. Sci. Food Agric.* 18, 269.

Gaillardon, P., Calvert, R., Terce, M. (1977) Adsorption and desorption of terbutryne by a calcium-montmorrillonite and humic acid alone or in combination. *Weed Res.* 17, 41-48.

Garten, Jr., C. T., Trablka, J.R. (1983) Evaluation of models for predicting terrestrial food chain behavior of xenobiotics. *Environ. Sci. Technol.* 17, 590-595.

Gear, J. R., Michel, J. G., Grover, R. (1982) Photochemical degradation of picloram. *Pest. Sci.* 13, 189.

Geller, A. (1980) Studies on degradation of atrazine by bacterial communities enriched from various biotypes. *Arch. Environ. Contam. Toxicol.* 9, 289.

GEMS (1986) Graphical Exposure Modeling Systems. Fate of atmospheric pollutants (FAP) database. Office of Toxic Substances, U.S. Environmental Protection Agency.

Gerstl, Z., Helling, C. S. (1987) Evaluation of molecular connectivity as a predictive method for the adsorption of pesticides by soils. *J. Environ. Sci. Health* B22, 55-69.

Gerstl, Z, Mingelgrin, U. (1984) Sorption of organic substances by soils and sediments. *J. Environ. Sci. Health* B19(3), 297-312.

Getzen, F. W., Ward, T. M. (1971) Influence of water structure on aqueous solubility. *Ind. Eng. Chem. Prod. Res. Develop.* 10, 122-132.

Geyer, H., Kraus, A, G., Klein, W., Richter, E., Korte, F. (1980) Relationship between water solubility and bioaccumulation potential of organic chemicals in rats. *Chemosphere* 9, 277-291.

Geyer, H., Politzki, G., Freitag, D. (1984) Prediction of ecotoxicological behaviour of chemicals: Relationship between n-octanol/water partition coefficient and bioaccumulation of organic chemicals by alga chlorella. *Chemosphere* 13, 269-284.

Geyer, H., Scheunert, I., Brüggemann, R., Steinberg, C., Korte, F., Kettrup, A. (1991) QSAR for organic chemical bioconcentration in *Daphnia*, algae, and mussels. *Sci. Total Environ.* 109/110, 387-394.

Geyer, H., Visvanathan, R., Freitag, D., Korte, F. (1981) Relationship between water solubility of organic chemicals and their bioaccumulation by the *Alga chlorella*. *Chemosphere* 10, 1307-1313.

Ghassemi, M., Fargo, L., Painter, P., Quinlivan, S., Scofield, R., Takata, A. (1981) Environmental fates and impacts of major forest use pesticides and toxic substances. A-149 (Citing glyphosate registration data). Washington DC.

Gingerich, L. L., Zimdahl, R. L. (1976) Soil persistence of isopropalin and oryzalin. *Weed Sci.* 24, 431-434.

Gish, T. J., Sadeghi, A., Wienhold, B. J. (1995) Volatilization of alachlor and atrazine as influenced by surface litter. *Chemosphere* 31, 2971-2982.

Glass, B. L. (1975) Photosensitization and luminescence of picloram. *J. Agric. Food Chem.* 23, 1109.

Glotfelty, D. E. (1981) Atmospheric dispersion of pesticides from treated fields. Ph.D. Thesis of Univ. of Maryland, College Park, Maryland.

Glotfelty, D. E., Taylor, A. W., Turner, B. C., Zoller, W. H. (1984) Volatilization of surface-applied pesticides from fallow soils. *J. Agric. Food Chem.* 32, 638-643.

Glotfelty, D. E., Leech, M. M., Jersey, J., Taylor, A. W. (1989) Volatilization and wind erosion of soil surface applied atrazine, simazine, alachlor, and toxaphene. *J. Agric. Food Chem.* 37, 546-551.

Görge, G., Nagel, R. (1990) Kinetics and mechanism of 14C-lindane and 14C-atrazine in early life stages of zebrafish (*Brachdanio rerio*). *Chemosphere* 21, 1125-1137.

Goring, C. A. I. (1967) Physical aspects of soil in relation to the action of soil fungicides. *Ann. Rev. Phytopath.* 5, 285-318.

Goswami, K. P., Green, R. E. (1971) Microbial degradation of the herbicide atrazine and its 2-hydroxy analog in submerged soils. *Environ. Sci. Technol.* 5, 426.

Grain, C. F. (1982) Chapter 14, Vapor pressure. In: *Handbook on Chemical Property Estimation Methods. Environmental Behavior of Organic Compounds*. Lyman, W. J., Reehl, W. F., Rosenblatt, D. H., Editors, McGraw-Hill Inc., New York.

Grayson, B. T., Fosbraey, L. A. (1982) Determination of the vapor pressure of pesticides. *Pest. Sci.* 13, 269-278.

Grover, R. (1974) Adsorption and desorption of trifluralin, triallate and diallate by various adsorbents. *Weed Sci.* 22(4), 405-408.

Grover, R. (1991) Chapter 2, Nature transport, and fate of airborne residues. In: *Environmental Chemistry of Herbicides*. Vol. II., Grover, R., Cessna, A. J., Editors, pp. 90-117, CRC Press, Inc., Boca Raton, Florida.

Grover, R., Editor (1988) *Environmental Chemistry of Herbicides*. Volume I, CRC Press, Inc., Boca Raton, Florida.

Grover, R., Cessna, A. J., Editors (1991) *Environmental Chemistry of Herbicides*. Volume II, CRC Press, Inc., Boca Raton, Florida.

Grover, R., Cessna, A. J., Banting, J. D., Morse, P. M. (1979) Adsorption and bioactivity of diallate, triallate and trifluralin. *Weed Res.* 19, 363-369.

Grover, R., Spencer, W. F., Farmer, W., Shoup, T. D. (1978) Triallate vapor pressure and volatilization from glass surfaces. *Weed Sci.* 26, 505-508.

Guenzi, W. D., Beard, W. E. (1974) In: *Pesticides in Soil and Water*. Guenzi, W. D., Editor, American Soil Science Society, Madison, Wisconsin. pp. 108-122.

Gunkel, G., Streit, B. (1980) Mechanisms of bioaccumulation of a herbicide (atrazine, s-triazine) in a freshwater molluse (*Ancylus fluviatilis* müll) and a fish (*Coregonus fera* jurine). *Water Res.* 14, 1573-1584.

Günther, F. A., Westlake, W. E., Jaglan, P. S. (1968) Reported solubilities of 738 pesticide chemicals in water. *Res. Rev.* 20, 1-148.

Gustafson, D. I. (1989) Groundwater ubiquity score: A simple method for assessing pesticide leachability. *Environ. Toxicol. Chem.* 8, 339-357.

Guzik, F. F. (1978) Photolysis of isopropyl 3-chlorocarbanilate in water. *J. Agric. Food. Chem.* 26, 53.

Gysin, H. (1962) Triazine herbicides - their chemistry, biological properties and mode of action. *Chem. Ind.* 31, 1393.

Haag, W. R., Yao, C. C. D. (1992) Rate constants for reaction of hydroxyl radicals with several drinking water contaminants. *Environ. Sci. Technol.* 26, 1005-1013.

Haderlein, S. B., Weissnahr, K. W., Schwarzenbach, R. P. (1996) Specific adsorption of nitroaromatic explosives and pesticides to clay minerals. *Environ. Sci. Technol.* 30, 612-622.

Hahan, R. R., Burnside, O. C., Lavy, T. L. (1969) Dissipation and phytotoxicity of dicamba. *Weed Sci.* 17, 3-8.

Halfon, E., Galassi, S., Brüggermann, R., Provini, A. (1996) Selection of priority properties to assess environmental hazard of pesticides. *Chemosphere* 33(8), 1543-1562.

Hall, R. C., Giam, C. S., Merkle, M. G. (1968) The photolytic degradation of picloram. *Weed Res.* 8, 292.

Hamaker, J. W. (1972) Decomposition: Quantitative aspects. In: *Organic Chemicals in the Soil Environment.* Goring, C. A. I., Hammaker, J. W., Editors, pp. 253-341, Marcel Dekker, New York, New York.

Hamaker, J. W. (1975) The interpretation of soil leaching experiments. In: *Environmental Dynamics of Pesticides.* Haque, R., Freed, V. H., Editors, pp. 115-133, Plenum Press, New York, New York.

Hamaker, J. W., Kerlinger, H. O. (1969) Vapor pressure of pesticides. In: *Pesticidal Formulation Research: Physical and Colloidal Chemical Aspects.* Gould, R.F., Ed., pp. 39-54, Adv. Chem. Ser. 86, Am. Chem. Soc., Washington, DC.

Hamaker, J. W., Thompson, J. M. (1972) Adsorption. In: *Organic Chemicals in the Soil Environment.* Volume I. Goring, C. A. I., Hamaker, J. W., Editors, pp. 49-143, Marcel Dekker Inc., New York, New York.

Hance, R. J. (1969) Decomposition of herbicides in soil. *J. Sci. Food Agric.* 20(3), 144-145.

Hance, R. J. (1969) Empirical relation between structure and the sorption of some herbicides by soils. *J. Agric. Food Chem.* 17(3), 667-668.

Hance, R. J. (1974) Soil organic matter and the adsorption and decomposition of the herbicides atrazine and linuron. *Soil Biol. Biochem.* 6, 39-42.

Hance, R. J. (1976) Adsorption of glyphosate by soils. *Pest. Sci.* 7, 363-366.

Hance, R. J. (1979) Effect of pH on the degradation of atrazine, dichlorprop, linuron and propyzamide in soil. *Pest. Sci.* 10, 83-86.

Hance, R. J., Haynes, R. A. (1981) The kinetics of linuron and metribuzin decomposition in soil using different laboratory systems. *Weed Res.* 21, 87-92.

Hansch, C., Leo, A. (1985) Medchem. Project Issue No. 26, Pomona College, Claremont, California.

Hansch, C., Leo, A. (1987) Medchem. Project Issue No. 28, Pomona College, Claremont, California.

Hansch, C., Leo, A., Hoekman, D. (1995) *Exploring QSAR. Hydrophobic, Electronic, and Steric Constants.* ACS Professional Reference Book, American Chemical Society, Washington, DC.

Harris, J. C. (1982) Chapter 8, Rate of aqueous photolysis. In: *Handbook on Chemical Property Estimation Methods. Environmental Behavior of Organic Compounds.* Lyman, W. J., Reehl, W. F., Rosenblatt, D. H., Editors, McGraw-Hill Inc., New York.

Harris, C. I., Warren, G. F. (1964) Detection of phosphorus fixation capacity in organic soil. *Weeds* 12, 120-126.

Hartley, D., Kidd, H. (1983) *The Agrochemicals Handbook.* Royal Society of Chemistry, Union Brothers Ltd., Old Working Surrey, England.

Hartley, D., Kidd, H. (1987) *The Agrochemicals Handbook.* 2nd Edition, Royal Society of Chemistry, Union Brothers Ltd., Old Working Surrey, England.

Hartley, G. S., Graham-Bryce, I. J. (1980) *Physical Principles of Pesticide Behaviour.* Academic Press, New York.

Harvey, Jr., J., Pease, H. L. (1973) Decomposition of methomyl in soil. *J. Agric. Food Chem.* 21, 784-786.

Harvey, R. G. (1974) Soil adsorption and volatility of dinitroaniline herbicides. *Weed Sci.* 22, 120-124.

Hedlund, R. T., Youngson, C. R. (1972) The rates of photodecomposition of picloram in aqueous systems. In: *Fate of Organic Pesticides in the Aquatic Environment.* Advances in Chemistry Series No. 111, Faust, S., Editor, p. 159, American Chemical Society, Washington, DC.

Heller, S. R., Scott, K., Bigwood, D. W. (1989) The need for data evaluation of physical and chemical properties of pesticides: The ARS pesticide properties database. *J. Chem. Inf. Comput. Sci.* 29, 159-162.

Helling, C. S. (1976) Dinitroaniline herbicides in soils. *J. Environ. Qual.* 5, 1-15.

Hemond, H. F., Fechner, E. J. (1994) *Chemical Fate and Transport in the Environment.* Academic Press, New York.

Herbicide Handbook (1974) *Herbicide Handbook.* 3rd Edition, Weed Science Society of America, Champaign, Illinois.

Herbicide Handbook (1978) *Herbicide Handbook.* 4th Edition, Weed Science Society of America, Champaign, Illinois.

Herbicide Handbook (1983) *Herbicide Handbook.* 5th Edition, Beste, C. E., Editor, Weed Science Society of America, Champaign, Illinois.

Herbicide Handbook (1989) *Herbicide Handbook.* 6th Edition, Weed Science Society of America, Champaign, Illinois.

Herrmann, M., Kotzias, D., Korte, F. (1985) Photochemical behaviour of chlorsulfuron in water and in adsorbed phase. *Chemosphere* 14, 3.

Hiltibran, R. C. (1972) Fate of diquat in the aquatic environment. Research Report No. 52, Water Resources Center, Univ. of Illinois, Urbana, Illinois.

Hine, J. Mookerjee, P. K. (1975) The intrinsic hydrophilic character of organic compounds. Correlations in terms of structural contributions. *J. Org. Chem.* 40, 292-298.

Hinman, M. L., Klaine, S. J. (1992) Uptake and translocation of selected organic pesticides by the rooted aquatic plant *Hydrilla verticillata* royale. *Environ. Sci. Technol.* 26, 609-613.

Hodge, H. C., Downs, W. L., Panner, B. S., Smith, D. W., Maynard, E. A. (1967) *Fed. Cosmet. Toxicol.* 5, 513-531.

Hodgman, C. R., Editor (1952) *Handbook of Chemistry and Physics.* 34th Edition, Chemical Rubber Publishing Co., Cleveland, Ohio.

Hodson, J., Williams, N. A. (1988) The estimation of the adsorption coefficient (K_{OC}) for soils by high performance liquid chromatography. *Chemosphere* 17, 67-77.

Hormann, W. D., Eberle, D. O. (1972) The aqueous solubility of 2-chloro-4-ethylamino-6-isopropylamino-1,3,5-triazine (atrazine) obtained by an improved analytical method. *Weeds Res.* 12, 199-202.

Hornsby, A. G., Wauchope, R. D., Herner, A. E. (1996) *Pesticide Properties in the Environment.* Springer-Verlag, New York.

Horowitz, M., Herzlinger, G. (1974) Soil conditions affecting the dissipation of diuron, flumeturon and propham from the soil surface. *Weed Res.* 14, 257.

Howard, P. H., Editor (1989) *Handbook of Environmental Fate and Exposure Data for Organic Chemicals. Volume I. Large Production and Primary Pollutants.* Lewis Publishers, Inc., Chelsea, Michigan.

Howard, P. H., Editor (1991) *Handbook of Environmental Fate and Exposure Data for Organic Chemicals. Volume III. Pesticides.* Lewis Publishers, Inc., Chelsea, Michigan.

Howard, P. H., Boethling, R. S., Jarvis, W. F., Meylan, W. M., Michalenko, E. M. (1991) *Handbook of Environmental Degradation Rates.* Lewis Publishers, Inc., Chelsea, Michigan.

Huber, G., Gemes, E. (1981) Decomposition of urea herbicide linuron (3-(3,4-dichlorophenyl)-1-methoxy-1-methylurea) in water of Lake Balaton. *Hungar. J. Ind. Chem.* 9, 113.

Hurle, R. J., Freed, V. H. (1972) Effect of electrolytes on the solubility of some 1,3,5-triazines and substituted ureas and their adsorption of soil. *Weeds Res.* 12, 1-10.

IARC (1983) Miscellaneous Pesticides. *IARC* 30, 235-244.

Iglesias-Jimenez, E., Sanchez-Martin, M. J., Sanchez-Camazano, M. (1996) Pesticide adsorption in a soilwater system in the presence of surfactants. *Chemosphere* 32(9), 1771-1782.

Ilchmann, A., Wienke, G., Meyer, T., Gmehling, J. (1993) Concurrent liquid/liquid chromatography - A reliable method for determination of partition coefficients. *Chem.-Ing.-Tech.* 65(1), 72-75.

Isensee, A. R. (1976) Variability of aquatic model ecosystem-derived data. *Inst. J. Environ. Studies* 10, 35.

Isensee, A. R. (1991) Chapter 5, Bioaccumulation and food chain accumulation. In: *Environmental Chemistry of Herbicides.* Vol. II, Grover, R., Cessna, A. J., Editors, pp. 188-198, CRC Press, Inc., Boca Raton, Florida.

Isnard, P., Lambert, S. (1988) Estimating bioconcentration factors from octanol-water partition coefficients and aqueous solubility. *Chemosphere* 17, 21-34.

Jacob, F., Neumann, S. (1983) Quantitative determination of mobility of xenobiotics of mobility and criteria of their phloem and xylem mobility. In: *International Union of Pure and Applied Chemistry. Pesticide Chemistry: Human Welfare and the Environment.* Vol. 4, Miyamoto, J., Kearney, P. C., Editors, Pergamon Press, Oxford, England.

Jafvert, C. T., Westall, J. C., Grieder, E., Schwarzenbach, R. P. (1990) Distribution of hydrophobic ionogenic organic compounds between octanol and water: organic acids. *Environ. Sci. Technol.* 24(12), 1795-1803.

Johnsen, Jr., T. N., Warskow, W. L. (1980) Picloram dissipation in a small southwestern stream. *Weed Sci.* 28, 612.

Johnson, W. W., Julin, A. M. (1974) A Review of the Literature on the Use of Diuron in Fisheries. Bureau of Sport Fish and Wildlife. PB 235446, U.S. Dept. of Interior, Columbia, Missouri.

Jones, T. W., Kemp, W. M., Stevenson, J. C., Means, J. C. (1982) Degradation of atrazine in estuarine water/sediments systems and soils. *J. Environ. Qual.* 11(4), 632-638.

Jorden, L. S., Farmer, W. J., Goodin, J. R., Day, B. E. (1970) Nonbiological detoxication of the *s*-triazine herbicides. *Res. Rev.* 32, 267-286.

Jury, W. A., Farmer, W. J., Spencer, W. F. (1984) Behavior assessment model for trace organics in soil: II. Chemical classification and parameter sensitivity. *J. Environ. Qual.* 13, 567-572.

Jury, W. A., Farmer, W. J., Spencer, W. F. (1984) Behavior assessment model for trace organics in soil: III. Application of screening model. *J. Environ. Qual.* 13, 573-579.

Jury, W. A., Grover, R., Spencer, W. F., Farmer, W. J. (1980) Modeling vapor losses of soil incoporated triallate. *Soil Sci. Soc. Am. J.* 44, 445-450.

Jury, W. A., Spencer, W. F., Farmer, W. J. (1982) Behavior assessment model for trace organic in soil: I. Model description. *J. Environ. Qual.* 12, 558-564.

Jury, W. A., Spencer, W. F., Farmer, W. J. (1983) Use of models for assessing relative volatility, mobility, and persistence of pesticides and other trace organics in soil systems. In: *Hazard Assessment of Chemicals: Recent Developments.* Vol. 2, Saxena, J., Editor, Academic Press, New York.

Jury, W. A., Focht, D. D., Farmer, W. J. (1987b) Evalution of pesticide groundwater pollution potential from standard indices of soil-chemical adsorption and biodegradation. *J. Environ. Qual.* 16, 422-428.

Jury, W. A., Ghodrati, M. (1989) Overview of organic chemical environmental fate and transport modeling approaches. In: *Reactions and Movement of Organic Chemicals in Soils.* SSSA Special Publication No. 22, Sawhney, B. L., Brown, K., Eds., pp.271-304, Soil Sci. Soc. of America and Soc. of Agronomy, Madison, Wisconsin.

Jury, W. A., Russo, D., Streile, G., El Abd, H. (1990) Evaluation of volatilization by organic chemicals residing below the soil surface. *Water Resources Res.* 26(1), 13-20.

Jury, W. A., Winer, A. M., Spencer, W. F., Focht, D. D. (1987a) Transport and transformations of organic chemicals in the soil-air-water ecosystem. *Rev. Environ. Contam. Toxicol.* 99, 120-164.

Kanazawa, J. (1981) Measurement of the bioconcentration factors of pesticides by fresh-water fish and their correlation with physicochemical properties of acute toxicities. *Pest. Sci.* 12, 417-424.

Kanazawa, J. (1989) Relationship between the soil sorption constants for pesticides and their physicochemical properties. *Environ. Toxicol. Chem.* 8, 477-484.

Karcher, W., Devillers, J. (1990) SAR and QSAR in environmental chemistry and toxicology: Scientific tool or wishful thinking? In: *Practical Applications of Quantitative-Structure Relationships (QSAR) in Environmental Chemistry and Toxicology, 1-12.* Karcher, W., Devillers, J., Editors, ECSC, EEC, EAEC, Brussels and Luxembourg.

Karickhoff, S. K. (1981) Semi-empirical estimation of sorption of hydrophobic pollutants on natural sediments and soils. *Chemosphere* 10, 833-846.

Karickhoff, S. K., Morris, K. R. (1985) Sorption dynamics of hydrophobic pollutants in sediment suspensions. *Environ. Toxicol. Chem.* 4, 469-479.

Kaufman, D. D. (1966) Microbile degradation of herbicide combinations: Amitrole and dalapon. *Weeds* 14, 130-134.

Kaufman, D. D. (1976) Soil degradation and persistence. In: A Literature Survey of Benchmark Pesticides. pp. 19-71. The George Washington University Medical Center, Dept. of Medical and Public Affairs, Science Communication Division, Washington DC.

Kaufman, D. D., Doyle, R. D. (1977) Biodegradation of organics. National Conf. Composting Municipal Residues Sludges. 75 pp.

Kearney, P. C., Kaufman, D. D. (1975) *Herbicides: Chemistry, Degradation and Mode of Action.* 2nd Edition, Vol. 2, Marcel Dekker, Inc., New York.

Kearney, P. C., Nash, R. G., Isensee, A. R. (1969) Persistence of pesticides in soil. In: *Chemical Fallout: Current Research on Persistence Pesticides.* Chapter 3, pp. 54-67, Miller, M. W., Berg, C. C., Editors, Charles C. Thomas, Springfield, Illinois.

Kearney, P. C., Plimmer, J. R., Wheeler, W. B., Konston, A. (1976) Persistence and metabolism of dinitroaniline herbicides in soils. *Pestic. Biochem. Physiol.* 6, 229-238.

Kenaga, E. E. (1974) Toxilogical and residue data useful in the environmental safety evaluation of dalapon. *Res. Rev.* 53, 109-151.

Kenaga, E. E. (1975) In: *Environmental Dynamics of Pesticides.* Haque, R., Freed, V.H., Editors, Plenum Press, New York. pp. 217-273.

Kenaga, E. E. (1980) Predicted bioconcentration factors and soil sorption coefficients of pesticides and other chemicals. *Ecotoxicol. Environ. Saf.* 4, 26-38.

Kenaga E. E., Goring, C. A. I. (1980) Relationship between water solubility, soil sorption, octanol-water partitioning, and concentration of chemicals in biota. In: *Aquatic Toxicology*. ASTM STP 707, Eaton, J. G., Parrish, P. R., Hendricks, A. C., Editors, pp. 78-115, American Soc. for Testing and Materials, Philadelphia, Pennsylvania.

Kerler, F., Schönherr, J. (1988) Accumulation of lipophilic chemicals across plant cuticles: prediction from octanol/water partition coefficients. *Arch. Environ. Contam. Toxicol.* 17, 1-6.

Khan, S. U. (1978) Kinetics of hydrolysis atrazine in aqueous fulvic acid solution. *Pest. Sci.* 9, 39-45.

Khan, S. U. (1980) *Pesticides in the Soil Environment, Fundamental Aspects of Pollution Control and Environmental Series 5*. Elsevier, Amsterdam, The Netherlands.

Khan, S. U., Gamble, D. S. (1983) Ultraviolet irradiation of an aqueous solution of prometryn in the presence of humic materials. *J. Agric. Food Chem.* 31, 1099-1108.

Khan, S. U., Schnitzer, M. (1978) UV irradiation of atrazine in aqueous fulvic acid solution. *J. Environ. Sci. Health* B13, 299-310.

Kirkland, K., Frayer, J. D. (1972) Degradation of several herbicides in soil previously treated with MCPA. *Weed Res.* 12, 90-95.

Klecka, G. M. (1985) Chapter 6, Biodegradation. In: *Environmental Exposure from Chemicals*. Volume I., Neely, W. B., Blau, G. E., Editors, pp. 110-155, CRC Press, Inc., Boca Raton, Florida.

Klein, W., Geyer, H., Freitag, D., Rohleder, H. (1984) Sensitivity of schemes for ecotoxicological hazard ranking of chemicals. *Chemosphere* 13(1), 203-211.

Klein, W., Kördel, W., Weiβ, M., Poremski, H. J. (1988) Updating of the OECD test guideline 107 "Partition coefficient n-octanol/water": OECD laboratory intercomparison test on HPLC method. *Chemosphere* 17, 361-386.

Kochany, J. (1992) Effects of carbonates on the aquatic photodegradation rate of bromoxynil (3,5-dibromo-4-hydroxybenzonitrile) herbicide. *Chemosphere* 24, 1119-1126.

Kollig, H. P., Editor (1993) Environmental Fate Constants for Organic Chemicals under consideration for EPA's Hazardous Waste Identification Projects. EPA/600/R-93/132. Environmental Research Laboratory, U.S. EPA, Athens, Georgia.

Kolpin, D. W., Kalkhoff, S. J. (1993) Atrazine degradation in a small stream in Iowa. *Environ. Sci. Technol.* 27, 134-139.

Kördel, W., Kotthoff, G., Müller, J. (1995a) HPLC-screening method for the determination of adsorption coefficient on soil - results of a ring-test. *Chemosphere* 30, 1373-1384.

Kördel, W., Stutte, J., Kotthoff, G. (1995b) HPLC-screening method to determine the adsorption coefficient in soil - comparison of immobilized humic acid and clay mineral phases for cyanopropyl columns. *Sci. Total Environ.* 162, 119-125.

Kotzias, D., Klein, W., Korte, F. (1974) Beiträge zur ökologischen chemie. LXXXIX. Reaktionen von buturon und monolinuron in fester und flüssiger phase bei UV-bestrahlung. *Chemosphere* 3, 161.

Kruger, E. L., Somasundaram, L., Kanwar, R. S. (1993) Persistence and degradation of [^{14}C]atrazine and [^{14}C]deisopropylatrazine as affected by soil depth and moisture conditions. *Environ. Toxicol. Chem.* 12, 1959-1967.

Kühne, R., Ebert, R.-U., Kleint, F., Schmidt, G., Schüürmann, G. (1995) Group contribution methods to estimate water solubility of organic chemicals. *Chemosphere* 30, 2061-2077.

Kulshrestha, G., Mukerjee, S. K. (1986) The photochemical decomposition of the herbicide isoproturon. *Pest. Sci.* 17, 489.

Kuwatsuka, S. (1972) Degradation of several herbicides in soils under different conditions. In: *Environmental Toxicology of Pesticides*. Matsumura, F., Boush, G. M., Misato, T., Editors, pp. 385, Academic Press, New York.

Kwok, E. S. C., Atkinson, R., Arey, J. (1992) Gas-phase atmospheric chemistry of selected thiocarbamates. *Environ. Sci. Technol.* 26(9), 1798-1807.

Lane, L. J., Morton, H. L., Wallace, D. E., Wilson, R. E., Martin, R. D. (1977) Nonpoint source pollutants to determine runoff source areas. *Hydrology and Water Resources in Arizona and the Southwest* 7, 89.

Lartiges, S. B., Garrigues, P. P. (1995) Degradation kinetics of organophosphorus and organonitrogen pesticides in diferent waters under various environmental conditions. *Environ. Sci. Technol.* 29, 1246-1254.

Lee, L. S., Bellin, C. A., Pinal, R., Rao, P. S. C. (1993) Cosolvent effects on sorption of organic acids by soils from mixed solvents. *Environ. Sci. Technol.* 27, 165-171.

Lee, R. F., Ryan, C. (1979) Microbial degradation of organochlorine compounds in esturine waters and sediments. In: *Proceedings of the Workshop: Microbial Degradation of Pollutants in Marine Environments.* EPA 600/9-79-012, Washington DC.

Lee, Y.-C., Myrdal, P. B., Yalkowsky, S. H. (1996) Aqueous functional group activity coefficients (AQUAFAC). 4: Applications to complex organic compounds. *Chemosphere* 33(11), 2129-2144.

Leistra, M., Smelt, J. H., Verlaat, J. G., Zandvoort, R. (1974) Measured and computed concentration patterns of propyzamide in field soils. *Weed Res.* 14, 87-95.

Leo, A., Hansch, C., Elkins, D. (1971) Partition coefficients and their uses. *Chem. Rev.* 71, 525-616.

Leonard, R. A., Bailey, G. W., Swank, Jr., R. R. (1976) Transport, detoxification, fate and effects of pesticides in soil and water environments in land application of waste materials. Soil Conservation Society of America, Ankeny, Iowa. 48 p.

Leopold, A. C., van Schaik, P., Neal, M. (1960) Molecular structure and herbicide adsorption. *Weeds* 8, 48.

Li, G. C., Felbeck, Jr., G. T. (1972) Atrazine hydrolysis as catalyzed by humic acids. *Soil Sci.* 114, 201-208.

Liu, D., Strachan, W. M. J., Thomson, K., Kwasniewska, K. (1981) Determination of the biodegradability of organic compounds. *Environ. Sci. Technol.* 15, 788-793.

Liu, J., Qian, C. (1995) Hydrophobic coefficients of s-trazine and phenylurea herbicides. *Chemosphere* 31(8), 3951-3959.

Lohninger, H. (1994) Estimation of soil partition coefficients of pesticides from their chemical structure. *Chemosphere* 29, 1611-1626.

Loos, M. A. (1975) Phenoxyalkanoic acids. In: *Herbicides, Chemistry, Degradation and Mode of Action.* Vol. I, 2nd Edition, Kearney, P. C., Kaufmann, D. D., Editors, Marcel Dekker, Inc., New York.

Lopez-Avila, V., Hirata, P., Kraska, S., Flanagan, M., Taylor, Jr., J. H., Hern, S. C. (1985) Determination of atrazine, lindane, pentachlorophenol, and diazinon in water and soil by isotope dilution gas chromatography/mass spectrometry. *Anal. Chem.* 57, 2797-2801.

Lopez-Avila, V., Hirata, P., Kraska, S., Flanagan, M., Taylor, J. H., Hern, S. C., Melanon, S., Pollard, J. P. (1989) Movement of selected pesticides and herbicides through columns of sandy loam. In: *Evaluation of Pesticides in Ground Water.* Garner, W. Y., Honeycutt, R. C., Editors, American Chemical Society, Washington DC.

Lord, K. A., Briggs, G. C., Neale, M. C., Manlove, R. (1980) Uptake of pesticides from water and soil by earthworms. *Pestic. Sci.* 11, 401-408.

Lund-Høie, K., Friestad, H. O. (1986) Photodegradation of the herbicide glyphosate in water. *Bull. Environ. Contam. Toxicol.* 36, 723-729.

Lyman, W. J. (1982) Chapter 2, Solubility in water. In: *Handbook on Chemical Property Estimation Methods. Environmental Behavior of Organic Compounds.* Lyman, W. J., Reehl, W. F., Rosenblatt, D. H., Editors, McGraw-Hill, New York, New York.

Lyman, W. J. (1985) Chapter 2, Estimation of physical properties. In: *Environmental Exposure from Chemicals.* Vol. 1, Neely, W. B., Blau, G. E., Editors, pp. 13-48, CRC Press, Inc., Boca Raton Florida.

Lyman, W. J., Reehl, W. F., Rosenblatt, D. H., Editors (1982) *Handbook on Chemical Property Estimation Methods. Environmental Behavior of Organic Compounds.* McGraw-Hill, Inc., New York, New York.

Lyman, W. J., Reehl, W. F., Rosenblatt, D. H., Editors (1990) *Handbook on Chemical Property Estimation Methods. Environmental Behavior of Organic Compounds.* 2nd printing, American Chemical Society, Washington DC.

Lynch, T. R., Johnson, H. E., Adams, W. J. (1982) The fate of atrazine and a hexachlorophenyl isomer in naturally-derived model stream ecosystems. *Environ. Toxicol. Chem.* 1, 179-192.

Mackay, D. (1982) Correlation of bioconcentration factors. *Environ. Sci. Technol.* 16, 274-278.

Mackay, D., Bobra, A., Shiu, W. Y., Yalkowsky, S. H. (1980) Relationships between aqueous solubility and octanol-water partition coefficient. *Chemosphere* 9, 701-711.

Mackay, D., Stiver, W. (1991) Chapter 8, Predicatability and environmental chemistry. In: *Environmental Chemistry of Herbicides.* Vol. II, Grover, R., Cessna, A. J., Editors, pp. 281-297, CRC Press, Inc., Boca Raton, Florida.

Madhun, Y. A., Freed, V. H. (1987) Degradation of herbicides bromacil, diuron, and chlortoluron in soil. *Chemosphere* 16, 1003-1011.

Madhun, Y. A., Freed, V. H., Young, J. L., Fang, S. C. (1986) Sorption of bromacil, chlortoluron, and diuron by soils. *Soil Sci. Soc. Am. J.* 50, 1467-1471.

MAFF (1992a) Food and Environment Protection Act 1985, Part III. Control of Pesticides Regulations 1986. Evaluation of Atrazine. Ministry of Agriculture Fisheries and Food, United Kingdom.

MAFF (1992b) Food and Environment Protection Act 1985, Part III. Control of Pesticides Regulations 1986. Evaluation of Simazine. Ministry of Agriculture Fisheries and Food, United Kingdom.

Magee, P. S. (1991) Complex factors in hydrocarbon/water, soil/water and fish/water partitioning. *Sci. Total Environ.* 109/110, 155-178.

Majewski, M. S., Capel, P. D. (1995) *Pesticides in the Atmosphere. Distribution, Trends, and Governing Factors.* Vol. 1 of the series *Pesticide in the Hydrologic System.* Gilliom, R. J., Editor, Ann Arbor Press Inc., Chelsea, Michigan.

Mansour, M., Feicht, E., Meallier, P. (1989) Improvement of the photostability of selected substances in aqueous medium. *Toxicol. Environ. Contam.* 20-21, 139-147.

Martin, H. (1961) *Guide to the Chemicals used in Crop Protection.* 4th Edition, Canadian Dept. of Agriculture Publication 1093, Ottawa, Ontario.

Martin, H. (1972) *Pesticide Manual.* 3rd Edition, British Crop Protection Council, United Kingdom.

Martin, H., Worthing, C. R., Editors (1977) *Pesticide Manual.* 5th Edition, British Crop Protection Council, United Kingdom.

Massini, P. (1961) Movement of 2,6-dichlorobenzonitrile in soils and in plants in relation to its physical properties. *Weed Res.* 1, 142-146.

McCall, P. J., Agin, G. L. (1985) Desorption kinetics of picloram as affected by residence time in the soil. *Environ. Toxicol. Chem.* 4, 37-44.

McCall, P. J., Swann, R. L., Laskowski, D. A., Unger, S. M., Vrona, S. A., Dishburger, H. J. (1980) Estimation of chemical mobility in soil from liquid chromatographic retention times. *Bull. Environ. Contam. Toxicol.* 24, 190-195.

McCall, P. J., Swann, R. L., Laskowski, D. A., Vrona, S. A., Unger, S. M., Dishburger, H. J. (1981) Prediction of chemical mobility in soil from sorption coefficients. In: *Aquatic Toxicology and Hazard Assessment.* ASTM STP 737, Branson, D. R., Dickson, K. L., Editors, pp. 49-58, Am. Soc. for Testing and Materials, Philadelphia, Pennsylvania.

McCall, P. J., Vrona, S. A., Kelley, S. S. (1981) Fate of uniformly carbon-14 ring labelled 2,4,5-trichlorophenoxyacetic acid and 2,4-dichlorophenoxyacetic acid. *J. Agric. Food Chem.* 29, 100-107.

McDuffie, B. (1981) Estimation of octanol/water partition coefficients for organic pollutants using reverse-phase HPLC. *Chemosphere* 10, 73-83.

Meakins, N. C., Bubb, J. M., Lester, J. N. (1994) The behaviour of the *s*-triazine herbicides, atrazine and simazine, during primary and secondary biological waste water treatment. *Chemosphere* 28, 1611-1622.

Means, J. C., Wijayaratne, R. D. (1982) Role of natural colloids in the transport of hydrophobic pollutants. *Science* 215, 968.

Means, J. C., Wijayaratne, R. D., Boynton, W. R. (1983) Fate and transport of selected herbicides in an estuarine environment. *Can. J. Fish Aquat. Sci.* 40 (Suppl. 2), 337-345.

Medchem (1988) Medcham Database, Release 3.54 of 1988. Daylight Chemical Information System Inc., California.

Melnikov, N. N. (1971) Chemistry of pesticides. *Res. Rev.* 36, 1-447.

The Merck Index (1983) *An Encyclopedia of Chemicals, Drugs and Biologicals.* 10th Edition, Windholz, M., Editor, Merck and Co., Inc., Rahway, New Jersey.

The Merck Index (1989) *An Encyclopedia of Chemicals, Drugs and Biologicals.* 11th Edition, Budavari, S., Editor, Merck and Co., Inc., Rahway, New Jersey.

Metcalf, R. L., Sanborn, J. R. (1975) Pesticides and environmental quality in Illinois. *Illinois Natural History Survey Bulletin* 31, 381-436.

Meylan, W., Howard, P. H. (1991) Bond contribution method for estimating Henry's law constants. *Environ. Toxicol. Chem.* 10, 1283-1293.

Meylan, W., Howard, P. H., Boethling, R. S. (1992) Molecular topology/fragment contribution method for predicting soil sorption coefficients. *Environ. Sci. Technol.* 26, 1560-1567.

Milne, G. W. A., Editor (1995) *CRC Handbook of Pesticides.* CRC Press, Inc., Boca Raton, Florida.

Mitsutake, K.-I., Iwasmura, H., Shimizu, R., Fujita, T. (1986) Quantitative structure-activity relationships of photosystem II inhibitors in chloroplasts and its link to herbicidal action. *J. Agric. Food Chem.* 34, 725-732.

Miyazaki, S., Sikka, H. C., Lynch, R. S. (1975) Metabolism of dichlobenil by microorganism in the aquatic environment. *J. Agric. Food Chem.* 23, 365.

Moilanen, K. W., Crosby, D. G. (1972) Photodecomposition of 3',4'-dichloropropionanilide (propanil). *J. Agric. Food Chem.* 20, 950-953.

Mongar, K., Miller, G. C. (1988) Vapor phase photolysis of trifluralin in an outdoor chamber. *Chemosphere* 17, 2183-2188.

Montgomery, J. H. (1993) *Agrochemicals Desk Reference. Environmental Data.* Lewis Publishers, Chelsea, Michigan.

Morrill, L. G., Mahilum, B. C., Mohiuddin, S. H. (1982) *Organic Compounds in Soils.* Ann Arbor Science Publishers, Inc., Ann Arbor, Michigan.

Mosier, A. R., Guenzi, W. D. (1973) Picloram photolytic decomposition. *J. Agric. Food Chem.* 21, 835-837.

Moyer, J. R. R., Hance, R. J., McKone, C. E. (1972) The effect of adsorption of adsorbents on the rate of degradation of herbicides incubated with soil. *Soil. Biol. Biochem.* 4, 307-311.

Muir, D. C. G. (1991) Chapter 1, Dissipation and transformation in water and sediment. In: *Environmental Chemistry of Herbicides*. Vol. II, Grover, R., Cessna, A. J., Editors, pp. 1-88, CRC Press, Inc., Boca Raton, Florida.

Muir, D. C. G., Grift, N. P. (1982) Fate of fluoridone in sediment and water in laboratory and field experiments. *J. Agric. Food Chem.* 30, 238.

Muir, D. C. G., Grift, N. P., Blouw, A. P., Lockhart, W. L. (1980) Persistence of fluridone in small ponds. *J. Environ. Qual.* 9, 151-156.

Muir, D. C. G., Yarechewski, A. L. (1982) Degradation of terbutryn in sediments and water under various redox conditions. *J. Environ. Sci. Health* B17, 363.

Müller, J. F., Hawker, D. W., Connell, D. W. (1994) Calculation of bioconcentration factors of persistent hydrophobic compounds in the air/vegetation system. *Chemosphere* 20(4), 623-640.

Muszkat, L., Halmann, M., Raucher, D., Bir, L. (1992) Solar photodegradation of xenobiotic contaminants in polluted well water. *J. Photochem. Photobiol. A: Chem.* 65, 409-417.

Nash, R. G. (1980) Dissipation rate of pesticides from soils. In: *CREAMS*. Vol. 3, Niesel, W. G., Editor, pp. 560-594, U.S. Dept. of Agriculture, Washington DC.

Nash, R. G. (1983) Distribution of butylate, heptachlor, lindane, and dieldrin emulsifiable concentrated and butyrated microencapsulated formulations in microagroecosystem chambers. *J. Agric. Food Chem.* 31, 1195.

Nash, R. G. (1983) Comparative volatilization and dissipation rates of several pesticides from soil. *J. Agric. Food Chem.* 31, 210.

Nash, R. G. (1983) Determining environmental fate of pesticides with microagroecosystems. *Res. Rev.* 85, 199-215.

Nash, R. G. (1988) Chapter 5, Dissipation from soil. In: *Environmental Chemistry of Herbicides*. Vol. I, Grover, R., Ed., pp. 131-169, CRC Press, Inc., Boca Raton, Florida.

Nash, R. G. (1989) Models for estimating pesticide dissipation from soil and vapor decline in air. *Chemosphere* 18(11/12), 2375-2381.

Nash, R. G., Beall, M. L. (1980) Distribution of silvex, 2,4-D, and TCDD applied to turf in chambers and field plots. *J. Agric. Food Chem.* 28, 614.

Neary, D. G., Bush, P. B., Michael, J. L. (1993) Fate, dissipation and environmental effects of pesticides in southern forests: A review of a decade of research progress. *Environ. Toxicol. Chem.* 12, 411-428.

Neely, W. B., Blau, G. E. (1985) Chapter 1, Introduction to environmental exposure from chemicals. In: *Environmental Exposure From Chemicals*. Volume I, Neely, B. W., Blau, G. E., Editors, pp. 1-11. CRC Press, Inc., Boca Raton, Florida.

Nelson, N. H., Faust, S. D. (1969) Acidic dissociation constants of selected aquatic herbicides. *Environ. Sci. Technol.* 3, 1186-1188.

Nesbitt, H. J., Watson, J. R. (1980a) Degradation of the herbicide 2,4-D in river water-I. Description of study area and survey of rate determining factors. *Water Res.* 14, 1683-1688.

Nesbitt, H. J., Watson, J. R. (1980b) Degradation of the herbicide 2,4-D in river water-II, the role of suspended sediment. Nutrients and water temperature. *Water Res.* 14, 1689-1694.

Newsom, H. C., Woods, W. G. (1973) Photolysis of the herbicide dinitramine (N^3, N^3-diethyl-2,4-dinitro-6-trifluromethyl-*m*-phenylenediamine). *J. Agric. Food Chem.* 21, 598.

Nex, R. W., Swezey, A. W. (1954) Some chemical and physical properties of weed killers. *Weeds* 3, 241-253.

Nilles, G. P., Zabik, M. J. (1974) Photochemistry of bioactive compounds. Multiphase photodegradation of basalin. *J. Agric. Food Chem.* 22, 684-688.

Nkedi-Kizza, P., Rao, P. S. C., Johnson, J. W. (1983) Adsorption of diuron and 2,4,5-T on soil particle-size separates. *J. Environ. Qual.* 12, 195-197.

Nkedi-Kizza, P., Rao, P. S. C., Hornsby, A. G. (1985) Influence of organic cosolvents on sorption of hydrophobic organic chemicals by soils. *Environ. Sci. Technol.* 19, 975-979.

Nomura, N. S., Hilton, H. W. (1977) The adsorption and degradation of glyphosate in five Hawaiian sugarcane soils. *Weed Res.* 17, 113-121.

Novick, N. J., Alexander, M. (1985) Cometabolism of low concentration of propachlor, alachlor and cycloate in sewage and lake water. *Appl. Environ. Microbiol.* 49, 737.

Nyholm, N., Jacobsen, B. N., Pedersen, B. M., Poulsen, O., Damboroj, A., Schultz, B. (1992) Removal of organic micro pollutants at ppb levels in laboratory activated sludge reactors under various operating conditions: Biodegradation. *Water Res.* 26, 339-353.

Nzengung, V. A., Voudrias, E. A., Nekdi-Kizza, P., Wampler, J. M., Weaver, C. E. (1996) Organic cosolvent effects on sorption equilibrium of hydrophobic organic chemicals by organoclays. *Environ. Sci. Technol.* 30, 89-96.

OECD (1981) *OECD Guidelines for the Testing of Chemicals*. Paris.

Oehler, D. D., Ivie, G. W. (1980) Metabolic fate of the herbicide dicamba in a lactating cow. *J. Agric. Food Chem.* 28, 685-689.

Ohyama, H., Kawatsuka, S. (1978) Degradation of bifenox, a diphenylether herbicide, methyl-5-(2,4-dichlorophenoxy)-2-nitrobenzoate, in soils. *J. Pest. Sci.* 3, 401.

Pacakova, V., Stulik, K., Prihoda, M. (1988) High performance liquid chromatography of s-triazines and their degradation products using ultraviolet photometric and amperometric detection. *J. Chromatogr.* 442, 147-156.

Pait, A. S., De Souza, A. E., Farrow, D. R. G. (1992) *Agricultural Pesticide Use in Coastal Areas: A National Summary*, National Oceanic and Atmospheric Administration, Rockville, Maryland.

Pape, B. E., Zabik, M. J. (1970) Photochemistry of bioactive compounds. Photochemistry of selected 2-chloro- and 2-methylthio-4,6-di-(alkylamino)-s-triazine herbicides. *J. Agric. Food Chem.* 18, 202-207.

Paris, D. F., Steen, W. C., Baughman, G. L. (1978) Prediction of microbial transformation of pesticides in natural waters. (unpublished), presented before the American Chemical Society, Division of Pesticide Chemistry, Anaheim, California, Environmental Research Laboratory, U.S. Environmental Protection Agency, Athens, Georgia.

Paris, D. F., Steen, W. C., Baughman, G. L., Barnett, Je., J. T. (1981) Second-order model to predict microbial degradation of organic compounds in natural waters. *Appl. Environ. Microbiol.* 41, 603-609.

Parr, J. F., Smith, S. (1973) Degradation of trifluralin under laboratory conditions and soil anaerobiosis. *Soil Sci.* 115, 55-63.

Patchett, G. G., Batchelder, G. H., Menn, J. J. (1964) In: *Analytical Methods for Pesticides and Plant Growth Regulators*. Vol. 4, Zweig, G., Editor, Academic Press, New York. pp. 117-123.

Patchett, G. G., Gray, R. A., Reed, A., Hyzak, D. L. (1983) Thiolcarbamate sulfoxides protected against dry soil deactivation. U.S. Patent 4,389,237.

Patil, G. S. (1994) Prediction of aqueous solubility and octanol-water partition coefficient for pesticides based on their molecular structure. *J. Hazard. Materials* 36, 35-43.

Peck, D. E., Corwin, D. L., Farmer, W. J. (1980) Adsorption-desorption of diuron by fresh water sediments. *J. Environ. Qual.* 9, 101-106.

Pillai, V. N. R. (1977) Role of singlet oxygen in the environmental degradation of chlorthiamid and dichlobenil. *Chemosphere* 6, 777.

Pinsuwan, S., Li, A., Yalkowsky, S. H. (1995) Correlation of octanol/water solubility ratios and partition coefficients. *J. Chem. Eng. Data* 40, 623-626.

Plato, C. (1972) Differential scanning calorimetry as a general method for the determining purity and heat of fusion of high-purity organic chemicals. Application to 64 compounds. *Anal. Chem.* 44(8), 1531-1534.

Plato, C., Glasgow, A. R., Jr. (1972) Differential scanning calorimetry as a general method for the determining purity and heat of fusion of high-purity organic chemicals. Application to 95 compounds. *Anal. Chem.* 41(2), 330-336.

Plimmer. J. R. (1970) The photochemistry of halogenated herbicides. *Res. Rev.* 33, 47.

Plimmer, J. R., Hummer, B. E. (1969) Photolysis of amiben (3-amino-2,5-dichlorobenzoic acid) and its methyl ester. *J. Agric. Food Chem.* 17, 83-85.

Plimmer, J. R., Kearney, P. C., Chisaka, H. (1970) Microbial conversion of ^{14}C labelled propanil in Japanese soils. *Weed Sci. Soc. Am.*, Abst. No. 167.

Plimmer, J. R., Klingebiel, U. I. (1974) Photochemistry of N-*sec*-butyl-4-*tert*-butyl-2,6-dinitroaniline. *J. Agric. Food Chem.* 22, 689-693.

Probst, G. W., Golab, T., Wright, W. L. (1975) Dinitroanilines. In: *Herbicides, Chemistry, Degradation and Mode of Action.* Vol. 1, Kearney, P. C., Kaufman, D. D., Editors, Marcel Dekker, Inc., New York.

Probst, G. W., Golab, T., Herberg, R. J., Holser, F. J., Parka, S. J., Van der Schans, C., Tepe, J. B. (1967) Fate of trifluralin in soils and plants. *J. Agric. Food Chem.* 15, 592-599.

Que Hee, S., Sutherland, R. G. (1981) *The Phenoxyalkanoic Herbicides. Volume 1: Chemistry, Analysis, and Environmental Pollution.* CRC Press, Inc., Boca Raton, Florida.

Quellette, R. P., King, J. A. (1977) *Chemical Week. Pesticide Register.* McGraw-Hill, Inc., New York.

Rao, P. S. C., Davidson, J. M. (1979) Adsorption and movement of selected pesticides at high concentrations in soils. *Water Res.* 13, 375-380.

Rao, P. S. C., Davidson, J. M. (1980) Estimation of pesticide retention and transformation parameters required in nonpoint source pollution models. In: *Environmental Impact of Nonpoint Source Pollution.* Overcash, M. R., Davidson, J. M., Editors, Ann Arbor Science Publishers, Inc., Ann Arbor, Michigan.

Rao, P. S. C., Davidson, J. M. (1982) Retention and Transformation of Selected Pesticides and Phosphorus in Soil Water System: A Critical Review. U.S. EPA, EPA-600/3-82-060.

Reinert, K. H. (1989) Environmental behavior of aquatic herbicides in sediments. In: *Reactions and Movement of Organic Chemicals in Soils.* SSSA Special Publ. No.22, pp. 335-348, Soil Sci. Soc. of Ameroca and Soc. of Argonomy, Madison, Wisconsin.

Reinert, K. H., Rogers, J. H. (1984) Influence of sediment types on the sorption of endothall. *Bull. Environ. Contam. Toxicol.* 32, 557-564.

Reinert, K. H., Rogers, J. H. (1987) Fate and persistence of aquatic herbicides. *Rev. Environ. Contam. Toxicol.* 98, 69-91.

Reinhold, K. A. et al. (1979) Adsorption of Energy Related Organic Pollutants. U.S. EPA, EPA-600/3-79-086. p. 103.

Rejto, M., Saltzman, S., Acher, A. J. (1984) Photodecomposition of propachlor. *J. Agric. Food Chem.* 32, 226.

Rejto, M., Saltzman, S., Acher, A. J., Muszkat, L. (1983) Identification of sensitized photooxidation products of *s*-triazine herbicides in water. *J. Agric. Food Chem.* 31, 138-142.

Rekker, R. F. (1977) *The Hydrophobic Constants; Its Derivation and Application; A Means of Characterizing Membrane Systems.* Nauta, W. T., Rekker, R. F., Editors, Elsevier Scientific Publishing Company, New York, New York.

Ribo, J. M. (1988) The octanol/water partition coefficient of the herbicide chlorsulfuron as a function of pH. *Chemosphere* 17, 709-715.

Richards, R. P., Baker, D. B. (1993) Pesticide concentration patterns in agricultural drainage networks in the Lake Erie basin. *Environ. Toxicol. Chem.* 12, 13-26.

Riederer, M. (1990) Estimating partitioning and transport of organic chemicals in the foliage/atmosphere system: Discussion of a fugacity-based model. *Environ. Sci. Technol.* 24, 829-837.

Roberts, T. R. (1974) The fate of WL-6361 in a static aquatic system. *Proc. European Weed Res. Council*, 4th Int'l Symp. Aquat. Weeds, Vienna, Austria. pp. 232.

Rosen, J. D., Strusz, R. F., Still, C. C. (1969) Photolysis of phenylurea herbicides. *J. Agric. Food Chem.* 17, 206-207.

Rueppel, M. L., Brightwell, B. B., Schaefer, J., Marvel, J. T. (1977) Metabolism and degradation of glyphosate in soil and water. *J. Agric. Food Chem.* 25, 517-528.

Ruzo, L. O., Casida, J. E. (1985) Photochemistry of thiocarbamate herbicides: Oxidative and free radical processes of thiobencarb and diallate. *J. Agric. Food Chem.* 33, 272.

Ruzo, L. O., Lee, J. K., Zabik, M. J. (1980) Solution-phase photodecomposition of several substituted diphenyl ether herbicides. *J. Agric. Food Chem.* 28, 1289.

Ruzo, L. O., Zabik, M. J., Schuetz, R. D. (1973) Photochemistry of bioactive compounds. Kinetics of selected s-triazines in solution. *J. Agric. Food Chem.* 21, 1047-1049.

Sabljic, A. (1984) Prediction of the nature and strength of soil sorption of organic pollutants by molecular topology. *J. Agric. Food Chem.* 32, 243-246.

Sabljic, A. (1987) On the prediction of soil sorption coefficients of organic pollutants from molecular structure: Application of molecular topology model. *Environ. Sci. Technol.* 21, 358-366.

Sagebiel, J. C., Seiber, J. N., Woodrow, J. E. (1992) Comparison of headspace and gas-stripping methods for determining the Henry's law constant (H) for organic compounds of low to intermediate H. *Chemosphere* 25(12), 1763-1768.

Saito, S., Tanoue, A., Matsuo, M. (1992) Applicability of the i/o-characters to a quantitative description of bioconcentration of organic chemicals in fish. *Chemosphere* 24(1), 81-87.

Saito, S., Koyasu, J., Yoshida, K., Shigeoka, T., Koike, S. (1993) Cytotoxicity of 109 chemicals to goldfish GFS cells and relationships with 1-octanol/water partition coefficients. *Chemosphere* 26(5), 1015-1028.

Sanborn, J. R., Francis, B. M., Metcalf, R. L. (1977) The Degradation of Selected Pesticides in Soil: A Review of the Published Literature. Prepared for the U.S. Environmental Protection Agency, Cincinnati, Ohio. Publication No. U.S. EPA-600/9-77-022.

Sanchez-Camazano, M., Arienzo, M., Sanchez-Martin, M. J., Crisanto, T. (1995) Effect of different surfactants on the mobility of selected non-ionic pesticides in soil. *Chemosphere* 31(8), 3793-3801.

Sanders, D. G., Mosier, J. W. (1983) Photolysis of the aquatic herbicide fluridone in aqueous solution. *J. Agric. Food Chem.* 31, 237-241.

Sangster, J. (1993) LOGKOW Data Bank. Sangster Research Laboratory, Montreal, Canada.

Sattar, M. A., Paasivirta, J. (1980) Fate of the herbicide MCPA in soil. Analysis of the residues of MCPA by an internal standard method. *Chemosphere* 9, 365-375.

Savage, K. E. (1978) Persistence of several dinitroaniline herbicides as affected by soil moisture. *Weed Sci.* 26, 465.

Savage, K. E., Jordan, T. N. (1980) Persistence of three dinitroaniline herbicides on the soil surface. *Weed Sci.* 28, 105-110.

Savage, K. E., Wauchope, R. D. (1974) Flumeturon adsorption-desorption equilibria in soil. *Weed Sci.* 22, 106-110.

Schliebe, K. A., Burnside, O. C., Lavy, T. L. (1965) Dissipation of amiben. *Weeds* 13, 321.

Schmidt, G. (1975) Von problematik der verhaltensprüfung von pflanzenschtzmitteln im oberflächenwasser, schriftenr. des Ver Wasser-, Boden-, Lufthyg. *Berlin-Dahlem* 46, 155.

Schnoor, J. L., Editor (1992) *Fate of Pesticides and Chemicals in the Environment*. John Wiley & Sons, Inc., New York.

Schnoor, J. L. (1992) Chemical fate and transport in the environment. In: *Fate of Pesticides and Chemicals in the Environment*. Schnoor, J. L., Editor, John Wiley & Sons, Inc., New York. pp. 1-24.

Schnoor, J. L., McAvoy, D. C. (1981) Pesticide transport and bioconcentration model. *J. Environ. Eng. Div. (Am. Soc. Civ. Eng.)* 107(EE6), 1229-1246.

Schnoor, J. L., Rao, N. B., Cartwright, K. J., Noll, R. M. (1982) In: *Modeling the Fate of Chemicals in the Aquatic Environment*. Dickson, K. L., Maki, A. W., Cairns, Jr., J., Editors, Ann Arbor Science, Ann Arbor, Michigan. pp. 145.

Schwartz, H. J. (1967) Microbial degradation of pesticides in aqueous solutions. *J. Water Pollut. Control Fed.* 39, 1701.

Scifres, C. J., Allen, T. J., Leinweber, C. L., Pearson, K. H. (1973) Dissipation and phototoxicity of dicamba residues in water. *J. Environ. Qual.* 2, 306.

Scow, K. M. (1982) Chapter 9, Rate of biodegradation. In: *Handbook on Chemical Property Estimation Methods. Environmental Behavior of Organic Compounds*. Lyman, W.J., Reehl, W.F., Rosenblatt, D.H., Editors, McGraw-Hill, New York.

Seiber, J. N., McChesney, M. M. (1987) Measurement and computer model simulation of the volatilization flux of molinate and methyl parathion from a flooded rice field. Final Report to the Department of Food and Agriculture, Sacramento, California.

Seiber, J. N., McChesney, M. M., Sanders, P. F., Woodrow, J. E. (1986) Models for assessing the volatilization of herbicides applied to flooded rice fields. *Chemosphere* 15, 127-138.

Seiber, J. N., McChesney, M. M., Sanders, P. F., Woodrow, J. E. (1989) Air borne residues resulting from use of methyl parathion, molinate and thiobencarb on rice in the Sacramento Valley, California. *Environ. Toxicol. Chem.* 8, 577-588.

Sheets, T. J. (1963) Photochemical alteration and inactivation of amiben. *Weeds* 11, 186.

Shirmohammadi, A., Magette, W. L., Brinsfield, R. B., Staver, K. (1989) Ground water loading of pesticides in the Atlantic Coastal Plain. *Ground Water Monitor Rev.* 9, 141-148.

Shiu, W. Y., Ma, K. C., Mackay, D. (1990) Solubilities of pesticides in water. Part 1, Environmental physical chemistry and Part 2, Data compilation. *Rev. Environ. Contam. Toxicol.* 115, 1-187.

Sicbaldi, F., Finizio, A. (1993) K_{ow} estimation by combination of RP-HPLC and molecular connectivity indexes for a heterogeneous set of pesticide. In: *Proceedings IX Symposium Pesticide Chemistry, Mobility and Degradation of Xenobiotics*. Oct. 1993, Piacenza, Italy.

Siebers, J., Gottschild, D., Nolting, H.-G. (1994) Pesticides in precipitation in Northern Germany. *Chemosphere* 28, 1559-1570.

Simsiman, G. V., Chesters, G. (1976) Persistence of diquat in the aquatic environment. *Water Res.* 10, 105.

Skurlatov, Y. I., Zepp, R. G., Baughman, G. L. (1983) Photolysis rates of (2,4,5-trichlorophenoxy)acetic acid and 4-amino-3,5,6-trichloropicolinic acid in natural waters. *J. Agric. Food Chem.* 31, 1065-1071.

Slade, P., Smith, A. E. (1967) Photochemical degradation of diquat. *Nature* 213, 919.

Smith, A. E. (1969) Factors affecting the loss of tri-allate from soils. *Weeds Res.* 9, 306.

Smith, A. E. (1970) Degradation, adsorption, and volatility of di-allate and tri-allate in prairie soils. *Weed Res.* 10, 331-339.

Smith, A. E. (1973) Degradation of dicamba in prairie soils. *Weed Res.* 13, 373-378.

Smith, A. E. (1978) Relative persistence of di- and tri-chlorophenoxyalkanoic acid herbicides in Saskatchewan soils. *Weed Res.* 18, 275-279.

Smith, A. E. (1979) Soil persistence experiments with (^{14}C) 2,4-D in herbicidal mixtures and field persistence studies with tri-allate and triflualin both singly and combined. *Weed Res.* 19, 165-170.

Smith, A. E., Biggs, G. G. (1978) The fate of the herbicide chlortoluron and its possible degradation products in soils. *Weed Res.* 18, 1-7.

Smith, A. E., Grove, J. (1969) Photochemical degradation of diquat in dilute aqueous solution and on silica gel. *J. Agric. Food Chem.* 17, 609-613.

Smith, A. E., Hayden, B. J. (1981) Relative persistence of MCPA and mecoprop in Saskatchewan soils and the identification of MCPA in MCPB-treated soil. *Weeds Sci.* 21, 179-183.

Soderquist, C. J., Bowers, J. B., Crosby, D. G. (1977) Dissipation of molinate in a rice field. *J. Agric. Food Chem.* 25, 940.

Soderquist, C. J., Crosby, D. G. (1975) Dissipation of 4-chloro-2-methyl-phneoxyacetic acid (MCPA) in a rice field. *Pest. Sci.* 6, 17.

Somasundaram, L., Coats, J. R., Racke, K. D. (1991) Mobility of pesticides and their hydrolysis metabolites in soil. *Environ. Toxicol. Chem.* 10, 185-194.

Spacie, A., Hamelink, J. L. (1979) Dynamics of trifluralin accumulation in river fishes. *Environ. Sci. Technol.* 13(7), 817-822.

Spain, J. C., Van Veld, P. A. (1983) Adaptation of natural microbial communities to degradation of xenobiotic compounds: Effects of concentration, exposure time, inoculum and chemical structure. *Appl. Environ. Microbiol.* 45, 428.

Spencer, E. Y., Editor (1973) *Guide to the Chemicals Used in Crop Protection.* 6th Edition, Research Branch Agriculture Canada, Ontario, Canada.

Spencer, E. Y., Editor (1981) *Guide to the Chemicals Used in Crop Protection.* 7th Edition, Research Branch Agriculture Canada, Ontario, Canada.

Spencer, E. Y., Editor (1982) *Guide to the Chemicals Used in Crop Protection.* 8th Edition, Research Branch Agriculture Canada, Ontario, Canada.

Spencer, W. F. (1976) Vapor pressure and vapor losses of benchmark pesticides. In: *A Literature Survey of Benchmark Pesticides.* pp. 72-165. The George Washington University Medical Center, Dept. of Medical and Public Affairs, Science Communication Division, Washington, DC.

Spencer, W. F., Cliath, M. M. (1990) Chapter 1, Movement of pesticides from soil to the atmosphere. In: *Long Range Transport of Pesticides.* Kurtz, D.A., Editor, Lewis Publishers, Inc., Ann Arbor, Michigan.

Spencer, W. F., Cliath, M. M. (1974) Factors affecting vapor loss of trifluralin from soil. *J. Agric. Food Chem.* 22, 987-991.

Spencer, W. F., Cliath, M. M. (1983) Measurement of pesticide vapor pressures. *Res. Rev.* 85, 57-71.

Spencer, W. F., Cliath, M. M., Jury, W. A., Zhang, L. Z. (1988) Volatilization of organic chemicals from soil as related to their Henry's law constants. *J. Environ. Qual.* 17(3), 504-509.

Spencer, W. F., Farmer, W. J., Cliath, M. M. (1973) Pesticide volatilization. *Res. Rev.* 49, 1-47.

Sprankle, P., Meggitt, W. F., Penner, D. (1975) Adsorption, mobility and microbial degradation of glyphosate in the soil. *Weed Sci.* 23, 229-234.

Spurlock, F. C. (1992) Thermodynamics of organic chemical partition in soils. Ph.D. Thesis, University of California at Davis, California.

Spurlock, F. C., Biggar, J. W. (1994a) Thermodynamics of organic chemical partition in soils. 2. Nonlinear partition of substituted phenylureas from aqueous solution. *Environ. Sci. Technol.* 28, 996-1002.

Spurlock, F. C., Biggar, J. W. (1994b) Thermodynamics of organic chemical partition in soils. 3. Nonlinear partition from water-miscible cosolvent solutions. *Environ. Sci. Technol.* 28, 1003-1009.

Steen, W. C., Collette, T. W. (1989) Microbial degradation of seven amides by suspended bacterial populations. *Appl. Environ. Microbiol.* 55, 2545-2549.

Steen, W. C., Paris, D. F., Baughman, G. L. (1979) Effects of sediment sorption on microbial degradation of toxic substances. in *Proc. 177th National Meeting of American Chemical Society*, April 1979, Honolulu, Hawaii.

Steen, W. C., Paris, D. F., Baughman, G. L. (1982) Effects of sediment sorption on microbial degradation of toxic substances. In: *Contaminants and Sediments: Fate and Transport, Case Studies, Modeling, Toxicity.* Vol. 1, Baker, R. A., Editor, p. 477, Ann Arbor Science, Ann Arbor, Michigan.

Stepp, T. A., Camper, N. D., Paynter, M. J. B. (1985) Anaerobic microbial degradation of selected 3,4-dihalogenated aromatic compounds. *Pest. Biochem. Physiol.* 23, 256.

Stewart, D. K. R., Gaul, S. O. (1977) Persistence of 2,4-D dichlorophenoxyacetic acid, 2,4,5-T, and dicamba in a Dykeland soil. *Bull. Environ. Contam. Toxicol.* 18, 210.

Sukop, M., Cogger, C. G. (1992) Adsorption of carbofuran, metalaxyl, and simazine: KOC evaluation and relation to soil transport. *J. Environ. Sci. Health* B27(5), 565-590.

Suntio, L. R., Shiu, W. Y., Mackay, D., Seiber, J. N., Glotfelty, D. (1988) Critical review of Henry's law constants. *Rev. Environ. Contam. Toxicol.* 103, 1-59.

Swann, R. L., Laskowski, D. A., McCall, P. J., Vanderkuy, K., Dishburger, H. J. (1983) A rapid method for the estimation of the environmental parameters octanol/water partition coefficient, soil sorption constant, water to air ratio, and water solubility. *Residue Rev.* 85, 17-28.

Swezey, A. W., Nex, R. W. (1961) Some physical and chemical properties of weed killers. *Suppl. I Weeds* 9, 209.

Tanaka, F. S., Wien, R. G., Mansager, E. R. (1979) Effects of nonionic surfactants on the photochemistry of 3-(4-chlorophenyl)-1,1-dimethylurea in aqueous solution. *J. Agric. Food Chem.* 27, 774-779.

Tanaka, F. S., Wien, R. G., Mansager, E. R. (1981) Survey for surfactant effects on the photodegradation of herbicides in aqueous media. *J. Agric. Food Chem.* 29, 227-230.

Tanaka, F. S., Wien, R. G., Mansager, E. R. (1982) Photolytic demethylation of monuron and demethylmonuron in aqueous solution. *Pest. Sci.* 13, 287.

Tanaka, F. S., Wien, R. G., Zaylskie, G. (1977) Photolysis of 3-(4-chlorophenyl)-1,1-dimethylurea in dilute aqueous solution. *J. Agric. Food Chem.* 25, 1068.

Taylor, A. W., Glotfelty, D. E. (1988) Chapter 4, Evaporation from soils and crops. In: *Environmental Chemistry of Herbicides.* Vol. I, Grover, R., Editor, pp. 89-130, CRC Press, Inc., Boca Raton, Florida.

Taylor, A. W., Spencer, W. F. (1990) Volatilization and vapor transport processes. In: *Pesticides in the Soil Environment: Processes, Impacts, and Modeling.* Cheng, H.H., Editor, Soil Science Society of America, Inc., Madison, Wisconsin.

Thomas, R. G. (1982) Chapter 15: Volatilization from water and Chapter 16: Volatilization from soil. In: *Handbook on Chemical Property Estimation Methods, Environmental Behavior of Organic Compounds.* Lyman, W. J., Reehl, W. F., Rosenblatt, D. H., Editors, McGraw-Hill, Inc., New York.

Thomas, V. M., Holt, C. L. (1980) The degradation of [^{14}C]-molinate in soil under flooded and non-flooded conditions. *J. Environ. Sci. Health* B15, 475.

Thor (1989) from *MedChem Release 3.54*, Daylight Chemical Information Systems Inc., Claremont, California.

Tomlin, C. (1994) *The Pesticide Manual. (A World Compendium)*. 10th Edition. The British Crop Protection Council, Surrey, England and The Royal Society of Chemistry, Cambridge, England.

Traub-Eberhard, U., Kördel, W., Klein, W. (1994) Pesticide movement into subsurface drains on a loamy silt soil. *Chemosphere* 28, 273-284.

Travis, C. C., Arms, A. D. (1988) Bioconcentration of organics in beef, milk, and vegetation. *Environ. Sci. Technol.* 22, 271-274.

Trevisan, M., Montepiani, C., Ghebbioni, C., Del Re, A. A. M. (1991) Evaluation of potential hazard of propanil to groundwater. *Chemosphere* 22, 637-643.

Tucker, C. S., Boyd, C. E. (1981) Relationships between pond sediments and simazine loss from waters of laboratory systems. *J. Aquat. Plant Manag.* 19, 55.

U. Oklahoma (1986) Univ. of Oklahoma Data Base, 1986.

Urosol, N. J., Hance, R. J. (1974) The effect of temperature and water contents on the rate of decomposition of the herbicide linuron. *Weed Sci.* 16, 19-21.

Ursin, C. (1985) Degradation of organic chemicals at trace levels in sea water and marine sediment. The effect of concentration on the initial fractional turnover rate. *Chemosphere* 14, 1539.

USDA (1989) Final environment impact statement, vegetation management in the Piedmont and Coastal Plain. Southern Region Management Bulletin R8-MB-23. U.S. Dept. of Agriculture, Forest Service, Atlanta, Georgia.

USEPA (1975) Substitute Chemical Program - Initial Scientific and Minieconomic Review of Bromocil. U.S. EPA-540/1-75-006. U.S. Government Printing Office, Washington DC.

USEPA (1988) Graphical Exposure Modeling System. (GEMS), CLOGP3, U.S. Environmental Protection Agency.

Veeh, R. H., Inskeep, W. P., Camper, A. K. (1996) Soil depth and temperature effects on microbial degradation of 2,4-D. *J. Environ. Qual.* 25, 5-12.

Veith, G. D., Defoe, D. L., Bergstedt, B. V. (1979) Measuring and estimating the bioconcentration factor of chemicals in fish. *J. Fish Res. Board Can.* 26, 1040-1048.

Veith, G. D., Kosian, P. (1982) In: *Physical Behavior of PCBs in the Great Lakes*. Chapter 15, pp. 269-282, Ann Arbor Science, Michigan.

Veith, G. D., Macek, K. J., Petrocelli, S. R., Caroll, J. (1980) An evaluation of using partition coefficient and water solubilities to estimate bioconcentration factors for organic chemicals in fish. In: *Aquatic Toxicology*. Eaton, J. G., Parrish, P. R., Hendricks, A. C., Editors, ASTM STP 707, American Society for Testing and Materials. pp. 116-129.

Verloop, A. (1972) Fate of the herbicide diclobenil in plants and soil in relation to its biological activity. *Res. Rev.* 43, 55-103.

Verschueren, K. (1977) *Handbook of Environmental Data on Organic Chemicals*. Van Nostrand Reinhold, New York, New York.

Verschueren, K. (1983) *Handbook of Environmental Data on Organic Chemicals*. 2nd. Edition, Van Nostrand Reinhold, New York, New York.

Virtanen, M., Hattula, M. L., Arstila, A. U. (1979) Behavior and fate of 4-chloro-2-methylphenoxyacetic acid (MCPA) and 2,6-dichloro-*o*-cresol as studied in an aquatic-terrestrial model ecosystem. *Chemosphere* 8, 431.

von Oepen, B., Kördel, W., Klein, W. (1991) Sorption of nonpolar and polar compounds to soils: Processes, measurements and experiences with the applicability of the modified OECD-guideline 106. *Chemosphere* 22, 285-304.

Walker, A. (1976) Simulation of herbicide persistence in soil. III Propyzamide in different soil types. *Pest. Sci.* 7, 59-64.

Walker, A. (1978) Simulation of the persistence of eight soil applied herbicides. *Weed Res.* 18, 305-313.

Walker, A., Cotterill, E. G., Welch, S. J. (1989) Adsorption and degradation of chlorsulfuron and metsulfuron-methyl in soils from different depths. *Weed Res.* 29, 281-287.

Walker, A., Thompson, J. A. (1977) The degradation of simazine, linuron, and propyzamide in different soils. *Weed Res.* 17, 399-405.

Walker, A., Welch, S. J. (1991) Enhanced degradation of some soil-applied herbicides. *Weed Res.* 31, 49-57.

Walker, A., Welch, S. J. (1992) Further studies of the enhanced biodegradation of some soil-applied herbicides. *Weed Res.* 32, 19-27.

Walker, A., Zimdahl, R. L. (1981) Simulation of the persistence of atrazine, linuron and metocholor in soil at different sites in U.S.A. *Weed Res.* 21, 255-265.

Walker, W. W. (1978) Insecticide persistence in natural seawater as affected by salinity, temperature and sterility. EPA-600/3-78-044. U.S. Environmental Protection Agency, Gulf Breeze, Florida.

Walker, W. W., Cripe, C. R., Pritchard, P. H., Bourquin, A. W. (1988) Biological and abiotic degradation of xenobiotic compounds in *in vitro* esturine water and sediment/water systems. *Chemosphere* 17, 2255-2270.

Wang, X., Harada, S., Watanabe, M., Koshikawa, H., Geyer, H. J. (1996) Modelling the bioconcentration of hydrophobic organic chemicals in aquatic organisms. *Chemosphere* 32(9), 1783-1793.

Wang, Y.-S., Jaw, C.-G., Tang, H.-C., Lin, T.-S., Chen, Y.-L. (1992) Accumulation and release of herbicides butachlor, thiobencarb, and chlomethoxyfen by fish, clam, and shrimp. *Bull. environ. Contam. Toxicol.* 48, 474-480.

Wang, Y.-S., Madsen, E. L., Alexander, M. (1985) Microbial degradation by mineralization or cometabolism determined by chemical concentration and environment. *J. Agric. Food Chem.* 33, 495.

Wang, Y.-S., Subba-Rao, R. V., Alexander, M. (1984) Effect of substrate concentration and organic and inorganic compounds on the occurence and rate of mineralization and cometabolism. *Appl. Environ. Microbiol.* 47, 1195.

Ward, T. M., Weber, J. B. (1968) Aqueous solubility of alkylamino-s-triazines as a function of pH and molecular structure. *J. Agric. Food Chem.* 16, 959-961.

Wauchope, R. D. (1978) The pesticide content of surface water draining from agricultural fields - A review. *J. Environ. Qual.* 7, 459-472.

Wauchope, R. D. (1989) *ARS/SCS Pesticides Properties Database.* Version 1.9, preprint, August, 1989.

Wauchope, R. D., Buttler, T. M., Hornsby, A. G., Augustijn-Beckers, P. W. M., Burt, J. P. (1992) The SCS/ARS/SCS Pesticides Properties Database for Environmental Decision-Making. *Rev. Environ. Contam. Toxicol.* 123, 1-164.

Wauchope, R. D., Hornsby, A. G., Goss, D. W., Burt, J. P. (1991) The SCS/ARS/SCS Pesticides Properties Database: A set of parameter values for first-tier comparative water pollution risk analysis. *Proceedings, National Pesticide Conference,* Brookfield, Virginia, November 8-9, 1990, pp. 455-470.

Wauchope, R. D., Meyers, R. S. (1985) Adsorption-desorption kinetics of atrazine and linuron in freshwater-sediment aqueous slurries. *J. Environ. Qual.* 14, 132-137.

Weber, J. B. (1970) Mechanisms of adsorption of s-trazines by clay colloids and factors affecting plant availability. *Res. Rev.* 32, 93-130.

Weber, J. B. (1972) Interaction of organic pesticides with particulate matter in aquatic and soil systems. In: *Fate of Organic Pesticides in the Aquatic Environment. Adv. Chem. Ser.* 111. American Chemical Society, Washington, DC. p. 55.

Weber, J. B., Peter, C. J. (1982) Adsorption, bioactivity, and evaluation of soil tests for alachlor, acetochlor, and metolachlor. *Weed Sci.* 30, 14-20.

Weber, J.B., Shea, P.J., Strek, H.J. (1980) An evaluation of nonpoint sources of pesticide pollution in runoff. In: *Environmental Impact of Nonpoint Source Pollution.* Overcash, M., Davidson, J., Editors, Ann Arbor Science Publishers, Ann Arbor, Michigan.

Weidner, C. W. (1974) Degradation in ground water and mobility of herbicides. Report prepared for the Office of Water Research and Technology, U.S. Environmental Protection Agency, Washington DC. PB 239242.

West, S. D., Burger, R. O., Poole, G. M., Mowrey, O. H. (1983) Bioconcentration and field dissipation of the aquatic herbicide fluridone and its degradation products in aquatic environments. *J. Agric. Food Chem.* 31, 579-585.

West, S. D., Day, E. W., Jr., Burger, R. O. (1979) Dissipation of the experimental aquatic herbicide fluridone from lakes and ponds. *J. Agric. Food Chem.* 27, 1067.

Wienhold, B. J., Gish, T. J. (1994) Chemical properties influencing rate of release of starch encapsulated herbicides: Implications for modifying environmental fate. *Chemosphere* 28(5), 1035-1046.

Willis, G. H., McDowell, L. L. (1982) Pesticides in agricultural runoff and their effects on downstream water quality. *Environ. Toxicol. Chem.* 1, 267-279.

Willis, G. H., Wander, R. C., Southwick, L. M. (1974) Degradation of trifluralin in soil suspensions as related to redox potential. *J. Environ. Qual.* 3, 262-265.

Wilson, R. G., Jr., Cheng, H. H. (1978) Fate of 2,4-D in a Naff silt loam soil. *J. Environ. Qual.* 7, 281.

Windholz, M., Editor (1983) *The Merck Index. An Encyclopedia of Chemicals, Drugs and Biologicals.* 10th Edition. The Merck & Co. Inc., Rahway, New Jersey.

Winkelmann, D. A., Klaine, S. J. (1991) Degradation and bound residue formation of atrazine in a western Tennessee soil. *Environ. Toxicol. Chem.* 10, 335-345.

Wolf, D. C., Jackson, R. L. (1982) Atrazine degradation, sorption, and bioaccumulation in water systems. Arkansas Water Resources Center. NTIS PB83-150151.

Wolfe, N. L., Zepp, R. G., Baughman, G. L., Fincher, R. C., Gordon, J. A. (1976) Chemical and photochemical transformation of selected pesticides in aquatic systems. U.S. Environmental Protection Agency, Athens, Georgia. EPA-600/3-76-067.

Wolfe, N. L., Zepp, R. G., Paris, D. F. (1978) Carbaryl, propham, and chloropropham: A comparison of the rates of hydrolysis and photolysis with the rate of biolysis. *Water Res.* 12, 565-571.

Wood, A. L., Davidson, J. M. (1975) *Soil Science Society of America Proceedings* 39, 820-825.

Wood, M. J. et al. (1991) In: *Pesticides in Soil and Water: Current Perspectives.* Walker, A., Editor, (BCPC Monograph), 47, 175-182.

Woodford, E. K., Evans, S. A., Editors (1963) *Weed Control Handbook: Properties of Herbicides.* Blackwell Scientific, Oxford, England.

Worthing, C. R., Editor (1983) *The Pesticide Manual. (A World Compendium).* 7th Edition, The British Crop Protection Council, Croydon, Emgland.

Worthing, C. R., Editor (1987) *The Pesticide Manual. (A World Compendium).* 8th Edition, The British Crop Protection Council, Croydon, England.

Worthing, C. R., Hance, R., Editors (1990) *The Pesticide Manual. (A World Compendium).* 9th Edition, The British Crop Protection Council, Croydon, England.

Worthing, C. R., Editor (1991) *The Pesticide Manual. (A World Compendium).* 10th Edition, The British Crop Protection Council, Croydon, England.

Yalkowsky, S. H. (1989) *Arizona Database of Aqueous Solubilities.* University of Arizona, Tucson, Arizona.

Yalkowsky, S. H., Banerjee, S. (1992) *Aqueous Solubility. Methods of Estimation for Organic Compounds.* Marcel Dekker, Inc., New York, New York.

Yalkowsky, S. H., Valvani, S. C., Kun, W.-Y., Dannenfelser, R. M., Editors (1987) *Arizona Database of Aqueous Solubility for Organic Compounds.* College of Pharmacy, University of Arizona, Tucson, Arizona.

Yao, C. C. D., Haag, W. R. (1991) Rate constants for direct reactions of ozone with several drinking water contaminants. *Water Res.* 25, 761-773.

Yoshioka, Y., Mizuno, T., Ose, Y., Sato, T. (1986) The estimation for toxicity of chemicals on fish by physico-chemical properties. *Chemosphere* 15(2), 195-203.

Yu, C-C., Hansen, D. J., Booth, G. M. (1975) Fate of dicamba in a model ecosystem. *Bull. Environ. Contam. Toxicol.* 13, 280-283.

Zamdahl, R. L., Clark, S. K. (1982) Degradation of three acetanilide herbicides in soil. *Weed Sci.* 30, 545-548.

Zandvoort, R., Van Dord, D. C., Leistra, M., Verlaat, J. G. (1979) The decline of propyzamide in soil under field conditions in the Netherlands. *Weed Res.* 19, 157.

Zepp, R. G. (1978) Quantum yields for reactions of pollutants in dilute aqueous solution. *Environ. Sci. Technol.* 12, 327-329.

Zepp, R. G. (1980) 9. Assessing the photochemistry of organic pollutants in aquatic environments. In: *Dynamics, Exposure and Hazard Assessment of Toxic Chemicals.* Haque, R., Editor, pp. 60-110, Ann Arbor Science Publishers, Ann Arbor, Michigan.

Zepp, R. G. (1991) Photochemical fate of agrochemicals in natural waters. In: *Pesticide Chemistry.* Advances in International Research, Development, and Legislation. Frehse, H., Editor, pp. 329-345, VCH, New York, New York.

Zepp, R. G., Baughman, G. L. (1978) Prediction of photochemical transformation of pollutants in aquatic environment. In: *Aquatic Pollutants: Transformation and Biological Effects.* Hutzinger, O., Van Lelyveld, I. H., Zoeteman, B. C. J., Editors, pp. 237-264, Pergamon Press, Oxford, England.

Zepp, R. G., Cline, D. M. (1977) Rates of direct photolysis in aqueous environment. *Environ. Sci. Technol.* 11, 359-366.

Zepp, R. G., Schlotzhauer, P. F., Simmons, M. S., Miller, G. C., Baughman, G. L., Wolfe, N. L. (1984) Dynamics of pollutant photoreactions in the hydrosphere. *Fresenius Z. Anal. Chem.* 319, 119-125.

Zepp, R. G., Wolfe, N. L., Gordon, J. A., Baughman, G. L. (1975) Dynamics of 2,4-D esters in surface waters. Hydrolysis, photolysis and vaporization. *Environ. Sci. Technol.* 9, 1144.

Zimdahl, R. L., Gwynn, S. M. (1977) Soil degradation of three dinitroanilines. *Weed Sci.* 25, 247-251.

Zitko, V., McLeese, D. W., Carson, W. G., Welch, H. E. (1976) Toxicity of alkyl-dinitrophenols to some aquatic organisms. *Bull. Environ. Contam. Toxicol.* 16, 508-515.

Chapter 3. Insecticides

3.1 List of Chemicals and Data Compilations:
  3.1.1 Organophosphorus compounds:
    a) Phosphates:
- Chlorfenvinphos 370
- Crotoxyphos 378
- Dichlorvos 409
- Dicrotophos 412
- Mevinphos 504
- Monocrotophos 510

    b) Phosphorothioates:
- Acephate 336
- Chlorpyrifos 372
- Demeton 402
- Diazinon 404
- Fenitrothion 442
- Fenthion 448
- Leptophos 476
- Parathion 514
- Parathion-methyl 520
- Ronnel (Fenchlorphos) 547
- Trichlorfon 556

    c) Phosphorodithioates (Phosphorothiolothionates*):
- Azinphos-methyl 349
- Dimethoate 423
- Disulfolton 427
- Ethion 439
- Fonofos 456
- Malathion 490
- Phenthoate* 536
- Phorate 538
- Phosmet 542
- Terbufos 550

  3.1.2 Carbamates:
- Aldicarb 338
- Aminocarb 347
- Bendiocarb 352
- Carbaryl 354
- Carbofuran 359
- Fenoxycarb 446
- Methiocarb 495
- Methomyl 497
- Oxamyl 512
- Propoxur 544

3.1.3 Organochlorines:
- Aldrin 342
- Chlordane 364
- DDD 383
- DDE 387
- DDT 391
- Dieldrin 414
- Endrin 435
- α HCH 459
- β HCH 462
- δ HCH 464
- Heptachlor 466
- Heptachlor epoxide 471
- Kepone 474
- Lindane (γ HCH) 479
- Methoxychlor 500
- Mirex 506
- Toxaphene 552

3.1.4 Phenols:
- Denoseb 132
- Pentachlorophenol (PCP) 525

3.1.5 Synthetic pyrethroids:
- Cypermethrin 380
- Fenvalerate 451
- Permethrin 533

3.1.6 Miscellaneous:
- Diflubenzuron (Benzoylphenylurea) 421
- Endosulfan (Organochloro sulfide) 431
- Flucythrinate (Fluorophenoxy benzoester) 454

3.2 Summary Tables 559
3.3 Illustrative Fugacity Calculations: Levels I, II and III 572
- Carbaryl 572
- Chlorpyrifos 576
- DDT 580
- Diazinon 584
- Lindane 588
- Malathion 592
- Methyl parathion 596

3.4 Commentary on Physical-Chemical Properties and Environmental Fate 600
3.5 References 602

Common Name: Acephate
Synonym: Chevron RE 12420, ENT 27822, Orthene, Ortho 12420, Ortran, Ortril, RE 12420, 75 SP, Tornado
Chemical Name: acetylphosphoramidothioic acid *O,S*-dimethyl ester; *O,S*-dimethyl acetylphosphoramidothioate; *N*-[methoxy(methylthio)phosphinoyl]acetamide
Uses: systemic insecticide with contact and stomach action to control a wide range of chewing and sucking insects in fruit, cotton, hops, vines, soybeans, olives, groundnuts, beet, brassicas, celery, potatoes, rice ornamentals, forestry and other crops; also used as cholinesterase inhibitor.
CAS Registry No: 30560-19-1
Molecular Formula: $C_4H_{10}NO_3PS$
Molecular Weight: 183.2
Melting Point (°C):
    82-89   (Shiu et al. 1990; Worthing 1991)
    64-68   (The Merck Index 1989; Montgomery 1993)
    88-90   (Tomlin 1994)
    85.50   (Kühne et al. 1995)
    82-93   (Milne 1995)
Boiling Point (°C):
Density (g/cm³ at 20°C):
    1.35   (Spencer 1982; Worthing 1991; Montgomery 1993; Tomlin 1994; Milne 1995)
Molar Volume (cm³/mol):
    135.7   (calculated from density)
Molecular Volume (Å³):
Total Surface Area, TSA (Å²):
Dissociation Constant pK_a:
Heat of Fusion, ΔH_{fus}, kcal/mol:
Entropy of Fusion, ΔS_{fus}, cal/mol·K (e.u.):
Fugacity Ratio at 25°C (assuming ΔS_{fus} = 13.5 e.u.), F: 0.252

Water Solubility (g/m³ or mg/L at 25°C):
    650000   (Spencer 1973, 1982; Worthing 1979, 1987, 1991; quoted, Bowman & Sans 1983a; Shiu et al. 1990; Majewski & Capel 1995)
    650000   (Martin & Worthing 1977; quoted, Kenaga 1980; Kenaga & Goring 1980)
    >5000   (20°C, shake flask-GC, Bowman & Sans 1983a; quoted, Shiu et al. 1990)
    790000   (20°C, Agrochemicals Handbook 1987)
    818000   (Wauchope 1989; quoted, Shiu et al. 1990; Lohninger 1994)
    818000   (20-25°C, selected, Wauchope et al. 1992; Hornsby et al. 1996)
    790000   (20°C, Montgomery 1993; Milne 1995)
    790000   (20°C, Tomlin 1994; quoted, Sanchez-Camazano et al. 1995)
    635220   (quoted, Kühne et al. 1995)
    36550   (calculated-group contribution fragmentation method, Kühne et al. 1995)
    650000   (selected, Iglesias-Jimenez et al. 1996)

Vapor Pressure (Pa at 25°C):
    2.26×10^{-4} (20°C, Agrochemicals Handbook 1987)
    2.26×10^{-4} (24°C, Worthing 1987, 1991; Tomlin 1994; Majewski & Capel 1995)
    2.27×10^{-4} (20-25°C, selected, Wauchope et al. 1992; Hornsby et al. 1996)
    2.27×10^{-4} (20°C, Montgomery 1993)

Henry's Law Constant (Pa·m³/mol):
    5.27×10^{-8} (20-25°C, calculated-P/C, Montgomery 1993)
    6.37×10^{-8} (20-25°C, calculated-P/C as per Worthing 1987, Majewski & Capel 1995)
    5.06×10^{-8} (calculated-P/C, this work)

Octanol/Water Partition Coefficient, log K_{OW}:
    −1.87 (calculated, Montgomery 1993)
    −0.886 (Tomlin 1994; quoted, Sanchez-Camazano et al. 1995)
    −0.886 (selected, Iglesias-Jimenez et al. 1996)
    −0.28 (quoted, calculated, Finizio et al. 1997)
    1.12, −0.89, 0.43 (RP-HPLC, ClogP, calculated-S, Finizio et al. 1997)

Bioconcentration Factor, log BCF:
    −0.523 (calculated-S, Kenaga 1980)
    0.053 (wet wt. basis, rainbow trout, Geen et al. 1984; quoted, De Bruijn & Hermens 1991)

Sorption Partition Coefficient, log K_{OC}:
    0.477 (calculated-S as per Kenaga & Goring 1978, Kenaga 1980)
    0.30 (soil, 20-25°C, selected, Wauchope et al. 1992; Dowd et al. 1993; Hornsby et al. 1996)
    0.48 (Montgomery 1993)
    0.30 (estimated-chemical structure, Lohninger 1994)

Half-Lives in the Environment:
    Air:
    Surface water:
    Groundwater:
    Sediment:
    Soil: selected field half-life of 3.0 d (Wauchope et al. 1992; Dowd et al. 1993; Hornsby et al. 1996); half-life of 7-10 d in soil (Tomlin 1994).
    Biota:

Environmental Fate Rate Constants or Half-Lives:
    Volatilization:
    Photolysis:
    Oxidation: calculated rate constant of $\sim 51 \times 10^{-12}$ cm³/molecule·s for the vapor phase reaction with hydroxyl radicals in air (Winer & Atkinson 1990).
    Hydrolysis: half-life of 60 h at pH 9 and 710 hours at pH 3 both at 40°C (Montgomery 1993).
    Biodegradation:
    Biotransformation:
    Bioconcentration, Uptake (k_1) and Elimination (k_2) Rate Constants:

Common Name: Aldicarb

Synonym: Ambush, Carbanolate, ENT 27093, NCI-C08640, matadan, OMS 771, Pounce, Temik, Union Carbide 21149

Chemical Name: 2-methyl-2-(methylthio)propionaldehyde O-(methylcarbamoyl) oxime; 2-methyl-2-(methylthio)propanal O-(methylamino)carbonyl) oxime

Uses: systemic insecticide, acaricide, and nematocide with contact and stomach action; also used as cholinesterase inhibitor.

CAS Registry No: 116-06-3
Molecular Formula: $C_7H_{14}N_2O_2S$
Molecular Weight: 190.25

Melting Point (°C):
    100   (Khan 1980)
    99-101   (Spencer 1982)
    99   (Bowman & Sans 1983b; Patil 1994)
    99-100   (Suntio et al. 1988; Howard 1991; Montgomery 1993; Milne 1995)
    98-100   (Worthing 1991; Tomlin 1994)

Boiling Point (°C):
    100   (decomposes above this temp., Howard 1991)

Density (g/cm³ at 20°C):
    1.195   (25°C, Agrochemicals Handbook 1987; Montgomery 1993; Tomlin 1994; Milne 1995)

Molar Volume (cm³/mol):
    224.3   (calculated-LeBas method, Suntio et al. 1988; Fisher et al. 1993)
    1.202   ($V_I/100$, Fisher et al. 1993)

Molecular Volume (Å³):

Total Surface Area, TSA (Å²):

Dissociation Constant pK_a:

Heat of Fusion, ΔH_{fus}, kcal/mol:
    6.20   (DSC method, Plato & Glasgow 1969)

Entropy of Fusion, ΔS_{fus}, cal/mol·K (e.u.):

Fugacity Ratio at 25°C (assuming ΔS_{fus} = 13.5 e.u.), F: 0.185

Water Solubility (g/m³ or mg/L at 25°C):
    4000   (24°C, shake flask-GC, Felsot & Dahm 1979; quoted, Shiu et al. 1990)
    7800   (Kenaga 1980a; Kenaga & Goring 1980; quoted, Suntio et al. 1988; Isnard & Lambert 1989; Shiu et al. 1990)
    6000   (Khan 1980; Agrochemicals Handbook 1987)
    6016, 6000   (exptl., corrected-M.P., Briggs 1981)
    6000   (20°C, shake flask-GC, Bowman & Sans 1983b; quoted, Suntio et al. 1988; Shiu et al. 1990; Patil 1994)
    7760   (Garten & Trabalka 1983; quoted, Shiu et al. 1990)
    6000   (Verschueren 1983; Worthing 1987, 1991; quoted, Yalkowsky 1987; quoted, Howard 1991)
    5730   (Seiber 1987; quoted, Shiu et al. 1990)
    6000   (20°C, selected, Suntio et al. 1988; quoted, Majewski & Capel 1995
    6000   (The Merck Index 1989; quoted, Milne 1995)
    6000   (20-25°C, selected, Wauchope et al. 1992; Hornsby et al. 1996)
    6000   (Montgomery 1993; Lohninger 1994)
    38840   (calculated, Patil 1994)
    4930   (20°C at pH 7, Tomlin 1994)

Vapor Pressure (Pa at 25°C):
    6.67   (20°C, Khan 1980)
    0.013   (selected, Suntio et al. 1988; quoted, Howard 1991; Majewski & Capel 1995)
    0.013   (20°C, Agrochemicals Handbook 1987)
    0.013   (Worthing 1991)
    0.004   (20-25°C, selected, Wauchope et al. 1992; Hornsby et al. 1996)
    0.0046   (Montgomery 1993)
    0.013   (20°C, Tomlin 1994)

Henry's Law Constant (Pa·m^3/mol):
    2.48×10^{-4} (Jury et al. 1987a, Jury & Ghodrati 1989)
    3.20×10^{-4} (calculated-P/C, Suntio et al. 1988; quoted, Howard 1991; Fisher et al. 1993; Majewski & Capel 1995)
    1.47×10^{-4} (20-25°C, calculated-P/C, Montgomery 1993)
    1.27×10^{-4} (calculated-P/C, this work)

Octanol/Water Partition Coefficient, log K_{OW}:
    0.85   (shake flask, Felsot & Dahm 1979; quoted, Bowman & Sans 1983b)
    1.10   (Hansch & Leo 1979; quoted, Fisher et al. 1993)
    0.70   (quoted, Rao & Davidson 1980; Bowman & Sans 1983b)
    1.57   (shake flask-UV, Briggs 1981; quoted, Bowman & Sans 1983b)
    1.00   (approx. = log BCF_{lipid}, Briggs 1981)
    1.13   (20°C, shake flask-GC, Bowman & Sans 1983b; Patil 1994)
    1.13   (Hansch & Leo 1985; quoted, Isnard & Lambert 1989; Howard 1991)
    1.10   (selected, Suntio et al. 1988; quoted, Bintein & Devillers 1994)
    1.15   (selected, Travis & Arms 1988)
    1.57   (Thor 1989; quoted, Connell & Markwell 1990)
    0.70, 1.13   (Montgomery 1993)
    1.13   (selected, Sangster 1993)
    0.50   (calculated, Patil 1994)
    1.13   (selected, Hansch et al. 1995)

Bioconcentration Factor, log BCF:
    1.62   (fish in static water, Metcalf & Sanborn 1975; quoted, Kenaga & Goring 1980)
    0.85   (vegetation, correlated-K_{OW}, Iwata et al. 1977; Maitlen & Power 1982; quoted, Travis & Arms 1988)
    0.602   (calculated-S, Kenaga 1980; quoted, Howard 1991)
    1.64   (earthworm, Lord et al. 1980; quoted, Connelly & Markwell 1990)
    1.00   (log BCF_{lipid}, Briggs 1981)

Sorption Partition Coefficient, log K_{OC}:
    1.36-1.57   (Felsot & Dahm 1979; quoted, Howard 1991)
    0.91, 1.20   (Bromilow & Leistra 1980; quoted, Howard 1991)
    1.51   (calculated-S as per Kenaga & Goring 1978, Kenaga 1980)
    1.39   (reported as log K_{OM}, Briggs 1981)
    1.18   (log $BCF_{protein}$, Briggs 1981)

1.51   (estimated, Kenaga 1980; quoted, Howard 1991)
1.30-1.40   (Bilkert & Rao 1985; quoted, Howard 1991)
1.56   (soil, screening model calculations, Jury et al. 1987a,b; Jury & Ghodrati 1989)
1.48   (soil, 20-25°C, selected, Wauchope et al. 1992; Dowd et al. 1993; Hornsby et al. 1996)
0.85-1.67 (Montgomery 1993)
1.48   (estimated-chemical structure, Lohninger 1994)

Half-Lives in the Environment:
  Air: 1-9.5 h, based on an estimated rate constant for vapor-phase reaction with hydroxyl radicals in air (Atkinson 1987; quoted, Howard 1991; Howard et al. 1991).
  Surface water: 480-8664 h, based on estimated aqueous aerobic biodegradation half-life (Howard et al. 1991).
  Groundwater: 960-15240 h, based on estimated aqueous aerobic biodegradation half-life and water grab sample data (Miles & Delfino 1985; quoted, Howard et al. 1991).
  Sediment:
  Soil: half-lives of 9, 7, and 12 d in clay, silty clay loam and fine sandy loam at an application rate of 20 ppm (Coppedge et al. 1967; quoted, Montgomery 1993); hydrolysis half-lives have been reported to be 23 d at pH 7.2 (Smelt et al. 1978; quoted, Howard 1991), 9.9 d at pH 6.3-7.0 at 15°C (Bromilow et al. 1980; Bromilow & Leistra 1980; quoted, Howard 1991; Montgomery 1993); both at 15°C and 0.4-3.2 d at pH 4.5-4.9 and 25°C (Rao et al. 1984; quoted, Howard 1991); degradation rate constants of 0.000222 h^{-1} for discharge rate of 30 cm/yr and 0.000233 h^{-1} for discharge rate of 61 cm/yr with half-life of 30 d (Jones & Back 1984); 480-8664 h, based on unacclimated aerobic soil grab sample data (Ou et al. 1985; quoted, Howard 1991; Howard et al. 1991; Montgomery 1993); reported half-life was 70 d from screening model calculations (Jury et al. 1987a,b; Jury & Ghodrati 1989; quoted, Montgomery 1993); oxidation half-life of 1.7-12 d for pH 1-10 with little change in rate between pH 4.4-10 (Lemley et al. 1988; quoted, Howard 1991); selected field half-life of 30 d (Wauchope et al. 1992; Dowd et al. 1993; Hornsby et al. 1996).
  Biota: biochemical half-life of 70 d from screening model calculations (Jury et al. 1987a,b; Jury & Ghodrati 1989).

Environmental Fate Rate Constants or Half-Lives:
  Volatilization:
  Photolysis:
  Oxidation: photooxidation half-life of 1-9.5 h, based on an estimated rate constant for vapor-phase reaction with hydroxyl radicals in air (Atkinson 1987; quoted, Howard et al. 1991); half-life of 1.7-12 d in soil for pH 1-10 with little change in rate between pH 4.4-10 (Lemley et al. 1988; quoted, Howard 1991); rate constant of 5.9×10^9 M^{-1} s^{-1} for the reaction (Fenton with reference to acetophenone) with hydroxyl radicals in aqueous solutions at pH 3.5 and at (24 ± 1)°C (Buxton et al. 1988; quoted, Faust & Hoigné 1990; Haag & Yao 1992); rate constant of $(8.1 \pm 1.1) \times 10^9$ M^{-1} s^{-1} for the reaction (Fenton with reference to acetophenone) with hydroxyl radicals in aqueous solutions at pH 3.5 and at (24 ± 1)°C (Haag & Yao 1992).

Hydrolysis: half-lives have been reported to be 23 d at pH 7.2 (Smelt et al. 1978; quoted, Howard 1991), 9.9 d at pH 6.3-7.0 (Bromilow & Leistra 1980; quoted, Howard 1991) both at 15°C and 0.4-3.2 d at pH 4.5-4.9 and 25°C (Rao et al. 1984; quoted, Howard 1991); first-order half-life of 4580 d, based on a first-order rate constant of 1.51×10^{-4} d^{-1} at pH 5.5 and 5°C (Hansen & Spiegel 1983; quoted, Howard et al. 1991); half-lives in soil have been reported to be 0.4-3.2 d at pH 4.5-4.9 and 25°C (Rao et al. 1984; quoted, Howard 1991); half-lives for pH buffered distilled water at 20°C: 131 d at pH 3.95, 559 d at pH 6.02, 324 d at pH 7.96, 55 days at pH 8.85, and 6 d at pH 9.85 (Montgomery 1993).

Biodegradation: aqueous aerobic half-life of 480-8664 h, based on unacclimated aerobic soil grab sample data (Ou et al. 1985; quoted, Howard 1991; Howard et al. 1991); aqueous anaerobic half-life of 1488-15240 h, based on anaerobic ground water grab sample data (Miles & Delfino 1985; quoted, Howard et al. 1991); half-life of 70 d in 0-10 cm depth of soil (Jury et al. 1987a,b; Jury & Ghodrati 1989).

Biotransformation:

Bioconcentration, Uptake (k_1) and Elimination (k_2) Rate Constants:

Common Name: Aldrin
Synonym: Aldrec, Aldrex, Aldrite, Aldrosol, Altox, Compound 118, Drinox, ENT 15949, HHDN, NA 2761, NA 2762, Octalene, Seedrin
Chemical Name: 1,2,3,4,10,10-hexachloro-1,4,4a,5,8,8a-hexahydro-1,4-endoexo-5,8-dimethano-naphthalene
Uses: Insecticide/Fumigant
CAS Registry No: 309-00-2
Molecular Formula: $C_{12}H_8Cl_6$
Molecular Weight: 364.93
Melting Point (°C):
    104-104.5 (Martin 1972; Callahan et al. 1979; Khan 1980; Spencer 1982; Milne 1995)
    104 (Suntio et al. 1988; Howard 1991; Liu et al. 1991; Noegrohati & Hammers 1992; Montgomery 1993)
    104.3 (Patil 1994; Kühne et al. 1995)
Boiling Point (°C):
    145 (at 2 mmHg, Agrochemicals Handbook 1987; Montgomery 1993; Milne 1995)
Density (g/cm³ at 20°C):
    1.70 (Montgomery 1993)
Molar Volume (cm³/mol):
    316.8 (calculated-LeBas method, Suntio et al. 1988)
    244.1 (Ruelle & Kesselring 1997)
    328.3 (calculated-LeBas method, this work)
    214.7 (calculated-density, this work)
Molecular Volume (Å³):
Total Surface Area, TSA (Å²):
Dissociation Constant pK_a:
Heat of Fusion, ΔH_{fus}, kcal/mol:
    3.87 (Ruelle & Kesselring 1997)
Entropy of Fusion, ΔS_{fus}, cal/mol·K (e.u.):
Fugacity Ratio at 25°C (assuming ΔS_{fus} = 13.5 e.u.), F: 0.165

Water Solubility (g/m³ or mg/L at 25°C):
    0.20 (shake flask-GC/UV, Richardson & Miller 1960; quoted, Shiu et al. 1990)
    0.20 (Stephen & Stephen 1963; quoted, Lu & Metcalf 1975)
    0.027 (25-29°C, shake flask-GC/ECD, Park & Bruce 1968; quoted, Callahan et al. 1979; Khan 1980; Suntio et al. 1988; Shiu et al. 1990)
    0.18 (particle size ≤5.0 μ, shake flask-GC/ECD, Biggar & Riggs 1974; quoted, Callahan et al. 1979; Suntio et al. 1988; Shiu et al. 1990)
    0.013, 0.14, 0.18 (particle size: 0.01, 0.05 & 5.0 μ, shake flask-GC/ECD, Biggar & Riggs 1974)
    0.017 (gen. col.-GC/ECD, Weil et al. 1974; quoted, Callahan et al. 1979; Geyer et al. 1980; Suntio et al. 1988; Shiu et al. 1990)
    0.027 (Martin & Worthing 1977; quoted, Briggs 1981)
    0.01-0.2 (20-25°C, Wauchope 1978; Willis & McDowell 1982)
    0.013 (Kenaga 1980a,b; Kenaga & Goring 1980; Garten & Trabalka 1983; quoted, Shiu et al. 1990)
    0.027 (27°C, Spencer 1982; Worthing 1987; quoted, Shiu et al. 1990)
    0.20 (quoted, Thomas 1982)
    0.025 (U.S. EPA 1984; quoted, McLean et al. 1988)

<0.05 (rm. temp., Agrochemicals Handbook 1987)
0.20 (quoted, Isnard & Lambert 1988, 1989)
0.02 (20°C, selected, Suntio et al. 1988; quoted, Howard 1991; Majewski & Capel 1995)
0.017-0.18 (Montgomery 1993)
0.027 (20-25°C, selected, Augustijn-Beckers et al. 1994; Hornsby et al. 1996)
0.20, 0.011 (quoted, calculated, Patil 1994)
0.017, 0.132 (quoted, calculated-group contribution fragmentation method, Kühne et al. 1995)
<0.05 (Milne 1995)
0.020; 0.070, 0.0024 (quoted; predicted-molar volume, Ruelle & Kesselring 1997)

Vapor Pressure (Pa at 25°C):
0.0008 (Günther & Günther 1971; quoted, Callahan et al. 1979)
0.0031 (20°C, Martin 1972; quoted, Callahan et al. 1979)
0.0008 (20°C, Mackay & Wolkoff 1973; quoted, Dobbs & Cull 1982)
0.075 (20°C, Khan 1980)
0.001 (20°C, estimated-rel. volatilization rate, Dobbs & Cull 1982)
0.0086 (20°C, extrapolated from gas saturation-GC measurement, ln P (Pa) = 32.9 − 11044/T for temp. range 35.5-70°C, Grayson & Fosbraey 1982)
0.0008 (20°C, quoted, Thomas 1982)
0.0031 (20°C, U.S. EPA 1984; quoted, McLean et al. 1988)
0.0071 (20°C, gas saturation-mixed bed, Kim 1985; quoted, Suntio et al. 1988)
0.0092 (20°C, gas saturation, Kim 1985; quoted, Suntio et al. 1988)
0.0086 (20°C, Agrochemicals Handbook 1987)
0.005 (20°C, selected, Suntio et al. 1988; quoted, Howard 1991; Majewski & Capel 1995)
0.001 (20°C, Budavari 1989)
0.0031 (20°C, Montgomery 1993)
0.0009 (20-25°C, selected, Augustijn-Beckers et al. 1994; Hornsby et al. 1996)

Henry's Law Constant (Pa·m^3/mol):
1.418 (calculated-P/C, Thomas 1982)
50.25 (20°C, exptl., Warner et al. 1987; quoted, Suntio et al. 1988; Howard 1991)
91.23 (20°C, calculated-P/C, Suntio et al. 1988; quoted, Majewski & Capel 1995)
49.47 (known LWAPC of Warner et al. 1987, Meylan & Howard 1991)
39.2 (bond-estimated LWAPC, Meylan & Howard 1991)
50.25 (calculated-P/C, Montgomery 1993)
91.23 (calculated-P/C, this work)

Octanol/Water Partition Coefficient, log K_{ow}:
3.01 (Lu & Metcalf 1975; quoted, Suntio et al. 1988)
5.67 (quoted, Callahan et al. 1979)
5.66 (calculated, Kenaga 1980a,b; quoted, Isnard & Lambert 1988, 1989)
7.50 (estimated-RP-TLC, Lord et al. 1980)
7.40 (extrapolated from RP-TLC, Briggs 1981)
5.68 (selected, Dao et al. 1983)

5.52   (Garten & Trabalka 1983; quoted, De Bruijn et al. 1989)
5.66   (shake flask, Geyer et al. 1984; quoted, Devillers et al. 1991; Liu et al. 1991; Noegrohati & Hammers 1992)
5.68   (U.S. EPA 1984; quoted, McLean et al. 1988)
5.52   (Travis & Arms 1988)
5.48   (selected, Chapman 1989)
6.496   (slow-stirring method, De Bruijn et al. 1989; quoted, Howard 1991)
5.52   (Thor 1989; quoted, Connell & Markwell 1990)
6.50   (selected, Geyer et al. 1991)
6.496   (estimated by QSAR & SPARC, Kollig 1993)
5.17-7.4 (Montgomery 1993)
5.66, 5.74 (quoted, RP-HPLC, Sicbaldi & Finizio 1993)
5.52, 6.53 (quoted, calculated, Patil 1994)
6.50   (selected, Hansch et al. 1995)
5.74, 5.49, 5.39 (RP-HPLC, ClogP, calculated-S, Finizio et al. 1997)

Bioconcentration Factor, log BCF:
0.398   (bioaccumulation factor log BF, adipose tissue in female Albino rats, Quaife et al. 1967; quoted, Geyer et al. 1980)
3.56-4.88   (earthworms, Wheatley & Hardman 1968; quoted, Connell & Markwell 1990)
2.80   (lake bacteria, Leshniowsky et al. 1970; quoted, Baughman & Paris 1981 1983)
4.36   (*Diptera*, nonsteady-state, Johnson et al. 1971; quoted, Biddinger & Gloss 1984)
4.50   (*Epemeoptera*, nonsteady-state, Johnson et al. 1971; quoted, Biddinger & Gloss 1984)
5.15   (*Cladocera*, nonsteady-state, Johnson et al. 1971; quoted, Biddinger & Gloss 1984)
4.55   (*Daphnia magna*, wet wt. basis, Johnson et al. 1971; quoted, Geyer et al. 1991)
3.56-4.60   (*Oedogonium sp.*, Metcalf et al. 1973; quoted, Baughman & Paris 1981)
3.50   (Metcalf 1974; quoted, Kenaga & Goring 1980)
3.11   (*Anabaena cylindrica*, Schauberger & Wildman 1977; quoted, Baughman & Paris 1981; Swackhamer & Skoglund 1991)
2.30   (*Acacystis nidulans*, Schauberger & Wildman 1977; quoted, Baughman & Paris 1981)
2.99   (*Acacystis nidulans*, Schauberger & Wildman 1977; quoted, Baughman & Paris 1981)
4.03, 3.50 (fish: flow water, static water; Kenaga 1980b)
3.85, 1.34 (calculated-S, K_{oc}, Kenaga 1980a)
0.431   (average beef fat diet, Kenaga 1980b)
4.10   (*Chlorella fusca*, Geyer et al. 1981; quoted, Swackhamer & Skoglund 1991)
3.59   (golden orfe, Freitag et al. 1982; quoted, Howard 1991)
4.10   (algae, Freitag et al. 1982)
4.26   (activated sludge, Freitag et al. 1982, 1984)
4.03   (Garten & Trabalka 1983; quoted, Howard 1991)
4.13   (clam fat, 60-d expt., Hartley & Johnson 1983)
4.09   (*Chlorella fusca*, Geyer et al. 1984; quoted, Howard 1991; Swackhamer & Skoglund 1991)
4.09, 3.44, 4.26 (algae, golden ide, activated sludge, Freitag et al. 1985)
3.66   (molluscs, Hawker & Connell 1986; quoted, Howard 1991)
3.12   (quoted, Isnard & Lambert 1988)

−1.07 (beef biotransfer factor log B_b, correlated-K_{OW} from Radeleff et al. 1952 & Kenaga 1980; Travis Arms 1988)

−1.62 (milk biotransfer factor log B_m, correlated-K_{OW} from Saha 1969; Travis Arms 1988)

−1.67 (vegetation, correlated-K_{OW} from Lichtenstein 1960 & Weisgerber et al. 1974; Travis Arms 1988)

Sorption Partition Coefficient, log K_{OC}:
- 2.61 (soil, Hamaker & Thompson 1972; quoted, Kenaga 1980a,b; Kenaga & Goring 1980)
- 4.68 (calculated-S as per Kenaga & Goring 1978, Kenaga 1980)
- 4.36 (calculated-K_{OW} as per Kenaga & goring 1980, Chapman 1989)
- 4.69 (derived from exptl., Meylan et al. 1992)
- 5.02 (calculated-χ, Meylan et al. 1992)
- 6.18 (estimated by QSAR & SPARC, Kollig 1993)
- 2.61, 4.69 (Montgomery 1993)
- 3.70 (20-25°C, selected, Augustijn-Beckers et al. 1994; Hornsby et al. 1996)

Sorption Partition Coefficient, log K_{OM}:
- 4.45 (Briggs 1981)

Half-Lives in the Environment:

Air: estimated half-life of 35.5 min for the vapor phase reaction with hydroxyl radicals in air (GEMS 1986; quoted, Howard 1991); 0.9-9.1 h, based on an estimated rate constant for vapor-phase reaction with hydroxyl radicals in air (Atkinson 1987; quoted, Howard et al. 1991).

Surface water: 504-14200 h, based on unacclimated aerobic river die-away test data (Eichelberger & Lichtenberg 1971; quoted, Howard et al. 1991) and soil field test data (Lichtenstein et al. 1971; quoted, Howard et al. 1991).

Groundwater: 24-28400 h, based on estimated aqueous aerobic and anaerobic biodegradation half-lives (Howard et al. 1991).

Sediment:

Soil: half-life of 5-10 yr persistence in soil (Nash & Woolson 1967); 504-14200 h, based on unacclimated aerobic river die-away test data (Eichelberger & Lichtenberg 1971; quoted, Howard et al. 1991) and soil field test data (Lichtenstein et al. 1971; quoted, Howard et al. 1991); persistence of 2 years (Edwards 1973; quoted, Morrill et al. 1982); more than 24 months of persistence in soil (Wauchope 1978); estimated first-order half-life of 53.3 d from biodegradation rate constant of 0.013 d^{-1} by soil incubation studies from die-away tests (Rao & Davidson 1980; quoted, Scow 1982); moderately persistent with a half-life of 20-100 d (Willis & McDowell 1982; quoted, Howard 1991); half-life in a sandy loam soil incubated in the dark of 43-63 d (McLean et al. 1988; quoted, Howard 1991); selected field half-life of 365 d (Augustijn-Beckers et al. 1994; Hornsby et al. 1996).

Environmental Fate Rate Constants or Half-Lives:
- Volatilization: half-life of a few hours to a few days (Callahan et al. 1979); calculated half-life of 68 h from water (Thomas 1982).
- Photolysis:
- Oxidation: photooxidation half-life of 0.9-9.1 h, based on an estimated rate constant for vapor-phase reaction with hydroxyl radicals in air (Atkinson 1987; quoted, Howard et al. 1991).
- Hydrolysis: not readily hydrolyzable with half-life more than 4 yr (Callahan et al. 1979); first-order half-life of 760 d, based on a first-order rate constant of 3.8×10^{-5} h^{-1} at pH 7.0 and 25°C (Ellington et al. 1987, 1988; quoted, Howard et al. 1991); no disappearance in sealed glass ampules after two weeks at pH 11 and 85°C (Kollig 1993); half-life of 760 d at pH 7 and 25°C (Montgomery 1993).
- Biodegradation: aqueous aerobic half-life of 504-14200 h, based on unacclimated aerobic river die-away test data (Eichelberger & Lichtenberg 1971; quoted, Howard et al. 1991) and soil field test data (Lichtenstein et al. 1971; quoted, Howard et al. 1991); rate constant of 0.013 d^{-1} by soil incubation studies from die-away tests (Rao & Davidson 1980; quoted, Scow 1982); aqueous anaerobic half-life of 24-168 h, based on soil and freshwater mud grab sample data (Maule et al. 1987; quoted, Howard et al. 1991); half-life in a sandy loam soil incubated in the dark of 43-63 d (McLean et al. 1988; quoted, Howard 1991).
- Biotransformation:
- Bioconcentration, Uptake (k_1) and Elimination (k_2) Rate Constants:

Common Name: Aminocarb
Synonym: A 363, Bay 44646, Bayer 5080, ENT 25784, Matacil, Mitacil
Chemical Name: 4-dimethylamino-m-tolyl methylcarbamate, 4-dimethylamino-3-methylphenol methylcarbamate
Uses: nonsystemic, broad-spectrum insecticide used to control the spruce budworm in forests and also as molluscicide.
CAS Registry No: 2032-59-9
Molecular Formula: $C_{11}H_{16}N_2O_2$
Molecular Weight: 208.25
Melting Point (°C):
    93-94   (Spencer 1982; Montgomery 1993)
    93.5   (Bowman & Sans 1983b; Patil 1994)
Boiling Point (°C):
Density (g/cm³ at 20°C):
Molar Volume (cm³/mol):
    250.0   (calculated-LeBas method, this work)
Molecular Volume (Å³):
Total Surface Area, TSA (Å²):
Dissociation Constant pK_a:
Heat of Fusion, ΔH_{fus}, kcal/mol:
Entropy of Fusion, ΔS_{fus}, cal/mol·K (e.u.):
Fugacity Ratio at 25°C (assuming ΔS_{fus} = 13.5 e.u.), F: 0.210

Water Solubility (g/m³ or mg/L at 25°C):
    915   (20°C, shake flask-GC, Bowman 1982 unpublished result, quoted, Geyer et al. 1982)
    915   (20°C, shake flask-GC, Bowman & Sans 1983a,b; quoted, Shiu et al. 1990; Patil 1994)
    915   (20°C, Montgomery 1993)
    1360   (30°C, Montgomery 1993)
    915   (20-25°C, selected, Augustijn-Beckers et al. 1994; Hornsby et al. 1996)
    690   (calculated, Patil 1994)

Vapor Pressure (Pa at 25°C):
    0.00227   (20-25°C, selected, Augustijn-Beckers et al. 1994; Hornsby et al. 1996)

Henry's Law Constant (Pa·m³/mol):
    5.17×10^{-4} (20-25°C, calculated-P/C, this work)

Octanol/Water Partition Coefficient, log K_{OW}:
    1.74   (Zitko & McLeese 1980; quoted, Geyer et al. 1982)
    1.734   (20°C, shake flask-GC, Bowman 1982 unpublished result, quoted, Geyer et al. 1982)
    1.734   (20°C, shake flask-GC, Bowman & Sans 1983b; Patil 1994)
    1.70   (Richardson & Qadri 1986; quoted, McCarty et al. 1991)
    1.73   (Montgomery 1993)
    1.93   (calculated, Patil 1994)
    1.90   (pH 9, selected, Hansch et al. 1995)

Bioconcentration Factor, log BCF:
    0.690   (mussel, McLeese et al. 1980; quoted, Geyer et al. 1982)

Sorption Partition Coefficient, log K_{OC}:
    1.92   (calculated, Montgomery 1993)
    2.00   (20-25°C, selected, Augustijn-Beckers et al. 1994; Hornsby et al. 1996)

Half-Lives in the Environment:
    Air:
    Surface water:
    Groundwater:
    Sediment:
    Soil: selected field half-life of 6 d (Augustijn-Beckers et al. 1994; Hornsby et al. 1996).

Environmental Fate Rate Constants or Half-Lives:
    Volatilization:
    Photolysis:
    Oxidation:
    Hydrolysis:
    Biodegradation:
    Biotransformation:
    Bioconcentration, Uptake (k_1) and Elimination (k_2) Rate Constants:

Common Name: Azinphos-methyl
Synonym: Bay or Bayer 9027, Bay 17147, Carfene, Cotnion, Cotnion methyl, Crysthion 21, DBD, ENT 23233, Gothnion, Guthion, Gusathion, Metiltriazotion, R 1582
Chemical Name: O,O-dimethyl-S-[-4-oxo-1,2,3-benzotriazin-3(4H)-yl)methyl]phosphorodithioate; O,O-dimethyl-S-[3,4-dihydro-4-keto-1,2,3-benzotriazinyl-3-methyl)dithiophosphate
Uses: nonsystemic insecticide and acaricide for control of insects and pests in blueberry, grape, maize, vegetable, cotton, and citrus crops.
CAS Registry No: 86-50-0
Molecular Formula: $C_{10}H_{12}N_3O_3PS_2$
Molecular Weight: 317.34
Melting Point (°C):
    73-74   (Khan 1980; Spencer 1982; Suntio et al. 1988; Milne 1995)
    73.5   (Bowman & Sans 1983b)
    72.4   (Montgomery 1993; Tomlin 1994)
Boiling Point (°C):
    >200   (dec., Montgomery 1993)
Density (g/cm³ at 20°C):
    1.518   (Tomlin 1994)
    1.44   (Milne 1995; Montgomery 1993)
Molar Volume (cm³/mol):
    270.4   (calculated-LeBas method, Suntio et al. 1988)
Molecular Volume (Å³):
Total Surface Area, TSA (Å²):
Dissociation Constant pK_a:
Heat of Fusion, ΔH_{fus}, kcal/mol:
    7.40   (DSC method, Plato & Glasgow 1969)
Entropy of Fusion, ΔS_{fus}, cal/mol·K (e.u.):
Fugacity Ratio at 25°C (assuming ΔS_{fus} = 13.5 e.u.), F: 0.331

Water Solubility (g/m³ or mg/L at 25°C):
    33   (rm. temp., Spencer 1973; Worthing 1979; Khan 1980; quoted, Bowman & Sans 1983; Cohen & Steinmetz 1986)
    30   (20°C, Melnikov 1971; Spencer 1973; quoted, Shiu et al. 1990)
    33   (20-25°C, Willis & McDowell 1982)
    20.9   (20°C, shake flask-GC, Bowman & Sans 1983a,b; quoted, Shiu et al. 1990; Howard 1991; Patil 1994)
    33   (Merck Index 1983,89; quoted, Suntio et al. 1988; Shiu et al. 1990)
    29   (Verschueren 1983; quoted, Suntio et al. 1988; Shiu et al. 1990)
    29   (Agrochemicals Handbook 1987; Lohninger 1994)
    30   (20°C, selected, Suntio et al. 1988; quoted, Majewski & Capel 1995)
    33   (selected, USDA 1989; quoted, Neary et al. 1993)
    28   (20°C, Worthing 1991; Tomlin 1994)
    29   (20-25°C, selected, Wauchope et al. 1992; Hornsby et al. 1996)
    15.5   (calculated, Patil 1994)
    30   (Milne 1995)
    29   (selected, Halfon et al. 1996)

Vapor Pressure (Pa at 25°C):
- 2.93×10^{-5} (20°C, Melnikov 1971; quoted, Kim et al. 1984; Kim 1985; Suntio et al. 1988)
- 0.051 (20°C, Khan 1980)
- 1.00×10^{-6} (20°C, Worthing 1983; quoted, Howard 1991)
- 1.11×10^{-5} (20°C, GC-calculated value, Kim et al. 1984; Kim 1985)
- 0.31×10^{-5} (20°C, GC-calculated value with M.P. correction, Kim et al. 1984; Kim 1985)
- <0.001 (20°C, Agrochemicals Handbook 1987)
- 3.00×10^{-5} (20°C, selected, Suntio et al. 1988; quoted, Majewski & Capel 1995)
- $<1.8 \times 10^{-4}$ (20°C, Worthing 1991)
- 1.8×10^{-4} (20°C, Tomlin 1994)
- 2.67×10^{-5} (20-25°C, selected, Wauchope et al. 1992; Hornsby et al. 1996)
- 2.13×10^{-4} (20°C, Montgomery 1993)
- 2.70×10^{-5} (selected, Halfon et al. 1996)

Henry's Law Constant (Pa·m^3/mol):
- 0.0032 (20°C, calculated-P/C, Suntio et al. 1988; quoted, Majewski & Capel 1995)
- 1.52×10^{-5} (calculated-P/C, Howard 1991)
- 3.17×10^{-4} (calculated-P/C, this work)

Octanol/Water Partition Coefficient, log K_{ow}:
- 2.69 (20°C, shake flask-GC, Bowman & Sans 1983b; quoted, Suntio et al. 1988; Patil 1994)
- 2.75 (Hansch & Leo 1985; quoted, Howard 1991; selected, Sangster 1993)
- 2.70 (selected, Suntio et al. 1988)
- 2.76 (calculated, De Bruijn & Hermens 1991; quoted, Verhaar et al. 1992)
- 2.75 (selected, Hulzebos et al. 1993)
- 2.69, 2.75 (Montgomery 1993)
- 3.18 (calculated, Patil 1994)
- 2.96 (Tomlin 1994)
- 2.75 (selected, Hansch et al. 1995)

Bioconcentration Factor, log BCF:
- 1.96 (calculated-S as per Kenaga 1980, this work)
- 1.86 (calculated-K_{ow}, Lyman et al. 1982; quoted, Howard 1991)

Sorption Partition Coefficient, log K_{oc}:
- 2.61 (calculated-S, Lyman et al. 1982; quoted, Howard 1991)
- 2.28 (Fröbe et al. 1989; quoted, Howard 1991)
- 1.30 (selected, USDA 1989; quoted, Neary et al. 1993)
- 2.28 (derived from exptl., Meylan et al. 1992)
- 1.84 (calculated-χ, Meylan et al. 1992)
- 3.00 (soil, 20-25°C, selected, Wauchope et al. 1992; Dowd et al. 1993; Lohninger 1994; Hornsby et al. 1996)
- 2.47-3.53 (Montgomery 1993)
- 2.95 (soil, HPLC-screening method, Kördel et al. 1995)

Half-Lives in the Environment:
- Air: 1.3 h, based on an estimated rate constant for the vapor-phase reaction with hydroxyl radicals in air (Atkinson 1987; quoted, Howard et al. 1991).
- Surface water: half-lives of 415 d at 6°C, 115 d at 22°C in darkness for Milli-Q water; 278 d at 6 °C, 42 d at 22°C in darkness, 8 d under sunlight conditions for river water at pH 7.3; 506 d at 6°C, 35 d at 22°C in darkness for filtered river water at pH 7.3; 26 d at 22°C in darkness, 11 d under sunlight conditions for seawater at pH 8.1 (Lartiges & Garrigues 1995).
- Groundwater:
- Sediment:
- Soil: for dry soil with 2-3% moisture, half-lives were 484, 88, and 32 d at 6, 25, and 40°C, respectively; while for moist soil with 50% moisture content, half-lives were much shorter: 64, 13, and 5 d at 6, 25, and 40°C, respectively (Yaron et al. 1974; quoted, Montgomery 1993); selected field half-life of 10 d (Wauchope et al. 1992; Hornsby et al. 1996); average half-life of 40 d (Dowd et al. 1993); half-life in soil ranges from a few days to many weeks, depending on soil type (Tomlin 1994); 10 d (selected, Halfon et al. 1996).
- Biota: average half-life of 30 d in forest (selected, USDA 1989; quoted, Neary et al. 1993).

Environmental Fate Rate Constants or Half-Lives:
- Volatilization:
- Photolysis:
- Oxidation: photooxidation half-life of 1.3 h, based on an estimated rate constant for the vapor-phase reaction with hydroxyl radicals in air (Atkinson 1987; quoted, Howard et al. 1991).
- Hydrolysis: first-order half-lives of 36.4, 27.9, 7.2 d in water at pH 8.6 and 6°C, 25°C and 40°C (Heuer et al. 1974; quoted, Howard 1991); half-life of 27.9 d at pH 8.6 and 25°C (Montgomery 1993); half-lives at 22°C: 87 days at pH 4, 50 d at pH 7, and 4 d at pH 9 (Tomlin 1994).
- Biodegradation: studies with aquatic water/sediment microorganisms at 5 mg/L and pH 6.7 indicate half-lives of 3.3 d in microcosms compared to 2.7 days in field studies (Portier 1985; quoted, Howard 1991).
- Biotransformation:
- Bioconcentration, Uptake (k_1) and Elimination (k_2) Rate Constants:

Common Name: Bendiocarb
Synonym: Bencarbate, Dycarb, Ficam, Garvox, Multamat, Multimet, NC 6897, Niomil, Rotate, Seedox, Tatto, Turcam
Chemical Name: 2,3-isopropylidenedioxyphenylmethylcarbamate; 2,2-dimethyl-1,3-benzodioxol-4-yl methylcarbamate
Uses: contact insecticide used to control beetles, wireworms, flies, wasps, and mosquitoes in beets and maize.
CAS Registry No: 22781-23-3
Molecular Formula: $C_{11}H_{13}NO_4$
Molecular Weight: 223.2
Melting Point (°C):
    129.0-130.0 (Spencer 1982; Montgomery 1993; Milne 1995)
    124.6-128.7 (Tomlin 1994)
Boiling Point (°C):
Density (g/cm³ at 20°C):
    1.25 (Worthing 1991; Montgomery 1993; Tomlin 1994; Milne 1995)
Molar Volume (cm³/mol):
    231.7 (calculated-LeBas method, this work)
    178.6 (calculated-density, this work)
Molecular Volume (Å³):
Total Surface Area, TSA (Å²):
Dissociation Constant pK_a:
    8.80 (Worthing 1991; Wauchope et al. 1992; Montgomery 1993; Tomlin 1994; Hornsby et al. 1996)
Heat of Fusion, ΔH_{fus}, kcal/mol:
Entropy of Fusion, ΔS_{fus}, cal/mol·K (e.u.):
Fugacity Ratio at 25°C (assuming ΔS_{fus} = 13.5 e.u.), F: 0.0915

Water Solubility (g/m³ or mg/L at 25°C):
    40 (Spencer 1973, 1982; quoted, Shiu et al. 1990)
    40 (Martin & Worthing 1977; quoted, Kenaga 1980)
    40 (20°C, Agrochemicals Handbook 1987)
    40 (Worthing 1987; quoted, Shiu et al. 1990)
    26000 (Worthing 1991)
    40 (20-25°C, selected, Wauchope et al. 1992; Hornsby et al. 1996)
    40 (20°C, Montgomery 1993)
    40 (Lohninger 1994)
    280 (20°C at pH 7, Tomlin 1994)
    40 (20°C, Milne 1995)

Vapor Pressure (Pa at 25°C):
    6.6×10^{-4} (Agrochemicals Handbook 1987)
    0.00467 (20-25°C, selected, Wauchope et al. 1992; Hornsby et al. 1996)
    6.6×10^{-4} (20°C, Montgomery 1993)
    0.0046 (gas saturation-GC, Tomlin 1994)

Henry's Law Constant (Pa·m³/mol):
    0.365   (20°C, calculated-P/C, Montgomery 1993)

Octanol/Water Partition Coefficient, log K_{OW}:
    5.29   (selected, Dao et al. 1983)
    1.70   (Worthing 1991)
    1.70   (Montgomery 1993)
    1.72   (at pH 6.55, Tomlin 1994)
    1.70   (Milne 1995)
    1.70   (selected, Hansch et al. 1997)

Bioconcentration Factor, log BCF:
    1.89   (calculated-S, Kenaga 1980)

Sorption Partition Coefficient, log K_{OC}:
    2.76   (calculated-S, Kenaga 1980)
    2.76   (soil, 20-25°C, selected, Wauchope et al. 1992; Hornsby et al. 1996)
    2.76   (Montgomery 1993)
    2.76   (estimated-chemical structure, Lohninger 1994)
    1.45-1.60   (Tomlin 1994)

Half-Lives in the Environment:
    Air:
    Surface water: hydrolysis half-life of 4 d at 25°C and pH 7 under EPA gudelines (Spencer 1982)
    Groundwater:
    Sediment:
    Soil: half-life of several days to a few weeks (Hartley & Kidd 1987; quoted, Montgomery 1993); selected field half-life of 5.0 d (Wauchope et al. 1992; Hornsby et al. 1996).
    Biota:

Environmental Fate Rate Constants or Half-Lives:
    Volatilization:
    Photolysis:
    Oxidation:
    Hydrolysis: half-life of 4 d at pH 7 and 25°C (Spencer 1982; Montgomery 1993; Tomlin 1994).
    Biodegradation:
    Biotransformation:
    Bioconcentration, Uptake (k_1) and Elimination (k_2) Rate Constants:

Common Name: Carbaryl
Synonym: Arylam, Atoxan, Caproline, Carbamine, Carbatox, Carpolin, Carylderm, Cekubaryl, Crag sevin, Denapon, Devicarb, Dicarbam, ENT 23969, Gamonil, Germain's, Hexavin, Karbaspray, Karbatox, Karbosep, OMS 29, naphthyl carbamate, Panam, Ravyon, Rylam, Seffein, Septene, Sevimol, Sevin, Sok, Tercyl, Toxan, Union Carbide 7744
Chemical Name: carbamic acid, methyl-, 1-naphthyl ester; 1-naphthalenol, methyl carbamate; 1-naphthyl-N-methyl carbamate; 1-naphthyl methylcarbamate; 1-naphthalenyl methylcarbamate
Uses: contact insecticide used to control most insects on fruits, vegetables, and ornamentals; also used as growth regulator for fruit thinning of apples.
CAS Registry No: 63-25-2
Molecular Formula: $C_{12}H_{11}NO_2$
Molecular Weight: 201.22
Melting Point (°C):
    142   (Khan 1980; Karickhoff 1981; Spencer 1982; Lyman 1982; Bowman & Sans 1983b; Swann et al. 1983; Suntio et al. 1988; Patil 1994; Tomlin 1994; Kühne et al. 1995; Milne 1995)
    142.2   (Montgomery 1993)
Boiling Point (°C): dec. on distillation
Density (g/cm³ at 20°C):
    1.232   (Spencer 1982; Agrochemicals Handbook 1987; Montgomery 1993; Tomlin 1994; Milne 1995)
    0.52-0.61   (Worthing 1991)
Molar Volume (cm³/mol):
    218.7   (calculated-LeBas method, Suntio et al. 1988; Fisher et al. 1993)
    1.183   ($V_I/100$, Fisher et al. 1993)
Molecular Volume (Å³):
Total Surface Area, TSA (Å²):
Dissociation Constant pK_a:
Heat of Fusion, ΔH_{fus}, kcal/mol:
    5.80   (DSC method, Plato & Glasgow 1969)
Entropy of Fusion, ΔS_{fus}, cal/mol·K (e.u.):
Fugacity Ratio at 25°C (assuming ΔS_{fus} = 13.5 e.u.), F: 0.0696

Water Solubility (g/m³ or mg/L at 25°C):
    40   (shake flask, David et al. 1960; quoted, Shiu et al. 1990)
    40   (30°C, Spencer 1973; quoted, Khan 1980; Sharom et al. 1980; Suntio et al. 1988; Shiu et al. 1990; Schomburg et al. 1991)
    40   (Martin & Worthing 1977; quoted, Kenaga 1980; Kenaga & Goring 1980; Briggs 1981; Karickhoff 1981; Garten & Tralbalka 1983; Shiu et al. 1990; Somasundaram et al. 1991)
    40   (quoted, Wauchope 1978; Briggs 1981; Lyman 1982)
    40   (Weber et al. 1980; quoted, Willis & McDowell 1982)
    34   (20-25°C, shake flask-GC, Kanazawa 1981; quoted, Shiu et al. 1990)
    50   (20°C, Spencer 1982)
    104   (20°C, shake flask-GC, Bowman & Sans 1983a,b; quoted, Shiu et al. 1990)
    82.6   (gen. col., Swann et al. 1983; quoted, Shiu et al. 1990)
    590   (RP-HPLC, Swann et al. 1983; quoted, Shiu et al. 1990)
    83   (Seiber 1987; quoted, Shiu et al. 1990)

120   (30°C, Agrochemicals Handbook 1987)
40   (selected, Gerstl & Helling 1987)
120   (30°C, Worthing 1987; quoted, Shiu et al. 1990)
32   (20°C, selected, Suntio et al. 1988; quoted, Majewski & Capel 1995)
114   (Wauchope 1989; quoted, Shiu et al. 1990)
40   (30°C, Worthing 1991; Milne 1995)
120   (20-25°C, selected, Wauchope et al. 1992; Lohninger 1994; Hornsby et al. 1996)
30   (quoted, Pait et al. 1992)
104, 130 (20°C, 30°C, Montgomery 1993)
104, 45 (quoted, calculated, Patil 1994)
120   (20°C, Tomlin 1994)
106, 36 (quoted, calculated-group contribution fragmentation method, Kühne et al. 1995)
1000   (quoted, Milne 1995)
40   (30°C, Milne 1995)
120   (selected, Halfon et al. 1996)

Vapor Pressure (Pa at 25°C):
    <0.665 (26°C, Melnikov 1971; quoted, Khan 1980; Suntio et al. 1988)
    2.80×10^{-3} (20°C, Hartley & Graham-Bryce 1980; quoted, Suntio et al. 1988)
    <0.133 (20-25°C, Weber et al. 1980; quoted, Willis & McDowell 1982)
    1.81×10^{-4} (Ferreira & Seiber 1981; quoted, Howard 1991)
    <0.665 (26°C, Agrochemicals Handbook 1987)
    2.00×10^{-4} (20°C, selected, Suntio et al. 1988; quoted, Majewski & Capel 1995)
    <0.0053 (Worthing 1991)
    1.60×10^{-4} (20-25°C, selected, Wauchope et al. 1992; Hornsby et al. 1996)
    8.77×10^{-4} (Montgomery 1993)
    2.00×10^{-4} (23.5°C, Tomlin 1994)
    1.60×10^{-4} (selected, Halfon et al. 1996)

Henry's Law Constant (Pa·m^3/mol):
    0.0013   (calculated-P/C, Suntio et al. 1988; quoted, Howard 1991; Schomburg et al. 1991; Fisher et al. 1993; Majewski & Capel 1995)
    4.41×10^{-4} (known LWAPC, Meylan & Howard 1991)
    3.18×10^{-4} (bond-estimated LWAPC, Meylan & Howard 1991)
    1.287   (20°C, calculated-P/C, Montgomery 1993)
    4.48×10^{-5} (calculated-P/C, this work)

Octanol/Water Partition Coefficient, log K_{ow}:
    2.36   (Freed et al. 1976; quoted, Kenaga & Goring 1980; Bowman & Sans 1983b; selected, Dao et al. 1983; Hodson & Williams 1988)
    2.81   (Hansch & Leo 1979; quoted, Fisher et al. 1993)
    2.81   (quoted, Rao & Davidson 1980; Karickhoff 1981; Bowman & Sans 1983b; Trapp & Pussemier 1991)
    2.32   (Briggs 1981; quoted, Bowman & Sans 1983b; Bintein & Devillers 1994)
    2.36   (extrapolated from RP-TLC, Briggs 1981)
    2.36   (quoted, Lyman et al. 1982; Magee 1991; Trapp & Pussemier 1991)

2.31   (20°C, shake flask-GC, Bowman & Sans 1983b; quoted, Somasundaram et al. 1991; Patil 1994)
2.36   (Hansch & Leo 1985; selected, Gerstl & Helling 1987; Howard 1991)
2.30   (selected, Suntio et al. 1988)
3.13   (quoted, Isnard & Lambert 1989)
2.30   (Thor 1989; quoted, Connell & Markwell 1990)
2.408 (calculated as per Broto et al. 1984, Karcher & Devillers 1990)
2.34   (Suzuki & Kudo 1990; quoted, Trapp & Pussemier 1991)
2.14   (RP-HPLC-RT, Trapp & Pussemier 1991)
2.31-2.81 (Montgomery 1993)
1.99   (RP-HPLC, Saito et al. 1993)
2.27   (calculated-S with regression, Patil 1994)
1.59   (Tomlin 1994)
2.36   (selected, Hansch et al. 1995)
2.31   (selected, Devillers et al. 1996)

Bioconcentration Factor, log BCF:
< 0.0 (fish in static water, Metcalf & Sanborn 1975; Freed et al. 1976; quoted, Kenaga & Goring 1980)
1.89   (calculated-S, Kenaga 1980)
1.08   (calculated-K_{OC}, Kenaga 1980)
1.64   (earthworm, Lord et al. 1980; quoted, Connell & Markwell 1990)
0.95   (*Pseudorasbora parva*, Kanazawa 1981; quoted, Devillers et al. 1996)
1.86   (algae, Freitag et al. 1982)
1.53   (golden orfe, Freitag et al. 1982; quoted, Howard 1991)
1.95   (activated sludge, Freitag et al. 1982, 1984)
0.954 (topmouth gudgeon, Kanazawa 1983; quoted, Howard 1991)
1.45   (golden ide, Freitag et al. 1984; quoted, Howard 1991)
2.15   (quoted, Pait et al. 1992)

Sorption Partition Coefficient, log K_{OC}:
2.36   (soil, Leenheer & Atrichs 1971; LaFleur 1976; quoted, Kenaga 1980; Kenaga & Goring 1980; Karickhoff 1981; Lyman 1982; Hodson & Williams 1988)
2.36   (Kenaga 1980; quoted, Howard 1991; Schomburg et al. 1991)
2.76   (calculated-S as per Kenaga & Goring 1978, Kenaga 1980; quoted, Schomburg et al. 1991)
2.49   (av. of 3 soils, McCall et al. 1980)
2.02   (Briggs 1981; quoted, Bahnick & Doucette 1988)
1.78   (reported as log K_{OM}, Briggs 1981)
2.49   (average of 3 soils, HPLC-RT, McCall et al. 1980)
2.02   (soil slurry & UV method, Briggs. 1981; quoted, Howard 1991)
3.04, 2.50 (estimated-S, S & M.P., Karickhoff 1981)
2.42   (estimated-K_{OW}, Karickhoff 1981)
2.76, 2.66 (estimated-S, K_{OW}, Lyman 1982)
2.59   (soil slurry method, Swann et al. 1983; quoted, Howard 1991)
2.57   (reverse phase HPLC, Swann et al. 1983; quoted, Howard 1991)
1.85, 1.48, 1.95 (algae, golden ide, activated sludge, Freitag et al. 1985)
2.36, 2.14 (quoted, calculated-χ, Gerstl & Helling 1987)

2.36 (soil, screening model calculations, Jury et al. 1987b)
2.23 (calculated-χ, Bahnick & Doucette 1988)
2.47, 2.04 (reported as log K_{OM}, estimated as log K_{OM}, Magee 1991)
2.48 (soil, 20-25°C, selected, Wauchope et al. 1992; Hornsby et al. 1996)
2.30 (soil, Dowd et al. 1993)
2.02-2.59 (Montgomery 1993)
2.71 (estimated-chemical structure, Lohninger 1994)

Half-Lives or Fate Rate Constants in the Environment:
  Air: 12.6 hours, based on estimated rate constant for the vapor-phase reaction with photochemically produced hydroxyl radicals in the atmosphere (GEMS 1985; quoted, Howard 1991); 4.5-7.4 h, based on estimated rate constant for the vapor-phase reaction with hydroxyl radicals in air (Atkinson 1987; quoted, Howard et al. 1991); atmospheric transformation lifetime was estimated to be <1 d (Kelly et al. 1994).
  Surface water: 3.2-200 h, based on aqueous hydrolysis half-life at pH 9 and 28°C and photolysis half-life for winter sunlight at 40°N (Wolfe et al. 1976; quoted, Harris 1982; Howard et al. 1991); biodegradation rate constant of 2.4×10^{-10} mL(cell)$^{-1}$ d^{-1} in aquatic system (Wolfe et al. 1978; quoted, Scow 1982); half-lives of 37 d at 22°C for Milli-Q water at pH 6.1; 31 d at 6°C, 11 d at 22°C in darkness, 9 d under sunlight conditions for river water at pH 7.3; 45 d at 6°C, <2 d at 22°C in darkness for filtered river water at pH 7.3; 22 d at 6°C, <2 d at 22°C in darkness and 13 d under sunlight conditions for seawater at pH 8.1 (Lartiges & Garrigues 1995).
  Groundwater: 3.2-1440 h, based on aqueous hydrolysis half-life at pH 9 and 28°C (Wolfe et al. 1976; quoted, Howard et al. 1991) and unacclimated aerobic biodegradation half-life (Howard et al. 1991).
  Sediment:
  Soil: 97-251 h in dry soil and 4458-688 h in wet or saturated soil (Hautala 1978; quoted, Howard 1991); persistence of less than one month (Wauchope 1978); 3.2-720 h, based on aqueous hydrolysis half-life at pH 9 and 28°C (Wolfe et al. 1976; quoted, Howard et al. 1991) and unacclimated aerobic biodegradation half-life (Howard et al. 1991); biodegradation rate constant of 0.037 d^{-1} in soil by die-away test (Rao & Davidson 1980; quoted, Scow 1982); moderately persistent in soils with half-life of 20-100 d (Willis & McDowell 1982); half-life of 22 d from screening model calculations (Jury et al. 1987b); selected field half-life of 10 d (Wauchope et al. 1992; Dowd et al. 1993; Hornsby et al. 1996); soil half-life of 8 d (Pait et al. 1992); under aerobic conditions for concn. at 1 ppm will be degraded with a half-life of 7-14 d in a sandy loam and 14-28 d in a clay loam (Tomlin 1994); 10 d (selected, Halfon et al. 1996).
  Biota: biochemical half-life of 22 d from screening model calculations (Jury et al. 1987b).

Environmental Fate Rate Constants or Half-Lives:
  Volatilization: half-life of 3000 d was estimated from Henry's law constant for a body of water 1 m deep, flowing at 1 m/s and with a wind speed of 3 m/sec (Lyman et al. 1982; quoted, Howard 1991).
  Photolysis: half-life of 52-200 h in the atmosphere, based on aqueous photolysis data (Wolfe et al. 1976; quoted, Howard et al. 1991); aqueous half-life of 52-200 h,

based on reported photolysis half-life for summer and winter sunlight at 40°N (Wolfe et al. 1976; quoted, Howard et al. 1991).

Oxidation: photooxidation half-life of 4.5-7.4 h, based on estimated rate constant for the vapor-phase reaction with hydroxyl radicals in air (Atkinson 1987; quoted, Howard et al. 1991).

Hydrolysis: first-order half-life of 312 h, based on base rate constant at pH 7 and 25°C (Wolfe et al. 1976; quoted, Howard et al. 1991); half-life of 0.15 day at pH 9 and 27°C (Wolfe et al. 1978); half-lives at 27°C: 1500 days at pH 5, 15 d at pH 7, and 0.15 d at pH 9 (Montgomery 1993); half-lives: 12 d at pH 7 and 3.2 h at pH 9 (Tomlin 1994).

Biodegradation: aqueous aerobic half-life of 40-720 h, based on unacclimated aerobic river die-away test data (Eichelberger & Lichtenberg 1971; quoted, Howard et al. 1991) and freshwater grab sample data (Wolfe et al. 1978; quoted, Scow 1982; Howard et al. 1991); aqueous anaerobic half-life of 160-2880 h, based on unacclimated aerobic biodegradation half-life (Howard et al. 1991).

Biotransformation:

Bioconcentration, Uptake (k_1) and Elimination (k_2) Rate Constants:

Common Name: Carbofuran
Synonym: Bay 70143, Curaterr, ENT 27164, Furadan, NIA 10242, Niagara 10242, Yaltox
Chemical Name: 2,3-dihydro-2,2-dimethylbenzofuran-7-yl methylcarbamate; 2,3-dihydro-2,2-dimethyl-7-benzofuranyl methylcarbamate
Uses: broad-spectrum systemic insecticide, nematocide and acaricide applied in soil to control insects and nematodes; also to control insects and mites on foliage.
CAS Registry No: 1563-66-2
Molecular Formula: $C_{12}H_{15}NO_3$
Molecular Weight: 221.30
Melting Point (°C):
- 150-152 (Khan 1980; Spencer 1982; Suntio et al. 1988)
- 151 (Karickhoff 1981; Lyman 1982; Bowman & Sans 1983b; Swann et al. 1983; Patil 1994)
- 153-154 (Montgomery 1993; Tomlin 1994)
- 150-153 (Milne 1995)

Boiling Point (°C):
Density (g/cm³ at 20°C):
- 1.18 (Agrochemicals Handbook 1987; Trotter et al. 1991; Montgomery 1993; Tomlin 1994; Milne 1995)

Molar Volume (cm³/mol):
- 240.8 (calculated-LeBas method, Suntio et al. 1988; Fisher et al. 1993)
- 1.303 ($V_I/100$, Fisher et al. 1993)
- 187.5 (calculated-density)

Molecular Volume (Å³):
Total Surface Area, TSA (Å²):
Dissociation Constant pK_a:
Heat of Fusion, ΔH_{fus}, kcal/mol:
Entropy of Fusion, ΔS_{fus}, cal/mol·K (e.u.):
Fugacity Ratio at 25°C (assuming ΔS_{fus} = 13.5 e.u.), F: 0.0567

Water Solubility (g/m³ or mg/L at 25°C):
- 700 (Spencer 1973, 1982; quoted, Wauchope 1978; Bowman & Sans 1979)
- 250 (Caro et al. 1976)
- 415 (Martin & Worthing 1977; quoted, Kenaga 1980)
- 415 (Herbicide Handbook 1978, 1983; quoted, Kenaga & Goring 1980; Karickhoff 1981; Isnard & Lambert 1989; Shiu et al. 1990; Pait et al. 1992)
- 320 (19°C, shake flask-GC, Bowman & Sans 1979, 1983b; quoted, Fuhremann & Lichtenstein 1980; Sharom et al. 1980; Belluck & Felsot 1981; Jury et al. 1983, 1984; Shiu et al. 1990)
- 700 (Khan 1980; Weber et al. 1980; quoted, Willis & McDowell 1982)
- 415 (Lyman 1982; Thomas 1982; quoted, Nash 1988)
- 320 (Jury et al. 1983; quoted, Taylor & Glotfelty 1988)
- 480 (gen. col.-HPLC-RI, Swann et al. 1983; quoted, Shiu et al. 1990)
- 670 (RP-HPLC, Swann et al. 1983; quoted, Shiu et al. 1990)
- 700 (Verschueren 1983; quoted, Suntio et al. 1988; Shiu et al. 1990)
- 700 (Merck Index 1983, 1989; quoted, Somasundaram et al. 1991)
- 700 (Agrochemicals Handbook 1987; Milne 1995)
- 700 (Worthing 1987; quoted, Shiu et al. 1990; quoted, Howard 1991)
- 600 (20°C, selected, Suntio et al. 1988; quoted, Majewski & Capel 1995)

700 (selected, USDA 1989; quoted, Neary et al. 1993)
351 (Wauchope 1989; quoted, Shiu et al. 1990; Lohninger 1994)
700 (quoted, Trotter et al. 1991; Behrendt & Brüggemann 1993)
351 (20-25°C, selected, Wauchope et al. 1992; Hornsby et al. 1996)
320 (20°C, Montgomery 1993; Milne 1995)
375 (30°C, Montgomery 1993)
32, 2723 (quoted, calculated, Patil 1994)
320 (20°C, Tomlin 1994)

Vapor Pressure (Pa at 25°C):
1.12×10^{-3} (Knudsen effusion method, Cook 1973; quoted, Jury et al. 1983)
1.11×10^{-3} (Caro et al. 1976; quoted, Fuhrmann & Lichtenstein 1980)
1.11×10^{-3} (Fuhrmann & Lichtenstein 1980; quoted, Howard 1991)
8.67×10^{-4} (20°C, Hartley & Graham-Bryce 1980; quoted, Suntio et al. 1988)
2.70×10^{-3} (33°C, Khan 1980)
2.67×10^{-3} (20-25°C, Weber et al. 1980; quoted, Willis & McDowell 1982)
2.70×10^{-4} (Thomas 1982; quoted, Nash 1988)
1.10×10^{-3} (Jury et al. 1983; quoted, Taylor & Glotfelty 1988; Taylor & Spencer 1990)
2.70×10^{-3} (20°C, Agrochemicals Handbook 1987)
1.50×10^{-3} (20°C, selected, Suntio et al. 1988; quoted, Majewski & Capel 1995)
1.10×10^{-3} (selected, Taylor & Spencer 1990)
8.00×10^{-5} (20-25°C, selected, Wauchope et al. 1992; Hornsby et al. 1996)
2.70×10^{-3} (quoted, Behrendt & Brüggemann 1993)
7.20×10^{-5} (Tomlin 1994)

Henry's Law Constant (Pa·m^3/mol):
3.95×10^{-4} (calculated-P/C, Lyman et al. 1982; quoted, Howard 1991)
7.69×10^{-4} (Jury et al. 1984)
9.42×10^{-6} (Jury et al. 1987a,b; Jury & Ghodrati 1989)
5.10×10^{-4} (calculated-P/C, Suntio et al. 1988; quoted, Fisher et al. 1993; Majewski & Capel 1995)
7.69×10^{-4} (calculated-P/C, Taylor & Glotfelty 1988)
5.04×10^{-5} (calculated-P/C, this work)

Octanol/Water Partition Coefficient, log K_{ow}:
2.32 (Hansch & Leo 1979,85; quoted, Howard 1991; Fisher et al. 1993)
1.60 (quoted from Dow Chemical data, Kenaga & Goring 1980; quoted, Bowman & Sans 1983b; Suntio et al. 1988; Somasundaram et al. 1991)
2.32 (Rao & Davidson 1980; quoted, Bowman & Sans 1983b; Suntio et al. 1988)
2.88 (calculated as per Chiou et al. 1977, Belluck & Felsot 1981)
2.07 (quoted, Karickhoff 1981; quoted, Isnard & Lambert 1989; Trapp & Pussemier 1991)
1.60 (calculated, Lyman 1982; quoted, Trapp & Pussemier 1991)
1.63 (Suntio et al. 1988; quoted, Bintein & Devillers 1994; Patil 1994)
1.82 (RP-HPLC-RT, Trapp & Pussemier 1991)
1.82 (quoted, Behrendt & Brüggemann 1993)
1.60-2.32 (Montgomery 1993)

1.60   (RP-HPLC, Saito et al. 1993)
0.83   (calculated, Patil 1994)
1.52   (20°C, Tomlin 1994)
1.23-1.42 (Milne 1995)
2.32   (selected, Hansch et al. 1995)

Bioconcentration Factor, log BCF:
1.00   (estimated-log K_{OW}, Neely et al. 1974; quoted, Trotter et al. 1991)
1.32   (calculated-S, Kenaga 1980; quoted, Pait et al. 1992)
0.60   (*Triaenodes tardus*, Belluck & Felsot 1981)
1.00   (selected, Schnoor & McAvoy 1981; quoted, Schnoor 1992)
1.53   (calculated-log K_{OW}, Lyman et al. 1982; quoted, Howard 1991)
1.18   (calculated-S, Lyman et al. 1982; quoted, Howard 1991)
2.07   (*Tilapia nilotica*, Tejada & Magallona 1985; quoted, Howard 1991)
1.00   (*Pila luzonica*, Tejada & Magallona 1985; quoted, Howard 1991)
2.07   (paddy field fish, Tejada 1995; quoted, Abdullah et al. 1997)

Sorption Partition Coefficient, log K_{OC}:
2.20   (calculated-S as per Kenaga & Goring 1978, Kenaga 1980)
1.67   (calculated values for 6 samples while high organic carbon >15% were omitted from calculation by Felsot & Wilson 1980; quoted, Sukop & Cogger 1992)
1.78-2.20 (3 soils of org. content 0.68-2.01, MaCall et al. 1980; quoted, Howard 1991)
2.02   (av. of 3 soils, McCall et al. 1980)
1.46   (soil/sediments, Rao & Davidson 1980; quoted, Karickhoff 1981; Lyman 1982; Sukop & Cogger 1992)
2.46   (estimated-S, Karickhoff 1981)
1.51   (estimated-S & M.P., Karickhoff 1981)
1.68   (estimated-K_{OW}, Karickhoff 1981)
2.70   (selected, sediment/water, Schnoor & McAvoy 1981; quoted, Schnoor 1992)
2.25   (calculated-S, Lyman 1982)
1.00   (Thomas 1982; quoted, Nash 1988)
1.47   (av. of 5 different soils, Rao & Davidson 1982; quoted, Howard 1991)
1.46   (quoted, Jury et al. 1983, 1984)
2.11   (retention times of RP-HPLC, Swann et al. 1983)
2.00   (soil slurry method, Swann et al. 1983; quoted, Howard 1991)
1.45   (soil, screening model calculations, Jury et al. 1987a,b; Jury & Ghodrati 1989)
1.73   (calculated-Freund isotherm linearized for 12 samples, Sukop & Cogger 1992)
0.903   (selected, USDA 1989; quoted, Neary et al. 1993)
1.34   (soil, 20-25°C, selected, Wauchope et al. 1992; Dowd et al. 1993; Lohninger 1994; Tomlin 1994; Hornsby et al. 1996)
1.98-2.32 (Montgomery 1993)

Half-Lives in the Environment:
Air: 4.6 h, based on estimated rate constant for the vapor-phase reaction with photochemically produced hydroxyl radicals in the atmosphere (Atkinson 1987; quoted, Howard 1991).

Surface water: average half-life in rice paddy water was 57 h, but pH dependent, e.g., half-life 1.2 h at pH 10 and 864 h at pH 7 (Seiber et al. 1978); the half-lives for degradation in river, lake and seawater from Greece which were irradiated with sunlight were approximately 2, 6 and 12 h, respectively (Samanidou et al. 1988; quoted, Howard 1991).

Groundwater:

Sediment:

Soil: persistence of less than one month (Wauchope 1978); half-life of 11-13 days at pH 6.5, and 60-75 d for a granular formulation (Ahmad et al. 1979; quoted, Montgomery 1993); estimated first-order half-lives of 15 d in soil from biodegradation rate constant of 0.047 d^{-1} by die-away test from soil incubation studies and of 26 d from biodegradation rate constant of 0.026 d^{-1} in anaerobic system from flooded soil incubation studies by die-away test (Rao & Davidson 1980; quoted, Scow 1982); moderately persistent in soils with half-life of 20-100 days (Willis & McDowell 1982); half-life of 1-2 months (Hartley & Kidd 1987; quoted, Montgomery 1993); half-life of 40 d from screening model calculations (Jury et al. 1987a,b; Jury & Ghodrati 1989); selected field half-life of 50 d (Wauchope et al. 1992; Dowd et al. 1993; Hornsby et al. 1996); soil half-life of 81 days (Pait et al. 1992); half-lives of 30 d for soil depth <5 cm, 60 d for soil depth 5-20 cm and 120 d for soil depth >20 cm (Dowd et al. 1993); half-life of 60 d in forest soil (Neary et al. 1993); half-lives of 42.4 d in loam and 95.5 d in sand (Behrendt & Brüggemann 1993); half-life of 30-60 d in soil (Tomlin 1994).

Biota: biochemical half-life of 40 d from screening model calculations (Jury et al. 1987a,b; Jury & Ghodrati 1989); 4 d in wheat/barley (Behrendt & Brüggemann 1993); average half-life of 60 d in the forest (USDA 1989; quoted, Neary et al. 1993).

Environmental Fate Rate Constants or Half-Lives:

Volatilization: initial rate constant of 1.2×10^{-3} h^{-1} and a predicted rate constant of 2.9×10^{-4} h^{-1} from soil with half-life of 2390 h (Thomas 1982).

Photolysis: near surface direct sunlight photolysis rate constant of 0.003 d^{-1} with half-life of about 200 d (Schnoor & McAvoy 1981; quoted, Schnoor 1992); the half-lives for degradation in river, lake and seawater from Greece which were irradiated with sunlight were approximately 2, 6, and 12 h, respectively (Samanidou et al. 1988; quoted, Howard 1991).

Oxidation: photooxidation half-life of 4.6 h, based on estimated rate constant for the vapor-phase reaction with hydroxyl radicals in air (Atkinson 1987; quoted, Howard et al. 1991); calculated rate constant of 7×10^9 M^{-1} s^{-1} for the reaction with hydroxyl radicals in aqueous solutions at $(24 \pm 1)°C$ (Haag & Yao 1992).

Hydrolysis: the aqueous hydrolysis half-life at 27°C was found to be 5.1 weeks at pH 7.0 and 1.2 h at pH 10 (Seiber et al. 1978; quoted, Howard 1991); alkaline chemical hydrolysis rate constant of 6×10^{-5} M^{-1} s^{-1} with half-life of >10,000 days (Schnoor & McAvoy 1981; quoted, Schnoor 1992); half-lives of 690, 8.2, and 1.0 weeks in water at 25°C and pH 6.0, 7.0 and 8.0 respectively (Chapman & Cole 1982; quoted, Howard 1991); hydrolysis rate constants: 30.6 ± 0.6 L/min/mol at 15°C, 67.0 ± 0.4 L/min/mol at 25°C and 163 ± 1.0 l/min/mol at 35°C (Trotter et al. 1991); half-lives at 25°C: 170 weeks at pH 4.5, 690 weeks at pH 5-6, 8.2 weeks at ph 7 and one week at pH 8.0 (Montgomery

1993); half-lives at 22°C: >1 yr at pH 4, 121 d at pH 7, and 31 d at pH 9 (Tomlin 1994).

Biodegradation: rate constants of 0.047 d^{-1} from soil incubation studies and 0.026 d^{-1} in anaerobic system from flooded soil incubation studies both by die-away test (Rao & Davidson 1980; quoted, Scow 1982); half-life of 40 d in 0 to 10 cm depth of soil (Rao & Davidson 1980; quoted, Jury et al. 1983, 1984, 1987a,b; Jury & Ghodrati 1989).

Biotransformation:

Bioconcentration, Uptake (k_1) and Elimination (k_2) Rate Constants:

Common Name: Chlordane

Synonym: A 1068, Aspon-chlordane, Belt, beta-chlordane, CD-68, Chlorindan, Chlor-Kill, Chlortox, Corodane, Cortilan-neu, Dichlorochlordene, Dowchlor, ENT 9932, ENT 25552, HCS 3260, Kypchlor, M 140, Octachlor, Octaterr, Orthoklor, Shell SD 5532, Synklor, Tat chlor 4, Topichlor, Toxichlor, Velsicol

Chemical Name: 1,2,4,5,6,7,8,8-octachloro-3a,4,7,7a-tetrahydro-4,7-methano-1H-indane; 1,2,4,5,6,7,8,8-octachloro-3a,4,7,7a-tetrahydro-4,7-methanoindane

Uses: nonsystemic insecticide with contact, stomach, and respiratory action and also used as fumigant.

CAS Registry No: 57-74-9 (nonstereospecific chlordane); 5103-71-9 (*cis*- or α-isomer); 5103-74-2 (*trans*- or β-isomer); 5564-34-7 (γ-isomer); 12789-03-6 (technical grade chlordane)

Molecular Formula: $C_{10}H_6Cl_8$

Molecular Weight: 409.8

Melting Point (°C):
    107-108.8   (*cis*-isomer, Callahan et al. 1979; Howard 1991)
    103-105   (*trans*-isomer, Callahan et al. 1979; Suntio et al. 1988; Howard 1991)
    106-107   (*cis*-isomer, Tomlin 1994)
    104-105   (*trans*-isomer, Tomlin 1994)

Boiling Point (°C):
    175   (at 2 mmHg, Roark 1951; Callahan et al. 1979; Howard 1991; Montgomery 1993)
    262, 363, 365 (estimated from structure, Tucker et al. 1983)
    175   (at 1 mmHg, Agrochemicals Handbook 1987; Tomlin 1994; Milne 1995)

Density (g/cm^3 at 20°C):
    1.59-1.63 (25°C, Worthing 1987, 1991; Tomlin 1994; Milne 1995)
    1.59-1.63 (Montgomery 1993)

Molar Volume (cm^3/mol):
    336.5 (calculated-LeBas method, Suntio et al. 1988)
    256.6 (Ruelle & Kesselring 1997)
    340.5 (calculated-LeBas method, this work)

Molecular Volume (Å^3):

Total Surface Area, TSA (Å^2):

Dissociation Constant pK_a:

Heat of Fusion, ΔH_{fus}, kcal/mol:
    6.70 (*cis*-isomer, DSC method, Plato 1972)
    6.80 (*trans*-isomer, DSC method, Plato 1972)

Entropy of Fusion, ΔS_{fus}, cal/mol·K (e.u.):

Fugacity Ratio at 25°C (assuming ΔS_{fus} = 13.5 e.u.), F:
    0.162 (Mackay et al. 1986)
    0.140 (20°C, Suntio et al. 1988)

Water Solubility (g/m^3 or mg/L at 25°C):
    1.850 (gen. col.-GC/ECD, Weil et al. 1974; quoted, Callahan et al. 1979; Warner et al. 1987; Suntio et al. 1988; Shiu et al. 1990)
    0.056 (shake flask-LSC, Sanborn et al. 1976; quoted, Callahan et al. 1979; Garten & Trabalka 1983; Suntio et al. 1988; Shiu et al. 1990; Howard 1991; Mortimer & Connell 1995)
    0.056 (Martin & Worthing 1977; quoted, Kenaga 1980a,b; Kenaga & Goring 1980; quoted, Bysshe 1982; Zaroogian et al. 1985; Mackay et al. 1986)

0.10   (Agrochemicals Handbook 1987)
0.10   (Worthing 1987, 1991; quoted, Shiu et al. 1990; Howard 1991)
0.56   (Agency for Toxic Substances and Disease Registry 1988; quoted, Burmaster et al. 1991)
0.056   (α-chlordane, quoted, Isnard & Lambert 1988)
0.05   (20°C, selected, Suntio et al. 1988; quoted, Majewski & Capel 1995)
0.06   (quoted, Hinman & Klaine 1992)
0.032; 0.009-0.056 (shake flask-LSC; quoted lit. range, Johnson-Logan et al. 1992)
1.83   (selected, Yalkowsky & Banerjee 1992)
0.056   (Montgomery 1993)
0.06   (20-25°C, selected, Augustijn-Beckers et al. 1994; Hornsby et al. 1996)
0.057   (quoted, Kühne et al. 1995)
0.127   (calculated-group contribution fragmentation method, Kühne et al. 1995)
0.10   (Tomlin 1994; Milne 1995)
0.050; 0.061, 0.002 (quoted; predicted-molar volume, Ruelle & Kesselring 1997)

Vapor Pressure (Pa at 25°C):
0.0013 (Martin 1972, Spencer 1973, 1982; quoted, Bidleman & Christensen 1979; Callahan et al. 1979; Khan 1980; Mackay et al. 1986; Suntio et al. 1988)
0.0013 (SRI International 1980; quoted, Tucker et al. 1983)
0.00227, 1.6×10^{-5}, 1.3×10^{-5} (estimated-B.P., Tucker et al. 1983)
2.9×10^{-3}, 3.86×10^{-3} (*cis*-, *trans*-chlordane, 20°C, supercooled liq. value, Bidleman et al. 1986)
0.0013 (Agrochemicals Handbook 1987)
0.0011 (20°C, selected, Suntio et al. 1988; quoted, Majewski & Capel 1995)
0.0613, 0.00133 (technical, refined, Worthing 1987, quoted, Howard 1991)
0.00293, 0.00040 (*cis*-isomer, GC-RT, supercooled liquid, solid crystal, Foreman & Bidleman 1987; quoted Howard 1991)
0.00387, 0.00052 (*trans*-isomer, GC-RT, supercooled liquid, solid crystal, Foreman & Bidleman 1987, quoted, Howard 1991)
0.00130 (Agency for Toxic Substances and Disease Registry 1988)
4.5×10^{-3}, 5.1×10^{-3}, 4.8×10^{-3} (*cis*-chlordane, GC-RT, supercooled liquid values, Hinckley et al. 1990)
6.3×10^{-3}, 6.9×10^{-3}, 6.7×10^{-3} (*trans*-chlordane, GC-RT, supercooled liquid values, Hinckley et al. 1990)
$\log P_L$ (Pa) = 12.04 − 4284/T (*cis*-chlordane, supercooled liquid values, Hinckley et al. 1990)
$\log P_L$ (Pa) = 11.95 − 4216/T (*trans*-chlordane, supercooled liquid values, Hinckley et al. 1990)
0.0013 (refined, Worthing 1991; Tomlin 1994)
0.0610 (technical grade, Worthing 1991)
0.00269, 0.00813 (*cis*-chlordane, supercooled liquid values at 20°C, 30°C, calculated from Hinckley et al. 1990; Cotham & Bidleman 1992)
0.000133 (20°C, Montgomery 1993)
0.00133 (20-25°C, selected, Augustijn-Beckers et al. 1994; Hornsby et al. 1996)
4.59×10^{-4} (quoted as mean of *cis* and *trans* forms from Howard 1991, Mortimer & Connell 1995)

Henry's Law Constant (Pa·m³/mol):
- 2.92-9.5 (Callahan et al. 1979, quoted, Suntio et al. 1988)
- 4.92 (exptl., Warner et al. 1980, quoted, Tucker et al. 1983; Suntio et al. 1988)
- 0.294 (calculated-P/C, Levins 1981; quoted, Tucker et al. 1983)
- 112 (batch stripping, average of *cis*- and *trans*- isomers, Atlas et al. 1982; quoted, Cotham & Bidleman 1991)
- 87.7, 134 (exptl.: α-chlordane, γ-chlordane, Atlas et al. 1982; quoted, Suntio et al. 1988, Howard 1991)
- 9.12 (Mabey et al. 1982; quoted, Suntio et al. 1988)
- 3.44 (estimated-group method of Hine & Mookerjee 1975, Tucker et al. 1983)
- 9.64 (calculated-P/C, Mackay et al. 1986)
- 0.248 (calculated-P/C, Jury et al. 1990)
- 4.86 (gas stripping-GC, Warner et al. 1987; quoted, Cotham & Bidleman 1991)
- 9.02 (20°C, calculated-P/C, Suntio et al. 1988; quoted, Cotham & Bidleman 1991; Majewski & Capel 1995)
- 9.66 (WERL Treatability Database, quoted, Ryan et al. 1988)
- 0.97 (Agency for Toxic Substances and Disease Registry 1988; quoted, Burmaster et al. 1991)
- 4.91 (technical chlordane, Howard 1991)
- 6.60 (wetted-wall column-GC, Fendinger et al. 1989; quoted, Cotham & Bidleman 1991)
- 0.87 (0°C, selected, Cotham & Bidleman 1991)
- 9.02 (20°C), 140, 570 (23°C), 9.64 (25°C) (*trans*-chlordane, quoted, Iwata et al. 1993)
- 9.02 (20°C), 89, 420 (23°C), 9.64, 11.2 (25°C) (*cis*-chlordane, quoted, Iwata et al. 1993)
- 8.42 (γ-chlorane, wetted-wall column-concn ratio-GC/ECD, Fendinger et al. 1989)
- 5.45 (γ-chlorane, fog chamber-concn ratio-GC/ECD, Fendinger et al. 1989)
- 8.11, 8.42 (calculated-bond contribution, quoted, Meylan et al. 1991)

Octanol/Water Partition Coefficient, log K_{OW}:
- 2.78 (shake flask-LSC, Sanborn et al. 1976; quoted, Callahan et al. 1979; Mackay et al. 1986; Suntio et al. 1988; Chapman 1989; Sicbaldi & Finizio 1993)
- 6.00 (Veith et al. 1979, 1980; Veith & Kosian 1983; quoted, Bysshe 1982; Oliver & Charlton 1984; Zaroogian et al. 1985; Hawker & Connell 1986; Isnard & Lambert 1988, 1989; Chapman 1989; Saito et al. 1992)
- 5.16 (calculated, Kenaga 1980a,b)
- 3.32 (quoted, Rao & Davidson 1980; Suntio et al. 1988)
- 5.89 (selected, Dao et al. 1983)
- 5.54 (CLOGP 1986; quoted, Howard 1991)
- 5.58 (selected, Yoshioka et al. 1986)
- 2.78, 5.48 (Schnoor et al. 1987; quoted, Wilcock et al. 1993)
- 3.30 (Agency for Toxic Substances & Disease Registry 1988; quoted, Burmaster et al. 1991)
- 3.0 (selected, Suntio et al. 1988; quoted, Findinger et al. 1989)
- 5.9, 6.1 (*cis*-, *trans*-chlordane, Kawano et al. 1988)
- 4.30 (Ryan et al. 1988)
- 6.00 (Travis & Arms 1988; Markwell et al. 1989; Thomann 1989)
- 6.00 (Endicott & Richardson 1989; quoted, Parkerton et al. 1993)
- 5.54 (Thor 1989; quoted, Connell & Markwell 1990)
- 5.58 (quoted, Hinman & Klaine 1992; Johnson-Logan et al. 1992)

6.21 (estimated by QSAR & SPARC, Kollig 1993)
6.00 (Montgomery 1993; Devillers et al. 1996)
5.08 (RP-HPLC, Sicbaldi & Finizio 1993)
5.08, 5.80, 4.75 (RP-HPLC, ClogP, calculated-S, Finizio et al. 1997)

Bioconcentration Factor, log BCF:
    3.51-3.92 (wet wt. basis 96-h test, eastern oysters, Parrish et al. 1976)
    3.60-3.78 (wet wt. basis 96-h test, pink shrimp, Parrish et al. 1976)
    3.28-3.36 (wet wt. basis 96-h test, grass shrimp, Parrish et al. 1976)
    4.10-4.27 (wet wt. basis 96-h test, pin fish, Parrish et al. 1976)
    4.69 (*Oedogonium cardiacum*, Sanborn et al. 1976; quoted, Baughman & Paris 1981)
    3.66 (spot fish, 24-h flow system, *trans*-chlordane, Schimmel et al. 1976, quoted, Howard 1991)
    3.57-4.23 (96 h exposures of *trans*-chlordane to whole marine fish, Schimmel et al. 1976)
    4.11-4.34 (186 days chronic exposures *trans*-chlordane to sheephead minnow, Parrish et al. 1978; quoted, Howard 1991)
    3.85-4.46 (exposures *trans*-chlordane to 28-day-old second generation sheephead minnow, Parrish et al. 1978; quoted, Howard 1991)
    4.58 (fathead minnows, 32-d exposure, Veith et al. 1979, 1980; Veith & Kosian 1983; quoted, Bysshe 1982; Zaroogian et al. 1985; Devillers et al. 1996)
    4.01 (green algae, Glooschenko et al. 1979; quoted, Howard 1991; Saito et al. 1992)
    4.06, 3.92 (fish: flowing water, static ecosystem, Kenaga 1980a,b; Kenaga & Goring 1980; quoted, Howard 1991)
    3.50 (calculated-S, Kenaga 1980a)
    -0.523 (average beef fat diet, Kenaga 1980b)
    2.08, 4.32 (estimated-S, K_{ow}, Bysshe 1982)
    3.68, 3.64 (α-, γ-chlordane, clam fat, 60-d expt., Hartley & Johnson 1983)
    4.06-4.58 (fish, Bysshe 1987)
    5.04-5.88 (earthworms, Gish & Hughes 1982; quoted, Connell & Markwell 1990)
    4.58; 4.33, 4.40 (measured for fathead minnows; calculated-K_{ow} for sheepshead minnows, Zaroogian et al. 1985)
    4.58; 3.90, 3.92 (measured for fathead minnows; calculated-K_{ow} for pinfish, Zaroogian et al. 1985)
    4.58; 4.33, 4.40 (measured for fathead minnows; calculated-K_{ow} for oyster, Zaroogian et al. 1985)
    3.70 (oyster, Hawker & Connell 1986)
    6.73, 6.99 (total chlordanes, zooplankton, thick-billed murre, Kawano et al. 1986)
    3.78, 4.30 (juvenile and adult sheepshead minnows, 28-129 d exposure, Parrish et al. 1978; quoted, Howard 1991)
    6.0-7.0 (zooplankton and Chum salmon, Kawano et al. 1988)
    5.57 (Markwell et al. 1989)
    3.52, 2.60 (large mouth bass, clams, 106-127 d exposure, NRC 1974, quoted, Howard 1991)
    3.86 (eastern oyster, 10-d exposure, NRC 1974; quoted, Howard 1991)
    3.74 (white sucker and redhorse, Roberts et al. 1977; quoted, Howard 1991)
    2.03, 2.51, 3.0 (frogs, bluegills, goldfish, Verschueren 1983)
    6.73, 7.89, 6.99 (total chlordane: zooplankton, Dall's porpoise, thick-billed murre, Kawano et al. 1986)
    4.58 (estimated-S and K_{ow}, Isnard & Lambert 1988)

-2.13 (beef biotransfer factor log B_b, correlated-K_{OW} from Kenaga 1980, Travis & Arms 1988)
-3.43 (milk biotransfer factor log B_m, correlated-K_{OW} from Dorough & Hemken 1973, Travis & Arms 1988)
-1.81 (vegetation, correlated-K_{OW} from Dorough & Pass 1973; Tafuri et al. 1977; Travis & Arms 1988)
3.03 (*Hydrilla*, Hinman & Klaine 1992)
4.01, 3.50; 2.80 (estimated: fish-based, duckweed-based, Hinman & Klaine 1992)
3.117, 3.098 (*cis*-, *trans*-chlordane, goldfish, Park & Erstfeld 1997)

Sorption Partition Coefficient, log K_{OC}:
4.33 (calculated-S as per Kenaga & Goring 1978, Kenaga 1980a,b)
1.58 (screening model calculations, Jury et al. 1987b)
4.64, 4.09 (calculated-K_{OW} as per Kenaga & Goring 1980, Chapman 1989)
5.50, 5.60 (α-chlordane: field sediment trap material, calculated-K_{OW}, Oliver & Charlton 1984)
5.40, 5.60 (γ-chlordane, field sediment trap material, calculated-K_{OW}, Oliver & Charlton 1984)
4.58 (soil, screening model calculations, Jury et al. 1987b, 1990)
4.39, 4.19 (calculated-K_{OW} and solubility, Howard 1991)
4.77, 4.94 (α-chlordane: quoted, calculated-χ, Meylan et al. 1992)
5.45, 4.40-4.86 (Aldrich humic acid, soil, Johnson-Logan 1992)
5.90 (estimated by QSAR & SPARC, Kollig 1993)
4.85-5.57 (Montgomery 1993)
4.30 (20-25°C, selected, Augustijn-Beckers et al. 1994; Hornsby et al. 1996)
4.33, 4.42 (log K_P, *cis*-, *trans*-chlordane, Park & Erstfeld 1997)

Half-Lives in the Environment:
Air: 5.2-51.7 h, based on estimated photooxidation half-life in air (Atkinson 1987; quoted, Howard et al. 1991; Mortimer & Connell 1995); atmospheric transformation lifetime was estimated to be >1 d (Kelly et al. 1994).
Surface water: 5712-33264 h, based on unacclimated aerobic river die-away test data (Eichelberger & Lichtenberg 1971; quoted, Howard et al. 1991; Mortimer & Connell 1995) and reported soil grab sample data (Castro & Yoshida 1971; quoted, Howard et al. 1991; Mortimer & Connell 1995).
Groundwater: 11424-66528 h, based on estimated aqueous aerobic biodegradation half-life (Howard et al. 1991).
Sediment: 20000 h (quoted mean value from Howard et al. 1991, Mortimer & Connell 1995).
Soil: half-life of about 6 yr persistence in soil (Nash & Woolson 1967); estimated persistence of 5 yr in soil (Kearney et al. 1969; Edwards 1973; quoted, Morrill et al. 1982; Jury et al. 1987); 5712-33264 h, based on unacclimated aerobic river die-away test data (Eichelberger & Lichtenberg 1971; quoted, Howard et al. 1991) and reported soil grab sample data (Castro & Yoshida 1971; quoted, Howard et al. 1991); rate constant of 0.0024 d^{-1} with a biodegradation half-life of 1214 d under field conditions (Rao & Davidson 1980); field half-life of 9 d in fallow soil (Glotfelty 1981; quoted, Nash 1983); persistent with half-life >100 d (Willis & McDowell 1982); microagroecosystem half-life of 10-13 d in moist fallow soil (Nash 1983); half-life in soil about 1 yr (Hartley & Kidd 1987; quoted, Montgomery 1993); half-life of 3500 d from screening model

calculations (Jury et al. 1987b); >50 d when subject to plant uptake via volatilization (Ryan et al. 1988); degradation half-life of 100 d in soil (Jury et al. 1990); mean half-life under field conditions is 3.3 yr (Howard 1991); estimated field half-life of 350 d (Augustijn-Beckers et al. 1994; Tomlin 1994; Hornsby et al. 1996).

Biota: 1 d for daphids and 60 d for fish (Callahan et al. 1979; quoted, Wilcock et al. 1993); biochemical half-life of 3500 d from screening model calculations (Jury et al. 1987b); depuration half-lives of 7.1 d for γ-chlordane, and 5.9 days for α-chlordane (rats, Dearth & Hites 1991); 12 d for elimination from *T. liliana* (Wilcock et al. 1993).

Environmental Fate Rate Constants or Half-Lives:

Volatilization: the volatilization half-lives of γ- and α- chlordane from a model river 1 m deep flowing 1 m/s with a wind velocity of 3 m/s is estimated to be 7.3 and 7.9 h respectively (Lyman et al. 1982, Howard 1991); and from a model environmental pond (2 m deep), river (3 m deep) and lake (5 m deep) are estimated to be 18-26, 3.6-5.2, 14.4-20.6 d respectively (Lyman et al. 1982; Howard 1991); based on the Henry's law constant of technical chlordane, estimated half-life of 43 h from a model river one meter deep flowing 1 m/s (Lyman et al. 1982, Howard 1991); measured rate constant of 0.3 d^{-1} (Glotfelty et al. 1984; quoted, Glotfelty et al. 1989); calculated rate constant of 1.0 d^{-1} (Glotfelty et al. 1989).

Photolysis:

Oxidation: half-life of 5.2-51.7 h in air, based on estimated photooxidation half-life in air (Atkinson 1987; quoted, Howard et al. 1991); rate constant of 8×10^8 $M^{-1} \cdot s^{-1}$ for the reaction (photo-Fenton with reference to lindane) with hydroxyl radicals in aqueous solutions at pH 3.3 and at (24 ± 1)°C (Buxton et al. 1988; quoted, Faust & Hoigné 1990; Haag & Yao 1992); rate constant of $(6-170) \times 10^8$ $M^{-1} \cdot s^{-1}$ for the reaction (photo-Fenton with reference to acetophenone) with hydroxyl radicals in aqueous solutions at pH 3.3 and at (24 ± 1)°C (Haag & Yao 1992).

Hydrolysis: half-life >4 yr (Callahan et al. 1979); first-order half-life >197000 yr, based on base rate constant of 4.3×10^{-3} $M^{-1} h^{-1}$ at pH 7.0 and 25°C (Ellington et al. 1987, 1988; quoted, Howard et al. 1991).

Biodegradation: aqueous aerobic half-life of 5712-33264 h, based on unacclimated aerobic river die-away test data (Eichelberger & Lichtenberg 1971; quoted, Howard et al. 1991) and reported soil grab sample data (Castro & Yoshida 1971; quoted, Howard et al. 1991); rate constant of 0.0024 d^{-1} with a biodegradation half-life of 1214 d under field conditions (Rao & Davidson 1980); aqueous anaerobic half-life of 24-168 h, based on soil and freshwater mud grab sample data for aldrin, dieldrin, endrin and heptachlor epoxide (Maule et al. 1987; quoted, Howard et al. 1991); half-life of 100 d in soil (Jury et al. 1990).

Biotransformation:

Bioconcentration, Uptake (k_1) and Elimination (k_2) Rate Constants:

k_2: 0.0974 d^{-1} (γ-chlordane from rats, Dearth & Hites 1991)
k_2: 0.1170 d^{-1} (α-chlordane from rats, Dearth & Hites 1991)

Common Name: Chlorfenvinphos
Synonym: Apachlor, Birlane, Clofenvenfos, GC 4092, Sapecron, SD 7859
Chemical Name: 2-chloro-1-(2,4-dichlorophenyl)vinyl diethyl phosphate; 2-chloro-1-(2,4-dichlorophenyl)ethenyl diethyl phosphate
Uses: soil application of insecticide to control root flies, root worms and other soil insects in vegetables; foliar application to control Colorado beetles on potatoes; scale insects and mite eggs on citrus fruit; stem borers and leafhoppers on rice, maize and sugar cane; and white flies on cotton; aside from control of mosquito larvae, it is also used as acaricide and animal ectoparasiticide.
CAS Registry No: 470-90-6 (Z)-isomer, 18708-87-7 (E)-isomer
Molecular Formula: $C_{12}H_{14}Cl_3O_4P$
Molecular Weight: 359.56
Melting Point (°C):
    −19.0 (Khan 1980; Suntio et al. 1988)
    −23 to −19 (Tomlin 1994)
    −22 to −16 (Milne 1995)
Boiling Point (°C):
    167-170 (at 0.5 mmHg, Agrochemicals Handbook 1987; Tomlin 1994)
    167-170 (at 0.05 mmHg, Milne 1995)
    120 (at 0.001 mmHg, Milne 1995)
Density (g/cm³ at 20°C):
    1.36 (Agrochemicals Handbook 1987; Tomlin 1994)
Molar Volume (cm³/mol):
    321.4 (calculated-LeBas method, Suntio et al. 1988)
Molecular Volume (Å³):
Total Surface Area, TSA (Å²):
Dissociation Constant pK_a:
Heat of Fusion, ΔH_{fus}, kcal/mol:
Entropy of Fusion, ΔS_{fus}, cal/mol·K (e.u.):
Fugacity Ratio at 25°C (assuming ΔS_{fus} = 13.5 e.u.), F: 1.0

Water Solubility (g/m³ or mg/L at 25°C):
    145 (20°C, Melnikov 1971; quoted, Shiu et al. 1990)
    145 (Martin & Worthing 1977; quoted, Kenaga 1980)
    145 (23°C, Khan 1980; quoted, Suntio et al. 1988; Shiu et al. 1990)
    146 (quoted, Briggs 1981)
    124 (20°C, shake flask-GC, Bowman & Sans 1983a,b; quoted, Shiu et al. 1990)
    145 (23°C, Agrochemicals Handbook 1987; Tomlin 1994)
    130 (20°C, selected, Suntio et al. 1988)
    145 (23°C, Worthing 1987, 1991; quoted, Shiu et al. 1990)
    145 (Milne 1995)

Vapor Pressure (Pa at 25°C):
    0.00053 (20°C, Khan 1980; quoted, Suntio et al. 1988)
    2.7×10^{-5} (Verschueren 1983; quoted, Suntio et al. 1988)
    0.00100 (Agrochemicals Handbook 1987; Tomlin 1994)
    0.00010 (20°C, selected, Suntio et al. 1988)
    0.00053 (20°C, Worthing 1991)

Henry's Law Constant (Pa·m^3/mol):
    0.00028   (20°C, calculated-P/C, Suntio et al. 1988)
    0.00029   (calculated-P/C, this work)

Octanol/Water Partition Coefficient, log K_{OW}:
    3.23   (shake flask, Lord et al. 1980; Thor 1989; quoted, Connell & Markwell 1990)
    3.10   (shake flask-GC, Briggs 1981; quoted, Bintein & Devillers 1994)
    3.81   (shake flask-GC, Bowman & Sans 1983; quoted, Suntio et al. 1988)
    3.84   (shake flask, Eadsforth & Moser 1983; quoted, Sicbaldi & Finizio 1993)
    3.79   (HPLC method, Eadsforth & Moser 1983; quoted, Sicbaldi & Finizio 1993)
    3.10   (quoted and selected, Suntio et al. 1988)
    3.10, 3.81, 3.82, 3.79; 3.56 (quoted values; RP-HPLC, Sicbaldi & Finizio 1993)
    3.81   (selected, Sangster 1993)
    3.85, 4.22   ((Z) isomer, (E) isomer, Tomlin 1994)
    3.10   (selected, Hansch et al. 1995)
    3.56, 3.27, 3.08 (RP-HPLC, ClogP, calculated-S, Finizio et al. 1997)

Bioconcentration Factor, log BCF:
    1.57   (calculated-S, Kenaga 1980)
    2.30   (earthworms, Lord et al. 1980; quoted, Connell & Markwell 1990)

Sorption Partition Coefficient, log K_{OC}:
    2.45   (calculated-S, Kenaga 1980)
    2.23   (reported as log K_{OM}, Briggs 1981)
    2.47   (soil, quoted exptl., Meylan et al. 1992)
    2.77   (soil, calculated-χ and fragment contribution, Meylan et al. 1992)

Half-Lives in the Environment:
    Air:
    Surface water:
    Groundwater:
    Sediment:
    Soil: > 24 weeks in sterile sandy loam and < 1.0 week in nonsterile sandy loam; > 24 weeks in sterile organic soil and 1.0 week in nonsterile organic soil (Miles et al. 1979).
    Biota:

Environmental Fate Rate Constants or Half-Lives:
    Volatilization:
    Photolysis:
    Oxidation:
    Hydrolysis: half-lives at 38°C: > 700 h at pH 1.1 and > 400 h at pH 9.1; half-life of 1.28 h at pH 13 and 20°C (Tomlin 1994).
    Biodegradation:
    Biotransformation:
    Bioconcentration, Uptake (k_1) and Elimination (k_2) Rate Constants:

Common Name: Chlorpyrifos
Synonym: Brodan, Chlorpyrifos-ethyl, Detmol UA, Dowco 179, Dursban, ENT 27311, Eradex, Killmaster, Lorsban, NA 2783, OMS 971, Pyrinex
Chemical Name: O,O-diethyl O-3,5,6-trichloro-2-pyridylphosphorothioate; O,O-diethyl O-(3,5,6-trichloro-2-pyridinyl) phosphorothioate
Uses: insecticide used to control insects on a wide variety of crops including fruits, vegetables, ornamentals and trees.
CAS Registry No: 2921-88-2
Molecular Formula: $C_9H_{11}Cl_3NO_3PS$
Molecular Weight: 350.6
Melting Point (°C):
- 42-43.5 (Melnikov 1971; Freed et al. 1977, 1979; Worthing 1991; Tomlin 1994; Milne 1995)
- 42.5-43 (Khan 1980)
- 42.0 (Karickhoff 1981; Yalkowsky & Banerjee 1992)
- 42.7 (Bowman & Sans 1983b; Patil 1994)
- 41-43.5 (Montgomery 1993)
- 43.0 (Swann et al. 1983)
- 41-42 (Suntio et al. 1988; Howard 1991; Milne 1995)
- 41.5 (Kühne et al. 1995)

Boiling Point (°C):
Density (g/cm³ at 20°C):
Molar Volume (cm³/mol):
- 298.8 (calculated-LeBas method, Suntio et al. 1988)

Molecular Volume (Å³):
Total Surface Area, TSA (Å²):
Dissociation Constant pK_a:
Heat of Fusion, ΔH_{fus}, kcal/mol:
- 6.20 (DSC method, Plato & Glasgow 1969)

Entropy of Fusion, ΔS_{fus}, cal/mol·K (e.u.):
Fugacity Ratio at 25°C (assuming $\Delta S_{fus} = 13.5$ e.u.), F: 0.679

Water Solubility (g/m³ or mg/L at 25°C):
- 0.47 (Brust 1966; quoted, Bowman & Sans 1979; Garten & Trabalka 1983; Shiu et al. 1990)
- 0.40 (23°C, Brust 1966; quoted, Chiou et al. 1977; Freed et al. 1979; Bowman & Sans 1983; Suntio et al. 1988)
- 2.0 (Spencer 1973; Worthing 1987, 1991; quoted, Bowman & Sans 1979, 1983; Shiu et al. 1990)
- 0.40 (23°C, NIEHS 1975; quoted, Freed et al. 1977)
- 2.0 (Branson 1978; quoted, Neely 1980; Karickhoff 1981)
- 1.12 (shake flask-GC, Felsot & Dahm 1979; quoted, Briggs 1981; Shiu et al. 1990; Howard 1991)
- 0.70 (19°C, shake flask-GC, Bowman & Sans 1979; quoted, Sharom et al. 1980; Shiu et al. 1990; Patil 1994)
- 0.73 (20°C, shake flask-GC, Bowman & Sans 1983)
- 0.30 (23°C, Kenaga 1980a,b; quoted, Shiu et al. 1990)
- 0.30 (quoted from Dow Chemical Data, Kenaga & Goring 1980; quoted, Suntio et al. 1988; Schomburg et al. 1991)

2.00   (35°C, Khan 1980)
0.30, 12.8 (quoted exptl., calculated-Parachor, Briggs 1981)
2.00   (Thomas 1982; quoted, Nash 1988)
2.00   (20-25°C, Willis & McDowell 1982)
0.40   (Verschueren 1983; quoted, Pait et al. 1992)
0.73   (20°C, shake flask-GC, Bowman & Sans 1983a,b; quoted, Shiu et al. 1990)
0.073   (quoted, Schimmel et al. 1983)
1.07   (gen. col., Swann et al. 1983)
0.42   (RP-HPLC, Swann et al. 1983)
2.00   (Windholz 1983; quoted, Somasundaram et al. 1991)
0.87   (selected, Neely & Blau 1985; quoted, Karickhoff 1985; Mackay 1985)
2.00   (Agrochemicals Handbook 1987)
0.90   (selected, Gerstl & Helling 1987)
0.50   (quoted, Isnard & Lambert 1988)
0.30   (20°C, selected, Suntio et al. 1988; quoted, Fendinger et al. 1990; Majewski & Capel 1995)
0.30   (Kanazawa 1989)
0.40   (20-25°C, selected, Wauchope et al. 1992; Lohninger 1994; Hornsby et al. 1996)
0.75   (selected, Yalkowsky & Banerjee 1992)
0.73, 1.30 (20°C, 30°C, Montgomery 1993)
2.32   (calculated, Patil 1994)
1.40   (Tomlin 1994)
1.14   (quoted, Kühne et al. 1995)
62.3   (calculated-group contribution fragmentation method, Kühne et al. 1995)
2.00   (Milne 1995)

Vapor Pressure (Pa at 25°C):
0.00145 (20°C, Eichler 1965; Melnikov 1971; quoted, Kim 1985)
0.00145 (20°C, Wolfdietrich 1965; quoted, Kim et al. 1984)
0.00088 (20°C, GC-calculated value, Kim et al. 1984; Kim 1985)
0.00052 (20°C, GC-calculated value with M.P. correction, Kim et al. 1984; Kim 1985)
0.0025 (Brust 1966; quoted, Suntio et al. 1988)
0.0037 (Hamaker 1975; quoted, Hinckley et al. 1990)
0.0025 (Melnikov 1971; quoted, Freed et al. 1977,79; Suntio et al. 1988; Howard 1991)
0.0104 (30°C, NIEHS 1975; quoted, Freed et al. 1977)
0.0025 (selected, Neely & Blau 1977; quoted, Karickhoff 1985; Lyman 1985; Mackay 1985)
0.0025 (Khan 1980; quoted, Neely 1980)
0.0025 (Thomas 1982; quoted, Nash 1988)
0.0039 (Kim et al. 1984; quoted, Hinckley et al. 1990)
0.0025 (Agrochemicals Handbook 1987; Worthing 1991; Montgomery 1993)
0.0015 (20°C, selected, Suntio et al. 1988; quoted, Findinger et al. 1990; Majewski & Capel 1995)
0.0067 (GC method, Hinckley et al. 1990)
0.0023 (20-25°C, selected, Wauchope et al. 1992; Hornsby et al. 1996)

Henry's Law Constant (Pa·m³/mol):
- 1.0   Mackay 1985)
- 1.75   (20°C, calculated-P/C, Suntio et al. 1988; quoted, Majewski & Capel 1995)
- 0.418   (calculated-P/C, Findinger & Glotfelty 1990; quoted, Findinger et al. 1990)
- 7.902   (calculated-P/C, Howard 1991)
- 0.418   (23°C, known LWAPC of Findinger & Glotfelty 1990, Meylan & Howard 1991)
- 4.06×10^{-3} (bond-estimated LWAPC, Meylan & Howard 1991)
- 0.421   (23°C, quoted, Schomburg et al. 1991)
- 0.421   (calculated-P/C, Montgomery 1993)
- 1.090   (calculated-P/C, this work)

Octanol/Water Partition Coefficient, log K_{ow}:
- 5.11   (NIEHS 1975; quoted, Freed et al. 1977, 1979)
- 5.11   (20°C, shake flask-GC, Chiou et al. 1977; quoted, Felsot & Dahm 1979; Karickhoff 1981; Bowman & Sans 1983b; Suntio et al. 1988; De Bruijn et al. 1989; Howard 1991)
- 4.99   (quoted, Kenaga 1980b; Kenaga & Goring 1980)
- 3.31, 4.82, 5.11 (quoted values, Rao & Davidson 1980)
- 4.99   (quoted exptl., Briggs 1981; quoted, Bowman & Sans 1983b)
- 4.27   (calculated-Parachor, Briggs 1981)
- 4.82   (Veith et al. 1979; quoted, Karickhoff 1981; Veith & Kosian 1983; Mackay 1982; Suntio et al. 1988; Saito et al. 1992)
- 5.04   (quoted from D.L. Macalady of U.S. EPA 1980 unpublished result, McDuffie 1981)
- 4.67, 4.77 (estimated-RP-HPLC, McDuffie 1981)
- 4.96   (Bowman & Sans 1983b; quoted, Suntio et al. 1988; De Bruijn et al. 1989; De Bruijn & Hermens 1991; Somasundaram et al. 1991; Patil 1994)
- 5.20   (quoted, Schimmel et al. 1983; Kanazawa 1989)
- 4.60   (selected, Neely & Blau 1985; quoted, Karickhoff 1985)
- 4.97   (selected, Yoshioka et al. 1986; Travis & Arms 1988)
- 4.41   (selected, Gerstl & Helling 1987)
- 4.99   (quoted, Isnard & Lambert 1988)
- 5.00   (selected, Suntio et al. 1988)
- 5.267   (slow-stirring method, De Bruijn et al. 1989; quoted, Debruijn & Hermens 1991)
- 5.20   (selected, Magee 1991)
- 4.70   (Worthing 1991; Tomlin 1994; Milne 1995)
- 3.31-5.27 (Montgomery 1993)
- 4.73   (RP-HPLC, Saito et al. 1993)
- 4.30, 3.99 (quoted, RP-HPLC, Sicbaldi & Finizio 1993)
- 4.96   (selected, Sangster 1993)
- 4.50   (calculated, Patil 1994)
- 5.27   (selected, Hansch et al. 1995)

Bioconcentration Factor, log BCF:
- 2.67   (rainbow trout, Neely & Blau 1977; quoted, McLeese et al. 1976)
- 2.67   (mosquito fish for 35 days exposure, Veith et al. 1979; quoted, Veith & Kosian 1983)

2.65, 2.51 (fish: flowing water, static water; quoted, Kenaga 1980b; Kenaga & Goring 1980)
3.09 (calculated-S, Kenaga 1980a)
3.04 (calculated-K_{OC}, Kenaga 1980)
−1.70 (average beef fat diet, Kenaga 1980b)
2.67 (mosquito fish for 30 days exposure, Veith et al. 1980; quoted, Bysshe 1982; Howard 1991)
2.65 (quoted exptl., Briggs 1981)
2.38 (calculated-K_{OW}, Briggs 1981)
3.54 (estimated-regression from log K_{OW}, Lyman et al. 1982; quoted, Howard 1991)
3.08 (estimated-regression from S, Lyman et al. 1982; quoted, Howard 1991)
3.50 (calculated-K_{OW}, Mackay 1982)
2.67 (mosquito fish, Veith & Kosian 1983; quoted, Saito et al. 1992)
2.67 (quoted, Isnard & Lambert 1988)
−3.55 (beef biotransfer factor log B_b, correlated-K_{OW} from Kenaga 1980, Travis & Arms 1988)
−4.73 (milk biotransfer factor log B_m, correlated-K_{OW} from McKellar et al. 1976, Travis & Arms 1988)
2.67 (rainbow trout, wet wt. basis, De Bruijn & Hermens 1991)
2.51 (mosquito fish, wet wt. basis, De Bruijn & Hermens 1991)
4.32 (stickleback, lipid-based lab data, Deneer 1994)
2.68 (Pait et al. 1992)

Sorption Partition Coefficient, log K_{OC}:
4.13 (soil, quoted from Dow Chemical Data, Kenaga 1980a,b; Kenaga & Goring 1980; quoted, Karickhoff 1981; Howard 1991; Schomburg et al. 1991)
3.93 (calculated-S as per Kenaga & Goring 1978, Kenaga 1980; quoted, Schomburg et al. 1991)
3.78 (average of 3 soils, HPLC-RT, McCall et al. 1980; quoted, Howard 1991)
3.96, 4.87 (estimated-S, S & M.P., Karickhoff 1981)
2.92, 4.43, 4.72 (estimated-K_{OW}, Karickhoff 1981)
4.11 (soil, Thomas 1982; quoted, Nash 1988)
3.79, 4.0 (soil slurry method, RP-HPLC, Swann et al. 1983)
1.61 (average value calculated from Freundlich coeffs. without Baldwin Lake site data, Corwin & Farmer 1984)
4.17, 3.35 (quoted, calculated-χ, Gerstl & Helling 1987)
3.78 (soil, screening model calculations, Jury et al. 1987b)
3.27 (average of 2 soils, Kanazawa 1989)
3.78 (soil, 20-25°C, selected, Wauchope et al. 1992; quoted, Dowd et al. 1993; Richards & Baker 1993)
3.77-4.13 (Montgomery 1993)
4.37 (selected, Lohninger 1994)
3.78 (soil, 20-25°C, selected, Hornsby et al. 1996)

Sorption Partition Coefficient, log K_{OM}:
3.42 (Felsot & Dahm 1979; quoted, Briggs 1981)
3.78 (av. of 3 soils, McCall et al. 1980)
3.90 (exptl., Briggs 1981)

2.90 (calculated-Parachor, Briggs 1981)
3.10-4.31 (Mingelgrin & Gerstl 1983)
4.24 (quoted, Karickhoff 1985; Neely & Blau 1985)
4.50 (best estimate at low sediment concn., Karickhoff 1985)
4.13, 3.74 (selected, estimated, Magee 1991)

Half-Lives in the Environment:
  Air: 6.34 h for the vapor phase reaction with hydroxyl radicals in air (Atkinson 1987; quoted, Howard 1991).
  Surface water: based on Henry's law constant, volatilization half-life was estimated to be 9.0 d for a model river 1 meter deep, flowing 1 m/s with a wind velocity of 3 m/s (Lyman et al. 1982; quoted, Howard 1991); half-lives of a 100 mL pesticide-seawater solution containing 10 g of sediment were: <2.0 d, indoor (25°C with 12-h photoperiod white fluorescent light), 4.6 d, outdoor-light (stoppered, Pyrex flasks exposed to ambient sunlight with temperature 22-45°C), 7.1 d, outdoor-dark (foil-covered flasks) and 24 d in an estuary (Schimmel et al. 1983; quoted, Montgomery 1993); half-life of 120 d in water at pH 6.1, 20°C (quoted, Lartiges & Garrigues 1995).
  Groundwater:
  Sediment: half-lives: 24 d in 10 g untreated sediment/100 mL of a pesticide-seawater solution and >28 d in 10 g sterile sediment/100 mL of a pesticide-seawater solution (Schimmel et al. 1983).
  Soil: 17.0 weeks in sterile sandy loam and <1.0 week in nonsterile sandy loam; >24 weeks in sterile organic soil and 2.5 weeks in nonsterile organic soil (Miles et al. 1979); half-lives in a silt loam and clay loam were 12 and 4 weeks, while in sterilized soils were 24 weeks; however, temperature also had noticeable effects on decomposition as half-lives were 25, 13, and 6 weeks for soil samples incubated at 15, 25, and 35°C, respectively (Getzin 1981a; quoted, Montgomery 1993); hydrolysis half-life of 8 d in Sultan silt loam (Getzin 1981b; quoted, Montgomery 1993); half-life of 60-100 d in soil (Hartley & Kidd 1987; quoted, Montgomery 1993); half-life of 63 d from screening model calculations (Jury 1987b); persists in soil for 60-120 d (Worthing 1991); selected field half-life of 30 d (Wauchope et al. 1992; quoted, Dowd et al. 1993; Richards & Baker 1993; Hornsby et al. 1996); half-life of 30 d (Pait et al. 1992); half-life of 60-120 d (Tomlin 1994).
  Biota: 335 hours clearance from fish (Neely 1980); biochemical half-life of 63 d from screening model calculations (Jury et al. 1987b); elimination half-life of about 3.3 d in channel catfish (Barron et al. 1991); uptake and elimination rate coefficients were 7000 ± 2000 L kg^{-1} d^{-1} and (0.40 ± 0.11) d^{-1} (guppy, lipid-based modeling data, Deneer 1993); uptake and elimination rate coefficients were $(26 \pm 8.0) \times 10^3$ L kg^{-1} d^{-1} and (1.2 ± 0.4) d^{-1} (stickleback, lipid-based lab data, Deneer 1994).

Environmental Fate Rate Constants or Half-Lives:
  Volatilization: based on Henry's law constant, half-life estimated to be 9.0 d for a model river 1 meter deep, flowing 1 m/s with a wind velocity of 3 m/s (Lyman et al. 1982; quoted, Howard 1991); initial rate constant of 8.8×10^{-2} h^{-1} and predicted rate constant of 1.3×10^{-3} h^{-1} from soil with half-life of 533 hours (Thomas 1982); half-life of 0.3-3.2 d for disappearance from an inert surface at 25°C (Meikle et al. 1983).

Photolysis: exptl. half-life of 22 d determined under midday summer sunlight in California (Meikle et al. 1983; quoted, Howard 1991); estimated half-lives based on exptl. data at 40°N latitude: 31 d under midsummer surface conditions, 345 d under midwinter surface conditions, 43 d for midsummer conditions all for 1 m depth pure water and 2.7 yr for midsummer 1 m depth river water with average light attenuation (Dilling et al. 1984; quoted, Howard 1991; Montgomery 1993).

Oxidation: photooxidation half-life of 6.34 h for the vapor phase reaction with hydroxyl radicals in air (Atkinson 1987; quoted, Howard 1991).

Hydrolysis: half-life of 53 d at pH 7.4 and 20°C (NIEHS 1975; quoted, Freed et al. 1977, 1979); half-life of 120 d at pH 6.1 and 53 d at pH 7.4 in water and soil at 20°C as per Ruzicka et al. 1967 using gas chromatographic-retention time method for hydrolysis rates determination (Freed et al. 1979; quoted, Montgomery 1993); half-life of 78 d relatively independent of pH from pH 1 to 7 (Macalady & Wolfe 1983; quoted, Howard 1991); half-life of 1.5 d in water at pH 8 and 25°C (Worthing 1991; Tomlin 1994).

Biodegradation: rate constant of 0.014 d^{-1} in soil at 28°C (Miles et al. 1979; quoted, Klečka 1985); rate constant of 0.008-0.025 d^{-1} in soil at 25°C (Getzin 1981; quoted, Klečka 1985); first-order rate constant of -0.000945 to -0.00243 h^{-1} in nonsterile sediment and -0.000562 to -0.00151 h^{-1} in sterile sediment by shake-tests at Range Point and also first-order rate constant of -0.00109 to -0.00231 h^{-1} in nonsterile water and -0.00144 to -0.00197 h^{-1} in sterile water by shake-tests at Range Point (Walker et al. 1988).

Biotransformation:

Bioconcentration, Uptake (k_1) and Elimination (k_2) Rate Constants: elimination half-life of about 3.3 d in channel catfish (Barron et al. 1991); uptake and elimination rate coefficients were $(26 \pm 8.0) \times 10^3$ L kg^{-1} d^{-1} and (1.2 ± 0.4) d^{-1} (stickleback, lipid-based lab data, Deneer 1994).

Common Name: Crotoxyphos
Synonym: Ciodrin, Ciovap, Cyodrin, Cypona EC, Decrotox, Duo-kill, Duravos, ENT 24717, Volfazol
Chemical Name: dimethyl(E)-1-methyl-2-(1-phenyl-ethoxycarbonyl)vinyl phosphate
Uses: insecticide
CAS Registry No: 7700-17-6
Molecular Formula: $C_{14}H_{19}O_6P$
Molecular Weight: 314.3
Melting Point (°C): liquid
Boiling Point (°C):
    135   (at 0.03 mmHg, Montgomery 1993)
Density (g/cm³ at 20°C):
    1.2   (Spencer 1982)
    1.19   (25°C, Montgomery 1993)
Molar Volume (cm³/mol):
    264.1   (calculated-density)
Molecular Volume (Å³):
Total Surface Area, TSA (Å²):
Dissociation Constant pK_a:
Heat of Fusion, ΔH_{fus}, kcal/mol:
Entropy of Fusion, ΔS_{fus}, cal/mol·K (e.u.):
Fugacity Ratio at 25°C (assuming ΔS_{fus} = 13.5 e.u.), F: 1.0

Water Solubility (g/m³ or mg/L at 25°C):
    1000   (Günther et al. 1968; quoted, Shiu et al. 1990)
    1000   (Melnikov 1971; quoted, Shiu et al. 1990)
    1000   (Spencer 1973, 1982; Worthing 1987; quoted, Shiu et al. 1990)
    1000   (Martin & Worthing 1977; quoted, Kenaga & Goring 1980)
    1000   (Montgomery 1993)

Vapor Pressure (Pa at 25°C):
    0.0019   (20°C, Khan 1980)
    0.00187, 0.0052, 0.013   (20, 30, 40°C, Spencer 1982)
    0.0019   (20°C, Montgomery 1993)

Henry's Law Constant (Pa·m³/mol):
    0.00063   (20-25°C, calculated-P/C, Montgomery 1993)
    0.00060   (calculated-P/C, this work)

Octanol/Water Partition Coefficient, log K_{ow}:
    2.23   (Kenaga 1980)
    1.28   (calculated, Montgomery 1993)
    3.0, 3.30; 3.30   (quoted values, selected, Sangster 1993)
    3.30   (selected, Hansch et al. 1995)

Bioconcentration Factor, log BCF:
    1.10   (calculated-S as per Kenaga 1980, this work)

Sorption Partition Coefficient, log K_{OC}:
    2.23   (soil, Hamaker & Thompson 1972; quoted, Kenaga & Goring 1980)
    2.00   (soil, quoted exptl., Meylan et al. 1992)
    1.70   (soil, calculated-χ and fragment contribution, Meylan et al. 1992)
    2.23   (Montgomery 1993)

Half-Lives in the Environment:
    Air:
    Surface water: biodegradation half-lives of 7.5 d at pH 9 and 22.5 d at pH 2 from river die-away tests (Konrad & Chester 1969; quoted, Scow 1982).
    Groundwater:
    Sediment:
    Soil:
    Biota:

Environmental Fate Rate Constants or Half-Lives:
    Volatilization:
    Photolysis:
    Oxidation:
    Hydrolysis:
    Biodegradation:
    Biotransformation:
    Bioconcentration, Uptake (k_1) and Elimination (k_2) Rate Constants:

Common Name: Cypermethrin
Synonym: Agrothrin, Ambush C, Barricade, CCN 52, Cymbush, Cyperkill, Demon, FMC 30980, Folcord, Imperator, Kafil Super, Polytrin, Ripcord, Sherpa, Stocade, Toppel
Chemical Name: cyano(3-phenoxyphenyl)methyl 3-(2,2-dichloroethenyl)-2,2-dimethylcyclopropanecarboxylate; (RS)-α-cyano-3-phenoxybenzyl($1RS,3RS;1RS,3SR$)-3(2,2-dichlorovinyl)-2,2-dimethylcyclopropanecarboxylate
Uses: nonsystemic insecticide with contact and stomach action to control a wide range of insects in fruits, vegetables, vines, potatoes, cucurbits, capsicums, cereals, maize, soybeans, cotton, coffee, coca, rice, pecans, ornamentals and forestry, etc.; also used to control flies in animal houses and mosquitoes, cockroaches, houseflies and other pests in public health.
CAS Registry No: 52315-07-8
Molecular Formula: $C_{22}H_{19}Cl_2NO_3$
Molecular Weight: 416.3
Melting Point (°C):
    80.5   (Tomlin 1994)
Boiling Point (°C):
Density (g/cm^3 at 20°C):
    1.23   (Tomlin 1994)
    1.25   (Milne 1995)
Molar Volume (cm^3/mol):
    457.7   (calculated-LeBas method, this work)
    335.7   (calculated-density, this work)
Molecular Volume (Å^3):
Total Surface Area, TSA (Å^2):
Dissociation Constant pK_a:
Heat of Fusion, ΔH_{fus}, kcal/mol:
Entropy of Fusion, ΔS_{fus}, cal/mol·K (e.u.):
Fugacity Ratio at 25°C (assuming ΔS_{fus} = 13.5 e.u.), F: 0.283

Water Solubility (g/m^3 or mg/L at 25°C):
    0.041   (shake flask-GC, Coats & O'Donnell-Jefferey 1979; quoted, Spencer 1982; Shiu et al. 1990)
    0.005-0.01 (Stephenson 1982; quoted, Clark et al. 1989)
    0.010   (20°C, Agrochemicals Handbook 1987)
    0.01-0.2 (21°C, Worthing 1987; quoted, Shiu et al. 1990)
    0.004   (Wauchope 1989; quoted, Shiu et al. 1990)
    0.004   (20-25°C, selected, Wauchope et al. 1992; quoted, Majewski & Capel 1995; Hornsby et al. 1996)
    0.004   (Montgomery 1993)
    0.004   (Lohninger 1994)
    0.004   (at pH 7, Tomlin 1994)
    0.009   (20°C, selected, Siebers & Mattusch 1996)

Vapor Pressure (Pa at 25°C):
    8.7×10^{-7} (Barlow 1978; quoted, Hinckley et al. 1990)
    4.3×10^{-7} (Grayson et al. 1982; quoted, Hinckley et al. 1990)
    $< 1.3 \times 10^{-5}$ (Spencer 1982)

5.0×10^{-10} (70°C, Agrochemicals Handbook 1987)
2.4×10^{-6} (GC-RT, supercooled value, Hinckley et al. 1990)
1.9×10^{-7} (20-25°C, selected, Wauchope et al. 1992; quoted, Majewski & Capel 1995; Hornsby et al. 1996)
1.9×10^{-7} (20°C, extrapolated, Montgomery 1993)
2.3×10^{-7} (20°C, Tomlin 1994)
2.3×10^{-6} (20°C, selected, Siebers & Mattusch 1996)

Henry's Law Constant (Pa·m^3/mol):
    0.0199 (20-25°C, calculated-P/C, Montgomery 1993)
    0.0194 (20-25°C, calculated-P/C as per Wauchope et al. 1992, Majewski & Capel 1995)
    0.080 (selected, Siebers & Mattusch 1996)
    0.0195 (calculated-P/C, this work)

Octanol/Water Partition Coefficient, log K_{OW}:
    4.47 (shake flask-GC, Coats & O'Donnell-Jefferey 1979; quoted, Shiu et al. 1990)
    2.44 (calculated, McLeese et al. 1980)
    5.90 (Schimmel et al. 1983; quoted, Clark et al. 1989)
    5.2 ± 0.6 (*cis*-form, correlated-HPLC-RT, Muir et al. 1985)
    5.0 ± 0.6 (*trans*-form, correlated-HPLC-RT, Muir et al. 1985)
    6.60 (Montgomery 1993)
    4.47, 6.0 (quoted, Sangster 1993)
    6.60 (Tomlin 1994)
    6.60 (Milne 1995)
    6.05 (selected, Devillers et al. 1996)
    6.05, 6.05 (α-, β-isomer, Hansch et al. 1995)
    5.56, 6.35, 5.60 (RP-HPLC, ClogP, calculated-S, Finizio et al. 1997)

Bioconcentration Factor, log BCF:
    2.99 (activated sludge, Freitag et al. 1984)
    3.52, 2.62, 2.99 (algae, golden ide, activated sludge, Freitag et al. 1985)
    1.73-2.34 (*trans*-form on sediment, 24 h BCF for chironomid larvae in water, Muir et al. 1985)
    1.63-2.39 (*trans*-form on sediment, 24 h BCF for chironomid larvae in sediment, Muir et al. 1985)
    1.49-2.05 (*trans*-form on sediment, 24 h BCF for chironomid larvae in sediment/pore water, Muir et al. 1985)
    1.53-2.38 (*cis*-form on sediment, 24 h BCF for chironomid larvae in water, Muir et al. 1985)
    1.84-2.59 (*cis*-form on sediment, 24 h BCF for chironomid larvae in sediment, Muir et al. 1985)
    1.68-2.02 (*cis*-form on sediment, 24 h BCF for chironomid larvae in sediment/pore water, Muir et al. 1985)
    2.89 (*Oncorhynchus mykiss*, Muir et al. 1994; quoted, Devillers et al. 1996)
    2.92 (*Oncorhynchus mykiss*, Muir et al. 1994; quoted, Devillers et al. 1996)

Sorption Partition Coefficient, log K_{OC}:
- 2.36 (*cis*-form, silt, K_P on 24% DOC, Muir et al. 1985)
- 2.57 (*cis*-form, clay, K_P on 56% DOC, Muir et al. 1985)
- 2.59 (*trans*-form, silt, K_P on 10% DOC, Muir et al. 1985)
- 5.00 (soil, 20-25°C, estimated, Wauchope et al. 1992; quoted, Lohninger 1994; Hornsby et al. 1996)
- 4.0-4.53 (Montgomery 1993)

Half-Lives in the Environment:
    Air:
    Surface water: half-life of 5 d in river water (Tomlin 1994).
    Groundwater:
    Sediment:
    Soil: estimated field half-life of 30 d (Wauchope et al. 1992; Hornsby et al. 1996).
    Biota:

Environmental Fate Rate Constants or Half-Lives:
    Volatilization:
    Photolysis:
    Oxidation:
    Hydrolysis:
    Biodegradation:
    Biotransformation:
    Bioconcentration, Uptake (k_1) and Elimination (k_2) Rate Constants:

Common Name: DDD
Synonym: dichloro diphenyl dichloroethane; p,p'-DDD; Dilene; ENT 4225; ME 1700; NCI-C00475; Rhothane; p,p'-TDE; TDE; tetrachlorodiphenylethane
Chemical Name: 1,1-dichloro-2,2-bis(4-chlorophenyl)ethane; 1,1'-(2,2-dichloroethylidene)bis[4-chlorobenzene
Uses: degradation product of DDT used as insecticide.
CAS Registry No: 72-54-8 (p,p'-DDD or DDD); 53-10-0 (o,p'-DDD)
Molecular Formula: $C_{14}H_{10}Cl_4$
Molecular Weight: 321.0
Melting Point (°C):
    112   (Martin 1972; Callahan et al. 1979)
    109.5   (Ballschmiter & Wittlinger 1991; Kühne et al. 1995)
    109   (Liu et al. 1991)
    109-110   (Milne 1995)
Boiling Point (°C):
Density (g/cm³ at 20°C):
Molar Volume (cm³/mol): 312.6   (LeBas method, Suntio et al. 1988)
Molecular Volume (Å³):
Total Surface Area, TSA (Å²):
Dissociation Constant pK_a:
Heat of Fusion, ΔH_{fus}, kcal/mol:
    7.40   (DSC method, Plato & Glasgow 1969)
    6.528   (Ruelle & Kesselring 1997)
Entropy of Fusion, ΔS_{fus}, cal/mol·K (e.u.):
    19.4   (Plato & Glasgow 1969; Hinckley et al. 1990)
Fugacity Ratio at 25°C (assuming ΔS_{fus} = 13.5 e.u.), F: 0.144

Water Solubility (g/m³ or mg/L at 25°C):
    0.002   (shake flask-LSC, Metcalf et al. 1973)
    0.005, 0.015, 0.09   (shake flask-GC with particle sizes: 0.01, 0.05, 5.0 micron, Biggar & Riggs 1974)
    0.24   (shake flask-GC, o, p'-DDD with particle sizes: 0.05 micron, Biggar & Riggs 1974)
    0.050, 0.090, 0.15, 0.24   (shake flask-GC, at 15, 25, 35, 45 °C with particle sizes: 5.0 micron or less, Biggar & Riggs 1974; quoted, Shiu et al. 1990)
    0.060, 0.10, 0.25, 0.315   (shake flask-GC, at 15, 25, 35, 45 °C with particle sizes: 5.0 micron or less, Biggar & Riggs 1974; quoted, Shiu et al. 1990)
    0.020   (gen. col.-GC/ECD, Weil et al. 1974; quoted, Callahan et al. 1979; Geyer et al. 1982; Shiu et al. 1990)
    0.005   (Martin & Worthing 1977; quoted, Kenaga 1980; Bruggeman et al. 1981; Zaroogian et al. 1985; Adams 1987)
    0.09, 0.10   (quoted, p,p'-, o,p'-, Callahan et al. 1979)
    0.004   (nephelometry, Hollifield 1979; quoted, Belluck & Felsot 1981)
    0.020   (selected, Yoshida et al. 1983)
    0.0048   (Isnard & Lambert 1988; quoted, Ballschmiter & Wittlinger 1991)
    0.050   (20°C, selected, Suntio et al. 1988; quoted, Majewski & Capel 1995)
    0.10   (20°C, selected, o,p'-DDD, Suntio et al. 1988)
    0.16   (Agency for Toxic Substances & Disease Registry 1988; quoted, Burmaster et al. 1991)
    0.020, 0.0127   (quoted, estimated-group contribution method, Kühne et al. 1995)

0.020 (20-25°C, selected, Hornsby et al. 1996)
0.050; 0.010, 0.0035 (quoted; predicted-molar volume, Ruelle & Kesselring 1997)

Vapor Pressure (Pa at 25°C):
   1.30×10^{-4} (30°C, Balson 1947; quoted, Bowery 1964; Spencer & Cliath 1970; Spencer 1972,75; Callahan et al. 1979; Ballschmiter & Wittlinger 1991)
   2.52×10^{-4} (30°C, calculated-vapor density, Spencer & Cliath 1972)
   1.36×10^{-4} (30°C, Spencer 1975; quoted, Callahan et al. 1979)
   2.52×10^{-4} (30°C, o,p'-DDD, Spencer 1975; quoted, Callahan et al. 1979)
   1.33×10^{-4} (selected, Yoshida et al. 1983)
   4.34×10^{-4}; 1.63×10^{-3}, 6.24×10^{-4} (supercooled liquid values, P_L: quoted; GC-RT, Bidleman 1984)
   1.40×10^{-4} (Agency for Toxic Subtances and Disease Registry 1988; quoted, Burmaster et al. 1991)
   1.00×10^{-4} (20°C, selected, Suntio et al. 1988; quoted, Majewski & Capel 1995)
   2.00×10^{-4} (20°C, o,p'-, selected, Suntio et al. 1988)
   9.72×10^{-4} (calculated from log $P°_L = 12.49 - 4622/T$ (Hinckley et al. 1990)
   4.35×10^{-4}, 9.84×10^{-4} (quoted $P°_L$-GC values, Hinckley et al. 1990)
   1.33×10^{-4} (20-25°C, estimated, Hornsby et al. 1996)

Henry's Law Constant (Pa·m^3/mol):
   2.18 (calculated-P/C, Yoshida et al. 1983)
   0.27 (Agency for Toxic Subtances & Disease Registry 1988; quoted, Burmaster et al. 1991)
   0.64 (20°C, calculated-P/C, Suntio et al. 1988; quoted, Majewski & Capel 1995)
   9.00 (calculated-P/C, Ballschmiter & Wittlinger 1991)
   8.51 (calculated-P/C, this work)

Octanol/Water Partition Coefficient, log K_{ow}:
   5.99 (O'Brien 1974; quoted, Callahan et al. 1979; Geyer et al. 1982; Chin et al. 1986; Suntio et al. 1988)
   6.02 (Ernst 1977; quoted, Hawker & Connell 1986)
   6.02 (Veith & Morris 1978; Veith et al. 1979; quoted, Zaroogian et al. 1985)
   6.26 (calculated as per Chiou 1977, Belluck & Felsot 1981)
   5.69 (Hansch & Leo 1979; quoted, Yoshida et al. 1983; De Kock & Lord 1987)
   5.99, 6.08 (quoted, p,p'-, o,p'-DDD, Callahan et al. 1979)
   6.00 (Kenaga & Goring 1980; quoted, De Kock & Lord 1987; De Bruijn et al. 1989)
   5.19 (predicted-RP-HPLC, Chin et al. 1986)
   5.90, 5.80 (lit. average, estimated, De Kock & Lord 1987)
   5.00 (predicted-RP-HPLC, De Kock & Lord 1987)
   6.20 (Agency for Toxic Subtances & Disease Registry 1988; quoted, Burmaster et al. 1991)
   6.02 (Isnard & Lambert 1988,89; quoted, Ballschmiter & Wittlinger 1991)
   5.50, 6.10 (20°C, p,p'-, o,p'-, selected, Suntio et al. 1988)
   6.217 ± 0.031 (slow-stirring method, De Bruijn et al. 1989; Sicbaldi & Finizio 1993)
   6.21 (quoted calculated value, De Bruijn et al. 1989)
   6.70 (estimated-SAR, Karickhoff et al. 1989; quoted, Hoke et al. 1994)
   6.02 (selected, Travis & Arms 1988; Liu et al. 1991)
   5.59 (calculated-f const., Noegrohati & Hammers 1992)
   6.02 (selected, Verhaar et al. 1992; Saito et al. 1993)

6.21 (estimated-QSAR & SPARC, Kollig 1993)
6.22, 4.82 (quoted, RP-HPLC, Sicbaldi & Finizio 1993)
4.87, 6.06, 5.34 (*o,p'*-, RP-HPLC, CLOGP, calculated-S, Finizio et al. 1997)

Bioconcentration Factor, log BCF:
    4.92, 3.92 (*Gambusia, Physa,* Metcalf et al. 1973)
    3.96 (mussel, Ernst 1977; quoted, Geyer et al. 1982; Hawker & Connell 1986)
    4.72 (fathead minnow, Veith et al. 1979; quoted, Zaroogian et al. 1985)
    4.09 (calculated-S, Kenaga 1980)
    4.11 (calculated-S or K_{ow}, Kenaga & Goring 1980; quoted, Yoshida et al. 1983)
    3.30 (*Triaenodes tardus*, Belluck & Felsot 1981)
    4.34, 4.42 (mussel, oyster; calculated-K_{ow}, Zaroogian et al. 1985)
    4.68 (oyster, Zaroogian et al. 1985; quoted, Hawker & Connell 1986)
    2.85-4.29 (benthic macroinvertebrates, Reich et al. 1986)
    4.81 (calculated-S and K_{ow}, Isnard & Lambert 1988)
    −1.90 (beef biotransfer factor log B_b, correlated-K_{ow} from Fries et al. 1969, Travis & Arms 1988)
    −2.52 (milk biotransfer factor log B_m, correlated-K_{ow} from Fries et al. 1969, Travis & Arms 1988)
    0.301 (earthworms, quoted, Menzie et al. 1992)
    −0.456, −0.745, −0.602 (earthworms, field/lab. estimated, field leaf litter, calculated-modeled, Menzie et al. 1992)
    4.68 (calculated-log K_{ow} as per Mackay 1982, this work)

Sorption Partition Coefficient, log K_{OC}:
    4.91 (calculated-S, Kenaga 1980)
    4.63 (calculated-S or K_{ow}, Kenaga & Goring 1980; quoted, Yoshida et al. 1983)
    5.86 (calculated-S, Mill et al. 1980; quoted, Adams 1987)
    5.89 (estimated-QSAR & SPARC, Kollig 1993)
    5.00 (20-25°C, estimated, Hornsby et al. 1996)

Half-Lives in the Environment:
    Air: 17.7-177 h, based on estimated photooxidation half-life in air (Howard et al. 1991).
    Surface water: 2-15.6 yr, based on observed rates of biodegradation of DDT in aerobic soils under field conditions (Lichtenstein & Schultz 1959; Stewart & Chisholm 1971; quoted, Howard et al. 1991); dehydrochlorination rate constant of 5.00×10^{-2} h^{-1} with a half-life of 13.9 h for 1.0 ppm *p,p'*-DDD and 0.76×10^{-2} h^{-1} with a half-life of 96.3 h for *o,p'*-DDD both at 21 ± 2°C and pH 12.8 (in 0.1 N NaOH solution) (Choi & Chen 1976).
    Groundwater: 1680-270,000 h, based on anaerobic flooded soil die-away study data for two flooded soils (Castro & Yoshida 1971; quoted, Howard et al. 1991) and observed rates of biodegradation of DDT in aerobic soils under field conditions (Lichtenstein & Schultz 1959; Stewart & Chisholm 1971; quoted, Howard et al. 1991).
    Sediment:
    Soil: 2-15.6 yr, based on observed rates of biodegradation of DDT in aerobic soils under field conditions (Lichtenstein & Schultz 1959; Stewart & Chisholm 1971; quoted, Howard et al. 1991); estimated field half-life of 1000 d (20-25°C, Hornsby et al. 1996).
    Biota: 119 hours in mussels (Ernst 1977).

Environmental Fate Rate Constants or Half-Lives:
    Volatilization: aquatic half-life of a few days to about a month (summarized data, Callahan et al. 1979).
    Photolysis: aquatic half-life of > 150 d (summarized data, Callahan et al. 1979).
    Oxidation: aquatic half-life of about 22 yr (summarized data, Callahan et al. 1979); photooxidation half-life of 13.3-133 h in air, based on estimated rate constant for reaction with hydroxyl radicals in air (Atkinson 1987; quoted, Howard et al. 1991).
    Hydrolysis: half-life of about 570 d at pH 9 and 190 yr at pH 5 (summarized data, Callahan et al. 1979); half-life of 28 yr at pH 7 and 25°C, calculated from measured neutral and base catalyzed hydrolysis constants of $(2.8 \pm 0.9) \times 10^{-6}$ h^{-1} and 5.2 M^{-1} h^{-1} (Ellington et al. 1987, 1988, 1989; quoted, Howard et al. 1991); rate constant of 2.5×10^{-2} y^{-1} at pH 7 and 25°C (Kollig 1993).
    Biodegradation: aqueous aerobic half-life of 2-15.6 yr, based on observed rates of biodegradation of DDT in aerobic soils under field conditions (Lichtenstein & Schultz 1959; Stewart & Chisholm 1971; quoted, Howard et al. 1991); aqueous anaerobic half-life of 70-294 d, based on anaerobic flooded soil die-away study data for two flooded soils (Castro & Yoshida 1971; quoted, Howard et al. 1991).
    Biotransformation:
    Bioconcentration, Uptake (k_1) and Elimination (k_2) Rate Constants:
        k_1: 52.9 h^{-1} (mussel, Ernst 1977; quoted, Hawker & Connell 1986)
        k_2: 0.0058 h^{-1} (mussel, Ernst 1977; quoted, Hawker & Connell 1986)

Common Name: DDE (*p,p'*-DDE; *o,p'*-DDE)
Synonym: 4,4'-DDE, DDE; 2,4-DDE
Chemical Name: 1,1-dichloro-2,2-bis(*p*-chlorophenyl)-ethylene
CAS Registry No: 72-55-9 (*p,p'*-DDE); 3424-82-6 (*o,p*-DDE)
Molecular Formula: $C_{14}H_8Cl_4$
Molecular Weight: 319.0
Melting Point (°C):
    88.5   (Ballschmiter & Wittlinger 1991)
Boiling Point (°C):
Density (g/cm^3 at 20°C):
Molar Volume (cm^3/mol):
    305.2   (calculated-LeBas method, Suntio et al. 1988)
    243.1   (Ruelle & Kesselring 1997)
Heat of fusion ΔH_{fus}, kcal/mol:
    7.30   (*o,p'*-DDE, DSC method, Plato & Glasgow 1969)
    5.80   (*p,p'*-DDE, DSC method, Plato & Glasgow 1969)
Entropy of fusion, ΔS_{fus}:
Fugacity Ratio at 25°C (assuming ΔS = 13.5 e.u.), F: 0.233

Water Solubility (g/m^3 or mg/L at 25°C):
p,p'-DDE
    0.0013   (shake flask-LSC, Metcalf et al. 1973,75; Hollifield 1979)
    0.055, 0.12, 0.235, 0.45   (15, 25, 35, 45°C, shake flask-GC for particles 5 μ or less, Biggar & Riggs 1974)
    0.014   (gen. col.-GC/ECD, Weil et al. 1974; quoted, Yoshida et al. 1983)
    0.040   (20°C, shake flask-GC, Chiou et al. 1977; Freed et al. 1977; Hollifield 1979)
    0.040   (quoted, Callahan et al. 1979, Mabey et al. 1982; Calamari et al. 1991)
    0.065   (shake flask-nephelometry, Hollifield 1979)
    0.0079   (Kenaga & Goring 1980; quoted, Isnard & Lambert 1988,89; Ballschmiter & Wittlinger 1991)
    0.0017   (30°C, semimicro gas-saturation method, Westcott et al. 1981)
    0.0011, 0.006   (gen. col., HPLC-RT, Swann et al. 1983)
    0.004   (selected, Suntio et al. 1988)
    0.018   (10°C, estimated, McLachlin et al. 1990)
    0.174   (supercooled value, quoted, Majewski & Capel 1995)
    0.040; 0.029, 0.001   (quoted; predicted-molar volume, Ruelle & Kesselring 1997)
o,p-DDE
    0.140   (shake flask-GC for particles 5 μ or less, Biggar & Riggs 1974)
    0.0013   (Zepp et al. 1978)
    0.140   (quoted, Callahan et al. 1979, Mabey et al. 1982)
    0.10   (selected, Suntio et al. 1988; Hornsby et al. 1996)

Vapor Pressure (Pa at 25°C):
p,p'-DDE
    8.60×10^{-4} (30°C, Balson 1947; quoted, Ballschmiter & Wittlinger 1991)
    7.87×10^{-4} (gas saturation-GC, Spencer & Cliath 1972)
    8.65×10^{-4} (30°C, calculated-vapor density, Spencer & Cliath 1972)
    8.67×10^{-4} (quoted, Callahan et al. 1979, Mabey et al. 1982)
    9.87×10^{-4} (GC-RT, Westcott & Bidleman 1981)
    1.733×10^{-4} (30°C, gas saturation-GC, Westcott et al. 1981)

8.66×10^{-4} (selected, Yoshida et al. 1983)
2.55×10^{-3}; 2.70×10^{-3}, 2.09×10^{-3} (supercooled liquid value P_L: quoted; GC-RT, Bidleman 1984)
1.73×10^{-3} (20°C, supercooled liquid value, Bidleman et al. 1986)
7.50×10^{-5} (10°C, estimated, McLachlin et al. 1990)
0.00225, 0.00334 (supercooled liquid values, GC-RT, Hinckley et al. 1990)
log P_L (Pa) = A_L + B_L/T, A_L = 12.79, B_L = −4554 (liquid, Hinckley et al. 1990)
5.13×10^{-4}, 1.82×10^{-3} (supercooled liquid values at 10°C, 20°C, Cothom & Bidleman 1992)
4.36×10^{-3} (supercooled liquid value, quoted, Majewski & Capel 1995)

o,p-DDE
8.21×10^{-4} (30°C, calculated-vapor density, Spencer & Cliath 1972)
8.27×10^{-4} (quoted, Callahan et al. 1979, Mabey et al. 1982)
8.67×10^{-4} (isomer unspecified, estimated, Hornsby et al. 1996)

Henry's Law Constant (Pa·m^3/mol):
27.4 (Levins 1981; quoted, Tucker et al. 1983)
0.78 (estimated-group method per Hine & Mookerjee 1975, Tucker et al. 1983)
6.89 (calculated-P/C, Mabey et al. 1982)
124   (gas stripping, Atlas et al. 1982; quoted, Müller et al. 1994)
19.59 (calculated-P/C, Yoshida et al. 1983)
7.95 (calculated-P/C, Suntio et al. 1988; quoted, Paterson et al. 1991)
1.25 (10°C, calculated-P/C, McLachlin et al. 1990)
34.0 (calculated-P/C, Ballschmiter & Wittlinger 1991)
120, 370 (23°C), 7.95 (20°C) (quoted, Iwata et al. 1993)
7.95 (20-25°C, calculated-P/C, Majewski & Capel 1995)

Octanol/Water Partition Coefficient, log K_{OW}:
5.80 (*o,p*-DDE, O'Brien 1974, quoted, Suntio et al. 1988)
4.28 (correlated, Metcalf et al. 1975)
5.69 (*p,p'*-DDE, O'Brien 1975, quoted, Chiou et al. 1977; Freed et al. 1977; De Bruijn et al. 1989; Isnard & Lambert 1988,89; Ballschmiter & Wittlinger 1991)
5.69 (*p,p'*-DDE, HPLC-RT, Veith et al. 1979, quoted, Rao & Davidson 1980, Mackay 1982; Swann et al. 1983; Veith & Kosian 1983; Burkhard et al. 1985; Oliver & Charlton 1984; Chin et al. 1986; Lydy et al. 1990, 1992)
5.77 (Kenaga & Goring 1980; quoted, De Bruijn et al. 1989)
6.96 (calculated-f const., Mabey et al. 1982)
5.83 (Garten & Trabalka 1983; quoted, De Bruijn et al. 1989)
5.63 (RP-HPLC, Swann et al. 1983)
5.69 (*p,p'*-DDE, selected, Yoshida et al. 1983; Thomann 1989)
5.89 (estimated-HPLC/MS, Burkhard et al. 1985)
6.29 (predicted-RP-HPLC, Chin et al. 1986)
5.79, 6.09 (lit. average, RP-HPLC, De Kock & Lord 1987)
6.89 (*p,p'*-DDE, Connell et al. 1988; Markwell et al. 1989)
6.51 (Medchem 1988; quoted, Müller et al. 1994)
6.956 ± 0.011 (*p,p'*-DDE, slow stirring method, De Bruijn et al. 1989; quoted, Sicbaldi & Finizio 1993; Hansch et al. 1995)
6.94 (quoted calculated value, De Bruijn et al. 1989)
6.36 (calculated, Karickhoff et al. 1989; quoted, Hoke et al. 1994)
5.70 (quoted, Thomann 1989; Paterson et al. 1991)

6.51 (Thor 1989; quoted, Connell & Markwell 1990)
5.00 (Calamari et al. 1991; quoted, Meakins et al. 1994)
6.72 (p,p'-DDE, calculated-f const., Noegrohati & Hammers 1992)
6.96, 5.78 (quoted, RP-HPLC, Sicbaldi & Finizio 1993)
5.43, 6.94, 5.45 (o,p'-, RP-HPLC, CLOGP, calculated-S, Finizio et al. 1997)

Bioconcentration Factor, log BCF:
4.44, 4.29 (*Gambusia, Physa*, Metcalf et al. 1973)
4.05, 4.56, 4.77, 4.08 (alga, snail, mosquito, fish, Metcalf et al. 1975)
4.71 (fathead minnows, 32 days exposure, Veith et al. 1979; quoted, Veith & Kosian 1983; Zaroogian et al. 1985)
3.80 (calculated-S or K_{ow}, Kenaga & Goring 1980; quoted, Yoshida et al. 1983)
4.71, 4.37 (quoted exptl, calculated-K_{ow}, Mackay 1982)
5.95 (microorganism-water: calculated-K_{ow}, Mabey et al. 1982)
4.91, 4.08 (rainbow trout: kinetic, steady-state, Oliver & Niimi 1985)
3.70-5.32 (p,p'-DDE, benthic macroinvertebrates, Reich et al. 1986)
3.70-5.32 (o,p'-DDE, benthic macroinvertebrates, Reich et al. 1986)
4.13 (azalea leaves, Bacci & Gaggi 1987)
6.01 (p,p-DDE, Connell et al. 1988)
4.71 (correlated, Isnard & Lambert 1988)
7.48 (Azalea leaves, Bacci et al. 1990; quoted, Paterson et al. 1991; Müller et al. 1994)
1.025 (earthworms, quoted, Menzie et al. 1992)
−0.824 (earthworms, field/lab. estimated, Menzie et al. 1992)
−0.602 (earthworms, calculated-modeled, Menzie et al. 1992)

Bioaccumulation Factor, log BAF:
8.35 (rainbow trout, Thomann 1989)

Sorption Partition Coefficient, log K_{OC}:
4.48 (calculated-S or K_{ow}, Kenaga & Goring 1980; quoted, Yoshida et al. 1983)
6.64 (sediment, calculated-K_{ow}, Mabey et al. 1982)
4.70, 5.17 (quoted, calculated-χ, Sabljic 1984)
6.0, 5.3 (sediment: field measurement, calculated-K_{ow}, Oliver & Charlton 1984)
3.70 (soil, estimated, Hornsby et al. 1996)
4.82 (av. lit. value, Gerstl 1991)

Half-Lives in the Environment:
Air: atmospheric transformation lifetime was estimated to be <1 d (Kelly et al. 1994).
Surface water: hydrolytic half-life of >120 yr in water at 27°C (Wolfe et al. 1977); estimated half-life of >300 d in lakes (Zoeteman et al. 1980).
Groundwater:
Sediment:
Soil: field half-life of 1000 d (estimated, Hornsby et al. 1996).
Biota: elimination half-life of 340 d (rainbow trout, Oliver & Niimi 1985); half-life of 264 d in herring gulls compared to lit. half-life of average 300 d for birds (Norstrom et al. 1986); elimination half-life of 2230 h (Azalea leaves, Bacci & Gaggi 1987); half-lives in the midge (*Chironomus riparius*) under varying

sediment conditions: 231 h (water only system with no sediment), 150 h (midge screened from the sediment), 87 h (midge screened from 3% organic carbon sediment), 99 h (midge screened from 3% organic carbon sediment) (Lydy et al. 1992).

Environmental Fate Rate Constants or Half-Lives:
    Volatilization:
    Photolysis:
    Oxidation:
    Hydrolysis: the first-order rate constant of 1.4×10^{-9} M^{-1} s^{-1} and the hydrolytic half-life at 27°C in water > 120 yr ((Wolfe et al. 1977); a hydrolytic half-life at pH 7 and 25°C of 120 yr and a rate constant of 6.6×10^{-7} h^{-1} (Callahan et al. 1979, Mabey et al. 1982).
    Biodegradation:
    Biotransformation:
    Bioconcentration, Uptake (k_1) and Elimination (k_2) Rate Constants:
        k_1: 170.0 d^{-1} (rainbow trout, Oliver & Niimi 1985)
        k_2: 0.021 d^{-1} (rainbow trout, Oliver & Niimi 1985)
        k_2: 0.950 y^{-1} (*Larus argentatus*, Norstrom et al. 1986)
        k_1: 20800 d^{-1} (*Oligochaetes*, Connell et al. 1988)
        k_2: 0.020 d^{-1} (*Oligochaetes*, Connell et al. 1988)
        k_2: 0.0004 h^{-1} (azalea leaves, Paterson et al. 1991)
        k_2: 0.0030 h^{-1} (midge *C. riparius*, water only system, Lydy et al. 1992)
        k_2: 0.0046 h^{-1} (midge *C. riparius*, screened, Lydy et al. 1992)
        k_2: 0.0080 h^{-1} (midge *C. riparius*, 3% organic carbon, Lydy et al. 1992)
        k_2: 0.0046 h^{-1} (midge *C. riparius*, 15% organic carbon, Lydy et al. 1992)

Common Name: DDT

Synonym: Agritan, Anofex, Arkotine, Azotox, Bosan supra, Bovidermol, Cesarex, chlorophenoethane, Chlorophenothanum, Chlorophenotoxum, Citox, Clofenotan, Dedelo, Deoval, Devol, Detox, Detoxan, Dibovan, Dichophane, dichlorodiphenyltrichloroethane, Didigam, Didimac, Dodat, Dykol, ENT 1506, Estonate, Genitox, Gesafid, Gesapon, Gesarex, Gesarol, Guesarol, Gyron, Havero-extra, Ivoran, Ixodex, Kopsol, Mutoxin, Neocid, Parachlorocidum, PEBI, Pentachlorin, Pentech, p,p'-DDT, 4,4'-DDT, Rukseam, Santobane, Zeidane, Zerdane

Chemical Name: 1,1,1-trichloro-2,2-bis-(4-chlorophenyl)-ethane;1,1'-(2,2,2-trichloroethylidene)-bis(4-chlorobenzene)

Uses: persistent nonsystemic insecticide with contact and stomach action to control mosquitoes for the eradication of malaria but is now prohibited and displaced with less persistent insecticides on crop application.

CAS Registry No: 50-29-3 (p,p'-DDT, DDT), 789-02-6 (o,p'-DDT)

Molecular Formula: $C_{14}H_9Cl_5$

Molecular Weight: 354.5

Melting Point (°C):
- 108.5-109 (Callahan et al. 1979; Suntio et al. 1988; Tomlin 1994; Milne 1995)
- 109 (Karickhoff 1981; Swann et al. 1983; Bidleman & Foreman 1987; Gobas et al. 1988; Ballschmiter & Wittlinger 1991; Noegrohati & Hammers 1992; Yalkowsky & Banerjee 1992)
- 108.5 (Kenaga 1972; Spencer 1982; Neely & Blau 1985; Mackay 1991; Montgomery 1993; Patil 1994; Kühne et al. 1995; Mortimer & Connell 1995)
- 107 (Liu et al. 1991)

Boiling Point (°C):
- 185 (Günther & Günther 1971; Callahan et al. 1979)
- 315, 369 (estimated from structure, Tucker et al. 1983)
- 185-187 (at 0.05 mmHg with dec., Agrochemicals Handbook 1987; Tomlin 1994)

Density (g/cm^3 at 20°C):
- 1.55 (Hadaway et al. 1970; Kenaga 1972)

Molar Volume (cm^3/mol):
- 250 (calculated-density, Chiou 1985)
- 363.5 (Gobas et al. 1988)
- 333.5 (calculated-LeBas method, Suntio et al. 1988)
- 261.3 (Ruelle & Kesselring 1997)

Molecular Volume (Å^3):

Total Surface Area, TSA (Å^2):

Dissociation Constant pK_a:

Heat of Fusion, ΔH_{fus}, kcal/mol:
- 6.50 (o,p'-DDT, DSC method, Plato & Glasgow 1969)
- 6.30 (p,p'-DDT, DSC method, Plato & Glasgow 1969)
- 6.282 (Ruelle & Kesselring 1997)

Entropy of Fusion, ΔS_{fus}, cal/mol·K (e.u.):
- 16.5 (Plato & Glasglow 1969)
- 16.8 (Hinckley et al. 1990)

Fugacity Ratio at 25°C (assuming ΔS_{fus} = 13.5 e.u.), F:
- 0.147 (Mackay et al. 1986)
- 0.130 (20°C, Suntio et al. 1988)

Water Solubility (g/m^3 or mg/L at 25°C):
- 0.0002-0.001 (15°C, shake flask-bioassay, Richards & Cutkomp 1946; quoted, Biggar & Riggs 1974; Shiu et al. 1990)
- 0.0020 (Balson 1947; quoted, Spencer & Cliath 1970; Ballschmiter & Wittlinger 1991)
- 0.0374 (shake flask-radiometric, Babers 1955; quoted, Richardson & Miller 1960; Biggar & Riggs 1974; Shiu et al. 1990)
- ≤0.0012 (shake flask-radiometric method, Bowman et al. 1960; quoted, Biggar & Riggs 1974; Callahan et al. 1979; Fuhremann & Lichtenstein 1980; Sharom et al. 1980; Belluck & Felsot 1981; Suntio et al. 1988)
- 0.035 (colorimetric, Lipke & Kearns 1960; quoted, Shiu et al. 1990)
- 0.0012 (Stephen & Stephen 1963; quoted, Lu & Metcalf 1975)
- 0.0016 (99% pure DDT isomers plus DDE at rm. temp., shake flask-GC, Robeck et al. 1965; quoted, Biggar & Riggs 1974)
- 0.0034 Biggar et al. 1966; quoted, Chiou et al. 1977; Freed et al. 1977; Shiu et al. 1990; Chiou 1985; Karickhoff 1985)
- 0.0017 (ultracentrifugation-GC, Biggar et al. 1967; quoted, Biggar & Riggs 1974; Kenaga & Goring 1980; Briggs 1981; Karickhoff 1981; Bysshe 1982; Zaroogian et al. 1985)
- 0.0012-0.0374 (quoted, Günther et al. 1968; Metcalf et al. 1973; Jury et al. 1983,84; Clark et al. 1988; Suntio et al. 1988)
- 0.0012 (quoted, Mackay & Wolkoff 1973; Mackay & Leinonen 1975; quoted, Geyer et al. 1980; Giam et al. 1980)
- 0.0017, 0.006, 0.025 (shake flask-GC, *p,p'*-DDT, particle size: 0.01, 0.05, 5.0 μ, Biggar & Riggs 1974)
- 0.004, 0.012, 0.085 (shake flask-GC, *o,p'*-DDT, particle size: 0.01, 0.05, 5.0μ size or less, Biggar & Riggs 1974)
- 0.017, 0.025, 0.037, 0.045 (shake flask-GC, *p,p'*-DDT, at 15, 25, 35, 45 °C, particle size 5μ or less, Biggar & Riggs 1974)
- 0.050, 0.085, 0.135, 0.200 (shake flask-GC, *o,p'*-DDT, at 15, 25, 35,45 °C, Biggar & Riggs 1974)
- 0.025 (shake flask-GC, 5.0 μ particle size, Biggar & Riggs 1974; quoted, Callahan et al. 1979; Davies & Dobbs 1984; Shiu et al. 1990)
- 0.0055 (gen. col.-GC/ECD, Weil et al. 1974; quoted, Callahan et al. 1979; Geyer et al. 1980, 1982; Chiou 1981; Chiou & Schmedding 1981; Chiou et al. 1986; Clark et al. 1988; Suntio et al. 1988; Shiu et al. 1990)
- 0.0031 (quoted from Bowman et al. 1960, Chiou et al. 1977; Freed et al. 1977; Chiou 1985; Mackay 1991; Mortimer & Connell 1995)
- 0.0017 (Martin & Worthing 1977; quoted, Kenaga 1980a,b)
- 0.001 (shake flask-GC, Paris et al. 1977; quoted, Shiu et al. 1990)
- 0.001 (quoted, Branson 1978, Wauchope 1978)
- 0.004 (shake flask-nephelometry, Hollifield 1979)
- 0.0055 (quoted, Callahan et al. 1979, Mabey et al. 1982; Chin et al. 1986)
- 0.003 (selected, Ellgehausen et al. 1980)
- 0.0012 (Hartley & Graham-Bryce 1980; quoted, Taylor & Glotfelty 1988)
- 0.0012 (Neely 1980; quoted, Zoeteman et al. 1981)
- 0.001 (Weber et al. 1980; quoted, Willis & McDowell 1982)
- 0.0017, 0.0039 (quoted exptl., calculated-Parachor, Briggs 1981)
- 0.0012 (quoted, Burkhard & Guth 1981)
- 0.040 (shake flask-HPLC, Ellgehausen et al. 1981)
- 0.0645 centrifuge-GC, Chiou et al. 1982)

0.0012 (quoted, Neely 1982; Thomas 1982; Nash 1988)
0.0030 (Jury et al. 1983, 1987; quoted, Taylor & Glotfelty 1988; Shiu et al. 1988; Mackay & Stiver 1991; Harner & Mackay 1995)
0.0023 (gen. col., Swann et al. 1983; quoted, Shiu et al. 1990)
0.020 (RP-HPLC, Swann et al. 1983; quoted, Shiu et al. 1990)
0.0031-0.0034 (quoted, Verschueren 1983)
0.0045 (shake flask-LSC, Gerstl & Mingelgrin 1984; quoted, Shiu et al. 1990)
0.0017 (Yashida et al. 1983)
0.0055 (quoted, Caron et al. 1985)
0.0033 (quoted, Neely & Blau 1985; Mackay 1985)
0.467, 0.030 (predicted-K_{ow}, RPHPLC, Chin et al. 1986)
0.0054 (24°C, shake flask-GC/ECD, Chiou et al. 1986; quoted, Wang & Brusseau 1993)
9.0×10^{-3} (quoted, Mackay et al. 1986)
0.0032 (selected, Gerstl & Helling 1987)
0.00316 (selected, Eadie & Robbins 1987)
0.0033 (selected, Elzerman & Coates 1987)
0.004-0.0000008 (quoted literature range, Sabljic 1987)
0.0034 (Agency for Toxic Substances and Disease Registry 1988; quoted, Burmaster et al. 1991)
0.002 (quoted, Isnard & Lambert 1988; 1989; Ballschmiter & Wittlinger 1991)
2.69×10^{-3} (quoted, Mackay et al. 1988)
0.0645 (Mailhot & Peters 1988; quoted, Chessells et al. 1992)
0.003 (20°C, selected, Suntio et al. 1988; quoted, Calamari et al. 1991; Majewski & Capel 1995; selected, Harner & Mackay 1995)
0.0055 (shake flask-GC, Chiou et al. 1991)
0.00295 (selected, Yalkowsky & Banerjee 1992)
0.001-0.0055 (Montgomery 1993)
0.0055 (20-25°C, selected, Augustijn-Beckers et al. 1994; Hornsby et al. 1996)
0.003, 0.001 (quoted, calculated, Patil 1994)
0.025, 0.0063 (quoted, calculated-group contribution fragmentation method, Kühne et al. 1995)
0.0044; 0.016, 0.085 (quoted; predicted-Aquafac, predicted, Myrdal et al. 1995)
0.0109 (quoted, Pinsuwan et al. 1995)
0.0030; 0.0035, 0.00012 (quoted; predicted-molar volume, Ruelle & Kesselring 1997)

Vapor Pressure (Pa at 25°C):

2.00×10^{-5} (20°C, effusion manometer, Balson 1947, quoted, Metcalf 1948; Kenaga 1972; Dobbs & Cull 1982; Windholz 1983; Suntio et al. 1988)
3.30×10^{-5} (Balson 1947; quoted, Spencer & Cliath 1970; Ballschmiter & Wittlinger 1991)
3.33×10^{-5} (20°C, partition coefficient, Atkins & Eggleton 1971; quoted, Bidleman & Christensen 1979; Suntio et al. 1988)
2.03×10^{-5}, 9.68×10^{-5}, 4.426×10^{-4} (20, 30, 40°C, gas saturation, Spencer & Cliath 1972; quoted, Suntio et al. 1988)
9.68×10^{-5} (30°C, calculated-vapor density, Spencer & Cliath 1972; Spencer 1975; quoted, Spencer et al. 1973; Callahan et al. 1979)
4.50×10^{-5} (gas saturation, Spencer & Cliath 1972; quoted, Jury et al. 1983)

2.53×10^{-5} (20°C, Spencer 1973; quoted, Callahan et al. 1979, Mabey et al. 1982; Fuhremann & Lichtenstein 1980; Khan 1980; Burkhard & Guth 1981)
1.33×10^{-5} (quoted, Mackay & Leinonen 1975; Branson 1978; Giam et al. 1980)
7.33×10^{-4} (*o,p'*-DDT, 30°C, Spencer 1975; quoted, Callahan et al. 1979)
2.50×10^{-5} (20°C, Hartley & Graham-Bryce 1980; quoted, Suntio et al. 1988; Taylor & Glotfelty 1988; Worthing 1991; Tomlin 1994)
2.00×10^{-5} (20-25°C, Weber et al. 1980; quoted, Willis & McDowell 1982)
1.87×10^{-3}, 1.73×10^{-3} (GC-RT, Westcott & Bidleman 1981)
1.15×10^{-5} (*o,p'*-DDT, 30°C, semimicro gas-saturation, Westcott et al. 1981)
4.30×10^{-5} (estimated-relative volatilization rate, Dobbs & Cull 1982)
2.67×10^{-5}; 2.67×10^{-3}, 2.67×10^{-5} (20°C, quoted; calculated values, Grain 1982)
1.33×10^{-5} (quoted, Neely 1980, 1982; quoted, Zoeteman et al. 1981)
2.93×10^{-5}, 5.73×10^{-5}, 1.24×10^{-4} (20°C, 25°C, 30°C, gas saturation, Rothman 1980)
1.50×10^{-4} (20°C, GC, Seiber et al, 1981; quoted, Suntio et al. 1988)
2.01×10^{-5} - 2.8×10^{-5} (gas saturation, Jaber et al. 1982; Suntio et al. 1988)
log P (mmHg) = $14.191 - 6160/T$ (Gückel et al. 1982)
4.31×10^{-5} (20°C, relative loss rate, Dobbs & Cull 1982; quoted, Suntio et al. 1988)
3.60×10^{-5} (Thomas 1982; quoted, Nash 1988)
4.50×10^{-5} (Jury et al. 1983, 1987; quoted, Taylor & Glotfelty 1988; Taylor & Spencer 1990)
1.33×10^{-5} (Yoshida et al. 1983)
0.00133; 0.00162, 0.00128 (*o,p'*-DDT, supercooled liquid values P_L: quoted; GC-RT, Bidleman 1984)
3.16×10^{-4}; 8.30×10^{-4}, 4.70×10^{-4} (*p,p'*-DDT, supercooled liquid values P_L: quoted; GC-RT, Bidleman 1984)
2.13×10^{-5}, 1.73×10^{-4} (20°C: quoted average, calculated-supercooled values from quoted average, Bidleman & Foreman 1987)
2.53×10^{-6} (Caron et al. 1984)
2.63×10^{-5} (quoted, Neely & Blau 1985; Lyman 1985; Mackay 1985)
3.33×10^{-5} (quoted, Mackay et al. 1986)
1.73×10^{-4} (20°C, supercooled liquid value, Bidleman et al. 1986)
2.50×10^{-5} (Agrochemicals Handbook 1987)
7.00×10^{-4} (Agency for Toxic Substances and Disease Registry 1988; quoted, Burmaster et al. 1991)
1.94×10^{-5} (quoted, Clark et al. 1988; Mackay et al. 1988)
2.00×10^{-5} (20°C, selected, Suntio et al. 1988; quoted, Majewski & Capel 1995)
2.50×10^{-5} (20°C, quoted, Calamari et al. 1991)
3.16×10^{-4}, 5.12×10^{-4} (supercooled liquid values, GC-RT, Hinckley et al. 1990)
log P_L (Pa) = $12.32 - 4640/T$ (liquid, Hinckley et al. 1990)
log P_L (Pa) = $13.02 - 4865/T$ (GC-RT, liquid, Hinckley et al. 1990)
3.30×10^{-5} (quoted, Ballschmiter & Wittlinger 1991)
4.20×10^{-5} (quoted, Mackay & Stiver 1991)
6.92×10^{-5}, 2.69×10^{-4}, 9.33×10^{-4} (supercooled liquid values at 10°C, 20°C, 30°C, calculated from Hinckley et al. 1990; Cotham & Bidleman 1992)
2.27×10^{-8} (20°C, Montgomery 1993)
2.53×10^{-5} (20-25°C, selected, Augustijn-Beckers et al. 1994; Hornsby et al. 1996)
1.715×10^{-5}, 8.180×10^{-5} (20°C, 30°C, gas saturation-GC/ECD, Wania et al. 1994)
2.00×10^{-5} (selected, Mackay 1991; quoted, Mortimer & Connell 1995)

Henry's Law Constant (Pa·m^3/mol):
- 7.29 (calculated-P/C, Levins 1981; quoted, Tucker et al. 1983)
- 3.94 (calculated-P/C, Mackay & Leinonen 1975; quoted, Tucker et al. 1983)
- 6.02 (20-25°C, calculated, Thibodeaux 1979; quoted, Suntio et al. 1988; Bacci et al. 1990)
- 5.30 (calculated-P/C, Mackay & Shiu 1981)
- 1.60 (calculated-P/C, Mabey et al. 1982)
- 3.85 (calculated-P/C, Thomas 1982)
- 0.466 (estimated-group method per Hine & Mookerjee 1975, Tucker et al. 1983)
- 2.73 (Yoshida et al. 1983)
- 4.96 (calculated-P/C, Jury et al. 1984, 1987a; Jury & Ghodrati 1989)
- 1.63 (calculated-P/C, Caron et al. 1984)
- 2.84 (estimated, Mackay 1985)
- 1.31 (calculated-P/C, Mackay et al. 1986; quoted, Iwata et al. 1993)
- 73.0 (Agency for Toxic Substances and Disease Registry 1988; quoted, Burmaster et al. 1991)
- 4.96 (WERL Treatability Database, quoted, Ryan et al. 1988)
- 2.36 (Suntio et al. 1988; quoted, Fendinger et al. 1989; Müller et al. 1994)
- 4.96, 8.18 (calculated-P/C, Taylor & Glotfelty 1988)
- 2.36 (20°C, calculated-P/C, Suntio et al. 1988; quoted, Cotham & Bidleman 1991; Majewski & Capel 1995)
- 1.31 (fog chamber of concn. ratio-GC/ECD, Fendinger et al. 1989)
- 1.20 (wetted-wall column, GC/ECD, Fendinger et al. 1989; quoted, Cotham & Bidleman 1991)
- 1.30 (measured, Atkins & Eggleton 1971; quoted, Cotham & Bidleman 1991)
- 0.16 (0°C, selected, Cotham & Bidleman 1991)
- 2.90 (calculated-P/C, Calamari et al. 1991)
- 6.0 (calculated-P/C, Ballschmiter & Wittlinger 1991)
- 1.31 (25°C), 0.86 (24°C) (quoted, Iwata et al. 1993)
- 1.31 (23°C, Montgomery 1993)

Octanol/Water Partition Coefficient, log K_{ow}:
- 3.98 (Kapoor et al. 1973; quoted, Callahan et al. 1979)
- 3.98 (Lu & Metcalf 1975)
- 6.19 (calculated, O'Brien 1975; quoted, Chiou et al. 1977; Freed et al. 1977; Callahan et al. 1979; Karickhoff 1981, 1985; Zaroogian et al. 1985; De Bruijn et al. 1989)
- 6.36 (shake flask-GC, Chiou et al. 1977, 1982; quoted, Chiou 1981; Chiou & Schmedding 1981, Chiou 1985; Neely & Blau 1985; Zaroogian et al. 1985; Brooke et al. 1986; Chiou et al. 1986; Chiou et al. 1987; De Bruijn et al. 1989; Wang & Brusseau 1993)
- 4.89 (quoted, Wolfe et al. 1977; Callahan et al. 1979)
- 5.98 (Callahan et al. 1979; quoted, Kenaga 1980b; Chin et al. 1986)
- 6.19 (shake flask-GC, Freed et al. 1979)
- 3.98-6.19 (Hansch & Leo 1979; quoted, Platford et al. 1982)
- 5.10 ± 0.1 (radioactive analysis method, Platford et al. 1982)
- 6.19 (quoted, Hansch & Leo 1979, quoted, Veith et al. 1979a; Mackay et al. 1980; Geyer et al. 1982; Harnisch et al. 1983; Klein et al. 1988)
- 6.36 (shake flask, Karickhoff et al. 1979; quoted, Sicbaldi & Finizio 1993)

5.75 (quoted, Veith et al. 1979b, 1980; quoted, Mackay 1982; Bysshe 1982; Veith & Kosian 1983)
6.19 (selected, Ellgehausen et al. 1980, 1981)
5.98 (quoted, Kenaga & Goring 1978, 1980; Callahan et al. 1979)
6.19 (quoted, Rao & Davidson 1980; Suntio et al. 1988; Geyer et al. 1991)
6.69 (calculated as per Chiou et al. 1977, Belluck & Felsot 1981)
5.98, 6.89 (quoted exptl., calculated-Parachor, Briggs 1981)
5.94, 6.19 (shake flask-GC/LC, quoted, Ellgehausen et al. 1981; quoted, Sicbaldi & Finizio 1993)
5.57, 5.55 (quoted, HPLC-k', McDuffie 1981)
6.38 (correlated-HPLC, Hammers et al. 1982; quoted, De Bruijn et al. 1989)
6.91 (calculated-f const., Mabey et al. 1982)
4.98 (Clement Associates 1983; quoted, Chapman 1989)
5.10 (Platford 1982, 1983)
6.00 (quoted, Neely 1982)
5.75 (Veith & Kosian 1982; quoted, Dao et al. 1983; Saito et al. 1992)
5.90 (average of shake flask values, Eadsforth & Moser 1983)
6.12 (average of HPLC method values, Eadsforth & Moser 1983)
6.20 (Elgar 1983)
7.35 (calculated-f const. as per Rekker 1977, Harnisch et al. 1983)
6.06, 5.84 (quoted HPLC method, Harnisch et al. 1983)
5.44 (shake flask-GC or LSC, Gerstl & Mingelgrin 1984)
5.98 (Pavlou & Weston 1983, 1984)
6.36 (selected, Davies & Dobbs 1984, Elzerman & Coates 1987)
5.44 (estimated-HPLC/MS, Burkhard et al. 1985)
4.98 (quoted from Hansch & Leo 1979, Burkhard et al. 1985)
6.91 (calculated-CLOGP, Burkhard et al. 1985)
6.36 (Brooke et al. 1986; quoted, Hodson & Williams 1988; Sicbaldi & Finizio 1993)
6.06 (RPHPLC, Chin et al. 1986)
6.19, 6.21, 6.91 (quoted shake flask, estimated-HPLC, calculated-structural data file MACCS, Eadsforth 1986)
6.19 (selected, Hawker & Connell 1986, Gerstl & Helling 1987)
6.01 (quoted, Mackay et al. 1986)
4.99 (selected, Yoshioka et al. 1986)
6.19, 6.19 (quoted, RP-HPLC, De Kock & Lord 1987)
3.98-6.36 (quoted literature range, Sabljic 1987)
6.20 (Agency for Toxic Substances and Disease Registry 1988; quoted, Burmaster et al. 1991)
5.97 (quoted, Clark et al. 1988; Mackay et al. 1988)
5.98 (quoted, Isnard & Lambert 1988; Ryan et al. 1988; Ballschmiter & Wittlinger 1991; Liu et al. 1991)
6.36 (Medchem 1988; quoted, Müller et al. 1994)
6.00 (selected, Suntio et al. 1988; quoted, Calamari et al. 1988; Bacci et al. 1990)
6.91 (slow-stirring method, De Bruijn et al. 1989; quoted, Sicbaldi & Finizio 1993; selected, Hansch et al. 1995)
5.98, 6.00 (quoted, Thomann 1989)
5.89 (correlated value, Isnard & Lambert 1988, 1989)
5.76 (selected, Travis & Arms 1988)
6.84 (estimated-SAR, Karickhoff et al. 1989; quoted, Hoke et al. 1994)
6.307 ± 0.045 (BRE value, Brooke et al. 1990)

6.914 ± 0.030 (RITOX value, Brooke et al. 1990)
5.90, 6.36 (quoted shake-flask method, Brooke et al. 1990)
6.12, 6.19, 6.21, 6.38, 6.40 (quoted HPLC method, Brooke et al. 1990)
6.37, 6.91, 7.48 (calculated values, Brooke et al. 1990)
6.36 (Thor 1989; quoted, Connell & Markwell 1990, Chessells et al. 1992)
6.00 (quoted from Isnard & Lambert 1988, Ballschmiter & Wittlinger 1991)
6.19 (quoted, Banerjee & Baughman 1991; Noegrohati & Hammers 1992)
6.36 (quoted from Chiou et al. 1987, Chin et al. 1991)
4.89-6.91 (quoted, Montgomery 1993)
6.91 (estimated-QSAR & SPARC, Kollig 1993)
6.613 (calculated-f. const., Rekker et al. 1993; quoted, Pinsuwan et al. 1995)
5.50 (RP-HPLC, Sicbaldi & Finizio 1993)
6.36 (selected, Sangster 1993)
6.19, 6.82 (quoted, calculated, Patil 1994)
6.083 (Yalkowsky & Dannenfelser 1994; quoted, Pinsuwan et al. 1995)
6.20 (selected, Harner & Mackay 1995)
6.00 (selected, Devillers et al. 1996)
5.65, 6.76, 5.53 (*o, p'*-, RP-HPLC, ClogP, calculated-S, Finizio et al. 1997)
4.82, 6.06, 5.53 (*p, p'*-, RP-HPLC, ClogP, calculated-S, Finizio et al. 1997)

Bioconcentration Factor, log BCF:
5.31-6.23 (earthworms, Wheatley & Hardman 1968; quoted, Connell & Markwell 1990)
2.42 (*Cylindrotheca closterium*, Keil & Priester 1969; quoted, Baughman & Paris 1981)
4.40 (*Syracosphaera carterae*, Cox 1970; quoted, Swackhamer & Skoglund 1991)
4.90 (*Amphidirium cartaria*, Cox 1970; quoted, Swackhamer & Skoglund 1991)
4.40 (*Tholassiosira fluviatilus*, Cox 1970; quoted, Swackhamer & Skoglund 1991)
4.00 (pinfish, Hansen & Wilson 1970; quoted, Biddinger & Gloss 1984)
4.58 (Atlantic croaker, Hansen & Wilson 1970; quoted, Biddinger & Gloss 1984)
3.94 (soft clam, Butler 1971; quoted, Hawker & Connell 1986)
4.20-4.36 (*Daphnia magna* over concn. gradient 8 µg/L to 1.1 mg/L, Crosby & Tucker 1971; quoted, Biddinger & Gloss 1984)
5.00 (*Daphnia magna* at water level 80 ng/L, Johnson et al. 1971; quoted, Biddinger & Gloss 1984)
4.27 (*Daphnia magna*, wet wt. basis, Crosby & Tucker 1971; quoted, Geyer et al. 1991)
4.45 (*Daphnia magna*, wet wt. basis, Johnson et al. 1971; quoted, Geyer et al. 1991)
4.08-4.60 (fishes, Menzie 1972)
4.93, 4.54 (*Gambusia, Physa,* Metcalf et al. 1973)
4.68 (oyster, Parrish 1974)
4.79 (*Ankistrodesmus*, Neudorf & Khan 1975; quoted, Swackhamer & Skoglund 1991)
1.27 (bioaccumulation factor log BF, adpose tissue in male Albino rats, Berdanier & de Dennis 1977; quoted, Geyer et al. 1980)
3.52-3.63, 3.11-3.43 (bacteria, algae, Wolfe et al. 1977)
3.14 (trout muscle, Branson 1978)
4.47 (fathead minnows, 32 days exposure, Veith et al. 1979b, 1980; quoted, Bysshe 1982; Veith & Kosian 1983; Zaroogian et al. 1985; Devillers et al. 1996)

4.72   (bluegill sunfish-kinetic value, Bishop & Maki 1980; quoted, Davies & Dobbs 1984)
4.20   (log BF-bioaccumulation factor of algae, Ellgehausen et al. 1980)
3.539   (log BF-bioaccumulation factor of catfish, Ellgehausen et al. 1980)
3.35   (log BF-bioaccumulation factor of daphnids, Ellgehausen et al. 1980)
4.79, 4.93 (fish: flowing water, static water; quoted, Kenaga 1980a,b; Kenega & Goring 1980)
4.35   (calculated-S, Kenaga 1980a)
4.43   (calculated-K_{OC}, Kenaga 1980a)
−0.045 (average beef fat diet, Kenaga 1980b)
4.15   (pulex, Kenaga & Goring 1980; quoted, Hawker & Connell 1986)
4.11   (algae, estimated, Baughman & Paris 1981)
2.95-3.03 (*Rhodotorulus solani*, Baughman & Paris 1981,83)
3.02-3.13 (*Alfafa tissue*, Baughman & Paris 1981)
2.10   (*Triaenodes tardus*, Belluck & Felsot 1981)
4.79, 5.38 (quoted exptl., calculated-K_{OW}, Briggs 1981)
5.11   (selected, Schnoor & McAvoy 1981; quoted, Schnoor 1992)
4.36, 4.15, 4.43 (estimated-S, K_{OW}, K_{OC}, Bysshe 1982)
4.47, 4.43 (fish: quoted, correlated, Mackay 1982)
4.37   (mussels, quoted average, Geyer et al. 1982)
6.90   (microorganism-water, Mabey et al. 1982)
4.47   (fathead minnows, Veith & Kosian 1982; quoted, Saito et al. 1992)
2.30, 4.08 (trout, pinfish, Verschueren 1983)
4.71   (15°C, rainbow trout, Davies & Dobbs 1984)
5.00   (25°C, fathead minnow-steady state, Davies & Dobbs 1984)
4.15   (activated sludge, Freitag et al. 1984)
3.97, 3.46, 4.15 (algae, fish, activated sludge, Klein et al. 1984)
3.97, 3.28, 4.15 (algae, golden ide, activated sludge, Freitag et al. 1985)
4.97   (*Oncorhynchus mykiss*, Muir et al. 1985; quoted, Devillers et al. 1996)
3.91, 3.08 (rainbow trout: kinetic, steady-state, Oliver & Niimi 1985)
4.47, 4.56 (oyster, calculated-K_{OW} & models, Zaroogian et al. 1985)
4.68   (oyster, quoted from Zaroogian et al. 1985; Hawker & Connell 1986)
3.24-5.00 (*p,p'*-DDT, benthic macroinvertebrates, Reich et al. 1986)
3.44-5.71 (*o,p'*-DDT, benthic macroinvertebrates, Reich et al. 1986)
4.08   (*Selenastrum capricornutum*, Mailhot 1987; quoted, Swackhamer & Skoglund 1991)
4.47   (quoted, Isnard & Lambert 1988)
6.50   (zooplankton, chum salmon; Kawano et al. 1988)
−1.55   (beef biotransfer factor log B_b, correlated-K_{OW} from Radeleff et al. 1952 & Kenaga 1980, Travis & Arms 1988)
−2.62   (milk biotransfer factor log B_m, correlated-K_{OW} from Fries et al. 1969; Saha 1969 & Whiting et al. 1973, Travis & Arms 1988)
−1.80   (vegetation, correlated-K_{OW} from Beall & Nash 1972 & Voerman & Besemer 1975, Travis & Arms 1988)
5.28, 7.64 (dry leaf, wet leaf, Bacci et al. 1990)
4.47, 4.30 (quoted, calculated, Banerjee & Baughman 1991)
4.72   (selected, Chessells et al. 1992)
−0.155, −1.0 (earthworms, quoted, field/lab., Menzie et al. 1992)
−1.0, −0.602 (earthworms, field leaf litter, calcd-model, Menzie et al. 1992)

4.81, 4.86, 4.95, 4.99 (*Oncorhynchus mykiss*, Muir et al. 1994; quoted, Devillers et al. 1996)
5.10 (fish, reported as log BAF_w, LeBlanc 1996)

Sorption Partition Coefficient, log K_{OC}:
5.38 (soil, Hamaker & Thompson 1972; quoted, Kenaga 1980; Kenaga & Goring 1980; Karickhoff 1981)
3.93 (sediment, Wolfe et al. 1977)
5.38 (calculated-K_{OW}, Kenaga 1980)
5.16 (soil, calculated-S as per Kenaga & Goring 1978, Kenaga 1980)
5.38 (Kenaga & Goring 1980; quoted, Hodson & Williams 1988)
5.18 (average 3 soils, HPLC-RT, McCall et al. 1980)
5.39 (average soils/sediments, Rao & Davidson 1980; quoted, Lyman 1982; Jury et al. 1983)
5.20, 5.18, 5.18; 5.18 (commerce soil, Tracy soil, Catlin soil; average soil, McCall et al. 1980)
5.00 (selected, sediment/water, Schnoor & McAvoy 1981; quoted, Schnoor 1992)
4.38 (quoted from Kenaga & Goring 1980, Byssche 1982)
5.62 (estimated-S, Karickhoff 1981)
6.81, 5.80 (estimated-S & M.P., K_{OW}, Karickhoff 1981)
6.59 (calculated-K_{OW}, Mabey et al. 1982)
5.38 (soil, Thomas 1982; quoted, Nash 1988)
5.20 (Pavlou & Weston 1983, 1984)
5.18, 4.64 (soil slurry method, HPLC-RT, Swann et al. 1983)
5.38 (soil, Jury et al. 1984; quoted, Mackay & Stiver 1991)
5.38, 5.33 (soil: quoted, calculated-χ, Sabljic 1984)
5.61 (Caron et al. 1984)
5.39 (soil, estimated, Karickhoff 1985; quoted, Neely & Blau 1985)
6.00 (best estimate at low sediment concn., Karickhoff 1985)
5.11 - 5.45 (Aldrich humics, Landrum et al. 1985)
4.28 - 4.66 (natural water, Landrum et al. 1985)
5.61 ± 0.11 (Chiou et al. 1987; quoted, Chin et al. 1991)
6.03 (predicted-K_{OW}, Chiou et al. 1987; quoted, Chin et al. 1991)
5.39 (selected, Elzerman & Coates 1987)
5.26, 3.94 (quoted, calculated-χ, Gerstl & Helling 1987)
5.38 (soil, screening model calculations, Jury et al. 1987a,b; Jury & Ghodrati 1989)
4.58 (sediment, Mackay et al. 1988)
5.38, 5.34 (quoted, calculated-χ, Bahnick & Doucette 1988)
4.09 (calculated-K_{OW} as per Kenaga & Goring 1980, Chapman 1989)
5.15 - 6.26 (Montgomery 1993)
6.59 (estimated-QSAR & SPARC, Kollig 1993)
6.30 (20-25°C, estimated, Augustijn-Beckers et al. 1994; Hornsby et al. 1996)
or log K_{OM}:
5.14 (exptl., Briggs 1981)
4.24 (calculated-Parachor, Briggs 1981)
4.88 - 5.41 (Mingelgrin & Gerstl 1983)
5.69, 5.59, 5.69 (average soil, sediment, soil and sediment, Gerstl & Mingelgrin 1984)

Half-Lives in the Environment:
- Air: 17.7-177 h, based on estimated rate constant for the reaction with hydroxyl radicals in air (Atkinson 1987; quoted, Howard et al. 1991; Mortimer & Connell 1995).
- Surface water: dehydrochlorination rate constant of 1.75×10^{-2} h^{-1} for 1 ppm p,p'-DDT and 1.65×10^{-2} h^{-1} for 1 ppm o,p'-DDT both at 21 ± 2°C and pH 12.8 (in 0.1 N NaOH solution) (Choi & Chen 1976); degradation half-life in water at 27°C is about 8 years (Wolfe et al. 1977); half-life of 73.9 h for a pond 1 m deep (Branson 1978); 168-8400 h, based on measured rate of photooxidation in two natural waters under sunlight for seven days and 56 d (Callahan et al. 1979; quoted, Howard et al. 1991; Mortimer & Connell 1995); degradation half-life of about 10 yr average from the loss rates in Lake Michigan (Bierman & Swain 1982).
- Groundwater: 16 d to 31.3 yr, based on anaerobic flooded soil die-away data for two flooded soils (Castro & Yoshida 1971; quoted, Howard et al. 1991) and observed rates of biodegradation in aerobic soils under field conditions (Lichtenstein & Schultz 1959; Stewart & Chisholm 1971; quoted, Howard et al. 1991).
- Sediment: 3 to 5 yr (Bierman & Swain 1982); 21 years (conversion of p,p'-DDT to p,p'-DDE in sediment, Oliver et al. 1989); 78800 h (quoted mean value from Howard et al. 1991, Mortimer & Connell 1991).
- Soil: 2-15.6 yr, based on observed rates of biodegradation in aerobic soils under field conditions (Lichtenstein & Schultz 1959; Stewart & Chisholm 1971; quoted, Howard et al. 1991); average half-life of ~12 yr in 3 different soils for ~50 ppm in soil (Nash & Woolson 1967); estimated persistence of 4 yr in soil (Kearney et al. 1969; Edwards 1973; quoted, Morrill et al. 1982; Jury et al. 1987a); field half-life of 173 d when incorporated into soil (Willis et al. 1971; quoted, Nash 1983); microagroecosystem half-life of >50 d with open cotton canopy (Nash & Harris 1977; quoted, Nash 1983); persistence of more than 36 months (Wauchope 1978); >50 d and subject to plant uptake via volatilization (Callahan et al. 1979; quoted, Ryan et al. 1988); estimated first-order half-lives of 14.6 yr from biodegradation rate constant of 0.00013 d^{-1} from soil incubation studies and 198 d from rate constant of 0.0035 d^{-1} from flooded soil incubation studies in anaerobic system both by die-away test (Rao & Davidson 1980; quoted, Scow 1982); very persistent in soils with half-life >100 d (Willis & McDowell 1982); microagroecosystem half-life of >50 d in moist fallow soil (Nash 1983); half-life of 3837 d from screening model calculations (Jury et al. 1984, 1987a,b; Jury & Ghorati 1989); 173 d from field study, >50 d from microagroecosystem, 116 d from laboratory data (Nash 1985); 3800 d (Jury et al. 1987; quoted, Montgomery 1993); reaction half-life of 3837 d and overall half-life in soil 9.4 yr (Mackay & Stiver 1991); estimated field half-life of 2000 d (Augustijn-Beckers et al. 1994; Hornsby et al. 1996).
- Biota: field half-life of 15 d in fruit leaves (Decker et al. 1950; quoted, Nash 1983); microagroecosystem half-life of 29 d in cotton leaves (Nash & Harris 1977; quoted, Nash 1983); 915 h from fish compared with calculated value of 517 hours from regression (Neely 1980); 0.70 h in algae, 3.65 d in catfish and 315 h in daphnids (Ellgehausen et al. 1980); 340 d in rainbow trout (Oliver & Niimi 1985); biochemical half-life of 3837 d (Jury et al. 1987a,b; Jury & Ghodrati 1989); biological half-lives for fishes: 77 d for trout, 31 d for salmon, 4 d for catfish (Niimi 1987).

Environmental Fate Rate Constants or Half-Lives:
: Volatilization: half-life of 3.7 d for water depth of 1 m (Mackay & Walfkoff 1972); 73.9 h (Mackay & Leinonen 1973, Branson 1978); initial rate constant of 6.9×10^{-4} h^{-1} and predicted rate constant of 1.2×10^{-3} h^{-1} from soil with half-life of 578 h (Thomas 1982); calculated half-life of 45 h from water (Thomas 1982); half-flux values times, 0.3 day from field study, 0.3-12 d from microagroecosystem, >80-1000 d from laboratory data (Nash 1985).
: Photolysis: using fungus and either 254 or 300 nm UV light, more than 97% initial added amounts were metabolized in 3 weeks of incubation (Katayama & Matsumura 1991).
: Oxidation: photooxidation half-life of 168-8400 h in water, based on measured rate of photooxidation in two natural waters under sunlight for 7 d and 56 d (Callahan et al. 1979; quoted, Howard et al. 1991); rate constants: for singlet oxygen, <3600 M^{-1} h^{-1} (Mabey et al., 1982) and for RO$_2$, 3600 M^{-1} h^{-1} (Mabey et al., 1982); photooxidation half-life of 17.7-177 h in air, based on estimated rate constant for the reaction with hydroxyl radicals in air (Atkinson 1987; quoted, Howard et al. 1991).
: Hydrolysis: estimated first-order half-life of 22 yr, based on a measured neutral hydrolysis rate constant of 1.9×10^{-9} s^{-1} and a base catalyzed constant of 9.90×10^{-3} M^{-1} s^{-1} at pH 7 and 27°C for 1×10^{-8} M in water (Wolfe et al. 1977; quoted, Callahan et al. 1979; Harris 1982; Howard et al. 1991); rate constant of 1.57×10^{-4} h^{-1} at pH 7 (Neely & Blau 1985); rate constant of 6.0×10^{-2} y^{-1} at pH 7.0 and 25°C (Kollig 1993).
: Biodegradation: aqueous aerobic half-life of 2-15.6 yr based on observed rates of biodegradation in aerobic soils under field conditions (Lichtenstein & Schultz 1959; Stewart & Chisholm 1971; quoted, Howard et al. 1991); aqueous anaerobic half-life of 16-100 d, based on anaerobic flooded soil die-away data for two flooded soils (Castro & Yoshida 1971; quoted, Howard et al. 1991); half-life of 3837 d (Hamaker 1972; quoted, Jury et al. 1983); rate constant of 0.00013 d^{-1} from soil incubation studies and 0.0035 d^{-1} from flooded soil incubation studies in anaerobic system both by die-away test (Rao & Davidson 1980; quoted, Scow 1982); half-life of 3837 d (Jury et al. 1984, 1987a,b; Jury & Ghodrati 1989).
: Biotransformation:
: Bioconcentration, Uptake (k_1) and Elimination (k_2) Rate Constants:
: k_2: 0.002, 0.0007 h^{-1} (algae, daphnids, Ellgehausen et al. 1980)
: k_2: 0.052 d^{-1} (catfish, Ellgehausen et al. 1980)
: k_1: 170 d^{-1} (rainbow trout, Oliver & Niimi 1985)
: k_2: 0.0021 d^{-1} (rainbow trout, Oliver & Niimi 1985)
: k_1: 818 d^{-1} (rainbow trout, Muir et al. 1985)
: k_2: 0.009 d^{-1} (rainbow trout, Muir et al. 1985)
: k_2: 8.60 y^{-1} (*P. hoyi*, Evans et al. 1991)
: k_2: 4.50 y^{-1} (*M. relicta*, Evans et al. 1991)

Common Name: Demeton
Synonym: Bayer 8169, Demeton-S, E-1059, mercaptophos, Systox
Chemical Name: O,O-diethyl-O-(2-ethylthioethyl)phosphorothioate mixture with O,O-diethyl-S-(2-ethylthioethyl)phosphorothioate
Uses: insecticide to control sucking insects and mites in a wide range of crops, including fruit, nuts, vegetables, ornamentals, and field crops; also used as acaricide.
CAS Registry No: 8065-48-3
Molecular Formula: $C_8H_{19}O_3PS_2$
Molecular Weight: 258.34
Melting Point (°C): liquid
Boiling Point (°C):
    123   (Khan 1980)
Density (g/cm^3 at 20°C):
    1.119   (25°C, Spencer 1982)
    1.119-1.132 (Agrochemicals Handbook 1987)
Molar Volume (cm^3/mol):
    264.8 (calculated-LeBas method, Suntio et al. 1988)
Molecular Volume (Å^3):
Total Surface Area, TSA (Å^2):
Dissociation Constant pK_a:
Heat of Fusion, ΔH_{fus}, kcal/mol:
Entropy of Fusion, ΔS_{fus}, cal/mol·K (e.u.):
Fugacity Ratio at 25°C (assuming ΔS_{fus} = 13.5 e.u.), F: 1.0

Water Solubility (g/m^3 or mg/L at 25°C):
    60   (20°C, Kenaga 1980a; quoted, Shiu et al. 1990)
    60   (22°C, Khan 1980; quoted, Suntio et al. 1988; Shiu et al. 1990)
    100   (20-25°C, Willis & McDowell 1982)
    60   (22°C, Worthing 1983; quoted, Shiu et al. 1990)
    60   (rm. temp., Spencer 1982; Agrochemicals Handbook 1987)
    60   (20°C, selected, Suntio et al. 1988)
    60   (20-25°C, selected, Augustijn-Beckers et al. 1994; Hornsby et al. 1996)

Vapor Pressure (Pa at 25°C):
    0.0347 (20°C, Melnikov 1971; quoted, Kim 1985; Suntio et al. 1988)
    0.0347 (20°C, Hartley Graham-Bryce 1980; quoted, Suntio et al. 1988)
    0.0331 (Khan 1980)
    0.033   (20°C, Spencer 1982)
    0.0167 (20°C, calculated-GC, Kim 1985)
    0.0340 (20°C, Agrochemicals Handbook 1987)
    0.030   (20°C, selected, Suntio et al. 1988)
    0.1333 (20-25°C, selected, Augustijn-Beckers et al. 1994; Hornsby et al. 1996)

Henry's Law Constant (Pa·m^3/mol):
    0.130 (20°C, calculated-P/C, Suntio et al. 1988)
    0.150 (calculated-P/C, this work)

Octanol/Water Partition Coefficient, log K_{OW}:
    1.20   (calculated-S as per Chiou et al. 1977 & Chiou 1981, this work)

Bioconcentration Factor, log BCF:
    1.79   (calculated-S, Kenaga 1980)

Sorption Partition Coefficient, log K_{OC}:
    2.66   (soil, calculated-S, Kenaga 1980)
    1.85   (20-25°C, estimated, Augustijn-Beckers et al. 1994; Hornsby et al. 1996)

Half-Lives in the Environment:
    Air:
    Surface water:
    Groundwater:
    Sediment:
    Soil: selected field half-life of 15 d (Augustijn-Beckers et al. 1994; Hornsby et al. 1996).
    Biota:

Environmental Fate Rate Constants or Half-Lives:
    Volatilization:
    Photolysis:
    Oxidation: calculated rate constant of 128×10^{-12} cm^3/molecule·s for the vapor phase reaction with hydroxyl radicals in air (Winer & Atkinson 1990).
    Hydrolysis:
    Biodegradation:
    Biotransformation:
    Bioconcentration, Uptake (k_1) and Elimination (k_2) Rate Constants:

Common Name: Diazinon
Synonym: Alfa-Tox, AG-500, Basudin, Bazinon, Bazuden, Ciazinon, Dacutox, Dassitox, Dazzel, Desapon, Dianon, Diater, Diaterr-fos, Diazitol, Diazide, Diazol, Dicid, Dimpylate, Dipofene, Dizinon, Dyzol, ENT 19507, Flytrol, G 301, Gardentox, Geigy 24480, Kayazinon, Kayazol, NA 2763, Nedicisol, Neocidol, Nipsan, Nucidol, Sarolex, Spectracide
Chemical Name: O,O-diethyl O-2-isopropyl-6-methylpyrimidin-4-yl phosphorothioate; O,O-diethyl-O-(2-isopropyl-6-methyl 4-pyrimidinyl) phosphorothioate; O,O-diethyl 2-isopropyl-4-methylpyrimidinyl-6-thiophosphate
Uses: nonsystemic insecticide to control flies, aphids and spider mites in soil, fruit, vegetables and ornamentals; also used as acaricide.
CAS Registry No: 333-41-5
Molecular Formula: $C_{12}H_{21}N_2O_3PS$
Molecular Weight: 304.36
Melting Point (°C): liquid
Boiling Point (°C):
    125   (at 1 mmHg, Agrochemicals Handbook 1987; Tomlin 1994; Milne 1995)
    83-84   (at 0.0002 mmHg, Montgomery 1993; Tomlin 1994)
    120   (Kühne et al. 1995)
Density (g/cm³ at 20°C):
    1.11   (Worthing 1991)
    1.116-1.118 (Montgomery 1993; Tomlin 1994; Milne 1995)
Molar Volume (cm³/mol):
    320.2   (calculated-LeBas method, Suntio et al. 1988)
Molecular Volume (Å³):
Total Surface Area, TSA (Å²):
Dissociation Constant pK_a:
    <2.5   (Albert 1963; Perrin 1989; Somasundaram et al. 1991; Montgomery 1993)
Heat of Fusion, ΔH_{fus}, kcal/mol:
Entropy of Fusion, ΔS_{fus}, cal/mol·K (e.u.):
Fugacity Ratio at 25°C (assuming ΔS_{fus} = 13.5 e.u.), F: 1.000

Water Solubility (g/m³ or mg/L at 25°C):
    40   (Spencer 1973, 1982; Martin & Worthing 1977; Bowman & Sans 1979, 1983; Worthing 1979, 1987; Sharom et al. 1980; Taylor & Glofelty 1988; quoted, Cohen & Steinmetz 1986; Kim et al. 1984; Shiu et al. 1990)
    40   (Martin & Worthing 1977; quoted, Kenaga 1980; Kenaga & Goring 1980; Jury et al. 1983, 1984; Zaroogian et al. 1985)
    40   (Wauchope 1978; quoted, Pait et al. 1992)
    68.8   (22°C, shake flask-GC, Bowman & Sans 1979, 1983a,b; quoted, Shiu et al. 1990; Howard 1991; Patil 1994)
    40   (Hartley & Graham-Bryce 1980; quoted, Taylor & Glotfelty 1988)
    40   (22°C, Khan 1980; quoted, Suntio et al. 1988; Schomberg et al. 1991)
    40   (quoted, Briggs 1981; Burkhard & Guth 1981; Kanazawa 1989)
    40.5   (20-25°C, shake flask-GC, Kanazawa 1981; quoted, Shiu et al. 1990)
    40   (20-25°C, selected, Willis & McDowell 1982)
    52.5   (Garten & Trabalka 1983; quoted, Shiu et al. 1990)
    40   (20°C, Windholz 1983; quoted, Somasundaram et al. 1991)
    40   (20°C, Agrochemicals Handbook 1987; Worthing 1991; Milne 1995)

52.5 (quoted, Isnard & Lambert 1988)
38 (20°C, selected, Suntio et al. 1988; quoted, Fendinger et al. 1990; Majewski & Capel 1995)
60 (20-25°C, selected, Wauchope et al. 1992; quoted, Lohninger 1994; Hornsby et al. 1996; Halfon et al. 1996)
52.9 (selected, Yalkowsky & Banerjee 1992)
53.5, 43.7 (20°C, 30°C, Montgomery 1993)
12.7 (calculated, Patil 1994)
60 (20°C, Tomlin 1994; quoted, Sanchez-Camazano et al. 1995)
70, 3.19 (quoted, calculated-group contribution fragmentation method, Kühne et al. 1995)
40 (selected, Iglesias-Jimenez et al. 1996)

Vapor Pressure (Pa at 25°C):
0.0111 (20°C, Wolfdietrich 1965; quoted, Kim et al. 1984)
0.0112 (20°C, Melnikov 1971; quoted, Suntio et al. 1988; Howard 1991)
0.0161 (gas saturation, Heiber & Szelagiewicz 1976; quoted, Jury et al. 1983; Taylor & Spencer 1990)
0.020 (Marti 1976; quoted, Hinckley et al. 1990)
0.0187 (Worthing 1979; quoted, Cohen & Steinmetz 1986)
0.019 (20°C, Hartley & Graham-Bryce 1980; quoted, Suntio et al. 1988; Taylor & Glotfelty 1988)
0.0187 (Khan 1980; quoted, Suntio et al. 1988)
0.0097 (quoted, Burkhard & Guth 1981)
0.00236-0.00469 (20°C, GC, Seiber et al. 1981; quoted, Suntio et al. 1988)
0.0113 (25.3°C, gas saturation method, Kim et al. 1984; quoted, Hinckley et al. 1990)
0.0064 (20°C, extrapolated-Clausius Clapeyron eqn., Kim et al. 1984; quoted, Suntio et al. 1988)
0.0024 (20°C, GC-calculated value, Kim et al. 1984; Kim 1985)
0.0064 (20°C, gas saturatiom method, Kim 1985)
9.7×10^{-5} (20°C, Agrochemicals Handbook 1987)
0.008 (20°C, selected, Suntio et al. 1988; quoted, Findinger et al. 1990; Majewski & Capel 1995; Halfon et al. 1996)
0.020 (GC-RT, supercooled value, Hinckley et al. 1990)
0.016 (selected, Taylor & Spencer 1990)
9.7×10^{-5} (20°C, Worthing 1991)
0.008 (20-25°C, selected, Wauchope et al. 1992; Hornsby et al. 1996)
0.0113 (20°C, Montgomery 1993)
0.012 (20°C, Tomlin 1994)

Henry's Law Constant (Pa·m³/mol):
0.0114 (calculated, Adachi et al. 1984; quoted, Howard 1991)
0.124 (calculated-P/C, Jury et al. 1984, 1987a; Jury & Ghodrati 1989)
0.0114 (wetted-wall col.-GC/ECD, Fendinger & Glotfelty 1988; quoted, Fendinger et al. 1989,90; Schomburg et al. 1991)
0.0669 (20°C, calculated-P/C, Suntio et al. 1988; quoted, Fendinger & Glotfelty 1988; Fendinger et al. 1989; Majewski & Capel 1995)
0.1438 (calculated-P/C, Taylor & Glotfelty 1988)

0.0119 (fog chamber-GC/ECD, Fendinger et al. 1989)
0.0114 (23°C, known LWAPC of Fendinger & Glotfelty 1990, Meylan & Howard 1991)
0.007 (bond-estimated LWAPC, Meylan & Howard 1991)
0.0114 (20°C, calculated-P/C, Montgomery 1993)
0.0406 (calculated-P/C, this work)

Octanol/Water Partition Coefficient, log K_{ow}:
   3.02   (Rao & Davidson 1980; quoted, Bowman & Sans 1983b; Kim et al. 1984; quoted, Suntio et al. 1988)
   3.11   (shake flask-GC, Briggs 1981; quoted, Bowman & Sans 1983b; Suntio et al. 1988; Sicbaldi & Finizio 1993; Bintein & Devillers 1994)
   3.14   (shake flask-GC, Kanazawa 1980,81; quoted, Zaroogian et al. 1985; Kanazawa 1989; Sicbaldi & Finizio 1993)
   3.81   (shake flask-GC, Bowman & Sans 1983b; quoted, Suntio et al. 1988; De Bruijn & Hermens 1991; Somasundaram et al. 1991; Sicbaldi & Finizio 1993; Patil 1994; Devillers et al. 1996)
   1.92   (Veith & Kosian 1983; quoted, Saito et al. 1992)
   3.81   (Hansch & Leo 1985; quoted, Howard 1991)
   3.54   (selected, Yoshioka et al. 1986)
   3.31   (quoted, Isnard & Lambert 1988; Travis & Arms 1988)
   3.30   (selected, Suntio et al. 1988)
   3.81   (Thor 1989; quoted, Connell & Markwell 1990; Magee 1991)
   3.02-3.81 (Montgomery 1993)
   3.70   (RP-HPLC, Saito et al. 1993)
   3.58   (RP-HPLC, Sicbaldi & Finizio 1993)
   4.16   (calculated, Patil 1994)
   3.30   (Tomlin 1994; quoted, Sanchez-Camazano et al. 1995)
   3.30   (selected, Iglesias-Jimenez et al. 1996)
   3.81   (selected, Hansch et al. 1995)
   3.58, 3.50, 3.42 (RP-HPLC, ClogP, calculated-S, Finizio et al. 1997)

Bioconcentration Factor, log BCF:
   2.39   (motsugo, Kanazawa 1975; quoted, McLeese et al. 1976)
   1.54   (fish in flowing water, Alison & Hermanutz 1977; quoted, Kenaga 1980; Kenaga & Goring 1980)
   2.18, 1.56 (topmouth gudugeon, silver crucian carp, Kanazawa 1978, 1981; quoted, Veith & Kosian 1983; Howard 1991; Tsuda et al. 1992; Devillers et al. 1996)
   1.81, 1.24 (carp, guppy, Kanazawa 1978; quoted, Veith & Kosian 1983; Howard 1991; Devillers et al. 1996)
   0.69, 1.23, 0.77 (crayfish, red snail, pond snail, Kanazawa 1978; quoted, Howard 1991)
   1.83   (fish, Kanazawa 1978; quoted, McLeese et al. 1976)
   0.954   (invertebrates, Kanazawa 1978; quoted, McLeese et al. 1976)
   1.89   (calculated-S, Kenaga 1980)
   2.75   (earthworms, Lord et al. 1980; quoted, Connell & Markwell 1990)
   2.08, 1.80 (carp, rainbow trout, Seguchi & Asaka 1981; quoted, Howard 1991; Devillers et al. 1996)
   1.41   (loach, Seguchi & Asaka 1981; quoted, Howard 1991; Devillers et al. 1996)
   0.477   (shrimp, Seguchi & Asaka 1981; quoted, Howard 1991)

1.81   (carp, Veith & Kosian 1982; quoted, Saito et al. 1992)
1.24   (guppy, Veith & Kosian 1982; quoted, Saito et al. 1992)
2.18   (topmouth gudgeon, Veith & Kosian 1982; quoted, Saito et al. 1992)
2.32   (topmouth gudgeon, Kanazawa 1983; quoted, Sancho et al. 1993)
2.15, 1.97 (sheesphead minnow, calculated-K_{OW} & models, Zaroogian et al. 1985)
2.30   (sheepshead minnow, Zaroogian et al. 1985; quoted, Howard 1991)
1.46   (quoted, Isnard & Lambert 1988)
−0.59   (vegetation, correlated-K_{OW}, Nash 1974; quoted, Travis & Arms 1988)
2.39   (willow shiner, Tsuda et al. 1989; quoted, Tsuda et al. 1992)
1.81, 2.08 (carp, De Bruijn & Hermens 1991)
1.38, 1.81, 1.81 (loach, motsugo, rainbow trout, De Bruijn & Hermens 1991)
2.16-2.33 (sheepshead minnow, De Bruijn & Hermens 1991)
0.477   (shrimp, De Bruijn & Hermens 1991)
1.56, 2.18 (silver crucian carp, topmouth gudgeon, De Bruijn & Hermens 1991)
2.18, 1.79 (pale chub, ayu sweetfish, calculated-field data, Tsuda et al. 1992)
1.93   (quoted, Pait et al. 1992)
3.20   (eel, Sancho et al. 1993)
1.34, 1.45 (*Oryzias latipes*, Tsuda et al. 1995; quoted, Devillers et al. 1996)

Sorption Partition Coefficient, log K_{OC}:
2.76   (calculated-S as per Kenaga & Goring 1978, Kenaga 1980; quoted, Howard 1991; Schomburg et al. 1991)
2.93   (Rao & Davidson 1980; quoted, Jury et al. 1983, 1984)
2.28   (average of 3 soils, Sharom et al. 1980; quoted, Howard 1991)
2.40   (one sediment, Sharom et al. 1980; quoted, Howard 1991)
2.12   (reported as log K_{OM}, Briggs 1981)
2.36   (estimated, Lyman et al. 1982; quoted, Howard 1991; Lohninger 1994)
2.93   (screening model calculations, Jury et al. 1987b; Jury & Ghodrati 1989)
2.40   (average of 2 soils, Kanazawa 1989)
2.12, 3.27 (reported, estimated as log K_{OM}, Magee 1991)
2.75   (soil, quoted exptl., Meylan et al. 1992)
3.13   (soil, calculated-χ and fragment contribution, Meylan et al. 1992)
3.00   (soil, 20-25°C, estimated, Wauchope et al. 1992; Hornsby et al. 1996)
2.76   (soil, average value, Dowd et al. 1993)
3.00-3.27 (Montgomery 1993)

Half-Lives in the Environment:
Air: photooxidation half-life of 4.1 h, estimated from the vapor-phase reaction with 5×10^5 hydroxyl radicals/m^3 in air at 25°C (Martin & Worthing 1977; quoted, Howard 1991).
Surface water: half-lives of 144 d at 6°C, 69 d at 22°C in darkness for Milli-Q water; 181 d at 6°C, 80 d at 22°C in darkness, 43 d under sunlight conditions for river water at pH 7.3; 132 d at 6°C, 52 d at 22°C in darkness for filtered river water at pH 7.3; 125 d at 6°C, 50 d at 22°C in darkness, 47 d under sunlight conditions for seawater, pH 8.1 (Lartiges & Garrigues 1995).
Groundwater:
Sediment:
Soil: half-life in sterile soil at pH 4.7 was 43.8 d (Sethunathan & MacRae 1969; quoted, Montgomery 1993); estimated persistence of 12 weeks in soil (Kearney et al. 1969; Edwards 1973; quoted, Morrill et al. 1982; Jury et al. 1987a); persistence

of 3 months (Wauchope 1978); 12.5 weeks in sterile sandy loam and <1.0 week in nonsterile sandy loam; 6.5 weeks in sterile organic soil and 2.0 weeks in nonsterile organic soil (Miles et al. 1979); estimated first-order half-life of 30 d in soil from biodegradation rate constant of 0.023 d^{-1} for soil incubation studies by soil die-away test (Rao & Davidson 1980; quoted, Scow 1982); moderate persistent in soil with half-life of 20-100 d (Willis & McDowell 1982); half-life of 32 d from screening model calculations (Jury et al. 1987a,b; Jury & Ghodrati 1989); dissipation half-life of 19 d in soil (Glotfelty et al. 1990); av. degradation rate constant of 0.0193 d^{-1} in silty clay with half-life of 36 days and average degradation rate constant of 0.0245 d^{-1} in sandy clay with half-life of 28 d (Sattar 1990); selected field half-life of 40 d (Wauchope et al. 1992; Dowd et al. 1993; Hornsby et al. 1996); soil half-life of 65 d (Pait et al. 1992); 40 d (selected, Halfon et al. 1996).

Biota: biochemical half-life of 32 d from screening model calculations (Jury et al. 1987a,b; Jury & Ghodrati 1989); excretion half-life of 9.9 h by willow shiner (Tsuda et al. 1989); 25 h in eel's liver and 26 h in eel's muscle (Sancho et al. 1993).

Environmental Fate Rate Constants or Half-Lives:
Volatilization: mostly dissipated through volatilization with half-life of 19 d from soil (Glotfelty et al. 1990).

Photolysis: calculated half-life of 15 d for photolysis in an aqueous buffer solution at pH 7 and 25°C under UV light for 24 h (Burkhard & Guth 1979; quoted, Montgomery 1993).

Oxidation: photooxidation half-life of 4.1 h in air, estimated from the vapor-phase reaction with 5×10^5 hydroxyl radicals/m^3 in air at 25°C (Martin & Worthing 1977; quoted, Howard 1991).

Hydrolysis: rate constants of 2.1×10^{-2} M^{-1} s^{-1} for acid catalyzed hydrolysis, 4.3×10^{-8} $M^{-1} \cdot s^{-1}$ for neutral hydrolysis and 5.3×10^{-3} M^{-1} s^{-1} for base catalyzed hydrolysis with 10^{-5} M in aqueous buffer (Faust & Gomaa 1972; quoted, Freed 1976; Harris 1982); half-lives of hydrolysis in water at 20°C of 11.77 hours at pH 3.1, 185 d at pH 7.4 and 6.0 d at pH 10.4 (Worthing 1991; Tomlin 1994); half-lives at 20°C: 11.77 h at pH 3.1, 185 d at pH 7.4, 136 days at pH 9.0, and 6 d at pH 10.4 (Montgomery 1993).

Biodegradation: half-lives of 4.91 d at pH 3.1 and 185 d at pH 7.4 from river die-away tests (Gomaa et al. 1969; quoted, Scow 1982); half-lives: 12.5 weeks in sterile soils and <1 week in nonsterile soils; 6.5 weeks in sterile sandy loam and 2 weeks in nonsterile sandy loam (Miles et al. 1979; quoted, Howard 1991); half-life of 32 d in 0-10 cm depth of soil by 100 d leaching screening test (Rao & Davidson 1980; quoted, Jury et al. 1983, 1984, 1987a,b; Jury & Ghodrati 1989).

Biotransformation:

Bioconcentration, Uptake (k_1) and Elimination (k_2) Rate Constants:
k_2: 0.070 h^{-1} (willow shiner, Tsuda et al. 1989)
k_2: 0.023 h^{-1} (eel's liver, Sancho et al. 1993)
k_2: 0.019 h^{-1} (eel's muscle, Sancho et al. 1993)

Common Name: Dichlorvos
Synonym: Apavap, Astrobot, Atgard, Bay 19149, Benfos, Bibesol, Brevinyl, Canogard, Cekusan, Chlorvinphos, Cyanophos, Cypona, DDVF, DDVP, Dedevap, Deriban, Derribante, Devikol, Dichlorman, Dichlorophos, Divipan, Duo-kill, Duravos, ENT 20738, Equigard, Equigel, Estrosel, Estrosol, Fecama, Fly-die, Fly fighter, Herkal, Herkol, Krecalvin, Lindan, Mafu, Mafu strip, Marvex, Mopari, NA 2783, Nerkol, Nogos, No-pest, Nuva, Nuvan, Oko, OMS 14, Phosvit, SD-1750, Szklarniak, Tap 9VP, Task, Tenac, Tetravos, UDVF, Unifos, Vapona, Vaponite, Vapora II, Verdican, Verdipor, Vinylofos, Vinylophos
Chemical Name: 2,2-dichlorovinyl-O,O-dimethyl phosphate; 2,2-dichloroethenyl-O,O-dimethyl phosphate
Uses: insecticide and fumigant to control flies, mosquitoes, and moths; also used as acaricide.
CAS Registry No: 62-73-7
Molecular Formula: $C_4H_7Cl_2O_4P$
Molecular Weight: 220.98
Melting Point (°C): liquid
Boiling Point (°C):
    35, 120, 140 (at 0.05, 14, 20 mmHg, Montgomery 1993)
    74, 117 (at 1, 10 mmHg, Agrochemicals Handbook 1987; Worthing 1991)
    234 (Tomlin 1994)
    84, 140 (at 1, 20 mmHg, Milne 1995)
Density (g/cm^3 at 20°C):
    1.415 (25°C, Spencer 1982; Agrochemicals Handbook 1987; Montgomery 1993; Milne 1995)
    1.420 (25°C, Worthing 1991)
    1.425 (Tomlin 1994)
    1.440 (Montgomery 1993)
Molar Volume (cm^3/mol):
    167.5 (calculated-LeBas method, Suntio et al. 1988; Fisher et al. 1993)
    0.861 ($V_I/100$, Fisher et al. 1993)
Molecular Volume (Å^3):
Total Surface Area, TSA (Å^2):
Dissociation Constant pK_a:
Heat of Fusion, ΔH_{fus}, kcal/mol:
Entropy of Fusion, ΔS_{fus}, cal/mol·K (e.u.):
Fugacity Ratio at 25°C (assuming ΔS_{fus} = 13.5 e.u.), F: 1.0

Water Solubility (g/m^3 or mg/L at 25°C):
    10000 (Günther et al. 1968)
    10000 (Melnikov 1971; Spencer 1982; quoted, Shiu et al. 1990)
    10000 (Martin & Worthing 1977; quoted, Kenaga 1980; Kenaga & Goring 1980; Kim et al. 1984)
    10000 (Worthing 1979; quoted, Bowman & Sans 1983b)
    10000 (Kenaga 1980a; Khan 1980; Worthing 1987; quoted, Suntio et al. 1988; Shiu et al. 1990)
    10000 (quoted, Lyman 1982)
    10000 (20°C, Agrochemicals Handbook 1987)
    8000 (20°C, selected, Suntio et al. 1988)
    16000 (Kawamoto & Urano 1989; quoted, Howard 1991)

10000 (20°C, Worthing 1991; Milne 1995)
16000 (20°C, Montgomery 1993)
10000 (20-25°C, estimated, Augustijn-Beckers et al. 1994; Hornsby et al. 1996)
10100, 10100 (quoted, calculated, Patil 1994)
8000 (Tomlin 1994)

Vapor Pressure (Pa at 25°C):
    1.60 (20°C, Wolfdietrich 1965; quoted, Kim et al. 1984)
    1.60 (20°C, Melnikov 1971; Hartley & Graham-Bryce 1980; Spencer 1982; quoted, Suntio et al. 1988; Montgomery 1993)
    1.60 (Khan 1980; Brouwer et al. 1994)
    4.011 (20°C, extrapolated-Clausius Clapeyron eqn., Kim et al. 1984)
    0.746 (20°C, GC-calculated value, Kim et al. 1984; Kim 1985)
    7.025 (gas saturation method, Kim et al. 1984; quoted, Suntio et al. 1988; Howard 1991)
    4.011 (20°C, gas saturation method, Kim 1985)
    1.60, 3.90 (20°C, 30°C, Agrochemicals Handbook 1987)
    7.0 (20°C, selected, Suntio et al. 1988)
    0.29 (20°C, Worthing 1991)
    0.267 (20-25°C, estimated, Augustijn-Beckers et al. 1994; Hornsby et al. 1996)
    2.10 (Tomlin 1994)

Henry's Law Constant (Pa·m^3/mol):
    0.190 (20°C, calculated-P/C, Suntio et al. 1988; quoted, Fisher et al. 1993)
    0.097 (calculated-P/C, Howard 1991)
    506.5 (Montgomery 1993)
    0.194 (calculated-P/C, this work)

Octanol/Water Partition Coefficient, log K_{OW}:
    1.40 (Leo et al. 1971; Hansch & Leo 1979; quoted, Kenaga & Goring 1980; Bowman & Sans 1983b; Suntio et al. 1988; Fisher et al. 1993)
    2.29 (Rao & Davidson 1980; quoted, Kim et al. 1984)
    1.40 (calculated, Lyman 1982)
    1.47 (shake flask-GC, Bowman & Sans 1983; quoted, Sicbaldi & Finizio 1993)
    1.45 (selected, Suntio et al. 1988)
    1.16 (Kawamoto & Urano 1989; quoted, Howard 1991)
    1.40-2.29 (Montgomery 1993)
    1.73 (RP-HPLC, Sicbaldi & Finizio 1993)
    1.47, 1.78 (quoted, calculated, Patil 1994)
    1.90 (Tomlin 1994)
    1.47 (selected, Devillers et al. 1996)
    1.42 (selected, Hansch et al. 1995)
    1.73, 1.89, 1.74 (RP-HPLC, ClogP, calculated-S, Finizio et al. 1997)

Bioconcentration Factor, log BCF:
    0.477 (calculated-S, Kenaga 1980a; quoted, Howard 1991)
    −0.097 (whole body willow shiner after 24-168 hr exposure, Tsuda et al. 1992; quoted, Devillers et al. 1996)
    < −0.30 (whole body carp, Tsuda et al. 1993)

Sorption Partition Coefficient, log K_{OC}:
- 1.45 (soil, calculated-S, Kenaga 1980a; quoted, Howard 1991)
- 1.67 (correlated, Kawamoto & Urano 1989)
- 1.70 (calculated, Montgomery 1993)
- 1.48 (20-25°C, estimated, Augustijn-Beckers et al. 1994; Hornsby et al. 1996)

Half-Lives in the Environment:
Air: half-life of 320 d, based on an estimated rate constant of 3.58×10^{-20} cm^3 molecule^{-1} s^{-1} at 25°C for the vapor-phase reaction with ozone of 7×10^{11}/cm^3 in air (Atkinson & Carter 1984; quoted, Howard 1991); atmospheric half-life of 2 d, based on an estimated rate constant of 9.24×10^{-12} cm^3 molecule^{-1} s^{-1} at 25°C for the vapor-phase reaction with hydroxyl radicals of 5×10^5/cm^3 in air (Atkinson 1987; quoted, Howard 1991); atmospheric transformation lifetime was estimated to be <1 d (Kelly et al. 1994).
Surface water: half-lives in lakes and rivers are reported to be approximately 4 d (Lamoreaux & Newland 1978; quoted, Howard 1991).
Groundwater:
Sediment:
Soil: average degradation rate constant of 0.0423 d^{-1} in silty clay with half-life of 16 days and average degradation rate constant of 0.0444 d^{-1} in sandy clay with half-life of 16 d (Sattar 1990); selected field half-life of 0.5 d (Augustijn-Beckers et al. 1994; Hornsby et al. 1996).
Biota:

Environmental Fate Rate Constants or Half-Lives:
Volatilization: based on the Henry's law constant, half-life from a model river has been estimated to be 57 d (Lyman et al. 1982; quoted, Howard 1991); half-life from an model pond, which considered the effect of adsorption, has been estimated to be over 400 yr (USEPA 1987; quoted, Howard 1991).
Photolysis:
Oxidation: photooxidation atmospheric half-life of 320 d, based on an estimated rate constant of 3.58×10^{-20} cm^3 molecule^{-1} s^{-1} at 25°C for the vapor-phase reaction with ozone of 7×10^{11}/cm^3 in air (Atkinson & Carter 1984; quoted, Howard 1991); photooxidation atmospheric half-life of 2 d, based on an estimated rate constant of 9.24×10^{-12} cm^3 molecule^{-1} s^{-1} at 25°C for the vapor-phase reaction with hydroxyl radicals of 5×10^5/cm^3 in air (Atkinson 1987; quoted, Howard 1991).
Hydrolysis: half-life of 462 min at pH 7 and 30 min at pH 8 (Montgomery 1993); estimated half-lives at 22°C: 31.9 d at pH 4, 2.9 d at pH 7, and 2.0 d at pH 9 (Tomlin 1994).
Biodegradation: the presence of active microorganisms reduced the half-life in autoclaved clay and calcareous soil from 0.9 to 0.75 d and 0.85 to 0.70 d, respectively (Guirguis & Shafik 1975; quoted, Howard 1991); rate constant of 0.99 d^{-1} with half-life of 0.70 d (Kawamoto & Urano 1990).
Biotransformation:
Bioconcentration, Uptake (k_1) and Elimination (k_2) Rate Constants

Common Name: Dicrotophos
Synonym: Bidirl, Bidrin, C 709, Cabicron, Carbomicron, CIBA 709, Diapadrin, Dicrotofos, Ektafos, ENT 24,482, Karbicron, Shell SD-3562
Chemical Name: (E)-2-dimethylcarbamoyl-1-methylvinyl dimethyl phosphate; (E)-3-(diethylamino)-1-methyl-3-oxo-1-propenyl dimethyl phosphate
Uses: contact and systemic insecticide and acaricide to control pests on rice, cotton, maize, soybeans, coffee, citrus, and potatoes.
CAS Registry No: 141-66-2
Molecular Formula: $C_8H_{16}NO_5P$
Molecular Weight: 237.20
Melting Point (°C): liquid
Boiling Point (°C):
    130   (at 0.1 mmHg, Worthing 1991; Montgomery 1993; Milne 1995)
    400   (technical grade, Worthing 1991; Montgomery 1993)
    400   (Tomlin 1994; Milne 1995)
Density (g/cm^3 at 20°C):
    1.216   (Agrochemicals Handbook 1987)
    1.216   (15°C, Merck Index 1989; Milne 1995)
    1.21   (technical grade, Worthing 1991)
    1.216   (Worthing 1991; Montgomery 1993; Tomlin 1994)
Molar Volume (cm^3/mol):
Molecular Volume (Å^3):
Total Surface Area, TSA (Å^2):
Dissociation Constant pK_a:
Heat of Fusion, ΔH_{fus}, kcal/mol:
Entropy of Fusion, ΔS_{fus}, cal/mol·K (e.u.):
Fugacity Ratio at 25°C (assuming ΔS_{fus} = 13.5 e.u.), F: 1.0

Water Solubility (g/m^3 or mg/L at 25°C):
    miscible (Spencer 1973; quoted, Shiu et al. 1990)
    miscible (Agrochemicals Handbook 1987; Merck Index 1989; Milne 1995)
    miscible (Worthing 1987; quoted, Shiu et al. 1990)
    1000000 (20-25°C, estimated, Wauchope et al. 1992; Hornsby et al. 1996)
    miscible (Montgomery 1993)
    miscible (Tomlin 1994)

Vapor Pressure (Pa at 25°C):
    0.0115   (20°C, extrapolated from gas saturation measurement, ln P(Pa) = 21.6 -7631/T for temp range 32.3-77°C, Grayson & Fosbraey 1982)
    0.0093   (20°C, Agrochemicals Handbook 1987)
    0.0093   (20°C, Worthing 1991; Tomlin 1994)
    0.0213   (20-25°C, estimated, Wauchope et al. 1992; Hornsby et al. 1996)
    0.0093   (20°C, Montgomery 1993)

Henry's Law Constant (Pa·m^3/mol):
    5.05x10^{-6} (20-25°C, calculated-P/C as per estimated values from Wauchope et al. 1992, this work)

Octanol/Water Partition Coefficient, log K_{OW}:
    −0.260 (calculated as per Broto et al. 1984, Karcher & Devillers 1990)
    −0.50   (Montgomery 1993)
    0.0   (Hansch et al. 1995)

Bioconcentration Factor, log BCF:
    −2.00 (calculated-K_{OW} as per Kenaga 1980, this work)

Sorption Partition Coefficient, log K_{OC}:
    1.88   (soil, 20-25°C, estimated, Wauchope et al. 1992; Hornsby et al. 1996)
    1.04-2.27 (Montgomery 1993)

Half-Lives in the Environment:
    Air:
    Surface water:
    Groundwater:
    Sediment:
    Soil: half-life of 3 d in sandy loam soil (Lee et al. 1989; quoted, Montgomery 1993); selected field half-life of 20 d (Wauchope et al. 1992; Hornsby et al. 1996).
    Biota:

Environmental Fate Rate Constants or Half-Lives:
    Volatilization:
    Photolysis:
    Oxidation:
    Hydrolysis: half-lives of 117, 72, and 28 d in buffer solutions of pH 5, 7, and 9 respectively at 25°C (Lee et al. 1989; quoted, Montgomery 1993); calculated half-lives in water at 20°C of 88 d at pH 5 and 23 d at pH 9 (Worthing 1991; Montgomery 1993; Tomlin 1994).
    Biodegradation:
    Biotransformation:
    Bioconcentration, Uptake (k_1) and Elimination (k_2) Rate Constants:

Common Name: Dieldrin
Synonym: Aldren, Alvit, Alyran, Compound 497, Dieldrite, Dieldrix, Dorytox, ENT 16225, HEOD, Illoxol, Insectlack, Kombi-Albertan, NA 2761, NCI-C00124, Octalox, Panoram D-31, Quintox
Chemical Name: 1,2,3,4,10,10-hexachloro-6,7-epoxy-1,4,4a,5,6,7,8,8a-octahydroendo-1,4-exo-5,8-dimethano-naphthalene; 3,4,5,6,9,9-hexachloro-1a,2,2a,3,6,6a,7,7a-octahydro-2,7:3,6-dimethanonaphth[2,3-b]oxirene
Uses: insecticide to control public health insect pests, termites, locusts, and tropical disease vectors.
CAS Registry No: 60-57-1
Molecular Formula: $C_{12}H_8Cl_6O$
Molecular Weight: 380.93
Melting Point (°C):
- 175-176 (Khan 1980; Howard 1991; Montgomery 1993)
- 176-177 (Suntio et al. 1988; Milne 1995)
- 176 (Liu et al. 1991; Noegrohati & Hammers 1992)
- 175.5 (Patil 1994)
- 175 (Kühne et al. 1995; Mortimer & Connell 1995)

Boiling Point (°C):
- 265, 352, 404 (estimated from structure, Tucker et al. 1983)

Density (g/cm³ at 20°C):
- 1.75 (Agrochemicals Handbook 1987; Montgomery 1993)

Molar Volume (cm³/mol):
- 318.2 (calculated-LeBas method, Suntio et al. 1988)
- 253.6 (Ruelle & Kesselring 1997)

Molecular Volume (Å³):
Total Surface Area, TSA (Å²):
Dissociation Constant pK_a:
Heat of Fusion, ΔH_{fus}, kcal/mol:
- 4.40 (Ruelle & Kesselring 1997)

Entropy of Fusion, ΔS_{fus}, cal/mol·K (e.u.):
- 9.88, 11.5 (Plato 1972)
- 10.7 (Hinckley et al. 1990)

Fugacity Ratio at 25°C (assuming ΔS_{fus} = 13.5 e.u.), F:
- 0.026 (20°C, Suntio et al. 1988)
- 0.033 (Mackay et al. 1986)

Water Solubility (g/m³ or mg/L at 25°C):
- 0.19 (colorimetric, Lipke & Kearns 1960; quoted, Shiu et al. 1990)
- 0.25 (shake flask-GC/UV, Richardson & Miller 1960; quoted, Spencer & Cliath 1969; Metcalf et al. 1973; Freed 1976; Nash 1988; Shiu et al. 1990)
- 0.14-0.18 (particle size of 0.04-5.0μ, shake flask-GC, Robeck et al. 1965)
- 0.15 (Eye 1968; quoted, Freed 1976; Jury et al. 1983,84)
- 0.20 (Günther et al. 1968; quoted, Clark et al. 1988; Suntio et al. 1988; Shiu et al. 1990)
- 0.186 (25-29°C, shake flask-GC/ECD, Park & Bruce 1968; quoted, Freed 1976; Khan 1980; Sharom et al. 1980; Belluck & Felsot 1981; Suntio et al. 1988; Shiu et al. 1990)
- 0.022 (Biggar & Riggs 1974; quoted, Kenaga 1980a,b; Kenaga & Goring 1980; Zaroogian et al. 1985)

0.195   (particle size of ≤ 5.0 μ, shake flask-GC/ECD, Biggar & Riggs 1974; quoted, Davies & Dobbs 1984; Suntio et al. 1988; Shiu et al. 1990)
0.022, 0.15, 0.195   (particle size: 0.01, 0.05 & 5.0 μ; shake flask-GC/ECD, Biggar & Riggs 1974)
0.20   (gen. col.-GC/ECD, Weil et al. 1974; quoted, Geyer et al. 1982; Clark et al. 1988; Suntio et al. 1988; Shiu et al. 1990)
0.186   (quoted, Caro et al. 1976)
0.187   (Martin & Worthing 1977; quoted, Briggs 1981)
0.1-0.25   (Wauchope 1978)
0.022   (Kenaga 1980a,b; Garten & Trabalka 1983; quoted, Shiu et al. 1990)
0.10   (Weber et al. 1980; quoted, Willis & McDowell 1982)
0.468   (20-25°C, shake flask-GC, Kanazawa 1981; quoted, Shiu et al. 1990)
2.0   (selected, Schnoor & McAvoy 1981; quoted, Schnoor 1992)
0.15   (quoted, Jury et al. 1984,87)
0.19   (20°C, Agrochemicals Handbook 1987)
0.10   (selected, Eadie & Robbins 1987)
0.186   (20°C, Worthing 1987; quoted, Shiu et al. 1990)
0.183   (quoted, Clark et al. 1988)
0.17   (20°C, selected, Suntio et al. 1988; quoted, Howard 1991; Majewski & Capel 1995; Mortimer & Connell 1995)
0.022   (quoted, Isnard & Lambert 1989)
0.14, 0.20   (20°C, 30°C, Montgomery 1993)
0.20   (20-25°C, selected, Augustijn-Beckers et al. 1994; Hornsby et al. 1996)
0.022, 0.009   (quoted, calculated, Patil 1994)
0.195, 1.26   (quoted, calculated-group contribution fragmentation method, Kühne et al. 1995)
0.215   (quoted, Pinsuwan et al. 1995)
0.17; 0.399, 0.00024   (quoted; predicted-molar volume, Ruelle & Kesselring 1997)

Vapor Pressure (Pa at 25°C):
1.04×10^{-4} (20°C, Porter 1964; quoted, Freed 1976; Suntio et al. 1988)
1.04×10^{-4}, 1.91×10^{-4}, 3.65×10^{-4} (20, 30, 40°C, effusion method, Porter 1964 as quoted in Spencer & Cliath 1969)
2.40×10^{-5} (Eichler 1965; Martin 1971, 1972; Melnikov 1971; quoted, SRI International 1980; Tucker et al. 1983; Suntio et al. 1988)
2.47×10^{-4}, 1.33×10^{-3}, 4.63×10^{-3} (20, 30, 40°C, gas saturation method, Spencer & Cliath 1969; quoted, Freed 1976)
6.59×10^{-4} (calculated from vapor pressure eqn. log P (mmHg) = 12.07 - (5178/T) for 20-40°C, apparent vapor pressure, Spencer & Cliath 1969)
6.77×10^{-4} (gas saturation, Spencer & Cliath 1969; quoted, Jury et al. 1983; Nash 1988, 1989; Taylor & Glotfelty 1988; Taylor & Spencer 1990)
3.87×10^{-4} (20°C, partition coeff., Atkins & Eggleton 1971; quoted, Freed 1976)
3.47×10^{-4} (quoted, Caro et al. 1976)
4.13×10^{-4} (20°C, Khan 1980)
2.40×10^{-5} (20-25°C, Weber et al. 1980; quoted, Willis & McDowell 1982)
3.60×10^{-4} (selected, Schnoor & McAvoy 1981; quoted, Schnoor 1992)
8.90×10^{-4} (20°C, GC, Seiber et al. 1981; quoted, Suntio et al. 1988)
4.2×10^{-4} (20°C, extrapolated from gas saturation measurement, ln P (Pa) = 30.7 - 11285/T for temp range 35 to 75.2°C, Grayson & Fosbraey 1982)
0.213, 5.2×10^{-5}, 4.7×10^{-7} (estimated-B.P., Tucker et al. 1983)

2.40×10⁻⁵ (Verschueren 1983)
3.30×10⁻⁴ (20°C, selected lit. value, Dobbs et al. 1984)
0.0215; 0.00532, 0.060 (supcooled liquid values P_L: quoted; GC-RT, Bidleman 1984)
3.80×10⁻⁴ (Mackay et al. 1986)
4.00×10⁻⁴ (20°C, Agrochemicals Handbook 1987)
6.77×10⁻⁴ (quoted, Jury et al. 1987)
5.00×10⁻⁴ (20°C, selected, Suntio et al. 1988; quoted, Howard 1991; Majewski & Capel 1995; Mortimer & Connell 1995)
2.57×10⁻⁴ (selected, Clark et al. 1988)
6.90×10⁻⁴ (quoted, Nash 1989)
0.0215, 0.0101 (subcooled liquid values, GC-RT, Hinckley et al. 1990)
$\log P_L$ (Pa) = 11.30 − (3965/T) (quoted, liquid, Hinckley et al. 1990)
$\log P_L$ (Pa) = 12.46 − (4310/T) (GC-RT, liquid, Hinckley et al. 1990)
2.37×10⁻⁵ (20°C, Montgomery 1993)
4.00×10⁻⁴ (20-25°C, selected, Augustijn-Beckers et al. 1994; Hornsby et al. 1996)

Henry's Law Constant (Pa·m³/mol):
- 4.59 (gas stripping, Atkins & Eggleton 1971)
- 0.02 (calculated-P/C, Mackay & Leinonen 1975; quoted, Suntio et al. 1988)
- 5.84 (exptl., Warner et al. 1980; quoted, Tucker et al. 1983; Suntio et al. 1988)
- 0.0456 (calculated-P/C, Levins 1981; quoted, Tucker et al. 1983)
- 1.10 (20°C, Mackay & Shiu 1981; quoted, Tucker et al. 1983; Suntio et al. 1988)
- 2.94 (20°C, measured, Slater & Spedding 1981; quoted, Suntio et al. 1988; Cotham & Bidleman 1991)
- 0.172 (estimated-group method per Hine & Mookerjee 1975, Tucker et al. 1983)
- 1.66 (calculated-P/C, Jury et al. 1984, 1987a; Jury & Ghodrati 1989)
- 0.78 (calculated-P/C, Mackay et al. 1986)
- 0.74 (WERL Treatability Database, quoted, Ryan et al. 1988)
- 1.12 (20°C, calculated-P/C, Suntio et al. 1988; quoted, Cotham & Bidleman 1991; Majewski & Capel 1995)
- 1.0 (calculated-P/C, Nash 1989)
- 5.9 (quoted from Warner et al. 1987, Cotham & Bidleman 1991; Howard 1991)
- 4.5 (quoted from Atkins & Eggleton 1971, Cotham & Bidleman 1991)
- 5.88 (Montgomery 1993)
- 3.28 (quoted average, Müller et al. 1994)
- 1.12 (calculated-P/C, this work)

Octanol/Water Partition Coefficient, $\log K_{OW}$:
- 2.60 (Hansch & Leo 1979)
- 5.48 (calculated, Kenaga 1980a,b; quoted, Chapman 1989)
- 3.69 (quoted, Rao & Davidson 1980; Suntio et al. 1988)
- 4.32 (shake flask-GC, Kanazawa 1981; quoted, Geyer et al. 1982; Davies & Dobbs 1984; Noegrohati & Hammers 1992; Sicbaldi & Finizio 1993)
- 5.21 (calculated as per Chiou et al. 1977, Belluck & Felsot 1981)
- 6.20 (extrapolated from RP-TLC, Briggs 1981; quoted, Clark et al. 1988)
- 5.30 (RP-HPLC from W.E. Hammers private communication, Hermens & Leeuwangh 1982; quoted, Hermens et al. 1985; De Bruijn et al. 1989; Verhaar et al. 1992)
- 5.48 (selected, Mackay 1982; Thomann 1989)
- 5.10 (Platford 1982)

5.11 (correlated, Hammers et al. 1982)
4.41 (selected, Dao et al. 1983)
5.16 (quoted, Garten & Trabalka 1983; De Bruijn et al. 1989)
4.32 (Hansch & Leo 1985; quoted, Howard 1991)
4.69 (quoted from G.D. Veith's personal communication, Zaroogian et al. 1985)
4.50 (quoted, Mackay et al. 1986)
4.51, 4.49, 4.60, 4.55 (Brooke et al. 1986; quoted, Sicbaldi & Finizio 1993)
5.48 (selected lit. average, Hawker & Connell 1986; Chessells et al. 1992)
5.34 (selected, Yoshioka et al. 1986)
4.51 (HPLC-RT, Kock & Lord 1987)
5.16 (quoted, Isnard & Lambert 1988, 1989; Travis & Arms 1988; Liu et al. 1991)
4.32 (Medchem Database 1988; quoted, Müller et al. 1994)
3.70 (selected, Suntio et al., 1988)
2.90 (quoted, Ryan et al. 1988)
5.40 (slow-stirring method, De Bruijn et al. 1989; quoted, Geyer et al. 1991; Parkerton et al. 1993; Sicbaldi & Finizio 1993; Devillers et al. 1996)
4.32 (Thor 1989; quoted, Connell & Markwell 1990)
4.60 (Devillers et al. 1991; quoted, Bintein & Devillers 1994)
4.32 (quoted exptl., Noegrohati & Hammers 1992)
5.40 (estimated-QSAR & SPARC, Kollig 1993)
3.69-6.20 (Montgomery 1993)
3.176 (calculated-f. const., Rekker et al. 1993; quoted, Pinsuwan et al. 1995)
4.76 (RP-HPLC, Sicbaldi & Finizio 1993)
5.16, 5.88 (quoted, calculated, Patil 1994)
5.109 (Yalkowsky & Dannenfelser 1994; quoted, Pinsuwan et al. 1995)
3.176 (calculated, Pinsuwan et al. 1995)
5.20 (selected, Hansch et al. 1995)
4.76, 3.63, 5.17 (RP-HPLC, ClogP, calculated, Finizio et al. 1997)

Bioconcentration Factor, log BCF:
3.08, 4.14, 4.69 (algae, daphnia, guppies, Reinert 1967; quoted, Biddinger & Gloss 1984)
3.65-4.69 (earthworms, Wheatley & Hardman 1968; quoted, Connell & Markwell 1990)
0.230 (bioaccumulation factor log BF, adipose tissue in male Albino rats, Robinson et al. 1969; quoted, Geyer et al. 1980)
0.322 (bioaccumulation factor log BF, adipose tissue in male Albino rats, Walker et al. 1969; quoted, Geyer et al. 1980)
3.04-3.66 (*Saccharomyces cerevisiae*, Voerman & Tammes 1969; quoted, Baughman & Paris 1981)
3.0-5.48 (benthic algae, Rose & McIntire 1970; quoted, Baughman & Paris 1981; Swackhamer & Skoglund 1991)
0.301 (bioaccumulation factor log BF, adipose tissue in male Albino rats, Baron & Walton 1971; quoted, Geyer et al. 1980)
3.24 (soft clam, Butler 1971; quoted, Hawker & Connell 1986)
3.11 (*Scenedemus obliquus*, Reinert 1972; quoted, Baughman & Paris 1981)
3.54 (*Daphania magna*, Reinert 1972; quoted, Geyer et al. 1991)
2.37 (wet-wt. basis, *scenedemus obliquus*, Reinert 1972; quoted, Swackhamer & Skoglund 1991)
3.43, 4.79 (*Gambusia, Physa*, Metcalf et al. 1973)
2.66-4.60 (*Oedogonium sp.*, Metcalf et al. 1973; quoted, Baughman & Paris 1981)

4.51 (wet-wt. basis, *Ankistrodesmus*, Neudorf & Khan 1975; quoted, Swackhamer & Skoglund 1991)
3.39 (oyster, Mason & Rowe 1976; quoted, Hawker & Connell 1986)
3.20 (mussel, steady state, Ernst 1977; quoted, Renberg & Sundström 1979; Geyer et al. 1982; Zaroogian et al. 1985; Hawker & Connell 1986)
2.30 (*Anabaena cylindrica*, Schauberger & Wildman 1977; quoted, Baughman & Paris 1981; Swackhamer & Skoglund 1991)
2.70 (*Anacystis nidulans*, Schauberger & Wildman 1977; quoted, Baughman & Paris 1981)
3.26 (*Nostoc muscorum*, Schauberger & Wildman 1977; quoted, Baughman & Paris 1981)
2.0-4.0 (Callahan et al. 1979; quoted, Howard 1991)
3.76, 3.65 (fish: flowing water, static water; quoted, Kenaga 1980a,b; Kenaga & Goring 1980)
3.72 (calculated-S, Kenaga 1980a)
0.362 (average beef fat diet, Kenaga 1980b)
3.54 (pulex, Kenaga & Goring 1980; quoted, Hawker & Connell 1986)
3.62 (earthworms, Lord et al. 1980; quoted, Connell & Markwell 1990)
2.00 (*Triaenodes tardus*, Belluck & Felsot 1981)
3.65 (*Pseudorasbora parva*, Kanazawa 1981; quoted, Devillers et al. 1996)
1.0-5.0 (selected, Schnoor & McAvoy 1981; quoted, Schnoor 1992)
3.37 (mussel, quoted average, Geyer et al. 1982)
4.23-4.98 (earthworms, Gish & Hughes 1982; quoted, Connell & Markwell 1990)
4.16 (fish, correlated, Mackay 1982)
3.52 (trout, Verschueren 1983)
3.55 (clam fat, 60-d expt., Hartley & Johnson 1983)
4.10 (guppy, Davies & Dobbs 1984)
4.25 (activated sludge, Freitag et al. 1984)
3.36, 3.48, 4.25 (algae, golden ide, activated sludge, Freitag et al. 1985)
3.33, 3.29 (mussel, calculated values, Zaroogian et al. 1985)
3.33, 3.29 (oyster, calculated values, Zaroogian et al. 1985)
3.70, 3.90 (oyster, quoted from Zaroogian et al. 1985; Hawker & Connell 1986)
1.72-1.95 (human fat lipid basis, Geyer et al., 1987)
1.56-1.78 (human fat wet wt. basis, Geyer et al., 1987)
4.10 (quoted, Isnard & Lambert 1988; Howard 1991)
−2.10 (beef biotransfer factor log B_b, correlated-K_{OW}, Potter et al. 1974; quoted, Travis & Arms 1988)
−1.97 (milk biotransfer factor log B_m, correlated-K_{OW}, Saha 1969; Wilson & Cook 1972; quoted, Travis & Arms 1988)
−1.01 (vegetation, correlated-K_{OW}, Beall & Nash 1972; quoted, Travis & Arms 1988)
2.96-4.11 (aquatic food web, Fordham & Reagan 1991)
3.81 (fish, Fordham & Reagan 1991)
3.88 (selected, Chessells et al. 1992)

Sorption Partition Coefficient, log K_{OC}:
4.55 (soil, calculated-S as per Kenaga & Goring 1978, Kenaga 1980; quoted, Nash 1988, 1989)
4.08 (calculated-K_{OW}, Rao & Davidson 1980; quoted, Jury et al. 1983)
3.87 (extrapolated from RP-TLC and reported as log K_{OM}, Briggs 1981; quoted, Howard 1991)

4.0 (selected, Schnoor & McAvoy 1981; quoted, Schnoor 1992)
3.36-3.85 (reported as log K_{OM}, Mingelgrin & Gerstl 1983)
4.08 (soil, screening model calculations, Jury et al. 1984, 1987a,b; Jury & Ghodrati 1989; quoted, Howard 1991)
4.36 (calculated-K_{OW} as per Kenaga & Goring 1980, Chapman 1989)
4.15 (soil, Kishi et al. 1990)
4.50 (sediment, Fordham & Reagan 1991)
4.10 (soil, quoted exptl., Meylan et al. 1992)
4.03 (soil, calculated-χ and fragment contribution, Meylan et al. 1992)
5.08 (estimated-QSAR & SPARC, Kollig 1993)
4.08-4.55 (Montgomery 1993)
4.08 (20-25°C, estimated, Augustijn-Beckers et al. 1994; Hornsby et al. 1996)

Half-Lives in the Environment:
Air: 4-40.5 h, based on an estimated rate constant for vapor-phase reaction with hydroxyl radicals in air (Atkinson 1987; quoted, Howard et al. 1991; Mortimer & Connell 1995); calculated life-time of 1.1 d in troposphere (Atkinson et al. 1992).
Surface water: estimated half-life of ≥300 d in lake waters (Zoeteman et al. 1980); 4200-25920 h, based on estimated aqueous aerobic biodegradation half-life (Howard et al. 1991; quoted, Mortimer & Connell 1995).
Groundwater: 24-51840 h, based on estimated aqueous aerobic and anaerobic biodegradation half-lives (Howard et al. 1991).
Sediment: 15100 h (mean value quoted from Howard et al. 1991).
Soil: field half-life of 49 d in nondisked soil (Lichtenstein & Schultz 1961; quoted, Nash 1983); half-life of about 7 yr persistence in soil (Nash & Woolson 1967); estimated persistence of 3 yr in soil (Kearney et al. 1969; Edwards 1973; quoted, Morrill et al. 1982; Jury et al. 1987); 4200-25920 h, based on unacclimated aerobic soil grab sample data (Castro & Yoshida 1971; quoted, Howard et al. 1991) and reported soil field test data (Kearney et al. 1969; quoted, Howard et al. 1991); persistence of more than 36 months (Wauchope 1978); estimated first-order half-life of 53.3 d from rate constant of 0.013 d^{-1} from soil incubation studies by die-away tests (Rao & Davidson 1980; quoted, Scow 1982); moderately persistent in soils with half-life of 20-100 d (Willis & McDowell 1982); microagroecosystem half-life of 19-26 d in moist fallow soil (Nash 1983); measured dissipation rate of 0.055 d^{-1} (Nash 1983; quoted, Nash 1988); estimated dissipation rate of 0.034 and 0.049 d^{-1} (Nash 1988); biodegradation half-life of 868 d (Jury et al. 1984, 1987a,b; Jury & Ghodrati 1989; quoted, Montgomery 1993); >50 d and subject to plant uptake via volatilization (Ryan et al. 1988); estimated field half-life of 1000 d (Augustijn-Beckers et al. 1994; Hornsby et al. 1996).
Biota: estimated half-lives in rat's liver were ~1.3 and 10.2 d and similar values estimated for the blood in rat and 10.3 and 3 d in adipose tissue of rat (Robinson et al. 1969); half-life of 53.1 h in mussels (Ernst 1977; quoted, Hawker & Connell 1986); biochemical half-life of 868 d from screening model calculations (Jury et al. 1987a,b; Jury & Ghodrati 1989).

Environmental Fate Rate Constants or Half-Lives:
Volatilization: half-life of a model river of depth 1 m flowing at 1 m/s with a wind velocity of 3 m/s was estimated to be 1.4 d by using Henry's law constant (Lyman et al. 1982; quoted, Howard 1991).

Photolysis: rate constant of 4.8×10^{-4} h^{-1} by direct sunlight at 40° latitude (Mabey et al. 1982); using fungus and 254 nm UV, more than 90% initial added amounts were degraded in 4 weeks of incubation (Katayama & Matsumura 1991).

Oxidation: rate constant for singlet oxygen, <3600 M^{-1} h^{-1} and for RO_2 radicals <30 M^{-1} h^{-1} (Mabey et al. 1982); photooxidation half-life of 4-40.5 h, based on an estimated rate constant for vapor-phase reaction with hydroxyl radicals in air (Atkinson 1987; quoted, Howard et al. 1991); calculated tropospheric lifetimes due to gas-phase reaction with OH radical are 1.1 d (Atkinson et al. 1992).

Hydrolysis: first-order half-life of 10.5 yr, based on a first-order rate constant of 7.5×10^{-6} h^{-1} at pH 7.0 and 25°C (Ellington et al. 1986, 1987, 1988; quoted, Howard et al. 1991; Montgomery 1993); rate constant of 6.3×10^{-2} y^{-1} at pH 7 and 25°C (Kollig 1993).

Biodegradation: aqueous aerobic half-life of 4200-25,920 h, based on unacclimated aerobic soil grab sample data (Castro & Yoshida 1971; quoted, Howard et al. 1991; Howard 1991) and reported soil field test data (Kearney et al. 1969; quoted, Howard et al. 1991); half-life of 868 d (Nash 1980; quoted, Jury et al. 1983); rate constant of 0.013 d^{-1} from soil incubation studies by die-away tests (Rao & Davidson 1980; quoted, Scow 1982); aqueous anaerobic half-life of 24-168 h, based on soil and freshwater mud grab sample data (Maule et al. 1987; quoted, Howard et al. 1991); half-life of 870 d in soil by 100-d leaching screening simulation in 0-10 cm depth of soil (Jury et al. 1984, 1987a,b; Jury & Ghodrati 1989).

Biotransformation:

Bioconcentration, Uptake (k_1) and Elimination (k_2) Rate Constants:

k_1: 20.40 h^{-1} (Ernst 1977; quoted, Hawker & Connell 1986)
k_2: 0.013 h^{-1} (Ernst 1977; quoted, Hawker & Connell 1986)
k_2: 0.017 d^{-1} (fish, Fordham & Reagan 1991)
k_2: 0.014 d^{-1} (birds, Fordham & Reagan 1991)

Common Name: Diflubenzuron
Synonym: Deflubenzon, difluron, Dimilin, DU 112307, Duphacid, ENT 29054, OMS 1804, Largon, Micromite, PDD 60401, PH 60-40, TH-6040
Chemical Name: 1-(4-chlorophenyl)-3-(2,6-difluorobenzol) urea; N-[[(4-chlorophenyl)-amino]carbonyl]-2,6-difluorobenzamide
Uses: nonsystemic insecticide to control leaf-eating larvae and leaf miners in forestry, woody ornamentals and fruit trees.
CAS Registry No: 35367-38-5
Molecular Formula: $C_{14}H_9ClF_2N_2O_2$
Molecular Weight: 310.7
Melting Point (°C):
    210-230 (Spencer 1982)
    230-232 (Montgomery 1993; Milne 1995)
    230-232 (dec., Tomlin 1994)
    239 (Milne 1995)
Boiling Point (°C):
    dec. on distillation (Montgomery 1993)
Density (g/cm³ at 20°C):
    288.3 (calculated-LeBas method, this work)
Molar Volume (cm³/mol):
Molecular Volume (Å³):
Total Surface Area, TSA (Å²):
Dissociation Constant pK_a:
Heat of Fusion, ΔH_{fus}, kcal/mol:
Entropy of Fusion, ΔS_{fus}, cal/mol·K (e.u.):
Fugacity Ratio at 25°C (assuming ΔS_{fus} = 13.5 e.u.), F: 0.00920

Water Solubility (g/m³ or mg/L at 25°C):
    0.25 (Ivie et al. 1980; quoted, Belluck & Felsot 1981)
    0.20 (Spencer 1982; Wauchope 1989; quoted, Shiu et al. 1990)
    0.14 (20°C, Agrochemicals Handbook 1987; Milne 1995)
    0.10 (20°C, Worthing 1987, 1991; quoted, Shiu et al. 1990)
    14.0 (Montgomery 1993)
    0.30 (Milne 1995)
    0.08 (selected, Lohninger 1994)
    0.08 (20-25°C, selected, Hornsby et al. 1996)

Vapor Pressure (Pa at 25°C):
    $<3.3 \times 10^{-5}$ (50°C, Agrochemicals Handbook 1987)
    $<1.3 \times 10^{-5}$ (Worthing 1991)
    3.33×10^{-5} (20°C, Montgomery 1993)
    1.20×10^{-7} (gas saturation, Tomlin 1994)
    1.20×10^{-7} (20-25°C, selected, Hornsby et al. 1996)

Henry's Law Constant (Pa·m³/mol):
    7.40×10^{-4} (20-25°C, calculated-P/C, Montgomery 1993)
    4.70×10^{-4} (20-25°C, calculated-P/C, this work)

Octanol/Water Partition Coefficient, log K_{OW}:
- 5.06 (calculated as per Chiou et al. 1977, Belluck & Felsot 1981)
- 3.10 (selected, Nendza 1991)
- 3.29 (calculated, Montgomery 1993)
- 3.89 (Tomlin 1994)
- 3.32 (calculated as per Chiou et al. 1977 & Chiou 1981, this work)
- 3.88 (selected, Hansch et al. 1995)

Bioconcentration Factor, log BCF:
- 2.88 (calculated-S as per Kenaga & Goring, this work)
- 2.44 (calculated-K_{OW} as per Kenaga & Goring, this work)

Sorption Partition Coefficient, log K_{OC}:
- 3.01 (calculated, Montgomery 1993)
- 4.00 (average value, Dowd et al. 1993)
- 4.00 (20-25°C, selected, Hornsby et al. 1996)
- 4.06 (estimated-chemical structure, Lohninger 1994)

Half-Lives in the Environment:
Air:
Surface water:
Groundwater:
Sediment:
Soil: half-life is less than one week (Hartley & Kidd 1987; quoted, Montgomery 1993; Tomlin 1994); half-life of 10 d in forest soil (Dowd et al. 1993); field half-life of 10 d (20-25°C, selected, Hornsby et al. 1996).
Biota:

Environmental Fate Rate Constants or Half-Lives:
Volatilization:
Photolysis:
Oxidation:
Hydrolysis: half-lives at 20°C: >150 d at pH 5 & 7 and 42 d at pH 9 (Tomlin 1994).
Biodegradation:
Biotransformation:
Bioconcentration, Uptake (k_1) and Elimination (k_2) Rate Constants:

Common Name: Dimethoate

Synonym: AC 12880, AC 18682, American Cynamid 12880, BI 58, Cekuthoate, Chemathoate, CL 12880, Cygon, Daphene, De-fend, Demos-L40, Devigon, Dimetate, Dimeton, Dimevur, ENT 24650, Ferkethion, Fip, Fortion NM, Fosfamid, Fosfotox, Fostion MM, L 395, Lurgo, NC 262, Perfekthion, Phosphamid, Rebelate, Rogodial, Rogor, Roxion, Sinoratox, Trimetion

Chemical Name: O,O-dimethyl S-methylcarbamoyl-methyl phosphorodithioate; O,O-dimethyl S-(N-monomethyl)carbamylmethyl dithiophosphate; 2-dimethoxyphosphinothioylthio-N-methylacetamide

Uses: systemic and contact insecticide to control thrips and red spider mites on many agricultural crops, sawflies on apples and plums, also wheat bulb and olive flies.

CAS Registry No: 60-51-5
Molecular Formula: $C_5H_{12}NO_3PS_2$
Molecular Weight: 229.28
Melting Point (°C):
    42-46   (NIEHS 1975; Freed et al. 1977)
    43.5-45.8 (Spencer 1973; Freed et al. 1977)
    51-52   (Khan 1980)
    51.5   (Bowman & Sans 1983b; Patil 1994)
    52-52.5 (Suntio et al. 1988; Montgomery 1993; Milne 1995)
    49.0   (Worthing 1991; Tomlin 1994; Milne 1995)

Boiling Point (°C):
    107   (at 0.05 mmHg, Melnikov 1971; Freed et al. 1977)
    117   (at 0.1 mmHg, Agrochemicals Handbook 1987; Tomlin 1994; Milne 1995)
    117   (tech. grade at 0.1 mmHg, Worthing 1991)

Density (g/cm³ at 20°C):
    1.277   (65°C, Agrochemicals Handbook 1987; Montgomery 1993; Tomlin 1994; Milne 1995)
    1.281   (50°C, Worthing 1991; Montgomery 1993)

Molar Volume (cm³/mol):
    205.6   (calculated-LeBas method, Suntio et al. 1988)

Molecular Volume (Å³):
Total Surface Area, TSA (Å²):
Dissociation Constant pK_a:
Heat of Fusion, ΔH_{fus}, kcal/mol:
    5.60   (DSC method, Plato & Galsgow 1969)
Entropy of Fusion, ΔS_{fus}, cal/mol·K (e.u.):
Fugacity Ratio at 25°C (assuming ΔS_{fus} = 13.5 e.u.), F: 0.5346

Water Solubility (g/m³ or mg/L at 25°C):
    39000   (Melnikov 1971; quoted, Freed et al. 1977; Shiu et al. 1990)
    25000   (Martin & Worthing 1977; Worthing 1979; quoted, Kenaga & Goring 1980; Khan 1980; Bowman & Sans 1983a; Kim et al. 1984; Suntio et al. 1988)
    25000   (Worthing 1979; quoted, Bowman & Sans 1983b; Patil 1994)
    25000   (Kenaga 1980a; quoted, Shiu et al. 1990)
    25140   (Briggs 1981)
    7000-30000 (20-25°C, selected, Willis & McDowell 1982)
    >5000   (20°C, shake flask-GC, Bowman & Sans 1983a)
    25020   (20°C, Bowman & Sans 1983b; quoted, Shiu et al. 1990)

25000 (22°C, Verschueren 1983; quoted, Suntio et al. 1988; Shiu et al. 1990)
25000 (21°C, Agrochemicals Handbook 1987)
25000 (21°C, Worthing 1987,91; quoted, Shiu et al. 1990; Howard 1991)
20000 (20°C, selected, Suntio et al. 1988; quoted, Majewski & Capel 1995)
25120 (Kanazawa 1989)
39800 (20-25°C, selected, Wauchope et al. 1992; Hornsby et al. 1996)
25000 (21°C, Montgomery 1993)
39800 (selected, Lohninger 1994)
25000, 9340 (quoted, calculated, Patil 1994)
23300, 23800, 25000 (20°C, at pH 5, 7, 9, Tomlin 1994)
21000 (21°C, Milne 1995)
39800 (selected, Halfon et al. 1996)

Vapor Pressure (Pa at 25°C):
11.3×10^{-4} (20°C, Wolfdietrich 1965; quoted, Kim et al. 1984)
11.3×10^{-4} (20°C, Melnikov 1971; quoted, Freed et al. 1977; Suntio et al. 1988)
3.73×10^{-4} (20°C, gravimetric method, Gückel et al. 1973; quoted, Suntio et al. 1988)
11.0×10^{-4} (Worthing 1979; quoted, Dobbs et al. 1984)
11.3×10^{-4} (Khan 1980)
8.90×10^{-4} (20°C, GC, Seiber et al. 1981; quoted, Suntio et al. 1988)
6.8×10^{-4} (gas saturation method, Kim et al. 1984; quoted, Suntio et al. 1988; Howard 1991)
3.87×10^{-4} (20°C, extrapolated value, Kim et al. 1984)
85.0×10^{-4} (20°C, GC-calculated value, Kim et al. 1984; Kim 1985)
41.0×10^{-4} (20°C, GC-calculated value with M.P. correction, Kim et al. 1984; Kim 1985)
3.87×10^{-4} (20°C, gas saturation method, Kim 1985)
11.0×10^{-4} (Agrochemicals Handbook 1987)
0.010 (20°C, selected, Suntio et al. 1988; quoted, Majewski & Capel 1995)
2.90×10^{-4} (20°C, Worthing 1991)
33.3×10^{-4} (20-25°C, selected, Wauchope et al. 1992; Hornsby et al. 1996; quoted, Halfon et al. 1996)
6.75×10^{-4} (20°C, Montgomery 1993)
11.0×10^{-4} (Tomlin 1994)

Henry's Law Constant (Pa·m^3/mol):
6.23×10^{-6} (calculated-P/C, Lyman et al. 1982; quoted, Howard 1991)
1.10×10^{-4} (20°C, calculated-P/C, Suntio et al. 1988; quoted, Majewski & Capel 1995)
2.66×10^{-6} (20-21°C, calculated-P/C, Montgomery 1993)
1.15×10^{-4} (calculated-P/C, this work)

Octanol/Water Partition Coefficient, log K_{ow}:
−0.29 (Hamaker 1975; quoted, Kenaga & Goring 1980; Bowman & Sans 1983b; Suntio et al. 1988)
−0.29 (NIEHS 1975; quoted, Freed et al. 1977; Dao et al. 1983)
−0.294 (shake flask-GC, Freed et al. 1979)

0.79   (20 ± 2°C, shake flask-UV, Briggs 1981; quoted, Bowman & Sans 1983b; Suntio et al. 1988; Bintein & Devillers 1994)
0.50, 0.78 (recommended, Hansch & Leo 1985; quoted, Howard 1991)
0.80   (selected, Suntio et al. 1988)
2.71   (Kanazawa 1989)
-0.197 (calculated as per Broto et al. 1984, Karcher & Devillers 1990)
0.78   (selected, Magee 1991)
0.699   (Worthing 1991; Milne 1995)
0.452   (estimated-QSAR & SPARC, Kollig 1993)
0.51-0.78 (Montgomery 1993)
0.77, 1.02 (quoted, calculated, Patil 1994)
0.704   (Tomlin 1994)
0.78   (selected, Hansch et al. 1995)
0.75, 1.44 (ClogP, calculated-S, Finizio et al. 1997)

Bioconcentration Factor, log BCF:
    2.00   (estimated-S, Howard 1991)

Sorption Partition Coefficient, log K_{OC}:
    1.23   (soils, calculated, Kenaga 1980a; quoted, Howard 1991)
    0.72   (20 ± 2°C, shake flask-UV and reported as log K_{OM}, Briggs 1981)
    1.43   (average of 2 soils, Kanazawa 1989)
    1.26, 1.56 (clay loam soil, Kanazawa 1989; quoted, Howard 1991)
    0.716   (clay soil, Kanazawa 1989; quoted, Howard 1991)
    0.72   (reported as log K_{OM}, Magee 1991)
    1.47   (estimated as log K_{OM}, Magee 1991)
    1.20   (soil, quoted exptl., Meylan et al. 1992)
    1.39   (soil, calculated-χ and fragment contribution, Meylan et al. 1992)
    1.30   (soil, 20-25°C, selected, Wauchope et al. 1992; Hornsby et al. 1996)
    0.132   (estimated-QSAR & SPARC, Kollig 1993)
    0.96   (Montgomery 1993)
    1.00   (estimated-chemical structure, Lohninger 1994)
    1.21   (sandy loam soil, Tomlin 1994)
    1.72   (sandy loam sand, Tomlin 1994)

Half-Lives in the Environment:
    Air: 0.469-4.69 h, based on estimated rate constant for the reaction with hydroxyl radicals in air (Atkinson 1987; quoted, Howard et al. 1991).
    Surface water: 264-1344 h, based on estimated unacclimated aqueous aerobic biodegradation half-life (Howard et al. 1991); half-lives of 423 d at 6°C, 193°C in darkness for Milli-Q water; 171 d at 6°C, 43 d at 22°C in darkness, 29 d under sunlight conditions for river water at pH 7.3; 173 d at 6°C, 29 d at 22°C in darkness for filtered river water, pH 7.3; 219 d at 6°C, 36 d at 22°C in darkness, 74 d under sunlight conditions for seawater, pH 8.1 (Lartiges & Garrigues 1995).
    Groundwater: 528-2688 h, based on estimated unacclimated aqueous aerobic biodegradation half-life (Howard et al. 1991).

Soil: 264-888 h, based on soil die-away test data for two soils (Bro-Rasmussen et al. 1970; quoted, Howard et al. 1991); selected half-life of 7.0 d (Wauchope et al. 1992; Hornsby et al. 1996); aerobic half-life of 2-4.1 d in soil and photolytic half-life of 7-16 d on soil surface (Tomlin 1994); 7.0 d (selected, Halfon et al. 1996).

Biota: disappearance rate and half-life from treated plants: 2.95 d for cabbage, 3.40 d for tomato leaves and 2.40 d for tomato fruits (Othman et al. 1987).

Environmental Fate Rate Constants or Half-Lives:
  Volatilization:
  Photolysis:
  Oxidation: photooxidation half-life of 0.469-4.69 h in air, based on estimated rate constant for the reaction with hydroxyl radicals in air (Atkinson 1987; quoted, Howard et al. 1987).
  Hydrolysis: half-life of 0.8 h at pH 9 and 21 h at pH 2 both at 70°C (Melnikov 1971; quoted, Freed et al. 1977); neutral rate constant of 1.7×10^{-4} h^{-1} with a calculated half-life of 118 h at pH 7 and 25°C (Ellington et al. 1987, 1988; quoted, Montgomery 1993); first-order half-life of 2822 h, based on measured neutral and base catalyzed hydrolysis rate constants (Ellington et al. 1987; quoted, Howard et al. 1991); rate constant of 1.68 y^{-1} at pH 7.0 and 25°C (Kollig 1993); half-life of 12 d at pH 9 (Tomlin 1994).
  Biodegradation: aqueous aerobic half-life of 264-1344 h, based on river die-away test data (Eichelberger & Lichtenburg 1971; quoted, Howard et al. 1991) and soil die-away test data for two soils (Bro-Rasmussen et al. 1970; quoted, Howard et al. 1991); aqueous anaerobic half-life of 1056-5376 h, based on estimated unacclimated aqueous aerobic biodegradation half-life (Howard et al. 1991).
  Biotransformation:
  Bioconcentration, Uptake (k_1) and Elimination (k_2) Rate Constants:

Common Name: Disulfoton
Synonym: Di-Syston, Dimaz, Disipton, Disystox, Dithiosystox, Frumin AL, Glebofos, Solvirex
Chemical Name: O,O-diethyl S-[2-(ethylthio)ethyl] phosphorodithioate; phosphorodithioic acid, O,O-diethyl S-[2-(ethylthio)ethyl] ester
Uses: insecticide to control aphids, thrips, mealybugs, and other sucking insects, and spider mites in potatoes, vegetables, cereals, maize, sorghum, rice, soybeans, groundnuts, lucerne, clover, sugar cane, sugar beet, hops, strawberries, cotton, coffee, pineapples, tobacco, ornamentals, fruit and nut crops, and forestry nurseries; also used as acaricide.
CAS Registry No: 298-04-4
Molecular Formula: $C_8H_{19}O_2PS_3$
Molecular Weight: 274.38
Melting Point (°C):
    < −25 (Tomlin 1994)
    −25 (Milne 1995)
Boiling Point (°C):
    62.0 (at 0.01 mmHg, Agrochemicals Handbook 1987)
    128 (at 1 mmHg, Worthing 1991; Tomlin 1994; Milne 1995)
    108 (at 0.01 mmHg, Milne 1995)
Density (g/cm³ at 20°C):
    1.144 (Agrochemicals Handbook 1987; Tomlin 1994)
    1.14 (Worthing 1991)
Molar Volume (cm³/mol):
    282.1 (calculated-LeBas method, Suntio et al. 1988)
Molecular Volume (Å³):
Total Surface Area, TSA (Å²):
Dissociation Constant pK_a:
Heat of Fusion, ΔH_{fus}, kcal/mol:
Entropy of Fusion, ΔS_{fus}, cal/mol·K (e.u.):
Fugacity Ratio at 25°C (assuming ΔS_{fus} = 13.5 e.u.), F: 0.151

Water Solubility (g/m³ or mg/L at 25°C):
    66 (Günther 1968; quoted, Suntio et al. 1988; Shiu et al. 1990)
    25 (20°C, Melnikov 1971; Spencer 1973; quoted, Bowman & Sans 1979; Suntio et al. 1988; Shiu et al. 1990)
    25 (Martin & Worthing 1977; quoted, Kenaga 1980; Kenaga & Goring 1980)
    16.3 (19.5°C, shake flask-GC, Bowman & Sans 1979,83b; quoted, Shiu et al. 1990; Patil 1994)
    25 (22°C, Khan 1980; Worthing 1983; quoted, Suntio et al. 1988; Shiu et al. 1990)
    15-66 (20-25°C, selected, Willis & McDowell 1982)
    25 (U.S. EPA 1984; quoted, McLean et al. 1988)
    25 (22°C, Agrochemicals Handbook 1987)
    25 (selected, Gerstl & Helling 1987)
    25 (20°C, selected, Suntio et al. 1988; quoted, Majewski & Capel 1995)
    12 (22°C, Worthing 1991)
    25 (20-25°C, selected, Wauchope et al. 1992; Hornsby et al. 1996)
    25 (quoted, Pait et al. 1992; Lohninger 1994)
    16.3, 112 (quoted, calculated, Patil 1994)
    12 (20°C, Tomlin 1994)
    16 (quoted, Kühne et al. 1995)

10.4   (calculated-group contribution fragmentation method, Kühne et al. 1995)
12   (22°C, Milne 1995)

Vapor Pressure (Pa at 25°C):
0.024   (20°C, vapor density, MacDougall 1964; quoted, Suntio et al. 1988)
0.024   (20°C, Eichler 1965; quoted, Kim 1985)
0.024   (20°C, Melnikov 1971; Khan 1980; quoted, Kim 1985; Suntio et al. 1988)
0.024   (Worthing 1983; quoted, Howard 1991)
0.024   (20°C, quoted exptl. value, Kim et al. 1984; Kim 1985)
0.0041 (20°C, GC-calculated value, Kim et al. 1984; Kim 1985)
0.024   (20°C, U.S. EPA 1984; quoted, McLean et al. 1988)
0.024   (20°C, Agrochemicals Handbook 1987)
0.020   (20°C, selected, Suntio et al. 1988; quoted, Majewski & Capel 1995)
0.024   (20°C, Worthing 1991)
0.020   (20-25°C, selected, Wauchope et al. 1992; Hornsby et al. 1996)
0.013   (Tomlin 1994)

Henry's Law Constant (Pa·m^3/mol):
0.404   (calculated-P/C, Lyman et al. 1982; quoted, Howard 1991)
0.22   (20°C, calculated-P/C, Suntio et al. 1988; quoted, Majewski & Capel 1995)
0.101, 0.253   (10°, 20°C, quoted, Wanner et al. 1989)
0.22   (calculated-P/C, this work)

Octanol/Water Partition Coefficient, log K_{OW}:
3.88   (shake flask-UV, Hermens & Leeuwangh 1982)
4.02   (shake flask-GC, Bowman & Sans 1983b; quoted, Gerstl & Helling 1987; Suntio et al. 1988; Patil 1994)
4.02   (recommended, Hansch & Leo 1985; quoted, Howard 1991)
4.00   (selected, Suntio et al. 1988)
2.671   (calculated as per Broto et al. 1984, Karcher & Devillers 1990)
4.02   (selected, Magee 1991)
3.26   (estimated-QSAR & SPARC, Kollig 1993)
3.84   (RP-HPLC, Saito et al. 1993)
3.00   (calculated, Patil 1994)
3.95   (Tomlin 1994)
4.02   (selected, Hansch et al. 1995)

Bioconcentration Factor, log BCF:
2.00   (calculated-S, Kenaga 1980; quoted, Pait et al. 1992)
2.04   (calculated-K_{OC}, Kenaga 1980)
2.00   (estimated-S, Lyman et al. 1982; quoted, Howard 1991)
2.83   (estimated-K_{OW}, Lyman et al. 1982; quoted, Howard 1991)
2.65   (carp, Takase & Oyama 1985; quoted, Howard 1991)

Sorption Partition Coefficient, log K_{OC}:
- 3.25 (soil, Hamaker & Thompson 1972; quoted, Kenaga 1980; Kenaga & Goring 1980)
- 2.81, 3.04, 3.72 (Hamaker & Thompson 1972; quoted, Howard 1991)
- 2.87 (soil, calculated-S as per Kenaga & Goring 1978, Kenaga 1980)
- 3.20 (av. soils/sediments, Rao & Davidson 1980; quoted, Lyman 1982)
- 2.67-3.70 (reported as log K_{OM}, Mingelgrin & Gerstl 1983)
- 3.61, 2.90 (quoted, calculated-χ, Gerstl & Helling 1987)
- 3.20 (soil, screening model calculations, Jury et al. 1987b)
- 3.25 (Sabljic 1987; quoted, Howard 1991)
- 3.25 (reported as log K_{OM}, Magee 1991)
- 3.36 (estimated as log K_{OM}, Magee 1991)
- 3.22 (soil, quoted exptl., Meylan et al. 1992)
- 2.91 (soil, calculated-χ and fragments contribution, Meylan et al. 1992)
- 2.78 (soil, 20-25°C, estimated, Wauchope et al. 1992; Hornsby et al. 1996)
- 2.94 (estimated-QSAR & SPARC, Kollig 1993)
- 3.49 (estimated-chemical structure, Lohninger 1994)
- 2.91 (soil, HPLC-screening method, Kördel et al. 1995)

Half-Lives in the Environment:
- Air: 0.50-4.80 h, based on estimated rate constant for the reaction with hydroxyl radicals in air (Atkinson 1987; quoted, Howard et al. 1991).
- Surface water: gas exchange half-lives of 900 d for winter, 360 d for summer; abiotic hydrolysis half-lives of 170 d for winter, 62 d for summer; photolytic transformation half-lives of 1000 d for winter, 200 d for summer and primary biodegradation half-lives of 7-41 d for winter, 8-28 d for summer in Rhine River under environmental conditions (Wanner et al. 1989); overall half-life of 72-504 h, based on estimated unacclimated aqueous aerobic biodegradation half-life (Howard et al. 1991).
- Groundwater: 144-1008 h, based on estimated unacclimated aqueous aerobic biodegradation half-life (Howard et al. 1991).
- Sediment:
- Soil: estimated persistence of 4 weeks in soil (Kearney et al. 1969; Edwards 1973; quoted, Morrill et al. 1982); 72-504 h, based on aerobic soil field data (Szeto et al. 1983; quoted, Howard et al. 1991) and reported half-lives for soil (Domsch 1984; quoted, Howard et al. 1991); half-life of 5 d from screening model calculations (Jury et al. 1987b); estimated half-life of 30 d (Wauchope et al. 1992; Hornsby et al. 1996); soil half-life of 9 d (Pait et al. 1992).
- Biota: biochemical half-life of 5 d from screening model calculations (Jury et al. 1987b).

Environmental Fate Rate Constants or Half-Lives:
- Volatilization: gas exchange half-lives of 900 d for winter and 360 d for summer in Rhine River (Wanner et al. 1989).
- Photolysis: photolytic half-lives of 1000 d for winter and 100 d for summmer in the Rhine River (Wanner et al. 1989); half-life of 1-4 d (Tomlin 1994).
- Oxidation: half-life ranged from about 5 h of midday sunlight during summer to 12 h during winter estimated from kinetic data for oxygenation reactions photosensitized by humic substances in water exposed to sunlight (Zepp et al.

1981); photooxidation half-life of 0.50-4.80 h in air, based on estimated rate constant for the reaction with hydroxyl radicals in air (Atkinson 1987; quoted, Howard et al. 1987).

Hydrolysis: first-order half-life of 103 d, based on measured overall rate constant of 2.8×10^{-4} h^{-1} at pH 7, 25°C (Ellington et al. 1986, 1987, 1988; quoted, Howard et al. 1991); abiotic hydrolysis rate constant of 1.3×10^{-7} s^{-1} under neutral condition, 2.0×10^{-3} s^{-1} under base-catalyzed condition at 20°C and hydrolysis half-lives of 170 d at 11°C, pH 9 and 62 d in summer were predicted in Rhine River (Wanner et al. 1989); half-lives in water of 3.04 yr at pH 1-5 and at 20°C; 1.2 d at pH 7 and 7.2 h at pH 9 both at 70°C (Worthing 1991); rate constant of 3.23 y^{-1} at pH 7.0 and 25°C (Kollig 1993); half-lives at 22°C: 133 days at pH 4, 169 d at pH 7, and 131 d at pH 9 (Tomlin 1994).

Biodegradation: primary biodegradation rate constant was 0.2 $\mu g^{\frac{1}{2}}$ $L^{\frac{1}{2}}$ d^{-1} with a half-life of 41 d, and the degradation half-lives were between 7-41 d for winter and 4-28 d for summer in Rhine River (Wanner et al. 1989); aqueous aerobic half-life of 72-504 h, based on aerobic soil field data (Szeto et al. 1983; quoted, Howard et al. 1991) and reported half-lives for soil (Domsch 1984; quoted, Howard et al. 1991); aqueous anaerobic half-life of 288-2016 h, based on estimated unacclimated aqueous aerobic biodegradation half-life (Howard et al. 1991).

Biotransformation:

Bioconcentration, Uptake (k_1) and Elimination (k_2) Rate Constants:

Common Name: Endosulfan
Synonym: Benzoepin, Beosit, Bio 5462, Chlorthiepin, Crissulfan, Cyclodan, Endocel, ENT 23979, FMC 5462, Hildan, Hoe 2671, Insectophene, KOP-thiodan, Malix, NCI-C00566, Niagara 5462, OMS-570, Thifor, Thimul, Thiodan, Thiofor, Thionex, Thiosulfan, Tionel, Tiovel
Chemical Name: 1,4,5,6,7,7-hexachloro-5-norbornene-2,3-dimethyl cyclic sulfite; 1,2,3,4,7,7-hexachlorobicyclo-2,2,1-hepten-5,6-bisoxymethylene sulfite; (1,4,5,6,7,7-hexachloro-8,9,10-trinorborn-5-en-2,3-ylenebismethylene)sulfite; 6,7,8,9,10,10-hexachloro-1,5,5a,6,9,9a-hexahydro-6,9-methano-2,4,3-benzodioxathiepine 3-oxide
Uses: insecticide for vegetable crops and also used as acaricide.
CAS Registry No: 115-29-7; 959-98-8 (α-Endosulfan); 33213-65-9 (β-Endosulfan)
Molecular Formula: $C_9H_6Cl_6O_3S$
Molecular Weight: 406.95
Melting Point (°C):
    70-100 (tech. grade, Khan 1980; Suntio et al. 1988; Worthing 1991; Milne 1995)
    70-100 (α-Endosulfan, Suntio et al. 1988)
    108-110 (β-Endosulfan, Suntio et al. 1988)
    106 (α-Endosulfan, Montgomery 1993)
    207-209 (β-Endosulfan, Montgomery 1993)
    109.2 (α-Endosulfan, Tomlin 1994)
    213.3 (β-Endosulfan, Tomlin 1994)
    106, 109.2 (Milne 1995)
Boiling Point (°C):
    106 (at 0.7 mmHg, Agrochemicals Handbook 1987; Milne 1995)
Density (g/cm³ at 20°C):
    1.80 (tech. grade, Tomlin 1994)
    1.745 (Milne 1995; Montgomery 1993)
Molar Volume (cm³/mol):
    312.8 (calculated-LeBas method, Suntio et al. 1988)
Molecular Volume (Å³):
Total Surface Area, TSA (Å²):
Dissociation Constant pK_a:
Heat of Fusion, ΔH_{fus}, kcal/mol:
Entropy of Fusion, ΔS_{fus}, cal/mol·K (e.u.):
Fugacity Ratio at 25°C (assuming ΔS_{fus} = 13.5 e.u.), F:
    0.22 (20°C, α-Endosulfan, Suntio et al. 1988)
    0.13 (20°C, β-Endosulfan, Suntio et al. 1988)

Water Solubility (g/m³ or mg/L at 25°C):
    0.53 (α-Endosulfan, gen. col.-GC, Weil et al. 1974; quoted, Geyer et al. 1980, 1982; Suntio et al. 1988; Shiu et al. 1990)
    0.286 (β-Endosulfan, gen. col.-GC, Weil et al. 1974; quoted, Geyer et al. 1980; Suntio et al. 1988; Shiu et al. 1990)
    <1.0 (Wauchope 1978)
    0.050 (Weber et al. 1980; quoted, Willis & McDowell 1982)
    0.510 (α-Endosulfan, 20°C, shake flask-GC, Bowman & Sans 1983a; quoted, Shiu et al. 1990; Howard 1991)
    0.45 (β-Endosulfan, 20°C, shake flask-GC, Bowman & Sans 1983a; quoted, Shiu et al. 1990; Howard 1991)

0.06-0.15 (U.S. EPA 1984; quoted, McLean et al. 1988)
0.32 (22°C, Agrochemicals Handbook 1987; quoted, Lohninger 1994)
0.15 (20°C, selected, Suntio et al. 1988; quoted, Majewski & Capel 1995)
0.32 (α-Endosulfan at 22°C, Worthing 1991; Tomlin 1994; Milne 1995)
0.33 (β-Endosulfan at 22°C, Worthing 1991; Tomlin 1994)
0.32 (20-25°C, selected, Wauchope et al. 1992; Hornsby et al. 1996)
very soluble (quoted for CAS 115-29-7, Pait et al. 1992)
0.53 (α-Endosulfan, Montgomery 1993)
0.28 (β-Endosulfan, Montgomery 1993)
0.32 (selected, Halfon et al. 1996)

Vapor Pressure (Pa at 25°C):
0.00133 (Martens 1972; Khan 1980; quoted, Suntio et al. 1988)
0.013 (Endosulfan I, Barlow 1978; quoted, Hinckley et al. 1990)
>0.00013 (20-25°C, Weber et al. 1980; quoted, Willis & McDowell 1982)
0.0013 (U.S. EPA 1984; quoted, McLean et al. 1988)
1.20 (80°C, Agrochemicals Handbook 1987)
0.0061 (Endosulfan I, GC-RT, supercooled value, Hinckley et al. 1990)
0.0032 (Endosulfan II, GC-RT, supercooled value, Hinckley et al. 1990)
0.00133 (Suntio et al. 1988; quoted, Howard 1991)
0.0011 (20°C, selected, Suntio et al. 1988; quoted, Majewski & Capel 1995)
1.20 (tech. grade at 80°C, Worthing 1991)
2.27×10^{-5} (20-25°C, selected, Wauchope et al. 1992; Hornsby et al. 1996)
0.00133 (Montgomery 1993)
8.3×10^{-5} (20°C, 2 to 1 mixture of α- & β-Endosulfan, Tomlin 1994)
2.3×10^{-5} (selected, Halfon et al. 1996)

Henry's Law Constant (Pa·m^3/mol):
1.09 (Mabey et al. 1982; quoted, Suntio et al. 1988)
0.033 (20°C, calculated-P/C, Suntio et al. 1988; quoted, Majewski & Capel 1995)
1.135 (calculated-P/C an av. of α- and β-Endosulfan, Howard 1991)
10.23 (α-Endosulfan, Montgomery 1993)
1.935 (β-Endosulfan, calculated-P/C, Montgomery 1993)
1.06 (calculated-P/C, this work)

Octanol/Water Partition Coefficient, log K_{ow}:
3.55 (α-Endosulfan, Ali 1978; quoted, Callahan et al. 1979; Geyer et al. 1982; Ryan et al. 1988; Suntio et al. 1988)
3.62 (β-Endosulfan, Ali 1978; quoted, Suntio et al. 1988)
3.83 (α-Endosulfan, shake flask-GC, Hermens & Leeuwangh 1982; quoted, Verhaar et al. 1992)
3.55 (U.S. EPA 1984; quoted, McLean et al. 1988)
3.83 (α-Endosulfan, Hansch & Leo 1985; quoted, Shiu et al. 1990; Howard 1991)
5.07 (α-Endosulfan, calculated-S, Hawker & Connell 1986)
3.60 (selected, Suntio et al. 1988)
2.23 (selected, Travis & Arms 1988)
4.74, 4.78 (α-, β-Endosulfan, calculated-f const., Noegrohati & Hammers 1992)

4.30 (α or β-Endosulfan, estimated-QSAR & SPARC, Kollig 1993)
3.55, 3.62 (α-, β-Endosulfan, Montgomery 1993)
4.74, 4.79 (α-, β-Endosulfan at pH 5, Tomlin 1994)
3.62, 3.83 (α-, β-Endosulfan, Hansch et al. 1995)

Bioconcentration Factor, log BCF:
- −3.66 (beef biotransfer factor log B_b, correlated-K_{OW}, Beck et al. 1966; quoted, Travis & Arms 1988)
- 2.78 (α-Endosulfan for mussel, Ernst 1977; quoted, Renberg & Sunström 1979; Geyer et al. 1982; Hawker & Connell 1986)
- −1.52, −1.22 (α-, β-Endosulfan, bioaccumulation factor log BF, adipose tissue in female Albino rats, Dorough et al. 1978; quoted, Geyer et al. 1980)
- 2.63, 2.44 (α-, β-Endosulfan, paddy field fish, Soon & Hock 1987; quoted, Abdullah et al. 1997)
- 1.91, 2.33 (α-, β-Endosulfan, paddy field fish, Tejada 1995; quoted, Abdullah et al. 1997)

Sorption Partition Coefficient, log K_{OC}:
- 3.46 (α-Endosulfan, estimated, Lyman et al. 1982; quoted, Howard 1991)
- 3.83 (β-Endosulfan, calculated-S, Lyman et al. 1982; quoted, Howard 1991)
- 3.11 (for CAS 112-29-7, Pait et al. 1992)
- 4.09 (soil, 20-25°C, selected, Wauchope et al. 1992; Hornsby et al. 1996)
- 4.00 (α- or β-Endosulfan, estimated-QSAR & SPARC, Kollig 1993)
- 3.31 (α-Endosulfan, calculated, Montgomery 1993)
- 3.37 (β-Endosulfan, calculated, Montgomery 1993)
- 4.09 (estimated-chemical structure, Lohninger 1994)
- 3.48-4.30 (Tomlin 1994)
- 4.09 (α-Endosulfan, HPLC-screening method, Kördel et al. 1995)

Half-Lives in the Environment:
Air: 2.5-24.8 h, based on an estimated rate constant for the vapor-phase reaction with hydroxyl radicals in air with a deoxygenated Endosulfan analog (Atkinson 1987; quoted, Howard et al. 1991).
Surface water: persistence of up to 4 weeks in river water (Eichelberger & Lichtenberg 1971); estimated half-life of 30-300 d in lakes (β-Endosulfan, Zoeteman et al. 1980); 4.5-218 h, based on aqueous hydrolysis half-lives for both α- and β-Endosulfan at pH 7 and 9 and 25°C respectively (Ellington et al. 1987; quoted, Howard et al. 1991); half-life of 1.3 d in rice paddy water (Tejada et al. 1993; quoted, Abdullah et al. 1997).
Groundwater: estimated half-life of 30-300 d in lakes and groundwater (β-Endosulfan, Zoeteman et al. 1980); 4.5-218 h, based on aqueous hydrolysis half-lives for both α- and β-Endosulfan at pH 7 and 9 and 25°C respectively (Ellington et al. 1987; quoted, Howard et al. 1991).
Sediment:
Soil: 4.5-218 h, based on aqueous hydrolysis half-lives for both α- and β-Endosulfan at pH 7 and 9 and 25°C respectively (Ellington et al. 1987; quoted, Howard et al. 1991); >50 d and subject to plant uptake via volatilization (Ryan et al. 1988);

selected half-life of 50 d (Wauchope et al. 1992; Hornsby et al. 1996); half-life of 1.2 d in rice soil (Tejada et al. 1993; quoted, Abdullah et al. 1997); soil half-life of 120 d (Pait et al. 1992); degraded in soil with half-life of 30-70 d (Tomlin 1994); 50 days (selected, Halfon et al. 1996).

Biota: 33.8 hours in mussels (α-Endosulfan, Ernst 1977); half-life of 1.0 d in rice leaves (Tejada et al. 1993; quoted, Abdullah et al. 1997).

Environmental Fate Rate Constants or Half-Lives:
  Volatilization:
  Photolysis:
  Oxidation: photooxidation half-life of 2.5-24.8 h, based on an estimated rate constant for the vapor-phase reaction with hydroxyl radicals in air with a deoxygenated endosulfan analog (Atkinson 1987; quoted, Howard et al. 1991).
  Hydrolysis: first-order half-life of 218 h, based on neutral aqueous hydrolysis rate constant of $(3.2 \pm 2.0) \times 10^{-3}$ h^{-1} for α-Endosulfan at pH 7 and 25°C (Ellington et al. 1986, 1987, 1988; quoted, Howard et al. 1991; Montgomery 1993); first-order half-life of 187 h, based on neutral aqueous hydrolysis rate constant of $(3.7 \pm 2.0) \times 10^{-3}$ h^{-1} for β-Endosulfan at pH 7 and 25°C (Ellington et al. 1987, 1988; quoted, Howard et al. 1991; Montgomery 1993); rate constant of 6.1×10^{-2} y^{-1} for α-Endosulfan at pH 7 and 25°C and rate constant of 8.9×10^{-2} y^{-1} for β-Endosulfan at pH 7 and 25°C (Kollig 1993).
  Biodegradation: aqueous aerobic half-life of 48-336 h, based on unacclimated aerobic river die-away test data (Eichelberger & Lichtenburg 1971; quoted, Howard et al. 1991) and reported soil grab sample data (Bowman et al. 1965; quoted, Howard et al. 1991); first-order rate constant of -0.00502 h^{-1} in nonsterile sediment and -0.00796 h^{-1} in sterile sediment by shake-tests at Range Point and first-order rate constant of -0.0157 h^{-1} in nonsterile water and -0.0325 h^{-1} in sterile water by shake-tests at Range Point (Walker et al. 1988); first-order rate constants of -0.00165 to -0.00296 h^{-1} in nonsterile sediment and -0.00426, -0.00545 h^{-1} in sterile sediment by shake-tests at Davis Bayou and first-order rate constants of -0.00335 to -0.00490 h^{-1} in nonsterile water and -0.0130, -0.00866 h^{-1} in sterile water by shake-tests at Davis Bayou (Walker et al. 1988).
  Biotransformation:
  Bioconcentration, Uptake (k_1) and Elimination (k_2) Rate Constants:
    k_1: 12.3 h^{-1} (mussel from α-Endosulfan, Ernst 1977; quoted, Hawker & Connell 1986)
    k_2: 0.0205 h^{-1} (mussel from α-Endosulfan, Ernst 1977; quoted, Hawker & Connell 1986)

Common Name: Endrin
Synonym: Endrex, ENT 17521, Hexadrin, Isodrin epoxidek, Mendrin, NA 2761, NCI-C00157, Nendrin, RCRA
Chemical Name: 1,2,3,4,10,10-hexachloro-6,7-epoxy-1,4,4a,5,6,7,8,8a-octahydro-exo-1,4-exo-5,8-dimethanonaphthalene
Uses: Insecticide/Avicide/Rodenticide
CAS Registry No: 72-20-8
Molecular Formula: $C_{12}H_8Cl_6O$
Molecular Weight: 380.92
Melting Point (°C):
    235 (dec. Callahan et al. 1979)
    200-210 (Suntio et al. 1988)
    226-230 (Howard 1991)
Boiling Point (°C):
    245 (dec. Montgomery 1993)
Density (g/cm³ at 20°C):
    1.70, 1.65 (pure, technical, at 25°C, Montgomery 1993)
Molar Volume (cm³/mol):
    318.2 (calculated-LeBas method, Suntio et al. 1988)
    249.9 (Ruelle & Kesselring 1997)
Molecular Volume (Å³):
Total Surface Area, TSA (Å²):
Dissociation Constant pK_a:
Heat of Fusion, ΔH_{fus}, kcal/mol:
    5.7075 (Ruelle & Kesselring 1997)
Entropy of Fusion, ΔS_{fus}, cal/mol·K (e.u.):
Fugacity Ratio at 25°C (assuming ΔS_{fus} = 13.5 e.u.), F: 0.015

Water Solubility (g/m³ or mg/L at 25°C):
    0.23 (shake flask-UV, Richardson & Miller 1960; quoted, Sharom et al. 1980; Isnard & Lambert 1988,89)
    0.26 (rm. temp., shake flask-GC, Robeck et al. 1965)
    0.23 (Günther et al. 1968; quoted, Metcalf et al. 1973)
    0.13, 0.25, 0.42, 0.625 (15, 25, 35, 45°C, shake flask-GC/ECD, Biggar & Riggs 1974; quoted, Callahan et al. 1979; Davies & Dobbs 1984; Suntio et al. 1988; Shiu et al. 1990)
    0.022, 0.15, 0.195 (particle size: 0.01, 0.05 and 5.0 μ, shake flask-GC/ECD, Biggar & Riggs 1974)
    0.26 (gen. col.-GC/ECD, Weil et al. 1974; quoted, Callahan et al. 1979; Geyer et al. 1980; Geyer et al. 1984; Suntio et al. 1988; Shiu et al. 1990)
    0.10 Weber et al. 1980; quoted, Willis & McDowell 1982)
    0.024 (Bruggeman et al. 1981; quoted, Adams 1987)
    0.23 (20°C, selected, Suntio et al. 1988)
    0.25 (misquoted as 0.25 μg/L from Biggar & Riggs, Howard 1991)
    0.22-0.26 (Montgomery 1993)
    0.23 (20-25°C, selected, Augustijn-Beckers et al. 1994)
    0.23 (20-25°C, selected, Hornsby et al. 1996)
    15.85 (20-25°C, supercooled liquid value, quoted, Majewski & Capel 1995)
    0.255 (selected, Pinsuwan et al. 1995)
    0.23; 0.105, 0.000065 (quoted; predicted-molar volume, Ruelle & Kesselring 1997)

Vapor Pressure (Pa at 25°C):
- 4.00×10^{-4} (20°C, Bowery 1964; quoted, Nash 1983; Howard 1991)
- 2.67×10^{-5} (Eichler 1965; Melnikov 1971; Martin 1972; Quellette & King 1977)
- 2.67×10^{-5} (20-25°C, Weber et al. 1980; quoted, Willis & McDowell 1982)
- 2.00×10^{-5} (20°C, selected, Suntio et al. 1988)
- 9.33×10^{-5} (25°C, Montgomery 1993)
- 2.67×10^{-5} (20-25°C, selected, Augustijn-Beckers et al. 1994; Hornsby et al. 1996)
- 1.38×10^{-3} (20-25°C, supercooled liquid value, Majewski & Capel 1995)

Henry's Law Constant (Pa·m^3/mol):
- 1.8×10^{-4} (calculated-P/C, Mabey et al. 1982)
- 0.042 (Ryan et al. 1988)
- 0.033 (20°C, calculated-P/C, Suntio et al. 1988; quoted, Majewski & Capel 1995)
- 0.762 (calculated, Howard 1991)
- 0.0507 (calculated-P/C, Montgomery 1993)
- 0.0331 (calculated-P/C, this work)

Octanol/Water Partition Coefficient, log K_{OW}:
- 5.60 (calculated, Neely et al. 1974)
- 4.56 (Veith et al. 1979; quoted, Mackay 1982; Veith & Kosian 1983; Geyer et al. 1984; Zaroogian et al. 1985; Noegrohati & Hammers 1992)
- 5.34 (Kenaga & Goring 1980; quoted, De Bruijn et al. 1989)
- 3.21 (Rao & Davidson 1980)
- 4.82 (Veith & Kosian 1982; quoted, Saito et al. 1992)
- 5.34 (selected, Dao et al. 1983)
- 5.16 (Garten & Trabalka 1983; quoted, De Bruijn et al. 1989)
- 5.43 (Patton et al. 1984; quoted, Hawker & Connell 1989)
- 4.56, 5.01 (quoted lit. shake-flask value, estimated-HPLC, Eadsforth 1986)
- 5.34 (selected, Hawker & Connell 1986)
- 4.56 (Isnard & Lambert 1988, 1989; quoted, Howard 1991; Saito et al. 1992)
- 5.195 ± 0.005 (slow-stirring method, De Bruijn et al. 1989; quoted, Sicbaldi & Finizio 1993)
- 4.50 (Ryan et al. 1988)
- 3.20 (selected, Suntio et al. 1988)
- 3.21-5.34 (Montgomery 1993)
- 4.71 (RP-HPLC, Sicbaldi & Finizio 1993)
- 5.20, 3.18 (selected, calculated, Pinsuwan et al. 1995)
- 5.20 (selected, Hansch et al. 1995)
- 4.71, 3.63, 5.05 (RP-HPLC, ClogP, calculated-S, Finizio et al. 1997)

Bioconcentration Factor, log BCF:
- 2.40-2.18 (bluegills, Bennett & Day 1970; quoted, Beddinger & Gloss 1984)
- 2.60-2.88 (channel catfish, Argyle et al. 1973; quoted, Beddinger & Gloss 1984)
- 3.21 (channel catfish, 55-d exposure, Argyle et al. 1973; quoted, Veith & Kosian 1983)
- 3.13, 4.69 (*Gambusia, Physa*, Metcalf et al. 1973)

3.11, 4.69, 3.66 (fish, snail, algae, Metcalf et al. 1973; quoted, Beddinger & Gloss 1984)
2.83, 2.49, 2.48 (fish, mosquitoes, Daphnia, 3-d expt. with no dietary routes, Metcalf et al. 1973; quoted, Callahan et al. 1979)
3.43 (oyster, Mason & Rowe 1976; quoted, Beddinger & Gloss 1984; Hawker & Connell 1986)
3.28 (mussel, Ernst 1977; quoted, Hawker & Connell 1986)
3.17, 3.24 (quoted, calculated-K_{OW}, Mackay 1982)
4.02 (flagfish, 30-d exposure, Hermanutz 1978; quoted, Veith & Kosian 1983)
4.18 (flagfish, 65-d exposure, Hermanutz 1978; quoted, Veith & Kosian 1983)
3.85 (flagfish, 110-d exposure, Hermanutz 1978; quoted, Veith & Kosian 1983)
3.70 (fathead minnow, Jarvinen & Tyo 1978; quoted, Davies & Dobbs 1984)
3.17 (mosquito fish, 35-d exposure, Veith et al. 1979; Veith & Kosian 1983; Saito et al. 1992)
3.66 (fathead minnow, 300-d exposure, Veith et al. 1979; Veith & Kosian 1983; quoted, Saito et al. 1992)
3.66 (*Oedogonium sp.*, Baughman & Paris 1981)
3.85-4.18 (flag fish, mosquito fish, Veith & Kosian 1983; quoted, Saito et al. 1992)
3.24, 3.18 (mussel, calculated-K_{OW} & models, Zaroogian et al. 1985)
3.17 (fathead minnow, quoted from Veith et al. 1979, Zaroogian et al. 1985)
3.40 (Isnard & Lambert 1988)
3.13-4.0 (fish, quoted, Howard 1991)
3.85 (fathead minnow, whole body, after 300 days, Howard 1991)
3.21-3.30 (channel catfish, after 41 and 55-d exposure, Howard 1991)
4.18 (flagfish, whole body after 65 days, Howard 1991)
3.52-3.68 (sheepshead minnow, 33 days for embryojuveniles, Howard 1991)
3.40-3.81 (sheepshead minnow, adults, after 28-161 days, Howard 1991)
2.70-3.10 (shellfish, Howard 1991)
4.69, 3.22-3.44, 3.48 (snail, oyster, grass shrimp, mussels, Howard 1991)
2.15-2.30 (algae, Howard 1991)
3.83 (fish, reported as log BAF_w, LeBanc 1996)

Sorption Partition Coefficient, log K_{OC}:
4.53 (calculated, Kenaga 1980, quoted, Howard 1991)
5.36 (calculated-S, Mill et al. 1980; quoted, Adams 1987)
4.00 (20-25°C, selected, Augustijn-Beckers et al. 1994; Hornsby et al. 1996)

Half-Lives in the Environment:
Air: a half-life of 1.45 h was predicted for reaction with hydroxyl radicals (Howard 1991).
Surface water: >8 weeks in river water (Eichelberger & Lichtenberg 1971).
Groundwater:
Sediment:
Soil: half-life of about 12 yr in Congaree sandy loam soil (Nash & Woolson 1967); field half-life of 63 d for sugar cane in soil (Willis & Hamilton 1973; quoted, Nash 1983); moderately persistent in soil with half-life of 20-100 d (Willis & McDowell 1982); microagroecosystem half-life of 33 d in moist fallow soil

(Nash 1983); half-life of >50 d in soil (Ryan et al. 1988); selected field half-life of 4300 d (Augustijn-Beckers et al. 1994; Hornsby et al. 1996).

Biota: elimination half-life was about 24 hours (Ernst 1977, quoted, Callahan et al. 1979).

Environmental Fate Rate Constants or Half-Lives:

Volatilization: half-life from a model river 1 m deep, flowing 1 m/s with a wind speed of 3 m/s was estimated to be 9.6 d, and greater than 14 yr from a model pond (Howard 1991).

Photolysis:

Oxidation: rate constant of $(2.7 \pm 0.7) \times 10^8$ M^{-1} s^{-1} for the reaction with OH radicals in aqueous solution (Fenton reaction) at $24 \pm 1°C$ and pH 2.8 (Haag & Yao 1992) with reference to 4.2×10^8 M^{-1} s^{-1} for the reaction of DPCP with OH radicals in aqueous solution (Buxton et al. 1988; quoted, Haag & Yao 1992); rate constants of $(1.3 \pm 0.4) \times 10^9$ M^{-1} s^{-1} (Fenton reaction) and $(1.1 \pm 0.2) \times 10^9$ M^{-1} s^{-1} (photo-Fenton reaction) for the reaction with OH radicals in aqueous solution at $24 \pm 1°C$ and pH 3.4 (Haag & Yao 1992) with reference to 8×10^8 M^{-1} s^{-1} for the reaction of lindane with OH radicals in aqueous solution (Buxton et al. 1988; quoted, Haag & Yao 1992).

Hydrolysis: half-life of at least 4 yr (Callahan et al. 1979).

Biodegradation: half-life in thick anaerobic sewage sludge is 5-14 d (Howard 1991).

Biotransformation:

Bioconcentration, Uptake (k_1) and Elimination (k_2) Rate Constants:

$-\log k_2$: 1.99 h (oyster, Mason & Rowe 1976; quoted, Hawker & Connell 1986)

$\log k_1$: 1.5 h^{-1} (mussel, Ernst 1977; quoted, Hawker & Connell 1986)

$-\log k_2$: 1.78 h (mussel, Ernst 1977; quoted, Hawker & Connell 1986)

Common Name: Ethion
Synonym: AC 3422, Bladan, diethion, Embathion, ENT 24105, Ethanox, Ethiol, Ethodan, Ethopaz, FMC 1240, Fosfono 50, Hylemax, Hylemox, Itopaz, KWIT, NA 2783, NIA 1240, Niagara 1240, Nialate, Vegfru fosmite
Chemical Name: bis(S-(dimethoxyphosphinothioyl)mercapto)methane; O,O,O',O'-tetraethyl-S,S'-methylene bis(phosphorodithioate); O,O,O',O'-tetraethyl-S,S'-methylene-bisphosphorothiolothionate
Uses: nonsystemic insecticide and acaricide used on apples.
CAS Registry No: 563-12-2
Molecular Formula: $C_9H_{22}O_4P_2S_4$
Molecular Weight: 384.48
Melting Point (°C):
    −12   (Suntio et al. 1988)
    −12 to −13 (Howard 1991)
    −12 to −15 (Montgomery 1993; Tomlin 1994; Milne 1995)
Boiling Point (°C):
    164-165 (at 0.3 mmHg, Agrochemicals Handbook 1987; Howard 1991; Tomlin 1994; Milne 1995)
Density (g/cm³ at 20°C):
    1.22   (Agrochemicals Handbook 1987; Worthing 1991; Montgomery 1993; Tomlin 1994; Milne 1995)
Molar Volume (cm³/mol):
    350.2   (calculated-LeBas method, Suntio et al. 1988; Fisher et al. 1993)
    1.80   ($V_I/100$, Fisher et al. 1993)
Molecular Volume (Å³):
Total Surface Area, TSA (Å²):
Dissociation Constant pK_a:
Heat of Fusion, ΔH_{fus}, kcal/mol:
Entropy of Fusion, ΔS_{fus}, cal/mol·K (e.u.):
Fugacity Ratio at 25°C (assuming ΔS_{fus} = 13.5 e.u.), F: 1.0

Water Solubility (g/m³ or mg/L at 25°C):
    2.0   (Metcalf 1971,74; quoted, Kenaga 1980a; Kenaga & Goring 1980; Suntio et al. 1988; Shiu et al. 1990)
    0.60   (Miles 1976; Miles & Harris 1978; quoted, Bowman & Sans 1979; Sharom et al. 1980)
    1.0   (20-25°C, selected, Willis & McDowell 1982; Gerstl & Helling 1987)
    1.1   (19.5°C, shake flask-GC, Bowman & Sans 1983b; quoted, Shiu et al. 1990; Howard 1991; Patil 1994)
    2.0   (Garten & Trabalka 1983; quoted, Shiu et al. 1990)
    1.8   (20°C, selected, Suntio et al. 1988; quoted, Majewski & Capel 1995)
    1.1   (20-25°C, selected, Wauchope et al. 1992; Lohninger 1994; Hornsby et al. 1996)
    0.68   (20°C, Montgomery 1993)
    0.76   (30°C, Montgomery 1993)
    3.43   (calculated, Patil 1994)
    2.00   (Tomlin 1994)
    1.10   (quoted, Kühne et al. 1995)
    0.279   (calculated-group contribution fragmentation method, Kühne et al. 1995)

Vapor Pressure (Pa at 25°C):
- 0.0002 (Khan 1980)
- 0.0002 (Merck Index 1983, 1989; quoted, Suntio et al. 1988)
- 0.0002 (Worthing 1983, 1991; quoted, Howard 1991)
- 0.0002 (Agrochemicals Handbook 1987)
- 0.00015 (20°C, selected, Suntio et al. 1988; quoted, Majewski & Capel 1995)
- 0.00032 (20-25°C, selected, Wauchope et al. 1992; Hornsby et al. 1996)
- 0.0002 (Montgomery 1993)
- 0.0002 (Tomlin 1994)

Henry's Law Constant (Pa·m^3/mol):
- 0.0699 (calculated-P/C, Lyman et al. 1982; quoted, Howard 1991)
- 0.032 (20°C, calculated-P/C, Suntio et al. 1988; quoted, Fisher et al. 1993; Majewski & Capel 1995)
- 0.0384 (calculated-P/C, Montgomery 1993)
- 0.0320 (calculated-P/C, this work)

Octanol/Water Partition Coefficient, log K_{OW}:
- 5.07 (Hansch & Leo 1979; quoted, Magee 1991; Fisher et al. 1993)
- 5.073 (shake flask-GC, Bowman & Sans 1983b; quoted, Suntio et al. 1988; Howard 1991; Patil 1994)
- 4.52 (selected, Yoshioka et al. 1986)
- 5.10 (selected, Suntio et al. 1988)
- 4.28, 5.07 (Montgomery 1993)
- 4.42 (calculated, Patil 1994)
- 5.07 (selected, Hansch et al. 1995)

Bioconcentration Factor, log BCF:
- 2.77 (estimated-log K_{OW}, Howard 1991)
- 2.77 (estimated-S, Howard 1991)

Sorption Partition Coefficient, log K_{OC}:
- 4.19 (av. 4 soils, King & McCarthy 1968; quoted, Howard 1991)
- 4.19 (soil, Hamaker & Thompson 1972; quoted, Kenaga & Goring 1980)
- 3.81 (organic soil, Sharom et al. 1980; quoted, Howard 1991)
- 3.94 (Beverley sandy loam, Sharom et al. 1980; quoted, Howard 1991)
- 4.00 (Plainsfield sand, Sharom et al. 1980; quoted, Howard 1991)
- 4.31, 3.66 (quoted, calculated-χ, Gerstl & Helling 1987)
- 4.19 (reported as log K_{OM}, Magee 1991)
- 4.28 (estimated as log K_{OM}, Magee 1991)
- 4.06 (soil, quoted exptl., Meylan et al. 1992)
- 4.12 (soil, calculated-χ and fragment contribution, Meylan et al. 1992)
- 4.00 (soil, 20-25°C, selected, Wauchope et al. 1992; Hornsby et al. 1996)
- 3.54-4.34 (Montgomery 1993)
- 4.43 (estimated-chemical structure, Lohninger 1994)

Half-Lives in the Environment:
- Air: half-life was estimated to be 6.95 h for the vapor-phase reaction with hydroxyl radicals in air (Howard 1991).
- Surface water: half-life of 4 weeks in river water (Eichelberger & Lichtenberg 1971).
- Groundwater:
- Sediment:
- Soil: >24 weeks in sterile sandy loam and 7 weeks in nonsterile sandy loam; >24 weeks in sterile organic soil and 8.0 weeks in nonsterile organic soil (Miles et al. 1979; quoted, Howard 1991); selected field half-life of 150 d (Wauchope et al. 1992; Hornsby et al. 1996); half-life of 90 d in soil (Tomlin 1994).
- Biota:

Environmental Fate Rate Constants or Half-Lives:
- Volatilization: using Henry's law constant, half-life from a model river 1 m deep, flowing 1 m/s with wind velocity of 3 m/s was estimated to be 102 d (Lyman et al. 1982; quoted, Howard 1991).
- Photolysis:
- Oxidation: photooxidation half-life was estimated to be 6.95 ho for the vapor-phase reaction with hydroxyl radicals in air (Howard 1991).
- Hydrolysis: half-lives in water at 25°C and pHs of 4.5, 5.0, 6.0, 7.0 and 8.0 were 99, 63, 58, 25, and 8.4 weeks, respectively (Chapman & Cole 1982; quoted, Montgomery 1993); exptl. half-life of 20.8 weeks was determined in buffered distilled water at 30°C between pH 4 and 7, 8.9 weeks at pH 8 and 1 day at pH 10 (Dierberg & Pfeuffer 1983; quoted, Howard 1991); half-life of 390 days at pH 9 (Tomlin 1994).
- Biodegradation: half-life of >24 weeks in sterile sandy loam and 7 weeks in nonsterile sandy loam; >24 weeks in sterile organic soil and 8.0 weeks in nonsterile organic soil (Miles et al. 1979; quoted, Howard 1991); half-life in both sterilized and unsterilized Florida canal water was found to be 24-26 days over 12 weeks observation (Dierberg & Pfeuffer 1983; quoted, Howard 1991).
- Biotransformation:
- Bioconcentration, Uptake (k_1) and Elimination (k_2) Rate Constants:

Common Name: Fenitrothion

Synonym: Accothion, Agria 1050, Agrothion, Arbogal, Cyfen, Cytel, Dybar, Falithion, Fenitox, Kotion, Sumithion

Chemical Name: *O,O*-dimethyl *O*-4-nitro-*m*-tolyl phosphorothioate; phosphorothioic acid *O,O*-dimethyl *O*-4-nitro-*m*-tolyl ester; *O,O*-dimethyl *O*-(3-methyl-4-nitrophenyl) phosphorothioate

Uses: insecticide to control boring, chewing and sucking insects in cereals, cotton, maize, sorghum, citrus fruit, pome fruit, stone fruit, soft fruit, vines, bananas, olives, rice, soybeans, beet, sugar cane, oilseed rape, vegetables, lucerne, coffee, cocoa, tea, tobacco, ornamentals and forestry; also used as a public health insecticide to control household insects, flies in animal houses, mosquito larvae, and locusts.

CAS Registry No: 122-14-5
Molecular Formula: $C_9H_{12}NO_5PS$
Molecular Weight: 277.25
Melting Point (°C):
    3.4   (Tomlin 1994)
Boiling Point (°C):
    95   (at 0.01 mmHg, Melnikov 1971; Freed et al. 1977)
    164   (at 1 mmHg, Worthing 1991; Milne 1995)
    140-145   (at 0.1 mmHg, dec., Tomlin 1994)
    118   (at 0.05 mmHg, Milne 1995)
Density (g/cm^3 at 20°C):
    1.328   (Worthing 1991; Tomlin 1994)
    1.3227   (25°C, Milne 1995)
Molar Volume (cm^3/mol):
    229.7   (calculated-LeBas method, this work)
Molecular Volume (Å^3):
Total Surface Area, TSA (Å^2):
Dissociation Constant pK_a:
    7.20   (Kortum et al. 1961; Wolfe 1980)
Heat of Fusion, ΔH_{fus}, kcal/mol:
Entropy of Fusion, ΔS_{fus}, cal/mol·K (e.u.):
Fugacity Ratio at 25°C (assuming ΔS_{fus} = 13.5 e.u.), F: 1.0

Water Solubility (g/m^3 or mg/L at 25°C):
    30   (Macy 1948; quoted, Chiou et al. 1977)
    30   (20°C, Bright et al. 1950; Melnikov 1971; Hamaker 1975; quoted, Shiu et al. 1990)
    30   (20°C, Melnikov 1971; quoted, Chiou et al. 1977; Freed et al. 1977,79; Bowman & Sans 1979; Metcalf et al. 1980; Geyer et al. 1982)
    30   (Hamaker 1975; quoted, Kenaga 1980; Kenaga & Goring 1980; Geyer et al. 1982; Zaroogian et al. 1985)
    25.2   (20°C, shake flask-GC, Bowman & Sans 1979; quoted, Shiu et al. 1990; Patil 1994)
    38.7   (20-25°C, shake flask-GC, Kanazawa 1981; quoted, Shiu et al. 1990)
    30   (quoted, Kanazawa 1989)
    21   (20°C, Worthing 1991; Tomlin 1994)
    30   (20-25°C, selected, Augustijn-Beckers et al. 1994; Hornsby et al. 1996)
    25, 31.8   (quoted, calculated, Patil 1994)

25.3   (quoted, Kühne et al. 1995)
21   (calculated-group contribution fragmentation method, Kühne et al. 1995)
30   (21°C, Milne 1995)

Vapor Pressure (Pa at 25°C):
- 8.0×10^{-3} (20°C, Melnikov 1971; quoted, Freed et al 1977)
- 7.2×10^{-3} (20°C, Freed et al. 1979; quoted, Metcalf et al. 1980)
- 8.0×10^{-4} (20°C, Hartley & Graham-Bryce 1980; quoted, Taylor & Glotfelty 1988; Taylor & Spencer 1990)
- 8.0×10^{-4} (20°C, Khan 1980)
- 5.5×10^{-3} (gas saturation method, Addison 1981; quoted, Hinckley et al. 1990)
- 5.4×10^{-3} (gas saturation-extrapolated, Addison 1981)
- 8.0×10^{-3} (Budavari 1989)
- 1.1×10^{-2} (GC-RT, supercooled value, Hinckley et al. 1990)
- 8.0×10^{-4} (selected, Taylor & Spencer 1990)
- 1.5×10^{-4} (20°C, Worthing 1991)
- 1.3×10^{-4} (20-25°C, estimated, Augustijn-Beckers et al. 1994; Hornsby et al. 1996)
- 0.0180   (20°C, Tomlin 1994)

Henry's Law Constant (Pa·m^3/mol):
- 0.0942 (exptl., Metcalf et al. 1980)
- 0.0669 (estimated, Metcalf et al. 1980)
- 0.0012 (calculated-P/C, this work)

Octanol/Water Partition Coefficient, log K_{OW}:
- 3.38   (20°C, shake flask-GC, Chiou et al. 1977; quoted, Kenaga & Goring 1980; Geyer et al. 1982; Bowman & Sans 1983b; Zaroogian et al. 1985; De Bruijn & Hermens 1991; Finizio et al. 1997)
- 3.38   (shake flask-GC, Freed et al. 1979)
- 3.36   (Rao & Davidson 1980; quoted, Bowman & Sans 1983b)
- 3.44   (shake flask-GC, Kanazawa 1981; quoted, Kanazawa 1989; Sicbaldi & Finizio 1993; Finizio et al. 1997)
- 3.397   (shake flask-GC, Bowman & Sans 1983b; quoted, De Bruijn & Hermens 1991; Patil 1994; Finizio et al. 1997)
- 3.38   (selected, Hawker & Connell 1986)
- 3.466 ± 0.003 (slow-stirring method, De Bruijn & Hermens 1991; quoted, Verhaar et al. 1992; Sicbaldi & Finizio 1993; Finizio et al. 1997)
- 3.43   (20°C, Worthing 1991; Tomlin 1994)
- 2.96   (RP-HPLC, Saito et al. 1993)
- 3.03   (RP-HPLC, Sicbaldi & Finizio 1993)
- 3.24   (calculated, Patil 1994)
- 3.43   (Milne 1995)
- 3.30   (selected, Hansch et al. 1995)
- 3.47   (selected, Devillers et al. 1996)
- 3.03, 3.21, 3.54 (RP-HPLC, CLOGP, Calculated-S, Finizio et al. 1997)

Bioconcentration Factor, log BCF:
- 1.00 (fish in static water, Leo et al. 1971; quoted, Kenaga & Goring 1980)
- 2.23 (motsugo, Kanazawa 1975; quoted, McLeese et al. 1976)
- 2.34 (rainbow trout, Takimoto & Miyamoto 1976; quoted, McLeese et al. 1976)
- 2.02 (mussel, McLeese et al. 1979; quoted, Geyer et al. 1982)
- 1.96 (calculated-S, Kenaga 1980)
- 2.39 (*Pseudorasbora parva*, Kanazawa 1981; quoted, Devillers et al. 1996)
- 2.34, 2.17 (mussel, calculated-K_{OW} & models, Zaroogian et al. 1985)
- 2.11 (mussel, Zaroogian et al. 1985; quoted, Hawker & Connell 1986)
- 2.74, 2.75 (*Oryzias latipes*, Takimoto et al. 1984; quoted, Devillers et al. 1996)
- 2.48 (*Oryzias latipes*, Takimoto et al. 1987; quoted, Devillers et al. 1996)
- 2.60 (willow shiner, Tsuda et al. 1989)
- 3.36 ± 0.04 (guppy, calculated on an extractable lipid wt. basis, De Bruijn & Hermens 1991; Devillers et al. 1996)
- 2.37, 2.72 (killifish, De Bruijn & Hermens 1991)
- 2.18, 2.31, 1.48 (minnow, motsugo, mullet, De Bruijn & Hermens 1991)
- 3.54 (*Poecilia reticulata*, De Bruijn & Hermens 1991; Devillers et al. 1996)
- 2.30, 2.39 (rainbow trout, topmouth gudgeon, De Bruijn & Hermens 1991)
- 1.65, 1.68 (*Oryzias latipes*, Tsuda et al. 1995; quoted, Devillers et al. 1996)

Sorption Partition Coefficient, log K_{OC}:
- 2.83 (soil, calculated-S as per Kenaga & Goring 1978, Kenaga 1980)
- 2.63 (average of 2 soils, Kanazawa 1989)
- 3.30 (20-25°C, selected, Augustijn-Beckers et al. 1994; Hornsby et al. 1996)

Half-Lives in the Environment:
Air:
Surface water: half-life of 15-168 h in summer, Palfrey Lake, Canada at pH 6.7, 11°C under sunlight conditions (Metcalfe et al. 1980); 36-48 h at pH 7.0-7.5, 19-23°C under sunlight conditions, 518-1188 h at pH 7.5, 23°C under dark conditions in Lac Bourgeous, Quebec (Greenhalgh et al. 1980); 13 h in winter, irrigation ditch from Ebre Delta, Spain under sunlight conditions, at pH 7.8, 11°C (Lacorte & Barcelo 1994); half-lives of 202 d at 6°C, 62 d at 22°C in darkness for Milli-Q water at pH 6.1; 103 d at 6°C, 31 d at 22°C in darkness, 4 d under sunlight conditions for river water at pH 7.3; 143 d at 6°C, 27 d at 22°C in darkness for filtered water at pH 7.3; 224 d at 6°C, 34 d at 22°C in darkness, 3 d under sunlight conditions in seawater (Arcachon Bay, France) at pH 8.1, 22-25°C (Lartiges & Garri gues 1995); 11-19.3 h at pH 7.8-8.2, 25-20°C under sunlight conditions in rice crop field; 70-74 h at pH 8.2, 15-18°C under dark conditions from Ebre Delta, Spain (Oubina et al. 1996).
Groundwater:
Sediment:
Soil: selected field half-life of 4 d (Augustijn-Beckers et al. 1994; Hornsby et al. 1996); half-life of 12-28 d under upland conditions and 4-20 d under submerged conditions (Tomlin 1994).
Biota: excretion half-life of 9.9 h (willow shiner, Tsuda et al. 1989); elimination rate constants of $(0.28 \pm 0.02) \times 10^3$ (NADPH) and $(0.15 \pm 0.02) \times x10^3$ (GSH)

min^{-1} · mg protein^{-1} (rainbow trout, De Bruijn et al. 1993); degraded half-life of 4 d in balsam fir and spruce foliage (Tomlin 1994).

Environmental Fate Rate Constants or Half-Lives:
- Volatilization: 6.3 d from the bottom of Palfrey Lake and 7.2 d from the surface of Palfrey Lake vs. a calculated half-life of 20.6 d; 0.9 d from Palfrey Brook vs. a calculated half-life of 5.40 d (Metcalf et al. 1980).
- Photolysis: disappearance rate constant of 0.053 h^{-1} with calculated first-order half-life of 13 h (Lacorte & Barcelo 1994).
- Oxidation:
- Hydrolysis: second-order alkaline hydrolysis rate constant of 4.2×10^{-3} M^{-1} s^{-1} at 27°C (Maquire & Hale 1980; quoted, Wolfe 1980); estimated half-lives at 22°C: 108.8 d at pH 4, 84.3 d at pH 7, and 75 d at pH 9 (Tomlin 1994).
- Biodegradation:
- Biotransformation:
- Bioconcentration, Uptake (k_1) and Elimination (k_2) Rate Constants: uptake rate constant of 88 d^{-1} and excretion rate constant of 0.4 d^{-1} (rainbow trout, Takimoto & Miyamoto 1976; quoted, McLeese et al. 1976); excretion rate constant of 0.070 h^{-1} (willow shiner, Tsuda et al. 1989); uptake and elimination rate constants of $(3.89 \pm 1.39) \times 10^3$ mL/g/d and (1.13 ± 0.07) d^{-1} (guppy, De Bruijn & Hermens 1991); elimination rate constants of $(0.28 \pm 0.02) \times 10^3$ (NADPH) and $(0.15 \pm 0.02) \times 10^3$ (GSH) min^{-1} · mg protein^{-1} (rainbow trout, De Bruijn et al. 1993).

Common Name: Fenoxycarb
Synonym: Insegar, Logic, Pictyl, Torus, Varikil
Chemical Name: ethyl2-(4-phenoxyphenoxy)ethylcarbamate; ethyl[2-(p-phenoxy)ethyl]-carbamate
Uses: insecticide to control lepidoptera, scale insects, and psyllids on fruit, cotton and ornamentals; and also cockroaches, fleas, mosquito larvae, and fire ants in public health situations.
CAS Registry No: 79127-80-3
Molecular Formula: $C_{17}H_{19}NO_4$
Molecular Weight: 301.3
Melting Point (°C):
    53-54   (Agrochemicals Handbook 1987; Tomlin 1994; Milne 1995)
    49-54   (Worthing 1991)
Boiling Point (°C):
Density (g/cm³ at 20°C):
    1.23   (Tomlin 1994)
Molar Volume (cm³/mol):
    344.2   (calculated-LeBas method, this work)
Molecular Volume (Å³):
Total Surface Area, TSA (Å²):
Dissociation Constant pK_a:
Heat of Fusion, ΔH_{fus}, kcal/mol:
Entropy of Fusion, ΔS_{fus}, cal/mol·K (e.u.):
Fugacity Ratio at 25°C (assuming ΔS_{fus} = 13.5 e.u.), F: 0.517

Water Solubility (g/m³ or mg/L at 25°C):
    6.0   (20°C, Agrochemicals Handbook 1987; Tomlin 1994; Milne 1995)
    5.7   (Worthing 1991)
    6.0   (20-25°C, selected, Hornsby et al. 1996)

Vapor Pressure (Pa at 25°C):
    1.7×10^{-6} (Agrochemicals Handbook 1987)
    7.8×10^{-6} (20°C, Worthing 1991)
    8.7×10^{-7} (Tomlin 1994)
    1.7×10^{-6} (20-25°C, selected, Hornsby et al. 1996)

Henry's Law Constant (Pa·m³/mol):
    8.5×10^{-5} (calculated-P/C, this work)

Octanol/Water Partition Coefficient, log K_{OW}:
    4.30   (Worthing 1991; Milne 1995)
    4.07   (Tomlin 1994)
    4.30   (selected, Hansch et al. 1995)

Bioconcentration Factor, log BCF:
    2.35   (calculated-S as per Kenaga 1980, this work)
    3.11   (calculated-K_{OW} as per Kenaga 1980, this work)

Sorption Partition Coefficient, log K_{OC}:
    3.00   (20-25°C, estimated, Hornsby et al. 1996)

Half-Lives in the Environment:
    Air:
    Surface water:
    Groundwater:
    Sediment:
    Soil: half-life of 1.7-2.5 months in laboratory soil and water and a few days to 31 days in field soil and water (Tomlin 1994); field half-life of 1 d (20-25°C, selected, Hornsby et al. 1996).

Environmental Fate Rate Constants or Half-Lives:
    Volatilization:
    Photolysis:
    Oxidation:
    Hydrolysis:
    Biodegradation:
    Biotransformation:
    Bioconcentration, Uptake (k_1) and Elimination (k_2) Rate Constants:

Common Name: Fenthion
Synonym: Bay 29493, Baycid, Bayer 9007, Baytex, Baycid, DMTP, Ekalux, ENT 25540, Entex, Lebacid, Lebaycid, Mercaptophos, MPP, NCI-C08651, OMS 2, Queletox, Spottan, Talodex, Tiquvon
Chemical Name: O,O-dimethyl O-(3-methyl-4-(methylthio)phenyl) phosphorothioate; O,O-dimethyl O-4-methylthio-m-tolyl phosphorothioate
Uses: insecticide with contact, stomach and respiratory action and also used as acaricide and cholinesterase inhibitor.
CAS Registry No: 55-38-9
Molecular Formula: $C_{10}H_{15}O_3PS_2$
Molecular Weight: 278.34
Melting Point (°C):
    7.0   (Montgomery 1993)
    7.5   (Tomlin 1994; Milne 1995)
Boiling Point (°C):
    87.0   (at 0.01 mmHg, Agrochemicals Handbook 1987; Worthing 1991; Montgomery 1993; Tomlin 1994; Milne 1995)
Density (g/cm³ at 20°C):
    1.246   (Agrochemicals Handbook 1987; Tomlin 1994; Milne 1995)
    1.25   (Worthing 1991; Montgomery 1993)
Molar Volume (cm³/mol):
    264.6   (calculated-LeBas method, Suntio et al. 1988)
Molecular Volume (Å³):
Total Surface Area, TSA (Å²):
Dissociation Constant pK_a:
Heat of Fusion, ΔH_{fus}, kcal/mol:
Entropy of Fusion, ΔS_{fus}, cal/mol·K (e.u.):
Fugacity Ratio at 25°C (assuming ΔS_{fus} = 13.5 e.u.), F: 1.0

Water Solubility (g/m³ or mg/L at 25°C):
    55   (Günther et al. 1968; Garten & Trabalka 1983; Budavari 1989; quoted, Suntio et al. 1988; Shiu et al. 1990)
    54-56   (rm. temp., Spencer 1973, 1980; quoted, Bowman & Sans 1983; Shiu et al. 1990)
    55   (Martin & Worthing 1977; quoted, Kenaga 1980)
    56   (22°C, Khan 1980; quoted, Suntio et al. 1988; Shiu et al. 1990)
    55   (22°C, Verschueren 1983; quoted, Suntio et al. 1988; Shiu et al. 1990)
    7.51   (20°C, shake flask-GC, Bowman & Sans 1983a,b; quoted, Heller et al. 1989; Shiu et al. 1990; Patil 1994)
    9.3   (20°C, Bowman & Sans 1985; quoted, Heller et al. 1985)
    54-56   (20°C, Agrochemicals Handbook 1987)
    2.0   (20°C, Worthing 1987,91; quoted, Shiu et al. 1990)
    2.0, 4.2, 7.51, 9.3, 50   (20°C, literature data variability, Heller et al. 1985)
    54-56   (rm. temp., literature data variability, Heller et al. 1985)
    50   (20°C, selected, Suntio et al. 1988)
    4.2   (20-25°C, selected, Wauchope et al. 1992; Lohninger 1994; Hornsby et al. 1996)
    9.30, 11.3   (20°C, 30°C, Montgomery 1993)
    7.50, 6.68   (quoted, calculated, Patil 1994)
    4.2   (20°C, Tomlin 1994)

7.49, 15.7 (quoted, calculated-group contribution, Kühne et al. 1995)
2.0 (20°C, Milne 1995)

Vapor Pressure (Pa at 25°C):
4.0×10^{-3} (20°C, Eichler 1965; quoted, Kim 1985; Suntio et al. 1988)
4.0×10^{-3} (20°C, Melnikov 1971; quoted, Kim 1985; Suntio et al. 1988)
4.0×10^{-3} (20°C, Hartley & Graham-Bryce 1980; quoted, Suntio et al. 1980)
4.0×10^{-3} (20°C, Khan 1980; Budavari 1989; Worthing 1991; Montgomery 1993)
4.0×10^{-3} (20°C, quoted exptl. value, Kim et al. 1984)
8.4×10^{-3} (20°C, GC-calculated value, Kim et al. 1984; Kim 1985)
4.0×10^{-3}, 10×10^{-3} (20°C, 30°C, Agrochemicals Handbook 1987)
4.0×10^{-3} (20°C, selected, Suntio et al. 1988)
3.7×10^{-4} (20-25°C, selected, Wauchope et al. 1992; Hornsby et al. 1996)
7.4×10^{-4} (Tomlin 1994)

Henry's Law Constant (Pa·m^3/mol):
0.022 (20°C, calculated-P/C, Suntio et al. 1988)
0.547 (Montgomery 1993)

Octanol/Water Partition Coefficient, log K_{OW}:
4.091 (shake flask-GC, Bowman & Sans 1983b; quoted, Suntio et al. 1988; Patil 1994; Finizio et al. 1997)
2.65 (selected, Dao et al. 1983)
4.10 (selected, Suntio et al. 1988)
3.16 (selected, Travis & Arms 1988)
3.432 (calculated as per Broto et al. 1984, Karcher & Devillers 1990)
4.167 ± 0.009 (slow-strring method, De Bruijn & Hermens 1991; quoted, Verhaar et al. 1992; Devillers et al. 1996; Finizio et al. 1997)
4.09, 4.84 (Montgomery 1993)
3.56 (RP-HPLC, Saito et al. 1993)
3.94 (calculated, Patil 1994)
4.84 (Tomlin 1994)
4.09 (selected, Hansch et al. 1995)
3.91, 3.51 (ClogP, calculated-S, Finizio et al. 1997)

Bioconcentration Factor, log BCF:
1.81 (calculated-S, Kenaga 1980)
−4.50 (beef biotransfer factor log B_b, correlated-K_{OW}, MacDougall 1972; quoted, Travis & Arms 1988)
−5.60 (milk biotransfer factor log B_m, correlated-K_{OW}, Johnson & Bowman 1972; quoted, Travis & Arms 1988)
4.22 ± 0.08 (guppy, calculated on an extractable lipid wt. basis, De Bruijn & Hermens 1991; quoted, Devillers et al. 1996)
4.17 (*Poecilia reticulata*, De Bruijn & Hermens 1991; quoted, Devillers et al. 1996)
2.68 (whole body willow shiner after 24-168 h exposure, Tsuda et al. 1992; quoted, Devillers et al. 1996)
1.34, 1.46, 1.43, 1.41 (whole body carp: 24 h, 72 h, 120 h, and 148 h; Tsuda et al. 1993)
1.96, 2.02 (*Oryzias latipes*, Tsuda et al. 1995; quoted, Devillers et al. 1996)

Sorption Partition Coefficient, log K_{OC}:
- 2.68 (calculated-S, Kenaga 1980)
- 3.18 (soil, 20-25°C, selected, Wauchope et al. 1992; Hornsby et al. 1996)
- 0.89-1.58 (Montgomery 1993)
- 3.18 (Tomlin 1994; Lohninger 1994)
- 3.31 (soil, HPLC-ring test, Kördel et al. 1995a)
- 3.31 (soil, HPLC-screening method, Kördel et al. 1995b)

Half-Lives in the Environment:
  Air:
  Surface water: persistence of up to 4 weeks in river water (Eichelberger & Lichtenberg 1971); half-lives of 189 d at 6°C, 71 d at 22°C in darkness for Mill-Q water at pH 6.1; 149 d at 6°C, 42 d at 22°C in darkness, 2 d under sunlight conditions for river water at pH 7.3; 104 d at 6°C, 33 d at 22°C in darkness for filtered river water, pH 7.3; 227 d at 6°C, 26 d at 22°C in darkness, 5 d under sunlight conditions for seawater at pH 8.1 (Lartiges & Garrigues 1995).
  Groundwater:
  Sediment:
  Soil: selected field half-life of 34 d (Wauchope et al. 1992; Hornsby et al. 1996); half-life is ca. 1 d in soil and water (Tomlin 1994).
  Biota: excretion rate constant of 0.07 h^{-1} from whole body willow shiner (Tsuda et al. 1992); elimination rate constants of $(0.64 \pm 0.09) \times 10^3$ (NADPH) and $(0.12 \pm 0.02) \times 10^3$ (GSH) $min^{-1} \cdot mg$ $protein^{-1}$ (rainbow trout, De Bruijn et al. 1993); excretion rate constant of 0.34 h^{-1} with half-life of 2.0 d from carp (Tsuda et al. 1993).

Environmental Fate Rate Constants or Half-Lives:
  Volatilization:
  Photolysis:
  Oxidation:
  Hydrolysis: half-lives at 22°C: 223 d at pH 4, 200 d at pH 7, and 151 d at pH 9 (Tomlin 1994).
  Biodegradation: rate constants of $-0.00745\, h^{-1}$ in nonsterile sediment and $-0.00199\, h^{-1}$ in sterile sediment by shake-tests at Range Point and $-0.00129\, h^{-1}$ in nonsterile water by shake-tests at Range Point (Walker et al. 1988).
  Biotransformation:
  Bioconcentration, Uptake (k_1) and Elimination (k_2) Rate Constants: uptake and elimination rate constants of $(8.81 \pm 0.72) \times 10^3$ mL/g/d and (0.60 ± 0.02) d^{-1} (guppy, De Bruijn & Hermens 1991); excretion rate constant of 0.07 h^{-1} from whole body willow shiner (Tsuda et al. 1992); elimination rate constants of $(0.64 \pm 0.09) \times 10^3$ (NADPH) $min^{-1} \cdot mg$ $protein^{-1}$ and $(0.12 \pm 0.02) \times 10^3$ (GSH) $min^{-1} \cdot mg$ $protein^{-1}$ (rainbow trout, De Bruijn et al. 1993); excretion rate constant of 0.34 h^{-1} from carp with half-life of 2.0 d (Tsuda et al. 1993).

Common Name: Fenvalerate
Synonym: Belmark, Ectrin, Pydrin, Pyrethroid, S 5602, Sanmarton, SD 43775, Sumicide, Sumicidin, Sumifly, Sumipower, WL 43775
Chemical Name: (RS)-α-cyano-3-phenoxybenzyl(RS)-2-(4-chlorophenyl)-3-methylbutyrate;cyano-(3-phenoxyphenyl)methyl 4-chloro-α-(-1-methylethyl)benzeneacetate
Uses: non-systemic insecticide to control a wide variety of pests and also used as acaricide.
CAS Registry No: 51630-58-1
Molecular Formula: $C_{25}H_{22}ClNO_3$
Molecular Weight: 419.9
Melting Point (°C): liquid
Boiling Point (°C):
Density (g/cm³ at 20°C):
    1.26   (22°C, Spencer 1982)
    1.17   (23°C, Agrochemicals Handbook 1987)
    1.175   (tech. grade at 25°C, Worthing 1991; Montgomery 1993; Tomlin 1994)
    1.17   (23°C, Milne 1995)
Molar Volume (cm³/mol):
    479.6   (calculated-LeBas method, this work)
Molecular Volume (Å³):
Total Surface Area, TSA (Å²):
Dissociation Constant pK_a:
Heat of Fusion, ΔH_{fus}, kcal/mol:
Entropy of Fusion, ΔS_{fus}, cal/mol·K (e.u.):
Fugacity Ratio at 25°C (assuming ΔS_{fus} = 13.5 e.u.), F: 1.0

Water Solubility (g/m³ or mg/L at 25°C):
    0.085   (shake flask-GC, Coats & O'Donnell-Jafferey 1979; quoted, Shiu et al. 1990)
    0.085   (Verschueren 1983; quoted, Pait et al. 1992)
    0.024   (in seawater, quoted, Schimmel et al. 1983; Zaroogian et al. 1985; Clark et al. 1989)
    <1.0   (20°C, Worthing 1979, 1987; Spencer 1982; quoted, Cohen & Steinmetz 1986; Shiu et al. 1990)
    <1.0   (20°C, Agrochemicals Handbook 1987)
    <0.02   (Davies & Lee 1987; quoted, Kawamoto & Urano 1989)
    <1.0   (selected, USDA 1989; quoted, Neary et al. 1993)
    0.002   (Wauchope 1989; quoted, Shiu et al. 1990; Lohninger 1994)
    <1.0   (tech. grade at 20°C, Worthing 1991)
    0.002   (20-25°C, selected, Wauchope et al. 1992; quoted, Majewski & Capel 1995; Hornsby et al. 1996)
    <1.0   (20°C, Montgomery 1993)
    <1.0   (20°C, Milne 1995)

Vapor Pressure (Pa at 25°C):
    4.90×10^{-7} (Barlow 1978; quoted, Hinckley et al. 1990)
    3.07×10^{-5} (Worthing 1979; quoted, Cohen & Steinmetz 1986)
    1.33×10^{-5} (22°C, Spencer 1982)
    3.70×10^{-5} (Agrochemicals Handbook 1987)
    1.47×10^{-6} (Budavari 1989)

    3.73×10⁻⁵ (Kawamoto & Urano 1989)
    8.10×10⁻⁷ (GC-RT, supercooled value, Hinckley et al. 1990)
    3.70×10⁻⁵ (tech. grade, Worthing 1991)
    1.47×10⁻⁶ (20-25°C, selected, Wauchope et al. 1992; quoted, Majewski & Capel 1995; Hornsby et al. 1996)
    1.92×10⁻⁷ (20°C, Tomlin 1994)

Henry's Law Constant (Pa·m³/mol):
    0.0152 (20-25°C, calculated-P/C, Montgomery 1993)
    0.308 (20-25°C, calculated-P/C as per Wauchope et al. 1992, Majewski & Capel 1995)
    0.0211 (calculated-P/C, this work)

Octanol/Water Partition Coefficient, log K_{OW}:
    4.42 (shake flask-GC, Coats & O'Donnell-Jafferey 1979; quoted, Shiu et al. 1990)
    6.20 (quoted, Schimmel et al. 1983; Zaroogian et al. 1985; Clark et al. 1989)
    5.2 ± 0.6 (correlated-HPLC-RT, Muir et al. 1985)
    4.08 (23°C, Milne 1995)
    6.20 (selected, Hawker & Connell 1986; Travis & Arms 1988)
    6.25 (quoted, Kawamoto & Urano 1989)
    4.09 (23°C, Worthing 1991)
    4.09-6.25 (Montgomery 1993)
    5.01 (23°C, Tomlin 1994)
    6.20 (selected, Hansch et al. 1995)
    6.20 (selected, Devillers et al. 1996)

Bioconcentration Factor, log BCF:
    3.67 (quoted, Schimmel et al. 1983)
    1.67-1.84 (sand, 24 h BCF for chironomid larvae in water, Muir et al. 1985)
    2.01-2.24 (sand, 24 h BCF for chironomid larvae in sediment, Muir et al. 1985)
    1.30-1.53 (sand, 24 h BCF for chironomid larvae in sediment/pore water, Muir et al. 1985)
    1.62-1.87 (silt, 24 h BCF for chironomid larvae in water, Muir et al. 1985)
    1.36-2.06 (silt, 24 h BCF for chironomid larvae in sediment, Muir et al. 1985)
    1.26-1.97 (silt, 24 h BCF for chironomid larvae in sediment/pore water, Muir et al. 1985)
    1.36-1.51 (clay, 24 h BCF for chironomid larvae in water, Muir et al. 1985)
    2.09-2.19 (clay, 24 h BCF for chironomid larvae in sediment, Muir et al. 1985)
    0.95-1.70 (clay, 24 h BCF for chironomid larvae in sediment/pore water, Muir et al. 1985)
    4.48, 4.57 (oyster, calculated-K_{OW} & models, Zaroogian et al. 1985
    4.48, 4.57 (sheepshead minnow, calculated-K_{OW} & models, Zaroogian et al. 1985)
    3.67 (oyster, Zaroogian et al. 1985; quoted, Hawker & Connell 1986)
    −3.09 (milk biotransfer factor log B_m, correlated-K_{OW}, Wszolek et al. 1980; quoted, Travis & Arms 1988)
    2.61, 2.96 (*Oncorhynchus mykiss*, Muir et al. 1994; quoted, Devillers et al. 1996)
    2.70 (calculated, Pait et al. 1992)

Sorption Partition Coefficient, log K_{OC}:
- 2.58 (silt, reported as K_P on 78% DOC, Muir et al. 1985)
- 2.61 (clay, reported as K_P on 61% DOC, Muir et al. 1985)
- 1.30 (selected, USDA 1989; quoted, Neary et al. 1993)
- 3.72 (soil, 20-25°C, selected, Wauchope et al. 1992; Dowd et al. 1993; Hornsby et al. 1996)
- 3.64 (calculated, Montgomery 1993)
- 3.72 (estimated-chemical structure, Lohninger 1994)

Half-Lives in the Environment:
Air:
Surface water: half-lives in 100 mL of a pesticide-seawater solution: 14 days under outdoor light, >14 d under outdoor dark condition and >28 d under indoor condition (Schimmel et al. 1983); 27-42 d in an estuary (Schimmel et al. 1983; quoted, Montgomery 1993).
Groundwater:
Sediment: half-lives in 10 grams of sediment/100 mL of a pesticide-seawater solution: 34 d in untreated condition and >28 d in sterile condition (Schimmel et al. 1983).
Soil: selected field half-life of 35 d (Wauchope et al. 1992; Dowd et al. 1993; Hornsby et al. 1996); soil half-life of 50 d (Pait et al. 1992).
Biota: average half-life of 35 d in the forest (USDA 1989; quoted, Neary et al. 1993).

Environmental Fate Rate Constants or Half-Lives:
Volatilization:
Photolysis:
Oxidation:
Hydrolysis:
Biodegradation: rate constant of 0.007 d^{-1} with half-life of 99 d (Kawamoto & Urano 1990).
Biotransformation:
Bioconcentration, Uptake (k_1) and Elimination (k_2) Rate Constants:

Common Name: Flucythrinate
Synonym: AC 222705, Cybolt, Cythrin, Pay-Off
Chemical Name: (RS)-α-cyano-3-phenoxybenzyl(S)-2-(4-difluoromethoxyphenyl)-3-methylbutyrate;
cyano(3-phenoxyphenyl)methyl 4-(difluoromethoxy)-α-(1-methylethyl)benzeneacetate
Uses: non-systemic insecticide with contact and stomach action to control a wide range of insect pests in cotton, fruit trees, strawberries, vines, fruits, olives, coffee, cocoa, hops, vegetables, soybeans, cereals, maize, alfalfa, sugar beet, sunflowers and ornamentals.
CAS Registry No: 70124-77-5
Molecular Formula: $C_{26}H_{23}F_2NO_4$
Molecular Weight: 451.48
Melting Point (°C):
    <25   (dark amber liquid, Montgomery 1993)
Boiling Point (°C):
    108.0   (at 0.35 mmHg, Agrochemicals Handbook 1987; Worthing 1991; Montgomery 1993; Tomlin 1994; Milne 1995)
Density (g/cm³ at 20°C):
    1.189   (22°C, Agrochemicals Handbook 1987; Montgomery 1993; Tomlin 1994; Milne 1995)
    1.190   (22°C, Worthing 1991)
Molar Volume (cm³/mol):
    499.9   (calculated-LeBas method, this work)
    379.4   (22°C, calculated-density, this work)
Molecular Volume (Å³):
Total Surface Area, TSA (Å²):
Dissociation Constant pK_a:
Heat of Fusion, ΔH_{fus}, kcal/mol:
Entropy of Fusion, ΔS_{fus}, cal/mol·K (e.u.):
Fugacity Ratio at 25°C (assuming ΔS_{fus} = 13.5 e.u.), F: 1.0

Water Solubility (g/m³ or mg/L at 25°C):
    0.049   (in seawater, Schimmel et al. 1983; quoted, Clark et al. 1989; Shiu et al. 1990)
    0.50   (21°C, Agrochemicals Handbook 1987; Worthing 1987,91; quoted, Shiu et al. 1990; Tomlin 1994; Milne 1995)
    0.06   (20-25°C, selected, Wauchope 1989; quoted, Shiu et al. 1990; Hornsby et al. 1996)
    0.50   (21°C, Montgomery 1993)

Vapor Pressure (Pa at 25°C):
    1.2×10^{-6} (Agrochemicals Handbook 1987; Worthing 1991; Tomlin 1994)
    9.066   (Montgomery 1993)
    1.2×10^{-6} (20-25°C, selected, Hornsby et al. 1996)

Henry's Law Constant (Pa·m³/mol):
    8187   (21-25°C, calculated-P/C, 8.08×10^{-2} atm·m³/mol, Montgomery 1993)
    0.0011   (calculated-P/C, this work)

Octanol/Water Partition Coefficient, log K_{ow}:
    6.28   (shake flask-GC, Schimmel et al. 1983; quoted, Sangster 1993)
    6.20   (quoted, Clark et al. 1989)
    2.08   (Worthing 1991; Tomlin 1994; Milne 1995)
    5.55   (shake flask, Huang & Leng 1993; quoted, Sangster 1993)
    4.70   (Montgomery 1993)
    6.20   (recommended, Sangster 1993)
    6.20   (selected, Hansch et al. 1995)

Bioconcentration Factor, log BCF:
    2.96   (calculated-S as per Kenaga 1980, this work)

Sorption Partition Coefficient, log K_{oc}:
    3.81   (calculated, Montgomery 1993)
    5.00   (20-25°C, selected, Hornsby et al. 1996)

Half-Lives in the Environment:
    Air:
    Surface water: half-life pf 34 d in an estuarine environment (Schimmel et al. 1983; quoted, Montgomery 1993).
    Groundwater:
    Sediment:
    Soil: half-life of ca. 2 months in soil (Tomlin 1994); field half-life of 21 d (20-25°C, selected, Hornsby et al. 1996).
    Biota:

Environmental Fate Rate Constants or Half-Lives:
    Volatilization:
    Photolysis: half-life of ca. 21 d for degradation on soil plates by simulated sunlight and 4.0 days in aqueous solutions (Tomlin 1994).
    Oxidation:
    Hydrolysis: half-lives of 40, 52, and 6.3 d at pH 3, 5, 9 all at 27°C (Hartley & Kidd 1987; quoted, Montgomery 1993; Tomlin 1994).
    Biodegradation:
    Biotransformation:
    Bioconcentration, Uptake (k_1) and Elimination (k_2) Rate Constants:

Common Name: Fonofos
Synonym: Difonate, Dyfonate, ENT-25796, Fonophos, N 2788, N-2790, Stauffer NA 2790
Chemical Name: O-ethyl S-phenyl (RS)-ethylphosphorodithioate; (±)-O-ethyl S-phenyl ethylphosphorodithioate
Uses: soil insecticide to control rootworms, wireworms, crickets and similar crop pests in vegetables, sorghum, ornamentals, cereals, maize, vines, olives, sugar beet, sugar cane, potatoes, groundnuts, tobacco, turf, and fruit crops.
CAS Registry No: 944-22-9 (unstated stereochemistry); 66767-39-3 (racemate); 62705-71-9 (R)-isomer; 62680-03-9 (S)-isomer
Molecular Formula: $C_{10}H_{15}OPS_2$
Molecular Weight: 246.3
Melting Point (°C): liquid
Boiling Point (°C):
    130   (at 0.1 mmHg, Agrochemicals Handbook 1987; Worthing 1991; Montgomery 1993; Tomlin 1994; Milne 1995)
Density (g/cm³ at 20°C):
    1.160   (25°C, Agrochemicals Handbook 1987; Tomlin 1994; Milne 1995)
    1.154   (Worthing 1991; Montgomery 1993)
Molar Volume (cm³/mol):
    213.4   (calculated-density, this work)
Molecular Volume (Å³):
Total Surface Area, TSA (Å²):
Dissociation Constant pK_a:
Heat of Fusion, ΔH_{fus}, kcal/mol:
Entropy of Fusion, ΔS_{fus}, cal/mol·K (e.u.):
Fugacity Ratio at 25°C (assuming ΔS_{fus} = 13.5 e.u.), F: 1.0

Water Solubility (g/m³ or mg/L at 25°C):
    13   (22°C, Spencer 1973; quoted, Bowman & Sans 1979; Shiu et al. 1990)
    13   (Wauchope 1978)
    15.7   (20°C, shake flask-GC, Bowman & Sans 1979, 1983b; quoted, Fuhrenmann & Lichtenstein 1980; Shiu et al. 1990; Patil 1994)
    13   (Agrochemicals Handbook 1987; Milne 1995)
    13   (20°C, Worthing 1987; quoted, Shiu et al. 1990; Schomburg et al. 1991; Majewski & Capel 1995)
    13   (Worthing 1991)
    16.9   (20-25°C, selected, Wauchope et al. 1992; Lohninger 1994; Hornsby et al. 1996)
    13   (rm. temp., Montgomery 1993)
    15.7, 8.35   (quoted, calculated, Patil 1994)
    13   (22°C, Tomlin 1994)

Vapor Pressure (Pa at 25°C):
    0.0267   (Menn 1969; quoted, Fuhrenmann & Lichtenstein 1980)
    0.028   (Khan 1980; Agrochemicals Handbook 1987)
    0.028   (Worthing 1987, 1991; quoted, Majewski & Capel 1995)
    0.0453   (20-25°C, selected, Wauchope et al. 1992; Hornsby et al. 1996)
    0.028   (Montgomery 1993; Tomlin 1994)

Henry's Law Constant (Pa·m^3/mol):
    0.5206 (calculated-P/C as per Worthing 1987, Schomburg et al. 1991)
    0.5268 (20-25°C, calculated-P/C, Montgomery 1993)
    0.530   (calculated-P/C as per Worthing 1987, Majewski & Capel 1995)
    0.693   (calculated-P/C, this work)

Octanol/Water Partition Coefficient, log K_{OW}:
    3.892 (shake flask-GC, Bowman & Sans 1983b; quoted, De Bruijn & Hermens 1991; Patil 1994)
    3.90   (20°C, Worthing 1991)
    3.89, 3.90 (Montgomery 1993)
    3.89, 3.79 (quoted, calculated, Patil 1994)
    3.94   (Tomlin 1994)
    3.90   (Milne 1995)
    3.94   (selected, Hansch et al. 1995)

Bioconcentration Factor, log BCF:
    2.11   (calculated-S as per Kenaga 1980, this work)
    2.79   (calculated-K_{OW} as per Kenaga 1980, this work)
    1.89   (mosquito fish, wet wt. basis, De Bruijn & Hermens 1991)

Sorption Partition Coefficient, log K_{OC}:
    2.3-2.7 (selected, sediment/water, Schnoor & McAvoy 1981; quoted, Schnoor 1992)
    1.83   (screening model calculations, Jury et al. 1987b)
    1.18   (loam soil, Worthing 1991)
    2.94   (soil, 20-25°C, selected, Wauchope et al. 1992; quoted, Richards & Baker 1993)
    3.03   (calculated, Montgomery 1993)
    2.94   (soil, 20-25°C, selected, Hornsby et al. 1996)
    2.94   (estimated-chemical structure, Lohninger 1994)

Half-Lives in the Environment:
    Air:
    Surface water:
    Groundwater:
    Sediment:
    Soil: persistence of less than one month in soil (Wauchope 1978); >24 weeks in sterile sandy loam and 3.0 weeks in nonsterile sandy loam; >24 weeks in sterile organic soil and 4.0 weeks in nonsterile organic soil (Miles et al. 1979); half-life of 60 d from screening model calculations (Jury 1987b); 16.5-28 d at 24°C (Worthing 1991); selected field half-life of 40 d (Wauchope et al. 1992; quoted, Richards & Baker 1993; Hornsby et al. 1996).
    Biota: biochemical half-life of 60 d from screening model calculations (Jury et al. 1987b).

Environmental Fate Rate Constants or Half-Lives:
    Volatilization:
    Photolysis: half-life of 12 d in water at pH 5 and 25°C (Worthing 1991; Tomlin 1994).
    Oxidation:
    Hydrolysis: alkaline chemical hydrolysis rate constant of 1×10^{-4} M^{-1} sec^{-1} with half-life of >365 d (selected, sediment/water, Schnoor & McAvoy 1981; quoted, Schnoor 1992); half-lives hydrolysis in water at 40°C of 74.0-127 d at pH 7 and 101 days at pH 4 (Worthing 1991; quoted, Montgomery 1993; Tomlin 1994).
    Biodegradation:
    Biotransformation:
    Bioconcentration, Uptake (k_1) and Elimination (k_2) Rate Constants:

Common Name: α-HCH
Synonym: α-BHC, α-Hexachlorocyclohexane
Chemical Name: α-1,2,3,4,5,6-hexachlorocyclohexane
CAS Registry No: 319-84-6
Molecular Formula: $C_6H_6Cl_6$
Molecular Weight: 290.85
Melting Point (°C):
    157.5-158 (Slade 1945)
    158 (Ballschmiter & Wittlinger 1991; Fischer et al. 1991)
Boiling Point (°C): 288
Density (g/cm^3 at 20 °C):
Molar Volume (cm^3/mol):
    243.6 (calculated-LeBas method, Suntio et al. 1988)
    179.5 (Ruelle & Kesselring 1997)
Enthalpy of fusion (kcal/mol), ΔH_{fus}:
    7.40 (Ruelle & Kesselring 1997)
Entropy of fusion, ΔS_{fus}:
Fugacity Ratio at 25°C, (assuming ΔS = 13.5 e.u.), F: 0.030

Water Solubility (g/m^3 or mg/L at 25°C):
    10 (20°C, Slade 1945, quoted, Günther et al. 1968; Horvath 1982)
    1.63 (shake flask-GC, Kanazawa et al. 1971; quoted, Horvath 1982)
    10 (Ulman 1972; quoted, Malaiyandi et al. 1982)
    2.03, 1.21 (28°C, shake flask-centrifuge, membrance filter-GC, max. 0.1μm particle size, Kurihara et al. 1973)
    1.77, 1.48 (28°C, shake flask-centrifuge, sonic and centrifuge-GC, max. 0.05μm particle size, Kurihara et al. 1973)
    1.21-2.03 (28°C, Kurihara et al. 1973; quoted, Callahan et al. 1979)
    1.63 (Brooks 1974; quoted, Callahan et al. 1979)
    2.0 (gen. col.-GC/ECD, Weil et al. 1974, quoted, Callahan et al. 1979; Geyer et al. 1982, Kucklick et al. 1991)
    4.34 (shake flask-GC/ECD, Malaiyandi et al. 1982)
    1.51 (20°C, Deutsche Forschungsgemeinschaft 1983; quoted, Ballschmiter & Wittlinger 1991; Fischer et al. 1991)
    2.0 (quoted, Isnard & Lambert 1988, 1989)
    1.0 (selected, Suntio et al. 1988; quoted, Schreitmüller & Ballschmiter 1995)
    2.01 (quoted, Calamari et al. 1991)
    21.6 (quoted, supercooled liquid value, Majewski & Capel 1995)
    10.1; 0.666, 0.023 (quoted; predicted-molar volume, Ruelle & Kesselring 1997)

Vapor Pressure (Pa at 25°C):
    2.67, 8.0, 44 (20, 40, 60°C, static method, Slade 1945)
    0.00172 (15°C, effusion menometer, Balson 1947, quoted, Kucklick et al. 1991)
    0.00333 (20°C, effusion menometer, Balson 1947, quoted, Kucklick et al. 1991)
    0.00631 (effusion menometer, Balson 1947, quoted, Kucklick et al. 1991)
    0.27 (supercooled liquid value, Balson 1947; quoted, Hinckley et al. 1990)
    0.0830 (Mackay & Shiu 1981; quoted, Fischer et al. 1991)

0.0073 (20°C, Deutsche Forschungsgemeinschaft 1983; quoted, Ballschmiter & Wittlinger 1991; Fischer et al. 1991; Schreitmüller & Ballschmiter 1995)

0.0840 (20°C, supercooled liquid value, Bidleman et al. 1986)

0.2270 (GC-RT, supercooled liquid value, Hinckley et al. 1990)

$\log P_L$ (Pa) = $A_L + B_L/T$, A_L = 10.49, B_L = -3301 (liquid, Hinckley et al. 1990)

$\log P_L$ (Pa) = $A_L + B_L/T$, A_L = 11.34, B_L = -3375 (liquid, Hinckley et al. 1990)

0.003 selected, Suntio et al. 1988, quoted, Calamari et al. 1991; Schreitmüller & Ballschmiter 1995)

0.0060 (quoted, Howard 1991)

0.00647 (quoted, supercooled liquid value, Majewski & Capel 1995)

Henry's Law Constant (Pa·m^3/mol):

0.47-0.792 (Callahan et al. 1979)

2.16 (gas stripping, Atlas et al. 1982)

0.55 (calculated-P/C, Mabey et al. 1982)

0.87 (calculated-P/C, Suntio et al. 1988; quoted, Paterson et al. 1991; Müller et al. 1994; Majewski & Capel 1995)

1.10 (calculated-P/C, Ballschmiter & Wittlinger 1991; Fischer et al. 1991)

1.07 (calculated-P/C, Howard 1991)

0.43 (calculated-P/C, Calamari et al. 1991)

0.215 at 8.5°C in Green Bay, 0.491 at 18.9°C in Lake Michigan, 0.473 at 18.5°C in Lake Huron, 0.630 at 22.3°C in Lake Erie and 0.630 at 22.3°C in Lake Ontario (concn. ratio-GC, McConnell et al. 1993)

$\log H$ (Pa·m^3 mol^{-1}) = $-2810/T + 9.31$ (temp. range 0.5-45°C, Kucklick et al. 1991, McConnell et al. 1993)

$\log H$ (Pa·m^3 mol^{-1}) = $-2969/T + 9.88$ (temp. range 0.5-45°C, artificial seawater, Kucklick et al. 1991)

0.87 (20°C), 2.40, 1.10, 0.677, 0.710 (23°C) (quoted, Iwata et al. 1993)

0.872 (calculated-P/C, this work)

Octanol/Water Partition Coefficient, $\log K_{OW}$:

3.81 (shake flask-GC, Kurihara et al. 1973, quoted, Callahan et al. 1979; Geyer et al. 1982; Noegrohati & Hammers 1992)

3.90 (calculated-π constant, Kurihara et al. 1973)

3.81 (HPLC-RT, Sugiura et al. 1979)

3.90 (Veith et al. 1979; quoted, Oliver & Charlton 1984)

3.98 (Hawker & Connell 1986)

3.80 (selected, Suntio et al. 1988; quoted, Thomann 1989; Howard 1991, Calamari et al. 1991, Paterson et al. 1991; Schreitmüller & Ballschmiter 1995)

3.89 (Isnard & Lambert 1988,89; quoted, Ballschmiter & Wittlinger 1991; Fischer et al. 1991)

3.81 (av. lit. value, Gerstl 1991)

3.85 (quoted, McConnell et al. 1993)

4.44 (Vigano et al. 1992)

3.80 (selected, Hansch et al. 1995)

Bioconcentration Factor, log BCF:
- 2.03 (mussels, Ernst 1977; quoted, Renberg & Sundström 1979; Hawker & Connell 1986)
- 3.08, 2.52, 2.78, 2.77 (golden orfe, carp, brown trout, guppy, Suguira et al. 1979)
- 2.20, 2.82 (mussels, Geyer et al. 1982)
- 3.20, 3.38 (rainbow trout, Oliver & Niimi 1985; quoted, Thomann 1989)
- 1.93 (paddy field fish, Soon & Hock 1987; quoted, Abdullah et al. 1997)
- 2.15 (calculated, Isnard & Lambert 1988)
- 6.01 (azalea leaves, Bacci et al. 1990; quoted, Peterson et al. 1991; Müller et al. 1994)
- 2.33 (early juvenile of rainbow trout, Vigano et al. 1992)
- 5.72 (azalea leaves, calculated, Müller et al. 1994)

Sorption Partition Coefficient, log K_{oc}:
- 3.81 (calculated-S, Lyman et al. 1982)
- 4.10, 3.5 (field sediment trap material, calculated-K_{ow}, Oliver & Charlton 1984)
- 3.25 (av. lit. value, Gerstl 1991)
- 3.32 (derived from exptl., Meylan et al. 1992)
- 3.53 (calculated-χ, Meylan et al. 1992)

Half-Lives in the Environment:
Air: atmospheric half-life was estimated to be 2.3 d based on reaction with OH radicals at 25°C (Atkinson 1987).
Surface water:
Groundwater:
Sediment:
Soil:
Biota: half-life of 19.2 h (mussels, Ernst 1977).

Environmental Fate Rate Constants or Half-Lives:
Volatilization: volatilization half-life from a model river of 1 m deep flowing 1 m/s with a wind speed of 3 m/s) was estimated to be 6 d (Saleh et al. 1982); from a model pond was estimated to be 500 days (Howard 1991).
Photolysis:
Hydrolysis: hydrolytic half-life of 26 yr at pH 8 and 5°C (Ngabe et al. 1993).
Oxidation: photooxidation half-life of 2.3 d for reaction with OH radicals in the gas phase (Atkinson 1987).
Biodegradation: half-lives were calculated from experiments S1-S3 of 35 ± 0.5 h for (+) enantiomer and 99 ± 3.5 h for (-) enantiomer in sewage sludge (Buser & Müller 1995).
Biotransformation:
Bioconcentration, Uptake (k_1) and Elimination (k_2) Rate Constants:
- k_1: 3.82 h^{-1} (mussels, Ernst 1977; quoted, Hawker & Connell 1986)
- k_2: 0.036 h^{-1} (mussels, Ernst 1977; quoted, Hawker & Connell 1986)
- k_1: 0.52 d^{-1}, 0.56 d^{-1}, 0.91 d^{-1}, and 0.42 d^{-1} (golden orfe, carp, brown trout, and guppy at steady state, Sugiura et al. 1979)
- k_2: 0.0009 h^{-1} (azalea leaves, Peterson et al. 1991)
- k_1: 27.6 h^{-1} (early juvenile of rainbow trout, Vigano et al. 1992)
- k_2: 0.13 h^{-1} (early juvenile of rainbow trout, Vigano et al. 1992)

Common Name: β-HCH
Synonym: β-BHC, β-Hexachlorocyclohexane
Chemical Name: β-1,2,3,4,5,6-hexachlorocyclohexane
CAS Registry No: 319-85-7
Molecular Formula: $C_6H_6Cl_6$
Molecular Weight: 290.85
Melting Point (°C):
    309   (Slade 1945; Ballschmiter & Wittlinger 1991)
Boiling Point (°C):
Density (g/cm³ at 20 °C):
Molar Volume (cm³/mol):
    243.6   (calculated-LeBas method, Suntio et al. 1988)
    179.5   (Ruelle & Kesselring 1997)
Enthalpy of fusion, ΔH_{fus}:
Entropy of fusion, ΔS_{fus}:
Fugacity Ratio at 25°C, (assuming ΔS = 13.5 e.u.), F: 0.00155

Water Solubility (g/m³ or mg/L at 25°C):
    5.0   (20°C, Slade 1945; quoted, Gunther et al. 1968; Horvath 1982)
    0.70   (20°C, shake flask-GC, Kanazawa et al. 1971; quoted, Horvath 1982)
    0.20, 0.13   (28°C, shake flask-centrifuge, membrance filter-GC, max. 0.1 μm particle size, Kurihara et al. 1973)
    0.70   (20°C, Brooks 1974)
    0.24   (gen. col.-GC/ECD, Weil et al. 1974)
    0.13-0.70   (Callahan et al. 1979)
    2.04   (20°C, Deutsche Forschungsgemeinschaft 1983; quoted, Ballschmiter & Wittlinger 1991; Fischer et al. 1991)
    7.0   (Worthing 1983)
    0.10   (selected, Suntio et al. 1988)
    1.60   (Isnard & Lambert 1988, 1989)
    69.5   (quoted, supercooled liquid value, Majewski & Capel 1995)
    0.10; 0.042, 0.0015   (quoted; predicted-molar volume, Ruelle & Kesselring 1997)

Vapor Pressure (Pa at 25°C):
    0.67, 22.7, 77.3   (20, 40, 60°C, static method, Slade 1945)
    3.73×10^{-5} (20°C, effusion menometer, Balson 1947, quoted, Kucklick et al. 1991)
    4.90×10^{-5} (20°C, Deutsche Forschungsgemeinschaft 1983; quoted, Ballschmiter & Wittlinger 1991; Fischer et al. 1991)
    4.00×10^{-5} (selected, Suntio et al. 1988, quoted, Calamari et al. 1991)
    0.02720   (quoted, supercooled liquid value, Majewski & Capel 1995)

Henry's Law Constant (Pa·m³/mol):
    0.055   (20-25 °C, Mabey et al. 1982)
    0.120   (calculated-P/C, Suntio et al. 1988)
    0.070   (calculated-P/C, Ballschmiter & Wittlinger 1991; Fischer et al. 1991)
    0.0727   (quoted, Majewski & Capel 1995)
    0.116   (calculated-P/C, this work)

Octanol/Water Partition Coefficient, log K_{OW}:
- 3.80 (shake flask-GC, Kurihara et al. 1973; quoted, Suntio et al. 1988; Noegrohati & Hammers 1992)
- 3.80 (calculated-π constant, Kurihara et al. 1973)
- 4.15 (HPLC-RT, Sugiura et al. 1979)
- 3.96 (Isnard & Lambert 1988,89; quoted, Ballschmiter & Wittlinger 1991)
- 4.00 (selected, Fischer et al. 1991)
- 3.80 (av. lit. value, Gerstl 1991)
- 3.78 (selected, Hansch et al. 1995)

Bioconcentration Factor, log BCF:
- 3.08, 2.26, 2.62 (activated sludge, algae, golden ide, reported as log BF, Freitag et al. 1985)
- 2.99, 2.44, 2.82, 3.17 (golden orfe, carp, brown trout, guppy, Sugiura et al. 1979)
- 2.66 (calculated, Isnard & Lambert 1988)
- 3.14 (calculated-S as per Kenaga 1980, this work)

Sorption Partition Coefficient, log K_{OC}:
- 3.36 (av. lit. value, Gerstl 1991)
- 3.98 (soil, calculated-S as per Kenaga 1980, this work)
- 3.50 (derived from exptl., Meylan et al. 1992)
- 3.53 (calculated-χ, Meylan et al. 1992)

Half-Lives in the Environment:
 Air:
 Surface water:
 Groundwater:
 Sediment:
 Soil:
 Biota:

Environmental Fate Rate Constants or Half-Lives:
 Volatilization:
 Photolysis:
 Hydrolysis:
 Oxidation:
 Biodegradation: calculated half-life of 178 h in sewage sludge from experiments S1-S3 (Buser & Müller 1995).
 Biotransformation:
 Bioconcentration, Uptake (k_1) and Elimination (k_2) Rate Constants:
  k_1: 0.46 d^{-1}, 0.33 d^{-1}, 0.53 d^{-1}, and 0.18 d^{-1} (golden orfe, carp, brown trout, and guppy at steady state, Sugiura et al. 1979)

Common Name: δ-HCH
Synonym: δ-BHC, δ-Hexachlorocyclohexane
Chemical Name: δ-1,2,3,4,5,6-hexachlorocyclohexane
CAS Registry No: 319-86-8
Molecular Formula: $C_6H_6Cl_6$
Molecular Weight: 290.85
Melting Point (°C):
    138-139 (Slade 1945)
    138.5   Ballschmiter & Wittlinger 1991)
    139.5   (Fischer et al. 1991)
Boiling Point (°C):
Density (g/cm³ at 20 °C):
Molar Volume (cm³/mol):
    243.6   (calculated-LeBas method, Suntio et al. 1988)
    179.5   (Ruelle & Kesselring 1997)
Heat of fusion (kcal/mol), ΔH_{fus}:
    5.10   (DSC method, Plato 1972)
    4.90   (Ruelle & Kesselring 1997)
Entropy of fusion, ΔS_{fus}:
Fugacity Ratio at 25°C (assuming ΔS = 13.5 e.u.), F: 0.0746

Water Solubility (g/m³ or mg/L at 25°C):
    10   (20°C, Slade 1945; quoted, Horvath 1982)
    15.7, 10.7   (28°C, shake flask-centrifuge, membrane filter-GC, max. 0.1 μm particle size, Kurihara et al. 1973)
    11.6, 8.64   (28°C, shake flask-centrifuge, sonic and centrifuge-GC, max. 0.05 μm particle size, Kurihara et al. 1973)
    8.64-31.4   (shake flask-GC, Kurihara et al. 1973; quoted, Callahan et al. 1979)
    10   (20°C, quoted, Günther et al. 1968)
    21.3   (20°C, shake flask-GC, Kanazawa et al. 1971; quoted, Horvath 1982)
    21.3   (20°C, Brooks 1974)
    31.4   (gen. col.-GC/ECD, Weil et al. 1974)
    9.01   (20°C, Deutsche Forschungsgemeinschaft 1983; quoted, Ballschmiter & Wittlinger 1991; Fischer et al. 1991)
    8.0   (selected, Suntio et al. 1988)
    109   (quoted, supercooled liquid value, Majewski & Capel 1995)
    8.0; 3.19, 0.113   (quoted; predicted-molar volume, Ruelle & Kesselring 1997)

Vapor Pressure (Pa at 25°C):
    2.67, 12, 45.33   (20, 40, 60°C, static method, Slade 1945)
    2.27×10^{-3} (20°C, effusion menometer, Balson 1947)
    2.00×10^{-3} (selected, Suntio et al. 1988)
    0.0309   (quoted, supercooled liquid value, Majewski & Capel 1995)

Henry's Law Constant (Pa·m³/mol):
    0.018   (20-25 °C, Mabey et al. 1982)
    0.073   (calculated-P/C, Suntio et al. 1988)
    0.0825   (calculated-P_L/C_L, Majewski & Capel 1995)

Octanol/Water Partition Coefficient, log K_{OW}:
- 4.14 (shake flask-GC, Kurihara et al. 1973, quoted, Callahan et al. 1979; Sangster 1993)
- 4.10 (calculated-π constant, Kurihara et al. 1973)
- 4.10 (selected, Suntio et al. 1988)
- 4.14 (recommended, Sangster 1993)
- 4.14 (selected, Hansch et al. 1995)

Bioconcentration Factor, log BCF:
- 1.95 (calculated-S as per Kenaga 1980, this work)

Sorption Partition Coefficient, log K_{OC}:
- 2.82 (soil, calculated-S as per Kenaga 1980, this work)

Half-Lives in the Environment:
Air:
Surface water:
Groundwater:
Sediment:
Soil:
Biota:

Environmental Fate Rate Constants or Half-Lives:
Volatilization:
Photolysis:
Hydrolysis:
Oxidation:
Biodegradation: calculated half-life of 126 h in sewage sludge from experiments S1-S3 (Buser & Müller 1995).
Biotransformation:
Bioconcentration, Uptake (k_1) and Elimination (k_2) Rate Constants:

Common Name: Heptachlor

Synonym: Aahepta, Aathepta, Agroceres, Basaklor, 3-Chlorochlordene, Drinox, ENT 15152, Hepta, Heptachlorane, Heptagran, Heptagranox, Heptamak, Heptamul, Heptasol, Heptox, methanoindene, NA 2761, NCI-C00180, Rhodiachlor, Soleptax, Velsicol

Chemical Name: 1,4,5,6,7,8,8-heptachloro-3a,4,7,7a-tetrahydro-4,7-methanoindene; 3-4,5,6,7,8,8a-heptachloro-dicyclopentadiene

Uses: non-systemic insecticide with contact, stomach, and some respiratory action to control termites, ants, and soil insects in cultivated and uncultivated soils; also used to control household insects.

CAS Registry No: 76-44-8

Molecular Formula: $C_{10}H_5Cl_7$

Molecular Weight: 373.3

Melting Point (°C):
- 95-96 (Martin 1972; Callahan et al. 1979; Khan 1980; Spencer 1982; Suntio et al. 1988; Worthing 1991; Montgomery 1993; Tomlin 1994; Milne 1995)
- 95.5 (Patil 1994; Kühne et al. 1995)

Boiling Point (°C):
- 135-145 (at 1-1.5 mmHg, Montgomery 1993; Tomlin 1994)

Density (g/cm³ at 20°C):
- 1.65-1.67 (25°C, Agrochemicals Handbook 1987; Tomlin 1994)
- 1.66 (Montgomery 1993)

Molar Volume (cm³/mol):
- 308.2 (calculated-LeBas method, Suntio et al. 1988)
- 238.5 (Ruelle & Kesselring 1997)

Molecular Volume (Å³):

Total Surface Area, TSA (Å²):

Dissociation Constant pK_a:

Heat of Fusion, ΔH_{fus}, kcal/mol:
- 5.49 (Ruelle & Kesselring 1997)

Entropy of Fusion, ΔS_{fus}, cal/mol·K (e.u.):

Fugacity Ratio at 25°C (assuming ΔS_{fus} = 13.5 e.u.), F: 0.1985

Water Solubility (g/m³ or mg/L at 25°C):
- 0.056 (25-29°C, shake flask-GC, Park & Bruce 1968; quoted, Callahan et al. 1979; Shiu et al. 1990)
- 0.10, 0.18, 0.315, 0.49 (shake flask-GC, at 15, 25, 35, 45°C, with particle size ≤5.0 µ, Biggar & Riggs 1974)
- 0.18 (particle size ≤5.0 µ, shake flask-GC, Biggar & Riggs 1974; quoted, Callahan et al. 1979; Suntio et al. 1988; Shiu et al. 1990; Howard 1991)
- 0.03, 0.125, 0.180 (particle size: 0.01, 0.05 & 5.0 µ, shake flask-GC, Biggar & Riggs 1974)
- 0.03 (Martin & Worthing 1977; quoted, Kenaga 1980a,b; Kenaga & Goring 1980; quoted, Bysshe 1982; Suntio et al. 1988; Shiu et al. 1990)
- <1.0 (Wauchope 1978)
- 0.0056 (Hartley & Graham-Bryce 1980; quoted, Taylor & Glotfelty 1988)
- 0.05 (Khan 1980)
- 0.30 (Herbicide Handbook 1983; quoted, Nash 1988)

0.056   (Garten & Trabalka 1983; Worthing 1987; Taylor & Glofelty 1988; quoted, Shiu et al. 1990)
0.03   (Schimmel et al. 1983; quoted, Zaroogian et al. 1985)
0.056   (U.S. EPA 1984; quoted, McLean et al. 1988)
0.056   (Agrochemicals Handbook 1987)
0.056   (quoted, Isnard & Lambert 1988, 1989)
0.10   (20°C, selected, Suntio et al. 1988; quoted, Majewski & Capel 1995)
0.056   (25-29°C, Worthing 1991; Tomlin 1994)
0.18   (Montgomery 1993)
0.056   (20-25°C, selected, Augustijn-Beckers et al. 1994; Hornsby et al. 1996)
0.056, 2.5×10^{-5} (quoted, calculated, Patil 1994)
0.18, 0.258 (quoted, calculated-group contribution fragmentation method, Kühne et al. 1995)
0.10; 0.049, 0.0017 (quoted; predicted-molar volume, Ruelle & Kesselring 1997)

Vapor Pressure (Pa at 25°C):
0.025   (Bowery 1964; quoted, Nash 1988)
0.040   (Eichler 1965; quoted, Suntio et al. 1988)
0.040   (Martin 1972; quoted, Callahan et al. 1979; Khan 1980)
0.021   (20°C, Hartley & Graham-Bryce 1980; quoted, Suntio et al. 1988; Taylor & Glotfelty 1988; Taylor & Spencer 1990)
0.053   (Worthing 1983, 1991; quoted, Suntio et al. 1988)
0.040   (U.S. EPA 1984; quoted, McLean et al. 1988)
0.053   (Agrochemicals Handbook 1987; Tomlin 1994)
0.0533 (20°C, selected, Suntio et al. 1988; quoted, Howard 1991; Majewski & Capel 1995)
0.031   (GC method, supercooled liq. value, Hinckley et al. 1990)
0.022   (20°C, selected, Taylor & Spencer 1990)
0.040   (20°C, Montgomery 1993)
0.0533 (20-25°C, selected, Augustijn-Beckers et al. 1994; Hornsby et al. 1996)

Henry's Law Constant (Pa·m³/mol):
150   (gas stripping-GC, Warner et al. 1987; quoted, Howard 1991)
154   (quoted from WERL Treatability Database, Ryan et al. 1988)
112   (20°C, calculated-P/C, Suntio et al. 1988; quoted, Majewski & Capel 1995)
845.4   (calculated-P/C, Jury et al. 1990)
150.2   (known LWAPC of Warner et al. 1987, Meylan & Howard 1991)
17.8   (bond-estimated LWAPC, Meylan & Howard 1991)
233   (Montgomery 1993)

Octanol/Water Partition Coefficient, log K_{OW}:
5.44   (Veith et al. 1979, 1980; Veith & Kosian 1983; quoted, Bysshe 1982; Zaroogian et al. 1985; Chapman 1989; Saito et al. 1992)
3.87   (quoted, Rao & Davidson 1980; Suntio et al. 1988)
3.90   (U.S. EPA 1984; quoted, McLean et al. 1988)
5.38   (selected, Hawker & Connell 1986; Thomann 1989)

5.73   (selected, Yoshioka et al. 1986)
3.90   (quoted, Ryan et al. 1988)
5.44   (quoted, Isnard & Lambert 1988, 1989)
5.27   (Schüürmann & Klein 1988; quoted, Howard 1991)
3.90   (selected, Suntio et al. 1988)
5.44   (selected, Travis & Arms 1988)
4.61   (Thor 1989; quoted, Connell & Markwell 1990)
5.27   (quoted Medchem value, Chessells et al. 1992)
5.34   (calculated-f const., Noegrohati & Hammers 1992)
5.53   (estimated-QSAR & SPARC, Kollig 1993)
4.40-5.50 (Montgomery 1993)
5.44, 8.25 (quoted, calculated, Patil 1994)
5.27, 5.58 (quoted, Hansch et al. 1995)
5.44   (selected, Devillers et al. 1996)
5.24, 4.92, 5.46 (RP-HPLC, ClogP, calculated-S, Finizio et al. 1997)

Bioconcentration Factor, log BCF:
  −1.81 (beef biotransfer factor log B_b, correlated-K_{ow}, Claborn et al. 1960; Kenaga 1980; quoted, Travis & Arms 1988)
  −1.48 (vegetation, correlated-K_{ow}, Lichtenstein 1960; Nash 1974; quoted, Travis & Arms 1988)
  −2.49 (milk biotransfer factor log B_m, correlated-K_{ow}, Saha 1969; quoted, Travis & Arms 1988)
  4.26   (oysters, wet wt. basis, Wilson 1963; quoted, Biddinger & Gloss 1984)
  3.26   (bluegill, field tests, Andrews et al. 1966; quoted, Biddinger & Gloss 1984)
  3.41   (soft clam, Butler 1971; quoted, Hawker & Connell 1986)
  3.45-4.33 (estuarine fish for 96-h exposure, Schimmel et al. 1976; quoted, Biddinger & Gloss 1984)
  3.76-3.92 (spot fish, whole body 24-d exposure, Schimmel et al. 1976; quoted, Howard 1991)
  3.67   (spot fish, edible tissue 24-d exposure, Schimmel et al. 1976; quoted, Howard 1991)
  3.58   (mosquito fish, Callahan et al. 1979; quoted, Howard 1991)
  3.56   (spot fish for 72-h test, Callahan et al. 1979; quoted, Howard 1991)
  3.87   (spot fish for 96-h test, Callahan et al. 1979; quoted, Howard 1991)
  4.57   (snails, Callahan et al. 1979; quoted, Howard 1991)
  4.32   (algae, Callahan et al. 1979; quoted, Howard 1991)
  3.98   (fathead minnows, 32-d exposure, Veith et al. 1979, 1980; quoted, Bysshe 1982; Veith & Kosian 1983; Howard 1991; Devillers et al. 1996)
  4.30   (fathead minnows, 276-d exposure, Veith et al. 1979; quoted, Veith & Kosian 1983; Zaroogian et al. 1985)
  4.30   (sheepshead minnows, Veith et al. 1979; quoted, Zaroogian et al. 1985)
  4.24, 3.33 (fish: flowing water, static water; quoted, Kenaga 1980; Kenaga & Goring 1980)
  3.65   (calculated-S, Kenaga 1980)
  4.30   (fathead minnows, 276-d exposure, Veith et al. 1980; quoted, Bysshe 1982; Zaroogian et al. 1985)
  3.65, 3.90 (estimated-S, K_{ow}, Bysshe 1982)

3.11-3.56 (earthworms, Gish & Hughes 1982; quoted, Connell & Markwell 1990)
3.98, 4.30 (fathead minnows, Veith & Kosian 1983; quoted, Saito et al. 1992)
4.03 (clam fat, 60-d expt., Hartley & Johnson 1983)
4.26 (oysters, Biddinger & Gloss 1984; quoted, Howard 1991)
3.90, 3.92 (oyster, calculated values, Zaroogian et al. 1985)
3.90, 3.92 (pinfish, calculated values, Zaroogian et al. 1985)
3.90, 3.92 (sheepshead minnow, calculated values, Zaroogian et al. 1985)
4.30, 4.33 (measured for fathead minnows, sheepshead minnows, Zaroogian et al. 1985; quoted, Howard 1991)
3.93 (oyster, Zaroogian et al. 1985; quoted, Hawker & Connell 1986)
3.98 (calculated, Isnard & Lambert 1988)
4.11 (selected, Chessells et al. 1992)

Sorption Partition Coefficient, log K_{oc}:
4.48 (soil, calculated-S as per Kenaga & Goring 1978, Kenaga 1980a; quoted, Nash 1988; Howard 1991)
4.38 (screening model calculations, Jury et al. 1987b)
4.34 (calculated-K_{ow} as per Kenaga & Goring 1980, Chapman 1989)
3.81 (Jury et al. 1990)
5.21 (estimated-QSAR & SPARC, Kollig 1993)
4.38 (Montgomery 1993)
4.38 (20-25°C, selected, Augustijn-Beckers et al. 1994; Hornsby et al. 1996)

Half-Lives in the Environment:
Air: 9.8-59.0 h, based on estimated photooxidation half-life in air (Atkinson 1987; quoted, Howard et al. 1991); atmospheric transformation lifetime was estimated to be <1 d (Kelly et al. 1994).
Surface water: persistence up to 2 weeks in river water (Eichelberger & Lichtenberg 1971); 23.1-129.4 h, based on hydrolysis half-lives (Kollig et al. 1987 and Chapman & Cole 1982; quoted, Howard et al. 1991).
Groundwater: 23.1-129.4 h, based on hydrolysis half-lives (Kollig et al. 1987 and Chapman & Cole 1982; quoted, Howard et al. 1991).
Sediment:
Soil: half-life of about 2-5 yr persistence in soil (Nash & Woolson 1967); estimated persistence of 2 yr in soil (Kearney et al. 1969; Edwards 1973; quoted, Morrill et al. 1982; Jury et al. 1987a); persistence of >24 months (Wauchope 1978); <10 d and subject to plant uptake via volatilization (Callahan et al. 1979; quoted, Ryan et al. 1988); first-order half-life of 63 d from biodegradation rate constant of 0.011 d^{-1} by die-away test in soil (Rao & Davidson 1980; quoted, Scow 1982); field half-life of 0.3 d in moist fallow soil (Glotfelty 1981; quoted, Nash 1983); microagroecosystem half-life of 3 d in moist fallow soil (Nash 1983); measured dissipation rate of 0.28 d^{-1} (Nash 1983; quoted, Nash 1988); estimated dissipation rate of 1.0 and 0.20 d^{-1} (Nash 1988); reported half-life of 9-10 months in soil (Hartley & Kidd 1987; quoted, Montgomery 1993); 23.1-129.4 h, based on hydrolysis half-lives (Kollig et al. 1987 and Chapman & Cole 1982; quoted, Howard et al. 1991); estimated biodegradation half-life of 220 d in soil (Jury et al. 1990); selected field half-life of 250 d (Augustijn-Beckers et

al. 1994; Hornsby et al. 1996); half-life of 9-10 months when used at agricultural rates (Tomlin 1994).
Biota: biochemical half-life of 2000 d from screening model calculations (Jury et al. 1987b).

Environmental Fate Rate Constants or Half-Lives:
Volatilization: measured rate constant of 3.0 d^{-1} (Glotfelty et al. 1984; quoted, Glotfelty et al. 1989); calculated rate constant of 5.0 d^{-1} (Glotfelty et al. 1989).
Photolysis:
Oxidation: half-life of 5.2-51.7 h in air, based on estimated rate constant for the vapor-phase reaction with hydroxyl radicals in air (Atkinson 1987; quoted, Howard et al. 1991).
Hydrolysis: first-order half-life of 23.1 h, based on rate constant of 2.97×10^{-2} h^{-1} at pH 7.0 and 25°C (Demayo 1972; quoted, Callahan et al. 1979; Kollig et al. 1987; Howard et al. 1991); rate constant of 61 y^{-1} at pH 7.0 and 25°C (Kollig 1993).
Biodegradation: aqueous aerobic half-life of 360-1567 h, based on unacclimated aerobic soil grab sample test data (Castro & Yoshida 1971; quoted, Howard et al. 1991); rate constant of 0.011 d^{-1} by die-away test in soil (Rao & Davidson 1980; quoted, Scow 1982); estimated half-life of 220 d in soil (Jury et al. 1990); aqueous anaerobic half-life of 1440-6268 h, based on unacclimated aerobic biodegradation half-life (Howard et al. 1991).
Biotransformation:
Bioconcentration, Uptake (k_1) and Elimination (k_2) Rate Constants:

Common Name: Heptachlor Epoxide
Synonym: β-Heptachlorepoxide, Epoxyheptachlor, HCE, Velsicol 53-CS-17
Chemical Name: 1,4,5,6,7,8,8-heptachloro-2,3-epoxy-3a,4,7,7a-tetrahydro-4,7-methanoindan;
2,3,4,5,7,8-hexahydro-2,5-methano-2H-indeno(1,2b)oxirene
Uses: a degradation product of heptachlor
CAS Registry No: 1024-57-3
Molecular Formula: $C_{10}H_5Cl_7O$
Molecular Weight: 389.2
Melting Point (°C):
- 157-160 (Callahan et al. 1979; Mabey et al. 1982; Montgomery 1993)
- 160-161.5 (Howard 1991)
- 157-159 (Milne 1995)

Boiling Point (°C):
Density (g/cm³ at 20°C):
Molar Volume (cm³/mol):
- 247.4 (Ruelle & Kesselring 1997)
- 317.2 (calculated-LeBas method, this work)

Molecular Volume (Å³):
Total Surface Area, TSA (Å²):
Dissociation Constant pK_a:
Heat of Fusion, ΔH_{fus}, kcal/mol:
- 5.14 (Ruelle & Kesselring 1997)

Entropy of Fusion, ΔS_{fus}, cal/mol·K (e.u.):
Fugacity Ratio at 25°C (assuming ΔS_{fus} = 13.5 e.u.), F: 0.0473

Water Solubility (g/m³ or mg/L at 25°C):
- 0.035 (25-29°C, shake flask-GC, Park & Bruce 1968; quoted, Callahan et al. 1979; Shiu et al. 1990)
- 0.025, 0.120, 0.20 (shake flask-GC, particle size: 0.01, 0.05 and 5.0 μ, Biggar & Riggs 1974)
- 0.11, 0.20, 0.35, 0.60 (shake-flask, 15, 25, 35, 45 °C, particle size 5.0μ or less, Biggar & Riggs 1974)
- 0.20 (particle size ≤5.0 μ, shake flask-GC, Biggar & Riggs 1974; quoted, Callahan et al. 1979; Shiu et al. 1990; Howard 1991)
- 0.35 (gen. col.-GC/ECD, Weil et al. 1974; quoted, Callahan et al. 1979; Geyer et al. 1980; Geyer et al. 1982; Zaroogian et al. 1985)
- 0.20-0.35 (quoted, Mills et al. 1982; Mabey et al. 1982)
- 0.90 (quoted, Zaroogian et al. 1985)
- 0.275 (quoted, Montgomery 1993)
- 5.91 (supercooled liquid value, 20-25°C, quoted, Majewski & Capel 1995)
- 0.35; 0.60, 0.0004 (quoted; predicted-molar volume, Ruelle & Kesselring 1997)

Vapor Pressure (Pa at 25°C):
- 0.045 (estimated, Mabey et al. 1982)
- 0.00256 (estimated, Howard 1991)
- 3.47×10^{-4} (20°C, Montgomery 1993)
- 0.0997 (supercooled liquid value, 20-25°C, quoted, Majewski & Capel 1995)

Henry's Law Constant (Pa m^3/mol):
- 395 (calculated-P/C, Mabey et al. 1982)
- 3.42 (gas-stripping, Warner et al. 1987; quoted, Howard 1991)
- 3.25 (Montgomery 1993)
- 65.5 (20-25°C, Majewski & Capel 1995)

Octanol/Water Partition Coefficient, log K_{OW}:
- 4.43 (Briggs 1981; quoted, Geyer et al. 1982)
- 5.40 (Veith et al. 1979; quoted, Zaroogian et al. 1985; Hawker & Connell 1986; Howard 1991)
- 2.65 (calculated-f const., Mabey et al. 1982)
- 4.56 ± 0.05 (calculated-f const., Noegrohati & Hammers 1992)
- 3.65 (Montgomery 1993)

Bioconcentration Factor, log BCF:
- 3.30 (algae-microcosm expt., Lu & Metcalf 1975)
- 4.90 (snail-microcosm expt., Lu & Metcalf 1975)
- 3.78 (mosquito fish-microcosm expt., Lu & Metcalf 1975)
- 3.23 (mussel, Ernst 1977; quoted, Geyer et al. 1982; Hawker & Connell 1986)
- 4.16 (fathead minnows, 32-d flow-through aquarium, Veith et al. 1979
- 2.03 (microorganism, calculated-K_{OW}, Mabey et al. 1982)
- 3.37 (clam fat, 60-d expt., Hartley & Johnson 1983)
- 2.93 (oyster, Zaroogian et al. 1985; quoted, Hawker & Connell 1986)
- 3.87, 3.89 (for sheepshead minnow, pinfish, mussel and oyster, calculated-K_{OW} & models, Zaroogian et al. 1985)
- −1.45 (beef biotransfer factor log B_b, correlated-K_{OW}, quoted, Travis & Arms 1988)
- 3.88 (calculated-K_{OW}, Howard 1991)

Sorption Partition Coefficient, log K_{OC}:
- 2.34 (sediment, calculated-K_{OW}, Mabey et al. 1982)
- 2.00 (bentonite clay, Hill & McCarty 1967; quoted, Howard 1991)
- 4.0-4.3 (suspended solids in river, Frank 1981; quoted, Howard 1991)
- 3.89 (calculated-S, Howard 1991)
- 4.32 (calculated, Montgomery 1993)
- 3.98 (activated carbon-water, Blum et al. 1994)

Half-Lives in the Environment:
- Air: estimated half-life of 1.5 d for vapor-phase reaction with photochemically produced hydroxyl radicals (Howard 1991).
- Surface water: half-life of 35 d in lower Rhine River (Zoeteman 1980; quoted, Howard 1991).
- Groundwater:
- Sediment:
- Soil:
- Biota:

Environmental Fate Rate Constants or Half-Lives:
    Volatilization: half-life of 60 h from a model river (Howard 1991).
    Photolysis:
    Oxidation: oxidation rate constants of < 3600 M^{-1} h^{-1} for reaction with singlet oxygen, and 20 M^{-1} h^{-1} for reaction with peroxy radical (Mabey et al. 1982).
    Hydrolysis:
    Biodegradation: half-life approximately 25 d under anaerobic conditions when incubated with thick digester sludge at 35°C (Howard 1991).
    Biotransformation: rate constant for bacterial transformation in water of 3×10^{-12} mL $cell^{-1}$ h^{-1} (Mabey et al. 1982).
    Bioconcentration, Uptake (k_1) and Elimination (k_2) Rate Constants:

Common Name: Kepone
Synonym: Chlordecone, CIBA 8514
Chemical Name: 1,2,3,4,5,5,6,7,9,10,10-dodecachlorooctahydro-1,3,4-metheno-2-cyclobuta-[c,d]-pentalone
CAS Registry No: 143-50-0
Uses: Insecticide/Fungicide
Molecular Formula: $C_{10}Cl_{10}O$
Molecular Weight: 490.68
Melting Point (°C): 350 (dec. Howard 1991; Montgomery 1993)
Boiling Point (°C):
Density (g/cm³ at 20°C):
Molar Volume (cm³/mol):
    369.9   (calculated-LeBas method, this work)
Molecular Volume (Å³):
Total Surface Area, TSA (Å²):
Dissociation Constant pK_a:
Heat of Fusion, ΔH_{fus}, kcal/mol:
Entropy of Fusion, ΔS_{fus}, cal/mol·K (e.u.):
Fugacity Ratio at 25°C (assuming ΔS_{fus} = 13.5 e.u.), F: 6.0×10^{-4}

Water Solubility (g/m³ or mg/L at 25°C):
    4.0   (100°C, Günther et al. 1968)
    2.7   (quoted Weil 1978 unpublished result, Kilzer et al. 1979)
    3.0   (20°C, Kenaga & Goring 1978,80; quoted, Kilzer et al. 1979)
    3.0   (Kenaga 1980; quoted, Zaroogian et al. 1985)
    2.7, 3.0   (quoted, Geyer et al. 1980; Kilzer et al. 1985)
    7.6   (24°C, shake flask-nephelometry/fluo., Hollifield 1979; quoted, Howard 1991; Montgomery 1993)

Vapor Pressure (Pa at 25°C):
    3.0×10^{-5} (Kilzer et al. 1979; quoted, Howard 1991; Montgomery 1993)

Henry's Law Constant (Pa·m³/mol):
    0.00153   (calculated-P/C, Howard 1991)
    0.00311   (calculated-P/C, Montgomery 1993, reported as 3.11×10^{-2} atm·m³/mol)
    0.00140   (calculated-P/C, this work)

Octanol/Water Partition Coefficient, log K_{OW}:
    5.50   (Di Toro 1985)
    6.08   (quoted from G.D. Veith personal communication, Zaroogian et al. 1985; quoted, Hawker & Connell 1986)
    5.41   (shake flask, Hansch & Leo 1987; quoted, Sangster 1993)
    3.45   (correlated, Hodson & Williams 1988)
    3.80   (selected, Thomann 1989)
    4.50   (quoted, Howard 1991)
    4.07   (calculated, Montgomery 1993)
    5.41   (recommended, Sangster 1993)
    5.44   (selected, Hansch et al. 1995)

Bioconcentration Factor, log BCF:
- 3.92   (Kenaga & Goring 1980, quoted, Howard 1991)
- 4.0, 2.65, 2.76 (sludge, algae, golden ide, Freitag et al. 1985)
- 3.84   (oyster, Zaroogian et al. 1985; quoted, Hawker & Connell 1986)
- 4.39, 4.46 (oyster, calculated-K_{OW} & models, Zaroogian et al. 1985)
- 4.39, 4.47 (sheephead minnow, calculated-K_{OW} & models, Zaroogian et al. 1985)
- 3.04-3.34 (fathead minnows, quoted, Howard 1991)
- 3.19, 3.09, 2.84, 0.91 (*Cyprinodon variegatus, Leiostomus xanthrus, Palaemonetes pugio, Callinetes sapidus*, quoted, Howard 1991)
- 3.36-3.99 (Atlantic menhaden, quoted, Howard 1991)
- 4.34-4.78 (Atlantic silversides, quoted, Howard 1991)

Sorption Partition Coefficient, log K_{OC}:
- 3.38-3.41 (calculated, Howard 1991)
- 4.74   (calculated, Montgomery 1993)

Half-Lives in the Environment:
- Air: estimated half-life of 438,000 to 4.2×10^7 h or 50-200 yr (Howard et al. 1991).
- Surface water: half-life of 7488 to 17,280 h or 312 d to 2 yr, based on aerobic aquatic microcosm study of soil and water grab samples (Howard et al. 1991).
- Groundwater: estimated half-life of 14,976 to 34,560 h (624 d to 4 yr) based on aqueous aerobic biodegradation (Howard et al. 1991).
- Sediment:
- Soil: estimated half-life of 7488 to 17,280 h (312 d to 2 yr) based on aerobic aquatic microcosm study (Howard et al. 1991).
- Biota:

Environmental Fate Rate Constants or Half-Lives:
- Volatilization: A half-life of 3.8-46 yr predicted for evaporation from a river 1 m deep, flowing at 1 m/s with a wind velocity of 3 m/s (Howard 1991).
- Photolysis: indefinite in air (Howard et al. 1991).
- Oxidation:
- Hydrolysis: no hydrolyzable group (Howard et al. 1991).
- Biodegradation: aerobic aqueous half-life of 7488 to 17,280 h (312 d to 2 years), based on aerobic aquatic microcosm study, anerobic half-life of 29,952 - 69,120 h (1248 d to 8 yr) based on unacclimated aerobic biodegradation half-life (Howard et al. 1991).
- Biotransformation:
- Bioconcentration, Uptake (k_1) and Elimination (k_2) Rate Constants:

Common Name: Leptophos
Synonym: Abar, Phosvel, VCS-506
Chemical Name: O-(4-bromo-2,5-dichlorophenyl) O-methyl phenylphosphorothioate
Uses: insecticide
CAS Registry No: 21609-90-5
Molecular Formula: $C_{13}H_{10}BrCl_2O_2PS$
Molecular Weight: 412.06
Melting Point (°C):
    70.2-70.6 (Spencer 1973, 1982; Freed et al. 1977, 1979; Khan 1980)
    71.5-72.0 (Freed et al. 1977)
    70.4 (Bowman & Sans 1983b; Patil 1994)
    55-67 (Suntio et al. 1988)
Boiling Point (°C):
Density (g/cm³ at 20°C):
    1.53 (25°C, Merck Index 1989)
Molar Volume (cm³/mol):
    317.8 (calculated-LeBas method, Suntio et al. 1988; Fisher et al. 1993)
    1.643 ($V_I/100$, Fisher et al. 1993)
    269.3 (calculated-density, this work)
Molecular Volume (Å³):
Total Surface Area, TSA (Å²):
Dissociation Constant pK_a:
Heat of Fusion, ΔH_{fus}, kcal/mol:
Entropy of Fusion, ΔS_{fus}, cal/mol·K (e.u.):
Fugacity Ratio at 25°C (assuming ΔS_{fus} = 13.5 e.u.), F: 0.494

Water Solubility (g/m³ or mg/L at 25°C):
    0.03 (Velsicol Chem. 1972; quoted, Freed et al. 1977)
    0.03 (shake flask-AS, Carringer et al. 1975)
    0.03 (20°C, GC, Freed 1976; quoted, Suntio et al. 1988; Shiu et al. 1990)
    0.0047 (20°C, shake flask-GC, Chiou et al. 1977; quoted, Freed et al. 1977; Bowman & Sans 1979, 1983; Suntio et al. 1988; Shiu et al. 1990)
    2.4 (Martin & Worthing 1977; quoted, Kenaga 1980; Kenaga & Goring 1980; Khan 1980; Shiu et al. 1990)
    0.07 (20°C, shake flask-GC, Bowman & Sans 1979; quoted, Bowman & Sans 1983; Sharom et al. 1980; Shiu et al. 1990)
    0.0047 (20-25°C, shake flask-GC, Freed et al. 1979)
    0.06 (corrected-MP, Briggs 1981)
    0.005 (20-25°C, shake flask-GC, Kanazawa 1981; quoted, Shiu et al. 1990)
    0.021 (20°C, shake flask-GC, Bowman & Sans 1983a,b; quoted, Shiu et al. 1990; Patil 1994)
    0.005 (20°C, selected, Suntio et al. 1988)
    0.03 (Merck Index 1989)
    0.021, 0.018 (quoted, calculated, Patil 1994)

Vapor Pressure (Pa at 25°C):
    3.07×10^{-6} (20°C, NIEHS 1975; quoted, Freed et al. 1977; Suntio et al. 1988)
    2.27×10^{-5} (30°C, NIEHS 1975; quoted, Freed et al. 1977)
    3.07×10^{-6} (20-25°C, Freed et al. 1979)
    3.00×10^{-6} (20°C, selected, Suntio et al. 1988)
    0.0002   (Merck Index 1989)

Henry's Law Constant (Pa·m^3/mol):
    0.27   (20°C, calculated-P/C, Mackay & Shiu 1981; quoted, Suntio et al. 1988)
    0.25   (20°C, calculated-P/C, Suntio et al. 1988; quoted, Fisher et al. 1993)
    0.247   (calculated-P/C, this work)

Octanol/Water Partition Coefficient, log K_{OW}:
    6.30   (NIEHS 1975; quoted, Freed et al. 1977)
    6.31   (shake flask-GC, Chiou et al. 1977; quoted, Kenaga & Goring 1980; Rao & Davidson 1980; Bowman & Sans 1983b; Suntio et al. 1988)
    6.31   (Freed et al. 1979)
    6.31   (Hansch & Leo 1979; quoted, Fisher et al. 1993)
    6.00   (appr. = BCF_{lipid}, Briggs 1981)
    4.32   (Kanazawa 1981; quoted, Suntio et al. 1988)
    5.88   (shake flask-GC, Bowman & Sans 1983b; quoted, De Bruijn & Hermens 1991)
    5.90   (selected, Suntio et al. 1988)
    5.88   (selected, Geyer et al. 1991)
    6.31   (selected, Magee 1991)
    5.88, 6.14 (quoted, calculated, Patil 1994)
    6.31   (selected, Hansch et al. 1995)
    5.88   (selected, Devillers et al. 1996)

Bioconcentration Factor, log BCF:
    2.81   (*Daphnia magna*, wet wt. basis, Macek et al. 1979; quoted, Geyer et al. 1991)
    2.88, 3.16 (fish: flowing water, static water; quoted, Kenaga 1980; Kenaga & Goring 1980)
    2.58   (calculated-S, Kenaga 1980)
    2.86   (calculated-K_{OC}, Kenaga 1980)
    6.00   (log BCF_{lipid}, Briggs 1981)
    3.78   (log $BCF_{protein}$, Briggs 1981)
    3.78   (*Pseudorasbora parva*, Kanazawa 1981; quoted, Devillers et al. 1996)
    3.16   (mosquito fish, wet wt. basis, De Bruijn & Hermens 1991)
    3.78   (topmouth gudgeon, wet wt. basis, De Bruijn & Hermens 1991)
    2.88   (fish, reported as log BAF_W, LeBanc 1995)

Sorption Partition Coefficient, log K_{OC}:
    3.97   (soil, Carringer et al. 1975; quoted, Kenaga 1980; Kenaga & Goring 1980)
    3.43   (soil, calculated-S as per Kenaga & Goring 1978, Kenaga 1980)
    3.78   (log K_{OM} approx. = log $BCF_{protein}$, Briggs 1981)
    3.97   (reported as log K_{OM}, Magee 1991)
    4.45   (estimated as log K_{OM}, Magee 1991)

Half-Lives in the Environment:
    Air:
    Surface water:
    Groundwater:
    Sediment:
    Soil:
    Biota:

Environmental Fate Rate Constants or Half-Lives:
    Volatilization:
    Photolysis:
    Oxidation:
    Hydrolysis:
    Biodegradation:
    Biotransformation:
    Bioconcentration, Uptake (k_1) and Elimination (k_2) Rate Constants:

Common Name: Lindane (γ-HCH)
Synonym: Aalindan, Aficide, Agrisol G-20, Agrocide, Agronexit, Ambocide, Ameisenatod, Ameisenmittelmerck, Aparacin, Aparasin, Aphtiria, Aplidal, Arbitex, BBX, Ben-hex, Bentox 10, Benzenehexachloride, Benzex, Bexol, BHC, γ-BHC, Celanex, Chloran, Chloresene, Codechine, DBH, Detmol-extrakt, Detox 25, Devoran, Dolmix, ENT 7796, Entomoxan, Exagama, Forlin, Gallogama, Gamacid, Gamaphex, Gamene, Gamiso, Gamahexa, Gamalin, Gammexane, Gammopaz, Gexane, HCCH, Gyben, HCCH, HCH, γ-HCH, Heclotox, Hexa, Hexachlor, γ-Hexachlor, Hexachloran, γ-Hexachloran, Hexachlorane, γ-Hexachlorane, γ-Hexachlorobenzene, Hexamul, Hexapurdre, Hexatox, Hexaverm, Hexdow, Hexicide, Hexyclan, HGI, Hortex, Inexit, Isaton, Isotox, Jacutin, Kokotine, Kotol, Kwell, Lendine, Lentox, Lidenal, Lindafor, Lindagam, Lindagrain, Lindagranox, γ-Lindine, Lindapoudre, Lindatox, Lindosep, Lintox, Lorexane, Milbol 49, Mszycol, NA 2761, NCI-C00204, Neo-scabicidol, Nexen FB, Nexit, Nexit-stark, Nexol-E, Nicochloran, Novigam, Omnitox, Ovadziak, Owadziak, Pedraczak, Pflanzol, Quellada, Silvanol, Soprocide, Spritz-rapidin, Spruehpflanzol, Streunex, Tap 85, TBH, Tri-6, Viton
Chemical Name: 1,2,3,4,5,6-hexachlorocyclohexane; γ-hexachlorocyclohexane; γ-1,2,3,4,5,6-hexachlorocyclohexane
Uses: insecticide and pesticide with contact, stomach, and respiratory action to control a broad spectrum of phytophagous and soil inhibiting insects, public health pests, and animal ectoparasites.
CAS Registry No: 58-89-9
Molecular Formula: $C_6H_6Cl_6$
Molecular Weight: 290.85
Melting Point (°C):
    112.5   (Slade 1945; Suntio et al. 1988; Howard 1991; Yalkowsky & Banerjee 1992; Montgomery 1993; Milne 1995)
    112.9   (Martin 1972; Spencer 1982; Callahan et al. 1979)
    113.0   (Noegrohati & Hammers 1992)
    112.5-113.5   (Tomlin 1994)
Boiling Point (°C):
    323.4   (Howard 1991; Montgomery 1993)
Density (g/cm³ at 20°C):
    1.87   (Montgomery 1993)
Molar Volume (cm³/mol):
    243.6   (LeBas method, Suntio et al. 1988)
    1.146   (intrinsic V_I/100, Luehrs et al. 1996)
    179.5   (Ruelle & Kesselring 1997)
Heat of Vaporization, kcal/mol:
    24.17   (Spencer & Cliath 1970)
Heat of Fusion, ΔH_{fus}, kcal/mol:
    5.29   (Ruelle & Kesselring 1997)
Entropy of Fusion, ΔS_{fus} :
    41.4   J/mol·K (Plato & Glasglow 1969)
    61.1   J/mol·K (Hinckley et al. 1990)
Fugacity Ratio at 25 °C (assuming ΔS = 13.5 e.u.), F:
    0.12   (20°C, Suntio et al. 1988)
    0.138   (Mackay et al. 1986)

Water Solubility (g/m^3 or mg/L at 25°C):
- 10 (20°C, Slade 1945; quoted, Günther et al. 1968; Suntio et al. 1988)
- 7.3 (shake flask-polarography, Richardson & Miller 1960; quoted, Biggar & Riggs 1974; Freed 1976)
- 0.50-6.60 (particle size of 0.04-5 μ, shake flask-GC, room temp., Robeck et al. 1965; quoted, Biggar & Riggs 1974)
- 7.3 (quoted, Günther et al. 1968, Mackay & Wolkoff 1973; Metcalf et al. 1973; Mackay & Leinonen 1975; Geyer et al. 1980; Suntio et al. 1988)
- 5.7 (partition coefficient, Atkins & Eggleton 1971)
- 7.52 ± 0.041 (shake flask-centrifuge/GC, Masterton & Lee 1972; quoted, Freed 1976; Callahan et al. 1979)
- 10 (Ulmann 1972; quoted, Malaiyandi et al. 1982)
- 7.40, 5.75 (28°C, shake flask-centrifuge, membrane filter-GC, max. 0.1 μm particle size, Kurihara et al. 1973)
- 6.61, 6.24 (28°C, shake flask-centrifuge, sonic and centrifuge-GC, max. 0.05 μm particle size, Kurihara et al. 1973)
- 5.75-7.40 (28°C, Kurihara et al. 1973; quoted, Callahan et al. 1979)
- 10 (Spencer 1973, Spencer 1982; quoted, Fuhrenmann & Lichtenstein 1980; Sharom et al. 1980)
- 12 (26.5°C, Bhavnagary & Jayaram 1974; quoted, Callahan et al. 1979)
- 2.15, 6.80, 11.4, 15.2 (shake flask-GC, at 15, 25, 35, 45 °C, for particle size ≤5μ, Biggar & Riggs 1974)
- 2.15 (15°C, shake flask-GC, Biggar & Riggs 1974, quoted, Callahan et al. 1979; Kucklick et al. 1991)
- 6.8 (particle size ≤5.0 μ, shake flask-GC, Biggar & Riggs 1974; quoted, Callahan et al. 1979; Suntio et al. 1988; Kucklick et al. 1991)
- 0.15, 0.60, 6.80 (shake flask-GC, at 25°C, for different particle sizes: 0.01μ, 0.05μ, 5.0μ, Biggar & Riggs 1974)
- 0.15 (Biggar & Riggs 1974; quoted, Kenaga & Goring 1980)
- 7.8 (gen. col.-GC/ECD, Weil et al. 1974; quoted, Callahan et al. 1979; Geyer et al. 1980,82; Chiou et al. 1986; Suntio et al. 1988)
- 7.5 (Freed 1976; quoted, Jury et al. 1983,84,87; Spencer et al. 1988; Spencer & Cliath 1990)
- 0.15 (Martin & Worthing 1977; quoted, Kenaga 1980a,b; Bruggeman et al. 1981; Zaroogian et al. 1985; Adams 1987; Suntio et al. 1988)
- 2.0 (shake flask-nephelometry, Hollifield 1979; quoted, Howard 1991)
- 10 (quoted, Burkhard & Guth 1981)
- 7.88 (20-25°C, shake flask-GC, Kanazawa 1981; quoted, Platford 1981)
- 9.12 (OECD 1981; quoted, Davies & Dobbs 1984)
- 10 (20-25°C, shake flask-GC, Platford 1981)
- 10.3 (shake flask-GC/ECD, Malaiyandi et al. 1982)
- 9.50, 7.9-8.2 (shake flask-GC/ECD: Milli-Q water, environmental surface waters, Saleh et al. 1982)
- 7.0 (quoted, Thomas 1982; Nash 1988)
- 6.11 (20°C, Deutsche Forschungsgemeinschaft 1983; quoted, Ballschmiter & Wittlinger 1991; Fischer et al. 1991)
- 7.5 (Jury et al. 1983; quoted, Taylor & Glotfelty 1988)
- 6.98 (Caron et al. 1985)
- 7.87 (24°C, shake flask-GC, Chiou et al. 1986)

7.52   (quoted, Mackay et al. 1986)
7.3   (Agrochemicals Handbook 1987; Tomlin 1994)
10   (selected, Gerstl & Helling 1987)
7.3   (Yalkowsky 1987; selected, Yalkowsky & Banerjee 1992)
7.08   (quoted, Isnard & Lambert 1988. 1989)
5.79   (Mailhot & Peters 1988; quoted, Chessells et al. 1992)
6.50   (20°C, selected, Suntio et al. 1988; Schreitmüller & Ballschmiter 1995)
7.94   (Kanazawa 1989)
10   (selected, USDA 1989; quoted, Neary et al. 1993)
10   (selected, Boehncke et al. 1990; Pinsuwan et al. 1995)
7.3   (20°C, Lyman et al. 1990; quoted, Hemond & Fechner 1994)
0.9   (10°C, estimated, Mclachlan et al. 1990)
6.107   (20°C, quoted, Ballschmiter & Wittlinger 1991)
7.0   (20°C, selected, Calamari et al. 1991)
7.5   (quoted, Mackay & Stiver 1991)
7.3, 2.0 (quoted, Howard 1991)
7.0   (20-25°C, selected, Wauchope et al. 1992; Lohninger 1994; Hornsby et al. 1996)
6.1   (quoted, Fischer et al. 1993)
7.52   (Montgomery 1993)
7.3   (Milne 1995)
6.5; 3.92, 0.14 (quoted; predicted-molar volume, Ruelle & Kesselring 1997)

Vapor Pressure (Pa at 25°C):
    4.0, 18.7, 64 (20, 40, 60°C, static method, Slade 1945)
    0.000545   (15°C, effusion manometer, Balson 1947, quoted, Kucklick et al. 1991)
    0.00124   (20°C, effusion manometer, Balson 1947, quoted, Freed 1976; Spencer 1982; Kucklick et al. 1991; Boehncke et al. 1996)
    log P (mmHg) = 15.515 − 6020/T,   (temp range 50-90°C, Balson 1947)
    0.00435   (20°C, gas-saturation, Spencer & Cliath 1970; quoted, Freed 1976; Dobbs & Cull 1982; Spencer & Cliath 1983; quoted, Suntio et al. 1988; Boehncke et al. 1996)
    0.0171   (30°C, gas-saturation, Spencer & Cliath 1970; quoted, Spencer et al. 1973; Spencer & Cliath 1983)
    $\log_{10}$ P (mmHg) = 13.544 − 5288/T, (20-60°C, Spencer & Cliath 1970)
    0.00413   (20°C, Partition coefficient, Atkins & Eggleton 1971; quoted, Freed 1976; Suntio et al. 1988; Boehncke et al. 1996)
    0.02133   (20°C, Demozay & Marechal 1972; quoted, Callahan et al. 1979)
    0.00125   (20°C, quoted, Martin 1972, Melnikov 1971, Quellette & King 1977, Callahan et al. 1979; Hartley & Graham-Bryce 1980; Suntio et al. 1988)
    0.00125   (quoted, Mackay & Wolkoff 1973, Mackay & Leinonen 1975)
    0.0028   (20°C, estimated from diffusion rate, Zimmerli & Marek 1974; quoted, Freed 1976)
    0.0026   (20°C, estimated, Dobbs & Grant 1980; quoted, Dobbs & Cull 1982)
    0.00125   (20°C, Spencer 1973; quoted, Fuhrenmann & Lichtenstein 1980)
    0.00426   (quoted, Burkhard & Guth 1981)
    log P (mmHg) = 15.515 − 6020/T (Gückel et al. 1982; quoted, Boehncke et al. 1996)
    0.011   (quoted, Thomas 1982; Nash 1988)

0.019 (20°C, Deutsche Forschungsgemeinschaft 1983; quoted, Ballschmiter & Wittlinger 1991; Fischer et al. 1991)
0.0086 (Jury et al. 1983, 1984; quoted, Taylor & Glotfelty 1988; Spencer & Cliath 1990; Taylor & Spencer 1990)
0.0056 (20°C, Worthing 1983, 1991; Tomlin 1994)
0.0552; 0.0107, 0.0654 (supercooled liquid values P_L: quoted; GC-RT, Bidleman 1984)
0.0359 (Caron et al. 1985)
0.00368 (gas saturation-mixed bed, Kim 1985)
0.0041 (quoted, Mackay et al. 1986)
0.0056 (20°C, Agrochemicals Handbook 1987)
0.00863 (quoted, Jury et al. 1987; Spencer et al. 1988; Spencer et al. 1988)
0.003 (20°C, selected, Suntio et al. 1988; Schreitmüller & Ballschmiter 1995)
8.1×10^{-4} (10°C, estimated, Mclachlan et al. 1990)
0.0552, 0.065 (subcooled liquid values, GC-RT, Hinckley et al. 1990)
$\log P_L$ (Pa) = $A_L + B_L/T$, $A_L = 13.63$, $B_L = -4416$ (liquid, Hinckley et al. 1990)
$\log P_L$ (Pa) = $A_L + B_L/T$, $A_L = 11.15$, $B_L = -3680$ (GC-RT, liquid, Hinckley et al. 1990; quoted, Boehncke et al. 1996))
0.00122 (20°C, Lyman et al. 1990; quoted, Hemond & Fechner 1994)
0.004 (20°C, selected, Calamari et al. 1991)
0.0019, 0.032 (quoted, Fischer et al. 1991)
0.0085 (quoted, Mackay & Stiver 1991)
7.426×10^{-4} (quoted, Howard 1991)
0.0044 (20-25°C, selected, Wauchope et al. 1992; Hornsby et al. 1996)
0.0145, 0.0398, 0.1035 (subcooled liquid values at 10°C, 20°C, 30°C, calculated from Hinckley et al. 1990; Cotham & Bildeman 1992)
0.00125 (20°C, Montgomery 1993)
0.0094 (20°C, gas saturation-GC/ECD, Wania et al. 1994; quoted, Boehncke et al. 1996)
0.04192 (30°C, gas saturation-GC/ECD, Wania et al. 1994)
0.00374, 0.00737, 0.0116, 0.0225, 0.0381 (20, 25, 28.4, 33.6, 37.9°C, Knudsen effusion method, Boehncke et al. 1996)
0.00383 (20°C, interpolated from vapor pressure eq. ln P (Pa) = $(34.53 \pm 0.21) - (11754 \pm 72)/T$, temp range 20-50°C, Boehncke et al. 1996)

Henry's Law Constant (Pa·m³/mol):
0.005 (calculated-P/C, Mackay & Leinonen 1975; quoted, Suntio et al. 1988)
0.22 (gas stripping, Atkins & Eggleton 1971)
0.32 (24°C, calculated-P/C, Chiou et al. 1980)
0.018-0.55 (calculated-P/C, Mabey et al. 1982)
0.27-0.32 (calculated-P/C, Mackay & Shiu 1981; quoted, Suntio et al. 1988)
0.05 (calculated-P/C, Lyman et al. 1982; quoted, Suntio et al. 1988)
0.0486 (calculated-P/C, Thomas 1982)
0.322 (calculated-P/C, Jury et al. 1984, 1987a; Jury & Ghodrati 1989; Spencer et al. 1988; Spencer & Cliath 1990)
1.496, 1.334 (Caron et al. 1985)
0.158 (calculated-P/C, Mackay et al. 1986; quoted, Iwata et al. 1993)
0.203 (wetted-wall column-GC/ECD, Fendinger & Glotfelty 1988)
1.49 (quoted from WERL Treatability Database, Ryan et al. 1988)

0.129 (20°C, calculated-P/C, Suntio et al. 1988; quoted, Fendinger & Glotfelty 1988; Bacci et al. 1990; Müller et al. 1994)
0.322 (calculated-P/C, Taylor & Glotfelty 1988)
0.206 (fog chamber-concentraion ratio-GC/ECD, Fendinger et al. 1989)
0.0486 (20°C, Lyman et al. 1990; quoted, Hemond & Fechner 1994)
0.26 (calculated, 10°C, Mclachlan et al. 1990)
0.10 (calculated-P/C, Ballshmiter & Wittlinger 1991; Fischer et al. 1991)
0.296 (calculated-P/C, Howard 1991)
0.187, 0.074-0.287 (15°C, gas stripping, calc-P/C, Kucklick et al. 1991)
0.353, 0.102-0.358 (25°C, gas stripping, calc-P/C, Kucklick et al. 1991)
0.31, 0.35 (gas stripping: distilled water, seawater, Kucklick et al. 1991)
0.17 (calculated-P/C, Calamari et al. 1991)
0.10, 1.50 (calculated-P/C, Fischer et al. 1991)
0.296 (calculated-P/C, Howard 1991)
0.13 (20°C), 0.20, 0.339, 0.363 (23°C), 0.158 (quoted, Itawa et al. 1993)
0.121 at 8°C in Green Bay, 0.242 at 18.9°C in Lake Michigan, 0.236 at 18.5°C in Lake Huron, 0.301 at 22.3°C in Lake Erie and 0.301 at 22.3°C in Lake Ontario (concentraion ratio-GC, MacConnell et al. 1993)
$\log H$ (Pa·m^3 mol^{-1}) = $-2382/T + 7.54$ (0.5-45°C, Kucklick et al. 1991, MacConnell et al. 1993)
$\log H$ (Pa·m^3 mol^{-1}) = $-2703/T + 8.68$ (0.5-45°C, artificial seawater, Kucklick et al. 1991)
0.0246 (20°C, Montgomery 1993)
0.149 (calculated-P/C, this work)

Octanol/Water Partition Coefficient, log K_{OW}:
3.72 (shake flask-GC, Kurihara et al. 1973, quoted, Callahan et al. 1979; McDuffie 1981; Mackay et al. 1986; Gerstl & Helling 1987; Suntio et al. 1988; Ryan et al. 1988; Sicbaldi & Finizio 1993; Finizio et al. 1997)
3.70 (calculated-π constant, Kurihara et al. 1973)
3.65 (HPLC-RT, Sugiura et al. 1979)
3.85 (HPLC-RT, Veith et al. 1979; quoted, Mackay 1982; Veith & Kosian 1983; Chapman 1989; Saito et al. 1992; MacConnell et al. 1993)
3.89 (Veith et al. 1979; quoted, Zaroogian et al. 1985)
2.81 (quoted, Rao & Davidson 1980)
3.66 (shake flask-GC, concn. ratio, Kanazawa 1981; quoted, Davies & Dobbs 1984; Kanazawa 1989; Sicbaldi & Finizio 1993; Finizio et al. 1997)
3.72, 3.62 (quoted, HPLC-k', McDuffie 1981)
3.25 (shake flask-GC, Platford 1982)
3.20, 3.29 (quoted, Geyer et al. 1982)
3.53 (shake flask-GC/FID, Hermens & Leeuwangh 1982; quoted, Hermens et al. 1985; Verhaar et al. 1992)
5.43 (selected, Dao et al. 1983)
4.81 (Hawker & Connell 1986)
3.90 (Elgar 1983; quoted, Suntio et al. 1988)
3.30, 3.20 (quoted, Geyer et al. 1984; quoted, Sicbaldi & Finizio 1993)
3.61 (Hansch & Leo 1985; quoted, Howard 1991; Müller et al. 1994)
3.70 (McKim et al. 1985, Thomann 1989; selected, Boehncke et al. 1990)

3.24   (selected, Carlberg et al. 1986)
3.85   (Isnard & Lambert 1988, 1989; Thomann 1989; quoted, Ballschmiter & Wittlinger 1991; Banerjee & Baughman 1991)
3.80   (selected, Suntio et al. 1988, quoted, Bacci et al. 1990; Calamari et al. 1991; Schreitmüller & Ballschmiter 1995)
3.66   (selected, Travis & Arms 1988)
3.688   (slow stirring, De Bruijn et al. 1989; quoted, Parkerton et al. 1993; Sicbaldi & Finizio 1993; Finizio et al. 1997)
3.69   (selected, Geyer et al. 1990)
3.80   (selected, Lydy et al. 1990,92)
3.70   (selected, Mclachlan et al. 1990)
3.61   (quoted, Howard 1991; Chessells et al. 1992)
3.51   (exptl., Noegrohati & Hammers 1992)
4.04   (quoted, Vigano et al. 1992)
3.90   (quoted, Fischer et al. 1993)
3.20-3.89 (Montgomery 1993)
5.32   (RP-HPLC, Sicbaldi & Finizio 1993)
3.517   (Yalkowsky & Dannenfelser 1994; quoted, Pinsuwan et al. 1995)
3.752   (calculated, Pinsuwan et al. 1995)
3.72   (selected, Hansch et al. 1995; Devillers et al. 1996)
3.52, 3.75, 3.26 (RP-HPLC, ClogP, calculated-S, Finizio et al. 1997)

Bioconcentration Factor, log BCF:
    −1.78   (beef biotransfer factor log B_b, correlated-K_{OW}, Radeleff et al. 1952; Kenaga 1980; quoted, Travis & Arms 1988)
    −0.41   (vegetation, correlated-K_{OW}, Lichtenstein 1959; Voerman & Besemer 1975; quoted, Travis & Arms 1988)
    −2.60   (milk biotransfer factor log B_m, correlated-K_{OW}, Saha 1969; quoted, Travis & Arms 1988)
    2.15, 2.34 (Voerman & Tammes 1969; quoted, Baughman & Paris 1981)
    1.83, 3.24 (brine shrimp, silverside fish, Matsumura et al. & Benezet 1973)
    1.98   (brine shrimp in water, Matsumura & Benezet 1973; quoted, Howard 1991)
    2.26   (brine shrimp in sand, Matsumura & Benezet 1973; quoted, Howard 1991)
    3.21   (northern brook silverside fish to lindane residues on sand, Matsumura & Benezet 1973; quoted, Howard 1991)
    2.75, 2.66 (fish, snail, Metcalf et al. 1973; quoted, Howard 1991)
    2.26 (fathead minnows, Canton et al. 1975; quoted, Davies & Dobbs 1984)
    2.23, 2.65 (zooplankton, Hamelink & Waybrant 1976; quoted, Baughman & Paris 1981)
    2.00   (mussels, steady state, Ernst 1977; quoted, Renberg & Sundström 1979; Hawker & Connell 1986)
    1.92, 2.34, 1.80, 2.69 (pink shrimp, pinfish, grass shrimp, sheepshead minnows, Schimmel et al. 1977; quoted, Howard 1991)
    2.88, 2.45, 2.65, 2.97 (golden orfe, carp, brown trout, guppy, Sugiura et al. 1979)
    2.68   (fathead minnows, Veith et al. 1979; quoted, Zaroogian et al. 1985)
    2.26   (fathead minnows, 32-d exposure, Veith et al. 1979; quoted, Veith & Kosian 1983; Devillers et al. 1996)
    2.51, 2.75 (fish: flowing water, static water; quoted, Kenaga 1980b; Kenaga & Goring 1980)

3.26, 1.73 (calculated-S, K_{OC}, Kenaga 1980)
−0.26 (average beef fat diet, Kenaga 1980b)
2.67 (fathead minnows, 30-d exposure, Veith et al. 1980; quoted, Bysshe 1982)
2.25 (fathead minnows, 32-d exposure, Veith et al. 1980; quoted, Bysshe 1982)
3.10 (topmouth gudgeon, Kanazawa 1981, quoted Howard 1991; Devillers et al. 1996)
2.19 (mussel, quoted average, Geyer et al. 1982)
2.26, 2.67 (fathead minnows, Veith & Kosian 1983; quoted, Saito et al. 1992)
3.10 (topmouth gudgeons, Kanazawa 1983; quoted, Howard 1991)
3.42 (clam fat, 60-d exptl., Hartley & Johnson 1983)
2.38, 2.46 (algae: exptl., calculated, Geyer et al. 1984)
2.26 (fathead minnow, Davies & Dobbs 1984)
2.38, 2.88, 2.91 (algae, fish, activated sludge, Klein et al. 1984)
2.38, 2.57, 2.91 (algae, golden ide, activated sludge, Freitag et al. 1985)
3.32, 3.20 (rainbow trout: kinetic, steady state, Oliver & Niimi 1985; quoted, Devillers et al. 1996)
2.50 (*Salmo gairdneri* Richardson fry, Ramamoorthy 1985; quoted, Howard 1991)
2.78, 2.61 (mussel, calculated-K_{OW} & models, Zaroogian et al. 1985)
2.73, 2.61 (pinfish, calculated-K_{OW} & models, Zaroogian et al. 1985)
2.78, 2.61 (sheepshead minnow, calculated-K_{OW} & models, Zaroogian et al. 1985)
2.38, 2.67 (quoted values: mussel, sheepshead minnow, Zarrogian et al. 1985; quoted, Hawker & Connell 1986)
2.76 (salmon fry in in Humic water April 1982 at steady state, Carlberg et al. 1986)
2.43 (salmon fry in in Humic water Oct. 1983 at steady state, Carlberg et al. 1986)
2.42, 2.84 (salmon fry in lake water, Carlberg et al. 1986)
2.45-3.18 (quoted values, Carlberg et al. 1986)
2.33 (*Daphnia magna*, wet wt. basis, Korte & Freitag 1986; quoted, Geyer et al. 1990)
3.53 (azalea leaves, Bacci & Gaggi 1987)
2.38 (paddy field fish, Soon & Hock 1987; quoted, Abdullah et al. 1997)
2.51 (quoted, Isnard & Lambert 1988)
2.51 (measured, Isnard & Lambert 1988; quoted, Banerjee & Baughman 1991)
4.30 (zooplankton, chum salmon, Kawano et al. 1988)
3.53, 5.88 (dry leaf, wet leaf, Bacci et al. 1990)
2.33 (*Daphnia magna*, Geyer et al. 1990)
2.09, 2.70, 2.29, 2.34 (zebrafish: egg, embryo, yolk sac fry, juvenile, Görge & Nagel 1990)
1.96 (calculated, Banerjee & Baughman 1991)
2.93, 2.96 (*Brachydanio rerio*, Butte et al. 1991; quoted, Devillers et al. 1996)
2.67 (selected, Chessells et al. 1992)
1.58 (*Hydrilla*, Hinman & Klaine 1992)
2.16-2.57 (rainbow trout in early life stages on wet wt. basis, Vigano et al. 1992; quoted, Devillers et al. 1996)
3.77-3.85 (rainbow trout in early life stages on lipid basis, Vigano et al. 1992)

Bioaccumulation Factor, log BAF:
- −0.398 (bioaccumulation factor log BF, adipose tissue in male Albino rats, Jacob et al. 1974; quoted, Geyer et al. 1980)
- 0.146 (bioaccumulation factor log BF, adipose tissue in male Albino rats, Baron et al. 1975; quoted, Geyer et al. 1980)
- 4.10 (rainbow trout, Thomann 1989)

Sorption Partition Coefficient, log K_{OC}:
- 2.96 (soil, Hamaker & Thompson 1972; quoted, Kenaga 1980a,b; Kenaga & Goring 1980)
- 4.09 (soil, calculated-S as per Kenaga & Goring 1978, Kenaga 1980)
- 3.40 (soil, Kenaga 1980; quoted, Thomas 1982; Nash 1988)
- 2.87 (average of 3 soils, HPLC-RT, McCall et al. 1980)
- 4.64 (calculated-S, Mill et al. 1980; quoted, Adams 1987)
- 3.03 (av. for 3 soils, Rao & Davidson 1982; quoted, Lyman 1982; Howard 1991)
- 2.88, 2.95, 2.74; 2.87 (Commerce soil, Tracy soil, Catlin soil; average soil, McCall et al. 1980)
- 4.07, 2.90 (estimated-S, K_{OW}, Lyman 1982)
- 3.11 (soil, screening model simulations, Jury et al. 1984, 1987a,b; Jury & Ghodrati 1989; Spencer et al. 1988; Spencer & Cliath 1990)
- 4.30, 3.50 (field data of river sediment, calculated-K_{OW}, Oliver & Charlton 1984)
- 3.03 (Rao & Davidson 1982, Howard 1991)
- 2.63-3.18 (reported as log K_{OM}, Mingelgrin & Gerstl 1983)
- 3.04 (Caron et al. 1985)
- 1.63 (log K_P with first-order adsorption rate 0.088 h^{-1}, Miller & Weber 1986; quoted, Brusseau & Rao 1989)
- 3.11, 2.82 (quoted, calculated-χ, Gerstl & Helling 1987)
- 0.114 (screening model calculations, Jury et al. 1987b)
- 3.47 (calculated-K_{OW} as per Kenaga & Goring 1980, Chapman 1989)
- 2.38 (average of 2 soils, Kanazawa 1989)
- 1.18 (selected, USDA 1989; quoted, Neary et al. 1993)
- 2.99 (average of 5 soils, Kishi et al. 1990)
- 3.11 (soil, Mackay & Stiver 1991)
- 3.04 (soil, 20-25°C, selected, Wauchope et al. 1992; quoted, Dowd et al. 1993; Lohninger 1994; Hornsby et al. 1996)
- 2.38-3.52 (Montgomery 1993)

Half-Lives in the Environment:
- Air: half-life of about 2.3 d was estimated, based on rate constant 6.94×10^{-12} cm^3/molecule·s for the vapor-phase reaction with hydroxyl radicals in air (Atkinson 1987; quoted, Howard 1991); calculated tropospheric lifetimes due to gas-phase reaction with OH radicals was estimated to be about 7 d (Atkinson et al. 1992); atmospheric transformation lifetime was estimated to be < 1 d (Kelly et al. 1994); lifetime of 13 d was estimated for atmospheric reaction with OH radicals in the tropics (Schreitmüller and Ballsmiter 1995).
- Surface water: half-lives of 3-30 d in rivers and 30-300 d in lakes (Zoeteman et al. 1980); hydrolysis half-life of 92 h (exptl.), 89 h (calcd.) for Roselawn Cemetery

Pond at pH 9.3; 771 h (exptl.), 578 h (calcd.) for Cross Lake at pH 7.3; 648 h (exptl.), 231 h (calcd.) for Indiana Quarry at pH 7.8; photolysis half-lives for direct sunlight during July and adjusted for mid-winter half-life of 779 h, 1560 h for Milli-Q water at pH 6.98, 169 h, 339 h for Roselawn Pond at pH 9.3, 1791 h, 3590 h for Cross Lake and 1540 h, 3090 h for Indiana Quarry (Saleh et al. 1982).

Groundwater: half-life of ≥ 300 d (Zoeteman et al. 1980).

Sediment:

Soil: half-life of about 2 yr persistence in soil (Nash & Woolson 1967; quoted, Kaufman 1976); persistence of 3 yr in soil (Edwards 1973; quoted, Morrill et al. 1982); >50 d and subject to plant uptake via volatilization (Callahan et al. 1979; quoted, Ryan et al. 1988); first-order half-life of 266 d in soil from biodegradation rate constant of 0.0026 d^{-1} by die-away test in soil (Rao & Davidson 1980; quoted, Scow 1982); field half-life of 0.3 d in moist fallow soil (Glotfelty 1981; quoted, Nash 1983); microagroecosystem half-life of 1-4 d in moist fallow soil (Nash 1983); measured dissipation rate of 0.16 d^{-1} (Nash 1983; quoted, Nash 1988); estimated dissipation rate of 0.20, 0.10 d^{-1} (Nash 1988); biodegradation half-life of 266 d (soil, Jury et al. 1984, 1987); first-order adsorption rate 0.088 h^{-1} (Miller & Weber 1986; quoted, Brusseau & Rao 1989); half-lives in soil surfaces at (20 $\pm$ 1)°C: 5.5 to 15.9 d in peat soil and 2.7 to 6.7 d in sandy soil (Dörfler et al. 1991); reported half-life of 266 d in soil (Jury et al. 1987a,b; Jury & Ghodrati 1989; quoted, Montgomery 1993); reaction half-life of 266 d (Mackay & Stiver 1991); selected field half-life of 400 d (Wauchope et al. 1992; Dowd et al. 1993; Hornsby et al. 1996); half-lives of 14 d for soil depth <5 cm, 90 days for 5-20 cm and 180 d for >20 cm (Dowd et al. 1993).

Biota: 22.1 h (mussels, Ernst 1977); biological half-lives for fishes: 11 d for trout muscle, 1 d for goldfish, <1 d for sunfish and 4 d for guppy (Niimi 1987); 678 h (azalea leaves, Bacci & Gaggi 1987); biochemical half-life of 266 d from screening model calculations (Jury et al. 1987b); half-lives in plant surfaces at (20 $\pm$ 1)°C: 0.56 d in bean, 0.40 d in turnips and 0.31 d in oats (Dörfler et al. 1991); elimination half-lives in the midge (*Chironomus riparius*) under varying sediment conditions: 11 h for water only system, 10 h for screened system, 9 hours for 3% organic carbon system and 6 h for 15% organic carbon system (Lydy et al. 1992).

Average half-life: 90 d (for pesticides used in conjunction with forest management, Neary et al. 1993).

Environmental Fate Rate Constants or Half-Lives:

Volatilization: half-life of 191 d was estimated from water (Mackay & Leinonen 1975, quoted, Howard 1991); estimated half-life of more than 200 d (Callahan et al. 1979); measured half-lives from 4.5 cm deep distilled water at 24°C: exptl. half-lives: 3.2 d in nonstirred water and 1.5 d in stirred water (Chiou et al. 1980; quoted, Howard 1991); estimated half-lives: 3.4 d in nonstirred water and 2.3 d in stirred water (Chiou et al. 1980); half-life of 22 d, estimated from a model river of 1 m deep flowing 1 m/s with a wind speed of 3 m/s (Lyman et al. 1982; quoted, Howard 1991); initial rate constant of 4.4×10^{-2} h^{-1} and predicted rate constant of 1.4×10^{-2} h^{-1} from soil with a half-life of 49.5 h (Thomas 1982); calculated half-life of 2760 h from water (Thomas 1982); measured rate constant of 3.0 d^{-1} (Glotfelty et al. 1984; quoted, Glotfelty et al. 1989); calculated rate constant of 0.01 d^{-1} (Glotfelty et al. 1989); half-life of 266 d from lab. and field experiments (Jury 1984; quoted, Spencer & Cliath 1990); half-lives in soil surfaces at (20 ± 1)°C: 5.5 to 15.9 d in peat soil and 2.7 to 6.7 d in sandy soil; half-lives in plant surfaces at (20 ± 1)°C: 0.56 d in bean, 0.40 d in turnips and 0.31 d in oats (Dörfler et al. 1991).

Photolysis: first-order rate constant of 1.429×10^{-2} d^{-1} for photolysis in natural waters (Malaiyandi et al. 1982); first-order aqueous photolysis rate constants were 8.9×10^{-4} h^{-1} for Mille-Q water, 4.1×10^{-3}, 3.9×10^{-4}, and 4.5×10^{-4} h^{-1} for natural surface water samples from eutrophic pond, dystrophic reservoir and oligotrophic rock quarry and the half-lives were 779, 169, 1791, and 1540 h under direct sunlight (Saleh et al. 1982; quoted, Howard 1991).

Oxidation: half-life of about 2.3 d was estimated, based on rate constant 6.94×10^{-12} cm^3/molecule·s for the vapor-phase reaction with hydroxyl radicals in air (Atkinson 1987; quoted, Howard 1991); calculated tropospheric lifetimes due to gas-phase reaction with OH radicals was estimated to be about 7 d (Atkinson et al. 1992); rate constant of 4.2×10^8 M^{-1} s^{-1} for the reaction (Fenton with reference to DBCP) with hydroxyl radicals in aqueous solutions at pH 2.9 and at (24 ± 1)°C (Buxton et al. 1988; quoted, Faust & Hoigné 1990; Haag & Yao 1992); rate constants of $(5.8 \pm 1.9) \times 10^8$ M^{-1} s^{-1} (Fenton with reference to DBCP) and $(5.2 \pm 0.9) \times 10^8$ M^{-1} s^{-1} (photo-Fenton with reference to DBCP) for the reaction with hydroxyl radicals in aqueous solutions at pH 2.9 and at (24 ± 1)°C (Haag & Yao 1992).

Hydrolysis: rate constant in neutral medium 1.6×10^{-4} h^{-1} indicating that neutral hydrolysis is unimportant, rate constants of 7.5×10^{-3}, 8.99×10^{-4}, and 1.07×10^{-3} h^{-1} corresponded to half-lives of 92, 771 and 648 h in natural surface water samples from eutrophic pond, dystrophic reservoir and oligotrophic rock quarry respectively (Saleh et al. 1982; quoted, Howard 1991); neutral hydrolysis rate constant of $(1.2 \pm 0.2) \times 10^{-4}$ h^{-1} with a calculated half-life of 206 d at pH 7 (Ellington et al. 1987, 1988; quoted, Montgomery 1993); half-life of 42 yr at pH 8 and 5°C (Ngabe et al. 1993); half-lives at 22°C: 191 d at pH 7, and 11 h at pH 9 (Tomlin 1994).

Biodegradation: rate constant of 0.0026 d^{-1} by die-away test in soil (Rao & Davidson 1980; quoted, Scow 1982); half-life of 266 d (soil, Jury et al. 1987); degradation half-lives were estimated to be 3 to 30, 30 to 300 d and >300 d for river lake and groundwater (Zoeteman et al. 1980; quoted, Howard 1991); half-life of 266 d for 100-d leaching screening test in 0-10 cm depth of soil (Jury et al. 1984, 1987a,b; Jury & Ghodrati 1989); calculated half-life in sewage sludge of 20.4 ± 0.1 h from experiments S1-S3 (Buser & Müller 1995).

Biotransformation:

Bioconcentration, Uptake (k_1) and Elimination (k_2) Rate Constants:
    k_1: 3.13 h^{-1} (mussels, Ernst 1977)
    k_2: 0.0313 h^{-1} (mussels, Ernst 1977)
    k_1: 130 d^{-1} (rainbow trout, Oliver & Niimi 1985)
    k_2: 0.063 d^{-1} (rainbow trout, Oliver & Niimi 1985)
    k_1: 14, 179, 196 h^{-1} (zebrafish: egg, yolk sac fry, juvenile, Görge & Nagel 1990)

k_2: 0.06 h^{-1} (*Chironomus riparius*-water only system, Lydy et al. 1992)
k_2: 0.0661 h^{-1} (*Chironomus riparius*-screened system, Lydy et al. 1992)
k_2: 0.08 h^{-1} (*Chironomus riparius*-3% organic carbon system, Lydy et al. 1992)
k_2: 0.0661 h^{-1} (*Chironomus riparius*-15% organic carbon system, Lydy et al. 1992)
k_1: 9.0-26.4 h^{-1} (rainbow trout in early life stages on wet wt. basis, Vigano et al. 1992)
k_2: 0.04-0.18 h^{-1} (rainbow trout in early life stages on wet wt. basis, Vigano et al. 1992)
k_1: 180-939 h^{-1} (rainbow trout in early life stages on lipid basis, Vigano et al. 1992)
k_2: 0.031-0.13 h^{-1} (rainbow trout in early life stages on lipid basis, Vigano et al. 1992)

Common Name: Malathion
Synonym: American Cyanamid 4049, Calmathion, Carbethoxy malathion, Carbetovur, Carbetox, Carbofos, Carbophos, Celthion, Chemathion, Cimexan, Cythion, Detmol MA, EL 4049, Emmatos, Emmatos extra, ENT 17034, Ethiolacar, Etio, Fog 3, Formal, Forthion, Fosfothion, Fyfanon, Hithion, Karbofos, Kop-thion, Kypfos, Malacide, Malafor, Malakill, Malagran, Malamar, Malaphele, Malaphos, Malasol, Malaspray, Malatol, Malatox, Maldison, Malmed, Malphos, Maltox, Mercaptothion, MLT, Moscardia, NA 2783, NCI-C00215, Oleophosphothion, Orthomalathion, Phosphothion, Prioderm, Sadofos, Sadophos, SF 60, Siptox I, Sumitox, Tak, TM-4049, Vegfru malatox, Vetiol, Zithiol
Chemical Name: S-[1,2-bis(ethoxycarbonyl)ethyl] O,O-dimethyl phosphorodithioate
Uses: as insecticide to control sucking and chewing insects and spider mites on vegetables, fruits, ornamentals, field crops in greenhouses, gardens and forestry; also used as acaricide.
CAS Registry No: 121-75-5
Molecular Formula: $C_{10}H_{19}O_6PS_2$
Molecular Weight: 330.36
Melting Point (°C):
    2.85   (NIEHS 1975; Freed et al. 1977; Khan 1980; Montgomery 1993; Tomlin 1994; Worthing 1991)
    2.90   (Suntio et al. 1988; Yalkowsky & Banerjee 1992)
Boiling Point (°C):
    120   (at 0.2 mmHg, Melnikov 1971; Freed et al. 1977)
    156-157   (at 0.7 mmHg, Agrochemicals Handbook 1987; Worthing 1991; Montgomery 1993; Tomlin 1994)
Density (g/cm³ at 20°C):
    1.23   (25°C, Spencer 1982; Agrochemicals Handbook 1987; Worthing 1991; Montgomery 1993)
Molar Volume (cm³/mol):
    319.1   (calculated-LeBas method, Suntio et al. 1988; Fisher et al. 1993)
    1.421   (intrinsic $V_I/100$, Fisher et al. 1993)
Molecular Volume (Å³):
Total Surface Area, TSA (Å²):
Dissociation Constant pK_a:
Heat of Fusion, ΔH_{fus}, kcal/mol:
Entropy of Fusion, ΔS_{fus}, cal/mol·K (e.u.):
Fugacity Ratio at 25°C (assuming ΔS_{fus} = 13.5 e.u.), F: 1.0

Water Solubility (g/m³ or mg/L at 25°C):
    145   (20°C, Macy 1948; quoted, Chiou et al. 1977)
    145   (Spiller 1961; quoted, Shiu et al. 1990)
    145   (20°C, Melnikov 1971; Spencer 1973; quoted, Chiou et al. 1977; Freed et al. 1977; Belluck & Felsot 1981; Shiu et al. 1990)
    145   (rm. temp., Spencer 1973; Martin & Worthing 1977; Worthing 1979; quoted, Kenaga 1980; Kenaga & Goring 1980; Karickhoff 1981; Bowman & Sans 1983a,b; Kim et al. 1984; Pait et al. 1992)
    150   (Hartley & Graham-Bryce 1980; Beste & Humburg 1983; quoted, Taylor & Glotfelty 1988; Shiu et al. 1990)

145   (Kenaga 1980a; quoted, Lyman 1982; Suntio et al. 1988; Shiu et al. 1990; Nash 1988; Schomburg et al. 1991)
145   (22°C, Khan 1980)
145   (Willis & McDowell 1982; Agrochemicals Handbook 1987; Tomlin 1994)
143   (20°C, shake flask-GC, Bowman & Sans 1983a,b; quoted, Shiu et al. 1990; Howard 1991; Patil 1994)
145   (rm. temp., Worthing 1987, 1991; quoted, Shiu et al. 1990)
145   (20°C, selected, Suntio et al. 1988; quoted, Majewski & Capel 1995)
145   (selected, USDA 1989; quoted, Neary et al. 1993)
130   (20-25°C, selected, Wauchope et al. 1992; Lohninger 1994; Hornsby et al. 1996)
144   (selected, Yalkowsky & Banerjee 1992)
145, 164 (20°C, 30°C, Montgomery 1993)
143, 166 (quoted, calculated, Patil 1994)
141, 109 (quoted, calcd-group contribution fragmentation method, Kühne et al. 1995)
130   (selected, Halfon et al. 1996)

Vapor Pressure (Pa at 25°C):
1.67×10^{-4} (20°C, Wolfdietrich 1965; quoted, Kim et al. 1984)
1.67×10^{-4} (20°C, Melnikov 1971; quoted, Freed et al. 1977)
7.33×10^{-4} (20°C, gravimetric method, Gückel et al. 1973; quoted, Freed 1976; Suntio et al. 1988)
2.90×10^{-3} (Woolford 1975; quoted, Hinckley et al. 1990)
1.30×10^{-3} (20°C, Hartley & Graham-Bryce 1980; quoted, Taylor & Glotfelty 1988; Taylor & Spencer 1990)
5.30×10^{-3} (30°C, Khan 1980)
9.20×10^{-4} (20°C, GC, Seiber et al. 1981; quoted, Suntio et al. 1988)
1.05×10^{-3} (20°C, Kim et al. 1984; quoted, Howard 1991)
0.60×10^{-3} (20°C, extrapolated value, Kim et al. 1984)
1.06×10^{-3} (gas saturation method, Kim et al. 1984; Kim 1985; quoted, Suntio et al. 1988; Hinckley et al. 1990)
0.67×10^{-3} (20°C, GC-calculated value, Kim et al. 1984; Kim 1985)
0.60×10^{-3} (20°C, gas saturation method, Kim 1985)
5.30×10^{-3} (30°C, Agrochemicals Handbook 1987; Tomlin 1994)
1.00×10^{-3} (20°C, selected, Suntio et al. 1988; quoted, Majewski & Capel 1995)
4.70×10^{-3} (GC-RT, supercooled value, Hinckley et al. 1990)
1.30×10^{-3} (20°C, selected, Taylor & Spencer 1990)
1.07×10^{-3} (20-25°C, selected, Wauchope et al. 1992; Hornsby et al. 1996)
1.67×10^{-4} (20°C, Montgomery 1993)
1.07×10^{-3} (selected, Halfon et al. 1996)

Henry's Law Constant (Pa·m^3/mol):
0.038   (calculated-P/C, Mackay & Shiu 1981; quoted, Suntio et al. 1988)
2.30×10^{-3} (20°C, calculated-P/C, Suntio et al. 1988; quoted, Fisher et al. 1993; Majewski & Capel 1995)
3.22×10^{-3} (calculated-P/C, Taylor & Glotfelty 1988)
2.03×10^{-3} (calculated-P/C, Howard 1991)

0.49×10^{-3} (23°C, quoted, Schomburg et al. 1991)
0.49×10^{-3} (Montgomery 1993)
2.28×10^{-3} (calculated-P/C, this work)

Octanol/Water Partition Coefficient, log K_{OW}:
- 2.89 (20°C, shake flask-GC, Chiou et al. 1977; quoted, Kenaga & Goring 1980; Karickhoff 1981; Yoshioka et al. 1986; Suntio et al. 1988; De Bruijn & Hermens 1991; Finizio et al. 1997)
- 2.89 (shake flask-GC, Freed et al. 1979; Yoshioka et al. 1986)
- 2.36 (Hansch & Leo 1979; quoted, Fisher et al. 1993)
- 2.36 (Rao & Davidson 1980; quoted, Kim et al. 1984; Suntio et al. 1988)
- 3.23 (calculated as per Chiou et al. 1977, Belluck & Felsot 1981)
- 2.82 (shake flask-GC/FID, Hermens & Leeuwangh 1982; quoted, Hermens et al. 1985; Verhaar et al. 1992)
- 2.89 (quoted, Lyman 1982)
- 2.838 (shake flask-GC, Bowman & Sans 1983b; quoted, Suntio et al. 1988; De Bruijn & Hermens 1991; Sicbaldi & Finizio 1993; Patil 1994; Finizio et al. 1997)
- 2.36 (Hansch & Leo 1985; quoted, Howard 1991)
- 2.80 (selected, Suntio et al. 1988)
- 2.89 (selected, Travis & Arms 1988)
- 2.90 (selected, Thomann 1989)
- 2.94 (slow-stirring method, De Bruijn et al. 1991; quoted, Verhaar et al. 1992)
- 2.75 (Worthing 1991; Tomlin 1994)
- 2.36-2.89 (Montgomery 1993)
- 2.68 (RP-HPLC, Saito et al. 1993)
- 2.18 (RP-HPLC, Sicbaldi & Finizio 1993)
- 2.84, 2.79 (quoted, calculated, Patil 1994)
- 2.36 (selected, Hansch et al. 1995)
- 2.18, 2.31, 3.06 (RP-HPLC, ClogP, calculated-S, Finizio et al. 1997)

Bioconcentration Factor, log BCF:
- 1.11 (carp, calcd. from k_1 of Bender 1969, McLeese et al. 1976)
- −4.74 (beef biotransfer factor logB_b, correlated-K_{OW}, Pasarela et al. 1962; quoted, Travis & Arms 1988)
- 0.867, 1.47 (lake trout, coho salmon, Walsh & Ribelin 1973; quoted, Howard 1991)
- 2.94, 2.98 (white shrimp, brown shrimp, Conte & Parker 1975; quoted, Howard 1991)
- 1.57 (calculated-S, Kenaga 1980a; quoted, Howard 1991)
- 0.40 (*Triaenodes tardus*, Belluck & Felsot 1981)
- 1.54 (willow shiner, Tsuda et al. 1989)
- 0.85 (carp, wet wt. basis, De Bruijn & Hermens 1991)
- 2.00 (topmouth gudgeon, wet wt. basis, De Bruijn & Hermens 1991)
- 1.57 (Pait et al. 1992)

Sorption Partition Coefficient, log K_{OC}:
- 2.45 (soil, calculated-S as per Kenaga & Goring 1978, Kenaga 1980a; quoted, Nash 1988; Howard 1991; Schomburg et al. 1991)
- 3.26 (av. soils/sediments, Rao & Davidson 1980; quoted, Lyman 1982)
- 3.25 (Rao & Davidson 1980; quoted, Howard 1991; Schomburg et al. 1991)
- 3.25 (Karickhoff 1981; quoted, Howard 1991)
- 2.83, 3.29 (estimated-S, S & M.P., Karickhoff 1981)
- 2.50 (estimated-K_{OW}, Karickhoff 1981)
- 2.36 (Bomberger et al. 1983; quoted, Howard 1991)
- 3.25 (screening model calculations, Jury et al. 1987b)
- 0.903 (selected, USDA 1989; quoted, Neary et al. 1993)
- 3.26 (soil, 20-25°C, selected, Wauchope et al. 1992; quoted, Dowd et al. 1993; Lohninger 1994; Hornsby et al. 1996)
- 2.61 (Montgomery 1993)

Half-Lives in the Environment:

Air: 1.0-9.8 h, based on an estimated rate constant for the vapor-phase reaction with hydroxyl radicals in air (Atkinson 1987; quoted, Howard et al. 1991); calculated lifetime of 3 h for the vapor-phase reaction with OH radicals in the troposphere (Atkinson et al. 1992).

Surface water: persistence of up to 4 weeks in river water (Eichelberger & Lichtenberg 1971); half-life of 100-1236 h, based on unacclimated aerobic river die-away test data (Eichelberger & Lichtenberg 1971; quoted, Howard et al. 1991) and estuarine water grab sample data (Walker 1978; quoted, Howard et al. 1991); half-life in Indian River water (24 ppt salinity; pH 8.16) was 1.65 d (Wang & Hoffman 1991); half-lives of 212 d at 6°C, 42 d at 22°C in darkness for Milli-Q water, pH 6.1; 55 d at 6°C, 19 d at 22°C in darkness, 8 d under sunlight conditions for river water at pH 7.3; 53 d at 6°C, 7 d at 22°C in darkness for filtered river water at pH 7.3; 41 d at 6°C, 6 d at 22°C in darkness, 14 d under sunlight conditions for seawater at pH 8.1 (Lartiges & Garrigues 1995).

Groundwater: 200-2472 h, based on estimated aqueous aerobic biodegradation half-life (Howard et al. 1991).

Sediment:

Soil: estimated persistence of one week (Kearney et al. 1969; Edwards 1973; quoted, Morrill et al. 1982; Jury et al. 1987); 72-168 h, based on unacclimated aerobic soil grab sample data (Walker & Stojanovic 1973; quoted, Howard et al. 1991); biodegradation rate constant of 1.4 d^{-1} in soil (Rao & Davidson 1980; quoted, Scow 1982); non-persistent in soil with half-life of <20 d (Willis & McDowell 1982); half-life of 1 d in screening model simulations (Jury et al. 1987b); selected field half-life of 1.0 d (Wauchope et al. 1992; Dowd et al. 1993; Halfon et al. 1996; Hornsby et al. 1996); soil half-life of 11 d (Pait et al. 1992); half-lives of 1 day for soil depth <5 cm, 7 d for soil depth 5-20 cm and 14 d for soil depth >20 cm (Dowd et al. 1993).

Biota: biochemical half-life of 1 d from screening model calculations (Jury et al. 1987b); excretion half-life of 1.4 h (willow shiner, Tsuda et al. 1989); average half-life of 20 d in the forest (USDA 1989; quoted, Neary et al. 1993).

Environmental Fate Rate Constants or Half-Lives:
- Volatilization: half-life of 131 d, based on volatilization rate from water with a wind speed of 0-2.5 m/s (Sanders & Seiber 1984; quoted, Howard 1991).
- Photolysis: half-life of 15 h for direct sunlight photolysis in aqueous media (Wolfe et al. 1976; quoted, Harris 1982); both atmospheric and aqueous photolysis half-life of 990-20000 h, based on experimental photolysis rate constant in aqueous solution at pH 6 exposure to >290 nm under summer sunlight at 30°N (Wolfe et al. 1976, 1977; quoted, Howard et al. 1991).
- Oxidation: photooxidation half-life of 1.0-9.8 h, based on an estimated rate constant for the vapor-phase reaction with hydroxyl radicals in air (Atkinson 1987; quoted, Howard et al. 1991); calculated rate constant of 64×10^{-12} cm^3/molecule·s for the vapor-phase reaction with hydroxyl radicals in air (Winer & Atkinson 1990); calculated lifetime of 3 h for the vapor-phase reaction with OH radicals in the troposphere (Atkinson et al. 1992).
- Hydrolysis: half-life of 10.5 d at pH 7.4 and 20°C (NIEHS 1975; quoted, Freed et al. 1977, 1979; Montgomery 1993); half-life of 120 d at pH 6.1 and 11 d at pH 7.4 in water and soil at 20°C as per Ruzicka et al. 1967 using gas chromatographic-retention time method for hydrolysis rates determination (Freed et al. 1979; quoted, Montgomery 1993); first-order half-life of 8.8 years, based on reported rate constant of 2.5×10^{-2} M^{-1} s^{-1} at pH 7 and 0°C (Wolfe et al. 1977; quoted, Howard et al. 1991); disappearance rate constants at 27°C: 4.8×10^{-5} M^{-1} s^{-1} (acid hydrolysis), 7.7×10^{-9} s^{-1} (neutral hydrolysis), 5.5×10^{0} M^{-1} s^{-1} (alkaline hydrolysis); all for 10^{-4} M in 1% acetonitrile and water (Wolfe et al. 1977; quoted, Harris 1982); half-life of 9 d at pH 6 (Montgomery 1993).
- Biodegradation: aqueous aerobic biodegradation half-life of 100-1236 h, based on unacclimated aerobic river die-away test data (Eichelberger & Lichtenberg 1971; quoted, Howard et al. 1991) and estuarine water grab sample data (Walker 1978; quoted, Howard et al. 1991); rate constant of 1.4 d^{-1} in soil (Rao & Davidson 1980; quoted, Scow 1982); biodegradation rate constants in aquatic systems: 1.9×10^{-1} mg(g fungi)$^{-1}$ d^{-1} (Lewis et al. 1975; quoted, Scow 1982); 6.2×10^{-8} mL(cell)$^{-1}$ d^{-1} (Paris et al. 1975; quoted, Scow 1982), 5.0×10^{-8} mL(cell)$^{-1}$ d^{-1} (Baughman & Lassiter 1978; quoted, Scow 1982), and $2.6-16.1 \times 10^{-7}$ mL(cell)$^{-1}$ d^{-1} (Paris et al. 1978; quoted, Scow 1982); primary biodegradation rate constant of $(4.5 \pm 0.74) \times 10^{-11}$ L(cell)$^{-1}$ h^{-1} in North American waters (Paris et al. 1981; quoted, Battersby 1990); aqueous aerobic biodegradation half-life of 200-2472 h, based on estimated aqueous aerobic biodegradation half-life (Howard et al. 1991).

Biotransformation:

Bioconcentration, Uptake (k_1) and Elimination (k_2) Rate Constants:
- k_2: 0.49 h^{-1} (willow shiner, Tsuda et al. 1989)
- k_1: 1.07 d^{-1} (carp, Bender 1969; quoted, McLeese et al. 1976)
- k_2: 0.08 d^{-1} (carp, calcd. from k_1 of Bender 1969, McLeese et al. 1976)

Common Name: Methiocarb
Synonym: Bayer 37344, Draza, Ensurol, Mercaptodimethur, Mesurol, Mesurol Phenol, metmercapturon
Chemical Name: 4-methylthio-3,5-xylyl methylcarbamate; 3,5-dimethyl-4-(methylthio)phenol methylcarbamate
Uses: Insecticide/Acaricide/Molluscicide/Repellent; to control slugs and snails in a wide range of agricultural situations; broad range control of lepidoptera, coleoptera, diptera, and homoptera and spider mites in pome fruit, stone fruit, citrus fruit, strawberries, hops, potatoes, beet, maize, oilseed rape, vegetables and ornamentals; also used as a bird repellent.
CAS Registry No: 2032-65-7
Molecular Formula: $C_{11}H_{15}NO_2S$
Molecular Weight: 225.3
Melting Point (°C):
    117-118 (Khan 1980)
    117.5 (Patil 1994)
    119.0 (Tomlin 1994)
    119, 121.5 (Milne 1995)
Boiling Point (°C):
Density (g/cm³ at 20°C):
    1.236 (Tomlin 1994)
Molar Volume (cm³/mol):
    261.4 (calculated-LeBas method, this work)
    182.3 (calculated-density, this work)
Molecular Volume (Å³):
Total Surface Area, TSA (Å²):
Dissociation Constant pK_a:
Heat of Fusion, ΔH_{fus}, kcal/mol:
Entropy of Fusion, ΔS_{fus}, cal/mol·K (e.u.):
Fugacity Ratio at 25°C (assuming ΔS_{fus} = 13.5 e.u.), F: 0.123

Water Solubility (g/m³ or mg/L at 25°C):
    30 (20°C, Agrochemicals Handbook 1987)
    30 (20°C, Worthing 1987; quoted, Shiu et al. 1990)
    30, 400 (quoted, calculated, Patil 1994)
    30 (selected, Lohninger 1994)
    27 (20°C, Tomlin 1994)
    30 (20°C, Milne 1995)
    24 (20-25°C, selected, Hornsby et al. 1996)

Vapor Pressure (Pa at 25°C):
    0.015 (60°C, Agrochemicals Handbook 1987)
    1.5×10^{-5} (20°C, Tomlin 1994)
    3.6×10^{-5} (Tomlin 1994)
    0.016 (20-25°C, selected, Hornsby et al. 1996)

Henry's Law Constant (Pa·m^3/mol):
    0.120   (calculated-P/C, this work)

Octanol/Water Partition Coefficient, log K_{OW}:
    2.92   (shake flask as per Fujita et al. 1964; Briggs 1981)
    2.92   (selected, Magee 1991)
    2.92, 1.86   (quoted, calculated, Patil 1994)
    3.34   (Tomlin 1994)
    2.92   (selected, Hansch et al. 1995)

Bioconcentration Factor, log BCF:
    1.96   (calculated-S as per Kenaga 1980, this work)
    1.69   (calculated-K_{OW} as per Kenaga 1980, this work)

Sorption Partition Coefficient, log K_{OC}:
    2.08   (20 ± 2 °C, reported as log K_{OM}, Briggs 1981)
    2.08   (reported as log K_{OM}, Magee 1991)
    2.33   (estimated as log K_{OM}, Magee 1991)
    2.32   (selected, Lohninger 1994)
    2.82   (soil, HPLC-screening method, Kördel et al. 1995)
    2.48   (20-25°C, estimated, Hornsby et al. 1996)

Half-Lives in the Environment:
    Air:
    Surface water:
    Groundwater:
    Sediment:
    Soil: field half-life of 30 d (20-25°C, estimated, Hornsby et al. 1996).
    Biota:

Environmental Fate Rate Constants or Half-Lives:
    Volatilization:
    Photolysis: photodegradation half-life of 6-16 d (Tomlin 1994).
    Oxidation:
    Hydrolysis:
    Biodegradation:
    Biotransformation:
    Bioconcentration, Uptake (k_1) and Elimination (k_2) Rate Constants:

Common Name: Methomyl

Synonym: Du Pont 1179, ENT 27341, Lannate, Mesomile, Nu-bait II, Nudrin, SD 14999, WL 18236

Chemical Name: S-methyl-N-(methylcarbamoyloxy) thioacetimidate; methyl-N-(((methylamino)-carbonyl)oxy)ethanimidothioate

Uses: insecticide/acaricide; control a wide range of insects and spider mites in fruit, vines, olives, hops, vegetables, ornamentals, field crops, cucurbits, flax, cotton, soya beans, etc.; also used for control of flies in animal and poultry houses and dairies.

CAS Registry No: 16752-77-5
Molecular Formula: $C_5H_{10}N_2O_2S$
Molecular Weight: 162.2
Melting Point (°C):
    78-79 (Khan 1980; Spencer 1982; Suntio et al. 1988; Worthing 1991; Montgomery 1993; Tomlin 1994; Milne 1995)

Boiling Point (°C):
Density (g/cm³ at 20°C):
    1.2946 (25°C, Spencer 1982; Worthing 1991; Tomlin 1994)
    1.2946 (24°C, Milne 1995; Montgomery 1993)

Molar Volume (cm³/mol):
    179.9 (calculated-LeBas method, Suntio et al. 1988)

Molecular Volume (Å³):
Total Surface Area, TSA (Å²):
Dissociation Constant pK_a:
Heat of Fusion, ΔH_{fus}, kcal/mol:
    5.80 (DSC method, Plato 1972)

Entropy of Fusion, ΔS_{fus}, cal/mol·K (e.u.):
Fugacity Ratio at 25°C (assuming ΔS_{fus} = 13.5 e.u.), F: 0.2924

Water Solubility (g/m³ or mg/L at 25°C):
    58000 (Worthing 1979; Khan 1980; quoted, Bowman & Sans 1983a; Suntio et al. 1988; Shiu et al. 1990)
    10000 (Kenaga 1980a; Kenaga & Goring 1980; quoted, Suntio et al. 1988; Shiu et al. 1990)
    soluble (Spencer 19982)
    >1000 (20°C, shake flask-GC, Bowman & Sans 1983a; quoted, Shiu et al. 1990)
    579520 (20°C, Bowman & Sans 1983b; quoted, Shiu et al. 1990)
    58000 (Worthing 1983, 1987, 1991; quoted, Shiu et al. 1990; Howard 1991)
    57900 (Agrochemicals Handbook 1987; Tomlin 1994)
    10000 (20°C, selected, Suntio et al. 1988; quoted, Majewski & Capel 1995)
    57900 (Montgomery 1993)
    58000 (selected, Lohninger 1994)
    57000 (Milne 1995)
    58000 (20-25°C, selected, Hornsby et al. 1996)

Vapor Pressure (Pa at 25°C):
    3.47×10^{-3} (20°C, Hartley & Graham-Bryce 1980; quoted, Suntio et al. 1988)
    6.66×10^{-3} (Khan 1980; Spencer 1982; quoted, Suntio et al. 1988)
    0.162 (30°C, GC, Seiber et al. 1981; quoted, Suntio et al. 1988)

6.67×10^{-3} (Worthing 1983; quoted, Howard 1991)
6.65×10^{-3} (Agrochemicals Handbook 1987; Tomlin 1994)
4.00×10^{-3} (20°C, selected, Suntio et al. 1988; quoted, Majewski & Capel 1995)
6.65×10^{-3} (Worthing 1991)
6.65×10^{-3} (Montgomery 1993)
6.67×10^{-3} (20-25°C, selected, Hornsby et al. 1996)

Henry's Law Constant (Pa·m^3/mol):
    1.82×10^{-5} (calculated, Lyman et al. 1982; quoted, Howard 1991)
    6.50×10^{-5} (20°C, calculated-P/C, Suntio et al. 1988; quoted, Majewski & Capel 1995)
    6.48×10^{-5} (calculated-P/C, Montgomery 1993)
    1.87×10^{-5} (calculated-P/C, this work)

Octanol/Water Partition Coefficient, log K_{OW}:
    0.30   (quoted from Dow Chemical data, Kenaga & Goring 1980)
    1.08   (Rao & Davidson 1980; quoted, Suntio et al. 1988)
    0.13   (Bowman & Sans 1983b; quoted, Suntio et al. 1988)
    0.60   (Hansch & Leo 1985; quoted, Howard 1991)
    0.20   (selected, Suntio et al. 1988)
    1.434   (calculated as per Broto et al. 1984, Karcher & Devillers 1990)
    0.13, 1.08   (Montgomery 1993)
    0.09   (Tomlin 1994)
    0.60   (selected, Hansch et al. 1995)

Bioconcentration Factor, log BCF:
    0.477   (calculated-S, Kenaga 1980)
    0.903   (calculated-K_{OC}, Kenaga 1980)
    0.23   (calculated-K_{OW}, Lyman et al. 1982; quoted, Howard 1991)
    0.11   (calculated-S, Lyman et al. 1982; quoted, Howard 1991)

Sorption Partition Coefficient, log K_{OC}:
    2.20   (soil, Fung & Uren 1977; quoted, Kenaga 1980; Kenaga & Goring 1980)
    1.45   (soil, calculated-S as per Kenaga & Goring 1978, Kenaga 1980)
    1.71   (calculated-K_{OW}, Lyman et al. 1982; quoted, Howard 1991)
    1.00   (calculated-S, Lyman et al. 1982; quoted, Howard 1991)
    2.20   (Worthing 1983; quoted, Howard 1991)
    1.30   (soil, quoted exptl., Meylan et al. 1992)
    1.08   (soil, calculated-χ and fragments contribution, Meylan et al. 1992)
    1.86, 2.20   (Montgomery 1993)
    1.86   (estimated-chemical structure, Lohninger 1994)
    1.86   (Tomlin 1994)
    1.86   (soil, 20-25°C, selected, Hornsby et al. 1996)

Half-Lives in the Environment:
- Air: estimated to be 1.14 months, based on rate constant of 2.919×10^{-13} cm^3/molecule·s for the vapor-phase reaction with 8×10^5/cm^3 hydroxyl radicals in air (GEMS 1986; quoted, Howard 1991).
- Surface water: experimental half-life of 262 d has been determined in pure water at 25°C (Ellington et al. 1988; quoted, Howard 1991).
- Groundwater: half-life of <0.2 d in groundwater samples (Smelt 1983; quoted, Tomlin 1994).
- Sediment:
- Soil: field half-life of 30 d (20-25°C, selected, Hornsby et al. 1996).
- Biota: half-life of ca. 3-5 d in plants following leaf application (Harvey & Reiser 1973; quoted, Tomlin 1994); half-life of 0.4-8.5 d on cotton plants, 0.8-1.2 d on mint plants and approximately 2.5 d on Bermuda grass (Willis & McDowell 1987; quoted, Howard 1991).

Environmental Fate Rate Constants or Half-Lives:
- Volatilization:
- Photolysis:
- Oxidation: photooxidation half-life estimated to be 1.14 months, based on vapor-phase reaction with hydroxyl radicals in air (GEMS 1986; quoted, Howard 1991).
- Hydrolysis: experimental half-life of 262 d from rate constant of 8.9×10^5 h^{-1} has been determined in pure water at pH 7 and 25°C (Ellington et al. 1987,88; quoted, Howard 1991; Montgomery 1993).
- Biodegradation: rate constants of -0.000215 h^{-1} in nonsterile sediment, -0.000747 h^{-1} in sterile sediment by shake-tests at Range Point and -0.000175 h^{-1} in nonsterile water and -0.000383 h^{-1} in sterile water by shake-tests at Range Point (Walker et al. 1988).
- Biotransformation:
- Bioconcentration, Uptake (k_1) and Elimination (k_2) Rate Constants:

Common Name: Methoxychlor
Synonym: Chemform, Dimethoxy-DDT, DMDT, DMTD, ENT 1716, Maralate, Marlate, Methorcide, Methoxo, Metox, Moxie NCI-C00497
Chemical Name: 1,1,1-trichloro-2,2-bis(4-methoxyphenyl)ethane; 1,1'-(2,2,2-trichloroethylidene)bis[4-methoxybenzene]
Uses: insecticide to control mosquito larvae, house flies, and other insect pests in field crops, fruits, and vegetables; also to control ectoparasites on cattle, sheep, and goats.
CAS Registry No: 72-43-5
Molecular Formula: $C_{16}H_{15}Cl_3O_2$
Molecular Weight: 345.65
Melting Point (°C):
    89.0 (Karickhoff 1981; Noegrohati & Hammers 1992; Tomlin 1994; Kühne et al. 1995; Milne 1995)
    86-88, 89-98 (Montgomery 1993)
    77.0 (Patil 1994)
    78-78.2 (Milne 1995)
Boiling Point (°C):
Density (g/cm³ at 20°C):
    1.41 (25°C, Agrochemicals Handbook 1987; Montgomery 1993; Tomlin 1994; Milne 1995)
Molar Volume (cm³/mol):
    279.5 (Ruelle & Kesselring 1997)
    354.3 (calculated-LeBas method, this work)
    245.1 (calculated-density, this work)
Molecular Volume (Å³):
Total Surface Area, TSA (Å²):
Dissociation Constant pK_a:
Heat of Fusion, ΔH_{fus}, kcal/mol:
    6.60 (DSC method, Plato & Glasgow 1969)
    5.708 (Ruelle & Kesselring 1997)
Entropy of Fusion, ΔS_{fus}, cal/mol·K (e.u.):
Fugacity Ratio at 25°C (assuming ΔS_{fus} = 13.5 e.u.), F: 0.244

Water Solubility (g/m³ or mg/L at 25°C):
    0.10 (shake flask-UV, Richardson & Miller 1960; quoted, Spencer 1973; Freed 1976; Shiu et al. 1990)
    0.62 (Karpoor et al. 1970; quoted, Belluck & Felsot 1981)
    0.045 (particle size of ≤5.0 µ, shake flask-GC, Biggar & Riggs 1974; quoted, Shiu et al. 1990; quoted, Howard 1991)
    0.02, 0.045, 0.095, 0.185 (shake flask-GC/ECD, at 15, 25, 35, 45°C, or particle size 5µ or less, Biggar & Riggs 1974)
    0.003, 0.01, 0.045 (particle size of 0.01, 0.05 & 5.0 µ; shake flask-GC, Biggar & Riggs 1974)
    0.003 (shake flask-GC, Biggar & Riggs 1974; quoted, Kenaga 1980a,b; Kenaga & Goring 1980; Bruggeman et al. 1981; Bysshe 1982; Adams 1987)
    0.10 (gen. col.-GC/ECD, Weil et al. 1974; quoted, Geyer et al. 1980; Shiu et al. 1990)
    0.12 (quoted from Zepp et al. 1976, Karickhoff et al. 1979; Karickhoff 1981)
    0.1-0.25 (Wauchope 1978)

0.10 (Weber et al. 1980; quoted, Willis & McDowell 1982)
0.10 (Worthing 1983, 1987, 1991; quoted, Shiu et al. 1990; Lohninger 1994)
0.10 (Agrochemicals Handbook 1987; Tomlin 1994)
0.10 (quoted, Isnard & Lambert 1988, 1989)
0.10 (20-25°C, selected, Wauchope et al. 1992; Hornsby et al. 1996)
0.04 (24°C, Montgomery 1993)
0.12, 0.097 (quoted, calculated, Patil 1994)
0.045, 0.025 (quoted, calculated-group contribution fragmentation method, Kühne et al. 1995)
0.10 (Milne 1995)
0.10; 0.077, 0.00009 (quoted; predicted-molar volume, Ruelle & Kesselring 1997)

Vapor Pressure (Pa at 25°C):
 $< 1.33 \times 10^{-4}$ (20-25°C, Weber et al. 1980; quoted, Willis & McDowell 1982)
 1.910×10^{-4} (estimated, USEPA 1988; quoted, Howard 1991)

Henry's Law Constant (Pa·m^3/mol):
 1.60 (estimated, Hine & Mookerjee 1975; quoted, Howard 1991)
 0.999 (calculated-P/C, this work)

Octanol/Water Partition Coefficient, log K_{OW}:
 4.68 (Veith & Morris 1978; quoted, Kenaga 1980b; Kenaga & Goring 1980)
 5.08 (shake flask-UV, Karickhoff et al. 1979; Karickhoff 1981; quoted, Rao & Davidson 1980; Baughman & Paris 1981; McDuffie 1981; Hodson & Williams 1988; Noegrohati & Hammers 1992; Bintein & Devillers 1994)
 4.30 (Veith et al. 1979,80; quoted, Bysshe 1982; Veith & Kosian 1983; Renberg et al. 1985)
 4.20 (Mackay et al. 1980; quoted, Renberg et al. 1985)
 4.83 (calculated as per Chiou et al. 1977, Belluck & Felsot 1981)
 4.51 (HPLC-k', McDuffie 1981)
 4.30 (Veith & Kosian 1983; quoted, Saito et al. 1992)
 4.68 (selected, Dao et al. 1983; Thomann 1989)
 4.68-5.08 (Hansch & Leo 1985; quoted, Howard 1991)
 4.26 (RPTLC, Renberg et al. 1985)
 4.30 (quoted, Isnard & Lambert 1988, 1989)
 5.08 (estimated-QSAR & SPARC, Kollig 1993)
 3.31, 5.08 (Montgomery 1993)
 5.08, 5.67 (quoted, calculated, Patil 1994)
 5.08 (selected, Hansch et al. 1995; Devillers et al. 1996)
 4.58, 5.17, 5.15 (RP-HPLC, ClogP, calculated-S, Finizio et al. 1997)

Bioconcentration Factor, log BCF:
 −1.70 (bioaccumulation factor logBF, adipose tissue in female Albino rats, Harris et al. 1974; quoted, Geyer et al. 1980)
 4.68 (*Bacillus subtilis*, Paris et al. 1975; Paris & Lewis 1976; quoted, Baughman & Paris 1981)

3.08 (*Flavobacterium harrisonii*, Paris et al. 1975; Paris & Lewis 1976; quoted, Baughman & Paris 1981)
3.72 (*Aspergillus sp.*, Paris et al. 1975; Paris & Lewis 1976; quoted, Baughman & Paris 1981)
3.92 (*Chlorella pyrenoidosa*, Paris et al. 1975; Paris & Lewis 1976; quoted, Baughman & Paris 1981)
4.40 (bacterial sorption, Paris & Lewis 1976; quoted, Baughman & Paris 1981)
2.14 (sheepshead minnow, Parrish et al. 1977; quoted, Howard 1991)
3.92, 3.72 (algae, fungi, Wolfe et al. 1977; quoted, Howard 1991)
3.92 (fathead minnows, 32 days exposure, Veith et al. 1979, 1980; quoted, Bysshe 1982; Veith & Kosian 1983; Howard 1991; Devillers et al. 1996)
3.70-3.93 (snail, Anderson & Defoe 1980; quoted, Howard 1991)
2.54-3.05 (stonefly, Anderson & Defoe 1980; quoted, Howard 1991)
2.27, 3.19 (fish: flowing water, static water; quoted, Kenaga 1980b; Kenaga & Goring 1980)
4.21, 3.91 (calculated-S, K_{OC}, Kenaga 1980)
1.15 (*Triaendoes tardus*, Belluck & Felsot 1981)
4.20, 3.04, 3.91 (estimated-S, K_{OW}, K_{OC}, Bysshe 1982)
3.92 (fathead minnows, Veith & Kosian 1983; quoted, Saito et al. 1992)
4.08 (mussel, Renberg et al. 1985; quoted, Howard 1991)
3.18 (soft clams, Hawker & Connell 1986; quoted, Howard 1991)
3.92 (calculated, Isnard & Lambert 1988)
5.40 (calculated field bioaccumulation, Thomann 1989)
5.29 (rainbow trout lipid base, estimated, Noegrohati & Hammers 1992)

Sorption Partition Coefficient, log K_{OC}:
2.79 (water-sediment, Wolfe et al. 1977; quoted, Howard 1991)
4.90 (av. for isotherms on sediments, Karickhoff et al. 1979; quoted, Karickhoff 1981)
3.99-4.61 (sand, Karickhoff et al. 1979; quoted, Howard 1991)
4.90-5.00 (fine silt, Karickhoff et al. 1979; quoted, Hodson & Williams 1988; Howard 1991)
4.86-4.96 (clay, Karickhoff et al. 1979; quoted, Hodson & Williams 1988; Howard 1991)
4.90 (soil, quoted, Kenaga 1980a,b; Kenaga & Goring 1980; Bysshe 1982)
5.03 (soil, calculated-S as per Kenaga & Goring 1978, Kenaga 1980)
6.04 (calculated-S, Mill et al. 1980; quoted, Adams 1987)
4.90 (av. soils/sediments, Rao & Davidson 1980; quoted, Lyman 1982)
4.67, 4.69, 5.54 (estimated-S, K_{OW}, S & M.P., Karickhoff 1981)
4.26 (soil, screening model calculations, Jury et al. 1987b)
4.90 (soil, quoted exptl., Meylan et al. 1992; Lohninger 1994)
4.63 (soil, calculated-χ and fragment contribution, Meylan et al. 1992)
4.90 (soil, 20-25°C, selected, Wauchope et al. 1992; Hornsby et al. 1996)
4.90 (estimated-QSAR & SPARC, Kollig 1993)
4.90, 4.95 (Montgomery 1993)

Half-Lives in the Environment:
- Air: estimated to be 1.12-11.2 h, based on rate constant for the vapor-phase reaction with hydroxyl radicals in air (Atkinson 1987; quoted, Howard et al. 1991); atmospheric transformation lifetime was estimated to be <1 d (Kelly et al. 1994).
- Surface water: 2.2-5.4 h, based on measured photooxidation in river water exposed to midday May sunlight (Zepp et al. 1976; quoted, Howard et al. 1991).
- Groundwater: 1200-8760 h, based on aerobic and anaerobic soil die-away test study data (Fogel et al. 1982; quoted, Howard et al. 1991).
- Sediment:
- Soil: half-lives: 1.5 week at pH 4.7 and 6.5 and 1.0 week at pH 7.8 (Carlo et al. 1952; quoted, Kaufman 1976); 4320-8760 hours, based on very slow biodegradation observed in an aerobic soil die-away test study data (Fogel et al. 1982; quoted, Howard et al. 1991); half-life of 42 d in scareening model calculations (Jury et al. 1987b); selected field half-life of 120 d (Wauchope et al. 1992; Hornsby et al. 1996).
- Biota: half-life of 0.4-8.5 d on cotton plants, 0.8-1.2 d on mint plants and approximately 2.5 d on Bermuda grass (Willis & McDowell 1987; quoted, Howard 1991).

Environmental Fate Rate Constants or Half-Lives:
- Volatilization: half-life of 4.5 d from water was estimated based on Henry's law constant for a model river 1 m deep with a current of 1 m/s and a wind speed of 3 m/sec (Lyman et al. 1982; quoted, Howard 1991).
- Photolysis: half-life of 300-2070 h in both air and water, based on measured photolysis rates in distilled water under midday sunlight (Wolfe et al. 1976; Zepp et al. 1976; quoted, Howard et al. 1991) and adjusted for approximate winter sunlight intensity (Lyman et al. 1982; quoted, Howard et al. 1991).
- Oxidation: photooxidation half-life of 2.2-5.4 h in water, based on measured photooxidation in river water exposed to midday May sunlight (Zepp et al. 1976; quoted, Howard et al. 1991); photooxidation half-life estimated to be 1.12-11.2 h in air, based on rate constant for the vapor-phase reaction with hydroxyl radicals in air (Atkinson 1987; quoted, Howard et al. 1991); calculated rate constant of 2×10^{10} M^{-1} s^{-1} for the reaction with hydroxyl radicals in aqueous solutions at $(24 \pm 1)°C$ (Haag & Yao 1992).
- Hydrolysis: first-order half-life of 1.05 years, based on neutral and base catalyzed hydrolysis rate constants (Kollig et al. 1987; quoted, Howard et al. 1991); rate constant of 0.60 y^{-1} at pH 7 and 25°C (Kollig 1993).
- Biodegradation: aqueous aerobic half-life of 4320-8760 h (6 months to 1 yr), based on very slow biodegradation observed in an aerobic soil die-away test study data (Fogel et al. 1982; quoted, Howard et al. 1991); aqueous anaerobic half-life of 1200-4320 h (50 d to 6 months), based on anaerobic soil die-away test study data (Fogel et al. 1982; quoted, Howard et al. 1991); first-order rate constant of -0.00236 h^{-1} in nonsterile sediment and -0.000639 h^{-1} in sterile sediment by shake-tests at Range Point and first-order rate constant of -0.000139 h^{-1} in nonsterile water and -0.00000327 h^{-1} in sterile water by shake-tests at Range Point (Walker et al. 1988).
- Biotransformation:
- Bioconcentration, Uptake (k_1) and Elimination (k_2) Rate Constants:

Common Name: Mevinphos
Synonym: Apavinfos, CMDP, Compound 2046, Duraphos, ENT 22374, Fosdrin, Gesfid, Gestid, Meniphos, Menite, NA 2783, OS 2046, PD 5, Phosdrin, Phosfene
Chemical Name: 2-carbomethoxy-1-methylvinyl dimethyl phosphate; 1-methoxycarbonyl-1-propen-2-yl dimethyl phosphate; methyl-3-(dimethoxyphosphinoyloxy)but-2-enoate 2-carbomethoxy-1-methylvinyl dimethyl phosphate
Uses: contact insecticide and acaricide to control chewing insects and spider mites in fruits, vegetables, and ornamentals.
CAS Registry No: 7786-34-7 [formerly 298-01-1 for (E) isomer & 338-45-4 for (Z) isomer]
Molecular Formula: $C_7H_{13}O_6P$
Molecular Weight: 224.1
Melting Point (°C):
    -56.1   (Montgomery 1993)
    21   ((E) and (Z) isomer, Milne 1995)
Boiling Point (°C):
    99-103   (at 0.03 mmHg, Martin 1971; Freed et al. 1977; Milne 1995)
    76.0   (at 0.2 mmHg, Melnikov 1971; Freed et al. 1979)
    110   (at 1.6 mmHg, Agrochemicals Handbook 1987)
    106-107.5   (at 1 mmHg, Montgomery 1993)
Density (g/cm³ at 20°C):
    1.24   (Agrochemicals Handbook 1987; Tomlin 1994; Milne 1995)
    1.25   (Montgomery 1993)
    1.235, 1.245   ((E) isomer, (Z) isomer, Tomlin 1995)
Molar Volume (cm³/mol):
    180.7   (calculated from density)
Molecular Volume (Å³):
Total Surface Area, TSA (Å²):
Dissociation Constant pK_a:
Heat of Fusion, ΔH_{fus}, kcal/mol:
Entropy of Fusion, ΔS_{fus}, cal/mol·K (e.u.):
Fugacity Ratio at 25°C (assuming ΔS_{fus} = 13.5 e.u.), F: 1.0

Water Solubility (g/m³ or mg/L at 25°C):
    miscible (Spencer 1973; Worthing 1979; Freed et al. 1979; quoted, Bowman & Sans 1983a; Sharom et al. 1980; Montgomery 1993)
    >2000   (shake flask-GC, Bowman & Sans 1983a; quoted, Shiu et al. 1990)
    miscible (Agrochemicals Handbook 1987; Tomlin 1994)
    miscible (Worthing 1987; quoted, Shiu et al. 1990)
    600000   (Wauchope 1989; quoted, Shiu et al. 1990; Lohninger 1994)
    >999   (selected, Boehncke et al. 1990)
    600000   (20-25°C, selected, Wauchope et al. 1992; Hornsby et al. 1996)

Vapor Pressure (Pa at 25°C):
    0.293   (20-25°C, Melnikov 1971; quoted, Freed et al. 1979)
    0.293   (20°C, quoted, experimental value, Kim et al. 1984; Kim 1985)
    0.0757   (20°C, GC-calculated value, Kim et al. 1984; Kim 1985)
    0.017   (20°C, Agrochemicals Handbook 1987; Tomlin 1994)
    0.0173   (20-25°C, selected, Wauchope et al. 1992; Hornsby et al. 1996)
    0.293   (20°C, Montgomery 1993)

Henry's Law Constant (Pa·m^3/mol):
  6.35×10^{-6} (calculated-P/C, this work)

Octanol/Water Partition Coefficient, log K_{OW}:
  0.845   (Melnikov 1971; quoted, Freed et al. 1977)
  0.954   (Velsicol Chem. 1972; quoted, Freed et al. 1977)
  0.550   (selected, Dao et al. 1983)
  0.200   (selected, Boehncke et al. 1990)
  0.130   (Tomlin 1994)
  1.20   (selected, Hansch et al. 1995)

Bioconcentration Factor, log BCF:

Sorption Partition Coefficient, log K_{OC}:
  1.64   (soil, 20-25°C, selected, Wauchope et al. 1992; Hornsby et al. 1996)
  1.64   (estimated-chemical structure, Lohninger 1994)

Half-Lives in the Environment:
  Air:
  Surface water:
  Groundwater:
  Sediment:
  Soil: selected field half-life of 3 d (Wauchope et al. 1992; Hornsby et al. 1996).
  Biota: estimated half-life of 19 ± 2 and 24 ± 7 h in lettuce in the summer and 20± 11 hours in the fall, 50 h in cauliflower in the summer and 18 ± 1 hours in the fall, 25 ± 2 h in celery in the summer and 16 hours in the fall (Spencer et al. 1992).

Environmental Fate Rate Constants or Half-Lives:
  Volatilization:
  Photolysis:
  Oxidation:
  Hydrolysis: half-lives: 1.8 h for *cis*- and 3.0 h for *trans*-isomer at pH 11.6 (Casida et al. 1956; quoted, Montgomery 1993); half-life of 30-35 d (Melnikov 1971; quoted, Freed et al. 1977); half-lives: 120 d at pH 6, 35 d at pH 7, 3 days at pH 9, and 1.4 h at pH 11 (Montgomery 1993; Tomlin 1994).
  Biodegradation:
  Biotransformation:
  Bioconcentration, Uptake (k_1) and Elimination (k_2) Rate Constants:

Common Name: Mirex
Synonym: Bichlorendo, Declorane, ENT 25719, Ferriamicide, Paramex, Perclordecone
Chemical Name: 1,1a,2,2,3,3a,4,5,5,5a,5b,6-dodecachloro-octahydro-1,3,4-methano-1H-cyclobuta(cd) pentalene; dodecachloro-pentacyclodecane
Uses: Insecticide.
CAS Registry No: 2385-85-5
Molecular Formula: $C_{10}Cl_{12}$
Molecular Weight: 545.59
Melting Point (°C):
    485   (dec., Smith et al. 1978; Spencer 1982; Suntio et al. 1988; Kühne et al. 1995; Milne 1995)
Boiling Point (°C):
Molar Volume (cm³/mol):
    403.2   (calculated-LeBas method, Suntio et al. 1988)
    291.6   (Ruelle & Kesselring 1997)
Molecular Volume (Å³):
Total Surface Area, TSA (Å²):
Dissociation Constant pK_a:
Heat of Fusion, ΔH_{fus}, kcal/mol:
Entropy of Fusion, ΔS_{fus}, cal/mol·K (e.u.):
Fugacity Ratio, at 25°C (assuming ΔS_{fus} = 13.5 e.u.), F: 2.82×10^{-5}

Water Solubility (g/m³ or mg/L at 25°C):
    0.001   (quoted from D. Dollar of Miss. State Chem. Lab. unpublished results, Alley 1973)
    0.085   (shake flask-LSC, Metcalf et al. 1973)
    0.60   (Neely 1978; quoted, Kenaga 1980; Kenaga & Goring 1980)
    7.0×10^{-5} (22°C, shake flask-GC, Smith et al. 1978; quoted, McKim et al. 1985; selected, Patterson & Mackay 1985; Clark et al. 1988; Suntio et al. 1988)
    7.0×10^{-5} (quoted, Geyer et al. 1980)
    0.005   (selected, Neely 1980)
    0.02   (24°C, Verschueren 1983; quoted, Clark et al. 1988)
    6.5×10^{-5} (20°C, selected, Suntio et al. 1988)
    0.001   (quoted, Isnard & Lambert 1988, 1989)
    3.0×10^{-5} (10°C, estimated, Mclachlan et al. 1990)
    7.0×10^{-5} (20-25°C, estimated, Augustijn-Beckers et al. 1994; Hornsby et al. 1996)
    0.086   (quoted, Kühne et al. 1995)
    0.0019   (calculated-group contribution fragmentation method, Kühne et al. 1995)
    7.0×10^{-5}; 3.1×10^{-5}, 1.1×10^{-6} (quoted; predicted-molar volume, Ruelle & Kesselring 1997)

Vapor Pressure (Pa at 25°C):
    8.0×10^{-4} (50°C, Smith, et al. 1978; quoted, Suntio et al. 1988)
    1.3×10^{-4} (20°C, Smith et al. 1978; selected, Patterson & Mackay 1985; Suntio et al. 1988)
    1.0×10^{-4} (20°C, selected, Suntio et al. 1988; quoted, Bacci et al. 1990)
    9.0×10^{-7} (10°C, estimated, Mclachlan et al. 1990)
    2.5×10^{-4}, 2.9×10^{-4}, 2.8×10^{-4} (GC-RT, supercooled liquid, Hinckley et al. 1990)
    5.2×10^{-5} (12°C, extrapolated supercooled liquid value, Hinckley et al. 1990)
    1.1×10^{-4} (20-25°C, selected, Augustijn-Beckers et al. 1994; Hornsby et al. 1996)

Henry's Law Constant (Pa·m³/mol):
- 1013 (20°C, calculated, Smith et al. 1978; quoted, Suntio et al. 1988)
- 53.2 (22°C, gas stripping, Yin & Hassett 1986, quoted, Suntio et al. 1988, Calamari et al. 1991; Comba et al. 1993; Müller et al. 1994)
- 840 (20°C, calculated-P/C, Suntio et al. 1988; quoted, Bacci et al. 1990)

Octanol/Water Partition Coefficient, log K_{OW}:
- 7.50 (Hansch & Leo 1979; quoted, McKim et al. 1985; Thomann 1989)
- 6.89 (Veith et al. 1979; Veith & Kosian 1983, quoted, Baughman & Paris 1981; Mackay 1982; Oliver 1987; Saito et al. 1992)
- 7.40 (quoted, McKim et al. 1985)
- 6.89 (estimated, Clark et al. 1988; quoted, Comba et al. 1993)
- 6.89 (selected, Connell et al. 1988; Travis & Arms 1988)
- 5.28 (Medchem Database 1988; quoted, Müller et al. 1994)
- 6.90 (selected, Suntio et al. 1988, quoted, Bacci et al. 1990; McLachlan et al. 1990; Calamari et al. 1991)
- 6.89 (quoted, Isnard & Lambert 1988, 1989; Markwell et al. 1989)
- 5.28 (Yalkowsky & Dannenfelser 1994; quoted, Pinsuwan et al. 1995)
- 4.65 (calculated, Pinsuwan et al. 1995)
- 5.28 (selected, Hansch et al. 1995)
- 6.89 (selected, Devillers et al. 1996)

Bioconcentration Factor, log BCF:
- 2.34, 3.07 (*Gambusia, Physa*, Metcalf et al. 1973)
- 2.78 (*Oedogonium sp.*, Metcalf et al. 1973; quoted, Baughman & Paris 1981)
- 2.18 (bioaccumulation factor log BF, adipose tissue in female Albino rats, Ivie et al. 1974; quoted, Geyer et al. 1980)
- −2.02 (milk biotransfer factor log B_m, correlated-K_{OW}, Dorough & Ivie 1974; quoted, Travis & Arms 1988)
- 2.34 (fish in static water, Metcalf 1974; quoted, Kenaga & Goring 1980)
- −1.25 (beef biotransfer factor log B_b, correlated-K_{OW}, Bond et al. 1975; quoted, Travis & Arms 1988)
- −1.14 (vegetation, correlated-K_{OW}, De La Cruz & Rajanna 1975; quoted, Travis & Arms 1988)
- 3.86 (*Chlorococcum sp.*, Hollister et al. 1975; quoted, Baughman & Paris 1981)
- 3.51 (*Chlamydomonas sp.*, Hollister et al. 1975; quoted, Baughman & Paris 1981)
- 3.61 (*Dunaliella tertiolecta*, Hollister et al. 1975; quoted, Baughman & Paris 1981)
- 3.70 (*Thallasidsira pseudomana*, Hollister et al. 1975; quoted, Baughman & Paris 1981)
- 5.60 (bacterial sorption, Smith et al. 1978; quoted, Baughman & Paris 81)
- 4.26 (fathead minnows, 32-d exposure, Veith et al. 1979, 1980; quoted, Bysshe 1982; Veith & Kosian 1983; Devillers et al. 1996)
- 2.91 (calculated-S, Kenaga 1980)
- 4.71 (fathead minnows to ^{14}C mirex, Huckins et al. 1982)
- 4.34 (fish, correlated, Mackay 1982)
- 4.26 (fathead minnows, Veith & Kosian 1983; quoted, Saito et al. 1992)
- 4.09, 3.41 (algae, fish, Verschueren 1983)

6.50   (fish, selected, Patterson & Mackay 1985)
6.08   (rainbow trout, calculated-K_{OW}, Oliver & Niimi 1985)
4.18   (Ontario rainbow trout, field data, Oliver & Niimi 1985)
5.08, 2.87   (rainbow trout, kinetic-k_1/k_2, steady-state, Oliver & Niimi 1985)
2.87   (fish, Oliver & Niimi 1985; Oliver 1987)
4.34   (worms, Oliver 1987)
6.17   (oligochaetes, Connell et al. 1988)
4.26   (quoted, Isnard & Lambert 1988)
6.41   (smelt, Oliver & Niimi 1988; quoted, Comba et al. 1993)
4.31   (*Poecilia reticulata*, Gobas et al. 1989; quoted, Devillers et al. 1996)
6.42, 7.16   (guppy, correlated, Gobas et al. 1989)
6.40   (Markwell et al. 1989)
8.27   (calculated field bioaccumulation, Thomann 1989)
4.72, 7.07   (dry leaf, wet leaf, Bacci et al. 1990)
7.07   (wet leaf, Bacci et al. 1990)

Sorption Partition Coefficient, log K_{OC}:
5.56   (natural sediment, Smith et al. 1978)
7.38   (av. soils/sediments, Smith et al. 1978; quoted, Lyman 1982)
3.76   (soil, quoted exptl., Kenaga 1980)
3.08   (soil, calculated-S as per Kenaga & Goring 1978, Kenaga 1980)
6.00   (derived from exptl., Meylan et al. 1992)
5.67   (calculated-χ, Meylan et al. 1992)
6.42 ± 0.39   (suspended particulate matter of the St. Lawrence River, Comba et al. 1993)
6.00   (20-25°C, estimated, Augustijn-Beckers et al. 1994; Hornsby et al. 1996)

Half-Lives in the Environments:
Air:
Surface water: overall half-life of 0.83 h in river or stream, 420 h in pond, and 1480 hours by sorption in both eutrophic lake and oligotrophic lake; with photolysis half-lives >8000 h and oxidation half-lives >1000 h in pond, river, eutrophic lake and oligotrophic lake (Smith et al. 1978); degradation rate constant of 1.93×10^{-4} h^{-1} (Mackay et al. 1985; quoted, Mackay & Paterson 1991); half-life of 7 days in sunlit, air-equilibrated humic acid solution, or natural water (Mudambi & Hassett 1988; Burns et al. 1996).
Ground water:
Soil: estimated field half-life of 3000 d (Augustijn-Beckers et al. 1994; Hornsby et al. 1996).
Biota: >1000 d (Skea et al. 1981; Oliver & Niimi 1985); >28 d in fathead minnows to ^{14}C mirex (Huckins et al. 1982); >500 d (4°C, rainbow trout, Niimi & Palazzo 1985); 114 d as observed and 495 d as adjusted (12°C, rainbow trout, Niimi & Palazzo 1985); 103 d as observed and >1000 d as adjusted (18°C, rainbow trout, Niimi & Palazzo 1985).

Rate Constants and Environmental Half-Lives:
- Volatilization: 5.37×10^{-2} h^{-1} (Hill et al. 1976) with half-life of 500 h from river, 700 h from pond, 1980 h from eutrophic lake, and 1980 hours from oligotrophic lake (Smith et al. 1978).
- Photolysis: rate constants of $<5.0 \times 10^{-8}$ s^{-1} (laboratory data, Smith et al. 1978); 4.2×10^{-3} d^{-1} (field data, Smith et al. 1978); 3.9×10^3 h (aquatic half-life, Haque et al. 1980); 0.123 d^{-1} (sunlight, distilled water containing 2.0 mg DOC/L humic acid, Mudambi & Hasset 1988); 0.033 d^{-1} (sunlight, distilled water, summer, Mudambi & Hassett 1988); 0.102 d^{-1} (sunlight, Lake Ontario water, Mudambi & Hassett 1988); 0.019 d^{-1} (sunlight, distilled water, fall, Mudambi & Hassett 1988).
- Oxidation: lab. data rate constant of <30 M^{-1} s^{-1} (Smith et al. 1978); half-life of $>>$ 0.7 yr (Smith et al. 1978; quoted, Cheung 1984).
- Hydrolysis: lab. data rate constant of 1×10^{-10} s^{-1} (Smith et al. 1978); 2×10^{-10} s^{-1} with half-life of 250 yr (Cheung 1984); degradation rate constant of 1.93×10^{-4} h^{-1} (Mackay et al. 1985; quoted, Mackay & Paterson 1991).
- Biodegradation: slow process (Cheung 1984).
- Biotransformation:
- Bioconcentration, Uptake (k_1) and Elimination (k_2) Rate Constants:
    - $k_1 > 8.5000$ d^{-1} (rainbow trout, Oliver & Niimi 1985)
    - $k_2 < 0.0007$ d^{-1} (rainbow trout, Oliver & Niimi 1985)

Common Name: Monocrotophos
Synonym: Apadrin, Azodrin, Bilobran, Crotos, ENT 27129, Monocron, Nuvacron
Chemical Name: dimethyl (E)-1-methyl-2-(2-methylcarbamoyl)vinyl phosphate
Uses: systemic insecticide and acaricide to control pests in cotton, sugar cane, coffee, tobacco, olives, rice hops, sorghum, maize, deciduous fruits, citrus fruits, potatoes, sugar beet, tomatoes, soya beans, and ornamentals.
CAS Registry No: 6923-22-4
Molecular Formula: $C_7H_{14}NO_5P$
Molecular Weight: 223.1
Melting Point (°C):
    54-55   (Worthing 1991; Montgomery 1993; Tomlin 1994; Milne 1995)
Boiling Point (°C):
    125   (at 0.0005 mmHg, Agrochemicals Handbook 1987; Worthing 1991; Montgomery 1993; Tomlin 1994; Milne 1995)
Density (g/cm³ at 20°C):
    1.33   (Agrochemicals Handbook 1987; Worthing 1991; Montgomery 1993; Milne 1995)
    1.22   (Tomlin 1994)
Molar Volume (cm³/mol):
Molecular Volume (Å³):
Total Surface Area, TSA (Å²):
Dissociation Constant pK_a:
Heat of Fusion, ΔH_{fus}, kcal/mol:
Entropy of Fusion, ΔS_{fus}, cal/mol·K (e.u.):
Fugacity Ratio at 25°C (assuming ΔS_{fus} = 13.5 e.u.), F: 0.505

Water Solubility (g/m³ or mg/L at 25°C):
    miscible   (Spencer 1973; quoted, Shiu et al. 1990)
    miscible   (Agrochemicals Handbook 1987; Tomlin 1994)
    1000000   (Worthing 1987,91; quoted, Shiu et al. 1990; Milne 1995)
    miscible   (Merck Index 1989; quoted, Shiu et al. 1990)
    1000000   (20-25°C, selected, Wauchope et al. 1992; Hornsby et al. 1996)
    miscible   (Montgomery 1993)

Vapor Pressure (Pa at 25°C):
    9.33×10^{-3} (20°C, Wolfdietrich 1965; Melnikov 1971; quoted, Kim et al. 1984)
    9.33×10^{-3} (20°C, quoted exptl. value, Kim et al. 1984; Kim 1985)
    5.09×10^{-3} (20°C, GC-calculated value, Kim et al. 1984; Kim 1985)
    2.30×10^{-3} (20°C, GC-calculated value with M.P. correction, Kim 1985)
    9.00×10^{-3} (Agrochemicals Handbook 1987)
    9.33×10^{-3} (Merck Index 1989; quoted, Shiu et al. 1990)
    2.90×10^{-4} (20°C, Worthing 1991; Tomlin 1994)
    9.33×10^{-3} (20-25°C, selected, Wauchope et al. 1992; Hornsby et al. 1996)
    9.00×10^{-3} (20°C, Montgomery 1993)

Henry's Law Constant (Pa·m³/mol):
    2.08×10^{-6} (20-25°C, calculated-P/C, this work)

Octanol/Water Partition Coefficient, log K_{OW}:
    −1.97   (calculated, Montgomery 1993)
    −0.22   (calculated, Tomlin 1994)
    −0.20   (selected, Hansch et al. 1995)

Bioconcentration Factor, log BCF:

Sorption Partition Coefficient, log K_{OC}:
    0.0   (soil, 20-25°C, estimated, Wauchope et al. 1992; Hornsby et al. 1996)

Half-Lives in the Environment:
    Air:
    Surface water:
    Groundwater:
    Sediment:
    Soil: selected field half-life of 30 d (Wauchope et al. 1992; Hornsby et al. 1996); half-life of 1-5 d in lab. soil (Tomlin 1994).

Environmental Fate Rate Constants or Half-Lives:
    Volatilization:
    Photolysis:
    Oxidation:
    Hydrolysis: calculated half-lives at 20°C of 96 d at pH 5, 66 d at pH 7 and 17 d at pH 9 (Worthing 1991; Montgomery 1993; Tomlin 1994).
    Biodegradation:
    Biotransformation:
    Bioconcentration, Uptake (k_1) and Elimination (k_2) Rate Constants:

Common Name: Oxamyl
Synonym: D 1410, Dioxamyl, Dupont 1410, Nematicide 1410, Thioxamyl, Vydate
Chemical Name: N,N-dimethylcarbamoyloxyimino-2-(methylthio)acetamide; ethanimidothioic acid, 2-(dimethylamino)-N-[[(methylamino)carbonyl]oxy]-2-oxo-, methyl ester
Uses: insecticide/acaricide/nematicide
CAS Registry No: 23135-22-0
Molecular Formula: $C_7H_{13}N_3O_3S$
Molecular Weight: 219.3
Melting Point (°C):
    100-102 (Spencer 1982; Suntio et al. 1988)
    100-102 (changes to a dimorphic form melts at 108-110°C, Montgomery 1993, Tomlin 1994)
    108-110 (phase change 100-102°C, Milne 1995)
Boiling Point (°C):
Density (g/cm³ at 20°C):
    0.97 (Montgomery 1993; Tomlin 1994; Milne 1995)
Molar Volume (cm³/mol):
    212.4 (calculated-LeBas method, Suntio et al. 1988)
    226.1 (calculated-density, this work)
    1.412 (intrinsic V_I/100, Luehrs et al. 1996)
Molecular Volume (Å³):
Total Surface Area, TSA (Å²):
Dissociation Constant pK_a:
Heat of Fusion, ΔH_{fus}, kcal/mol:
Entropy of Fusion, ΔS_{fus}, cal/mol·K (e.u.):
Fugacity Ratio at 25°C (assuming ΔS_{fus} = 13.5 e.u.), F: 0.177
    0.15 (20°C, Suntio et al. 1988)

Water Solubility (g/m³ or mg/L at 25°C):
    281000 (Martin & Worthing 1977; quoted, Gerstl 1984)
    280000 (Khan 1980, Spencer 1982; quoted, Suntio et al. 1988)
    282513 (Briggs 1981; quoted, Gerstl & Helling 1987)
    280000 (Montgomery 1993; Milne 1995)
    282000 (20-25°C, selected, Wauchope et al. 1992)
    282000 (20-25°C, selected, Hornsby et al. 1996)
    280000 (Tomlin 1994)

Vapor Pressure (Pa at 25°C):
    0.0306 (Khan 1980; Spencer 1982; selected, Suntio et al. 1988)
    0.0306 (20-25°C, selected, Wauchope et al. 1992; Hornsby et al. 1996)
    0.0311 (Montgomery 1993)
    0.0310 (Tomlin 1994)

Henry's Law Constant (Pa·m³/mol):
    0.00026 (calculated-P/C, Suntio et al. 1988)
    0.260 (calculated-P/C, Montgomery 1993)

Octanol/Water Partition Coefficient, log K_{OW}:
- −0.432 (Briggs 1973; quoted, Gerstl 1984; Gerstl & Mingelgrin 1984)
- −0.470 (20-25°C, shake flask-RC, Briggs 1981; quoted, Sangster 1993)
- −0.432 (shake flask-centrifuge, Gerstl 1984; Gerstl & Helling 1987)
- −0.400 (Montgomery 1993)
- −0.440 (pH 5, Tomlin 1994)
- −0.470 (selected, Hansch et al. 1995)

Bioconcentration Factor, log BCF:

Sorption Partition Coefficient, log K_{OC}:
- 0.47 (reported as log K_{OM}, Briggs 1981)
- 1.66, 1.07, 1.20, 1.32, 1.84 (reported as K_{OM}, Israeli soils, Gerstl 1984)
- 0.176 - 1.16 (reported as K_{OM}, estimated-S, Gerstl 1984)
- −0.886 - 0.38 (reported K_{OM}, estimated-K_{OW}, Gerstl 1984)
- 0.778 (soil, screening model calculations, Jury et al. 1987b)
- 1.20, 2.47 (quoted, calculated- χ, Gerstl & Helling 1987)
- 0.70 (soil, Carsel 1989)
- 1.40 (soil, Wauchope et al. 1992; Hornsby et al. 1996)
- −0.70 - 1.40 (Montgomery 1993)

Half-Lives in the Environment:
Air:
Surface water:
Groundwater:
Sediment:
Soil: decomposition in soil was as a function of moisture content, and followed first-order kinetics with reported soil half-lives of 4-13 d at 25°C and, 33 d at 15°C in Bet Dagan soil (Gerstl 1984); half-life of 6 d in screening model calculations (Jury et al. 1987b); half-lives reported as 7 d (Worthing 1991); 8-50 d (Ou & Rao 1986) and 10.2-13.1, 6.2, 7.1 and 17.8 d in Pitstone, Devizes, Sutton, Veany soils respectively (quoted, Montgomery 1993); field half-life is 4 d (selected, Wauchope et al. 1992; Hornsby et al. 1996).
Biota:

Environmental Fate Rate Constants or Half-Lives:
Volatilization:
Photolysis: half-life of 55.4 h (absorbance wavelength 223 nm) (Montgomery 1993).
Oxidation:
Hydrolysis: hydrolysis half-lives >31 d (pH 5), 8 d (pH 7) and 3 h (pH 9) (Tomlin 1994).
Biodegradation: decomposition rate constants range from 0.182 d^{-1} to 0.021 d^{-1} corresponding to half-lives of 4 to 33 d in Bet Dagan soil depending on moisture, and decomposition rate constant ranges from 0.23 to 0.11 d^{-1} corresponding to half-lives of 3.1 to 6.5 d in five Israeli soils (Gerstl 1984); half-life of 6 d in screening model calculations (Jury et al. 1987b); .
Biotransformation:
Bioconcentration, Uptake (k_1) and Elimination (k_2) Rate Constants:

Common Name: Parathion

Synonym: AAT, AATP, AC 3422, Alkron, Alleron, American Cyanamid 3422, Aphamite, Aralo, B 404, Bay E-605, Bladan, Corothion, Corthion, Corthione, Danthion, DDP, Diethyl parathion, DNDP, DPP, E 605, Ecatox, Ekatox, ENT 15108, Ethlon, Ethyl parathion, Etilon, Folidol, Fosfermo, Fosferno, Fosfex, Fosfive, Fosova, Fostern, Fostox, Gearphos, Genithion, Kolphos, Kypthion, Lethalaire G54, Lirothion, Murfos, NA 2783, NCI-C00226, Niran, Nitrostigmine, Orthophos, Pac, Panthion, Paradust, Paraflow, Paramar, Paraphos, Paraspray, Parathene, Parathionethyl, Parawet, Penphos, Pestox plus, Pethion, Phoskil, Phosphemol, Phosphenol, Phosphostigmine, RB, Rhodiasol, Rhodiatox, Rhodiatrox, Selephos, SNP, Soprathion, Stathion, Strathion, Sulphos, Super rodiatox, T-47, Thiofos, Tiophos, Tox 47, Vapophos, Vitrex

Chemical Name: *O,O*-diethyl *O*-4-(nitrophenyl) phosphorothioate; diethyl 4-nitrophenyl phosphorothioate; phosphorothioc acid *O,O*-diethyl *O*-(4-nitrophenyl) ester

Uses: insecticide and acaricide to control chewing and sucking insects and mites in fruits, vegetables, ornamentals and field crops.

CAS Registry No: 56-38-2

Molecular Formula: $C_{10}H_{14}NO_5PS$

Molecular Weight: 291.27

Melting Point (°C):
    6.1   (Melnikov 1971; Freed et al. 1977; Montgomery 1993; Tomlin 1994; Milne 1995)
    6.0   (Spencer 1982; Suntio et al. 1988)

Boiling Point (°C):
    115   (at 0.05 mmHg, Melnikov 1971; Freed et al. 1977)
    375   (Montgomery 1993; Milne 1995)
    105   (at 80 Pa, Tomlin 1994)

Density (g/cm³ at 20°C):
    1.265   (25°C, Spencer 1982; Agrochemicals Handbook 1987; Milne 1995)
    1.26   (25°C, Merck Index 1989; Montgomery 1993)
    1.2694   (25°C, Worthing 1991; Tomlin 1994)

Molar Volume (cm³/mol):
    251.9   (calculated-LeBas method, Suntio et al. 1988; Fisher et al. 1993)
    1.41   ($V_I/100$, Fisher et al. 1993)
    230.3   (calculated from density)

Molecular Volume (Å³):

Total Surface Area, TSA (Å²):

Dissociation Constant pK_a:
    7.14   (Kortum et al. 1961; Wolfe 1980)

Heat of Fusion, ΔH_{fus}, kcal/mol:
    4.75   (DSC method, Plato & Glasgow 1969)

Entropy of Fusion, ΔS_{fus}, cal/mol·K (e.u.):

Fugacity Ratio at 25°C (assuming ΔS_{fus} = 13.5 e.u.), F: 1.0

Water Solubility (g/m³ or mg/L at 25°C):
    24   (Macy 1948; quoted, Chiou et al. 1977)
    18-31   (rm temp., >95% purity with max. particle size 0.07-5.0 µ, shake flask-GC, Robeck et al. 1965)

24	(Günther et al. 1968; Melnikov 1971; Spencer 1973; quoted, Bowman & Sans 1979; Worthing 1987; Taylor & Glofelty 1988; Suntio et al. 1988; Shiu et al. 1990)
11.9	(20°C, NIEHS 1975; quoted, Freed et al. 1977)
11.9	(20°C, O'Brien 1975; quoted, Suntio et al. 1988; Schomburg et al. 1991)
24	(Martin & Worthing 1977; quoted, Kenaga 1980; Kenaga & Goring 1980; Karickhoff 1981; Jury et al. 1983, 1984; Kim et al. 1984)
24	(Wauchope 1978; Khan 1980; Lyman 1982; Willis & McDowell 1982)
12.4	(20°C, shake flask-GC, Bowman & Sans 1979,83b; quoted, Fuhremann & Lichtenstein 1980; Sharom et al. 1980; Shiu et al. 1990; Patil 1994)
6.54	(shake flask-GC, Felsot & Dahm 1979; quoted, Shiu et al. 1990; Howard 1991)
24, 6	(quoted, corrected-MP, Briggs 1981)
24	(Bruggeman et al. 1981; quoted, Adams 1987)
20	(quoted, Burkhard & Guth 1981)
24	(Jury et al. 1983; quoted, Taylor & Glotfelty 1988)
20	(Merck Index 1983,89; quoted, Somasundaram et al. 1991)
14	(Gerstl & Mingelgrin 1984; quoted, Shiu et al. 1990)
24	(U.S. EPA 1984; quoted, McLean et al 1988)
24	(Agrochemicals Handbook 1987; Isnard & Lambert 1988)
15	(20°C, selected, Suntio et al. 1988; quoted, Findinger et al. 1990; Majewski & Capel 1995)
11	(20°C, Worthing 1991; Tomlin 1994)
24	(quoted, Pait et al. 1992)
12.9, 15.2	(20°C, 30°C, Montgomery 1993)
12.4, 37.5	(quoted, calculated, Patil 1994)
6.37, 7.5	(quoted, calculated-group contribution fragmentation method, Kühne et al. 1995)
24	(Milne 1995)
24	(selected, Halfon et al. 1996)
12	(20°C, selected, Siebers & Mattusch 1996)
24	(20°C, selected, Hornsby et al. 1996)

Vapor Pressure (Pa at 25°C):

5.04×10^{-3}	(20°C, Bright et al. 1950; quoted, Freed et al. 1977; Khan 1980; Spencer 1982; Suntio et al. 1988)
12.3×10^{-3}	(30°C, Bright et al. 1950; quoted, Spencer et al. 1973)
0.76×10^{-3}	(20°C, Wolfdietrich 1965; quoted, Kim et al. 1984)
5.07×10^{-3}	(20°C, Spencer 1973; quoted, Fuhremann & Lichtenstein 1980)
5.85×10^{-4}	(20°C, Gückel et al. 1973, 1974; quoted, Suntio et al. 1988)
0.76×10^{-3}	(20°C, Gückel et al. 1973, 1974; quoted, Suntio et al. 1988)
0.63×10^{-3}	(20°C, gas saturation method, Spencer et al. 1979; quoted, Kim et al. 1984; Suntio et al. 1988
1.26×10^{-3}	(gas saturation, Spencer et al. 1979; quoted, Jury et al. 1983; Taylor & Glotfelty 1988; Taylor & Spencer 1990)
5.05×10^{-3}	(quoted, Burkhard & Guth 1981)
1.29×10^{-3}	(Spencer 1983; quoted, Howard 1991)
1.30×10^{-3}	(25.3°C, gas saturation method, Kim et al. 1984)
0.69×10^{-3}	(20°C, extrapolated-Clausius Clapeyron eqn., Kim et al. 1984; quoted, Suntio et al. 1988)

0.81×10^{-3}	(20°C, GC-calculated value, Kim et al. 1984; Kim 1985)
5.00×10^{-3}	(20°C, U.S. EPA 1984; quoted, McLean et al 1988)
0.69×10^{-3}	(20°C, gas saturation method, Kim 1985)
5.00×10^{-3}	(20°C, Agrochemicals Handbook 1987)
6.00×10^{-4}	(20°C, selected, Suntio et al. 1988; quoted, Findinger et al. 1990; Majewski & Capel 1995)
1.30×10^{-3}	(selected, Taylor & Spencer 1990)
8.90×10^{-4}	(20°C, Worthing 1991; Tomlin 1994)
0.0533	(20°C, Montgomery 1993)
6.67×10^{-4}	(20°C, selected, Hornsby et al. 1996)
5.00×10^{-3}	(selected, Halfon et al. 1996)
1.30×10^{-3}	(20°C, selected, Siebers & Mattusch 1996)

Henry's Law Constant (Pa·m³/mol):
0.120	(20°C, calculated-P/C, Mackay & Shiu 1981; quoted, Suntio et al. 1988)
0.096	(24°C, calculated-P/C, Chiou et al. 1980)
0.015	(calculated-P/C, Jury et al. 1984, 1987a; Jury & Ghodrati 1989)
0.012	(20°C, calculated-P/C, Suntio et al. 1988; quoted, Fisher et al. 1993; Majewski & Capel 1995)
0.015	(calculated-P/C, Taylor & Glotfelty 1988)
0.0087	(23°C, Findinger & Glotfelty 1990; quoted, Findinger et al. 1990; Schomburg et al. 1991)
0.057	(calculated-P/C, Howard 1991)
0.0087	(calculated-P/C, Montgomery 1993)
0.020	(selected, Siebers & Mattusch 1996)
0.0141	(calculated-P/C, this work)

Octanol/Water Partition Coefficient, log K_{OW}:
3.81	(20°C, NIEHS 1975; quoted, Freed et al. 1977)
3.81	(shake flask-GC, Chiou et al. 1977; quoted, Kenaga & Goring 1980; Rao & Davidson 1980; Karickhoff 1981; Lyman 1982; Bowman & Sans 1983b; Suntio et al. 1988; De Bruijn & Hermens 1991)
3.40	(Felsot & Dahm 1979; quoted, Bowman & Sans 1983b; Gerstl & Mingelgrin 1984; Suntio et al. 1988)
3.81	(shake flask-GC, Freed et al. 1979)
3.80	(Hansch & Leo 1979; quoted, Isnard & Lambert 1988; Fisher et al. 1993)
3.81	(Rao & Davidson 1980; quoted, Kim et al. 1984)
3.93	(Briggs 1981; quoted, Bowman & Sans 1983b; Suntio et al. 1988)
4.00	(appr. = log BCF_{lipid}, Briggs 1981)
3.76	(shake flask-GC, Bowman & Sans 1983b; quoted, Suntio et al. 1988; De Bruijn & Hermens 1991; Somasundaram et al. 1991)
3.40	(selected, Dao et al. 1983)
3.81	(U.S. EPA 1984; quoted, McLean et al 1988)
3.83	(Hansch & Leo 1985; quoted, Howard 1991)
2.98	(selected, Yoshioka et al. 1986)
3.90	(selected, Gerstl & Helling 1987)
3.80	(selected, Suntio et al. 1988; quoted, Bintein & Devillers 1994)
3.93	(Thor 1989; quoted, Connell & Markwell 1990)
2.609	(calculated as per Broto et al. 1984, Karcher & Devillers 1990)
3.83	(selected, Magee 1991)

3.47   (estimated-QSAR & SPARC, Kollig 1993)
2.15-3.93 (Montgomery 1993)
3.76, 3.26 (quoted, calculated, Patil 1994)
3.83   (Tomlin 1994)
3.83   (selected, Hansch et al. 1995)
3.45, 3.47, 3.81 (RP-HPLC, ClogP, calculated-S, Finizio et al. 1997)

Bioconcentration Factor, log BCF:
    2.53   (fish in static water, Metcalf & Sanborn 1975; quoted, Kenaga & Goring 1980; De Bruijn & Hermens 1991)
    1.81   (tadpoles, Hall & Kolbe 1980; quoted, Howard 1991)
    2.00, 2.54 (calculated-S, K_{OC}, Kenaga 1980)
    3.14   (earthworms, Lord et al. 1980; quoted, Connell & Markwell 1990)
    4.00   (log BCF_{lipid}, Briggs 1981)
    2.68   (calculated-K_{OW}, Hansch & Leo 1985; quoted, Howard 1991)
    2.48   (Am. oysters after 84 d, USEPA 1986; quoted, Howard 1991)
    1.48, 2.34 (av., fathead minnow after 70 d, 820,138 d, USEPA 1986; quoted, Howard 1991)
    2.34   (av., fathead minnow after 82-138 d, USEPA 1986; quoted, Howard 1991)
    1.91, 2.27, 2.40, 1.43   (av., bluegill after 12 h, 29 h, 46 h, 504 d, USEPA 1986; quoted, Howard 1991)
    1.95, 2.39, 2.50 (av., brook trout muscle after 8 h, 6 d, 180 d, USEPA 1986; quoted, Howard 1991)
    1.90   (quoted, Isnard & Lambert 1988)
    2.53   (Pait et al. 1992)

Sorption Partition Coefficient, log K_{OC}:
    3.68   (soil, Swoboda & Thomas 1968; quoted, Kenaga 1980; Kenaga & Goring 1980; Karickhoff 1981)
    4.02   (av., 4 soils, Hamaker & Thompson 1972; quoted, Howard 1991)
    3.30   (av., soils, Chiou et al. 1979; quoted, Howard 1991)
    2.88   (soil, calculated-S as per Kenaga & Goring 1978, Kenaga 1980)
    2.90   (calculated-S, Mill et al. 1980; quoted, Adams 1987)
    4.03   (av. soils/sediments, Rao & Davidson 1980; quoted, Karickhoff 1981; Lyman 1982; Jury et al. 1983,84)
    2.78   (log K_{OM} appr. = log BCF_{lipid}, Briggs 1981)
    3.25, 3.95, 3.42 (estimated-S, S. & M.P., K_{OW}, Karickhoff 1981)
    2.26-3.96   (reported as log K_{OM}, Mingelgrin & Gerstl 1983)
    2.83   (av., 8 Israeli soils, Gerstl & Mingelgrin 1984; quoted, Howard 1991)
    3.19   (av., 4 Israeli sediments, Gerstl & Mingelgrin 1984; quoted, Howard 1991)
    3.52, 2.58 (quoted, calculated- χ, Gerstl & Helling 1987)
    3.04   (screening model calculations, Jury et al. 1987a,b; Jury & Ghodrati 1989)
    3.68, 3.33 (reported, estimated as log K_{OM}, Magee 1991)
    3.15   (estimated-QSAR & SPARC, Kollig 1993)
    2.50-4.20 (Montgomery 1993)
    3.70   (20°C, selected, Hornsby et al. 1996)

Adsorption Coefficient, K_d (L kg^{-1}):
    9.6   (homoionic K^+-montmorillonite clay minerals, Haderlein et al. 1996)

Half-Lives in the Environment:
> Air: atmospheric transformation lifetime was estimated to be <1 d (Kelly et al. 1994).
> Surface water: persistence of up to 8 weeks in river water (Eichelberger & Lichtenberg 1971); estimated half-life >4250 d from biodegradation rate constant in aquatic systems from river water samples (Williams 1977; quoted, Scow 1982); half-life in the Indian River water (24 ppt salinity; pH 8.16) was 7.84 d (Wang & Hoffman 1991); half-lives of 120 d at 6°C, 84 d at 22°C in darkness for Milli-Q water at pH 6.1; 120 d at 6°C, 86 d at 22°C in darkness, 8 d under sunlight conditions for river water at pH 7.3; 122 d at 6°C, 33 d at 22°C in darkness for filtered river water at pH 7.3; 542 d at 6°C, 44 d at 22°C in darkness, 18 d under sunlight conditions for seawater at pH 8.1 (Lartiges & Garrigues 1995).
> Groundwater:
> Sediment:
> Soil: persistence of one week (Edwards 1973; quoted, Morrill et al. 1982); persistence of less than one month (Wauchope 1978); >24 weeks in sterile sandy loam and <1.0 week in nonsterile sandy loam; >24 weeks in sterile organic soil and 1.5 week in nonsterile organic soil (Miles et al. 1979); estimated first-order half-life of 23.9 d from biodegradation rate constant of 0.029 d^{-1} in soil by die-away tests from soil incubation studies (Rao & Davidson 1980; quoted, Scow 1982); moderately persistent in soil with half-life of 20-100 d (Willis & McDowell 1982); reported half-life of 18 d calculated using screeening model calculations (Jury et al. 1987a,b; Jury & Ghodrati 1989; quoted, Montgomery 1993); av. degradation rate constant of 0.030 d^{-1} in silty clay with half-life of 23 d and av. degradation rate constant of 0.0315 d^{-1} in sandy clay with half-life of 22 d (Sattar 1990); 14 d (selected, Halfon et al. 1996); field half-life of 14 d (20-25°C, selected, Hornsby et al. 1996); soil half-life of 6 d (Pait et al. 1992).
> Biota: biochemical half-life of 15 d from screening model calculations (Jury et al. 1987a,b; Jury & Ghodrati 1989).

Environmental Fate Rate Constants or Half-Lives:
> Volatilization: exptl. half-lives: 14 d in nonstirred aqueous solutions and 9.3 days in stirred aqueous solutions, and estimated half-lives: 13 d in nonstirred aqueous solutions and 8.7 d in stirred aqueous solutions (Chiou et al. 1980).
> Photolysis: photoreacted 390 times more rapidly when sorbed by algae than in distilled water (Zepp & Schlotzhauer 1983); direct photolysis has a half-life of <1 d to 10 d in surface waters, the presence of photosensitizers, free radicals, hydrogen peroxide, or algae which are found in eutropic waters may accelerate degradation considerably (GEMS 1986; quoted, Howard 1991); photodegradation half-life of 88 h (Hazardous Substances Data Bank 1989; quoted, Montgomery 1993).
> Oxidation: calculated rate constant of 92×10^{-12} cm^3/molecule·s for the vapor-phase reaction with hydroxyl radicals in air (Winer & Atkinson 1990).
> Hydrolysis: second-order alkaline hydrolysis rate constant of 1.2×10^{-3} M^{-1} s^{-1} at 27°C (Ketelaar 1950; quoted, Wolfe 1980); half-life of 130 d at pH 7.4 and 20°C (NIEHS 1975; quoted, Freed et al. 1977,79; Montgomery 1993); reported half-lives of 24, 19 weeks at pH 6, 7.4 and 20°C (Freed et al. 1979; quoted, Howard 1991; Montgomery 1993); half-lives of 43, 24, and 15 weeks at pH 5, 7, 8 and 20°C (Chapman & Cole 1982; quoted, Howard 1991; Montgomery 1993); rate constant of 2.4 y^{-1} at pH 7.0 and 25°C (Kollig 1993); half-life of 3.5 weeks at

pH 6 (Montgomery 1993); half-lives at 22°C: 272 d at pH 4, 260 d at pH 7, and 130 d at pH 9 (Tomlin 1994).

Biodegradation: generally occurs with a half-life of several weeks but in well acclimated water, complete degradation may occur in two weeks (Eichelberger & Lichtenberg 1971; Sharom et al. 1980; quoted, Howard 1991); estimated half-life >4250 d from biodegradation rate constant in aquatic systems from river water samples (Williams 1977; quoted, Scow 1982); rate constant of 0.029 d^{-1} in soil by die-away tests from soil incubation studies (Rao & Davidson 1980; quoted, Scow 1982); half-life of 18 d for a 100 d leaching and screening test in 0-10 cm depth of soil (Rao & Davidson 1980; quoted, Jury et al. 1983, 1984, 1987a,b; Jury & Ghodrati 1989); <0.00016 d^{-1} of aerobic degradation observed in incubations of river water samples (Lyman et al. 1990; quoted, Hemond & Fechner 1994).

Biotransformation:

Bioconcentration, Uptake (k_1) and Elimination (k_2) Rate Constants:

Common Name: Parathion-methyl
Synonym: Bladan M, Folidol-M, Metacide, Nitrox 80
Chemical Name: O,O-dimethyl O-4-(nitrophenyl) phosphorothioate; dimethyl 4-nitrophenyl phosphorothioate
Uses: insecticide to control chewing and sucking insects, and mites in a wide range of crops, including fruits, vines, vegetables, ornamentals, cotton, and also used as acaricide.
CAS Registry No: 298-00-0
Molecular Formula: $C_8H_{10}NO_5PS$
Molecular Weight: 263.5
Melting Point (°C):
    36.0   (Smith et al. 1978; Karickhoff 1981)
    35-36   (Khan 1980; Worthing 1991; Tomlin 1994)
    35.5   (Bowman & Sans 1983b; Patil 1994)
    37-38   (Howard 1991)
    35.0   (Kühne et al. 1995)
Boiling Point (°C):
    109   (at 0.05 mmHg, Freed et al. 1977)
    119   (at 0.1 mmHg, Agrochemicals Handbook 1987)
    154   (at 1.0 mmHg, Agrochemicals Handbook 1987; Tomlin 1994)
    143   (Howard 1991)
Density (g/cm³ at 20°C):
    1.358   (Agrochemicals Handbook 1987; Worthing 1991; Tomlin 1994)
Molar Volume (cm³/mol):
    207.5   (calculated-LeBas method, Suntio et al. 1988; Fisher et al. 1993)
    1.214   (V_I/100, Fisher et al. 1993)
    194.0   (calculated from density)
Molecular Volume (Å³):
Total Surface Area, TSA (Å²):
Dissociation Constant pK_a:
    7.15   (Kortum et al. 1961; Wolfe 1980)
Heat of Fusion, ΔH_{fus}, kcal/mol:
    5.75   (Plato & Glasgow 1969)
Entropy of Fusion, ΔS_{fus}, cal/mol·K (e.u.):
Fugacity Ratio at 25°C (assuming ΔS_{fus} = 13.5 e.u.), F: 0.752

Water Solubility (g/m³ or mg/L at 25°C):
    55   (Melnikov 1971; quoted, Shiu et al. 1990)
    60   (Leonard et al. 1976; quoted, Nash 1988)
    57   (Martin & Worthing 1977; quoted, Kenaga 1980; Kenaga & Goring 1980; Karickhoff 1981; Jury et al. 1983; Kim et al. 1984; Pait et al. 1992)
    50   (quoted, Smith et al. 1978; Wauchope 1978)
    37.7   (19.5°C, shake flask-GC, Bowman & Sans 1979, 1983b; quoted, Shiu et al. 1990)
    55   (20°C, Freed et al. 1979; quoted, Metcalf et al. 1980)
    55-60   (Worthing 1979, 1983; quoted, Cohen & Steinmetz 1986; Shiu et al. 1990)
    60   (Khan 1980)
    53   (Weber et al. 1980; quoted, Willis & McDowell 1982)
    >1.0   (quoted, Schimmel et al. 1983)
    55-60   (U.S. EPA 1984; quoted, McLean et al. 1988)

55-60   (Agrochemicals Handbook 1987)
58   (selected, Gerstl & Helling 1987)
25   (20°C, Suntio et al. 1988; quoted, Howard 1991; Schomburg et al. 1991; Majewski & Capel 1995)
37.7   (22°C, selected, Seiber et al. 1989)
55   (20°C, Worthing 1991; Tomlin 1994)
60   (20-25°C, selected, Wauchope et al. 1992; Hornsby et al. 1996; quoted, Halfon et al. 1996)
37.7, 75.9   (quoted, calculated, Patil 1994)
37.7, 50   (quoted, calculated-group contribution fragmentation method, Kühne et al. 1995)

Vapor Pressure (Pa at 25°C):
0.0013   (20°C, Wolfdietrich 1965; quoted, Kim et al. 1984)
0.0013   (20°C, von Rümker & Horay 1972; quoted, Freed et al. 1977)
0.0013   (20°C, Gückel et al. 1973; quoted, Freed 1976)
0.0011   (20°C, gas saturation method, Spencer et al. 1979; quoted, Dobbs et al. 1984; Kim et al. 1984)
0.0024   (gas saturation, Spencer et al. 1979; quoted, Jury et al. 1983; Spencer & Cliath 1983; Nash 1988; Taylor & Spencer 1990)
0.0023   (calculated-vapor density, Spencer et al. 1979; Hinckley et al. 1990)
0.0013   (Worthing 1979; quoted, Cohen & Steinmetz 1986)
0.0013   (20°C, Khan 1980)
0.0013   (20°C, Metcalf et al. 1980; quoted, Howard 1991)
>0.0133   (20-25°C, Weber et al. 1980; quoted, Willis & McDowell 1982)
0.0020   (gas saturation method, Kim et al. 1984; quoted, Hinckley et al. 1990)
0.0008   (20°C, extrapolated-Clausius Clapeyron eqn., Kim et al. 1984)
0.0013   (20°C, U.S. EPA 1984; quoted, McLean et al. 1988)
0.00084   (20°C, gas saturation method, Kim 1985)
0.0013   (20°C, Agrochemicals Handbook 1987)
0.0020   (20°C, selected, Suntio et al. 1988; quoted, Majewski & Capel 1995)
0.0015   (22°C, selected, Seiber et al. 1989)
0.023   (GC method, supercooled liq., Hinckley et al. 1990)
0.0024   (selected, Taylor & Spencer 1990)
0.0002   (20°C, Worthing 1991; Tomlin 1994)
0.002   (20-25°C, selected, Wauchope et al. 1992; Hornsby et al. 1996; quoted, Halfon et al. 1996)
0.00041   (Tomlin 1994)

Henry's Law Constant (Pa·m³/mol):
0.0101   (Metcalf et al. 1980; quoted, Howard 1991)
0.0061   (estimated, Metcalf et al. 1980)
0.0109   (calculated-P/C, Jury et al. 1987a; Jury & Ghodrati 1989)
0.021   (20°C, calculated-P/C, Suntio et al. 1988; quoted, Fisher et al. 1993; Majewski & Capel 1995)
0.0101   (22°C, selected, Seiber et al. 1989)
0.0062   (23°C, quoted, Schomburg et al. 1991)
0.0211   (calculated-P/C, this work)

Octanol/Water Partition Coefficient, log K_{OW}:
- 2.04 (shake flask, Leo et al. 1971; quoted, Sicbaldi & Finizio 1993)
- 1.91 (Freed et al. 1976; quoted, Kenaga & Goring 1980; Bowman & Sans 1983b)
- 3.32 (Hansch & Leo 1979; quoted, Fisher et al. 1993)
- 3.32 (Rao & Davidson 1980; quoted, Karickhoff 1981; Bowman & Sans 1983b; Kim et al. 1984)
- 2.94 (shake flask-GC, Bowman & Sans 1983b; quoted, Sicbaldi & Finizio 1993; Patil 1994)
- 1.80 (quoted, Schimmel et al. 1983)
- 1.91 (U.S. EPA 1984; quoted, McLean et al. 1988)
- 2.86 (Hansch & Leo 1985; quoted, Howard 1991)
- 2.52 (selected, Yoshioka et al. 1986)
- 3.32 (selected, Gerstl & Helling 1987)
- 1.887 (calculated as per Broto et al. 1984, Karcher & Devillers 1990)
- 3.039 ± 0.005 (guppy, calculated on an extractable lipid wt. basis, De Bruijn & Hermens 1991; quoted, Verhaar et al. 1992; Sicbaldi & Finizio 1993)
- 2.71 (RP-HPLC, Sicbaldi & Finizio 1993)
- 2.94, 2.64 (quoted, calculated, Patil 1994)
- 3.00 (Tomlin 1994)
- 2.86 (selected, Hansch et al. 1995)
- 2.94 (selected, Devillers et al. 1996)
- 2.71, 2.79, 3.40 (RP-HPLC, ClogP, calculated-S, Finizio et al. 1997)

Bioconcentration Factor, log BCF:
- 0.778 (carp/lipids, Chigareva 1973; quoted, McLeese et al. 1976)
- 0.0 (carp/muscle, Chigareva 1973; quoted, McLeese et al. 1976)
- 1.98 (fish in static water, Metcalf 1974; quoted, Kenaga & Goring 1980)
- 2.69 (bacteria, Smith et al. 1978; quoted, Baughman & Paris 1981)
- 1.80 (calculated-S, Kenaga 1980)
- 2.89 (calculated-K_{OC}, Kenaga 1980)
- 2.98 (guppy, calculated on an extractable lipid wt. basis, De Bruijn & Hermens 1991; quoted, Devillers et al. 1996)
- 3.04 (*Poecilia reticulata*, De Bruijn & Hermens 1991; quoted, Devillers et al. 1996)
- 1.85 (Pait et al. 1992)
- 1.92 (paddy field fish, Tejada 1995; quoted, Abdullah et al. 1997)

Sorption Partition Coefficient, log K_{OC}:
- 3.99 (soil, Hamaker & Thompson 1972; quoted, Kenaga 1980; Kenaga & Goring 1980; Karickhoff 1981)
- 1.699 (av. all sediments, Smith et al. 1978)
- 2.63 (av. of 3 soils, Rao & Davidson 1979; quoted, Nash 1988)
- 2.67 (soil, calculated-S as per Kenaga & Goring 1978, Kenaga 1980)
- 3.71 (Rao & Davidson 1980; quoted, Karickhoff 1981; Jury et al. 1983)
- 3.02, 3.47, 2.93 (estimated-S, S & M.P., K_{OW}, Karickhoff 1981)
- 3.71 (screening model calculations, Jury et al. 1987a,b; Jury & Ghodrati 1989)
- 3.84, 1.97 (quoted, calculated-χ, Gerstl & Helling 1987)
- 3.71-3.99 (soil, Carsel 1989)
- 3.71 (soil, 20-25°C, selected, Wauchope et al. 1992; Hornsby et al. 1996)

Adsorption Coefficient, K_d (L kg^{-1}):
    6.20   (homoionic K$^+$-montmorillonite clay minerals, Haderlein et al. 1996)

Half-Lives in the Environment:
- Air: estimated half-life of 3.6 d for the vapor-phase reaction with hydroxyl radicals in air (Atkinson 1985; quoted, Howard 1991); photooxidation half-life of 1.0-10.5 h in air based on estimated rate constant for the vapor-phase reaction with hydroxyl radicals in air (Atkinson 1987; quoted, Howard et al. 1991).
- Surface water: persistence up to 4.0 weeks in river water (Eichelberger & Lichtenberg 1971); overall half-lives of 0.6 h in river, 15 h in eutrophic pond, 28.3 h in eutrophic lake and 157.5 h oligotrophic lake (Smith et al. 1978); half-life of 8 d in summer and 38 d in winter for direct sunlight photolysis in natural water (Howard 1991; Howard et al. 1991); half-lives in 100 mL pesticide-seawater solution: >28 d under indoor conditions, 6.3 d under outdoor light conditions and 18 d under outdoor dark conditions (Schimmel et al. 1983); first-order biodegradation rate constant of 0.30 d^{-1} in river sediment and 0.02 d^{-1} in river water (Cripe et al. 1987; quoted, Battersby 1990); 44 h of dissipation from rice field water (Seiber & McChesney 1987; quoted, Seiber et al. 1989); half-lives of 237 d at 6°C, 46 d at 22°C in darkness for Milli-Q water at pH 6.1; 95 d at 0°C, 23 d at 22°C in darkness, 11 d under sunlight conditions for river water at pH 7.3; 173 d at 6°C, 18 d at 22°C in darkness for filtererd river water at pH 7.3; 233 d at 6°C, 30 d at 22°C in darkness, 34 d under sunlight conditions for seawater at pH 8.1 (Lartiges & Garrigues 1995).
- Groundwater: 24-1680 h based on estimated aqueous aerobic and anaerobic biodegradation half-life (Howard et al. 1991).
- Sediment: half-lives in 10 grams sediment/100 mL pesticide-seawater solution: <1.2 d under untreated conditions and >28 d under sterile conditions (Schimmel et al. 1983); disappearance rate constants: $(3.5 \pm 0.6) \times 10^{-3}$ min^{-1} in Beaver Dam sediments samples at pH 6.7, $(2.9 \pm 1.2) \times 10^{-3}$ min^{-1} in Memorial Park sediments samples at pH 6.5 and $(2.8 \pm 2.4) \times 10^{-3}$ min^{-1} in Hickory Hills sediments samples at pH 6.9 near Athens, Georgia (Wolfe et al. 1986).
- Soil: 2,408,640 h, based on unacclimated aerobic soil grab sample data (Davidson et al. 1980; Butler et al. 1981; quoted, Howard et al. 1991); measured dissipation rate of 0.010-0.034 d^{-1} (Baker & Applegate 1970; quoted, Nash 1988); estimated dissipation rate of 0.029, 0.042 d^{-1} (Nash 1988); persistence of less than one month (Wauchope 1978); non-persistent in soils with half-life <20 d (Willis & McDowell 1982); rate constant of 0.16 d^{-1} with half-life of 4 d under lab. conditions and rate constant of 0.046 d^{-1} with half-life of 15 d under field conditions (Rao & Davidson 1980); half-life of 15 d in screening model calculations (Jury et al. 1987a,b; Jury & Ghodrati 1989); selected field half-life of 5.0 d (Wauchope et al. 1992; quoted, Halfon et al. 1996; Hornsby et al. 1996); soil half-life of 44 d (Pait et al. 1992).
- Biota: biochemical half-life of 15 d from screening model calculations (Jury et al. 1987a,b; Jury & Ghodrati 1989).

Environmental Fate Rate Constants or Half-Lives:
- Volatilization: volatilization rate of <0.01 kg/ha/d in a flooded rice field (Seiber et al. 1986; Seiber & McChesney 1987; quoted, Seiber et al. 1989)
- Photolysis: lab. rate constant of 2.7×10^{-7} s^{-1} in early January with photolysis half-lives of 240 h, 850 h, 850 h and 170 h in river, pond, eutrophic lake and oligotropic lake predicted by the one-compartment model (Smith et al. 1978; quoted, Howard et al. 1991); half-life of 8 d in summer and 38 days in winter for direct sunlight photolysis in natural water (Smith et al. 1978; quoted, Howard 1991; Howard et al. 1991); photolytic half-life of 200 h in aquatics (Haque et al. 1980); photoreacted 390 times more rapidly when sorbed by algae than in distilled water (Zepp & Schlotzhauer 1983).
- Oxidation: lab. rate constant of 3.0 M^{-1} s^{-1} (Smith et al. 1978); estimated photooxidation half-life of 3.6 d in air for the vapor-phase reaction with hydroxyl radicals in the atmosphere (Atkinson 1985; quoted, Howard 1991); photooxidation half-life of 1.0-10.5 h in air based on estimated rate constant for the vapor-phase reaction with hydroxyl radicals in air (Atkinson 1987; quoted, Howard et al. 1991).
- Hydrolysis: half-life of 125 h for pH <11.0 at 15°C and rate constant was found to be 9.4×10^{-2} mol/min (Ketelaar & Gersmann 1958; quoted, Freed 1976); pseudo first-order hydrolysis half-life of 8.4 h with pH 6 buffer at 70°C in 20% ethanol aqueous solution (Ruzicka et al. 1967; quoted, Freed 1976; Smith et al. 1978); half-life of 1.7 weeks at pH about 6 and room temperature (Cowart et al. 1971; quoted, Smith et al. 1978); first-order hydrolysis half-life of 72 d based on rate constant of 1.1×10^{-7} s^{-1} at pH 7 and 25°C (Mabey & Mill 1978; quoted, Howard et al. 1991); lab. rate constant of 9×10^{-7} s^{-1} (Smith et al. 1978); second-order alkaline hydrolysis rate constant of 5.3×10^{-3} M^{-1} s^{-1} at 27°C and pH 10 (Smith et al. 1978; quoted, Wolfe 1980); half-lives at 25°C: 68 d at pH 5, 40 d at pH 7, 33 d at pH 9 (Tomlin 1994).
- Biodegradation: lab. rate constant of 1.7×10^{-7} μg $cell^{-1}$ h^{-1} (Smith et al. 1978); aqueous aerobic half-life of 360-1680 h, based on an unacclimated aerobic river die-away test data (Bourquin et al. 1979; Spain et al. 1980; quoted, Howard et al. 1991); aqueous anaerobic half-life of 24-168 h, based on unacclimated anaerobic soil and sediment grab sample data (Adhya et al. 1981; Wolfe et al. 1986; quoted, Howard et al. 1991); half-life of 15 d for a 100 days leaching and screening test in 0-10 cm depth of soil (Rao & Davidson 1980; quoted, Jury et al. 1983, 1987a,b; Jury & Ghodrati 1989); rate constants of 0.003 ± 0.0003 h^{-1} with half-life of 220.9 h in surface aerobic soils at Williamsburg and 0.0017 ± 0.00009 h^{-1} with half-life of 410 h in subsurface aerobic soils at Sault Ste. Marie (Ward 1985); first-order biodegradation rate constant of 0.30 d^{-1} in river sediment and 0.02 d^{-1} in river water (Cripe et al. 1987; quoted, Battersby 1990).
- Biotransformation:
- Bioconcentration, Uptake (k_1) and Elimination (k_2) Rate Constants: uptake and elimination rate constants of $(2.59 \pm 0.88) \times 10^3$ mL g^{-1} d^{-1} and (2.38 ± 0.14) d^{-1} (guppy, De Bruijn & Hermens 1991); elimination rate constants of $(0.12 \pm 0.02) \times 10^3$ (NADPH) and $(0.11 \pm 0.03) \times 10^3$ (GSH) mg·min^{-1} $protein^{-1}$ (rainbow trout, De Bruijn et al. 1993).

Common Name: Pentachlorophenol
Synonym: chlorophen, PCP, penchlorol
Chemical Name: pentachlorophenol
Uses: insecticide/fungicide/herbicide; control of termites; as wood preservatives to protect against fungal rots and wood boring insects; as a pre-harvest defoliant in cotton; and also as a general pre-emergence herbicide.
CAS Registry No: 87-86-5
Molecular Formula: C_6Cl_5OH
Molecular Weight: 266.34
Melting Point (°C):
- 190 (Verschueren 1977,83; Callahan et al. 1979; Crosby 1981; Lee et al. 1991)
- 191 (Firestone 1977; Weast 1978; Kaiser et al. 1984; Agrochemicals Handbook 1987)
- 187 (Schmidt-Bleek et al. 1982)

Boiling Point (°C):
- 210 (dec., Firestone 1977)
- 310 (Verschueren 1977, 1983; Callahan et al. 1979; Renner 1990)
- 300.6 (Crosby 1981)
- 309-310 (Agrochemicals Handbook 1987)

Density (20°C, g/cm^3):
- 1.987 (Firestone 1977)
- 1.978 (Schmidt-Bleek et al. 1982; Verschueren 1983)
- 1.980 (22°C, Agrochemicals Handbook 1987)

Dissociation Constants, pK_a:
- 4.80 (Blackman et al. 1955; Sillén & Martell 1971; McLeese et al. 1979; Kaiser et al. 1984)
- 5.0 (Farquharson et al. 1958; Renner 1990)
- 4.92 (Doedens 1967; Jones 1981; Bintein & Devillers 1994)
- 4.74 (Drahonovsky & Vacek 1971; Callahan et al. 1979; Ugland et al. 1981; Könemann 1981; Könemann & Musch 1981; Dean 1985; Westall et al. 1985; Lagas 1988; Renner 1990; Lee et al. 1990,91)
- 4.71 (Cessna & Grover 1978; Saarikoski & Viluksela 1982; Saarikoski et al. 1986; Tratnyek & Hoigné 1991)
- 5.30 (Gebefügi et al. 1979; Xie 1983; Schellenberg et al. 1984)
- 4.70 (Crosby 1981; Hoigné & Bader 1983)
- 5.20 (Renberg 1981; Renner 1990; Larsson et al. 1993)
- 4.90 (Xie & Dryssen 1984; Xie et al. 1986; Shigeoka et sl. 1988; Söderström et al. 1994)
- 4.75 (Leuenberger et al. 1985)
- 4.60 (Nendza & Seydel 1988)

Molar Volume (cm^3/mol):
- 207.9 (calculated-LeBas method)
- 134.3 (calculated-density)

Molecular Volume, TMV (Å^3):
Total Surface Area, TSA (Å^2):
- 199.7 (calculated-χ, Sabljic 1987b)

Heat of Fusion, ΔH_{fus}, kcal/mol:
- 4.10 (Plato & Glasgow 1969)

Entropy of Fusion, ΔS_{fus}, kcal/mol·K (e.u.):
Fugacity Ratio at 25°C, (assuming ΔS_{fus} = 13.5 e.u.) F: 0.0336

Water Solubility (g/m^3 or mg/L at 25°C):

- 15.4 (gravimetric, Carswell & Nason 1938; quoted, Ma et al. 1993)
- 18 (27°C, gravimetric, Carswell & Nason 1938)
- 9.59 (shake flask-UV, at pH 5.1, Blackman et al. 1955; quoted, Ma et al. 1993)
- 14 (20°C, shake flask-UV, at pH 3.0, Bevenue & Beckman 1967; quoted, Valsaraj et al. 1991)
- 10 (shake flask-gravimetric, at pH 5.0, Toyota & Kuwahara 1967; quoted, Ma et al. 1993)
- 14 (gravimetric at pH 5.0, Toyota & Kuwahara 1967; quoted, Ma et al. 1993)
- 20-25 (selected, Günther et al. 1968; quoted, Gebefügi et al. 1979; Crossland & Wolff 1985)
- 20 (30°C, Firestone 1977)
- 14 (20°C, Verschueren 1977, 1983; quoted, Callahan et al. 1979; Kilzer et al. 1979; LeBlanc 1984; Shigeoka et al. 1988; Howard 1991; Lee et al. 1991)
- 20 (20°C, Körte et al. 1978)
- 14 (Kenaga & Goring 1980; Geyer et al. 1982; McKim et al. 1985)
- 14, 20-25 (selected lit. values, Geyer et al. 1980, Geyer et al. 1984)
- 14 (20°C, quoted, Crosby 1981)
- 14 (selected, Geyer et al. 1981)
- 15 (23°C, Klöpffer et al. 1982)
- 20 (20°C, Schmidt-Bleek et al. 1982; quoted, Geyer et al. 1982; Hattemer-Frey & Travis 1989)
- 15.4 (selected, Cheung 1984)
- 5-10 (at pH 5-6 in contaminated water, Goerlitz et al. 1985)
- 14 (recommended at pH 4.5-5.5, IUPAC 1985)
- 80 (20°C, Agrochemicals Handbook 1987)
- 14.1 (quoted, Isnard & Lambert 1988)
- 41 (predicted-χ, Nirmalakhandan & Speece 1988; quoted, Valsaraj et al. 1991)
- 14 (estimated, Suntio et al. 1988)
- 8 ± 2 (shake flask-UV at pH 2.5, Valsaraj et al. 1991)
- 32 ± 3 (shake flask-UV at pH 5.0, Valsaraj et al. 1991)
- 19 (quoted, Müller & Klein 1992)
- 18.4 (shake flask-HPLC/UV, at pH 4.8, Ma et al. 1993)

Vapor Pressure (Pa at 25°C):

- 0.0227 (20°C, static method, Carswell & Nason 1938)
- 0.0147 (20°C, Bevenue & Beckman 1967; quoted, Callahan et al. 1979)
- 0.231 (subcooled liq. extrapolated-Antoine eqn., Weast 1976-77; quoted, Bidleman & Renberg 1985)
- 0.10 (Weast 1972-73)
- 0.0211 (quoted, Chiou & Freed 1977)
- 0.0213 (Firestone 1977)
- 0.0147-0.0227 (20°C, Goll 1954; Bevenue & Beckman 1967; Neumüller 1974; quoted, Gebefügi et al. 1979; Kilzer et al. 1979)
- 0.0147 (20°C, quoted, Klizer et al. 1979)
- 0.0956 (subcooled liq., Hamilton 1980; quoted, Bidleman & Renberg 1985)
- 0.0227 (20°C, quoted, Crosby 1981)
- 0.00415 (23°C, OECD, Klöpffer et al. 1982)
- 0.0093 (20°C, Schmidt-Bleek et al. 1982; quoted, Hattemer-Frey & Travis 1989)

0.0147   (20°C, quoted, Verschueren 1983; Howard 1991)
0.1153   (extrapolated-Antoine eqn., Boublik et al. 1984)
0.50   (20°C, quoted, Crossland & Wolff 1985)
0.115   (capillary GC-RT, Bidleman & Renberg 1985)
0.127   (extrapolated-Antoine eqn., Stephenson & Malanowski 1987)

Henry's Law Constant (Pa m^3/mol):
0.00248 (calculated-P/C, Hellmann 1987; quoted, Meylan & Howard 1991)
0.0127   (estimated-bond contribution, Hellmann 1987; quoted, Meylan & Howard 1991)
0.277   (calculated-P/C, Howard 1991)
0.079   (calculated-P/C, this work)

Octanol/Water Partition Coefficient, log K_{ow}:
5.01   (quoted unpublished result, Leo et al. 1971; quoted, Chiou & Freed 1977; Callahan et al. 1979; Beltrame et al. 1984; Crossland & Wolff 1985; Westall et al. 1985; Gerstl & Helling 1987)
5.01   (Firestone 1977)
5.01, 5.12, 5.86, 3.81 (Hansch & Leo 1979)
5.01   (quoted, Veith et al. 1979; Mackay 1982; Chin et al. 1986)
5.01   (calculated, Veith et al. 1979b; McLeese et al. 1979)
2.97   (Veith et al. 1979; quoted, Veith & Kosian 1983; Saito et al. 1992)
3.69   (quoted from Kotzias 1980 unpublished result, Geyer et al. 1982)
4.16   (quoted, Rao & Davidson 1980)
5.10   (calculated-HPLC-k', Butte et al. 1981; quoted, Butte et al. 1987)
3.25   (quoted, Crosby 1981)
5.19   (calculated-f const., Könemann 1981; Könemann & Musch 1981)
4.00, 0.0 (at pH 4, 8, Renberg 1981; quoted, Larsson et al. 1993)
5.12   (quoted from Hansch & Leo 1979, Miyake & Terada 1982)
5.08   (RP-HPLC-k', Miyake & Terada 1982; quoted, Thomann 1989)
5.15   (shake flask-GC, Saarikoski & Viluksela 1982; Saarikoski et al. 1986)
5.05   (quoted, Kaiser & Valdmanis 1982)
4.84   (shake flask-GC, apparent value at pH 1.2, Kaiser & Valdmanis 1982)
1.30   (shake flask-GC, apparent value at pH 10.5, Kaiser & Valdmanis 1982)
3.69   (Geyer et al. 1982, Schmidt-Bleek et al. 1982; quoted, Hattemer-Frey & Travis 1989)
3.29   (shake flask average, OECD/EEC Lab. comparison tests, Harnish et al. 1983)
5.01   (quoted from Hansch & Leo 1979, Kaiser 1983)
5.01   (Verschueren 1983; quoted, Lee et al. 1991)
5.85   (calculated as per Leo et al. 1971, Xie 1983)
5.86   (quoted from Hansch & Leo 1982, Banerjee et al. 1984)
5.01   (quoted from Hansch & Leo 1979, Davies & Dobbs 1984; Kaiser et al. 1984)
5.12   (selected best lit. value, Garst & Wilson 1984)
5.11 ± 0.07 (exptl.-ALPM, Garst & Wilson 1984)
3.69, 3.81 (shake flask, OECD 1981 guidelines, Geyer et al. 1984)
5.24   (shake flask-HPLC/UV, Schellenberg et al. 1984)
5.04, 5.08, 5.85, 5.22 (shake flask-GC, HPLC-k', calculated-π const., calculated-f const., Xie et al. 1984; Bintein & Devillers 1994)
5.05   (calculated, Xie & Dryssen 1984; quoted, Lagas 1988)

3.81   (quoted, Freitag et al. 1985)
5.24   (OECD 1981 guidelines, Leuenberger et al. 1985)
5.01, 4.71 (quoted, RPHPLC-RT, Chin et al. 1986)
5.04   (quoted from Xie et al. 1984, Xie et al. 1986; Söderström et al. 1994)
2.50   (at pH 4.7, Geyer et al. 1987)
5.10   (quoted, Butte et al. 1987)
4.47   (CPC-RV, Tereda et al. 1987)
5.12   (quoted, Tereda et al. 1987; Hodson & Williams 1988)
4.07   (OECD 81 method, Kerler & Schönherr 1988)
5.04   (HPLC-RP, Shigeoka et al. 1988; quoted, Saito et al. 1993)
5.05   (selected, Suntio et al. 1988)
5.00   (batch equilibration-UV, Beltreme et al. 1988)
5.01   (quoted, Isnard & Lambert 1988, 1989)
5.24   (THOR 1989; quoted, Connell & Markwell 1990)
5.01   (quoted, Liu et al. 1991)
5.12   (quoted, Howard 1991)
5.06   (calculated-CLOGP, Müller & Klein 1992)
5.01, 5.38 (quoted, calculated-original UNIFAC, Chen et al. 1993)
5.24   (EPA CLOGP Data Base, Hulzebos et al. 1993)
5.18   (recommended, LOGKOW databank, Sangster 1993)
5.06, 5.12 (COMPUTOX databank, Kaiser 1993)
5.12   (selected, Hansch et al. 1995)

Bioconcentration Factor, log BCF:
3.75   (fish, Statham et al. 1976; quoted, Hattemer-Frey & Travis 1989)
3.04   (fish, Körte et al. 1978)
2.89   (fathead minnow, 32 d exposure, Veith et al. 1979; quoted, Veith & Kosian 1983; Davies & Dobbs 1984)
2.89   (fathead minnow, calculated value, Veith et al. 1979b)
3.09, 2.64 (algae: exptl, calculated, Geyer et al. 1981)
2.00   (trout, Hattula et al. 1981)
3.04, 3.10, 3.02 (activated sludge, algae, golden orfe, Freitag et al. 1982)
2.54   (mussel *Mytilus edulis*, quoted average, Geyer et al. 1982)
2.89, 3.69 (quoted, calculated-K_{ow}, Mackay 1982)
1.60   (killifish, Trujillo et al. 1982)
1.86, 1.72, 1.60 (low-PCP flowing, high-PCP flowing, high-PCP static soft water; Brockway et al. 1984)
1.66, 1.62, 1.26 (low-PCP flowing, high-PCP flowing, high-PCP static hard water; Brockway et al. 1984)
3.10   (*alga chlorella fusca* in culture flasks, Geyer et al. 1984; quoted, Brockway et al. 1984)
3.10, 2.72 (algae: exptl, calculated-K_{ow}, Geyer et al. 1984)
3.10, 3.02, 3.04 (algae, fish, sludge, Klein et al. 1984)
3.00   (quoted, LeBlanc 1984)
3.04, 3.10, 2.42 (activated sludge, algae, golden ide, Freitag et al. 1985)
0.57   (human fat, Geyer et al. 1987)
2.99   (zebrafish, Butte et al. 1987)
0.46   (15°C, initial concn. 1.0 mg/L uptake by *Allolobophora caliginosa* at 24 hours, Haque & Ebing 1988)

0.38   (15°C, initial concn. 10.0 mg/L uptake by *Allolobophora caliginosa* at 24 h, Haque & Ebing 1988)
0.80   (whole *Allolobophora caliginosa*/soil, uptake from soil after 131 d-exposure in outdoor lysimeters, Haque & Ebing 1988)
1.35   (whole *Lumbricus terrestris*/soil, uptake from soil after 131 d-exposure in outdoor lysimeters, Haque & Ebing 1988)
2.89   (quoted, Isnard & Lambert 1988)
2.80, 2.63   (earthworm *E. fetida andrei*: in Kooyenburg soil, Holten soil, van Gestel & Ma 1988)
−2.66   (daily intake/cow adipose tissue, Travis & Arms 1988; quoted, Hattemer-Frey & Travis 1989)
4.10   (rainbow trout, field bioaccumulation, Thomann 1989)
4.38, 4.50, 4.53, 4.90   (earthworm system, quoted, Connell & Markwell 1990)
4.00, 5.30, 3.40, 8.00   (earthworm system, derived data, Connell & Markwell 1990)
2.97   (*P. hoyi*, Landrum & Dupuis 1990)
2.11   (*M. relicta*, Landrum & Dupuis 1990)
2.16-2.53   (soft tissue of freshwater mussel, Mäkelä & Oikari 1990)
2.33; 3.21   (flagfish: whole fish; fish lipid, Smith et al. 1990)
2.78, 2.11, 1.72   (goldfish at pH 7, pH 8, pH 9, Stehly & Hayton 1990)
2.89, 1.11   (fathead minnow, bluegill; Saito et al. 1992)
3.0, 3.4, 3.9, 4.0   (perch bile to water, Söderström et al. 1994)

Sorption Partition Coefficient, log K_{OC}:
2.95   (soil, calculated-K_{OW}, Kenaga & Goring 1980)
3.11-5.65   (soil, calculated-K_{OW}, model of Karickhoff et al. 1979, Sabljic 1987a,b)
3.17-4.54   (soil, calculated-K_{OW}, model of Kenaga & Goring 1980, Sabljic 1987a,b)
3.37-3.69   (soil, calculated-K_{OW}, model of Briggs 1981, Sabljic 1987a,b)
3.00-5.54   (soil, calculated-K_{OW}, model of Means et al. 1982, Sabljic 1987a,b)
2.21-4.49   (soil, calculated-K_{OW}, model of Chiou et al. 1983, Sabljic 1987a,b)
4.52   (sediment, Schellenberg et al. 1984)
2.95, 2.41   (quoted, calculated-χ, Gerstl & Helling 1987)
3.73   (quoted average of Kenaga & Goring 1980 and Schellenberg et al. 1984 values, Sabljic 1987a,b)
3.46   (soil, calculated-χ, Sabljic 1987a,b)
2.95   (soil, calculated-χ, Bahnick & Doucette 1988)
4.04   (estimated, HPLC-k', Hodson & Williams 1988)
4.40   (calculated, Lagas 1988)
3.10, 3.26   (totally dissociated as phenolate-calculated, Lagas 1988)
5.27, 5.71   (Bluepoint soil at pH 7.8, pH 7.4, Bellin et al. 1990)
5.58, 5.52   (Glendale soil at pH 7.3, pH 4.3, Bellin et al. 1990)
3.49, 3.57   (Norfolk soil at pH 4.3, pH 4.4, Bellin et al. 1990)
4.32-4.65   (Norfolk + lime soil at pH 6.9, Bellin et al. 1990)

Half-Lives in the Environment:
   Air: 139.2-1392 h, based on an estimated rate constant for the vapor-phase reaction with hydroxyl radicals in air (Howard et al. 1991); photolysis half-life of 6.5 h in noonday summer sunshine (Howard 1991); half-lives at latitude of 43.70° N is

216 h at noon in January to less than 24 h in mid-summer for reaction with hydroxyl radicals (Bunce et al. 1991).

Surface water: calculated photolysis half-life of 4.75 h from a determined rate of 3.4×10^{-4} s^{-1} for a depth of 300 cm at pH 7 with light intensity of 0.04 watts/cm^2 between 290 and 330 nm on a midsummer day at the latitude of Cleveland, Ohio (Hiatt et al. 1960; quoted, Callahan et al. 1979); photolysis half-life of 1.5 d was estimated from photolytic destruction by sunlight in an aqueous solution at Davis, California (Wong & Crosby 1978; quoted, Callahan et al. 1979); photolytic half-life of 10-15 d (Brockway et al. 1984); rate constant of $> > 3.0 \times 10^5$ M^{-1} s^{-1} for the reaction with ozone at pH 2.0 (Hoigné & Bader 1983); 1.5 to 3.0 d for direct photo-transformation from outdoor ponds (Crossland & Wolff 1985); half-lives of 1 h (summer), 2 h (winter) for distilled water; 2 h (summer), 3 h (winter) for estuarine water; 2 h (summer), 3 hours (winter) for poisoned estuarine water, based on photo-transformation rate constants (Hwang et al. 1986); 6 d (summer), 14 days (winter) for distilled water; 3 d (summer), 7 d (winter) for estuarine water; 6 d (summer), 10 days (winter) for poisoned estuarine water, based on photo-mineralization rate constants (Hwang et al. 1986); 0.75 h and 0.96 h, based on photochemical transformation in Xenotest 1200 (Svenson & Björndal 1988); half-life of 1-110 hours, based on aqueous photolysis half-life (Howard et al. 1991); photodegradation half-lives ranging from hours to days, more rapid at the surface (Howard 1991); photodegradation half-lives of 1.0 h (summer), 2.0 hours (winter) in distilled water and 2.0 h (summer), 3.0 h (winter) in estuarine water under irradiation by natural sunlight (quoted from Hwang et al. 1987, Sanders et al. 1993).

Groundwater: 1104-36,480 h, based on estimated unacclimated aqueous aerobic sediment grab sample data (Delaune et at. 1983; selected, Howard et al. 1991) and unacclimated anaerobic grab sample data for groundwater (Baker & Mayfield 1980; selected, Howard et al. 1991).

Sediment:

Soil: disappearance half-lives of 23.2 d from Kooyenburg soil, 47.9 d from Holten soil with earthworm *E. fetida andrei* and 27.4 d from Kooyenburg soil, 31.8 d from Holten soil with earthworm *L. rubellus* (van Gestel & Ma 1988); half-life 552-4272 h, based on estimated unacclimated aqueous aerobic biodegradation half-life (Howard et al. 1991); 12.0 d in an acidic clay soil with <1.0% organic matter and 6.7 d in a slightly basic sandy loam soil with 3.25% organic matter, based on aerobic batch lab. microcosm experiments (Loehr & Matthews 1992).

Biota: biological half-life in guppy *Lebistes reticulatus* was estimated at about 30 d (Landner et al. 1977); elimination half-lives 23, 9.3, 6.9, and 6.2 h for fat, liver muscle, and blood respectively (rainbow trout, Call et al. 1980); estimated half-life of 7.0 d in trout (Niimi & Cho 1983; quoted, Niimi & Palazzo 1985); clearance from flagfish: 0.68 d from whole fish and 0.68 d from fish lipid (Smith et al. 1990).

Environmental Fate Rate Constants and Half-Lives:

Volatilization/evaporation: half-life of 84 h from the rate of loss experiment on watch glass for an exposure period of 192 h (Dobbs & Grant 1980); stripping loss rate constant of 0.0076 d^{-1} (Moos et al. 1983); 0.028 d^{-1} for nondissociated PCP, assuming diffusion coefficient in air to be 7×10^{-6} m^2/s and in water 7×10^{-10}

m²/s with wind speed 0.1 m above the pond is 2 m/s and the average temperature is 15°C for water depth of 1 m (Crossland & Wolff 1985); calculated rate constant of 5×10^{-4} d^{-1} to 1×10^{-7} d^{-1} for total PCP (Crossland & Wolff 1985).

Photolysis: calculated photolysis half-life of 4.75 h from observed rate of 3.4×10^{-4} s^{-1} for a depth of 300 cm at pH 7 with light intensity of 0.04 watts/cm² between 290 and 330 nm on a midsummer day at the latitude of Cleveland, Ohio (Hiatt et al. 1960; quoted, Callahan et al. 1979); photolysis half-life of 1.5 d was estimated from photolytic destruction by sunlight in an aqueous solution at Davis, California (Wong & Crosby 1978; quoted, Callahan et al. 1979); exposure of aqueous PCP solutions to either sunlight or laboratory ultraviolet light resulted in rapid degradation at pH 7.3 and slower degradation at pH 3.3 (Wong & Crosby 1981); photolytic half-life of 10-15 d (Brockway et al. 1984); 0.23 to 0.46 d^{-1} for direct photo-transformation, is the main loss process for PCP from ponds, with half-life of 1.5 to 3.0 d (Crossland & Wolff 1985); photo-transformation rate constants of 0.6 h^{-1} with half-life of 1 h for distilled water in summer (mean temperature 25°C) and 0.37 h^{-1} with half-life of 2 hours in winter (mean temperature 11°C); 0.37 h^{-1} with half-life of 2 h for both poisoned estuarine water and estuarine water in summer and 0.27 h^{-1} with half-life of 3 h in winter during days when exposed to full sunlight and microbes (Hwang et al. 1986); photo-mineralization rate constants of 0.11 h^{-1} with half-life of 6 d for distilled water in summer (mean temperature 25°C) and 0.049 h^{-1} with half-life of 14 d in winter (mean temperature 11°C); 0.12 h^{-1} with half-life of 6 d for poisoned estuarine water in summer and 0.07 h^{-1} with half-life of 10 d in winter; 0.25 h^{-1} with half-life of 10 d for estuarine water in summer and 0.10 h^{-1} with half-life of 7 d for winter during days when exposed to full sunlight and microbes (Hwang et al. 1986); phototransformation half-life of 0.75 h in Xenotest 1200 (Svenson & Björndal 1988); aqueous photolysis half-life of 1-110 h (Hwang et al. 1986; Sugiura et al. 1984; selected, Howard et al. 1991); assuming a linear rate of photolysis during 96-h period, they determined the half-life of PCP in their systems to be 7.43 d (Smith et al. 1987); photodegradation rate constants of 0.60 h^{-1} corresponding to a half-life of 1.0 h (summer), 0.37 h^{-1} corresponding to a half-life of 2 h (winter) in distilled water; and 0.37 h^{-1} corresponding to a half-life of 2 h (summer), 0.27 h^{-1} corresponding to a half-life of 3.0 h (winter) in estuarine water under irradiation by natural sunlight (quoted from Hwang et al. 1987, Sanders et al. 1993).

Oxidation: rate constant of $>> 3.0 \times 10^5$ M$^{-1} \cdot$s^{-1} for the reaction with ozone in water at pH 2.0 (Hoigné & Bader 1983); photooxidation half-life of 66-3480 h in water, based on reported reaction rate constants for reaction of OH and RO$_2$ radicals with phenol class in aqueous solution (Mill & Mabey 1985; Guesten et al. 1981; quoted, Howard et al. 1991); photooxidation half-life of 139.2-1392 h, based on an estimated rate constant for the vapor-phase reaction with hydroxyl radicals in air (Atkinson 1987; quoted, Howard et al. 1991); rate constant of $(0.2 \pm 5.5) \times 10^6$ M$^{-1} \cdot$s^{-1} for the reaction with singlet oxygen in aqueous phosphate buffer at (27 ± 1)°C (Tratnyek & Hoigné 1991); atmospheric half-lives vary from less than 24 h at noon in mid-summer to 216 h in January at latitude of 41.79°N for reaction with OH radicals (Bunce et al. 1991).

Hydrolysis: is not expected to occur (Crossland & Wolff 1985).

Biodegradation: 1800-2160 h and 480-∞ h to obtain 75% degradation in mineral medium and seawater respectively (De Kreuk & Hanstveit 1981); aqueous aerobic half-life of 552-4272 h, based on unacclimated and acclimated aerobic sediment grab sample data (Delaune et al. 1983; Baker & Mayfield 1980; quoted, Howard et al. 1991); aqueous anaerobic half-life of 1008-36480 h, based on unacclimated anaerobic grab sample data for soil and ground water (Ide et al. 1972; Baker & Mayfield 1980; quoted, Howard et al. 1991); aerobic degradation rate constant of 0.0017 L $\mu g^{-1} \cdot d^{-1}$ (Moos et al. 1983); microbial degradation negligible in darkness (Hwang et al. 1986); degradation rate constant of 0.12 ± 0.01 h^{-1} in the absence of light (Minero et al. 1993).

Biotransformation: degradation rate of 3×10^{-14} mol·cell^{-1}·h^{-1} with microorganisms in Seneca River waters (Banerjee et al. 1984).

Bioconcentration Uptake (k_1) and Elimination (k_2) Rate Constants:

k_1: 18.3 h^{-1}, 19 h^{-1} (at 1 mM buffer concn), 18.5 h^{-1} (at 10 mM buffer concn) at pH 8 (guppy *P. reticulata* Peters, Saarikoski et al. 1986)

k_1: 222 d^{-1}, 1677 d^{-1} (flagfish: whole fish; fish lipid, Smith et al. 1990)

k_2: 1.03 d^{-1}, 1.03 d^{-1} (flagfish: whole fish; fish lipid, Smith et al. 1990)

k_2: 1.03 d^{-1}, 0.95 d^{-1} (flagfish: BCF based, toxicity based, Smith et al. 1990)

k_2: 0.00195 ± 0.00063 h^{-1} (*M. relicta*, Landrum & Dupuis 1990)

k_2: 0.00330 ± 0.00140 h^{-1} (*P. hoyi*, Landrum & Dupuis 1990)

Common Name: Permethrin

Synonym: Ambush, Dragnet, Ectiban, Exmin, FMC 33297, FMC 41665, ICI-PP 557, Kafil, Kestrel, NDRC-143, NIA 33297, Niagara 33297, Outflank, Outflank-stockade, Perthrine, Picket, Pounce, Pramex, S 3151, SBP-1513, Talcord, WL 43479

Chemical Name: 3-phenoxybenzyl (1RS, 3RS; 1RS, 3SR)-3(2,2-dichlorovinyl)-2,2-dimethylcyclopropanecarboxylate; 3-(2,2-dichloroethenyl)-2,2-dimethylcyclopropanecarboxylic acid (3-phenoxyphenyl)methyl ester

Uses: insecticide to control overwintering forms of spider mites, aphids, and scale insects on fruit trees, vines, olives, bananas and ornamentals; used as herbicides to control grass and broadleaf weeds in umbelliferous crops, and in tree nurseries; also used as acaricide and surfactant.

CAS Registry No: 52645-53-1
Molecular Formula: $C_{21}H_{20}Cl_2O_3$
Molecular Weight: 391.3
Melting Point (°C):
    liquid   (tech. grade, Worthing 1991)
    34-39   (pure, Worthing 1991; Montgomery 1993)
    63-65   (*cis* isomer, Worthing 1991)
    44-47   (*trans* isomer, Worthing 1991)
    34-35   (Tomlin 1994; Milne 1995)

Boiling Point (°C):
    200   (at 0.01 mmHg, Agrochemicals Handbook 1987; Milne 1995)
    200   (tech. grade at 0.1 mmHg, Worthing 1991; Tomlin 1994)
    220   (at 0.05 mmHg, Montgomery 1993)
    >290   (Tomlin 1994)

Density (g/cm³ at 20°C):
    1.19-1.27   (Agrochemicals Handbook 1987; Montgomery 1993; Tomlin 1994; Milne 1995)
    1.214   (tech. grade at 25°C, Worthing 1991)

Molar Volume (cm³/mol):
    431   (calculated-LeBas method, this work)
    318.1   (calculated-density, this work)

Molecular Volume (Å³):
Total Surface Area, TSA (Å²):
Dissociation Constant pK_a:
Heat of Fusion, ΔH_{fus}, kcal/mol:
Entropy of Fusion, ΔS_{fus}, cal/mol·K (e.u.):
Fugacity Ratio at 25°C (assuming ΔS_{fus} = 13.5 e.u.), F: 0.770

Water Solubility (g/m³ or mg/L at 25°C):
    0.20   (Martin & Worthing 1977; quoted, Kenaga 1980)
    0.04   (shake flask-GC, Coats & O'Donnell-Jefferey 1979; quoted, Spener 1982; Shiu et al. 1990)
    ~0.2   (Spencer 1982)
    0.05   (in seawater, quoted, Schimmel et al. 1983; Zaroogian et al. 1985; Clark et al. 1989)
    0.20   (20°, Agrochemicals Handbook 1987; Tomlin 1994)
    0.20   (30°, Worthing 1987, 1991; quoted, Shiu et al. 1990)
    0.006   (20-25°, selected, Wauchope et al. 1992; quoted, Lohninger 1994; Majewski & Capel 1995; Hornsby et al. 1996)

1.0 (quoted, Pait et al. 1992)
0.2 (20°, Montgomery 1993; Milne 1995)

Vapor Pressure (Pa at 25°C):
4.8×10^{-6} (*cis* isomer, Barlow 1978; quoted, Hinckley et al. 1990)
4.9×10^{-6} (*cis* isomer, Wells et al. 1986; quoted, Hinckley et al. 1990)
3.7×10^{-6} (*trans* isomer, Barlow 1978; quoted, Hinckley et al. 1990)
3.1×10^{-6} (*trans* isomer, Wells et al. 1986; quoted, Hinckley et al. 1990)
4.5×10^{-5} (Agrochemicals Handbook 1987; Tomlin 1994)
1.0×10^{-5} (*cis* isomer, GC-RT, supercooled liq., Hinckley et al. 1990)
8.1×10^{-6} (*trans* isomer, GC-RT, supercooled liq., Hinckley et al. 1990)
1.3×10^{-6} (tech. grade at 20°C, Worthing 1991)
2.5×10^{-6} (pure *cis* isomer at 20°C, Worthing 1991; Montgomery 1993; Tomlin 1994)
1.5×10^{-6} (pre *trans* isomer at 20°C, Worthing 1991; Tomlin 1994)
1.7×10^{-6} (20-25°, selected, Wauchope et al. 1992; quoted, Majewski & Capel 1995; Hornsby et al. 1996)

Henry's Law Constant (Pa·m³/mol):
0.00486 (20°, calculated-P/C, Montgomery 1993)
0.0157 (20-25°, calculated-P/C, Majewski & Capel 1995)
0.111 (calculated-P/C, this work)

Octanol/Water Partition Coefficient, log K_{OW}:
3.49 (shake flask-GC, Coats & O'Donnell-Jefferey 1979; quoted, Shiu et al. 1990)
3.95 (calculated, McLeese et al. 1980)
2.88 (quoted, Rao & Davidson 1980)
6.60 (calculated, Briggs 1981)
6.50 (quoted, Schimmel et al. 1983; Zaroogian et al. 1985; Clark et al. 1989)
6.2 ± 0.9 (*cis*-form, correlated-HPLC-RT, Muir et al. 1985)
5.7 ± 0.7 (*trans*-form, correlated-HPLC-RT, Muir et al. 1985)
6.10 (tech. grade at 20°C, Worthing 1991; Tomlin 1994)
2.88-6.10 (Montgomery 1993)
6.10 (Milne 1995)
6.50 (selected, Devillers et al. 1996)
5.85, 7.12, 5.63 (RP-HPLC, ClogP, calculated-S, Finizio et al. 1997)

Bioconcentration Factor, log BCF:
3.18 (calculated-S, Kenaga 1980)
3.28 (quoted, Schimmel et al. 1983)
3.23, 3.49, 3.52 (*Pimephales promelas*, Spehar et al. 1983; quoted, Devillers et al. 1996)
1.49-1.84 (*trans*-form on sediment, 24 h BCF for chironomid larvae in water, Muir et al. 1985)
1.08-2.13 (*trans*-form on sediment, 24 h BCF for chironomid larvae in sediment, Muir et al. 1985)
0.95-1.70 (*trans*-form on sediment, 24 h BCF for chironomid larvae in sediment/pore water, Muir et al. 1985)
0.90-2.22 (*cis*-form on sediment, 24 h BCF for chironomid larvae in water, Muir et al. 1985)

1.46-2.62 (*cis*-form on sediment, 24 h BCF for chironomid larvae in sediment, Muir et al. 1985)
1.32-2.47 (*cis*-form on sediment, 24 h BCF for chironomid larvae in sediment/pore water, Muir et al. 1985)
4.71, 4.83 (oyster, calculated-K_{OW} & models, Zaroogian et al. 1985)
4.71, 4.83 (sheepshead minnow, calculated-K_{OW} & models, Zaroogian et al. 1985)
3.29, 3.39 (*Oncorhynchus mykiss*, Muir et al. 1994; quoted, Devillers et al. 1996)
2.79 (quoted, Pait et al. 1992)

Sorption Partition Coefficient, log K_{OC}:
4.03 (calculated-S, Kenaga 1980)
2.76 (*cis*-form, silt, K_P on 34% DOC, Muir et al. 1985)
2.64 (*cis*-form, clay, K_P on 77% DOC, Muir et al. 1985)
2.64 (*trans*-form, silt, K_P on 23% DOC, Muir et al. 1985)
2.64 (*trans*-form, clay, K_P on 0% DOC, Muir et al. 1985)
4.80 (soil, quoted exptl., Meylan et al. 1992)
5.25 (soil, calculated-χ and fragment contribution, Meylan et al. 1992)
5.00 (soil, 20-25°, selected, Wauchope et al. 1992; Lohninger 1994; Hornsby et al. 1996)
1.32-2.79 (Montgomery 1993)

Half-Lives in the Environment:
Air:
Surface water: half-lives in 100 mL pesticide-seawater solution: >21 d under indoor conditions, 14 d under outdoor light conditions and >14 d under outdoor dark conditions (Schimmel et al. 1983).
Groundwater:
Sediment: half-lives in 10 grams sediment/100 mL pesticide-seawater solution: < 2.5 days for untreated sediment and >28 d for sterile sediment (Schimmel et al. 1983).
Soil: reported half-life in soil containing 1.3-51.3% organic matter at pH 4.2-7.7 was <38 days (Holmstead et al. 1978; quoted, Worthing 1991; Montgomery 1993; Tomlin 1994); selected field half-life of 30 d (Wauchope et al. 1992; Hornsby et al. 1996); soil half-life of 30 d (Pait et al. 1992); half-life of 32 d for forest soil (Dowd et al. 1993).
Biota:

Environmental Fate Rate Constants or Half-Lives:
Volatilization:
Photolysis: photodegradation rate constant of 1.73×10^{-3} min^{-1} and half-life of 400 min with TiO_2 as catalyst after 20 h irradiation at 222 nm (Hidaka et al. 1992).
Oxidation:
Hydrolysis:
Biodegradation:
Biotransformation:
Bioconcentration, Uptake (k_1) and Elimination (k_2) Rate Constants:

Common Name: Phenthoate
Synonym: Cidial, Elsan
Chemical Name: ethyl 2-dimethoxyphosphinothioylthio(phenyl)acetate; ethyl 2-dimethoxy-thiophosphorylthio-2-phenylacetate; S-α-ethoxycarbonylbenzyl O,O-dimethyl phosphorodithioate; ethyl α-[(dimethoxyphosphinothioyl)thio]benzeneacetate
Uses: insecticide to control aphids, scale insects, jassids, lepidopterous larvae, bollworms, mealybugs, psyllids, thrips, spider mites, etc. in citrus fruit, pome fruit, olives, cotton, cereals, rice, coffee, tea, sunflower, sugar cane, tobacco, ornamentals, and vegetables; also used as acaricide and for control of mosquito larvae.
CAS Registry No: 2597-03-7
Molecular Formula: $C_{12}H_{17}O_4PS_2$
Molecular Weight: 320.4
Melting Point (°C):
    17.5   (Khan 1980; Spencer 1982)
    17-18   (Worthing 1991; Tomlin 1994)
Boiling Point (°C): 78-80 (Spencer 1982)
Density (g/cm³ at 20°C):
    1.226   (Agrochemicals Handbook 1987; Worthing 1991; Tomlin 1994)
Molar Volume (cm³/mol):
    261.3   (calculated from density)
Molecular Volume (Å³):
Total Surface Area, TSA (Å²):
Dissociation Constant pK_a:
Heat of Fusion, ΔH_{fus}, kcal/mol:
Entropy of Fusion, ΔS_{fus}, cal/mol·K (e.u.):
Fugacity Ratio at 25°C (assuming ΔS_{fus} = 13.5 e.u.), F: 1.0

Water Solubility (g/m³ or mg/L at 25°C):
    200   (Martin & Worthing 1977; quoted, Kenaga 1980; Spencer 1982)
    11   (20-25°C, shake flask-GC, Kanazawa 1981; quoted, Shiu et al. 1990)
    11   (20°C, Khan 1980; Agrochemicals Handbook 1987)
    11   (24°C, Worthing 1987,91; quoted, Shiu et al. 1990; Tomlin 1994)
    11   (20-25°C, selected, Augustijn-Beckers et al. 1994; Hornsby et al. 1996)

Vapor Pressure (Pa at 25°C):
    0.005   (40°C, Agrochemicals Handbook 1987)
    0.0053   (40°C, Worthing 1991; Tomlin 1994)
    3.5×10^{-4} (20-25°C, selected, Augustijn-Beckers et al. 1994; Hornsby et al. 1996)

Henry's Law Constant (Pa·m³/mol):
    0.01019   (calculated-P/C, this work)

Octanol/Water Partition Coefficient, log K_{OW}:
    2.89   (shake flask-GC, Kanazawa 1981; quoted, Sangster 1993)
    3.96   (calculated, De Bruijn et al. 1991; quoted, Verhaar et al. 1992)
    3.69   (Worthing 1991; Tomlin 1994)
    3.32   (RP-HPLC, Saito et al. 1993)
    3.69   (selected, Hansch et al. 1995)
    3.69   (selected, Devillers et al. 1996)

Bioconcentration Factor, log BCF:
- 1.49 (calculated-S, Kenaga 1980)
- 1.56 (*Pseudorasbora parva*, Kanazawa 1981; quoted, De Bruijn & Hermens 1991; Devillers et al. 1996)
- 2.85 (whole body willow shiner after 168 h exposure, Tsuda et al. 1992)
- 1.57, 1.43, 1.30, 1.51 (whole body carp: 24 h, 72 h, 120 h and 168 h; Tsuda et al. 1993)

Sorption Partition Coefficient, log K_{OC}:
- 2.38 (soil, calculated-S, Kenaga 1980)
- 3.00 (20-25°C, estimated, Augustijn-Beckers et al. 1994; Hornsby et al. 1996)

Half-Lives in the Environment:
Air:
Surface water:
Groundwater:
Sediment:
Soil: selected field half-life of 11 d (Augustijn-Beckers et al. 1994; Hornsby et al. 1996); half-life of 10 d in silty clay loam and other soils (Tomlin 1994).
Biota: excretion rate constant of 0.05 h^{-1} from whole body willow shiner (Tsuda et al. 1992); excretion rate constant of 0.52 h^{-1} with half-life of 1.3 h (Tsuda et al. 1993).

Environmental Fate Rate Constants or Half-Lives:
Volatilization:
Photolysis:
Oxidation:
Hydrolysis:
Biodegradation:
Biotransformation:
Bioconcentration, Uptake (k_1) and Elimination (k_2) Rate Constants: excretion rate constant of 0.05 h^{-1} from whole body willow shiner (Tsuda et al. 1992); excretion rate constant of 0.52 h^{-1} with half-life of 1.3 h (Tsuda et al. 1993).

Common Name: Phorate
Synonym: AC 3911, American Cyanamid 3911, ENT 24042, Foraat, Gramitox, Granutox, Rampart, Thimate, Thimet, Timet, Vegfu, Vergfru foratox
Chemical Name: O,O-diethyl-S-(ethylthio)methyl phosphorodithioate; O,O-diethyl-S-ethylmercaptomethyl dithiophosphate; phosphorodithioic acid O,O-diethyl S-((ethylthio)methyl) ester
Uses: insecticide to control mites, chewing and sucking insects in fruits and vegetables, cotton, and some ornamentals; also used as acaricide and nematicide.
CAS Registry No: 298-02-2
Molecular Formula: $C_7H_{17}O_2PS_3$
Molecular Weight: 260.4
Melting Point (°C):
- -42.9 (Spencer 1982; Suntio et al. 1988)
- <-15 (Montgomery 1993)

Boiling Point (°C):
- 118-120 (at 0.8 mmHg, Agrochemicals Handbook 1987; Montgomery 1993; Milne 1995)
- 125-127 (at 2 mmHg, Merck Index 1989; Milne 1995)
- 118-120 (tech. grade at 0.8 mmHg, Worthing 1991)

Density (g/cm³ at 20°C):
- 1.156 (25°C, Merck Index 1989; Montgomery 1993; Milne 1995)
- 1.167 (tech. grade at 25°C, Spencer 1982; Worthing 1991)

Molar Volume (cm³/mol):
- 259.9 (calculated-LeBas method, Suntio et al. 1988)

Molecular Volume ($Å^3$):
Total Surface Area, TSA ($Å^2$):
Dissociation Constant pK_a:
Heat of Fusion, ΔH_{fus}, kcal/mol:
Entropy of Fusion, ΔS_{fus}, cal/mol·K (e.u.):
Fugacity Ratio at 25°C (assuming ΔS_{fus} = 13.5 e.u.), F: 1.0

Water Solubility (g/m³ or mg/L at 25°C):
- 19 (26°C, 95% pure, shake flask-GC, Lord & Burt 1964; quoted, Freed 1976; Suntio et al. 1988; Shiu et al. 1990)
- 14 (15°C, shake flask-GC, Lord & Burt 1964; quoted, Shiu et al. 1990)
- 85 (Günther et al. 1968; quoted, Suntio et al. 1988)
- 70 (Melnikov 1971; quoted, Shiu et al. 1990)
- 50 (Spencer 1973, 1982; ; quoted, Bowman & Sans 1979; Khan 1980)
- 50 (Martin & Worthing 1977; quoted, Kenaga 1980a; Kenaga & Goring 1980; Jury et al. 1983,84; Kim et al. 1984; Suntio et al. 1988)
- 80-85 (Wauchope 1978)
- 20 (shake flask-GC, Felsot & Dahm 1979; quoted, Shiu et al. 1990)
- 17.9 (20°C, shake flask-GC, Bowman & Sans 1979, 1983b; quoted, Fuhremann & Lichtenstein 1980; Shiu et al. 1990)
- 50 (Briggs 1981)
- 50-85 (20-25°C, selected, Willis & McDowell 1982)
- 50 (U.S. EPA 1984; quoted, McLean et al. 1988)
- 50 (Agrochemicals Handbook 1987)
- 50 (rm. temp., Worthing 1987; Merck Index 1989; quoted, Shiu et al. 1990)

40	(20°C, selected, Suntio et al. 1988; quoted, Majewski & Capel 1995)
22	(Wauchope 1989; quoted, Shiu et al. 1990)
50	(tech. grade at rm. temp., Worthing 1991)
22	(20-25°C, selected, Wauchope et al. 1992; Hornsby et al. 1996)
17.9, 50	(quoted, calculated, Pait et al. 1992)
20	(24°C, Montgomery 1993)
22.0	(selected, Lohninger 1994; Halfon et al. 1996)
17.9, 217	(quoted, calculated, Patil 1994)
20, 29,2	(quoted, calculated-group contribution fragmentation method, Kühne et al. 1995)
50	(Milne 1995)

Vapor Pressure (Pa at 25°C):
- 0.112 (20°C, Wolfdietrich 1965; quoted, Kim et al. 1984)
- 0.112 (20°C, Spencer 1973, 1982; quoted, Fuhremann & Lichtenstein 1980; Khan 1980)
- 0.25 (Woolford 1975; quoted, Hinckley et al. 1990)
- 0.148 (gas saturation method, Sutherland et al. 1980; quoted, Jury et al. 1983)
- 0.074 (gas saturation method, Kim et al. 1984; Suntio et al. 1988)
- 0.042 (20°C, extrapolated value, Kim et al. 1984)
- 0.011 (20°C, GC-calculated value, Kim et al. 1984; Kim 1985)
- 0.11 (20°C, U.S. EPA 1984; quoted, McLean et al. 1988)
- 0.042 (20°C, gas saturation, Kim 1985)
- 0.11 (20°C, Agrochemicals Handbook 1987)
- 0.10 (20°C, selected, Suntio et al. 1988; quoted, Majewski & Capel 1995)
- 0.112 (20°C, Merck Index 1989)
- 0.11 (GC method, Hinckley et al. 1990)
- 0.085 (tech. grade, Worthing 1991)
- 0.0853 (20-25°C, selected, Wauchope et al. 1992; Hornsby et al. 1996)
- 0.112 (20°C, Montgomery 1993)
- 0.085 (selected, Halfon et al. 1996)

Henry's Law Constant (Pa·m^3/mol):
- 0.769 (calculated-P/C, Jury et al. 1984, 1987a, 1990; Jury & Ghodrati 1989)
- 0.65 (20°C, calculated-P/C, Suntio et al. 1988; quoted, Majewski & Capel 1995)
- 0.648 (20-24°C, calculated-P/C, Montgomery 1993)
- 1.010 (calculated-P/C, this work)

Octanol/Water Partition Coefficient, log K_{ow}:
- 3.33 (shake flask-GC, Felsot & Dahm 1979; quoted, Bowman & Sans 1983b; Suntio et al. 1988)
- 2.92 (Rao & Davidson 1980; quoted, Bowman & Sans 1983b; Kim et al. 1984; Suntio et al. 1988)
- 4.26 (Briggs 1981; quoted, Bowman & Sans 1983b; Suntio et al. 1988)
- 3.33 (selected, Dao et al. 1983)
- 1.26 (U.S. EPA 1984; quoted, McLean et al. 1988)
- 3.80 (selected, Suntio et al. 1988; quoted, Bintein & Devillers 1994)

4.70   (selected, Travis & Arms 1988)
4.26   (Thor 1989; quoted, Connell & Markwell 1990)
2.92   (selected, Magee 1991)
3.92   (Worthing 1991)
2.96   (estimated-QSAR & SPARC, Kollig 1993)
2.91-3.92 (Montgomery 1993)
3.83, 2.70 (quoted, calculated, Patil 1994)
3.92   (Milne 1995)
3.56   (selected, Hansch et al. 1995)
4.25, 3.46, 3.44 (RP-HPLC, ClogP, calculated-S, Finizio et al. 1997)

Bioconcentration Factor, log BCF:
    1.83   (calculated-S, Kenaga 1980; quoted, Pait et al. 1992)
    2.34   (calculated-K_{OC}, Kenaga 1980)
    3.34   (earthworms, Lord et al. 1980; quoted, Connell & Markwell 1990)
    −1.70   (vegetation, correlated-K_{OW}, U. Oklahoma 1986; quoted, Travis & Arms 1988)

Sorption Partition Coefficient, log K_{OC}:
    3.51   (soil, Hamaker & Thompson 1972; quoted, Kenaga 1980; Kenaga & Goring 1980)
    2.71   (soil, calculated-S as per Kenaga & Goring 1978, Kenaga 1980)
    2.82   (Rao & Davidson 1980; quoted, Jury et al. 1983, 1984, 1990)
    2.58   (reported as log K_{OM}, Briggs 1981)
    2.32-3.60 (reported as log K_{OM}, Mingelgrin & Gerstl 1983)
    2.82   (screening model calculations, Jury et al. 1987a,b; Jury & Ghodrati 1989)
    2.58   (reported as log K_{OM}, Magee 1991)
    2.88   (estimated as log K_{OM}, Magee 1991)
    2.73   (soil, Worthing 1991)
    3.00   (soil, 20-25°C, estimated, Wauchope et al. 1992; quoted, Richards & Baker 1993; Hornsby et al. 1996)
    2.64   (estimated-QSAR & SPARC, Kollig 1993)
    2.51-2.80 (Montgomery 1993)
    2.82   (selected, Lohninger 1994)

Half-Lives in the Environment:
    Air:
    Surface water:
    Groundwater:
    Sediment:
    Soil: half-life of 68 d in a sandy soil (Way & Scopes 1968; quoted, Montgomery 1993); estimated persistence of 2 weeks (Kearney et al. 1969; Edwards 1973; quoted, Morrill et al. 1982; Jury et al. 1987a); persistence of less than one month (Wauchope 1978); biodegradation half-life of 82 d in soil (Jury et al. 1984, 1987a,b, 1990; Jury & Ghodrati 1989; quoted, Montgomery 1993); 2-14 d (Worthing 1991); estimated field half-life of 60 d (Wauchope et al. 1992; quoted, Richards & Baker 1993; selected, Halfon et al. 1996; Hornsby et al. 1996); soil half-life of 25 d (Pait et al. 1992).

Biota: half-lives in coastal Bermuda grass and alfalfa are 1.4 d (Leuck & Bowman 1970; quoted, Montgomery 1993) and 3.6 d (Dobson et al. 1960; quoted, Montgomery 1993); biochemical half-life of 82 d from screening model calculations (Jury et al. 1987a,b; Jury & Ghodrati 1989).

Environmental Fate Rate Constants or Half-Lives:
Volatilization:
Photolysis:
Oxidation:
Hydrolysis: neutral hydrolysis rate constant of 7.2×10^{-3} h^{-1} with a calculated half-life of 96 hours at pH 7 (Ellington et al. 1987, 1988; quoted, Montgomery 1993); calculated rate constant of $\geq 100 \times 10^{-12}$ cm^3/molecule·s for the vapor-phase reaction with hydroxyl radicals in air (Winer & Atkinson 1990); half-lives of 3.2 d at pH 7 and 3.9 d at pH 9 (Worthing 1991); rate constant of 62 y^{-1} at pH 7.0 and 25°C (Kollig 1993).
Biodegradation: half-life of 82 d for a 100 d leaching and screening test in 0-10 cm depth of soil (Rao & Davidson 1980; quoted, Jury et al. 1983); half-life of 82 d in soil (Jury et al. 1984, 1987a,b, 1990; Jury & Ghorati 1989); first-order rate constant of -0.0403 h^{-1} in nonsterile sediment and -0.0209 h^{-1} in sterile sediment by shake-tests at Range Point and first-order rate constant of -0.0206 h^{-1} in nonsterile water and -0.0186 h^{-1} in sterile water by shake-tests at Range Point (Walker et al. 1988); first-order rate constants of -0.0241 h^{-1} in nonsterile sediment and -0.0185 h^{-1} in sterile sediment by shake-tests at Davis Bayou and first-order rate constants of -0.0262 h^{-1} in nonsterile water and -0.0185 h^{-1} in sterile water by shake-tests at Davis Bayou (Walker et al. 1988).
Biotransformation:
Bioconcentration, Uptake (k_1) and Elimination (k_2) Rate Constants:

Common Name: Phosmet
Synonym: APPA, Decemthion, Decemthion p-6, ENT 25,705, Ftalophos, Imidan, Percolate, Phthalophos, Prolate, R 1504, Safidon, Smidan, Stauffer R-1504
Chemical Name: O,O-dimethyl S-phthalimidomethyl phosphorodithioate; N-dimethoxyphosphino-thioylthiomethyl)phthalimide; S-[(1,3-dihydro-1,3-dioxo-$2H$-isoindol-2-yl)methyl] O,O-dimethyl phosphorodithioate; phosphorodithioic acid, S-[(1,3-dihydro-1,3-dioxo-$2H$-isoindol-2-yl)methyl] O,O-dimethyl ester
Uses: nonsystemic acaricide and insecticide.
CAS Registry No: 732-11-6
Molecular Formula: $C_{11}H_{12}NO_4PS_2$
Molecular Weight: 317.32
Melting Point (°C):
    71.9   (Spencer 1982; Suntio et al. 1988; Milne 1995)
    72.0-72.7, 66.5-69.5 (pure, technical grade, Montgomery 1993; Tomlin 1994)
Boiling Point (°C):
    decompose rapidly > 100°C (Montgomery 1993)
Density (g/cm³ at 20°C):
Molar Volume (cm³/mol):
    263.3   (calculated-LeBas method, Suntio et al. 1988)
Molecular Volume (Å³):
Total Surface Area, TSA (Å²):
Dissociation Constant pK_a:
Heat of Fusion, ΔH_{fus}, kcal/mol:
    7.40   (Plato & Glasgow 1969)
Entropy of Fusion, ΔS_{fus}, cal/mol·K (e.u.):
Fugacity Ratio at 25°C (assuming ΔS_{fus} = 13.5 e.u.), F: 0.343
    0.3   (20°C, Suntio et al. 1988)

Water Solubility (g/m³ or mg/L at 25°C):
    25   (Bright et al. 1950, quoted, Suntio et al. 1988)
    25   (Melnikov 1971; Spencer 1982; quoted, Chiou et al. 1977)
    25   (Agrochemicals Handbook 1987)
    20   (20-25°C, selected, Wauchope et al. 1992; Hornsby et al. 1996)
    22-25 (quoted, Montgomery 1993)
    25   (Tomlin 1994; Milne 1995)

Vapor Pressure (Pa at 25°C):
    6.03×10^{-4} (20°C, Freed et al. 1977, quoted, Suntio et al. 1988)
    0.133   (50°C, Spencer 1982; Agrochemicals Handbook 1987)
    6.53×10^{-5} (20-25°C, selected, Wauchope et al. 1992; Hornsby et al. 1996)
    6.03×10^{-5}, 0.133 (30, 50°C, Montgomery 1993)
    6.50×10^{-5} (Tomlin 1994)

Henry's Law Constant (Pa·m³/mol):
    9.50×10^{-4} (calculated-P/C, Suntio et al. 1988)
    9.53×10^{-4} (calculated-P/C, Montgomery 1993)
    7.62×10^{-4} (calculated-P/C, this work)

Octanol/Water Partition Coefficient, log K_{OW}:
- 2.83 (20°C, shake flask-GC, Chiou et al. 1977; quoted, Sangster 1993)
- 2.83 (Rao & Davidson 1980)
- 2.78 (22°C, shake flask-GC, Bowman & Sans 1983; quoted, Sangster 1993)
- 2.80 (selected, Suntio et al. 1988)
- 2.83 (selected, Saito et al. 1992)
- 2.78-3.04 (Motgomery 1993)
- 2.95 (Tomlin 1994)
- 3.40 (Milne 1995)
- 2.78 (selected, Hansch et al. 1995)

Bioconcentration Factor, log BCF:
- 0.90 (bluegill sunfish/fathead minnows, Saito et al. 1992)
- 1.04 (channel catfish, Saito et al. 1992)
- 1.56 (av. whole body willow shiner after 24-168 h exposure, Tsuda et al. 1992)
- 0.23 (av. whole body carp after 24-168 h exposure, Tsuda et al. 1993)

Sorption Partition Coefficient, log K_{OC}:
- 2.91 (soil, Wauchope et al. 1992, Hornsby et al. 1996)

Half-Lives in the Environment:
Air:
Surface water: half-lives of 33 d at 6°C, 5 d at 22°C in darkness for Milli-Q water at pH 6.1 (Lartiges & Garrigues 1995).
Groundwater:
Sediment:
Soil: field half-life of 10 d (Wauchope et al. 1992; Hornsby et al. 1996).
Biota: half-life in Bermuda grass was 6.5 d (Montgomery 1993).

Environmental Fate Rate Constants or Half-Lives:
Volatilization:
Photolysis: half-life of 53.25 h for absorbance wavelength at 243 nm (Montgomery 1993).
Oxidation:
Hydrolysis: the hydrolysis half-lives at 20°C were 7.2 d at pH 6.1 and 7.1 h at pH 7.4; 1.1 h at 37.5°C (Freed et al. 1979; quoted, Montgomery 1993); half-lives of 13 d at pH 4.5, < 12 h at pH 7 and < 4 h at pH 8.3 in buffered aqueous solution at 20°C (Montgomery 1993); half-lives of 7.0 d at pH 6.1, and 7.1 h at pH 7.4 at 20°C (Lartiges & Garrigues 1995).
Biodegradation:
Biotransformation:
Bioconcentration, Uptake (k_1) and Elimination (k_2) Rate Constants:
k_2: 0.28 h^{-1} (whole body willow shiner, Tsuda et al. 1992)

Common Name: Propoxur
Synonym: Baygon, Blattanex, Under, arprocarb, PHC, Sendran, Suncide, Aracarb, Tugon Fliegendugel
Chemical Name: 2-(1-methylethoxy)phenol methyl carbamate
CAS Registry No: 114-26-1
Uses: insecticide to control cockroaches, flies, fleas, mosquitoes, bugs, ants, millepedes and other insect pests in food storage areas, houses, animal houses, etc.; also to control sucking and chewing insects in fruits, vegetables, ornamentals, vines, maize, lucerne, soya beans, cotton, sugar cane, rice cocoa, forestry, etc.
Molecular Formula: $C_{11}H_{15}NO_3$
Molecular Weight: 209.24
Melting Point (°C)
  91.50 (Spencer 1982; Howard 1991; Kühne et al. 1995)
  84-87 (Montgomery 1993)
  85.50 (Patil 1994)
Boiling Point (°C):
Density (g/cm³ at 20°C):
Molar Volume (cm³/mol):
  244.7 (calculated-LeBas method, Suntio et al. 1988; Fisher et al. 1993)
  1.245 ($V_I/100$, Fisher et al. 1993)
Molecular Volume (Å³):
Total Surface Area, TSA (Å²):
Dissociation Constant pK_a:
Heat of Fusion, ΔH_{fus}, kcal/mol:
Entropy of Fusion, ΔS_{fus}, cal/mol·K (e.u.):
Fugacity Ratio at 25°C (assuming ΔS_{fus} = 13.5 e.u.), F: 0.2199

Water Solubility (g/m³ or mg/L at 25°C):
  2000 (20°C, Spencer 1973; 1982)
  2000 (Kenaga 1980; Kanazawa 1981)
  2000 (20°C, Worthing 1983, 1987, 1991)
  1860 (20°C, shake flask-GC, Bowman & Sans 1983; quoted, Patil 1994)
  1600 (20°C, selected, Suntio et al. 1988)
  1750 (quoted, Howard et al. 1991)
  1800 (20-25°C, selected, Wauchope et al. 1992; Lohninger 1994; Hornsby et al. 1996)
  1740, 1930, 2440 (10, 20, 30°C, Montgomery 1993)
  1900 (selected, Siebers et al. 1994)
  550 (calculated, Patil 1994)
  1865, 550 (selected, calculated, Kühne et al. 1995)

Vapor Pressure (Pa at 25°C):
  1.333 (120°C, Melnikov 1971)
  1.333 (120°C, Spencer 1973, 1982)
  4.13×10^{-4} (20°C, Hartley & Graham-Bryce 1980)
  1.0 (20°C, selected, Suntio et al. 1988)
  4.00×10^{-4} (20°C, quoted, Howard 1991)
  1.69×10^{-3} (20-25°C, selected, Wauchope et al. 1992, Hornsby et al. 1996)
  1.30×10^{-3} (20°C, Montgomery 1993; Siebers et al. 1994)

Henry's Law Constant (Pa·m^3/mol):
- 0.1308   (20°C, calculated-P/C, Suntio et al. 1988; Fisher et al. 1993)
- 4.46×10^{-5} (calculated-P/C, Howard 1991)
- 1.32×10^{-4} (calculated-P/C, Montgomery 1993)
- 1.40×10^{-4} (calculated-P/C, Siebers et al. 1994)
- 1.98×10^{-6} (calculated-P/C, this work)

Octanol/Water Partition Coefficient, log K_{OW}:
- 1.50   (Hansch & Leo 1979; quoted, Fisher et al. 1993)
- 1.45   (Rao & Davidson 1980)
- 1.52   (Kenaga & Goring 1980; Kanazawa 1981)
- 1.552   (shake flask-GC, Bowman & Sans 1983)
- 1.50   (selected, Suntio et al. 1988)
- 1.52   (Hansch & Leo 1985; quoted, Howard 1991)
- 1.75   (RP-HPLC-RT, Trapp & Pussemier 1991)
- 1.45-1.56   (Montgomery 1993)
- 1.552, 2.01   (quoted, calculated, Patil 1994)
- 1.56   (selected, Siebers et al. 1994)
- 1.52   (selected, Hansch et al. 1995)

Bioconcentration Factor, log BCF:
- 0.924   (calculated, Howard 1991)

Sorption Partition Coefficient, log K_{OC}:
- 1.67   (measurements for average of 2 soils, Kanazawa 1981, 1989)
- 1.86   (calculated, Howard 1991)
- 1.48   (soil, Wauchope et al. 1992; Hornsby et al. 1996)
- 0.48-1.97   (Montgomery 1993)
- 1.48   (estimated-chemical structure, Lohninger 1994)

Half-Lives in the Environment:
- Air: 0.71-7.1 h, based on estimated rate constant for the vapor-phase reaction with hydroxyl radicals in air (Atkinson 1987; quoted, Howard et al. 1991); half-life of about 4 h reacting with photochemically produced hydroxyl radicals in air (Howard 1991).
- Surface water: 38-672 h, based on estimated hydrolysis half-life at pH 9 (Aly & El-Dib 1971; quoted, Howard et al. 1991) and estimated unacclimated aqueous aerobic biodegradation half-life (Howard et al. 1991); half-life from 1 day to 1 week by degradation, photolyze rapidly with a half-life of 13 to 88 h (Howard 1991).
- Groundwater: 38-1344 hours, based on estimated hydrolysis half-life at pH 9 (Aly & El-Dib 1971; quoted, Howard et al. 1991) and estimated unacclimated aqueous aerobic biodegradation half-life (Howard et al. 1991).
- Sediment:
- Soil: 38-672 h, based on estimated hydrolysis half-life at pH 9 (Aly & El-Dib 1971; quoted, Howard et al. 1991) and estimated unacclimated aqueous aerobic biodegradation half-life (Howard et al. 1991).
- Biota:

Environmental Fate Rate Constants or Half-Lives:
   Volatilization:
   Photolysis: atmospheric and/or aqueous photolysis half-life of 62.5-87.9 h, based on measured rate of photolysis on bean leaves in sunlight (Ivie & Casida 1971; quoted, Howard et al. 1991) and in aqueous solution under simulated sunlight (Jensen-Korte et al. 1987; quoted, Howard et al. 1991); photolyze in water with a half-life of 88 h and decreased with humic material to 13-41 h; half-life of 87.9 h in water when irradiated with light > 290 nm (Howard 1991).
   Oxidation: photooxidation half-life of 0.71-7.1 h in air, based on estimated rate constant for the vapor-phase reaction with hydroxyl radicals in air (Atkinson 1987; quoted, Howard et al. 1991); vapor-phase photooxidation half-life of 4.3 h for reaction with ambient hydroxyl radical (Howard 1991).
   Hydrolysis: half-life at pH 10 at 20°C of 40 min, hydrolyzes at a rate of 1.5% d^{-1} in 1% aqueous solution at pH 7 (Spencer 1982); half-lives of 16, 1.6 and 0.17 days at pH 8, 9, 10, but stable between pH 3-7, with a half-life of 40 min at pH 10 (Howard 1991); hydrolysis half-life 290 d at pH 7, 17.9 d at pH 8 and 48 minutes at pH 10 (Montgomery et al. 1993); hydrolysis half-lives in water at pH 8, 9 and 10 were 16 d, 1.6 d and 4.2 h and 3.0 d (Aly & El-Dib 1971, 1976; quoted, Montgomery 1993).
   Biodegradation: aqueous aerobic half-life of 168-672 h, based on unacclimated aqueous aerobic screening test data (Gummer 1979; Kanazawa 1987; quoted, Howard et al. 1991); aqueous anaerobic half-life of 672-2688 h, based on estimated unacclimated aqueous aerobic biodegradation half-life (Howard et al. 1991); half-life of 44 d under aerobic conditions and 59 d under anaerobic conditions in water used a combination of activated sludge, silt loam soil and sediment as an inoculum; a half-life of 78 d under aerobic conditions and 125 d under anaerobic conditions at pH 6.9 (Howard 1991).
   Biotransformation:
   Bioconcentration, Uptake (k_1) and Elimination (k_2) Rate Constants:

Common Name: Ronnel
Synonym: Blitex, Dermafos, Dermaphos, dimethyl trichlorophenyl thiophosphate, Dow ET 14, Dow ET 57, Ectoral, ENT 23284, Etrolene, Fenchlorfos, Fenchlorphos, Gesektin K, Karlan, Korlan, Nanchor, Nanker, Nankor, OMS 123, Phenchlorfos, Remelt, Rovan, trichlorometafos, Trolen, Trolene, Viozene
Chemical Name: O,O-dimethyl O-(2,4,5-trichlorophenyl)thiophosphate; O,O-dimethyl O-2,4,5-trichlorophenyl phosphorothioate; phosphoric acid O,O-dimethyl O-(2,4,5-trichlorophenyl)ester
Uses: insecticide.
CAS Registry No: 299-84-3
Molecular Formula: $C_8H_8Cl_3O_3PS$
Molecular Weight: 321.57
Melting Point (°C):
    40-42 (Spencer 1982)
    41 (Bowman & Sans 1983; Suntio et al. 1988; Montgomery 1993; Patil 1994; Kühne et al. 1995; Milne 1995)
Boiling Point (°C):
Density (g/cm^3 at 20°C):
    1.48 (25°, Montgomery 1993)
Molar Volume (cm^3/mol):
    257.3 (calculated-LeBas method, Suntio et al. 1988)
Molecular Volume (Å^3):
Total Surface Area, TSA (Å^2):
Dissociation Constant pK_a:
Heat of Fusion, ΔH_{fus}, kcal/mol:
    5.70 (Plato & Glasgow 1969)
Entropy of Fusion, ΔS_{fus}, cal/mol·K (e.u.):
Fugacity Ratio at 25°C (assuming ΔS_{fus} = 13.5 e.u.), F: 0.695

Water Solubility (g/m^3 or mg/L at 25°C):
    44 (Günther et al. 1968; Melnikov 1971; quoted, Shiu et al. 1990)
    1.08 (20°, shake flask-GC, Chiou et al. 1977; quoted, Shiu et al. 1990)
    1.08 (20-25°, shake flask-GC/ECD, Freed et al. 1979)
    2.5 (20°C, Spencer 1982)
    0.60 (20°, shake flask-GC, Bowman & Sans 1979, 1983; quoted, Shiu et al. 1990; Patil 1994)
    0.98 (20°, corrected supercooled liq. value, shake flask-GC, Bowman & Sans 1979,83)
    6.0 (quoted from Dow Chemical unpublished data, Kenaga 1980a,b; Kenaga & Goring 1980)
    40 (22°, Khan 1980; quoted, Shiu et al. 1990)
    1.0 (20°, shake flask-HPLC, Ellgehausen et al. 1981; quoted, Shiu et al. 1990)
    1.61 (20°, selected, Suntio et al. 1988)
    3.07 (calculated, Patil 1994)
    40 (Montgomery 1993; Milne 1995)
    0.61 (quoted exptl., Kühne et al. 1995)
    1.47 (calculated, Kühne et al. 1995)

Vapor Pressure (Pa at 25°C):
- 0.0533 (20°, Eichler 1965; Melnikov 1971; quoted, Kim et al. 1984; Kim 1985)
- 0.0071 (20-25°, Freed et al. 1979)
- 1.067 (Spencer 1982)
- 0.0017 (20°, GC-calculated value, Kim et al. 1984; Kim 1985)
- 0.0011 (20°, GC-calculated value with M.P. correction, Kim et al. 1984; Kim 1985)
- 0.016 (20°, selected, Suntio et al. 1988)
- 0.0045 (20°, Montgomery 1993)

Henry's Law Constant (Pa·m^3/mol):
- 3.22 (20°C, calculated-P/C, Suntio et al. 1988)
- 0.857 (20-25°C, calculated-P/C, Montgomery 1993)

Octanol/Water Partition Coefficient, log K_{OW}:
- 4.88 (20°C, shake flask-GC, Chiou et al. 1977; quoted, Ellgehausen et al. 1981; Bowman & Sans 1983; Zepp et al. 1984; De Bruijn et al. 1989)
- 4.67 (quoted from Chiou et al. 1977, Kenaga 1980b; Kenaga & Goring 1980; quoted, De Bruijn et al. 1989)
- 4.88 (20-25°, shake flask-GC/ECD, Freed et al. 1979)
- 5.34 (calculated-S, Ellgehausen et al. 1981)
- 4.81 (20°, shake flask-GC, Bowman & Sans 1983; quoted, De Bruijn et al. 1989)
- 4.054 (calculated-f const., Broto et al. 1984; quoted, Karcher & Devillers 1990)
- 4.88 (quoted, Gerstl & Helling 1987)
- 4.80 (selected, Suntio et al. 1988)
- 5.068 ± 0.004 (slow-stirring method, De Bruijn et al. 1989; quoted, De Bruijn & Hermens 1991; Verhaar et al. 1992)
- 4.67-5.068 (Montgomery 1993)
- 4.81, 4.25 (quoted, calculated, Patil 1994)
- 5.07 (selected, Hansch et al. 1995)
- 4.81 (selected, Devillers et al. 1996)

Bioconcentration Factor, log BCF:
- 2.35 (calculated-S, Kenaga 1980a,b)
- −1.38 (average beef fat diet, Kenaga 1980b)
- 4.55, 4.64 (*Poecilia reticulata*, De Bruijn & Hermens 1991; quoted, Devillers et al. 1996)

Sorption Partition Coefficient, log K_{OC}:
- 3.20 (soil, calculated-S as per Kenaga & Goring 1978, Kenaga 1980a,b)
- 2.90 (soil, calculated-χ, Gerstl & Helling 1987)
- 2.76 (calculated, Montgomery 1993)

Half-Lives in the Environment:
    Air:
    Surface water:
    Groundwater:
    Sediment:
    Soil:
    Biota:

Environmental Fate Rate Constants or Half-Lives:
    Volatilization:
    Photolysis:
    Oxidation:
    Hydrolysis: estimated half-life of 3 d at pH 6 (Montgomery 1993).
    Biodegradation:
    Biotransformation:
    Bioconcentration, Uptake (k_1) and Elimination (k_2) Rate Constants:
        k_2:   0.38 d^{-1} (observed, De Bruijn & Hermens 1991)
        k_2:   0.14 d^{-1} (calculated-K_{ow}, De Bruijn & Hermens 1991)

Common Name: Terbufos
Synonym: AC 92100, Counter, ST-100
Chemical Name: S-(*tert*-butylthio)methyl O,O-diethyl phosphorodithioate; S-[[(1,1-dimethylethyl)thio]methyl] O,O-diethyl phosphorodithioate; phosphorodithioic acid S-((*tert*-butylthio)methyl) O,O-diethyl ester
Uses: insecticide in soil to control insects and also used as nematocide to control nematodes in beet, maize, cotton, sorghum, onions, cabbage, and bananas.
CAS Registry No: 13071-79-9
Molecular Formula: $C_9H_{21}O_2PS_3$
Molecular Weight: 288.4
Melting Point (°C):
    -29.2   (Worthing 1991; Brecken-Folse et al. 1994; Howe et al. 1994; Montgomery 1993; Tomlin 1994; Milne 1995)
Boiling Point (°C):
    69   (at 0.01 mmHg, Worthing 1991; Montgomery 1993; Tomlin 1994; Milne 1995)
    312   (Brecken-Folse et al. 1994; Howe et al. 1994)
Density (g/cm³ at 20°C):
    1.105   (24°C, Worthing 1991; Montgomery 1993; Tomlin 1994; Milne 1995)
Molar Volume (cm³/mol):
    261   (24°C, calculated from density)
Molecular Volume (Å³):
Total Surface Area, TSA (Å²):
Dissociation Constant pK_a:
Heat of Fusion, ΔH_{fus}, kcal/mol·
Entropy of Fusion, ΔS_{fus}, cal/mol·K (e.u.):
Fugacity Ratio at 25°C (assuming ΔS_{fus} = 13.5 e.u.), F: 1.0

Water Solubility (g/m³ or mg/L at 25°C):
    12   (Martin & Worthing 1977; quoted, Kenaga 1980; Kenaga & Goring 1980)
    5.07   (shake flask-GC, Felsot & Dahm 1979; quoted, Belluck & Felsot 1981)
    ≤ 10   (Spencer 1982)
    5.5   (19°C, shake flask-GC, Bowman & Sans 1983a,b)
    10-15   (Worthing 1991)
    5   (20-25°C, selected, Wauchope et al. 1992; Lohninger 1994; Hornsby et al. 1996)
    5.5, 15   (quoted, calculated, Pait et al. 1992)
    4.5   (27°C, Montgomery 1993; quoted, Tomlin 1994; Majewski & Capel 1995)
    0.10   (Howe et al. 1994)
    5.50, 67.6   (quoted, calculated, Patil 1994)
    4500   (4.5 mg/mL reported, Milne 1995)

Vapor Pressure (Pa at 25°C):
    0.0346   (Worthing 1991; Tomlin 1994)
    0.0427   (20-25°C, selected, Wauchope et al. 1992; Hornsby et al. 1996)
    0.0351   (20°C, Montgomery 1993; quoted, Majewski & Capel 1995)

Henry's Law Constant (Pa·m³/mol):
    2.229   (20-27°C, calculated-P/C, Montgomery 1993; quoted, Majewski & Capel 1995)
    2.463   (calculated-P/C, this work)

Octanol/Water Partition Coefficient, log K_{OW}:
- 3.68 (Felsot & Dahm 1979; quoted, Bowman & Sans 1983b)
- 2.22 (quoted, Rao & Davidson 1980; Bowman & Sans 1983b)
- 4.17 (calculated as per Chiou et al. 1977, Belluck & Felsot 1981)
- 4.477 (shake flask-GC, Bowman & Sans 1983b; quoted, De Bruijn & Hermens 1991; Patil 1994)
- 3.67 (selected, Gerstl & Helling 1987)
- 4.52 (Worthing 1991; Tomlin 1994)
- 2.22-4.70 (Montgomery 1993)
- 3.54 (Brecken-Folse et al. 1994)
- 0.832 (12°C in reconstituted test water at pH 7.5, Howe et al. 1994)
- 4.48, 3.30 (quoted, calculated, Patil 1994)
- 4.52 (Milne 1995)
- 4.48 (selected, Hansch et al. 1995)
- 4.86, 4.17, 3.98 (RP-HPLC, ClogP, calculated-S, Finizio et al. 1997)

Bioconcentration Factor, log BCF:
- 2.73 (topmouth gudgeon, Metcalf & Sanborn 1975; quoted, Kenaga & Goring 1980; De Bruijn & Hermens 1991)
- 2.18 (calculated-S, Kenaga 1980; quoted, Pait et al. 1992)
- 1.0 (*Triaenodes tardus*, Belluck & Felsot 1981)

Sorption Partition Coefficient, log K_{OC}:
- 3.04 (soil, calculated-S per Kenaga & Goring 1978, Kenaga 1980)
- 2.76, 3.29 (quoted, calculated-χ, Gerstl & Helling 1987)
- 2.70 (soil, 20-25°C, selected, Wauchope et al. 1992; quoted, Richards & Baker 1993; Lohninger 1994; Hornsby et al. 1996)
- 2.46-3.03 (Montgomery 1993)

Half-Lives in the Environment:
  Air:
  Surface water:
  Groundwater:
  Sediment:
  Soil: half-life of 9-27 d in soil (Worthing 1991; quoted, Montgomery 1993; Tomlin 1994); selected field half-life of 5.0 d (Wauchope et al. 1992; quoted, Richards & Baker 1993; Hornsby et al. 1996); soil half-life of 5 d (Pait et al. 1992).
  Biota:

Environmental Fate Rate Constants or Half-Lives:
  Volatilization:
  Photolysis:
  Oxidation:
  Hydrolysis:
  Biodegradation:
  Biotransformation:
  Bioconcentration, Uptake (k_1) and Elimination (k_2) Rate Constants:

Common Name: Toxaphene
Synonym: Agricide maggot killer, Alltex, Alltox, Camphochlor, Chem-Phene, chlorinated Camphene, Chloro-camphene, Coopertox, Crestoxo, Cristoxo, ENT 9735, Estonox, Fasco terpene, Geniphene, Gy-phene, Hercules 3956, Huilex, Kamfochlor, Melipax, Motox, NA 2761, NCI-C00259, Octachlorcamphene, Polychlorocamphene, Strobane-T, Texadust, Toxakil, Toxon 63
Chemical Name: mixtures of chlorinated camphene
Uses: pesticide used primarily on lettuce, cotton, corn, tomatoes, peanuts, wheat and soybean.
CAS Registry No: 8001-35-2
Molecular Formula: $C_{10}H_{10}Cl_8$
Molecular Weight: 414
Melting Point (°C):
    65-90   (Howard 1991; Montgomery 1993; Milne 1995)
    35   (dec., Milne 1995)
Boiling Point (°C):
    246, 351, 360 (estimated from structure, Tucker et al. 1983)
Density (g/cm^3 at 20°C):
    1.65   (25°C, Montgomery 1993)
Molar Volume (cm^3/mol):
    358.8   (calculated-LeBas method, Suntio et al. 1988)
    366.8   (calculated-LeBas method, this work)
Heat of Fusion, ΔH_{fus}, kcal/mol:
Fugacity Ratio at 20 °C (assuming ΔS = 13.5 e.u.), F: 0.20-0.35
    0.30   (Mackay et al. 1986)

Water Solubility (g/m^3 or mg/L at 25°C):
    3.0   (Brooks 1974)
    0.74   (gen. col.-GC/ECD, Weil et al. 1974)
    0.40   (Leonard et al. 1976; quoted, Nash 1988)
    0.40   (Sanborn et al. 1976; quoted, Kenaga & Goring 1980; Zaroogian et al. 1985)
    0.40   (Martin & Worthing 1977; quoted, Kenaga 1980)
    0.50   (shake flask-GC, Paris et al. 1977)
    0.40   (Wauchope 1978)
    3.0   (22°C, Khan 1980)
    0.40   (Weber et al. 1980; quoted, Willis & McDowell 1982)
    0.40   (Graten & Trabalka 1983)
    0.3-3.0 (U.S. EPA 1984; quoted, McLean et al 1988)
    0.74   (quoted, Mackay et al. 1986)
    0.56   (20°C, calculated-estimated M.W., Murphy et al. 1987)
    0.55   (20°C, Murphy et al. 1987; quoted, Howard 1991)
    3.0   (Worthing 1987; quoted, Shiu et al. 1990)
    0.40   (quoted, Isnard & Lambert 1988)
    0.50   (20°C, selected, Suntio et al. 1988; quoted, Majewski & Capel 1995)
    0.55   (20°C, Montgomery 1993)
    0.63   (calculated from vapor pressure & HLC, Wania & Mackay 1993)
    log C (mol/m^3) = 0.77 − 1071/T (Wania & Mackay 1993)
    insoluble (Milne 1995)
    3.0   (20-25°C, selected, Hornsby et al. 1996)

Vapour Pressure (Pa at 25°C):
- 4.0×10^{-5} (20°C, Spencer 1973; quoted, Bidleman & Christensen 1979; Tucker et al. 1983)
- 27-53 (Brooks 1974; Khan 1980)
- 1.3×10^{-4} (Leonard et al. 1976; quoted, Nash 1988)
- 1.3×10^{-4} (20-25°C, Weber et al. 1980; quoted, Willis & McDowell 1982)
- 1.3×10^{-4} (30°C, Seiber et al. 1981; quoted, Tucker et al. 1983)
- 2.0×10^{-5}, 4.5×10^{-5}, 0.667 (estimated-B.P., Tucker et al. 1983)
- 27.0 (U.S. EPA 1984; quoted, McLean et al. 1988)
- 4.0×10^{-5} (quoted, Mackay et al. 1986)
- 0.00089 (20°C, Murphy et al. 1987; quoted, Howard 1991)
- 0.0005 (20°C, selected, Suntio et al. 1988; quoted, Majewski & Capel 1995)
- 27-54 (20°C, Montgomery 1993)
- 0.0016 (calculated from the eqn. below, Wania & Mackay 1993)
- $\log P (Pa) = 12.25 - 4487/T$ (Wania & Mackay 1993)
- 5.3×10^{-4} (20-25°C, selected, Hornsby et al. 1996)

Henry's Law Constant (Pa·m^3/mol):
- 490 (Warner et al. 1980; quoted, Tucker et al. 1983)
- 6380 (Kavanaugh & Trussel 1980)
- 45.59 (estimated-group method per Hine & Mookerjee 1975, Tucker et al. 1983)
- 0.0238 (calculated-P/C, Mackay et al. 1986)
- 496 (Warner et al. 1987; quoted, Cothram & Bidleman 1991)
- 520 (quoted from WERL Treatability Database, Ryan et al. 1988)
- 0.42 (20°C, calculated-P/C, Suntio et al. 1988; quoted, Majewski & Capel 1995)
- 3097 (calculated-P/C, Jury et al. 1990)
- 0.6078 (20°C, Murphy et al. 1987, quoted, Howard 1991)
- 0.62 (20°C, average value for toxaphene complex mixture, Murphy et al. 1987; quoted, Cotham & Bidleman 1991)
- 0.067 (0°C, selected, Cotham & Bidleman 1991)
- 6382 (Montgomery 1993)
- 1.054 (calculated from the eqn. below, Wania & Mackay 1993)
- $\log H (Pa \cdot m^3/mol) = 11.48 - 3416/T$ (Wania & Mackay 1993)

Octanol/Water Partition Coefficient, $\log K_{OW}$:
- 3.52 (shake flask-GC, Paris et al. 1977)
- 5.30 (HPLC-RT, Veith et al. 1979; quoted, Parkerton et al. 1993)
- 3.23 (quoted, Rao & Davidson 1980)
- 5.28 (Veith & Kosian 1983; quoted, Saito et al. 1992)
- 3.30 (Clement Associates 1983; quoted, Chapman 1989)
- 2.92 (U.S. EPA 1984; quoted, McLean et al. 1988)
- 4.83 (quoted from Veith's personal communication, Zaroogian et al. 1985)
- 3.30 (quoted, Mackay et al. 1986)
- 3.30 (selected, Suntio et al. 1988)
- 4.82 (calculated-solubility, Lyman et al. 1982; quoted, Howard 1991)
- 5.50 (Garten & Trabalka 1983)
- 4.83 (quoted from G.D. Veith's communication, Zaroogian et al. 1985)
- 3.85 (quoted, Ryan et al. 1988)
- 5.50 (Isnard & Lambert 1988,89; quoted, Wania & Mackay 1993)

5.50   (selected, Travis & Arms 1988)
4.63   (estimated-QSAR & SPARC, Kollig 1993)
3.23-5.50   (Montgomery 1993)

Bioconcentration Factor, log BCF:
- −2.79   (beef biotransfer factor log B_b, correlated-K_{OW}, Radeleff et al. 1952; Claborn et al. 1953,60; quoted, Travis & Arms 1988)
- −3.20   (milk biotransfer factor log B_m, correlated-K_{OW}, Saha 1969; quoted, Travis & Arms 1988)
- 3.53   (*Bacillus subtilis*, Paris et al. 1975, 1977; quoted, Baughman & Paris 1981)
- 3.72   (*Flavobacterium harrisonii*, Paris et al. 1975, 1977; quoted, Baughman & Paris 1981)
- 4.23   (*Aspergillus sp.*, Paris et al. 1975, 1977; quoted, Baughman & Paris 1981)
- 4.04   (*Chlorella prenoidosa*, Paris et al. 1975, 1977; quoted, Baughman & Paris 1981; Swackhamer & Skoglund 1991)
- 3.63   (*Gambusia*, Sanborn et al. 1976; quoted, Howard 1991)
- 4.84   (fathead minnows, Mayer et al. 1977; quoted, Davies & Dobbs 1984)
- 3.51-4.23   (microorganisms, Paris et al. 1977)
- 3.59   (pinfish, 4-d exposure, Schimmel et al. 1977; quoted, Veith & Kosian 1983)
- 3.64   (sheepshead minnow, 4-d exposure, Schimmel et al. 1977; quoted, Veith & Kosian 1983)
- 3.49-4.52   (fish, Reish et al. 1978; quoted, Howard 1991)
- 2.60-3.08   (shrimp, Reish et al. 1978; quoted, Howard 1991)
- 4.42, 3.63   (fish: flowing water, static water; quoted, Kenaga & Goring 1980)
- 4.42   (fish, Kenaga 1980, quoted, Howard 1991)
- 3.02   (calculated-S, Kenaga 1980)
- 3.59   (pinfish, Veith & Kosian 1983; quoted, Saito et al. 1992)
- 3.64   (sheepshead minnow, Veith & Kosian 1983; quoted, Saito et al. 1992)
- 3.81, 3.72   (fish: flowing system, microcosm, Garten & Trabalka 1983; quoted, Howard 1991)
- 3.84, 3.98   (algae: snail, Garten & Trabalka 1983)
- 3.44, 3.41   (oyster, calculated-K_{OW} & models, Zaroogian et al. 1985)
- 3.44, 3.41   (pinfish, calculated-K_{OW} & models, Zaroogian et al. 1985)
- 3.44, 3.41   (sheepshead minnow, calculated-K_{OW} & models, Zaroogian et al. 1985)
- 3.81   (quoted, Isnard & Lambert 1988)

Sorption Partition Coefficient, log K_{OC}:
- 3.86   (calculated-S, Kenaga 1980; quoted, Howard 1991)
- 4.99   (soil, McDowell et al. 1981; quoted, Nash 1988)
- 4.32   (soil, screening model calculations, Jury et al. 1987a,b, 1990; Jury & Ghodrati 1989)
- 5.32   (sediment, Bomberger et al. 1983; quoted, Howard 1991)
- 3.17   (calculated-K_{OW} as per Kenaga & Goring 1980, Chapman 1989)
- 4.31   (estimated-QSAR & SPARC, Kollig 1993)
- 3.18   (calculated, Montgomery 1993)
- 5.00   (20-25°C, selected, Hornsby et al. 1996)

Half-Lives in the Environment:
- Air: 4-5 d for the vapor-phase reaction with hydroxyl radicals (GEMS 1986; quoted, Howard 1991).
- Surface water:
- Groundwater:
- Sediment:
- Soil: very persistent with reported half-life from 0.8 yr (Adams 1967; quoted, Howard 1991) to 14 yr (Nash & Woolson 1967; quoted, Howard 1991); >50 d when subject to plant uptake via volatilization (Callahan et al. 1979; quoted, Ryan et al. 1988); measured dissipation rate of 0.010 d^{-1} (Seiber et al. 1979; quoted, Nash 1988); half-life of 9 d in screening model calculations (Jury et al 1987b); estimated dissipation rate of 0.0011 and 0.013 d^{-1} (Nash 1988); 3650 d, half-life of volatilization to atmosphere from chemical below soil surface (Jury et al. 1990); field half-life of 9 d (20-25°C, selected, Hornsby et al. 1996).
- Biota: field half-life of 15.6 d in fruit tree leaves (Decker et al. 1950; quoted, Nash 1983); microagroecosystem half-life of 19 d in cotton leaves (Nash & Harris 1977; quoted, Nash 1983); field half-life of ca. 6.3 d in cotton canopy (Willis et al. 1980; quoted, Nash 1983); half-life 524 d for white suckers, and 232 to 322 d for lake trout (total toxaphene, Delorme et al. 1993).

Environmental Fate Rate Constants or Half-Lives:
- Volatilization: volatilization half-life from chemical below soil surface of 2650 d (Jury et al. 1990).
- Photolysis:
- Hydrolysis: estimated half-life of >10 yr at pH 5-8 and 25°C (Callahan et al. 1979; quoted, Howard 1991); hydrolysis rate constant of $(8.0 \pm 2.2) \times 10^{-6}$ h^{-1} at pH 7 with a calculated half-life of 10 yr (Ellington et al. 1987, 1988, quoted, Montgomery 1993); rate constant of 7.0×10^{-2} y^{-1} at pH 7.0 and 25°C (Kollig 1993).
- Oxidation: photooxidation half-life of 4-5 d for the vapor-phase reaction with hydroxyl radicals in air (GEMS 1986; quoted, Howard 1991); rate constant of 8×10^8 M^{-1} s^{-1} for the reaction (Fenton with reference to lindane) with hydroxyl radicals in aqueous solutions at pH 1.9 ± 0.1 and at (24 ± 1)°C (Buxton et al. 1988; quoted, Faust & Hoigné 1990; Haag & Yao 1992); rate constant of $(1.2$-$8.1) \times 10^8$ M^{-1} s^{-1} for the reaction (Fenton with reference to lindane) with hydroxyl radicals in aqueous solutions at pH 1.9 ± 0.1 and at (24 ± 1)°C (Haag & Yao 1992).
- Biodegradation: very resistent to degradation in soils with reported half-life from 0.8 year (Adams 1967; quoted, Howard 1991) to 14 yr (Nash & Woolson 1967; quoted, Howard 1991).
- Biotransformation:
- Bioconcentration, Uptake (k_1) and Elimination (k_2) Rate Constants:

Common Name: Trichlorfon

Synonym: Aerol 1, Agroforotox, Anthion, Bay 15922, Bayer 15922, Bilarcil, Bovinox, Britten, Britton, Cekufon, Chlorak, Chlorfos, Chlorphos, Chloroftalm, Chloroxyphos, Ciclosom, Combat, Combot, Danex, DEP, Depthon, DETF, Dimetox, Dipterax, Diptevur, Ditrifon, Dylox, Dyrex, Dyvon, ENT 19763, Equino-acid, Flibol E, Forotox, Foschlor, Hypodermacid, Leivasom, Loisol, Masoten, Mazoten, Methyl chlorophos, Metifonate, Metrifonate, Metriphonate, NA 2783, NCI-C54831, Neguvon, Phoschlor, Proxol, Ricifon, Ritsifon, Soldep, Sotipox, Trichlorphon, Trichlorphene, Trinex, Tugon, Volfartol, Votexit, Wotexit

Chemical Name: dimethyl 2,2,2-trichloro-hydroxyethylphosphorate; 2,2,2-trichloro-hydroxyethylphosphoric acid dimethyl ester

Uses: insecticide to control flies and roaches.

CAS Registry No: 52-68-6

Molecular Formula: $C_4H_8Cl_3O_4P$

Molecular Weight: 257.4

Melting Point (°C):
- 78-80 (Burchfield & Johnson 1965; Freed et al. 1977)
- 83-84 (Spencer 1973, 1982; Freed et al. 1977; Suntio et al. 1988; Montgomery 1993; Milne 1995)
- 83.50 (Bowman & Sans 1983b; Patil 1994)
- 83 (Kawamoto & Urano 1989)
- 75-79 (Worthing 1991; Milne 1995)
- 81-82 (Spencer 1982; Brecken-Folse et al. 1994; Howe et al. 1994)
- 78.5-84 (Tomlin 1994)
- 83.5 (Kühne et al. 1995)

Boiling Point (°C):
- 100 (at 0.1 mmHg, Spencer 1973; Freed et al. 1977; Montgomery 1993; Brecken-Folse et al. 1994; Howe et al. 1994; Milne 1995)

Density (g/cm³ at 20°C):
- 1.73 (Spencer 1982; Worthing 1991; Montgomery 1993; Tomlin 1994; Milne 1995)

Molar Volume (cm³/mol):
- 194.9 (calculated-LeBas method, Suntio et al. 1988)

Molecular Volume (Å³):

Total Surface Area, TSA (Å²):

Dissociation Constant pK_a:

Heat of Fusion, ΔH_{fus}, kcal/mol:

Entropy of Fusion, ΔS_{fus}, cal/mol·K (e.u.):

Fugacity Ratio at 25°C (assuming ΔS_{fus} = 13.5 e.u.), F: 0.261

Water Solubility (g/m³ or mg/L at 25°C):
- 154000 (Spencer 1973, 1982; quoted, Freed et al. 1977; Bowman & Sans 1983a,b)
- 154000 (Martin & Worthing 1977; quoted, Kenaga 1980a; Kenaga & Goring 1980; Suntio et al. 1988)
- 154000 (Worthing 1979; quoted, Bowman & Sans 1983a,b; Patil 1994)
- >5000 (20°C, shake flask-GC, Bowman & Sans 1983a)
- 154000 (20°C, selected, Suntio et al. 1988)
- 150000 (Davies & Lee 1987; quoted, Kawamoto & Urano 1989)
- 120000 (20°C, Worthing 1991; Tomlin 1994)

120000 (20-25°C, selected, Wauchope et al. 1992; Lohninger 1994; Hornsby et al. 1996)
154000 (Montgomery 1993)
90000 (Brecken-Folse et al. 1994)
9000 (Howe et al. 1994)
154000, 45778 (quoted, calculated, Patil 1994)
155100, 34722 (quoted, calculated-group contribution fragmentation method, Kühne et al. 1995)
154000 (Milne 1995)

Vapor Pressure (Pa at 25°C):
0.00095 (20°C, vapor density, MacDougall 1964; quoted, Suntio et al. 1988)
0.00104 (20°C, Eichler 1965; quoted, Kim 1985; Suntio et al. 1988)
0.00104 (20°C, Melnikov 1971; quoted, Kim et al. 1984; Kim 1985; Suntio et al. 1988)
0.0640 (20°C, GC-calculated value, Kim et al. 1984; Kim 1985)
0.0187 (20°C, GC-calculated value with M.P. correction, Kim et al. 1984; Kim 1985)
0.00104 (20°C, Spencer 1973; quoted, Freed et al. 1977)
0.00104 (20°C, Hartley & Graham-Bryce 1980; quoted, Suntio et al. 1988)
0.0010 (20°C, selected, Suntio et al. 1988)
0.00104 (20°C, quoted, Kawamoto & Urano 1989)
0.00021 (20°C, Worthing 1991; Tomlin 1994)
0.00027 (20-25°C, selected, Wauchope et al. 1992; Hornsby et al. 1996)
0.00104 (20°C, Montgomery 1993)
0.00051 (Tomlin 1994)

Henry's Law Constant (Pa·m^3/mol):
1.7×10^{-6} (20°C, calculated-P/C, Suntio et al. 1988)
1.7×10^{-6} (calculated-P/C, Montgomery 1993)

Octanol/Water Partition Coefficient, log K_{OW}:
0.48 (quoted from Dow Chemical data, Kenaga & Goring 1980; quoted, Bowman & Sans 1983b; Suntio et al. 1988)
0.431 (shake flask-GC, Bowman & Sans 1983b; quoted, Suntio et al. 1988; Sicbaldi & Finizio 1993; Sangster 1993; Patil 1994)
0.40 (selected, Suntio et al. 1988)
0.76 (HPLC-RT, Kawamoto & Urano 1989; quoted, Sangster 1993)
0.43-0.76 (Montgomery 1993)
0.51 (shake flask, Sicbaldi & Finizio 1993)
0.72 (RP-HPLC, Sicbaldi & Finizio 1993)
1.70 (shake flask, Brecken-Folse et al. 1994)
0.304 (12°C in reconstituted test water at pH 7.5, Howe et al. 1994)
0.43, 0.90 (quoted, calculated, Patil 1994)
0.43 (Tomlin 1994)
0.51 (selected, Hansch et al. 1995)
0.72, 0.30, 0.93 (RP-HPLC, ClogP, calculated-S, Finizio et al. 1997)

Bioconcentration Factor, log BCF:
    −0.155 (calculated-S, Kenaga 1980)

Sorption Partition Coefficient, log K_{OC}:
    0.778   (calculated-S as per Kenaga & Goring 1978, Kenaga 1980)
    1.90   (correlated, Kawamoto & Urano 1989)
    1.90   (soil, quoted exptl., Meylan et al. 1992)
    1.73   (soil, calculated-χ and fragments contribution, Meylan et al. 1992)
    1.00   (soil, 20-25°C, selected, Wauchope et al. 1992; quoted, Dowd et al. 1993; Lohninger 1994; Hornsby et al. 1996)
    0.99-1.58   (Montgomery 1993)

Half-Lives in the Environment:
    Air: photooxidation half-life of 1-101 h, based on an estimated rate constant for the vapor-phase reaction with hydroxyl radicals in air (Atkinson 1987; quoted, Howard et al. 1991).
    Surface water: 22-588 h, based on aqueous hydrolysis half-lives at pH 6 and 8 and 25°C (Chapman & Cole 1982; quoted, Howard et al. 1991).
    Groundwater: 22-588 h, based on aqueous hydrolysis half-lives at pH 6 and 8 and 25°C (Chapman & Cole 1982; quoted, Howard et al. 1991).
    Sediment:
    Soil: 24-1080 h, based on unacclimated soil grab sample data (Guirguis & Shafik 1975; Kostovetskii et al. 1976; quoted, Howard et al. 1991); selected field half-life of 10 d (Wauchope et al. 1992; Dowd et al. 1993; Hornsby et al. 1996).
    Biota:

Environmental Fate Rate Constants or Half-Lives:
    Volatilization:
    Photolysis:
    Oxidation: photooxidation half-life of 1-101 h, based on an estimated rate constant for the vapor-phase reaction with hydroxyl radicals in air (Atkinson 1987; quoted, Howard et al. 1991).
    Hydrolysis: first-order hydrolysis half-life of 68 h, based on first-order rate constant at pH 7 and 25°C (Chapman & Cole 1982; quoted, Howard et al. 1991); half-lives at 22°C: 510 d at pH 4, 46 h at pH 7, and < 30 min at pH 9 (Tomlin 1994).
    Biodegradation: aqueous aerobic half-life of 24-1080 h, based on unacclimated soil grab sample data (Guirguis & Shafik 1975; Kostovetskii et al. 1976; quoted, Howard et al. 1991); aqueous anaerobic half-life of 96-4320 h, based on unacclimated aerobic biodegradation half-life (Howard et al. 1991).
    Biotransformation:
    Bioconcentration, Uptake (k_1) and Elimination (k_2) Rate Constants:

Table 3.2.1 Common names, chemical names and physical-chemical properties of insecticides

Name	Synonym	Chemical Name	Formula	MW	mp, °C	Fug. ratio F at 25°C	pK_a
Acephate [30560-19-1]	Orthene	O,S-dimethyl acetylphosphoramidothioate	$C_4H_{10}NO_3PS$	183.2	82-89	0.252	
Aldicarb [116-06-3]	Temik	2-methyl-2-(methylthio)-propionaldehyde O-(methylcarbamoyl) oxime	$C_7H_{14}N_2O_2S$	190.25	99-100	0.185	
Aldrin [309-00-2]	Aldrec, Aldrex, Aldrite, Octalene	1,2,3,4,10,10-hexachloro-1,4,4a,5,8,8a-hexahydro-1,4-endoexo-5,8-dimethano-naphthalene	$C_{12}H_8Cl_6$	364.93	104	0.165	
Aminocarb [2032-59-9]	Matacil	4-dimethylamino-m-tolyl methylcarbamate	$C_{11}H_{16}N_2O_2$	208.3	93-94	0.208	
Azinphos-methyl [86-50-0]	Guthion	O,O-dimethyl-S-[4-oxo-1,2,3-benzotriazin-3(4H)-yl)methyl]phosphorodithioate	$C_{10}H_{12}N_3O_3PS_2$	317.34	73-74	0.331	
Bendiocarb [22781-23-3]	Bencarbate, Dycarb, Garvox, Multamat	2,2-dimethyl-1,3-benzodioxol-4-yl-methylcarbamate	$C_{11}H_{13}NO_4$	223.2	129-130	0.0915	8.8
Carbaryl [63-25-2]	Sevin	1-naphthyl-N-methyl carbamate	$C_{12}H_{11}NO_2$	201.22	142	0.0696	
Carbofuran [1563-66-2]	Furadan, Yaltox	2,3-dihydro-2,2-dimethylbenzofuran-7-yl methylcarbamate	$C_{12}H_{15}NO_3$	221.3	151	0.0567	
Chlordane [57-74-9]	Aspon-chlordane, Chlorindan, Octachlor	1,2,4,5,7,8,8-octachloro-3a,4,7,7a-tetrahydro-4,7-methanoindane	$C_{10}H_6Cl_8$	409.8	103-105	0.165	
cis- or α-chlordane [5103-71-9]			$C_{10}H_6Cl_8$	409.8	107-109	0.151	
trans- or β-chlordane [5103-74-2]			$C_{10}H_6Cl_8$	409.8	103-105	0.165	
γ-chlordane [5564-34-7]			$C_{10}H_6Cl_8$	409.8	131	0.0895	
technical grade [12789-03-6]							
Chlorfenvinphos [470-90-6]	Birlane, Sapecron	2-chloro-1-(2,4-dichlorophenyl) vinyl diethyl phosphate	$C_{12}H_{14}Cl_3O_4P$	359.56	−19	1	

Name	Synonym	Chemical Name	Formula	MW	mp, °C	Fug. ratio F at 25°C	pK$_a$
Chlorpyrifos [2921-88-2]	Brodan, Dursban, Dowco 179	O,O-diethyl O-3,5,6-trichloro 2-pyridyl phosphorothioate	C$_9$H$_{11}$Cl$_3$NO$_3$PS	350.6	41-42	0.679	
Chlorpyrifos-methyl [5598-13-0]	Reldan, Dowco 214	O,O-dimethyl O-3,5,6-trichloro-2-pyridyl phosphorothioate	C$_7$H$_7$Cl$_3$NO$_3$PS	322.5	45.5-46.5	0.620	
Crotoxyphos [7700-17-6]	Ciodrin	dimethyl(E)-1-methyl-2-(1-phenyl-ethoxycarbonyl)vinyl phosphate	C$_{14}$H$_{19}$O$_6$P	314.3	liquid	1	
Cypermethrin [52315-07-8]	Polytrin, Ambush C, Kakfil Super, BSI, draft E-ISO	(RS)-α-cyano-3-phenoxybenzyl (1RS,3RS; 1RS,3RS)-3-(2,2-dichlorovinyl)-2,2-= dimethylcyclopropanecarboxylate	C$_{22}$H$_{19}$Cl$_2$NO$_3$	416.3	80.5	0.283	
α-cypermethrin [67375-30-8]			C$_{22}$H$_{19}$Cl$_2$NO$_3$	416.3	78-81	0.286	
β-cypermethrin [65731-84-2]			C$_{22}$H$_{19}$Cl$_2$NO$_3$	416.3	64-71		
ζ-cypermethrin [52315-07-8]			C$_{22}$H$_{19}$Cl$_2$NO$_3$	416.3	−22.4	1	
DDD							
p,p'-DDD [72-54-8]	p,p'-TDE	1,1-Dichloro-2,2-bis (4-chlorophenyl)ethane	C$_{14}$H$_{10}$Cl$_4$	320	109-110	0.144	
o,p'-DDD [53-10-0]		1,1-dichloro-(2-chlorophenyl)-2-(4-chlorophenyl)ethane	C$_{14}$H$_{10}$Cl$_4$	320	112	0.138	
DDE							
p,p'-DDE [72-55-9]	p,p'-DDE	1,1-dichloro-2,2-bis-(p-chlorophenyl)-ethylene	C$_{14}$H$_8$Cl$_4$	319	88-90	0.233	
o,p'-DDE [3424-82-6]	o,p'-DDE	1,1-Dichloro-2(2-chlorophenyl)-2-(4-chlorophenyl)ethylene	C$_{14}$H$_8$Cl$_4$	318	88-90	0.233	
DDT							
p,p'-DDT [50-29-3]	Agritan	1,1,1-trichloro-2,2-bis-(4-chlorophenyl)-ethane	C$_{14}$H$_9$Cl$_5$	354.5	108.5-109	0.148	
o,p'-DDT [789-02-6]		1,1,1-trichloro-2-(chlorophenyl)-2-(4-chlorophenyl)-ethane	C$_{14}$H$_9$Cl$_5$	354.5			

Name	Synonym	Chemical Name	Formula	MW	mp, °C	Fug. ratio F at 25°C	pK$_a$
Deltamethrin [62918-63-5]	Decis, K-Othrine Butox, Butoflin	(S)-α-cyano-3-phenoxybenyl(1R,3R)-3-(2,2-dibromvinyl)-2,2-dimethycyclopropanecarboxylate	$C_{22}H_{19}Br_3NO_3$	505.2	98-101	0.181	
Demeton [8065-48-3]	Systox	O,O-diethyl O-2-ethylthioethyl phosphorothioate	$C_8H_{19}O_3PS_2$	258.3	liquid	1.0	
Demeton-S-methyl [919-86-8]	Metasystoxi	S-2-ethylthioethyl O,O-dimethyl phosphorothioate	$C_6H_{15}O_3PS_2$	230.3	liquid	1	
Dialifor [10311-84-9]	Torak	S-2-chloro-1-phthalimidoethyl O,O-diethyl phosphorodithioate	$C_{14}H_{17}ClNO_4PS_2$	393.84	67-69	0.376	
Diamidaphos [1754-58-1]	Nillite	phenyl N,N-dimethylphosphorodiamidate					
Diazinon [333-41-5]	Basudin, Diazide, Spectracide	O,O-diethyl O-2-isopropyl-6-methylpyrimidin-4-yl phosphorothioate	$C_{12}H_{21}N_2O_3PS$	304.36	liquid	1.0	
Dicapthon [2463-84-5]	Dicaptan	O-(2-chloro-4-nitrophenyl)-O,O-dimethyl phosphorothioate	$C_8H_9ClNO_5PS$	297.68	62-63	0.421	
Dichlofenthion [97-17-6]	Mobilawn	O-2,4-dichlorophenyl O,O-=diethyl phosphorothioate	$C_{10}H_{13}Cl_2O_3PS$	315.17	liquid	1.0	
Dichlorvos [62-73-7]	Vapona, Nuvan, DDVP, Dedevap	2,2-dichlorovinyl-O,O-dimethyl phosphate	$C_4H_7Cl_2O_4P$	220.98	liquid	1.0	
Dicrotophos [141-66-2]	Carbicron, Ektafos, Bidrin	(E)-2-dimethylcarbamoyl-1-methylvinyl dimethyl phosphate	$C_8H_{16}NO_5P$	237.2	liquid	1.0	
Dieldrin [60-57-1]	HEOD	1,2,3,4,10,10-hexachloro-6,7-epoxy-1,4,4a,5,6,7,8,8a-octahydro-exo-1,4-endo-5,8-dimethanonaphthalene	$C_{12}H_8Cl_6O$	380.93	176-177	0.0314	
Diflubenzuron [35367-38-5]	Deflubenzon, Dimilin	1-(4-chlorophenyl)-3-(2,6-difluorobenzol) urea	$C_{14}H_9ClF_2N_2O_2$	310.7	230-232	0.0092	

Name	Synonym	Chemical Name	Formula	MW	mp, °C	Fug. ratio F at 25°C	pK$_a$
Dimethoate [60-51-5]	Cygon	O,O-dimethyl-S-(N-methyl-carbamoyl-methyl phosphorodithioate	C$_5$H$_{12}$NO$_3$PS$_2$	229.28	52-52.5	0.535	
Dinoseb [88-85-7]	Antox, Aretit, BNP 30, DNBP	2-sec-butyl-4,6-dinitrophenol	C$_{10}$H$_{12}$N$_2$O$_5$	240.22	38-42	0.711	
Disulfoton [298-04-4]	Di-Syston, Dithiosystox	O,O-diethyl-S-(ethylthio)-ethyl phosphorodithioate	C$_8$H$_{19}$O$_2$PS$_3$	274.38	108	0.151	
Disulfoton sulfone							
Disulfoton sulfoxide							
DNOC [534-52-1]	Elgetol DNC, Antinonnin	2-methyl-4,6-dinitrophenol sodium salt	C$_7$H$_5$N$_2$O$_5$Na	148.1	86	0.249	
Endosulfan [115-29-7]	Thiodan, Cyclodan, Malix, Thifor	5-norbornene-2,3-dimethanol-1,4,5,6,7,7-hexachlorocyclic sulfite	C$_9$H$_6$Cl$_6$O$_3$S	406.92	70-100		
α-Endosulfan [959-98-8]			C$_9$H$_6$Cl$_6$O$_3$S	406.92	106	0.158	
β-Endosulfan [33213-65-9]			C$_9$H$_6$Cl$_6$O$_3$S	406.92	207-209	0.0155	
Endosulfan-salfate							
Endrin [72-20-8]	endrine, nendrin	1,2,3,4,10,10-hexachloro-6,7-epoxy-1,4,4a,5,6,7,8,8a-octahydro-=exo-1,4-exo-5,8-dimethanonaphthalene	C$_{12}$H$_8$Cl$_6$O	380.93	208-210	0.0151	
Ethion [563-12-2]	Nialate, diethion	O,O,O',O'-tetraethyl-S,S'-methylene bis(phosphorodithioate)	C$_9$H$_{22}$O$_4$P$_2$S$_4$	384.48	−12 to −15	1.0	
Fenitrothion [122-14-5]	Sumithion, Folithion Cyfen	O,O-dimethyl O-4-nitro-m-tolyl phosphorothioate	C$_9$H$_{12}$NO$_5$PS	277.25	liquid	1.0	
Fenoxycarb [79127-80-3]	Logic, Pictyl, Varodo	ethyl 2-(4-phenoxyphenoxy)ethyl-carbamate	C$_{17}$H$_{19}$NO$_4$	301.3	53-54	0.517	
Fenthion [55-38-9]	Baytex, Baycid, Mercaptophos,	O,O-dimethyl-O-(3-methyl-4-(methylthio)phenyl)phosphorothioate	C$_{10}$H$_{15}$O$_3$PS$_2$	278.34	liquid	1.0	

Name	Synonym	Chemical Name	Formula	MW	mp, °C	Fug. ratio F at 25°C	pK$_a$
Fenvalerate [51630-58-1]	Sumicidin, Belmark, Pydrin	(RS)-α-cyano-3-phenoxybenzyl (RS)-2-(4-chlorophenyl)-3-methylbutyrate	$C_{25}H_{22}ClNO_2$	419.9	liquid	1.0	
Flucythrinate [70124-77-5]	Cybolt, Cythrin, Pay-Off	(RS)-α-cyano-3-phenoxybenzyl (S)-2-(4-difluoromethoxyphenyl)-3-methylbutyrate	$C_{26}H_{23}F_2NO_4$	451.48	liquid	1	
Fonofos [944-22-9]	Dyfonate, Fonophos	O-ethyl-S-phenyl (RS)-ethyl-phosphorodithioate	$C_{10}H_{15}OPS_2$	246.3	liquid	1.0	
Heptachlor [76-44-8]	Methanoindene	1,4,5,6,7,8,8-heptachloro-3a-4,7-7a-tetrahydro-4,7-endo-methanoindene	$C_{10}H_5Cl_7$	373.4	95-96	0.199	
Heptachlor epoxide [1024-57-3]		1,4,5,6,7,8,8-heptachloro-2,3-epoxy-2,3,3a-4,7,7a-tetrahydro-4,7-methanoindene	$C_{10}H_5Cl_7O$	389.4	160-161.5	0.0473	
Hexachlorocyclohexane	BHC, HCH	1,2,3,4,5,6-Hexachlorocyclohexane	$C_6H_6Cl_6$	290.8			
α-HCH [319-84-6]	α-BHC	1,2,3,4,5,6-Hexachlorocyclohexane		290.8	157-160	0.030	
β-HCH [319-85-7]	β-BHC	1,2,3,4,5,6-Hexachlorocyclohexane		290.8	309-310	0.0015	
γ-HCH (Lindane) [319-86-8]	Gammexane	1,2,3,4,5,6-hexachlorocyclohexane	$C_6H_6Cl_6$	290.8	112	0.138	
δ-HCH	δ-BHC	1,2,3,4,5,6-hexachlorocyclohexane	$C_6H_6Cl_6$	290.8	138-139	0.074	
Iodofenphos [25311-71-1]	Nuvanol N	O-2,5-dichloro-4-iodophenyl O,O-dimethyl phosphorothioate	$C_8H_8Cl_2IO_3PS$	345.3	oil	1	
Isophorone	Isooctaphenone	3,5,5-trimethyl-2-cyclohexene-1-one	$C_9H_{14}O$	138.2	−8	1.0	
Kepone [143-50-0]	Chlordecone	decachlorooctahydro-1,3,4-metheno-2H-cyclobuta[cd]pentalen-2-one	$C_{10}H_{10}O$	490.68	350 dec.	0.0006	
Leptophos [21609-90-5]	Phosvel	O-(4-bromo-2,5-dichlorophenyl) O-methyl phenylphosphorothioate	$C_{13}H_{10}BrCl_2O_2PS$	412.06	55-67	0.494	
Lindane [58-89-9]	γ-BHC, γ-HCH	1,2,3,4,5,6-hexachlorocyclohexane	$C_6H_6Cl_6$	290.85	112.5	0.136	
Malathion [121-75-5]	Karbofos, Cythion, mercaptothion	S-[1,2-bis(ethoxycarbenyl)ethyl]-O,O-dimethyl phosphorodithioate	$C_{10}H_{19}O_6PS$	330.36	2.9	1.0	

Name	Synonym	Chemical Name	Formula	MW	mp, °C	Fug. ratio F at 25°C	pK$_a$
Mecarbam [2595-54-2]	Afos	S-(N-ethyoxycarbonyl-N-methyl-carbamoylmethyl) O,O-diethyl	C$_{10}$H$_{20}$NO$_5$PS$_2$	329.4	oil	1.0	
Methamidophos [10265-92-6]	Monitor, Tamaron	O,S-dimethylphosphoramidothioate	C$_2$H$_8$NO$_2$PS	141.1	44.5	0.641	
Methiocarb [2032-65-7]	Mesurol, Draza	4-methylthio-3,5-xylyl methylcarbamate	C$_{11}$H$_{15}$NO$_2$S	225.3	117-118	0.123	
Methomyl [16752-77-5]	Lannate	S-methyl-N-(methylcarbamoyl-oxy)-thioaceticimidate	C$_5$H$_{10}$N$_2$O$_2$S	162.2	78-79	0.292	
Methoxychlor [72-43-5]	Marlate	1,1,1-trichloro-2,2-bis(4-methoxy-phenyl)ethane	C$_{16}$H$_{15}$Cl$_3$O$_2$	345.7	86-88	0.244	
Mevinphos [7786-34-7]	Apavinfos, Duraphos	2-carbomethoxy-1-methylvinyl dimethyl phosphate	C$_7$H$_{13}$O$_6$P	224.1	−56.1	1	
Mirex [2385-85-5]	Dechlorane	1,1a,2,2,3a,4,5,5,5a,5b,6-dodeca-chlorooctahydro-1,3,4-metheno-1H-cyclobuta(cd)pentalene	C$_{10}$Cl$_{12}$	545.59	485	2.82×10^{-5}	
Monocrotophos [6923-22-4]	Nuvacron, Azodrin	dimethyl (E)-1-methyl-2-(methyl-carbamoyl)vinyl phosphate	C$_7$H$_{14}$NO$_5$P	223.2	54-55	0.505	
Oxamyl [23135-22-0]	Vydate	N'N'-dimethyl-2-methylcarbamoyloxyimino-2-(methylthio)acetamine	C$_7$H$_{13}$N$_3$O$_3$S	219.25	100-102	0.177	
Parathion [56-38-2]	Folidol, Bladan, Niran	O,O-diethyl O-4-nitrophenyl phosphorothioate	C$_{10}$H$_{14}$NO$_5$PS	291.27	6	1	
Parathion-methyl [298-00-0]	Dalf, Nitrox	O,O-dimethyl O-(p-nitrophenyl) phosphorothioate	C$_8$H$_{10}$NO$_5$PS	263.5	37-38	0.752	
Pentachlorophenol [87-86-5]	PCP	pentachlorophenol	C$_6$H$_5$OH	266.34	190	0.0336	4.74

Pentachlorophenol sodium salt (Pentacon)

Name	Synonym	Chemical Name	Formula	MW	mp, °C	Fug. ratio F at 25°C	pK$_a$
Permethrin [52645-53-1]	Ambush, Kafil, Picket, Pramex	3-phenoxybenzyl(1*RS*,3*RS*;1*RS*,3*RS*)-3(2,2-dichlorovinyl)-2,2-=dimethylcyclo-prapanecarboxylate	C$_{21}$H$_{20}$Cl$_2$O$_3$	391.3	34-39	0.770	
cis-Permethrin					63-65	0.411	
trans-Permethrin					44-47	0.620	
technical grade							
Phenthoate [2597-03-7]	Cidial, Elsan	ethyl 2-dimethoxythiophosphorythio-2-phenylacetate	C$_{12}$H$_{17}$O$_4$PS$_2$	320.4	17-18	1	
Phorate [298-02-2]	Forsaat, Gramitox	*O*,*O*-diethyl-*S*-(ethylthio)methyl phosphordithioate	C$_7$H$_{17}$O$_2$PS$_3$	260.4	−43	1	
Phorate-sulfone				292.3			
Phorate-sulfoxide				276.4			
Phosmet [732-11-6]	Imidan	*S*-[(1,3-dihydro-1,3-dioxo-2*H*-isoindol-2-yl)methyl]	C$_{11}$H$_{12}$NO$_4$PS$_2$	317.32	71.9	0.344	
Phosphamidon [13171-21-6]	Dimecron	2-chloro-2-diethylcarbamoyl-1-methylvinyl dimethyl phosphate	C$_{10}$H$_{19}$ClNO$_5$P	299.69	−45	1	
Pirimor [23103-98-2]	Pirimicarb, Aphox	2-dimethylamino-5,6-dimethyl-pyrimidin-4-yl dimethylcarbamate	C$_{11}$H$_{18}$N$_4$O$_2$	238.29	90.5	0.225	
Profenofos [41198-08-7]	Selecron	*O*-4-bromo-2-chlorophenyl *O*-ethyl *S*-propyl phosphorothioate	C$_{11}$H$_{15}$BrClO$_3$PS	373.6	liquid	1	
Propoxur [114-26-1]	Baygon	2-(1-Methylethoxy)phenol methyl carbamate	C$_{11}$H$_{15}$NO$_3$	209.24	91.5	0.220	
Ronnel [299-84-3]	Fenchlorphos, Korlan, Etrolene, Trolene	*O*,*O*-dimethyl *O*-2,4,5-trichlorophenyl-phosphorothioate	C$_8$H$_8$Cl$_3$O$_3$PS	321.57	41	0.695	
Sulfotep [3689-24-5]	dithio, thiotep, ENT, Bladafum	*O*,*O*,*O'*,*O'*-tetraethyl dithiopyrophosphate	C$_8$H$_{20}$O$_5$P$_2$S$_2$	322.3	liquid	1.0	

Name	Synonym	Chemical Name	Formula	MW	mp, °C	Fug. ratio F at 25°C	pKa
Terbacil [5902-51-2]	Sinbar, Turbacil	5-chloro-3-(1,1-dimethyl)-6-methyl-2,4-(1H,3H)-pyrimidine-dione	$C_9H_{13}Cl_2N_2O_3$	216.7	175-177	0.0321	9.0
Terbufos [13071-79-9]	Contraven, Counter	S-$tert$-butylthiomethyl O,O-diethyl phosphorodithioate	$C_9H_{21}O_2PS_3$	288.4	liquid	1	
Terbufos sulfone				320.41			
Terbufos sulfoxide				304.41			
Tetramethrin [7696-12-0]	Neo-Pynamin, phthalthrin	cyclohex-1-ene-1,2-dicarboximidomethyl (1RS,3RS;1RS,3SR)-2,2-dimethyl-3-methylprop-1-enyl)cyclopropanecarboxylate	$C_{19}H_{25}NO_4$	331.4	60-80		
Thiobencarb [28249-77-6]	Benthiocarb, Bolero, Saturno	S-(4-chloropheny)methyl diethyl-carbamothioate	$C_{12}H_{16}ClNOS$	257.8			
Thiodicarb [59669-26-0]	Cicarbosulf, Larvin, Lepicron	dimethyl N,N'-thiobis(methylimino)-carbonyloxy bisethanimidothioate	$C_{10}H_{18}N_4O_4S_3$	354.47	168-172	0.0368	
Toxaphene [8001-35-2]	Camphechlor	chlorinated camphene (67-69% Cl content) - mixture	$C_{10}H_{10}Cl_8$	413.8	65-90		
Trichlorfon [52-68-6]	Tugon, Chlorophos, Dipterex, Neguvon	dimethyl 2,2,2-trichloro-1-hydroxy-ethylphosphonate	$C_4H_8Cl_3O_4P$	257.45	83-84	0.2609	
Zinophos [297-92-2]	Thionazin, Namafos Cynem	O,O-diethyl-O-pyrazin-2-yl phosphorothioate	$C_8H_{13}N_2O_3PS$	248.2	−1.69	1	

Table 3.2.2 Summary of physical-chemical properties insecticides at 25°C

Name	Selected properties							
	vapor pressure		solubility			$\log K_{ow}$	H, Pa·m³/mol calcd P/C at 25 °C	$\log K_{oc}$
	P^S, Pa	P_L, Pa	g/m³	C^S, mol/m³	C_L, mol/m³			
Acephate	2.26×10^{-4}	8.96×10^{-4}	818000	4465	17710	-1	5.06×10^{-8}	0.301
Aldicarb	0.004	0.0216	6000	31.54	170	1.1	1.27×10^{-4}	1.48
Aldrin	0.005	0.0302	0.02	5.48×10^{-5}	3.31×10^{-4}	3.01	91.23	2.61
Aminocarb	0.00227	0.0109	915	4.39	21.1	1.73	5.17×10^{-4}	2.00
Azinphos-methyl	3.0×10^{-5}	9.05×10^{-5}	30	0.0945	0.285	2.7	3.17×10^{-4}	2.61
Bendiocarb	6.6×10^{-4}	7.21×10^{-3}	40	0.179	1.96		3.68×10^{-3}	2.76
Carbaryl	2.67×10^{-5}	3.83×10^{-4}	120	0.596	8.56	2.36	4.48×10^{-5}	2.36
Carbofuran	8.0×10^{-5}	1.41×10^{-3}	351	1.59	28.0	2.32	5.04×10^{-5}	2.02
Chlordane								
cis- or α-chlordane	4.0×10^{-4}	2.65×10^{-3}	0.056	1.37×10^{-4}	9.07×10^{-4}	6.0	0.342	5.5
trans- or β-chlordane	5.2×10^{-4}	3.15×10^{-3}	0.056	1.37×10^{-4}	8.30×10^{-4}	6.0	0.262	5.5
γ-chlordane								
Chlorfenvinphos	1.0×10^{-4}	1.0×10^{-4}	124	0.345	0.345	3.82	2.90×10^{-4}	2.47
Chlorpyrifos	0.00227	3.34×10^{-3}	0.73	2.08×10^{-3}	3.07×10^{-3}	4.92	1.09	3.78
Chlorpyrifos-methyl	0.006	9.68×10^{-3}	4.76	0.0148	0.0238		0.407	3.48
Crotoxyphos	0.0019	1.90×10^{-3}	1000	3.18	3.18	2.23	5.97×10^{-4}	2.23
Cypermethrin#	1.87×10^{-7}	6.62×10^{-7}	0.004	9.61×10^{-6}	3.40×10^{-5}	6.6	0.0195	2.59
α-cypermethrin	2.30×10^{-7}	8.21×10^{-7}	0.01	2.40×10^{-5}	8.41×10^{-5}	6.94*	0.0098	
β-cypermethrin	1.80×10^{-7}	5.13×10^{-7}	0.0934	2.24×10^{-4}	6.4×10^{-4}	4.70*	8.02×10^{-7}	
ζ-cypermethrin	2.50×10^{-7}	2.5×10^{-7}	0.045	1.08×10^{-4}	1.08×10^{-4}		2.31×10^{-3}	
DDD								
p,p'-DDD	1.30×10^{-4}	6.93×10^{-4}	0.05	1.56×10^{-4}	1.08×10^{-3}	5.5	0.640	5.0
o,p'-DDD	2.0×10^{-4}*	1.39×10^{-3}	0.10*			6.0		

Name	Selected properties								
	vapor pressure			solubility			$\log K_{ow}$	H, Pa·m³/mol calcd P/C at 25 °C	$\log K_{oc}$
	P^s, Pa	P_L, Pa	g/m³	C^s, mol/m³	C_L, mol/m³				
DDE									
p,p'-DDE	8.66×10^{-4}	3.72×10^{-3}	0.04	1.26×10^{-4}	5.40×10^{-4}	5.7	7.95		5.0
o,p'-DDE	8.0×10^{-4}	3.44×10^{-3}	0.1	3.14×10^{-4}	1.35×10^{-3}	5.8	2.54		
DDT									
p,p'-DDT	2.0×10^{-5}	1.35×10^{-4}	0.0055	1.55×10^{-5}	1.11×10^{-4}	6.19	2.36		5.4
o,p'-DDT	2.53×10^{-5}	1.72×10^{-4}	0.026	7.33×10^{-5}	4.96×10^{-4}		0.347		
Deltamethrin	1.0×10^{-5}	5.52×10^{-5}	0.002	3.96×10^{-6}	2.18×10^{-5}		2.53		5.66
Demeton	0.0347	0.0347	60	0.232	0.232		0.15		1.85
Demeton-S-methyl	0.04	0.040	3300	14.3	14.3		2.79×10^{-3}		
Dialifor	6.50×10^{-5}	1.73×10^{-4}	0.18	4.57×10^{-4}	1.22×10^{-3}	4.7	0.14		
Diamidaphos			50000						
Diazinon	0.008	8.0×10^{-3}	60	0.197	0.197	3.3	0.0406		2.76
Dicapthon	5.0×10^{-4}	1.19×10^{-3}	6.25	0.021	0.05	3.6	0.0238		
Dichlofenthion	25	25.0	0.25	7.93×10^{-4}	7.93×10^{-4}	5.1	31646		
Dichlorvos	7.02	7.02	8000	36.20	36.20	1.45	0.194		1.45
Dicrotophos	0.0213	0.0213	1000000	4216	4216		5.05×10^{-6}		1.88
Dieldrin	0.0005	0.016	0.17	4.46×10^{-4}	0.0142	5.20	1.120		4.08
Diflubenzuron	1.20×10^{-7}	1.31×10^{-5}	0.08	2.57×10^{-4}	0.0281	0.78	4.66×10^{-4}		3.01
Dimethoate	0.01	0.019	20000	87.23	163.2	0.8	1.15×10^{-4}		1.3
Dinoseb	10	14.07	47	0.196	0.275		51.11		2.09
Disulfoton	0.02	0.132	25	0.0911	0.603	4.02	0.220		3.25
DNOC	0.011	0.044	150	1.013	4.063		0.0109		
Endosulfan	0.0013		0.5	1.23×10^{-3}		3.6	1.06		4.09
α-Endosulfan	0.0013	0.008	0.5	1.23×10^{-3}	0.008	3.62			3.4
β-Endosulfan	0.0061	0.394	0.45	1.11×10^{-3}	0.071	3.83			3.5
Endrin	2.0×10^{-5}	1.32×10^{-3}	0.23	6.04×10^{-4}	0.0399	5.2	0.0331		4

Name	Selected properties								
	vapor pressure			solubility				H, Pa·m³/mol	
	P^s, Pa	P_L, Pa	g/m³	C^s, mol/m³	C_L, mol/m³	log K_{OW}	calcd P/C at 25 °C	log K_{OC}
---	---	---	---	---	---	---	---	---
Ethion	1.5×10^{-4}	1.50×10^{-4}	1.8	4.68×10^{-3}	4.68×10^{-3}	5.7	0.0320	4.19
Fenitrothion	1.3×10^{-4}	1.30×10^{-4}	30	0.108	0.108	3.4	1.20×10^{-3}	3.3
Fenoxycarb	1.70×10^{-6}	3.29×10^{-6}	6	0.0199	0.039	4.3	8.54×10^{-5}	3.0
Fenthion	0.004	4.0×10^{-3}	50	0.180	0.180	4.1	0.0223	3.18
Fenvalerate	4.27×10^{-6}	4.27×10^{-6}	0.085	2.02×10^{-4}	2.02×10^{-4}	6.2	0.0211	4.0
Flucythrinate	1.20×10^{-6}*	1.20×10^{-6}	0.5*	1.11×10^{-3}	1.11×10^{-3}	6.2	1.08×10^{-3}	5.0
Fonofos	0.045	0.045	16	0.0650	0.065	3.9	0.693	2.94
Heptachlor	0.053	0.267	0.056	1.50×10^{-4}	7.56×10^{-4}	5.27	353.4	4.38
Heptachlor epoxide			0.35	8.99×10^{-4}	0.0190	5.0		4.0
Hexachlorocyclohexane								
α-BHC	0.003	0.10	1	3.44×10^{-3}	0.115	3.81	0.872	3.81
β-BHC	4.0×10^{-5}	0.0264	0.1	3.44×10^{-4}	0.227	3.8	0.116	3.36
δ-BHC	0.002	0.0268	8	0.0275	0.369	4.14	0.0727	
Iodofenphos	4.4×10^{-4}	4.4×10^{-4}	18	0.0521	0.0521	4.04	8.45×10^{-3}	
Isophorone	50	50.0	12000	86.83	86.83	1.7	0.576	
Kepone	2.93×10^{-5}	0.05	3	0.0061	10.02	5.4	0.005	4.74
Leptophos	3.0×10^{-6}	6.08×10^{-6}	0.005	1.21×10^{-5}	2.46×10^{-5}	5.9	0.247	3.97
Lindane	0.00374	0.0274	7.3	0.0251	0.184	3.7	0.149	3.0
Malathion	0.001	0.001	145	0.439	0.439	2.8	2.28×10^{-3}	3.26
Mecarbam	negligible		< 1000					
Methamidophos	0.0023	3.59×10^{-3}	200000	142	2210			
Methiocarb	0.016*	0.130	30	0.133	1.082	2.92	0.120	2.48
Methomyl	0.0067	0.0229	58000	358	1223	0.60	1.87×10^{-5}	
Methoxychlor	0.00013	5.46×10^{-4}	0.045*	1.30×10^{-4}	5.47×10^{-4}	5.08	0.999	4.9
Mevinphos	0.017	0.0170	600000	268	2677	0.5	6.35×10^{-6}	1.64
Mirex	0.0001	3.545	6.5×10^{-5}	1.19×10^{-7}	4.22×10^{-3}	6.9	839.4	6.0

Name	Selected properties								
	vapor pressure			solubility			$\log K_{ow}$	H, Pa·m³/mol calcd P/C at 25 °C	$\log K_{oc}$
	P^s, Pa	P_L, Pa	g/m³	C^s, mol/m³	C_L, mol/m³				
Monocrotophos	0.00933	0.0185	1000000	448	8870	−0.20		2.08×10^{-6}	1.4
Oxamyl	0.0306	0.173	282000	1290	7261	−0.47		2.38×10^{-5}	4.02
Parathion	6.0×10^{-4}	6.0×10^{-4}	12.4	0.0426	0.0426	3.8		0.0141	3.7
Parathion methyl	0.002	2.69×10^{-3}	25	0.095	0.128	3.0		0.0211	4
Pentachlorophenol	0.00415	0.12	14	0.053	1.565	5.05		0.79	4.8
Permethrin	1.70×10^{-6}	2.34×10^{-6}	0.006	1.53×10^{-5}	2.11×10^{-5}	6.1		0.111	3.00
Phenthoate	3.5×10^{-4}	3.50×10^{-4}	11	0.0343	0.034	3.69		0.0102	2.82
Phorate	0.085	0.085	22	0.0845	0.084	3.56		1.01	2.8
Phosmet	6.0×10^{-5}	1.75×10^{-4}	25	0.0788	0.229	2.8		7.62×10^{-4}	0.845
Phosphamidon	0.003	0.003	2.5	8.34×10^{-3}	0.0083			0.360	3.34
Pirimor	0.003	0.0133	2200	9.232	41.03			3.25×10^{-4}	1.48
Profenofos	1.2×10^{-4}	1.20×10^{-4}	28	0.0749	0.075			1.60×10^{-3}	5
Propoxur	1.70×10^{-5}	7.73×10^{-5}	1800	8.603	39.12	1.5		1.98×10^{-6}	2.9
Pyrethrins	1.33×10^{-6}		0.001	3.05×10^{-6}				0.437	
Ronnel (Fenchlorofos)	0.107	0.154	0.6	1.87×10^{-3}	2.69×10^{-3}	5.07		57.35	1.74
Sulfotep	0.0227	0.0227	25	0.0776	0.0776			0.293	2.70
Terbacil	4.13×10^{-5}	1.29×10^{-3}	710	3.276	102.06			1.26×10^{-5}	2.95
Terbufos	0.0427	0.0427	5	0.0173	0.017	4.48		2.463	
Thiobencarb	0.00293		19.2	0.0745				0.0393	
Thiodicarb	0.00431	0.117	35	0.0987	2.68			0.0437	5
Toxaphene	0.0009		0.5	1.21×10^{-3}		5.50		0.745	
Trichlorfon	0.001	3.83×10^{-3}	154000	598	2290	0.51		1.67×10^{-6}	1.00
Zinophos	0.4	0.40	1000	4.029	4.03			0.0993	

Note: # isomer not specified

* The reported values for this quantity vary considerably; whereas this selected value represents the best judgement of the authors, the reader is cautioned that it may be subject to a large error.

Table 3.2.3 Suggested half-life classes of insecticides in various environmental compartments

Compounds	Air class	Water class	Soil class	Sediment class
Aldcarb	1	5	6	8
Aldrin	1	8	8	9
Carbaryl	3	4	5	6
Carbofuran	1	4	5	6
Chloropyrifos	2	4	4	6
Chlordane	3	8	8	9
DDE	4	9	9	9
p,p'-DDT	4	7	8	9
Dieldrin	3	8	8	9
Diazinon	5	6	6	7
γ-HCH (lindane)	4	8	8	9
Heptachlor	3	5	6	7
Malathion	2	3	3	5
Methoxychlor	2	4	6	7
Mirex	4	4	9	9
Parathion	2	5	5	6
Parathion-methyl	2	5	5	6
Propoxur	1	5	5	6
Toxaphene	4	9	9	9

where,

Class	Mean half-life (hours)	Range (hours)
1	5	< 10
2	17 ($\sim$ 1 day)	10-30
3	55 ($\sim$ 2 days)	30-100
4	170 ($\sim$ 1 week)	100-300
5	550 ($\sim$ 3 weeks)	300-1,000
6	1700 ($\sim$ 2 months)	1,000-3,000
7	5500 ($\sim$ 8 months)	3,000-10,000
8	17000 ($\sim$ 2 years)	10,000-30,000
9	55000 ($\sim$ 6 years)	> 30,000

Chemical name: Carbaryl

<u>Level I calculation:</u> (six-compartment model)

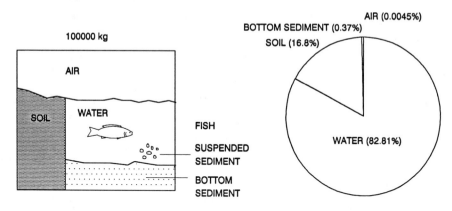

Distribution of mass

Physical-chemical properties:
molecular wt., g/mol	201.22
melting point, °C	142
solubility, g/m³	120
vapor pressure, Pa	1.60E−04
log K_{OW}	2.35
fugacity ratio, F	0.0696
dissoc. const. pK_a	

Partition coefficients:
H, Pa·m³/mol	0.0003
K_{AW}	1.08E−07
K_{OC}	93.93
BCF	11.45
K_{SW}	4.51
$K_{SD/W}$	9.02
$K_{SSD/W}$	28.18
$K_{AR/W}$	2.61E+09

COMPARTMENT	Z	CONCENTRATION				AMOUNT	AMOUNT
	mol/m³·Pa	mol/m³	mg/L, or g/m³		µg/g	kg	%
AIR	4.03E−04	2.23E−13	4.48E−11		3.78E−08	4.481	4.48E−03
WATER	3.73E+03	2.06E−06	4.14E−04		4.14E−04	82809	82.809
SOIL	1.68E+04	9.28E−06	1.87E−03		7.78E−04	16800	16.80
BIOTA (FISH)	4.27E+04	2.36E−05	4.74E−03		4.74E−03	0.949	9.49E−04
SUSPENDED SEDIMENT	1.05E+05	5.80E−05	1.17E−02		7.78E−03	11.67	1.17E−02
BOTTOM SEDIMENT	3.36E+04	1.86E−05	3.73E−03		1.56E−03	374.34	0.373
Total						100000	100

Fugacity, f = 5.521E−10 Pa

Chemical name: Carbaryl
Level II calculation: (six-compartment model)

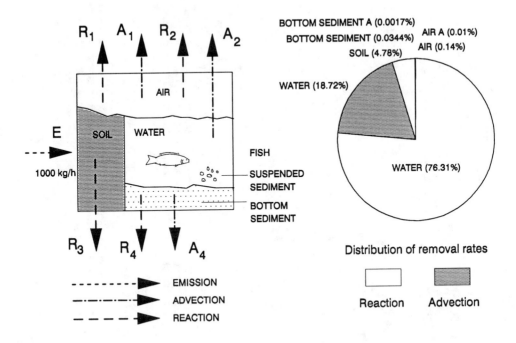

COMPARTMENT	Half-Life h	D VALUES Reaction mol/Pa·h	D VALUES Advection mol/Pa·h	CONC'N mol/m³	LOSS Reaction kg/h	LOSS Advection kg/h	REMOVAL %
AIR	5	5.59E+09	4.03E+08	5.03E−13	1.404	0.101	0.151
WATER	170	3.04E+12	7.45E+11	4.65E−06	763.09	187.19	90.028
SOIL	550	1.91E+11		2.10E−05	47.852		4.785
BIOTA (FISH)				5.33E−05			
SUSPENDED SEDIMENT				1.31E−04			
BOTTOM SEDIMENT	1700	1.37E+09	6.72E+07	4.19E−05	0.344	0.0169	0.0361
Total		3.23E+12	7.46E+11		812.69	187.31	100
Reaction + Advection			3.98E+12			1000	

Fugacity, f = 1.248E−09 Pa

Total amount = 226054 kg

Overall residence time = 226.05 h

Reaction time = 278.16 h

Advection time = 1206.83 h

Fugacity Level III calculations: (four-compartment model)
Chemical name: Carbaryl

Phase Properties and Rates:

Compartment	Bulk Z mol/m³ Pa	Half-life h	D Values Reaction mol/Pa h	D Values Advection mol/Pa h
Air (1)	4.245E-04	5	5.88E+09	4.24E+08
Water (2)	3.728E+03	170	3.04E+12	7.46E+11
Soil (3)	9.520E+03	550	2.16E+11	
Sediment (4)	9.703E+03	1700	1.98E+09	9.70E+07

$E(1)=1000$ $E(2)=1000$ $E(3)=1000$ $E(1,2,3)$

Overall residence time =	596.19	197.84	604.66 h
Reaction time =	619.42	246.35	643.98 h
Advection time =	15895.99	1004.64	9902.08 h

- - - - → EMISSION (E)
- - - - → REACTION (R)
- - - - → ADVECTION (A)
──→ TRANSFER D VALUE mol/Pa h

Diagram:

E_1 → AIR (1), R_1, A_1 up; 1.722E08, D_{13}, D_{31} 3.377E10, D_{12} 2.017E07, D_{21}

E_3 → SOIL (3); R_3, R_4, A_4

E_2 → WATER (2) 3.754E09; R_2, A_2; D_{24} 3.794E09, D_{42} 4.252E09; SEDIMENT (4); D_{32} 1.679E10

Phase Properties, Compositions, Transport and Transformation Rates:

Emission, kg/h

E(1)	E(2)	E(3)		f(1)	f(2)	f(3)	f(4)		C(1)	C(2)	C(3)	C(4)
1000	0	0		1.134E-07	1.854E-10	1.645E-08	1.343E-10		9.689E-09	1.391E-04	3.152E-02	2.623E-04
0	1000	0		6.043E-13	1.313E-09	8.764E-14	9.510E-07		5.162E-14	9.846E-04	1.679E-07	1.857E-03
0	0	1000		8.393E-11	9.475E-11	2.135E-08	6.865E-11		7.169E-12	7.108E-05	4.090E-02	1.340E-04
50	250	700		5.731E-09	4.037E-10	1.577E-08	2.925E-10		4.895E-10	3.028E-04	3.021E-02	5.711E-04

Fugacity, Pa; Concentration, g/m³

Emission, kg/h; Loss, Reaction, kg/h; Loss, Advection, kg/h

E(1)	E(2)	E(3)		R(1)	R(2)	R(3)	R(4)		A(1)	A(2)	A(3)	A(4)
1000	0	0		1.343E+02	1.134E+02	7.15E+02	5.346E-02		9.689E+00	2.781E+01		2.623E-03
0	1000	0		7.154E-04	8.027E+02	3.81E+01	3.785E-01		5.162E-05	1.969E+02		1.857E-02
0	0	1000		9.936E+00	5.795E+01	9.28E+02	2.732E-02		7.169E-03	1.422E+01		1.340E-03
50	250	700		6.784E+00	2.469E+02	6.85E+02	1.164E-01		4.895E-01	6.057E+01		5.711E-04

Amounts, kg

m(1)	m(2)	m(3)	m(4)	Total Amount, kg
9.689E+02	2.781E+04	5.673E+05	1.311E+02	5.962E+05
5.162E-03	1.969E+05	3.022E+00	9.284E+02	1.978E+05
7.169E-01	1.422E+04	7.363E+05	6.702E+01	7.506E+05
4.895E+01	6.057E+04	5.438E+05	2.856E+02	6.047E+05

Intermedia Rate of Transport, kg/h

T12	T21	T13	T31	T32	T24	T42
air-water	water-air	air-soil	soil-air	soil-water	water-sed	sed-water
8.568E+01	7.525E-04	7.709E+02	5.701E-01	5.557E+01	1.586E-01	1.026E-01
4.564E-04	5.327E-03	4.107E-03	3.037E-06	2.961E-04	1.123E+00	7.261E-01
6.340E-02	5.327E-03	5.704E-01	7.399E-01	7.213E+01	8.108E-02	5.242E-02
4.329E+00	1.639E-03	3.895E+01	5.465E-01	5.327E+01	3.455E-01	2.235E-01

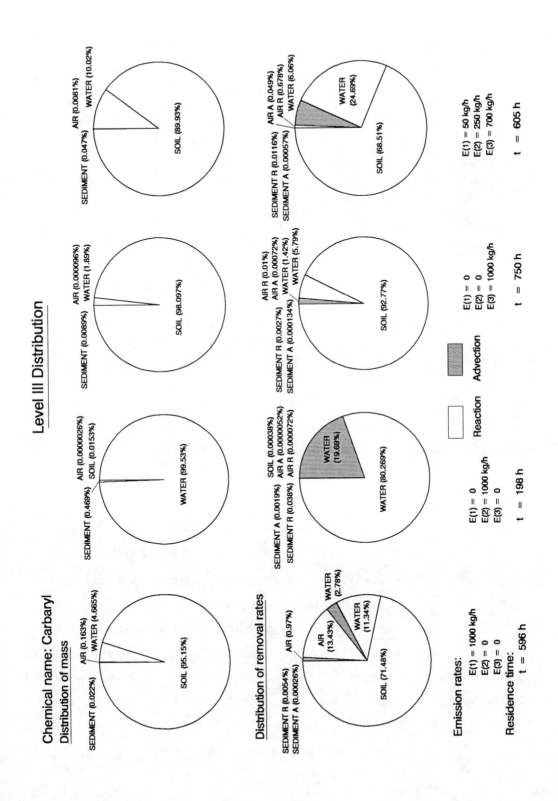

Chemical name: Chlorpyrifos

Level I calculation: (six-compartment model)

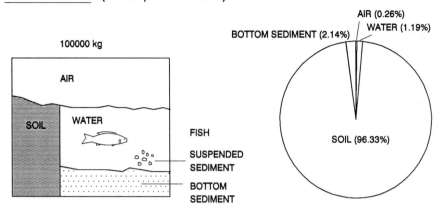

Distribution of mass

Physical-chemical properties:
molecular wt., g/mol	350.60
melting point, °C	41
solubility, g/m³	0.73
vapor pressure, Pa	2.27E−03
log K_{OW}	4.96
fugacity ratio, F	0.6946
dissoc. const. pK_a	

Partition coefficients:
H, Pa·m³/mol	1.0902
K_{AW}	4.40E−04
K_{OC}	37392
BCF	4560
K_{SW}	1795
$K_{SD/W}$	3590
$K_{SSD/W}$	11218
$K_{AR/W}$	1.84E+09

COMPARTMENT	Z	CONCENTRATION				AMOUNT	AMOUNT
	mol/m³·Pa	mol/m³	mg/L, or g/m³	µg/g		kg	%
AIR	4.03E−04	7.48E−12	2.62E−09	2.21E−06		262.285	0.2623
WATER	9.17E−01	1.70E−08	5.96E−06	5.96E−06		1192.7	1.1927
SOIL	1.65E+03	3.05E−05	1.07E−02	4.46E−03		96332	96.332
BIOTA (FISH)	4.18E+03	7.76E−05	2.72E−02	2.72E−02		5.439	5.44E−03
SUSPENDED SEDIMENT	1.03E+04	1.91E−04	6.69E−02	4.46E−02		66.90	6.69E−02
BOTTOM SEDIMENT	3.29E+03	6.11E−05	2.14E−02	8.92E−03		2140.7	2.141
Total						100000	100

Fugacity, f = 1.854E−08 Pa

Chemical name: Chlorpyrifos
Level II calculation: (six-compartment model)

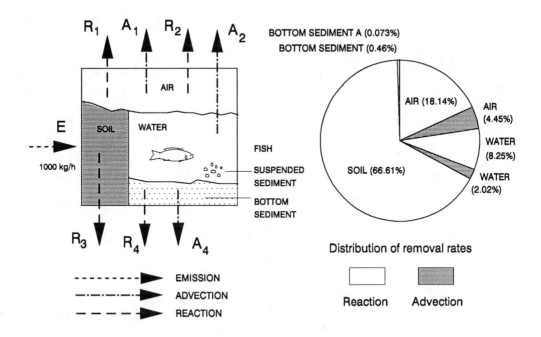

COMPARTMENT	Half-Life	D VALUES		CONC'N	LOSS	LOSS	REMOVAL
		Reaction	Advection		Reaction	Advection	%
	h	mol/Pa·h	mol/Pa·h	mol/m³	kg/h	kg/h	
AIR	17	1.64E+09	4.03E+08	1.27E−10	181.37	44.492	22.586
WATER	170	7.48E+08	1.83E+08	2.89E−07	82.475	20.232	10.271
SOIL	1700	6.04E+09		5.18E−04	666.13		66.613
BIOTA (FISH)				1.32E−03			
SUSPENDED SEDIMENT				3.24E−03			
BOTTOM SEDIMENT	5500	4.15E+07	6.59E+06	1.04E−03	4.5754	0.7263	0.5302
Total		8.43E+09	5.87E+08		934.55	65.45	100
Reaction + Advection			9.02E+09			1000	

Fugacity, $f = 3.146E-07$ Pa

Total amount = 1696309 kg

Overall residence time = 1696.31 h

Reaction time = 1815.11 h

Advection time = 25918 h

Fugacity Level III calculations: (four-compartment model)
Chemical name: Chlorpyrifos

Phase Properties and Rates:

Compartment	Bulk Z mol/m3 Pa	Half-life h	D Values Reaction mol/Pa h	D Values Advection mol/Pa h
Air (1)	4.182E-04	17	1.70E+09	4.18E+08
Water (2)	9.729E-01	170	7.93E+08	1.95E+08
Soil (3)	8.234E+02	1700	6.04E+09	
Sediment (4)	6.593E+02	5500	4.15E+07	6.59E+06

	E(1)=1000	E(2)=1000	E(3)=1000	E(1,2,3)
Overall residence time =	79.96	481.20	2450.67	1839.77 h
Reaction time =	99.02	597.27	2451.23	1953.80 h
Advection time =	415.38	2476.32	10694590	31521.30 h

- - - - ▲ EMISSION (E)
- - - ▲ REACTION (R)
- - ▲ ADVECTION (A)
——▲ TRANSFER D VALUE mol/Pa h

Diagram:
- E_1 → AIR (1); R_1, A_1 outflows; 1.539E06, D_{31}; 1.932E07; D_{13} 4.979E07; D_{12}
- E_3, E_2 → WATER (2) 2.468E07; D_{21}; R_2, A_2
- 5.609E06; D_{32}; D_{24} 7.502E06
- SOIL (3); SEDIMENT (4) D_{42} 5.236E07
- R_3, R_4, A_4

Phase Properties, Compositions, Transport and Transformation Rates:

Emission, kg/h

E(1)	E(2)	E(3)	Fugacity, Pa				Concentration, g/m3				Loss, Advection, kg/h			
			f(1)	f(2)	f(3)	f(4)	C(1)	C(2)	C(3)	C(4)	A(1)	A(2)	A(3)	A(4)
1000	0	0	1.298E-06	3.050E-08	1.069E-08	2.871E-08	1.904E-07	1.040E-05	3.085E-03	6.637E-03	1.904E+02	2.081E+00		6.637E-02
0	1000	0	2.383E-08	2.711E-06	1.962E-10	2.552E-06	3.495E-09	9.246E-04	5.663E-05	5.898E-01	3.495E+00	1.849E+02		5.898E+00
0	0	1000	3.523E-10	2.521E-09	4.715E-07	2.374E-09	5.165E-11	8.601E-07	1.361E-01	5.486E-04	5.165E-02	1.720E-01		5.486E-03
50	250	700	7.111E-08	6.810E-07	3.306E-07	6.410E-07	1.043E-08	2.323E-04	9.545E-02	1.482E-01	1.043E+01	4.646E+01		1.482E+00

Emission, kg/h

E(1)	E(2)	E(3)	Loss, Reaction, kg/h				Amounts, kg				Total Amount, kg
			R(1)	R(2)	R(3)	R(4)	m(1)	m(2)	m(3)	m(4)	
1000	0	0	7.760E+02	8.483E+00	2.26E+01	4.181E-01	1.904E+04	2.081E+03	5.553E+04	3.318E+03	7.996E+04
0	1000	0	1.425E+01	7.539E+02	4.16E-01	3.716E+01	3.495E+02	1.849E+05	1.019E+03	2.949E+05	4.812E+05
0	0	1000	2.106E-01	7.012E-01	9.99E+02	3.456E-02	5.165E+00	1.720E+02	2.450E+06	2.743E+02	2.451E+06
50	250	700	4.251E+01	1.894E+02	7.00E+02	9.335E+00	1.043E+03	4.646E+04	1.718E+06	7.408E+04	1.840E+06

Intermedia Rate of Transport, kg/h

T12	T21	T13	T31	T32	T24	T42
air-water	water-air	air-soil	soil-air	soil-water	water-sed	sed-water
1.123E+01	2.066E-01	2.266E+01	5.764E-03	2.101E-01	5.600E-01	7.553E-02
2.062E-01	1.836E+01	4.161E-01	1.058E-04	3.858E-04	4.977E+01	6.712E+00
3.048E-03	1.708E-02	6.149E-03	2.543E-01	9.273E-01	4.629E-02	6.243E-03
6.154E-01	4.613E+00	1.241E+00	1.784E-01	6.502E-01	1.250E+01	1.686E+00

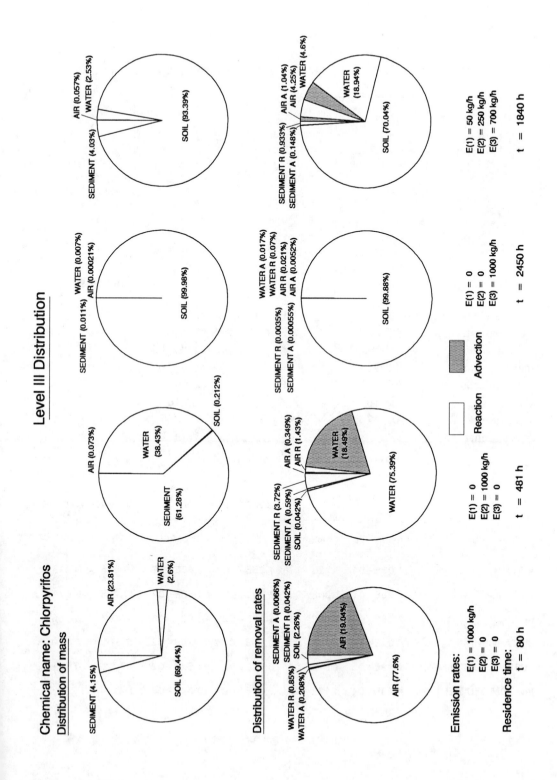

Chemical name: p,p'-DDT

<u>Level I calculation</u>: (six-compartment model)

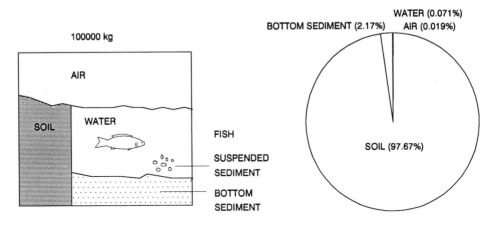

Distribution of mass

Physical-chemical properties:
molecular wt., g/mol	354.50
melting point, °C	109
solubility, g/m^3	0.0055
vapor pressure, Pa	2.0E−05
log K_{OW}	6.19
fugacity ratio, F	0.1476
dissoc. const. pK_a	

Partition coefficients:
H, Pa·m^3/mol	1.2891
K_{AW}	5.20E−04
K_{OC}	635015
BCF	77441
K_{SW}	30481
$K_{SD/W}$	60961
$K_{SSD/W}$	190504
$K_{AR/W}$	4.43E+10

COMPARTMENT	Z	CONCENTRATION			AMOUNT	AMOUNT
	mol/m^3·Pa	mol/m^3	mg/L, or g/m^3	µg/g	kg	%
AIR	4.03E−04	5.22E−13	1.85E−10	1.56E−07	18.515	0.0185
WATER	7.76E−01	1.00E−09	3.56E−07	3.56E−07	71.205	0.0712
SOIL	2.36E+04	3.06E−05	1.09E−02	4.52E−03	97667	97.667
BIOTA (FISH)	6.01E+04	7.78E−05	2.76E−02	2.76E−02	5.514	5.51E−03
SUSPENDED SEDIMENT	1.48E+05	1.91E−04	6.78E−02	4.52E−02	67.82	6.78E−02
BOTTOM SEDIMENT	4.73E+04	6.12E−05	2.17E−02	9.04E−03	2170.4	2.170
Total					100000	100

Fugacity, f = 2.816E−12 Pa

Chemical name: p,p'-DDT
Level II calculation: (six-compartment model)

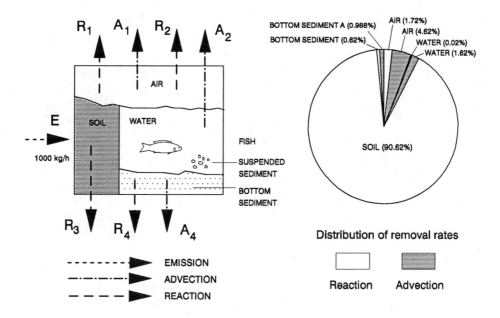

COMPARTMENT	Half-Life h	D VALUES Reaction mol/Pa·h	D VALUES Advection mol/Pa·h	CONC'N mol/m^3	LOSS Reaction kg/h	LOSS Advection kg/h	REMOVAL %
AIR	170	1.64E+08	4.03E+08	1.19E−10	17.181	42.147	5.933
WATER	5500	1.95E+07	1.55E+08	2.29E−07	2.042	16.209	1.825
SOIL	17000	8.67E+09		6.97E−03	906.31		90.631
BIOTA (FISH)				1.77E−02			
SUSPENDED SEDIMENT				4.36E−02			
BOTTOM SEDIMENT	55000	5.96E+07	9.46E+07	1.39E−02	6.225	9.881	1.611
Total		8.86E+09	5.59E+08		931.76	68.24	100
Reaction + Advection			9.42E+09			1000	

Fugacity, f = 2.947E−07 Pa

Total amount = 22763998 kg

Overall residence time = 22764 h

Reaction time = 24431 h

Advection time = 333601 h

Fugacity Level III calculations: (four-compartment model)
Chemical name: p,p'-DDT

Phase Properties and Rates:

Compartment	Bulk Z mol/m3 Pa	Half-life h	D Values Reaction mol/Pa h	Advection mol/Pa h
Air (1)	7.608E-04	170	3.10E+08	7.61E+08
Water (2)	1.575E+00	5500	3.97E+07	3.15E+08
Soil (3)	1.182E-04	17000	8.68E+09	
Sediment (4)	9.459E+03	55000	5.96E+07	9.46E+07

	E(1)=1000	E(2)=1000	E(3)=1000	E(1,2,3)
Overall residence time =	12047.48	17530.83	24509.07	22141.43 h
Reaction time =	19874.56	63896.78	24561.38	27766.74 h
Advection time =	30591.01	24159.18	11507952	109290.95 h

- - - - EMISSION (E)
- · - · REACTION (R)
- · · - ADVECTION (A)
────► TRANSFER D VALUE mol/Pa h

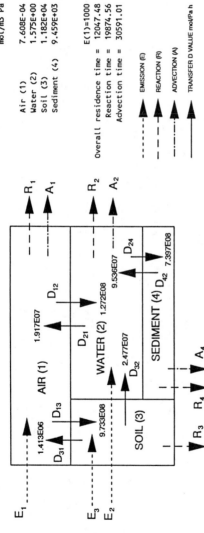

Phase Properties, Compositions, Transport and Transformation Rates:

Emission, kg/h

E(1)	E(2)	E(3)		f(1)	f(2)	f(3)	f(4)
1000	0	0		1.301E-06	2.035E-07	1.455E-07	6.031E-07
0	1000	0		3.005E-08	3.400E-06	3.358E-09	1.008E-05
0	0	1000		2.968E-10	9.712E-09	3.242E-07	2.879E-08
50	250	700		7.277E-08	8.670E-07	2.351E-07	2.570E-06

Fugacity, Pa

Concentration, g/m3

				C(1)	C(2)	C(3)	C(4)
				3.509E-07	1.136E-04	6.099E-01	2.022E+00
				8.098E-09	1.898E-03	1.408E-02	3.379E+01
				8.004E-11	5.422E-06	1.359E+00	9.654E-02
				1.962E-08	4.840E-04	9.852E-01	8.617E-01

Emission, kg/h

E(1)	E(2)	E(3)		R(1)	R(2)	R(3)	R(4)
1000	0	0		1.430E+02	2.862E+00	4.48E+02	1.274E+01
0	1000	0		3.301E+00	4.783E+01	1.05E+01	2.129E+02
0	0	1000		3.263E-02	1.366E-01	9.97E+02	6.082E-01
50	250	700		8.000E+00	1.220E+01	7.23E+02	5.429E+01

Loss, Reaction, kg/h

Loss, Advection, kg/h

				A(1)	A(2)	A(3)	A(4)
				3.509E+02	2.272E+01		2.022E+01
				8.098E+00	3.796E+02		3.379E+02
				8.004E-02	1.084E+00		9.654E-01
				1.962E+01	9.679E+01		8.617E+00

Amounts, kg

	m(1)	m(2)	m(3)	m(4)
	3.509E+04	2.272E+04	1.098E+07	1.011E+06
	8.098E+02	3.796E+05	2.534E+05	1.690E+07
	8.004E+00	1.084E+03	2.446E+07	4.827E+04
	1.962E+03	9.679E+04	1.773E+07	4.309E+06

Total Amount, kg

	1.205E+07
	1.753E+07
	2.451E+07
	2.214E+07

Intermedia Rate of Transport, kg/h

T12	T21	T13	T31	T32	T24	T42
air-water	water-air	air-soil	soil-air	soil-water	water-sed	sed-water
5.865E+01	1.383E+00	4.489E+02	7.291E-02	1.278E+00	5.335E+01	2.039E+01
1.353E+00	2.311E+01	1.036E+01	1.683E-03	2.949E-02	8.915E+02	3.407E+02
1.338E-02	6.602E-02	1.024E-01	1.624E+01	2.847E+01	2.547E+00	9.732E-01
3.280E+00	5.893E+00	2.511E+01	1.178E-01	2.064E+00	2.273E+02	8.687E+01

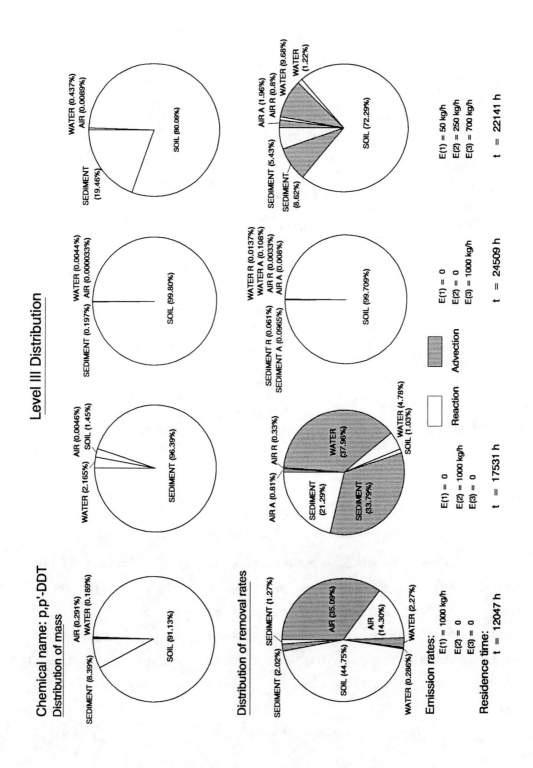

Chemical name: Diazinon

Level I calculation: (six-compartment model)

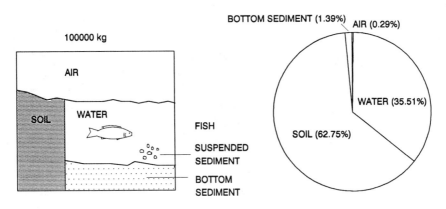

Distribution of mass

Physical-chemical properties:
molecular wt., g/mol	304.40
melting point, °C	84
solubility, g/m³	60.0
vapor pressure, Pa	8.0E−03
log K_{OW}	3.30
fugacity ratio, F	0.2609
dissoc. const. pK_a	

Partition coefficients:
H, Pa·m³/mol	0.0406
K_{AW}	1.64E−05
K_{OC}	818.06
BCF	99.763
K_{SW}	39.267
$K_{SD/W}$	78.534
$K_{SSD/W}$	245.417
$K_{AR/W}$	1.96E+08

COMPARTMENT	Z	CONCENTRATION			AMOUNT	AMOUNT
	mol/m³·Pa	mol/m³	mg/L, or g/m³	µg/g	kg	%
AIR	4.03E−04	9.55E−12	2.91E−09	2.45E−06	290.74	0.2907
WATER	2.46E+01	5.83E−07	1.78E−04	1.78E−04	35514	35.514
SOIL	9.67E+02	2.29E−05	6.97E−03	2.91E−03	62754	62.754
BIOTA (FISH)	2.46E+03	5.82E−05	1.77E−02	1.77E−02	3.543	3.54E−03
SUSPENDED SEDIMENT	6.05E+03	1.43E−04	4.36E−02	2.91E−02	43.58	4.36E−02
BOTTOM SEDIMENT	1.93E+03	4.58E−05	1.39E−02	5.81E−03	1394.5	1.395
Total					100000	100

Fugacity, f = 2.368E−08 Pa

Chemical name: Diazinon
Level II calculation: (six-compartment model)

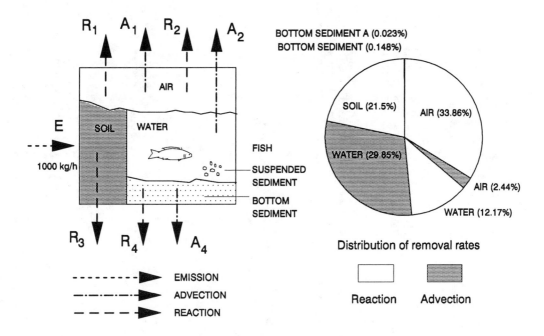

COMPARTMENT	Half-Life h	D VALUES Reaction mol/Pa·h	D VALUES Advection mol/Pa·h	CONC'N mol/m³	LOSS Reaction kg/h	LOSS Advection kg/h	REMOVAL %
AIR	5	5.59E+09	4.03E+08	8.03E−11	338.69	24.436	36.312
WATER	1700	2.01E+09	4.93E+09	4.90E−06	121.68	298.486	42.016
SOIL	1700	3.55E+09		1.93E−04	215.0		21.50
BIOTA (FISH)				4.89E−04			
SUSPENDED SEDIMENT				1.20E−03			
BOTTOM SEDIMENT	5500	2.44E+07	3.87E+06	3.85E−04	1.4768	0.2344	0.1711
Total		1.11E+10	5.33E+09		676.84	323.16	100
Reaction + Advection			1.65E+10			1000	

Fugacity, f = 1.990E−07 Pa

Total amount = 840473 kg

Overall residence time = 840.5 h

Reaction time = 1241.8 h

Advection time = 2600.82 h

Fugacity Level III calculations: (four-compartment model)
Chemical name: Diazinon

Phase Properties and Rates:

Compartment	Bulk Z mol/m3 Pa	Half-life h	D Values Reaction mol/Pa h	D Values Advection mol/Pa h
Air (1)	4.050E-04	5	5.61E-09	4.05E+08
Water (2)	2.468E+01	1700	2.01E+09	4.94E+09
Soil (3)	4.913E+02	1700	3.60E+09	
Sediment (4)	4.068E+02	5500	2.56E+07	4.07E+06

	E(1)=1000	E(2)=1000	E(3)=1000	E(1,2,3)
Overall residence time =	104.97	733.23	2388.88	1860.77 h
Reaction time =	112.88	2496.26	2441.38	2311.88 h
Advection time =	1497.67	1038.18	111074	9536.24 h

▲ EMISSION (E)
▲ REACTION (R)
▲ ADVECTION (A)
▲ TRANSFER D VALUE mol/Pa h

Phase Properties, Compositions, Transport and Transformation Rates:

Emission, kg/h

E(1)	E(2)	E(3)	Fugacity, Pa f(1)	f(2)	f(3)	f(4)	Concentration, g/m3 C(1)	C(2)	C(3)	C(4)
1000	0	0	5.209E-07	3.919E-09	3.435E-08	3.695E-09	6.420E-08	2.943E-05	5.135E-03	4.574E-04
0	1000	0	1.499E-09	4.697E-07	9.888E-11	4.429E-07	1.848E-10	3.528E-03	1.478E-05	5.483E-02
0	0	1000	2.880E-09	1.407E-08	8.796E-07	1.327E-08	3.550E-10	1.057E-04	1.315E-01	1.643E-03
50	250	700	2.844E-08	1.275E-07	6.174E-07	1.202E-07	3.505E-09	9.574E-04	9.231E-02	1.488E-02

Emission, kg/h

E(1)	E(2)	E(3)	Loss, Reaction, kg/h R(1)	R(2)	R(3)	R(4)	Loss, Advection, kg/h A(1)	A(2)	A(3)	A(4)
1000	0	0	8.898E+02	2.399E+00	3.77E+01	2.882E-02	6.420E+01	5.886E+00		4.574E-03
0	1000	0	2.561E+00	2.876E+02	1.08E-01	3.454E+00	1.848E-01	7.055E+02		5.483E-01
0	0	1000	4.920E+00	8.616E+00	9.65E+02	1.035E-01	3.550E-01	2.114E+01		1.643E-02
50	250	700	4.857E+01	7.805E+01	6.77E+02	9.375E-01	3.505E+00	1.915E+02		1.488E-01

			Amounts, kg					Total
			m(1)	m(2)	m(3)	m(4)		Amount, kg
			6.420E+03	5.886E+03	9.244E+04	2.287E+02		1.050E+05
			1.848E+01	7.055E+05	2.661E+02	2.742E+04		7.332E+05
			3.550E+01	2.114E+04	2.367E+06	8.213E+02		2.389E+06
			3.505E+02	1.915E+05	1.662E+06	7.440E+03		1.861E+06

Intermedia Rate of Transport, kg/h

T12	T13	T21	T23	T24	T31	T32	T34	T42
air-water	air-soil	water-air	water-soil	water-sed	soil-air	soil-water	sed-water	
7.174E+00	3.906E+01	2.401E-02		6.546E-02	2.126E-01	1.168E+00	3.206E-02	
2.065E-02	1.124E-01	2.879E+00		7.846E+00	6.120E-04	3.363E-03	3.843E+00	
3.967E-02	2.160E-01	8.623E-02		2.350E-01	5.446E+00	2.992E+01	1.151E-01	
3.916E-01	2.132E+00	7.812E-01		2.129E+00	3.822E+00	2.100E+01	1.043E+00	

D values from figure:
- D_{31} = 2.034E07
- D_{13} = 2.464E08
- D_{21} = 1.014E07
- D_{12} = 4.526E07
- D_{32} = 1.118E08
- D_{24} = 2.852E07
- D_{42} = 5.489E07

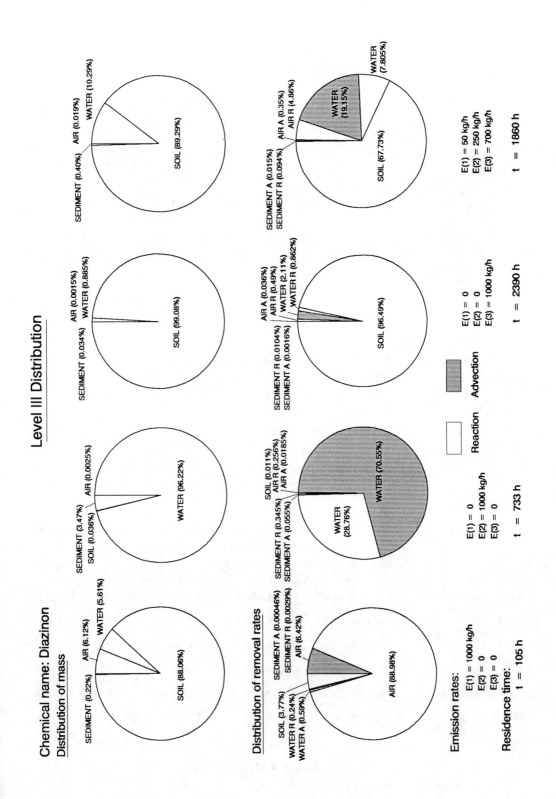

Chemical name: Lindane

Level I calculation: (six-compartment model)

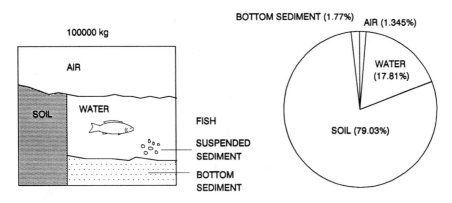

Distribution of mass

Physical-chemical properties:
molecular wt., g/mol	354.50
melting point, °C	109
solubility, g/m³	7.0
vapor pressure, Pa	7.40E−03
log K_{OW}	3.70
fugacity ratio, F	0.1476
dissoc. const. pK_a	

Partition coefficients:
H, Pa·m³/mol	0.3748
K_{AW}	1.51E−04
K_{OC}	2054.87
BCF	250.59
K_{SW}	98.63
$K_{SD/W}$	197.27
$K_{SSD/W}$	616.46
$K_{AR/W}$	1.20E+08

COMPARTMENT	Z	CONCENTRATION			AMOUNT	AMOUNT
	mol/m³·Pa	mol/m³	mg/L, or g/m³	μg/g	kg	%
AIR	4.03E−04	3.80E−11	1.35E−08	1.14E−05	1345.99	1.346
WATER	2.67E+00	2.51E−07	8.90E−05	8.90E−05	17806	17.806
SOIL	2.63E+02	2.48E−05	3.66E−03	3.66E−03	79032	79.032
BIOTA (FISH)	6.69E+02	6.29E−05	2.23E−02	2.23E−02	4.462	4.46E−03
SUSPENDED SEDIMENT	1.64E+03	1.55E−04	3.66E−02	3.66E−02	54.88	5.49E−02
BOTTOM SEDIMENT	5.26E+02	4.95E−05	7.32E−03	7.23E−03	1756.27	1.756
Total					100000	100

Fugacity, f = 9.412E−08 Pa

Chemical name: Lindane
Level II calculation: (six-compartment model)

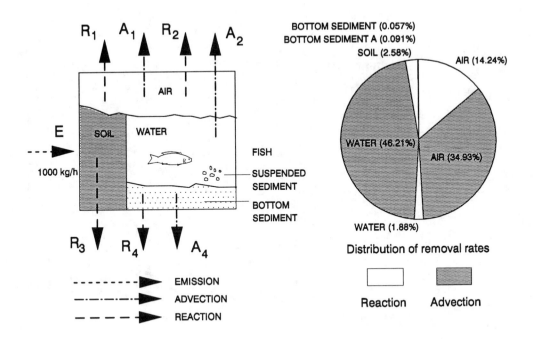

COMPARTMENT	Half-Life	D VALUES		CONC'N	LOSS	LOSS	REMOVAL
		Reaction	Advection		Reaction	Advection	%
	h	mol/Pa·h	mol/Pa·h	mol/m^3	kg/h	kg/h	
AIR	170	1.64E+08	4.03E+08	9.85E−10	142.40	349.32	49.172
WATER	17000	2.18E+07	5.34E+08	6.52E−06	18.838	462.11	48.095
SOIL	55000	2.98E+07		6.43E−04	25.844		2.584
BIOTA (FISH)				1.63E−03			
SUSPENDED SEDIMENT				4.02E−03			
BOTTOM SEDIMENT	55000	6.63E+05	1.05E+06	1.29E−03	0.574	0.912	0.1486
Total		2.16E+08	9.37E+08		187.66	812.35	100
Reaction + Advection			1.15E+09			1000	

Fugacity, f = 2.443E−06 Pa

Total amount = 2595261 kg

Overall residence time = 2595.26 h

Reaction time = 13830 h

Advection time = 3194.8 h

Fugacity Level III calculations: (four-compartment model)
Chemical name: Lindane

Phase Properties and Rates:

Compartment	Bulk Z mol/m3 Pa	Half-life h	D Values Reaction mol/Pa h	Advection mol/Pa h
Air (1)	4.044E-04	170	1.65E+08	4.04E+08
Water (2)	2.677E+00	17000	2.18E+07	5.35E+08
Soil (3)	1.324E+02	17000	9.71E+07	
Sediment (4)	1.074E+02	5500	6.77E+06	1.07E+06

	E(1)=1000	E(2)=1000	E(3)=1000	E(1,2,3)
Overall residence time =	1122.25	1040.74	21333.61	15249.83
Reaction time =	3628.87	17711.42	24280.24	23635.75
Advection time =	1624.69	1105.71	175789	42981.66

▲ EMISSION (E)
▲ REACTION (R)
▲ ADVECTION (A)
▲ TRANSFER D VALUE mol/Pa h

Phase Properties, Compositions, Transport and Transformation Rates:

Emission, kg/h | Fugacity, Pa | Concentration, g/m3

E(1)	E(2)	E(3)	f(1)	f(2)	f(3)	f(4)	C(1)	C(2)	C(3)	C(4)
1000	0	0	4.549E-06	2.028E-07	1.201E-06	1.911E-06	6.522E-07	1.925E-04	5.638E-02	7.277E-03
0	1000	0	1.547E-07	4.833E-06	4.084E-08	4.553E-06	2.217E-08	4.587E-03	1.917E-03	1.734E-01
0	0	1000	1.412E-07	5.317E-07	2.512E-05	5.009E-07	2.024E-08	5.046E-04	1.179E+00	1.907E-02
50	250	700	3.650E-07	1.590E-06	1.765E-05	1.498E-06	5.232E-08	1.510E-03	8.286E-01	5.706E-02

Loss, Reaction, kg/h | Loss, Advection, kg/h

E(1)	E(2)	E(3)	R(1)	R(2)	R(3)	R(4)	A(1)	A(2)	A(3)	A(4)
1000	0	0	2.659E+02	1.570E+01	4.14E+01	4.584E-01	6.522E+02	3.850E+01		7.277E-02
0	1000	0	9.038E+00	3.740E+02	1.41E+00	1.092E+01	2.217E+01	9.174E+02		1.734E+00
0	0	1000	8.252E+00	4.114E+01	8.65E+02	1.202E+00	2.024E+01	1.009E+02		1.907E-01
50	250	700	2.133E+01	1.231E+02	6.08E+02	3.595E+00	5.232E+01	3.019E+02		5.706E-01

Amounts, kg | Total Amount, kg

E(1)	E(2)	E(3)	m(1)	m(2)	m(3)	m(4)	
1000	0	0	6.522E+04	3.850E+04	1.015E+06	3.638E+03	1.122E+06
0	1000	0	2.217E+03	9.174E+05	3.450E+04	8.669E+04	1.041E+06
0	0	1000	2.024E+03	1.009E+05	2.122E+07	9.537E+03	2.133E+07
50	250	700	5.232E+03	3.019E+05	1.491E+07	2.853E+04	1.525E+07

Intermedia Rate of Transport, kg/h

	T12	T21	T13	T31	T32	T24	T42
	air-water	water-air	air-soil	soil-air	soil-water	water-sed	sed-water
	3.682E+01	1.429E+00	4.790E+01	5.215E+00	1.773E-01	7.833E-01	2.521E+00
	1.252E+00	3.404E+01	1.628E+00	4.451E-02	1.090E+02	1.866E+01	6.006E+00
	1.143E+00	3.745E+00	1.487E+00	2.738E+01	7.663E+01	2.053E+02	6.608E-01
	2.954E+00	1.120E+01	3.842E+00	1.924E+01	6.142E+00	1.977E+00	

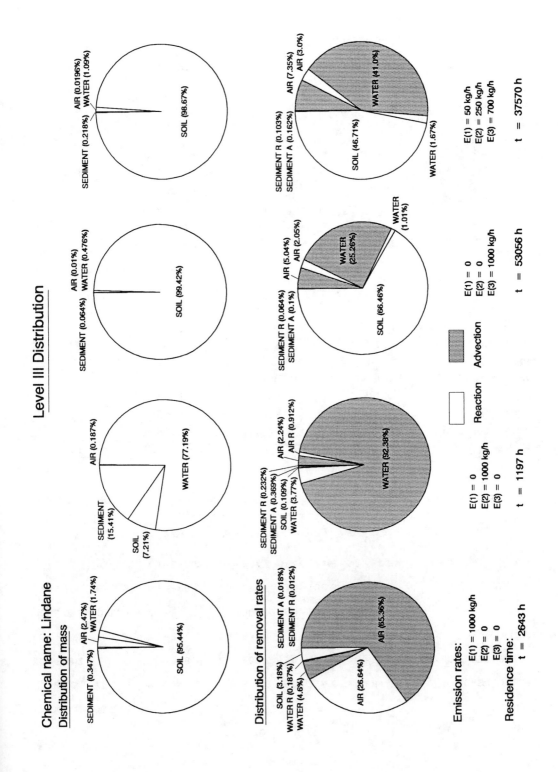

Chemical name: Malathion

Level I calculation: (six-compartment model)

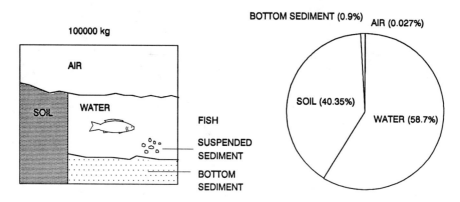

Distribution of mass

Physical-chemical properties:
molecular wt., g/mol	330.36
melting point, °C	2.90
solubility, g/m³	145
vapor pressure, Pa	1.00E−03
log K_{OW}	2.89
fugacity ratio, F	1.0
dissoc. const. pK_a	

Partition coefficients:
H, Pa·m³/mol	2.28E−03
K_{AW}	9.19E−07
K_{OC}	318.26
BCF	38.81
K_{SW}	15.28
$K_{SD/W}$	30.55
$K_{SSD/W}$	95.48
$K_{AR/W}$	6.00E+09

COMPARTMENT	Z	CONCENTRATION				AMOUNT	AMOUNT
	mol/m³·Pa	mol/m³	mg/L, or g/m³		µg/g	kg	%
AIR	4.03E−04	8.17E−13	2.70E−10		2.28E−07	26.97	0.0270
WATER	4.39E+01	8.88E−07	2.93E−04		2.93E−04	58696	58.696
SOIL	6.71E+03	1.36E−05	4.48E−03		1.87E−03	40350	40.350
BIOTA (FISH)	1.70E+04	3.45E−05	1.14E−02		1.14E−02	2.278	2.28E−03
SUSPENDED SEDIMENT	4.19E+04	8.48E−05	2.80E−02		1.87E−02	28.02	2.80E−02
BOTTOM SEDIMENT	1.34E+04	2.71E−05	8.97E−03		3.74E−03	896.67	0.8967
Total						100000	100

Fugacity, f = 2.024E−09 Pa

Chemical name: Malathion
Level II calculation: (six-compartment model)

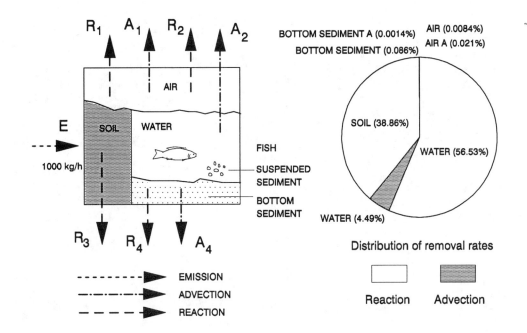

COMPARTMENT	Half-Life	D VALUES		CONC'N	LOSS	LOSS	REMOVAL
		Reaction	Advection		Reaction	Advection	%
	h	mol/Pa·h	mol/Pa·h	mol/m³	kg/h	kg/h	
AIR	170	1.64E+08	4.03E+08	6.24E−13	0.0841	0.2062	0.0290
WATER	55	1.11E+12	8.78E+11	6.79E−07	565.33	44.868	61.02
SOIL	55	7.60E+11		1.04E−05	388.63		38.863
BIOTA (FISH)				2.64E−05			
SUSPENDED SEDIMENT				6.48E−05			
BOTTOM SEDIMENT	550	1.69E+09	2.68E+07	2.07E−05	0.864	0.0137	0.0877
Total		1.87E+12	8.82E+10		954.91	45.087	100
Reaction + Advection		1.95E+12			1000		

Fugacity, $f = 1.547E-09$ Pa

Total amount = 76441 kg

Overall residence time = 76.44 h

Reaction time = 80.05 h

Advection time = 31695.4 h

Fugacity Level III calculations: (four-compartment model)

Chemical name: Malathion

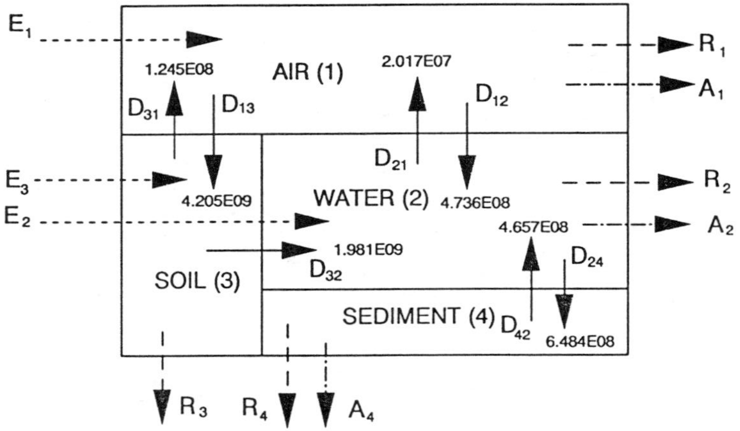

Phase Properties and Rates:

Compartment	Bulk Z mol/m3 Pa	Half-life h	D Values Reaction mol/Pa h	D Values Advection mol/Pa h
Air (1)	4.518E-04	170	1.84E+08	4.52E+08
Water (2)	4.391E+02	55	1.11E+12	8.78E+10
Soil (3)	3.484E+03	55	7.90E+11	
Sediment (4)	3.033E+03	550	1.91E+09	3.03E+07

	E(1)=1000	E(2)=1000	E(3)=1000	E(1,2,3)
Overall residence time =	77.87	73.84	79.35	77.90 h
Reaction time =	85.73	79.70	79.37	79.74 h
Advection time =	848.98	1004.56	400363	3372.15 h

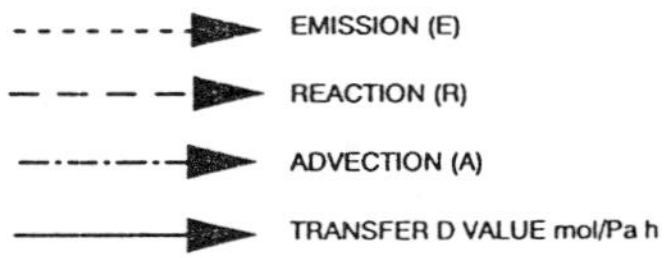

Phase Properties, Compositions, Transport and Transformation Rates:

Emission, kg/h			Fugacity, Pa				Concentration, g/m3				Amounts, kg				Total Amount, kg
E(1)	E(2)	E(3)	f(1)	f(2)	f(3)	f(4)	C(1)	C(2)	C(3)	C(4)	m(1)	m(2)	m(3)	m(4)	
1000	0	0	5.696E-07	2.308E-10	3.023E-09	6.217E-11	8.502E-08	3.348E-05	3.480E-03	6.229E-05	8.502E+03	6.695E+03	6.264E+04	3.115E+01	7.787E+04
0	1000	0	9.613E-12	2.533E-09	5.102E-14	6.824E-10	1.435E-12	3.675E-04	5.873E-08	6.838E-04	1.435E-01	7.350E+04	1.057E+00	3.419E+02	7.384E+04
0	0	1000	8.949E-11	6.370E-12	3.821E-09	1.716E-12	1.336E-11	9.241E-07	4.398E-03	1.720E-06	1.336E+00	1.848E+02	7.916E+04	8.598E-01	7.935E+04
50	250	700	2.854E-08	6.493E-10	2.826E-09	1.749E-10	4.261E-09	9.419E-05	3.253E-03	1.753E-04	4.261E+02	1.884E+04	5.855E+04	8.763E+01	7.790E+04

Emission, kg/h			Loss, Reaction, kg/h				Loss, Advection, kg/h			Intermedia Rate of Transport, kg/h T12	T21	T13	T31	T32	T24	T42
E(1)	E(2)	E(3)	R(1)	R(2)	R(3)	R(4)	A(1)	A(2)	A(4)	air-water	water-air	air-soil	soil-air	soil-water	water-sed	sed-water
1000	0	0	3.466E+01	8.436E+01	7.89E+02	3.924E-02	8.502E+01	6.695E+00	6.229E-04	8.912E+01	1.538E-03	7.913E+02	1.243E-01	1.979E+00	4.943E-02	9.565E-03
0	1000	0	5.850E-04	9.261E+02	1.33E-02	4.308E-01	1.435E-03	7.350E+01	6.838E-03	1.504E-03	1.688E-02	1.336E-02	2.098E-06	3.340E-05	5.426E-01	1.050E-01
0	0	1000	5.445E-03	2.329E+00	9.97E+02	1.083E-03	1.336E-02	1.848E-01	1.720E-05	1.400E-02	4.244E-05	1.243E-01	1.571E-01	2.501E+00	1.365E-03	2.640E-04
50	250	700	1.737E+00	2.374E+02	7.38E+02	1.104E-01	4.261E+00	1.884E+01	1.753E-03	4.466E+00	4.326E-03	3.966E+01	1.162E-01	1.849E+00	1.391E-01	2.691E-02

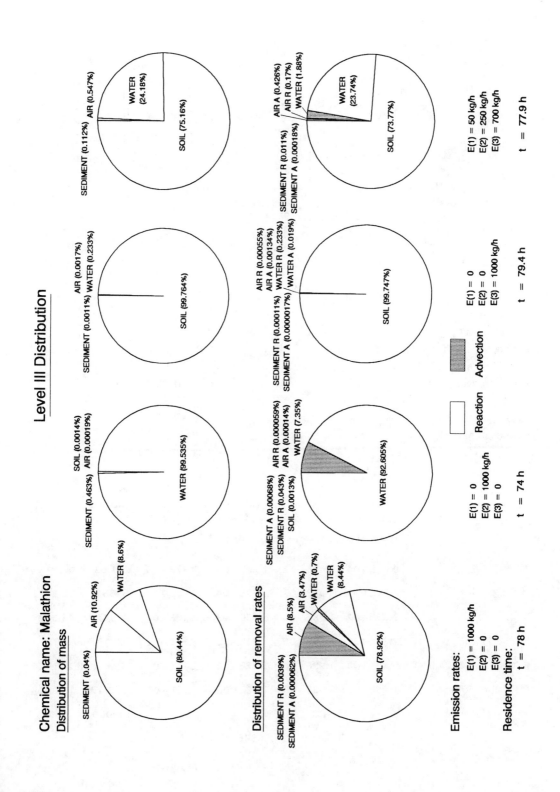

Chemical name: Methyl parathion

Level I calculation: (six-compartment model)

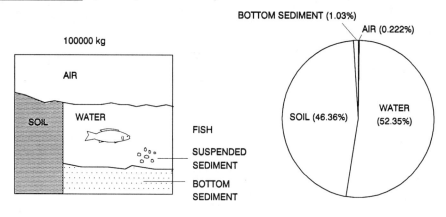

Distribution of mass

Physical-chemical properties:
molecular wt., g/mol	262.23
melting point, °C	37
solubility, g/m³	25
vapor pressure, Pa	2.0E−03
log K_{OW}	3.0
fugacity ratio, F	0.7609
dissoc. const. pK_a	

Partition coefficients:
H, Pa·m³/mol	0.0210
K_{AW}	8.46E−06
K_{OC}	410.0
BCF	50.00
K_{SW}	19.68
$K_{SD/W}$	39.36
$K_{SSD/W}$	123.00
$K_{AR/W}$	2.28E+09

COMPARTMENT	Z	CONCENTRATION			AMOUNT	AMOUNT
	mol/m³·Pa	mol/m³	mg/L, or g/m³	µg/g	kg	%
AIR	4.03E−04	8.45E−12	2.22E−09	1.87E−06	221.53	0.2215
WATER	4.77E+01	9.98E−07	2.62E−04	2.62E−04	52351	52.351
SOIL	9.38E+02	1.96E−05	5.15E−03	2.15E−03	46362	46.362
BIOTA (FISH)	2.38E+03	4.99E−05	1.31E−02	1.31E−02	2.618	nn2.62E−
SUSPENDED SEDIMENT	5.86E+03	1.23E−04	3.22E−02	2.15E−02	32.30	3.22E−02
BOTTOM SEDIMENT	1.88E+03	3.93E−05	1.03E−02	4.29E−03	1030.27	1.0303
Total					100000	100

Fugacity, $f = 2.094\text{E}-08$ Pa

Chemical name: Methyl parathion
Level II calculation: (six-compartment model)

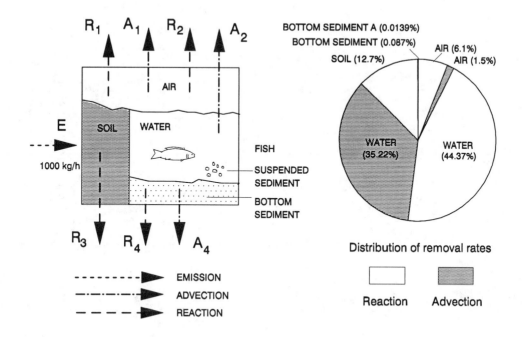

COMPARTMENT	Half-Life	D VALUES		CONC'N	LOSS	LOSS	REMOVAL
		Reaction	Advection		Reaction	Advection	%
	h	mol/Pa·h	mol/Pa·h	mol/m^3	kg/h	kg/h	
AIR	17	1.64E+08	4.03E+08	5.68E−11	60.766	14.907	7.567
WATER	550	1.20E+10	9.53E+09	6.79E−06	443.87	352.27	79.61
SOIL	1700	3.44E+09		1.32E−04	127.18		12.718
BIOTA (FISH)				3.36E−04			
SUSPENDED SEDIMENT				8.26E−04			
BOTTOM SEDIMENT	5500	2.36E+07	3.75E+06	2.64E−04	0.874	0.1387	0.1012
Total		1.71E+10	9.94E+09		632.68	367.32	100
Reaction + Advection			2.70E+10			1000	

Fugacity, f = 1.409E−07 Pa

Total amount = 672906 kg

Overall residence time = 672.91 h

Reaction time = 1063.58 h

Advection time = 1831.94 h

Fugacity Level III calculations: (four-compartment model)
Chemical name: Methyl parathion

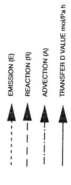

Phase Properties and Rates:

Compartment	Bulk Z mol/m3 Pa	Half-life h	D Values Reaction mol/Pa h	Advection mol/Pa h
Air (1)	4.218E-04	17	1.72E+09	4.22E+08
Water (2)	4.752E-01	550	1.20E+10	9.50E+09
Soil (3)	4.815E+02	1700	3.53E+09	
Sediment (4)	4.118E+02	5500	2.59E+07	4.12E+06

	E(1)=1000	E(2)=1000	E(3)=1000	E(1,2,3)
Overall residence time =	463.55	450.94	2321.11	1760.69 h
Reaction time =	559.55	807.88	2384.66	2041.82 h
Advection time =	2701.95	1020.64	87105	12787.66 h

▲ ---- EMISSION (E)
▲ —— REACTION (R)
▲ —·— ADVECTION (A)
▲ —··— TRANSFER D VALUE mol/Pa h

Phase Properties, Compositions, Transport and Transformation Rates:

Emission, kg/h			Fugacity, Pa				Concentration, g/m3				Amounts, kg				Total Amount, kg
E(1)	E(2)	E(3)	f(1)	f(2)	f(3)	f(4)	C(1)	C(2)	C(3)	C(4)	m(1)	m(2)	m(3)	m(4)	
1000	0	0	1.396E-06	6.627E-09	1.890E-07	6.252E-09	1.550E-07	8.289E-05	2.395E-02	6.777E-04	1.550E+04	1.658E+04	4.311E+05	3.389E+02	4.635E+05
0	1000	0	1.307E-09	1.765E-07	1.769E-10	1.665E-07	1.451E-10	2.208E-03	2.242E-05	1.805E-02	1.451E+01	4.415E+05	4.036E+02	9.025E+03	4.509E+05
0	0	1000	1.301E-08	1.007E-08	1.006E-06	9.501E-09	1.445E-09	2.519E-04	1.275E-01	1.030E-03	1.445E+02	2.519E+04	2.295E+06	5.149E+02	2.321E+06
50	250	700	7.922E-08	5.150E-08	7.137E-07	4.859E-08	8.797E-09	6.442E-04	9.046E-02	5.267E-03	8.797E+02	1.288E+05	1.628E+06	2.634E+03	1.761E+06

Emission, kg/h			Loss, Reaction, kg/h				Loss, Advection, kg/h				Intermedia Rate of Transport, kg/h						
E(1)	E(2)	E(3)	R(1)	R(2)	R(3)	R(4)	A(1)	A(2)	A(3)	A(4)	T12	T21	T13	T31	T32	T24	T42
											air-water	water-air	air-soil	soil-air	soil-water	water-sed	sed-water
1000	0	0	6.318E+02	2.089E+01	1.76E+02	4.270E-02	1.550E+02	1.658E+01	1.882E+02	6.777E-03	2.688E+01	3.516E-02	1.882E+02	1.744E+00	1.067E-01	1.338E-01	8.431E-02
0	1000	0	5.915E-01	5.563E+02	1.65E-01	1.137E+00	1.451E-01	4.415E+02	1.762E-01	1.805E-02	2.517E-02	9.363E-01	1.762E-02	1.633E-03	9.991E-03	3.563E+00	2.245E+00
0	0	1000	5.890E+00	3.174E+01	9.36E+02	6.488E-02	1.445E+00	2.519E+01	2.519E+01	1.030E-02	2.506E-01	5.342E-02	1.754E+00	9.287E+00	5.681E+01	2.033E-01	1.281E-01
50	250	700	3.586E+01	1.623E+02	6.64E+02	3.318E-01	8.797E+01	1.288E+02	1.288E+02	5.267E-02	1.526E+00	2.732E-01	1.068E+01	6.588E+00	4.030E+01	1.040E+00	6.552E-01

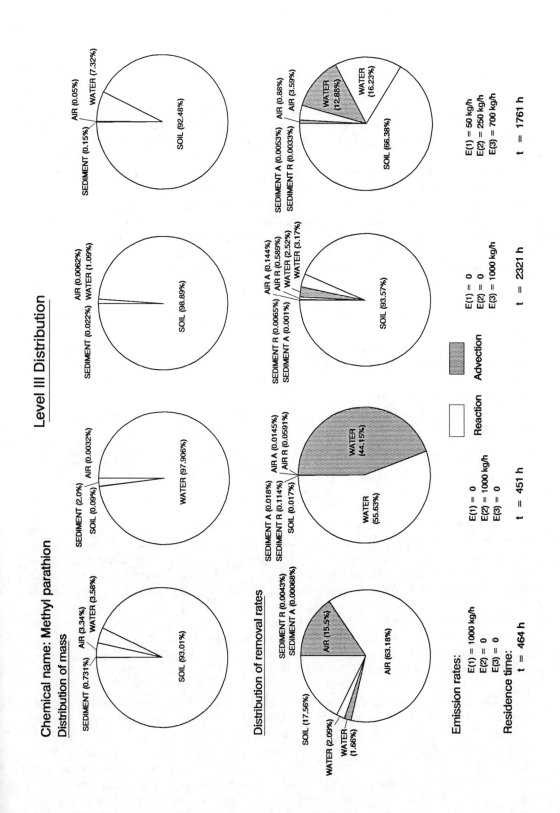

3.4 COMMENTARY ON PHYSICAL-CHEMICAL PROPERTIES AND ENVIRONMENTAL FATE

Properties and Reactivity

Selected physical-chemical properties of insecticides are given in Tables 3.1 and 3.2. Table 3.3 gives the estimated half-lives. These half-lives should be regarded as only tentative, average values or ranges. The insecticides range in structure from relatively simple chlorinated hydrocarbons to complex organo-phosphorus and nitrogen compounds, and to mixtures of uncertain composition such as toxaphene. Accordingly, their physical-chemical properties vary greatly. The requirement for absorption or uptake by insects tends to result in relatively high hydrophobicity, i.e., log K_{OW} values usually exceeding 3.0. As a result, solubilities in water are often low. Often, both vapor pressure and solubility are low thus their ratio, the Henry's law constants, can be appreciable and highly variable and uncertain. Chlorinated substances such as mirex, toxaphene, DDE, chlordane and heptachlor thus have relatively high Henry's law constants and may be subject to appreciable evaporation and subsequent long-range transport. Vapor pressure is not a good indicator of volatility, the air-octanol and air-water partition coefficients being more descriptive of the tendency to evaporate from soil and water.

Some of the insecticides, such as toxaphene are complex mixtures and the assignment of a set of properties is inherently unreliable and possibly misleading. It is only recently that the dominant chemical species have been identified. As in the case of herbicides, it is essential to understand the fundamental chemistry of these substances in aqueous solution because dissociation can have a profound effect on fate.

Evaluative Calculations

Illustrative calculations are given for DDT (which was also addressed in Chapter 1), carbaryl, chlorpyrifos, diazinon, lindane, malathion and methylparathion.

The Level I distributions are primarily between soil and water, the relative proportion being controlled by K_{OW}. For example, malathion, lindane, chlorpyrifos and DDT distribute 40%, 80%, 96% and 98% to soil respectively because of their log K_{OW} values of 2.9, 3.7, 5.0 and 6.2.

The Level II calculations show the most likely significant removal processes as dictated primarily by soil-water partitioning and the relative half-lives in these media. Even these very simple calculations reveal a wide range of overall residence times or persistencies and suggest how rates of loss by degradation and advection compare. For example, chlorpyrifos and DDT have overall residence times of 1700 h (2 months) and 23,000 h (2.6 years). Others such as malathion and carbaryl have residence times of days to weeks. Although only approximate, these calculations provide an early indication of potential for transport and subsequent impact on non-target organisms. Any substance with a residence time in excess of 2000 h (12 weeks), corresponding to a half-life of 1400 h or 8 weeks, is likely to be subject to greater regulatory scrutiny because of its environmental persistence.

The Level III calculations give a more accurate estimate of persistence because they take into account how the chemical is introduced into the environment and its susceptibility to intermedia transport. Of particular interest are the relative rates of loss by competing processes and the ultimate distribution. Comparing, for example, chlorpyrifos and DDT, inspection of the final scenario with emissions to air, water and soil suggests that 4% of the chlorpyrifos and 20% of the DDT will reside in sediments, despite neither having been discharged to sediments. DDT is more readily deposited from the atmosphere at a total rate of 28 kg/h compared with chlorpyrifos for which the corresponding figure is only 3 kg/h. Despite its much lower vapor pressure, DDT evaporates from water somewhat more rapidly than chlorpyrifos (5.9 versus 4.6 kg/h). The total advective loss of DDT is 202 kg/h, i.e., 20% of the input suggesting a potential for global transport. For chlorpyrifos the corresponding figure is 58 kg/h or 6% of the input.

It is hoped that these calculations will help to identify the combinations of properties which result in undesirable environmental persistence, susceptibility to long range transport and thus impact non-target species locally, regionally and even globally. Again the reader is urged to repeat these calculations, modifying the input data to reflect more closely the local conditions and to identify the key sensitivities in the evaluation.

3.5 REFERENCES:

Abdullah, A. R., Bajet, C. M., Matin, M. A., Nhan, D. D., Sulaiman, A. H. (1997) Ecotoxicology of pesticides in the tropical paddy field ecosystem. *Environ. Toxicol. Chem.* 16(1), 59-70.

Albert, A. (1963) Ionization constants. In: *Physical Methods in Heterocyclic Chemistry.* Katritzky, A. R., Editor, Academic Press, New York.

Adachi, K., Mitsuhashi, T., Ohkuni, N. (1984) Pesticides and trialkyl phosphates in tap water. *Hyogo-Ken Eisei Kenkyusho Kenkyu Hokoku* 19, 1-6.

Adams, Jr., R. S. (1967) The fate of pesticide residues in soil. *J. Minn. Acad. Sci.* 34, 44-48.

Adams, W. J. (1987) Chapter 16, Bioavailability of neutral lipophilic organic chemicals contained on sediments: A review. In: *Fate and Effects of Sediment-Bound Chemicals in Aquatic Systems.* Dickson, K. L., Maki, A. W., Brungs, W. A., Eds., Pergamon Press, New York.

Addison, J. B. (1981) Measurement of of vapour pressures of fenitrothion and metacil. *Chemosphere* 10, 355-364.

Adhya, T. K., Sudhakar-Barik, R., Sethunathan, N. (1981) Fate of fenitrothion, methyl parathion and parathion in anoxic sulfur-containing soil systems. *Pest. Biochem. Phys.* 16, 14.

Agency for Toxic Subtances and Disease Registry (1988) Toxicological profile for chlordane. Agency for Toxic Subtances and Disease Registry, Atlanta, Georgia.

Agency for Toxic Subtances and Disease Registry (1988) Toxicological profile for DDT, DDE, and DDD. Agency for Toxic Subtances and Disease Registry, Atlanta, Georgia.

Agrochemicals Handbook (1987) *The Agrochemicals Handbook.* 2nd Edition, Hartley, D., Kidd, H., Eds., The Royal Society of Chemistry, The University, Nottingham, England.

Ahmad, A., Walgenbach, D. D., Sutter, G. R. (1979) Degradation rates of technical carbofuran and a granular formulation in four soils with known insecticide use history. *Bull. Environ. Contam. Toxicol.* 23(4/5), 572-574.

Alexander, M. (1973) Nonbiodegradable and other recalcitrant molecules. *Biotech. Bioeng.* 15, 611-647.

Ali, S. (1978) Degradation and Environmental Fate of Endosulfan Isomers and Endosulfan Sulfate in Mouse, Insect and Laboratory Ecosystem. *Diss. Abstr. Int. B.* 39(5), 2117., Ph.D. Thesis, University of Illinois.

Alison, D. T., Hermanutz, R. O. (1977) Toxicity of Diazinon to Brook Trout and Fathead Minnows. EPA 600/3-77-060, U.S. EPA, Duluth, Minnesota.

Alley, E. G. (1973) The use of mirex in control of the imported fire ant. *J. Environ. Quality* 2(1), 52-61.

Aly, O. M., El-Dib, M. A. (1971) Studies on the persistence of some carbamate insecticides in the aquatic environment. I. Hydrolysis of sevin, baygon, pyrolan, and dimetilan in waters. *Water Res.* 5, 1191-1205.

Anderson, R. L., Defoe, D. L. (1980) Toxicity and bioaccumulation of endrin and methoxychlor in aquatic invertebrates and fish. *Environ. Pollut. Ser.* A22, 111.

Andrews, A. K., Van Valin, C. C., Stebbings, B. E. (1966) Some effects of heptchlor on bluegills (*Lepomis macrochirus*). *Trans. Amer. Fish Soc.* 95, 297.

Argyle, R. L., Williams, G. C., Dupree, H. K. (1973) Endrin uptake and release by fingerling channel catfish (*Ictalurus puncatus*). *J. Fish Res. Board Can.* 30, 1743.

Ashton, F. M., Crafts, A. S. (1981) *Mode of Action of Herbicides.* John Wiley & Sons, New York.

Atkins, D. H. F., Eggleton, A. E. J. (1971) Studies of atmospheric wash-out and deposition of γ-BHC, dieldrin and *p,p'*-DDT using radio-labelled pesticides. In: *Proc. Symp. on Nucl. Tech. Environ. Pollut.*, pp. 521-533, Vienna.

Atkinson, R. (1985) Kinetics and mechanisms of the gas-phase reactions of hydroxyl radicals with organic compounds under atmospheric conditions. *Chem. Rev.* 85, 69-201.

Atkinson, R. (1987) Structure-activity relationship for the estimation of rate constants for the gas-phase reactions of OH radicals with organic compounds. *Int'l. J. Chem. Kinetics* 19, 799-828.

Atkinson, R., Carter, W. P. L. (1984) Kinetics and mechanisms of the gas-phase reactions of ozone with organic compounds under atmospheric conditions. *Chem. Rev.* 84, 437-470.

Atkinson, R., Kwok, E. S. C., Arey, J. (1992) Photochemical processes affecting the fate of pesticides in the atmosphere. *Brighton Crop Prot. Conf. - Pests Dis.* 2, 469-476.

Atlas, E. Foster, R., Giam, C. S. (1982) Air-sea exchange of high molecular weight organic pollutants: laboratory studies. *Environ. Sci. Technol.* 16, 283-286.

Augustijn-Beckers, P. W. M., Hornsby, A. G., Wauchope, R. D. (1994) The SCS/ARS/CES pesticide-properties database for environmental decision - making. II. Additional compounds. *Rev. Environ. Contam. Toxicol.* 137, 1-82.

Babers, F. H. (1955) The solubility of DDT in water determined radiometrically. *J. Am. Chem. Soc.* 77, 4666.

Bacci, E., Calamari, D., Gaggi, C., Vighi, M. (1990) Bioconcentration of organic chemical vapors in plant leaves: Experimental measurements and correlation. *Environ. Sci. Technol.* 24, 885-889.

Bacci, E., Gaggi, C. (1987) Chlorinated hydrocarbon vapours and plant foliage: Kinetics and applications. *Chemosphere* 16, 2515-2522.

Bahnick, D. A., Doucette, W. J. (1988) Use of molecular connectivity indices to estimate soil sorption coefficients for organic chemicals. *Chemosphere* 17, 1703-1715.

Baker, M. D., Mayfield, C. I. (1980) Microbial and non-biological decomposition of chlorophenols and phenols in soil. *Water Air Soil Pollut.* 13, 411.

Baker, R. A., Editor (1991) *Organic Substances and Sediments in Water*. Vol. 1 and 2. Lewis Publishers, Inc., Chelsea, Michigan.

Baker, R. D., Applegate, H. G. (1970) Effect of temperature and ultraviolet radiation on the persistence of methyl parathion and DDT in soils. *Argon. J.* 62, 509.

Ballschmiter, K., Wittlinger, R. (1991) Interhemisphere exchange of hexachlorohexanes, hexachlorobenzene, polychlorobiphenyls and 1,1,1-trichloro-2,2-bis(*p*-chlorophenyl)-ethane in the lower troposphere. *Environ. Sci. Technol.* 25, 1103-1111.

Balson, E. W. (1947) Studies in vapour pressure measurement. III. An infusion manometer sensitive to 5×10^{-4} mmHg. Vapour pressures of DDT and other slightly volatile substances. *Trans. Faraday Soc.* 43, 54-60.

Banerjee, S., Baughman, G. L. (1991) Bioconcentration factors and lipid solubility. *Environ. Sci. Technol.* 25, 536-539.

Banerjee, S., Howard, P. H., Rosenberg, A. M., Dombrowski, A. E., Sikka, H., Tullis, D. L. (1984) Development of general kinetic model for biodegradation and its application to chlorophenols and related compounds. *Environ. Sci. Technol.* 18, 416-422.

Banson, D.R. (1978) Predicting the fate of chemicals in the aquatic environment from laboratory data. In: *Estimating the Hazard of Chemical Sbstances to Aquatic Life*. ASTM STP 657, Cairns, Jr., J., Dickson, K. L., Maki, A. W., Eds., pp. 55-70, American Society for Testing and Materials, Philadelphia, Pennsylvania.

Barlow, F. (1978) Presented at the 4th International Congress of Pesticide Chemistry. Zurich, Switzerland, July 24-28, 1978.

Baron, R. L., Copeland, F., Walton, M. S. (1975) In: *Environment Quality and Safety*. Supplement Vol. III., Coulston, F., Korte, F., Editors., G. Thieme Publishers, Stuttgart. Pesticides Lectures held at the IUPAC 3rd International Congress of Pesticide Chemistry, Helsinki, July 3-9, 1974.

Baron, R. L., Walton, M. S. (1971) Dynamics of HEOD (dieldrin) in adipose tissue of the rats. *Toxicol. Appl. Pharmacol.* 18, 958-963.

Barron, M. G., Plakas, S. M., Wilga, P. C. (1991) Chlorpyrifos pharmacokinetics and metabolism following intravascular and dietary administration in channel catfish. *Toxicol. Appl. Pharmacol.* 108, 474-482.

Battersby, N. S. (1990) A review of biodegradation kinetics in the aquatic environment. *Chemosphere* 21(10-11), 1243-1284.

Baughman, G. L., Lassiter, R. R. (1978) Prediction of enviromental pollutant concentration. In: *Estimating the Hazard of Chemical Substances to Aquatic Life.* ASTM STP 657, Cairns, Jr., J., Dickson, K. L., Maki, A. W., Editors, American Society for Testing and Materials, Philadelphia, Pennsylvania.

Baughman, G. L., Paris, D. F. (1981) Microbial bioconcentration of organic pollutants from aquatic systems-A critical review. *CRC Critical Review in Microbiology.* CRC Press, Inc., Boca Raton, Florida.

Beall, M. L., Nash, R. G. (1972) Insecticide depth in soil - Effect on soyabean uptake in the greenhouse. *J. Environ. Qual.* 1, 283-288.

Beck, E. W., Johnson, Jr., J. C., Woodham, D. W., Leuck, D. B., Dawsey, L. H., Robbins, J. E., Bowman, M. C. (1966) *J. Econ. Entomol.* 59, 1444.

Behrendt, H., Brüggemann, R. (1993) Modeling the fate of organic chemicals in the soil plant environment: Model study of root uptake of pesticides. *Chemosphere* 27(12), 2325-2332.

Bellin, C. A., O'Connor, G. A., Yin, Y. (1990) Sorption and degradation of pentachlorophenol in sludge-amended soils. *J. Environ. Qual.* 19, 603-608.

Belluck, D., Felsot, A. (1981) Bioconcentration of pesticides by egg masses of the caddisfly, *Triaenodes tardus* milne. *Bull. Environ. Contam. Toxicol.* 26, 299-306.

Beltrame, P., Beltrame, P. L., Cartini, P. (1984) Inhibiting action of chloro- and nitro-phenols on biodegradation of phenols: A structure-toxicity relationship. *Chemosphere* 13, 3-9.

Beltrame, P., Beltrame, P. L., Cartini, P. (1984) Influence of feed concentration on the kinetics of biodegradation of phenol in a continuous stirred reactor. *Water Res.* 18, 403-407.

Beltrame, P., Beltrame, P. L., Cartini, P., Guardione, D., Lanzetta, C. (1988) Inhibiting action of chlorophenols on biodegradation of phenol and its correlation with structured structural properties of inhibitors. *Biotechn. Bioeng.* 31, 821-828.

Beltrame, P., Beltrame, P. L., Cartini, P., Lanzetta, C. (1988) New results on the inhibiting action of chloro- and nitro-substituted phenols on the biodegradation of phenol. *Chemosphere* 17, 235-242.

Bender, M. E. (1969) Uptake and retention of malathion by the carp. *Prog. Fish-Cult.* 31, 155-159.

Bennett, H. J., Day, J. W. (1970) Absorption of endrin by the bluegill sunfish, *Lepomis macrochirus*. *Pest. Monit. J.* 3, 201.

Berdanier, C. D., de Dennis, S. K. (1977) Effect of exercise on the responses of rats to DDT. *J. Toxicol. Environ. Health* 2, 651-656.

Beste, C. E., Humburg, N. E. (1983) *Herbicide Handbook of the Weed Science Society of America.* 5th Edition, Weed Science Society, Champaign, Illinois.

Bevenue, A., Beckman, H. (1967) Pentachlorophenol: A discussion of its properties and its occurrence as a residue in human and animal tissues. *Res. Rev.* 19, 83-134.

Bhavnagary, H. M., Jayaram, M. (1974) Determination of water solubilities of lindane and dieldrin at different temperatures. *Bull. Grain Technol.* 12(2), 95-99.

Biddinger, G. R., Gloss, S. P. (1984) The importance of trophic transfer in the bioaccumulation of chemical contaminants in organic ecosystem. *Res. Rev.* 91, 103-145.

Bidleman, T. F. (1984) Estimation of vapor pressures for nonpolar organic compounds by capillary gas chromatography. *Anal. Chem.* 56, 2490-2496.

Bidleman, T. F., Billings, W. N., Foreman, W. T. (1986) Vapor-particle partitioning of semivolatile organic compounds-Estimation from field collections. *Environ. Sci. Technol.* 20(10), 1038-1043.

Bidleman, T. F., Christensen, E. J. (1979) Atmospheric removal processes for high molecular weight organochlorines. *J. Geophys. Res.* 84(C12), 7857-7862.

Bidleman, T. F., Foreman, W. T. (1987) 2. Vapor-particle partitioning of semivolatile organic compounds. In: *Sources and Fates of Aquatic Pollutants.* Hites, R. A., Eisenreich, S. J., Editors, pp. 27-56, Advances Chemistry Series 216, American Chemical Society, Washington DC.

Bidleman, T. F., Renberg, L. (1985) Determination of vapor pressures for chloroguaiacols, chloroveratrols, and nonylphenol by gas chromatography. *Chemosphere* 14, 1475-1481.

Bierman, V. Swain, W. (1982) Mass balance modeling of DDT dynamics in Lakes Michigan and Superior. *Environ. Sci. Technol.* 16, 572-579.

Biggar, J. W., Riggs, I. R. (1974) Apparent solubility of organochlorine insecticides in water at various temperatures. *Hilgardia* 42(10), 383-391.

Biggar, J. W., Doneen, L. D., Riggs, I. R. (1966) Soil interaction with organically polluted water. Summary Report, Dept. of Water Science and Engineering, University of California, Davis, California.

Biggar, J. W., Dutt, G. R., Riggs, I. R. (1967) Predicting and measuring the solubility of p,p'-DDT in water. *Bull. Environ. Contam. Toxicol.* 2(3), 90.

Bilkert, J. N., Rao, P. S. C. (1985) Sorption and leaching of three nonfumigant nematocides in soils. *J. Environ. Sci. Health* B20, 1-26.

Bintein, S., Devillers, J. (1994) QSAR for organic chemical sorption in soils and sediments. *Chemosphere* 28, 1171-1188.

Bishop, W. E., Maki, A. W. (1980) A critical comparison of two bioconcentration test methods. In: *Aquatic Toxicology.* ASTM STP 707, pp. 61-77, American Society for Testing and Materials, Philadelphia, Pennsylvania.

Blackman, G. E., Parke, M. H., Garton, G. (1955) The physiological activity of substituted phenols. I. Relationships between chemical structure and physiological activity. *Arch. Biochem. Biophys.* 54(1), 55-71.

Blum, D. J. W., Suffet, I. H., Duguet, J. P. (1994) Quantitative structure-activity relationship using molecular connectivity for the activated carbon adsorption of organic chemicals in water. *Water Res.* 28, 687-699.

Boehncke, A., Seiber, J., Nolting, H.-G. (1990) Investigations of the evaporation of selected pesticides from natural and model surfaces in field and laboratory. *Chemosphere* 21(9), 1109-1124.

Boehncke, A., Martin, K., Müller, M. G., Cammenga, H. K. (1996) The vapor pressure of lindane (γ-1,2,3,4,5,6-hexachlorocyclohexane) - A comparison of Knudsen effusion measurements with data from other techniques. *J. Chem. Eng. Data* 41, 543-545.

Bomberger, D. C., Gwinn, J. L., Mabey, W. R., Tus, D., Chou, T. W. (1983) Environmental fate and transport at the terrestrial-atmospheric interface. *ACS Symp. Ser.* 225, 197-214.

Bond, C. A., Woodham, D. W., Ahrens, E. H., Medley, J. G. (1975) The accumulation and disappearance of mirex residues. II. In milk and tissues of cows fed two concentrations of the insecticide in their diet. *Bull. Environ. Contam. Toxicol.* 14, 25-31.

Bottoni, P., Funari, E. (1992) Criteria for evaluating the impact of pesticides on groundwater quality. *Sci. Total Environ.* 123/124, 581-590.

Boublik, T., Fried, V., Hala, E. (1984) *The Vapor Pressures of Pure Substances.* 2nd Edition, Elsevier, Amsterdam, The Netherlands.

Bourquin, A. W., Garnas, R. I., Pritchard, P. H., Wilkes, F. G., Cripe, C. R., Rubinstein, N. I. (1979) Interdependent microcosms for the assessment of pollutants in the marine environment. *Int. J. Environ. Studies* 13, 131-140.

Bowery, T. G. (1964) In: *Analytical Methods for Pesticides, Plant Growth Regulators, and Food Additives.* Vol. 2, Zweig, G., Editor, Academic Press, New York.

Bowman, B. T., Sans, W. W. (1979) The aqueous solubility of twenty-seven insecticides and related compounds. *J. Environ. Sci. Health* B14(6), 625-634.

Bowman, B. T., Sans, W. W. (1983a) Further water solubility determination of insecticidal compounds. *J. Environ. Sci. Health* B18(2), 221-227.

Bowman, B. T., Sans, W. W. (1983b) Determination of octanol-water partitioning coefficients (K_{ow}) of 61 organophosphorous and carbamate insecticides and their relationship to respective water solubility (S) values. *J. Environ. Sci. Health* B18(6), 667-683.

Bowman, M. C., Acree, Jr., F., Corbett, M. K. (1960) Solubility of carbon-14 DDT in water. *J. Agric. Food Chem.* 8(5), 406-408.

Bowman, M. C., Schechter, M. S., Carter, R. L. (1965) Behavior of chlorinated insecticides in a broad spectrum of soil types. *J. Agric. Food Chem.* 13 360-365.

Branson, D. R. (1978) Predicting the fate of chemicals in the aquatic environment from laboratory data. pp. 55-70. In: *Estimating the Hazard of Chemical Substances to Aquatic Life.* ASTM STP 657, Cairns, Jr., J., Dickson, K. L., Maki, A. W., Editors, American Society for Testing and Materials, Philadelphia, Pennsylvania.

Brecken-Folse, J. A., Mayer, F. L., Pedigo, L. E., Marking, L. L. (1994) Acute toxicity of 4-nitrophenol, 2,4-dinitrophenol, terbufos and trichlorfon to grass shrimp (*Palaemonetes* spp.) and sheepshead minnows (*Cyprinodon variegatus*) as affected by salinity and temperature. *Environ. Toxicol. Chem.* 13, 67-77.

Briggs, G. G. (1973) A simple relationship between soil adsorption of organic chemicals and their octanol/water partitioning coefficients. *Proc. 7th British Insecticide and Fungicide Conference* 1, 83-86.

Briggs, G. G. (1981) Theoretical and experimental relationships between soil adsorption, octanol-water partition coefficients, water solubilities, bioconcentration factors, and the Parachor. *J. Agric. Food Chem.* 29, 1050-1059.

Bright, N. F. H., Cuthill, J. C., Woodbury, N. H. (1950) The vapor pressure of parathion and related compounds. *J. Sci. Food Agric.* 1, 344.

Brockway, D. L., Smith, P. D., Stancil, F. E. (1984) Fates and effects of pentachlorophenol in hard- and soft-water microcosms. *Chemosphere* 13(12), 1363-1377.

Bromilow, R. H., Baker, R. J., Freeman, M. A. H., Görög, K. (1980) The degradation of aldicarb and oxamyl in soil. *Pest. Sci.* 11(4), 389-395.

Bromilow, R. H., Leistra, M. (1980) Measured and simulated behavior of aldicarb and its oxidation products in fallow soils. *Pest. Sci.* 11, 389-395.

Brooke, D., Nielsen, I., De Bruijn, J., Hermens, J. (1990) An interlaboratory evaluation of the stir-flask method for the determination of octanol water partition coefficients. (LOG POW). *Chemosphere* 21, 119-133.

Brooke, D. N., Dobbs, A. J., Williams, N. (1986) Octanol/water partition coefficients (P): Measurement, estimation, and interpretation, particularly for chemicals with $P > 10^5$. *Ecotox. Environ. Saf.* 11, 251-260.

Brooks, G. T. (1974) *Chlorinated Insecticides*: Volume I: Technology and Applications. CRC Press, Cleveland, Ohio.

Bro-Rasmussen, F., Noddegaard, E., Voldum-Claussen, K. (1970) Comparison of the disappearance of eight organophosphorus insecticides from soil in laboratory and in outdoor experiments. *Pest. Sci.* 1, 179-182.

Broto, P., Moreau, G., Vandycke, C. (1984) Molecular structures: Perception, autocorrelation descriptor and SAR studies. System of atomic contribution for the calculation of the n-octanol/water partition coefficients. *Eur. J. Med. Chem. Chim. Term.* 19, 71-78.

Bruggeman, W. A., Martron, L. B. J. M., Kooiman, D., Hutzinger, O. (1981) Accumulation and elimination kinetics of di-, tri-, and tetrachlorobiphenyls by goldfish after dietary and aqueous exposure. *Chemosphere* 10, 811-832.

Brusseau, M. L., Rao, P. S. C. (1989) The influence of sorbate-organic matter interactions on sorption nonequilibrium. *Chemosphere* 18, 1691-1706.

Brust, H. F. (1966) A summary of chemical and physical properties of Dursban. *Down to Earth* 22(3), 21-22.

Budavari, S., Editor (1989) *The Merck Index. An Encyclopedia of Chemicals, Drugs and Biologicals.* 11th Edition, Merck and Co., Inc., Rahway, New Jersey.

Bunce, N. J., Nakai, J. S., Yawching, M. (1991) A model for estimating the rate of chemical transformation of a VOC in the troposphere by two pathways: Photolysis by sunlight and hydroxyl radical attack. *Chemosphere* 22, 305-315.

Burchfield, H. P., Johnson, D. F. (1965) *Guide to the Analysis of Pesticide Residues.* Vol. II, U.S. HEW, Washington DC.

Burkhard, L. P., Kuehl, D. W., Veith, G. D. (1985) Evaluation of reverse phase liquid chromatography/mass spectrometry for estimation of N-octanol/water partition coefficients. *Chemosphere* 14, 1551-1560.

Burkhard, N., Guth, J. A. (1979) Photolysis of organophosphorous insecticides on soil surfaces. *Pest. Sci.* 10(4), 313-319.

Burkhard, N., Guth, J. A. (1981) Rate of volatilisation of pesticides from soil surfaces: Comparison of calculated results with those determined in a laboratory model system. *Pest. Sci.* 12(1), 37-44.

Burmaster, D. E., Menzie, C. A., Freshman, J. S., Burris, J. A., Maxwell, N. I., Drew, S. R. (1991) Assessment of methods for estimating aquatic hazards at superfund-type sites: A cautionary tale. *Environ. Toxicol. Chem.* 10, 827-842.

Burns, S. E., Hassett, J. P., Rossi, M. V. (1996) Binding effects on jumic-mindiated photoreaction: intrahumic dechlorination of mirex in water. *Environ. Sci. Technol.* 30, 2934-2941.

Buser, H-R., Müller, M. D. (1995) Isomer and enantioselective degradation of hexachlorocyclohexane isomers in sewage sludge under anaerobic conditions. *Environ. Sci. Technol.* 29, 664-672.

Butler, P. A. (1971) Influence of pesticides on marine ecosystems. *Proc. Roy. Soc. London* (Ser. B) 177, 321-329.

Butler, L. C., Stauff, D. C., Davis, R. L. (1981) Methyl parathion persistence in soil following simulated spillage. *Arch. Environ. Contam. Toxicol.* 10, 451-458.

Butte, W., Fox, K., Zauke, G. P. (1991) Kinetics of bioaccumulation and clearance of isomeric hexachlorocyclohexanes. *Sci. Total Environ.* 109/110, 377-382.

Butte, W., Fooken, C., Klussman, R., Schuller, D. (1981) Evaluation of lipophilic properties for a series of phenols, using reversed-phase high performance liquid chromatography and high-performance thin-layer chromatography. *J. Chromatogr.* 214, 59-67.

Butte, W., Willing, A., Zanke, G. P. (1987) Bioaccumulation of phenols in zebrafish determined by a dynamic flow through test. In: *QSAR in Environmental Toxicology II.* Kaiser, K. L. E., pp. 43-53, Editor, D. Reidel Publishing Company, Dordrecht, The Netherlands.

Buxton, G. V., Greenstock, C. L., Helman, W. P., Ross, A. B. (1988) Critical review of rate constants for reactions of hydrated electrons, hydrogen atoms and hydroxyl radicals ($\cdot$OH/$\cdot$O$^-$) in aqueous solution. *J. Phys. Chem. Ref. Data* 17, 513-886.

Bysshe, S.E. (1982) Chapter 5, Bioconcentration factor in aquatic organisms. In: *Handbook on Chemical Property Estimation Methods, Environmental Behavior of Organic Compounds.* Lyman, W. J., Reehl, W. F., Rosenblatt, D. H., Editors, McGraw-Hill, New York.

Calamari, D., Bacci, E., Forcardi, S., Gaggi, C., Morosini, M., Vighi, M. (1991) Role of plant biomass in the global environmental partitioning of chlorinated hydrocarbons. *Environ. Sci. Technol.* 25, 1489-1495.

Call, D. J., Brooke, L. T., Lu, P. Y. (1980) Uptake, elimination and metabolism of three phenols by fathead minnows. *Arch. Environ. Contam. Toxicol.* 9, 699-714.

Callahan, M. A., Slimak, M. W., Gabel, N. W., May, I. P., Fowler, C. F., Freed, J. R., Jennings, P., Durfee, R. L., Whitmore, F. C., Maestri, B., Mabey, W. R., Holt, B. R., Gould, C. (1979) *Water-Related Environmental Fate of 129 Priority Pollutants.* Vol.1, EPA Report No. 440/4-79-029a, Versar, Inc., Springfield, Virginia.

Canton, J. H., Greve, P. A., Sloof, W., van Esch, G. J. (1975) Toxicity accumulation and elimination studies of alpha-hexacyclohexane (alpha-HCH) with fresh water organisms of different trophic levels. *Water Res.* 9, 1163-1169.

Carlberg, G. E., Martinsen, K., Kringstad, A., Gjessing, E., Grande, M., Källqvist, T., Skåre, J. U. (1986) Influence of aquatic humus on the bioavailability of chlorinated micropollutants in atlantic salmon. *Arch. Environ. Contam. Toxicol.* 15, 543-548.

Carlo, C. P., Ashdown, D., Heller, V. G. (1952) The persistence of parathion, toxaphene and methoxychlor in soil. *Okla. Agric. Exp. Stn. Tech. Bull.* No. T-42, 3-11.

Caro, J. H., Taylor, A. W., Freeman, H. P. (1976) Comparative behaviour of dieldrin and carbofuran in the field. *Arch. Environ. Contam. Toxicol.* 3, 437-447.

Caron, G., Suffet, I. H., Belton, T. (1985) Effect of dissolved organic carbon on the environmental distribution of nonpolar organic compounds. *Chemosphere* 14, 993-1000.

Carringer, R. D., Weber, J. B., Monaco, T. J. (1975) Adsorption-desorption of selected pesticides by organic matter and Montmorillonite. *J. Agric. Food Chem.* 23(3), 568-572.

Carsel, R. F. (1989) Hydrologic processes affecting the movement of organic chemicals in soils. In: *Reactions and Movement of Organic Chemicals in Soils.* SSSA Special Publication No. 22, Sawhney, B. L., Brown, K., Ed., pp. 439-445, Soil Sci. Soc. of America and Soc. of Agronomy, Madison, Wisconsin.

Carswell, T, G., Nason, H. K. (1938) Properties and uses of pentachlorophenol. *Ind. Eng. Chem.* 30, 622-626.

Casida, J. E., Gatterdam, P. E., Getzin, Jr., L. W., Chapman, R. K. (1956) Residual properties of the systemic insecticide O,O-dimethyl 1-carbomethoxy-1-propen-2-yl phosphate. *J. Agric. Food Chem.* 4(3), 236-243.

Castro, T. F., Yoshida, T. (1974) Effect of organic matter on the biodegradation of some organochlorine insecticides in submerged soils. *Soil Sci. Plant Nutri.* 20, 363-370.

Castro, T. F., Yoshida, T. (1971) Degradation of organochlorine insecticides in flooded soils in the Philippines. *J. Agric. Food Chem.* 19, 1168-1170.

Cessna, A. J., Grover, R. (1978) Spectrophotometric determination of dissociation constants of selected acidic herbicides. *J. Agric. Food Chem.* 26, 289-292.

Chapman, P. M. (1989) Review of current approaches to developing sediment quality criteria. *Environ. Toxicol. Chem.* 8, 589-599.

Chapman, R. A., Cole, C. M. (1982) Observations on the influence of water and soil pH on the persistence of insecticides. *J. Environ. Sci. Health* B17, 487-504.

Chen, F., Holten-Andersen, J., Tyle, H. (1993) New developments of the UNIFAC model for environmental application. *Chemosphere* 26, 1325-1354.

Cheng, H. H., Editor (1990) *Pesticides in the Soil Environment: Processes, Impacts, and Modeling.* Soil Science Society of America, Inc., Madison, Wisconsin.

Chessells, M., Hawker, D. W., Connell, D. W. (1992) Influence of solubility on bioconcentration of hydrophobic compounds. *Ecotoxicol. Environ. Saf.* 23, 260-273.

Cheung, B. (1984) Environmental modelling studies of hazardous substances in Ontario. M. A. Sc. Thesis, University of Toronto, Toronto, Canada.

Chigareva, O. I. (1973) Metaphos distribution in fish organs and tissues. *Tr., Vses. Nauchno-Issled. Inst. Vet. Sanit.* 46, 102.

Chin, Y. P., Weber, Jr., W. J., Chiou, C. T. (1991) Chapter 14, A thermodynamic partition model for binding of nonpolar organic compounds by organic colloids and implications for their sorption to soils and sediments. In: *Organic Substances and Sediments in Water.* Vol. 1. Baker, R. A., Editor, pp. 251-273, Lewis Publishers, Inc., Chelsea, Michigan.

Chin, Y. P., Weber, Jr., W. J., Voice, T. C. (1986) Determination of partition coefficients and aqueous solubilities by reverse phase chromatography-II. *Water Res.* 20(11), 1443-1450.

Chiou, C. T. (1981) Partition coefficient and water solubility in environmental chemistry. In: *Hazard Assessment of Chemicals. Current Development.* Vol. 1, Saxena, J., Fisher, F., Editors, pp. 117-153, Academic Press, Inc., New York.

Chiou, C. T. (1985) Partition coefficients of organic compounds in lipid-water systems and correlations with fish concentration factors. *Environ. Sci. Technol.* 19, 57-62.

Chiou, C. T., Freed, V. H. (1977) *Chemodynamic Studies on Bench Mark Industrial Chemicals.* NSF/RA-770286 prepared for Research Applied to National Needs, National Science Foundation, Washington DC.

Chiou, C. T., Freed, V. H., Schmedding, D. W., Kohnert, R. (1977) Partition coefficient and bioaccumulation of selected organic chemicals. *Environ. Sci. Technol.* 11(5), 475-478.

Chiou, C. T., Freed, V. H., Peters, L. J., Kohnert, R. L. (1980) Evaporation of solutes from water. *Environ. Internat.* 3, 231-236.

Chiou, C. T., Kile, D. E., Brinton, T. I., Malcolm, R. L., Leenheer, J. A., MacCarthy, P. (1987) A comparison of water solubility enhancements of organic solutes by aquatic humic materials and commercial humic acids. *Environ. Sci. Technol.* 21, 1231.

Chiou, C. T., Kile, D. E., Rutherford, D. W. (1991) The natural oil in commercial linear alkylbenzenesulfonate and its effect on organic solute solubility in water. *Environ. Sci. Technol.* 25, 660-665.

Chiou, C. T., Malcolm, R. L., Brinton, T. I., Kile, D. E. (1986) Water solubility enhancement of some organic pollutants and pesticides by dissolved humic and fulvic acids. *Environ. Sci. Technol.* 20, 502-508.

Chiou, C. T., Peters, L. J., Freed, V. H. (1979) A physical concept of soil-water equilibria for nonionic organic compounds. *Science* 206, 831-832.

Chiou, C. T., Porter, P. E., Schmedding, D. W. (1983) Partition equilibria of nonionic organic compounds between soil organic matter and water. *Environ. Sci. Technol.* 17, 227-231.

Chiou, C. T., Schmedding, D. W. (1981) Measurement and interrelation of octanol-water partition coefficient and water solubility of organic chemicals. In: *Test Protocols for Environmental Fate and Movement of Toxicants.* J. Assoc. Anal. Chem., Arlington, Virginia.

Chiou, C. T., Schmedding, D. W., Manes, M. (1982) Partitioning of organic compounds in octanol-water system. *Environ. Sci. Technol.* 16, 4-10.

Choi, W.-W., Chen, K. Y. (1976) Associations of chlorinated hydrocarbons with fine particles and humic substances in nearshore surficial sediments. *Environ. Sci. Technol.* 10(8), 782-786.

Claborn, H. W., Bowers, J. W., Wells, R. W., Redeleff, R. D., Nickerson, W. J. (1953) Meat contamination from pesticides. *Agric. Chem.* 8, 37-39, 119, 121.

Claborn, H. W., Redeleff, R. D., Bushland, R. C. (1953) *Pesticide Residue in Meat and Milk.* Agriculture Research Service, U.S. Dept. of Agriculture, Washington DC.

Clark, J. R., Goodman, L. R., Borthwick, P. W., Patrick, Jr., J. M., Cripe, G. M., Moody, P. M., Moore, J. C., Lores, E. M. (1989) Toxicity of pyrethroids to marine invertebrates and fish: A literature review and test results with sediment-sorbed chemicals. *Environ. Toxicol. Chem.* 8, 393-401.

Clark, T., Clark, K., Patterson, S., Norstrom, R., Mackay, D. (1988) Wildlife monitoring, modeling, and fugacity. *Environ. Sci. Technol.* 22, 120-127.

Clement Associates (1983) Chemical physical, and biological properties of compounds presented at hazardous waste sites. Report prepared for the U. S. Environmental Protection Agency, Washington DC.

Cliath, M. M., Spencer, W. F. (1972) Dissipation of pesticides from soil by volatilization of degradation products. I. Lindane and DDT. *Environ. Sci. Technol.* 6, 910-914.

CLOGP (1986) Medchem Project of Pomona College, Claremont, California.

Coats, J. R., O'Donnell-Jeffery, N. L. (1979) Toxicity of four synthetic pyrethroid insecticides to rainbow trout. *Bull. Environ. Contam. Toxicol.* 23, 250-258.

Cohen, M. L., Steinmetz W. D. (1986) Foliar washoff of pesticides by rainfall. *Environ. Sci. Technol.* 20, 521-523.

Comba, M. E., Nostrom, R. J., Macdonald, C. R., Kaiser, K. L. E. (1993) A Lake Ontario-Gulf of St. Lawrence dynamic mass budget for mirex. *Environ. Sci. Technol.* 27, 2198-2206.

Connell, D. W., Bowman, M., Hawker, D. W. (1988) Bioconcentration of chlorinated hydrocarbons from sediment by oligochaetes. *Ecotoxicol. Environ. Saf.* 16, 293-302.

Connell, D. W., Hawker, D. W. (1986) Bioconcentration of lipophilic compounds by some aquatic organisms. *Ecotoxicol. Environ. Saf.* 11, 184-197.

Connell, D. W., Markwell, R. D. (1990) Bioaccumulation in the soil to earthworm system. *Chemosphere* 20(1-2), 91-100.

Conte, F. S., Parker, J. C. (1975) Effect of aerially-applied malathion on juvenile brown and white shrimp *Penaeus aztecus* and *Penaeus setiferus*. *Trans. Am. Fish Soc.* 104, 793-799.

Cook, R. F. (1973) Carbofuron. In: *Analytical Methods for Pesticides and Plant Growth Regulators*. Vol. 7, Zweig, G., Editor, pp. 187-210, Academic Press, New York.

Coppedge, J. R., Lindquist, D. A., Bull, D. L., Dorough, H. W. (1967) Fate of 2-methyl-2-(methylthio)propinaldehyde O-(methylcarbamoyl) oxime (Temik) in cotton plants and soil. *J. Agric. Food Chem.* 15(5), 902-910.

Corwin, D. L., Farmer, W. J. (1984) Non-single-valued adsorption-desorption of bromacil and diquat by freshwater sediments. *Environ. Sci. Technol.* 18, 507-514.

Cotham, W. E., Bidleman, T. F. (1991) Estimating the deposition of organic contaminants to the Arctic. *Chemosphere* 22, 165-188.

Cowart, R. P., Bonner, F. L., Epps, Jr., E. A. (1971) Rate of hydrolysis of seven organophosphate pesticides. *Bull. Environ. Contam. Toxicol.* 6, 231-234.

Cox, J. L. (1970) Low ambient level uptake of ^{14}C-DDT by three species of marine phytoplankton. *Bull. Environ. Contam. Toxicol.* 5, 218.

Cripe, C. R., Walker, W. W., Pritchard, P. H., Bourquin, A. W. (1987) A shake-flask test for estimation of biodegradability of toxic organic substances in the aquatic environment. *Ecotoxicol. Environ. Saf.* 14, 239-251.

Crosby, D. G., Tucker, R. K. (1971) Accumulation of DDT by *Daphnia magna*. *Environ. Sci. Technol.* 5, 714-716.

Crosby, T. (1981) Environmental chemistry of pentachlorophenol. *Pure Appl. Chem.* 53, 1051-1080.

Crossland, N. O., Wolff, C. J. M. (1985) Fate and biological effects of pentachlorophenol in outdoor ponds. *Environ. Toxicol. Chem.* 4, 73-86.

Dao, T. H., Lavy, T. L., Dragun, J. (1983) Rationale of the solvent selection for soil extraction of pesticide residues. *Res. Rev.* 87, 91-104.

David, W. A. L., Metcalf, R. L., Winton, M. (1960) The systematic insecticidal properties of certain carbamates. *J. Econ. Entmol.* 53, 1021-1025.

Davidson, J. M., Ou, L. T., Rao, P. S. C. (1980) Adsorption, movement, and biological degradation of high concentration of selected pesticides in soils. EPA-600-/2-80-124. U. S. Environmental Protection Agency, Cincinnati, Ohio.

Davies, J. E., Lee, J. A. (1987) Changing profiles in human health effects of pesticides. *Pestic. Sci. Biotechnol.* 53.

Davies, R. P., Dobbs, A. J. (1984) The prediction of bioconcentration in fish. *Water Res.* 18(10), 1253-1262.

Dean, J., Editor (1985) *Lange's Handbook of Chemistry*. 13th Edition, McGraw-Hill, Inc., New York.

Dearth, M. A., Hites, R. A. (1991) Depuration rates of chlordane compounds from rat fat. *Environ. Sci. Technol.* 25(6), 1125-1128.

De Bruijn, J., Busser, F., Seinen, W., Hermens, J. (1989) Determination of octanol/water partition coefficients for hydrophobic organic chemicals with the "slow-stirring" method. *Environ. Toxicol. Chem.* 8, 499-512.

De Bruijn, J., Hermens, J. (1991) Uptake and elimination kinetics of organophosphorus pesticides in the guppy (*Poecilia reticulata*): Correlations with the octanol/water partition coefficient. *Environ. Toxicol. Chem.* 10, 791-804.

De Bruijn, J., Seinen, W., Hermens, J. (1993) Biotransformation of organophosphorus compounds by rainbow trout (*Oncorhynchus mykiss*) liver in relation to bioconcentration. *Environ. Toxicol. Chem.* 12, 1041-1050.

De Kock, A. C., Lord, D. A. (1987) A simple procedure for determining octanol-water partition coefficients using reverse phase high performance liquid chromatography (RPHPLC). *Chemosphere* 16(1), 133-142.

De Kreuk, J. F., Hanstveit, A. O. (1981) Determination of the biodegradability of the organic fraction of chemical wastes. *Chemosphere* 10, 561-575.

De La Cruz, A. A., Rajanna, B. (1975) Mirex incorporation in the environment: Uptake and distribution in crop seedlings. *Bull. Environ. Contam. Toxicol.* 14, 38-42.

Decker, G. C., Weinman, C. J., Bann, J. M. (1950) A preliminary report on the rate of insecticide residue loss from treated plants. *J. Econ. Entomol.* 43, 919.

Delaune, R. D., Gambrell, R. P., Reddy, K. S. (1983) Fate of pentachlorophenol in estuarine sediment. *Environ. Pollut. Series* B6, 297-308.

Delorme, P. D., Muir, D. C. G., Lockhart, W. L., Mills, K. H., Ward, F. J. (1993) Depuration of toxaphene in lake trout and white suckers in a natural ecosystem following a single I. P. dose. *Chemosphere* 27(10), 1965-1973.

Demayo, A. (1972) Gas chromatographic determination of the rate constant for the hydrolysis of heptachlor. *Bull. Environ. Contam. Toxicol.* 8(4), 234-237.

Demozay, D., Marechal, G. (1972) Physical and chemical properties in lindane: Monograph of an insecticide, E. Ulman, pp. 15-21, K. Schiller, Freiburg im Breisgau.

Deneer, J. W. (1993) Uptake and elimination of chlorpyrifos in the guppy at sublethal and lethal aqueous concentrations. *Chemosphere* 26(9), 1607-1616.

Deneer, J. W. (1994) Bioconcentration of chlorpyrifos by the three-spined stickleback under laboratory and field conditions. *Chemosphere* 29(7), 1561-1575.

Deutsche Forschungsgemeinschaft (1983) *Hexachlorcyclohexan als Schadstoff in Lebensmitteln*. Verlag Chemie, Weinheim, Germany. 13p.

Devillers, J., Bintein, S., Domine, D. (1996) Comparison of BCF models based on log P. *Chemosphere* 33(6), 1047-1065.

Devillers, J., Thioulouse, J., Domine, D., Chastrette, M., Karcher, W. (1991) Multivariate analysis of the input and output data in the fugacity model level I. In: *Applied Multivariate Analysis in SAR and Environmental Studies*. Devillers, J., Karcher, W. Editors, Kluwer Academic Publishers, Dordrecht, The Netherlands.

Dierberg, F. E., Pfeuffer, R. J. (1983) Fate of ethion in canals draining a Florida citrus grove. *J. Agric. Food Chem.* 31, 704-709.

Dilling, W. L., Lickly, L. C., Lickly, T. D., Murphy, P. G., McKellar, R. L. (1984) Organic photochemistry. 19. Quantum yields for O,O-diethyl O-(3,5,6-trichloro-2-pyridinal) phosphorothiaoate and 3,5,6-trichloro-2-pyridinol in dilute aqueous solutions and their environmental transformation rates. *Environ. Sci. Technol.* 18, 540-543.

Di Toro, D. M. (1985) A particle interaction model of reversible organic chemical sorption. *Chemosphere* 14, 1503-1538.

Di Toro, D. M., Jeris, J. S., Ciarcia, D. (1985) Diffusion and partitioning of hexachlorobiphenyl in sediments. *Environ. Sci. Technol.* 19, 1169-1176.

Dobbs, A. J., Cull, M. R. (1982) Volatilization of chemicals-relative loss rates and the estimation of vapor pressures. *Environ. Pollut.* (series B) 3, 289-298.

Dobbs, A. J., Grant, C. (1980) Pesticide volatilisation rates-A new measure of the vapour pressure of pentachlorophenol at room temperature. *Pest. Sci.* 11, 29-32.

Dobbs, A. J., Hart, G. F., Parsons, A. H. (1984) The determinations of vapour pressures from relative volatilization rates. *Chemosphere* 13(5/6), 687-692.

Dobson, R. C., Throneberry, G. O., Belling, T. E. (1960) Residues of established alfalfa treated with granulated phorate (Thimet) and their effect on cattle fed the hay. *J. Econ. Entomol.* 53, 306-310.

Doedens, J. D., Editor (1967) *Lange's Handbook of Chemistry*. McGraw-Hill, New York.

Domsch, K. H. (1984) Effects of pesticides and heavy metals on biological processes in soil. *Plant Soil* 76, 367-378.

Dörfler, U., Adler-Köhler, R., Schneider, P., Scheunert, I., Korte, F. (1991) A laboratory model system for determining the volatility of pesticides from soil and plant surfaces. *Chemosphere* 23, 485-496.

Dorough, H. W., Hemken, R. W. (1973) Chlordane residues in milk and fat of cows fed HCS 3260 (High Purity Chlordane) in the diet. *Bull. Environ. Contam. Toxicol.* 10, 208-216.

Dorough, H. W., Ivie, G. W. (1974) Fate of mirex-carbon-14 during and after a 28-day feeding period to a lactating cow. *J. Environ. Qual.* 3(1), 65-67.

Dorough, H. W., Huhtanen, K., Marshall, T. C., Bryant, H. E. (1978) Fate of endosulfan in rats and toxicological conditions of apolar metabolites. *Pest. Biochem. Physio.* 8, 241-252.

Dorough, H. W., Pass, B. C. (1973) Residues in corn and soils treated with technical chlordane and high-purity chlordane (HS3260). *J. Econ. Entomol.* 65, 976-979.

Dowd, J. F., Bush, P. B., Neary, D. G., Taylor, J. W., Berisford, Y. C. (1993) Modeling pesticide movement in forested watersheds: Use of PRSM for evaluating pesticide options in loblolly pine stand management. *Environ. Toxicol. Chem.* 12, 429-439.

Doyle, R. C., Kaufman, D. D., Burt, G. W. (1978) Effect of dairy manure and sewage sludge on ^{14}C-pesticide degradation in soil. *J. Agric. Food Chem.* 26, 987-989.

Drahonovsky, J., Vacek, Z. (1971) Dissoziations konstanten und austauscherchromatographie chlorieter phenole. *Coll. Czech. Chem. Commun.* 36(10), 3431-3440.

Eadie, B. J., Robbins, J. A. (1987) 11. The role of particulate matter in the movement of contaminants in the Great Lakes. In: *Sources and Fates of Aquatic Pollutants*. Hites, R. A., Eisenreich, S. J., Editors, pp. 318-364, Advances Chemistry Series 216, American Chemical Society, Washington DC.

Eadsforth, C. V. (1986) Application of reverse-phase HPLC for determination of partition coefficients. *Pest. Sci.* 17, 311-325.

Eadsforth, C. V., Moser, P. (1983) Assessment of reverse phase chromatographic methods for determining partition coefficients. *Chemosphere* 12, 1459-1475.

Edwards, C. A. (1973) *Persistence Pesticides in the Environment.* 2nd edition, CRC Press, Cleveland, Ohio.

Eichelberger, J. W., Lichtenberg, J. J. (1971) Persistence of pesticides in river water. *Environ. Sci. Technol.* 5, 541-544.

Eichler, W., Editor (1965) *Hanbuch der Insectizidkunde.* Veb. Verlag Volk. Gesundheit, Berlin.

Elgar, K. E. (1983) Pesticides residues in water - an appraisal. In: *Pesticide Chemistry: Human Welfare and The Environment.* Vol. 4, Miyamoto, J., Kearney, P. C., Editors, International Union of Pure and Applied Chemistry, Pergamon Press, Oxford, England.

Ellgehausen, H., D'Hondt, C., Fuerer, R. (1981) Reversed-phase chromatography as a general method for determining octanol/water partition coefficients. *Pest. Sci.* 12, 219-227.

Ellgehausen, H., Guth, J. A., Esser, H. O. (1980) Factors determining bioaccumulation potential of pesticides in the individual compartments of aquatic food chains. *Ecotoxicol. Environ. Saf.* 4, 134-157.

Ellington, J. J., Stancil, F. E., Payne, W. D. (1987) Measurement of Hydrolysis Rate Constants for Evaluation of Hazardous Waste Land Disposal. Volume 1, Data on 32 chemicals. U.S. EPA-600/3-86/043, Washington DC.

Ellington, J. J., Stancil, F. E., Payne, W. D. (1987) Measurement of Hydrolysis Rate Constants for Evaluation of Hazardous Waste Land Disposal. Volume 2, Data on 54 chemicals. U.S. EPA-600/53-87/019, Washington DC.

Ellington, J. J. et al. (1988) Measurement of Hydrolysis Rate Constants for Evaluation of Hazardous Waste Land Disposal. Volume 3. U.S. EPA 600/3-88/028, Washington DC.

Elzerman, A. W., Coates, J. T. (1987) 10. Hydrophobic organic compounds on sediments: Equilibria and kinetics of sorption. In: *Sources and Fates of Aquatic Pollutants.* Hites, R. A., Eisenreich, S. J., Editors, pp. 263-317, Advances Chemistry Series 216, American Chemical Society, Washington DC.

Ernst, W. (1977) Determination of the bioconcentration potential of a marine organisms-A steady state approach. *Chemosphere* 6, 731-740.

Ernst, W. (1979) Factors affecting the evaluation of chemicals in laboratory experiments using marine organisms. *Ecotox. Environ. Saf.* 3(1), 90-98.

Evans, M. S., Noguchi, G. E., Rice, C. P. (1991) The biomagnification of polychlorinated biphenyls, toxaphene, and DDT compounds in a Lake Michigan offshore food web. *Arch. Environ. Contam. Toxicol.* 20, 87-93.

Eye, J. D. (1968) Aqueous transport of dieldrin residues in soils. *J. Water Pollut. Control Fed.* 40, R316-R332.

Farquharson, M. E., Gage, J. C., Northover, J. (1958) The biological action of chlorophenols. *Brit. J. Pharmacol.* 13, 20.

Faust, B. C., Hoigné, J. (1990) Photolysis of Fe (III)-hydroxy complexes as sources of OH radicals in clouds, fog and rain. *Atmos. Environ.* 24A, 79-89.

Faust, S. D., Gomaa, H. M. (1972) Chemical hydrolysis of some organic phosphorous and carbamate pesticides in aquatic environments. *Environ. Lett.* 3, 171-201.

Felsot, A., Dahm, P. A. (1979) Sorption of organophosphorous and carbamate insecticides by soil. *J. Agric. Food Chem.* 27, 557-563.

Felsot, A., Wilson, J. (1980) Adsorption of carbofuran and movement on soil thin layers. *Bull. Environ. Contam. Toxicol.* 24, 778-782.

Ferreira, G. A., Seiber, J. N. (1981) Volatilization and exudation losses of three N-methyl-carbamate insecticides applied systemically to rice. *J. Agric. Food Chem.* 29, 93-99.

Findinger, N. J., Glotfelty, D. E. (1988) A laboratory method for the experimental determination of air/water Henry's law constants for several pesticides. *Environ. Sci. Technol.* 22, 1289-1293.

Findinger, N. J., Glotfelty, D. E. (1990) Henry's law constants for selected pesticides, PAHs and PCBs. *Environ. Toxicol. Chem.* 9, 731-735.

Findinger, N. J., Glotfelty, D. E., Freeman, H. P. (1989) Comparison of two experimental techniques for determining air/water Henry's law constants. *Environ. Sci. Technol.* 23(12), 1528-1531.

Finizio, A., Vighi, M., Sandroni, D. (1997) Determination of *n*-octanol/water partition coefficient (K_{ow}) of pesticide, critical review and comparison of methods. *Chemosphere* 34, 131-161.

Firestone, D. (1977) Chemistry and analysis of pentachlorophenol and its contaminants. Division of Chemistry and Physics, Bureau of Foods. FDA By-Lines No. 2, September, 1977.

Fischer, R. C., Krämer, W., Ballschmiter, K. (1991) Hexachlorocyclohexane isomers as markers in the water flow of Atlantic Ocean. *Chemosphere* 23, 889-900.

Fisher, S. W., Lydy, M. J., Barger, J., Landrum, P. F. (1993) Quantitative structure-activity relationships for predicting the toxicity of pesticides in aquatic systems with sediment. *Environ. Toxicol. Chem.* 12, 1307-1318.

Fogel, S., Lancione, R., Sewall, A., Boethling, R. S. (1982) Enhanced biodegradation of methoxychlor in soil under enhanced environmental conditions. *Appl. Environ. Microbiol.* 44, 113-120.

Fordham, C. L., Reagan, D. P. (1991) Pathways analysis method for estimating water and sediment criteria at hazardous waster sites. *Environ. Toxicol. Chem.* 10, 949-960.

Foreman, W. T., Bidleman, T. F. (1987) An experimental system for investigating vapor-particle partitioning of trace organic pollutants. *Environ. Sci. Technol.* 21, 869-875.

Frank, R. (1981) Pesticides and PCB in the Grand and Saugeen River Basins. *J. Great Lakes Res.* 7, 440-454.

Freed, V. H. (1976) Solubility, hydrolysis, dissolution constants and other constants of benchmark pesticides. In: *A Literature Survey of Benchmark Pesticides*. George Washington University Medical Center, Washington, DC.

Freed, V. H., Chiou, C. T., Haque, R. (1977) Chemodynamics: Transport and behavior of chemicals in the environment - A problem in environmental health. *Environ. Health Perspect.* 20, 55-70.

Freed, V. H., Chiou, C. T., Schmedding, D. W. (1979) Degradation of selected organophosphorous pesticides in water and soil. *J. Agric. Food Chem.* 27, 706-708.

Freed, V. H., Kaufman, D. D., Metcalf, R. L., Farmer, W. J., Crosby, D. G., Spencer, W. (1976) Chemodynamics: Transport and Behavior of Chemicals in the Environment - A Problem in Environmental Health. George Washington University Medical Center, Washington, DC.

Freed, V. H., Schmedding, D. W., Kohnert, R., Haque, R. (1979) Physical chemical properties of several organophosphates: Some implications in environmental and biological behavior. *Pest. Biochem. Physiol.* 10, 203-211.

Freitag, D., Geyer, H., Kraus, A., Viswanathan, R., Kozias, D., Attar, A., Klein, W., Korte, F. (1982) Ecotoxicological profile analysis. VII. Screening chemicals for their environmental behavior by comparative evaluation. *Ecotox. Environ. Saf.* 6, 60-81.

Freitag, D., Balhorn, L., Geyer, H., Körte, F. (1985) Environmental hazard profile of organic chemicals. An experimental method for the assessment of the behaviour of chemicals in the ecosphere by simple laboratory tests with C-14 labelled chemicals. *Chemosphere* 14, 1589-1616.

Freitag, D., Lay, J. P., Körte, F. (1984) Environmental hazard profile - Test results as related to structures and translation into the environment. In: *QSAR in Environmental Toxicology.* Kaiser, K. L. E., Editor, pp. 111-136, D. Reidel Publishing Co., Dordrecht, The Netherlands.

Fries, G. F., Marrow, G. S., Gordon, C. H. (1969) Comparative excretion and retention of DDT analogs by dairy cows. *J. Dairy Sci.* 52, 1801-1805.

Fröbe, Z., Drevenkar, V., Štengl, B. (1989) Sorption behaviour of some organophosphorus pesticides in natural sediments. *Toxicol. Environ. Chem.* 19, 69-82.

Fujita, T., Iwasa, J., Hansch, C. (1964) A new substituent constant 'pi' derived from partition coefficients. *J. Am. Chem. Soc.* 86, 5175-5180.

Fuhremann, T. W., Lichtenstein, E. P. (1980) A comparative study of the persistence, movement, and metabolism of six carbon-14 insecticides in soils and plants. *J. Agric. Food Chem.* 28, 446-452.

Fung, K. K. H., Uren, N. C. (1977) Microbial transformation of S-methyl N-[(methylcarbamoyl)oxy]thioacitimidate. *J. Agric. Food Chem.* 25, 966-969.

Garst, J. E., Wilson, W. C. (1984) Accurate, wide-range, automated, high-performance liquid chromatographic method for the estimation of octanol/water partition coefficients. I: Effect of chromatographic conditions and procedure variables on accuracy and reproducibility of the method. *J. Pharm. Sci.* 73(11), 1616-1623.

Garten, Jr., C. T., Trabalka, J. R. (1983) Evaluation of models for predicting terrestrial food chain behavior of xenobiotics. *Environ. Sci. Tecnol.* 17, 590-595.

Gebefügi, I., Parlar, H., Körte, F. (1979) Occurrence of pentachlorophenol in enclosed environments. *Ecotox. Environ. Saf.* 3, 269-300.

Geer, R. D. (1978) *Predicting the Anaerobic Degradation of Organic Chemical Pollutants in Waste Water Treatment Plants from their Electrochemical Reduction Behavior.* Waste Resources Research Center, Montana State University, MUJWRRC-95, W79-01-OWRT-A-097-MONT(1), Pb-289 22478WP, Bozeman, Montana.

GEMS (1985) Graphical Exposure Modeling Systems. Fate of Atmosphere Pollutants (FAP). Office of Toxic Substances, U.S. EPA, Washington, DC.

GEMS (1986) Graphical Exposure Modeling Systems. Fate of Atmosphere Pollutants (FAP). Office of Toxic Substances, U.S. EPA, Washington, DC.

Gerstl, Z. (1984) Adsorption, decomposition and movement of oxamyl in soil. *Pestic. Sci.* 15, 9-17.

Gerstl, Z., Helling, C. S. (1987) Evaluation of molecular connectivity as a predictive method for the adsorption of pesticides by soils. *J. Environ. Sci. Health* B22, 55-69.

Gerstl, Z., Mingelgrin, U. (1984) Sorption of organic substances by soils and sediments. *J. Environ. Sci. Health* B19(3), 297-312.

Getzin, L. W. (1981a) Degradation of chlorpyrifos in soil: Influence of autoclaving, soil moisture, and temperature. *J. Econ. Entomol.* 74, 158-162.

Getzin, L. W. (1981b) Dissipation of chlorpyrifos from dry soil surfaces. *J. Econ. Entomol.* 74(6), 707-713.

Geyer, H., Kraus, A. G., Klein, W., Richter, E. Körte, F. (1980) Relationship between water solubility and bioaccumulation potential of organic chemicals in rats. *Chemosphere* 9, 277-291.

Geyer, H., Politzki, G., Freitag, D. (1984) Prediction of ecotoxiological behaviour of chemicals: Relationship between n-octanol/water partition coefficient and bioaccumulation of organic chemicals by alga chlorella. *Chemosphere* 13, 269-284.

Geyer, H., Scheunert, I., Brüggemann, R., Steinberg, C., Körte, F., Kettrup, A. (1991) QSAR for organic chemical bioconcentration in *Daphnia,* algae, and mussels. *Sci. Total Environ.* 109/110, 387-394.

Geyer, H., Scheunert, I., Körte, F. (1987) Correlation between the bioconcentration potential of organic environmental chemicals in humans and their *n*-octanol/water partition coefficients. *Chemosphere* 16, 239-252.

Geyer, H., Sheehan, P., Kotzias, D., Freitag, F., Körte, F. (1982) Prediction of ecotoxicological behaviour of chemicals: Relationship between physico-chemical properties and bioaccumulation of organic chemicals in the mussel *Mytilus edulis*. *Chemosphere* 11(11), 1121-1134.

Geyer, H., Viswanathan, R., Freitag, D., Korte, F. (1981) Relationship between water solubility of organic chemicals and their bioaccumulation by alga chlorella. *Chemosphere* 10, 1307-1313.

Giam, C. S., Atlas, E., Chan, H. S., Neff, G. S. (1980) Phthalate esters, PCB, and DDT residues in the Gulf of Mexico atmosphere. *Atoms. Environ.* 14, 65-69.

Gingerich, L. L., Zimdahl, R. L. (1976) Soil persistence of isopropalin and oryzalin. *Weed Sci.* 24, 431-434.

Gish, C. D., Hughes, D. L. (1982) Residues of DDT, dieldrin and heptachlor in earthworms during two years following application. *U.S. Fish Wildl. Serv. Spec. Sci. Rep.:* Wildl. 241.

Glooschenko, V. et al. (1979) Bioconcentration of chlordane by the green *Alga scenedesmus quadricauda*. *Bull. Environ. Contam. Toxicol.* 21, 515-520.

Glotfelty, D. E. (1981) Atmospheric dispersion of pesticides from treated fields. Ph.D. Thesis, pp. 94-187, University of Maryland, College Park, Maryland.

Glotfelty, D. E., Taylor, A. W., Turner, B. C., Zoller, W. H. (1984) Volatilization of surface-applied pesticides from fallow soils. *J. Agric. Food Chem.* 32, 638-643.

Glotfelty, D. E., Leech, M. M., Jersey, J., Taylor, A. W. (1989) Volatilization and wind erosion of soil surface applied atrazine, simazine, alachlor, and toxaphene. *J. Agric. Food Chem.* 37, 546-551.

Glotfelty, D. E., Schomburg, C. J., McChesney, M. M., Sagebiel, J. C., Seiber, J. N. (1990) Studies of the distribution, drift, and volatilization of diazinon resulting from spray application to a dormant peach orchard. *Chemosphere* 21(10-11), 1303-1314.

Gobas, F. A. P. C., Clark, K. E., Shiu, W. Y., Mackay, D. (1989) Bioconcentration of polybrominated benzenes and biphenyls and related superhydrophobic chemicals in fish: Role of bioavailability and elimination into the feces. *Environ. Toxicol. Chem.* 8, 231-245.

Gobas, F. A. P. C., Lahittete, J. M., Garofalo, G., Shiu, W. Y., Mackay, D. (1988) A novel method for measuring membrane-water partition coefficients of hydrophobic organic chemicals: Comparison with 1-octanol-water partitioning. *J. Pharma. Sci.* 77(3), 265-272.

Goerlitz, D. F., Troutman, D. E., Godsy, E. M., Franks, B. J. (1985) Migration of wood-preserving chemicals in contaminanted groundwater in sand aquifer at Pensacola, Florida. *Environ. Sci. Technol.* 19, 955-961.

Goll, O. (1954) Chlorophenol. In: *Ullmans Encyklopädie der Technischen Cheme*. Foerst, W., Ed., pp. 494-499, Urban and Schwarzenberg, Munich/Berlin.

Gomaa, H. M., Suffert, I. H., Faust, S. D. (1969) Kinetics of hydrolysis of diazinon. *Residue Rev.* 29, 171.

Görge, G., Nagel, R. (1990) Kinetics and metabolism of 14-C-lindane and 14-C-atrazine in early life stages of zebrafish (*Brachdanio rerio*). *Chemosphere* 21, 1125-1137.

Grain, C. F. (1982) Chapter 14, Vapor pressure. In: *Handbook on Chemical Property Estimation Methods, Environmental Behavior of Organic Compounds*. Lyman, W. J., Reehl, W. F., Rosenblatt, D. H., Editors, McGraw-Hill, Inc., New York.

Goring, C. A. I. (1967) *Ann. Rev. Phytopathol.* 5, 285-318.

Grayson, B. T., Fasbraey, L. A. (1982) Determination of the vapor pressure of pesticides. *Pest. Sci.* 13, 269-278.
Grayson, B. T., Langner, E., Wells, D. (1982) Comparison of two gas saturation methods for the determination of the vapor pressure of cypermethrin. *Pest. Sci.* 13, 552-556
Green, G. H., McKeown, B. A., Oloffs, P. C. (1984) Acephate in rainbow trout (*Salmo gairdneri*); Acute toxicity, uptake, elimination. *J. Environ. Sci. Health* B19, 131-155.
Greenhalgh, R., Dhawan, K., Weinberger, P. (1980) *J. Agric. Food Chem.* 28, 263-272.
Grover, R., Editor (1988) *Environmental Chemistry of Herbicides.* Volume I, CRC Press, Inc., Boca Raton, Florida.
Gückel, W., Kästel, R., Lewerenz, J., Synnatschke, G. (1982) A method for determining the volatility of active ingredients used in plant protection. Part III. The temperature relationship between vapor pressure and evaporation rate. *Pest. Sci.* 13, 161-168.
Gückel, W., Synnatsche, G., Rittig, R. (1973) A method for determining the volatility of active ingredients used in plant protection. *Pest. Sci.* 4, 137-147.
Gückel, W., Rittig, R., Synnatsche, G. (1974) A method for determining the volatility of active ingredients used in plant protection. II. Application to formulated products. *Pest. Sci.* 5, 393.
Güesten, H., Filby, W. G., Schoop, S. (1981) Prediction of hydroxyl radical reaction rates with organic compounds in the gas-phase. *Atom. Environ.* 15, 1763-1765.
Guinee, J., Heijungs, R. (1993) A proposal for the classification of toxic substances within the framework of life cycle assessment of products. *Chemosphere* 26(1), 1925-1944.
Guirguis, M. W., Shafik, M. T. (1975) Persistence of trichlorfon and dichlorvos in two different autoclaved and non-autoclaved soils. *Bull. Entmol. Soc. Egypt Econ. Ser.* 8, 29-32.
Gummer, W. D. (1979) Pesticide monitoring in the prairies of western Canada. In: Water Quality Interpretive Report No. 4., Inland Waters Directorate, Regina, Saskatchewan, Canada.
Günther, F. A., Günther, J. D. (1971) Residue of pesticides and other foreign chemicals in foods and feeds. *Res. Rev.* 36, 69-77.
Günther, F. A., Westlake, W. E., Jaglan, P. S. (1968) Reported solubilities of 738 pesticide chemicals in water. *Res. Rev.* 20, 1-148.
Haag, W. R., Yao, C. C. D. (1992) Rate constants for the reaction of hydroxyl radicals with several drinking water contaminants. *Environ. Sci. Technol.* 26, 1005-1013.
Hadaway, A. B., Barlow, F., Turner, C. R. (1970) The effect of particle size on the contact toxicity of insecticides to adult mosquitoes. *Bull. Entomol. Res.* 60, 17.
Haderlein, S. B., Weissnahr, K. W., Schwarzenbach, R. P. (1996) Specific adsorptin of nitroaromatic explosives and pesticides to clay minerals. *Environ. Sci. Technol.* 30, 612-622.
Halfon, E., Galassi, S., Brüggermann, R., Provini, A. (1996) Selection of priority properties to assess environmental hazard of pesticides. *Chemosphere* 33(8), 1543-1562.
Hall, R. J., Kolbe, E. (1980) Bioconcentration of organophosphorous pesticides to hazardous levels by amphibians. *J. Toxicol. Environ. Health* 6, 853-868.
Hamaker, J. W. (1972) Decomposition: Quantitative aspects. In: *Organic Chemicals in the Soil Environment.* Vol. 1, Goring, C. A. I., Hamaker, J. W., Editors, pp. 253-341, Marcel Dekker, Inc., New York.
Hamaker, J. W. (1975) The interpretation of soil leaching experiments. In: *Environmental Dynamics of Pesticides.* Haque, R., Freed, V. H., Editors, pp. 115-133, Plenum Press, New York.
Hamaker, J. W., Thompson, J. M. (1972) Adsorption. In: *Organic Chemistry in the Soil Environment.* Vol. 1, Goring, C. A. I., Hamaker, J. W., Editors, pp. 51-145, Marcel Dekker, Inc., New York.

Hamelink, J. L., Waybrant, R. C. (1976) DDE and lindane in a large-scale model lentic ecosystem. *Trans. Am. Fish Soc.* 105, 124.

Hamilton, D. J. (1980) Gas chromatographic measurement of volatility of herbicide esters. *J. Chromatogr.* 195, 75-83.

Hammers, W. E., Meurs, G. J., De Ligny, C. L. (1982) Correlations between liquid chromatographic capacity ratio data on lichrosorb RP-18 and partition coefficients in the octanol-water system. *J. Chromatogr.* 247, 1-13.

Hance, R. J. (1976) Adsorption of glyphosate by soils. *Pest. Sci.* 7, 363-366.

Hansch, C., Leo, A. (1979) *Substituent Constants for Correlation Analysis in Chemistry and Biology.* Wiley, New York.

Hansch, C., Leo, A. (1982) *Medchem. Project Issue No. 26,* Pomona College, Claremont, California.

Hansch, C., Leo, A. (1985) *Medchem. Project Issue No. 26,* Pomona College, Claremont, California.

Hansch, C., Leo, A., Hoekman, D. (1995) *Exploring QSAR, Hydrophobic, Electronic, and Steric Constants.* ACS Professional Reference Book, Am. Chem. Soc., Washington, DC.

Hansen, D. J., Wilson, A. J. (1970) Residues in fish, wildlife and esturies. *Pest. Monit. J.* 4, 51.

Hansen, J. L., Spiegel, M. H. (1983) Hydrolysis studies of aldicarb, aldicarb sulfoxide and aldicarb sulfone. *Environ. Toxicol. Chem.* 2, 147-153.

Haque, R., Ebing, W. (1988) Uptake and accumulation of pentachlorophenol and sodium pentachlorophenate by earth worms from water and soil. *Sci. Total Environ.* 68, 113-125.

Haque, R., Falco, J., Cohen, S., Riordan, C. (1980) 8. Role of transport and fate studies in the exposure, assessment and screening of toxic chemicals. In: *Dynamics, Exposure and Hazard Assessment of Toxic Chemicals.* Haque, R., Editor, pp. 47-67, Ann Arbor Science Publishers, Ann Arbor, Michigan.

Harner, T., Mackay, D. (1995) Measurement of octanol-air partition coefficients for chlorobenzenes, PCBs, and DDT. *Environ. Sci. Technol.* 29, 1599-1606.

Harnisch, M., Möckel, H. J., Schultze, G. (1983) Relationship between LOG P_{ow} shake-flask values and capacity factors derived from reversed-phase high-performance liquid chromatography for *n*-alkylbenzenes and some OECD reference substances. *J. Chromatogr.* 282, 315-332.

Harris, J. C. (1982) Chapter 7, Rate of hydrolysis and Chapter 8, Rate of aqueous photolysis. In: *Handbook on Chemical Property Estimation Methods, Environmental Behavior of Organic Compounds.* Lyman, W. J., Reehl, W. F., Rosenblatt, D. H., Editors, McGraw-Hill, New York.

Harris, S. J., Cecil, H. C., Bitman, J. (1974) Effect of several dietary levels of technical methoxychlor on reproduction in rats. *J. Agric. Food Chem.* 22(6), 969-973.

Hartley, D., Kidd, H., Editors (1987) *The Agrochemicals Handbook.* 2nd Edition, The Royal Society of Chemistry, Nottingham, England.

Hartley, D. M., Johnson, J. B. (1983) Use of freshwater clam *Corbicula manilensis* as a monitor for organochlorine pesticieds. *Bull. Environ. Contam. Toxicol.* 31, 33-40.

Hartley, G. S., Graham-Bryce, I. J. (1980) *Physical Principles and Pesticide Behavior.* Volume 2, Academic Press, New York.

Harvey, J., Reiser, R. W. (1973) Metabolism of methomyl in tobacco, corn and cabbage. *J. Agric. Food Chem.* 21, 775-783.

Hattemer-Frey, H. A., Travis, C. C. (1989) Pentachlorophenol: Environmental partitioning and human exposure. *Arch. Environ. Contam. Toxicol.* 18, 482-489.

Hattula, M. L., Wasenius, V.-M., Reunanen, H., Arstila, A. U. (1981) Acute toxicity of some chlorinated phenols, catechols and cresols in trout. *Bull. Environ. Contam. Toxicol.* 26, 295-298.

Hautala, R. P. (1978) Surfactant Effects on Pesticide Photochemistry in Water and Soil. EPA-600/3-78-060, U.S. EPA.

Hawker, D. W., Connell, D. W. (1986) Bioconcentration of lipophilic compounds by some aquatic organisms. *Ecotoxicol. Environ. Saf.* 11, 184-197.

Hawker, D. W., Connell, D. W. (1989) A simple water/octanol partition system for bioconcentration investigations. *Environ. Sci. Technol.* 23, 961-965.

Hazardous Substances Data Bank (1989) National Library of Medicine, Toxicology Information Program.

Heiber, O., Szelagiewicz, H. O. (1976) Ciba-Geigy Ltd., Personal Communication.

Heller, S. R., Scott, K., Bigwood, D. W. (1989) The need for data evaluation of physical and chemical properties of pesticides: The ARS pesticide properties database. *J. Chem. Inf. Comput. Sci.* 29, 159-162.

Hellmann, H. (1987) Model tests on volatilization of organic trace substances in surface waters. *Frensenius Z. Anal. Chem.* 328, 475-479.

Hemond, H. F., Fechner, E. J. (1994) *Chemical Fate and Transport in the Environment.* Academic Press, New York.

Herbicide Handbook (1974) *Herbicide Handbook.* 3rd Edition, Weed Science Society of America, Champaign, Illinois.

Herbicide Handbook (1978) *Herbicide Handbook.* 4th Edition, Weed Science Society of America, Champaign, Illinois.

Herbicide Handbook (1983) *Herbicide Handbook.* 5th Edition, Weed Science Society of America, Champaign, Illinois.

Hermanutz, R. O. (1978) Endrin and malathion toxicity to flagfish (*Jordanella floridae*). *Arch. Environ. Contam. Toxicol.* 7, 159-168.

Hermens, J., Leeuwangh, P. (1982) Joint toxicity of mixture of 8 and 24 chemicals to the guppy (*Poecilia reticulata*). *Ecotoxicol. Environ. Saf.* 6, 302-310.

Hermens, J., Könemann, H., Leeuwangh, P., Musch, A. (1985) Quantitative structure-activity relationships in aquatic toxicity studies of chemicals and complex mixtures of chemicals. *Environ. Toxicol. Chem.* 4, 273-279.

Heuer, B., Yaron, B., Birk, Y. (1974) Guthion half-life in aqueous solutions and on glass surfaces. *Bull. Environ. Contam. Toxicol.* 11, 532-537.

Hiatt, C. W., Haskins, W. T., Olivier, L. (1960) The action of sunlight on sodium pentachlorophenate. *Am. J. Trop. Med. Hyg.* 9, 527-531.

Hidaka, H., Nohara, K., Zhao, J., Serpone, N., Pelizzetti, E. (1992) Photo-oxidative degradation of the pesticide permethrin catalyzed by irradiated TiO_2 semiconductor slurries in aqueous media. *J. Photochem. Photobiol. A: Chem.* 64, 247-254.

Hill, D. W., McCarty, P. L. (1967) Anaerobic degradation of selected chlorinated hydrocarbon pesticides. *J. Water Pollut. Control Fed.* 39, 1259-1277.

Hill, J. C., Kolling, H. P., Paris, D. F., Wolfe, N. L., Zepp, R. G. (1976) Dynamic Behavior of Vinylchloride in Aquatic Ecosystems. U.S. EPA-600/3-76-001.

Hinckley, D. A., Bidleman, T. F., Foreman, W. T. (1990) Determination of vapor pressures for nonpolar and semipolar organic compounds from gas chromatographic retention data. *J. Chem. Eng. Data* 35, 232-237.

Hine, J. Mookerjee, P. K. (1975) The intrinsic hydrophilic character of organic compounds. Correlations in terms of structural contributions. *J. Org. Chem.* 40, 292-298.

Hinman, M. L., Klaine, S. J. (1992) Uptake and translocation of selected organic pesticides by the rooted aquatic plant *Hydrilla verticillata* royale. *Environ. Sci. Technol.* 26, 609-613.

Hodgman, C. R., Editor (1952) *Handbook of Chemistry and Physics*. 34th Edition, Chemical Rubber Publishing Co., Cleveland, Ohio.

Hodson, J., Williams, N. A. (1988) The estimation of the adsorption coefficient (K_{OC}) for soils by high performance liquid chromatography. *Chemosphere* 17, 67-77.

Hoigné, J., Bader, H. (1983) Rate constants of reactions of ozone with organic and inorganic compounds in water - I. Non-dissociating organic compounds. *Water Res.* 17, 173-183.

Hoigné, J., Bader, H. (1983) Rate constants of reactions of ozone with organic and inorganic compounds in water - II. Dissociating organic compounds. *Water Res.* 17, 185-194.

Hoke, R. A., Ankley, G. T., Cotter, A. M., Goldenstein, T., Kosian, P. A., Phipps, G. L., VanderMeiden, F. M. (1994) Evaluation of equilibrium partitioning theory for predicting acute toxicity of field-collected sediments contaminated with DDT, DDE and DDD to the amphipod (*Hyalella azteca*). *Environ. Toxicol. Chem.* 13, 157-166.

Hollifield, H. C. (1979) Rapid nephelometric estimate of water solubility of highly insoluble organic chemicals of environmental interests. *Bull. Environ. Contam. Toxicol.* 23, 579-586.

Hollister, T. A., Walsh, G. E., Forester, J. (1975) Mirex and marine unicellular algae: Accumulation, population growth, and oxygen evolution. *Bull. Environ. Contam. Toxicol.* 10, 753.

Holmstead, R. L., Casida, J. E., Luzo, L. O., Fullmer, D. G. (1978) Pyrethroid photodecomposition: Permethrin. *J. Agric. Food Chem.* 26, 590-595.

Hornsby, A. G., Wauchope, R. D., Herner, A. E. (1996) *Pesticide Properties in the Environment*. Springer-Verlag, Inc., New York, New York.

Horvath, A. L., Editor (1991) *Halogenated Hydrocarbons. Solubility-Miscibility with Water.* Marcel Dekker, Inc., New York.

Howard, P. H., Editor (1991) *Handbook of Environmental Fate and Exposure Data for Organic Chemicals. Pesticides.* Vol. III. Lewis Publishers Inc., Chelsea, Michigan.

Howard, P. H., Boethling, R. S., Jarvis, W. F., Meylan, W. M., Michalenko, E. M., Editors (1991) *Handbook of Environmental Degradation Rates*. Lewis Publishers, Inc., Chelsea, Michigan.

Howe, G. E., Marking, L. L., Bills, T. D., Rach, J. J., Mayer, Jr., F. L. (1994) Effects of water temperature and pH on toxicity of terbufos, trichlorfon, 4-nitrophenol and 2,4-dinitrophenol to the amphipod *Gammarus pseudolimnaeus* and rainbow trout (*Oncorhynchus mykiss*). *Environ. Toxicol. Chem.* 13, 51-66.

Huckins, J. N., Stalling, D. L., Petty, J. D., Buckler, D. R., Johnson, B. T. (1982) Fate of kepone and mirex in the aquatic environment. *J. Agric. Food Chem.* 30, 1020-1027.

Hulzebos, E. M., Adema, D. M. M., van Breemen, D., Henzen, L., van Dis, W. A., Herbold, H. A., Hoekstra, J. A., Baerselman, R., van Gestel, C. A. M. (1993) Phototoxicity studies with *Lactuca sativa* in soil and nutrient solution. *Environ. Toxicol. Chem.* 12, 1079-1094.

Huang, J.-Y., Leng, X.-F. (1993) Interaction of rat hepatocyte and cytochrome P450 with pyrethroids in vitro. *Dongwu Xuebao* 39(4), 418-423.

Hwang, H.-M., Hodson, R. E., Lee, R. F. (1986) Degradation of phenol and chlorophenols by sunlight and microbes in estuarine water. *Environ. Sci. Technol.* 20, 1002-1007.

Hwang, H.-M., Hodson, R. E., Lee, R. F. (1987) Photolysis of phenol and chlorophenols in estuarine water. In: *Photochemistry of Environmental Aquatic Systems*. American Chemical Society, Washington DC.

Ide, A., Niki, Y., Sakamoto, F., Watanabe, I. (1972) Decomposition of pentachlorophenol in paddy soil. *Agric. Biol. Chem.* 36, 1937-1944.

Iglesias-Jimenez, E., Sanchez-Martin, M. J., Sanchez-Camazano, M. (1996) Pesticide adsorption in a soil-water system in the presence of surfactants. *Chemosphere* 32(9), 1771-1782.

Isnard, P., Lambert, S. (1988) Estimating bioconcentration factors from octanol-water partition coefficient and aqueous solubility. *Chemosphere* 17, 21-34.

Isnard, P., Lambert, S. (1989) Aqueous solubility and octanol-water partition coefficient correlations. *Chemosphere* 18, 1837-1853.

IUPAC (1985) *Halogenated Benzenes, Toluenes and Phenols with Water, Solubility Data Series.* Vol. 20, Horvath, A. L., Getzen, F. W., Editors, Pergamon Press, Oxford.

Ivie, G. W., Bull, D. L., Veech, J. A. (1980) Fate of diflubenzuron in water. *J. Agric. Food Chem.* 28, 330-337.

Ivie, G. W., Casida, J. E. (1971) Photosensitizers for the accelerated degradation of chlorinated cyclodienes and other insecticide chemicals exposed to sunlight on bean leaves. *J. Agric. Food Chem.* 19, 410-416.

Ivie, G. W., Gibson, J. R., Bryant, H. E., Begin, J. J., Barnett, J. R., Dorough, H. W. (1974) Accumulation, distribution and excretion of mirex-^{14}C in animals exposed for long periods to the insecticide in the diet. *J. Agric. Food Chem.* 22(4), 646-653.

Iwata, H., Tanabe, S., Sakai, N., Tatsukawa, R. (1993) Distribution of persistent organochlorines in the oceceanic air and surface seawater and the role of ocean on their global transport and fate. *Environ. Sci. Technol.* 27, 1080-1098.

Iwata, Y., Westlake, W. E., Berkley, J. H., Carman, G. R., Gunther, F. A. (1977) Aldicarb residues in oranges, citrus by-products, orange leaves, and soil after an aldicarb soil-application in an orange grove. *J. Agric. Food Chem.* 25, 933-937.

Jaber, H. M., Smith, J. H., Cwirla, A. N. (1982) Evaluation of gas saturation methods to measure vapor pressure. (EPA Contract No. 68-01-5117), SRI International, Menlo Park, California.

Jacobs, A., Blangetti, M., Hellmund, E. (1974) Accumulation of noxious chlorinated substances from Rhine River water in the fatty tissue of rats. *Vom Wasser* 43, 259-274.

Jafvert, C. T., Weber, E. J. (1991) Sorption of ionizable organic compounds to sediments and soils. U.S. EPA Report EPA/600/3-91/017, Environmental Research Lab., U.S. EPA, Athens, Georgia.

Jarvinen, A. W., Tyo, R. M. (1978) Toxicity to fathead minnows of endrin in food and water. *Arch. Environ. Contam. Toxicol.* 7, 409-421.

Jensen-Korte, U., Anderson, C., Spiteller, M. (1987) Photodegradation of pesticides in the presence of humic substances. *Sci. Total Environ.* 62, 335-340.

Johnson, B. T., Saunders, C. R., Sanders, H. O. (1971) Biological magnification and degradation of DDT and aldrin by freshwater invertebrates. *J. Fish Res. Board Can.* 28, 705-709.

Johnson, Jr., J. C., Bowman, M. C. (1972) Responses from cows fed diets containing fenthion or fenitrothion. *J. Diary Sci.* 55, 777.

Johnson-Logan, L. R., Broshears, R. E., Klaine, S. J. (1992) Partitioning behavior and the mobility of chlordane in groundwater. *Environ. Sci. Technol.* 26, 2234-2239.

Jones, P. A. (1981) Chlorophenols and their impurities in the Canadian environment. Environment Canada, Report SPE 3-EC-81-2F. p. 322.

Jones, R. L., Back, R. C. (1984) Monitoring aldicarb in Florida soil and water. *Environ. Toxicol. Chem.* 3, 9-20.

Jury, W. A., Farmer, W. J., Spencer, W. F. (1984) Behavior assessment model for trace organics in soil: II. Chemical classification and parameter sensitivity. *J. Environ. Qual.* 13, 567-572.

Jury, W. A., Focht, D. D., Farmer, W. J. (1987b) Evaluation of pesticide groundwater pollution potential from standard indices of soil-chemical adsorption and biodegradation. *J. Environ. Qual.* 16(4), 422-428.

Jury, W. A., Ghodrati, M. (1989) Overview of organic chemical environmental fate and transport modeling approaches. In: *Reactions and Movement of Organic Chemicals in Soils*. SSSA Special Publication No. 22, pp. 271-304, Soil Sci. Soc. of America and Soc. of Agronomy, Madison, Wisconsin.

Jury, W. A., Russo, D., Streile, G., El Abd, H. (1990) Evaluation of volatilization by organic chemicals residing below the soil surface. *Water Resources Res.* 26, 13-20.

Jury, W. A., Spencer, W. F., Farmer, W. J. (1983) Use of models for assessing relative volatility, mobility, and persistence of pesticides and other trace organics in soil systems. In: *Hazard Assessments of Chemicals: Recent Developments*. Vol. 2, Saxena, J., Editor, Academic Press, New York.

Jury, W. A., Spencer, W. F., Farmer, W. J. (1984) Behavior assessment model for trace organics in soil: III. Application of screening model. *J. Environ. Qual.* 13, 573-579.

Jury, W. A., Winer, A. M., Spencer, W. F., Focht, D. D. (1987a) Transport and transformations of organic chemicals in the soil-air water ecosystem. *Rev. Environ. Contam. Toxicol.* 99, 120-164.

Kaiser, K. L. E. (1983) A non-linear function for the calculation of partition coefficients of aromatic compounds with multiple chlorine substitution. *Chemosphere* 12, 1159-1165.

Kaiser, K. L. E. (1993) COMUTOX database, National Water Research Institute, Burlington, Ontario, Canada.

Kaiser, K. L. E., Dixon, D. G., Hodson, P. V. (1984) QSAR studies on chlorophenols, chlorobenzenes, and para-substituted phenols. In: *QSAR in Experimental Toxicology*. Kaiser, K. L. E., Editor, pp.189-206, D. Reidel Publishing Company, Dordrecht, the Netherlands.

Kaiser, K. L. E., Valdmanis, I. (1982) Apparent octanol/water partition coefficients of pentachlorophenol as a function of pH. *Can. J. Chem.* 60, 2104-2106.

Kanazawa, J. (1975) Uptake and excretion of organophosphorus and carbamate insecticides by fresh water, Motsugo (*Pseudorasbora parva*). *Bull. Environ. Contam. Toxicol.* 14, 346-352.

Kanazawa, J. (1978) Bioconcentration ratio of diazinon by freshwater fish and snail. *Bull. Environ. Contam. Toxicol.* 20, 613-617.

Kanazawa, J. (1980) Prediction of biological concentration potential of pesticides in aquatic organisms. *Rev. Plant Protection Res.* (Japan) 13, 27-36.

Kanazawa, J. (1981) Measurement of the bioconcentration factors of pesticides by fresh-water fish and their correlation with physicochemical properties of acute toxicities. *Pest. Sci.* 12, 417-424.

Kanazawa, J. (1983) A method of predicting the bioconcentration potential of pesticides by using fish. *JARQ* 17(3), 173-179.

Kanazawa, J. (1987) Biodegradability of pesticides in water by microbes in activated sludge. *Environ. Monit. Assess.* 9, 57-70.

Kanazawa, J. (1989) Relationship between the soil sorption constants for pesticides and their physicochemical properties. *Environ. Toxicol. Chem.* 8, 477-484.

Kanazawa, J., Yushima, T., Kiritani, K. (1971) Pollution of the ecosystem by insecticides. II. Environmental pollution by organochlorine insecticides. *Kagaku* 41(7), 384-391.

Kapoor, I. P., Metcalf, R. L., Hirwe, A. S., Coats, J. R., Khaisa, M. S. (1973) Structure activity correlations of biodegradability of DDT analogs. *J. Agric. Food Chem.* 21(2), 310-315.

Kapoor, I. P., Metcalf, R. L., Nystrom, R. F., Sangha, G. K. (1970) Comparative metabolism of methoxychlor, methiochlor, and DDT in mouse, insects, and in a model ecosystem. *J. Agric. Food Chem.* 18, 1145-1152.

Karcher, W., Devillers, J. (1990) SAR and QSAR in environmental chemistry and toxicology: Scientific tool or wishful thinking? In: *Practical Applications of Quantitative Structure-Activity Relationships (QSAR) in Environmental Chemistry & Toxicology*. Karcher, W., Devillers, J., Editors, ECSC, EEC, EAEC, Brussels and Luxemburg.

Karickhoff, S. W. (1981) Semi-empirical estimation of sorption of hydrophobic pollutants on natural sediments and soils. *Chemosphere* 10, 833-846.

Karickhoff, S. W. (1985) Pollutant sorption in environmental systems. In: *Environmental Exposure from Chemicals*. Neely, W. B., Blau, G. E., Editors, pp. 49-64, CRC Press, Inc., Boca Raton, Florida.

Karickhoff, S. W., Brown, D. S., Scott, T. A. (1979) Sorption of hydrophilic pollutants on natural water sediments. *Water Res.* 13, 241-248.

Karickhoff, S. W., Carreira, L. A., Melton, C., McDaniel, V. K., Vellino, A. N., Nate, D. E. (1989) Computer prediction of chemical reactivity. The ultimate SAR. EPA 600/M-89-017. U.S. Environmental Protection Agency, Athens, Georgia.

Katayama, A., Matsumura, F. (1991) Photochemically enhanced microbial degradation of environmental pollutants. *Environ. Sci. Technol.* 25(7), 1329-1333.

Kaufman, D. D. (1976) Soil degradation and persistence of benchmark pesticides. In: *A Literature Survey of Benchmark Pesticides*. pp. 19-71. The George Washington University Medical Center, Dept. of Medical and Public Affairs, Science Communication Division, Washington DC.

Kavanaugh, M. C., Trussell, R. R. (1980) Design of aeration towers to strip volatile contaminants from drinking water. *J. Am. Water Works Assoc.* 72, 684-692.

Kawamoto, K., Urano, K. (1989) Parameters for predicting fate of organochlorine pesticides in the environment. (I) Octanol-water and air-water partition coefficients. *Chemosphere* 18, 1987-1996.

Kawamoto, K., Urano, K. (1989) Parameters for predicting fate of organochlorine pesticides in the environment. (II) Adsorption constant to soil. *Chemosphere* 19(8/9), 1223-1231.

Kawamoto, K., Urano, K. (1990) Parameters for predicting fate of organochlorine pesticides in the environment. (III) Biodegradation rate constants. *Chemosphere* 21(10-11), 1141-1152.

Kawano, M., Inoue, T., Hidaka, H., Tatsukawa, R. (1986) Chlordane residues in krill, fish and Weddell seal from the Antarctic. *Toxicol. Environ. Chem.* 11, 137.

Kawano, M., Inoue, T., Wada, T., Hidaka, H., Tatsukawa, R. (1988) Bioconcentration and residue of chlordane compounds in marine animals: Invertebrates, fish, mammals, and seabirds. *Environ. Sci. Technol.* 22, 792-797.

Kawano, M., Matsushita, S., Inoue, T., Tanaka, H. Tatsukawa, R. (1986) Biological accumulation of chlordane compounds in marine organisms from the morthern North Pacific and Bering Sea. *Mar. Pollut. Bull.* 17, 512-516.

Kearney, P. C., Kaufman, D. D. (1975) *Herbicides: Chemistry, Degradation and Mode of Action*. 2nd Edition, Vol. 2, Marcel Dekker, Inc., New York.

Kearney, P. C., Nash, R. G., Isensee, A. R. (1969) Persistence of pesticides in soil. Chapter 3, pp. 54-67, In: *Chemical Fallout: Current Research on Persistence Pesticides*. Miller, M. W., Berg, C. C., Editors, Charles C. Thomas, Springfield, Illinois.

Keil, J. E., Priester, L. E. (1969) DDT uptake and metabolism by a marine diatom. *Bull. Environ. Contam. Toxicol.* 4, 169.

Kelly, T. J., Mukund, R., Spicer, C. W., Pollack, A. J. (1994) Concentrations and transformations of hazardous air pollutants. *Environ. Sci. Technol.* 28, 378A-387A.

Kenaga, E. E. (1972) Factors related to bioconcentration of pesticides. In: *Environmental Toxicology of Pesticides*. Matsumura, F., Boush, G. M., Misato, T., Editors, pp. 193-228, Academic Press, New York.

Kenaga, E. E. (1972) Guidelines for environmental study of pesticides: Determination of bioconcentration potential. *Res. Rev.* 44, 73-111.

Kenaga, E. E. (1980a) Predicted bioconcentration factors and soil sorption coefficients of pesticides and other chemicals. *Ecotoxicol. Environ. Saf.* 4, 24-38.

Kenaga, E. E. (1980b) Correlation of bioconcentration factors of chemicals in aquatic and terrestrial organisms with their physical and chemical properties. *Environ. Sci. Technol.* 14, 553-556.

Kenaga E. E., Goring, C. A. I. (1978) Relationship between water solubility, soil-sorption, octanol-water partitioning, and bioconcentration of chemicals in biota. In: *Am. Soc. Test. Mat. 3rd. Aquatic Toxicology Sym.*, New Orleans, Louisiana. 63pp.

Kenaga E. E., Goring, C. A. I. (1980) Relationship between water solubility, soil sorption, octanol-water partitioning, and concentration of chemicals in biota. In: *Aquatic Toxicology.* ASTM STP 707, Eaton, J. G., Parrish, P. R., Hendricks, A. C., Editors, pp. 78-115, American Society for Testing and Materials, Philadelphia, Pennsylvania.

Kerler, F., Schönherr, J. (1988) Accumulation of lipophilic chemicals across plant cuticles: Prediction from octanol/water partition coefficients. *Arch. Environ. Contam. Toxicol.* 17, 1-6.

Ketelaar, J. A. A. (1950) Chemical studies of insecticides. II. The hydrolysis of O,O-diethyl- and O,O-dimethyl O-p-nitrophenylthiophosphates. *Rev. Trav. Chim.* 69, 649.

Ketelaar, J. A. A., Gersmann, H. R. (1958) Chemical studies on insecticides. VI. The rate of hydrolysis of some phosphorus acid esters. *Rev. Trav. Chim.* 77, 973-981.

Khan, S. U. (1980) *Pesticides in the Soil Environment, Fundamental Aspects of Pollution Control and Environmental Series 5*, Elsevier, Amsterdam, The Netherlands.

Kilzer, L., Scheunert, I., Geyer, H., Klein, W., Körte, F. (1979) Laboratory screening of the volatilization rates of organic chemicals from water and soil. *Chemosphere* 10, 751-761.

Kim, Y. H. (1985) Evaluation of a gas chromatographic method for estimating vapor pressures with organic pollutants. Ph.D. Thesis, University of California, Davis, California.

Kim, Y. H., Woodrow, J. E., Seiber, J. N. (1984) Evaluation of a gas chromatographic method for calculating vapor pressures with organophosphorous pesticides. *J. Chromatogr.* 314, 37-53.

King, P. H., McCarthy, P. L. (1968) A chromatographic model for predicting pesticide migration in soils. *Soil Sci. Soc. Am. Proc.* 106, 248-261.

Kishi, H., Kogure, N., Hashimoto, Y. (1990) Contribution of soil constitutents in adsorption coefficient of aromatic compounds, halogenated alicyclic and aromatic compounds to soil. *Chemosphere* 21(7), 867-876.

Kishino, T., Kobayashi, K. (1994) Relation between the chemical structures of chlorophenols and their dissociation constants and partition coefficients in several solvent-water systems. *Water Res.* 7, 1547-1552.

Klečka, G. M. (1985) Chapter 6, Biodegradation. In: *Environmental Exposure from Chemicals.* Neely, W. B., Blau, G. E., Editors, pp. 109-156, CRC Press, Inc., Boca Raton, Florida.

Klein, W., Geyer, H., Freitag, D., Rohleder, H. (1984) Sensitivity of schemes for ecotoxicological hazard ranking of chemicals. *Chemosphere* 13, 203-211.

Klein, W., Kördel, W., Weiss, M., Poremski, H. J. (1988) Updating of the OECD test guideline 107 "Partition Coefficient n-Octanol/Water": OECD laboratory intercomparison test of the HPLC method. *Chemosphere* 17, 361-386.

Klöpffer, W., Rippen, G., Frische, R. (1982) Physicochemical properties as useful tools for predicting the environmental fate of organic chemicals *Ecotox. Environ. Saf.* 6, 294-301.

Klopman, G., Namboodiri, K., Schochet, M. (1985) Simple method of computing the partition coefficient. *J. Comput. Chem.* 6, 28-38.

Kollig, H. P., Editor (1993) Environmental Rate Constants for Organic Chemicals under Consideration for EPA's Hazardous Waste Identification Projects. EPA/600/R-93/132. Environmental Research Laboratory, U.S. Environmental Protection Agency, Athens, Georgia.

Kollig, H. P., Ellington, J. J., Hamrick, K. J., Jafverts, C. T., Weber, E. J., Wolfe, N. L. (1987) *Hydrolysis Rate Constants, Partition Coefficients, and Water Solubilities for 129 Chemicals*. A Summary of Fate Constants Provided for the Concentration-Based Listing Program. U.S. EPA, Environmental Research Lab., Office of Research and Development, Athens, Georgia.

Könemann, W. H. (1981) Quantity structure-activity relationships in fish toxicity studies. Part 1: Realtionship for 50 industrial pollutants. *Toxicology* 19, 209-221.

Könemann, W. H., Musch, A. (1981) Quantitative structure-activity relationships in fish toxicity studies. Part 2: The influence of pH on the QSAR of chlorophenols. *Toxicology* 19, 223-228.

Konrad, J. G., Chesters, G. (1969) Degradation in soils of ciodrin, an organophosphate insecticide. *J. Agric. Food Chem.* 17, 226.

Kördel, W., Kotthoff, G., Müller, J. (1995a) HPLC-screening method for the determination of adsorption coefficient on soil-Results of a ring-test. *Sci. Total Environ.* 162, 119-125.

Kördel, W., Stutte, J., Kotthoff, G. (1995b) HPLC-screening method to determine the adsorption coefficient in soil-comparison of immobilized humic acid and clay mineral phases for cyanopropyl columns. *Sci. Total Environ.* 162, 119-125.

Körte, F., Freitag, D. (1986) Kriterien zur auswahl umweltgefährlicher alter stoffe. Mobilität einschliesslich abbaubarkeit und akkumulation. Umweltforschungsplan des Bundesministeriums des Innern. Forschungsbericht 106 05 25. GSF im Auftrag des Umweltbundesamtes.

Körte, F., Freitag, D., Geyer, H., Klein, W., Kraus, A. G., Lahaniatis, E. (1978) Ecotoxicologic profile analysis: A concept for establishing ecotoxicologic priority lists for chemicals. *Chemosphere* 1, 79-102.

Kortum, G., Vogel, W., Andrussow, K. (1961) *Dissociation Constants for Organic Acids in Aqueous Solutions*. Butterworths, London.

Kostovetskii, Y. I., Nasishten, S. Y., Tolstopyatova, G. V., Chegrinets, G. Y. (1976) Hygiene aspects of pesticide use in the catchment areas of water bodies. *Vodn. Resur.* 1, 67-72.

Kucklick, J. R., Hinckley, D. A., Bidleman, T. F. (1991) Determination of Henry's law constants for hexachlorocyclohexane in distilled water and artificial seawater as a function of temperature. *Marine Chem.* 34, 197-209.

Kühne, R., Ebert, R. -U., Kleint, F., Schmidt, G., Schüürmann, G. (1995) Group contribution methods to estimate water solubility of organic chemicals. *Chemosphere* 30(11), 2061-2077.

Kurihara, N., Uchida, M., Fujita, T., Nakajima, M. (1973) Studies on BHC isomers and related compounds. V. Some physicochemical properties of BHC isomers. *Pestic. Biochem. Physiol.* 2(4), 383-390.

Lacorte, S., Barcelo, D. (1994) Rapid degradation of fenitrothion in estuarine waters. *Environ. Sci. Technol.* 28, 1159-1163.

Lafleur, K. S. (1976) Carbaryl desorption and movement in soil columns. *Soil Sci.* 121, 212-216.

Lagas, P. (1988) Sorptions of chlorophenols in soil. *Chemosphere* 17(2), 205-216.

Lamoreaux, R. J., Newland, L. W. (1978) The fate of dichlorvos in soil. *Chemosphere* 10, 807-814.

Landner, L., Lindström, K., Karlsson, M., Nordin, J., Sörensen, L. (1977) Bioaccumulation in fish of chlorinated phenols from Kraft pulp mill bleachery effluents. *Bull. Environ. Contam. Toxicol.* 18, 663-673.

Landrum, R. F., Dupuis, W. S. (1990) Toxicity and toxicokinetics of pentachlorophenol and carbaryl to *Pontoporeia hoyi* and *Mysis relicta*. In: *Aquatic Toxicology and Risk Assessment.* 13th Volume, *ASTM STP* 1096, Landis, W. G., van der Schalie, W. H., Editors, American Society for Testing and Materials, Phildelphia.

Landrum, R. F., Nihart, S. R., Edie, B. J., Gardner, W. S. (1984) Reverse-phase separation method for determining pollutant binding to Aldrich humic acid and dissolved organic carbon of natural waters. *Environ. Sci. Technol.* 18, 187-192.

Larsson, P., Bremle, G., Okla, L. (1993) Uptake of pentachlorophenol in fish of acidified and non-acidified lakes. *Bull. Environ. Contam. Toxicol.* 50, 653-658.

Lartiges, S. B., Garrigues, P. P. (1995) Degradation kinetics of organophosphorous and organonitrogen pesticides in different waters under various environmental conditions. *Environ. Sci. Technol.* 29, 1246-1254.

LeBlanc, G. A. (1984) Interspecies relationships in acute toxicity of chemicals to aquatic organisms. *Environ. Toxicol. Chem.* 3, 47-60.

LeBlanc, G. A. (1996) Trophic-level defferences in the bioconcentration of chemicals: inplication in assessing environmental biomagnification. *Environ. Sci. Technol.* 29, 154-160.

Lee, L. S., Rao, P. S. C., Brusseau, M. L. (1991) Nonequilibrium sorption and transport of neutral and ionized chlorophenols. *Environ. Sci. Technol.* 25(4), 722-729.

Lee, L. S., Rao, P. S. C., Nkedl-Kizza, P., Delfino, J. J. (1990) Influence of solvent and sorbent characteristics on distribution of pentachlorophenol in octanol-water and soil-water systems. *Environ. Sci. Technol.* 24, 654-661.

Lee, P. W., Stearns, S. M., Hernandez, H., Powell, W. R., Naidu, M. V. (1989) Fate of dicrotophos in the soil environment. *J. Agric. Food Chem.* 37(4), 1169-1174.

Leenheer, J. A., Atrichs, J. L. (1971) *Soil Science Society of America Proceedings* 35, 700-705.

Lemley, A. T., Wagenet, R. J., Zhong, W. Z. (1988) Sorption of degradation of aldicarb and its oxidation products in a soil-water flow system as a function of pH and temperature. *J. Environ. Qual.* 17, 408-414.

Leo, A., Hansch, C., Elkins, D. (1971) Partition coefficients and their uses. *Chem. Rev.* 71, 525-616.

Leonard, R. A., Bailey, G. W., Swank, Jr., R. R. (1976) Transport, detoxification, fate and effects of pesticides in soil and water environments in land application of waste materials. Soil Conservation Society of America, Ankeny, Iowa. 48pp.

Leshniowsky, W. O., Dugan, P. R., Pfister, R. M., Frea, J. I., Randers, C. I. (1970) Aldrin: removal from lake water by flocculent bacteria. *Science* 169, 993.

Leuck, D. G., Bowman, M. C. (1970) Residues of phorate and five of its metabolites. Their persistence in forage corn and grass. *J. Econ. Entomol.* 63, 1838-1842.

Leuenberger, C., Giger, W., Coney, R., Graydon, J. W., Molnar-Kubica, E. (1985) Persistence chemicals in pulp mill effluents, occurrence and behavior in an activated sludge treatment plant. *Water Res.* 19, 885-894.

Leuenberger, C., Ligocki, M. P., Pankow, P. F. (1985) Trace organic compounds in rain. 4. Identities, concentrations, and scavenging mechanisms for phenols in urban air and rain. *Environ. Sci. Technol.* 19, 1053-1058.

Lewis, D. L., Paris, D. F., Baughman, G. L. (1975) Transformation of malathion by fungus, *Aspergillus oryzae*, isolated from a freshwater pond. *Bull. Environ. Contam. Toxicol.* 13, 596-601.

Lichtenstein, E. P. (1959) Absorption of some chlorinated hydrocarbon insecticides from soils into various crops. *J. Agric. Food Chem.* 7, 430-433.

Lichtenstein, E. P. (1960) Insecticidal residues in various crops grown in soils treated with an abnormal rate of aldrin and heptachlor. *J. Agric. Food Chem.* 8, 448-451.

Lichtenstein, E. P., Fuhremann, T. W., Schultz, K. R. (1971) Persistence and vertical distribution of DDT, lindane and aldrin residues, ten and fifteen years after a single soil application. *J. Agric. Food Chem.* 19, 718-721.

Lichtenstein, E. P., Schultz, K. R. (1959) Persistence of some chlorinated hydrocarbon insecticides influenced by soil types, rates of application and temperature. *J. Econ. Entomol.* 52, 124-131.

Lichtenstein, E. P., Schultz, K. R. (1959) Effect of soil cultivation, soil surface, and water on the persistence of insecticidal residues in soils. *J. Econ. Entomol.* 54, 517.

Lipke, H., Kearns, C. W. (1960) DDT-Dehydrochlorinase III. Solubilization of insecticides by lipoprotein. *J. Econ. Entomol.* 53, 31-35.

Liu, M. H., Kapila, S., Yanders, A. F., Clevenger, T. E., Elseewi, A. A. (1991) Role of entrainers in supercritical fluid extraction of chlorinated aromatics from soils. *Chemosphere* 23, 1085-1095.

Loehr, R. C., Matthews, J. E. (1992) Loss of organic chemicals in soil: Pure compound treatability studies. *J. Soil Contam.* 1(4), 339-360.

Lohninger, H. (1994) Estimation of soil partition coefficients of pesticides from their chemical structure. *Chemosphere* 29, 1611-1626.

Lokke, H. (1984) Sorption of selected organic pollutants in Danish soils. *Ecotoxicol. Environ. Safety* 8(5), 395-409.

Lord, K. A., Briggs, G. C., Nearle, M. C., Manlove, R. (1980) Uptake of pesticides from water and soil by earthworms. *Pest. Sci.* 11, 401-408.

Lord, K. A., Burt, P. E. (1964) Effect of temperature on water solubility of phorate and disulfoton. *Chem. Ind.* (London) July 11, 1262-1263.

Lu, P. Y., Metcalf, R. L. (1975) Environmental fate and biodegradability of benzene derivatives as studied in a model aquatic ecosystem. *Environ. Health Perspt.* 10, 269-284.

Lydy, M. J., Bruner, K. A., Fry, D. M., Fisher, S. W. (1990) Effects of sediment and the route of exposure on the toxicity and accumulation of neutral lipophilic and moderately water soluble metabolizable compounds in the midge, *Chironomus riparus*. In: *Aquatic Toxicology and Risk Assessment*. 12th Volume, ASTM STP 1096, pp. 104-164, American Society for Testing and Materials, Philadelphia, Pennsylvania.

Lydy, M. J., Oris, J. T., Baumann, P. C., Fisher, S. W. (1990) Effects of sediment organic carbon content on the elimination rates on neutral lipophilic compounds in the midge *(Chironomus riparus)*. *Environ. Toxicol. Chem.* 11, 347-356.

Lyman, W. J. (1982) Chapter 2, Solubility in water and Chapter 4, Adsorption coefficient for soils and sediments. In: *Handbook on Chemical Property Estimation Methods, Environmental Behavior of Organic Compounds*. Lyman, W. J., Reehl, W. F., Rosenblatt, D. H., Editors, McGraw-Hill, New York.

Lyman, W. J. (1985) Chapter 2, Estimation of physical properties. In: *Environmental Exposure from Chemicals*. Neely, W. B., Blau, G. E., Editors, pp. 13-48, CRC Press, Inc., Boca Raton, Florida.

Lyman, W. J., Reehl, W. F., Rosenblatt, D. H., Editors (1982) *Handbook on Chemical Property Estimation Methods, Environmental Behavior of Organic Compounds*. McGraw-Hill, Inc., New York.

Lyman, W. J., Reehl, W. F., Rosenblatt, D. H., Editors (1990) *Handbook on Chemical Property Estimation Methods, Environmental Behavior of Organic Compounds*. 2nd printing, American Chemical Society, Washington DC.

Ma, K. C., Shiu, W. Y., Mackay, D. (1993) Aqueous solubility of chlorophenols at 25°C. *J. Chem. Eng. Data* 38, 364-366.

Mabey, W. R., Mill, T. (1978) Critical review of hydrolysis of organic compounds in water under environmental conditions. *J. Phys. Chem. Ref. Data* 7, 383-415.

Mabey, W. R., Smith, J. H., Podoll, R. T., Johnson, H. L., Mill, T., Chou, T. W., Gates, J., Waight-Partridge, I., Jaber, H., Vanderberg, D. (1982) Aquatic Fate Process for Organic Priority Pollutants. EPA Report No. 440/4-81-014, U.S. EPA, Washington, DC.

Macalady, D. L., Wolfe, N. L. (1983) New perspectives on the hydrolytic degradation of the organophosphorothioate insecticide chlorpyrifos. *J. Agric. Food Chem.* 31, 1139.

MacDougall, D. (1964) Dylox. In: *Analytical Methods for Pesticides, Plant Growth Regulators, and Food Additives.* Vol. 2, Zweig, G., Editor, Academic Press, New York.

MacDougall, D. (1972) *Toxicity, Biodegradation.* Swets-Zeitlinger: Lisse, The Netherlands.

Macek, K. J., Petrocelli, S. R., Sleight, B. H. (1979) Consideration in assessing the potential for, and significance of, biomagnification of chemical residues in aquatic food chains. pp. 251-268. In: *Aquatic Toxicology.* ASTM STP 667, Marking, L. L., Kimerle, R. A., Editors, American Society for Testing and Materials, Philadelphia, Pennsylvania.

Mackay, D. (1982) Correlation of bioconcentration factors. *Environ. Sci. Technol.* 16, 274-278.

Mackay, D. (1985) Chapter 5, Air/water exchange coefficients. In: *Environmental Exposure from Chemicals.* Neely, W. B., Blau, G. E., Editors, pp. 91-108, CRC Press, Inc., Boca Raton, Florida.

Mackay, D. (1991) *Multimedia Environmental Models. The Fugacity Approach.* Lewis Publishers, Inc., Chelsea, Michigan.

Mackay, D., Bobra, A., Shiu, W. Y., Yalkowsky, S. H. (1980) Relationships between aqueous water solubility and octanol-water partition coefficient. *Chemosphere* 9, 701-711.

Mackay, D., Leinonen, P. (1975) Rate of evaporation of low-solubility contaminants from water bodies to atmosphere. *Environ. Sci. Technol.* 9, 1178-1180.

Mackay, D., Paterson, S. (1991) Evaluating the multimedia fate of organic chemicals: A level III fugacity model. *Environ. Sci. Technol.* 25, 427-436.

Mackay, D., Paterson, S., Chung, B., Neely, W. B. (1985) Evaluation of the environmental behavior of chemicals with a level III fugacity model. *Chemosphere* 14, 335-374.

Mackay, D., Paterson, S., Schroeder, W. H. (1986) Model describing the rates of transfer processes of organic chemicals between atmosphere and water. *Environ. Sci. Technol.* 20, 810-816.

Mackay, D., Shiu, W. Y. (1981) A critical review of Henry's law constants for chemicals of environmental interest. *J. Phys. Chem. Ref. Data* 10, 1175-1199.

Mackay, D., Stiver, W. (1991) Chapter 8, Predictability and environmental chemistry. In: *Environmental Chemistry of Herbicides.* Vol. II, Grover, R., Cessna, A. J., Editors, pp. 281-297, CRC Press, Inc., Boca Raton, Florida.

Mackay, D., Wolkoff, A. W. (1973) Rate of evaporation of low-solubility contaminants from water bodies to atmosphere. *Environ. Sci. Technol.* 7, 611-614.

Macy, R. (1948) Partition coefficients of fifty compounds between olive oil and water at 20°C. *J. Ind. Hyg. Toxicol.* 30, 140.

Magee, P. S. (1991) Complex factors in hydrocarbon/water, soil/water, and fish/water partitioning. *Sci. Total Environ.* 109/110, 155-178.

Mailhot, H. (1987) Prediction of algae bioaccumulation and uptake rate of nine organic compounds by ten physicochemical properties. *Environ. Sci. Technol.* 21, 1009-1013.

Mailhot, H., Peters, R. H. (1988) Empirical relationships between the 1-octanol/water partition coefficient and nine physicochemical properties. *Environ. Sci. Technol.* 22, 1479-1488.

Maitlen, J. C., Powell, D. M. (1982) Persistance of aldicarb in soil relative to the carry-over of residues into crops. *J. Agric. Food Chem.* 30, 589-592.

Majewski, M. S., Capel, P. D. (1995) *Pesticides in the Atmosphere. Distribution, Trends, and Governing Factors.* Vol. 1 of the series *Pesticides in the Hydrologic System.* Gilliom, R. J., Editor, Ann Arbor Press, Inc., Chelsea, Michigan.

Mäkelä, P., Oikari, O. J. (1990) Uptake and body distribution of chlorinated phenols in the freshwater mussel, *Anodonta anatina* L. *Ecotoxicol. Environ. Saf.* 20, 354-362.

Malaiyandi, M., Shah, S. M., Lee, P. (1982) Fate of α- and γ-hexachlorocyclohexane isomers under simulated environmental conditions. *J. Environ. Sci. Health* A17(3), 283-297.

Maquire, R. J., Hale, E. J. (1980) Fenitrothion sprayed on a pond: Kinetics of its distribution and transformation in water and sediment. *J. Agric. Food Chem.* 28, 372-378.

Markwell, R. D., Connell, D. W., Gabric, A. J. (1989) Bioaccumulation of lipophilic compounds from sediments by oligochaetes. *Water Res.* 23(11), 1443-1450.

Martens, R. (1972) Decomposition of endosulfan by soil microorganisms. *Schrifter Ver Wasser-, Bodden-, Luftig, Berlin-Dahlem* 37, 167-173.

Martin, H. (1972) *Pesticide Manual*, 3rd Edition, British Crop Protection Council, Worcester, England.

Martin, H., Worthing, C. R., Editors (1977) *Pesticide Manual.* 5th Edition, British Crop Protection Council. Thornton Heath, United Kingdom.

Mason, J. W., Rowe, D. R. (1976) The accumulation and loss of dieldrin and endrin in the eastern oyster. *Arch. Environ. Contam. Toxicol.* 4, 349-360.

Masterton, W. L., Lee, T. P. (1972) Effects of dissolved salts on water solubility of lindane. *Environ. Sci. Technol.* 6, 919-921.

Matsumura, F., Benezet, H. J. (1973) Studies on the bioaccumulation and microbial degradation of 2,3,7,8-tetrachlorodibenzo-p-dioxin. *Environ. Health Perspt.* 253-258.

Maule, A., Plyte, S., Quick, A. V. (1987) Dehalogenation of organochlorine insecticides by mixed anaerobic microbial populations. *Pesti. Biochem. Physiol.* 277, 229-236.

Mayer, F. L., Mehrle, P. M., Dwyer, W. P. (1977) Toxaphene: Chronic Toxicity to Fathead Minnows and Channel Catfish. EPA-600/3-77-069, U.S. Environmental Protection Agency.

McCall, P. J., Swann, R. L., Laskowski, D. A., Unger, S. M., Vrona, S. A., Dishburger, H. J. (1980) Estimation of chemical mobility in soil from liquid chromatographic retention times. *Bull. Environ. Contam. Toxicol.* 24, 190-195.

McCarty, L. S., Mackay, D., Smith, A. D., Ozburn, G. W., Dixon, D. G. (1991) Interpreting aquatic toxicity QSARs: The significance of toxicant body residues at the pharmacologic endpoint. *Sci. Total Environ.* 109/110, 515-525.

McConnell, L. L., Cotham, W. E., Bildleman, T. F. (1993) Gas exchange of hexachlorocyclohexane in the Great Lakes. *Environ. Sci. Technol.* 27, 1304-1311.

McDowell, L. L., Willis, G. H., Murphree, C. E., Southwick, L. M., Smith, S. (1981) Toxaphene and sediments yields in runoff from a Mississippi delta watershed. *J. Environ. Qual.* 10, 120.

McDuffie, B. (1981) Estimation of octanol/water partition coefficients for organic pollutants using reverse-phase HPLC. *Chemosphere* 10, 73-83.

McKellar, R. L., Dishburger, H. J., Rice, J. R., Craig, L. F., Pennington, J. J. (1976) Residues of chlorpyrifos, its oxygen analogue, and 3,5,6-trichloro-2-pyridinol in milk and cream from cows fed chlorpyrifos. *J. Agric. Food Chem.* 24, 283-286.

McKim, J., Schnieder, P., Veith, G. (1985) Absorption dynamics of organic chemical transport across trout gills as related to octanol-water partition coefficients. *Toxicol. Appl. Pharmacol.* 77, 1-10.

McLachlan, M., Mackay, D., Jones, P. H. (1990) A conceptual model of organic chemical volatilization at waterfalls. *Environ. Sci. Technol.* 24, 252-257.

McLean, J. E., Sims, R. C., Doucette, W. J., Caupp, C. R., Grenney, W. J. (1988) Evaluation of mobility of pesticides in soil using U.S. EPA methodology. *J. Environ. Eng.* 114, 689-703.

McLeese, D. W., Metcalf, C. D., Zitko, V. (1980) Lethality of permethrin, cypermethrin and fenvalerate to salmon, lobster and shrimp. *Bull. Environ. Contam. Toxicol.* 25, 950-955.

McLeese, D. W., Sergent, D. B., Metcalf, C. D., Zitko, V., Burridge, L. E. (1979) Uptake and excretion of aminocarb, nonylphenol and pesticide diluent 585 by mussels (*Mytilus edulis*). *Bull. Environ. Contam. Toxicol.* 24, 575-581.

McLeese, D. W., Zetko, V., Sergent, D. B. (1976) Uptake and excretion of fenitrothion by clams and mussel. *Bull. Environ. Contam. Toxicol.* 16, 508-515.

Means, J. C., Woods, S. G., Hassett, J. J., Banwart, W. L. (1982) Sorption of amino- and carboxy-substituted polynuclear aromatic hydrocarbons by sediments and soils. *Environ. Sci. Technol.* 16, 93-98.

Medchem (1988) Medchem Database, Release 3.54 of 1988. Daylight Chemical Information System Inc., California.

Meikle, R. W., Youngson, C. R. (1978) The hydrolysis rate of chlorpyrifos, O-O-diethyl-o-(3,5,6-trichloro-2-pyridyl)phosphorothioate,and its dimethyl analog, chlorpyrifos-methyl, in dilute aqueous solution. *Arch. Environ. Contam. Toxicol.* 7, 13-22.

Meikle, R. W., Kurihara, N. H., DeVries, D. H. (1983) Chlorpyrifos: The photodecomposition rates in dilute aqueous solution and on a surface, and the volatilization rate from a surface. *Arch. Environ. Contam. Toxicol.* 12, 189.

Melnikov, N. N. (1971) Chemistry of pesticides. *Res. Rev.* 36, 1-447.

Menn, J. J. (1969) Personal communication, Stauffer Chemical Co., Mountain View, California.

Menzie, C. A., Burmaster, D. E., Freshman, J. S., Callahan, C. A. (1992) Assessment of methods for estimating ecological risk in the terrestrial component: A case study at the Baird and McGuire Superund site in Holbrook, Massachusetts. *Environ. Toxicol. Chem.* 11, 245-260.

Menzie, C. M. (1972) Fate of pesticides in the environment. *Ann. Rev. Entomol.* 17, 199.

The Merck Index (1983) *An Encyclopedia of Chemicals, Drugs and Biologicals.* 10th Edition, Widholz, M., Editor, Merck and Co., Inc., Rahway, New Jersey.

The Merck Index (1989) *An Encyclopedia of Chemicals, Drugs and Biologicals.* 11th Edition, Budavari, S., Editor, Merck and Co., Inc., Rahway, New Jersey.

Metcalf, C. D., McLeese, D. W., Zitko, V. (1980) Rate of volatilization of fenitrothion from fresh water. *Chemosphere* 9, 151-155.

Metcalf, R. L. (1948) *The Mode of Action of Organic Insecticides.* Chemical-Biological Coordination Review No. 1. National Research Council, Washington DC.

Metcalf, R. L. (1971) The chemistry and biology of pesticides. In: *Pesticides in the Environment*. White-Stevens, J., Ed., Part I, Vol. 1, p. 50, Marcel Dekker, New York.

Metcalf, R. L. (1974) In: Comparative Studies of Food and Environmental Contaminants. Proceedings of the FAO/IAEA/WHO Symposium, Otaniemi, International Atomic Energy Agency, Vienna. pp. 3-22.

Metcalf, R. L., Kapoor, I. P., Lu, P-Y., Schuth, C. K., Sherman, P. (1973) Model ecosystem studies of the environemntal fate of six organochlorine pesticides. *Environ. Health Perspect.* 4, 35-44.

Metcalf, R. L., Sanborn, J. R. (1975) *Illinois Natural History Survey Bulletin* 31, 381-436.

Metcalf, R. L., Sanborn, J. R., Lu, P.-Y., Nye, D. (1975) Laboratory model ecosystem studies of the degradation and fate of radiolabeled tri-, tetra-, and pentachlorobiphenyl compared with DDE. *Arch. Environ. Contam. Toxicol.* 3, 151-165.

Meylan, W., Howard, P. H. (1991) Bond contribution method for estimating Henry's law constants. *Environ. Toxicol. Chem.* 10, 1283-1293.

Meylan, W., Howard, P. H., Boethling, R. S. (1992) Molecular topology/fragment contribution method for predicting soil sorption coefficients. *Environ. Sci. Technol.* 26, 1560-1567.

Miles, J. R. W. (1976) Fates of insecticides applied to lands and crops. *Pest. Monit. J.* 10, 87-91.

Miles, J. R. W., Delfino, J. J. (1985) Fate of aldicarb, aldicarb sulfoxide, and aldicarb sulfone in Floridan groundwater. *J. Agric. Food Chem.* 33(3), 455-460.

Miles, J. R. W., Harris, C. R. (1978) Insecticide residues in water, sediment, and fish of the drainage system of the Holland Marsh, Ontario, Canada. *J. Econ. Ent.* 71, 125-131.

Miles, J. R. W., Tu, C. M., Harris, C. R. (1979) Persistence of eight organophosphorous insecticides in sterile and non-sterile mineral and organic soils. *Bull. Environ. Contam. Toxicol.* 22, 312-318.

Mill, T., Hendry, D. M., Mabey, W. E., Johnson, D. J. (1980) Laboratory protocols for evaluating fate of organic chemicals in air and water. EPA-600/3-80-069. U.S. Enveirnmental Protection Agency, Washington DC.

Mill, T., Mabey, W. E. (1985) Photochemical transformations. In: *Environmental Exposure from Chemicals.* Neely, W.B., Blau, G.E., Editors, pp. 175-213, CRC Press, Inc., Boca Raton, Florida.

Miller, C. T., Weber, W. J., Jr. (1986) Sorptions of hydrophobic organic pollutants in saturated soil systems. *J. Contam. Hydrol.* 1, 243.

Mills, W. B., Dean, J. D., Porcella, D. B., Gherini, S. A., Hudson, R. J. M., Frick, W. E., Rupp, G. L. (1982) Water quality assessment: A screening procedure for toxic and conventional pollutants. Part 1, U.S. EPA Report No. EPA-600/6-82-004a, Environmental Research Lab., U. S. Environmental Protection Agency, Athens, Georgia.

Milne, G. W. A., Editor (1995) *CRC Handbook of Pesticides.* CRC Press, Boca Raton, Florida.

Minero, C., Pelizzetti, E., Malato, S., Blanco, J. (1993) Large solar plant photocatalytic water decontamination: degradation of pentachlorophenol. *Chemosphere* 26, 2103-2119.

Mingelgrin, U., Gerstl, Z. (1983) Reevaluation of partitioning as a mechanism of nonionic chemicals adsorption in soils. *J. Environ. Qual.* 12(1), 1-11.

Miyake, K., Terada, H. (1982) Determination of partition coefficients of very hydrophobic compounds by high-performance liquid chromatography on glycerol-coated controlled-pore glass. *J. Chromatogr.* 240, 9-20.

Montgomery, J. H. (1993) *Agrochemicals Desk Reference. Environmental Data.* Lewis Publishers, Chelsea, Michigan.

Moos, L. P., Kiesch, E. J., Wukasch, R. F., Grady, C. P. L., Jr. (1983) Pentachlorophenol biodegradation - I. Aerobic. *Water Res.* 17, 1575-1584.

Morrill, L. G., Mahilum, B. C., Mohiuddin, S. H. (1982) *Organic Compounds in Soils.* Ann Arbor Science Publishers, Inc., Ann Arbor, Michigan.

Mortimer, M. R., Connell, D. W. (1995) A model of the environmental fate of chloro-hydrocarbon contaminants associated with Sydney sewage discharge. *Chemosphere* 30, 2021-2038.

Mudami, A. R., Hassett, J. P. (1988) Photochemical activity of mirex associated with dissolved organic matter. *Chemosphere* 17, 1133-1146.

Muir, D. C. G., Hobdem, B. R., Servos, M. R. (1994) Bioconcentration of pyrethroid insecticides and DDT by rainbow trout: Uptake, depuration, and effect of dissolved organic carbon. *Aquatic Toxicol.* 29, 223-240.

Muir, D. C. G., Rawn, G. P., Townsend, B. E., Lockhart, W. L. (1985) Bioconcentration of cypermethrein, deltamethrin, fenvalerate and permethrin by *Chironomus tentans* larvae in sediment and water. *Environ. Toxicol. Chem.* 4, 51-61.

Mulla, M. S., Milan, L. S., Kawecki, J. A. (1981) Distribution, transport, and fate of the insecticides malathion and parathion in the environment. *Res. Rev.* 81, 1-172.

Müller, J. F., Hawker, D. W., Connell, D. W. (1994) Calculation of bioconcentration factors of persistent hydrophobic compounds in the air/vegetation system. *Chemosphere* 29, 623-640.

Müller, M., Klein, W. (1992) Comparative evaluation of methods predicting water solubility for organic compounds. *Chemosphere* 25, 769-782.

Murphy, T. J., Mullin, M. D., Meyer, J. A. (1987) Equilibration of polychlorinated biphenyls and toxaphene with air and water. *Environ. Sci. Technol.* 21, 155-162.

Myrdal, P. B., Manka, A. M., Yalkowsky, S. H. (1995) Aquafac 3: Aqueous functional group activity coefficients; Application to the estimation of aqueous solubility. *Chemosphere* 30, 1619-1637.

Nash, R. G. (1974) In: *Pesticides in Soil and Water.* Guenzi, W. D., Editor. Soil Science Society of America, Madison, Wisconsin.

Nash, R. G. (1980) Dissipation rrate of pesticides from soils. In: CREAMS: A field scale model for chemical, runoff, and erosion from agricultural management systems. Vol. 3, Knisel, W. G., Editor, pp. 560-594, USDA Conserv. Res. Rep. 26, U.S. Government Printing Office, Washington, DC.

Nash, R. G. (1983) Comparative volatilization and dissipation rates of several pesticides from soil. *J. Agric. Food Chem.* 31, 210-217.

Nash, R. G. (1983) Determining environmental fate of pesticides with microagroecosystems. *Res. Rev.* 85, 199-215.

Nash, R. G. (1988) Chapter 5. Dissipation from soil. In: *Environmental Chemistry of Herbicides.* Volume I, Grover, R., Editor, pp. 131-169, CRC Press, Inc., Boca Raton, Florida.

Nash, R. G. (1989) Models for estimating pesticide dissipation from soil and vapor decline in air. *Chemosphere* 18, 2375-2381.

Nash, R. G., Harris, W. G. (1983) Toxaphene and 1,1,1-trichloro-2,2-bis(*p*-chlorophenyl)ethane (DDT) losses from cotton in an agroecosystem chamber. *J. Agric. Food Chem.* 25, 336.

Nash, R. G., Woolson, E. A. (1967) Persistence of chlorinated hydrocarbon insecticides. *Science* 157, 924-927.

Neary, D. G., Bush, P. B., Michael, J. L. (1993) Fate, dissipation and environmental effects of pesticides in sourthern forests: A review of a decade of research progress. *Environ. Toxicol. Chem.* 12, 411-428.

Neely, W. B. (1976) Predicting the flux of organics across the air/water interface. National Conference on Control of Hazardous Materials Spills, New Orleans, Louisiana, 1976.

Neely, W. B. (1978) Personal communication. Dow Chemical Company, Midland, Michigan.

Neely, W. B. (1980) Chapter 20. A method for selecting the most appropriate environmental experiments on a new chemical. In: *Dynamics, Exposure and Hazard Assessment of Toxic Chemicals.* Haque, R., Ed., pp. 287-196, Ann Arbor Science Publishers, Ann Arbor, Michigan.

Neely, W. B. (1982) Review. Organizing data for environmental studies. *Environ. Toxicol. Chem.* 1, 259-266.

Neely, W. B., Blau, G. E. (1977) The use of laboratory data to predict the distribution of chlorpyrifos in a fish pond. In: *Pesticides in Aquatic Environments.* Khan, M. A. Q., Editor, Plenum Press, New York.

Neely, W. B., Blau, G. E. (1985) Chapter 1. Introduction to Environmental exposure from chemicals. In: *Environmental Exposure from Chemicals.* Neely, W. B., Blau, G. E., Editors, pp. 1-12, CRC Press, Inc., Boca Raton, Florida.

Neely, W. B., Blau, G. E., Editors (1985) *Environmental Exposure from Chemicals.* CRC Press, Inc., Boca Raton, Florida.

Neely, W. B., Branson, D. R., Blau, G. E. (1974) Partition coefficient to measure bioconcentration potential of organic chemicals in fish. *Environ. Sci. Technol.* 8, 1113-1115.

Nendza, M. (1991) Predictive QSAR models estimating ecotoxic hazard of phenylureas: Aquatic toxicity. *Chemosphere* 23, 497-506.

Nendza, M., Seydel, J. K. (1988) Quantitative structure-toxicity relationship for ecotoxicologically relevant biotest systems and chemicals. *Chemosphere* 17, 1585-1602.

Neudorf, S., Khan, M. A. Q. (1975) Pick-up and metabolism of DDt, dieldrin and photodieldrin by fresh water algae (*Ankistrodesmus*) and a micro-crustacean (*Daphnia pulex*). *Bull. Environ. Contam. Toxicol.* 13, 443-450.

Neumüller, O. A. (1974) *Römpp's Chemie-Lexikon.* p. 2538, Frank'sche Verlagsbuchhandlung, Stuttgart.

Ngabe, B., Bidleman, T. F., Falconer, R. L. (1993) Base hydrolysis of α- and γ-hexachlorocyclohexanes. *Environ. Sci. Technol.* 27, 1930-1933.

NIEHS (1975) National Institute of Environmental Health Services Grant No. ES 00040-10 Annual Progress Report.

Niimi, A. J. (1987) Biological half-life of chemicals in fishes. *Rev. Environ. Contam. Toxicol.* 99, 1-46.

Niimi, A. J., Cho, C. Y. (1983) Laboratory and field analysis of pentachlorophenol (PCP) accumulation by salmonids. *Water Res.* 17, 1791-1795.

Niimi, A. J., Palazzo, V. (1985) Temperature effect on the elimination of pentachlorophenol, hexachlorobenzene and mirex by rainbow trout (*Salmo gairdneri*). *Water Res.* 19(2), 205-207.

Nirmalakhandan, N. N., Speece, R. E. (1988) QSAR model for predicting Henry's law constant. *Environ. Sci. Technol.* 22, 1349-1357.

Norstrom, R. J., Clark, T. P., Jeffrey, D. A., Won, H. T., Gilman, A. P. (1986) Dynamics of organochlorine compounds in herring gulls (*Larus argentatus*): I. Distribution and clearance of [^{14}C]DDE in free-living herring gulls *(Larus argentatus). Environ. Toxicol. Chem.* 5, 41-48.

Noegrohati, S., Hammers, W. E. (1992) Regression models for octanol-water partition coefficients, and for bioconcentration in fish. *Toxicol. Environ. Chem.* 34, 155-173.

NRC (1974) Chlordane; Its effects on Canadian ecosystems and its chemistry. NRCC No. 14094, National Research Council, Ottawa, Canada.

O'Brien, R. D. (1974) Nonenzymic effects of pesticides on membranes. In: *Environmental Dynamics of Pesticides.* Haque, R., Freed, V. H., Editors, pp. 331-342, Plenum Press, New York.

OECD (1981) OECD Guidelines for testing of Chemicals, Paris.

Oliver, B. G. (1987) Biouptake of chlorinated hydrocarbons from laboratory-spiked and field sediments by oligochaete worms. *Environ. Sci. Technol.* 21, 785-790.

Oliver, B. G., Charlton, M. N. (1984) Chlorinated organic contaminants on settling particulates in the Niagara River vicinity of Lake Ontario. *Environ. Sci. Technol.* 18, 903-908.

Oliver, B. G., Charlton, M. N., Durham, R. W. (1989) Distribution, redistribution, and geochronology of polychlorinated biphenyl congeners and other chlorinated hydrocarbons in Lake Ontario sediments. *Environ. Sci. Technol.* 23, 200-208.

Oliver, B. G., Niimi, A. J. (1985) Bioconcentration factors of some halogenated organics for rainbow trout: Limitations in their use for prediction of environmental residues. *Environ. Sci. Technol.* 19, 842-849.

Oliver, B. G., Niimi, A. J. (1988) Trophodynamic analysis of polychlorinated biphenyl congeners and other chlorinated hydrocarbons in the Lake Ontario ecosystems. *Environ. Sci. Technol.* 22, 388-397.

Othman, M. A., Antonious, G. F., Khattab, M. M., Abdel-All, A., Khamis, A. E. (1987) Residues of dimethioate and methomyl on tomato and cabbage in relation to their effect on quality-related properties. *Environ. Toxicol. Chem.* 6, 947-952.

Ou, L. T., Sture, K., Edvardsson, V., Suresh, P., Rao, C. (1985) Aerobic and anaerobic degradation of aldicarb in soils. *J. Agric. Food Chem.* 33, 72-78.

Oubina, A., Ferrer, I, Gascon, J., Barcelo, D. (1996) Disappearance of aerially applied fenitrothion in rice crop waters. *Environ. Sci. Technol.* 30, 3551-3557.

Pait, A. W., De Souza, A. E., Farrow, D. R. G. (1992) *Agricultural Pesticide Use in Coastal Areas: A National Summary.* National Oceanic and Atmospheric Administration, Rockville, Maryland.

Pangrekar, J., Klopman, G., Rosenkranz, H. S. (1994) Expert system comparison of structural determinants of chemical toxicity to environmental bacteria. *Environ. Toxicol. Chem.* 13(6), 979-1001.

Park, S.S., Erstfeld, K.M. (1997) A numerical kinetic medel for bioaccumulatin of organic chemicals in sediment-water system. *Chemosphere* 34, 419-427.

Paris, D. F., Lewis, D. L. (1976) Accumulation of methoxychlor by microorganisms. *Bull. Environ. Contam. Toxicol.* 15, 24.

Paris, D. F., Lewis, D. L., Barnett, J. T. (1977) Bioconcentration of toxaphene by microorganisms. *Bull. Environ. Contam. Toxicol.* 17, 564-573.

Paris, D. F., Lewis, D. L., Barnett, J. T., Baughman, G. L. (1975) Microbial Degradation and Accumulation of Pesticides in Aquatic Systems. Report No. U.S. EPA-660/375-007. U.S. Environmental Protection Agency, Athens, Georgia.

Paris, D. F., Lewis, D. L., Barnett, J. T., Baughman, G. L. (1975) Microbial Degradation and Accumulation of Pesticides in Aquatic Systems. Report No. EPA-660/3-75-007, U.S. Environmental Protection Agency, Athens, Georgia.

Paris, D. F., Lewis, D. L., Wolfe, N. L. (1975) Rates of degradation of malathion by bacteria isolated from aquatic system. *Environ. Sci. Technol.* 9, 135.

Paris, D. F., Steen, W. C., Baughman, G. L. (1978) Prediction of microbial transformation of pesticides in natural waters. (unpublished), presented before the American Chemical Society, Division of Pesticide Chemistry, Anaheim, Calif., Environmental Research Laboratory, U.S. EPA, Athens, Georgia.

Paris, D. F., Steen, W. C., Baughman, G. L., Barnett, J. T. (1981) Second-order model to predict microbial degradation of organic compounds in natural waters. *Appl. Environ. Microbiol.* 41, 603-609.

Park, K. S., Bruce, W. N. (1968) The determination of the water solubility of aldrin, dieldrin, heptochlor and heptachlor epoxide. *J. Econ. Entomol.* 61(3), 770-774.

Parkerton, T. F., Connolly, J. P., Thomann, R. V., Uchrin, C. G. (1993) Do aquatic effects of human health end points govern the development of sediment-quality criteria for nonionic organic chemicals? *Environ. Toxicol. Chem.* 12, 507-523.

Parrish, P. R. (1974) Aroclor 1254, DDT and DDD, and dieldrin: Accumulation and loss by American oysters (*Crassostrea virginica*) exposed continuously for 56 weeks. *Proc. Natl. Shellfish Assoc.* 64, 7.

Parrish, P. R., Dyar, E. E., Enos, J. M., Wilson, W. G. (1978) Chronic toxicity of chlordane, trifuralin, and pentachlorophenol to sheepshead minnows (*Cyprinodon variegatus*). *EPA Ecol. Res. Ser.* EPA-600/3-78-010. U.S. Environmental Protection Agency, Gulf Breeze, Florida.

Parrish, P. R., Dyar, E. E., Lindberg, M. A., Shanika, C. M., Enos, J. M. (1977) Chronic toxicity of methoxychlor, malathion and carbofuran to sheepshead minnows (*Cyprinodon variegatus*). NTIS PB-272101.

Parrish, P. R., Schimmel, S. C., Hansen, D. J., Patrick, J. M., Jr., Forester, J. (1976) Chlordane: Effects on several estuarine organisms. *J. Toxicol. Environ. Health* 1, 485.

Pasarela, N. R., Brown, R. G., Shaffer, C. B. (1962) Feeding of malathion to cattle; residue analyses of milk and tissue. *J. Agric. Food Chem.* 10, 7.

Pass, B. C., Dorough, H. W. (1973) Insecticidal and residual properties of EC and encapsulated formulations of methyl parathion sprayed on alfalfa. *J. Econ. Entomol.* 66(5), 1117-1119.

Patil, G. S. (1994) Prediction of aqueous solubility and octanol-water partition coefficient for pesticides based on their molecular structure. *J. Hazard. Materials* 36, 35-43.

Paterson, S., Mackay, D. (1985) The fugacity concept in environmental modelling. In: *The Handbook of Environmental Chemistry.* Vol. 2, Part C, Hutzinger, O., Editor, pp. 121-140, Springer-Verlag, Heidelberg, Germany.

Patton, J. S., Stone, B., Papa, C., Abramowitz, R., Yalkowsky, S. H. (1984) Solubility of fatty acids and other hydrophobic molecules in liquid trioleoylglycerol. *J. Lipid Res.* 25, 189-197.

Pavlou, S. P., Weston, D. P. (1983, 1984) Initial Evaluation of Alternatives for Development of Sediment Related Criteria for Toxic Contaminants in Marine Waters (Puget Sound), Phase I and II. EPA Contract No. 68-01-6388.

Perrin, D. D. (1989) *pK_a Prediction for Organic Acids and Bases.* Chapman & Hall, New York.

Pinsuwan, S., Li, A., Yalkowsky, S. H. (1995) Correlation of octanol/water solubility ratios and partition coefficients. *J. Chem. Eng. Data* 40, 623-626.

Platford, R. F. (1981) The environmental significance of surface films. II. Enhanced partitioning of lindane in thin films of octanol on the surface of water. *Chemosphere* 10(7), 719-722.

Platford, R. F. (1982) Pesticide partitioning in artificial surface films. *J. Great Lakes Res.* 8, 307-309.

Platford, R. F. (1983) The octanol-water partitioning of some hydrophobic and hydrophilic compounds. *Chemosphere* 12, 1107-1111.

Platford, R. F., Carey, J. H., Hale, E. J. (1982) The environmental significance of surface films: Part 1. Octanol-water partition coefficients for DDT and hexachlorobenzene. *Environ. Pollution* (series B), 125-128.

Plato, C. (1972) Differential scanning calorimetry as a general method for determining the purity and heat of fusion of high purity organic chemicals. Application to 64 compounds. *Anal. Chem.* 44, 1531-1534.

Plato, C., Glasglow, A. R., Jr. (1969) Differential scanning calorimetry as a general method for determining the purity and heat of fusion of high purity organic chemicals. Application to 95 compounds. *Anal. Chem.* 41, 330-336.

Porter, P. E. (1964) Dieldrin. In: *Analytical Methods for Pesticides, Plant Growth Regulators and Food Additives.* Vol. 2, Zweig, G., Editor, pp. 143-163, Academic Press, New York.

Portier, R. J. (1985) In: *ASTM STP 865.* (Validat. Predict. Lab Methods Assess. Fate Eff. Contam. Aquat. Ecosyst.), pp. 14-30.

Potter, J. C., Marxmiller, R. L., Barber, G. F., Young, R., Loefller, J. E., Burton, W. B., Dixon, L. D. (1974) Total ^{14}C residues and dieldrin residues in milk and tissues of cows fed dieldrin-^{14}C. *J. Agric. Food Chem.* 22, 889-899.

Quaife, M. L., Winbush, J. S., Fitzhugh, O. G. (1967) Survey of quantitative relations between ingestion and storage of aldrin and dieldrin in animals and man. *Fed. Cosmet. Toxicol.* 5, 39-50.

Quellette, R. P., King, J. A. (1977) *Chemical Week. Pesticide Register.* McGraw-Hill Inc., New York.

Radleleff, R. D., Bushland, R. C., Claborn, H. V. (1952) *Insects: The Yearbook of Agriculture.* U.S. Dept. of Agriculture, Washington DC.

Ramamoorthy, S. (1985) Competition of fate processes in the bioconcentration of lindane. *Bull. Environ. Contam. Toxicol.* 34, 349-358.

Rao, P. S. C., Davidson, J. M. (1979) Adsorption and movement of selected pesticides at high concentrations in soils. *Water Res.* 13, 375-380.

Rao, P. S. C., Davidson, J. M. (1980) Estimation of pesticide retention and transformation parameters required in nonpoint source pollutant models. In: *Environmetal Impact of Nonpoint Pollution.* Overcash, M. R., Davidson, J. M., Editors, Ann Arbor Science Publishers Inc., Ann Arbor, Michigan.

Rao, P. S. C., Davidson, J. M. (1982) Retention and Transformation of Selected Pesticides and Phosphorus in Soil Water System: A Critical Review. U.S. EPA-600/S3-82-060.

Rao, P. S. C. et al. (1984) Degradation and sorption of aldicarb and metolachlor in Dougherty Plains soils. Progress report to U.S. EPA of EPA Co-operative Agreement CR-810464.

Reich, A. R., Perkins, J. L., Cutter, G. (1986) DDT contamination of a North Alabama aquatic ecosystem. *Environ. Toxicol. Chem.* 5, 725-736.

Reinert, R. E. (1967) The accumulation of deldrin in an algal (*Scenedesmus obliquus*), daphnia (*Daphnia magna*), guppy (*Lebistes reticulatus*) food chain. *Diss. Abstr.* 28, 2210-B.

Reinert, R. E. (1972) The accumulation of deldrin in an alga (*Scenedesmus obliquus*), daphnia (*Daphnia magna*), guppy (*Lebistes reticulatus*). *J. Fish Res. Board Can.* 29, 1413-1418.

Reish, D. J., Kauwling, T. J., Mearns, A. J., Oshida, P. S., Rossi, S. S., Wilkes, F. G., Ray, M. J. (1978) Marine and estuarine pollution. *J. Water Pollut. Control Fed.* 50, 1424-1469.

Rekker, R. F. (1977) *The Hydrophobic Constants: Its Derivation and Application; A means of Characterizing Membrane Systems.* Nauta, W. T., Rekker, R. F., Editors, Elsevier Scientific Publishing Company, New York.

Rekker, R. F., ter Laak, A. M., Mannhold, R. (1993) On the reliability of calculated Log P-values: Rekker, Hansch/Leo and Suzuki approach. *Quant. Struct.-Act. Relat.* 12, 152-157.

Renberg, L. (1981) Gas chromatographic determination of chlorophenols in environmental samples. *National Swedish Environment Protection Board Report* 1410, 135pp.

Renberg, L., Sundström, G. (1979) Prediction of bioconcentration potential of organic compounds using partition coefficients derived from reversed phase thin layer chromatography. *Chemosphere* 7, 449-459.

Renberg, L., Sundström, G., Rosen-Olofsson, S. (1985) The determination of partition coefficients of organic compounds in technical products and waste waters for the estimation of their bioaccumulation potential using reversed phase thin layer chromatography. *Toxicol. Environ. Chem.* 10, 333-349.

Renberg, L., Tarkpea, M., Linden, E. (1985) The use of the *Bivalve Mystilus edulis* as a test organism for bioconcentration studies. *Ecotox. Environ. Safety* 9, 171.

Renner, G. (1990) Gas chromatographic studies of chlorinated phenols, chlorinated anisoles, and chlorinated phenylacetates. *Toxicol. Environ. Chem.* 27, 217-224.

Richards, A. G., Cutkomp, L. K. (1946) Correlation between the possession of a chitinous cuticle and sensitivity to DDT. *Biol. Bull.* 90, 97-108.

Richards, R. P., Baker, D. B. (1993) Pesticide concentration patterns in agricultural drainage networks in the Lake Erie basin. *Environ. Toxicol. Chem.* 12, 13-26.

Richardson, G. M., Qadri, S. U. (1986) Tissue distribution of ^{14}C-labeled residues of aminocarb in brownhead (*Ictalurus nebulosus Le Sueur*) following acute exposure. *Ecotoxicol. Environ. Saf.* 12, 180-186.

Richardson, L. T., Miller, D. M. (1960) Fungitoxicity of chlorinated hydrocarbon insecticides in relation to water solubility and vapor pressure. *Can. J. Botany* 38, 163-175.

Roark, R. C. (1951) A digest of information on chlordane. U.S. Dept. of Agriculture, Bureau Entomol. and Plant Quarantine E-817. 132pp.

Robeck, G. G., Dostal, K. K., Cohen, J. M., Dreissal, J. F. (1965) Effectiveness of water treatment processes in pesticide removal. *J. Am. Water Works Assn.* 57, 181-200.

Roberts, J. R., De Frietas, A. S. W., Gidney, M. A. J. (1977) Influence of lipid pool size on bioaccumulation of the insecticide chlordane by northern redhorse suckers (*Moxostoma macrolepidotum*). *J. Fish Res. Board Can.* 34, 89.

Robinson, J., Roberts, M., Baldwin, M., Walker, A. I. T. (1969) Pharmacokinetics of HEOD (dieldrin) in the rat. *Fed. Cosmet. Toxicol.* 7, 317-332.

Rose, F. L., McIntire, C. D. (1970) Accumulation of dieldrin by benthic algae in laboratory streams. *Hydrobiologia* 35, 481.

Rothman, A. M. (1980) Low vapor pressure determination by the radiotracer transpiration method. *J. Agric. Food Chem.* 28, 1225-1228.

Ruelle, P., Kesselring, U. W. (1997) Aqueous solubility prediction of environmentally important chemicals from the mobile order thermodynamics. *Chemosphere* 34(2), 275-298.

Ruzicka, J. H., Thomson, J., Wheals, B. B. (1967) The gas chromatographic determination of organophosphorous pesticides. Part II. A comparative study of hydrolysis rates. *J. Chromatogr.* 31, 37-47.

Ryan, J. A., Bell, R. M., Davidson, J. M., O'Connor, G. A. (1988) Plant uptake of non-ionic organic chemicals from soils. *Chemosphere* 17, 2299-2322.

Saarikoski, J., Viluksela, M. (1982) Relation between physicochemical properties of phenols and their toxicity and accumulation in fish. *Ecotoxicol. Environ. Saf.* 6, 501-512.

Saarikoski, J., Lindström, R., Tyynelä, M., Viluksela, M. (1986) Factors affecting the absorption of phenolics and carboxylic acids in the guppy (*Poecilia reticulata*). *Ecotoxicol. Environ. Saf.* 11, 158-173.

Sabljic, A. (1984) Prediction of the nature and strength of soil sorption of organic pollutants by molecular topology. *J. Agric. Food Chem.* 32, 243-246.

Sabljic, A. (1987a) On the prediction of soil sorption coefficients of organic pollutants from molecular structure: Application of molecular topology model. *Environ. Sci. Technol.* 21, 358-366.

Sabljic, A. (1987b) Nonempirical modeling of environmental distribution and toxicity of major organic pollutants. In: *QSAR in Environmental Toxicology - II*. Kaiser, K. L. E., Editor, pp. 309-322, D. Reidel Publ. Co., Dordrecht, The Netherlands.

Saha, J. G. (1969) Significance of organochlorine insecticide residues in fresh plants as possible contaminants and beef products. *Res. Rev.* 26, 89-126.

Saito, S., Tanoue, A., Matsuo, M. (1992) Applicability of the i/o-characters to a quantitative description of bioconcentration of organic chemicals in fish. *Chemosphere* 24(1), 81-87.

Saito, S., Koyasu, J., Yoshida, K., Shigeoka, T., Koike, S. (1993) Cytotoxicity of 109 chemicals to goldfish GFS cells and relationships with 1-octanol/water partition coefficients. *Chemosphere* 26, 1015-1028.

Saleh, F. Y., Dickson, K. L., Rodgers, Jr. J. H. (1982) Fate of lindane in the aquatic environment: Rate constants of physical and chemical processes. *Environ. Toxicol. Chem.* 1, 289-297.

Samanidou, V., Fytianos, K., Pfister, G., Bahadir, M. (1988) Photochemical decomposition of waters of Northern Greece. *Sci. Total Environ.* 76, 85-92.

Sanborn, J. R., Metcalf, W. N. B., Bruce, W. N., Lu, P. Y. (1976) The fate of chlordane and toxaphene in a terrestrial-aquatic model ecosystem. *Environ. Entomol.* 5(3), 533-538.

Sanborn, T. R., Yu, C. (1973) The fate of dieldrin in a model ecosystem. *Bull. Environ. Contam. Toxicol.* 10, 340-346.

Sanchez-Camazano, M., Arienzo, M., Sanchez-Martin, M. J., Crisanto, T. (1995) Effect of different surfactants on the mobility of selected non-ionic pesticides in soil. *Chemosphere* 31(8), 37-93-3801.

Sancho, E., Ferrando, M. D., Andreu, E., Gamon, M. (1993) Bioconcentration and excretion of diazinon by eel. *Bull. Environ. Contam. Toxicol.* 50, 578-585.

Sanders, P. F., Jones, K. C., Hamilton-Taylor, J. (1993) A simple method to assess the susceptibility of polynuclear aromatic hydrocarbons to photolytic decomposition. *Atmos. Environ.* 27A, 139-144.

Sanders, P. F., Seiber, J. N. (1984) Organophosphorus pesticides volatilization. Model soil pits and evaporation ponds. In: *Treatment and Disposal of Pesticide Wastes*. Krueger, R.F., Seiber, J. N. Editors, Am. Chem. Soc. Sym. Series 259, 279-295.

Sangster, J. (1993) LOGKOW databank, Sangster Research Laboratory, Montreal, Quebec.

Sattar, M. A. (1990) Fate of organophosphorus pesticides in soils. *Chemosphere* 20, 387-396.

Schauberger, C. W., Wildman, R. B. (1977) Accumulation of aldrin and dieldrin by blue-green algae and related effects on photosynthetic pigments. *Bull. Environ. Contam. Toxicol.* 17, 534-541.

Schimmel, S. C., Patrick, Jr., J. M., Forester, J. (1976) Heptachlor: Uptake, depuration, retention, and metabolism by spot (*Leiostromus xanthurus*). *J. Toxicol. Environ. Health* 2, 169.

Schimmel, S. C., Patrick, Jr., J. M., Forester, J. (1977) Toxicity and bioconcentration of BHC and lindane in selected estuarine animals. *Arch. Environ. Contam. Toxicol.* 6, 355-363.

Schimmel, S. C., Garnas, R. L., Patrick, Jr., J. M., Moore, J. C. (1983) Acute toxicity, bioconcentration, and persistence of AC 222,705, benthiocarb, chlorpyrifos, fenvalerate, methyl parathion, and permethrin in the esturine environment. *J. Agric. Food Chem.* 31, 104-113.

Schellenberg, K., Leuenberger, C., Schwarzenbach, R. P. (1984) Sorption of chlorinated phenols by natural sediments and acquifer materials. *Environ. Sci. Technol.* 18, 652-657.

Schmidt-Bleek, F., Haberland, W., Klein, A. W., Caroli, S. (1982) Steps toward environmental hazard assessment of new chemicals (including a hazard ranking scheme, based upon directive 78/831/EEC). *Chemosphere* 11, 383-415.

Schnoor, J. L. (1992) 1. Chemical fate and transport in the environment. In: *Fate of Pesticides and Chemicals in the Environment*. Schnoor, J. L., Editor, pp. 1-24, John Wiley & Sons, Inc., New York.

Schnoor, J. L., Editor (1992) *Fate of Pesticides and Chemicals in the Environment*. John Wiley & Sons, Inc., New York.

Schnoor, J. L., McAvoy, D. C. (1981) Pesticide transport and bioconcentration model. *J. Environ. Eng. Div. (Am. Soc. Civ. Eng.)* 107(EE6), 1229-1246.

Schnoor, J. L., Sato, C., McKechnie, D., Sahoo, D. (1987) Processes, Coefficients, and Models for Simulating Toxic Organics and Heavy Metals in Surface Waters. EPA 600/3-87-015. U.S. Environmental Protection Agency, Athens, Georgia.

Schomburg, C. J., Glotfelty, D. E., Seiber, J. N. (1991) Pesticide occurrence and distribution in fog collected near Monterey, California. *Environ. Sci. Technol.* 25, 155-160.

Schreitmüller, J., Ballschmiter, K. (1995) Air-water equilibrium of hexachlorocyclohexanes and chloromethoxybenzenes in North and South Atlantic. *Environ. Sci. Technol.* 29, 207-215.

Schüürmann, G., Klein, W. (1988) Advances in bioconcentration prediction. *Chemosphere* 17, 1551-1574.

Scow, K. M. (1982) Rate of biodegradation. In: *Handbook of Chemical Property Estimation Methods*. Lyman, W. J., Rechl, W. F., Rosenblatt, D. H., Editors, McGraw-Hill, Inc., New York.

Seguchi, K., Asaka, S. (1981) Intake and excretion of diazinon in freshwater fishes. *Bull. Environ. Contam. Toxicol.* 27, 244-249.

Seiber, J. N. (1987) Solubility, partition coefficient and bioconcentration factor. In: *Fate of Pesticides in the Environment*. Biggar, J.W., Seiber, J.N., Editors, publication 3320 of the Agricultural Experiment Station, Division of Agriculture and Nature Resources, University of California, Oakland, California. pp. 53-59.

Seiber, J. N., Catahan, M. P., Barril, C. R. (1978) Loss of carbofuran from rice paddy water: Chemical and physical factors. *J. Environ. Sci. Health* B13, 131.

Seiber, J. N., Madden, S. C., McChesney, M. M., Winterlin, W. N. (1979) Toxaphene dissipation from treated cotton field environments: Component residual behaviour on leaves and in air, soil, and sediments determined by capillary gas chromatography. *J. Agric. Food Chem.* 27, 284.

Seiber, J. N., McChesney, M. M. (1987) Measurement and computer model simulation of the volatilization flux of molinate and methyl parathion from a flooded rice field. Final Report to Department of Food and Agriculture, Sacramento, California.

Seiber, J. N., McChesney, M. M., Woodrow, J. E. (1989) Airborne residues resulting from use of methyl parathion, molinate and thiobencarb on rice in the Sacramento Valley, California. *Environ. Toxicol. Chem.* 8, 577-588.

Seiber, J. N., McChesney, M. M., Sanders, P. F., Woodrow, J. E. (1986) Models for assessing the volatilization of herbicides applied to flooded rice fields. *Chemosphere* 15, 127-138.

Seiber, J. N., Woodrow, J. E., Sanders, P. F. (1981) Estimation of ambient vapor pressures of pesticides from gas chromatographic retention data. Abstract, 183rd Am. Chem. Soc. Meeting, New York.

Sethunathan, N., MacRae, I. C. (1969) Persistence and biodegradation of diazinon in submerged soils. *J. Agric. Food Chem.* 17(2), 221-225.

Sharom, M. S., Miles, J. R. W., Harris, C. R., McEwen, F. L. (1980) Persistence of 12 insecticides in water. *Water Res.* 14, 1089-1093.

Sharom, M. S., Miles, J. R. W., Harris, C. R., McEwen, F. L. (1980) Behaviour of 12 insecticides in soil and aqueous suspensions of soil and sediment. *Water Res.* 14, 1095-1100.

Shigeoka, T., Yamagata, T., Minoda, T., Yamauchi, F. (1988) *Jpn. J. Toxicol. Environ. Health* 34, 343-349.

Shiu, W. Y., Ma, K. C., Mackay, D. (1990) Solubilities of pesticides in water. Part 1, Environmental physical chemistry and Part 2, Data compilation. *Reviews Environ. Contam. Toxicol.* 115, 1-187.

Shiu, W. Y., Ma, K. C., Varhanickova, D., Mackay, D. (1994) Chlorophenols and alkylphenols: A review and correlation of environmentally relevant properties and fate in an evaluative environment. *Chemosphere* 29, 1155-1224.

Sicbaldi, F., Finizio, A. (1993) K_{ow} estimation by combination of RP-HPLC and molecular indexes for a heterogeneous set of pesticide. In: *Proceedings IX Synposium Pesticide Chemistry, Mobility and Degradation of Xenobiotics.* 11-13, Oct 1993, Piacenza, Italy.

Siebers, J., Gottschild, D., Nolting, H.-G. (1994) Pesticides in precipitation in Northern Germany. *Chemosphere* 28(8), 1559-1570.

Siebers, J., Mattusch, P. (1996) Determination of airborne residues in greenhouses after application of pesticides. *Chemosphere* 33(8), 1597-1607.

Sillén, L. G., Martell, A. E. (1971) *Stability Constants of Metal-Ion Complexes*. Supplement No.1, Spec. Publ. No. 25, The Chemical Society, London, England.

Skea, J. C., Simonin, H. J., Symula, J. (1981) Accumulation and retention of mirex by brook trout fed a contaminated diet. *Bull. Environ. Contam. Toxicol.* 27, 79-83.

Slade, R. E. (1945) The γ-isomer of hexachlorocyclohexane (Gammexane). An insecticide with outstanding properties. *Chem. Ind.* 40, 314-319.

Slater, R. M., Spedding, D. J. (1981) Transport of dieldrin between air and water. *Arch. Environ. Contam. Toxicol.* 10, 25-33.

Smelt, J. H., Dekker, A., Leistra, M., Houx, N. W. H. (1983) Conversion from carbamoyloximes in soil samples from above and below the soil water table. *Pest. Sci.* 14, 173-181.

Smelt, J.H., Leistra, M., Houx, N. W. H., Dekker, A. (1978) Conversion rates of aldicarb and its oxidation products in soils. III. Aldicarb. *Pest. Sci.* 9, 293-300.

Smith, A. D., Bharath, A., Mallard, C., Orr, D., McCarty, L. S., Ozbum, G. W. (1990) Bioconcentration kinetics of some chlorinated benzenes and chlorinated phenols in American flagfish, *Jordanella floridae* (Goode and Bean). *Chemosphere* 20, 379-386.

Smith, J. H., Mabey, W. R., Bahonos, N., Holt, B. R., Lee, S. S., Chou, T. W., Venberger, D. C., Mill, T. (1978) Environmental pathways of selected chemicals in freshwater systems: Part II. Laboratory Studies. Interagency Energy-Environmental Research Program Report. EPA-600/7-78-074. Environmental Research Laboratory Office of Research and Development. U.S. EPA, Athens, Georgia.

Smith, P. D., Brockway, D. L., Stencil, F. E., Jr. (1987) Effect of hardness, alklinity and pH on toxicity of petachlorophenol to *Selenastrum capricornutum* (printz). *Environ. Toxicol. Chem.* 6, 891-900.

Söderström, M., Wachtmeister, C. A., Förlin, L. (1994) Analysis of chlorophenolics from bleach Kraft Mill effluents (BKME) in bile of perch (*Perca fluviatilis*) from the Baltic Sea and development of an analytical procedure also measuring chlorocatechols. *Chemosphere* 28, 1701-1719.

Somasundaram, L., Coats, J. R., Racke, K. D. (1991) Mobility of pesticides and their hydrolysis metabolites in soil. *Environ. Toxicol. Chem.* 10, 185-194.

Soon, L. G., Hock, O. S. (1987) Environmental problems of pesticide usage in Malaysian rice-fields. In: *Management of Pests and Pesticides*. Tait, J., Napompeth, B., Editors, pp. 10-21, Westview, London, United Kingdom.

Spain, J. C., Pritchard, P., Bourquin, A. W. (1980) Effects of adaptation on biodegradation rates in sediment water cores from estuarine and freshwater environments. *Appl. Environ. Microbiol.* 40, 726-734.

Spehar, R. L., Tanner, D. K., Nordling, B. R. (1983) Toxicity of the synthetic pyrethroids, permethrin and AC 222, 705 and their accumulation in early life stages of fathead minnows and snails. *Aquatic Toxicol.* 3, 171-182.

Spencer, E. Y., Editor (1973) *Guide to the Chemicals Used in Crop Protection*. 6th Edition, Research Branch Agriculture Canada, Ontario, Canada.

Spencer, E. Y. (1976) Vapor pressure and vapor loss. In: *A Literature Survey of Bench-mark Pesticides*. Medical Center, Dept. of Medical and Public Affairs, Science Communication Division, The George Washington University, Washington DC.

Spencer, E. Y., Editor (1982) *Guide to the Chemicals Used in Crop Protection*. 7th Edition, Research Branch Agriculture Canada, Ontario, Canada.

Spencer, J. R., Hermandez, B. Z., Schneider, F. A., Gonzales, M., Begum, S., Krieger, R. I. (1992) Seasonal mevinphos degradation on row crops in Monterey county, 1990. *Chemosphere* 24, 773-777.

Spencer, W. F. (1975) Movement of DDT and its derivatives into the atmosphere. *Res. Rev.* 59, 91-117.

Spencer, W. F., Cliath, M. M. (1969) Vapor density of dieldrin. *Environ. Sci. Technol.* 3, 670-674.

Spencer, W. F., Cliath, M. M. (1970) Vapor density and apparent vapor pressure of lindane (γBHC). *J. Agric. Food Chem.* 18(3), 529-530.

Spencer, W. F., Cliath, M. M. (1972) Volatility of DDT and related compounds. *J. Agric. Food Chem.* 20, 645-649.

Spencer, W. F., Cliath, M. M. (1983) Measurement of pesticide vapor pressures. *Res. Rev.* 85, 57-71.

Spencer, W. F., Cliath, M. M. (1990) Chapter 1, Movement of pesticides from soil to the atmosphere. In: *Long Range Transport of Pesticides.* Kurtz, D. A., Editor, Lewis Publishers, Inc., Ann Arbor, Michigan.

Spencer, W. F., Cliath, M. M., Jury, W. A., Zhang, L. Z. (1988) Volatilization of organic chemicals from soil as their Henry's law constants. *J. Environ. Qual.* 17, 504-509.

Spencer, W. F., Farmer, W. J., Cliath, M. M. (1973) Pesticide volatilization. *Res. Rev.* 49, 1-47.

Spencer, W. F., Shoup, T. D., Cliath, M. M., Farmer, W. J., Haque, R. (1979) Vapor pressure and relative volatility of ethyl and methyl parathion. *J. Agric. Food Chem.* 27, 273-278.

Spiller, D. (1961) A digest of available information on insecticide malathion. *Adv. Pest. Control Res.* 4, 249.

SRI International (1980) Interim Report on Task No. 11, Contract No. 68-01-3867, U.S. EPA Monitoring and Data Support Div., Office of Water Regulations and Standards, Washington, DC.

Statham, C. N., Melancon, Jr., M. J., Leck, J. L. (1976) Bioconcentration of xenobiotics in trout bile: A proposed monitoring aid for some water borne chemicals. *Science* 193, 680-681.

Stehly, G. R., Hayton, W. L. (1990) Effect of pH on the accumulation kinetics of pentachlorophenol in goldfish. *Arch. Environ. Contam. Toxicol.* 19, 464-470.

Stephen, H., Stephen, T. (1963) *Solubilities of Inorganic and Organic Compounds.* Vols. I and II, MacMillan Co., New York.

Stephenson, R. M., Malanowski, A. (1987) *Handbook of the Thermodynamics of Organic Compounds.* Elsevier, New York.

Stephenson, R. R. (1982) Aquatic toxicity of cypermethrin. I. Acute toxicity to some freshwater fish and invertebrates in laboratory tests. *Aquatic Toxicology* 2, 175-185.

Stewart, D. K. R., Chisholm, D. (1971) Long term persistence of BHC, DDT and chlordane in a sandy loam clay. *Can. J. Soil Sci.* 61, 379-383.

Sugiura, K., Aoki, M., Kaneko, S., Daisaku, I., Komatsu, Y., Shibuya, H., Suzuki, H., Goto, M. (1984) Fate of 2,4,6-trichlorophenol, pentachlorophenol, *p*-chlorobiphenyl, and hexachlorobenzene in an outdoor experimental pond: Comparison between observations and predictions based on laboratory data. *Arch. Environ. Contam. Toxicol.* 13, 745-758.

Sugiura, K., Washino, T., Hattori, M., Sato, E., Goto, M. (1979) Accumulation of organochlorines in fishes-Difference of accumulation factors by fishes. *Chemosphere* 8(6), 359-364.

Sukop, M., Cogger, C. G. (1992) Adsorption of carbofuran, metalaxyl, and simazine: KOC evaluation and relation to soil transport. *J. Environ. Sci. Health* B27(5), 565-590.

Suntio, L. R., Shiu, W. Y., Mackay, D., Seiber, J. N., Glotfelty, D. (1988) Critical review of Henry's law constants. *Rev. Environ. Contam. Toxicol.* 103, 1-59.

Sutherland, G. L., Giang, P. A., Archer, T. E. (1980) In: *Analytical Methods for Pesticides and Plant Growth Regulators.* Vol. 11, Zweig, G., Editor, pp. 487-505, Academic Press, New York.

Svenson, S., Björndal, H. (1988) A convenient test method for photochemical transformation of pollutants in the aquatic environment. *Chemosphere* 17, 2397-2405.

Suzuki, T., Kudo, Y. (1990) Automatic log P estimation based on combined additive modeling methods. *J. Computer-Aided Design*, 4, 155-198.

Swackhamer, D. L., Skoglund, R. S. (1991) The role of phytoplankton in the partition of hydrophobic organic contaminants in water. In: *Organic Substances and Sediments in Water*. Vol. 2, Baker, R. A., Editor, pp. 95-106, Lewis Publishers, Inc., Chelsea, Michigan.

Swann, R. L., Laskowski, D. A., McCall, P. J., Vander, Kuy K., Dishburger, H. J. (1983) A rapid method for estimation of the environmental parameters octanol/water partition coefficient, soil sorption constant, water to air ratio, and water solubility. *Res. Rev.* 85, 17-28.

Swoboda, A. R., Thomas, G. W. (1968) Movement of parathion in soil columns. *J. Agric. Food Chem.* 16, 923-927.

Szeto, S. Y., Vernon, R. S., Brown, M. J. (1983) Degradation of disulfoton in soil and its translocation into asparagus. *J. Agric. Food Chem.* 31, 217-220.

Tafuri, F., Businelli, M., Scarponi, L., Marucchini, C. J. (1977) Decline and movement of AG chlordane in soil and its residues in alfalfa. *J. Agric. Food Chem.* 25, 353-356.

Takase, I., Oyama, H. (1985) Uptake and bioconcentration of disulfoton and its oxidation compounds in carp, *Cyprinus carpio L. Nippon Noyaku Gakkaishi* 10, 47-53.

Takimoto, Y., Miyamoto, J. (1976) Studies on the accumulation and metabolism of sumithion in fish. *J. Pest. Sci.* 1, 261-271.

Takimoto, Y., Ohshima, M., Miyamoto, J. (1987) Comparative metabolism of fenitrothion in aquatic organisms. I. Metabolism in the euryhaline fish, *Oryzalias latipes* and *Mugil cephalus*. *Ecotoxicol. Environ. Saf.* 13, 104-117.

Takimoto, Y., Ohshima, M., Yamada, H., Miyamoto, J. (1984) Fate of fenitrothion in several developmental stages of the killifish (*Oryzalias latipes*). *Arch. Environ. Contam. Toxicol.* 13, 579-587.

Taylor, A. W., Glotfelty, D. E. (1988) Evaporation from soils and crops. In: *Environmental Chemistry of Herbicides*. Vol. I, Grover, R., Editor, pp. 89-130, CRC Press, Inc., Boca Raton, Florida.

Taylor, A. W., Spencer, W. F. (1990) Volatilization and vapor transport processes. In: *Pesticides in the Soil Environment: Processes, Impacts, and Modeling*. Cheng, H. H., Editor, pp. 213-269, Soil Science Society of America, Inc., Madison, Wisconsin.

Tejada, A. W. (1995) Pesticide residues in foods and the environment as a consequence of crop protection. *Philipp. J. Agric.* 78, 63-79.

Tejada, A. W., Magallona, E. D. (1985) Fate of carbosulfan in a rice paddy environment. *Philipp. Entomol.* 6, 255-273.

Tejada, A. W., Varca, L. M., Ocampo, P., Bajet, C. M., Magallona, E. D. (1993) Fate and residues of pesticides in rice production. *Int. J. Pest. Manage.* 39, 281-287.

Terada, H., Kosuge, Y., Murayama, W., Nakaya, N., Nunogaki, Y., Nunogaki, K.-I. (1987) Correlation of hydrophobic parameters of organic compounds determined by centrifugal partition chromatography with partition coefficients between octanol and water. *J. Chromatogr.* 400, 343-351.

Thibodeaux, L. J. (1979) *Chemodynamics*. John Wiley & Sons, New York.

Thomann, R. V. (1989) Bioaccumulation model of organic chemical distribution in aquatic food chains. *Environ. Sci. Technol.* 23, 699-707.

Thomas, R. G. (1982) Chapter 15: Volatilization from water, and Chapter 16: Volatilization from soil. In: *Handbook of Chemical Property Estimation Methods*. Lyman, W. J., Rechl, W. F., Rosenblatt, D. H., Editors, McGraw-Hill, New York.

Thor (1989) from Medchem Release 3.54, Daylight Chemical Information Systems Inc., Claremont, California.

Tomlin, C. (1994) *The Pesticide Manual (A World Compendium)*. 10th Edition, Incorporating the Agrochemicals Handbook. The British Crop Protection Council, Surrey, UK and The Royal Society of Chemistry, Cambridge, United Kingdom.

Toyota, H., Kuwahara, M. (1967) The study on production of PCP chemical fertilizer and its effect as herbicide and fertilizer, the solubility in water of PCP in PCP chemical fertilizer. *Nippon Dojohiryogaku Zasshi* 38, 93-97.

Trapp, St., Pussemier, L. (1991) Model calculations and measurements of uptake and translocation of carbamates by bean plants. *Chemosphere* 22, 327-339.

Tratnyek, P. G., Hoigné, J. (1991) Oxidation of substituted phenols in the environment: A QSAR analysis of rate constants for reaction with singlet oxygen. *Environ. Sci. Technol.* 25, 626-631.

Travis, C. C., Arms, A. D. (1988) Bioconcentration of organics in beef, milk, and vegetation. *Environ. Sci. Technol.* 22, 271-274.

Trotter, D. M., Kent, R. A., Wong, M. P. (1991) Aquatic fate and effect of carbofuran. *Critical Reviews in Environ. Control* 21(2), 137-176.

Trujillo, D. A., Ray, L. E., Murray, H. E., Giam, C. S. (1982) Bioaccumulation of pentachlorophenol by killifish (*Fundulus similus*). *Chemosphere* 11, 25-31.

Tsuda, T., Aoki, S., Inoue, T., Kojima, M. (1995) Accumulation and excretion of diazinon, fenthion and fenitrothion by killifish: Comparison of individual and mixed pesticides. *Water Res.* 29, 455-458.

Tsuda, T., Aoki, S., Kojima, M., Fujita, T. (1992) Pesticides in water and fish from rivers flowing into Lake Biwa (II). *Chemosphere* 24, 1523-1531.

Tsuda, T., Aoki, S., Kojima, M., Fujita, T. (1992) Accumulation and excretion of organophosphorus pesticides by willow shiner. *Chemosphere* 25(12), 1945-1951.

Tsuda, T., Aoki, S., Kojima, M., Fujita, T. (1993) Accumulation and excretion of organophosphorus pesticides by carp *Cyprinus carpio*. *Comp. Biochem. Physiol.* 104C(2), 275-278.

Tsuda, T., Aoki, S., Kojima, M., Harada, H. (1989) Bioconcentration and excretion of diazinon, IBP, malathion and fenitrothion by willow shiner. *Toxicol. Environ. Chem.* 24, 185-190.

Tucker, W. A., Lyman, W. J., Preston, A. L. (1983) Estimation of the dry deposition velocity and scavenging ratio for organic chemicals. In: *Precipitation Scavenging, Dry Deposition, and Resuspension.* Pruppacher, et al., Editors, pp. 1242-1256, Elsevier Science Publishing Co., Inc., New York, New York.

Ugland, K., Lundanes, E., Greibrok, T., Bjoseth, A. (1981) Determination of chlorinated phenols by high-performance liquid chromatography. *J. Chromatogr.* 213, 83-90.

Ulmann, E. (1972) Lindane, Monograph of an Insecticide. Verlag K. Schillinger-Freiburg in Breisgau. p.16.

U. Oklahoma (1986) Univ. of Oklahoma Data Base 1986.

USDA (1989) Final environmental impact statement, vegetation management in the Piedmont and Coastal Plain. Southern Region Management Bulletin R8-MB-23, USDA Forest Service, Atlanta, Georgia.

USEPA (1980) EXAMS: An Exposure Analysis Modeling System. Preliminary draft document (February 1980). Available through R. R. Lassiter, Environmental Research Laboratory, U.S. Environmental Protection Agency, College Station Road, Athens, Georgia.

USEPA (1984) Review of In-place Treatment for Contaminated Surface Soils. Vol. 1 & 2, U.S. EPA 540/2-84-003. U.S. Environmental Protection Agency, Cincinnati, Ohio.

USEPA (1986) CIS Aquire Data Base. U.S. Environmental Protection Agency.

USEPA (1987) EXAMS II Computer Simulation. U.S. Environmental Protection Agency.

USEPA (1988) PCCHEM. Version 1.81. Computer estimation model based on U.S. EPA AUTOCHEM. U.S. Environmental Protection Agency.

Valsaraj, K. T., Thibodeaux, L. J., Lu, X.-Y. (1991) Studies in batch and continuous solvent sublation. III. Solubility of pentachlorophenol in alcohol-water mixtures and its effects on solvent sublation. *Sep. Sci. Technol.* 26(4), 529-538.

Van Gestel, C. A. M., Ma, W.-C. (1988) Toxicity and bioaccumulation of chlorophenols in earthworms, in relation to bioavailability in soil. *Ecotoxicol. Environ. Saf.* 16, 289-297.

Veith, G. D., Austin, N. M., Morris, R. T. (1979a) A rapid method for estimation log P for organic chemicals. *Water Res.* 13, 43-47.

Veith, G. D., Defoe, D. L., Bergstedt, B. V. (1979b) Measuring and estimating the bioconcentration factor of chemicals in fish. *J. Fish Res. Board Can.* 26, 1040-1048.

Veith, G. D., Kosian, P. (1983) Estimating bioconcentration potential from octanol/water partition coefficients. In: *Physical Behavior of PCBs in the Great Lakes*. Mackay, D., Paterson, S., Eosemreocj. S. J., Simmons, M. S., Editors, Chapter 15, pp. 269-282, Ann Arbor Science, Ann Arbor, Michigan.

Veith, G. D., Macek, K. J., Petrocelli, S. R., Caroll, J. (1980) An evaluation of using partition coefficient and water solubilities to estimate bioconcentration factors for organic chemicals in fish. In: *Aquatic Toxicology*. ASTM STP 707, Eaton, J. G., Parrish, P. R., Hendricks, A. C., Editors, p.p. 116-129, American Society for Testing and Materials, Philadelphia, Pennsylvania.

Veith, G. D., Morris, R. T. (1978) A Rapid Method for Estimating Log P for Organic Chemists. U.S. Environmental Protection Agency, ERL, Duluth, Minnesota.

Velsicol Chem. (1972) Phosvel insecticide. Velsicol Chemical Corp. General Bulletin #09-070-601A, Chicago, Illinois.

Ventullo, R.M., Larson, R.J. (1985) Metabolic diversity and activity of heterotrophic bacteria in ground water. *Environ. Toxicol. Chem.* 4, 759-711.

Verhaar, H. J. M., van Leeuwen, C. J., Hermens, J. L. M. (1992) Classifying environmental pollutants. 1: Structure-activity relationships for prediction of aquatic toxicity. *Chemosphere* 25, 471-491.

Verschueren, K. (1977) *Handbook of Environmental Data on Organic Chemicals*. Van Nostrand Reinhold, New York.

Verschueren, K. (1983) *Handbook of Environmental Data on Organic Chemicals*. 2nd Edition, Van Nostrand Reinhold, New York.

Vigano, L., Galassi, S., Gatto, M. (1992) Factors affecting the bioconcentration of hexachlorocyclohexanes in early life stages of *Oncorhynchus mykiss*. *Environ. Toxicol. Chem.* 11, 535-540.

Voerman, S., Besemer, A. F. H. (1975) Persistence of dieldrin, lindane, and DDT in a light sandy soil and their uptake by grass. *Bull. Environ. Contam. Toxicol.* 13, 501-505.

Voerman, S., Tammes, P. M. L. (1969) Adsorption and desorption of lindane and dieldrin by yeast. *Bull. Environ. Contam. Toxicol.* 4, 271.

von Rümker, R., Horay, F. (1972) *Basic Information on Thirty-five Pesticide Chemicals. Pesticide Manual*. Part II, U.S. Agency for International Development.

Walsh, A. H., Ribelin, W. E. (1973) In: *The Pathology of Fish*. Ribelin, W. E., Migaki, G., Editors, University of Wisconsin Press, Madison, Wisconsin.

Walker, W. W. (1978) Insecticide Persistence in Natural Seawater as Affected by Salinity, Temperature and Sterility. EPA-600/3-78-044. U.S. EPA, Gulf Breeze, Florida.

Walker, W. W., Cripe, C. R., Pritchard, P. H., Bourquin, A. W. (1988) Biological and abiotic degradation of xenobiotic compounds in *in vitro* esturine water and sediment/water system. *Chemosphere* 17(12), 2255-2270.

Walker, W. W., Stojanovic, B. J. (1973) Microbial versus chemical degradation of malathion in soil. *J. Environ. Qual.* 2, 229-232.

Wang, T. C., Hoffman, M. E. (1991) Degradation of organophosphorus pesticides in coastal water. *J. Assoc. Off. Anal. Chem.* 74(5), 883-886.

Wang, X., Brusseau, M. L. (1993) Solubilization of some low-polarity organic compounds by hydroxypropyl-β-cyclodextrin. *Environ. Sci. Technol.* 27, 2821-2825.

Wania, F., Mackay, D. (1993) Global fractionation and cold condensation of low volatility organochlorine compounds in polar regions. *Ambio* 22, 10-18.

Wania, F., Mackay, D. (1993) Modelling the global distribution of toxaphene: A discussion of feasibility and desirability. *Chemosphere* 27, 2079-2094.

Wania, F., Shiu, W. Y., Mackay, D. (1994) Measurements of the vapor pressure of several low-volatility organochlorine chemicals at low temperatures with a gas saturation method. *J. Chem. Eng. Data* 39, 572-577.

Wanner, O., Egli, T., Fleischmann, T., Lanz, K., Reichert, P., Schwazenbach, R. P. (1989) Behavior of the insecticides disulfolton and thiometon in the Rhine River: A chemodynamic study. *Environ. Sci. Technol.* 23, 1232-1242.

Ward, T. E. (1985) Characterizing the aerobic and anaerobic microbial activities in surface and subsurface soils. *Environ. Toxicol. Chem.* 4, 727-737.

Warner, H. P., Cohen, J. M., Ireland, J. C. (1980) Determination of Henry's Law Constants of Selected Priority Pollutants. *MERL*, Cincinnati, Ohio.

Warner, H. P., Cohen, J. M., Ireland, J. C. (1987) Determination of Henry's Law Constants of Selected Priority Pollutants. EPA/600/D-87/229; NTIS PB87-212684. U.S. Environmental Protection Agency, Cincinnati, Ohio.

Warner, H. P., Cohen, J. M., Ireland, J. C. (1980) In-house report of U.S. EPA, Municipal Environmental Research Laboratory, Wastewater Research Division, Cincinnati, Ohio.

Wauchope, R. D. (1978) The pesticide content of surface water draining from agricultural fields- A review. *J. Environ. Quality.* 7(4), 459-472.

Wauchope, R. D. (1989) *ARS/SCS Pesticide Properties Database*. Version 1.9, preprint, August, 1989.

Wauchope, R. D., Buttler, T. M., Hornsby, A. G., Augustijn-Beckers, P. W. M., Burt, J. P. (1992) The SCS/ARS/CES Pesticide Properties Database for Environmental Decision-Making. *Rev. Environ. Contam. Toxicol.* 123, 1-164.

Way, M. J., Scopes, N. E. A. (1968) Studies on the persistence and effects on soil fauna and some soil-applied systemic insecticides. *Ann. Appl. Biol.* 62, 199-214.

Weast, R. C., Editor (1972-73) *Handbook of Chemistry and Physics*. 53rd Edition, CRC Press, Cleveland, Ohio.

Weast, R. C., Editor (1976-77) *Handbook of Chemistry and Physics*. 57th Edition, CRC Press, Cleveland, Ohio.

Weast, R. C., Editor (1978) *Handbook of Chemistry and Physics*. 58th Edition, CRC Press, West Palm Beach, Florida.

Weber, J. B., Shea, P. J., Strek, H. J. (1980) An evaluation of nonpoint sources of pesticide pollution in runoff. In: *Environmental Impact of Nonpoint Source Pollution*. Overcash, M., Davidson, J., Editors, Ann Arbor Science Publishers, Ann Arbor, Michigan.

Weil, V. G., Dure, G., Quentin, K. E. (1974) Solubility in water of insecticide chlorinated hydrocarbons and polychlorinated biphenyls in view of water pollution. *Z. Wasser Abwasser Forsch* 7(6), 169-175.

Weisgerber, I., Kohli, J., Kaul, R., Klein, W., Korte, F. (1974) Fate of aldrin-^{14}C in maize, wheat, and soils under outdoor conditions. *J. Agric. Food Chem.* 22, 609-612.

Wells, D., Grayson, B. T., Langner, E. (1986) Vapor pressure of permethrin. *Pest. Sci.* 17, 473-476.

Westall, J. C., Leuenberger, C., Swarzenbach, R. P. (1985) Influence of pH and ionic strength on the aqueous-nonaqueous distribution of chlorinated phenols. *Environ. Sci. Technol.* 19, 193-198.

Westcott, J. W., Bidleman, T. F. (1981) Detemination of polychlorinated biphenyl vapor pressures by capillary gas chromatography. *J. Chromatogr.* 210, 331-336.

Westcott, J. W., Simon, C. G., Bidleman, T. F. (1981) Detemination of polychlorinated biphenyl vapor pressures by a semimicro gas saturation method. *Environ. Sci. Technol.* 15, 1375-1378.

Wheatley, G. A., Hardman, J. A. (1968) Organochlorine insecticide residues in earthworms from arable soils. *J. Sci. Food Chem. Agric.* 19, 219-225.

Whiting, F. M., Brown, W. H., Stull, J. W. (1973) Pesticide residues in milk and in tissues following long, low 2,2-bis(p-chlorophenyl)-1,1,1-trichloroethane intake. *J. Dairy Sci.* 56, 1324.

Wilcock, R. J., Smith, T. J., Pridmore, R. D., Thrush, S. F., Cummings, V. J., Hewitt, J. E. (1993) Bioaccumulation and elimination of chlordane by selected intertidal benthic fauna. *Environ. Toxicol. Chem.* 12, 733-742.

Williams, P. P. (1977) Metabolism of synthetic organic pesticides by anaerobic microorganisms. *Res. Rev.* 66, 63.

Willis, G. H., Hamilton, R. A. (1973) Agricultural chemicals in surface runoff, ground water, and soil: 1. Emdrin. *J. Environ. Qual.* 2, 463.

Willis, G. H., McDowell, L. L. (1982) Pesticides in agricultural runoff and their effects on downstream water quality. *Environ. Toxicol. Chem.* 1, 267-279.

Willis, G. H., McDowell, L. L. (1987) Pesticide persistence on foliage. *Rev. Environ. Contam. Toxicol.* 100, 23-73.

Willis, G. H., McDowell, L. L., Smith, S., Southwick, L. M., Lemon, E. R. (1980) Toxaphene volatilization from a mature cotton canopy. *Agron. J.* 72, 627.

Willis, G. H., Parr, J. F., Smith, S. (1971) Volatilization of soil-applied DDT and DDD from flooded and nonflooded plots. *Pest. Monit. J.* 4, 204.

Wilson, A. J. (1963) Chemical assays. In: *Annual Report of the Bureau of Commercial Fisheries*, U.S. Bureau of Commercial Fisheries Circ. #247. Biology Lab., Gulf Breeze, Florida.

Wilson, K. A., Cook, R. M. (1972) Metabolism of xenobiotics in remnants. IV. Storage and excretion of HEOD in Holstein cows. *J. Agric. Food Chem.* 20, 391.

Windholz, M., Editor (1983) *The Merck Index. An Encyclopedia of Chemicals, Drugs and Biologicals.* 11th Edition, Merck and Co., Inc., Rahway, New Jersey.

Winer, A. M., Atkinson, R. (1990) Chapter 9, Atmospheric reaction pathways and lifetimes for organophosphorous compounds. In: *Long Range Transport of Pesticides.* Kurtz, D. A., Editor, Lewis Publishers, Inc., Ann Arbor, Michigan.

Wolfdietrich, E., Editor (1965) *Handbuch der Insektizidkunde.* Veb Verlag Volk und Gesundheit, Berlin.

Wolfe, N. L. (1980) Organophosphate and organophosphorothionate esters: Application of linear free energy relationships to estimate hydrolysis rate constants for use in environmental fate assessment. *Chemosphere* 9, 571-579.

Wolfe, N. L., Kitchens, B. E., Macalady, D. L., Grundl, T. J. (1986) Physical and chemical factors that influence the anaerobic degradation of methyl parathion in sediment systems. *Environ. Toxicol. Chem.* 5, 1019-1026.

Wolfe, N. L., Paris, D. F., Steen, W. C., Baughman, G. L. (1980) Correlation of microbial degradation rates with chemical structure. *Environ. Sci. Technol.* 14, 1143.

Wolfe, N.L., Zepp, R.G., Gordon, T.A., Baughman, G.L., Cline, D.M. (1977) Kinetics of chemical degradation of malathion in water. *Environ. Sci. Technol.* 11, 88-93.

Wolfe, N. L., Zepp, R. G., Paris, D. F. (1978) Carbaryl, propham, and chloropropham: A comparison of the rates of hydrolysis and photolysis with the rate of biolysis. *Water Res.* 12, 565.

Wolfe, N. L., Zepp, R. G., Paris, D. F. (1978) Use of structure-reactivity relationships to estimate hydrolytic persistence of carbamate pesticides. *Water Res.* 12, 561-563.

Wolfe, N. L., Zepp, R. G., Paris, D. F., Baughman, G. L., Hollis, R. C. (1977) Methoxychlor and DDT degradation in water: Rates and products. *Environ. Sci. Technol.* 11, 1077-1081.

Wolfe, N. L., Zepp, R. G., Baughman, G. L., Fincher, R. C., Gordon, T. A. (1976) Chemical and Photochemical Transformation of Selected Pesticides in Aquatic Environments. U.S. EPA-600/3-76-067. U.S. EPA, Athens, Georgia.

Wong, A. S., Crosby, D. B. (1978) Photolysis of pentachlorophenol in water. In: *Pentachlorophenol: Chemistry, Pharmacology, and Environmental Toxicology.* Rao, K. R., Editor, pp. 19-25, Plenum Press, New York.

Wong, A. S., Crosby, D. B. (1981) Photodecomposition of pentachlorophenol in water. *J. Agric. Food Chem.* 29, 125-130.

Woolford, M. H., Jr. (1975) American Cyanamid Co. Letter to W.F. Spencer, Sept. 3, 1975.

Worthing, C. R., Editor (1979) *The Pesticide Manual (A World Compendium).* 6th Edition, The British Crop Protection Council, Croydon, England.

Worthing, C. R., Editor (1983) *The Pesticide Manual (A World Compendium).* 7th Edition, The British Crop Protection Council, Croydon, England.

Worthing, C. R., Editor (1987) *The Pesticide Manual (A World Compendium).* 8th Edition, The British Crop Protection Council, Croydon, England.

Worthing, C. R., Editor (1991) *The Pesticide Manual (A World Compendium).* 9th Edition, The British Crop Protection Council, Surrey, England.

Wszolek, P. C., Lein, D. H., Lisk, D. J. (1980) Excretion of fenvalerate insecticide in the milk of dairy cows. *Bull. Environ. Contam. Toxicol.* 24, 296.

Xie, T. M. (1983) Determination of trace amounts of chlorophenols and chloroguaicols in sediment. *Chemosphere* 12, 1183-1191.

Xie, T. M., Abrahamsson, K., Fogelqvist, E., Josefsson, B. (1986) Distribution of chlorophenols in a marine environment. *Environ. Sci. Technol.* 20, 457-463.

Xie, T. M., Dyrssen, D. (1984) Simultaneous determination of partition coefficients and acidity constants of chlorinated phenols and guaicols by gas chromatography. *Anal. Chem. Acta* 160, 21-30.

Xie, T. M., Hulthe, B., Folestad, S. (1984) Determination partition coefficients of chlorinated phenols guaicols and catecols by shake-flask GC and HPLC. *Chemosphere* 13, 445-459.

Yalkowsky, S. H., Editor (1987) *Arizona Data Base of Water Solubility.* University of Arizona, Tucson, Arizona.

Yalkowsky, S. H., Banerjee, S. (1994) *Aqueous Solubility. Methods of Estimation for Organic Compounds.* Marcel Dekker, Inc., New York.

Yalkowsky, S. H., Dannenfelser, R. M. (1994) *AQUASOL DATABASE*, 5th Edition, University of Arizona.

Yao, C. C. D., Haag, W. R. (1991) Rate constants for direct reactions of ozone with several drinking water contaminants. *Water Res.* 25, 761-773.

Yaron, B., Heuer, B., Birk, Y. (1974) Kinetics of azinphosmethyl losses in the soil environment. *J. Agric. Food Chem.* 22(3), 439-441.

Yin, C., Hassett, J. P. (1986) Gas partition approach for laboratory and field studies of mirex fugacity in water. *Environ. Sci. Technol.* 20, 1213-1217.

Yoshida, K., Shigeoka, T., Yamauchi, F. (1983) Non-steady state equilibrium model for the preliminary prediction of the fate of chemicals in the environment. *Ecotoxicol. Environ. Saf.* 7, 179-190.

Yoshida, K., Shigeoka, T., Yamauchi, F. (1983) Relationship between mole fraction and n-octanol/water partition coefficient. *Ecotoxicol. Environ. Saf.* 7, 558-565.

Yoshioka, Y., Mizuno, T., Ose, Y., Sato, T. (1986) The estimation for toxicity of chemicals on fish by physico-chemical properties. *Chemosphere* 15, 195-203.

Zaroogian, G. E., Hertshe, J. F., Johnson, M. (1985) Estimation of bioconcentration in marine species using structure-activity models. *Environ. Toxicol. Chem.* 4, 3-12.

Zepp, R. G., Baughman, G. L. (1978) Prediction of photochemical transformation of pollutants in the aquatic environment. In: *Aquatic Pollutants: Transformation and Biological Effects.* Hutzinger, O., Van Lelyveld, I. H., Zoeteman, B. C. J., Editors, pp. 237-164, Pergamon Press, Oxford, England.

Zepp, R. G., Baughman, G. L., Schlotzhauer, P. F. (1981) Comparison of photochemical behavior of various humic substances in water: I. Sunlight induced reactions of aquatic pollutants photosensitized by humic substances. *Chemosphere* 10, 109-117.

Zepp, R. G., Schlotzhauer, P. F. (1983) Influence of algae on photolysis rates of chemicals in water. *Environ. Sci. Technol.* 17, 462-468.

Zepp, R. G., Schlotzhauer, P. F., Simmons, M. S., Miller, G. C., Baughman, G. L., Wolfe, N. L. (1984) Dynamics of pollutant photoreactions in the hydrosphere. *Fresenius Z. Anal. Chem.* 319, 119-125.

Zepp, R. G., Wolfe, N. L., Gordon, J. A., Fincher, R. C. (1976) Light-induced transformation of methoxychlor in aquatic systems. *J. Agric. Food Chem.* 24, 727-733.

Zimmerli, B., Marek, B. (1974) Modellversuche zur kontamination von lebensmitteln mit pestiziden via gasphase. *Mitt Gebiete Lebensm Hyg.* 65, 55-64.

Zitko, V., McLeese, D. W. (1980) Canadian Technical Report, Fish Aquatic Science No. 985, Dec. 1980.

Zoeteman, B. C. J., de Greef, E., Brinkmann, F. J. J. (1981) Persistency of organic contaminants in groundwater. Lessons learned from soil pollution incidents in the Netherlands. *Sci. Total Environ.* 21, 187-202.

Zoeteman, B. C. J., Harmsen, K., Linders, J. B. H. J., Morra, C. F. H., Slooff, W. (1980) Persistent organic pollutants in river water and groundwater of the Netherlands. *Chemosphere* 9, 231-249.

Chapter 4. Fungicides

4.1 List of Chemicals and Data Compilations:
    4.1.1 Alanine Derivatives:
- Benalaxyl 653
- Metalaxyl 699

    4.1.2 Amides:
- Carboxin 666
- Fenfuram 684
- Thiram 713
- Tolylfluanid 717

    4.1.3 Carbamates:
- Benomyl 655
- Carbendazim 664
- Thiophanate-methyl 711

    4.1.4 Chlorobenzenes:
- Chloroneb 668
- Hexachlorobenzene 690
- Quintozene 709

    4.1.5 Imidazoles:
- Imazalil 697
- Triflumizole 721

    4.1.6 Phophrothioates:
- Edifenphos 678
- Tolclofos-methyl 715

    4.1.7 Quinones:
- Dichlone 674
- Dithianon 676
- Procymidone 705

    4.1.8 Thioimides:
- Captan 661
- Folpet 686

    4.1.9 Triazoles:
- Bitertanol 657
- Penconazole 703
- Propiconazole 707
- Triadimefon 719

    4.1.10 Miscellaneous:
- Anilazine (Triazine) 651
- Bupirimate (Pyridine Derivative) 659
- Chloropicrin (Chloronitromethane) 670
- Chlorothalonil (Benzonitrile) 672
- Etridiazole (Terrazole) 680
- Fenarimol (Chlorobenzoalcohol) 682
- Formaldehyde (Aldehyde) 688
- Oxycarboxin (Oxathiin dioxide) 701
- Triforine (Piperazine) 723
- Vinclozolin (Oxazolone) 725

4.2 Summary Tables 727
4.3 Illustrative Fugacity Calculations: Levels I, II and III 736
        Benomyl 736
        Captan 740
        Chloropicrin 744
        Chlorothalonil 748
4.4 Commentary on the Physical-Chemical Properties and Environmental Fate 752
4.5 References 753

Common Name: Anilazine
Synonym: Botrysan, Direz, Dyrene, Kemate, Triasyn, triazine, Zinochlor
Chemical Name: 2-chloro-N-(4,6-dichloro-1,3,5-triazin-2-yl)aniline; 2,4-dichloro-6-(o-chloro-anilino)-s-triazine; 4,6-dichloro-N-(2-chlorophenyl)-1,3,5-triazin-2-amine
Uses: as fungicide to control early and late blights of potatoes and tomatoes; anthracnose in cucurbits; leaf spot diseases in many crops; glume blotch of wheat; also used on vegetables, ornaments, berry fruits, melons, coffee and tobacco, etc.
CAS Registry No: 101-05-3
Molecular Formula: $C_9H_5Cl_3N_4$
Molecular Weight: 275.5
Melting Point (°C):
    159-160 (Spencer 1982; Agrochemicals Handbook 1987; Milne 1995)
    159 (Worthing 1991; Tomlin 1994)
Boiling Point (°C):
Density (g/cm³ at 20°C): 1.80 (Agrochemicals Handbook 1987; Tomlin 1994)
Molar Volume (cm³/mol):
    252.8 (calculated-LeBas method, this work)
    153.1 (calculated-density, this work)
Molecular Volume (Å³):
Total Surface Area, TSA (Å²):
Dissociation Constant pK_a:
Heat of Fusion, ΔH_{fus}, kcal/mol:
Entropy of Fusion, ΔS_{fus}, cal/mol·K (e.u.):
Fugacity Ratio at 25°C (assuming ΔS_{fus} = 13.5 e.u.), F: 0.0462

Water Solubility (g/m³ or mg/L at 25°C):
    8.00 (20°C, Agrochemicals Handbook 1987; Worthing 1991; Tomlin 1994)
    8.00 (20-25°C, selected, Wauchope et al. 1992; Hornsby et al. 1996)
    8.00 (selected, Lohninger 1994)
    8.00 (Milne 1995)

Vapor Pressure (Pa at 25°C):
    negligible (20°C, Agrochemicals Handbook 1987)
    8.20×10^{-7} (20°C, Worthing 1991; Tomlin 1994)
    8.26×10^{-7} (20-25°C, selected, Wauchope et al. 1992; Hornsby et al. 1996)

Henry's Law Constant (Pa·m³/mol):
    2.82×10^{-5} (20°C, calculated-P/C, this work)

Octanol/Water Partition Coefficient, log K_{OW}:
    4.39 (calculated, Chiou 1981)
    3.79 (calculated-CLOG program, Biagi et al. 1991)
    3.01 (20°C, Worthing 1991; Tomlin 1994)
    3.88 (RP-HPLC, Saito et al. 1993)
    1.91 (at pH 7, Milne 1995)
    3.00 (selected, Hansch et al. 1995)

Bioconcentration Factor, log BCF:
    2.28   (calculated-S as per Kenaga 1980, this work)

Sorption Partition Coefficient, log K_{OC}:
    3.00   (20-25°C, estimated, Wauchope et al. 1992; Hornsby et al. 1996)
    3.00   (estimated-chemical structure, Lohninger 1994)
    3.14   (calculated-S as per Kenaga 1980, this work)

Half-Lives in the Environment:
    Air:
    Surface water:
    Groundwater:
    Sediment:
    Soil: half-life in damp soil ca. 12 h (Agrochemicals Handbook 1987; Tomlin 1994); field half-life of 1 d (20-25°C, selected, Wauchope et al. 1992; Hornsby et al. 1996).
    Biota:

Environmental Fate Rate Constants or Half-Lives:
    Volatilization:
    Photolysis:
    Oxidation:
    Hydrolysis: stable in neutral and slightly acidic media, half-lives 730 h at pH 4, 790 h at pH 7, 22 hours at pH 9, 22°C (Tomlin 1994).
    Biodegradation:
    Biotransformation:
    Bioconcentration, Uptake (k_1) and Elimination (k_2) Rate Constants:

Common Name: Benalaxyl
Synonym: Galben, M 9834, Tairel
Chemical Name: methyl N-phenylacetyl-N-2,6-xylyl-DL-alaninate; methyl N-(2,6-dimethylphenyl)-N-(phenylacetyl)-DL-alaninate
CAS Registry No: 71626-11-4
Uses: as fungicide to control late blights of potatoes and tomatoes; downy mildews of hops, vines, lettuce, onions, soybeans and other crops; many diseases in flowers and ornamentals; and often used in combination with other fungicides, etc.
Molecular Formula: $C_{20}H_{23}NO_3$
Molecular Weight: 325.4
Melting Point (°C):
    78-80   (Agrochemicals Handbook 1987; Worthing 1991; Tomlin 1994; Milne 1995)
Boiling Point (°C):
Density (g/cm^3 at 20°C):
    1.27   (25°C, Agrochemicals Handbook 1987; Milne 1995)
Molar Volume (cm^3/mol):
    390.8   (calculated-LeBas method, this work)
    256.2   (calculated-density, this work)
Molecular Volume (Å^3):
Total Surface Area, TSA (Å^2):
Dissociation Constant pK_a:
Heat of Fusion, ΔH_{fus}, kcal/mol:
Entropy of Fusion, ΔS_{fus}, cal/mol·K (e.u.):
Fugacity Ratio at 25°C (assuming ΔS_{fus} = 13.5 e.u.), F: 0.286

Water Solubility (g/m^3 or mg/L at 25°C):
    37.0   (Agrochemicals Handbook 1987; Worthing 1991; Milne 1995)
    37.0   (20-25°C, Augustijn-Becker et al. 1994; Hornsby et al. 1996)
    37.0   (Tomlin 1994)

Vapor Pressure (Pa at 25°C):
    6.7×10^{-4} (Agrochemicals Handbook 1987; Worthing 1991)
    6.7×10^{-4} (Tomlin 1994)
    1.33×10^{-3} (20-25°C, Augustijn-Becker et al. 1994; Hornsby et al. 1996)

Henry's Law Constant (Pa·m^3/mol):
    0.0117   (calculated-P/C, this work)

Octanol/Water Partition Coefficient, log K_{OW}:
    3.40   (Worthing 1991; Milne 1995)
    3.40   (Tomlin 1994)
    3.40   (selected, Hansch et al. 1995)

Bioconcentration Factor, log BCF:
    1.91   (calculated-S as per Kenaga 1980, this work)

Sorption Partition Coefficient, log K_{OC}:
    3.44-3.86 (soil, Tomlin 1994)
    3.00   (soil, estimated, Augustin-Becker et al. 1994; Hornsby et al. 1996)

Half-Lives in the Environment:
    Air:
    Surface water:
    Groundwater:
    Sediment:
    Soil: half-life in soil is 20-71 d (Tomlin 1994); field half-life of 30 d (Augustijn-Becker et al. 1994; Hornsby et al. 1996).
    Biota:

Environmental Fate Rate Constants or Half-Lives:
    Volatilization:
    Photolysis:
    Oxidation:
    Hydrolysis: half-life of 86 d at pH 9, 25°C, but stable in aqueous solutions at pH 4-9 (Tomlin 1994).
    Biodegradation:
    Biotransformation:
    Bioconcentration, Uptake (k_1) and Elimination (k_2) Rate Constants:

Common Name: Benomyl
Synonym: Arilate, BBC, Benex, Benlate, Benosan, Fibenzo, Fundazol
Chemical Name: methyl N-(1-butylcarbamoyl-2-benzimidazole)carbamate; methyl 1-(butylcarbamoyl)benzimidazol-2-ylcarbamate; methyl 1-[(butylamino)carbonyl]-1H-benzimidazol-2-ylcarbamate
Uses: as fungicide to control a wide range of diseases of fruit, nuts, vegetables, mushrooms, field crops, ornamentals, turf and trees; also provides secondary acaricidal control, principally as an ovicide, etc.
CAS Registry No: 17804-35-2
Molecular Formula: $C_{14}H_{18}N_4O_3$
Molecular Weight: 290.62
Melting Point (°C):
    137   (Agrochemicals Handbook 1987)
    140   (dec., Tomlin 1994)
Boiling Point (°C):
Density (g/cm³ at 20°C):
Molar Volume (cm³/mol):
    320.0   (calculated-LeBas method, this work)
Molecular Volume (Å³):
Total Surface Area, TSA (Å²):
Dissociation Constant pK_a:
Heat of Fusion, ΔH_{fus}, kcal/mol:
Entropy of Fusion, ΔS_{fus}, cal/mol·K (e.u.):
Fugacity Ratio at 25°C (assuming ΔS_{fus} = 13.5 e.u.), F: 0.0729

Water Solubility (g/m³ or mg/L at 25°C):
    3.8   (Kenaga 1980; quoted, Howard 1991)
    2.8   (at pH 7, Singh & Chiba 1985; quoted, Howard 1991)
    2.0   (Agrochemicals Handbook 1987; Milne 1995)
    4.0   (pH 3-10, Worthing 1991)
    2.0   (20-25°C, selected, Wauchope et al. 1992; Hornsby et al. 1996)
    2.0   (stable only at pH 7, Montgomery 1993)
    4.0   (selected, Lohninger 1994)
    4.0   (pH 3-10, very soluble at pH 1, decomposes at pH 13, Tomlin 1994)

Vapor Pressure (Pa at 25°C):
    $< 1.00 \times 10^{-5}$ (20°C, Agrochemicals Handbook 1987)
    $< 1.33 \times 10^{-8}$ (20-25°C, selected, Wauchope et al. 1992; Hornsby et al. 1996)
    $< 4.90 \times 10^{-6}$ (Tomlin 1994)

Henry's Law Constant (Pa·m³/mol):
    $< 1.93 \times 10^{-6}$ (calculated-P/C)

Octanol/Water Partition Coefficient, log K_{OW}:
    2.12   (shake flask-AS, Austin & Briggs 1976; quoted, Sangster 1993)
    2.42   (Rao & Davidson 1982; Hansch & Leo 1985; quoted, Howard 1991)

3.11 (Garten & Trabalka 1983; Travis & Arms 1988; quoted, Howard 1991)
2.12 (Hansch & Leo 1985; quoted, Howard 1991)
2.42 (Hansch & Leo 1987; quoted, Sangster 1993)
1.40-3.11 (quoted, Montgomery 1993)
2.12 (selected, Hansch et al. 1995)

Bioconcentration Factor, log BCF:
    2.46 (estimated-S, Kenaga 1980; quoted, Howard 1991)
    −0.47 (vegetation, Popov & Sboeva 1974; Jalali & Anderson 1976; quoted, Travis & Arms 1988)

Sorption Partition Coefficient, log K_{OC}:
    3.32 (estimated-S, Kenaga 1980; quoted, Howard 1991)
    3.28 (soil, 20-25°C, selected, Wauchope et al. 1992; Hornsby et al. 1996)
    3.28 (soil, calculated, Montgomery 1993)
    3.28 (selected, Lohninger 1994)
    3.28 (soil, Tomlin 1994)

Half-Lives in the Environment:
    Air: estimated half-life of 1.6 h, based on the vapor-phase reaction with hydroxyl radicals in air (Atkinson 1987; quoted, Howard 1991).
    Surface water: half-life of 2 h (Tomlin 1994).
    Groundwater:
    Sediment:
    Soil: degradation occurred within 15 d in unsterilized soil (Hine et al. 1969; quoted, Howard 1991); half-life of 6-12 months (Hartley & Kidd 1987; quoted, Montgomery 1993); field half-life of 67 d (20-25°C, selected, Wauchope et al. 1992; Hornsby et al. 1996); half-life of 19 h in soil (Tomlin 1994).
    Biota: half-life on foliage was 3-7 d (quoted, Montgomery 1993).

Environmental Fate Rate Constants or Half-Lives:
    Volatilization:
    Photolysis:
    Oxidation: photooxidation half-life of 1.6 h in air, based on estimated rate constant for the vapor-phase reaction with hydroxyl radicals in air (Atkinson 1987; quoted, Howard 1991).
    Hydrolysis: very significant in water with half-life less than one week (Howard 1991).
    Biodegradation:
    Biotransformation:
    Bioconcentration, Uptake (k_1) and Elimination (k_2) Rate Constants:

Common Name: Bitertanol
Synonym: Baycor, Baymat, Biloxazol, Sibutol
Chemical Name: 1-(biphenyl-4-yloxy)-3,3-dimethyl-1-(1H-1,2,4-triazol-1-yl)butan-2-ol;β-([1,1'-biphenyl]-4-yloxy)-α-(1,1-dimethylethyl)-1H-1,2,4-triazole-1-ethanol
Uses: as fungicide to control scab on apples and pears; rusts and powdery mildews on ornamentals; black spot on roses; and leaf spot and other diseases of vegetables, cucurbits, cereals, deciduous fruit, bananas, groundnuts, soy beans, etc.
CAS Registry No: 70585-38-5 (diastereoisomer A), 55179-31-2 (diastereoisomer B)
Molecular Formula: $C_{20}H_{23}N_3O_2$
Molecular Weight: 337.4
Melting Point (°C):
    139.8   (diastereoisomer A, Agrochemicals Handbook 1987)
    146.3   (diastereoisomer B, Agrochemicals Handbook 1987)
    118.0   (eutectic mixture of the two diastereoisomers, Agrochemicals Handbook 1987; Worthing 1991)
    136.7   (diastereoisomer A, Worthing 1991; Tomlin 1994)
    145.2   (diastereoisomer B, Worthing 1991; Tomlin 1994)
Boiling Point (°C):
Density (g/cm³ at 20°C):
Molar Volume (cm³/mol):
    399.7   (calculated-LeBas method, this work)
Molecular Volume (Å³):
Total Surface Area, TSA (Å²):
Dissociation Constant pK_a:
Heat of Fusion, ΔH_{fus}, kcal/mol:
Entropy of Fusion, ΔS_{fus}, cal/mol·K (e.u.):
Fugacity Ratio at 25°C (assuming ΔS_{fus} = 13.5 e.u.), F: 0.120 (eutectic mixture)

Water Solubility (g/m³ or mg/L at 25°C):
    5.0   (20°C, eutectic mixture; Agrochemicals Handbook 1987; Worthing 1991)
    2.9   (20°C, diastereoisomer A, Worthing 1991; Tomlin 1994)
    1.6   (20°C, diastereoisomer B, Worthing 1991; Tomlin 1994)

Vapor Pressure (Pa at 25°C):
    1.0×10^{-6} (20°C, Agrochemicals Handbook 1987)
    0.0038   (100°C, diastereoisomer A, Worthing 1991)
    0.0032   (100°C, diastereoisomer B, Worthing 1991)
    2.2×10^{-10} (20°C, diastereoisomer A, Tomlin 1994)
    2.5×10^{-10} (20°C, diastereoisomer B, Tomlin 1994)

Henry's Law Constant (Pa·m³/mol):
    8.45×10^{-5} (20°C, eutectic mixture, calculated-P/C, this work)

Octanol/Water Partition Coefficient, log K_{OW}:
    4.10   (20°C, diastereoisomer A, Worthing 1991; Tomlin 1994)
    4.40   (20°C, diastereoisomer B, Worthing 1991; Tomlin 1994)
    4.16   (quoted, Schreiber & Schönherr 1992)
    4.16   (selected, Hansch et al. 1995)

Bioconcentration Factor, log BCF:
    2.40   (20°C, eutectic mixture, calculated-S as per Kenaga 1980, this work)

Sorption Partition Coefficient, log K_{OC}:
    2.25   (20°C, eutectic mixture, calculated-S as per Kenaga 1980, this work)

Half-Lives in the Environment:
    Air:
    Surface water: environmental half-life is 1 month to 1 year (Tomlin 1994).
    Groundwater:
    Sediment:
    Soil:
    Biota:

Environmental Fate Rate Constants or Half-Lives:
    Volatilization:
    Photolysis:
    Oxidation:
    Hydrolysis: stable in neutral, acidic and alkaline media, hydrolytic half-life at 25°C > 1 year, pH 4, 7 and 9 (Tomlin 1994).
    Biodegradation: degradation in soil is rapid (Tomlin 1994).
    Biotransformation:
    Bioconcentration, Uptake (k_1) and Elimination (k_2) Rate Constants:

Common Name: Bupirimate
Synonym: Nimrod, PP 588
Chemical Name: 5-butyl-2-ethylamino-6-methylpyrimidin-4-yl dimethylsulfamate; 5-butyl-2-(ethylamino)-6-methyl-4-pyrimidinyl dimethylsulfamate
Uses: as fungicide to control powdery mildews of apples and pears, stone fruit, strawberries, gooseberries, vines, roses and other ornamentals, cucurbits, hops, beet, and other crops, etc.
CAS Registry No: 41483-43-6
Molecular Formula: $C_{13}H_{24}N_4O_3S$
Molecular Weight: 316.4
Melting Point (°C):
    50-51   (Agrochemicals Handbook 1987; Worthing 1991; Tomlin 1994)
Boiling Point (°C):
Density (g/cm^3 at 20°C):
Molar Volume (cm^3/mol):
    368.9   (calculated-LeBas method, this work)
Molecular Volume (Å^3):
Total Surface Area, TSA (Å^2):
Dissociation Constant pK_a:
Heat of Fusion, ΔH_{fus}, kcal/mol:
Entropy of Fusion, ΔS_{fus}, cal/mol·K (e.u.):
Fugacity Ratio at 25°C (assuming ΔS_{fus} = 13.5 e.u.), F: 0.559

Water Solubility (g/m^3 or mg/L at 25°C):
    22.0   (Martin & Worthing 1977; quoted, Kenaga 1980)
    22.0   (Agrochemicals Handbook 1987; Worthing 1991; Tomlin 1994)
    23.0   (at pH 5.2, Worthing 1991)

Vapor Pressure (Pa at 25°C):
    6.7×10^{-5} (20°C, Agrochemicals Handbook 1987; Worthing 1991)
    1.0×10^{-4} (Tomlin 1994)

Henry's Law Constant (Pa·m^3/mol):
    9.64×10^{-3} (calculated-P/C, this work)

Octanol/Water Partition Coefficient, log K_{OW}:
    2.70   (shake flask, pH 7, Stevenson et al. 1988; quoted, Sangster 1993)
    3.70   (Worthing 1991)
    3.90   (Tomlin 1990)
    2.70   (selected, Hansch et al. 1995)

Bioconcentration Factor, log BCF:
    2.02   (calculated-S, Kenaga 1980)
    2.56   (calculated-K_{OW} as per Kenaga 1980, this work)

Sorption Partition Coefficient, log K_{OC}:
    2.90   (calculated-S, Kenaga 1980)

Half-Lives in the Environment:
    Air:
    Surface water:
    Groundwater:
    Sediment:
    Soil: half-life 35-90 d for nonsterile flooded or non-flooded soil, pH 5.1 to pH 7.3 (Tomlin 1994).
    Biota:

Environmental Fate Rate Constants or Half-Lives:
    Volatilization:
    Photolysis: rapidly decomposed by ultraviolet irradiation in aqueous solutions (Tomlin 1994).
    Oxidation:
    Hydrolysis: stable in dilute alkalis, but readily hydrolysed by dilute acids (Tomlin 1994).
    Biodegradation:
    Biotransformation:
    Bioconcentration, Uptake (k_1) and Elimination (k_2) Rate Constants:

Common Name: Captan
Synonym: Aacaptan, Amercide, Captab, Captaf, Captane, Captex, Flit 406, Glyodex 37-22, Malipur, Merpan, Orthocide, Pillarcap, Vondcaptan
Chemical Name: N-(trichloromethylthio)cyclohex-4-ene-1,2-dicarboximide; 1,2,3,6-tetrahydro-N-(trichloromethylthio)phthalimide; 3a,4,7,7a-tetrahydro-[(trichloromethyl)thio]-1H-isoindole-1,3(2H)-dione
CAS Registry No: 133-06-2
Uses: as fungicide to control a wide range of fungal diseases; also used as seed treatment on maize, ornamentals, vegetables, oilseed rape, and other crops.
Molecular Formula: $C_9H_8Cl_3NO_2S$
Molecular Weight: 300.6
Melting Point (°C):
    178   (Argrochemicals Handbook 1987; Suntio et al. 1988; Howard 1991; Worthing 1991; Patil 1994; Tomlin 1994; Milne 1995)
    160-170 (technical grade, Tomlin 1994)
Boiling Point (°C):
Density (g/cm^3 at 20°C):
    1.74   (Argrochemicals Handbook 1987; Tomlin 1994; Milne 1995)
Molar Volume (cm^3/mol):
    250.5   (calculated-LeBas method, Suntio et al. 1988)
    172.8   (calculated-density, this work)
Molecular Volume (Å^3):
Total Surface Area, TSA (Å^2):
Dissociation Constant pK_a:
Heat of Fusion, ΔH_{fus}, kcal/mol:
    10.60   (DSC method, Plato 1972)
Entropy of Fusion, ΔS_{fus}, cal/mol·K (e.u.):
Fugacity Ratio at 25°C (assuming ΔS_{fus} = 13.5 e.u.), F: 0.031

Water Solubility (g/m^3 or mg/L at 25°C):
    8.70   (colorimetric, Burchfield 1959; quoted, Freed 1976; quoted, Suntio et al. 1988)
    <0.5   (Martin & Worthing 1977; quoted, Kenaga 1980; Kenaga & Goring 1980; Khan 1980; Suntio et al. 1988)
    0.50   (quoted, Briggs 1981)
    3.30   (Argrochemicals Handbook 1987; Worthing 1991; Tomlin 1994)
    0.50   (20°C, selected, Suntio et al. 1988; quoted, Howard 1991; Majewski & Capel 1995)
    5.10   (20-25°C, selected, Wauchope et al. 1992; Hornsby et al. 1996)
    5.10   (selected, Lohninger 1994)
    0.50, 1.44 (quoted, calculated, Patil 1994)
    5.10   (Halfon et al. 1996)

Vapor Pressure (Pa at 25°C):
    <0.0013   (Khan 1980; quoted, Suntio et al. 1988)
    <0.0013   (Argrochemicals Handbook 1987; Worthing 1991; Tomlin 1994)
    0.0010   (20°C, selected, Suntio et al. 1988; quoted, Howard 1991; Majewski & Capel 1995)
    1.1×10^{-5} (20-25°C, selected, Wauchope et al. 1992; Hornsby et al. 1996)
    1.1×10^{-5} (Halfon et al. 1996)

Henry's Law Constant (Pa·m³/mol):
- 0.60 (20°C, calculated-P/C, Suntio et al. 1988; quoted, Howard 1991; Majewski & Capel 1995)

Octanol/Water Partition Coefficient, log K_{ow}:
- 2.35 (Leo et al. 1971; quoted, Kenaga & Goring 1980; Suntio et al. 1988; Sicbaldi & Finizio 1993)
- 1.52 (quoted, Rao & Davidson 1980; Suntio et al. 1988)
- 2.54 (shake flask-UV, Lord et al. 1980; Briggs 1981; quoted, Sicbaldi & Finizio 1993; Bintein & Devillers 1994)
- 2.35 (Hansch & Leo 1985; quoted, Howard 1991)
- 1.80 (selected, Suntio et al. 1988)
- 2.54 (Thor 1989; quoted, Connell & Markwell 1990)
- 2.79 (Worthing 1991; Milne 1995)
- 2.35 (RP-HPLC, Saito et al. 1993)
- 2.60 (RP-HPLC, Sicbaldi & Finizio 1993)
- 2.35 (selected, Sangster 1993)
- 2.54, 3.40 (quoted, calculated, Patil 1994)
- 2.35 (selected, Hansch et al. 1995)
- 2.60, 2.35, 4.10 (RP-HPLC, CLOGP, calculated-S, Finizio et al. 1997)

Bioconcentration Factor, log BCF:
- >2.96 (estimated-S, Kenaga 1980a; quoted, Howard 1991)
- 2.67 (earthworms, Lord et al. 1980; quoted, Connell & Markwell 1990)
- 1.30 (activated sludge, Freitag et al. 1984,85)
- 1.30 (algae, Freitag et al. 1984,85)
- 1.00 (golden ide, Freitag et al. 1985; quoted, Howard 1991)
- 1.56 (regression-log K_{ow}, Hansch & Leo 1985; quoted, Howard 1991)

Sorption Partition Coefficient, log K_{OC}:
- 2.06 (reported as log K_{OM}, Briggs 1981)
- 2.29 (Lyman et al. 1982; quoted, Howard 1991)
- 1.52 (estimated, Jury et al. 1987; quoted, Howard 1991)
- 1.52 (screening model calculations, Jury et al. 1987b)
- 2.30 (soil, quoted exptl., Meylan et al. 1992)
- 2.94 (soil, calculated-χ and fragments contribution, Meylan et al. 1992)
- 2.30 (soil, 20-25°C, selected, Wauchope et al. 1992; Hornsby et al. 1996)
- 2.30 (selected, Lohninger 1994)

Half-Lives in the Environment:
- Air: half-lives of 2.6 h and 1.4 h for the vapor-phase reaction with photochemically produced hydroxyl radicals and ozone (Atkinson 1985; quoted, Howard 1991); photooxidation half-life of 3.2-32 h in air, based on estimated rate constant for the vapor-phase reaction with hydroxyl radicals in air (Atkinson 1987; quoted, Howard et al. 1991); atmospheric transformation lifetime was estimated to be <1 day (Kelly et al. 1994).

Surface water: hydrolysis half-life of 170 min in a river water sample at pH 7 and 28°C (Wolfe et al. 1976; quoted, Howard 1991); half-lives in Lake Superior water were 7 hours at pH 7.6 and 12°C, 1 h at pH 7.6 and 25°C, 40 h at pH 6.7 and 12°C, and 8 h at pH 6.7 and 23°C (Wolfe et al. 1976; quoted, Howard 1991).

Groundwater: 10.5 min at pH 8.3 to 10.3 h at pH 5.2, based on first-order hydrolysis rate constants in surface waters (Wolfe et al. 1976; quoted, Howard et al. 1991).

Sediment:

Soil: 48-1440 h, based on unacclimated and acclimated soil grab sample data (Agnihotri 1970; Foschi et al. 1970; quoted, Howard et al. 1991); rate constant of 0.231 d^{-1} with a biodegradation half-life of 3 d (Rao & Davidson 1980); 2.5 days in soil (Halfon et al. 1996); field half-life of 2.5 d (20-25°C, selected, Wauchope et al. 1992; Hornsby et al. 1996); half-life of 1 d at pH 7.2 (Tomlin 1994).

Biota: biochemical half-life of 3 d from screening model calculations (Jury et al. 1987b).

Environmental Fate Rate Constants or Half-Lives:

Volatilization:

Photolysis: photolysis half-lives by UV-irradiation ($\lambda > 280$ nm): 37 min in isopropanol, 420 min in cyclohexene and 380 minutes in cyclohexane (Schwack & Flößer-Müller 1990).

Oxidation: photooxidation half-life of 3.2-32 h in air, based on estimated rate constant for the vapor-phase reaction with hydroxyl radicals in air (Atkinson 1987; quoted, Howard et al. 1991)

Hydrolysis: pseudo first-order hydrolysis half-life of 0.1 d (Burchfield 1959; quoted, Freed 1976); half-life of 1.8 h, based on first-order rate constant of 6.5×10^{-3} s^{-1} at pH 7.1 and 28°C (Wolfe et al. 1976; quoted, Howard et al. 1991); half-life of 10.3 h, based on first-order rate constant of $1.87 \times 10^{-5} \cdot s^{-1}$ at pH 5.2 and 28°C (Wolfe et al. 1976; quoted, Howard et al. 1991); half-life of 10.5 minutes, based on first-order rate constant of $1.10 \times 10^{-3} \cdot s^{-1}$ at pH 8.3 and 28°C (Wolfe et al. 1976; quoted, Howard et al. 1991); half-life of 170 min in a river water sample at pH 7 and 28°C (Wolfe et al. 1976; quoted, Howard 1991); half-lives in Lake Superior water were 7 h at pH 7.6 and 12°C, 1 h at pH 7.6 and 25°C, 40 h at pH 6.7 and 12°C, and 8 h at pH 6.7 and 23°C (Wolfe et al. 1976; quoted, Howard 1991).

Biodegradation: unacclimated aqueous aerobic degradation half-life of 48-1440 h, based on unacclimated and acclimated soil grab sample data (Agnihotri 1970; Foschi et al. 1970; quoted, Howard et al. 1991); unacclimated aqueous anaerobic degradation half-life of 192-5760 h, based on unacclimated aqueous aerobic half-life (Howard et al. 1991); rate constant of 0.231 d^{-1} with a biodegradation half-life of 3 d in soil (Rao & Davidson 1980).

Biotransformation:

Bioconcentration, Uptake (k_1) and Elimination (k_2) Rate Constants:

Common Name: Carbendazim
Synonym: Bavistin, BCM, BMK, Carbendazime, Carbendazol, Carbendazym, G 665, Kemdazin, Mecarzole
Chemical Name: carbamic acid, methyl 1H-benzimidazol-2-yl, methyl ester; methyl benzimidazole-2-ylcarbamate; methyl 1H-benzimidazol-2-ylcarbamate
Uses: as fungicide for control of a wide range of fungal diseases in cereals, fruit, vines, hops, oranmentals, vegetables, rice coffee, cotton, mushrooms, and other crops; also used by trunk injection to give some control of Dutch elm disease.
CAS Registry No: 10605-21-7
Molecular Formula: $C_9H_9N_3O_2$
Molecular Weight: 191.19
Melting Point (°C):
    302-307 (with dec., Agrochemicals Handbook 1987; Worthing 1991; Tomlin 1994; Milne 1995)
Boiling Point (°C):
Density (g/cm³ at 20°C):
    1.45 (Agrochemicals Handbook 1987; Tomlin 1994; Milne 1995)
Molar Volume (cm³/mol):
    194.8 (calculated-LeBas method, this work)
    131.9 (calculated-density, this work)
Molecular Volume (Å³):
Total Surface Area, TSA (Å²):
Dissociation Constant pK_a:
    4.48 (Austin & Briggs 1976)
    4.24 (Sangster 1993)
    4.20 (Tomlin 1994)
Heat of Fusion, ΔH_{fus}, kcal/mol:
Entropy of Fusion, ΔS_{fus}, cal/mol·K (e.u.):
Fugacity Ratio at 25°C (assuming ΔS_{fus} = 13.5 e.u.), F: 0.00172

Water Solubility (g/m³ or mg/L at 25°C):
    8.0 (24°C at pH 7, Agrochemicals Handbook 1987; Worthing 1991; Milne 1995)
    29, 8.0, 7.0 (24°C, at pH 4, 7, 8, Tomlin 1994)
    8.0 (20-25°C at pH 7, selected, Augustijn-Becker et al. 1994; Hornsby et al. 1996)

Vapor Pressure (Pa at 25°C):
    6.50×10^{-8} (20°C, Agrochemicals Handbook 1987)
    $<9 \times 10^{-5}$ (20°C, Worthing 1991)
    9.0×10^{-5}, 1.5×10^{-4}, 0.0013 (20, 25, 50°C, quoted, Tomlin 1994)
    $<1 \times 10^{-7}$ (20°C, quoted, Tomlin 1994)
    6.50×10^{-8} (20-25°C, selected, Augustijn-Becker et al. 1994; Hornsby et al. 1996)

Henry's Law Constant (Pa·m³/mol):
    1.55×10^{-6} (calculated-P/C, this work)

Octanol/Water Partition Coefficient, log K_{OW}:
- 1.52 (Austin & Briggs 1976; quoted, Sangster 1993; Bintein & Devillers 1994)
- 1.40 (shake flask-UV, Lord et al. 1980; quoted, Thor 1989; Connell & Markwell 1990; Sangster 1993)
- 1.34 (shake flask at pH 5, Barak et al. 1983; quoted, Sangster 1993)
- 1.56 (Worthing 1991; Milne 1995)
- 1.43 (selected, Sangster 1993)
- 1.38, 1.505, 1.49 (pH 5, 7, 9, quoted, Tomlin 1994)
- 1.56, 1.77 (pH 6, 7, quoted, Tomlin 1994)
- 1.52 (selected, Hansch et al. 1995)

Bioconcentration Factor, log BCF:
- 2.28 (calculated-S, Kenaga 1980)
- 1.57 (earthworms, Lord et al. 1980; quoted, Connell & Markwell 1990)

Sorption Partition Coefficient, log K_{OC}:
- 3.14 (soil, calculated-S, Kenaga 1980)
- 2.35 (soil, HPLC-screening method, Kördel et al. 1993,95)
- 2.30-2.40 (soil, Tomlin 1994)
- 2.69 (soil, 20-25°C at pH 7, selected, Augustijn-Becker et al. 1994; Hornsby et al. 1996)

Half-Lives in the Environment:
Air:
Surface water: half-lives of 2 and 25 months in water under aerobic and anaerobic conditions, respectively (Tomlin 1994).
Groundwater:
Sediment:
Soil: half-life 8-32 d under outdoor conditions, decomposes with half-lives of 6-12 months on bare soil, 3 to 6 months on turf (Tomlin 1994); field half-life of 120 d (20-25°C, selected, Augustijn-Becker et al. 1994; Hornsby et al. 1996).
Biota:

Environmental Fate Rate Constants or Half-Lives:
Volatilization:
Photolysis:
Oxidation:
Hydrolysis: half-life of more than 35 d (pH 5 and 7 at 22°C, Worthing 1991); slowly decomposed in alkaline solution, half-lives >350 d at pH 5, 7, 124 days at pH 9 (Tomlin 1994).
Biodegradation:
Biotransformation:
Bioconcentration, Uptake (k_1) and Elimination (k_2) Rate Constants:

Common Name: Carboxin
Synonym: Carbathiin, D 735, Kemikar, Kisvax, Vitavax
Chemical Name: 5,6-dihydro-2-methyl-1,4-oxathi-ine-3-carboxanilide; 2,3-dihydro-6-methyl-5-phenylcarbamoyl-1,4-oxathi-ine
CAS Registry No: 5234-68-4
Uses: as fungicide in seed treatment for control of seed diseases of barley, wheat, oats, rice, groundnuts, soybeans, cotton, vegetables, maize, and other crops, etc.
Molecular Formula: $C_{12}H_{13}NO_2S$
Molecular Weight: 235.3
Melting Point (°C):
- 91.5-92.5 (Spencer 1982; Agrochemicals Handbook 1987; Worthing 1991; Tomlin 1994)
- 98.0-100 (dimorphic, Spencer 1982; Agrochemicals Handbook 1987; Worthing 1991; Tomlin 1994)
- 94.0 (Kühne et al. 1995)
- 93-95 (Milne 1995)

Boiling Point (°C):
Density (g/cm³ at 20°C):
- 1.30 (Worthing 1991; Montgomery 1993; Tomlin 1994)

Molar Volume (cm³/mol):
- 246.6 (calculated-LeBas method, this work)
- 173.0 (calculated-density, this work)

Molecular Volume (Å³):
Total Surface Area, TSA (Å²):
Dissociation Constant pK_a:
Heat of Fusion, ΔH_{fus}, kcal/mol:
- 5.300 (DSC method, Plato 1972)

Entropy of Fusion, ΔS_{fus}, cal/mol·K (e.u.):
Fugacity Ratio at 25°C (assuming ΔS_{fus} = 13.5 e.u.), F: 0.217

Water Solubility (g/m³ or mg/L at 25°C):
- 170 (Martin & Worthing 1977; quoted, Kenaga 1980)
- 170 (Spencer 1982; Agrochemicals Handbook 1987; Milne 1995)
- 199 (Worthing 1991; Tomlin 1994)
- 170 (quoted, Kühne et al. 1995)
- 215 (calculated-group contribution fragmentation method, Kühne et al. 1995)
- 195 (20-25°C, selected, Wauchope et al. 1992; Hornsby et al. 1996)
- 195 (selected, Lohninger 1994)
- 170 (Montgomery 1993)

Vapor Pressure (Pa at 25°C):
- $<1 \times 10^{-3}$ (20°C, Agrochemicals Handbook 1987)
- 2.5×10^{-5} (Worthing 1991; Tomlin 1994)
- 1.3×10^{-5} (20-25°C, selected, Wauchope et al. 1992; Hornsby et al. 1996)
- 2.5×10^{-5} (20°C, Montgomery 1993)

Henry's Law Constant (Pa·m³/mol):
- 3.45×10^{-5} (calculated-P/C, Montgomery 1993)
- 1.57×10^{-5} (calculated-P/C, this work)

Octanol/Water Partition Coefficient, log K_{OW}:
    2.17   (Worthing 1991; Montgomery 1993; Milne 1995)
    2.18   (Tomlin 1994)
    2.14   (selected, Hansch et al. 1995)

Bioconcentration Factor, log BCF:
    1.53   (calculated-S, Kenaga 1980)

Sorption Partition Coefficient, log K_{OC}:
    2.41   (soil, calculated-S, Kenaga 1980)
    2.41   (soil, 20-25°C, selected, Wauchope et al. 1992; Hornsby et al. 1996)
    2.41   (calculated, Montgomery 1993)
    2.41   (estimated-chemical structure, Lohninger 1994)
    2.57   (soil, Tomlin 1994)

Half-Lives in the Environment:
    Air:
    Surface water:
    Groundwater:
    Sediment:
    Soil: half-life of about 24 h (Worthing 1991; quoted, Montgomery 1993; Tomlin 1994); field half-life of 3 d (20-25°C, selected, Wauchope et al. 1992; Hornsby et al. 1996).
    Biota:

Environmental Fate Rate Constants or Half-Lives:
    Volatilization:
    Photolysis: half-life <3 hours when exposed to light in aqueous solutions at pH 7 (Tomlin 1994).
    Oxidation:
    Hydrolysis: hydrolysis half-life <3 d when exposed to light (Montgomery 1993).
    Biodegradation:
    Biotransformation:
    Bioconcentration, Uptake (k_1) and Elimination (k_2) Rate Constants:

Common Name: Chloroneb
Synonym: Demosan; Tersan SP
Chemical Name: 1,4-dichloro-2,5-dimethoxybenzene
CAS Registry No: 2675-77-6
Uses: as fungicide applied to soil or used as seed treatment for control of seedling diseases of beans, cotton, soybeans, and beet; also used for control of snow mold (*Typhula spp.*) and Pythium blight on turf grass.
Molecular Formula: $C_8H_8Cl_2O_2$
Molecular Weight: 207.1
Melting Point (°C):
    133-135 (Spencer 1982; Agrochemicals Handbook 1987; Worthing 1991; Tomlin 1994; Milne 1995)
Boiling Point (°C):
    268 (Spencer 1982; Agrochemicals Handbook 1987; Worthing 1991; Tomlin 1994; Milne 1995)
Density (g/cm³ at 20°C):
    1.66 (Spencer 1982)
Molar Volume (cm³/mol):
    200.4 (calculated-LeBas method, this work)
    124.8 (calculated-density, this work)
Molecular Volume (Å³):
Total Surface Area, TSA (Å²):
Dissociation Constant pK_a:
Heat of Fusion, ΔH_{fus}, kcal/mol:
    7.30 (DSC method, Plato & Glasgow 1969)
Entropy of Fusion, ΔS_{fus}, cal/mol·K (e.u.):
Fugacity Ratio at 25°C (assuming ΔS_{fus} = 13.5 e.u.), F: 0.0835

Water Solubility (g/m³ or mg/L at 25°C):
    8 (Martin & Worthing 1977; Spencer 1982; quoted, Kenaga & Goring 1980)
    8 (Agrochemicals Handbook 1987; Worthing 1991; Milne 1995)
    8 (20-25°C, selected, Wauchope et al. 1992; Hornsby et al. 1996)
    8 (selected, Lohninger 1994)
    8 (Tomlin 1994)

Vapor Pressure (Pa at 25°C):
    0.40 (Spencer 1982)
    0.40 (Agrochemicals Handbook 1987; Worthing 1991; Tomlin 1994)
    0.40 (20-25°C, selected, Wauchope et al. 1992; Hornsby et al. 1996)

Henry's Law Constant (Pa·m³/mol):
    10.36 (calculated-P/C, this work)

Octanol/Water Partition Coefficient, log K_{OW}:
    1.93 (calculated-s as per Chiou et al. 1977; Chiou 1981, this work)

Bioconcentration Factor, log BCF:
    2.28   (calculated-S as per Kenaga 1980, this work)

Sorption Partition Coefficient, log K_{OC}:
    3.06   (soil, Hamaker & Thompson 1972; quoted, Kenaga & Goring 1980)
    3.10   (soil, quoted exptl., Meylan et al. 1992)
    2.36   (calculated-χ and fragments contribution, Meylan et al. 1992)
    3.22   (soil, 20-25°C, selected, Wauchope et al. 1992; Hornsby et al. 1996)
    3.22   (selected, Lohninger 1994)

Half-Lives in the Environment:
    Air:
    Surface water:
    Groundwater:
    Sediment:
    Soil: half-life of about 24 h (Worthing 1991); field half-life of 130 d (20-25°C, selected, Wauchope et al. 1992; Hornsby et al. 1996).
    Biota:

Environmental Fate Rate Constants or Half-Lives:
    Volatilization:
    Photolysis:
    Oxidation:
    Hydrolysis:
    Biodegradation:
    Biotransformation:
    Bioconcentration, Uptake (k_1) and Elimination (k_2) Rate Constants:

Common Name: Chloropicrin
Synonym: Acquinite, Nemax, Nitrochloroform, Picfume
Chemical Name: Trichloronitromethane
CAS Registry No: 76-06-2
Uses: fungicide/herbicide/insecticide/nematicide/rodenticide; used as a soil disinfectant for control of nematodes, soil insects, soil fungi, and weed seeds; also used for fumigation of stored grain to control insects and rodents, for glasshouse and mushroom-house fumigation, etc.
Molecular Formula: CCl_3NO_2
Molecular Weight: 164.38
Melting Point (°C):
  −64.0 (Spencer 1982; Agrochemicals Handbook 1987; Kawamoto & Urano 1989; Tomlin 1994)
  −69.2 (Howard 1991)
  −64.5, −69.2 (Montgomery 1993)
Boiling Point (°C):
  112.4 (Spencer 1982; Agrochemicals Handbook 1987; Tomlin 1994)
  112.0 (Kawamoto & Urano 1989)
  111.8 (Montgomery 1993)
Density (g/cm³ at 20°C):
  1.656 (20°C, Spencer 1982; Tomlin 1994)
  1.6558, 1.6483 (20, 25°C, Montgomery 1993)
Molar Volume (cm³/mol):
  113.9 (calculated-LeBas method, this work)
  99.3 (calculated-density, this work)
Molecular Volume ($Å^3$):
Total Surface Area, TSA ($Å^2$):
Dissociation Constant pK_a:
Heat of Fusion, ΔH_{fus}, kcal/mol:
Entropy of Fusion, ΔS_{fus}, cal/mol·K (e.u.):
Fugacity Ratio at 25°C (assuming ΔS_{fus} = 13.5 e.u.), F: 1.0

Water Solubility (g/m³ or mg/L at 25°C):
  2270 (Martin & Worthing 1977; Kenaga 1980)
  2270 (0°C, Spencer 1982; quoted, Howard 1991; Tomlin 1994)
  2270, 1620 (0, 25°, Agrochemicals Handbook 1987)
  2300 (Davies & Lee 1987; quoted, Kawamoto & Urano 1989)
  1621 (quoted, Howard 1991)
  2270 (20-25°C, selected, Wauchope et al. 1992; Hornsby et al. 1996)
  2270 (selected, Lohninger 1994)
  2270 (Montgomery 1993)
  1620 (Tomlin 1994)

Vapor Pressure (Pa at 25°C):
  760, 3173 (0, 25°C, Spencer 1982)
  3200 (Agrochemicals Handbook 1987)
  3173 (quoted, Howard 1991)
  2253, 3173, 4400 (20, 25, 30°C, Montgomery 1993)

Henry's Law Constant (Pa·m^3/mol):
    208.0   (Kawamoto & Urano 1989; quoted, Howard 1991; Montgomery 1993)
    197.3   (calculated-P/C, this work)

Octanol/Water Partition Coefficient, log K_{OW}:
    1.03   (HPLC-RT, Kawamoto & Urano 1989; quoted, Sangster 1993)
    2.09   (shake flask, Hansch & Leo 1987; quoted, Sangster 1993)
    2.07   (quoted, Howard 1991)
    1.03, 2.09 (Montgomery 1993)
    2.09   (selected, Hansch et al. 1995)

Bioconcentration Factor, log BCF:
    0.90   (calculated, Kenaga 1980; quoted, Howard 1991)

Sorption Partition Coefficient, log K_{OC}:
    1.79   (calculated, Kenaga 1980; quoted, Howard 1991)
    1.91   (soil, correlated-Freundlich Isotherm, Kawamoto & Urano 1989)
    1.79   (soil, Wauchope et al. 1992; Hornsby et al. 1996)
    1.79   (selected, Lohninger 1994)

Half-Lives in the Environment:
    Air: half-life of 20 d by photodegradation (Howard 1991).
    Surface water: volatilization half-life of 4.3 h from a model river and photodegradation half-life of 3 d in the surface layer of water (Howard 1991).
    Groundwater:
    Sediment:
    Soil: field half-life was estimated to be 1 d (Wauchope et al. 1992; Hornsby et al. 1996).
    Biota:

Environmental Fate Rate Constants or Half-Lives:
    Volatilization: half-life for evaporation from a body of water 1 m deep with a current of 1 m/s and a wind of 3 m/s is 4.3 h (Howard 1991).
    Photolysis: half-life of 20 d in simulated atmosphere, 3 d in aqueous solution with sunlight irradiation (Montgomery 1993).
    Oxidation:
    Hydrolysis: anaerobic biodegradation rate constant 12 d^{-1} and half-life 0.058 d (Kawamoto & Urano 1990, 1991); stable in neutral aqueous solution and with a minimum half-life of 11 yr (Howard 1991).
    Biodegradation:
    Biotransformation:
    Bioconcentration, Uptake (k_1) and Elimination (k_2) Rate Constants:

Common Name: Chlorothalonil
Synonym: Bravo, chlorthalonil, Daconil, DAC 2787, Exotherm, Forturf, Nopcocide N 96, TPN
Chemical Name: tetrachloroisophthalonitrile; 2,4,5,6-tetrachloro-1,3-benzenedicarbonitrile; 2,4,5,6-tetrachloro-1,3-dicyanobenzene
CAS Registry No: 1897-45-6
Uses: fungicide, fumigant, soil insecticide
Molecular Formula: $C_8Cl_4N_2$
Molecular Weight: 265.9
Melting Point (°C):
- 250-251 (Spencer 1982; Agrochemicals Handbook 1987; Worthing 1991; Montgomery 1993; Tomlin 1994; Milne 1995)
- 250.0 (Kawamoto & Urano 1989)
- 250.5 (Kühne et al. 1995)

Boiling Point (°C):
- 350 (Spencer 1982; Agrochemicals Handbook 1987; Kawamoto & Urano 1989; Worthing 1991; Montgomery 1993; Tomlin 1994; Milne 1995)

Density (g/cm^3 at 20°C): 1.80 (Montgomery 1993; Tomlin 1994)
Molar Volume (cm^3/mol):
- 233.0 (calculated-LeBas method, this work)
- 147.7 (calculated-density, this work)

Molecular Volume (Å^3):
Total Surface Area, TSA (Å^2):
Dissociation Constant pK_a:
Heat of Fusion, ΔH_{fus}, kcal/mol:
Entropy of Fusion, ΔS_{fus}, cal/mol·K (e.u.):
Fugacity Ratio at 25°C (assuming ΔS_{fus} = 13.5 e.u.), F: 0.0058

Water Solubility (g/m^3 or mg/L at 25°C):
- 0.60 (Martin & Worthing 1977; Kenaga 1980; Spencer 1982; Agrochemicals Handbook 1987; Worthing 1987,91; quoted, Gustafson et al. 1989; Montgomery 1993; Lohninger 1994; Kühne et al. 1995; Majewski & Capel 1995; Milne 1995)
- 0.30 (Davies & Lee 1987; quoted, Kawamoto & Urano 1989)
- 0.50 (calculated-group contribution fragmentation method, Kühne et al. 1995)
- 0.60 (20-25°C, selected, Wauchope et al. 1992; Hornsby et al. 1996)
- 0.90 (Tomlin 1994)

Vapor Pressure (Pa at 25°C):
- < 1.30 (40°C, Agrochemicals Handbook 1987; Worthing 1991)
- 232 (Worthing 1987; quoted, Majewski & Capel 1995)
- 0.133 (20-25°C, estimated, Wauchope et al. 1992; Hornsby et al. 1996)
- 1.3×10^{-3} (40°C, Montgomery 1993)
- 8.1×10^{-3} (selected, Brouwer et al. 1994)
- 7.6×10^{-5} (Tomlin 1994)

Henry's Law Constant (Pa·m^3/mol):
- 576 (calculated-P/C as per Worthing 1987, Majewski & Capel 1995)
- 0.02 (Montgomery 1993)
- 58.94 (calculated-P/C, this work)

Octanol/Water Partition Coefficient, log K_{OW}:
- 0.14 (screening model calculations, Jury et al. 1987b)
- 2.64 (HPLC-RT, Kawamoto & Urano 1989; quoted, Montgomery 1993)
- 2.89 (RP-HPLC, Saito et al. 1993)
- 2.90 (selected, Hansch et al. 1995)

Bioconcentration Factor, log BCF:
- 1.92 (calculated-S, Kenaga 1980)
- 1.66 (calculated-K_{OW} as per Kenaga 1980, this work)

Sorption Partition Coefficient, log K_{OC}:
- 3.76 (soil, calculated, Kenaga 1980)
- 3.14 (soil, screening model calculations, Jury et al. 1987b)
- 3.14 (soil, Gustafson et al. 1989)
- 3.26 (soil, correlated-Freundlich Isotherm, Kawamoto & Urano 1989)
- 3.14 (soil, 20-25°C, selected, Wauchope et al. 1992; Hornsby et al. 1996)
- 2.76, 3.14 (soil, Montgomery 1993)
- 3.00 (sand, quoted, Montgomery 1993)
- 3.14 (estimated-chemical structure, Lohninger 1994)
- 3.20, 4.15 (sand, silt, Tomlin 1994)

Half-Lives in the Environment:
- Air:
- Surface water:
- Groundwater:
- Sediment:
- Soil: half-life of 70 d from screening model calculations (Jury et al. 1987b); half-life of about 1.5-3 months (Agrochemicals Handbook 1987; Worthing 1991); soil half-life 68 d (Gustafson et al. 1989); field half-life of 30 days (20-25°C, selected, Wauchope et al. 1992; Hornsby et al. 1996); half-lives reported as 4.1 d and 1.5-3 months (Montgomery 1993); half-lives 5-35 d in aerobic and anaerobic soil studies and from a few hours to a few days in aerobic and anaerobic aquatic soil studies (Tomlin 1994).
- Biota: biochemical half-life of 70 d from screening model calculations (Jury et al. 1987b).

Environmental Fate Rate Constants or Half-Lives:
- Volatilization:
- Photolysis:
- Oxidation:
- Hydrolysis:
- Biodegradation: biochemical half-life of 70 d (Jury et al. 1987b); first-order rate constants in biotic and anbiotic shake-flask tests of -0.0161 and -0.0155 d^{-1} in nonsterile sediment/esturine water and -0.00574 d^{-1} in sterile sediment/esturaine water and -0.00355 and -0.00329 d^{-1} in nonsterile esturine water and -0.00283 d^{-1} in sterile esturine water both at Davis Bayou (Walker et al. 1988); aerobic degradation rate constant of 1.7 d^{-1} with half-life of 0.41 d (Kawamoto & Urano 1990, 1991).
- Biotransformation:
- Bioconcentration, Uptake (k_1) and Elimination (k_2) Rate Constants:

Common Name: Dichlone
Synonym: Algistat, Compound 604, Ent 3776, Phygon, Quintar, Sanquinon
Chemical Name: 2,3-dichloro-1,4-naphthoquinone; 2,3-dichloro-1,4-naphthalenedione
CAS Registry No: 117-80-6
Uses: fungicide/algicide; as fungicide for control of blossom blights, scab on apples and pears and brown spot on stone fruit, etc.; also used to control blue-green algae in ponds, lakes, and swimming pools.
Molecular Formula: $C_{10}H_4Cl_2O_2$
Molecular Weight: 227.06
Melting Point (°C):
    193   (Agrochemicals Handbook 1987; Worthing 1991; Montgomery 1993; Tomlin 1994; Milne 1995)
    195   (Howard 1991)
Boiling Point (°C):
    275   (at 2 mmHg, Agrochemicals Handbook 1987; Howard 1991; Montgomery 1993)
Density (g/cm³ at 20°C):
Molar Volume (cm³/mol):
    196.8   (calculated-LeBas method, this work)
Molecular Volume (Å³):
Total Surface Area, TSA (Å²):
Dissociation Constant pK_a:
Heat of Fusion, ΔH_{fus}, kcal/mol:
Entropy of Fusion, ΔS_{fus}, cal/mol·K (e.u.):
Fugacity Ratio at 25°C (assuming ΔS_{fus} = 13.5 e.u.), F: 0.0213

Water Solubility (g/m³ or mg/L at 25°C):
    0.10   (Kenaga 1980; quoted, Howard 1991)
    8.00   (20°C, Hodnett et al. 1983; quoted, Howard 1991)
    0.10   (Agrochemicals Handbook 1987; Worthing 1991; Tomlin 1994; Milne 1995)
    1.00   (Montgomery 1993)
    0.10   (20-25°C, selected, Augustijn-Becker et al. 1994; Hornsby et al. 1996)

Vapor Pressure (Pa at 25°C):
    1.47×10^{-4} (calculated from S & Henry's Law Constant, Howard 1991)
    10930   (20-25°C, estimated, Augustijn-Becker et al. 1994; Hornsby et al. 1996)

Henry's Law Constant (Pa·m³/mol):
    6.51×10^{-5} (Hine & Mookerjee 1975; quoted, Howard 1991)

Octanol/Water Partition Coefficient, log K_{ow}:
    3.16   (estimated, Hodnett et al. 1983; quoted, Howard 1991)
    5.62   (calculated, Montgomery 1993)

Bioconcentration Factor, log BCF:
    3.35   (estimated-S, Kenaga 1980; quoted, Howard 1991)

Sorption Partition Coefficient, log K_{OC}:
    4.19   (estimated-S, Kenaga 1980; quoted, Howard 1991)
    4.00   (20-25°C, estimated, Augustijn-Becker et al. 1994; Hornsby et al. 1996)
    4.19   (calculated, Montgomery 1993)

Half-Lives in the Environment:
    Air: estimated half-life of 3.87 d, based on the vapor-phase reaction with hydroxyl radicals in air (Atkinson 1987; quoted, Howard 1991).
    Surface water:
    Groundwater:
    Sediment:
    Soil: half-life in moist and dry silt loam soil at pH 6.2-6.4 and 29°C is 1 d and slightly under three months, respectively (Burchfield 1959; quoted, Howard 1991); field half-life of 10 d (20-25°C, estimated, Augustijn-Becker et al. 1994; Hornsby et al. 1996).
    Biota:

Environmental Fate Rate Constants or Half-Lives:
    Volatilization:
    Photolysis:
    Oxidation: estimated photooxidation half-life of 3.87 d in air, based on the vapor-phase reaction with hydroxyl radicals in air (Atkinson 1987; quoted, Howard 1991).
    Hydrolysis: half-life of 5 d at pH 7 (Howard 1991).
    Biodegradation:
    Biotransformation:
    Bioconcentration, Uptake (k_1) and Elimination (k_2) Rate Constants:

Common Name: Dithianon
Synonym: Delan, Delan-Col
Chemical Name: 2,3-dicyano-1,4-dithia-anthraquinone; 5,10-dihydro-5,10-dioxonaphtho[2,3-b]-
  p-dithin-2,3-dicarbonitrile
CAS Registry No: 3347-22-6
Uses: as fungicide for control of many foliar diseases.
Molecular Formula: $C_{14}H_4N_2O_2S_2$
Molecular Weight: 296.33
Melting Point (°C):
    230   (Agrochemicals Handbook 1987)
    225   (Worthing 1991; Tomlin 1994)
Boiling Point (°C):
Density (g/cm³ at 20°C):
    1.580   (Tomlin 1994)
Molar Volume (cm³/mol):
    264.0   (calculated-LeBas method, this work)
    187.6   (calculated-density, this work)
Molecular Volume (Å³):
Total Surface Area, TSA (Å²):
Dissociation Constant pK_a:
Heat of Fusion, ΔH_{fus}, kcal/mol:
Entropy of Fusion, ΔS_{fus}, cal/mol·K (e.u.):
Fugacity Ratio at 25°C (assuming ΔS_{fus} = 13.5 e.u.), F: 0.0110

Water Solubility (g/m³ or mg/L at 25°C):
    0.50   (20°C, Agrochemicals Handbook 1987; Worthing 1991; Tomlin 1994)

Vapor Pressure (Pa at 25°C):
    6.6×10^{-5} (Agrochemicals Handbook 1987; Worthing 1991; Tomlin 1994)

Henry's Law Constant (Pa·m³/mol):
    0.0391   (calculated-P/C, this work)

Octanol/Water Partition Coefficient, log K_{OW}:
    2.84   (Worthing 1991)
    3.20   (Tomlin 1994)
    2.84   (selected, Hansch et al. 1995)

Bioconcentration Factor, log BCF:
    2.96   (calculated-S per Kenaga 1980, this work)

Sorption Partition Coefficient, log K_{OC}:
    3.81   (soil, calculated-S per Kenaga 1980, this work)

Half-Lives in the Environment:
    Air:
    Surface water: hydrolysis half-life of 12.2 h at pH 7 and photolytic half-life of 19 hours when exposed to artificial sunlight in 0.1 mg/L aqueous solutions (Tomlin 1994).
    Groundwater:
    Sediment:
    Soil:
    Biota:

Environmental Fate Rate Constants or Half-Lives:
    Volatilization:
    Photolysis: half-life of 19 h when exposed to artificial sunlight in 0.1 mg/L aqueous solution (Tomlin 1994).
    Oxidation:
    Hydrolysis: half-life of 12.2 h at pH 7 (Tomlin 1994).
    Biodegradation:
    Biotransformation:
    Bioconcentration, Uptake (k_1) and Elimination (k_2) Rate Constants:

Common Name: Edifenphos
Synonym: EDDP, Hinosan
Chemical Name: *O*-ethyl *S,S*-diphenyl phosphorodithioate
CAS Registry No: 17109-49-8
Uses: as fungicide for control of blast and blight diseases in rice, etc.
Molecular Formula: $C_{14}H_{15}O_2PS_2$
Molecular Weight: 310.66
Melting Point (°C):
    −25   (Tomlin 1994)
Boiling Point (°C):
    154   (at 0.01 mmHg, Agrochemicals Handbook 1987; Worthing 1991)
Density (g/cm³ at 20°C):
    1.230   (Agrochemicals Handbook 1987; Worthing 1991)
    1.251   (Tomlin 1994)
Molar Volume (cm³/mol):
    250.5   (calculated from density)
Molecular Volume (Å³):
Total Surface Area, TSA (Å²):
Dissociation Constant pK_a:
Heat of Fusion, ΔH_{fus}, kcal/mol:
Entropy of Fusion, ΔS_{fus}, cal/mol·K (e.u.):
Fugacity Ratio at 25°C (assuming ΔS_{fus} = 13.5 e.u.), F: 1.0

Water Solubility (g/m³ or mg/L at 25°C):
    56.0   (20°C, Agrochemicals Handbook 1987)
    insoluble (Worthing 1991)
    56.0   (20°C, Tomlin 1994)

Vapor Pressure (Pa at 25°C):
    0.013   (20°C, Agrochemicals Handbook 1987)
    0.013   (20°C, Tomlin 1994)

Henry's Law Constant (Pa·m³/mol):
    0.0721   (20°C, calculated-P/C, this work)

Octanol/Water Partition Coefficient, log K_{OW}:
    3.48   (RP-HPLC, Saito et al. 1993)

Bioconcentration Factor, log BCF:
    1.81   (calculated-S as per Kenaga 1980, this work)

Sorption Partition Coefficient, log K_{OC}:
    2.68   (calculated-S as per Kenaga 1980, this work)

Half-Lives in the Environment:
    Air:
    Surface water: hydrolysis half-life of 19 d at pH 7 and 2 d at pH 9 (Tomlin 1994).
    Groundwater:
    Sediment:
    Soil: half-life in soil in the range of few days to a few weeks (Tomlin 1994).
    Biota:

Environmental Fate Rate Constants or Half-Lives:
    Volatilization:
    Photolysis:
    Oxidation:
    Hydrolysis: hydrolyzed by strong acids and alkalis, at 25°C, half-life of 19 d at pH 7 and 2 d at pH 9 (Tomlin 1994).
    Biodegradation:
    Biotransformation:
    Bioconcentration, Uptake (k_1) and Elimination (k_2) Rate Constants:

Common Name: Etridiazole
Synonym: Aaterra, Banrot, Dwell, Echlomezol, ETCMTD, Ethazole, ETMT, Koban, MF-344, OM 2425, Pansoil, Terracoat, Terrazole, Truban
Chemical Name: 5-ethoxy-3-(trichloromethyl)-1,2,4-thiadiazole; ethyl 3-trichloromethyl-1,2,4-thiadiazolyl ether
Uses: as fungicide for control of *Phytophthora* and *Pythium* spp. in cotton, ornamentals, vegetables, groundnuts, cucurbits, tomatoes, and other crops; also used as a nitrification inhibitor in maize, cotton and wheat.
CAS Registry No: 2593-15-9
Molecular Formula: $C_5H_5Cl_3N_2OS$
Molecular Weight: 247.53
Melting Point (°C):
    19.1   (Agrochemicals Handbook 1987)
    19.9   (Tomlin 1994; Milne 1995)
Boiling Point (°C):
    95.0   (at 1 mmHg, Agrochemicals Handbook 1987; Tomlin 1994; Milne 1995)
Density (g/cm³ at 20°C):
    1.503   (25°C, Agrochemicals Handbook 1987; Tomlin 1994)
Molar Volume (cm³/mol):
    219.0   (calculated-LeBas method, this work)
    164.7   (calculated-density, this work)
Molecular Volume (Å³):
Total Surface Area, TSA (Å²):
Dissociation Constant pK_a:
    2.77   (Tomlin 1994)
Heat of Fusion, ΔH_{fus}, kcal/mol:
Entropy of Fusion, ΔS_{fus}, cal/mol·K (e.u.):
Fugacity Ratio at 25°C (assuming ΔS_{fus} = 13.5 e.u.), F: 1.00

Water Solubility (g/m³ or mg/L at 25°C):
    50.0   (Agrochemicals Handbook 1987; Worthing 1991; Milne 1995)
    50.0   (20-25°C, selected, Wauchope et al. 1992; Hornsby et al. 1996)
    50.0   (selected, Lohninger 1994)

Vapor Pressure (Pa at 25°C):
    0.013   (20°C, Agrochemicals Handbook 1987)
    0.013   (rm. temp., Worthing 1991)
    0.013   (20-25°C, selected, Wauchope et al. 1992; Hornsby et al. 1996)

Henry's Law Constant (Pa·m³/mol):
    0.0644   (calculated-P/C, this work)

Octanol/Water Partition Coefficient, log K_{OW}:
    2.48-2.60   (Worthing 1991; Milne 1995)
    3.36   (Tomlin 1994)
    2.55   (selected, Hansch et al. 1995)

Bioconcentration Factor, log BCF:
    1.83   (calculated-S as per Kenaga 1980, this work)
    1.22   (calculated-K_{OW} as per Kenaga 1980, this work)

Sorption Partition Coefficient, log K_{OC}:
    0.725   (sandy soil, Worthing 1991)
    0.149   (silt loam, Worthing 1991)
    3.00   (soil, 20-25°C, estimated, Wauchope et al. 1992; Hornsby et al. 1996)
    3.00   (selected, Lohninger 1994)

Half-Lives in the Environment:
    Air:
    Surface water:
    Groundwater:
    Sediment:
    Soil: half-life of 9.5 d under aerobic, 3 d under anaerobic conditions, field dissipation half-life of 1 week in sandy clay loam (Tomlin 1994); field half-life of 103 d (20-25°C, selected, Hornsby et al. 1996).
    Biota:

Environmental Fate Rate Constants or Half-Lives:
    Volatilization:
    Photolysis:
    Oxidation:
    Hydrolysis: half-life of 103 d at pH 6 (Worthing 1991); half-lives of 12 d at pH 6, 45°C, 103 d at pH 6, 25°C (Tomlin 1994).
    Biodegradation: soil half-lives of 9.5 d under aerobic conditions and 3 d under anaerobic conditions (Tomlin 1994).
    Biotransformation:
    Bioconcentration, Uptake (k_1) and Elimination (k_2) Rate Constants:

Common Name: Fenarimol
Synonym: Bloc, EL-222, Rimidin, Rubigan
Chemical Name: ($\pm$)-2,4'-dichloro-α-(pyrimidin-5-yl)benzhydryl alcohol; α-(2-chlorophenyl)-α-(4-chlorophenyl)-5-pyrimidinemethanol
CAS Registry No: 60168-88-9
Uses: as fungicide for control of powdery mildews in pome fruit, strawberries, vines, cucurbits, roses, and beet; also for control of scab on pome fruit, brown patch and snow mold of turf.
Molecular Formula: $C_{17}H_{12}Cl_2N_2O$
Molecular Weight: 331.20
Melting Point (°C):
    117-119 (Agrochemicals Handbook 1987; Worthing 1991; Tomlin 1994; Milne 1995)
Boiling Point (°C):
Density (g/cm^3 at 20°C):
Molar Volume (cm^3/mol):
    338.8 (calculated-LeBas method, this work)
Molecular Volume (Å^3):
Total Surface Area, TSA (Å^2):
Dissociation Constant pK_a:
    2.58 (Sangster 1993)
Heat of Fusion, ΔH_{fus}, kcal/mol:
Entropy of Fusion, ΔS_{fus}, cal/mol·K (e.u.):
Fugacity Ratio at 25°C (assuming ΔS_{fus} = 13.5 e.u.), F: 0.120

Water Solubility (g/m^3 or mg/L at 25°C):
    13.7 (Martin & Worthing 1977; quoted, Kenaga 1980)
    13.7 (at pH 7, Agrochemicals Handbook 1987; Worthing 1991; Tomlin 1994; Milne 1995)
    14.0 (20-25°C, selected, Wauchope et al. 1992; Hornsby et al. 1996)
    14.0 (selected, Lohninger 1994)

Vapor Pressure (Pa at 25°C):
    < 1.3×10^{-5} (Agrochemicals Handbook 1987)
    1.30×10^{-5} (Worthing 1991)
    2.93×10^{-5} (20-25°C, selected, Wauchope et al. 1992; Hornsby et al. 1996)
    6.5×10^{-5} (vapor pressure balance, Tomlin 1994)

Henry's Law Constant (Pa·m^3/mol):
    6.93×10^{-4} (20-25°C, calculated-P/C, this work)

Octanol/Water Partition Coefficient, log K_{OW}:
    0.67 (shake flask-RC at pH 5.3, Martin & Edgington 1981; quoted, Sangster 1993)
    −1.59 (shake flask-AS at pH 5, Barak et al. 1983; quoted, Sangster 1993)
    3.70 (Stevens et al. 1988; quoted, Sangster 1993)
    3.60 (shake flask-HPLC, Bateman et al. 1990; quoted, Sangster 1993)
    3.69 (pH 7, Worthing 1991; Tomlin 1994; Milne 1995)
    3.60 (selected, Hansch et al. 1995)

Bioconcentration Factor, log BCF:
    2.16   (calculated-S, Kenaga 1980)

Sorption Partition Coefficient, log K_{OC}:
    3.01   (calculated-S, Kenaga 1980)
    2.78   (soil, 20-25°C, selected, Wauchope et al. 1992; Hornsby et al. 1996)
    0.176-1.08 (soil, Tomlin 1994)
    2.78   (estimated-chemical structure, Lohninger 1994)

Half-Lives in the Environment:
    Air:
    Surface water:
    Groundwater:
    Sediment:
    Soil: half-life >365 d under aerobic conditions in soil (28% sand, 14.7% clay, 57.3% silt and pH 6 (Tomlin 1994); field half-life of 360 d (20-25°C, selected, Wauchope et al. 1992; Hornsby et al. 1996).
    Biota:

Environmental Fate Rate Constants or Half-Lives:
    Volatilization:
    Photolysis: decomposed readily by sunlight (Tomlin 1994).
    Oxidation:
    Hydrolysis: half-life of 28 d at 52°C and pH 3, 6 and 9 (Tomlin 1994).
    Biodegradation: half-life >365 d under aerobic conditions in soil, and microbial degradation is accelerated by light (Tomlin 1994).
    Biotransformation:
    Bioconcentration, Uptake (k_1) and Elimination (k_2) Rate Constants:

Common Name: Fenfuram
Synonym: Panoram
Chemical Name: 2-methylfuran-3-carboxanilide; 2-methyl-3-furanilide; 2-methyl-N-phenyl-3-furancarboxamide
CAS Registry No: 24691-80-3
Uses: as fungicide for control of bunts and smuts (*Tilletie* and *Ustilago spp.*) in cereals, when applied as a seed treatment.
Molecular Formula: $C_{12}H_{11}NO_2$
Molecular Weight: 201.22
Melting Point (°C):
    109-110   (technical, Agrochemicals Handbook 1987; Tomlin 1994)
    109-110   (Worthing 1991)
Boiling Point (°C):
Density (g/cm³ at 20°C):
    1.36   (Worthing 1991)
Molar Volume (cm³/mol):
    217.1   (calculated-LeBas method, this work)
    148.0   (calculated-density)
Molecular Volume (Å³):
Total Surface Area, TSA (Å²):
Dissociation Constant pK_a:
Heat of Fusion, ΔH_{fus}, kcal/mol:
Entropy of Fusion, ΔS_{fus}, cal/mol·K (e.u.):
Fugacity Ratio at 25°C (assuming ΔS_{fus} = 13.5 e.u.), F:

Water Solubility (g/m³ or mg/L at 25°C):
    100   (Martin & Worthing 1977; Kenaga 1980)
    100   (20°C, Agrochemicals Handbook 1987)
    100   (20°C, Worthing 1991; Tomlin 1994)
    100   (20-25°C, selected, Hornsby et al. 1996)

Vapor Pressure (Pa at 25°C):
    2.0×10^{-5} (20°C, Agrochemicals Handbook 1987)
    2.0×10^{-5} (extrapolated to 20°C, Worthing 1991; Tomlin 1994)
    2.0×10^{-5} (20-25°C, selected, Hornsby et al. 1996)

Henry's Law Constant (Pa·m³/mol):
    4.02×10^{-5} (calculated-P/C, this work)

Octanol/Water Partition Coefficient, log K_{OW}:

Bioconcentration Factor, log BCF:
    1.66   (calculated-S, Kenaga 1980)

Sorption Partition Coefficient, log K_{OC}:
    2.54   (calculated-S, Kenaga 1980)
    2.48   (20-25°C, estimated, Hornsby et al. 1996)

Half-Lives in the Environment:
    Air:
    Surface water:
    Groundwater:
    Sediment:
    Soil: half-life of about 42 d (Agrochemicals Handbook 1987; Tomlin 1994); field half-life of 42 d (20-25°C, selected, Hornsby et al. 1996).
    Biota:

Environmental Fate Rate Constants or Half-Lives:
    Volatilization:
    Photolysis:
    Oxidation:
    Hydrolysis: stable in neutral media, but hydrolyzed by strong acids and alkalis (Tomlin 1994).
    Biodegradation:
    Biotransformation:
    Bioconcentration, Uptake (k_1) and Elimination (k_2) Rate Constants:

Common Name: Folpet
Synonym: ENT-26539, Faltan, Folpan, Folpel, Ftalan, Fungitrol, Orthophaltan, Phaltan, Spolacid, Thiophal, Vinicoll
Chemical Name: N-(trichloromethylthio)phthalimide; 2-[(trichloromethylthio]-1H-isoindole-1,3(2H)-dione
CAS Registry No: 133-07-3
Uses: fungicide for control of downy/powdery mildews, leaf spot diseases, etc.
Molecular Formula: $C_9H_4Cl_3NO_2S$
Molecular Weight: 296.56
Melting Point (°C):
    177   (dec., Agrochemicals Handbook 1987; Patil 1994)
    177   (Worthing 1991; Tomlin 1994; Milne 1995)
Boiling Point (°C):
Density (g/cm³ at 20°C):
Molar Volume (cm³/mol):
    246.2   (calculated-LeBas method, this work)
Molecular Volume (Å³):
Total Surface Area, TSA (Å²):
Dissociation Constant pK_a:
Heat of Fusion, ΔH_{fus}, kcal/mol:
    8.50   (DSC method, Plato 1972)
Entropy of Fusion, ΔS_{fus}, cal/mol·K (e.u.):
Fugacity Ratio at 25°C (assuming ΔS_{fus} = 13.5 e.u.), F: 0.0314

Water Solubility (g/m³ or mg/L at 25°C):
    1.0   (Martin & Worthing 1977; quoted, Kenaga 1980; Briggs 1981)
    1.0   (rm. temp., Agrochemicals Handbook 1987; Worthing 1991; Tomlin 1994)
    1.0, 4.4×10^{-5} (quoted, calculated, Patil 1994)
    1.0   (Milne 1995)

Vapor Pressure (Pa at 25°C):
    <0.0013   (20°C, Agrochemicals Handbook 1987)
    0.0013   (20°C, Worthing 1991; Tomlin 1994)

Henry's Law Constant (Pa·m³/mol):
    0.386   (20°C, calculated-P/C, this work)

Octanol/Water Partition Coefficient, log K_{OW}:
    3.63   (shake flask-UV, Briggs 1981; quoted, Bintein & Devillers 1994)
    2.85   (selected, Yoshioka et al. 1986)
    3.63   (Thor 1989; quoted, Connell & Markwell 1990)
    3.63, 6.94   (quoted, calculated, Patil 1994)
    3.11   (Tomlin 1994)
    2.85   (selected, Hansch et al. 1995)

Bioconcentration Factor, log BCF:
- 1.91 (calculated-S, Kenaga 1980)
- 3.32 (earthworms, Lord et al. 1980; quoted, Connell & Markwell 1990)

Sorption Partition Coefficient, log K_{OC}:
- 1.78 (calculated-S, Kenaga 1980)
- 3.03 (reported as log K_{OM}, Briggs 1981)
- 3.27 (soil, quoted exptl., Meylan et al. 1992)
- 2.16 (calculated-fragment contribution method, Meylan et al. 1992)

Half-Lives in the Environment:
   Air:
   Surface water: half-life of 4.3 d (Tomlin 1994).
   Groundwater:
   Sediment:
   Soil: half-life of 4.3 d (Tomlin 1994).
   Biota:

Environmental Fate Rate Constants or Half-Lives:
   Volatilization:
   Photolysis: half-lives by UV-irradiation (λ >280 nm): 101 min in isopropanol, 144 min in cyclohexene and 1620 min in cyclohexane (Schwack & Flöβer-Müller 1990).
   Oxidation:
   Hydrolysis: hydrolyze at pH 7 with rates similar to captan, i.e., first-order rate constant of 6.5×10^{-5} s^{-1} with half-life of 2.96 h in a phosphate buffer solution at pH 7.07 and 28°C (Wolfe et al. 1976).
   Biodegradation:
   Biotransformation:
   Bioconcentration, Uptake (k_1) and Elimination (k_2) Rate Constants:

Common Name: Formaldehyde
Synonym: formalin, methanal, oxomethane
Chemical Name: formaldehyde
Uses: fungicide/bactericide; used as soil sterilant in mushroom houses and other areas; also used as a silage preservative.
CAS Registry No: 50-00-0
Molecular Formula: HCHO
Molecular Weight: 30.03
Melting Point (°C):
    −92   (Weast 1982-83; Dean 1985)
    −118/−92 (Verschueren 1983)
Boiling Point (°C):
    −21   (Weast 1982-83)
    −21/−19 (Verschueren 1983)
    −19.5 (Dean 1985)
Density (g/cm^3):
    0.815   (Weast 1982-83)
    0.815   (−20°C, Verschueren 1983; Dean 1985)
Acid Dissociation Constants, pK_a:
Molar Volume (cm^3/mol):
    29.6   (calculated-LeBas method)
Molecular Volume, TMV (Å^3):
Total Surface Area, TSA (Å^2):
Heat of Fusion, ΔH_{fus}, kcal/mol:
Entropy of Fusion, ΔS_{fus}, kcal/mol·K (e.u.):
Fugacity Ratio at 25°C (assuming ΔS_{fus} = 13.5 e.u.), F: 1.0

Water Solubility (g/m^3 or mg/L at 25°C):
    1,220,000   (Dean 1985)
    very soluble, up to 55% (Howard 1989)

Vapor Pressure (Pa at 25°C):
    1333   (−88°C, Verschueren 1983)
    451030 (>1 atmospheric pressure, Howard 1989)

Henry's Law Constant (Pa·m^3/mol):
    0.0331 (Dong et al. 1986, quoted, Howard 1989)
    0.0169 (quoted, Gaffney et al. 1987)
    0.0298 (gas stripping-HPLC, Zhou & Mopper 1990)

Octanol/Water Partition Coefficient, log K_{OW}:
    −0.75 (calculated-f const. per Rekker 1977, Deneer et al. 1988; quoted, Verhaar et al. 1992; Saito et al. 1993)
    0.00   (calculated, Verschueren 1983)
    0.35   (quoted, Howard 1989)
    0.35   (recommended, Sangster 1989)

Bioconcentration Factor, log BCF:
   no bioconcn. in fish and shrimp observed (Howard 1989)

Sorption Partition Coefficient, log K_{OC}:
   0.365   (estimated-S as per Kenaga 1980, this work)

Half-Lives in the Environmental Compartments:
   Air: photooxidation half-life of 7.13-71.3 h, based on measured rate constant for the vapor-phase reaction with hydroxyl radicals in air (Atkinson 1985; quoted, Howard et al. 1991); half-life of 1.26-6.0 h, based on photolysis half-life in air (Howard et al. 1991).
   Surface water: 24-168 h, based on unacclimated aqueous aerobic biodegradation half-life (Howard et al. 1991).
   Ground water: 48-336 h, based on unacclimated aqueous aerobic biodegradation half-life (Howard et al. 1991).
   Sediment:
   Soil: 24-168 h, based on unacclimated aqueous aerobic biodegradation half-life (Howard et al. 1991).
   Biota:

Environmental Fate Rate Constants and Half-Lives:
   Volatilization:
   Photolysis: sunlight photolysis half-life of 1.25-6.0 h, based on measured gas-phase photolysis by simulated sunlight (Calvert et al. 1972; Su et al. 1979; quoted, Howard et al. 1991).
   Oxidation: rate constant of 3.2×10^{-16} cm^3 molecule^{-1} s^{-1} for the vapor-phase reaction with NO$_3$ radicals in the atmosphere at 298 ± 1 K (Atkinson & Lloyd 1984; quoted, Carlier et al. 1986); rate constant of 4.50×10^{-14} cm^3 molecule^{-1} s^{-1} for the vapor-phase reaction with HO$_2$ radicals in the atmosphere at 298 K (Baulch et al. 1984; quoted, Carlier et al. 1986); rate constant of 111.1×10^{-11} cm^3 molecule^{-1} s^{-1} for the vapor-phase reaction with OH radicals in the atmosphere at 298 K (Baulch et al. 1984; quoted, Carlier et al. 1986); atmospheric photooxidation half-life of 7.13-71.3 h, based on measured rate constant for the vapor-phase reaction with OH radicals in air (Atkinson 1985; quoted, Howard et al. 1991); aqueous photooxidation half-life of 4,813-190,000 h, based on measured rate constant for the reaction with OH radicals in water (Dorfman & Adams 1973; quoted, Howard et al. 1991).
   Hydrolysis: no hydrolyzable group (Howard et al. 1991).
   Biodegradation: aqueous aerobic half-life of 24-168 h, based on unacclimated aqueous aerobic biodegradation screening test data (Gellman & Heukelekian 1950; Heukelekian & Rand 1955; quoted, Howard et al. 1991); aqueous anaerobic half-life of 96-672 h, based on unacclimated aqueous aerobic biodegradation half-life (Howard et al. 1991).
   Biotransformation:
   Bioconcentration Uptake (k_1) and Elimination (k_2) Rate Constants:

Common Name: Hexachlorobenzene
Synonym: HCB, perchlorobenzene, anticarie, Bunt-cure, Bunt-no-more, Julin's carbon chloride
Chemical Name: hexachlorobenzene
Uses: as fungicide for seed treatment to control common bunt and dwarf bunt of wheat.
CAS Registry No: 118-74-1
Molecular Formula: C_6Cl_6
Molecular Weight: 284.79
Melting Point (°C):
    227.0   (Verschueren 1977,83)
    230.0   (Mackay & Shiu 1981; Yalkowsky et al. 1983; Kishii et al. 1987)
    228.0   (Schmidt-Bleek et al. 1982)
    227.9   (Miller et al. 1984)
    228.3   (Miller et al. 1985)
Boiling Point (°C):
    322   (Verschueren 1977, 1983; Mackay & Shiu 1981; Miller et al. 1984)
    318   (Schmidt-Bleek et al. 1982)
Density (g/cm^3):
    1.569   (23.6°C, Weast 1972-73; Horvath 1982)
Molar Volume (cm^3/mol):
    181.5   (23.6°C, calculated-density, Weast 1972-73; Horvath 1982)
    182.0   (calculated-density, Lande & Banerjee 1981; Mailhot 1987)
    221.4   (LeBas method, Miller et al. 1985; Shiu et al. 1987)
    1.031   (intrinsic volume: V_I/100, Kamlet et al. 1988; Hawker 1989, 1990)
Molecular Volume (Å^3):
    199.5   (De Bruijn & Hermens 1990)
Total Surface Area, TSA (Å^2):
    203.0   (Yalkowsky et al. 1979)
    202.2   (Kishii et al. 1987)
    209.4   (Sabljic 1987)
    202.1   (planar, Doucette & Andren 1988)
    207.9   (De Bruijn & Hermens 1990)
Heat of Fusion, kcal/mol:
    6.87   (Tsonopoulos & Prausnitz 1971)
    5.354   (Miller et al. 1984)
Entropy of Fusion, cal/mol K (e.u.):
    13.7   (Tsonopoulos & Prausnitz 1971)
    10.7   (Miller et al. 1984)
Fugacity Ratio, F (assuming ΔS_{fusion} = 13.5 e.u.):
    0.0090   (25°C, Miller et al. 1985)
    0.0075, 0.0094   (20°C, 25°C, Suntio et al. 1988)

Water Solubility (g/m^3 or mg/L at 25 °C):
    0.005   (generator column-GC/ECD, Weil et al. 1974; selected, Korte, et al. 1978; Kilzer et al. 1979; Geyer et al. 1980, 1981; Chiou et al. 1982; Pereira et al. 1988; Shiu et al. 1990)
    0.006   (shake flask-LSC/^{14}C, Lu & Metcalf 1975; selected, Callahan et al. 1979; Niimi & Cho 1980; Mackay et al. 1985; Isnard & Lambert 1988, 1989)
    0.110   (shake flask-nephelometry, Hollifield 1979)
    0.005   (shake flask-UV, Yalkowsky et al. 1979; selected, Miller et al. 1984; Shiu et al. 1990)

0.0350   (selected, Kenaga & Goring 1980; Kenaga 1980a; selected, Calamari et al. 1983)
0.0034   (calculated-K_{OW}, Yalkowsky et al. 1979; Yalkowsky & Valvani 1980; Valvani & Yalkowsky 1980)
0.0035   (selected, Neely 1980)
0.036   (selected, Briggs 1981)
0.0039   (shake flask-GC, Könemann 1981)
0.0054   (gen. col.-GC/ECD, Hashimoto et al. 1982; selected, McKim et al. 1985)
0.0012-0.014 (shake flask-GC/ECD, Hashimoto et al. 1982)
0.005   (recommended, Horvath 1982; selected, Wong et al. 1984; Oliver 1987b)
0.0084   (selected lit. average, Yalkowsky et al. 1983)
0.0051   (Deutsche Forschungsgemeinschaft 1983; Fischer et al. 1991)
0.0066   (selected, Yoshida et al. 1983b)
0.0019   (OECD 1981; selected, Dobbs et al. 1984)
0.047   (gen. col.-GC/ECD, Miller et al. 1984; 1985; selected, Abernethy et al. 1986; Doucette & Andren 1988; Shiu et al. 1990; Mackay & Paterson 1991)
0.0084, 0.0162 (selected, calculated-UNIFAC, Banerjee 1985)
0.005   (recommended, IUPAC 1985; selected, Bobra et al. 1985; Eadie & Robbins 1987; Suntio et al. 1988; quoted, Ballschmiter & Wittlinger 1991)
0.00495, 0.0146 (selected, calculated-K_{OW} & HPLC-RT, Chin et al. 1986)
0.0124   (selected, Mailhot 1987)
0.005-0.05, 0.006-0.2 (selected exptl., calculated-K_{OW}, Anliker & Moser 1987)
0.86   (selected, supercooled liquid value, Hawker 1989)
0.0024, 0.00537 (selected, calculated-UNIFAC, Banerjee et al. 1990)
0.0063   (selected, Figueroa & Simmons 1991)
0.005   (selected, Mackay et al. 1992; quoted, Mortimer & Connell 1995)

Vapor Pressure (Pa at 25°C):
0.00028 (Sears & Hopke 1949)
0.00150 (selected, Callahan et al. 1979; Neuhauser et al. 1985)
0.00260 (selected, OECD 1979)
0.00145 (20°C, Kiltzer et al. 1979)
0.00230 (gas saturation, Farmer et al. 1980)
0.00130 (selected, Neely 1980; Nash 1989)
0.000453; 0.000167 (Klein et al. 1981)
0.00150 (Mackay & Shiu 1981; Mackay & Paterson 1991)
0.00046 (evaporation rate, Dobbs & Cull 1982)
0.00121 (extrapolated, Antoine eqn., Gückel et al. 1982)
0.00060 (20°C, evaporation rate & gravimetric, Gückel et al. 1982)
0.00240 (20°C, Deutsche Forschungsgemeinschaft 1983; Fischer et al. 1991)
0.00147 (selected, Yoshida et al. 1983b)
0.303; 0.159; 0.121 (subcooled liquid, selected; GC-RT, Bidleman 1984)
0.00310 (selected, Mackay et al. 1985)
0.00147, 0.187 (20°C, selected, solid, subcooled liquid, Bidleman & Foreman 1987)
0.130   (selected, Suntio et al. 1988)
0.2450   (selected, Suntio et al. 1988; quoted, Ballschmiter & Wittlinger 1991)
0.303, 0.127 (subcooled liquid, selected, Hinckley et al. 1990)
0.0023   (selected from Mackay et al. 1992, Mortimer & Connell 1995)

Henry's Law Constant (Pa·m^3/mol):
- 68.2 (20°C, Callahan et al. 1979)
- 5.07 (calculated-P/C, Mackay & Shiu 1981; selected, Pankow et al. 1984)
- 131.3 (batch stripping, Atlas et al. 1982)
- 68.9 (20°C, calculated, Mabey et al. 1982)
- 12.16 (calculated-P/C, Calamari et al. 1983)
- 62.0 (calculated-P/C, Yoshida et al. 1983b)
- 139 (calculated-P/C, Bobra et al. 1985)
- 48.6 (20°C, batch stripping, Oliver 1985)
- 133, 115.9 (observed, calculated-QSAR, Nirmalakhandan & Speece 1988)
- 7.12 (20°C, calculated-P/C, Suntio et al. 1988)
- 11.0 (calculated, Nash 1989)
- 139.0 (calculated-P/C, Fischer et al. 1991)

Octanol/Water Partition Coefficient, log K_{ow}:
- 6.18 (Neely et al. 1974; selected, McKim et al. 1985)
- 4.13 (radioisotope tracer-^{14}C, Lu & Metcalf 1975)
- 6.51 (calculated-f const., Rekker 1977; quoted, Harnish et al. 1983)
- 6.18 (selected, Callahan et al. 1979; Neuhauser et al. 1985)
- 4.13 (Hansch & Leo 1979)
- 5.0, 6.27 (shake flask-GC, HPLC-k', Könemann et al. 1979; selected, Figueroa & Simmons 1991)
- 6.44 (calculated-f constant, Könemann et al. 1979; Könemann 1981; selected, Opperhuizen 1986)
- 5.23 (HPLC-RT, Veith et al. 1979a; selected, Mackay 1982; Freitag et al. 1985)
- 6.18 (HPLC-RT, Veith et al. 1979b; quoted, Veith & Kosian 1982; Saito et al. 1992)
- 6.53 (calculated-f const., Yalkowsky et al. 1979,1982,1983; Yalkowsky & Valvani 1980; Valvani & Yalkowsky 1980; selected, Miller et al. 1984)
- 5.23 (selected, Kenaga & Goring 1980; selected, Yoshida et al. 1983b)
- 5.44 (selected, Briggs 1981)
- 6.22 (HPLC-RT, McDuffie 1981)
- 5.50 (shake flask-GC, Chiou et al. 1982; Chiou 1985; selected, Oliver & Niimi 1983; Oliver & Charlton 1984; Oliver 1987a,b & c)
- 5.66 (HPLC-RT, Hammers et al. 1982)
- 5.40 (shake flask-GC, Wateral et al. 1982; quoted, Suntio et al. 1988)
- 6.13-6.27, 5.66 (range, mean, shake flask method, Eadsforth & Moser 1983)
- 6.27-6.48, 6.38 (range, mean, HPLC method, Eadsforth & Moser 1983)
- 5.0, 5.19 (selected, calculated, Kaiser 1983; Kaiser et al. 1984)
- 5.89 (selected, Calamari et al. 1983)
- 6.42 (calculated-f const., Veith et al. 1983)
- 5.23, 4.61 (selected, calculated-molar refraction, Yoshida et al. 1983)
- 5.47 (generator column-GC/ECD, Miller et al. 1984, 1985; Kerler & Schönherr 1988; Mackay & Paterson 1991)
- 5.75 (selected, Garst 1984)
- 5.70-5.79 (HPLC-RV, Garst & Wilson 1984; Garst 1984)
- 5.20, 5.23, 5.44, 5.50, 5.55 (reported lit., Geyer et al. 1984)
- 5.47 (Sarna et al. 1984)
- 5.50 (selected, Bobra et al. 1985; Suntio et al. 1988; quoted, Ballschmiter & Wittlinger 1991)

5.47, 6.86, 6.42 (selected, HPLC/MS, calculated-π, Burkhard et al. 1985)
5.50 (selected, Hawker & Connell 1985; Connell & Hawker 1988; Hawker 1990)
5.61 (selected, Mackay et al. 1985)
5.75 (selected OECD value, Brooke et al. 1986)
5.6, 5.9 (HPLC-RV, Brooke et al. 1986)
6.51, 6.18 (selected, calculated-K_{OW} & HPLC-RT, Chin et al. 1986)
5.50 (selected, Geyer et al. 1987)
6.92 (HPLC-k', De Kock & Lord 1987)
5.64 (HPLC-k', Mailhot 1987)
5.45 (selected, Gobas et al. 1987, 1989; Travis & Arms 1988)
5.47 (selected exptl., Doucette & Andren 1988)
6.42, 6.55, 6.22, 5.34, 4.86, 4.75 (calculated-π, f const., HPLC-RT, MW, χ, TSA, Doucette & Andren 1988)
5.47, 5.37 (selected, calculated-V_I, solvatochromic parameters, Kamlet et al. 1988)
6.18 (selected, Ryan et al. 1988)
5.50 (shake flask-GC, Pereira et al. 1988)
5.31, 6.58 (selected, calculated-UNIFAC, Banerjee & Howard 1988)
6.68 (calculated-f const., De Bruijn et al. 1989)
5.73 (slow stirring-GC, De Bruijn et al. 1989; De Bruijn & Hermens 1990; quoted, Bintein & Devillers 1994; Sijm et al. 1995; Hansch et al. 1995)
5.66 (correlated, Isnard & Lambert 1988, 1989)
5.50 (selected, Thomann 1989; Fischer et al. 1991)

Bioconcentration Factor, log BCF:
3.89 (rainbow trout, calculated-k_1/k_2, Neely et al. 1974)
3.09 (fish, Körte et al. 1978)
4.27, 3.73, 4.34 (fathead minnow, rainbow trout, green sunfish, Veith et al. 1979)
5.46 (guppy, lipid basis, Könemann & van Leeuwen 1980; selected, Chiou 1985)
4.27 (fish, Giam et al. 1980)
1.20 (rats, adipose tissue, Geyer et al. 1980)
3.93, 2.46 (fish, flowing water, static water, Kenaga & Goring 1980; Kenaga 1980a)
3.61, 2.45 (calculated from water solubility, K_{OC}, Kenaga 1980a)
4.39, 4.20 (algae, calculated, Geyer et al. 1981)
3.91 (fish, correlated, Mackay 1982)
4.27, 3.89 (fathead minnow, rainbow trout, selected, Bysshe 1982)
4.60 (guppy, calculated-χ, Koch 1983)
4.08-4.30 (rainbow trout, Oliver & Niimi 1983)
5.16-5.37 (rainbow trout, lipid basis, Oliver & Niimi 1983; selected, Chiou 1985)
4.31 (calculated-K_{OW}, Calamari et al. 1983)
3.93 (calculated-K_{OW}, Yoshida et al. 1983b)
4.39, 3.36, 4.54 (algae, fish, activated sludge, Klein et al. 1984)
4.39, 3.83 (algae: exptl., calculated, Geyer et al. 1984; quoted, Wang et al. 1996)
4.27 (fathead minnow, 25°C, calculated, Davis & Dobbs 1984; Anliker & Moser 1987)
4.34, 3.74 (green sunfish, rainbow trout, 15°C, calculated, Davis & Dobbs 1984)
4.39, 3.36, 4.54 (algae, fish, sludge, Klein et al. 1984)
4.54 (activated sludge, Freitag et al. 1984; Halfon & Reggiani 1986)
4.39, 3.41, 4.54 (algae, fish, activated sludge, Freitag et al. 1985)
3.05 (fish, selected, Hawker & Connell 1986)

2.62-2.97 (human fat, lipid basis, Geyer et al. 1987)
2.44-2.79 (human fat, wet weight, Geyer et al. 1987)
4.41 (algae, Mailhot 1987)
4.34 (fathead minnow, Carlson & Kosian 1987)
4.38, 4.30 (worms, fish, Oliver 1987a)
3.48 (fish-normalized, Tadokoro & Tomita 1987)
4.19 (guppy, calculated, Gobas et al. 1987)
5.46 (guppy-lipid phase, calculated-K_{OW}, Gobas et al. 1987, 1989)
6.42, 6.71, 5.96, 5.98 (field data-lipid base: Atlantic croakers, blue crabs, spotted sea trout, blue catfish, Pereira et al. 1988)
−1.35 (beef, reported as biotransfer factor log B_b, Travis & Arms 1988)
−2.07 (milk, reported as biotransfer factor log B_m, Travis & Arms 1988)
−0.32 (vegetable, reported as biotransfer factor log B_v, Travis & Arms 1988)
5.30 (guppy-lipid phase, calculated-K_{OW}, Gobas et al. 1989)
3.90, 4.19 (fish, selected, Connell & Hawker 1988; Hawker 1990)
5.30 (guppy, correlated, Gobas et al. 1989)
3.53 (picea omorika, Reischl et al. 1989)
3.57 (fish, calculated, Figueroa & Simmons 1991)
4.37, 4.16 (rainbow trout, guppy, Saito et al. 1992)
4.27, 4.37 (fathead minnows, Saito et al. 1992)
4.25 (*Chlorella pyrenoidosa*, Sijm et al. 1995)
4.39, 3.18 (*Chlorella fusca, Myriophyllum spicatum*, Wang et al. 1996)

Sorption Partition Coefficient, log K_{OC}:
3.59 (Kenaga & Goring 1980; Kenaga 1980a; selected, Lyman 1982; Yoshida 1983b; Nash 1989)
4.45 (Kenaga 1980a)
4.44, 4.21, 3.59 (estimated-S, K_{OW}, BCF, Lyman 1982)
6.08 (calculated, Mabey et al. 1982)
3.59 (selected, Bysshe 1982; Lyman et al. 1982)
2.56 (shake flask-GC/ECD, Speyer soil, Freundlich isotherm, Rippen et al. 1982)
2.70 (shake flask-GC/ECD, Alfisol, Freundlich isotherm, Rippen et al. 1982)
4.58 (calculated-K_{OW}, Calamari et al. 1983)
5.90 (field data, Oliver & Charlton 1984)
4.90 (bottom sediment, Karickhoff & Morris 1985a)
5.10 (calculated-K_{OW}, Oliver & Charlton 1984)
5.2-6.7, 6.1 (suspended sediment, average, Oliver 1987c)
5.80 (algae >50 μm, Oliver 1987c)
6.0-6.5, 6.3; 5.1 (Niagara River plume, range, mean; calculated-K_{OW}, Oliver 1987b)
4.77 (HPLC-k', Hodson & Williams 1988)

Sorption Partition Coefficient, log K_{OM}:
4.25 (shake flask-GC, soil-organic matter, Briggs 1981)
5.50 (Niagara River-organic matter, Oliver & Charlton 1984)

Sorption Partition Coefficient, log K_P:
3.04-4.51 (sediment suspensions, Karickhoff & Morris 1985b; selected, Brusseau & Rao 1989)
5.11 (simulation of Oliver 1985, Brusseau & Rao 1989)

Half-Lives in the Environment:
- Air: degradation rate constant of 0.0144 h^{-1} (Mackay et al. 1985; quoted, Mackay & Paterson 1991); 3753-37530 h, based on estimated photooxidation half-life (Atkinson 1987); 17000 h (selected from Mackay et al. 1992, Mortimer & Connell 1995).
- Surface Water: 23256-50136 h, based on estimated unacclimated aqueous aerobic biodegradation half-life (Beck & Hansen 1974); 1.4-50 d estimated, 0.3-3 d for river water and 30-300 days for lakes, estimated from persistence (Zoeteman et al. 1980); 55000 h (selected from Mackay et al. 1992, Mortimer & Connell 1995).
- Ground Water: 46512-100272 h, based on unacclimated aqueous aerobic biodegradation half-life (Beck & Hansen 1974); 30-300 d, estimated from persistence (Zoeteman et al. 1980).
- Soil: 23256-50136 h, based on unacclimated aerobic soil grab sample data (Beck & Hansen, 1974); >50 d (Ryan et al. 1988).
- Sediment: 55000 h (selected from Mackay et al. 1992, Mortimer & Connell 1995).
- Biota: half-life in rainbow trout, >224 d (Niimi & Cho 1981); in subadult rainbow trout- calculated to be 210 d at 4°C, 80 d at 12°C and 70 d at 18°C (Niimi & Palazzo 1985); in worms at 8°C, 27 d (Oliver 1987a); picea omorika, 30 d (Reischl et al. 1989); 163 h, clearance from fish (Neely 1980).

Environmental Fate Rate Constants or Half Lives:
- Volatilization/Evaporation: 3.45×10^{-10} mol/m^2·h (Gückel et al. 1982).
- Photolysis:
- Oxidation: rate constant in air, 1.44×10^{-2} h^{-1} (Brown et al. 1975; selected, Mackay et al. 1985); photooxidation half-life in air: 3753-37530 h, based on estimated rate constant for the vapor-phase reaction with hydroxyl radicals in air (Atkinson 1987).
- Hydrolysis: not expected to be important, based on k_h = 0, observed after 13 d at pH 3, 7, 11 and 85°C (Ellington et al. 1987).
- Biodegradation: aqueous aerobic biodegradation half-life: 23256-50136 h, based on unacclimated aerobic soil grab sample data (Beck & Hansen 1974); anaerobic aqueous biodegradation half-life: 93024-200544 h, based on estimated unacclimated aqueous aerobic biodegradation half-life (Beck & Hansen 1974) and degradation rate constant in soil, 1.9×10^{-5} h^{-1} (Beck & Hansen 1974; selected, Mackay et al. 1985; Mackay & Paterson 1991); not significant in an aerobic environment, and no significant degradation rate (Tabak et al. 1981; Mills et al. 1982).
- Bioconcentration Uptake (k_1) and Elimination (k_2) Rate Contants:
    - k_1: 18.76 h^{-1} (trout mussele, Neely et al. 1974)
    - k_2: 0.00238 h^{-1} (trout mussele, Neely et al. 1974)
    - k_1: 10000 d^{-1} (guppy, Könemann & van Leeuwen 1980)
    - k_1: 22.5 h^{-1} (guppy, selected, Hawker & Connell 1985)
    - k_1: 18.8 h^{-1} (trout, selected, Hawker & Connell 1985)
    - k_1: 540.0 d^{-1} (fish, selected, Opperhuizen 1986)
    - k_2: 0.00510, 0.00818, 0.00640, 0.0047 d^{-1} (rainbow trout, calculated-fish mean body weight, Barber et al. 1988)
    - 1/k_2: 420 h (trout, selected, Hawker & Connell 1985)
    - log k_1: 2.73 d^{-1} (fish, selected, Connell & Hawker 1988)
    - log k_1: 2.65 d^{-1} (fish, selected, Connell & Hawker 1988)

  log $1/k_2$: 1.24 d^{-1} (fish, selected, Connell & Hawker 1988)
  log k_2: -1.24 d^{-1} (fish, calculated-K_{OW}, Thomann 1989)
  k_1: 0.049 h^{-1} (uptake of mayfly-sediment model II, Gobas et al. 1989b)
  k_2: 0.023 h^{-1} (depuration of mayfly-sediment model II, Gobas et al. 1989b)
  k_1 10489 h^{-1} (*Chlorella fusca*, Wang et al. 1996)
  k_2: 0.424 h^{-1} (*Chlorella fusca*, Wang et al. 1996)
  k_1: 6.558 h^{-1} (*Myriophyllum spicatum*, Wang et al. 1996)
  k_2: 0.00429 h^{-1} (*Myriophyllum spicatum*, Wang et al. 1996)

Sediment Exchange Rate Constant:
  0.026-1.2 d^{-1} (natural sediment, Karickhoff & Morris 1985).

Sediment Burial Rate Constant:
  4.6×10^{-6} h^{-1} (Di Toro et al. 1981; selected, Mackay et al. 1985)

Stratospheric Diffusion Rate Constant:
  1.7×10^{-6} h^{-1} (Mackay et al. 1985)

Common Name: Imazalil
Synonym: Bromazil, Deccozil, Enilconazole, Fecundal, Freshgard, Fungaflor, Fungazil, R 23979
Chemical Name: 1-(β-allyloxy-2,4-dichlorophenylethyl)imidazole;1-[2-(2,4-dichlorophenyl)-2-(2-propenyloxy)ethyl]-1H-imidazole
CAS Registry No: 35554-44-0
Uses: as fungicide for control of a wide range of fungal diseases on fruit, vegetables, and ornamentals; also used as a seed dressing, for control of diseases of cereal and cotton, etc.
Molecular Formula: $C_{14}H_{14}Cl_2N_2O$
Molecular Weight: 297.18
Melting Point (°C):
    50.0   (Agrochemicals Handbook 1987; Milne 1995)
    52.6   (Worthing 1991; Tomlin 1994)
Boiling Point (°C):
    >340   (Worthing 1991; Tomlin 1994)
Density (g/cm^3 at 20°C):
    1.243   (23°C, Agrochemicals Handbook 1987; Worthing 1991; Milne 1995)
    1.348   (26°C, Tomlin 1994)
Molar Volume (cm^3/mol):
    318.8   (calculated-LeBas method, this work)
    239.1   (calculated-density, this work)
Molecular Volume (Å^3):
Total Surface Area, TSA (Å^2):
Dissociation Constant pK_a:
    6.53   (Worthing 1991)
    7.47   (pK_b, Tomlin 1994)
Heat of Fusion, ΔH_{fus}, kcal/mol:
Entropy of Fusion, ΔS_{fus}, cal/mol·K (e.u.):
Fugacity Ratio at 25°C (assuming ΔS_{fus} = 13.5 e.u.), F: 0.533

Water Solubility (g/m^3 or mg/L at 25°C):
    1400   (20°C, Agrochemicals Handbook 1987; Milne 1995)
    180   (pH 7.6, Worthing 1991; Tomlin 1994)
    1400   (20-25°C, selected, Augustijn-Becker et al. 1994; Hornsby et al. 1996)

Vapor Pressure (Pa at 25°C):
    9.30×10^{-6} (20°C, Agrochemicals Handbook 1987)
    1.60×10^{-4} (20°C, Worthing 1991)
    1.58×10^{-4} (20°C, Tomlin 1994)
    9.30×10^{-6} (20-25°C, selected, Augustijn-Becker et al. 1994; Hornsby et al. 1996)

Henry's Law Constant (Pa·m^3/mol):
    1.97×10^{-6} (20-25°C, calculated-P/C, this work)

Octanol/Water Partition Coefficient, log K_{OW}:
    3.82   (Worthing 1991; Milne 1995)
    3.82   (pH 9.2, Tomlin 1994)
    3.82   (selected, Hansch et al. 1995)

Bioconcentration Factor, log BCF:
    4.57   (calculated-S as per Kenaga 1980, this work)
    2.70   (calculated-K_{OW} as per Kenaga 1980, this work)

Sorption Partition Coefficient, log K_{OC}:
    2.26   (clay loam, Worthing 1991; Tomlin 1994)
    2.32   (sandy loam, Worthing 1991; Tomlin 1994)
    1.83   (sandy soil, Worthing 1991; Tomlin 1994)
    3.60   (soil, 20-25°C, selected, Augustijn-Becker et al. 1994; Hornsby et al. 1996)

Half-Lives in the Environment:
    Air:
    Surface water:
    Groundwater:
    Sediment:
    Soil: half-life of 30-170 d (Tomlin 1994); field half-life of 150 d (20-25°C, selected, Augustijn-Becker et al. 1994; Hornsby et al. 1996).
    Biota:

Environmental Fate Rate Constants or Half-Lives:
    Volatilization:
    Photolysis: stable to light under normal storage conditions (Tomlin 1994).
    Oxidation:
    Hydrolysis: very stable to hydrolysis in dilute acids and alkalis at room temperature (Tomlin 1994).
    Biodegradation:
    Biotransformation:
    Bioconcentration, Uptake (k_1) and Elimination (k_2) Rate Constants:

Common Name: Metalaxyl
Synonym: Apron, CGA 48988, Ridomil, Subdue
Chemical Name: methyl N-(2-methoxyacetyl)-N-(2,6-xylyl)-DL-alaninate; methyl-N-(2,6-dimethylphenyl)-N-(methoxyacetyl)-DL-alaninate
CAS Registry No: 57837-19-1
Uses: fungicide to control of foliar and soil-borne diseases caused by *Peronosporates* on a wide range of crops; also used to treat seeds, etc.
Molecular Formula: $C_{15}H_{21}NO_4$
Molecular Weight: 279.34
Melting Point (°C):
    71.8-72.3   (Agrochemicals Handbook 1987; Worthing 1991; Tomlin 1994)
    71.0-72.0   (Milne 1995)
Boiling Point (°C):
Density (g/cm³ at 20°C):
    1.21   (Agrochemicals Handbook 1987; Worthing 1991)
Molar Volume (cm³/mol):
    328.2   (calculated-LeBas method, this work)
    230.9   (calculated-density, this work)
Molecular Volume (Å³):
Total Surface Area, TSA (Å²):
Dissociation Constant pK_a: << 0 (Tomlin 1994)
Heat of Fusion, ΔH_{fus}, kcal/mol:
Entropy of Fusion, ΔS_{fus}, cal/mol·K (e.u.):
Fugacity Ratio at 25°C (assuming ΔS_{fus} = 13.5 e.u.), F: 0.343

Water Solubility (g/m³ or mg/L at 25°C):
    7100   (quoted, Burkhard & Guth 1981)
    7000   (shake flask-HPLC, Ellgehausen et al. 1981)
    7100   (20°C, Agrochemicals Handbook 1987; Worthing 1991)
    7000   (quoted-Yalkowsky & Dannenfelser 1994, Pinsuwan et al. 1995)
    8400   (20-25°C, selected, Wauchope et al. 1992; Hornsby et al. 1996)
    8400   (selected, Lohninger 1994)
    8400   (22°C, Tomlin 1994)

Vapor Pressure (Pa at 25°C):
    2.9×10^{-4} (quoted, Burkhard & Guth 1981)
    2.9×10^{-4} (20°C, Agrochemicals Handbook 1987; Worthing 1991)
    7.5×10^{-4} (20-25°C, selected, Wauchope et al. 1992; Hornsby et al. 1996)
    7.5×10^{-4} (Tomlin 1994)

Henry's Law Constant (Pa·m³/mol):
    2.48×10^{-5} (calculated-P/C, this work)

Octanol/Water Partition Coefficient, log K_{OW}:
    1.65   (shake flask, Ellgehausen et al. 1980, 1981; quoted, Geyer et al. 1991; Sangster 1993)

1.27   (calculated-S, Ellgehausen et al. 1981; quoted, Sangster 1993)
1.53   (shake flask, Hansch & Leo 1987; quoted, Sangster 1993)
1.60   (shake flask at pH 7, Stevens et al. 1988; quoted, Sangster 1993)
1.70   (shake flask at pH 7, Baker et al. 1992; quoted, Sangster 1993)
1.59   (recommended value, Sangster 1993)
1.75   (Tomlin 1994)
1.707, 1.693   (quoted-Yalkowsky & Dannenfelser 1994, calculated-f. const., Pinsuwan et al. 1995)
1.65   (selected, Hansch et al. 1995)

Bioconcentration Factor, log BCF:
    0.03   (*Daphnia magna*, wet wt. basis, Ellgehausen et al. 1980; quoted, Geyer et al. 1991)

Sorption Partition Coefficient, log K_{OC}:
    1.59   (av. of 3 soils, Sharom & Edgington 1982; quoted, Sukop & Cogger 1992)
    2.26   (av. of 7 soils, Carris 1983; quoted, Sukop & Cogger 1992)
    3.22   (av. of 12 soils, calculated-linearized Freundlich Isotherm, Sukop & Cogger 1992)
    1.70   (soil, 20-25°C, estimated, Wauchope et al. 1992; Hornsby et al. 1996)
    1.70   (estimated-chemical structure, Lohninger 1994)

Half-Lives in the Environment:
    Air:
    Surface water:
    Groundwater:
    Sediment:
    Soil: residual activity in soil is about 70-90 ds (Tomlin 1994); field half-life of 70 days (20-25°C, selected, Wauchope et al. 1992; Hornsby et al. 1996).
    Biota:

Environmental Fate Rate Constants or Half-Lives:
    Volatilization: rate of 0.71 ng cm^{-2} h^{-1} (calculated) and 0.35 ng cm^{-2} h^{-1} (determined) from moist soils at 20°C (Burkhard & Guth 1981).
    Photolysis: irradiated by UV at 290 nm in the presence of hydrogen peroxide and titanium dioxide respectively in aqueous solution resulted in 29% and 84% transformation in 2.5 h (Moza et al. 1994).
    Oxidation:
    Hydrolysis: stable in neutral and acidic media at room temp., calculated half-lives at 20°C of >200 days at pH 1, 115 d at pH 9 and 12 d at pH 10 (Worthing 1991; Tomlin 1994).
    Biodegradation:
    Biotransformation:
    Bioconcentration, Uptake (k_1) and Elimination (k_2) Rate Constants:

Common Name: Oxycarboxin
Synonym: DCMOD, Oxycarboxine
Chemical Name: 5,6-dihydro-2-methyl-1,4-oxathi-ine-3-carboxanilide4,4-dioxide; 5,6-dihydro-2-methyl-N-phenyl-1,4-oxathin-3-carboxamide 4,4-dioxide
CAS Registry No: 5259-88-1
Uses: as fungicide for control of rust diseases on ornamentals, cereals, and nursery trees, etc.
Molecular Formula: $C_{12}H_{13}NO_4S$
Molecular Weight: 267.30
Melting Point (°C):
    127.5-130.0 (Agrochemicals Handbook 1987; Worthing 1991)
    119.5-121.5 (Tomlin 1994)
Boiling Point (°C):
Density (g/cm³ at 20°C): 1.14 (Tomlin 1994)
Molar Volume (cm³/mol):
    261.4   (calculated-LeBas method, this work)
    234.5   (calculated-density, this work)
Molecular Volume (Å³):
Total Surface Area, TSA (Å²):
Dissociation Constant pK_a:
Heat of Fusion, ΔH_{fus}, kcal/mol:
Entropy of Fusion, ΔS_{fus}, cal/mol·K (e.u.):
Fugacity Ratio at 25°C (assuming ΔS_{fus} = 13.5 e.u.), F: 0.096

Water Solubility (g/m³ or mg/L at 25°C):
    1000   (Martin & Worthing 1977; quoted, Kenaga 1980)
    1000   (Agrochemicals Handbook 1987; Worthing 1991; Tomlin 1994)
    1000   (20-25°C, selected, Wauchope et al. 1992; Hornsby et al. 1996)
    1000   (selected, Lohninger 1994)

Vapor Pressure (Pa at 25°C):
    0.0010   (20°C, Agrochemicals Handbook 1987)
    < 133   (20°C, Worthing 1991)
    5.60×10^{-6} (Tomlin 1994)
    1.33×10^{-3} (20-25°C, selected, Waudhope et al. 1992; Hornsby et al. 1996)

Henry's Law Constant (Pa·m³/mol):
    3.56×10^{-4} (20-25°C, calculated-P/C, this work)

Octanol/Water Partition Coefficient, log K_{OW}:
    0.740   (shake flask-AS, Mathre 1971; quoted, Sangster 1993)
    0.740   (recommended, Sangster 1993)
    0.772   (Tomlin 1994)
    0.740   (selected, Hansch et al. 1995)

Bioconcentration Factor, log BCF:
    1.11   (calculated-S, Kenaga 1980)

Sorption Partition Coefficient, log K_{OC}:
    1.99   (calculated-S, Kenaga 1980)
    1.98   (soil, estimated, Wauchope et al. 1992; Hornsby et al. 1996)
    1.98   (selected, Lohninger 1994)

Half-Lives in the Environment:
    Air:
    Surface water:
    Groundwater:
    Sediment:
    Soil: half-life of 2.5-8 weeks in sandy loam by aerobic soil metabolism (Tomlin 1994); field half-life is 20 d (Wauchope et al. 1992; Hornsby et al. 1996).
    Biota:

Environmental Fate Rate Constants or Half-Lives:
    Volatilization:
    Photolysis:
    Oxidation:
    Hydrolysis: half-life of 44 d at pH 6, 25°C (Tomlin 1994).
    Biodegradation:
    Biotransformation:
    Bioconcentration, Uptake (k_1) and Elimination (k_2) Rate Constants:

Common Name: Penconazole
Synonym: Award, CGA 71818, Topas, Topaz, Topaze
Chemical Name: 1-(2,4-dichloro-β-propylphenylethyl)-1H-1,2,4-triazole; 1-[2-(2,4-dichlorophenyl)pentyl]-1H-1,2,4-triazole
CAS Registry No: 66246-88-6
Uses: as fungicide for control of pathogenic *Ascomycetes, Basidiomycetes* and *Deuteromycetes* (especially powdery mildews) on vines, cucurbits, pome fruit, ornamentals and vegetables..
Molecular Formula: $C_{13}H_{15}Cl_2N_3$
Molecular Weight: 284.2
Melting Point (°C):
    60.0   (Agrochemicals Handbook 1987; Worthing 1991)
    57.6-60.3 (Tomlin 1994)
Boiling Point (°C):
Density (g/cm³ at 20°C): 1.30 (Tomlin 1994)
Molar Volume (cm³/mol):
    312.3   (calculated-LeBas method, this work)
    218.6   (calculated-density, this work)
Molecular Volume (Å³):
Total Surface Area, TSA (Å²):
Dissociation Constant pK_a:
    1.51   (Tomlin 1994)
Heat of Fusion, ΔH_{fus}, kcal/mol:
Entropy of Fusion, ΔS_{fus}, cal/mol·K (e.u.):
Fugacity Ratio at 25°C (assuming ΔS_{fus} = 13.5 e.u.), F: 0.451

Water Solubility (g/m³ or mg/L at 25°C):
    70   (20°C, Agrochemicals Handbook 1987; Worthing 1991)
    73   (20°C, Tomlin 1994)

Vapor Pressure (Pa at 25°C):
    0.00021 (20°C, Agrochemicals Handbook 1987; Worthing 1991; Tomlin 1994)

Henry's Law Constant (Pa·m³/mol):
    0.00082 (20°C, calculated-P/C, this work)

Octanol/Water Partition Coefficient, log K_{OW}:
    3.40   (shake flask-HPLC, Bateman et al. 1990; quoted, Sangster 1993)
    3.20   (shake flask-HPLC, Chamberlain et al. 1991; quoted, Sangster 1993)
    3.72   (pH 5.7, Tomlin 1994)
    3.40, 3.50 (quoted, Hansch et al. 1995)

Bioconcentration Factor, log BCF:
    1.75   (20°C, calculated-S as per Kenaga 1980, this work)

Sorption Partition Coefficient, log K_{OC}:
    2.62   (20°C, calculated-S as per Kenaga 1980, this work)

Half-Lives in the Environment:
    Air:
    Surface water:
    Groundwater:
    Sediment:
    Soil: half-life is several months (Tomlin 1994).
    Biota:

Environmental Fate Rate Constants or Half-Lives:
    Volatilization:
    Photolysis:
    Oxidation:
    Hydrolysis: stable to hydrolysis pH 1-13, and to temperature up to 350°C (Tomlin 1994).
    Biodegradation:
    Biotransformation:
    Bioconcentration, Uptake (k_1) and Elimination (k_2) Rate Constants:

Common Name: Procymidone
Synonym: S-7131, Sialex, Sumiboto, Sumilex, Sumisclex
Chemical Name: N-(3,5-dichlorophenyl)-1,2-dimethylcyclopropane-1,2-dicarboximide; 3-(3,5-dichlorophenyl)-1,5-dimethyl-3-azabicyclo[3,1,0]hexane-2,4-dione
CAS Registry No: 32809-16-8
Uses: as fungicide for control of *Botrytis, Sclerotinia, Monilia,* and *Helminthosporium spp.* on fruit, vines, vegetables, cereals and ornamentals, etc.
Molecular Formula: $C_{13}H_{11}Cl_2NO_2$
Molecular Weight: 284.1
Melting Point (°C):
    166-166.5 (tech., Agrochemicals Handbook 1987; Tomlin 1994; Milne 1995)
    164-166.0 (Worthing 1991; Tomlin 1994)
Boiling Point (°C):
Density (g/cm³ at 20°C):
    1.452 (25°C, Agrochemicals Handbook 1987; Tomlin 1994; Milne 1995)
Molar Volume (cm³/mol):
    225.9 (calculated-LeBas method, this work)
    195.7 (calculated-density, this work)
Molecular Volume (Å³):
Total Surface Area, TSA (Å²):
Dissociation Constant pK_a:
Heat of Fusion, ΔH_{fus}, kcal/mol:
Entropy of Fusion, ΔS_{fus}, cal/mol·K (e.u.):
Fugacity Ratio at 25°C (assuming ΔS_{fus} = 13.5 e.u.), F: 0.041

Water Solubility (g/m³ or mg/L at 25°C):
    4.50 (Agrochemicals Handbook 1987; Worthing 1991; Tomlin 1994; Milne 1995)
    4.50 (20-25°C, selected, Augustijn-Becker et al. 1994; Hornsby et al. 1996)

Vapor Pressure (Pa at 25°C):
    0.018 (Agrochemicals Handbook 1987)
    0.011 (20°C, Worthing 1991)
    0.018, 0.0105 (20, 25°C, Tomlin 1994)
    0.0187 (20-25°C, selected, Augustijn-Becker et al. 1994; Hornsby et al. 1996)

Henry's Law Constant (Pa·m³/mol):
    1.181 (20-25°C, calculated-P/C, this work)

Octanol/Water Partition Coefficient, log K_{OW}:
    3.14 (26°C, Worthing 1991; Tomlin 1994; Milne 1995)
    3.0 (selected, Hansch et al. 1995)

Bioconcentration Factor, log BCF:
    2.42 (calculated-S as per Kenaga 1980, this work)

Sorption Partition Coefficient, log K_{OC}:
    3.18   (soil, estimated, Augustijn-Becker et al. 1994; Hornsby et al. 1996)
    3.28   (soil, calculated-S as per Kenaga 1980, this work)

Half-Lives in the Environment:
    Air:
    Surface water:
    Groundwater:
    Sediment:
    Soil: persists for ca. 4-12 weeks (Agrochemicals Handbook 1987; Tomlin 1994); field half-life of 7 d (Augustijn-Becker et al. 1994; Hornsby et al. 1996).
    Biota:

Environmental Fate Rate Constants or Half-Lives:
    Volatilization:
    Photolysis:
    Oxidation:
    Hydrolysis:
    Biodegradation:
    Biotransformation:
    Bioconcentration, Uptake (k_1) and Elimination (k_2) Rate Constants:

Common Name: Propiconazole
Synonym: Alamo, Banner, CGA 64250, Desmel, Orbit, Practis, Radar, Spire, Tilt
Chemical Name: (±)-1-[2-(2,4-dichlorophenyl)-4-propyl-1,3-dioxalan-2-ylmethyl]-1H-1,2,4-triazole; 1-[2-(2,4-dichlorophenyl)-4-propyl-1,3-dioxalan-2-ylmethyl]-1H-1,2,4-triazole
CAS Registry No: 60207-90-1
Uses: as fungicide for control of mildews, rusts on cereals, ornamentals, fruits and other crops; and also used for other diseases of turf and grass seed crops, etc.
Molecular Formula: $C_{15}H_{17}Cl_2N_3O_2$
Molecular Weight: 342.2
Melting Point (°C): liquid
Boiling Point (°C):
    180   (at 0.1 mmHg, Agrochemicals Handbook 1987; Worthing 1991; Tomlin 1994; Milne 1995)
Density (g/cm³ at 20°C):
    1.27   (Agrochemicals Handbook 1987; Worthing 1991; Milne 1995)
    1.29   (20°C, Tomlin 1994)
Molar Volume (cm³/mol):
    358.6   (calculated-LeBas method, this method)
    267.3   (calculated-density, this method)
Molecular Volume (Å³):
Total Surface Area, TSA (Å²):
Dissociation Constant pK_a:
    1.09   (Tomlin 1994)
Heat of Fusion, ΔH_{fus}, kcal/mol:
Entropy of Fusion, ΔS_{fus}, cal/mol·K (e.u.):
Fugacity Ratio at 25°C (assuming ΔS_{fus} = 13.5 e.u.), F: 1.00

Water Solubility (g/m³ or mg/L at 25°C):
    110   (20°C, Agrochemicals Handbook 1987; Worthing 1991; Milne 1995)
    100   (20°C, Tomlin 1994)
    110   (20-25°C, selected, Wauchope et al. 1992; Hornsby et al. 1996)
    110   (selected, Lohninger 1994)
    110   (20°C, quoted, Siebers et al. 1994)

Vapor Pressure (Pa at 25°C):
    0.00013   (20°C, Agrochemicals Handbook 1987)
    0.000133   (20°C, Worthing 1991)
    5.6×10^{-5} (20-25°C, Wauchope et al. 1992; Hornsby et al. 1996)
    0.00013   (20°C, quoted, Siebers et al. 1994)
    5.6×10^{-5} (Tomlin 1994)

Henry's Law Constant (Pa·m³/mol):
    4.0×10^{-4} (20°C, calculated-P/C, Siebers et al. 1994)
    0.00017   (20-25°C, calculated-P/C, this work)

Octanol/Water Partition Coefficient, log K_{OW}:
    3.50   (Bateman et al. 1990; quoted, Sangster 1993)
    3.72   (quoted, Siebers et al. 1994)
    3.72   (pH 6.6, Tomlin 1994)
    3.50   (selected, Hansch et al. 1995)

Bioconcentration Factor, log BCF:
    1.64   (calculated-S as per Kenaga 1980, this work)

Sorption Partition Coefficient, log K_{OC}:
    2.81   (soil, selected, Wauchope et al. 1992; Hornsby et al. 1996)
    2.52   (soil, calculated-S as per Kenaga 1980, this work)
    2.81   (selected, Lohninger 1994)

Half-Lives in the Environment:
    Air:
    Surface water: half-life in aerobic aquatic systems at 25°C id 25-85 d (Tomlin 1994).
    Groundwater:
    Sediment:
    Soil: field half-life of 110 d (Wauchope et al. 1992; Hornsby et al. 1996); half-life in aerobic soils at 25°C is 40-70 d (Tomlin 1994).
    Biota:

Environmental Fate Rate Constants or Half-Lives:
    Volatilization:
    Photolysis:
    Oxidation:
    Hydrolysis: no significant hydrolysis (Tomlin 1994).
    Biodegradation:
    Biotransformation:
    Bioconcentration, Uptake (k_1) and Elimination (k_2) Rate Constants:

Common Name: Quintozene
Synonym: Avicol, Batrilex, Brassicol, Chinozan, earthcide, Fartox, Folosan, Fomac 2, Fungiclor, Kobutol, KOBU, KP 2, Marisan forte, Olpisan, PCNB, Pentagen, Phomasan, PKhNB, Quinosan, Quinocene, saniclor 30, Terraclor, Terrafun
Chemical Name: pentachloronitrobenzene
CAS Registry No: 82-68-8
Uses: as fungicide for seed and soil treatment, for control of *Botrytis*, *Rhizoctonia*, and *Sclerotinia spp. on brassicas*, vegetables, ornamentals and other crops, and *Telletia caries* of wheat.
Molecular Formula: $C_6Cl_5NO_2$
Molecular Weight: 295.34
Melting Point (°C):
    143-144 (Agrochemicals Handbook 1987; Tomlin 1994; Milne 1995)
    146 (Spencer 1982; Worthing 1991)
Boiling Point (°C):
Density (g/cm³ at 20°C):
    1.718 (25°C, Spencer 1982; Agrochemicals Handbook 1987; Worthing 1991; Milne 1995)
    1.907 (21°C, Tomlin 1994)
Molar Volume (cm³/mol):
    207.3 (calculated-LeBas method, this work)
    154.9 (calculated-density, this work)
Molecular Volume (Å³):
Total Surface Area, TSA (Å²):
Dissociation Constant pK_a:
Heat of Fusion, ΔH_{fus}, kcal/mol:
Entropy of Fusion, ΔS_{fus}, cal/mol·K (e.u.):
Fugacity Ratio at 25°C (assuming ΔS_{fus} = 13.5 e.u.), F: 0.064

Water Solubility (g/m³ or mg/L at 25°C):
    practically insoluble (Spencer 1982; Worthing 1991; Milne 1995)
    0.55 (20-25°C, shake flask-GC, Kanazawa 1981)
    0.44 (20°C, Agrochemicals Handbook 1987; Pait et al. 1992; Milne 1995)
    0.40 (Davies & Lee 1987; quoted, Kawamoto & Urano 1989)
    0.44 (20-25°C, selected, Wauchope et al. 1992; Hornsby et al. 1996)
    0.10 (selected, Lohninger 1994)
    0.10 (20°C, Tomlin 1994)

Vapor Pressure (Pa at 25°C):
    0.0151 (Spencer 1982)
    6.6×10^{-3} (20°C, Agrochemicals Handbook 1987)
    1.80 (Worthing 1991)
    0.0147 (20-25°C, selected, Wauchope et al. 1992; Hornsby et al. 1996)
    0.0127 (Tomlin 1994)

Henry's Law Constant (Pa·m³/mol):
    0.3718 (known LWAPC of Kawamoto & Urano 1989, Meylan & Howard 1991)
    0.4812 (bond-estimated LWAPC, Meylan & Howard 1991)

Octanol/Water Partition Coefficient, log K_{OW}:
- 4.22 (20°C, shake flask-GC, Kanazawa 1981; quoted, Sangster 1993)
- 5.21 (HPLC-RT, McDuffie 1981; quoted, Sangster 1993)
- 5.00 (HPLC-RT, Ohori & Ihashi 1987; quoted, Sangster 1993)
- 5.18 (HPLC-RT, Kawamoto & Urano 1989; quoted, Sangster 1993)
- 4.77 (shake flask-GC, Niimi et al. 1989; quoted, Sangster 1993)
- 5.40 (calculated-f. const., Niimi et al. 1989)
- 5.02 (RP-HPLC, Saito et al. 1993)
- 5.0-6.0 (Tomlin 1994)
- 4.22 (selected, Hansch et al. 1995)

Bioconcentration Factor, log BCF:
- 2.38 (topmouth gudgeon, Kanazawa 1981)
- 3.49, 3.65, 3.06 (algae, activated sludge, fish, Freitag et al. 1985)
- 2.23 (rainbow trout, Niimi et al. 1989)
- 2.91 (quoted, Pait et al. 1992)

Sorption Partition Coefficient, log K_{OC}:
- 4.30 (correlated-Freundlich Isotherm, Kawamoto & Urano 1989)
- 4.30 (soil, quoted exptl., Meylan et al. 1992)
- 3.38 (soil, calculated-χ and fragments contribution, Meylan et al. 1992)
- 3.70 (soil, estimated, Wauchope et al. 1992; Hornsby et al. 1996)
- 3.78, 3.47 (for adsorption: silt loam, sand, Tomlin 1994)
- 3.98, 3.52 (for desorption: silt loam, sand, Tomlin 1994)
- 4.34 (soil, HPLC-screening method, Kördel et al. 1993,95)
- 4.30 (estimated-chemical structure, Lohninger 1994)

Half-Lives in the Environment:
Air:
Surface water:
Groundwater:
Sediment:
Soil: half-life of ca. 4-10 months (Agrochemicals Handbook 1987; Tomlin 1994), half-life 4 days (Pait et al. 1992); field half-life of 21 d (Wacuchope et al. 1992; Hornsby et al. 1996).
Biota:

Environmental Fate Rate Constants or Half-Lives:
Volatilization:
Photolysis:
Oxidation:
Hydrolysis:
Biodegradation: rate constant of 0.16 d^{-1} with both aerobic and anaerobic half-life of 4.3 d (Kawamoto & Urano 1989, 1991; rate constant of 6.5 d^{-1} with half-life of 0.11 d (Kawamoto & Urano 1990, 1991).
Bioconcentration, Uptake (k_1) and Elimination (k_2) Rate Constants:

Common Name: Thiophanate-methyl
Synonym: Cerobin, Enovit, Fumidor, Fungitox, Fungo, Fungus Fighter, Labilite, Mildothane, Neotosin, NF-44, Pelt 44, Seal 7 Heal, Sigma, Sipcaplant, Sipcasan, Topsin M, Trevin
Chemical Name: dimethyl 4,4'-(o-phenylene)bis(3-thioallophanate; dimethyl [1,2-phenylenebis(minocarbonothioyl)]biscarbamate
Uses: Fungicide/Wound Protectant
CAS Registry No: 23564-05-8
Molecular Formula: $C_{12}H_{14}N_4O_4S_2$
Molecular Weight: 342.40
Melting Point (°C):
    178   (dec., Agrochemicals Handbook 1987)
    172   (dec., Worthing 1991; Tomlin 1994; Milne 1995)
Boiling Point (°C):
Density (g/cm³ at 20°C):
Molar Volume (cm³/mol):
    344.0   (calculated-LeBas method, this work)
Molecular Volume (Å³):
Total Surface Area, TSA (Å²):
Dissociation Constant pK_a:
    7.28   (Worthing 1991; Tomlin 1994)
Heat of Fusion, ΔH_{fus}, kcal/mol:
Entropy of Fusion, ΔS_{fus}, cal/mol·K (e.u.):
Fugacity Ratio at 25°C (assuming ΔS_{fus} = 13.5 e.u.), F: 0.0350

Water Solubility (g/m³ or mg/L at 25°C):
    3.50   (20°C, Agrochemicals Handbook 1987)
    26.6   (Worthing 1991)
    insolubile (Tomlin 1994)
    26.6   (20°C, Milne 1995)
    3.50   (20-25°C, selected, Wauchope et al. 1992; Hornsby et al. 1996)
    3.50   (selected, Lohninger 1994)

Vapor Pressure (Pa at 25°C):
    $<1.0 \times 10^{-5}$ (20°C, Agrochemicals Handbook 1987)
    $<1.33 \times 10^{-5}$ (20-25°C, selected, Wauchope et al. 1992; Hornsby et al. 1996)
    9.50×10^{-6} (Tomlin 1994)

Henry's Law Constant (Pa·m³/mol):
    0.0013 (calculated-P/C, this work)

Octanol/Water Partition Coefficient, log K_{OW}:
    1.40   (Worthing 1991; Milne 1995)
    1.50   (Tomlin 1994)
    1.40   (selected, Hansch et al. 1995)

Bioconcentration Factor, log BCF:
    2.48   (calculated-S as per Kenaga 1980, this work)
    0.044   (calculated-K_{OW} as per Kenaga 1980, this work)

Sorption Partition Coefficient, log K_{OC}:
    0.079   (Worthing 1991)
    3.26   (20-25°C, estimated, Wauchope et al. 1992; Hornsby et al. 1996)
    3.26   (selected, Lohninger 1994)

Half-Lives in the Environment:
    Air:
    Surface water:
    Groundwater:
    Sediment:
    Soil: field half-life of 10 d (Wauchope et al. 1992; Hornsby et al. 1996); persistence ca. 3-4 weeks (Tomlin 1994).
    Biota:

Environmental Fate Rate Constants or Half-Lives:
    Volatilization:
    Photolysis:
    Oxidation:
    Hydrolysis: stable neutral, aqueous solution, half-life of 24.5 h at pH 9, 22°C (Tomlin 1994).
    Biodegradation:
    Biotransformation:
    Bioconcentration, Uptake (k_1) and Elimination (k_2) Rate Constants:

Common Name: Thiram
Synonym: Aapirol, Aatiram, Accel TMT, Accelerator T, Aceto TETD, Arasan, Atiram, Cyuram, Delsan, Ekagom TB, ENT-987, Falitiram, Fermide, Fernacol, Fernasan, Fernide, Thiuram, TMTD
Chemical Name: tetramethylthiuram disulphide; bis(dimethylthiocarbomoyl) disulfide
CAS Registry No: 137-26-8
Uses: fungicide and also as seed disinfectant.
Molecular Formula: $C_6H_{12}N_2S_4$
Molecular Weight: 240.44
Melting Point (°C):
    155-156 (Spencer 1982; Agrochemicals Handbook 1987; Howard 1991; Tomlin 1994; Milne 1995)
    146 (technical, Agrochemicals Handbook 1987; Milne 1995)
    146 (Worthing 1991)
    155.6 (Montgomery 1993)
Boiling Point (°C):
    129 (20 mmHg, Howard 1991)
    310-315 (15 mmHg, Montgomery 1993)
Density (g/cm^3 at 20°C):
    1.29 (Spencer 1982; Worthing 1991; Montgomery 1993; Tomlin 1994; Milne 1995)
Molar Volume (cm^3/mol):
    256.6 (calculated-LeBas method, this work)
    186.4 (calculated-density, this work)
Molecular Volume (Å^3):
Total Surface Area, TSA (Å^2):
Dissociation Constant pK_a:
Heat of Fusion, ΔH_{fus}, kcal/mol:
Entropy of Fusion, ΔS_{fus}, cal/mol·K (e.u.):
Fugacity Ratio at 25°C (assuming ΔS_{fus} = 13.5 e.u.), F: 0.064

Water Solubility (g/m^3 or mg/L at 25°C):
    17.4 (22°C, Spencer 1973, 1982; quoted, Shiu et al. 1990)
    30 (Martin & Worthing 1977; quoted, Kenaga 1980)
    30 (Agrochemicals Handbook 1987; Worthing 1991; Milne 1995)
    30 (Worthing 1987; quoted, Shiu et al. 1990)
    30 (20-25°C, selected, Wauchope et al. 1992; Hornsby et al. 1996)
    30 (Montgomery 1993)
    30 (selected, Lohninger 1994)
    18 (room temp., Tomlin 1994)
    30 (Halfon et al. 1996)

Vapor Pressure (Pa at 25°C):
    negligible (Agrochemicals Handbook 1987; Worthing 1991)
    0.00133 (Halfon et al. 1996)
    <0.00133 (20-25°C, selected, Wauchope et al. 1992; Hornsby et al. 1996)
    0.307 (Tomlin 1994)

Henry's Law Constant (Pa·m^3/mol):
- < 0.008 (calculated-P/C, Lyman et al. 1982; quoted, Howard 1991)
- 0.0107 (calculated-P/C, this work)

Octanol/Water Partition Coefficient, log K_{ow}:
- 1.73 (Tomlin 1994)

Bioconcentration Factor, log BCF:
- 1.96 (calculated-S, Kenaga 1980)
- 1.96 (calculated-S, Lyman et al. 1982; quoted, Howard 1991)

Sorption Partition Coefficient, log K_{OC}:
- 2.83 (calculated-S, Kenaga 1980)
- 2.83 (calculated-S, Lyman et al. 1982; quoted, Howard 1991)
- 2.83 (20-25°C, selected, Wauchope et al. 1992; Hornsby et al. 1996)
- 2.82-3.39 (soil, Montgomery 1993)
- 2.83 (estimated-chemical structure, Lohninger 1994)

Half-Lives in the Environment:
- Air: assuming an ambient hydroxyl radical concn. of 8×10^5 mol/cm^3, the photooxidation reaction half-life in atmosphere is estimated to be 26.6 d at 25°C (GEMS 1986; quoted, Howard 1991).
- Surface water: calculated hydrolysis half-life of 5.3 d at pH 7 (Ellington et al. 1988); hydrolysis half-lives of 128 d, 18 d and 9 h at pH 4, 7 and 9 (Tomlin 1994).
- Groundwater:
- Sediment:
- Soil: 15 days in soil (Halfon et al. 1996); field half-life of 15 d (20-25°C, selected, Wauchope et al. 1992; Hornsby et al. 1996); degradation half-life in sandy soil was 0.5 d at pH 6.7 (Tomlin 1994).
- Biota:

Environmental Fate Rate Constants or Half-Lives:
- Volatilization:
- Photolysis:
- Oxidation: assuming an ambient hydroxyl radical concn. of 8×10^5 mol/cm^3, the photooxidation reaction half-life is estimated to be 26.6 d at 25°C (GEMS 1986; quoted, Howard 1991).
- Hydrolysis: half-life of 5.3 d was estimated based on exptl. rate of 5.0×10^{-3} h^{-1} (Ellington et al. 1988; quoted, Howard 1991; Montgomery 1993); half-lives of 128 d at pH 4, 18 d at pH 7 and 9 h at pH 9 (Tomlin 1994).
- Biodegradation:
- Biotransformation:
- Bioconcentration, Uptake (k_1) and Elimination (k_2) Rate Constants:

Common Name: Tolclofos-methyl
Synonym: Risolex, Rizolex, S-3349
Chemical Name: O-2,6-dichloro-p-tolyl O,O-dimethyl phosphorothioate; O-(2,6-dichloro-4-methylphenyl) O,O-dimethyl phosphorothioate
CAS Registry No: 57018-04-9
Uses: as fungicide for control of soil-borne diseases caused by *Rhizoctonia, Sclerotium* and *Typhula spp.*; also used as a seed, bulb or tuber treatment, soil drench, foliar spray, or by soil incorporation.
Molecular Formula: $C_9H_{11}Cl_2O_3PS$
Molecular Weight: 301.17
Melting Point (°C):
    78-80 (Agrochemicals Handbook 1987; Worthing 1991; Tomlin 1994; Milne 1995)
Boiling Point (°C):
Density (g/cm^3 at 20°C):
Molar Volume (cm^3/mol):
Molecular Volume (Å^3):
Total Surface Area, TSA (Å^2):
Dissociation Constant pK_a:
Heat of Fusion, ΔH_{fus}, kcal/mol:
Entropy of Fusion, ΔS_{fus}, cal/mol·K (e.u.):
Fugacity Ratio at 25°C (assuming ΔS_{fus} = 13.5 e.u.), F: 0.292

Water Solubility (g/m^3 or mg/L at 25°C):
    0.3-0.4 (23°C, Agrochemicals Handbook 1987)
    0.3-0.4 (23°C, Worthing 1991)
    1.10 (Tomlin 1994)
    0.30 (20-25°C, selected, Augustiijn-Becker et al. 1994; Hornsby et al. 1996)

Vapor Pressure (Pa at 25°C):
    0.057 (20°C, Agrochemicals Handbook 1987)
    0.057 (20°C, Worthing 1991; Tomlin 1995)
    0.0573 (20-25°C, selected, Augustijn-Becker et al. 1994; Hornsby et al. 1996)

Henry's Law Constant (Pa·m^3/mol):
    57.5 (calculated-P/C, this work)

Octanol/Water Partition Coefficient, log K_{OW}:
    4.56 (Worthing 1991; Tomlin 1994)

Bioconcentration Factor, log BCF:
    3.09 (calculated-S as per Kenaga 1980, this work)

Sorption Partition Coefficient, log K_{OC}:
    3.30 (soil, estimated, Augustijn-Becker et al. 1994; Hornsby et al. 1996)

Half-Lives in the Environment:
- Air:
- Surface water: photodegradable with 8 hours of sunlight in water with half-life of 44 d, 15-28 d in lake and river water, and less than 2 d on soil surface (Agrochemicals Handbook 1987); half-lives of 44 d in water, 15-28 days in lake for photodegradation (Tomlin 1994).
- Groundwater:
- Sediment:
- Soil: half-life <2 d from soil surface by photodegradation (Tomlin 1994); field half-life is 30 d (20-25°C, Augustijn-Becker et al. 1994; Hornsby et al. 1996)
- Biota:

Environmental Fate Rate Constants or Half-Lives:
- Volatilization:
- Photolysis: photodegradable with 8 h of sunlight in water with half-life of 44 days, 15-28 d in lake and river water, and less than 2 d on soil surface (Agrochemicals Handbook 1987; Tomlin 1994).
- Oxidation:
- Hydrolysis:
- Biodegradation:
- Biotransformation:
- Bioconcentration, Uptake (k_1) and Elimination (k_2) Rate Constants:

Common Name: Tolylfluanid
Synonym: Tolylfluanide
Chemical Name: N-dichlorofluoromethylthio-$N'N'$-dimethyl-N-p-tolylsulphamide;1,1-dichloro-N-[(dimethylamino)sulfonyl]-1-fluoro-N-(4-methylphenyl)methane-sulfenamide
Uses: fungicide/acaricide; to control scab on apples and pears; *Botrytis* on strawberries, raspberries, blackberries, currants, grapes, ornamentals, etc.
CAS Registry No: 731-27-1
Molecular Formula: $C_{10}H_{13}Cl_2FN_2O_2S_2$
Molecular Weight: 347.26
Melting Point (°C):
    95-97   (Agrochemicals Handbook 1987; Worthing 1991; Milne 1995)
    96   (Tomlin 1994)
Boiling Point (°C): dec. on distillation (Tomlin 1994)
Density (g/cm³ at 20°C): 1.52 (Tomlin 1994)
Molar Volume (cm³/mol):
    326.0   (calculated-LeBas method, this work)
Molecular Volume (Å³):
Total Surface Area, TSA (Å²):
Dissociation Constant pK_a:
Heat of Fusion, ΔH_{fus}, kcal/mol:
Entropy of Fusion, ΔS_{fus}, cal/mol·K (e.u.):
Fugacity Ratio at 25°C (assuming ΔS_{fus} = 13.5 e.u.), F: 0.199

Water Solubility (g/m³ or mg/L at 25°C):
    4000   (Martin & Worthing 1977; quoted, Kenaga 1980)
    4000   (room temp., Agrochemicals Handbook 1987; Worthing 1991)
    0.90   (room temp., Tomlin 1994)

Vapor Pressure (Pa at 25°C):
    <0.001   (20°C, Agrochemicals Handbook 1987)
    1.3×10^{-5} (45°C, Worthing 1991)
    1.6×10^{-5} (20°C, Tomlin 1994)

Henry's Law Constant (Pa·m³/mol):

Octanol/Water Partition Coefficient, log K_{OW}:
    3.95   (20°C, Worthing 1991)
    3.90   (Tomlin 1994)
    3.95   (selected, Hansch et al. 1995)

Bioconcentration Factor, log BCF:
    0.778   (calculated, Kenaga 1980)

Sorption Partition Coefficient, log K_{OC}:
    1.66   (soil, calculated-S, Kenaga 1980)

Half-Lives in the Environment:
    Air:
    Surface water:
    Groundwater:
    Sediment:
    Soil: half-life of a few days (Tomlin 1994).
    Biota:

Environmental Fate Rate Constants or Half-Lives:
    Volatilization:
    Photolysis:
    Oxidation:
    Hydrolysis: half-lives at 22°C of 12 d at pH 4, 29 h at pH 7 and < 10 min at pH 9 (Worthing 1991; Tomlin 1994).
    Biodegradation:
    Biotransformation:
    Bioconcentration, Uptake (k_1) and Elimination (k_2) Rate Constants:

Common Name: Triadimefon
Synonym: Amiral, Bayleton, MEB 6447, Triadimefone
Chemical Name: 1-(4-chlorophenoxy)-3,3-dimethyl-1-(1H-1,2,4-triazol-1-yl)butanone; 1-(4-chlorophenoxy)-3,3-dimethyl-1-(1H-1,2,4-triazol-1-yl)-2-butanone
CAS Registry No: 43121-43-3
Uses: as fungicide for control of powdery mildews, rusts in cereals and *Rhynchosporium* in cereals and control of bunt, smuts, *Typhula spp.*, seedling blight, leaf stripe, net blotch, and other cereal diseases when used for seed treatment, etc.
Molecular Formula: $C_{14}H_{16}ClN_3O_2$
Molecular Weight: 293.76
Melting Point (°C):
    82.3   (Agrochemicals Handbook 1987; Worthing 1991; Tomlin 1994; Kühne et al. 1995; Milne 1995)
Boiling Point (°C):
Density (g/cm³ at 20°C):
    1.22   (Agrochemicals Handbook 1987; Tomlin 1994; Milne 1995)
Molar Volume (cm³/mol):
    321   (calculated-LeBas method, this work)
    240.8   (calculated-density, this work)
Molecular Volume (Å³):
Total Surface Area, TSA (Å²):
Dissociation Constant pK_a:
Heat of Fusion, ΔH_{fus}, kcal/mol:
Entropy of Fusion, ΔS_{fus}, cal/mol·K (e.u.):
Fugacity Ratio at 25°C (assuming ΔS_{fus} = 13.5 e.u.), F: 0.271

Water Solubility (g/m³ or mg/L at 25°C):
    260   (Martin & Worthing 1977; quoted, Kenaga 1980)
    260   (20°C, Agrochemicals Handbook 1987; Worthing 1991; Milne 1995)
    71.5   (20-25°C, selected, Wauchope et al. 1992)
    71.5   (20-25°C, selected, Hornsby et al. 1996)
    71.5   (selected, Lohninger 1994)
    64   (20°C, Tomlin 1994)
    72   (quoted, Kühne et al. 1995)
    69   (calculated-group contribution fragmentation method, Kühne et al. 1995)

Vapor Pressure (Pa at 25°C):
    <1.0×10^{-4} (20°C, Agrochemicals Handbook 1987; Worthing 1991)
    2.00×10^{-6} (20-25°C, selected, Wauchope et al. 1992; Hornsby et al. 1996)
    2.00×10^{-5}, 6×10^{-5} (20, 25°C, Tomlin 1994)

Henry's Law Constant (Pa·m³/mol):

Octanol/Water Partition Coefficient, log K_{OW}:
    1.80   (shake flask-AS at pH 5, Barak et al. 1983; quoted, Sangster 1993)
    3.26   (shake flask, Hansch & Leo 1987; quoted, Sangster 1993)

2.77 (shake flask-LC, Patil et al. 1988; quoted, Sangster 1993)
3.18 (Worthing 1991; Milne 1995)
2.90 (shake flask at pH 7, Baker et al. 1992; quoted, Sangster 1993)
3.26 (recommended, Sangster 1993)
2.77 (selected, Hansch et al. 1995)

Bioconcentration Factor, log BCF:
1.43 (calculated, Kenaga 1980)

Sorption Partition Coefficient, log K_{OC}:
2.41 (soil, calculated-S, Kenaga 1980)
2.48 (20-25°C, selected, Wauchope et al. 1992; Hornsby et al. 1996)
2.48 (estimated-chemical structure, Lohninger 1994)
2.48 (soil, Tomlin 1994)

Half-Lives in the Environment:
Air:
Surface water:
Groundwater:
Sediment:
Soil: field half-life of 26 d (20-25°C, selected, Wauchope et al. 1992; Hornsby et al. 1996).
Biota:

Environmental Fate Rate Constants or Half-Lives:
Volatilization:
Photolysis:
Oxidation:
Hydrolysis: half-lives at 22°C of >1 yr at pH 3, 6, and 9 (Worthing 1991; Tomlin 1994).
Biodegradation:
Biotransformation:
Bioconcentration, Uptake (k_1) and Elimination (k_2) Rate Constants:

Common Name: Triflumizole
Synonym: NF-114, Triflumizol, Trifmine
Chemical Name: (E)-4-chloro-α,α,α,-trifluoro-N-(1-imidazol-1-yl-2-propoxyethylidene)-o-toluidine; 1-[1-[[4-chloro-2-(trifluoromethyl)phenyl]imino]-2-propoxyethyl]-1H-imidazole
CAS Registry No: 68694-11-1
Uses: as fungicide for control of powdery mildews in fruit, vines, and vegetables; scab and rust in apples and pears; also used as seed treatment for barley, etc.
Molecular Formula: $C_{15}H_{15}ClF_3N_3O$
Molecular Weight: 345.5
Melting Point (°C):
    63.5   (Agrochemicals Handbook 1987; Worthing 1991; Tomlin 1994; Milne 1995)
Boiling Point (°C):
Density (g/cm³ at 20°C):
Molar Volume (cm³/mol):
    359.5   (calculated-LeBas method, this work)
Molecular Volume (Å³):
Total Surface Area, TSA (Å²):
Dissociation Constant pK_a:
    3.70   (Augustijn-Becker et al. 1994; Tomlin 1994; Hornsby et al. 1996)
Heat of Fusion, ΔH_{fus}, kcal/mol:
Entropy of Fusion, ΔS_{fus}, cal/mol·K (e.u.):
Fugacity Ratio at 25°C (assuming ΔS_{fus} = 13.5 e.u.), F: 0.416

Water Solubility (g/m³ or mg/L at 25°C):
    12500   (20°C, Agrochemicals Handbook 1987; Worthing 1991; Tomlin 1994; Milne 1995)
    12500   (20-25°C, selected, Augustijn-Becker et al. 1994; Hornsby et al. 1996)

Vapor Pressure (Pa at 25°C):
    1.40×10^{-6} (Worthing 1991)
    1.47×10^{-6} (20-25°C, selected, Augustijn-Becker et al. 1992; Hornsby et al. 1996)
    1.86×10^{-4} (Tomlin 1994)

Henry's Law Constant (Pa·m³/mol):
    4.07×10^{-8} (20-25°C, calculated-P/C, this work)

Octanol/Water Partition Coefficient, log K_{OW}:
    1.40   (Worthing 1991; Milne 1995; Tomlin 1994)
    1.40   (selected, Hansch et al. 1995)

Bioconcentration Factor, log BCF:
    0.48   (calculated-S per Kenaga 1980, this work)

Sorption Partition Coefficient, log K_{OC}:
    3.03-3.22 (Tomlin 1994)
    1.60   (20-25°C, estimated, Augustijn-Becker et al. 1994; Hornsby et al. 1996)

Half-Lives in the Environment:
    Air:
    Surface water:
    Groundwater:
    Sediment:
    Soil: half-life of 14 d on clay (Worthing 1991); field half-life of 14 d (20-25°C, selected, Augustijn-Becker et al. 1994; Hornsby et al. 1996).
    Biota:

Environmental Fate Rate Constants or Half-Lives:
    Volatilization:
    Photolysis: aqueous solutions degraded by sunlight with half-life of 29 h (Worthing 1991; Tomlin 1994).
    Oxidation:
    Hydrolysis:
    Biodegradation:
    Biotransformation:
    Bioconcentration, Uptake (k_1) and Elimination (k_2) Rate Constants:

Common Name: Triforine
Synonym: Biformychloazin, Cela W524, Compound W, Denarin, FMC, Funginex, Saprol, W 524
Chemical Name: 1.1'-piperazine-1,4-diyldi-[N-(2,2,2-trichloroethyl)formamide]; 1,4-bis(2,2,2-trichloro-1-formamidoethyl)piperazine; N,N'-[1,4-piperazinediylbis(2,2,2-trichloroethylidene)]bisformamide
Uses: systemic fungicide to control powdery mildews on cereals, fruit, vines, hops, cucurbits, vegetables, and ornamentals, etc.; also used to suppress spider mite activity.
CAS Registry No: 26644-46-2
Molecular Formula: $C_{10}H_{14}Cl_6N_4O_2$
Molecular Weight: 434.97
Melting Point (°C):
    155   (dec., Agrochemicals Handbook 1987; Worthing 1991; Tomlin 1994; Milne 1995)
Boiling Point (°C):
Density (g/cm³ at 20°C): 1.554 (Tomlin 1994)
Molar Volume (cm³/mol):
    389.2   (calculated-LeBas method, this work)
    279.9   (calculated-density, this work)
Molecular Volume (Å³):
Total Surface Area, TSA (Å²):
Dissociation Constant pK_a:
Heat of Fusion, ΔH_{fus}, kcal/mol:
Entropy of Fusion, ΔS_{fus}, cal/mol·K (e.u.):
Fugacity Ratio at 25°C (assuming ΔS_{fus} = 13.5 e.u.), F: 0.052

Water Solubility (g/m³ or mg/L at 25°C):
    30   (rm. temp., Agrochemicals Handbook 1987; Worthing 1991)
    6.0   (rm. temp., Worthing 1991; Milne 1995)
    30   (20-25°C, selected, Wauchope et al. 1992; Hornsby et al. 1996)
    30   (selected, Lohninger 1994)
    9.0   (20°C, Tomlin 1994)

Vapor Pressure (Pa at 25°C):
    2.6×10^{-5} (Agrochemicals Handbook 1987)
    2.7×10^{-5} (Worthing 1991; Tomlin 1994)
    2.7×10^{-5} (20-25°C, selected, Wauchope et al. 1992; Hornsby et al. 1996)

Henry's Law Constant (Pa·m³/mol):

Octanol/Water Partition Coefficient, log K_{OW}:
    2.20   (Worthing 1991; Tomlin 1994; Milne 1995)

Bioconcentration Factor, log BCF:
    1.96   (calculated-S per Kenaga 1980, this work)

Sorption Partition Coefficient, log K_{OC}:
    2.73   (20-25°C, estimated, Wauchope et al. 1992; Hornsby et al. 1996)
    2.30   (estimated-chemical structure, Lohninger 1994)

Half-Lives in the Environment:
    Air:
    Surface water:
    Groundwater:
    Sediment:
    Soil: half-life in soil ca. 3 weeks (Agrochemicals Handbook 1987; Tomlin 1994); field half-life of 21 d (20-25°C, estimated, Wauchope et al. 1992; Hornsby et al. 1996).
    Biota:

Environmental Fate Rate Constants or Half-Lives:
    Volatilization:
    Photolysis:
    Oxidation:
    Hydrolysis: half-life of 3.5 d at pH 5, 25°C in aqeuous solutions (Tomlin 1994).
    Biodegradation:
    Biotransformation:
    Bioconcentration, Uptake (k_1) and Elimination (k_2) Rate Constants:

Common Name: Vinclozolin
Synonym: BAS 352F, Ronilan, Vorlan
Chemical Name: (RS)-3-(3,5-dichlorophenyl)-5-vinyl-1,3-oxazolidine-2,4-dione; 3-(3,5-dichlorophenyl)-5-ethenyl-5-methyl-2,4-oxazolidinedione
CAS Registry No: 50471-44-8
Uses: fungicide to control *Botrytis/Sclerotinie spp.* in vines, oilseed rape, vegetables, fruit, and ornamentals, etc.
Molecular Formula: $C_{12}H_9Cl_2NO_3$
Molecular Weight: 286.11
Melting Point (°C):
    108   (Agrochemicals Handbook 1987; Worthing 1991; Tomlin 1994; Milne 1995)
Boiling Point (°C):
    131   (at 0.05 mmHg, Agrochemicals Handbook 1987; Tomlin 1994; Milne 1995)
Density (g/cm³ at 20°C):
    1.51   (Worthing 1991; Tomlin 1994; Milne 1995)
Molar Volume (cm³/mol):
    266.3   (calculated-LeBas method, this work)
    189.5   (calculated-density, this work)
Molecular Volume (Å³):
Total Surface Area, TSA (Å²):
Dissociation Constant pK_a:
Heat of Fusion, ΔH_{fus}, kcal/mol:
Entropy of Fusion, ΔS_{fus}, cal/mol·K (e.u.):
Fugacity Ratio at 25°C (assuming ΔS_{fus} = 13.5 e.u.), F: 0.1510

Water Solubility (g/m³ or mg/L at 25°C):
    1000   (Martin & Worthing 1977; quoted, Kenaga 1980)
    1000   (20°C, Agrochemicals Handbook 1987)
    3.40   (20°C, Worthing 1991; Tomlin 1994; Milne 1995)
    1000   (20-25°C, estimated, Augustijn-Becker et al. 1994; Hornsby et al. 1996)
    3.40   (20°C, quoted, Siebers et al. 1994)

Vapor Pressure (Pa at 25°C):
    <0.010   (20°C, Agrochemicals Handbook 1987)
    1.6×10^{-5} (20°C, Worthing 1991; Tomlin 1994)
    1.6×10^{-5} (20-25°C, selected, Augustijn-Becker et al. 1994; Hornsby et al. 1996)
    1.3×10^{-4} (20°C, quoted, Siebers et al. 1994)

Henry's Law Constant (Pa·³/mol):
    0.011   (20°C, quoted, Siebers et al. 1994)

Octanol/Water Partition Coefficient, log K_{OW}:
    3.00   (Stevens et al. 1988; quoted, Sangster 1993)
    3.00   (pH 7, Worthing 1991; Tomlin 1994; Milne 1995)
    2.47   (shake flask-HPLC at pH 6, Nielsen et al. 1992; quoted, Sangster 1993)
    3.00   (quoted, Siebers et al. 1994)
    3.10   (selected, Hansch et al. 1995)

Bioconcentration Factor, log BCF:
    1.26   (calcualted, Kenaga 1980)

Sorption Partition Coefficient, log K_{OC}:
    1.99   (soil, calculated-S, Kenaga 1980)
    2.0-2.87 (soil, Tomlin 1994)
    2.0   (soil, 20-25°C, estimated, Augustijn-Becker et al. 1994; Hornsby et al. 1996)

Half-Lives in the Environment:
    Air:
    Surface water:
    Groundwater:
    Sediment:
    Soil: field half-life of 20 d (20-25°C, selected, Augustijn-Becker et al. 1994; Hornsby et al. 1996).
    Biota:

Environmental Fate Rate Constants or Half-Lives:
    Volatilization:
    Photolysis:
    Oxidation:
    Hydrolysis: stable in neutral and weakly acidic media, in 0.1 N NaOH, 50% hydrolysis occurs in 3.8 h (Agrochemicals Handbook 1987; Tomlin 1994).
    Biodegradation:
    Biotransformation:
    Bioconcentration, Uptake (k_1) and Elimination (k_2) Rate Constants:

Table 4.2.1 Common names, chemical names and physical-chemical properties of fungicides

Name	Synonym	Chemcial Name	Formula	MW	mp, °C	Fug. ratio at 25°C	pK_a
Anilazine [101-05-3]	Botrysan, Direz, Dyren	2-chloro-N-(4,6-dichloro-1,3,5-triazin-2-yl)aniline	$C_9H_5Cl_3N_4$	275.5	159-160	0.0462	
Banalaxyl [71626-11-4]	Galben	methyl N-phenylacetyl-N-2,6-xylyl-DL-alaninate	$C_{20}H_{23}NO_3$	325.4	78-80	0.286	
Benodanil [15310-01-8]	Calirux	2-iodo-N-phenylbenzamide	$C_{13}H_{10}INO$	323.1	137	0.078	
Benomyl [17804-35-2]	Benlate	methyl 1-(butylcarbamoyl)benzimidazol-2-ylcarbamate	$C_{14}H_{18}N_4O_3$	290.3	dec.	0.0729	
Bitertanol [70585-36-3]	Baycor, Baymat, Biloxa, Sibutol	1-(biphenyl-4-yloxy)-3,3-dimethyl-1-1H-1,2,4-triazol-1-yl)butan-2-ol	$C_{17}H_{25}N_3O_5$	422.4	118 eutectic	0.1203	
diastereoisomer A [70585-38-5]				422.4	136.7	0.0786	
diastereoisomer B [55179-31-2]				422.4	145.2	0.0647	
Bupirmate [41483-43-6]	Nimrod, Nimrod T	5-butyl-2-ethylamino-6-methyl-pyrimidinyl dimethylsulfamate	$C_{13}H_{24}N_4O_3S$	316.4	50-51	0.553	
Captan [133-06-2]	Aacaptan, Amercide, Captab, orthocide	N-trichloromethylthio-4-cyclohexene-1,2-dicarboximide	$C_9H_8Cl_3NO_2S$	300.6	178	0.0307	
Carbendazim [10605-21-7]	Bavistin, BCM, Carbendazol	carbamic acid, methyl-1H-benzimidazol-2-yl	$C_9H_9N_3O_2$	191.19	302-307	0.0017	4.48
Carboxin [5234-68-4]	Vitavax, carbathiin	5,6-dihydro-2-methy-1,4-oxathi-ine-3-carboxanilide	$C_{12}H_{13}NO_2S$	235.3	91.5-92.5	0.2174	4.2
Chloroneb [2675-77-6]	Tersan SP	1,4-dichloro-2,5-dimethoxybenzene	$C_8H_8Cl_2O_2$	207.1	133-135	0.0835	
Chloropicrin [76-06-2]	Nitrochloroform	trichloronitromethane	CCl_3NO_2	164.39	−64	1	
Chlorothalonil [1897-45-6]	Bravo, Daconil	2,4,6-tetrachloro-1,3-benzene-dicarbonitrile	$C_8Cl_4N_2$	265.89	250-251	0.0058	

727

Name	Synonym	Chemical Name	Formula	MW	mp, °C	Fug. ratio at 25°C	pK$_a$
Dazomet (Fum.) [533-74-4]	Salvo, Mylone, Basamid	3,5-dimethyl-1,3,5-thiadiazinane-2-thione	$C_5H_{10}N_2S_2$	162.3	104-105	0.162	
Dichlofluanid [1085-98-9]	Euaren, Elvaron	N-dichlorofluoromethylthio-N,N'-dimethyl-N-phenylsuphamide	$C_9H_{11}Cl_2FN_2O_2S_2$	333.2	105-105.6	0.160	
Dichlone [117-80-6]	Phygon	2,3-dichloro-1,4-naphthoquinone	$C_{10}H_4Cl_2O_2$	227.0	197 subl.		
Dithianon [3347-22-6]	Delan	5,10-dihydro-5,10-dioxonaphtho[2,3-b]-]1,4-dithi-in-2,3-dicarbonitrile	$C_{13}H_4N_2O_2S_2$	296.3	225	0.011	
Edifenphos [17109-49-8]	EDDP, Hinosan	O-ethyl S,S-diphenyl-phosphoradithioate	$C_{14}H_{15}O_2PS_2$	310.66	−25	1	
Ethirimol [23947-60-6]		5-butyl-2-ethylamino-6-methyl-pyrimidim-4-ol	$C_{11}H_{19}N_3O$	209.3	159-160	0.046	
Ethoprophos (N,I) [13194-48-4]	Mocap, ethoprop	O-ethyl-S,S-dipropyl-phosphorodithioate	$C_8H_{19}O_2PS_2$	242.3	liquid	1	
Etridiazole [2593-15-9]	ethazol, ethazole, Terrazole	ethyl 3-trichloromethyl-1,2,4-thiadiazolyl ether	$C_5H_5Cl_3N_2OS$	247.5	19.9	1	2.27
Fenarimol [60168-88-9]	Bloc, Rimidin, Rubigan	(±)-2,4-dichloro-α-(pyrimidin-5-yl)benzhydryl alcohol	$C_{17}H_{12}Cl_2N_2O$	331.2	117-119	0.120	
Fenfuram [24691-80-3]	fenfurame	2-methyl-3-fruanilide	$C_{12}H_{11}NO_2$	201.2	109-110	0.144	
Fenpropimorph [67564-91-4]	Corbel, Mistral	(±)-cis-4-[3-(4-tert-butylphenyl)-2-methylpropyl]-2,6-dimethylmorpholine	$C_{20}H_{33}NO$	303.5	oil	1	6.98
Folpet [133-07-3]	Foltan, Folpan, Folpel, Spolacid	N-(trichloromethylthio)phthalimide	$C_9H_4Cl_3NO_2S$	296.56	177	0.031	
Formaldehyde [50-00-0]		formaldehyde	HCHO	30.03	−92	1	
Furalaxyl [57646-30-7]	Fongarid	methyl-N-(2-furoyl)-N-(2,6-xylyl)-DL-alaninate	$C_{17}H_{19}NO_4$	301.3	70	0.359	

Name	Synonym	Chemical Name	Formula	MW	mp, °C	Fug. ratio at 25°C	pK$_a$
Hexachlorobenzene [118-74-1]	HCB	hexachlorobenzene	C$_6$Cl$_6$	284.79	228	0.0098	
Imazalil [35554-44-0]	Bromazil, Deccozil	(±)-1-(β-allyloxy-2,4-dichlorophenyl-ethyl)imidazole	C$_{14}$H$_{14}$Cl$_2$N$_2$O	297.18	50	0.533	
Iprobenfos [26087-47-8]	Kitazin	S-benzy O,O-di-isopropyl phosphoro-thioate	C$_{13}$H$_{21}$O$_3$PS	288.3	oil	1	
Metalaxyl [57837-19-1]	Ridomil, Apron Fubol	methyl N-(2-methoxyacetyl)-2,6-xylyl)-DL-alaninate	C$_{15}$H$_{21}$NO$_4$	279.3	72	0.343	<< 0
Metiram [9006-42-4]	Carbatene, Polyram	zinc ammoniate ethylenebisdithio-carbamate-poly(ethylenethiurmdisulfide)					
Oxycarboxin [5259-88-1]	Plantvax	5,6-dihydro-2-methyl-1,4-oxathi-ine-3-carboxanilide 4,4-dioxide	C$_{12}$H$_{13}$NO$_4$S	267.3	127-130	0.096	
Penconazole [66246-88-6]	Topas, Topaz, Topaze	1-(2,4-dichloro-β-propylphenyl-ethyl)-1H-1,2,4-trizaole	C$_{13}$H$_{15}$Cl$_2$N$_3$	284.2	60	0.451	1.51
Procymidone [32809-16-8]	Sumisclex, Sumilex	N-(3,5-dichlorophenyl)-1,2-dimethyl-cyclopropane-1,2-dicarboximide	C$_{13}$H$_{11}$Cl$_2$NO$_2$	284.1	164-166	0.041	
Propiconazole [60207-90-1]		(±)-1-[2,4-(dichlorophenyl)-4-propyl-1,3-dioxolan-2-methyl]-	C$_{15}$H$_{17}$Cl$_2$N$_3$O$_2$	342.2	liquid	1	1.09
Quintozene [82-68-8]	Tritisan, Botrilex, Terrachlor	pentachloronitrobenzene	C$_6$Cl$_5$NO$_2$	295.3	146	0.064	
Tecnazene [117-18-0]	Folosan, Fusarex	1,2,4,5-tetrachloro-3-nitrobenzene	C$_6$HCl$_4$NO$_2$	260.9	99	0.185	
Thiabendazole [148-79-8]	Mertect, Storite	2-(thiazol-4-yl)benzimidazole	C$_{10}$H$_7$N$_3$S	201.2	304-305	0.0017	
Thiophanate-methyl [23564-05-8]	Topsin M, Mildothane	dimethyl 4,4'-(o-phenylene)bis(3-thioallphanate)	C$_{12}$H$_{14}$N$_4$O$_4$S$_2$	342.4	172	0.035	7.28

Name	Synonym	Chemical Name	Formula	MW	mp, °C	Fug. ratio at 25°C	pK$_a$
Thiram [137-26-8]	Arasan, Tersan, Fernasan	tetramethylthiuram disulphide	$C_6H_{12}N_2S_4$	240.4	145	0.064	
Tolclofos-methyl [57018-04-9]	Rizolex	O-2,6-dichloro-p-tolyl O,O-dimethyl phosphorothioate	$C_9H_{11}Cl_2O_3PS$	301.1	78-80	0.292	
Tolylfluanid [731-27-1]	Euparen M	N-dichlorofluoromethylthio-N,N′-dimethyl-N-p-tolysulphamide	$C_{10}H_{13}Cl_2FN_2O_2S_2$	347.2	96	0.199	
Triadimefon [43121-43-3]	Amiral, Bayeton	1-(chlorophenoxy)-3,3-dimethyl-1-1H-1,2,4-triazol-1-yl)butanone	$C_{14}H_{16}ClN_3O_2$	293.8	82.3	0.271	
Triadimenol [55219-65-3]	Baytan	1-(4-chlorophenoxy)-3,3-dimethyl-1-(1H-1,2,4-triazol-1-yl)butan-2-ol	$C_{14}H_{18}ClN_3O_2$	295.8	121-127	0.103	
Tricyclazole [41814-78-2]	Beam, Bim, Blascide	5-methyl-1,2,4-triazolo[3,4-b]-benzothiazole	$C_9H_7N_3S$	189.2	187-188	0.024	
Triflumizole [99387-89-0]	Trifmine	(E)-4-chloro-α,α,α-trifluoro-N-(1-imidazol-1-yl-propoxyethylidene)-o-toluidine	$C_{15}H_{15}ClF_3N_3O$	345.7	63.5	0.416	3.7
Triforine [26644-46-2]		1,4-bis(2,2,2-trichloro-1-formamido-ethyl)piperazine	$C_{10}H_{14}Cl_6N_4O_2$	435	155	0.052	
Vinclozolin [50471-44-8]	Ronilan	(RS)-3-(3,5-dichlorophenyl)-5-methyl-5-vinyl-1,3-oxazolidine-2,4-dione	$C_{12}H_9Cl_2NO_3$	286.1	108	0.151	
Warfarin (R.) [81-18-2]	Coumadin, Dethmor	4-hydroxy-3-(3-oxo-1-phenylbutyl)-2H-1-benzopyran-2-one	$C_{19}H_{16}O_4$	308.3	161	0.045	
Zineb [12122-67-7]	Zinebe	zinc ethylenebis(dithiocarbamate)	$C_4H_6N_2S_4Zn$	275.8	157 dec.		
Ziram [137-30-4]	Zirame	zinc bis(dimethyldithiocarbamate)	$C_6H_{12}N_2S_4Zn$	305.8	246	0.0065	

Note: Fum. - fumigant, I - insecticide, R. - rodenticide

Table 4.2.2 Summary of physical-chemical properties of fungicides at 25°C

Name	Selected Properties at 25°C				Solubility		$\log K_{ow}$	Henry's law const.		$\log K_{oc}$
	vapor pressure							calcd P_L/C_L		reported
	P^S, Pa	P_L, Pa	g/m³		C^S, mol/m³	C_L, mol/m³		H, Pa·m³/mol		
Anilazine	8.20×10^{-7}	1.77×10^{-5}	8		0.0290	0.628	3.80	2.82×10^{-5}		3.0
Banalaxyl	0.00133	4.65×10^{-3}	37		0.1137	0.398	3.40	0.012		3.44-3.86
Benodanil	1.00×10^{-8}	1.28×10^{-7}	2		0.0062	0.0793		1.62×10^{-6}		2.85
Benomyl	1.33×10^{-8}*	1.82×10^{-7}	2.0*		0.0069	0.0945	2.3	1.93×10^{-6}		3.28
Bitertanol	1.00×10^{-6}	8.31×10^{-6}	5		0.0118	0.0984		8.45×10^{-5}		
diastereoisomer A	2.20×10^{-9}	2.80×10^{-8}	2.9		0.0069	0.0874	4.1	3.20×10^{-7}		
diastereoisomer B	2.50×10^{-9}	3.86×10^{-8}	1.6		0.0038	0.0585	4.4	6.60×10^{-7}		
Buprirmate	0.00067	1.21×10^{-3}	22		0.0695	0.126	3.9	9.64×10^{-3}		2.9
Captan	1.10×10^{-5}	4.23×10^{-4}	5.1		0.0170	0.653	2.30	6.48×10^{-4}		2.29
Carbendazim	6.50×10^{-8}	3.82×10^{-5}	8		0.0418	24.60	1.52	1.55×10^{-6}		2.35
Carboxin	1.30×10^{-5}	5.98×10^{-5}	195		0.829	3.811	2.17	1.57×10^{-5}		2.41
Chloroneb	0.40	4.788	8		0.0386	0.462		173.8		3.06
Chloropicrin (I,Fum.)	2400	2400	2270		13.81	13.81	2.07	197.3		1.79
Chlorothalonil	0.133*	22.86	0.6		0.0023	0.388	2.64	58.94		3.2
Dazomet (Fum.)	4.0×10^{-4}	2.47×10^{-3}	3000		18.48	114.3	0.15	2.16×10^{-5}		0.48
Dichlofluanid	2.10×10^{-5}	1.32×10^{-4}	1.3		0.0039	0.0245	3.70	5.38×10^{-3}		
Dithianon	6.60×10^{-5}	6.28×10^{-4}	0.5		0.0017	0.1605	3.2	0.0391		
Edifenphos	0.013	0.013	56		0.180	0.180	3.48	0.072		
Ethirimol	2.67×10^{-4}	5.78×10^{-3}	200		0.956	20.68	2.3	2.79×10^{-4}		

Selected Properties at 25°C

Name	vapor pressure			Solubility			$\log K_{ow}$	Henry's law const. calcd P_L/C_L	$\log K_{oc}$
	P^s, Pa	P_L, Pa	g/m³	C^s, mol/m³	C_L, mol/m³		H, Pa·m³/mol	reported
Ethoprophos (N,I)	0.0507	0.0507	750	3.095	3.095	3.59	0.0164	1.85
Etridiazole	0.013	0.013	50	0.202	0.202	3.37	0.0644	
Fenarimol	2.93×10^{-5}	2.44×10^{-4}	14	0.0423	0.351	3.69	6.93×10^{-4}	2.78
Fenfuram	2.0×10^{-5}	1.39×10^{-4}	100	0.497	3.444		4.02×10^{-5}	2.48
Fenpropimorph	0.0023	2.30×10^{-3}	4.3	0.0142	0.014		0.162	2.94-3.65
Folpet	0.0013	0.0414	1	0.0034	0.107	3.63	0.386	3.27
Formaldehyde	>1 atm		miscible			0.35		
Furalaxyl			230	0.763	2.127	2.61		
Imazalil	9.30×10^{-6}	1.75×10^{-5}	1400	4.71	8.853	3.82	1.97×10^{-6}	3.60
Metalaxyl	7.47×10^{-4}	2.18×10^{-3}	8400	30.08	87.71	1.75	2.48×10^{-5}	1.7
Metiram	< 0.00001		0.1			0.3		5.7
Oxycarboxin	0.00133	0.0139	1000	3.741	39.06	0.74	3.56×10^{-4}	1.98
Penconazole	0.00021	4.66×10^{-4}	73	0.257	0.570	3.72	8.18×10^{-4}	2.62
Procymidone	0.0187	0.4534	4.5	0.0158	0.384	3.14	1.181	3.18
Propiconazole	5.60×10^{-5}	5.60×10^{-5}	110	0.321	0.321	3.72	1.74×10^{-4}	2.82
Quintozene	0.0066	0.104	0.44	0.0015	0.023	4.64	4.430	4.3, 3.38
Tecnazene			0.44	0.0017	0.0091			
Thiabendazole	5.33×10^{-7}	3.13×10^{-4}	50	0.249	146.1	2.69	2.14×10^{-6}	3.4
Thiophanate-methyl	1.30×10^{-5}	3.70×10^{-4}	3.5	0.0102	0.291	1.5	1.27×10^{-3}	3.26
Thiram	0.00133	0.0209	30	0.125	1.963	1.73	0.0107	2.83

Selected Properties at 25°C

Name	vapor pressure			Solubility			log K_{ow}	Henry's law const.		log K_{oc} reported
	P^s, Pa	P_L, Pa	g/m³	C^s, mol/m³	C_L, mol/m³			calcd P_L/C_L H, Pa·m³/mol		
Tolclofos-methyl	0.0573	0.196	0.3	0.001	0.0034		4.56	57.51		3.3
Tolylfluanid	1.6×10^{-5}	8.06×10^{-5}	0.9	0.0026	0.0131		3.90	6.17×10^{-3}		1.66
Triadimefon	2.0×10^{-6}	7.37×10^{-6}	71.5	0.243	0.897		3.26	8.22×10^{-6}		
Triadimenol	4.13×10^{-8}	4.03×10^{-7}	47	0.159	1.549		3.08	2.60×10^{-7}		3.00
diastereoisomer A	< 0.001		62	0.210	2.761		3.08			
diastereoisomer B	4.10×10^{-8}	4.85×10^{-7}	32	0.108	1.280		3.28			
Tricyclazole	2.67×10^{-5}	1.09×10^{-3}	1600	8.46	346.2		1.40	3.16×10^{-6}		3.00
Triflumizole	1.47×10^{-6}	3.53×10^{-6}	12500	36.16	86.90			4.07×10^{-8}		
Triforine	2.67×10^{-5}	5.16×10^{-4}	30	0.069	1.332		2.20	3.87×10^{-4}		2.3
Vinclozolin	1.33×10^{-5}	8.81×10^{-5}	1000	3.495	23.14		3.00	3.81×10^{-6}		2.6
Warfarin (R.)	1.55×10^{-4}	3.43×10^{-3}	17	0.0551	1.221		3.20	2.81×10^{-3}		2.96
Zineb	1.33×10^{-5}		10	0.0363			1.30	3.67×10^{-4}		3.00
Ziram	1.0×10^{-6}	1.53×10^{-4}	65	0.213	32.61		1.086	4.70×10^{-6}		2.60

Note: * The reported values for this quantity vary considerably; whereas this selected value represents the best judgement of the authors, the reader is cautioned that it may be subject to a large error.

Table 4.2.3 Suggested half-life classes of fungicides in various environmental compartments

Compounds	Air class	Water class	Soil class	Sediment class
Benomyl	1	4	6	7
Captan	2	2	5	5
Chlorothalonil	4	4	5	6
Chloropicrin	4	3	3	4
Thirm	4	4	5	6

where,

Class	Mean half-life (hours)	Range (hours)
1	5	< 10
2	17 (~ 1 day)	10-30
3	55 (~ 2 days)	30-100
4	170 (~ 1 week)	100-300
5	550 (~ 3 weeks)	300-1,000
6	1700 (~ 2 months)	1,000-3,000
7	5500 (~ 8 months)	3,000-10,000
8	17000 (~ 2 years)	10,000-30,000
9	55000 (~ 6 years)	> 30,000

4.3 Illustrative Fugacity Calculations: Level I, II, III

Chemical name: Benomyl

Level I calculation: (six-compartment model)

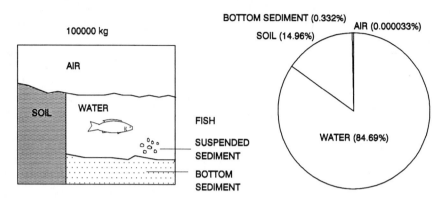

Distribution of mass

Physical-chemical properties:
molecular wt., g/mol	290.3
melting point, °C	140 dec.
solubility, g/m^3	2.0
vapor pressure, Pa	1.33E−08
log K_{OW}	2.30
fugacity ratio, F	0.0729
dissoc. const. pK_a	

Partition coefficients:
H, Pa·m^3/mol	1.93E−06
K_{AW}	7.79E−10
K_{OC}	81.086
BCF	9.976
K_{SW}	3.927
$K_{SD/W}$	7.853
$K_{SSD/W}$	24.542
$K_{AR/W}$	3.29E+13

COMPARTMENT	Z	CONCENTRATION			AMOUNT	AMOUNT
	mol/m^3·Pa	mol/m^3	mg/L, or g/m^3	μg/g	kg	%
AIR	4.03E−04	1.14E−13	3.30E−13	2.78E−10	0.033	3.30E−05
WATER	5.18E+05	1.46E−04	4.24E−04	4.24E−04	94691	84.69
SOIL	2.03E+06	5.73E−06	1.66E−03	6.93E−04	14965	14.97
BIOTA (FISH)	5.19E+06	1.46E−05	4.23E−03	4.23E−03	0.845	8.45E−03
SUSPENDED SEDIMENT	1.27E+07	3.58E−05	1.04E−02	6.93E−03	10.39	1.04E−02
BOTTOM SEDIMENT	4.07E+06	1.15E−05	3.33E−03	1.39E−03	332.55	0.333
Total					100000	100

Fugacity, f = 2.816E−12 Pa

Chemical name: Benomyl
Level II calculation: (six-compartment model)

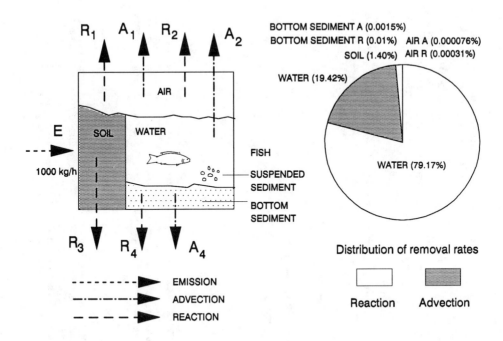

COMPARTMENT	Half-Life	D VALUES		CONC'N	LOSS	LOSS	REMOVAL
		Reaction	Advection		Reaction	Advection	%
	h	mol/Pa·h	mol/Pa·h	mol/m³	kg/h	kg/h	
AIR	17	1.64E+09	4.03E+08	2.61E−15	3.08E−03	7.56E−04	3.84E−04
WATER	170	4.22E+14	1.04E+14	3.34E−06	791.7	194.21	90.59
SOIL	1700	7.46E+12		1.31E−05	13.99		1.399
BIOTA (FISH)				3.34E−05			
SUSPENDED SEDIMENT				8.21E−05			
BOTTOM SEDIMENT	5500	5.13E+10	8.14E+09	2.63E−05	0.0961	0.0153	0.0111
Total		4.30E+14	1.04E+14		805.78	194.225	100
Reaction + Advection			5.33E+14			1000	

Fugacity, f = 6.457E−12 Pa

Total amount = 229314 kg

Overall residence time = 229.31 h

Reaction time = 284.59 h

Advection time = 1180.66 h

Fugacity Level III calculations: (four-compartment model)
Chemical name: Benomyl

Phase Properties and Rates:

Compartment	Bulk Z mol/m3 Pa	Half-life h	D Values Reaction mol/Pa h	Advection mol/Pa h
Air (1)	2.657E-01	17	1.08E+12	2.66E+11
Water (2)	5.181E+05	170	4.22E+14	1.04E+14
Soil (3)	1.172E+06	1700	8.60E+12	
Sediment (4)	1.228E+06	5500	7.74E+10	1.23E+10

	E(1)=1000	E(2)=1000	E(3)=1000	E(1,2,3)
Overall residence time =	1467.83	198.06	1972.04	1503.33 h
Reaction time =	1601.03	246.64	2058.54	1639.08 h
Advection time =	17642.93	1005.48	46930	18152.03 h

EMISSION (E)
REACTION (R)
ADVECTION (A)
TRANSFER D VALUE mol/Pa h

Phase Properties, Compositions, Transport and Transformation Rates:

Emission, kg/h				Fugacity, Pa				Concentration, g/m3			
E(1)	E(2)	E(3)	f(1)	f(2)	f(3)	f(4)	C(1)	C(2)	C(3)	C(4)
1000	0	0	4.703E-10	1.560E-12	2.313E-10	1.473E-12	3.627E-08	2.346E-04	7.872E-02	5.252E-04
0	1000	0	1.803E-17	6.548E-12	8.869E-18	6.184E-12	1.391E-15	9.848E-04	3.019E-09	2.205E-03
0	0	1000	7.808E-15	1.397E-12	3.150E-10	1.319E-12	6.022E-13	2.101E-04	1.072E-01	4.703E-04
50	250	700	2.352E-11	2.693E-12	2.321E-10	2.543E-12	1.814E-09	4.050E-04	7.898E-02	9.066E-04

Emission, kg/h			Loss, Reaction, kg/h				Loss, Advection, kg/h			
E(1)	E(2)	E(3)	R(1)	R(2)	R(3)	R(4)	A(1)	A(2)	A(3)	A(4)
1000	0	0	1.479E+02	1.913E+02	5.78E+02	3.309E-02	3.627E+01	4.692E+01		5.252E-03
0	1000	0	5.669E-06	8.029E+02	2.21E-05	1.389E-01	1.391E-06	1.970E+02		2.205E-02
0	0	1000	2.455E-03	1.713E+02	7.87E+02	2.963E-02	6.022E-04	4.202E+01		4.703E-03
50	250	700	7.394E+00	3.302E+02	5.80E+02	5.712E-02	1.814E+00	8.100E+01		9.066E-03

Emission, kg/h			Amounts, kg				Total Amount, kg
E(1)	E(2)	E(3)	m(1)	m(2)	m(3)	m(4)	
1000	0	0	3.627E+03	4.692E+04	1.417E+06	2.626E+02	1.468E+06
0	1000	0	1.391E-04	1.970E+05	5.433E+02	1.102E+03	1.981E+05
0	0	1000	6.022E-02	4.202E+04	1.930E+06	2.351E+02	1.972E+06
50	250	700	1.814E+02	8.100E+04	1.422E+06	4.533E+02	1.503E+06

Emission, kg/h			Intermedia Rate of Transport, kg/h						
E(1)	E(2)	E(3)	T12 air-water	T13 air-soil	T21 water-air	T24 water-sed	T31 soil-air	T32 soil-water	T42 sed-water
1000	0	0	8.159E+01	7.343E+02	9.134E-06	2.634E-01	1.218E-02	1.566E+02	2.250E-01
0	1000	0	3.128E-06	2.816E-05	3.834E-05	1.106E+00	4.672E-10	6.006E-06	9.446E-01
0	0	1000	1.355E-03	1.219E-02	8.179E-06	2.358E-01	1.659E-02	2.133E+02	2.015E-01
50	250	700	4.080E+00	3.672E+01	1.577E-05	4.546E-01	1.223E-02	1.572E+02	3.884E-01

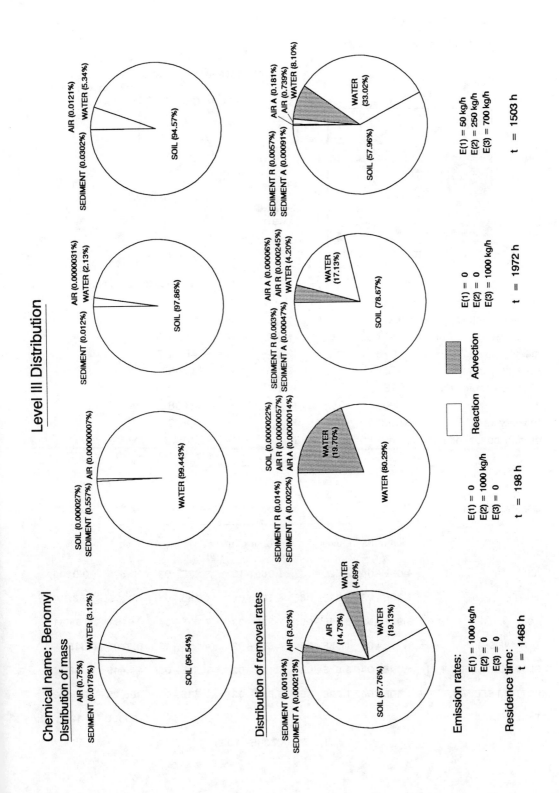

Chemical name: Captan

Level I calculation: (six-compartment model)

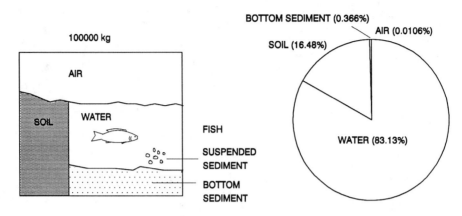

Distribution of mass

Physical-chemical properties:
molecular wt., g/mol	300.60
melting point, °C	178
solubility, g/m³	5.1
vapor pressure, Pa	1.07E−05
log K_{OW}	2.35
fugacity ratio, F	0.0307
dissoc. const. pK_a	

Partition coefficients:
H, Pa·m³/mol	6.31E−04
K_{AW}	2.54E−07
K_{OC}	91.788
BCF	11.194
K_{SW}	4.406
$K_{SD/W}$	8.812
$K_{SSD/W}$	27.546
$K_{AR/W}$	1.72E+10

COMPARTMENT	Z		CONCENTRATION			AMOUNT	AMOUNT
	mol/m³·Pa	mol/m³	mg/L, or g/m³	µg/g		kg	%
AIR	4.03E−04	3.52E−13	1.06E−10	8.92E−08		10.575	0.0106
WATER	1.59E+03	1.38E−06	4.16E−04	4.16E−04		83129	83.13
SOIL	6.99E+03	6.09E−06	1.83E−03	7.63E−04		16481	16.481
BIOTA (FISH)	1.77E+04	1.55E−05	4.65E−03	4.65E−03		0.9305	9.31E−04
SUSPENDED SEDIMENT	4.37E+04	3.81E−05	1.14E−02	7.63E−03		11.445	1.14E−02
BOTTOM SEDIMENT	1.40E+04	1.22E−05	3.66E−03	1.53E−03		366.25	0.3663
Total						100000	100

Fugacity, f = 8.720E−10 Pa

Chemical name: Captan
Level II calculation: (six-compartment model)

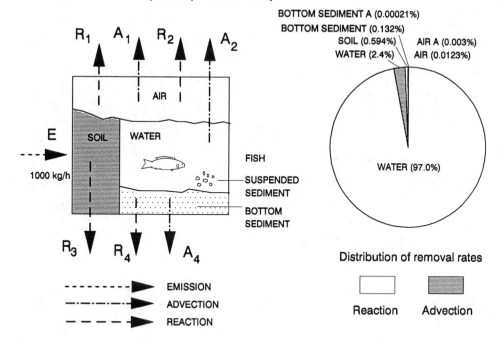

COMPARTMENT	Half-Life	D VALUES		CONC'N	LOSS	LOSS	REMOVAL
		Reaction	Advection		Reaction	Advection	%
	h	mol/Pa·h	mol/Pa·h	mol/m³	kg/h	kg/h	
AIR	17	1.64E+09	4.03E+08	1.01E−13	0.1234	0.0303	0.0154
WATER	17	1.29E+13	3.17E+11	3.96E−07	969.97	23.79	99.38
SOIL	550	7.92E+10		1.74E−06	5.944		0.594
BIOTA (FISH)				4.43E−06			
SUSPENDED SEDIMENT				1.09E−05			
BOTTOM SEDIMENT	550	1.76E+09	2.79E+07	3.49E−06	0.1321	0.0021	0.0134
Total		1.30E+13	3.18E+11		976.17	23.827	100
Reaction + Advection			1.33E+13			1000	

Fugacity, $f = 2.496E-10$ Pa

Total amount = 28623 kg

Overall residence time = 28.62 h

Reaction time = 29.32 h

Advection time = 1201.31 h

Fugacity Level III calculations: (four-compartment model)
Chemical name: Captan

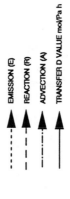

Phase Properties and Rates:

Compartment	Bulk Z mol/m3 Pa	Half-life h	D Values Reaction mol/Pa h	Advection mol/Pa h
Air (1)	5.422E-04	17	2.21E+09	5.42E-08
Water (2)	1.586E+03	17	1.29E+13	3.17E+11
Soil (3)	3.969E+03	550	9.00E+10	
Sediment (4)	4.063E+03	550	2.56E+09	4.06E+07

	E(1)=1000	E(2)=1000	E(3)=1000	E(1,2,3)
Overall residence time =	572.89	24.01	736.80	550.41 h
Reaction time =	591.64	24.60	738.14	555.32 h
Advection time =	18084.44	1002.72	407168	62288.83 h

▲ EMISSION (E)
▲ REACTION (R)
▲ ADVECTION (A)
▲ TRANSFER D VALUE mol/Pa h

Phase Properties, Compositions, Transport and Transformation Rates:

Emission, kg/h

E(1)	E(2)	E(3)	Fugacity, Pa				Concentration, g/m3				Loss, Advection, kg/h			
			f(1)	f(2)	f(3)	f(4)	C(1)	C(2)	C(3)	C(4)	A(1)	A(2)	A(3)	A(4)
1000	0	0	1.734E-07	3.579E-11	2.639E-08	1.532E-11	2.827E-08	1.706E-05	3.148E-02	1.871E-05	2.827E+01	3.412E+00		1.871E-04
0	1000	0	2.641E-13	2.511E-10	4.018E-14	1.075E-10	4.304E-14	1.197E-04	4.793E-08	1.313E-04	4.304E-05	2.394E+01		1.313E-03
0	0	1000	2.871E-10	1.849E-11	3.423E-08	7.915E-12	4.679E-11	8.813E-06	4.084E-02	9.666E-06	4.679E-02	1.763E+00		9.666E-05
50	250	700	8.873E-09	7.751E-11	2.528E-08	3.318E-11	1.446E-09	3.695E-05	3.016E-02	4.053E-05	1.446E+00	7.390E+00		4.053E-04

Emission, kg/h

E(1)	E(2)	E(3)	Loss, Reaction, kg/h				Amounts, kg				Total Amount, kg
			R(1)	R(2)	R(3)	R(4)	m(1)	m(2)	m(3)	m(4)	
1000	0	0	1.152E+02	1.391E+02	7.14E+02	1.179E-02	2.827E+03	3.412E+03	5.666E+05	9.357E+00	5.729E+05
0	1000	0	1.754E-04	9.760E+02	1.09E-03	8.272E-02	4.304E-03	2.394E+04	8.628E-01	6.565E+01	2.401E+04
0	0	1000	1.907E-01	7.186E+01	9.26E+02	6.090E-03	4.679E+00	1.763E+03	7.350E+05	4.833E+00	7.368E+05
50	250	700	5.895E+00	3.012E+02	6.84E+02	2.553E-02	1.446E+02	7.390E+03	5.429E+05	2.026E+01	5.504E+05

Intermedia Rate of Transport, kg/h

T12	T21	T13	T31	T32	T24	T42
air-water	water-air	air-soil	soil-air	soil-water	water-sed	sed-water
8.589E+01	2.170E-04	7.719E+02	1.278E+02	5.665E+01	1.941E-02	7.432E-03
1.308E-04	1.523E-03	1.175E-03	1.945E-06	8.625E-05	1.362E-01	5.214E-02
1.422E-01	1.121E-04	1.278E+00	1.657E+00	7.348E+01	1.003E-02	3.839E-03
4.394E+00	4.700E-04	3.949E+01	1.224E+00	5.427E+01	4.203E-02	1.609E-02

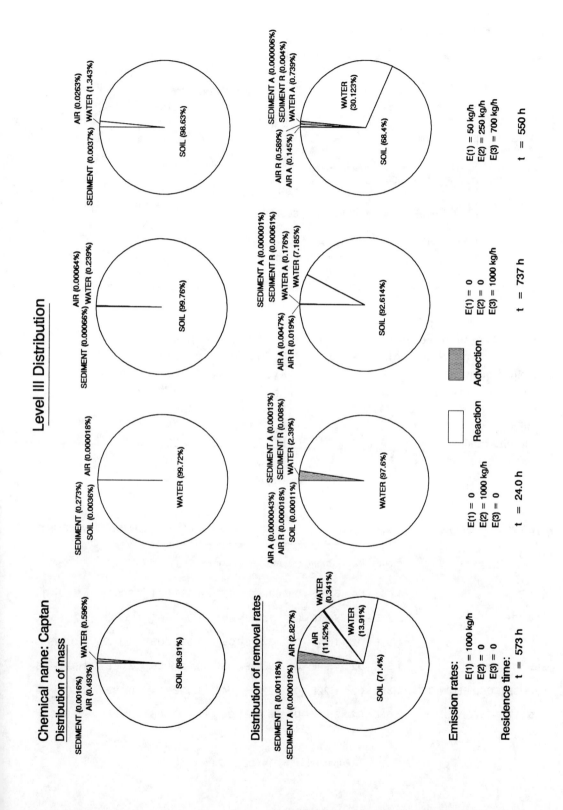

Chemical name: Chloropicrin

Level I calculation: (six-compartment model)

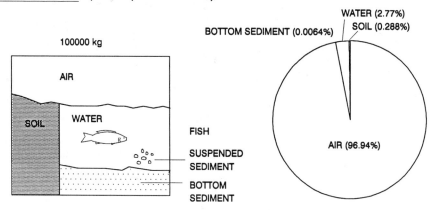

Distribution of mass

Physical-chemical properties:
molecular wt., g/mol	164.39
melting point, °C	−64
solubility, g/m³	2270
vapor pressure, Pa	2400
log K_{OW}	2.07
fugacity ratio, F	1.0
dissoc. const. pK_a	

Partition coefficients:
H, Pa·m³/mol	173.80
K_{AW}	0.0701
K_{OC}	48.17
BCF	5.87
K_{SW}	2.31
$K_{SD/W}$	4.62
$K_{SSD/W}$	14.45
$K_{AR/W}$	2500

COMPARTMENT	Z	CONCENTRATION				AMOUNT	AMOUNT
	mol/m³·Pa	mol/m³	mg/L, or g/m³	µg/g		kg	%
AIR	4.03E−04	5.90E−09	9.69E−07	8.18E−04		96941	96.941
WATER	5.75E−03	8.41E−08	1.38E−05	1.38E−05		2765	2.7652
SOIL	1.33E−02	1.94E−07	3.20E−05	1.33E−05		287.7	0.288
BIOTA (FISH)	3.38E−02	4.94E−07	8.12E−05	8.12E−05		0.0162	1.62E−05
SUSPENDED SEDIMENT	8.31E−02	1.22E−06	2.00E−04	1.33E−04		0.1998	2.00E−04
BOTTOM SEDIMENT	2.66E−02	3.89E−07	6.39E−05	2.66E−05		6.394	0.0064
Total						100000	100

Fugacity, f = 1.462E−05 Pa

Chemical name: Chloropicrin
Level II calculation: (six-compartment model)

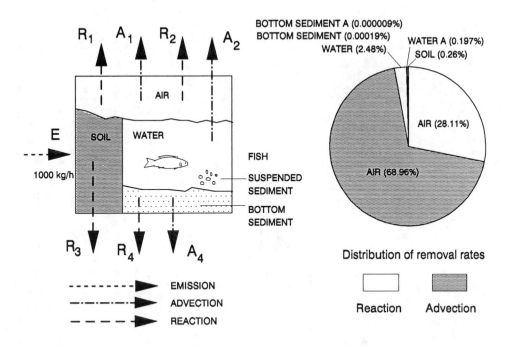

COMPARTMENT	Half-Life	D VALUES		CONC'N	LOSS	LOSS	REMOVAL
		Reaction	Advection		Reaction	Advection	%
	h	mol/Pa·h	mol/Pa·h	mol/m³	kg/h	kg/h	
AIR	170	1.64E+08	4.03E+08	4.19E−09	281.101	689.57	97.067
WATER	55	1.45E+07	1.15E+06	5.98E−08	24.783	1.967	2.675
SOIL	55	1.51E+06		1.38E−07	2.579		0.258
BIOTA (FISH)				3.51E−07			
SUSPENDED SEDIMENT				8.65E−07			
BOTTOM SEDIMENT	1700	1.08E+03	5.32E+01	2.77E−07	1.85E−03	9.10E−05	1.94E−04
Total		1.80E+08	4.05E+08		931.76	691.535	100
Reaction + Advection			5.85E+08			1000	

Fugacity, f = 1.040E−05 Pa

Total amount = 71133 kg

Overall residence time = 71.13 h

Reaction time = 230.60 h

Advection time = 102.86 h

Fugacity Level III calculations: (four-compartment model)
Chemical name: Chloropicrin

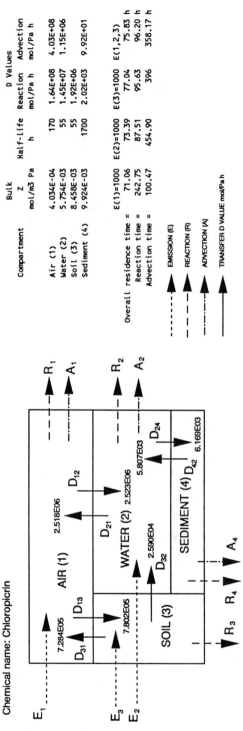

Phase Properties and Rates:

Compartment	Bulk Z mol/m3 Pa	Half-life h	D Values Reaction mol/Pa h	Advection mol/Pa h
Air (1)	4.034E-04	170	1.64E+08	4.03E+08
Water (2)	5.754E-03	55	1.45E+07	1.15E+06
Soil (3)	8.458E-03	55	1.92E+06	
Sediment (4)	9.924E-03	1700	2.02E+03	9.92E+01

	E(1)=1000	E(2)=1000	E(3)=1000	E(1,2,3)
Overall residence time =	71.06	73.39	77.04	75.83 h
Reaction time =	242.75	87.51	95.63	96.20 h
Advection time =	100.47	454.90	396	358.17 h

EMISSION (E)
REACTION (R)
ADVECTION (A)
TRANSFER D VALUE mol/Pa h

Phase Properties, Compositions, Transport and Transformation Rates:

Emission, kg/h

E(1)	E(2)	E(3)	f(1)	f(2)	f(3)	f(4)	C(1)	C(2)	C(3)	C(4)
1000	0	0	1.066E-05	1.485E-06	3.112E-06	1.155E-06	7.070E-07	1.405E-06	4.327E-06	1.885E-06
0	1000	0	1.477E-06	3.350E-04	4.312E-07	2.606E-04	9.796E-08	3.169E-04	5.996E-07	4.252E-04
0	0	1000	2.920E-06	3.651E-06	2.277E-03	2.841E-06	1.936E-07	3.454E-06	3.166E-03	4.635E-06
50	250	700	2.946E-06	8.638E-05	1.594E-03	6.721E-05	1.954E-07	8.170E-05	2.217E-03	1.096E-04

Emission, kg/h

E(1)	E(2)	E(3)	R(1)	R(2)	R(3)	R(4)	A(1)	A(2)	A(3)	A(4)
1000	0	0	2.882E+02	3.540E+00	9.81E-01	3.842E-04	7.070E+02	2.809E+01		1.885E-05
0	1000	0	3.993E+01	7.985E+02	1.36E-01	8.667E-02	9.796E+01	6.337E+03		4.252E-03
0	0	1000	7.893E+01	8.704E+00	7.18E+02	9.447E-04	1.936E+02	6.908E+01		4.635E-05
50	250	700	7.965E+01	2.059E+02	5.03E+02	2.235E-02	1.954E+02	1.634E+02		1.096E-03

Amounts, kg

m(1)	m(2)	m(3)	m(4)	Total Amount, kg
7.070E+04	2.809E+02	7.789E+01	9.425E-01	7.106E+04
9.796E+03	6.337E+04	1.079E+01	2.126E+02	7.339E+04
1.936E+04	6.908E+02	5.699E+04	2.318E+00	7.704E+04
1.954E+04	1.634E+04	3.990E+04	5.482E+01	7.583E+04

Intermedia Rate of Transport, kg/h

T12	air-water	water-air	air-soil	soil-air	soil-water	water-sed	sed-water
T12	T21	T13	T31	T32	T24	T42	
4.422E+00	6.146E-01	1.367E+00	3.727E-01	1.325E-02	1.506E-03	1.103E-03	
6.128E-01	1.386E+02	1.895E-01	5.164E-02	1.836E-03	3.397E-02	2.488E-01	
1.211E+00	1.511E+00	3.745E-01	5.699E+04	9.695E+00	3.703E-03	2.712E-03	
1.222E+00	3.575E+01	3.779E-01	1.909E+02	6.788E+00	8.760E-02	6.416E-02	

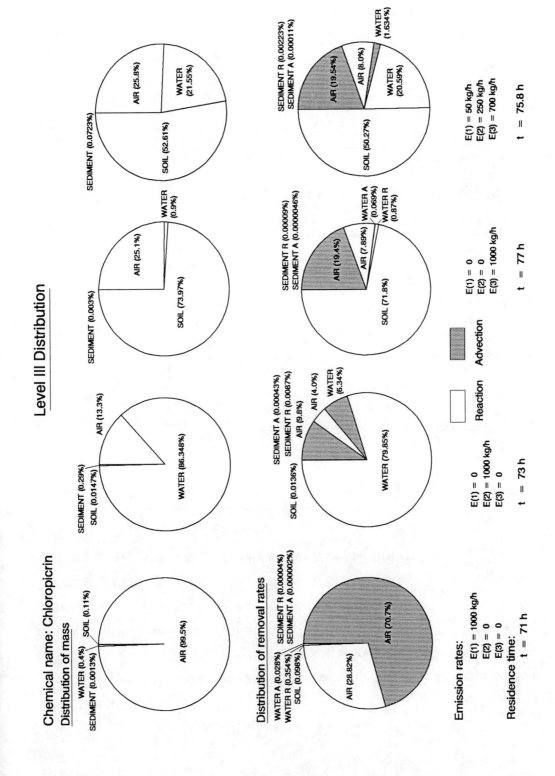

Chemical name: Chlorothalonil

<u>Level I calculation:</u> (six-compartment model)

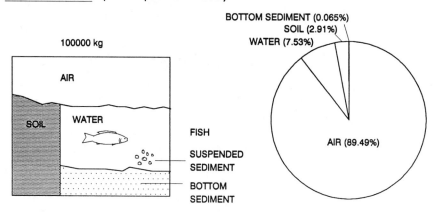

Distribution of mass

Physical-chemical properties:
molecular wt., g/mol	265.86
melting point, °C	250
solubility, g/m³	0.60
vapor pressure, Pa	0.133
log K_{OW}	2.64
fugacity ratio, F	0.006
dissoc. const. pK_a	

Partition coefficients:
H, Pa·m³/mol	58.9523
K_{AW}	2.38E−02
K_{OC}	178.971
BCF	21.826
K_{SW}	8.591
$K_{SD/W}$	17.181
$K_{SSD/W}$	53.691
$K_{AR/W}$	2.58E+05

COMPARTMENT	Z	CONCENTRATION			AMOUNT	AMOUNT
	mol/m³·Pa	mol/m³	mg/L, or g/m³	µg/g	kg	%
AIR	4.03E−04	3.37E−09	8.95E−07	7.55E−04	89494	89.494
WATER	1.70E−02	1.42E−07	3.76E−05	3.76E−05	7529	7.529
SOIL	1.46E−01	1.22E−06	3.23E−04	1.35E−04	2910	2.910
BIOTA (FISH)	3.70E−01	3.09E−06	8.22E−04	8.22E−04	0.164	1.64E−04
SUSPENDED SEDIMENT	9.11E−01	7.60E−06	2.02E−03	1.35E−03	2.021	2.02E−03
BOTTOM SEDIMENT	2.92E−01	2.43E−06	6.47E−04	2.69E−04	64.68	0.0647
Total					100000	100

Fugacity, f = 8.344E−06 Pa

Chemical name: Chlorothalonil
Level II calculation: (six-compartment model)

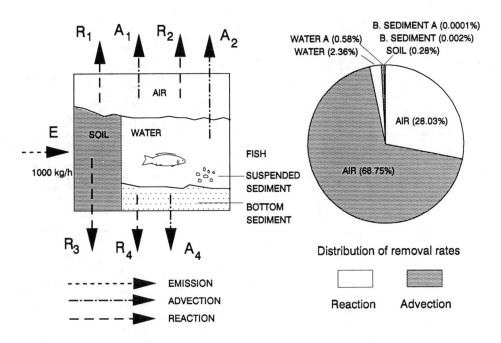

| COMPARTMENT | Half-Life | D VALUES | | CONC'N | LOSS | LOSS | REMOVAL |
| | | Reaction | Advection | | Reaction | Advection | % |
	h	mol/Pa·h	mol/Pa·h	mol/m^3	kg/h	kg/h	
AIR	170	1.64E+08	4.03E+08	2.59E−09	280.27	687.53	96.78
WATER	170	1.38E+07	3.39E+06	1.09E−07	23.577	5.784	2.936
SOIL	550	1.65E+06		9.34E−07	2.817		0.282
BIOTA (FISH)				2.37E−06			
SUSPENDED SEDIMENT				5.84E−06			
BOTTOM SEDIMENT	1700	1.19E+04	5.83E+02	1.87E−06	0.0203	0.0010	0.002
Total		1.80E+08	4.07E+08		306.8	693.315	100
Reaction + Advection			6.87E+08			1000	

Fugacity, f = 6.410E−06 Pa

Total amount = 76824 kg

Overall residence time = 76.82 h

Reaction time = 250.5 h

Advection time = 110.81 h

Fugacity Level III calculations: (four-compartment model)
Chemical name: Chlorothalonil

Phase Properties and Rates:

Compartment	Bulk Z mol/m3 Pa	Half-life h	D Values Reaction mol/Pa h	Advection mol/Pa h
Air (1)	4.034E-04	170	1.64E+08	4.03E+08
Water (2)	1.697E-02	170	1.38E+07	3.39E+06
Soil (3)	7.806E-02	550	1.77E+06	
Sediment (4)	7.189E-02	5500	4.53E+03	7.19E+02

	E(1)=1000	E(2)=1000	E(3)=1000	E(1,2,3)
Overall residence time =	72.81	166.46	569.21	443.70 h
Reaction time =	247.36	247.65	721.65	603.77 h
Advection time =	103.19	507.71	2695	1673.61 h

– – – ▲ EMISSION (E)
– – – ▲ REACTION (R)
– – – ▲ ADVECTION (A)
▲ TRANSFER D VALUE mol/Pa h

Phase Properties, Compositions, Transport and Transformation Rates:

Emission, kg/h

E(1)	E(2)	E(3)		f(1)	f(2)	f(3)	f(4)		C(1)	C(2)	C(3)	C(4)
1000	0	0		6.565E-06	1.702E-06	2.278E-06	1.607E-06		7.041E-07	7.679E-06	4.727E-05	3.070E-05
0	1000	0		1.689E-06	1.625E-04	5.861E-07	1.534E-04		1.812E-07	7.333E-04	1.216E-05	2.932E-03
0	0	1000		1.925E-06	5.294E-06	1.456E-03	4.998E-06		2.065E-07	2.389E-05	3.021E-02	9.551E-05
50	250	700		2.098E-06	4.441E-05	1.019E-03	4.193E-05		2.250E-07	2.004E-04	2.115E-02	8.013E-04

Fugacity, Pa — Concentration, g/m3

Emission, kg/h

E(1)	E(2)	E(3)		R(1)	R(2)	R(3)	R(4)		A(1)	A(2)	A(4)
1000	0	0		2.870E+02	6.261E+00	1.07E+00	1.934E-03		7.041E+02	1.536E+00	3.070E-04
0	1000	0		7.385E+01	5.978E+02	2.76E-01	1.847E-01		1.812E+02	1.467E+02	2.932E-02
0	0	1000		8.416E+01	1.948E+01	6.85E+02	6.017E-03		2.065E+02	4.778E+00	9.551E-04
50	250	700		9.173E+01	1.634E+02	4.80E+02	5.048E-02		2.250E+02	4.008E+01	8.013E-03

Loss, Reaction, kg/h — Loss, Advection, kg/h

Amounts, kg

m(1)	m(2)	m(3)	m(4)	Total Amount, kg
7.041E+04	1.536E+03	8.508E+02	1.535E+01	7.281E+04
1.812E+04	1.467E+05	2.189E+02	1.466E+03	1.665E+05
2.065E+04	4.778E+03	5.437E+05	4.776E+01	5.692E+05
2.250E+04	4.008E+04	3.807E+05	4.007E+02	4.437E+05

Intermedia Rate of Transport, kg/h

	T12	T21	T13	T31	T32	T24	T42
	air-water	water-air	air-soil	soil-air	soil-water	water-sed	sed-water
	1.045E+01	2.702E+00	1.566E+00	4.471E-01	4.632E-02	9.738E-03	7.497E-03
	2.690E+00	2.580E+02	4.028E-01	1.151E-01	1.192E-02	9.299E-01	7.158E-01
	3.066E+00	8.405E+00	4.591E-01	2.857E+02	2.960E+01	3.029E-02	2.332E-02
	3.341E+00	7.052E+01	5.003E-01	2.001E+02	2.073E+01	2.542E-01	1.957E-01

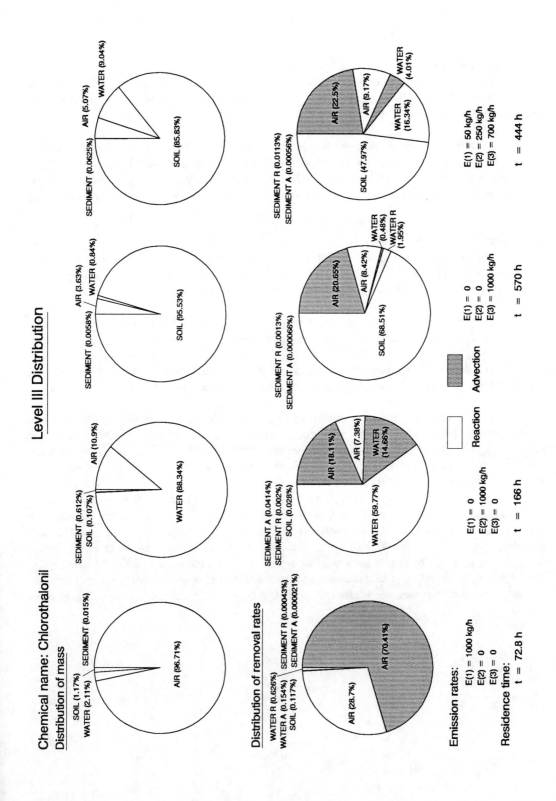

4.4 COMMENTARY ON THE PHYSICAL-CHEMICAL PROPERTIES AND ENVIRONMENTAL FATE

Properties and Reactivity

The selected physical-chemical properties of the fungicides are given in Tables 4.2.1 and 4.2.2. Table 4.2.3 gives the estimated half-lives. These half-lives should be regarded as only tentative, average values or ranges.

As with the insecticides there is a tendency for these substances to have relative high values of log K_{OW}, most exceeding 3.0. Some such as chloropicrin are highly volatile, but most have low vapor pressures and low Henry's law constants. Most are fairly short lived in the environment with half-lives in air and water less than 3 weeks, although they tend to persist longer in soils.

Evaluative Calculations

Illustrative calculations are given for four fungicides: captan which is slightly hydrophobic and is involatile; chloropicrin which is slightly hydrophobic but volatile and is used as a fumigant; chlorothalonil and benomyl which are similar to captan but more persistent.

The Level I distributions show the strong tendency of chloropicrin to volatilize, while the others partition primarily water and soil, as dictated by their K_{OW} values.

The Level II diagrams have the same distribution between media. Captan and chlorothalonil show an order of magnitude difference in persistence, but neither is persistent.

The Level III diagrams are most revealing, especially the final scenario with emissions to air, water and soil. Chloropicrin is primarily removed by reaction in soil (50%), reaction in water (21%) and advective loss in air (20%). Of the 700 kg/h applied to soil, 191 kg/h evaporate. The overall residence time is 76 h, attributable mainly to reaction losses. Captan and chlorothalonil both have persistences of about 600 h (25 days) primarily as a result of longer retention in soil. The apparent discrepancy between Level II and Level III persistencies arises because these compounds have similar half-lives in soil but captan is more reactive in water. In the Level III equilibrium calculation, most of both chemicals is in water, whereas in Level III with discharge primarily to soil, 90% or more remains in soil. In neither case is there significant evaporation, although there is some runoff from soil to water amounting to 30 to 55 kg/h or 4 to 8% of emissions to soil. It is thus important to evaluate how the chemical enters the environment, the extent to which there is intermedia transport, and the potential for reaction in all the affected media.

4.5 REFERENCES:

Abernethy, S. Bobra, A. M., Shiu, W. Y., Wells, P. G., Mackay, D. (1986) Acute lethal toxicity of hydrocarbons and chlorinated hydrocarbons to two planktonic crustaceans: The key role of organisms-water partitioning. *Aqua. Toxicol.* 8, 163-174.

Agnihotri, V. P. (1970) Persistent of captan and its effects on microflora, respiration, and nitrification of a forest nursery soil. *Can. J. Microbiol.* 17, 377-383.

Agrochemicals Handbook (1987) *The Agrochemicals Handbook.* 2nd Edition, Hartley, D., Kidd, H., Eds., The Royal Society of Chemistry, Nottingham, England.

Anliker, R., Moser, P. (1987) The limits of bioaccumulation of organic pigments in fish: their relation to the partition coefficient and the solubility in water and octanol. *Ecotoxicol. Environ. Saf.* 13, 43-52.

Atlas, E., Foster, R., Giam, C. S. (1982) Air-sea exchange of high molecular weight organic pollutants: laboratory studies. *Environ. Sci. Technol.* 16, 283-286.

Atkinson, R. (1985) Kinetics and mechanisms of the gas-phase reactions of OH radicals with organic compounds under atmospheric conditions. *Chem. Rev.* 85, 69-201.

Atkinson, R. (1987) Structure-activity relationship for the estimation of rate constants for the gas-phase reactions of OH radicals with organic compounds. *Int'l J. Chem. Kinetics* 19, 799-828.

Atkinson, R., Carter, W. P. L. (1984) Kinetics and mechanisms of the gas-phase reactions of ozone with organic compounds under atmospheric conditions. *Chem. Rev.* 84, 437-470.

Atkinson, R., Lloyd, A. C. (1984) Evaluation of kinetic and mechanistic data for modeling of photochemical smog. *J. Phys. Chem. Ref. Data* 13, 315-444.

Augustijn-Beckers, P. W. M., Hornsby, A. G., Wauchope, R. D. (1994) The SCS/ARS/CES pesticide-properties database for environmental decision making. II. Additional compounds. *Rev. Environ. Contam. Toxicol.* 137, 1-82.

Austin, D. J., Briggs, G. G. (1976) A new extraction method for benomyl residues in soil and its application in movement and persistence studies. *Pest. Sci.* 7, 201-210.

Baker, E. A., Hayes, A. L., Butler, R. C. (1992) Physicochemical properties of agrochemicals: Their effects on foliar penetration. *Pest. Sci.* 34(2), 167-182.

Ballschmiter, K., Wittlinger, R. (1991) Interhemisphere exchange of hexachlorohexanes, hexachlorobenzene, polychlorobiphenyl and 1,1,1-trichloro-2,2-bis(p-chloro-phenyl)-ethane in the lower atmosphere. *Environ. Sci. Technol.* 25, 1103-1111.

Banerjee, S. (1985) Calculation of water solubility of organic compounds with UNIFAC-derived parameters. *Environ. Sci. Technol.* 19, 369-370.

Banerjee, S., Howard, P. H. (1988) Improved estimation of solubility and partitioning through correction of UNIFAC-derived activity coefficients. *Environ. Sci. Technol.* 22, 839-841.

Banerjee, S., Howard, P. H., Lande, S. S. (1990) General structure vapor pressure relationship for organics. *Chemosphere* 21, 1173-1180.

Barak, E., Dinoor, A., Jacoby, B. (1983) Adsorption of systemic fungicides and a herbicide by some components of plant tissues, in relation to some physicochemical properties of the pesticides. *Pest. Sci.* 14(3), 213-219.

Bateman, G. L., Nicholls, P. H., Chamberlain, K. (1990) The effectiveness of eleven sterol biosysthesis-inhibiting fungicides against the take-all fungus (*Gaeumannomyces grammminis, var. tritici*) in relation to their physical properties. *Pest. Sci.* 29, 109-122.

Baulch, D. L., Cox, R. A., Hampson, R. F., Jr., Kerr, J. A., Troe, J., Watson, R. T. (1984) Evaluated kinetic and photochemical data for atmospheric chemistry (Supplement II). *J. Phys. Chem. Ref. Data* 13, 1255-1380.

Beck, J., Hansen, K. E. (1974) The degradation of quintozene, pentachlorobenzene, hexachloro-benzene and pentachloroaniline in soil. *Pest. Sci.* 5, 41-48.

Biagi, G. L., Guerra, M. C., Barbaro, A. M., Recanatini, M., Borea, P. A. (1991) Lipophilicity indices of triazine herbicides. *Sci. Total Environ.* 109/110, 33-40.

Bidleman, T. F. (1984) Estimation of vapor pressures for nonpolar organic compounds by capillary gas chromatography. *Anal. Chem.* 56, 2490-2496.

Bidleman, T. F., Christensen, E. J. (1979) Atmospheric removal processes for high molecular weight organochlorines. *J. Geophys. Res.* 84(C12), 7857-7862.

Bidleman, T. F., Foreman, W. T. (1987) Vapor-particle partitioning of semivolatile organic compounds. In: *Sources and Fate of Aquatic Pollutants.* Hite, R. A., Eisenreich, S. J., Editors, pp. 127-56, Advances in Chemistry Series 216, American Chemical Society, Washington, DC.

Bintein, S., Devillers, J. (1994) QSAR for organic chemical sorption in soils and sediments. *Chemosphere* 28(6), 1171-1188.

Bobra, A. M., Shiu, W. Y., Mackay, D. (1985) Quantitative structure-activity relationships for the acute toxicity of chlorobenzenes to *Daphnia magna. Environ. Toxicol. Chem.* 4, 297-305.

Brecken-Folse, J. A., Mayer, F. L., Pedigo, L. E., Marking, L. L. (1994) Acute toxicity of 4-nitrophenol, 2,4-dinitrophenol, terbufos and trichlorfon to grass shrimp (*Palaemonetes* spp.) and sheepshead minnows (*Cyprinodon variegatus*) as affected by salinity and temperature. *Environ. Toxicol. Chem.* 13, 67-77.

Briggs, G. G. (1981) Theoretical and experimental relationships between soil adsorption, octanol-water partition coefficients, water solubilities, bioconcentration factors, and the Parachor. *J. Agric. Food Chem.* 29, 1050-1059.

Brooke, D. N., Dobbs, A. J., Williams, N. (1986) Octanol/water partition coefficients (P): Measurement, estimation, and interpretation, particularly for chemicals with $P > 10^5$. *Ecotoxicol. Environ. Saf.* 11, 251-260.

Brouwer, D. H., Ravensberg, J. C., De Kort, W. L. A. M., Van Hemmen, J. J. (1994) A personal sampler for inhalable mixed-phase aerosols: Modification to an existing sampler and validation test with three pesticides. *Chemosphere* 28(6), 1135-1146.

Brown, S., Chan, F., Jones, J., Liu, D., MaCalab, K., Mill, T., Supios, K., Schendel, D. (1975) Research program on hazard priority ranking of manufactured chemicals: Phase II. Final report, Stanford Research Institute, Menlo Park, California.

Brusseau, M. L., Rao, P. S. C. (1989) The influence of sorbate-organic matter interactions on sorption nonequilibrium. *Chemosphere* 18, 1691-1706.

Budavari, S., Editor (1989) *The Merck Index. An Encyclopedia of Chemicals, Drugs and Biologicals.* 11th Edition, Merck and Co., Inc., Rahway, New Jersey.

Burchfield, H. P. (1959) Comparative stabilities of dyrene, 1-fluoro-2,4-dinitrobenzene, dichlone and captan in a silt loam soil. *Contrib. D. Boyce Thompson Inst.* 20, 205-215.

Burkhard, L. P., Kuehl, D. W., Veith, G. D. (1985) Evaluation of reversed phase LC/MS for estimation of *n*-octanol/water partition coefficients of organic chemicals. *Chemosphere* 14, 1551-1560.

Burkhard, N., Guth, J. A. (1981) Rate of volatilisation of pesticides from soil surfaces; Comparison of calculated results with those determined in a laboratory model system. *Pest. Sci.* 12(1), 37-44.

Bysshe, S. E. (1982) Chapter 5, Bioconcentration factor in aquatic organisms. In: *Handbook on Chemical Property Estimation Methods, Environmental Behavior of Organic Compounds.* Lyman, W. J., Reehl, W. F., Rosenblatt, D. H., Editors, McGraw-Hill, Inc., New York.

Calamari, D., Galassi, S., Sette, F., Vighi, M. (1983) Toxicity of selected chlorobenzenes to aquatic organisms. *Chemosphere* 12, 253-262.

Callahan, M. A., Slimak, M. W., Gabel, N. W., May, I. P., Fowler, C. F., Freed, J. R., Jennings, P., Durfee, R. L., Whitmore, F. C., Maestri, B., Mabey, W. R., Holt, B. R., Gould, C. (1979) *Water-Related Environmental Fate of 129 Priority Pollutants.* Vol. 1, EPA Report No. 440/4-79-029a, Versar, Inc., Springfield, Virginia.

Calvert, J. G., Demeyan, K. L., Kerr, J. A., McQuigg, R. D. (1972) Photolysis of formaldehyde as a hydrogen atom source in the lower atmosphere. *Science* 175, 751-752.

Carlier, P., Hannachi, H., Mouvier, G. (1986) The chemistry of carbonyl compounds in the atmosphere-A review. *Atmos. Environ.* 20(11), 2079-2099.

Carlson, A. R., Kosian, P. A. (1987) Toxicity of chlorinated benzenes to fathead minnows (*Pimephales promelas*). *Arch. Environ. Contam. Toxicol.* 16, 129-135.

Carris, L. M. (1983) Movement of the systemic fungicide metalaxyl in soils and its translocation in plants. M.S. Thesis, Washington State University, Pullman, Washington.

Chamberlain, K., Bateman, G. L., Nicholls, P. H. (1991) Volatile analogoues of penconazole and their activity against the take-all fungus (*Gaeumannomyces gramminis, var. tritici*). *Pest. Sci.* 31(2), 185-196.

Chin, Y. P., Weber, Jr., W. J., Voice, T. C. (1986) Determination of partition coefficients and aqueous solubilities by reverse phase chromatography-II. *Water Res.* 20, 1443-1450.

Chiou, C. T. (1981) Partition coefficient and water solubility in environmental chemistry. In: *Hazard Assessment of Chemicals. Current Development.* Vol. 1, Saxena, J., Fisher, S, Editors, pp. 117-153, Academic Press, Inc., New York.

Chiou, C. T. (1985) Partition coefficients of organic compounds in lipid-water systems and correlations with fish concentration factors. *Environ. Sci. Technol.* 19, 57-62.

Chiou, C. T., Schmedding, D. W. (1981) Measurement and interrelation of octanol-water partition coefficient and water solubility of organic chemicals. In: *Test Protocols for Environmental Fate and Movement of Toxicants. J. Assoc. Anal. Chem.*, Arlington, Virginia.

Chiou, C. T., Schmedding, D. W., Manes, M. (1982) Partitioning of organic compounds in octanol-water system. *Environ. Sci. Technol.* 16, 4-10.

Chiou, C. T., Freed, V. H., Schmedding, D. W., Kohnert, R. (1977) Partition coefficient and bioaccumulation of selected organic chemicals. *Environ. Sci. Technol.* 11(5), 475-478.

Connell, D. W., Hawker, D. W. (1988) Use of polynomial expressions to describe the bioconcentration of hydrophobic chemicals by fish. *Ecotoxicol. Environ. Saf.* 16, 242-257.

Connell, D. W., Markwell, R. D. (1990) Bioaccumulation in the soil to earthworm system. *Chemosphere* 20, 91-100.

Davies, J. E., Lee, J. A. (1987) Changing profiles in human health effects of pesticides. *Pest. Sci. Biotechnol. Proc. 6th Int'l Congr. Pesticide Chem.* 533-538.

Davies, R. P., Dobbs, A. J. (1984) The prediction of bioconcentration in fish. *Water Res.* 18, 1253-1262.

Dean, J., Editor (1985) *Lange's Handbook of Chemistry.* 13th Edition, McGraw-Hill, New York.

De Bruijn, J., Busser, F., Seinen, W., Hermens, J. (1989) Determination of octanol/water partition coefficients for hydrophobic organic chemicals with the "slow-stirring" method. *Environ. Toxicol. Chem.* 8, 499-512.

De Bruijn, J., Hermens, J. (1990) Relationships between octanol/water partition coefficients and total molecular surface area and total molecular volume of hydrophobic organic chemicals. *Quant. Struct.-Act. Relat.* 9, 11-21.

De Kock, A. C., Lord, D. A. (1987) A simple procedure for determining octanol-water partition coefficients using reversed phase high performance liquid chromatography (RPHPLC). *Chemosphere* 16(1), 133-142.

Deneer, J. W., Seinen, W., Hermens, J. L. M. (1988) The acute toxicity of aldehydes to the guppy. *Aqua. Toxicol.* 12, 185-192.

Deutsche Forschungsgemeinschaft (1983) *Hexachlorcyclohexan als Schadstoff in Lebensmitteln.* Verlag Chemie, Weinheim, Germany. pp.11-17.

DiToro, D. M., O'Conner, D. J., Thomann, R. V., St. John, J. P. (1981) Analysis of fate of chemicals in receiving waters. Phase I, Hydroqual, Inc., Prepared for the Chemical Manufacturing Association, Washington, DC., May, 1981.

Dobbs, A. J., Cull, M. R. (1982) Volatilization of chemicals-relative loss rates and the estimation of vapor pressures. *Environ. Pollut.* B3, 289-298.

Dobbs, A. J., Hart, G. F., Parsons, A. H. (1984) The determinations of vapour pressures from relative volatilization rates. *Chemosphere* 13, 687-692.

Dong, S., Dasgupta, P. K. (1986) Solubility of gaseous formaldehyde in liquid water and generation of trace standard gaseous formaldehyde. *Environ. Sci. Technol.* 6(20), 637-640.

Dorfman, L. M., Adams, G. E. (1973) Reactivity of the Hydroxyl Radicals in Aqueous Solution. NSRD-NDB-46. NTIS COM-73-50623. National Bureau of Standards, Washington, DC., 51 pp.

Doucette, W. J., Andren, A. W. (1988) Estimation of octanol/water partition coefficients: Evaluation of six methods for highly hydrophobic aromatic hydrocarbons. *Chemosphere* 17, 345-359.

Eadie, B. J., Robbins, J. A. (1987) The role of particulate matter in the movement of contaminants in the Great Lakes. In: *Sources and Fate of Aquatic Pollutants.* Hite, R. A., Eisenreich, S. J., Eds., pp. 320-364, Advances in Chemistry Series 216, American Chemical Society, Washington, DC.

Eadsforth, C. V., Moser, P. (1983) Assessment of reverse phase chromatographic methods for determining partition coefficients. *Chemosphere* 12, 1459-1475.

Ellgehausen, H., D'Hondt, C., Fuerer, R. (1981) Reversed-phase chromatography as a general method for determining octanol/water partition coefficients. *Pest. Sci.* 12, 219-227.

Ellgehausen, H., Guth, J. A., Esser, H. O. (1980) Factors determining bioaccumulation potential of pesticides in the individual compartments of aquatic food chains. *Ecotoxicol. Environ. Saf.* 4, 134-157.

Ellington, J. J., Stancil, F. E., Payne, W. D. (1987) *Measurement of Hydrolysis Rate Constants for Evaluation of Hazardous Waste Land Disposal.* Vol. 1, Data on 32 chemicals. EPA-600/3-86/043, U.S. EPA, Washington, DC.

Ellington, J. J., Stancil, F. E., Payne, W. D. (1987) *Measurement of Hydrolysis Rate Constants for Evaluation of Hazardous Waste Land Disposal.* Vol. 2, Data on 54 chemicals. EPA-600/53-87/019, U.S. EPA, Washington, DC.

Ellington, J. J., Stancil, F. E., Payne, W. D. (1988) *Measurement of Hydrolysis Rate Constants for Evaluation of Hazardous Waste Land Disposal.* Vol. 3, EPA 600/3-88/028, U.S. EPA, Washington, DC.

Farmer, W. J., Yang, M. S., Spencer, W. F. (1980) Hexachlorobenzene: Its vapor pressure and vapor phase diffusion in soil. *Soil Sci. Soc. Am. J.* 44, 676-680.

Figueroa, I. del C., Simmons, M. S. (1991) Structure-activity relationships of chlorobenzenes using DNA measurement as a toxicity parameter in algae. *Environ. Toxicol. Chem.* 10, 323-329.

Finizio, A., Vighi, M., Sandroni, D. (1997) Determination of N-octanolwater partition coefficient (Kow) of pesticide critical review and comparison of methods. *Chemosphere* 34, 131-161.

Fischer, R. C., Krämer, W., Ballschmiter, K. (1991) Hexachlorocyclohexane isomers as markers in the water flow of the Atlantic Ocean. *Chemosphere* 23, 889-900.

Foschi, S., Cesari, A., Ponti, I., Bentivogli, P. G., Bencivelli, A. (1970) Degradation and vertical movement of pesticides in the soil. *Notiz. Mal. Piante.* 82-83, 37-49.

Freed, V. H. (1976) Solubility, hydrolysis, dissolution constants and other constants of benchmark pesticides. In: *Literature Survey of Benchmark Pesticides*. George Washington University Medical Center, Washington, DC.

Freed, V. H., Chiou, C. T., Haque, R. (1977) Chemodynamics: Transport and behavior of chemicals in the environement - A problem in environmental health. *Environ. Health Perspect.* 20, 55-70.

Freed, V. H., Chiou, C. T., Schmedding, D. W. (1979) Degradation of selected organophosphorous pesticides in water and soil. *J. Agric. Food Chem.* 27, 706-708.

Freitag, D., Balhorn, L., Geyer, H., Körte, F. (1985) Environmental hazard profile of organic chemicals. An experimental method for the assessment of the behaviour of chemicals in the ecosphere by simple laboratory tests with C-14 labelled chemicals. *Chemosphere* 14, 1589-1616.

Freitag, D., Lay, J. P., Körte, F. (1984) Environmental hazard profile - Test results as related to structure and translation into the environment. In: *QSAR in Environmental Toxicology*. Kaiser, K. L. E., Ed., D. Reidel Publishing Co., Dordrecht, The Netherlands.

Fröbe, Z., Drevenkar, V., Stengle B. (1989) Sorption behaviour of some organophosphorus pesticides in natural sediments. *Toxicol. Environ. Chem.* 19, 69-82.

Gaffney, J. S., Streit, G. E., Spall, W. D., Hall, J. H. (1987) Beyond acid rain. Do soluble oxidants and organic toxins interact with SO_2 and NO_2 to increase ecosystem effects? *Environ. Sci. Technol.* 21(6), 519-524.

Garst, J. E. (1984) Accurate wide-range, automated, high-performance liquid chromatographic method for the estimation of octanol/water partition coefficients. II. Equilibrium in partition coefficient measurements, additivity of subsequent constants, and correlation of biological data. *J. Pharm. Sci.* 73, 1623-1629.

Garst, J. E., Wilson, W. C. (1984) Accurate wide-range, automated, high-performance liquid chromatographic method for the estimation of octanol/water partition coefficients. I. Effects of chromatographic conditions and procedure variables on accuracy and reproducibility of the method. *J. Pharm. Sci.* 73, 1616-1622.

Garten, C. T., Jr., Trabalka, J. R. (1983) Evaluation of models for predicting terrestrial food chain behavior of xenobiotics. *Environ. Sci. Technol.* 17, 590-595.

Gellman, I., Heukelekian, H. (1950) Biological oxidation of formaldehyde. *Sew. Indus. Waste* 22, 13-21.

GEMS (1986) *Graphical Exposure Modeling Systems*. Fate of Atmosphere Pollutants (FAP). Office of Toxic Substances, U.S. EPA, Washington, DC.

Geyer, H., Kraus, A. G., Klein, W., Richter, E., Körte, F. (1980) Relationship between water solubility and bioaccumulation potential of organic chemicals in rats. *Chemosphere* 9, 277-291.

Geyer, H., Politzki, G. R., Freitag, D. (1984) Prediction of ecotoxicological behaviour of chemicals: relationships between *n*-octanol/water partition coefficient and bioaccumulation of organic chemicals by *Alga chlorella*. *Chemosphere* 13, 269-284.

Geyer, H., Scheunert, I., Brüggemann, R., Steinberg, C., Korte, F., Kettrup, A. (1991) QSAR for organic chemical bioconcentration in *Daphnia*, algae, and mussels. *Sci. Total Environ.* 109/110, 387-394.

Geyer, H., Scheunert, I., Korte, F. (1987) Correlation between the bioconcentration potential of organic environmental chemicals in humans and their n-octanol/water partition coefficients. *Chemosphere* 16(1), 239-252.

Geyer, H., Visvanathan, R., Freitag, D., Körte, F. (1981) Relationship between water solubility of organic chemicals and their bioaccumulation by the *Alga chlorella*. *Chemosphere* 10, 1307-1313.

Giam, C. S., Atlas, E., Chan, H. S., Neff, G. S. (1980) Phthalate esters, PCB and DDT residues in the Gulf of Mexico atmosphere. *Atmos. Environ.* 14, 65-69.

Gobas, F. A. P. C., Bedard, D. C., Ciborowski, J. J. H. (1989b) Bioconcentration of chlorinated hydrocarbons by the mayfly (*Hexgenia limbata*) in Lake St. Clair. *J. Great Lakes Res.* 15(4), 581-588.

Gobas, F. A. P. C., Clark, K., Shiu, W. Y., Mackay, D. (1989a) Bioconcentration of polybrominated benzenes and biphenyls and related superhydrophobic chemicals in fish: Role of bioavailability and elimination into feces. *Environ. Toxicol Chem.* 8, 231-245.

Gobas, F. A. P. C., Shiu, W. Y., Mackay, D. (1987) Factors determining partitioning of hydrophobic organic chemicals in aquatic organisms. In: *QSAR in Environmental Toxicology II.*, Kaiser, K. L. E., Ed., pp. 107-124, D. Reidel Publishing Company, Dordrecht, The Netherlands.

Grayson, B. T., Fosbracy, L. A. (1982) Determination of the vapour pressure of pesticides. *Pest. Sci.* 13(3), 269-278.

Gückel, W., Kistel, R., Lewerenz, J., Synnatschke, G. (1982) A method for determining the volatility of active ingredients used in plant protection. Part III: the temperature relationship between vapor pressure and evaporation rate. *Pest. Sci.* 13, 161-168.

Günther, F. A., Westlake, W. E., Jaglan, P. S. (1968) Reported solubilities of 738 pesticide chemicals in water. *Res. Rev.* 20, 1-148.

Gustafson, D. I. (1989) Groundwater ubiquity score: A simple method for assessing pesticide leachability. *Environ. Toxicol. Chem.* 8, 339-357.

Halfon, E., Galassi, S., Brüggemann, R., Provini, A. (1996) Selection of priority properties to assess environmental hazard of pesticides. *Chemosphere* 33(8), 1543-1562.

Halfon, E., Reggiani, M. G. (1986) On ranking chemicals for environmental hazard. *Environ. Sci. Technol.* 20, 1173-1179.

Hamaker, J. W., Thompson, J. M. (1972) Adsorption. In: *Organic Chemistry in Soil Environment*. Goring, C. A. I., Hamaker, J. W., Eds., pp. 49-143, Vol. 1, Marcel Dekker, New York.

Hammers, W. E., Meurs, G. J., De Ligny, C. L. (1982) Correlations between liquid chromatographic capacity ratio data on lichrosorb RP-18 and partition coefficients in the octanol-water system. *J. Chromatogr.* 247, 1-13.

Hansch, C., Leo, A. (1979) *Substituent Constants for Correlation Analysis in Chemistry and Biology*. John Wiley & Sons, Inc., New York.

Hansch, C., Leo, A. (1985) *Medchem. Project Issue No. 26*, Pomona College, Claremont, California.

Hansch, C., Leo, A. (1987) *Medchem. Project Issue No. 28*, Pomona College, Claremont, California.

Hansch, C., Leo, A., Hoekman, D. (1995) *Exploring QSAR. Hydrophobic, Electronic, and Steric Constants*. ACS Professional Reference Book, Am. Chem. Soc., Washington, DC.

Harnish, M., Möckel, H. J., Schulze, G. (1983) Relationship between log P_{ow} shake-flask values and capacity factors derived from reversed-phase HPLC for *n*-alkylbenzenes and some OECD reference substances. *J. Chromatogr.* 282, 315-332.

Hartley, D., Kidd, H., Eds. (1987) *The Agrochemicals Handbook*. 2nd Edition, The Royal Society of Chemistry, London, England.

Hartley, G. S., Graham-Bryce, I. J. (1980) *Physical Principles and Pesticide Behavior*. Volume 2, Academic Press, New York.

Hashimoto, Y., Tokura, K., Ozaki, K., Strachan, W. M. J. (1982) A comparison of water solubility by the flask and micro-column methods. *Chemosphere* 11, 991-1001.

Hawker, D. W. (1989) The relationship between octan-1-ol/water partition coefficients and aqueous solubility in terms of solvatochromic parameters. *Chemosphere* 19(10/11), 1585-1593.

Hawker, D. W. (1990) Description of fish bioconcentration factors in terms of solvatochromic parameters. *Chemosphere* 20, 467-477.

Hawker, D. W., Connell, D. W. (1985) Relationships between partion coefficient and uptake rate constant, clearance rate constant, and time to equilibration for bioaccumulation. *Chemosphere* 14, 1205-1219.

Hawker, D. W., Connell, D. W. (1986) Bioconcentration of lipophilic compounds by some aquatic organisms. *Ecotoxicol. Environ. Saf.* 11, 184-197.

Hawker, D. W., Connell, D. W. (1988a) Octanol/water partition coefficient of polychlorinated biphenyl congeners. *Environ. Sci. Technol.* 22, 382-387.

Hawker, D. W., Connell, D. W. (1988b) Influence of partition coefficient of lipophilic compounds on bioconcentration kinetics with fish. *Water Res.* 22, 701-707.

Hermens, J., Leeuwangh, P. (1982) Joint toxicity of mixture of 8 and 24 chemicals to the guppy (*Poecilia reticulata*). *Ecotoxicol. Environ. Saf.* 6, 302-310.

Heukelekian, H., Rand, M. C. (1955) Biochemical oxygen demand for pure organic compounds. *J. Water Pollut. Control Assoc.* 29, 1040-1053.

Hinckley, D. A., Bidleman, T. F., Foreman, W. T. (1990) Determination of vapor pressures for nonpolar and semipolar organic compounds from gas chromatographic retention data. *J. Chem. Eng. Data* 35, 232-237.

Hine, J., Mookerjee, P. K. (1975) The intrinsic hydrophilic character of organic compounds. Correlations in terms of structural contributions. *J. Org. Chem.* 40, 292-298.

Hine, R. B., Johnson, D. L., Wenger, C. J. (1969) Persistence of two benzimidazole fungicides in soil and their fungistatic activity against *Phymatotrichum omnivorum*. *Phytopathology* 59, 798-801.

Hodnett, E. M., Wongwiechintana, C., Dunn, III, W. J., Marrs, P. (1983) Substituted 1,4-naphthoquinones vs. the ascitic Sarcoma 180 of mice. *J. Med. Chem.* 26, 570-574.

Hodson, J., Williams, N. A. (1988) The estimation of the adsorption coefficient (K_{OC}) for soils by high performance liquid chromatography. *Chemosphere* 17, 67-77.

Hollifield, H. C. (1979) Rapid nephelometric estimate of water solubility of highly insoluble organic chemicals of environmental interests. *Bull. Environ. Contam. Toxicol.* 23, 579-586.

Hornsby, A. G., Wauchope, R. D., Herner, A. E. (1996) *Pesticide Properties in The Environment.* Springer-Verlag, Inc., New York.

Horvath, A. L., Editor (1982) *Halogenated Hydrocarbons, Solubility-Miscibility with Water.* Marcel Dekker, Inc., New York.

Howard, P. H., Editor (1989) *Handbook of Fate Exposure Data for Organic Chemicals. Vol. I. Large Production and Priority Pollutants.* Lewis Publishers, Inc., Chelsea, Michigan.

Howard, P. H., Editor (1991) *Handbook of Fate Exposure Data for Organic Chemicals. Vol. III. Pesticides.* Lewis Publishers, Inc., Chelsea, Michigan.

Howard, P. H., Boethling, R. S., Jarvis, W. F., Meylan, W. M., Michalenko, E. M., Eds. (1991) *Handbook of Environmental Degradation Rates.* Lewis Publishers, Inc., Chelsea, Michigan.

Isnard, P., Lambert, S. (1988) Estimating bioconcentration factors from octanol-water partition coefficient and aqueous solubility. *Chemosphere* 17, 21-34.

Isnard, P., Lambert, S. (1989) Aqueous solubility and octanol-water partition coefficient correlations. *Chemosphere* 18, 1837-1853.

IUPAC Solubility Data series (1985) Volume 20: *Halogenated Benzenes, Toluenes and Phenols with Water.* Horvath, A. L, Getzen, F. W., Eds., Pergamon Press, Oxford, England.

Jalali, L., Anderson, J. P. E. (1976) Uptake of benomyl by the cultivated mushroom, *Agricus bisporus*. *J. Agric. Food Chem.* 24, 431-432.

Jury, W. A., Focht, D. D., Farmer, W. J. (1987b) Evaluation of pesticide groundwater pollution potential from standard indices of soil-chemical adsorption and biodegradation. *J. Environ. Qual.* 16(4), 422-428.

Jury, W. A., Winer, A. M., Spencer, W. F., Focht, D. D. (1987a) Transport and transformations of organic chemicals in the soil-air water ecosystem. *Rev. Environ. Contam. Toxicol.* 99, 120-164.

Kaiser, K. L. E. (1983) A non-linear function for the calculation of partition coefficients of aromatic compounds with multiple chlorine substitution. *Chemosphere* 12, 1159-1165.

Kaiser, K. L. E., Dixon, D. G., Hodson, P. V. (1984) QSAR studies on chlorophenols, chlorobenzenes and para-substituted phenols. In: *QSAR in Environmental Toxicology*. Kaiser, K. L. E., Ed., pp. 189-206, D. Reidel Publishing Co., Dordrecht, The Netherlands.

Kamlet, M. J., Doherty, R. M., Carr, P. W., Mackay, D., Abraham, M. H., Taft, R. W. (1988) Linear solvation energy relationship. 44. Parameter estimation rules that allow accurate prediction of octanol/water partition coefficients and other solubility and toxicity properties of polychlorinated biphenyls and polycyclic aromatic hydrocarbons. *Environ. Sci. Technol.* 22, 503-509.

Kanazawa, J. (1981) Measurement of the bioconcentration factors of pesticides by fresh-water fish and their correlation with physicochemical properties of acute toxicities. *Pest. Sci.* 12, 417-424.

Kanazawa, J. (1983) A method of predicting the bioconcentration potential of pesticides by using fish. *JARQ* 17(3), 173-179.

Kanazawa, J. (1989) Relationship between the soil sorption constants for pesticides and their physicochemical properties. *Environ. Toxicol. Chem.* 8, 477-484.

Karickhoff, S. W. (1981) Semi-empirical estimation of sorption of hydrophobic pollutants on natural sediments and soils. *Chemosphere* 10, 833-846.

Karickhoff, S. W. (1985) Chapter 3, Pollutant sorption in environmental systems. pp. 49-64. In: *Environmental Exposure from Chemicals*. Neely, W. B., Blau, G. E., Eds., CRC Press, Inc., Boca Raton, Florida.

Karickhoff, S. W., Brown, D. S., Scott, T. A. (1979) Sorption of hydrophilic pollutants on natural water sediments. *Water Res.* 13, 241-248.

Karickhoff, S. W., Morris, K. R. (1985a) Impact of tubificid oligochaetes on pollutant transport in bottom sediments. *Environ. Sci. Technol.* 19(1), 51-56.

Karickhoff, S. W., Morris, K. R. (1985b) Sorption dynamics of hydrophobic pollutants in sediment suspensions. *Environ. Toxicol. Chem.* 4, 469-479.

Kawamoto, K., Urano, K. (1989) Parameters for predicting fate of organochlorine pesticides in the environment. (I) Octanol-water and air-water partition coefficients. *Chemosphere* 18, 1987-1996.

Kawamoto, K., Urano, K. (1989) Parameters for predicting fate of organochlorine pesticides in the environment. (II) Adsorption constant to soil. *Chemosphere* 18, 1987-1996.

Kawamoto, K., Urano, K. (1990) Parameters for predicting fate of organochlorine pesticides in the environment. (III) Biodegradation rate constants. *Chemosphere* 21, 1141-1152.

Kawanoto, K., Urano, K. (1991) Corregendum. *Chemosphere* 23, 813.

Kearney, P. C., Kaufman, D. D. (1975) *Herbicides: Chemistry, Degradation and Mode of Action*. 2nd Edition, Vol. 2, Marcel Dekker, Inc., New York.

Kelly, T. J., Mukund, R., Spicer, C. W., Pollack, A. J. (1994) Concentrations and transformations of hazardous air pollutants. *Environ. Sci. Technol.* 28, 378A-387A.

Kenaga, E. E. (1980a) Predicted bioconcentration factors and soil sorption coefficients of pesticides and other chemicals. *Ecotoxicol. Environ. Saf.* 4, 26-38.

Kenaga, E. E. (1980b) Correlation of bioconcentration factors of chemicals in aquatic and terrestrial organisms with their physical and chemical properties. *Environ. Sci. Technol.* 14, 553-556.

Kenaga E. E., Goring, C. A. I. (1980) Relationship between water solubility, soil sorption, octanol-water partitioning, and bioconcentration of chemicals in biota. In: *Aquatic Toxicology*. ASTM STP 707, Eaton, J. G., Parrish, P. R., Hendricks, A. C., Eds., pp. 78-115, American Society for Testing and Materials, Philadelphia, Pennsylvania.

Kerler, F., Schönherr, J. (1988) Accumulation of lipophilic chemicals in plant cuticles: Prediction from octanol/water partition coefficients. *Arch. Environ. Contam. Toxicol.* 17, 1-6.

Khan, U. K. (1980) *Pesticides in the Soil Environment, Fundamental Aspects of Pollution Control and Environmental Series 5*. Elsevier, Amsterdam, The Netherlands.

Kilzer, L., Scheunert, I., Geyer, H., Klein, W., Korte, F. (1979) Laboratory screening of the volatilization rates of organic chemicals from water and soil. *Chemosphere* 10, 751-761.

Kim, Y. H., Woodrow, J. E., Seiber, J. N. (1984) Evaluation of a gas chromatographic method for calculating vapor pressures with organophosphorous pesticides. *J. Chromatogr.* 314, 37-53.

Kishii, H., Nakamura, M., Hashimoto, Y. (1987) Prediction of solubility of aromatic compounds in water by using total molecular surface area. *Nippon Kagaku Kaishi* 8, 1615-1622.

Klečka, G. M. (1985) Chapter 6, Biodegradation. In: *Environmental Exposure from Chemicals*. Neely, W. B., Blau, G. E., Eds., pp. 109-156, CRC Press, Inc., Boca Raton, Florida.

Klein, A. W., Harnish, M., Porenski, H. J., Schmidt-Bleek, F. (1981) OECD chemicals testing program physico-chemical tests. *Chemosphere* 10, 153-207.

Koch, R. (1983) Molecular connectivity index for assessing ecotoxicological behaviour of organic compounds. *Toxicol. Environ. Chem.* 6, 87-96.

Kollig, H. P., Editor (1993) *Environmental Rate Constants for Organic Chemicals under Consideration for EPA's Hazardous Waste Identification Projects*. EPA/600/R-93/132. Environmental Research Laboratory, U.S. EPA, Athens, Georgia.

Kollig, H. P., Ellington, J. J., Hamrick, K. J., Jafverts, C. T., Weber, E. J., Wolfe, N. L. (1987) *Hydrolysis Rate Constants, Partition Coefficients, and Water Solubilities for 129 Chemicals. A Summary of Fate Constants Provided for the Concentration-Based Listing Program*. Environmental Research Lab., Office of Research and Development, U.S. EPA, Athens, Georgia.

Könemann, H. (1981) Quantitative structure-activity relationships in fish toxicity studies. Part 1: Relationship for 50 industrial pollutants. *Toxicology* 19, 209-221.

Könemann, H., van Leeuwen, K. (1980) Toxicokinetics in fish: Accumulation and elimination of six chlorobenzenes by guppies. *Chemosphere* 9, 3-19.

Könemann, H., Zelle, R., Busser, F. (1979) Determination of log P_{oct} values of chloro-substituted benzenes, toluenes and anilines by high-performance liquid chromatography on ODS-silica. *J. Chromatogr.* 178, 559-565.

Kördel, W., Stutte, J., Kotthoff, G. (1993) HPLC-screening method for the determination of the adsorption coefficient on soil-Comparison of different stationary phases. *Chemosphere* 27(12), 2341-2352.

Kördel, W., Stutte, J., Kotthoff, G. (1995) HPLC-screening method to determine the adsorption coefficient in soil-Comparison of immobilized humic acid and clay mineral phases for cyanopropyl columns. *Sci. Total Environ.* 162, 119-125.

Körte, F., Freitag, D., Geyer, H., Klein, W., Kraus, A. G., Lahaniatus, E. (1978) Ecotoxicologic profile analysis-A concept for establishing ecotoxicologic priority lists for chemicals. *Chemosphere* No.1, 79-102.

Kühne, R., Ebert, R.-U., Kleint, F., Schmidt, G., Schüürmann, G. (1995) Group contribution methods to estimate water solubility of organic chemicals. *Chemosphere* 30, 2061-2077.

Lande, S. S., Banerjee, S. (1981) Predicting aqueous solubility of organic nonelectrolytes from molar volume. *Chemosphere* 10, 751-759.

Leo, A., Hansch, C., Elkins, D. (1971) Partition coefficients and their uses. *Chem. Rev.* 71, 525-616.

Lohninger, H. (1994) Estimation of soil partition coefficients of pesticides from their chemical structure. *Chemosphere* 29(8), 1611-1626.

Lord, K. A., Briggs, G. C., Nearle, M. C., Manlove, R. (1980) Uptake of pesticides from water and soil by earthworms. *Pest. Sci.* 11, 401-408.

Lu, P. Y., Metcalf, R. L. (1975) Environmental fate and biodegradability of benzene derivatives as studied in a model aquatic ecosystem. *Environ. Health Perspect.* 10, 269-284.

Lyman, W. J. (1982) Chapter 2, Solubility in water and Chapter 4, Adsorption coefficient for soils and sediments. In: *Handbook on Chemical Property Estimation Methods, Environmental Behavior of Organic Compounds.* Lyman, W. J., Reehl, W. F., Rosenblatt, D. H., Eds., McGraw-Hill, New York.

Lyman, W. J. (1985) Chapter 2, Estimation of physical properties. In: *Environmental Exposure from Chemicals.* Neely, W. B., Blau, G. E., Eds., pp. 13-48, CRC Press, Inc., Boca Raton, Florida.

Lyman, W. J., Reehl, W. F., Rosenblatt, D. H., Eds. (1982) *Handbook on Chemical Property Estimation Methods, Environmental Behavior of Organic Compounds.* McGraw-Hill, Inc., New York.

Mabey, W. R., Smith, J. H., Podoll, R. T., Johnson, H. L., Mill, T., Chou, T. W., Gates, J., Waight-Partridge, I., Jaber, H., Vanderberg, D. (1982) *Aquatic Fate Process for Organic Priority Pollutants.* EPA Report No. 440/4-81-014, U.S. EPA, Washington, DC.

Mackay, D. (1982) Correlation of bioconcentration factors. *Environ. Sci. Technol.* 16, 274-278.

Mackay, D. (1985) Chapter 5, Air/water exchange coefficients. In: *Environmental Exposure from Chemicals.* Neely, W. B., Blau, G. E., Eds., pp. 91-108, CRC Press, Inc., Boca Raton, Florida.

Mackay, D., Paterson, S. (1991) Evaluating the multimedia fate of organic chemicals. A level III fugacity model. *Environ. Sci. Technol.* 25, 427-436.

Mackay, D., Paterson, S., Chung, B., Neely, W. B. (1985) Evaluation of the environmental behaviour of chemicals with level III fugacity model. *Chemosphere* 13, 335-374.

Mackay, D., Shiu, W. Y. (1981) A critical review of Henry's law constants for chemicals of environmental interest. *J. Phys. Chem. Ref. Data* 10, 1175-1199.

Mackay, D., Shiu, W. Y., Ma, K. C. (1992) *Illustrated Handbook of Physical-Chemical Properties and Environmental Fate for Organic Chemicals, Vol. I. Monoaromatic-Hydrocarbons, Chlorobenzenes, and PCBs.* Lewis Publishers, Inc., Chelsea, Michigan.

Magee, P. S. (1991) Complex factors in hydrocarbon/water, soil/water and fish/water partitioning. *Sci. Total Environ.* 109/110, 155-178.

Mailhot, H. (1987) Prediction of algae bioaccumulation and uptake rate of nine organic compounds by ten physico-chemical properties. *Environ. Sci. Technol.* 21, 1009-1013.

Mailhot, H., Peters, R. H. (1988) Empirical relationships between the 1-octanol/water partition coefficient and nine physicochemical properties. *Environ. Sci. Technol.* 22, 1479-1488.

Majewski, M. S., Capel, P. D. (1995) *Pesticides in the Atmosphere. Distribution, Trends, and Governing Factors.* Vol. one in the Series of *Pesticides in the Hydrologic System.* Gilliom, R. J., Series Editor., Ann Arbor Press, Inc., Chelsea, Michigan.

Martin, H. (1972) *Pesticide Manual*, 3rd Edition, British Crop Protection Council, Worcester, England.

Martin, H., Worthing, C. R., Eds. (1977) *Pesticide Manual.* 5th Edition, British Crop Protection Council, Thornton, United Kingdom.

Martin, R. A., Edgington, L. V. (1981) Comparative systemic translocation of several xenobiotics and sucrose. *Pest. Biochem. Physiol.* 16(2), 87-96.

Mathre, D. E. (1971) Mode of action of oxathiin systemic fungicides. Structure-activity relations. *J. Agric. Food Chem.* 19(5), 872-874.

McDuffie, B. (1981) Estimation of octanol/water partition coefficients for organic pollutants using reverse-phase HPLC. *Chemosphere* 10(1), 73-83.

McKim, J., Schnieder, P., Veith, G. (1985) Absorption dynamics of organic chemical transport across trout gills as related to octanol-water partition coefficients. *Toxicol. Appl. Pharmacol.* 77, 1-10.

Melnikov, N. N. (1971) Chemistry of pesticides. *Res. Rev.* 36, 1-447.

The Merck Index (1983) *An Encyclopedia of Chemicals, Drugs and Biologicals.* 10th Editon, Windholz, M., Ed., Merck and Co., Inc., Rahway, New Jersey.

The Merck Index (1989) *An Encyclopedia of Chemicals, Drugs and Biologicals.* 11th Edition, Budavari, S., Ed., Merck and Co., Inc., Rahway, New Jersey.

Meylan, W., Howard, P. H. (1991) Bond contribution method for estimating Henry's law constants. *Environ. Toxicol. Chem.* 10, 1283-1293.

Meylan, W., Howard, P. H., Boethling, R. S. (1992) Molecular topology/fragment contribution method for predicting soil sorption coefficients. *Environ. Sci. Technol.* 26, 1560-1567.

Miller, M. M., Ghodbane, S., Wasik, S. P., Tewari, Y. B., Martire, D. E. (1984) Aqueous solubilities, octanol/water partition coefficients and entropies of melting of chlorinated benzenes and biphenyls. *J. Chem. Eng. Data* 29, 184-190.

Miller, M. M., Wasik, S. P., Huang, G. L., Shiu, W. Y., Mackay, D. (1985) Relationship between octanol-water partition coefficient and aqueous solubility. *Environ. Sci. Technol.* 19, 522-529.

Milne, G. W. A., Editor (1995) *CRC Handbook of Pesticides.* CRC Press, Inc., Boca Raton, Florida.

Montgomery, J. H., Editor (1993) *Agrochemicals Desk Reference. Environmental Data.* Lewis Publishers, Inc., Chelsea, Michigan.

Mortimer, M. R., Connell, D. W. (1995) A model of the environmental fate of chlorohydrocarbon contaminants associated with Sydney sewage discharge. *Chemosphere* 30, 2021-2038.

Moza, P. N., Sukul, P., Hustert, K., Kettrup, A. (1994) Photooxidation of metalaxyl in aqueous solution in the presence of hydrogen peroxide and titanium dioxide. *Chemosphere* 28, 341-347.

Neely, W. B. (1980) A method for selecting the most appropriate environmental experiments on a new chemical. In: *Dynamic, Exposure and Hazard Assessment of Toxic Chemicals.* Haque, R., Ed., pp. 287-298, Ann Arbor Science Publishers, Ann Arbor, Michigan.

Neely, W. B., Branson, D. R., Blau, G. E. (1974) Partition coefficient to measure bioconcentration potential of organic chemicals in fish. *Environ. Sci. Technol.* 8, 1113-1115.

Neuhauser, E. F., Loehr, R. C., Malecki, M. R., Milligan, D. l., Durin, P. R. (1985) The toxicity of selected organic chemicals to earthworm *Eisenia fetida*. *J. Environ. Qual.* 14(3), 383-388.

Nielsen, L. S., Bundgaard, H., Falch, E. (1992) Prodrugs of thiabendazole with increased water-solubility. *Acta Pharm. Nord.* 4(1), 43-49.

Niimi, A. J., Cho, C. Y. (1980) Uptake of hexachlorobenzene (HCB) from feed by rainbow trout (*Salmo gairdneri*). *Bull. Environ. Toxicol.* 24, 834-837.

Niimi, A. J., Lee, H. B., Kisson, G. P. (1989) Octanol/water partition coefficients and bioconcentration factors of chloronitrobenzenes in rainbow trout (*Salmo gairdneri*). *Environ. Toxicol. Chem.* 8, 817-823.

Niimi, A. J., Palazzo, V. (1985) Temperature effect on the elimination of pentachlorophenol, hexachlorobenzene and mirex by rainbow trout (*Salmo gairdneri*). *Water Res.* 19, 205-207.

Nirmalakhandan, N. N., Speece, R. E. (1988) QSAR model for predicting Henry's law constant. *Environ. Sci. Technol.* 22, 1349-1357.

OECD (1979) OECD Environmental Committee Chemicals Group, OECD Chemical Testing Programme Expert Group, Physical Chemical Final Report Vol. 1, Part 1 and Part 2, Summary of the OECD Laboratory Intercomparison Testing Programme Part 1-On the Physico-Chemical Properties. p. 33, Dec., 1979, Berlin, Germany.

OECD (1981) *OECD Guidelines for Testing of Chemicals.* Organization for Economic Cooperation and Development, OECD, Paris, France.

Ohori, Y., Ihashi, Y. (1987) Foliar absorption of pesticides using an isolated cucumber membrane. *Res. Dev. Rev.-Mitsubishi Chem.* 1(2), 22-26.

Oliver, B. G. (1985) Desorption of chlorinated hydrocarbons from spiked and anthropogenically contaminated sediments. *Chemosphere* 14, 1087-1106.

Oliver, B. G. (1987a) Biouptake of chlorinated hydrocarbons from laboratory-spiked and field sediments by oligochaete worms. *Environ. Sci. Technol.* 21, 785-790.

Oliver, B. G. (1987b) Fate of some chlorobenzenes from the Niagara River in Lake Ontario. In: *Sources and Fates of Aquatic Pollutants.* Hite, R. A., Eisenreich, S. J., Eds., pp. 471-489, Advances in Chemistry Series 216, American Chemical Society, Washington, DC.

Oliver, B. G. (1987c) Partitioning relationships for chlorinated organics between water and particulates in the St. Clair, Detroit and Niagara Rivers. In: *QSAR in Environmental Toxicology II.* Kaiser, K. L. E., Ed., pp. 251-260, D. Reidel Publishing Co., Dordrecht, The Netherlands.

Oliver, B. G., Charlton, M. N. (1984) Chlorinated organic contaminants on settling particulates in the Niagara River vicinity of Lake Ontario. *Environ. Sci. Technol.* 18, 903-908.

Oliver, B. G., Niimi, A. J. (1985) Bioconcentration factors of some halogenated organics for rainbow trout: Limitations in their use for prediction of environmental residues. *Environ. Sci. Technol.* 19, 842-849.

Opperhuizen, A. (1986) Bioconcentration of hydrophobic chemicals in fish. In: *Aquatic Toxicology and Environmental Fate.* 9th Vol. ASTM STP 921, Poston, T. M., Purdy, R., Eds., pp. 304-315, American Society for Testing and Materials, Philadelphia, Pennsylvania.

Pait, A. S., De Souza, A. E., Farrow, D. R. D. (1992) *Agriculture Pesticide Use in Coastal Areas: A National Summary.* National Oceanic and Atmospheric Adminstration (NOAA) Rockville, Maryland.

Pankow, J. F., Isabelle, L. M., Asher, W. E. (1984) Trace organic compounds in rain. 1. Sample design and analysis by adsorption/thermal desorption (ATD). *Environ. Sci. Technol.* 18, 310-318.

Patil, G. S. (1994) Prediction of aqueous solubility and octanol-water coefficient for pesticides based on their molecular structure. *J. Hazard. Materials* 36, 35-43.

Patil, G. S., Nicholls, P. H., Chamberlain, K., Briggs, G. G., Bromilow, R. H. (1988) Degradation rates in soil of 1-benzyltriazoles and two triazole fungicides. *Pest. Sci.* 22(4), 333-342.

Pereira, W. E., Rostad, C. E., Chiou, C. T., Brinton, T. I., Barber, I., L. B., Demcheck, D. K., Demas, C. R. (1988) Contamination of esturine water, biota and sediment by halogenated organic compounds: A field study. *Environ. Sci. Technol.* 22, 772-778.

Pinsuwan, S., Li, A., Yalkowsky, S. H. (1994) Correlation of octanol/water solubility and partition coefficients. *J. Chem. Eng. Data* 40, 623-626.

Plato, C. (1972) Differential scanning calorimetry as a general method for determining the purity and heat of fusion of high-purity organic chemicals. Application to 64 compounds. *Anal. Chem.* 44(8), 1531-1534.

Popov, V. I., Sboeva, J. N. (1974) Determination of benomyl in cotton leaves. *J. Environ. Qual. Saf.* 3.

Rao, P. S. C., Davidson, J. M. (1980) Estimation of pesticide retention and transformation parameters required in nonpoint source pollutant models. In: *Environmetal Impact of Nonpoint Pollution.* Overcash, M. R., Davidson, J. M., Eds., Ann Arbor Science Publishers Inc., Ann Arbor, Michigan.

Rao, P. S. C., Davidson, J. M. (1982) Retention and Transformation of Selected Pesticides and Phosphorus in Soil Water System: A Critical Review. U.S. EPA-600/3-82-060.

Reischl, A., Reissinger, M., Thoma, H., Hutzinger, O. (1989) Uptake and accumulation of PCDD/F in terrestrial plants: Basic considerations. *Chemosphere* 19, 467-474.

Rekker, R. F. (1977) *The Hydrophobic Fragmental Constants. Its Derivation and Application, A Means of Characterizing Membrane Systems.* Elsevier Science Publishing Co., Oxford, England.

Rippen, G., Ilgenstein, M., Klöpffer, W., Poreniski, H. J. (1982) Screening of the adsorption behavior of new chemicals: natural soils and model adsorbents. *Ecotoxicol. Environ. Saf.* 6, 236-245.

Ryan, J. A., Bell, R. M., Davidson, J. M., O'Connor, G. A. (1988) Plant uptake of non-ionic organic chemicals from soils. *Chemosphere* 17, 2299-2323.

Sabljic, A. (1987a) On the prediction of soil sorption coefficients of organic pollutants from molecular structure: Application of molecular topology model. *Environ. Sci. Technol.* 21, 358-366.

Sabljic, A. (1987b) Nonempirical modeling of environmental distribution and toxicity of major organic pollutants. In: *QSAR in Environmental Toxicology II.* Kaiser, K. L. E., Ed., pp. 309-322, D. Reidel Publ. Co., Dordrecht, The Netherlands.

Saito, S., Tanoue, A., Matsuo, M. (1992) Applicability of the i/o-characters to a quantitative description of bioconcentration of organic chemicals in fish. *Chemosphere* 24, 81-87.

Saito, S., Koyasu, J., Yoshida, K., Shigeoka, T., Koike, S. (1993) Cytotoxicity of 109 chemicals to goldfish GFS cells and relationships with 1-octanol/water partition coefficients. *Chemosphere* 26(5), 1015-1028.

Sangster, J. (1989) Octanol-water partition coefficients of simple organic compounds. *J. Phy. Chem. Ref. Data* 18, 1111-1230.

Sangster, J. (1993) LOGK_{ow} Databank. Sangster Research Labs., Montreal, Canada.

Sarna, I. P., Hodge, P. E., Webster, G. R. B. (1984) Octanol-water partition coefficients of chlorinated dioxins and dibenzofurans by reversed-phase HPLC using several C_{18} columns. *Chemosphere* 13, 975-983.

Schreiber, L., Schönherr, J. (1992) Uptake of organic chemicals in conifer needles: Surface adsorption and permeability of cuticles. *Environ. Sci. Technol.* 26, 153-159.

Schwack, W., Flößer-Müller, H. (1990) Fungicides and photochemistry. Photodehalogenation of captan. *Chemosphere* 21(7), 905-912.

Sharom, M. S., Edgington, L. V. (1982) The adsorption, mobility, and persistence of metalaxyl in soil and aqueous systems. *Can. J. Plant Pathol.* 4, 334-340.

Shiu, W. Y., Ma, K. C., Mackay, D (1990) Solubilities of pesticides in water. Part 1, Environmental physical chemistry and Part 2, Data compilation. *Rev. Environ. Contam. Toxicol.* 115, 1-187.

Shiu, W. Y., Gobas, F. P. A. C., Mackay, D (1987) Physical-chemical properties of three congeneric series of chlorinated aromatic hydrocarbons. In: *QSAR in Environmental Toxicology II.* Kaiser, K. L. E., Editor, pp. 347-362, D. Reidel Publishing Co., Dordrecht, The Netherlands.

Sicbaldi, F., Finizio, A. (1993) K_{ow} estimation by combination of RP-HPLC and molecular indexes for a heterogeneous set of pesticide. In: *Proceedings IX Symposium Pesticide Chemistry, Mobility and Degradation of Xenobiotics.* 11-13, Oct. 1993, Picenza, Italy.

Siebers, J., Gottschild, D., Nolting, H.-G. (1994) Pesticides in precipitation in Northern Germany. *Chemosphere* 28(8), 1559-1570.

Sijm, D. T. H. M., Middlekoop, J., Vrisekoop, K. (1995) Algal density dependent bioconcentration factors of hydrophobic chemicals. *Chemosphere* 31(9), 4001-4012.

Singh, R. P., Chiba, M. (1985) Solubility of benomyl in water at different pHs and its conversion to methyl 2-benzimidazolecarbamate, 3-butyl-2,4-dioxo[1,2-α]-s-triazinobenzimidazole, and 1-(2-benzimidazolyl)-3-n-butylurea. *J. Agric. Food Chem.* 33(1), 63-67.

Smith, J. H., Mabey, W. R., Bahonos, N., Holt, B. R., Lee, S. S., Chou, T. W., Venberger, D. C., Mill, T. (1978) Environmental pathways of selected chemicals in freshwater systems: Part II. Laboratory Studies. Interagency Energy-Environmental Research Program Report. EPA-600/7-78-074. Environmental Research Laboratory Office of Research and Development. U.S. Environmental Protection Agency, Athens, Georgia.

Spencer, E. Y., Editor (1973) *Guide to the Chemicals Used in Crop Protection.* 6th Edition, Research Branch Agriculture Canada, Ontario, Canada.

Spencer, E. Y., Editor (1981) *Guide to the Chemicals Used in Crop Protection.* 7th Edition, Research Branch Agriculture Canada, Ontario, Canada.

Spencer, W. F., Cliath, M. M. (1983) Measurement of pesticide vapor pressures. *Res. Rev.* 85, 57-71.

Spencer, W. F., Cliath, M. M. (1990) Chapter 1, Movement of pesticides from soil to the atmosphere. In: *Long Range Transport of Pesticides.* Kurtz, D. A., Editor, Lewis Publishers, Inc., Chelsea, Michigan.

Spencer, W. F., Cliath, M. M., Jury, W. A., Zhang, L. Z. (1988) Volatilization of organic chemicals from soil as their Henry's law constants. *J. Environ. Qual.* 17, 504-509.

SRI International (1980) Interim Report on Task No. 11, Contract No. 68-01-3867, U.S. EPA Monitoring and Data Support Div., Office of Water Regulations and Standards, Washington, DC.

Stevens, P. J. G., Baker, E. A., Anderson, N. H. (1988) Factors affecting the foliar absorption and redistribution of pesticides. 2. Physicochemical properties of the active ingredient and the role of surfactant. *Pest. Sci.* 24(1), 31-53.

Su, F., Calvert, J. G., Shaw, J. H. (1979) Mechanism of the photooxidation of gaseous formaldehyde. *J. Phys. Chem.* 83, 3185-3191.

Sukop, M., Cogger, C. G. (1992) Adsorption of carbofuran, metalaxyl, and simazine: KOC evaluation and relation to soil transport. *J. Environ. Sci. Health* B27, 565-590.

Suntio, L. R., Shiu, W. Y., Mackay, D. (1988) A review of the nature and properties of chemicals present in pulp mill effluents. *Chemosphere* 17, 1249-1290.

Suntio, L. R., Shiu, W. Y., Mackay, D., Seiber, J. N., Glotfelty, D. (1988) Critical review of Henry's law constants. *Rev. Environ. Contam. Toxicol.* 103, 1-59.

Swann, R. L., Laskowski, D. A., McCall, P. J., Vander, Kuy K., Dishburger, H. J. (1983) A rapid method for estimation of the environmental parameters octanol/water partition coefficient, soil sorption constant, water to air ratio, and water solubility. *Res. Rev.* 85, 17-28.

Tabak, H. H., Quave, S. A., Mahni, C. I., Barth, E. F. (1981) Biodegradability studies with organic priority pollutant compound. *J. Water Pollut. Control Fed.* 53, 1503-1518.

Tadokoro, H., Tomita, Y. (1987) The relationship between bioaccumulation and lipid content of fish. In: *QSAR in Environmental Toxicology II*. Kaiser, K. L. E., Editor, pp. 363-373, D. Reidel Publishing Co., Dordrecht, The Netherlands.

Thomann, R. V. (1989) Bioconcentration model of organic chemical distribution in aquatic food chains. *Environ. Sci. Technol.* 23, 699-707.

Thor (1989) from *Medchem Release* 3.54, Daylight Chemical Information Systems Inc., Claremont, California.

Tomlin, C. (1994) *The Pesticide Manual (A World Compendium)*. 10th Ed., Incorporating the Agrochemicals Handbook. The British Crop Protection Council, Surrey, UK and The Royal Society of Chemistry, Cambridge, UK.

Travis, C. C., Arms, A. D. (1988) Bioconcentration of organics in beef, milk, and vegetation. *Environ. Sci. Technol.* 22, 271-273.

Tsonopoulos, C., Prausnitz, J. M. (1971) Activity coefficients of aromatic solutes in dilute aqueous solutions. *Ind. Eng. Chem. Fundam.* 10, 593-600.

Valvani, S. C., Yalkowsky, S. H. (1980) Solubility and partitioning in drug design. In: *Physical Chemical Properties of Drug*. Medical Research Series, Vol. 10, Yalkowsky, S. H., Sinkinla, A. A., Valvani, S., Eds., pp. 201-229, Marcel Dekker, Inc., New York, New York.

Veith, G. D., Austin, N. M., Morris, R. T. (1979a) A rapid method for estimation log P for organic chemicals. *Water Res.* 13, 43-47.

Veith, G. D., Defoe, D. L., Bergstedt, B. V. (1979b) Measuring and estimating the bioconcentration factor of chemicals in fish. *J. Fish Res. Board Can.* 26, 1040-1048.

Verhaar, H. J. M., van Leeuwen, C. J., Hermens, J. L. M. (1992) Classifying environmental pollutants. 1: Structure-activity relationships for prediction of aquatic toxicity. *Chemosphere* 25(4), 471-491.

Verschueren, K. (1977) *Handbook of Environmental Data on Organic Chemicals*. Van Nostrand Reinhold, New York.

Verschueren, K. (1983) *Handbook of Environmental Data on Organic Chemicals*. 2nd Edition, Van Nostrand Reinhold, New York.

Walker, W. W., Cripe, C. R., Pritchard, P. H., Bourquin, A. W. (1988) Biological and abiotic degradation of xenobiotic compounds in vitro esturine water and sediment/water systems. *Chemosphere* 17(12), 2255-2270.

Wang, X., Harada, S., Wantanabe, M., Koshikawa, H., Geyer, H. J. (1996) Modelling the bioconcentration of hydrophobic organic chemicals in aquatic organisms. *Chemosphere* 32(9), 1783-1793.

Watarai, H., Tanaka, M., Suzuki, N. (1982) Determination of water partition coefficients of halobenzenes in heptane/water and 1-octanol/water systems and comparison with the scaled particle calculation. *Anal. Chem.* 54, 702-705.

Wauchope, R. D. (1989) *ARS/SCS Pesticide Properties Database*. Ver. 1.9, preprint, August, 1989.

Wauchope, R. D., Buttler, T. M., Hornsby, A. G., Augustijn-Beckers, P. W. M., Burt, J. P. (1992) The SCS/ARS/SCS Pesticide Properties Database for Environmental Decision Making. *Rev. Environ. Contam. Toxicol.* 123, 1-164.

Weast, R. C., Ed. (1972-73) *Handbook of Chemistry and Physics*. 53rd Edition, CRC Press, Inc., Cleveland, Ohio.

Weast, R. C., Ed. (1982-83) *Handbook of Chemistry and Physics*. 62th Edition, CRC Press, Inc., Boca Raton, Florida.

Weil, V. G., Dure, G., Quentin, K. E. (1974) Solubility in water of insecticide chlorinated hydrocarbons and polychlorinated biphenyls in view of water pollution. *Z. Wasser Abwasser Forsch* 7(6), 169-175.

Windholz, M., Ed. (1983) *The Merck Index. An Encyclopedia of Chemicals, Drugs and Biologicals*. 11th Edition, Merck and Co., Inc., Rahway, New Jersey.

Wolfe, N. L., Zepp, R. G., Baughman, G. L., Fincher, R. C., Gordon, T. A. (1976) *Chemical and Photochemical Transformation of Selected Pesticides in Aquatic Environments*. U.S. EPA-600/3-76-067, U.S. EPA, Athens, Georgia.

Wong, P. T. S., Chau, Y. K., Rhamey, J. S., Docker, M. (1984) Relationship between water solubility of chlorobenzenes and their effects on a fresh water green algae. *Chemosphere* 13, 991-996.

Worthing, C. R., Ed. (1979) *The Pesticide Manual (A World Compendium)*. 6th Edition, The British Crop Protection Council, Croydon, England.

Worthing, C. R., Ed. (1983) *The Pesticide Manual (A World Compendium)*. 7th Edition, The British Crop Protection Council, Croydon, England.

Worthing, C. R., Ed. (1987) *The Pesticide Manual (A World Compendium)*. 8th Edition, The British Crop Protection Council, Croydon, England.

Worthing, C. R., Ed. (1991) *The Pesticide Manual (A World Compendium)*. 9th Edition, The British Crop Protection Council, Croydon, England.

Yalkowsky, S. H., Dannenfelser, R. M. (1994) *AQUASOL DATABASE*. 5th Edition, University of Arizona, Tucson, Arizona.

Yalkowsky, S. H., Orr, R. J., Valvani, S. C. (1979) Solubility and partitioning. 3. The solubilities of halobenzenes in water. *Ind. Eng. Chem. Fundam.* 18, 351-353.

Yalkowsky, S. H., Valvani, S. C. (1979) Solubility and partitioning. 2. Relationships between aqueous solubilities, partition coefficients, and molecular surface areas of rigid aromatic hydrocarbons. *J. Chem. Eng. Data* 24, 127-129.

Yalkowsky, S. H., Valvani, S. C. (1980) Solubility and partitioning. 1. Solubility of nonelectrolytes in water. *J. Pharm. Sci.* 69, 912-922.

Yalkowsky, S. H., Valvani, S. C., Mackay, D. (1983) Estimation of the aqueous solubility of some aromatic compounds. *Res. Rev.* 85, 43-55.

Yoshioka, Y., Mizuno, T., Ose, Y., Sato, T. (1986) Estimation of toxicity on fish by physico-chemical properties. *Chemosphere* 15(2), 195-203.

Zaroogian, G. E., Hertshe, J. F., Johnson, M. (1985) Estimation of bioconcentration in marine species using structure-activity models. *Environ. Toxicol. Chem.* 4, 3-12.

Zepp, R. G., Baughman, G. L. (1978) Prediction of photochemical transformation of pollutants in the aquatic environment. In: *Aquatic Pollutants: Transformation and Biological Effects*. Hutzinger, O., Van Lelyveld, I. H., Zoeteman, B. C. J., Eds., pp. 237-264, Pergamon Press, Oxford.

Zhou, X., Mopper, K. (1990) Apparent partition coefficients of 15 carbonyl compounds between air and seawater and between air and freshwater; implications for air-sea exchange. *Environ. Sci. Technol.* 24, 1864-1869.

List of Symbols and Abbreviations:

A_i	area of phase i, m^2
ALPM	Automated Log-P Measurement
BCF	bioconcentration factor
B.P.	boiling point, °C
C	molar concentration, mol/L or mmol/m^3
C^S	saturated aqueous concentration, mol/L or mmol/m^3
C_L	liquid or supercooled liquid concentration, mol/L or mmol/m^3
C_S	solid molar concentration, mol/L or mmol/m^3
C_A	concentration in air phase, mol/L or mmol/m^3
C_W	concentration in water phase, mol/L or mmol/m^3
^{14}C	radioactive labelled carbon-14 compound
CC	Countercurrent Chromatography
COD	chemical oxygen demand
CPC	Centrifugal Partition Chromatography
D	D values, mol/Pa·h
D_A	D values for advection, mol/Pa·h
D_{Ai}	D values for advective loss in phase i, mol/Pa·h
D_R	D value for reaction, mol/Pa·h
D_{Ri}	D value for reaction loss in phase i, mol/Pa·h
D_{ij}	intermedia D values, mol/Pa·h
D_{VW}	intermedia D value for air-water diffusion (absorption), mol/Pa·h
D_{RW}	intermedia D value for air-water dissolution, mol/Pa·h
D_{QW}	D value for total particle transport (dry and wet), mol/Pa·h
D_{RS}	D value for rain dissolution (air-soil), mol/Pa·h
D_{QS}	D value for wet and dry deposition (air-soil), mol/Pa·h
D_{VS}	D value for total soil-air transport, mol/Pa·h
D_S	D value for air-soil boundary layer diffusion, mol/Pa·h
D_{SW}	D value for water transport in soil, mol/Pa·h
D_{SA}	D value for air transport in soil, mol/Pa·h
D_{Ti}	total transport D value in bulk phase i, mol/Pa·h
DOC	dissolved organic carbon
DOM	dissolved organic matter
DSC	Differential Scanning Calorimetry
DTA	Differential Thermal Analyzer
E	emission rate, mol/h or kg/h
EPICS	Equilibrium Partitioning In Closed System
F	Fugacity ratio
f	fugacity, Pa
f_i	fugacity in pure phase i, Pa
f-const.	fragmental constants

fluo.	fluorescence method
G	advective inflow, m^3/h
G_B	advective inflow to bottom sediment m^3/h
ΔG_v	Gibbs's free energy of vaporization kJ/mol or kcal/mol
GC	gas chromatography
GC/FID	GC analysis with flame ionization detector
GC/ECD	GC analysis with electron capture detector
GC-RT	GC retention time
gen. col.	generator-column
H, HLC	Henry's law constant, $Pa \cdot m^3/mol$
ΔH_{fus}	enthalpy of fusion, kcal/mol
ΔH_v	enthalpy of vaporization, kJ/mol or kcal/mol
HPLC	high pressure liquid chromatography
HPLC/UV	HPLC analysis with UV detector
HPLC/fluo.	HPLC analysis with fluorescence detector
HPLC-k'	HPLC-capacity factor correlation
HPLC-RI	HPLC-retention index correlation
HPLC-RT	HPLC-retention time correlation
HPLC-RV	HPLC-retention volume correlation
IP	ionization potential
J	intermediate quantities for fugacity calculation
K	Kjeldahl method
k	first-order rate constant, h^{-1} ($hour^{-1}$)
k_i	first-order rate constant in phase i, h^{-1}
k_A	air-water mass transfer coefficient, air-side, m/h
k_W	air-water mass transfer coefficient, water-side, m/h
$K_{AR/W}$	aerosol-water partition coefficient
K_{AW}	dimensionless air-water partition coefficient
K_B	bioconcentration factor
K_h	association coefficient
K_{OC}	organic-carbon sorption partition coefficient
K_{OM}	organic-matter sorption partition coefficient
K_{OW}	octanol-water partition coefficient
$K_{SD/W}$	sediment-water partition coefficient
$K_{SSD/W}$	suspended sediment-water partition coefficient
K_{SW}	soil-water partition coefficient
K_p	sorption coefficient
k_1	uptake/accumulation rate constant, d^{-1} (day^{-1})
k_2	elimination/clearance/depuration rate constant, d^{-1}
k_b	biodegradation rate constant, d^{-1}
k_h	hydrolysis rate constant, d^{-1}
k_p	photolysis rate constant, d^{-1}

L	lipid content of fish
LSC	Liquid Scintillation Counting
LSS	Liquid Scintillation Spectrometry
m_i	amount of chemical in phase i, mol or kg
M	total amount of chemical, mol or kg
MCI	molecular connectivity indices
MO	molecular orbital calculation
M.P.	melting point, °C
MR	molar refraction
MS	Mass Spectrometry
MW	molecular weight, g/mol
n_C	number of carbon atoms
n_{Cl}	number of chlorine atoms
P	vapor pressure, Pa (Pascal)
P_L	liquid or supercooled liquid vapor pressure, Pa
P_S	solid vapor pressure, Pa
Q	scavenging ratio
QSAR	Quantitative Structure-Activity Relationship
QSPR	Quantitative Structure-Property Relationship
RC	Radiochemical method
RP-HPLC	Reversed Phase High Pressure Liquid Chromatography
RP-TLC	Reversed Phase Thin Layer Chromatography
S	water solubility, mg/L or g/m^3
ΔS_{fus}	entropy of fusion, J/mol·K or cal/mol·K (e.u.)
$S_{octanol}$	solubility in octanol
SD	standard deviation
SPARC	a computational expert system that predicts chemical reactivity
t	residence time, h (hour)
t_o	overall residence time, h
t_A	advection persistence time, h
t_B	sediment burial residence time, h
t_R	reaction persistence time, h
$t_{1/2}$	half-life, h
T_{ij}	intermedia transport rate, mol/h or kg/h
T	system temperature, K
T_B	boiling point, K
T_M	melting point, K
TLC	thin-layer chromatography
TMV	total molecular volume per molecule, Å^3 (Angstrom^3)
TN	titration method
TSA	total surface area per molecule, Å^2
U_1	air side, air-water MTC (same as k_A), m/h

U_2	water side, air-water MTC (same as k_W), m/h	
U_3	rain rate (same as U_R), m/h	
U_4	aerosol deposition rate, m/h	
U_5	soil-air phase diffusion MTC, m/h	
U_6	soil-water phase diffusion MTC, m/h	
U_7	soil-air boundary layer MTC, m/h	
U_8	sediment-water MTC, m/h	
U_9	sediment deposition rate, m/h	
U_{10}	sediment resuspension rate, m/h	
U_{11}	soil-water run-off rate, m/h	
U_{12}	soil-solids run-off rate, m/h	
U_R	rain rate, m/h	
U_Q	dry deposition velocity, m/h	
U_B	sediment burial rate, m/h	
UV	UV spectrometry	
UNIFAC	UNIQUAC Functional Group Activity Coefficients	
V_i	volume of pure phase i, m³	
V_S	volume of bottom sediment, m³	
V_{Bi}	volume of bulk phase i, m³	
V_I	intrinsic molar volume, cm³/mol	
V_M	molar volume, cm³/mol	
v_i	volume fraction of phase i	
v_Q	volume fraction of aerosol	
VOC	volatile organic chemicals	
W	molecular mass, g/mol	
Z_i	fugacity capacity of phase i, mol/m³ Pa	
Z_{Bi}	fugacity capacity of bulk phase i, mol/m³ Pa	

Greek characters:

π-const.	substituent constants
γ	solute activity coefficient
γ_o	solute activity coefficient in octanol phase
γ_W	solute activity coefficient in water phase
ρ_i	density of pure phase i, kg/m³
ρ_{Bi}	density of bulk phase i, kg/m³
χ	molecular connectivity indices
ϕ_{OC}	organic carbon fraction
ϕ_i	organic carbon fraction in phase i

A.1 BASIC COMPUTER PROGRAM FOR FUGACITY CALCULATIONS

```
10 REM Fugacity Level I,II and III program, 6 compartments,(GENDISS)
20 REM Select condensed print
30 WIDTH "lpt1:",250
40 LPRINT CHR$(15)
50 DIM N$(9),V(9),Z(9),C(9),F(9),M(9),P(9),CG(9),CU(9),DEN(9),ORG(9),VZ(9),DR(9),
DA(9),CB(9),A(9),PA(9),PR(9),RK(9),GA(9)
60 DIM NR(9),NA(9),I(9),GD(9,9),D(9,9),N(9,9),GRA(9),TD(9,9),HL(9),U(20),UY(20),
U$(20)
70 REM N$   = six phases : air, water, soil, sediment, susp sedt and fish
80 REM V    = volume of the six phases (m3)
90 REM DEN  = density of the six phases (kg/m3)
100 REM HT  = depth of air, water, soil and sediment (m)
110 REM AR  = area of air, water, soil and sediment (m2)
120 REM ORG = the fraction of organic carbons in sediment and susp sedt
130 REM Z   = Z values for each phase (mol/m3.Pa)
140 REM VZ  = VZ values for each phase (mol/Pa)
150 REM F   = fugacity values for each phase (Pa)
160 REM C   = concentration of chemical in each phase (mol/m3)
170 REM CG  = concentration of chemical in each phase (g/m3)
180 REM CU  = concentration of chemical in each phase (ug/g)
190 REM M   = the total amount of chemical in each phase (mol)
200 REM MK  = the total amount of chemical in each phase (kg)
210 REM P   = the mole percent of chemical in each phase (%)
220 N$(1) = "Air        "
230 N$(2) = "Water      "
240 N$(3) = "Soil       "
250 N$(4) = "Sediment   "
260 N$(5) = "Susp sedt"
270 N$(6) = "Fish       "
280 ART = 100000!*1000000!:FAR(2) = .1:FAR(3) = 1-FAR(2)
290 AR(2) = FAR(2)*ART:AR(3) = FAR(3)*ART:AR(1) = ART:AR(4) = AR(2)'areas m2
300 HT(1) = 1000:HT(2) = 20:HT(3) = .1:HT(4) = .01
310 V(1) = AR(1)*HT(1):V(2) = AR(2)*HT(2):V(3) = AR(3)*HT(3):V(4) = AR(4)*HT(4)
320 V(5) = .000005*V(2):V(6) = .000001*V(2)
330 REM input properties
340 PRINT "Select chemical, chemical to be user-specified =1, benzene = 2, HCB = 3, Hexa-CB = 4,
BAP = 5,TriCB = 6, TCE = 7, p-cresol = 8, test = 9, to exit type 10 "
350 INPUT QC
360 ON QC GOTO 370,510,520,530,540,550,560,570,580,4360
370 INPUT "Name of chemical ",CHEM$
380 INPUT "Temperature eg 25 deg C";TC
390 INPUT "Melting point temperature or data temperature if chemical is liquid eg 80 deg C";TM
400 INPUT "Molecular mass eg 200 g/mol";WM
401 PRINT "For a non-dissociating chemical input a pKa of zero"
402 INPUT "Dissociation constant pKa eg 5.0";PKA
403 IF PKA = 0 GOTO 410
404 INPUT "pH of data determination eg 7.0";PHD
406 INPUT "pH of environment eg 6.0";PHE
410 INPUT "Vapor pressure eg 2 Pa ";P
420 INPUT "Water solubility eg 50 g/m3 ";SG
430 INPUT "Log octanol water coefficient eg 4.0 ";LKOW
440 PRINT "Input overall reaction rate half-lives eg 100 h "
450 PRINT "For zero reaction rate enter a fictitiously long half life eg 1E11 h"
460 INPUT "Half life in air       ";HL(1)
470 INPUT "Half life in water     ";HL(2)
480 INPUT "Half life in soil      ";HL(3)
490 INPUT "Half life in sediment  ";HL(4)
500 GOTO 590
```

```
510
CHEM$ = "Benzene":TC = 25!:WM = 78.11:P = 12700:SG = 1780:LKOW = 2.13:TM = 5.53:HL(1) = 17:HL(2
) = 170:HL(3) = 550:HL(4) = 1700:GOTO 590
520
CHEM$ = "Hexachlorobenzene(HCB)":TC = 25!:WM = 284.8:P = .0023:SG = .005:LKOW = 5.5:TM = 230!:H
L(1) = 17000:
HL(2) = 55000!:HL(3) = 55000!:HL(4) = 55000!:GOTO 590
530 CHEM$ = "Hexachlorobiphenyl
Hexa-CB)":TC = 25!:WM = 350:P = .0005:SG = .0035:LKOW = 6.8:TM = 106.85:
HL(1) = 5500:HL(2) = 55000!:HL(3) = 55000!:HL(4) = 55000!:GOTO 590
540
CHEM$ = "Benzo[a]pyrene(BAP)":TC = 25!:WM = 252.3:P = 7.3E-07:SG = .0038:LKOW = 5.98:TM = 175!:
HL(1) = 5:HL(2) = 5:HL(3) = 5500:HL(4) = 17000:GOTO 590
550 CHEM$ = "1,2,3-
Trichlorobenzene":TC = 25!:WM = 181.45:P = 28!:SG = 21!:LKOW = 4.1:TM = 53!:HL(1) = 550:
HL(2) = 1700:HL(3) = 5500:HL(4) = 17000!:GOTO 590
560
CHEM$ = "Trichloroethylene(TCE)":TC = 25!:WM = 131.5:P = 9870:SG = 1100:LKOW = 2.29:TM = -73!:HL(
1) = 170:HL(2) = 5500:HL(3) = 5500:HL(4) = 17000:GOTO 590
570 CHEM$ = "p-Cresol":TC = 25:WM = 108.14:P = 14.4:SG = 1.8:LKOW = 1.92:TM = 34.8:HL(1) = 5:
HL(2) = 17:HL(3) = 55:HL(4) = 170:GOTO 590
580 CHEM$ = "Test
phenolic":TC = 27.5:WM = 100!:P = .00001:SG = .001:LKOW = 6!:TM = 127.5:HL(1) = 100:
HL(2) = 1000:HL(3) = 5000:HL(4) = 50000!:PKA = 5:PHD = 6:PHE = 7:GOTO 590
590 MTK = 100000!
600 MT = MTK*1000/WM
610 REM    Input for Fugacity Level II program
620 EK = 1000
630 E = EK*1000/WM
635 PRINT "The emission rates for Level III are 1000 kg/h to each of air, water, and soil in turn "
640 PRINT "Input desired emission rates of chemical for fourth Level III calculation kg/h"
650 INPUT "Emission into air      ";IK4(1)
660 INPUT "Emission into water    ";IK4(2)
670 INPUT "Emission into soil     ";IK4(3)
680 GRA(1) = 100
690 GRA(2) = 1000
700 GRA(4) = 50000!
710 S = SG/WM  'solubility mol/m3
720 H = P/S    'Henry's law constant Pa.m3/mol
730 KOW = 10^LKOW  'Octanol-water partition coefficient
740 KOC = .41*KOW  'Organic carbon-water partition coefficient
750 KFW = .05*KOW  'Fish-water bioconcentration factor
760 TK = TC + 273.15 'Temperature K
770 RG = 8.314     'Gas constant
780 IF TM > TC GOTO 790 ELSE GOTO 810
790 FR = EXP(6.79*(1-(TM + 273.15)/TK))
800 GOTO 820
810 FR = 1
820 PL = P/FR
830 ORG(3) = .02:ORG(4) = .04:ORG(5) = .2'Organic carbon contents g/g
840 DEN(1) = .029*101325!/RG/TK:DEN(2) = 1000:DEN(3) = 2400 'Densities kg/m3
850 DEN(4) = 2400:DEN(5) = 1500:DEN(6) = 1000:DEN(7) = 2000 'Densities kg/m3
851 IF PKA = 0 GOTO 860
852 ID = 10^(PHD-PKA) 'Ratio ionic to non-ionic species at data pH
854 IE = 10^(PHE-PKA) 'Ratio ionic to non ionic species at environmental pH
860 REM calculate Z values, E and D refer to data and environment
861 REM N, I, and T refer to non-ionic , ionic and total
870 Z(1) = 1/RG/TK
880 ZTD(2) = 1/H: ZID(2) = ZTD(2)*ID/(1 + ID):ZND(2) = ZTD(2)/(1 + ID)
882 ZNE(2) = ZND(2):ZIE(2) = ZNE(2)*IE:ZTE(2) = ZNE(2) + ZIE(2)
```

```
885 Z(2) = ZTE(2)
890 Z(3) = ZTD(2)*DEN(3)*ORG(3)*KOC/1000
900 Z(4) = ZTD(2)*DEN(4)*ORG(4)*KOC/1000
910 Z(5) = ZTD(2)*DEN(5)*ORG(5)*KOC/1000
920 Z(6) = ZTD(2)*DEN(6)*KFW/1000
930 K71 = 6000000!/PL
940 Z(7) = Z(1)*K71
950 K12 = Z(1)/Z(2) 'Partition coefficients
960 K32 = Z(3)/Z(2)
970 K42 = Z(4)/Z(2)
980 K52 = Z(5)/Z(2)
990 K62 = Z(6)/Z(2)
1000 REM calculate distribution
1010 VZT = 0
1020 FOR N = 1 TO 6
1030 VZ(N) = V(N)*Z(N)
1040 VZT = VZT + VZ(N)
1050 NEXT N
1060 F1 = MT/VZT 'fugacity
1070 FOR N = 1 TO 6
1080 F(N) = F1
1090 C(N) = F(N)*Z(N) 'concentration mol/m3
1100 M(N) = C(N)*V(N) 'amount mol
1110 MK(N) = M(N)*WM/1000
1120 P(N) = 100*M(N)/MT 'percentages
1130 CG(N) = C(N)*WM 'concentration g/m3
1140 CU(N) = CG(N)*1000/DEN(N)'concentration ug/g
1150 NEXT N
1160 REM print out results
1170 LPRINT " PROGRAM 'GENDISS':SIX COMPARTMENT FUGACITY LEVEL I CALCULATION "
1180 LPRINT " "
1190 LPRINT "Properties of "CHEM$
1200 LPRINT " "
1210 LPRINT "Temperature deg C          ";TC
1220 LPRINT "Molecular mass g/mol       ";WM
1230 LPRINT "Melting point deg C        ";TM
1240 LPRINT "Fugacity ratio             ";FR
1241 IF PKA = 0 GOTO 1250
1242 LPRINT "Dissociation constant pKa     ";PKA
1244 LPRINT "pH of data determination      ";PHD
1246 LPRINT "pH of environment             ";PHE
1247 LPRINT "Ratio ionic to non-ionic (data) ";ID
1248 LPRINT "Ratio ionic to non-ionic (envir) ";IE
1250 LPRINT "Vapor pressure Pa          ";P
1260 LPRINT "Sub-cooled liquid vapor press Pa ";PL
1270 LPRINT "Solubility g/m3            ";SG
1280 LPRINT "Solubility mol/m3          ";S
1290 LPRINT "Henry's law constant Pa.m3/mol  ";H
1300 LPRINT "Log octanol-water p-coefficient ";LKOW
1310 LPRINT "Octanol-water partn-coefficient ";KOW
1320 LPRINT "Organic C-water ptn-coefficient ";KOC
1330 LPRINT "Fish-water partition coefficient ";KFW
1340 LPRINT "Air-water partition coefficient ";K12
1350 LPRINT "Soil-water partition coefficient ";K32
1360 LPRINT "Sedt-water partition coefficient ";K42
1370 LPRINT "Susp sedt-water partn coeffnt   ";K52
1380 LPRINT "Aerosol-air partition coeff     ";K71
1390 LPRINT "Aerosol Z value            ";Z(7)
1400 LPRINT "Aerosol density kg/m3         ";DEN(7)
1401 LPRINT " "
```

```
1402 IF PKA=0 GOTO 1410
1403 LPRINT " Z values in water "
1404 LPRINT "        At data pH              At environmental pH"
1405 LPRINT "    Neutral   Ionic    Total      Neutral    Ionic    Total"
1406 LPRINT USING " ##.###^^^^  ##.###^^^^  ##.###^^^^  ##.###^^^^
";ZND(2),ZID(2),ZTD(2),ZNE(2),ZIE(2),ZTE(2)
1408 LPRINT " "
1410 LPRINT " "
1420 LPRINT "Amount of chemical moles      ";MT
1430 LPRINT "Amount of chemical kilograms  ";MTK
1440 LPRINT "Fugacity Pa                   ";F1
1450 LPRINT "Total of VZ products          ";VZT
1460 LPRINT " "
1470 LPRINT "Phase properties and compositions"
1480 LPRINT " "
1490 LPRINT "Phase      "TAB(15) N$(1) TAB(30) N$(2) TAB(45) N$(3) TAB(60) N$(4) TAB(75) N$(5) TAB(90) N$(6)
1500 LPRINT "Volume m3  "TAB(15) V(1) TAB(30) V(2) TAB(45) V(3) TAB(60) V(4) TAB(75) V(5) TAB(90) V(6)
1510 LPRINT "Density kg/m3"TAB(15) DEN(1) TAB(30) DEN(2) TAB(45) DEN(3) TAB(60) DEN(4) TAB(75) DEN(5) TAB(90) DEN(6)
1520 LPRINT "Depth m    "TAB(15) HT(1) TAB(30) HT(2) TAB(45) HT(3) TAB(60) HT(4)
1530 LPRINT "Area  m2   "TAB(15) AR(1) TAB(30) AR(2) TAB(45) AR(3) TAB(60) AR(4)
1540 LPRINT "Frn org carb " TAB(45) ORG(3) TAB(60) ORG(4) TAB(75) ORG(5)
1550 LPRINT "Z mol/m3.Pa "TAB(15) Z(1) TAB(30) Z(2) TAB(45) Z(3) TAB(60) Z(4) TAB(75) Z(5) TAB(90) Z(6)
1560 LPRINT "VZ mol/Pa   "TAB(15) VZ(1) TAB(30) VZ(2) TAB(45) VZ(3) TAB(60) VZ(4 ) TAB(75) VZ(5) TAB(90) VZ(6)
1570 LPRINT "Fugacity Pa "TAB(15) F(1) TAB(30) F(2) TAB(45) F(3) TAB(60) F(4) TAB(75) F(5) TAB(90) F(6)
1580 LPRINT "Conc mol/m3 "TAB(15) C(1) TAB(30) C(2) TAB(45) C(3) TAB(60) C(4) TAB(75) C(5) TAB(90) C(6)
1590 LPRINT "Conc g/m3   "TAB(15) CG(1) TAB(30) CG(2) TAB(45) CG(3) TAB(60) CG(4 ) TAB(75) CG(5) TAB(90) CG(6)
1600 LPRINT "Conc ug/g   "TAB(15) CU(1) TAB(30) CU(2) TAB(45) CU(3) TAB(60) CU(4) TAB(75) CU(5) TAB(90) CU(6)
1610 LPRINT "Amount mol  "TAB(15) M(1) TAB(30) M(2) TAB(45) M(3) TAB(60) M(4) TAB(75) M(5) TAB(90) M(6)
1620 LPRINT "Amount kg   "TAB(15) MK(1) TAB(30) MK(2) TAB(45) MK(3) TAB(60) MK(4) TAB(75) MK(5) TAB(90) MK(6)
1630 LPRINT "Amount %    "TAB(15) P(1) TAB(30) P(2) TAB(45) P(3) TAB(60) P(4) TAB(75) P(5) TAB(90) P(6)
1640 LPRINT CHR$(12)
1650 REM Fugacity Level II program, 6 compartments
1660 REM calculate total inflows
1670 GA(1)=V(1)/GRA(1)
1680 GA(2)=V(2)/GRA(2)
1690 GA(4)=V(4)/GRA(4)
1700 REM calculate D values
1710 NRT=0:NRTK=0:NAT=0:NATK=0:VZT=0:MT=0:DT=0:DTA=0:DTR=0 'set totals to zero
1720 FOR N= 1 TO 4
1730 RK(N)=.693/HL(N) 'rate constants from half lives
1740 DR(N)=V(N)*Z(N)*RK(N):DA(N)=GA(N)*Z(N) 'reaction and advection D values
1750 DTR=DTR+DR(N):DTA=DTA+DA(N) 'total D values
1760 NEXT N
1770 DT=DTR+DTA 'total D value
1780 F2=E/DT 'fugacity
1790 FOR N=1 TO 6
1800 F(N)=F2
1810 C(N)=F(N)*Z(N) 'concentration mol/m3
```

```
1820 M(N) = C(N)*V(N) 'amount mol
1830 MK(N) = M(N)*WM/1000
1840 MT = MT + M(N) 'total amount
1850 CG(N) = C(N)*WM 'concentration g/m3
1860 CU(N) = CG(N)*1000/DEN(N) 'concentration ug/g
1870 NR(N) = V(N)*C(N)*RK(N):NRK(N) = NR(N)*WM/1000 'reaction rates mol/h and kg/h
1880 NA(N) = GA(N)*C(N):NAK(N) = NA(N)*WM/1000 'advection rates mol/h and kg/h
1890 NRT = NRT + NR(N):NAT = NAT + NA(N) 'total rates mol/h
1900 NRTK = NRTK + NRK(N):NATK = NATK + NAK(N) 'total rates kg/h
1910 NEXT N
1920 NT = NRT + NAT:NTK = NRTK + NATK
1930 MTK = MT*WM/1000
1940 FOR N = 1 TO 6
1950 P(N) = 100*M(N)/MT 'percentages of amount
1960 PR(N) = 100*NR(N)/NT 'percentages of reaction rate
1970 PA(N) = 100*NA(N)/NT 'percentages of advection rate
1980 NEXT N
1990 IF NRT = 0 THEN TR = 0 ELSE TR = MT/NRT
2000 IF NAT = 0 THEN TA = 0 ELSE TA = MT/NAT
2010 TOV = MT/NT 'overall residence time h
2020 REM print out results
2030 LPRINT
2040 LPRINT "SIX COMPARTMENT FUGACITY LEVEL II CALCULATION ";CHEM$
2050 LPRINT " "
2060 LPRINT "Emission rate of chemical mol/h ";E
2070 LPRINT "Emission rate of chemical  kg/h ";EK
2080 LPRINT "Fugacity Pa                     ";F2
2090 LPRINT "Total amount of chemical mol    ";MT
2100 LPRINT "Total amount of chemical kg     ";MTK
2110 LPRINT " "
2120 LPRINT "Phase properties,compositions and rates"
2130 LPRINT " "
2140 LPRINT "Phase        "TAB(15) N$(1) TAB(30) N$(2) TAB(45) N$(3) TAB(60) N$(4) TAB(75) N$(5) TAB(90) N$(6)
2150 LPRINT "Adv.flow m3/h"TAB(15) GA(1) TAB(30) GA(2) TAB(45) GA(3) TAB(60) GA(4)
2160 LPRINT "Adv.restime h"TAB(15) GRA(1) TAB(30) GRA(2) TAB(45) GRA(3) TAB(60) GRA(4)
2170 LPRINT "Rct halflife h"TAB(15) HL(1) TAB(30) HL(2) TAB(45) HL(3) TAB(60) HL(4)
2180 LPRINT "Rct rate c.h-1"TAB(15) RK(1) TAB(30) RK(2) TAB(45) RK(3) TAB(60) RK(4)
2190 LPRINT "Fugacity Pa "TAB(15) F(1) TAB(30) F(2) TAB(45) F(3) TAB(60) F(4) TAB(75) F(5) TAB(90) F(6)
2200 LPRINT "Conc mol/m3 "TAB(15) C(1) TAB(30) C(2) TAB(45) C(3) TAB(60) C(4) TAB(75) C(5) TAB(90) C(6)
2210 LPRINT "Conc g/m3   "TAB(15) CG(1) TAB(30) CG(2) TAB(45) CG(3) TAB(60) CG(4 ) TAB(75) CG(5) TAB(90) CG(6)
2220 LPRINT "Conc ug/g   "TAB(15) CU(1) TAB(30) CU(2) TAB(45) CU(3) TAB(60) CU(4) TAB(75) CU(5) TAB(90) CU(6)
2230 LPRINT "Amount mol  "TAB(15) M(1) TAB(30) M(2) TAB(45) M(3) TAB(60) M(4) TAB(75) M(5) TAB(90) M(6)
2240 LPRINT "Amount kg   "TAB(15) MK(1) TAB(30) MK(2) TAB(45) MK(3) TAB(60) MK(4) TAB(75) MK(5) TAB(90) MK(6)
2250 LPRINT "Amount %    "TAB(15) P(1) TAB(30) P(2) TAB(45) P(3) TAB(60) P(4) TAB(75) P(5) TAB(90) P(6)
2260 LPRINT "D rct mol/Pa.h"TAB(15) DR(1) TAB(30) DR(2) TAB(45) DR(3) TAB(60) DR(4)
2270 LPRINT "D adv mol/Pa.h"TAB(15) DA(1) TAB(30) DA(2) TAB(45) DA(3) TAB(60) DA(4)
2280 LPRINT "Rct rate mol/h"TAB(15) NR(1) TAB(30) NR(2) TAB(45) NR(3) TAB(60) NR(4)
2290 LPRINT "Adv rate mol/h"TAB(15) NA(1) TAB(30) NA(2) TAB(45) NA(3) TAB(60) NA(4)
2300 LPRINT "Rct rate kg/h "TAB(15) NRK(1) TAB(30) NRK(2) TAB(45) NRK(3) TAB(60) NRK(4)
2310 LPRINT "Adv rate kg/h "TAB(15) NAK(1) TAB(30) NAK(2) TAB(45) NAK(3) TAB(60) NAK(4)
2320 LPRINT "Reaction %  "TAB(15) PR(1) TAB(30) PR(2) TAB(45) PR(3) TAB(60) PR(4)
2330 LPRINT "Advection % "TAB(15) PA(1) TAB(30) PA(2) TAB(45) PA(3) TAB(60) PA(4)
```

```
2340 LPRINT " "
2350 LPRINT "Total advection D value     ";DTA
2360 LPRINT "Total reaction D value      ";DTR
2370 LPRINT "Total D value               ";DT
2380 LPRINT " "
2390 LPRINT "Output by reaction    mol/h ";NRT
2400 LPRINT "Output by advection   mol/h ";NAT
2410 LPRINT "Total output by reaction and advection mol/h ";NT
2420 LPRINT" "
2430 LPRINT "Output by reaction    kg/h ";NRTK
2440 LPRINT "Output by advection   kg/h ";NATK
2450 LPRINT "Total output by reaction and advection kg/h ";NTK
2460 LPRINT" "
2470 LPRINT "Overall residence time  h ";TOV
2480 LPRINT "Reaction residence time h ";TR
2490 LPRINT "Advection residence time h ";TA
2500 LPRINT CHR$(12)
2510 LPRINT
2520 REM Fugacity Level III Program
2530 REM Set bulk phase volumes, densities and Z values
2540 VB(1)=V(1):VB(2)=V(2):VB(3)=1.8E+10:VB(4)=5E+08
2550 VA(1)=1:VQ(1)=2E-11 'volume fractions
2560 VW(2)=1:VP(2)=.000005:VF(2)=.000001
2570 VA(3)=.2:VW(3)=.3:VE(3)=.5
2580 VW(4)=.8:VS(4)=.2
2590 DENB(1)=VA(1)*DEN(1)+VQ(1)*DEN(7)
2600 DENB(2)=VW(2)*DEN(2)+VP(2)*DEN(5)+VF(2)*DEN(6)
2610 DENB(3)=VW(3)*DEN(2)+VA(3)*DEN(1)+VE(3)*DEN(3)
2620 DENB(4)=VW(4)*DEN(2)+VS(4)*DEN(4)
2630 ZB(1)=VA(1)*Z(1)+VQ(1)*Z(7)
2640 ZB(2)=VW(2)*Z(2)+VP(2)*Z(5)+VF(2)*Z(6)
2650 ZB(3)=VA(3)*Z(1)+VW(3)*Z(2)+VE(3)*Z(3)
2660 ZB(4)=VW(4)*Z(2)+VS(4)*Z(4)
2670 KB12=ZB(1)/ZB(2)
2680 KB32=ZB(3)/ZB(2)
2690 KB42=ZB(4)/ZB(2)
2700 REM Parameters
2710 U(1)=5        :U$(1)="air side air-water MTC         "
2720 U(2)=.05      :U$(2)="water side air-water MTC       "
2730 U(3)=.0001    :U$(3)="rain rate                      "
2740 U(4)=6E-10    :U$(4)="aerosol deposition velocity    "
2750 U(5)=.02      :U$(5)="soil air phase diffusion MTC   "
2760 U(6)=.00001   :U$(6)="soil water phase diffusion MTC "
2770 U(7)=5        :U$(7)="soil air boundary layer MTC    "
2780 U(8)=.0001    :U$(8)="sediment-water diffusion MTC   "
2790 U(9)=.0000005 :U$(9)="sediment deposition velocity   "
2800 U(10)=.0000002:U$(10)="sediment resuspension velocity "
2810 U(11)=.00005  :U$(11)="soil water runoff rate         "
2820 U(12)=1E-08   :U$(12)="soil solids runoff rate        "
2830 'Calculate D values
2840 DRW=AR(2)*U(3)*Z(2)
2850 DQW=AR(2)*U(4)*Z(7)
2860 DVWA=AR(2)*U(1)*Z(1)
2870 DVWW=AR(2)*U(2)*Z(2)
2880 DVW=1/(1/DVWA+1/DVWW)
2890 D(2,1)=DVW
2900 D(1,2)=DVW+DQW+DRW
2910 DVSB=AR(3)*U(7)*Z(1)
2920 DVSA=AR(3)*U(5)*Z(1)
2930 DVSW=AR(3)*U(6)*Z(2)
```

```
2940 DRS = AR(3)*U(3)*Z(2)
2950 DQS = AR(3)*U(4)*Z(7)
2960 DVS = 1/(1/DVSB + 1/(DVSW + DVSA))
2970 D(3,1) = DVS
2980 D(1,3) = DVS + DRS + DQS
2990 DSWD = AR(2)*U(8)*Z(2)
3000 DSD = AR(2)*U(9)*Z(5)
3010 DSR = AR(4)*U(10)*Z(4)
3020 D(2,4) = DSWD + DSD
3030 D(4,2) = DSWD + DSR
3040 DSWW = AR(3)*U(11)*Z(2)
3050 DSWS = AR(3)*U(12)*Z(3)
3060 D(3,2) = DSWW + DSWS
3070 D(2,3) = 0
3075 IK(1) = 1000:IK(2) = 0:IK(3) = 0:QQ = 1   'first Level III emissions
3080 REM calculate total chemical inflows
3090 IN = 0:INK = 0
3100 FOR N = 1 TO 4
3110 I(N) = IK(N)*1000/WM
3120 IN = IN + I(N):INK = INK + IK(N)
3130 NEXT N
3140 REM calculate reaction and advection D values for bulk phases
3150 GAB(1) = VB(1)/GRA(1)
3160 GAB(2) = VB(2)/GRA(2)
3170 GAB(4) = VB(4)/GRA(4)
3180 VZBT = 0
3190 FOR N = 1 TO 4
3200 RK(N) = .693/HL(N)
3210 DR(N) = VB(N)*ZB(N)*RK(N):DA(N) = GAB(N)*ZB(N)
3220 VZB(N) = VB(N)*ZB(N)
3230 VZBT = VZBT + VZB(N)
3240 NEXT N
3250 FOR N = 1 TO 4
3260 FOR NN = 1 TO 4
3270 GD(N,NN) = D(N,NN)/ZB(N)
3280 IF GD(N,NN) = 0 GOTO 3300 ELSE GOTO 3290
3290 TD(N,NN) = .693*VB(N)/GD(N,NN)
3300 NEXT NN
3310 NEXT N
3320 DT(1) = DR(1) + DA(1) + D(1,2) + D(1,3)
3330 DT(2) = DR(2) + DA(2) + D(2,1) + D(2,3) + D(2,4)
3340 DT(3) = DR(3) + DA(3) + D(3,1) + D(3,2)
3350 DT(4) = DR(4) + DA(4) + D(4,2)
3360 J1 = I(1)/DT(1) + I(3)*D(3,1)/DT(3)/DT(1)
3370 J2 = D(2,1)/DT(1)
3380 J3 = 1-D(3,1)*D(1,3)/DT(1)/DT(3)
3390 J4 = D(1,2) + D(3,2)*D(1,3)/DT(3)
3400
F(2) = (I(2) + J1*J4/J3 + I(3)*D(3,2)/DT(3) + I(4)*D(4,2)/DT(4))/(DT(2)-J2*J4/J3-D(2,4)*D(4,2)/DT(4))
3410 F(1) = (J1 + F(2)*J2)/J3
3420 F(3) = (I(3) + F(1)*D(1,3))/DT(3)
3430 F(4) =  (I(4) + F(2)*D(2,4))/DT(4)
3440 NRT = 0:NAT = 0:MT = 0
3450 FOR N = 1 TO 4
3460 C(N) = F(N)*ZB(N)
3470 M(N) = C(N)*VB(N)
3480 MK(N) = M(N)*WM/1000
3490 MT = MT + M(N)
3500 CG(N) = C(N)*WM
3510 CU(N) = CG(N)*1000/DENB(N)
```

```
3520 NR(N)=F(N)*DR(N):NRK(N)=NR(N)*WM/1000
3530 NA(N)=F(N)*DA(N):NAK(N)=NA(N)*WM/1000
3540 NRT=NRT+NR(N):NAT=NAT+NA(N)
3550 NEXT N
3555 F(6)=F(2):C(6)=F(6)*Z(6):CG(6)=C(6)*WM:CU(6)=CG(6)*1000/DEN(6)
3560 MTK=MT*WM/1000
3570 NRTK=NRT*WM/1000:NATK=NAT*WM/1000
3580 NT=NRT+NAT:NTK=NT*WM/1000
3590 FOR N=1 TO 4
3600 P(N)=100*M(N)/MT
3610 PR(N)=100*NR(N)/NT
3620 PA(N)=100*NA(N)/NT
3630 NEXT N
3640 N(1,2)=D(1,2)*F(1):NK(1,2)=N(1,2)*WM/1000
3650 N(1,3)=D(1,3)*F(1):NK(1,3)=N(1,3)*WM/1000
3660 N(2,1)=D(2,1)*F(2):NK(2,1)=N(2,1)*WM/1000
3670 N(2,4)=D(2,4)*F(2):NK(2,4)=N(2,4)*WM/1000
3680 N(3,1)=D(3,1)*F(3):NK(3,1)=N(3,1)*WM/1000
3690 N(3,2)=D(3,2)*F(3):NK(3,2)=N(3,2)*WM/1000
3700 N(4,2)=D(4,2)*F(4):NK(4,2)=N(4,2)*WM/1000
3710 TR=MT/(NRT+.0000001)
3720 TA=MT/(NAT+.0000001)
3730 TOV=MT/NT
3740 TOVD=TOV/24
3750 REM print out results
3760 LPRINT " FOUR COMPARTMENT FUGACITY LEVEL III CALCULATION",CHEM$
3770 LPRINT
3780 LPRINT "Bulk phase properties,compositions and rates"
3790 LPRINT " "
3800 LPRINT "Phase       "TAB(15) N$(1) TAB(30) N$(2) TAB(45) N$(3) TAB(60) N$(4) TAB(75) "Total" TAB(90) N$(6)
3810 LPRINT "Bulk vol m3 "TAB(15) VB(1) TAB(30) VB(2) TAB(45) VB(3) TAB(60) VB(4)
3820 LPRINT "Density kg/m3"TAB(15) DENB(1) TAB(30) DENB(2) TAB(45) DENB(3) TAB(60) DENB(4)
3830 LPRINT "Bulk Z value"TAB(15) ZB(1) TAB(30) ZB(2) TAB(45) ZB(3) TAB(60) ZB(4 ) TAB(90) Z(6)
3840 LPRINT "Bulk VZ     "TAB(15) VZB(1) TAB(30) VZB(2) TAB(45) VZB(3) TAB(60) VZB(4) TAB(75) VZBT
3850 LPRINT "Emission mol/h"TAB(15) I(1) TAB(30) I(2) TAB(45) I(3) TAB(60) I(4) TAB(75) IN
3860 LPRINT "Emission kg/h "TAB(15) IK(1) TAB(30) IK(2) TAB(45) IK(3) TAB(60) IK(4) TAB(75) INK
3870 LPRINT "Fugacity Pa "TAB(15) F(1) TAB(30) F(2) TAB(45) F(3) TAB(60) F(4) TAB(90) F(6)
3880 LPRINT "Conc mol/m3 "TAB(15) C(1) TAB(30) C(2) TAB(45) C(3) TAB(60) C(4) TAB(90) C(6)
3890 LPRINT "Conc g/m3   "TAB(15) CG(1) TAB(30) CG(2) TAB(45) CG(3) TAB(60) CG(4 ) TAB(90) CG(6)
3900 LPRINT "Conc ug/g   "TAB(15) CU(1) TAB(30) CU(2) TAB(45) CU(3) TAB(60) CU(4) TAB(90) CU(6)
3910 LPRINT "Amount mol  "TAB(15) M(1) TAB(30) M(2) TAB(45) M(3) TAB(60) M(4) TAB(75) MT
3920 LPRINT "Amount kg   "TAB(15) MK(1) TAB(30) MK(2) TAB(45) MK(3) TAB(60) MK(4) TAB(75) MTK
3930 LPRINT "Amount %    "TAB(15) P(1) TAB(30) P(2) TAB(45) P(3) TAB(60) P(4)
3940 LPRINT "Adv.flow m3/h"TAB(15) GAB(1) TAB(30) GAB(2) TAB(45) GAB(3) TAB(60) GAB(4)
3950 LPRINT "D rct mol/Pa.h"TAB(15) DR(1) TAB(30) DR(2) TAB(45) DR(3) TAB(60) DR(4 )
3960 LPRINT "D adv mol/Pa.h"TAB(15) DA(1) TAB(30) DA(2) TAB(45) DA(3) TAB(60) DA(4 )
3970 LPRINT "Rct rate mol/h"TAB(15) NR(1) TAB(30) NR(2) TAB(45) NR(3) TAB(60) NR(4 ) TAB(75) NRT
3980 LPRINT "Rct rate kg/h "TAB(15) NRK(1) TAB(30) NRK(2) TAB(45) NRK(3) TAB(60) NRK(4 ) TAB(75) NRTK
3990 LPRINT "Adv rate mol/h"TAB(15) NA(1) TAB(30) NA(2) TAB(45) NA(3) TAB(60) NA(4 ) TAB(75) NAT
4000 LPRINT "Adv rate kg/h "TAB(15) NAK(1) TAB(30) NAK(2) TAB(45) NAK(3) TAB(60) NAK(4 ) TAB(75) NATK
4010 LPRINT "Reaction %   "TAB(15) PR(1) TAB(30) PR(2) TAB(45) PR(3) TAB(60) PR(4 )
```

```
4020 LPRINT "Advection %   "TAB(15) PA(1) TAB(30) PA(2) TAB(45) PA(3) TAB(60) PA(4 )
4030 LPRINT " "
4040 LPRINT "Overall residence time   h ";TOV
4050 LPRINT "Reaction residence time  h ";TR;
4060 LPRINT "   Advection residence time h ";TA
4070 LPRINT
4080 LPRINT "Intermedia Data.  Half times  Equiv flows  D values   Rates of transport "
4090 LPRINT "                   h         m3/h       mol/Pa.h    mol/h      kg/h"
4100 LPRINT "Air to water     ";:LPRINT USING " ##.####^^^^ ";TD(1,2);GD(1,2);D(1,2) ;N(1,2);NK(1,2)
4110 LPRINT "Air to soil      ";:LPRINT USING " ##.####^^^^ ";TD(1,3);GD(1,3);D(1,3) ;N(1,3);NK(1,3)
4120 LPRINT "Water to air     ";:LPRINT USING " ##.####^^^^ ";TD(2,1);GD(2,1);D(2,1) ;N(2,1);NK(2,1)
4130 LPRINT "Water to sediment";:LPRINT USING " ##.####^^^^ ";TD(2,4);GD(2,4);D(2,4)
;N(2,4);NK(2,4)
4140 LPRINT "Soil to air      ";:LPRINT USING " ##.####^^^^ ";TD(3,1);GD(3,1),D(3,1) ,N(3,1);NK(3,1)
4150 LPRINT "Soil to water    ";:LPRINT USING " ##.####^^^^ ";TD(3,2);GD(3,2);D(3,2) ;N(3,2);NK(3,2)
4160 LPRINT "Sediment to water";:LPRINT USING " ##.####^^^^ ";TD(4,2);GD(4,2),D(4,2)
,N(4,2);NK(4,2)
4170 LPRINT "  Transport velocity parameters          m/h         m/year "
4180 FOR I = 1 TO 12
4190 UY(I) = U(I)*8760
4200 LPRINT TAB(5) I TAB(10) U$(I) TAB(45) U(I) TAB(60) UY(I)
4210 NEXT I
4220 LPRINT "Individual process D values "
4230 LPRINT "Air-water diffusion (air-side)   ";DVWA TAB(50);
4240 LPRINT "Air-water diffusion (water-side) ";DVWW
4250 LPRINT "Air-water diffusion (overall)    ";DVW
4260 LPRINT "Rain dissolution to water        ";DRW TAB(50);
4270 LPRINT "Aerosol deposition to water      ";DQW
4280 LPRINT "Rain dissolution to soil         ";DRS TAB(50);
4290 LPRINT "Aerosol deposition to soil       ";DQS
4300 LPRINT "Soil-air diffusion (air-phase)   ";DVSA TAB(50);
4310 LPRINT "Soil-air diffusion (water-phase) ";DVSW
4320 LPRINT "Soil-air diffusion (bndry layer) ";DVSB TAB(50);
4330 LPRINT "Soil-air diffusion (overall)     ";DVS
4340 LPRINT "Water-sediment diffusion         ";DSWD
4350 LPRINT "Water-sediment deposition        ";DSD TAB(50);
4360 LPRINT "Sediment-water resuspension      ";DSR
4370 LPRINT "Soil-water runoff (water)        ";DSWW TAB(50);
4380 LPRINT "Soil-water runoff (solids)       ";DSWS
4385 LPRINT CHR$(12)
4390 ON QQ GOTO 4412,4414,4416,4418
4412 IK(1) =0:IK(2) = 1000:IK(3) =0:QQ = QQ + 1 :GOTO 3080
4414 IK(1) =0:IK(2) = 0:IK(3) = 1000:QQ = QQ + 1 :GOTO 3080
4416 IK(1) = IK4(1):IK(2) = IK4(2):IK(3) = IK4(3):QQ = QQ + 1 :GOTO 3080
4418 PRINT "If you wish to run another Level III calculation input 1 ,otherwise input 2."
4419 INPUT QQQ
4420 ON QQQ GOTO 4422,4500
4422 PRINT "Input emission rates of chemical for Level III calculation kg/h"
4430 INPUT "Emission into air     ";IK(1)
4440 INPUT "Emission into water   ";IK(2)
4450 INPUT "Emission into soil    ";IK(3)
4460 GOTO 3080
4500 END
```

PROGRAM 'GENDISS':SIX COMPARTMENT FUGACITY LEVEL I CALCULATION

Properties of DDT

Temperature deg C	25
Molecular mass g/mol	354.5
Melting point deg C	109
Fugacity ratio	.1476374
Vapor pressure Pa	.00002
Sub-cooled liquid vapor press Pa	1.354671E-04
Solubility g/m3	.0055
Solubility mol/m3	1.551481E-05
Henry's law constant Pa.m3/mol	1.289091
Log octanol-water p-coefficient	6.19
Octanol-water partn-coefficient	1548816
Organic C-water ptn-coefficient	635014.6
Fish-water partition coefficient	77440.81
Air-water partition coefficient	5.200423E-04
Soil-water partition coefficient	30480.7
Sedt-water partition coefficient	60961.4
Susp sedt-water partn coeffnt	190504.4
Aerosol-air partition coeff	4.429121E+10
Aerosol Z value	1.786786E+07
Aerosol density kg/m3	2000

Amount of chemical moles	282087.5
Amount of chemical kilograms	100000
Fugacity Pa	1.294631E-09
Total of VZ products	2.178903E+14

Phase properties and compositions

Phase	Air	Water	Soil	Sediment	Susp sedt	Fish
Volume m3	1E+14	2E+11	8.999999E+09	1E+08	999999.9	200000
Density kg/m3	1.185413	1000	2400	2400	1500	1000
Depth m	1000	20	.1	.01		
Area m2	1E+11	1E+10	9E+10	1E+10		
Frn org carb			.02	.04	.2	
Z mol/m3.Pa	4.034179E-04	.7757405	23645.11	47290.23	147782	60073.98
VZ mol/Pa	4.034179E+10	1.551481E+11	2.12806E+14	4.729023E+12	1.47782E+11	1.20148E+10
Fugacity Pa	1.294631E-09	1.294631E-09	1.294631E-09	1.294631E-09	1.294631E-09	1.294631E-09
Conc mol/m3	5.222771E-13	1.004297E-09	3.061169E-05	6.122337E-05	1.913231E-04	7.77736E-05
Conc g/m3	1.851472E-10	3.560234E-07	1.085184E-02	2.170369E-02	6.782402E-02	2.757074E-02
Conc ug/g	1.561879E-07	3.560234E-07	4.521601E-03	9.043202E-03	4.521602E-02	2.757074E-02
Amount mol	52.22771	200.8595	275505.2	6122.337	191.3231	15.55472
Amount kg	18.51472	71.20468	97666.58	2170.369	67.82401	5.514148
Amount %	1.851473E-02	7.120468E-02	97.66658	2.170369	6.782402E-02	5.514148E-03

SIX COMPARTMENT FUGACITY LEVEL II CALCULATION DDT

Emission rate of chemical mol/h   2820.875
Emission rate of chemical kg/h   1000
Fugacity Pa   2.947097E-07
Total amount of chemical mol   6.421439E+07
Total amount of chemical kg   2.2764E+07

Phase properties, compositions and rates

Phase	Air	Water	Soil	Sediment	Susp sedt	Fish
Adv.flow m3/h	1E+12	2E+08	0	2000		
Adv.restime h	100	1000	0	50000		
Rct halflife h	170	5500	17000	55000		
Rct rate c.h-1	4.076471E-03	.000126	4.076471E-05	.0000126		
Fugacity Pa	2.947097E-07	2.947097E-07	2.947097E-07	2.947097E-07	2.947097E-07	2.947097E-07
Conc mol/m3	1.188912E-10	2.286183E-07	6.968444E-03	1.393689E-02	4.355278E-02	1.770438E-02
Conc g/m3	4.214692E-08	8.104518E-05	2.470314	4.940627	15.43946	6.276204
Conc ug/g	3.555462E-05	8.104518E-05	1.029297	2.058595	10.29297	6.276204
Amount mol	11889.12	45723.65	6.271599E+07	1393689	43552.78	3540.877
Amount kg	4214.692	16209.04	2.223282E+07	494062.7	15439.46	1255.241
Amount %	1.851473E-02	7.120469E-02	97.66658	2.170369	6.782402E-02	5.514149E-03
D rct mol/Pa.h	1.644521E+08	1.954866E+07	8.674974E+09	5.958569E+07		
D adv mol/Pa.h	4.034179E+08	1.551481E+08	0	9.458046E+07		
Rct rate mol/h	48.46564	5.76118	2556.599	17.56048		
Adv rate mol/h	118.8912	45.72365	0	27.87378		
Rct rate kg/h	17.18107	2.042338	906.3143	6.225191		
Adv rate kg/h	42.14692	16.20903	0	9.881254		
Reaction %	1.718107	.2042338	90.63142	.622519		
Advection %	4.214691	1.620903	0	.9881252		

Total advection D value   6.531465E+08
Total reaction D value   8.91856E+09
Total D value   9.571706E+09

Output by reaction   mol/h   2628.386
Output by advection   mol/h   192.4886
Total output by reaction and advection mol/h   2820.875

Output by reaction   kg/h   931.7629
Output by advection   kg/h   68.23721
Total output by reaction and advection kg/h   1000

Overall residence time   h   22764
Reaction residence time   h   24431.11
Advection residence time   h   333601

FOUR COMPARTMENT FUGACITY LEVEL III CALCULATION   DDT

Bulk phase properties, compositions and rates

Phase	Air	Water	Soil	Sediment	Total	Fish
Bulk vol m3	1E+14	2E+11	1.8E+10	5E+08		
Density kg/m3	1.185413	1000.009	1500.237	1280		
Bulk Z value	7.607752E-04	1.574724	11822.79	9458.665		60073.98
Bulk VZ	7.607751E+10	3.149449E+11	2.128102E+14	4.729333E+12	2.179306E+14	
Emission mol/h	2820.875	0	0	0	2820.875	
Emission kg/h	1000	0	0	0	1000	
Fugacity Pa	1.301044E-06	2.034622E-07	1.455241E-07	6.031201E-07		2.034622E-07
Conc mol/m3	9.898019E-10	3.203968E-07	1.720501E-03	5.704711E-03		1.222278E-02
Conc g/m3	3.508848E-07	1.135807E-04	.6099175	2.02232		4.332976
Conc ug/g	2.960021E-04	1.135797E-04	.4065474	1.579937		4.332976
Amount mol	98980.19	64079.36	3.096902E+07	2852355	3.398443E+07	
Amount kg	35088.48	22716.13	1.097852E+07	1011160	1.204748E+07	
Amount %	.2912516	.1885551	91.12707	8.393122		
Adv.flow m3/h	1E+12	2E+08	0	10000		
D rct mol/Pa.h	3.101278E+08	3.968305E+07	8.675146E+09	5.958959E+07		
D adv mol/Pa.h	7.607752E+08	3.149449E+08	0	9.458665E+07		
Rct rate mol/h	403.4898	8.074	1262.443	35.93968	1709.946	
Rct rate kg/h	143.0371	2.862233	447.536	12.74062	606.176	
Adv rate mol/h	989.8019	64.07936	0	57.04711	1110.928	
Adv rate kg/h	350.8847	22.71613	0	20.2232	393.8241	
Reaction %	14.30371	.2862233	44.7536	1.274062		
Advection %	35.08848	2.271613	0	2.02232		

Overall residence time   h   12047.48
Reaction residence time   h   19874.56   Advection residence time h   30591.02

Intermedia Data.	Half times	Equiv flows	D values	Rates	of transport
	h	m3/h	mol/Pa.h	mol/h	kg/h
Air to water	4.1462E+02	1.6714E+11	1.2716E+08	1.6544E+02	5.8647E+01
Air to soil	5.4170E+01	1.2793E+12	9.7326E+08	1.2663E+03	4.4889E+02
Water to air	1.1383E+04	1.2176E+07	1.9174E+07	3.9011E+00	1.3830E+00
Water to sediment	2.9507E+02	4.6972E+08	7.3969E+08	1.5050E+02	5.3352E+01
Soil to air	1.0435E+08	1.1953E+02	1.4132E+06	2.0566E-01	7.2906E-02
Soil to water	5.9535E+06	2.0952E+03	2.4771E+07	3.6048E+00	1.2779E+00
Sediment to water	3.4370E+04	1.0081E+04	9.5356E+07	5.7511E+01	2.0388E+01

	Transport velocity parameters	m/h	m/year
1	air side air-water MTC	5	43800
2	water side air-water MTC	.05	438
3	rain rate	.0001	.876
4	aerosol deposition velocity	6E-10	5.256E-06
5	soil air phase diffusion MTC	.02	175.2
6	soil water phase diffusion MTC	.00001	.0876
7	soil air boundary layer MTC	5	43800
8	sediment-water diffusion MTC	.0001	.876
9	sediment deposition velocity	.0000005	.00438
10	sediment resuspension velocity	.0000002	.001752
11	soil water runoff rate	.00005	.438
12	soil solids runoff rate	1E-08	.0000876

Individual process D values

Air-water diffusion (air-side)	2.017089E+07	Air-water diffusion (water-side)	3.878703E+08
Air-water diffusion (overall)	1.917378E+07		
Rain dissolution to water	775740.5	Aerosol deposition to water	1.072072E+08
Rain dissolution to soil	6981665	Aerosol deposition to soil	9.648646E+08
Soil-air diffusion (air-phase)	726152.2	Soil-air diffusion (water-phase)	698166.4
Soil-air diffusion (bndry layer)	1.815381E+08	Soil-air diffusion (overall)	1413231
Water-sediment diffusion	775740.5		
Water-sediment deposition	7.389098E+08	Sediment-water resuspension	9.458046E+07
Soil-water runoff (water)	3490832	Soil-water runoff (solids)	2.12806E+07

FOUR COMPARTMENT FUGACITY LEVEL III CALCULATION   DDT

Bulk phase properties, compositions and rates

Phase	Air	Water	Soil	Sediment	Total	Fish
Bulk vol m3	1E+14	2E+11	1.8E+10	5E+08		
Density kg/m3	1.185413	1000.009	1500.237	1280		
Bulk Z value	7.607752E-04	1.574724	11822.79	9458.665		60073.98
Bulk VZ	7.607751E+10	3.149449E+11	2.128102E+14	4.729333E+12	2.179306E+14	
Emission mol/h	0	2820.875	0	0	2820.875	
Emission kg/h	0	1000	0	0	1000	
Fugacity Pa	3.002551E-08	3.399968E-06	3.358408E-09	1.007848E-05		3.399968E-06
Conc mol/m3	2.284266E-11	5.354012E-06	3.970575E-05	9.532894E-02		.2042496
Conc g/m3	8.097725E-09	1.897997E-03	1.407569E-02	33.79411		72.40647
Conc ug/g	6.831141E-06	1.897981E-03	9.382309E-03	26.40165		72.40647
Amount mol	2284.266	1070802	714703.5	4.766447E+07	4.945227E+07	
Amount kg	809.7725	379599.5	253362.4	1.689705E+07	1.753083E+07	
Amount %	4.619134E-03	2.165326	1.445239	96.38481		
Adv.flow m3/h	1E+12	2E+08	0	10000		
D rct mol/Pa.h	3.101278E+08	3.968305E+07	8.675146E+09	5.958959E+07		
D adv mol/Pa.h	7.607752E+08	3.149449E+08	0	9.458665E+07		
Rct rate mol/h	9.311744	134.9211	29.13468	600.5723	773.9398	
Rct rate kg/h	3.301013	47.82953	10.32824	212.9029	274.3617	
Adv rate mol/h	22.84266	1070.803	0	953.2894	2046.935	
Adv rate kg/h	8.097724	379.5995	0	337.9411	725.6383	
Reaction %	.3301013	4.782953	1.032824	21.29029		
Advection %	.8097725	37.95995	0	33.79411		

Overall residence time   h   17530.83
Reaction residence time   h   63896.78   Advection residence time h   24159.18

Intermedia Data.	Half times	Equiv flows	D values	Rates of transport	
	h	m3/h	mol/Pa.h	mol/h	kg/h
Air to water	4.1462E+02	1.6714E+11	1.2716E+08	3.8179E+00	1.3535E+00
Air to soil	5.4170E+01	1.2793E+12	9.7326E+08	2.9223E+01	1.0359E+01
Water to air	1.1383E+04	1.2176E+07	1.9174E+07	6.5190E+01	2.3110E+01
Water to sediment	2.9507E+02	4.6972E+08	7.3969E+08	2.5149E+03	8.9153E+02
Soil to air	1.0435E+08	1.1953E+02	1.4132E+06	4.7462E-03	1.6825E-03
Soil to water	5.9535E+06	2.0952E+03	2.4771E+07	8.3193E-02	2.9492E-02
Sediment to water	3.4370E+04	1.0081E+04	9.5356E+07	9.6105E+02	3.4069E+02

	Transport velocity parameters	m/h	m/year
1	air side air-water MTC	5	43800
2	water side air-water MTC	.05	438
3	rain rate	.0001	.876
4	aerosol deposition velocity	6E-10	5.256E-06
5	soil air phase diffusion MTC	.02	175.2
6	soil water phase diffusion MTC	.00001	.0876
7	soil air boundary layer MTC	5	43800
8	sediment-water diffusion MTC	.0001	.876
9	sediment deposition velocity	.0000005	.00438
10	sediment resuspension velocity	.0000002	.001752
11	soil water runoff rate	.00005	.438
12	soil solids runoff rate	1E-08	.0000876

Individual process D values

Air-water diffusion (air-side)	2.017089E+07	Air-water diffusion (water-side)	3.878703E+08
Air-water diffusion (overall)	1.917378E+07		
Rain dissolution to water	775740.5	Aerosol deposition to water	1.072072E+08
Rain dissolution to soil	6981665	Aerosol deposition to soil	9.648646E+08
Soil-air diffusion (air-phase)	726152.2	Soil-air diffusion (water-phase)	698166.4
Soil-air diffusion (bndry layer)	1.815381E+08	Soil-air diffusion (overall)	1413231
Water-sediment diffusion	775740.5		
Water-sediment deposition	7.389098E+08	Sediment-water resuspension	9.458046E+07
Soil-water runoff (water)	3490832	Soil-water runoff (solids)	2.12806E+07

FOUR COMPARTMENT FUGACITY LEVEL III CALCULATION   DDT

Bulk phase properties, compositions and rates

Phase	Air	Water	Soil	Sediment	Total	Fish
Bulk vol m3	1E+14	2E+11	1.8E+10	5E+08		
Density kg/m3	1.185413	1000.009	1500.237	1280		
Bulk Z value	7.607752E-04	1.574724	11822.79	9458.665		60073.98
Bulk VZ	7.607751E+10	3.149449E+11	2.128102E+14	4.729333E+12	2.179306E+14	
Emission mol/h	0	0	2820.875	0	2820.875	
Emission kg/h	0	0	1000	0	1000	
Fugacity Pa	2.967879E-10	9.712265E-09	3.242221E-07	2.878993E-08		9.712265E-09
Conc mol/m3	2.257889E-13	1.529414E-08	3.833209E-03	2.723143E-04		5.834544E-04
Conc g/m3	8.004216E-11	5.421773E-06	1.358873	9.653541E-02		.2068346
Conc ug/g	6.752258E-08	5.421727E-06	.905772	.0754183		.2068346
Amount mol	22.57889	3058.828	6.899777E+07	136157.2	6.913701E+07	
Amount kg	8.004216	1084.355	2.445971E+07	48267.71	2.450907E+07	
Amount %	3.265818E-05	4.424299E-03	99.7986	.1969382		
Adv.flow m3/h	1E+12	2E+08	0	10000		
D rct mol/Pa.h	3.101278E+08	3.968305E+07	8.675146E+09	5.958959E+07		
D adv mol/Pa.h	7.607752E+08	3.149449E+08	0	9.458665E+07		
Rct rate mol/h	9.204217E-02	.3854123	2812.674	1.71558	2814.867	
Rct rate kg/h	3.262895E-02	.1366287	997.0929	.6081732	997.8703	
Adv rate mol/h	.2257889	3.058828	0	2.723143	6.00776	
Adv rate kg/h	8.004216E-02	1.084355	0	.9653542	2.129751	
Reaction %	3.262895E-03	1.366287E-02	99.70928	6.081731E-02		
Advection %	8.004216E-03	.1084354	0	9.653541E-02		

Overall residence time   h   24509.07
Reaction residence time   h   24561.38   Advection residence time h   1.150795E+07

Intermedia Data.	Half times	Equiv flows	D values	Rates of transport	
	h	m3/h	mol/Pa.h	mol/h	kg/h
Air to water	4.1462E+02	1.6714E+11	1.2716E+08	3.7739E-02	1.3378E-02
Air to soil	5.4170E+01	1.2793E+12	9.7326E+08	2.8885E-01	1.0240E-01
Water to air	1.1383E+04	1.2176E+07	1.9174E+07	1.8622E-01	6.6015E-02
Water to sediment	2.9507E+02	4.6972E+08	7.3969E+08	7.1840E+00	2.5467E+00
Soil to air	1.0435E+08	1.1953E+02	1.4132E+06	4.5820E-01	1.6243E-01
Soil to water	5.9535E+06	2.0952E+03	2.4771E+07	8.0314E+00	2.8471E+00
Sediment to water	3.4370E+04	1.0081E+04	9.5356E+07	2.7453E+00	9.7321E-01

	Transport velocity parameters	m/h	m/year
1	air side air-water MTC	5	43800
2	water side air-water MTC	.05	438
3	rain rate	.0001	.876
4	aerosol deposition velocity	6E-10	5.256E-06
5	soil air phase diffusion MTC	.02	175.2
6	soil water phase diffusion MTC	.00001	.0876
7	soil air boundary layer MTC	5	43800
8	sediment-water diffusion MTC	.0001	.876
9	sediment deposition velocity	.0000005	.00438
10	sediment resuspension velocity	.0000002	.001752
11	soil water runoff rate	.00005	.438
12	soil solids runoff rate	1E-08	.0000876

Individual process D values

Air-water diffusion (air-side)	2.017089E+07	Air-water diffusion (water-side)	3.878703E+08
Air-water diffusion (overall)	1.917378E+07		
Rain dissolution to water	775740.5	Aerosol deposition to water	1.072072E+08
Rain dissolution to soil	6981665	Aerosol deposition to soil	9.648646E+08
Soil-air diffusion (air-phase)	726152.2	Soil-air diffusion (water-phase)	698166.4
Soil-air diffusion (bndry layer)	1.815381E+08	Soil-air diffusion (overall)	1413231
Water-sediment diffusion	775740.5		
Water-sediment deposition	7.389098E+08	Sediment-water resuspension	9.458046E+07
Soil-water runoff (water)	3490832	Soil-water runoff (solids)	2.12806E+07

FOUR COMPARTMENT FUGACITY LEVEL III CALCULATION   DDT

Bulk phase properties, compositions and rates

Phase	Air	Water	Soil	Sediment	Total	Fish
Bulk vol m3	1E+14	2E+11	1.8E+10	5E+08		
Density kg/m3	1.185413	1000.009	1500.237	1280		
Bulk Z value	7.607752E-04	1.574724	11822.79	9458.665		60073.98
Bulk VZ	7.607751E+10	3.149449E+11	2.128102E+14	4.729333E+12	2.179306E+14	
Emission mol/h	141.0437	705.2186	1974.612	0	2820.875	
Emission kg/h	50	250	700	0	1000	
Fugacity Pa	7.276631E-08	8.669637E-07	2.350713E-07	2.569929E-06		8.669637E-07
Conc mol/m3	5.535881E-11	1.365229E-06	2.779198E-03	2.430809E-02		5.208196E-02
Conc g/m3	1.96247E-08	4.839736E-04	.9852258	8.617219		18.46305
Conc ug/g	1.655515E-05	4.839695E-04	.6567134	6.732202		18.46305
Amount mol	5535.881	273045.8	5.002557E+07	1.215405E+07	6.245819E+07	
Amount kg	1962.47	96794.72	1.773406E+07	4308609	2.214143E+07	
Amount %	8.863338E-03	.4371657	80.09449	19.45949		
Adv.flow m3/h	1E+12	2E+08	0	10000		
D rct mol/Pa.h	3.101278E+08	3.968305E+07	8.675146E+09	5.958959E+07		
D adv mol/Pa.h	7.607752E+08	3.149449E+08	0	9.458665E+07		
Rct rate mol/h	22.56685	34.40377	2039.278	153.141	2249.389	
Rct rate kg/h	7.99995	12.19614	722.9239	54.28848	797.4085	
Adv rate mol/h	55.35881	273.0458	0	243.0809	571.4855	
Adv rate kg/h	19.6247	96.79472	0	86.17219	202.5916	
Reaction %	.799995	1.219614	72.29239	5.428849		
Advection %	1.96247	9.679473	0	8.617218		

Overall residence time   h   22141.43
Reaction residence time   h   27766.74   Advection residence time h   109291

Intermedia Data.	Half times	Equiv flows	D values	Rates	of transport
	h	m3/h	mol/Pa.h	mol/h	kg/h
Air to water	4.1462E+02	1.6714E+11	1.2716E+08	9.2527E+00	3.2801E+00
Air to soil	5.4170E+01	1.2793E+12	9.7326E+08	7.0821E+01	2.5106E+01
Water to air	1.1383E+04	1.2176E+07	1.9174E+07	1.6623E+01	5.8928E+00
Water to sediment	2.9507E+02	4.6972E+08	7.3969E+08	6.4128E+02	2.2733E+02
Soil to air	1.0435E+08	1.1953E+02	1.4132E+06	3.3221E-01	1.1777E-01
Soil to water	5.9535E+06	2.0952E+03	2.4771E+07	5.8231E+00	2.0643E+00
Sediment to water	3.4370E+04	1.0081E+04	9.5356E+07	2.4506E+02	8.6873E+01

Transport velocity parameters		m/h	m/year
1	air side air-water MTC	5	43800
2	water side air-water MTC	.05	438
3	rain rate	.0001	.876
4	aerosol deposition velocity	6E-10	5.256E-06
5	soil air phase diffusion MTC	.02	175.2
6	soil water phase diffusion MTC	.00001	.0876
7	soil air boundary layer MTC	5	43800
8	sediment-water diffusion MTC	.0001	.876
9	sediment deposition velocity	.0000005	.00438
10	sediment resuspension velocity	.0000002	.001752
11	soil water runoff rate	.00005	.438
12	soil solids runoff rate	1E-08	.0000876

Individual process D values

Air-water diffusion (air-side)	2.017089E+07	Air-water diffusion (water-side)	3.878703E+08
Air-water diffusion (overall)	1.917378E+07		
Rain dissolution to water	775740.5	Aerosol deposition to water	1.072072E+08
Rain dissolution to soil	6981665	Aerosol deposition to soil	9.648646E+08
Soil-air diffusion (air-phase)	726152.2	Soil-air diffusion (water-phase)	698166.4
Soil-air diffusion (bndry layer)	1.815381E+08	Soil-air diffusion (overall)	1413231
Water-sediment diffusion	775740.5		
Water-sediment deposition	7.389098E+08	Sediment-water resuspension	9.458046E+07
Soil-water runoff (water)	3490832	Soil-water runoff (solids)	2.12806E+07

Appendix 2. Fugacity calculations using Lotus 123 spreadsheet program.

```
Input parameters in Column C:      Column C
Chemical name:                     p,p'-DDT
system, C                          25
molecular weight, g/mol            354.5
melting point, C                   109
dissociation constant, log pKa     12   WHEN none available enter 12
pH of data determination           6    WHEN none available enter 6
pH of environment                  6
solubility, g/m3                   0.0055
vapor pressure, Pa                 2.00E-05
log Kow                            6.19
emission rate, kg/h
  - to air compartment only        1000    0       0       0
  - to water compartment only      0       1000    0       0
  - to soil compartment only       0       0       25      1000
  - to all compartments            0       0       5       700
degradation half-lives, h
  - in air compartment             170
  - in water compartment           5500
  - in soil compartment            17000
  - in sediment compartment        55000
Go To A288 to check both input and output results.
```

```
Fugacity Level I, II and III Calculations:
*  Amount of chemicals, moles                                       282087    moles      100000 kg
*  Emission rate of chemicals, E =                                  2820.87   mol/h      1000 kg/h
   Gas constant, Pa m3/mol K, R=                                    8.314
   System temperature, in C                                         25        * input data
   System temperature, in K, T = (t + 273.15)                       298.15
*  Molecular weight, g/mol                              MW =        354.5     * input data
*  Melting point, t in C                                M.P.=       109       * input data
   M.P. or Tm = (mp+273.15) in degree Kelvin                        382.15
   Fugacity ratio = exp(6.79(1-Tm/T)) for solid comp'ds, F          0.1476
*  Dissociation constant pKa, WHEN none available, enter 12         12        * input data
*  pH of data determination, pHd, WHEN none available, enter 6      6         * input data
*  pH of environment, pHe                                           6         * input data
   Ratio ionic to non-ionic (data deter'n), ID = 10^(pHd-pKa)       0.000001
   Ratio ionic to non-ionic (environ.), IE = 10^(pHe-pKa)           0.000001
*  Solubility, g/m3 or mg/L                             S =         0.0055    * input data
   molar solubility, mol/m3, c=S/MW                                 1.55E-05
*  Vapor pressure, Pa                                   P =         0.00002   * input data
   Vapor pressure, subcooled liquid value, Pa                       1.355E-04
   Henry's law constant, Pa m3/mol, H=p/c                           1.2891
*  Octanol/water partition coefficient,                 log Kow     6.19      * input data
   Kow =                                                            1548817
   Partition coefficient, organic C, Koc = 0.41*Kow*y               635014.81
     for soil (mole fraction organic C), y(3) =                     0.02
     suspended sediment, y(5) =                                     0.2
     bottom sediment, y(6) =                                        0.04
   Kp(3) = 0.41*Kow*y(3)                                            12700
   Kp(4) = 0.41*Kow*y(4)                                            127003
   Kp(5) = 0.41*Kow*y(5)                                            25401
   Bioconcentraion factor, BCF or Kb, K(6) = 0.050*Kow              77441
```

```
                                           * input
   Adv. flow, G(air)                       1.000E+12  mol/hr
   Adv. flow, G(water)                     5.00E+08   mol/hr
   Adv. flow, G(sed.)                      10000      mol/hr

   Half-lives, hours
*  t(air)                                  170     h
*  t(water)                                5500    h
*  t(soil)                                 17000   h
*  t(sediment)                             55000   h

   Emission, kg/h :
                         air           water         soil
   E                     50            250           700
   E(A)                  1000          0             0
   E(B)                  0             1000          0
   E(C)                  0             0             1000
```

```
If ionization ratio ID<=0.001, Z(2)=ZTD(2) otherwise Z(2)=ZTE(2)
Air/water partition coeff., Kaw, Z(1)/Z(2)                        5.200E-04
Soil/water partition coeff., Ksw, Z(3)/Z(2)                       3.048E+04
Sediment/water partition coeff., Ksd/w, Z(4)/Z(2)                 6.096E+04
Sus. sediment/water partition coeff., Kssd/w, Z(5)/Z(2)           1.905E+05
Aerosol/water partition coeff., Kar/w, Z(aerosol)/Z(air)          4.429E+10
Densities g/cm3 or kg/L
  air, d(1) = (0.029*101325/RT)                                   0.0011854
  water, d(2) =                                                   1
  soil, d(3) =                                                    2.4
  bottom sediment, d(4) =                                         2.4
  suspended sediment, d(5) =                                      1.5
  biota, d(6) =                                                   1
  aerosol density, g/m3, d(7) =                                   2
Fugacity capacities, Z:
  Z(1) or Z(air) = 1/RT                                           4.034E-04
  Z(2) = ZTD(2) or Z(water) = 1/H = c/p WHEN pKa >> pH             7.757E-01
  ZTD(2) = ZTD(2)*ID/(1+ID)                                       0.0000007
  ZND(2) = ZTD(2)/(1+ID)                                          0.7757397
  ZNE(2) = ZND(2), Z for neutral forms                            0.7757397
  ZIE(2) = ZNE(2)*IE                                              0.0000007
  ZTE(2) = ZNE(2)+ZIE(2)                                          0.7757404
  Z(2) = ZTE(2) WHEN dissociation is appreciable, see TEXT        0.7757404
  Z(3) or Z(soil) = Kp(s)*ZTD(2)*d(s)                             2.365E+04
  Z(4) or Z(bottom sediment) = Kp(bs)*ZTD(2)*d(bs)                4.729E+04
  Z(5) or Z(suspended sediment) = Kp(ss)*ZTD(2)*d(ss)             1.478E+05
  Z(6) or Z(biota) = K *ZTD(2)*d(B)                               6.007E+04
  Z(7) or Z(aerosol) = Z(1)*6*E6/p(L)RT                           1.787E+07
Fugacity (Level I), f = total no. of moles/sum(ViZi)              1.295E-09
Fugacity (Level II), f = emission/sum(D values)                   2.947E-07
```

Tansport parameters:

```
                                                    m/h         m/yr
                                                     k          k/8760
air-water MTC, air side, U(1)                        5          43800
air-water MTC, water side, U(2)                      0.05       438
rain rate, U(3) = 0.85/8760                          0.0001     0.876
aerosol deposition velocity, U(4)                    6.00E-10   0.0000052
soil air phase diffusion MTC, U(5)                   0.02       175.2
soil water phase diffusion MTC, U(6)                 1.00E-05   0.0876
soil air boundary layer MTC, U(7)                    5          43800
sediment-water MTC, U(8)                             0.0001     0.876
sediment deposition rate, U(9)                       5.00E-07   0.00438
sediment resuspended rate, U(10)                     2.00E-07   0.001752
soil water runoff rate, U(11)                        5.00E-05   0.438
soil solids runoff rate, U(12)                       1.00E-08   0.0000876
```

Define dimension of the generic environment (unit world):

Compartment	Volume, Vi m³	Depth, h m	Area, A m²	Density d, kg/m³	Fugacity Cap., Zi mol/Pa m³	ViZi	*input Advective flow, G mol/h	Residence time, V/G t(R),h	*input Reaction half-life t(1/2),h	Rate const. 0.693/t k, 1/hr	Emission rate, E mol/h	E(A) mol/h	E(B) mol/h	E(C) mol/h
Air (1)	1.00E+14	1000	1.000E+11	1.1854132	4.034E-04	4.03E+10	1.000E+12	100	170	4.076E-03	141.04	2820.87	0	0
Water (2)	2.00E+11	20	1.000E+10	1000	7.757E-01	1.55E+11	2.000E+08	1000	5500	1.260E-04	705.22	0	2820.87	0
Soil (3)	9.00E+09	0.1	9.000E+10	2400	2.365E+04	2.13E+14	0		17000	4.076E-05	1974.61	0	0	2820.87
Bottom sediment (4)	1.00E+08	0.01	1.000E+10	2400	4.729E+04	4.73E+12	2000	50000	55000	1.260E-05	0.00	0	0	0
Sus. sediment (5)	1.00E+06			1500	1.478E+05	1.48E+11			1.00E+11	6.930E-12	2820.87			
Biota (fish) (6)	2.00E+05			1000	6.007E+04	1.20E+10			1.00E+11	6.930E-12				
Aerosol (7)	2000			2000	1.787E+07	2.18E+14								

Level I calculation:

Compartment	Volume, Vi m³	Fugacity capacity Zi	ViZi	Conc'n c = f*Zi mol/m³	Amount m = ciVi mol	Amount w=m*MW/1E3 kg	Amount %	Conc'n, S mg/L (or g/m³)	Conc'n (S/d)*1000 ug/g
Air (1)	1.00E+14	4.034E-04	4.034E+10	5.223E-13	5.22E+01	1.851E+01	1.85E-02	1.85E-10	1.56E-07
Water (2)	2.00E+11	7.757E-01	1.551E+11	1.004E-09	2.01E+02	7.120E+01	7.12E-02	3.56E-07	3.56E-07
Soil (3)	9.00E+09	2.365E+04	2.128E+14	3.061E-05	2.76E+05	9.767E+04	9.77E+01	1.09E-02	4.52E-03
Bottom sediment (4)	1.00E+08	4.729E+04	4.729E+12	6.122E-05	6.12E+03	2.170E+03	2.17E+00	2.17E-02	9.04E-03
Sus. sediment (5)	1.00E+06	1.478E+05	1.478E+11	1.913E-04	1.91E+02	6.782E+01	6.78E-02	6.78E-02	4.52E-02
Biota (fish) (6)	2.00E+05	6.007E+04	1.201E+10	7.777E-05	1.56E+01	5.514E+00	5.51E-03	2.76E-02	2.76E-02
			2.179E+14		282087.44	100000	100		

Level II phase properties and rates:

Compartment	Rate const. k, 1/hr	D(reaction) VZk	D(advec'n) GZ	Conc'n c = f*Z mol/m³	Amount m = ciVi mol	Amount m*MW/1000 kg	Conc'n mg/L (or g/m³)	Conc'n ug/g	Conc'n (S/d)*1000	Loss Reaction mol/h Vck	Loss Advection mol/h Gc	% Loss reaction	% Loss advection	Removal %
Air (1)	0.004076470	1.645E+08	4.034E+08	1.189E-10	1.19E+04	4.21E+03	4.215E-08	3.56E-05		4.847E+01	1.189E+02	1.72E+00	4.21	5.93
Water (2)	0.000126	1.955E+07	1.551E+08	2.286E-07	1.62E+04	5.761E+00	8.105E-05	8.10E-05		5.761E+00	4.572E+01	2.04E-01	1.621	1.825
Soil (3)	0.000040764	8.675E+09		6.968E-03	2.22E+07	2.557E+03	1.03E+00	1.03E-05		2.557E+03		9.06E+01		9.06E+01
Bottom sediment (4)	0.0000126	5.959E+07	9.458E+07	1.394E-02	1.39E+06	4.941E+05	4.941E+00	2.06E+00		1.756E+01	2.787E+01	6.23E-01	9.88E-01	1.61E+00
Sus. sediment (5)	6.9300E-12	1.024E+00		4.355E-02	1.54E+04	1.544E+01	1.544E+01	1.03E+01		3.018E-07		1.07E-08		1.07E-08
Biota (fish) (6)	6.9300E-12	8.326E-02		1.770E-02	3.54E+03	1.26E+03	6.276E+00	6.28E+00		2.454E-08		8.70E-10		8.70E-10
Total R + A		8.919E+09	6.531E+08		22763997.	22763997				2628.39	192.49	93.18	6.82	100
		9.572E+09									2820.87447			

Total amount of chemicals, 64214380. moles 22763998 kg
Total reaction D value 8.92E+09
Total advection D value 653146483
Total D value 9.57E+09
Fugacity, E/sum D values 2.947E-07
Output by reaction, mol/h 2628.3859 931.76282 kg/h
Output by advection, mol/h 192.48849 20.981246 kg/h
Total output, mol/h 2820.8744 1077.9971 kg/h
Overall resistence time, h 22763.997
Reaction resistence time, h 24431.107
Advection resistence time, h 333601.12

Compartment	Subcomp't	Volume fraction v_i	Fugacity capacity Z_i	Bulk vol. V_B, m3	Bulk $ZB(i)$ sum(v_iZ_i)	Partition coeff. (Z_i/Z_w)	Bulk $VZ=VB*Z_i$	Bulk den. sum(v_id_i) kg/m3	Bulk Adv. flow GAB=VB/Gi
Air (1)	Air		4.034E-04	1.00E+14	0.0007607	4.83E-04	7.608E+10	1.18541328	1.00E+12
	Aerosol	2.0000E-11	1.787E+07						
Water (2)	Water	1	7.757E-01	2.00E+11	1.5747245	30480.711	3.149E+11	1000.0085	2.00E+08
	Particulate	0.000005	1.478E+05						
	Biota(fish)	0.000001	6.007E+04						
Soil (3)	Air	0.2	4.034E-04	1.80E+10	1.18E+04	60961.422	2.128E+14	1500.23708	
	Water	0.3	7.757E-01						
	Solids	0.5	23645.121						
Bottom sediment (4)	Water	0.8	0.7757404	5.00E+08	9458.6691	77440.830	4.729E+12	1280	10000
	Solids	0.2	47290.242		21283.038		2.179E+14		

Level III Intermedia Data:

	D values D_{ij} mol/Pa h	Eq. flows $D_{ij}/Z(i)$ GD$_{ij}$, m3/h	Half-life .693Vi/G $t(1/2)$, h	Rate of transport $D_{ij}*f(i)$ N, mol/h	N*MW/1000 Nk, kg/h
Air to water (D12)	1.2716E+08	1.671E+11	4.146E+02	9.253E+00	3.28E+00
Air to soil (D13)	9.7326E+08	1.279E+12	5.417E+01	7.082E+01	2.51E+01
Water to air (D21)	1.917E+07	1.218E+07	1.138E+04	1.662E+01	5.89E+00
Water to sed. (D24)	7.3969E+08	4.697E+08	2.951E+02	6.413E+02	2.27E+02
Soil to air (D31)	1.4132E+06	1.195E+02	5.218E+07	3.322E+01	1.18E-01
Soil to water (D32)	2.4771E+07	2.095E+03	2.977E+06	5.823E+00	2.06E+00
Sed. to water (D42)	9.5356E+07	1.008E+04	6.874E+03	2.451E+02	8.69E+01

Equations for Dij values:
D12 = A(2)*(1/(1/U(1)+1/(U(2)*Z(2)+U(4)*Z(7))
D13 = A(3)*(1/(1/(U(5)*Z(1)+U(6)*Z(2))+U(3)*Z(2)+U(4)*Z(7))
D21 = A(2)*(1/(1/(U(1)*Z(1)+1/(U(2)*Z(2)))
D24 = A(2)*(U(8)*Z(2)+U(9)*Z(5))
D31 = A(3)*(1/(1/(U(5)*Z(1)+U(6)*Z(2)+1/(U(7)*Z(1))+U(3)*Z(2)+U(4)*Z(7))
D32 = A(3)*(1/(U(11)*Z(2)+U(12)*Z(3))
D42 = A(4)*(U(8)*Z(2)+U(10)*Z(4))

					E(A)			E(B)			E(C)	
	D values D_{ij} mol/Pa h	Eq. flows $D_{ij}/Z(i)$ GD$_{ij}$, m3/h	Half-life .693Vi/G $t(1/2)$, h	Rate of transport $D_{ij}*f(i)$ N, mol/h	N*MW/1000 Nk, kg/h		Rate of transport $D_{ij}*f(i)$ N, mol/h	N*MW/1000 Nk, kg/h		Rate of transport $D_{ij}*f(i)$ N, mol/h	N*MW/1000 Nk, kg/h	
Air to water (D12)	1.2716E+08	1.671E+11	4.146E+02	1.654E+02	5.86E+01		3.818E+00	1.35E+00		3.774E-02	1.338E-02	
Air to soil (D13)	9.7326E+08	1.279E+12	5.417E+01	1.266E+03	4.49E+02		2.922E+01	1.04E+01		2.889E-01	1.024E-01	
Water to air (D21)	1.917E+07	1.218E+07	1.138E+04	3.901E+01	1.38E+01		6.519E+01	2.31E+01		1.862E-01	6.602E-02	
Water to sed. (D24)	7.3969E+08	4.697E+08	2.951E+02	1.505E+02	5.34E+01		2.515E+03	8.92E+02		7.184E+00	2.547E+00	
Soil to air (D31)	1.4132E+06	1.195E+02	5.218E+07	2.057E+01	7.29E+00		4.746E-01	1.68E-01		4.582E-01	1.624E-01	
Soil to water (D32)	2.4771E+07	2.095E+03	2.98E+06	3.605E+00	1.28E+00		8.319E-02	2.95E-02		8.031E+00	2.847E+00	
Sed. to water (D42)	9.5356E+07	1.008E+04	6.874E+03	5.751E+01	2.04E+01		9.610E+02	3.41E+02		2.745E+02	9.732E-01	

Phase properties and rates:

Compartment	Rate const. k, 1/hr	D Values Reaction VB*ZB*k mol/Pa h	Advection GAB*ZB mol/Pa h	DTs	Js	Js, E(A)	Js, E(B)	Js, E(C)
Air (1)	0.004076470	3.101E+08	7.608E+08	2.171E+09	6.511E-08	1.299E-06	0.000E+00	2.110E-10
Water (2)	0.000126	3.968E+07	3.149E+08	1.113E+09	8.830E-03	8.830E-03	8.830E-03	8.830E-03
Soil (3)	0.000040764	8.675E+09	0	8.701E+09	9.999E-01	9.999E-01	9.999E-01	9.999E-01
Bottom sediment (4)	0.000126	5.959E+07	9.459E+07	2.495E+08	1.299E+08	1.299E+08	1.299E+08	1.299E+08
Total R + A		9.085E+09	1.170E+09 1.025E+10					

For E(1,2,3), i.e., E(1)=1, E(2)=1 and E(3)=1 kg/h

Compartment	Fugacity f's Pa	Concentration C mol/m3	S mg/L	ug/g	Amount m, mol	Amount %	Loss Reaction mol/h (Ci*DRi)	Loss Advection mol/h (Ci*DAi)	Loss rate mol/h (Ci*DL)	Reaction %	Advection %	Removal %
Air (1)	7.277E-08	5.536E-11	1.962E-08	1.656E-05	5.54E+03	0.00886	2.26E+01	55.3588	0.00E+00	0.8000	1.9625	2.76
Water (2)	8.670E-07	1.365E-06	4.840E-04	4.840E-04	2.73E+05	0.43717	3.44E+01	273.0457	0	1.2196	9.6795	10.90
Soil (3)	2.351E-07	2.779E-03	9.852E-01	6.567E-01	5.00E+07	80.09448	2.04E+03	0.0000	0	72.2924	0.0000	72.29
Bottom sediment (4)	2.570E-06	2.431E-02	8.617E-03	6.732E+00	1.22E+07	19.45949	1.53E+02	243.0809	0.00E+00	5.4288	8.6172	1.40E+01
Total					6.25E+07	100	2.25E+03	571.4855	0.0000			100
								2.82E+03				

For E(A), i.e., E(1) = 1000 kg/h conditions:

Compartment	Fugacity f, Pa	Concentration mol/m3	mg/L	ug/g	Amount m, mol	Amount %	Loss Reaction mol/h	Loss Advection mol/h	Loss rate mol/h	Reaction %	Advection %	Removal %
Air (1)	1.301E-06	9.898E-10	3.509E-07	2.960E-04	9.90E+04	0.29125	403.49	989.8019	0.00E+00	14.3037	35.0885	49.39
Water (2)	2.035E-07	3.204E-07	1.136E-04	1.136E-04	6.41E+04	0.18856	8.0740	64.0793	0	0.2862	2.2716	2.56
Soil (3)	1.455E-07	1.721E-03	6.099E-01	4.065E-01	3.10E+07	91.12707	1262.4426	0.0000	0	44.7536	0.0000	44.7536
Bottom sediment (4)	6.031E-07	5.705E-03	2.022E+00	1.580E+00	2.85E+06	8.393E+00	35.9397	5.705E+01	0.00E+00	1.27E+01	2.02E+00	3.30E+00
Total					3.40E+07	100	1.71E+03	1110.9283	0.0000			100
								2821				

For E(B), i.e., E(2) = 1000 kg/h conditions:

Compartment	Fugacity f, Pa	Concentration mol/m3	mg/L	ug/g	Amount m, mol	Amount %	Reaction mol/h	Advection mol/h	Loss rate mol/h	Reaction %	Advection %	Removal %
Air (1)	3.003E-08	2.284E-11	8.098E-09	6.831E-06	2284	0.00462	9.3118	22.8427	0.00E+00	0.3301	0.8098	1.14
Water (2)	3.400E-06	5.354E-06	1.898E-03	1.898E-03	1.07E+06	2.16532	134.92	1070.8034	0	4.7830	37.9599	42.74
Soil (3)	3.358E-07	3.971E-05	1.408E-02	9.382E-03	714704.00	1.44524	29.1347	0.0000	0	1.0328	0.0000	1.0328
Bottom sediment (4)	1.008E-05	9.533E-02	3.379E+01	2.640E+01	47664527	96.38482	600.5730	2046.9366	0.00E+00	21.2903	3.38E+01	55.0844
Total					49452318.		773.94		0.0000			100
								2821				

For E(C) i.e., E(3) = 1000 kg/h conditions:

Compartment	Fugacity f, Pa	Concentration mol/m3	mg/L	ug/g	Amount m, mol	Amount %	Reaction mol/h	Advection mol/h	Loss rate mol/h	Reaction %	Advection %	Removal %
Air (1)	2.968E-10	2.258E-13	8.004E-11	6.752E-08	2.26E+01	0.00003	0.09	0.2258	0.00E+00	0.0033	0.0080	0.01
Water (2)	9.712E-09	1.529E-08	5.422E-06	5.422E-06	3.06E+03	0.00442	0.3854	3.0588	0	0.0137	0.1084	0.12
Soil (3)	3.242E-07	3.833E-03	1.36E+00	9.058E-01	6.90E+07	99.79860	2812.67	0.0000	0	99.7093	0.0000	99.71
Bottom sediment (4)	2.879E-08	2.723E-04	9.654E-02	7.542E-02	1.36E+05	0.19694	1.7156	2.723E+00	0.00E+00	6.08E-02	9.65E-02	1.57E-01
Total					69137005.		2814.87	6.0078	0.0000			100
								2821				

Level III summary

	E	E(A)	E(B)	E(C)
Total emission rate mol/h	2820.8744	2820.8744	2820.8744	2820.8744
Total VZ products	2.179E+14	2.179E+14	2.179E+14	2.179E+14
Total amount of chemicals	62458192.	33984425.	49452318.	69137005.
Total advection D value	9.085E+09	9.08E+09	9.08E+09	9.08E+09
Total reaction D value	1.17E+09	1.17E+09	1.17E+09	1.17E+09
Total D value	1.025E+10	1.03E+10	1.03E+10	1.03E+10
Output by reaction, mol/h	2249.3889	1709.9461	773.94071	2814.8667
Output by advection, mol/h	571.48548	1110.9283	2046.9366	6.0077594
Output by losses	0	0	0	0
Overall residence time, h	22141.429	12047.478	17530.829	24509.068
Reaction residence time, h	27766.736	19874.559	63896.778	24561.378
Advection residence time, h	109290.95	30591.014	24159.184	11507951.

p,p'-DDT
Level I calculation:

Physical-chemical properties:
		Partition coeff.:	
molecular wt., g/mol	354.50	HLC	1.289
melting point, C	109	Kaw	5.200E-04
solubility, g/m3	0.0055	Koc	635015
vapor pressure, Pa	0.00002	BCF	77441
log Kow	6.19	Ksw	30481
fugacity ratio, F	0.1476	Ksd/w	60961
dissoc. const. pKa	12	Kssd/w	190504
		Kar/w	4.429E+10

Compartment	Z mol/m3 Pa	Concentration mol/m3	mg/L (or g/m3)	ug/g	Amount kg	Amount %
Air	4.034E-04	5.223E-13	1.851E-10	1.562E-07	18.51	0.0185
Water	7.757E-01	1.004E-09	3.560E-07	3.560E-07	71.20	0.071
Soil	2.365E+04	3.061E-05	1.085E-02	4.522E-03	97667	97.667
Biota (fish)	6.007E+04	7.777E-05	2.757E-02	2.757E-02	5.514	5.51E-03
Suspended sediment	1.478E+05	1.913E-04	6.782E-02	4.522E-02	67.82	6.78E-02
Bottom sediment	4.729E+04	6.122E-05	2.170E-02	9.043E-03	2170.37	2.170
Total					100000	100

f = 1.295E-09 Pa

Level II Calculation:

Compartment	Half-life h	D(reaction) mol/Pa h	D Values D(advec'n) mol/Pa h	Conc'n mol/m3	Loss Reaction kg/h	Loss Advection kg/h	Removal %
Air	170	1.64E+08	4.03E+08	1.19E-10	17.181	42.147	5.933
Water	5500	1.95E+07	1.55E+08	2.29E-07	2.042	16.209	1.825
Soil	17000	8.67E+09		6.97E-03	906.31		90.631
Biota (fish)				1.77E-02			
Suspended sediment				4.36E-02			
Bottom sediment	55000	5.96E+07	9.46E+07	1.39E-02	6.225	9.881	1.611
Total R + A		8.86E+09	5.59E+08 9.42E+09		931.76	68.237 1000	100

f = 2.947E-07 Pa
Total amount = 22763998 kg

Overall residence time = 22764.00 h
Reaction time = 24431.11 h
Advection time = 333601.13 h

Level III Calculation: p,p'-DDT

Phase Properties and Rates:

Compartment	Bulk Z mol/m3 Pa	Half-life h	D Values Reaction mol/Pa h	Advection mol/Pa h
Air (1)	7.608E-04	170	3.10E+08	7.61E+08
Water (2)	1.575E+00	5500	3.97E+07	3.15E+08
Soil (3)	1.183E+04	17000	8.68E+09	
Bottom sediment (4)	9.461E+03	55000	5.96E+07	9.46E+07

	E(1)=1000	E(2)=1000	E(3)=1000	E(1,2,3)
Overall residence time =	12047.49	17530.86	24509.07	22141.44 h
Reaction time =	19874.58	63897.38	24561.38	27766.77 h
Advection time =	30591.07	24159.16	11507970	109290.73 h

Intermedia D values:

	mol/Pa h
air/water D12	1.272E+08
air/soil, D13	9.733E+08
water/air, D21	1.917E+07
water/sed., D24	7.399E+08
soil/air, D31	1.413E+06
soil/water, D32	2.478E+07
sed./water, D42	9.538E+07

Phase Properties, Compositions, Transport and Transformation Rates:

Emission, kg/h			Fugacity, Pa				Concentration, g/m3			
E(1)	E(2)	E(3)	f(1)	f(2)	f(3)	f(4)	C(1)	C(2)	C(3)	C(4)
1000	0	0	1.301E-06	2.035E-07	1.455E-07	6.031E-07	3.509E-07	1.136E-04	6.099E-01	2.022E+00
0	1000	0	3.003E-08	3.400E-06	3.358E-09	1.008E-05	8.096E-09	1.898E-03	1.407E-02	3.379E+01
0	0	1000	2.968E-10	9.712E-09	3.242E-07	2.879E-08	8.003E-11	5.422E-06	1.359E+00	9.654E-02
50	250	700	7.278E-08	8.670E-07	2.351E-07	2.570E-06	1.962E-08	4.840E-04	9.852E-01	8.617E+00

Emission, kg/h			Loss, Reaction, kg/h				Loss, Advection, kg/h			
E(1)	E(2)	E(3)	R(1)	R(2)	R(3)	R(4)	A(1)	A(2)	A(3)	A(4)
1000	0	0	1.430E+02	2.862E+00	4.48E+02	1.274E+01	3.509E+02	2.272E+01	3.509E+01	2.022E+01
0	1000	0	3.300E+00	4.783E+01	1.03E+01	2.129E+02	8.096E+00	3.796E+02	3.379E+02	3.379E+02
0	0	1000	3.262E-02	1.366E-01	9.97E+02	6.082E-01	8.003E-02	1.084E+00	1.084E+02	9.654E-01
50	250	700	8.000E+00	1.220E+01	7.23E+02	5.429E+01	1.962E+01	9.680E+01	9.680E+01	8.617E+01

Amounts, kg				Total amount, kg
m(1)	m(2)	m(3)	m(4)	
3.509E+04	2.272E+04	1.098E+07	1.011E+06	1.205E+07
8.096E+02	3.796E+05	2.533E+05	1.690E+07	1.753E+07
8.003E+00	1.084E+03	2.446E+07	4.827E+04	2.451E+07
1.962E+03	9.680E+04	1.773E+07	4.309E+06	2.214E+07

Intermedia Rate of Transport, kg/h						
T12	T21	T13	T31	T32	T24	T42
air-water	water-air	air-soil	soil-air	soil-water	water-sed	sed-water
5.865E+01	1.383E+00	4.489E+02	7.290E-02	1.278E+00	5.335E+01	2.039E+01
1.353E+02	2.310E+01	1.036E+02	1.682E-03	2.948E-02	8.915E+02	3.407E+02
1.338E-02	6.600E-02	1.024E-01	1.624E-01	2.847E+00	2.547E+00	9.732E-01
3.280E+00	5.891E+00	2.511E+01	1.178E-01	2.064E+00	2.273E+02	8.687E+01

Alphabetical Index (Volume V)

Acephate 338
Alachlor 57
Aldicarb 336
Aldrin 342
Ametryn 61
Aminocarb 347
Amitrole 67
Anilazine 651
Atrazine 67
Azinphos-methyl 349
Barban 76
Benalaxyl 653
Bendiocarb 352
Benefin 78
Benomyl 655
Bifenox 81
Bitertanol 657
Bromacil 82
Bromoxynil 85
Bupirimate 659
Butachlor 85
Butralin 89
Butylate 91
Captan 661
Carbaryl 352
Carbendazim 664
Carbofuran 359
Carboxin 666
Chloramben 93
Chlordane 364
Chlorfenvinphos 370
Chlorothalonil 672
Chloroneb 668
Chloropicrin 670
Chlorpropham 95
Chlorpyrifos 372
Chlorsulfuron 98
Chlortoluron 101
Crotoxyphos 378
Cyanazine 103
Cypermethrin 380
2,4-D 106
2,4-DB 114
Dalapon 111
DDD 383
DDE 387

DDT	391
Demeton	402
Diallate	116
Diazinon	404
Dicamba	119
Dichlobenil	123
Dichlone	674
Dichlorprop	126
Dichlorvos	409
Diclofop-methyl	128
Dicrotophos	412
Dieldrin	414
Diflubenzuron	421
Dimethoate	423
Dinitramine	130
Dinoseb	132
Diphenamid	135
Diquat	137
Disulfolton	427
Dithianon	673
Diuron	139
Edifenphos	675
Endosulfan	431
Endrin	435
EPTC	144
Ethion	439
Etridiazole	680
Fenarimol	682
Fenfuram	684
Fenitrothion	442
Fenoxycarb	446
Fenthion	448
Fenuron	147
Fenvalerate	451
Fluchloralin	150
Flucythrinate	454
Fluometuron	152
Fluorodifen	155
Fluridone	157
Folpet	686
Fonofos	456
Formaldehyde	688
Glyphosate	160
α-HCH	459
β-HCH	462
δ-HCH	464
Heptachlor	466
Heptachlor epoxide	471
Hexachlorobenzene	690
Imazalil	695

Isopropalin	163
Isoproturon	165
Kepone	474
Leptophos	476
Lindane (γ-HCH)	479
Linuron	167
Malathion	490
MCPA	171
MCPB	174
Mecoprop	176
Metalaxyl	699
Methiocarb	493
Methomyl	495
Methoxychlor	500
Metolachlor	178
Mevinphos	504
Mirex	506
Molinate	181
Monocrotophos	508
Monolinuron	184
Monuron	186
Neburon	190
Oryzalin	192
Oxamyl	512
Oxycarboxin	701
Parathion	514
Parathion-methyl	520
Pebulate	194
Penconazole	703
Pentachlorophenal (PCP)	525
Permethrin	533
Phenthoate	536
Phorate	538
Phosmet	540
Picloram	196
Procymidone	705
Profluralin	200
Prometon	203
Prometryn	206
Pronamide	209
Propachlor	212
Propanil	215
Propazine	228
Propham	221
Propiconazole	707
Propoxur	544
Pyrazon	224

Quintozene	709
Ronnel	547
Simazine	226
2,4,5-T	231
Terbacil	235
Terbufos	550
Terbutryn	237
Thiophanate-methyl	711
Thiram	710
Tolclofos-methyl	715
Tolylfluanid	717
Toxaphene	552
Triadimefon	719
Triallate	240
Trichlorfon	558
Triflumizole	721
Trifluralin	243
Triforine	723
Vernolate	249
Vinclozolin	725

Index for Volume I

1. Monoaromatics:
    - Benzene 55
    - Toluene 63
    - Ethylbenzene 72
    - *o*-Xylene 76
    - *m*-Xylene 81
    - *p*-Xylene 85
    - 1,2,3-Trimethylbenzene 90
    - 1,2,4-Trimethylbenzene 93
    - 1,3,5-Trimethylbenzene 96
    - *n*-Propylbenzene 100
    - *iso*-Propylbenzene 104
    - 1-Ethyl-2-methylbenzene 107
    - 1-Ethyl-4-methylbenzene 109
    - *iso*-Propyl-4-methylbenzene 111
    - *n*-Butylbenzene 113
    - *iso*-Butylbenzene 116
    - *sec*-Butylbenzene 118
    - *tert*-Butylbenzene 120
    - 1,2,3,4-Tetramethylbenzene 122
    - 1,2,3,5-Tetramethylbenzene 124
    - 1,2,4,5-Tetramethylbenzene 126
    - Pentamethylbenzene 128
    - Pentylbenzene 130
    - Hexamethylbenzene 132
    - Hexylbenzene 134
2. Chlorobenzenes:
    - Monochlorobenzene 193
    - 1,2-Dichlorobenzene 200
    - 1,3-Dichlorobenzene 206
    - 1,4-Dichlorobenzene 212
    - 1,2,3-Trichlorobenzene 220
    - 1,2,4-Trichlorobenzene 225
    - 1,3,5-Trichlorobenzene 232
    - 1,2,3,4-Tetrachlorobenzene 237
    - 1,2,3,5-Tetrachlorobenzene 241
    - 1,2,4,5-Tetrachlorobenzene 246
    - Pentachlorobenzene 250
    - Hexachlorobenzene 256
3. Polychlorinated Biphenyls (PCBs)
   PCB Congerers
    - Biphenyl (PCB-0) 328
    - 2-Chlorobiphenyl (PCB-1) 334
    - 3-Chlorobiphenyl (PCB-2) 338
    - 4-Chlorobiphenyl (PCB-3) 342
    - 2,2'-Dichlorobiphenyl (PCB-4) 346
    - 2,3 -Dichlorobiphenyl (PCB-5) 351
    - 2,4 -Dichlorobiphenyl (PCB-7) 353

2,4'-Dichlorobiphenyl (PCB-8) 356
2,5 -Dichlorobiphenyl (PCB-9) 359
2,6 -Dichlorobiphenyl (PCB-10) 363
3,3'-Dichlorobiphenyl (PCB-11) 366
3,4 -Dichlorobiphenyl (PCB-12) 369
3,5 -Dichlorobiphenyl (PCB-14) 372
4,4'-Dichlorobiphenyl (PCB-15) 374
2,2'3,-Trichlorobiphenyl (PCB-16) 378
2,2',5-Trichlorobiphenyl (PCB-18) 380
2,3,3'-Trichlorobiphenyl (PCB-20) 385
2,3,4 -Trichlorobiphenyl (PCB-21) 387
2,3,'5-Trichlorobiphenyl (PCB-26) 389
2,4,4'-Trichlorobiphenyl (PCB-28) 391
2,4,5 -Trichlorobiphenyl (PCB-29) 395
2,4,6 -Trichlorobiphenyl (PCB-30) 398
2,4',5-Trichlorobiphenyl (PCB-31) 401
2',3,4-Trichlorobiphenyl (PCB-33) 405
3,3',4-Trichlorobiphenyl (PCB-35) 408
3,4,4'-Trichlorobiphenyl (PCB-37) 410
2,2',3,3'-Tetrachlorobiphenyl (PCB-40) 413
2,2',3,5'-Tetrachlorobiphenyl (PCB-44) 417
2,2',4,4'-Tetrachlorobiphenyl (PCB-47) 420
2,2',4,5'-Tetrachlorobiphenyl (PCB-49) 423
2,2'4,6 -Tetrachlorobiphenyl (PCB-50) 426
2,2',4,6'-Tetrachlorobiphenyl (PCB-51) 428
2,2',5,5'-Tetrachlorobiphenyl (PCB-52) 430
2,2',5,6'-Tetrachlorobiphenyl (PCB-53) 436
2,2',6,6'-Tetrachlorobiphenyl (PCB-54) 438
2,3,4,4' -Tetrachlorobiphenyl (PCB-60) 441
2,3,4,5 -Tetrachlorobiphenyl (PCB-61) 443
2,3',4,4'-Tetrachlorobiphenyl (PCB-66) 446
2,3',4',5-Tetrachlorobiphenyl (PCB-70) 449
2,4,4',6 -Tetrachlorobiphenyl (PCB-75) 453
3,3',4,4'-Tetrachlorobiphenyl (PCB-77) 455
3,3',5,5'-Tetrachlorobiphenyl (PCB-80) 458
2,2',3,3',5-Pentachlorobiphenyl (PCB-83) 460
2,2',3,4,5 -Pentachlorobiphenyl (PCB-86) 462
2,2',3,4,5'-Pentachlorobiphenyl (PCB-87) 465
2,2',3,4,6 -Pentachlorobiphenyl (PCB-88) 468
2,2',3,5',6-Pentachlorobiphenyl (PCB-95) 470
2,2',4,4',5-Pentachlorobiphenyl (PCB-99) 472
2,2',4,4',6-Pentachlorobiphenyl (PCB-100) 474
2,2',4,5,5'-Pentachlorobiphenyl (PCB-101) 476
2,2',4,6,6'-Pentachlorobiphenyl (PCB-104) 482
2,3,3',4',6-Pentachlorobiphenyl (PCB-110) 484
2,3,4,5,6 -Pentachlorobiphenyl (PCB-116) 486
2,2'3,3',4,4' -Hexachlorobiphenyl (PCB-128) 489
2,2',3,3',4,5 -Hexachlorobiphenyl (PCB-129) 493
2,2',3,3',5,6 -Hexachlorobiphenyl (PCB-134) 496
2,2',3,3',6,6'-Hexachlorobiphenyl (PCB-136) 499

    2,2',3,4,4',5'-Hexachlorobiphenyl (PCB-138) 502
    2,2',4,4',5,5'-Hexachlorobiphenyl (PCB-153) 505
    2,2',4,4',6,6'-Hexachlorobiphenyl (PCB-155) 511
    3,3',4,4',5,5'-Hexachlorobiphenyl (PCB-169) 515
    2,2',3,3',4,4',5-Heptachlorobiphenyl (PCB-170) 517
    2,2',3,3',4,4',6-Heptachlorobiphenyl (PCB-171) 519
    2,2',3,4,4',5,5'-Heptachlorobiphenyl (PCB-180) 522
    2,2',3,4,5,5',6 -Heptachlorobiphenyl (PCB-185) 525
    2,2',3,4',5,5',6-Heptachlorobiphenyl (PCB-187) 528
    2,2',3,3',4,4',5,5'-Octachlorobiphenyl (PCB-194) 530
    2,2',3,3',5,5',6,6'-Octachlorobiphenyl (PCB-202) 533
    2,2',3,3',4,4',5,5',6-Nonachlorobiphenyl (PCB-206) 536
    2,2',3,3',4,4',5,6,6'-Nonachlorobiphenyl (PCB-207) 539
    2,2',3,3',4,5,5',6,6'-Nonachlorobiphenyl (PCB-208) 541
    2,2',3,3',4,4',5,5',6,6'-Decachlorobiphenyl (PCB-209) 543
PCB Isomer groups
    Monochlorobiphenyls 547
    Dichlorobiphenyls 549
    Trichlorobiphenyls 552
    Tetrachlorobiphenyls 554
    Pentachlorobiphenyls 557
    Hexachlorobiphenyls 560
    Heptachlorobiphenyls 562
    Octachlorobiphenyls 564
    Nonachlorobiphenyls 566
Aroclor mixtures
    Aroclor 1016 568
    Aroclor 1221 571
    Aroclor 1232 573
    Aroclor 1242 575
    Aroclor 1248 579
    Aroclor 1254 582
    Aroclor 1260 587

Index for Volume II

1. Polynuclear Aromatic Hydrocarbons (PAHs)

    Indan 60
    Naphthalene 62
    1-Methylnaphthalene 72
    2-Methylnaphthalene 76
    1,3-Dimethylnaphthalene 80
    1,4-Dimethylnaphthalene 82
    1,5-Dimethylnaphthalene 85
    2,3-Dimethylnaphthalene 87
    2,6-Dimethylnaphthalene 90
    1-Ethylnaphthalene 93
    2-Ethylnaphthalene 96
    1,4,5-Trimethylnaphthalene 98
    Biphenyl (also see Vol. I) 100
    4-Methylbiphenyl 107
    4,4'-Dimethylbiphenyl 109
    Diphenylmethane 111
    Bibenzyl 113
    trans-Stilbene 115
    Acenaphthene 117
    Acenaphthylene 122
    Fluorene 125
    1-Methylfluorene 130
    Phenanthrene 132
    1-Methylphenanthrene 141
    Anthracene 143
    2-Methylanthracene 153
    9-Methylanthracene 156
    9,10-Dimethylanthracene 160
    Pyrene 163
    Fluoranthene 172
    Benzo[*a*]fluorene 178
    Benzo[*b*]fluorene 180
    Chrysene 182
    Triphenylene 187
    p-Terphenyl 190
    Naphthacene 192
    Benz[*a*]anthracene 195
    Benzo[*b*]fluoranthene 201
    Benzo[*j*]fluoranthene 204
    Benzo[*k*]fluoranthene 206
    Benzo[*a*]pyrene 209
    Benzo[*e*]pyrene 217
    Perylene 219
    7,12-Dimethylbenz[*a*]anthracene 222
    9,10-Dimethylbenz[*a*]anthracene 225
    3-Methylcholanthrene 227
    Benzo[*ghi*]perylene 230

    Dibenz[*a,c*]anthracene 234
    Dibenz[*a,h*]anthracene 236
    Dibenz[*a,j*]anthracene 239
    Pentacene 241
    Coronene 243

2 Polychlorinated Dibenzo-*p*-Dioxins (PCDDs)
    Dibenzo-*p*-dioxin 370
    1-Chloro-dibenzo-*p*-dioxin 373
    2-Chloro-dibenzo-*p*-dioxin 375
    2,3-Dichloro-dibenzo-*p*-dioxin 378
    2,7-Dichloro-dibenzo-*p*-dioxin 380
    2,8-Dichloro-dibenzo-*p*-dioxin 383
    1,2,4-Trichloro-dibenzo-*p*-dioxin 385
    1,2,3,4-Tetrachloro-dibenzo-*p*-dioxin 388
    1,2,3,7-Tetrachloro-dibenzo-*p*-dioxin 391
    1,3,6,8-Tetrachloro-dibenzo-*p*-dioxin 395
    2,3,7,8-Tetrachloro-dibenzo-*p*-dioxin 400
    1,2,3,4,7-Pentachloro-dibenzo-*p*-dioxin 410
    1,2,3,4,7,8-Hexachloro-dibenzo-*p*-dioxin 414
    1,2,3,4,6,7,8-Heptachloro-dibenzo-*p*-dioxin 418
    Octachloro-dibenzo-*p*-dioxin 422

3. Polychlorinated Dibenzofurans (PCDFs)
    Dibenzofuran 501
    2,8-Dichloro-dibenzofuran 504
    2,3,7,8-Tetrachloro-dibenzofuran 506
    2,3,4,7,8-Pentachloro-dibenzofuran 509
    1,2,3,4,7,8-Hexachloro-dibenzofuran 512
    1,2,3,4,6,7,8-Heptachloro-dibenzofuran 514
    1,2,3,4,7,8,9-Heptachloro-dibenzofuran 517
    Octachlorodibenzofuran 519

Index for Volume III

1. Saturated Hydrocarbons
   Alkanes
   - Isobutane (2-methylpropane) 71
   - 2,2-Dimethylpropane (neopentane) 74
   - n-Butane 77
   - 2-Methylbutane (isopentane) 81
   - 2,2-Dimethylbutane 84
   - 2,3-Dimethylbutane 87
   - 2,2,3-Trimethylbutane 90
   - n-Pentane 92
   - 2-Methylpentane (isohexane) 97
   - 3-Methylpentane 100
   - 2,2-Dimethylpentane 103
   - 2,4-Dimethylpentane 105
   - 3,3-Dimethylpentane 108
   - 2,2,4-Trimethylpentane (isooctane) 110
   - 2,3,4-Trimethylpentane 113
   - n-Hexane 115
   - 2-Methylhexane (isoheptane) 120
   - 3-Methylhexane 123
   - 2,2,5-Trimethylhexane 126
   - n-Heptane 129
   - 2-Methylheptane 134
   - 3-Methylheptane 136
   - n-Octane 138
   - 4-Methyloctane 143
   - n-Nonane 145
   - n-Decane 148
   - n-Undecane 151
   - n-Dodecane 153

   Cycloalkanes
   - Cyclopentane 155
   - Methylcyclopentane 159
   - 1,1,3-Trimethylcyclopentane 162
   - Propylcyclopentane 164
   - Pentylcyclopentane 166
   - Cyclohexane 168
   - Methylcyclohexane 173
   - 1,2-cis-Dimethylcyclohexane 176
   - 1,4-trans-Dimethylcyclohexane 179
   - 1,1,3-Trimethylcyclohexane 181
   - Cycloheptane 183
   - Cyclooctane 185
   - Decalin 187

1.2 Unsaturated Hydrocarbons
   Alkenes
   - 2-Methylpropene 189
   - 1-Butene 192
   - 2-Methyl-1-butene 195

3-Methyl-1-butene	197
2-Methyl-2-butene	199
1-Pentene	201
cis-2-Pentene	204
2-Methyl-1-pentene	206
4-Methyl-1-pentene	208
1-Hexene	210
1-Heptene	213
1-Octene	216
1-Nonene	219
1-Decene	221

Dienes
1,3-Butadiene	223
2-Methyl-1,3-butadiene (isoprene)	227
2,3-Dimethyl-1,3-butadiene	230
1,4-Pentadiene	232
1,5-Hexadiene	235
1,6-Heptadiene	238

Alkynes
1-Butyne	240
1-Pentyne	242
1-Hexyne	244
1-Heptyne	246
1-Octyne	248
1-Nonyne	250

Cycloalkenes
Cyclopentene	252
Cyclohexene	255
1-Methylcyclohexene	258
Cycloheptene	260
1,4-Cyclohexadiene	262
Cycloheptatriene	265
dextro-Limonene [(*R*)-(+)-limonene]	267

1.3 Aromatic Hydrocarbons
Styrene	269
Methyl styrenes	272
Tetralin	274
Others (see Volume 1 and 2)	

2. Halogenated Hydrocarbons:
Methyl chloride	395
Dichloromethane	400
Chloroform	407
Carbon tetrachloride	416
Chloroethane (Ethyl chloride)	426
1,1-Dichloroethane	431
1,2-Dichloroethane	436
1,1,1-Trichloroethane	443
1,1,2-Trichloroethane	451
1,1,1,2-Tetrachloroethane	456
1,1,2,2-Tetrachloroethane	459

Pentachloroethane 465
Hexachloroethane 469
1-Chloropropane (*n*-Propyl chloride) 473
2-Chloropropane 476
1,2-Dichloropropane 479
1,2,3-Trichloropropane 483
1-Chlorobutane (*n*-Butyl chloride) 486
2-Chlorobutane 489
1-Chloropentane 491
Vinyl chloride 493
1,1-Dichloroethene 498
cis-1,2-Dichloroethene 503
trans-1,2-Dichloroethene 507
Trichloroethylene 512
Tetrachloroethylene 522
1,3-Dichloropropene 531
Chloroprene 534
Hexachlorobutadiene 536
Hexachlorocyclopentadiene 538
Bromomethane 541
Dibromomethane 545
Tribromomethane 548
Bromoethane (Ethyl bromide) 552
1,2-Dibromoethane 555
1-Bromopropane (*n*-Propyl bromide) 558
2-Bromopropane 561
1,2-Dibromopropane 564
1-Bromobutane 566
1-Bromopentane 569
Vinyl bromide 571
Methyl iodide 573
Iodoethane (Ethyl iodide) 576
1-Iodopropane (*n*-Propyl iodide) 579
2-Iodopropane 581
1-Iodobutane 583
Bromochloromethane 585
Bromodichloromethane 588
Dibromochloromethane 591
Chlorodifluoromethane 594
Dichlorodifluoromethane 597
Trichlorofluoromethane 600
1,1,2-Trichloro-1,2,2-trifluoroethane 604
1,1,2,2-Tetrachloro-1,2-difluoroethane 607
Fluorobenzene 609
Bromobenzene 611
Iodobenzene 614
Chlorobenzenes (see Volume I)
3. Ethers
    Aliphatic Ethers
        Dimethyl ether (Methyl ether) 750

Diethyl ether (Ethyl ether) 752
Methyl *t*-butyl ether 755
Di-*n*-propyl ether 757
Di-isopropyl ether 760
Butyl ethyl ether 763
Di-*n*-butyl ether 765
1,2-Propylene oxide 768
Furan 771
2-Methylfuran 774
Tetrahydrofuran 776
Tetrahydropyran 779
1,4-Dioxane 781
Aromatic Ethers
   Anisole 784
   Phenetole 787
   Benzyl ethyl ether 789
   Diphenyl ether 791
   Styrene oxide 794
Halogenated Ethers
   Epichlorohydrin 796
   Chloromethyl methyl ether 799
   Bis(chloromethyl)ether 801
   Bis(2-chloroethyl)ether 804
   Bis(2-chloroisopropyl)ether 808
   2-Chloroethyl vinyl ether 811
   4-Chlorophenyl phenyl ether 813
   4-Bromophenyl phenyl ether 815
   Bis(2-chloroethoxy)methane 817

Index for Volume IV

2. Alcohols:
    - Methanol 78
    - Ethanol 82
    - Propanol 86
    - Isopropanol 90
    - 1-Butanol 94
    - Isobutanol 100
    - *sec*-Butyl alcohol 104
    - *tert*-Butyl alcohol 107
    - 1-Pentanol 110
    - 1-Hexanol 114
    - 1-Heptanol 118
    - 1-Octanol 121
    - Ethylene Glycol 125
    - Allyl Alcohol 128
    - Cyclohexanol 130
    - Benzyl Alcohol 133
3. Aldehydes and Ketones:
    - 3.1.1 Aldehydes:
        - Methanal (formaldehyde) 181
        - Ethanal (acetaldehyde) 183
        - Propanal (propionaldehyde) 186
        - Butanal (*n*-butyraldehyde) 189
        - 2-Propenal (acrolein) 192
        - Furfural (2-furaldehyde) 195
        - Benzaldehyde 197
    - 3.1.2 Ketones:
        - Acetone 201
        - 2-Butanone (methyl ethyl ketone) 205
        - 2-Pentanone 210
        - 3-Pentanone 213
        - 4-Methyl-2-pentanone (methyl isobutyl ketone) 216
        - 2-Hexaonone 219
        - 2-Heptanone 222
        - Cyclohexanone 223
        - Acetophenone 226
        - Benzophenone 230
4. Phenolic compounds:
    - 4.1.1 Alkylphenols and other substituted phenols
        - Phenol 284
        - *o*-Cresol 293
        - *m*-Cresol 298
        - *p*-Cresol 303
        - 2,4-Dimethylphenol 309
        - 2,6-Dimethylphenol 313
        - 3,4-Dimethylphenol 316
        - 2,3,5-Trimethylphenol 318
        - 2,4,6-Trimethylphenol 319

    o-Ethylphenol 320
    p-Ethylphenol 321
    p-tert-Butylphenol 323
    4-Octylphenol 325
    4-Nonylphenol 327
    1-Naphthol 329
    2-Naphthol 331
    2-Phenylphenol (2-hydroxybiphenyl) 333
    4-phenylphenol (4-hydroxybiphenyl) 335
    4-Chloro-*m*-cresol 336
 4.1.2 Chlorophenols
    2-Chlorophenol 338
    3-Chlorophenol 342
    4-Chlorophenol 346
    2,4-Dichlorophenol 351
    2,6-Dichlorophenol 356
    3,4-Dichlorophneol 358
    2,3,4-Trichlorophenol 360
    2,4,5-Trichlorophenol 362
    2,4,6-Trichlorophenol 367
    2,3,4,5-Tetrachlorophenol 372
    2,3,4,6-Tetrachlorophenol 374
    Pentachlorophenol 377
 4.1.3 Nitrophenols
    2-Nitrophenol 385
    3-Nitrophenol 389
    4-Nitrophenol 391
    2,4-Dinitrophenol 396
    2,4,6-Trinitrophenol (picric acid) 399
    4,6-Dinitro-*o*-cresol 401
 4.1.4 Dihydroxybenzenes, Methoxyphenols and Chloroguaiacols
    Catechol (1,2-dihydroxybenzene) 404
    Resorcinol (1,3-dihydroxybenzene) 407
    Hydroquinone (1,4-dihydroxybenzene) 409
    2-Methoxyphenol (guaiacol) 412
    3-Methoxyphenol 414
    4-Methoxyphenol 415
    4-Chloroguaiacol 416
    4,5-Dichloroguaiacol 417
    3,4,5-Trichloroguaiacol 418
    4,5,6-Trichloroguaiacol 420
    Tetrachloroguaiacol 422
5.1 Carboxylic Acids:
 5.1.1 Aliphatic Acids:
    Formic acid 504
    Acetic acid 506
    Propionic acid 509
    Butyric acid 512
    Isobutyric acid 514
    n-Valeric acid 516

    Hexanoic acid (caproic acid) 518
    Stearic (octandecanoic) acid 520
    Oleic acie 521
    Acrylic acid (2-propenoic acid) 522
    Chloroacetic acid 524
    Dichloroacetic acid 526
    Trichloroacetic acid 528
  5.1.2 Aromatic Acids:
    Benzoic acid 530
    2-Methylbenzoic acid (o-toluic acid) 533
    3-Methylbenzoic acid (m-toluic acid) 535
    4-Methylbenzoic acid (p-toluic acid) 538
    Phenylacetic acid 539
    Phthalic acid 541
    2-Chlorobenzoic acid 544
    3-Chlorobenzoic acid 546
    4-Chlorobenzoic acid 548
    Salicylic acid 550
    2,4-Dichlorophenoxyacetic acid (2,4-D) 552
6.1 Esters:
  6.1.1 Aliphatic esters:
    Methyl formate 608
    Ethyl formate 610
    Propyl formate 612
    Methyl acetate 614
    Vinyl acetate 617
    Ethyl acetate 619
    Propyl acetate 623
    Butyl acetate 626
    Pentyl acetate 629
    Methyl acrylate 632
    Ethyl acrylate 634
    Methyl methacrylate 636
  6.1.2 Aromatic esters:
    Methyl benzoate 638
    Ethyl benzoate 639
    Propyl benzoate 641
    Benzyl benzoate 642
  6.1.3. Phthalate esters:
    Dimethyl phthalate (DMP) 643
    Diethyl phthalate (DEP) 646
    Di-*n*-butyl phthalate (DBP) 653
    Di-*n*-octyl phthalate (DOP) 658
    bis(2-Ethylhexyl) phthalate (DEHP) 662
    Butylbenzyl phthalate 668
7. Nitrogen and Sulfur compounds:
  7.1.1 Nitriles (organic cyanides):
    Acetonitrile 728
    Propionitrile 731
    Acrylonitrile (2-propenitrile) 734

Benzonitrile 737
7.1.2 Aliphatic amines:
    Dimethylamine 740
    Trimethylamine 743
    Ethylamine 746
    Diethylamine 749
    n-Propyl amine 752
    n-Butyl amine 755
    Ethanolamine 758
    Diethanolamine 760
    Triethanolamine 762
7.1.3 Aromatic amines:
    Aniline 764
    2-Chloroaniline 770
    3-Chloroaniline 773
    4-Chloroaniline 776
    3,4-Dichloroaniline 780
    o-Toluidine (2-methylbenzeneamine) 782
    m-Toluidine (3-methylbenzeneamine) 785
    p-Toluidine (4-methylbenzeneamine) 787
    N,N'-Dimethylaniline 789
    2,6-Xylidine (2,6-dimethylbenzamine) 792
    Diphenylamine 794
    Benzidine 797
    3,3'-Dichlorobenzidine 800
    N,N'-bianiline 802
    α-Naphthylamine (1-aminonaphthalene) 804
    β-Naphthalamine (2-aminonaphthalene) 806
    2-Nitroaniline 808
    4-Nitroaniline 810
7.1.4 Nitroaromatic Compounds:
    Nitrobenzene 812
    Nitrophenols (see Chapter 4: Phenolic Compounds) . . . .
    2-Nitrotoluene 817
    4-Nitrotoluene 820
    2,4-Dinitrotoluene (DNT) 823
    2,6-Dinitrotoluene 826
    1-Nitronaphthalene (α-Nitronaphthalene) 828
7.1.5 Amides and Urea:
    Acetamide 830
    Acryamide 832
    Benzamide 833
    Urea 835
7.1.6 Nitrosoamines:
    n-Nitrosodimethylamine 836
    n-Nitrosodipropylamine 838
    Diphenyl nitrosamine 840
7.1.7 Heterocyclic compounds:
    Pyrrole 842
    Indole 844

	Pyridine	846
	2-Methylpyridine	850
	3-Methylpyridine	852
	2,3-Dimethylpyridine	854
	Quinoline	855
	Isoquinoline	859
	Benzo(f)quinoline	861
	Carbazole	863
	7H-Dibenzo(c,g)carbazole	865
	Acridine	867
7.1.8	Sulfur compounds:	
	Dimethyl sulfide	870
	Dimethylsulfoxide (DMSO)	872
	Dimethyl sulfate	874
	Ethanethiol	876
	1-Butanethiol (butyl mercaptan)	878
	Thiophene	889
	Benzo(b)thiophene	882
	Dibenzothiophene	884
	Thiourea	886
	Thioacetamide	888